MULTICOMPONENT REACTIONS

MULTICOMPONENT REACTIONS

Concepts and Applications for Design and Synthesis

Edited By

RAQUEL P. HERRERA
EUGENIA MARQUÉS-LÓPEZ

Departamento de Química Orgánica
Instituto de Síntesis Química y Catálisis Homogénea (ISQCH)
CSIC-Universidad de Zaragoza
Zaragoza, Spain

WILEY

Published by John Wiley & Sons, Inc., Hoboken, New Jersey
Published simultaneously in Canada

Library of Congress Cataloging-in-Publication Data:

Multicomponent reactions : concepts and applications for design and synthesis / edited by Raquel P. Herrera, Eugenia Marques-Lopez.
pages cm
Includes bibliographical references and index.
ISBN 978-1-118-01600-8 (cloth)
1. Chemical reactions. 2. Catalysis. 3. Chemistry, Organic. I. Herrera, Raquel P., 1977– editor. II. Marques-Lopez, Eugenia, 1978– editor.
QD501.M868 2015
547′.2–dc23

2015000748

Set in 10/12pt Times LT Std by SPi Publisher Services, Pondicherry, India

10 9 8 7 6 5 4 3 2 1

1 2015

To our families

CONTENTS

LIST OF CONTRIBUTORS

Juan V. Alegre-Requena, Departamento de Química Orgánica, Instituto de Síntesis Química y Catálisis Homogénea (ISQCH), CSIC-Universidad de Zaragoza, Zaragoza, Spain.

José Alemán, Departamento de Química Orgánica (módulo-1), Universidad Autónoma de Madrid, Cantoblanco, Madrid, Spain.

Cuauhtémoc Alvarado, División Académica de Ciencias Básicas, Universidad Juárez Autónoma de Tabasco, Cunduacán, Tabasco, Mexico.

Silvia Cabrera, Departamento de Química Inorgánica (módulo-7), Universidad Autónoma de Madrid, Cantoblanco, Madrid, Spain.

Renata Marcia de Figueiredo, Institut Charles Gerhardt Montpellier ICGM (UMR 5253-CNRS-UM2-UM1-ENSCM), Ecole Nationale Supérieure de Chimie, Montpellier Cedex 5, France.

Tecla Gasperi, Dipartimento di Scienze, Università degli Studi Roma Tre, Roma, Italy.

Raquel P. Herrera, Departamento de Química Orgánica, Instituto de Síntesis Química y Catálisis Homogénea (ISQCH), CSIC-Universidad de Zaragoza, Zaragoza, Spain.

Javier Iglesias-Sigüenza, Departamento de Química Orgánica, Universidad de Sevilla, Sevilla, Spain.

Daniel Könning, Department of Medicinal Chemistry, Helmholtz Centre for Infection Research, Braunschweig, Germany.

Eugenia Marqués-López, Departamento de Química Orgánica, Instituto de Síntesis Química y Catálisis Homogénea (ISQCH), CSIC-Universidad de Zaragoza, Zaragoza, Spain.

Elena Mas-Marzá, Departamento de Física, Area Física Aplicada, Universidad Jaume I, Grupo de Dispositivos Fotovoltaicos y Optoelectrónicos, Castellón, Spain.

Inmaculada Matas, Instituto de Investigaciones Químicas (Consejo Superior de Investigaciones Científicas—Universidad de Sevilla), Sevilla, Spain.

David Monge, Departamento de Química Orgánica, Universidad de Sevilla, Sevilla, Spain.

Monica Orsini, Dipartimento di Ingegneria, Università degli Studi Roma Tre, Roma, Italy.

Lea Radtke, Institute of Pharmaceutical Sciences, Department of Chemistry and Applied Biosciences, Swiss Federal Institute of Technology (ETH) Zürich, Zurich, Switzerland.

Ramon Rios, Department of Organic Chemistry, University of Southampton, Highfield Campus, Southampton, UK.

Abel Ros, Instituto de Investigaciones Químicas (Consejo Superior de Investigaciones Científicas—Universidad de Sevilla), Sevilla, Spain.

Jörg Swatschek, TU Dortmund, Fakultät für Chemie und Chemische Biologie, Lehrbereich Anorganische Chemie, Dortmund, Germany.

Fabrizio Vetica, Dipartimento di Scienze, Università degli Studi Roma Tre, Roma, Italy.

PREFACE

It is not surprising to find that the social perception of chemistry has gradually gone worse over the past few decades. With an increased concern on sustainability, we should agree that many of today's environmental problems are caused, among others, by pollutants linked to chemical industries, but it is the same perception that makes us more difficult to recognize that the best solutions to many of these and other transcendental problems lie in *chemical science*.

Sustainable chemistry based on "the 12 green principles" was established as a new discipline by Anastas and Warner in the 1990s. Thus, chemistry has evolved with new ideas and main objectives such as (i) reducing the use of exhaustible resources, (ii) improving product design focused on their safety and nonpersistence after their life cycle, and (iii) improving productive processes by reducing the generation of pollutant hazardous chemical by-products.

Today, there is a growing interest in developing efficient and environmentally friendly chemical processes. In this context, multicomponent methods play a key role. Multicomponent reactions (MCRs) are commonly defined as reactions in which three or more compounds are added to a single vessel at the same time to obtain a final product that contains most of the atoms from the starting materials. Hence, these reactions encompass a sequence of more than one chemical transformation without the necessity of changing the reaction media after each conversion. This involves, among other advantages, an important energy, time, and materials saving. In addition, in-between purification procedures are avoided, and thereby waste generation is also minimized.

Multicomponent Reactions: Concepts and Applications for Design and Synthesis mainly deals with these contexts. Herein, we provide a comprehensive overview of the most remarkable examples of MCR published between 2005 and 2014 that certainly will appeal the interest of a wide range of scientists.

This book has been written by a group of young researchers internationally recognized in their respective areas of work. We have collected 13 appealing contributions covering a broad spectrum of this avant-garde field from 2005 to the present, providing modern and useful examples, including all types of reactivities, such as organocatalytic strategies (Chapter 2), metal-catalyzed processes (Chapter 3), the key use of organoboron compounds in MCRs (Chapter 4), and carbene-promoted MCRs (Chapter 5). In addition, specific chapters are dedicated to the Ugi (Chapter 7), Passerini (Chapter 8), Biginelli (Chapter 9), Bucherer–Bergs and Strecker MCRs (Chapter 10), and two more chapters focus on other essential MCRs (Chapters 12 and 13), such as Radziszewski, Sakurai, Gewald, Kabachnik–Fields, Yonemitsu, and Povarov MCRs, among others. Other two chapters complete this book: one devoted to unusual approaches followed in this area (Chapter 11) and the other to a wide array of examples covering the synthesis of target molecules (Chapter 6), such as natural products or biologically active compounds.

In particular, this book represents a significant tool for chemists working in this research area. In addition, the increasing necessity and interest on efficient, versatile, and potential synthetic procedures to achieve complex molecules and libraries, quickly and economically, makes this book an important reference instrument in organic synthesis. This wide-ranging collection can inspire both the academics and industrialists and can help in their future developments. We hope to fulfill the expectations of a great part of the scientific community, as well as organic chemistry faculties and students.

A word of gratitude to all those who helped us in our passionate endeavor. Firstly, we thank Jonathan T. Rose—editor of Wiley—for his incredible support and encouragement, and without him this book would not have been possible. We also thank our colleagues, who with their invaluable and interesting reports helped us to build this work. The huge and tireless effort of all the contributors has made this challenging assignment possible, which started some years ago in San Francisco, CA, and has ended up in print that the reader holds in their hands today.

Zaragoza
November 8, 2014

RAQUEL P. HERRERA AND
EUGENIA MARQUÉS-LÓPEZ

LIST OF ABBREVIATIONS

4CR four-component reaction
AAIL amino acid ionic liquid
AFPs amine-functional polysiloxanes
AHCC amine and heterocyclic carbene catalysis
AIBN azobisisobutyronitrile
APIs active pharmaceutical ingredients
AORC asymmetric organocatalytic relay cascade
BDMS bromodimethylsulfonium bromide
BIOS biology-oriented synthesis
BSA bovine serum albumin
BSMP bis(salicylaldehyde)-4-methyl-1,2-phenylenediimine
BTSC bis(trimethylsilyl) chromate
CAN cerammonium nitrate
CAN ceric ammonium nitrate
CBD condition-based divergence
CEYM cross-enyne metathesis
CM chorismate mutase
CoMSIA comparative molecular similarity indices analysis
CPN cupreine
CRP controlled radical polymerization
CSA camphorsulfonic acid
CSA cellulose sulfuric acid
CTA cetyltrimethylammonium
CuAAC copper-catalyzed azide–alkyne cycloaddition
DCPCC 2,6-dicarboxypyridinum chlorochromate
DDQ 2,3-dichloro-5,6-dicyano-1,4-benzoquinone
DHPMs dihydropyrimidines
DHPMs dihydropyrimidones
DMAD dimethylacetalenedicarboxylate
DNBA dinitrobenzenesulfonic acid
DTS divergent total synthesis
DOS diversity-oriented synthesis
EAN ethylammonium nitrate
EDDA ethylenediammonium diacetate
FAAH fatty acid amide hydrolase
FCCs Fischer carbene complexes
GAAS gluconic acid aqueous solution
G-3CR three-component Gewald reaction
HMDS hexamethyldisilazane
HPLC high-performance liquid chromatography
HSnP-2 hybrid tin phosphonate nanoparticles
HSVM high-speed vibration milling conditions
IBX 2-iodoxybenzoic acid
IEDDA inverse electron demand Diels–Alder
IEDHDA inverse electron demand hetero-Diels–Alder
ILs ionic liquids
IMCRs isocyanide-based multicomponent reactions
IR infrared
MACOS microwave-assisted continuous-flow organic synthesis
MAO-N monoaminooxidase N
MC multicomponent
MCAR multicomponent allylation reaction
MCR^2 combination of MCRs
MCRs multicomponent reactions
MDR multidrug-resistant
MIC minimum inhibitory concentration
MK-10 montmorillonite K-10
MMA methyl methacrylate

MMO	mixed metal oxide
MRS	modular reaction sequences
MS	mass spectroscopy
MTB	*Mycobacterium tuberculosis*
MW	microwave
NBS	*N*-bromosuccinimide
NCS	*N*-chlorosuccinimide
NHC	N-heterocyclic carbene
NHPI	*N*-hydroxyphthalimide
NMM	*N*-methylmorpholine
NMP	*N*-methylpyrrolidinone)/*n*-pentanol
NMR	nuclear magnetic resonance
OHS	organometallic hollow sphere
PADAM	Passerini reaction–amine deprotection–acyl migration
PBM	Petasis-borono–Mannich
PDE4B	phosphodiesterase 4B
PMA	phosphomolybdic acid
PTSA	*p*-toluenesulfonic acid
P-3CR	Passerini three-component reaction
RAFT	reversible addition fragmentation chain transfer
SAA	Strecker–allylic alkylation
SAAC	Strecker–allylic alkylation–cyclization
SAR	structure–activity relationship
SAW	Staudinger–aza-Wittig
SET	single-electron transfer
SILP	supported ionic liquid phase
SMS	silyl-modified Sakurai
SRR	single reactant replacement
SSA	silica sulfuric acid
STA	silica tungstic acid
TBAB	tetrabutylammonium bromide
TBHP	*tert*-butyl hydroperoxide
TCRA	three-component reductive alkylation
TCT	2,4,6-trichloro[1,3,5]triazine
TFA	trifluoroacetic acid
TFE	trifluoroethanol
TfOH	trifluoromethanesulfonic acid
TMG	tetramethylguanidine
THBCs	tetrahydro-β-carbolines
THF	tetrahydrofuran
TMSCN	trimethylsilyl cyanide
TPAP	tetra-*N*-propylammonium perruthenate
(TPyP)-CMP	tetrapyridylporphyrin supported on chloromethylated polystyrene
TS	transition state
TSA	tungstate sulfuric acid
UHP	hydrogen peroxide–urea complex
UMAM	Ugi/Michael/aza-Michael
U-4CR	Ugi four-component reaction
WD	Wells–Dawson

1

INTRODUCTION: MULTICOMPONENT STRATEGIES

JUAN V. ALEGRE-REQUENA, EUGENIA MARQUÉS-LÓPEZ AND RAQUEL P. HERRERA

Departamento de Química Orgánica, Instituto de Síntesis Química y Catálisis Homogénea (ISQCH), CSIC-Universidad de Zaragoza, Zaragoza, Spain

GENERAL INTRODUCTION

The goal of this book is to provide an overview of the most useful and noteworthy examples of multicomponent reactions (MCRs) published in this field between 2005 and 2014, in order to attract the attention of a wide range of readers. Previous examples are collected in an exceptional book edited by Zhu and Bienaymé, published in 2005 [1]. Since then, a great number of interesting and important reviews have also been written, and they will be cited throughout this book. For this reason, only the most pivotal examples will be reported and commented on in order to avoid repetitions.

MCRs are widely defined as reactions in which three or more components are added to a single vessel at the same time to lead to a final product that contains most of the atoms from the starting reagents. Therefore, these reactions encompass a sequence of more than one chemical transformation without the necessity of changing the reaction media after each transformation. It is not surprising then that MCRs lead to great molecular diversity and allow for the creation of libraries of small organic molecules while requiring less time and effort when compared with step-by-step procedures [2]. This is especially attractive for the pharmaceutical industry, for which the easy creation of large libraries of compounds with possible biological activity is a priority.

The significance of these processes can be observed in the large number of publications that have appeared in this field over the last decade. Also, the biological utility of compounds synthesized with MCRs has been confirmed by the discovery of many molecules with remarkable biological activity (Fig. 1.1) [7].

Over the last decade, interest in performing sustainable chemistry has drastically increased [8]. The application of ingenious strategies to synthesize complex scaffolds and highly substituted molecules, combining molecular diversity [9] with ecocompatibility [10], has been the main focus of many scientific groups. In effect, the rational design of reactions that transform simple and readily available substrates into complex structures in a single reaction is one of the current major challenges in organic synthesis. In this context, MCRs have become one of the best established approaches for reaching this goal, since these strategies imply atom economy [11] and bond-forming efficiency [12].

There are some authors that consider the reaction between bitter almond oil and ammonia, carried out by Laurent in 1838, as the first MCR [13]. This mixture could promote a condensation of ammonia, hydrogen cyanide, and benzaldehyde, resulting in an α-aminonitrile intermediate that, once formed, reacts with another molecule of benzaldehyde to give its corresponding Schiff base. However, in the compositions reported by the authors, none of the examined products lined up with the MCR's possible products, neither the α-aminonitrile nor its subsequent Schiff base. Therefore, the Strecker reaction could be considered the first reported MCR, due to the fact that it was the first time that an author was able to determine the structure of a product formed in a MCR.

Since the development of the Strecker reaction in 1850 [14], a great number of interesting MCRs have been reported, and amidst them, some of the most significant reactions are displayed in Table 1.1. In the following chapters, these pioneering reactions will be extensively discussed.

Multicomponent Reactions: Concepts and Applications for Design and Synthesis, First Edition. Edited by Raquel P. Herrera and Eugenia Marqués-López.

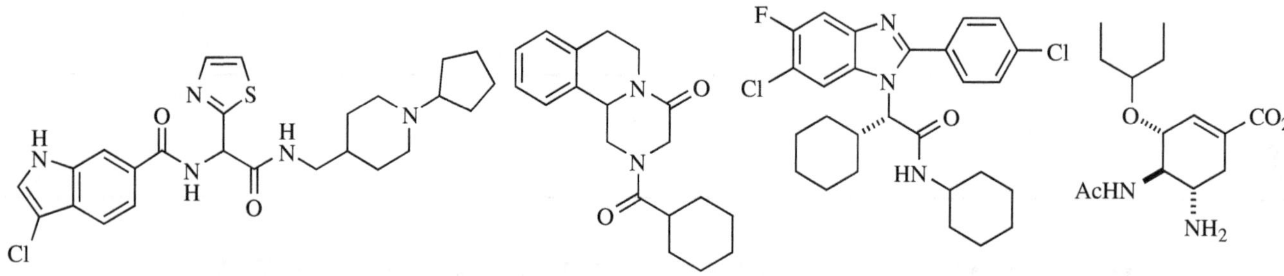

Factor Xa inhibitor — Praziquantel — Farnesoid X receptor agonist — (–)-Oseltamivir

FIGURE 1.1 Examples of drugs synthesized with MCRs: factor Xa inhibitors [3], praziquantel [4], farnesoid X receptor agonists [5], and (–)-oseltamivir [6].

TABLE 1.1 Some historically significant MCRs

Strecker [14], 1850	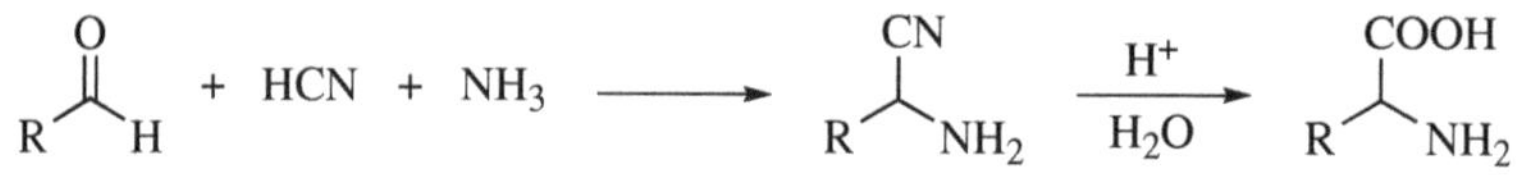 Synthesis of α-amino acids
Debus-Radziszewski [15], 1858, 1882	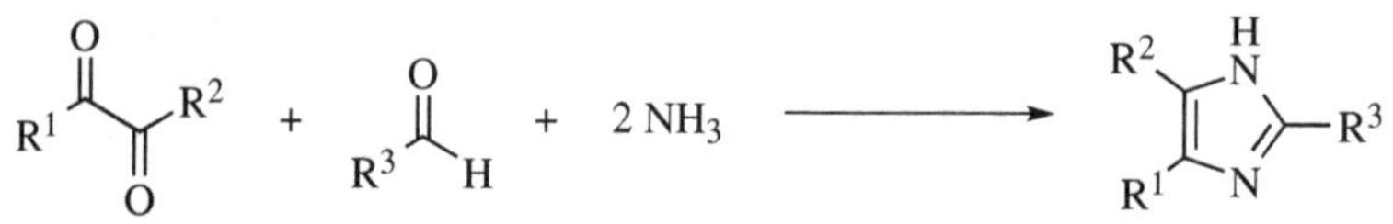 Synthesis of imidazoles
Hantzsch [16], 1881	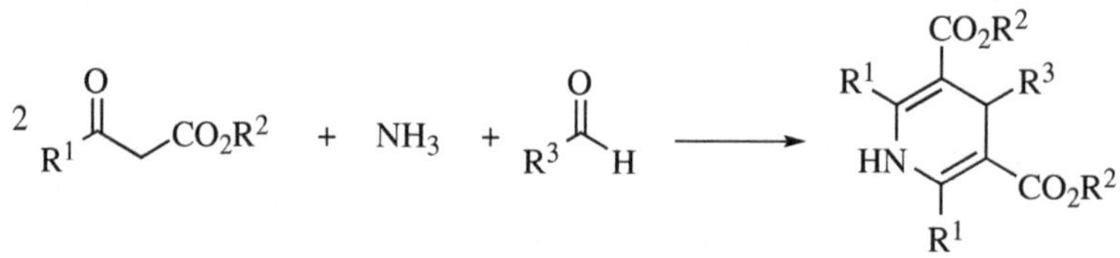 Synthesis of 1,4-dihydropyridines
Hantzsch [17], 1890	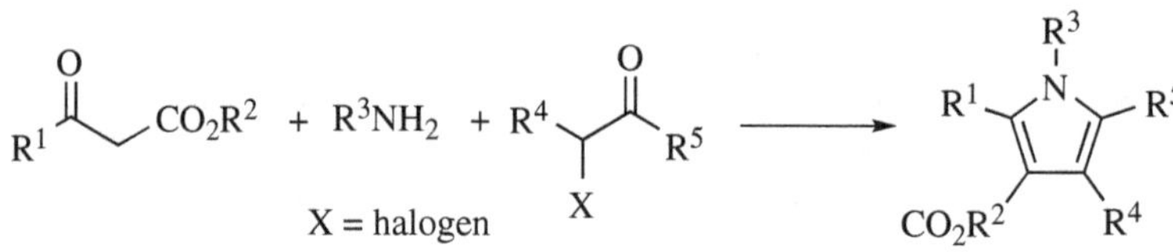 Synthesis of pyrroles
Biginelli [18], 1891	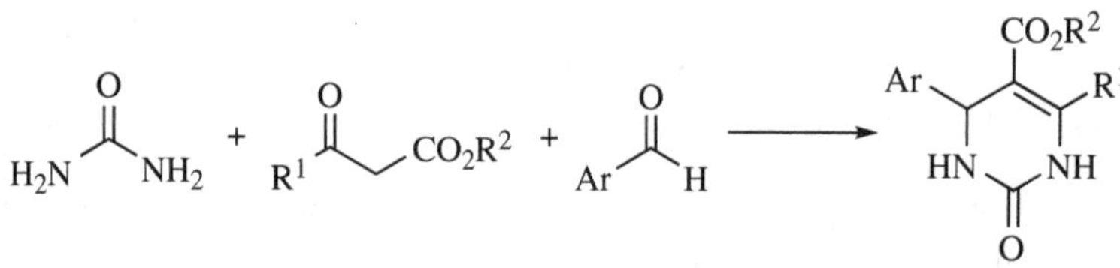 Synthesis of dihydropyrimidones
Mannich [19], 1912	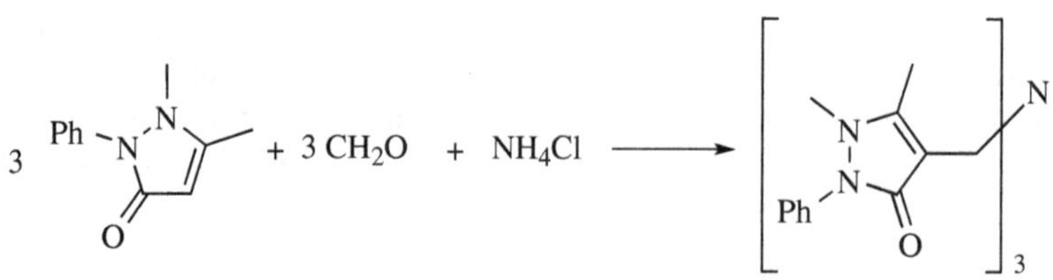 Synthesis of an antipyrine-derived tertiary amine

Table 1.1 *(continued)*

Mannich [20], 1917	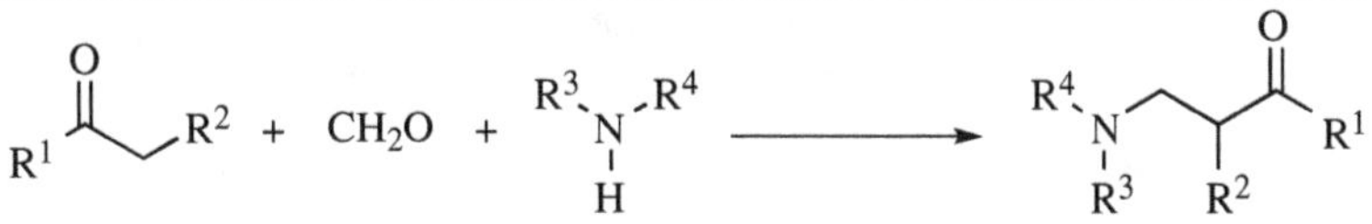 Synthesis of β-amino carbonyl compounds
Robinson [21], 1917	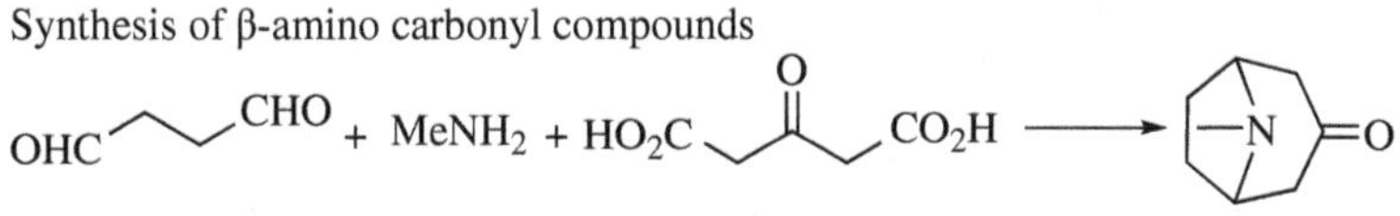 Synthesis of tropinone
Passerini [22], 1921	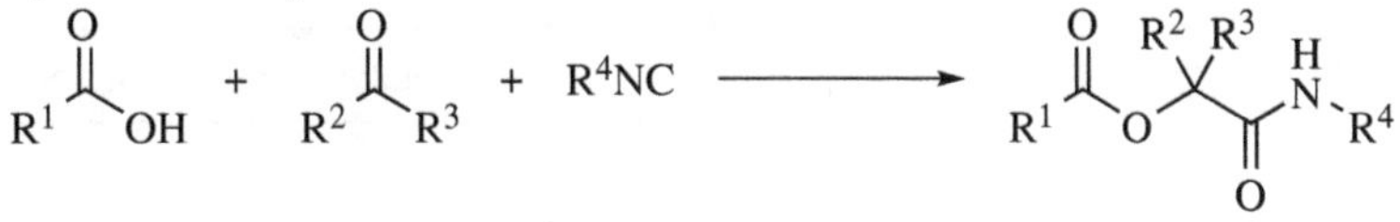 Synthesis of α-acyloxy amides
Bucherer–Bergs [23], 1929–1934	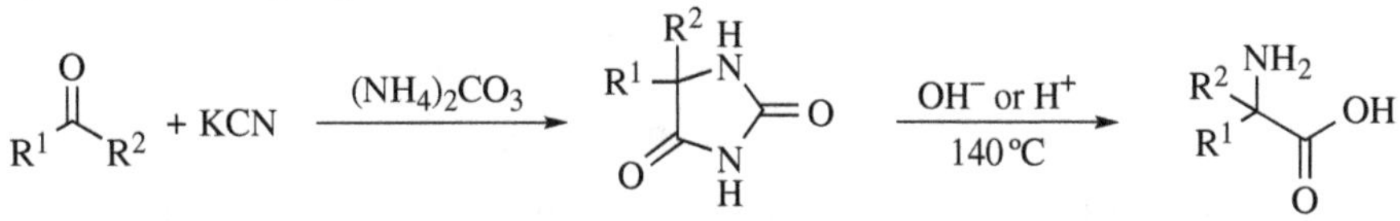 Synthesis of hydantoins and α-amino acids
Ugi [24], 1959	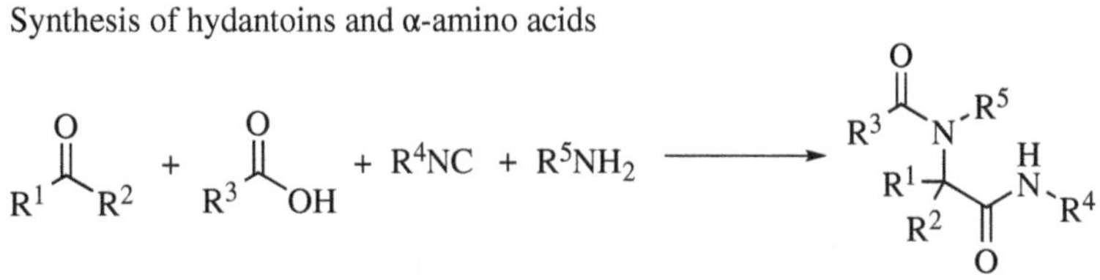 Synthesis of α-*N*-acylamino amides

1.1 BASIC CONCEPTS

Some of the basic concepts related to MCRs are briefly described in the following text in order to familiarize the reader with this field and its characteristics.

1.1.1 Clarifying Terminology: One-Pot, Domino/Cascade, Tandem, and MCRs

The previous terms are probably familiar for most chemists, but they have crucial differences that are important to know in order to distinguish each term from the others. The term one-pot reaction includes reactions that involve multiple chemical transformations between reagents that are carried out in a single reactor. Thus, MCRs fall into the category of one-pot reactions due to the sole reactor required for carrying out the reaction and that there are multiple chemical transformations involved.

Furthermore, Fogg and dos Santos categorized the different types of multicatalyzed one-pot reactions in 2004 [25], some years after Tietze set the definition of domino reactions [12]. In this categorization, domino/cascade catalysis, tandem catalysis, and multicatalytic one-pot reactions were distinguished depending on certain factors, such as the moment when the (pre)catalysts are added and the number of catalytic mechanisms involved (Fig. 1.2). Generally speaking, domino/cascade and tandem catalyses are one-pot reactions where all the components are introduced at the same time at the beginning of the reaction, while in multicatalytic one-pot reactions, all of the reaction's components are not added at the same time. Another requirement for domino/cascade and tandem catalyses is that all successive transformations must occur as a consequence of the intermediate generated in the previous reaction step. In Fogg's classification, domino/cascade and tandem catalyses are differentiated by the number of catalytic mechanisms present in the reaction.

With all the aforementioned concepts defined, it has been made clear that MCRs are one-pot reactions that might also fall under the category of domino/cascade or tandem reactions. A reaction is a domino/cascade or tandem MCR when it has the characteristics of one of these types of reactions in addition to including three or more reagents that react to form a final product.

1.1.2 Using Rational Design to Discover New MCRs

Designing new multicomponent approaches in a less haphazard and more rational manner is vital for increasing the limited scaffold diversity obtained by the MCRs reported

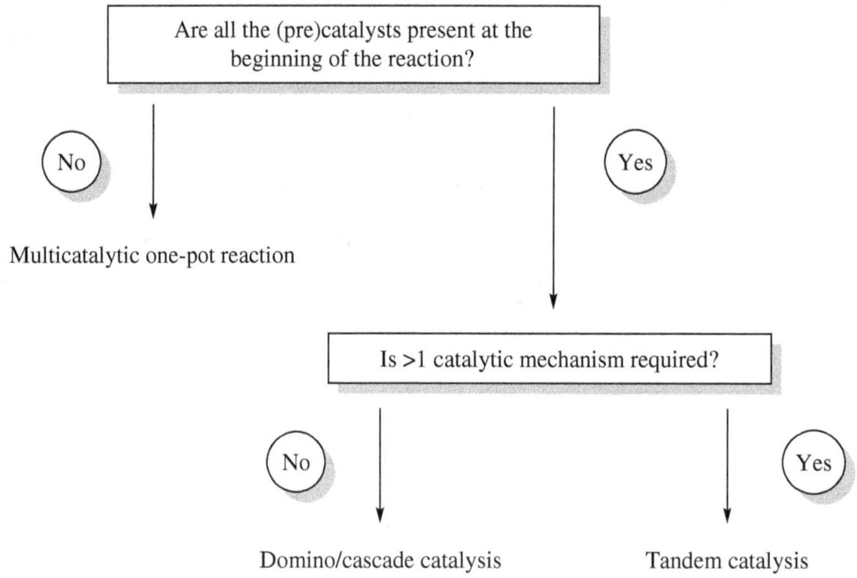

FIGURE 1.2 Fogg's simple classification of one-pot processes involving multiple catalytic transformations.

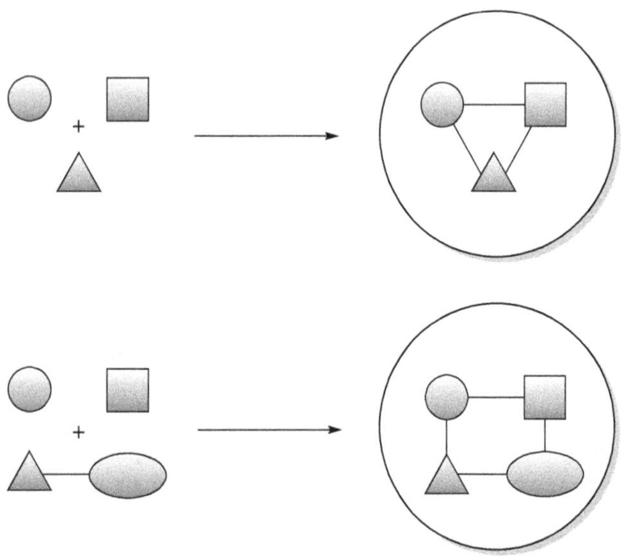

FIGURE 1.3 Single reactant replacement method for MCRs.

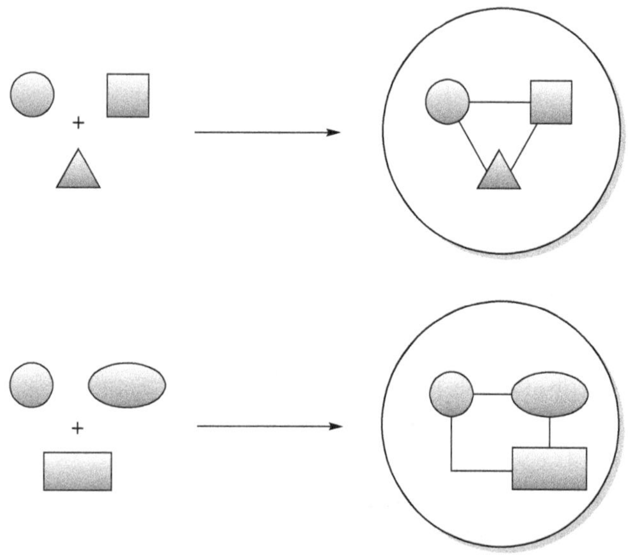

FIGURE 1.4 Example of a reaction-operator strategy carried out by changing two substrates.

until now. To do so, five different methods, most of them excellently explained by Orru and coworkers in their review [2b], have been developed to discover new MCRs: single reactant replacement (SRR), reaction-operator strategy, modular reaction sequences (MRS), condition-based divergence (CBD), and combination of MCRs (MCR^2).

1.1.2.1 SRR This strategy was first proposed by Ganem [26] and involves the replacement of one reactant with a different reactant that shows the same essential reactivity with other reagents, carrying out the same role in the reaction mechanism (Fig. 1.3). This approach has been demonstrated to be a valuable tool, providing different final adducts by incorporating additional functionalities in the reactants.

1.1.2.2 Reaction-Operator Strategy In this approach, defined by Mironov [27], there is a simultaneous replacement of two or more reagents with different reagents that show the same essential reactivity (Fig. 1.4). The name of this strategy comes from the comparison of chemical reactions with mathematical functions: in reactions, a reaction operator would be the equivalent to a function operator in mathematics. This reaction operator is introduced as an algorithm in a computer-controlled system, whose function is to find new reactions by using preexisting reactions as a starting point with the help of reaction preparation and analytical automated systems.

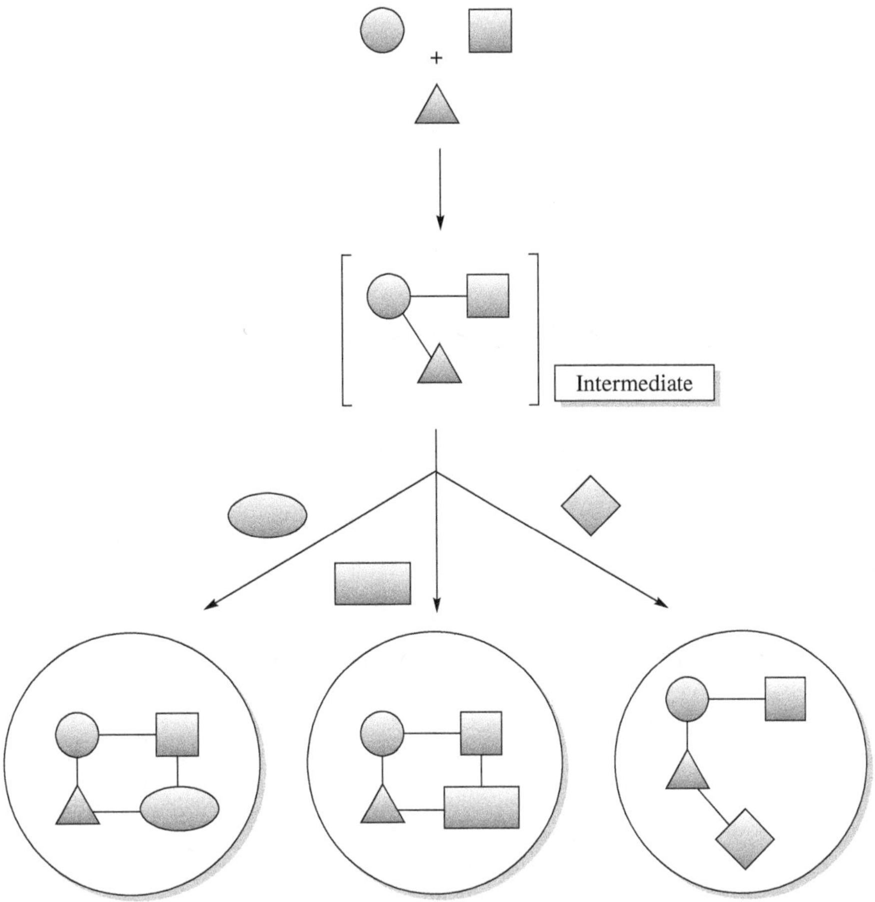

FIGURE 1.5 Modular reaction sequence approach in MCRs.

1.1.2.3 MRS This third approach involves a versatile reactive intermediate that is initially generated though a MCR from different substrates [28]. This compound is further treated *in situ* with a range of different compounds to produce a diverse set of more complex structures (Fig. 1.5). This divergent synthesis approach is very useful for rapidly generating scaffold diversity, creating large compound libraries.

1.1.2.4 CBD The use of specific catalysts, solvents, or additives could guide a reaction along different pathways, producing distinct final adducts (Fig. 1.6). There are some examples of MCRs that have different major products based on their reaction conditions [29]; however, it is uncommon to achieve a wide variety of adducts through this method. Many of these examples were discovered serendipitously, which is reflected in the limited number of reported examples. Although this approach is not frequently used, it is an efficient strategy for obtaining products with an attractive variety of scaffolds by simply varying reaction conditions.

1.1.2.5 MCR² The final strategy for the rational design of innovative MCRs is the combination of two or more different types of MCRs (Fig. 1.7) [30]. In this combinatorial chemistry approach, a certain compound with different functionalities that participates in a MCR reacts not only with the reagents of this MCR but also reacts with a molecule from a different MCR. When the results of each individual MCR are known, it might be possible to predict what products would be obtained from mixing both MCRs *via* a linker molecule. However, scientists should always take into consideration whether reactions that involve these linkers are reversible or not. In fact, this is a crucial parameter to be studied in order to make any MCR² work properly and to attain the best possible results.

1.1.3 Discovering New MCRs with Automated Combinatorial Reaction Finding

An appealing approach to discover innovative MCRs is to use the automated combinatorial reaction finding method, due to the large number of reactions that it studies simultaneously. This combinatorial chemistry approach combines mechanized analytical and reaction preparation systems in order to study the products from a large number of MCRs that are being carried out at the same time in separate well plates [31].

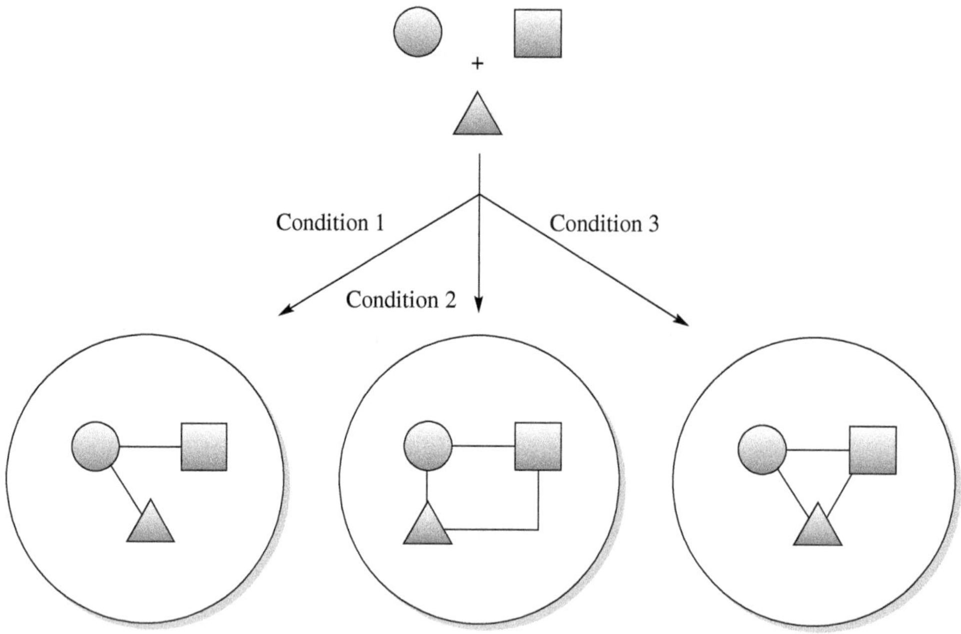

FIGURE 1.6 Divergence in MCRs achieved through changing reaction conditions.

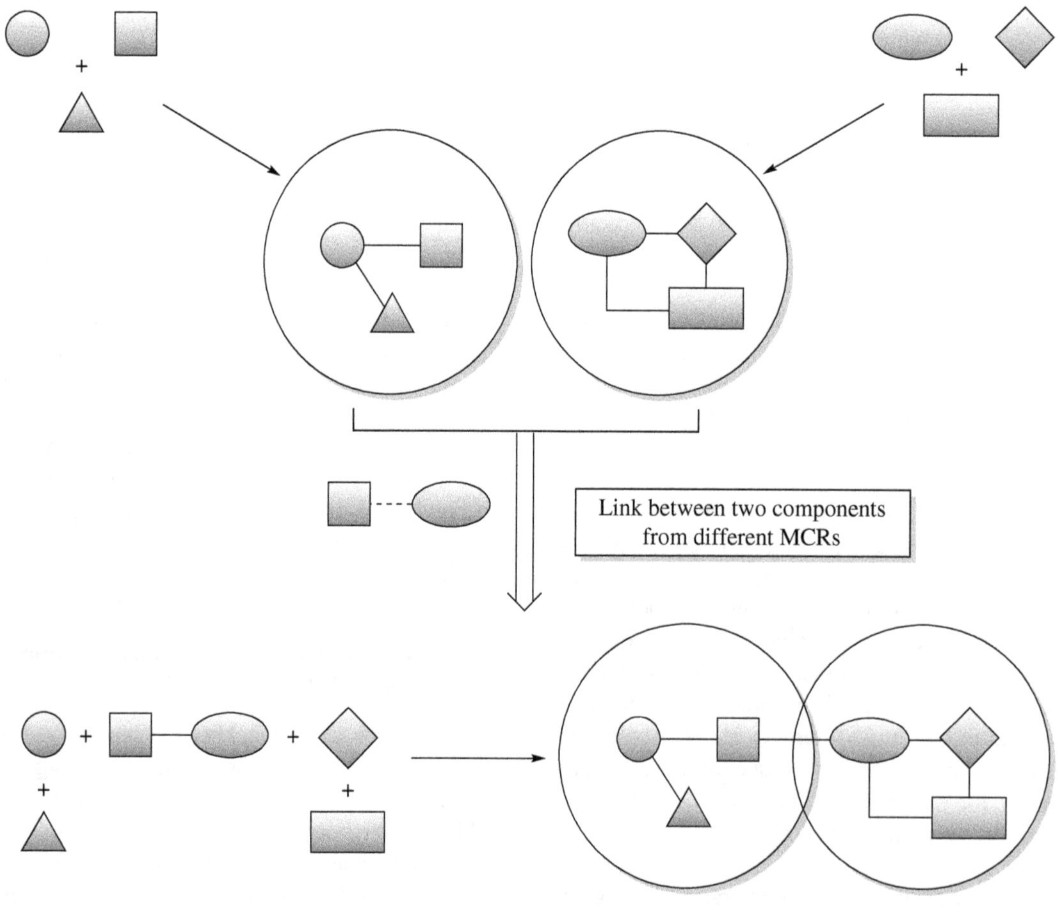

FIGURE 1.7 Combination of two MCRs.

In this approach, a robotic dispensing system prepares the reactions, then they are studied *via* automated high-performance liquid chromatography (HPLC) and/or mass spectroscopy (MS) systems, and finally, the data is collected and evaluated by a computer. This may save a considerable amount of time, since the exploration of such large number of reactions without using computer-controlled processes would require an enormous amount of time.

1.1.4 Computational and Analytical Tools to Study MCRs

Even though serendipity has always played a pivotal role in the discovery of new types of MCRs, the steady increase of publications in the last decade indicates an emergence of new ways to design MCRs in a more rational way. There are some tools that can help researchers to make decisions when performing MCRs. For example, there are computational programs that predict the main product when different starting reagents are put together in a reaction, enabling the creation of huge virtual compound libraries [31].

However, designing an enormous amount of products to synthesize is not very practical, since preparing and testing every product would require a considerable amount of time. Fortunately, there are computational programs that deal with this challenge. These programs sift through the virtual libraries that compile the resultant products from specific MCRs, selecting those products that may have biological activity. The ability to use these programs to target specific biological activities makes them especially appealing for drug discovery [32]. The results obtained after these filtering steps must always be verified through experimental work; however, doing these computational studies first can save time and effort.

Another strategy is using analytical techniques to study MCRs. There are diverse techniques that have appeared recently and are still being developed, such as coupling liquid chromatography, MS, and nuclear magnetic resonance (LC–MS–NMR) [31, 33]. These techniques allow having a better knowledge of the crude reaction mixture, making it possible in many cases to identify by-products, intermediates, and main products. This is helpful for interpreting the mechanisms that govern these MCRs, which helps researchers to predict whether or not certain changes in the reactants would result in variations in the final structures. Also, knowing different MCR mechanisms might help to predict whether a MCR[2] will have favorable results or not.

1.1.5 Diversity-Oriented Synthesis and Biology-Oriented Synthesis

Diversity-oriented synthesis (DOS) was defined by Schreiber in 2000 [34]. He referred to synthetic processes that contain only a few steps (~3–5) and are planned with a forward planning strategy rather than designed with a retrosynthetic analysis. The reason behind this is that products obtained using this approach are not aimed at one specific biological target, which makes a retrosynthetic analysis to create products that bind to specific sites useless. This approach proposes that in order to address biological targets that are innocuous to drugs currently being used, researchers should synthesize new designs that are not based in natural products, since natural products only interact with specific biological targets [35].

In 2006, Waldmann and coworkers introduced the concept of biology-oriented synthesis (BIOS) [36]. This approach's goal is to enable small organic molecules to bind to proteins by mimicking certain core structures found in natural products [37]. This is accomplished by adding the part of a natural product's structure that commonly binds to a certain protein to a small organic molecule, thus allowing it to bind to that protein.

Both of the aforementioned concepts are interesting because they could be useful for discovering new molecules with promising biological activity. However, in order to be successful, they require the synthesis of a great number of different compounds, since there are a huge number of variations among the synthesized structures that might drastically change their biological activity. MCRs have proven to be useful in DOS and BIOS due to the fact that libraries with a large amount of drug candidates can be created through easy approaches [38]. Synthesizing many different compounds through MCRs is especially appealing when the reactions are combined with high-throughput screening [5, 39]. This is an easy, fast, and effective way to generate a large group of diverse compounds which properties, biological or other, can be readily tested.

For this reason, MCRs are crucial tools along with DOS and BIOS. The development of these two promising areas will undoubtedly depend on the development of MCRs within these fields.

1.1.6 Optimization of MCRs

As previously stated, MCRs are reactions where more than two starting materials react in one single vessel. Since the collision of three or more independent molecules is highly unlikely, MCRs typically involve a number of basic reactions with individual mechanisms, each one normally requiring different reaction conditions.

In 1997, Ugi proposed a classification system for MCRs based on the reversibility of their individual reactions [40]. These reversibility parameters play an important role in the results obtained in MCRs, and they change depending on the reaction's conditions. Optimizing these conditions is a really challenging aspect because the best conditions for a specific individual reaction are not usually the best conditions for the developed MCR, since optimizing one of the chemical transformations generally results in changes in the efficiency of the other transformations. For this reason, a compromise must be found through the exploration and optimization of different reaction parameters, such as solvent, concentration, and temperature.

This optimization step may represent one of the most difficult tasks for developing new MCRs. However, recent developments in MCRs within other areas of chemistry, such as computational chemistry or analytical chemistry, have provided valuable tools for saving time and resources in the MCR optimization process. For example, certain computational programs, used in combination with the mechanized analytical and reaction preparation systems mentioned before, can expedite the optimization process of a MCR and drastically increase the yields obtained [31]. These computational programs use genetic algorithms, whose function is to change reaction parameters over successive sets of reactions depending on the results obtained in the previous set. This combination of genetic algorithms with automated systems has also proven to be very useful in drug discovery, since they can be used to optimize the values of a specific biological activity found in a MCR's products [41].

1.2 CATALYSIS IN MCRs AND VARIOUS SYNTHETIC APPROACHES

Many MCRs require a catalyst or a source of radiation to promote the formation of the desired products. The catalysts employed may have different functions, such as giving the desired stereocontrol [42], leading to the formation of a certain compound over the others, or simply ensuring that the reaction can be performed. In this section, different types of catalysis employed in MCRs will be discussed briefly, as well as different methods that improve the efficiency of these reactions.

1.2.1 Organocatalysis in MCRs

Chiral organocatalysis is a really attractive tool for performing effective reactions while avoiding the use of metals. Normally, organocatalysts contaminate less and are less toxic than organometallic catalysts because they do not include metals within their structures, which is beneficial for industries that try to avoid using metals, such as the pharmaceutical industry.

There are many different organocatalysts that have been used efficiently in asymmetric MCRs [43]. These organocatalysts have different ways of interacting with molecules, from forming covalent bonds with substrates, such as prolines, to interacting through hydrogen bonds, such as (thio) ureas. Many examples of how these types of organocatalysts are used in MCRs will be discussed in detail in Chapter 2.

1.2.2 Organometallic Catalysis in MCRs

Organometallic catalysis has many advantages over the other types of catalysis. Metal catalysts display good results in an extensive range of reactions, achieving an alluring combination of short reaction times and excellent results. This area of catalysis has been more developed than the others and, therefore, has a bigger number of examples within the field of MCRs [44]. There are many cases of metals that have been effectively employed in MCRs as catalysts, such as palladium, rhodium, and ruthenium, among many others. Numerous examples of interest will be disclosed in Chapter 3.

1.2.3 Biocatalysis in MCRs

Biocatalysis refers to the use of enzymes or whole cells as catalysts in chemical reactions. This type of catalysis does not have as many publications as the other types of catalysis, but there are noteworthy examples where biocatalysts have been used in MCRs with very good results [45]. Even though only a small number of examples have been discovered so far, their promising results make biocatalysis an area of great interest among the scientific community. Due to the high specificity of enzymes for certain substrates, a high stereocontrol can be achieved in these reactions. Also, it is possible to obtain a specific product out of all the possible final products, even in reactions in which none of the other types of catalysis obtain acceptable results (Scheme 1.1).

1.2.4 Combining Different Types of Catalysis

MCRs are not restricted to the use of only one type of catalysis. Combining different types of activation facilitates the synthesis of compounds that are difficult to obtain when

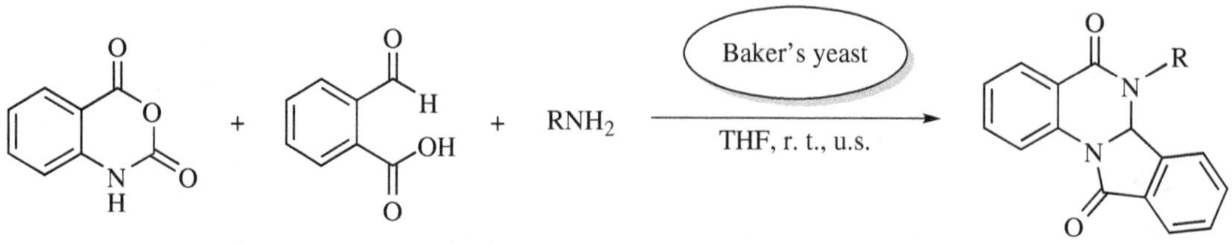

SCHEME 1.1 Biocatalyzed synthesis of isoindolo[2,1-*a*]quinazolines carried out by Raval and coworkers using baker's yeast as the catalyst [45a].

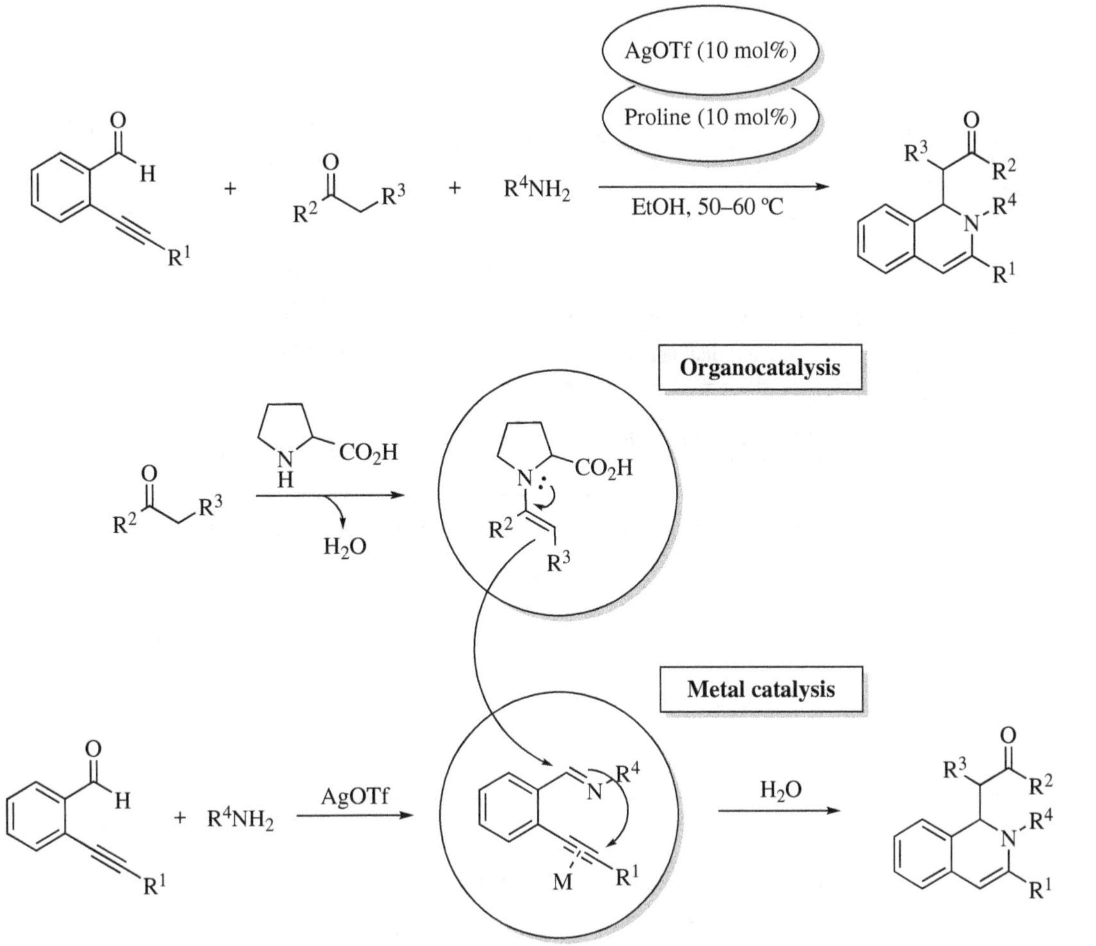

SCHEME 1.2 Cocatalyzed MCR using a metal catalyst and an organocatalyst along with the proposed mechanism [46].

using just one type of catalysis (Scheme 1.2) [46]. Furthermore, results of certain MCRs may be improved by shortening reaction times and by procuring higher yields and more stereocontrol, making this a promising approach that calls for further research.

1.2.5 Other Methods

There are a large number of MCRs that use alternative approaches to enhance their results; some remarkable examples will be covered in Chapter 11. One of the most common strategies is using microwave (MW) radiation in the reactions. Heating reactions with this source of radiation has enormous advantages over normal heating methods because, when using a polar solvent, the reaction media is heated up more efficiently, causing shorter reaction times [47]. For this reason, MW radiation can sometimes substitute a catalyst that would otherwise be necessary in certain MCRs to achieve good results in short reaction times (Scheme 1.3).

Another strategy employed to obtain better results is performing reactions with ultrasonication (Scheme 1.4). Ultrasound irradiation is especially useful in reactions where the starting reactants are nearly insoluble in the reaction media (multiphasic systems), when volatile gases are generated in the reaction, and in reactions that involve radical or ionic species [50].

Solid-phase synthesis is another useful method for the synthesis of products that is now being applied to MCRs [51]. Solid-phase synthesis is the synthesis of compounds through reactions in which the final products are attached to a solid support. Once the products are obtained, these compounds are separated from the support after the by-products have been removed from the reaction media through simple washing steps (Scheme 1.5). This feature can simplify the reaction's work-up process, requiring only filtration and further treatment to free the product from the support, avoiding the necessity of other more tedious purification processes while saving time and resources. This approach is very appealing for the easy and rapid creation of compound libraries when combined with automated reaction preparation systems [2a, 52].

Other methods that can be effectively employed in MCRs are infrared (IR) and photochemical irradiations. These approaches are useful in specific circumstances where reactants

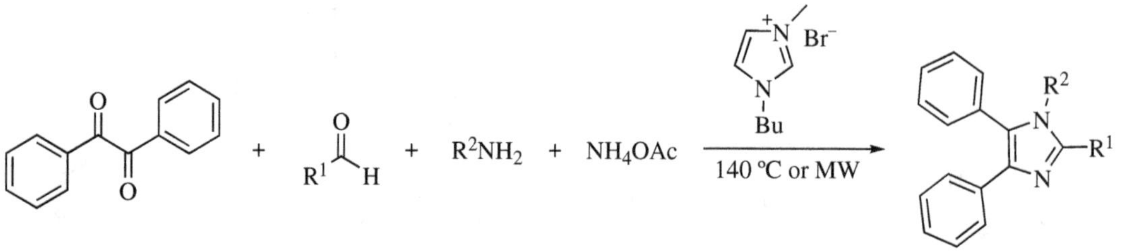

	Time	Yield
Conventional heating	1.5–5 h	82–93%
Microwave irradiation	3–8 min	82–93%

SCHEME 1.3 MCR using either conventional heating or microwave irradiation and the results observed by Hasaninejad and coworkers [48].

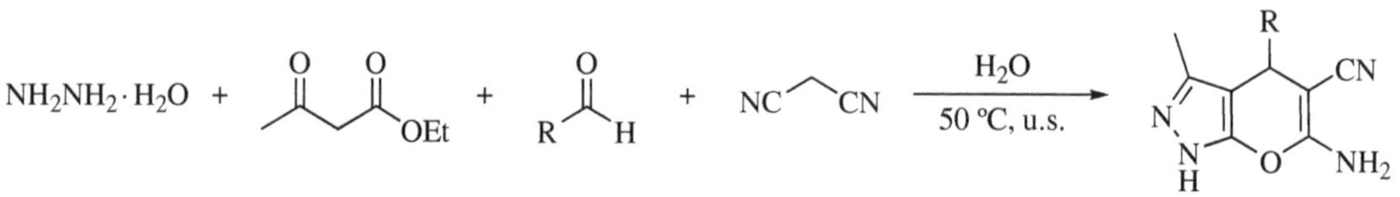

	Time	Yield
Normal stirring conditions	1–5 h	70–86%
Ultrasonication	15–40 min	79–95%

SCHEME 1.4 Synthesis of dihydropyrano[2,3-*c*]pyrazoles under ultrasound irradiation [49].

absorb those types of radiation, promoting their reaction with other reagents. There are only several examples of MCRs using these techniques [53], but further research may reveal new potential ways of synthesizing new complex molecules.

1.3 GREEN CHEMISTRY

MCRs are in agreement with the basic concepts of green chemistry set by Anastas and Warner in 1998 [54], providing final complex products in one step through novel synthetic strategies that are environmentally friendly. Some appealing advantages of using MCRs are the generation of a smaller amount of waste, the conservation of resources, and the reduction in the energy required. All these advantages have brought MCRs into the focus of researchers whose goal is developing pioneering green chemistry processes. Throughout this section, some techniques recently used in MCRs that adhere to the standards set by green chemistry will be discussed.

1.3.1 Atom Economy

Atom economy, a concept that was introduced by Trost in 1991, measures the efficiency of a reaction by comparing the amount of the targeted final product to the amount of other products generated [11]. This concept presents the need to design reactions where the majority of the reactants are incorporated into the desired product. This approach is preferred over others, such as breaking down a complex reactant to obtain a product, due to the fact that in the latter approach, even when reaction yields are 100%, the rest of the starting material normally is wasted.

MCRs are clear examples of successfully applied atom economy. In these reactions, different molecules are converted into a complex product in an efficient way. Thus, the development of MCRs also implies the development of atom-economical reactions.

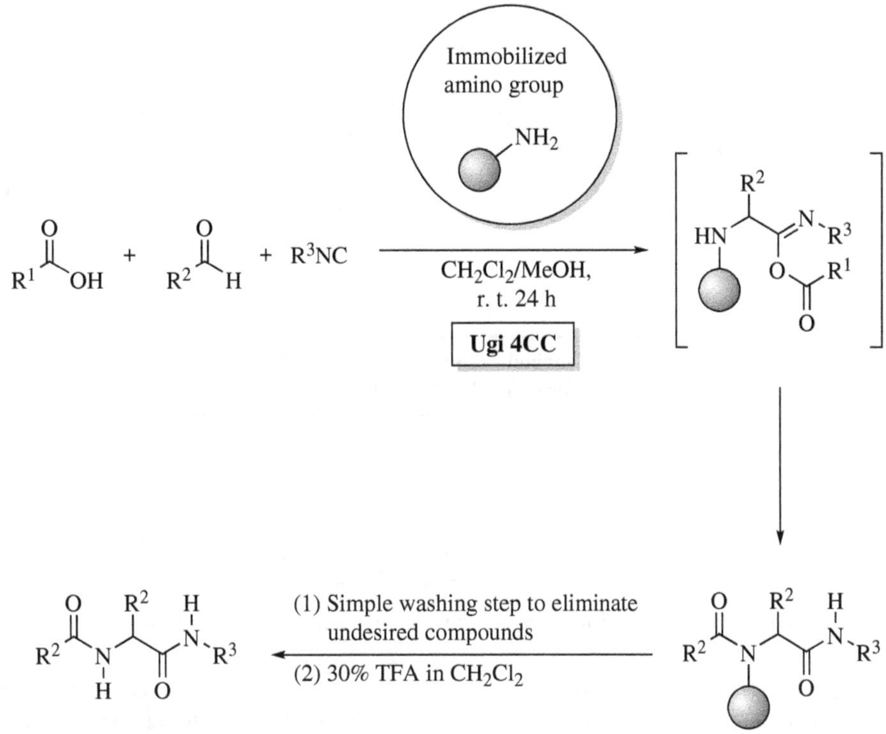

SCHEME 1.5 Syntheses of a large number of compounds having the same base structure through Ugi four-component condensations carried out by Armstrong and coworkers [51b].

1.3.2 Using Green Solvents

The use of green solvents in chemical reactions is a topic that has been concerning chemists throughout history [55]. There are many publications and ongoing research projects regarding this issue. In fact, researchers have always aimed to perform reactions with environmentally friendly solvents, especially if they are reusable and/or recyclable. Some examples of green solvents used in MCRs are water [56], ionic liquids (ILs) [57, 58], and bio-based solvents [56]. These solvents are not only environmentally friendly, but they also present fascinating properties that sometimes lead to unexpected results that would be difficult to obtain using other common solvents. Many innovative and interesting cases will be explored in Chapter 11 and throughout the rest of the book.

1.3.3 Solventless MCRs

Performing effective solventless reactions has been a challenging topic in many fields of chemistry, and there is a lot of ongoing research on these reactions. Implementing reactions in the absence of solvents is a textbook example of a green synthetic strategy in which the elimination process for waste generated by the solvent is no longer necessary. However, working without solvents can complicate the mixing process and make it more difficult for the substrates to react.

Many examples of MCRs have been performed without incorporating a solvent [59, 60]. The techniques that have had the best results in these reactions are mechanochemical procedures [61], MW irradiation [62], and IR irradiation [53].

1.3.4 Heterogeneous Catalysis in MCRs

Heterogeneous catalysis is really interesting not only because of multiphasic systems' intriguing characteristics but also because the catalysts are easy to recover and reuse. Reusable catalysts bring economic advantages to the table; they can reduce the costs of the reaction since they can be reutilized in subsequent reactions.

The most common strategy for synthesizing reusable catalysts is attaching the desired catalyst to a specific support, such as resin or silica, which makes them insoluble in many solvents. These supports do not degrade in solutions, allowing them to be separated easily by filtration at the end of the reaction and then reused in a subsequent reaction. A similar method is to use catalysts that are not soluble in the reaction media. Where, again, a simple filtration process is all that is required to separate the catalysts.

Also, another emerging approach used in MCRs that involves easily recoverable catalysts is the use of magnetic particles as catalysts or as supports for catalysts (Scheme 1.6). The enormous interest in this type of particle comes mainly from how effortless it is to recover these particles from the

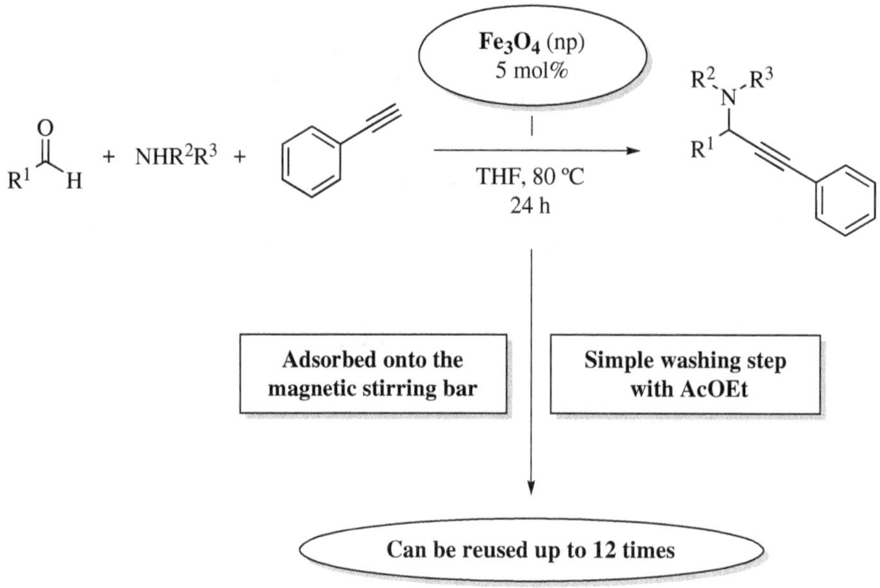

SCHEME 1.6 MCR using reusable magnetic particles as the catalyst [63a].

reaction media using magnets. There are many examples of MCRs that use magnetic particles. In some of these examples, the catalysts are compatible with water, and in others, they show promising results in solventless reactions [63].

1.4 IMPORTANCE AND EVOLUTION

MCRs represent a pivotal step in the development of modern chemistry. Unfortunately, many of the aforementioned reactions have only been extensively explored in their racemic version. With all the development that asymmetric catalysis has experienced over the years, new strategies have come to light, which will enable researchers to perform these previously racemic MCRs with stereocontrol.

Moreover, with the growing interest in using green chemistry to design cleaner organic reactions, mild, energy-efficient, and atom-economical processes will be the standards for chemistry in the near future. Also, other methods, already in use, will be improved in order to follow this trend in the pharmaceutical sector, aiming to synthesize drugs through more environmentally friendly reactions. MCRs have already been used for the synthesis of a handful of biologically active products with successful results (see Chapter 6). Further efforts are required to convert these procedures into competitive processes that can be implemented in the synthesis of a wide range of biologically active compounds.

The future of this area will most likely rely on the development of techniques that save time, energy, effort, and resources, as well as decrease the amount of waste generated in reactions. The development of the aforementioned computational and analytical techniques will play a vital role in future advances in MCRs. When using these techniques, it becomes easier to focus on the most valuable products of a certain MCR, shortening the time required for obtaining products with specific characteristics. Furthermore, these techniques expedite the study of products that result from a set of MCRs and the creation of novel reactions.

Also, there are already many examples of MCRs that employ methodologies from green chemistry, such as using reusable catalysts, recyclable solvents, and mechanochemical processes. The increase in the use of these techniques and green solvents in MCRs shows that these reactions will evolve along with green chemistry.

All the reasons mentioned previously make MCRs a valuable and competitive source for synthesizing biologically active products and other compounds with interesting uses in both minor scale and industrial research.

REFERENCES

1 J. Zhu, H. Bienaymé (Eds.) *Multicomponent reactions*. Wiley-VCH Verlag GmbH, Weinheim, Germany, 2005.

2 (a) H. Bienaymé, C. Hulme, G. Oddon, P. Schmitt, *Chem. Eur. J.* 2000, 6, 3321–3329. Maximizing synthetic efficiency: multi-component transformations lead the way. (b) E. Ruijter, R. Scheffelaar, R. V. A. Orru, *Angew. Chem. Int. Ed.* 2011, 50, 6234–6246. Multicomponent reaction design in the quest for molecular complexity and diversity.

3 S. M. Sheehan, J. J. Masters, M. R. Wiley, S. C. Young, J. W. Liebeschuetz, S. D. Jones, C. W. Murray, J. B. Franciskovich, D. B. Engel, W. W. Weber II, J. Marimuthu, J. A. Kyle, J. K. Smallwood, M. W. Farmen, G. F. Smith, *Bioorg. Med. Chem. Lett.* 2003, 13, 2255–2259. A four component coupling strategy for the synthesis of D-phenylglycinamide-derived non-covalent factor Xa inhibitors.

4 H. Cao, H. Liu, A. Dömling, *Chem. Eur. J.* 2010, 16, 12296–12298. Efficient multicomponent reaction synthesis of the schistosomiasis drug praziquantel.

5 H. G. F. Richter, G. M. Benson, D. Blum, E. Chaput, S. Feng, C. Gardes, U. Grether, P. Hartman, B. Kuhn, R. E. Martin, J.-M. Plancher, M. G. Rudolph, F. Schuler, S. Taylor, K. H. Bleicher, *Bioorg. Med. Chem. Lett.* 2011, 21, 191–194. Discovery of novel and orally active FXR agonists for the potential treatment of dyslipidemia & diabetes.

6 H. Ishikawa, T. Suzuki, Y. Hayashi, *Angew. Chem. Int. Ed.* 2009, 48, 1304–1307. High-yielding synthesis of the anti-influenza neuramidase inhibitor (–)-oseltamivir by three "one-pot" operations.

7 (a) L. Weber, *Curr. Med. Chem.* 2002, 9, 2085–2093. The application of multi-component reactions in drug discovery. (b) C. Hulme, V. Gore, *Curr. Med. Chem.* 2003, 10, 51–80. Multi-component reactions: emerging chemistry in drug discovery from xylocain to crixivan. (c) A. Dömling, *Chem. Rev.* 2006, 106, 17–89. Recent developments in isocyanide based multicomponent reactions in applied chemistry. (d) I. Akritopoulou-Zanze, *Curr. Opin. Chem. Biol.* 2008, 12, 324–331. Isocyanide-based multicomponent reactions in drug discovery. (e) A. Dömling, W. Wang, K. Wang, *Chem. Rev.* 2012, 112, 3083–3135. Chemistry and biology of multicomponent reactions.

8 For a special issue on green chemistry, see: P. Anastas, N. Eghbali, *Chem. Soc. Rev.* 2010, 39, 301–312. Green chemistry: principles and practice.

9 T. E. Nielsen, S. L. Schreiber, *Angew. Chem. Int. Ed.* 2008, 47, 48–56. Towards the optimal screening collection. A synthesis strategy.

10 Eco-compatible is defined as meaning both economically and ecologically compatible: Y. Coquerel, T. Boddaert, M. Presset, D. Mailhol, J. Rodriguez in *Ideas in chemistry and molecular sciences: advances in synthetic chemistry*. B. Pignataro (Ed.) Wiley-VCH Verlag GmbH, Weinheim, Germany, 2010, pp. 187–202. Multiple bond-forming transformations: the key concept toward eco-compatible synthetic organic chemistry.

11 B. M. Trost, *Science* 1991, 254, 1471–1477. The atom economy—a search for synthetic efficiency.

12 L. F. Tietze, *Chem. Rev.* 1996, 96, 115–136. Domino reactions in organic synthesis.

13 (a) A. Laurent, *Ann. Chim. Phys.* 1837, 66, 181–195. Sur diverses combinaisons azotées du benzoïle. (b) *Ann. Pharm.* 1838, 28, 265–269. Ueber einige Stickstoffverbindungen des Benzoyls.

14 A. Strecker, *Justus Liebigs Ann. Chem.* 1850, 75, 27–45. Ueber die künstliche Bildung der Milchsäure und einen neuen, dem Glycocoll homologen Körper.

15 (a) H. Debus, *Justus Liebigs Ann. Chem.* 1858, 107, 199–208. Ueber die Einwirkung des Ammoniaks auf Glyoxal. (b) Br. Radzisewski, *Ber. Dtsch. Chem. Ges.* 1882, 15, 2706–2708. Ueber Glyoxalin und seine Homologe.

16 A. Hantzsch, *Ber. Dtsch. Chem. Ges.* 1881, 14, 1637–1638. Condensationsprodukte aus Aldehydammoniak und ketonartigen Verbindungen.

17 A. Hantzsch, *Ber. Dtsch. Chem. Ges.* 1890, 23, 1474–1476. Neue Bildungsweise von Pyrrolderivaten.

18 (a) P. Biginelli, *Ber. Dtsch. Chem. Ges.* 1891, 24, 1317–1319. Ueber Aldehyduramide des Acetessigäthers. (b) P. Biginelli, *Ber. Dtsch. Chem. Ges.* 1891, 24, 2962–2967. Ueber Aldehyduramide des Acetessigäthers. II.

19 C. Mannich, W. Krösche, *Arch. Pharm.* 1912, 250, 647–667. Ueber ein Kondensationsprodukt aus Formaldehyd, Ammoniak und Antipyrin.

20 (a) C. Mannich, *J. Chem. Soc. Abstracts* 1917, 112, 634–635. Synthesis of β-ketonic bases. (b) C. Mannich, *Arch. Pharm.* 1917, 255, 261–276. Eine Synthese von β-Ketonbasen.

21 R. Robinson, *J. Chem. Soc. Trans.* 1917, 111, 762–768. A synthesis of tropinone.

22 M. Passerini, *Gazz. Chim. Ital.* 1921, 51, 126–129. Sopra gli isonitrili (I). Composto del p-isonitril-azobenzolo con acetone ed acido acetico.

23 (a) H. Bergs, Ger. Pat. 1929, DE566094. Hydantoin derivatives. (b) H. T. Bucherer, W. Steiner, *J. Prakt. Chem.* 1934, 140, 291–316. Syntheses of hydantoins. I. Reactions of α-hydroxy and α-amino nitriles. (c) H. T. Bucherer, V. A. Lieb, *J. Prakt. Chem.* 1934, 141, 5–43. Syntheses of hydantoins. II. Formation of substituted hydantoins from aldehydes and ketones.

24 (a) I. Ugi, R. Meyr, U. Fetzer, C. Steinbrückner, *Angew. Chem.* 1959, 71, 386–388. Versuche mit Isonitrilen. (b) I. Ugi, *Angew. Chem. Int. Ed.* 1962, 1, 8–21. The α-addition of immonium ions and anions to isonitriles accompanied by secondary reactions.

25 D. E. Fogg, E. N. dos Santos, *Coord. Chem. Rev.* 2004, 248, 2365–2379. Tandem catalysis: a taxonomy and illustrative review.

26 B. Ganem, *Acc. Chem. Res.* 2009, 42, 463–472. Strategies for innovation in multicomponent reaction design.

27 M. A. Mironov, *QSAR Comb. Sci.* 2006, 25, 423–431. Design of multi-component reactions: from libraries of compounds to libraries of reactions.

28 B. Groenendaal, E. Ruijter, R. V. A. Orru, *Chem. Commun.* 2008, 5474–5489. 1-Azadienes in cycloaddition and multi-component reactions towards *N*-heterocycles.

29 (a) N. Elders, E. Ruijter, F. J. J. de Kanter, M. B. Groen, R. V. A. Orru, *Chem. Eur. J.* 2008, 14, 4961–4973. Selective formation of 2-imidazolines and 2-substituted oxazoles by using a three-component reaction. (b) X.-S. Wang, J.-R. Wu, J. Zhou, M.-M. Zhang, *J. Heterocyclic Chem.* 2011, 48, 1056–1060. A green method for the synthesis of thiochromene derivatives in ionic liquids.

30 (a) A. Dömling, I. Ugi, *Angew. Chem. Int. Ed. Engl.* 1993, 32, 563–564. The seven-component reaction. (b) A. Dömling, *Curr. Opin. Chem. Biol.* 2000, 4, 318–323. The discovery of new isocyanide-based multi-component reactions.

31 L. Weber, K. Illgen, M. Almstetter, *Synlett* 1999, 366–374. Discovery of new multi component reactions with combinatorial methods.

32 L. Weber, *Drug Discov. Today* 2002, 7, 143–147. Multi-component reactions and evolutionary chemistry.

33 For a more extended view of such analytical techniques, see: (a) P. A. Keifer, *Curr. Opin. Chem. Biol.* 2003, 7, 388–394.

Flow NMR applications in combinatorial chemistry. (b) O. Corcoran, M. Spraul, *Drug Discov. Today* 2003, 8, 624–631. LC–NMR–MS in drug discovery.

34 S. L. Schreiber, *Science* 2000, 287, 1964–1969. Target-oriented and diversity-oriented organic synthesis in drug discovery.

35 S. L. Schreiber, *Nature* 2009, 457, 153–154. Organic chemistry: molecular diversity by design.

36 A. Nören-Müller, I. Reis-Corrêa Jr., H. Prinz, C. Rosenbaum, K. Saxena, H. J. Schwalbe, D. Vestweber, G. Cagna, S. Schunk, O. Schwarz, H. Schiewe, H. Waldmann, *Proc. Natl. Acad. Sci. U. S. A.* 2006, 103, 10606–10611. Discovery of protein phosphatase inhibitor classes by biology-oriented synthesis.

37 R. Breinbauer, I. R. Vetter, H. Waldmann, *Angew. Chem. Int. Ed.* 2002, 41, 2878–2890. From protein domains to drug candidates—natural products as guiding principles in the design and synthesis of compound libraries.

38 (a) A. Ulaczyk-Lesanko, D. G. Hall, *Curr. Opin. Chem. Biol.* 2005, 9, 266–276. Wanted: new multicomponent reactions for generating libraries of polycyclic natural products. (b) R. Ghahremanzadeh, M. Sayyafi, S. Ahadi, A. Bazgir, *J. Comb. Chem.* 2009, 11, 393–396. Novel one-pot, three-component synthesis of spiro[indoline-pyrazolo[4′,3′:5,6]pyrido[2,3-*d*] pyrimidine]trione library. (c) Y. Zou, Y. Hu, H. Liu, D. Shi, *ACS Comb. Sci.* 2012, 14, 38–43. Rapid and efficient ultrasound-assisted method for the combinatorial synthesis of spiro[indoline-3,4′-pyrano[2,3-c]pyrazole] derivatives.

39 (a) E. Brahmachary, F. H. Ling, F. Svec, J. M. J. Fréchet, *J. Comb. Chem.* 2003, 5, 441–450. Chiral recognition: design and preparation of chiral stationary phases using selectors derived from Ugi multicomponent condensation reactions and a combinatorial approach. (b) M. J. Chow, C. Licona, D. Y. Q. Wong, G. Pastorin, C. Gaiddon, W. H. Ang, *J. Med. Chem.* 2014, 57, 6043–6059. Discovery and investigation of anticancer ruthenium-arene Schiff-base complexes *via* water-promoted combinatorial three-component assembly.

40 (a) I. Ugi, *J. Prakt. Chem.* 1997, 339, 499–516. Multikomponentenreaktionen (MCR). I. Perspektiven von Multikomponentenreaktionen und deren Bibliotheken. (b) A. Dömling, I. Ugi, *Angew. Chem. Int. Ed.* 2000, 39, 3168–3210. Multicomponent reactions with isocyanides.

41 L. Weber, S. Wallbaum, C. Broger, K. Gubernator, *Angew. Chem. Int. Ed. Engl.* 1995, 34, 2280–2282. Optimization of the biological activity of combinatorial compound libraries by a genetic algorithm.

42 For selected general reviews of asymmetric MCRs, see: (a) D. J. Ramón, M. Yus, *Angew. Chem. Int. Ed.* 2005, 44, 1602–1634. Asymmetric multicomponent reactions (AMCRs): the new frontier. (b) C. de Graaff, E. Ruijter, R. V. A. Orru, *Chem. Soc. Rev.* 2012, 41, 3969–4009. Recent developments in asymmetric multicomponent reactions. (c) H. Pellissier, *Chem. Rev.* 2013, 113, 442–524. Stereocontrolled domino reactions.

43 For selected reviews on organocatalyzed MCRs, see: (a) G. Guillena, D. J. Ramón, M. Yus, *Tetrahedron: Asymmetry* 2007, 18, 693–700. Organocatalytic enantioselective multicomponent reactions (OEMCRs). (b) H. Pellissier, *Adv. Synth. Catal.* 2012, 354, 237–294. Recent developments in asymmetric organocatalytic domino reactions. (c) C. M. R. Volla, I. Atodiresei, M. Rueping, *Chem. Rev.* 2014, 114, 2390–2431. Catalytic C–C bond-forming multi-component cascade or domino reactions: pushing the boundaries of complexity in asymmetric organocatalysis.

44 For representative reviews on metal-catalyzed MCRs, see: (a) D. M. D'Souza, T. J. J. Müller, *Chem. Soc. Rev.* 2007, 36, 1095–1108. Multi-component syntheses of heterocycles by transition-metal catalysis. (b) J. A. Varela, C. Saá, *Synlett* 2008, 2571–2578. Recent advances in the synthesis of pyridines by transition-metal-catalyzed [2+2+2] cycloaddition. (c) H. Clavier, H. Pellissier, *Adv. Synth. Catal.* 2012, 354, 3347–3403. Recent developments in enantioselective metal-catalyzed domino reactions.

45 (a) J. R. Avalani, D. S. Patel, D. K. Raval, *J. Mol. Cat. B: Enzym.* 2013, 90, 70–75. Saccharomyces cerevisiae catalyzed one pot synthesis of isoindolo[2,1-*a*]quinazoline performed under ultrasonication. (b) P. P. Bora, M. Bihani, G. Bez, *J. Mol. Cat. B: Enzym.* 2013, 92, 24–33. Multicomponent synthesis of dihydropyrano[2,3-*c*]pyrazoles catalyzed by lipase from Aspergillus niger. (c) S. Kłossowski, B. Wiraszka, S. Berłożecki, R. Ostaszewski, *Org. Lett.* 2013, 15, 566–569. Model studies on the first enzyme-catalyzed Ugi reaction.

46 Q. Ding, J. Wu, *Org. Lett.* 2007, 9, 4959–4962. Lewis acid- and organocatalyst-cocatalyzed multicomponent reactions of 2-alkynylbenzaldehydes, amines, and ketones.

47 S. A. Galema, *Chem. Soc. Rev.* 1997, 26, 233–238. Microwave chemistry.

48 A. Hasaninejad, A. Zare, M. Shekouhy, J. A. Rad, *J. Comb. Chem.* 2010, 12, 844–849. Catalyst-free one-pot four component synthesis of polysubstituted imidazoles in neutral ionic liquid 1-butyl-3-methylimidazolium bromide.

49 Y. Zou, H. Wu, Y. Hu, H. Liu, X. Zhao, H. Ji, D. Shi, *Ultrason. Sonochem.* 2011, 18, 708–712. A novel and environment-friendly method for preparing dihydropyrano[2,3-*c*]pyrazoles in water under ultrasound irradiation.

50 For publications that discuss sonochemistry, see: (a) L. H. Thompson, L. K. Doraiswamy, *Ind. Eng. Chem. Res.* 1999, 38, 1215–1249. Sonochemistry: science and engineering. (b) T. J. Mason, *Ultrason. Sonochem.* 2003, 10, 175–179. Sonochemistry and sonoprocessing: the link, the trends and (probably) the future. (c) V. K. Patel, D. J. Sen, H. U. Patel, C. N. Patel, *J. Chem. Pharm. Res.* 2010, 2, 573–580. Sonochemistry: the effect of sonic waves on chemical systems.

51 (a) P. Wipf, A. Cunningham, *Tetrahedron Lett.* 1995, 36, 7819–7822. A solid phase protocol of the Biginelli dihydropyrimidine synthesis suitable for combinatorial chemistry. (b) P. A. Tempest, S. D. Brown, R. W. Armstrong, *Angew. Chem. Int. Ed. Engl.* 1996, 35, 640–642. Solid-phase, parallel syntheses by Ugi multicomponent condensation. (c) L. F. Tietze, M. E. Lieb, *Curr. Opin. Chem. Biol.* 1998, 2, 363–371. Domino reactions for library synthesis of small molecules in combinatorial chemistry.

52 C. O. Kappe, *QSAR Comb. Sci.* 2003, 22, 630–645. The generation of dihydropyrimidine libraries utilizing Biginelli multicomponent chemistry.

53 (a) M. O. Noguez, V. Marcelino, H. Rodríguez, O. Martín, J. O. Martínez, G. A. Arroyo, F. J. Pérez, M. Suárez, R. Miranda, *Int. J. Mol. Sci.* 2011, 12, 2641–2649. Infrared assisted production of 3,4-dihydro-2(1*H*)-pyridones in solvent-free conditions. (b) M. I. Flores-Conde, L. Reyes, R. Herrera, H. Rios, M. A. Vazquez, R. Miranda, J. Tamariz, F. Delgado, *Int. J. Mol. Sci.* 2012, 13, 2590–2617. Highly regio- and stereoselective Diels-Alder cycloadditions *via* two-step and multicomponent reactions promoted by infrared irradiation under solvent-free conditions.

54 P. T. Anastas, J. C. Warner (Eds.) *Green chemistry: theory and practice*. Oxford University Press, New York, 1998, p. 30.

55 For special issues on green solvents, see: (a) J. M. DeSimone, *Science* 2002, 297, 799–803. Practical approaches to green solvents. (b) R. A. Sheldon, *Green Chem.* 2005, 7, 267–278. Green solvents for sustainable organic synthesis: state of the art. (c) C. Capello, U. Fischer, K. Hungerbühler, *Green Chem.* 2007, 9, 927–934. What is a green solvent? A comprehensive framework for the environmental assessment of solvents.

56 Y. Gu, *Green Chem.* 2012, 14, 2091–2128. Multicomponent reactions in unconventional solvents: state of the art.

57 There is currently some controversy about how green ionic liquids really are. For several discussions on this topic, see: (a) M. J. Earle, K. R. Seddon, *Pure Appl. Chem.* 2000, 72, 1391–1398. Ionic liquids. Green solvents for the future. (b) R. P. Swatloski, J. D. Holbrey, R. D. Rogers, *Green Chem.* 2003, 5, 361–363. Ionic liquids are not always green: hydrolysis of 1-butyl-3-methylimidazolium hexafluorophosphate.

58 N. Isambert, M. del M. S. Duque, J.-C. Plaquevent, Y. Génisson, J. Rodriguez, T. Constantieux, *Chem. Soc. Rev.* 2011, 40, 1347–1357. Multicomponent reactions and ionic liquids: a perfect synergy for eco-compatible heterocyclic synthesis.

59 M. S. Singh, S. Chowdhury, *RSC Adv.* 2012, 2, 4547–4592. Recent developments in solvent-free multicomponent reactions: a perfect synergy for eco-compatible organic synthesis.

60 (a) A. Alizadeh, M. Babaki, N. Zohreh, *Tetrahedron* 2009, 65, 1704–1707. Solvent-free synthesis of penta-substituted pyrroles: one-pot reaction of amine, alkyl acetoacetate, and fumaryl chloride. (b) O. A. Attanasi, G. Favi, F. Mantellini, G. Moscatelli, S. Santeusanio, *J. Org. Chem.* 2011, 76, 2860–2866. Synthesis of functionalized pyrroles *via* catalyst- and solvent-free sequential three-component enamine-azoene annulation. (c) A. Kumar, S. Sharma, *Green Chem.* 2011, 13, 2017–2020. A grinding-induced catalyst- and solvent-free synthesis of highly functionalized 1,4-dihydropyridines *via* a domino multicomponent reaction.

61 (a) S. Mashkouri, M. R. Naimi-Jamal, *Molecules* 2009, 14, 474–479. Mechanochemical solvent-free and catalyst-free one-pot synthesis of pyrano[2,3-*d*]pyrimidine-2,4(1*H*,3*H*)-diones with quantitative yields. (b) M. Pascu, A. Ruggi, R. Scopelliti, K. Severin, *Chem. Commun.* 2013, 49, 45–47. Synthesis of borasiloxane-based macrocycles by multicomponent condensation reactions in solution or in a ball mill.

62 S.-L. Cui, X.-F. Lin, Y.-G. Wang, *J. Org. Chem.* 2005, 70, 2866–2869. Parallel synthesis of strongly fluorescent polysubstituted 2,6-dicyanoanilines *via* microwave-promoted multicomponent reaction.

63 (a) T. Zeng, W.-W. Chen, C. M. Cirtiu, A. Moores, G. Song, C.-J. Li, *Green Chem.* 2010, 12, 570–573. Fe_3O_4 nanoparticles: a robust and magnetically recoverable catalyst for three-component coupling of aldehyde, alkyne and amine. (b) Z.-H. Zhang, H.-Y. Lü, S.-H. Yang, J.-W. Gao, *J. Comb. Chem.* 2010, 12, 643–646. Synthesis of 2,3-dihydroquinazolin-4(1*H*)-ones by three-component coupling of isatoic anhydride, amines, and aldehydes catalyzed by magnetic Fe_3O_4 nanoparticles in water. (c) M. J. Aliaga, D. J. Ramón, M. Yus, *Org. Biomol. Chem.* 2010, 8, 43–46. Impregnated copper on magnetite: an efficient and green catalyst for the multicomponent preparation of propargylamines under solvent free conditions. (d) R. Mohammadi, M. Z. Kassaee, *J. Mol. Catal. A: Chem.* 2013, 380, 152–158. Sulfochitosan encapsulated nano-Fe_3O_4 as an efficient and reusable magnetic catalyst for green synthesis of 2-amino-4*H*-chromen-4-yl phosphonates.

2

ORGANOCATALYTIC ASYMMETRIC MULTICOMPONENT REACTIONS

Tecla Gasperi[1], Monica Orsini[2], Fabrizio Vetica[1] and Renata Marcia de Figueiredo[3]

[1] *Dipartimento di Scienze, Università degli Studi Roma Tre, Roma, Italy*

[2] *Dipartimento di Ingegneria, Università degli Studi Roma Tre, Roma, Italy*

[3] *Institut Charles Gerhardt Montpellier ICGM (UMR 5253-CNRS-UM2-UM1-ENSCM), Ecole Nationale Supérieure de Chimie, Montpellier Cedex 5, France*

2.1 INTRODUCTION

Asymmetric organocatalysis, the use of small organic molecules as catalysts [1], is becoming an increasingly effective alternative to the metal-based and bioorganic methods as clearly stated by the growing amount of publications [2] appeared since the first reported efforts [3]. Undeniably, many advantages are associated with the use of organocatalytic systems: (i) catalysts are usually quite cheaper than transition metal complexes, more stable, easier to be synthesized, and readily available; (ii) the reactions can be generally performed under low demanding conditions without the regular need for inert atmosphere or dry solvents; and, most important, (iii) different activation modes are possible with respect to substrates, reagents, and reactions (i.e. covalent and noncovalent interactions could be alternatively exploited). Within this context, one of the biggest attractiveness of organocatalysis relies on these organic molecules abilities of variously interacting with dramatically different substrates according to the accurate choice of the organocatalyst properties and chemical structure. Basically, these organic molecules could act either (i) by the reversible formation of iminiums/enamines (covalent activation), above all when aldehydes and ketones are employed with secondary or primary amines, or (ii) via hydrogen bond formation (noncovalent activation) when chiral catalysts bearing hydrogen bond donors/acceptors are used. Moreover, dual and cooperative catalysis, which allows the catalyst to bind to more than one substrate by using different activation modes, has paved the way to original reactions giving access to novel useful scaffolds [4].

Among the extensive realm of organocatalytic applications, the very last years have witnessed a golden rush to the development and implementation of multicomponent reactions (MCRs) that, coupled with the just disclosed features of organocatalysis, represent an incomparable breakthrough in total synthesis. Indeed, these reactions, which basically avoid time-consuming and costly processes, furnish a straightforward access to advanced key building blocks as well as complex final natural products in high yields and excellent stereoselectivities [5].

Within this context, this chapter will outline the application of organocatalysis to the improvement of asymmetric MCRs. Starting with the vast field of Mannich reaction (Section 2.2) and cycloaddition process (Section 2.3), a huge segment is devoted to the domino reaction (Section 2.4), wherein a remarkable amount of sequences involving a Michael addition plays a key role. Additionally, it will be highlighted how the organocatalytic MCRs find significant applications in the total synthesis of both natural products and their analogues having notorious pharmaceutical, clinical, and medicinal employment as well as biocidal activities (Section 2.5). Finally, the last section takes into account few reactions, which cannot be inserted in the previous sections, but merit the citation in this chapter (Section 2.6). Conversely, we are not evaluating the most famous MCRs that deserve a comprehensive and detailed description in the subsequent chapters.

Multicomponent Reactions: Concepts and Applications for Design and Synthesis, First Edition. Edited by Raquel P. Herrera and Eugenia Marqués-López.

2.2 THREE-COMPONENT MANNICH REACTION

Since its invention, the Mannich reaction has been one of the most important and studied strategies for C–C bond formation to achieve β-amino carbonyl compounds [6]. This aminoalkylation is accomplished by the interaction among an aldehyde, ammonia or an amine derivative, and a carbonyl compound. Particularly, the direct three-component Mannich reaction has been widely investigated as a source of alkaloid building blocks, essential for the preparation of natural products and biologically active compounds [7]. Therefore, the enhancement of such useful methodology in terms of diastereo- and enantioselectivity has been of substantial interest as well as a challenging target over the years. Within this context, the exceptional features owned by organocatalysis have been employed in the direct asymmetric multicomponent Mannich reaction, which relies on the ability of chiral secondary amines as powerful and distinguished organocatalysts in carbonyl-dependent reactions [8]. Particularly, (*S*)-proline (**1**) establishes a remarkable stereogenic control in an enamine-dependent pathway, as demonstrated by List in the first direct asymmetric organocatalytic three-component Mannich reaction (Scheme 2.1) [9]. Indeed, the condensation among acetone (**2a**, R^1 = H), *p*-anisidine (**3**), and several aldehydes **4** has brought to the expected product **5a** in high yield and impressive enantiomeric excess (up to 95% ee). After this pioneering work, List et al. have further improved the reaction scope employing variously decorated carbonyl compounds as well as different proline derivatives as catalysts [10]. They have exceptionally achieved the enantioenriched 1,2-amino alcohols **5b** using hydroxyacetone (**2b**, R^1 = OH) as synthon (up to >99% ee).

Moreover, possible side reactions such as aldol condensation or self-Mannich reactions have been never observed, and the mechanistic pathway, capable to explain the experimental evidences, has been broadly established, as depicted in Scheme 2.2 [11]. The (*S*)-proline (**1**) partakes in an enamine activation of the enolizable carbonyl compound **6** producing the enamine **7** and favoring the subsequent nucleophilic attack on the imine **8**, simultaneously generated by the interaction of the aldehydes **9** and the amine **3**.

In the proposed transition state **A**, the carboxylic moiety coordinates the nitrogen of imine **8** with a noncovalent interaction, directing the addition of the enamine double bond. The following hydrolysis of iminium **10** furnishes the wanted β-amino ketone **11** and regenerates the catalyst (*S*)-**1**.

Subsequently, a plethora of proline-catalyzed multicomponent Mannich reactions have been copiously and successfully developed as epitomized in Table 2.1.

All reactions usually proceed with extremely high enantio- and diastereoselectivity, forming as major product the *syn*-isomer. Moreover, quite often, the ee and dr values have been detected after reduction of the carbonyl function with $NaBH_4$ (Entries 1–3 and 6), which not only furnished more stable compounds but gave access to the corresponding γ-amino alcohols **12**, derivatives of great biological and synthetic interest. Enders et al., instead, have employed a stereoselective reduction with L-selectride in order to accomplish the stereosynthesis of amino sugars (Entry 5). Moreover, the temperature values are generally restricted to low degrees (down to −20 °C), even though a study, reported by Bolm's group in 2006, showed that analogous outcomes in terms of yield and enantiocontrol could be easily obtained by microwave-assisted organocatalytic Mannich reaction with proline (1) at high temperature [18]. The German group highlighted how the reaction is thermally accelerated reducing both time and catalyst loading (Entry 7). Meanwhile, Hayashi's group has employed high pressure as accelerating agent in the direct Mannich reaction. Unusually, they developed an efficient new method for the generation of high pressure by freezing water in a sealed autoclave, a protocol that effectively accelerated the reaction without any decrease in the enantioselectivity (Entry 8) [19]. Zhong and coworkers have observed

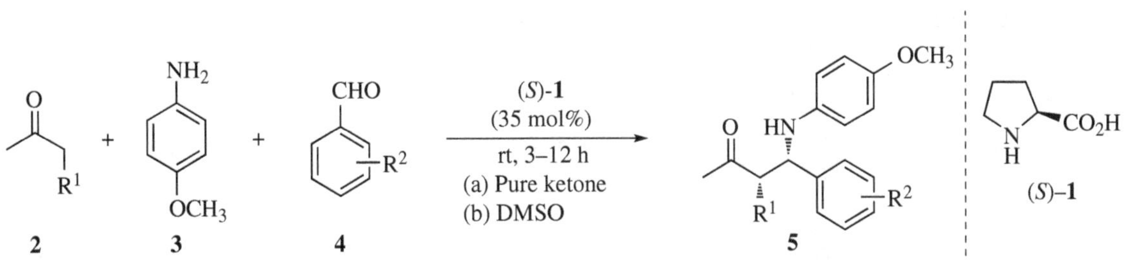

5a R^1 = H	5b R^1 = OH
6 examples	8 examples
35–90% yield	52–92% yield
70–95% ee	61–>99% ee
	dr 3:1–20:1

SCHEME 2.1 First three-component Mannich reaction by List et al.

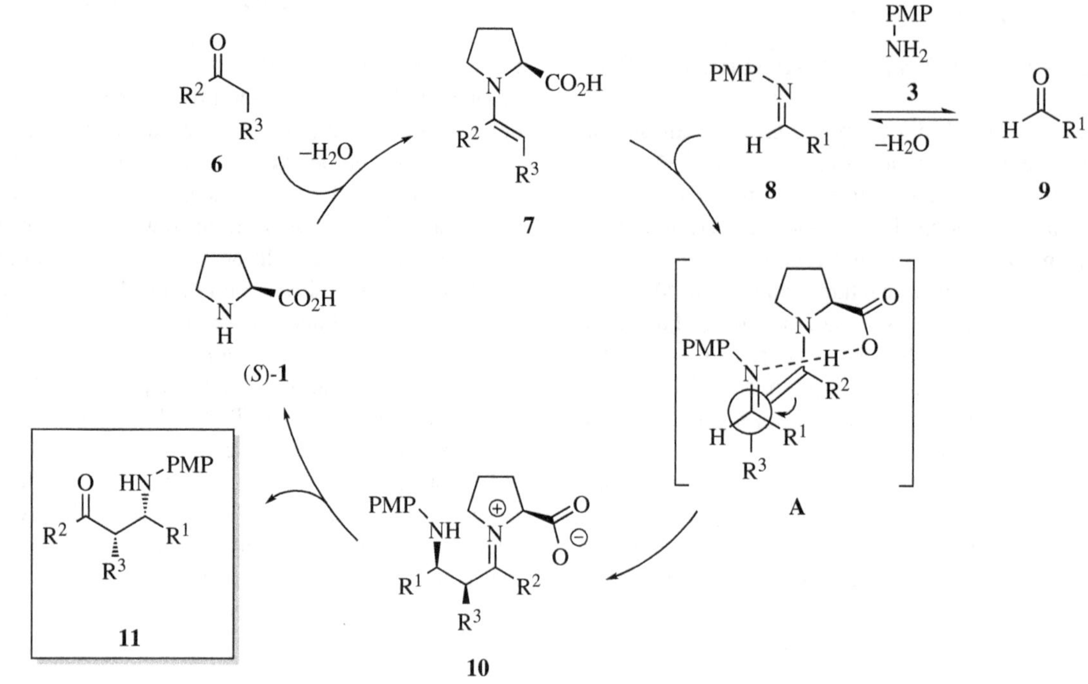

SCHEME 2.2 Plausible mechanism for the proline-catalyzed Mannich reaction.

TABLE 2.1 Several proline-catalyzed multicomponent Mannich reactions

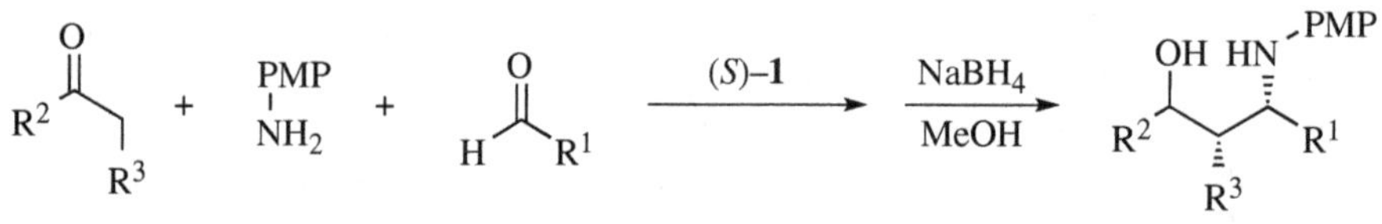

6 3 9 12

Entry[a]	R^1	R^2	R^3	Condition	Yield (%)	dr	ee (%)	Ref
1	*p*-NO_2-C_6H_4	H	Me	(*S*)-**1** (35 mol%) DMF, 4 °C	78	>10:1	99	[12]
2	*p*-C_5H_4N	H	Me	(*S*)-**1** (10 mol%) NMP, –20 °C	84	95:5	>99	[13]
3	*o*-C_5H_4N	H	Me	(*S*)-**1** (20 mol%) DMF, 0 °C	85		>99	[14]
4[b]	*p*-NO_2-C_6H_4	H	Me	(*S*)-**1** (20 mol%) DMSO, rt	66	19:1	98	[15]
5[c]	$CH(OMe)_2$	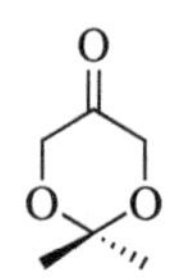		(*S*)-**1** (30 mol%) DMF, H_2O (4 eq) 2 °C	91	199:1	98	[16]
6	H	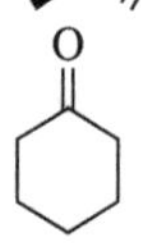		(*S*)-**1** (10 mol%) DMSO, rt	94		>99	[17]
7	H	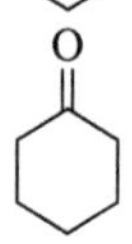		(*S*)-**1** (10 mol%) DMSO, 45 °C Microwave	81		95	[18]
8	2-Furanyl	Me	Me	(*S*)-**1** (30 mol%) DMSO, 200 MPa –20 °C	95		>99	[19]

[a] Only the best result of each work has been reported.
[b] The authors did not proceed to the reduction, obtaining only the Mannich product.
[c] Stereoselective reduction with L-selectride.

interesting effects, when an inverse electron demand ketone was employed in presence of both (*S*)-proline (**1**) and 4-siloxyproline **14** (Scheme 2.3). Indeed, using fluoroacetone **13**, they have observed as main product the isomer **15**, consequence of a different enamine attack **A** compared with the previously cited studies.

The presence of electron-withdrawing fluorine should direct the generation of the less-substituted enamine, which leads to the linear product **15** [20].

Besides the widely investigated proline catalysis, other pyrrolidine derivatives have been considered in multicomponent Mannich reaction, as depicted in Scheme 2.4. In these works, the effects of either a carboxylic **17** or ester **18** moiety in position 5 on the pyrrolidine ring have been examined, achieving the desired class of products **19** with comparable enantiomeric and diastereomeric excesses (up to 98% ee, dr up to 94:6) [21].

Alternatively, Garter's group has worked on the synthesis of (*S*)-*N*-(*p*-dodecylphenylsulfonyl)-2-pyrrolidinecarboxamide ((*S*)-**20**) and its use as organocatalyst for the direct Mannich reaction (Scheme 2.5) [22], obtaining the β-amino carbonyl compounds **21** and **22** as well as the γ-amino alcohol **23**, after a reduction with $NaBH_4$.

Likewise, Hayashi et al. have successfully developed an efficient strategy using aqueous medium, which starting from dimethoxyacetaldehyde (**24**), PMP-protected amine **3**, and various ketones **6** and employing their designed siloxy-tetrazole hybrid proline catalyst **25** furnished the titled β-amino ketones **26** in interesting yields (53–93% yield) and good to high optical purity (83–97% ee) (Scheme 2.6) [23].

During their studies on Diels–Alder cycloaddition, Kotsuki's group has developed the synthesis of a new class of anthracene-fused proline catalysts. Among them, the amino acid (*S*)-**27** resulted to be the most efficient when the

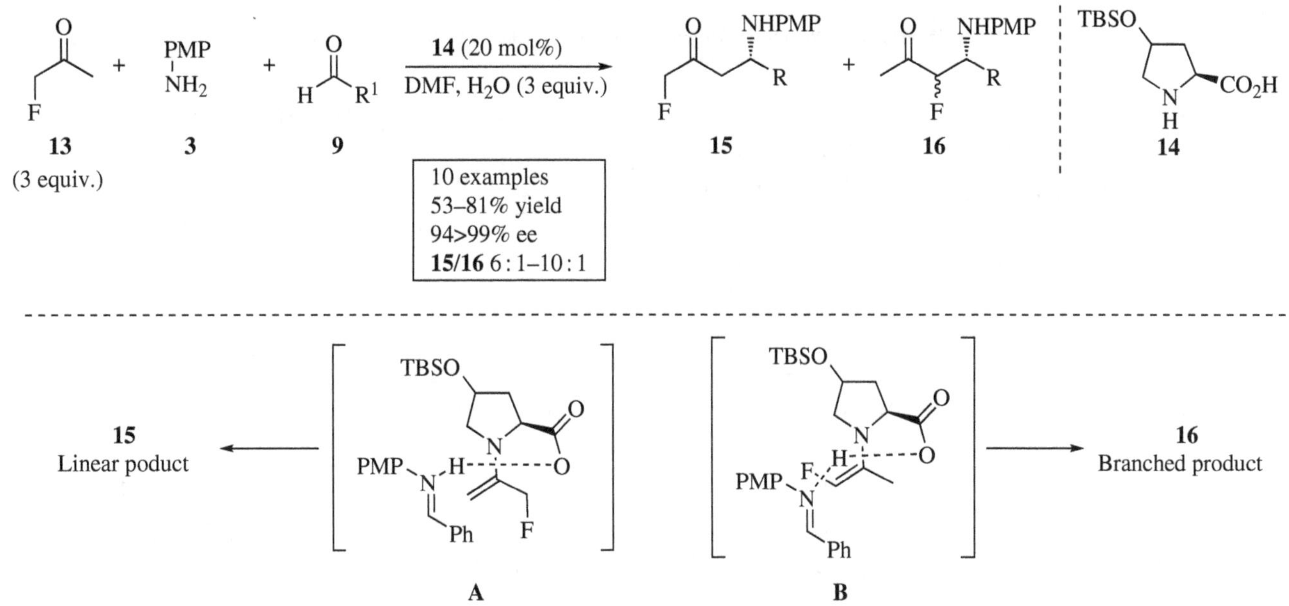

SCHEME 2.3 Reversal regioselectivity using fluoroacetone.

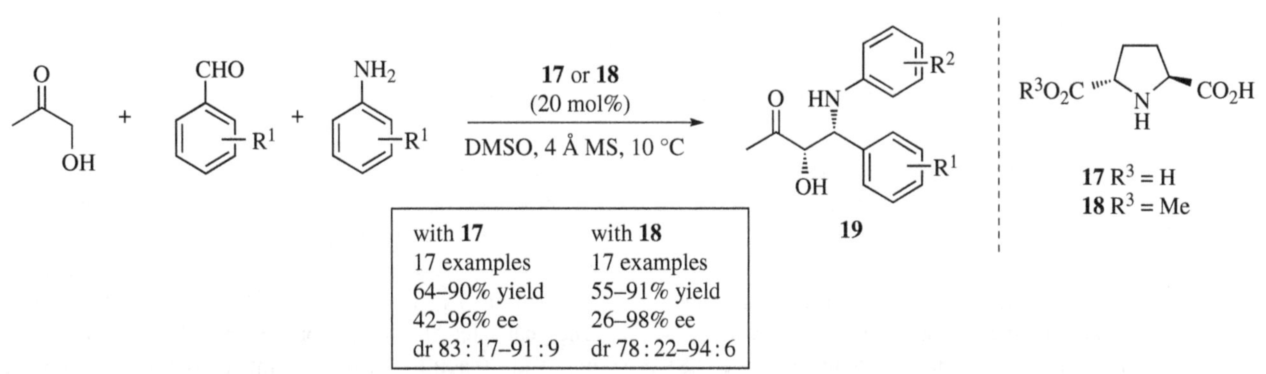

SCHEME 2.4 Selected examples of proline derivative catalysis in Mannich reaction.

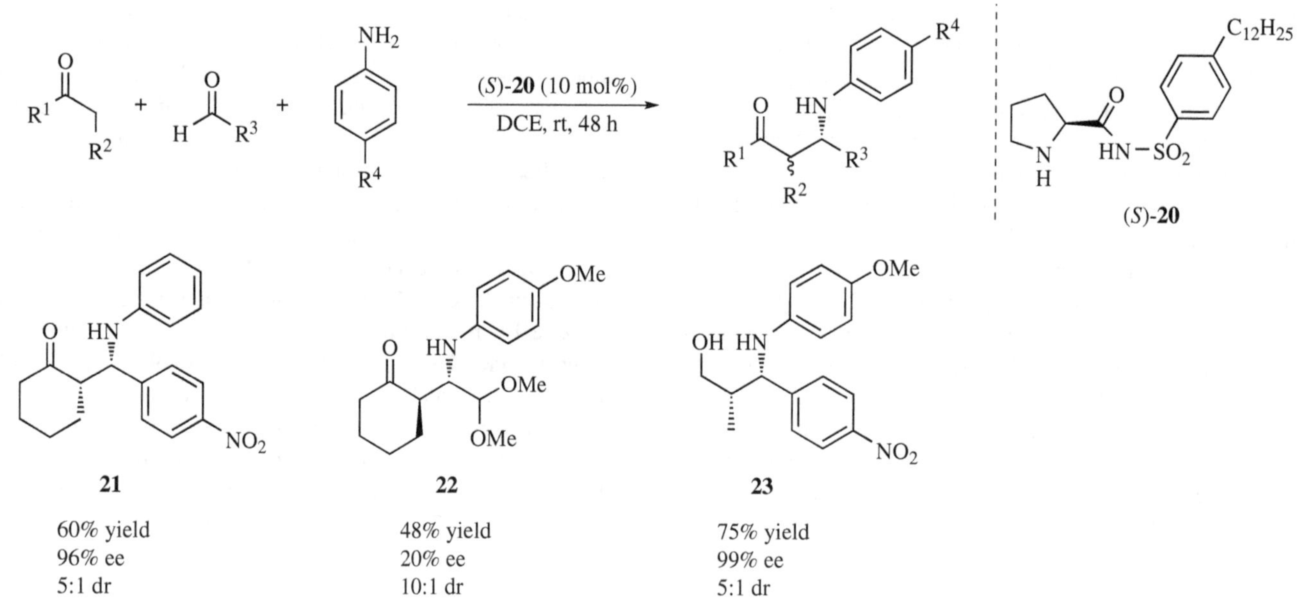

SCHEME 2.5 *N-p*-dodecylphenylsulfonamide-based proline catalysis.

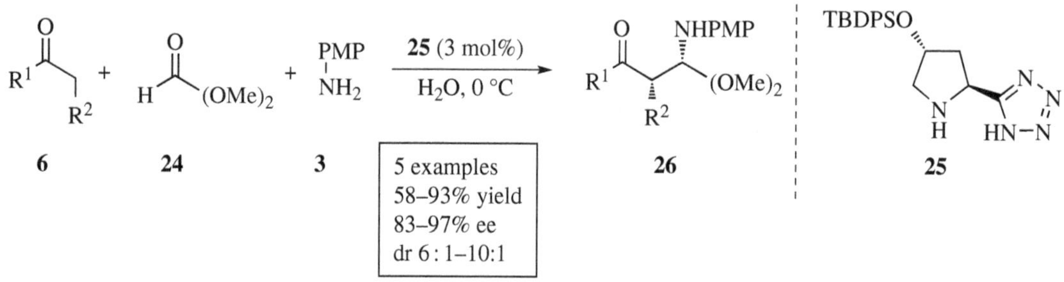

SCHEME 2.6 Asymmetric Mannich reaction under wet conditions.

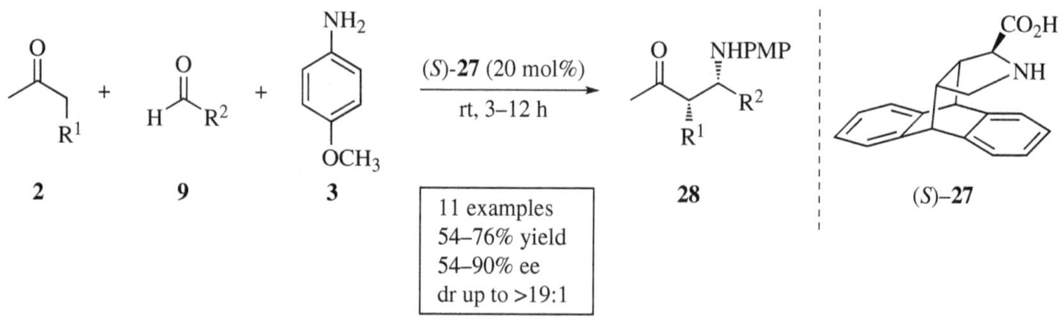

SCHEME 2.7 Anthracene-fused proline catalysis.

three-component Mannich reaction was performed in pure ketone **2**, achieving the amino carbonyl compounds **28** in 54–76% yields and up to 90% of ee (Scheme 2.7) [24].

On the other hand, in 2011, An and coworkers settled a series of isosteviol–amino acids, in which the steric hindrance and the hydrophobic interactions allow to fix the configuration of the common enamine intermediate (Scheme 2.8). Indeed, even though various amphiphilic catalysts were screened, only the amino acid–isosteviol derivatives demonstrated powerful features, displaying higher activities than the long-chain substituted proline analogues. Initially, the best results were achieved with threonine–isosteviol derivative **29** (10mol%) in DMF (Scheme 2.8a, up to >99% ee) [25]. One year later, the same group improved such results studying the reaction with proline–isosteviol catalyst **30**. Comparing the observed outcomes, An stated that, in presence

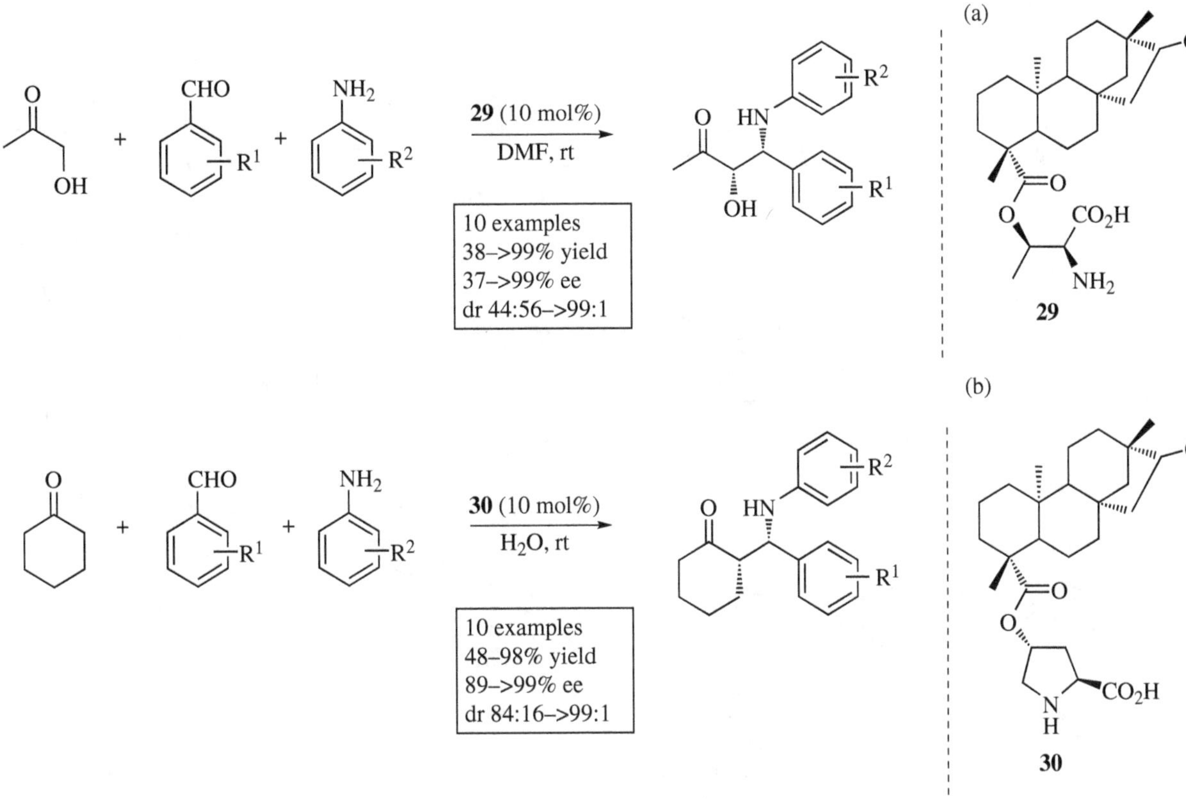

SCHEME 2.8 Isosteviol–amino acids organocatalysis by Tao's group.

of water, catalyst **30** was more suitable than **29** for asymmetric three-component Mannich reaction due to the rational matching of proline and isosteviol cavity. Additionally, the presence of water appeared necessary to reach high level of diastereo- and enantioselectivities (Scheme 2.8b, up to >99% ee) [26].

In this area, considerable efforts have been devoted to the synthesis of other amino acid derivatives with particular attention to threonine and serine scaffolds as consequence of their outstanding activities as organocatalysts. For instance, Barbas III et al. have explored the possibility to generate several *anti*-1,2-amino alcohols via direct asymmetric *anti*-Mannich procedure, which offered a valid alternative to the *syn*-approach previously implemented. In order to accomplish this aim, the American group designed a library of primary chiral amines and amino acids, including threonine derivative **32** and (*S*)-tryptophan **33** (Scheme 2.9a), which, starting from α-hydroxyacetone **31**, exhibited the preferential formation of *anti*-1,2-amino alcohols **34** [27]. The reason of this inverse diastereoselectivity probably relies on the favored formation of the (*Z*)-enamine in the C–C bond-forming transition state **A**, involving a stabilizing intramolecular H bond (Scheme 2.9); the subsequent nucleophilic attack leads to the *anti*-products **34** with elevated stereoselectivity (dr up to 15:1, up to 98% ee). Subsequently, Fu et al. enhanced the one-pot synthesis of novel threonine derivatives and engaged them in a direct three-component reaction involving various aromatic aldehydes **4** (Scheme 2.9b) [28]. The reaction, performed in NMP at 0 °C, afforded the expected *anti*-product **36** in very high yields (75–95%) and amazing enantioselectivity (up to >99% ee) when only 10 mol% of catalyst **35** was employed.

Additionally, the Chinese group realized a 100 mmol-scale example of this asymmetric synthesis, starting from *p*-NO_2-benzaldehyde achieving comparable ee (97% ee) and yield (90%), despite a decrease in the diastereomeric ratio (dr 11:1) [29].

Moreover, Teo's group worked on the development and application of siloxy serine organocatalysts [30]. Particularly, when (*S*)-**37** was used, the reaction could be performed in water with a consequent low environmental impact (Scheme 2.10a), and the *anti*-Mannich products **38** were isolated in interesting ee (up to 90%). Pursuing this research, in 2011, the same group has highlighted the remarkable properties of ionic liquids **39** as solvents (Scheme 2.10b, up to 99% ee) [31]. The developed strategy ensured the recyclability of the catalyst (*S*)-**37** in Mannich reaction: after three cycles, they obtained an identical stereoselectivity and reactivity as well as the ionic liquid preserved its biphasic properties, whereas the next three cycles displayed a decrease in ee and yield.

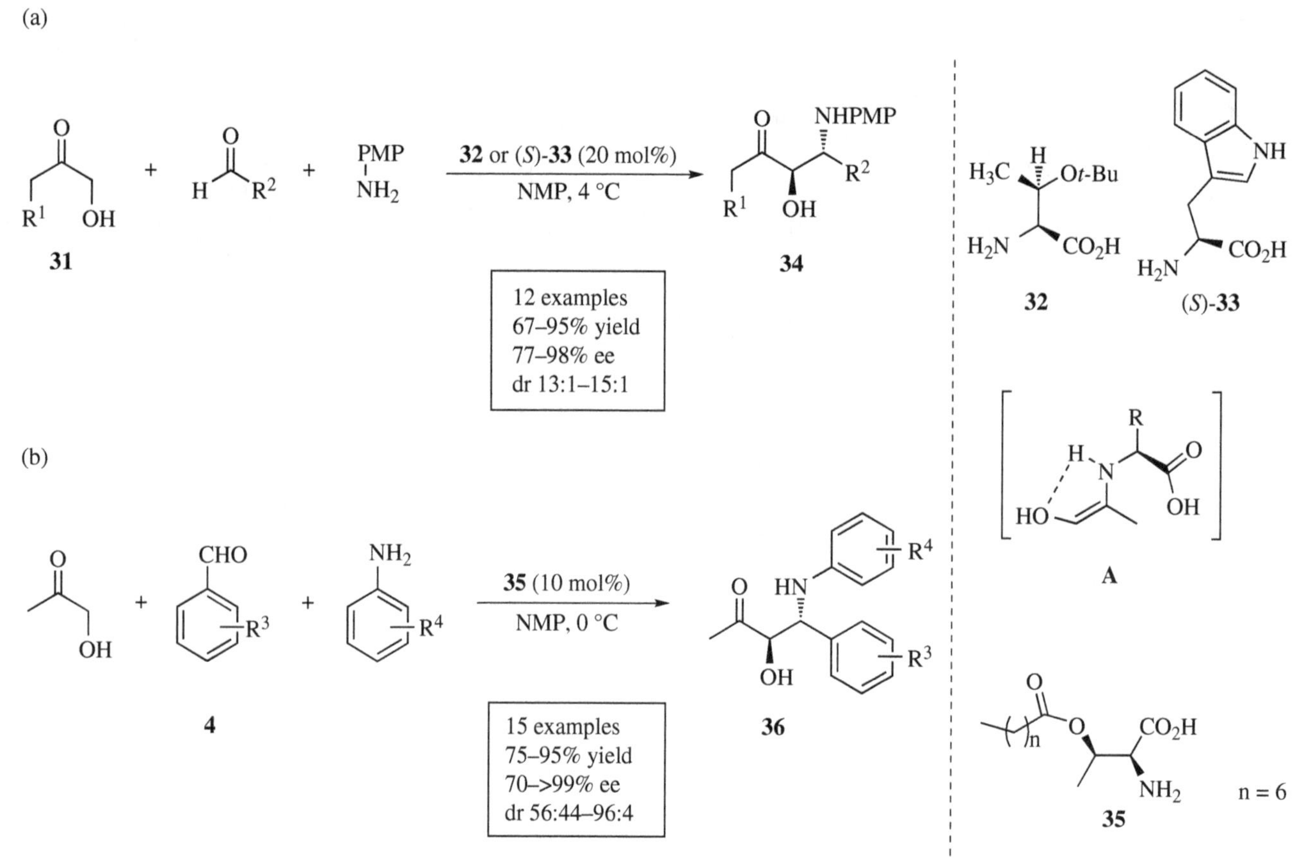

SCHEME 2.9 *Anti*-Mannich product by threonine derivative catalysis.

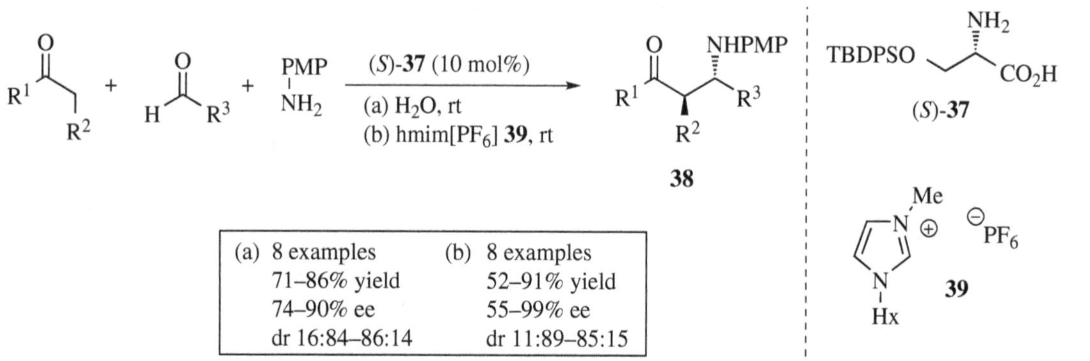

SCHEME 2.10 Siloxy serine catalysis in water and ionic liquid.

Comparably, Barbas III et al. observed interesting results employing the quite similar ionic liquid **40** as solvent in a proline-catalyzed Mannich reaction (Scheme 2.11a, up to 95% ee) [32], whereas, Wang's group examined the properties of more complex ionic liquids as enamine activation catalysts (Scheme 2.11b). The best outcome in terms of yield (up to 99%) and stereoselectivities (91–99% ee) was observed performing the reaction in DMF at −20 °C and using 30 mol% of [EMIm][Pro] **41** [33].

In addition to the classical amino acid derivative catalysis, a wide group of miscellaneous small organic molecules have been involved in a direct asymmetric Mannich reaction. For instance, Córdova et al. synthetized a large group of acyclic chiral primary amines [34]. In particular, chiral amines (*S*)-**43** and (*S*)-**44** were adopted to catalyze the reaction among a series of cyclic and acyclic ketones **42**, *p*-anisidine (**3**), and α-glyoxylate or aromatic aldehydes **9**, affording to the *syn*-γ-amino ketone **45** (Scheme 2.12, up to >99% ee). The authors further highlighted how the designed catalysts furnished particular high level of enantiocontrol when cyclic ketones were used.

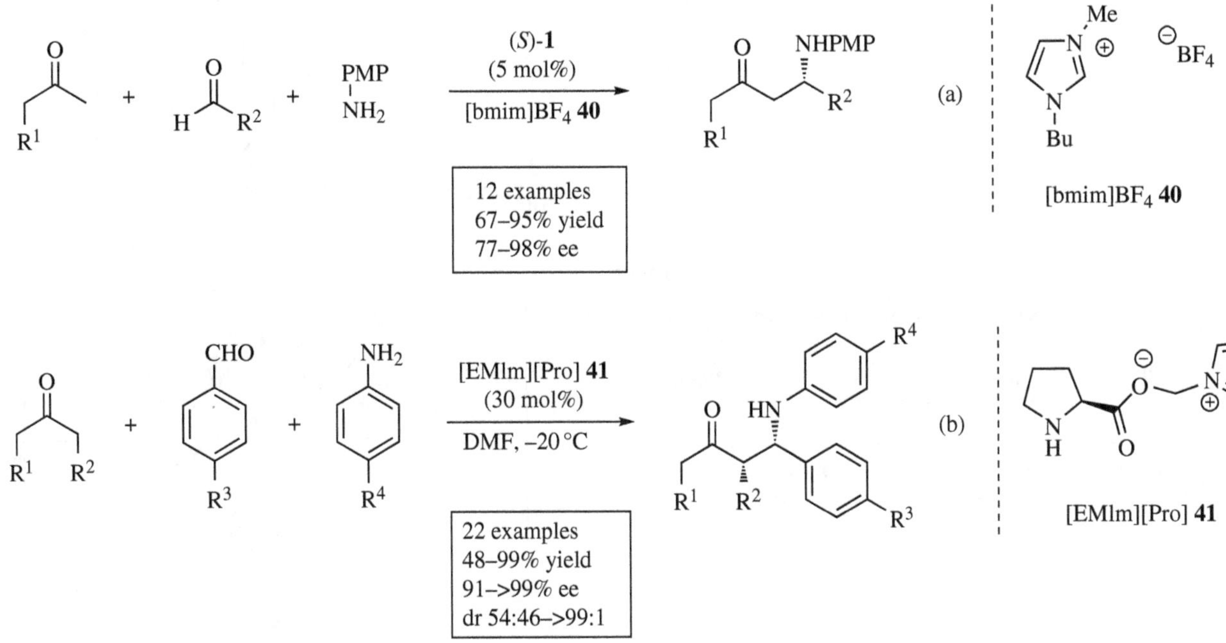

SCHEME 2.11 (a) (*S*)-Proline (**1**) catalysis in ionic liquid. (b) Ionic liquid **41** as organocatalyst.

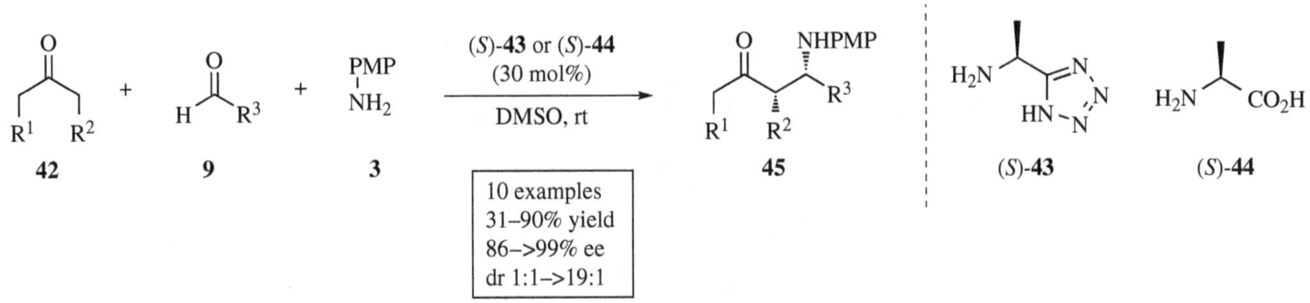

SCHEME 2.12 Asymmetric organocatalytic MCR developed by Córdova.

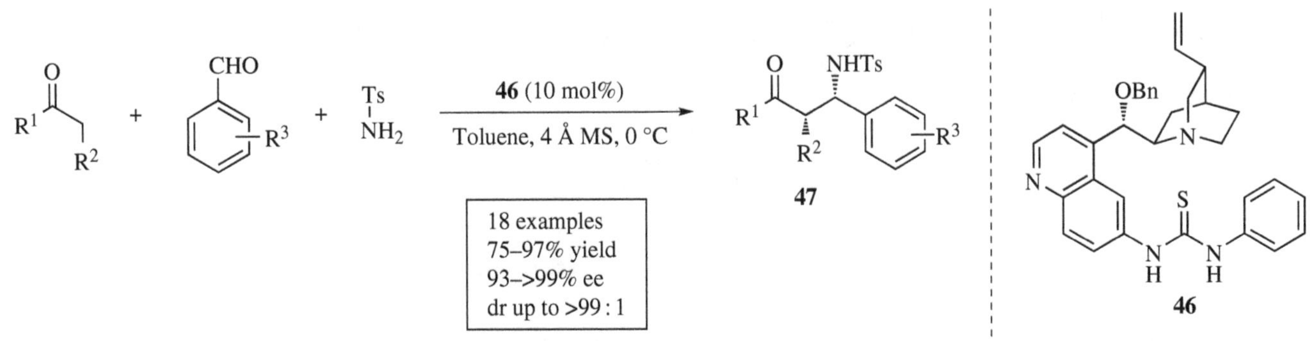

SCHEME 2.13 Bifunctional thiourea catalysis.

On the other hand, Zhao's group achieved remarkable results with bifunctional cinchona alkaloid and thiourea catalysts [35], which, over the years, have been broadly employed as catalysts in Mannich reaction. Indeed, as Brønsted bases, the preferred bifunctional catalyst **46** can lead to the Mannich three-component product **47** with extremely high diastereo- and enantioselectivity (Scheme 2.13, up to >99% ee, dr up to >99:1). The proposed pathway involves an enolate intermediate rather than the enamine activation previously depicted.

Likewise, bifunctional thiourea catalysts have been applied in MCRs in a host–guest complex, largely increasing the reaction stereocontrol [36]. Thus, aliphatic aldehydes **9**, acetone

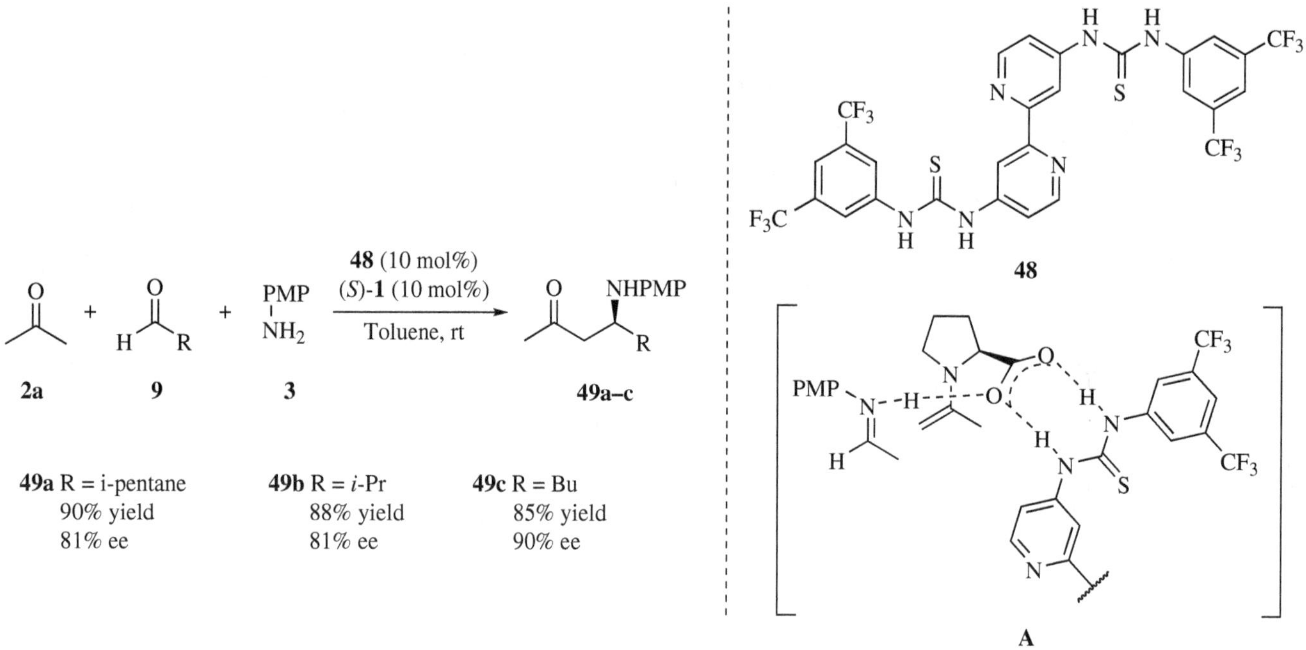

SCHEME 2.14 Mannich reaction catalyzed by a proline–thiourea host–guest complex.

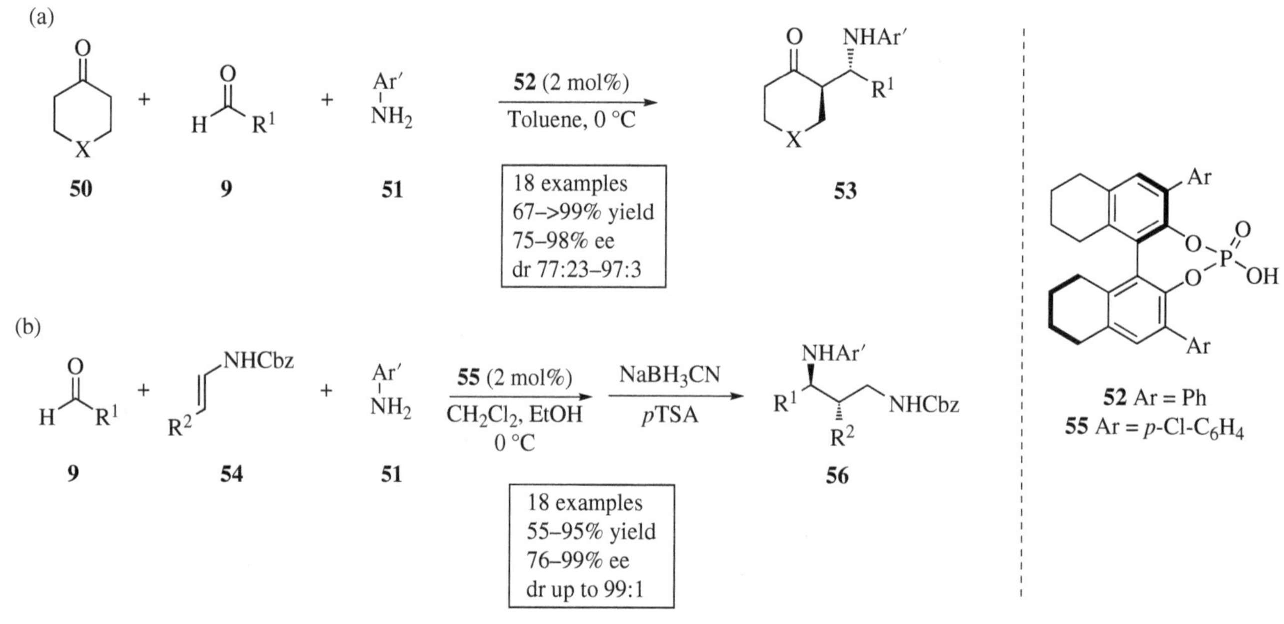

SCHEME 2.15 Phosphoric acid catalysis.

(**2a**), and PMP amine **3** were used for a Mannich reaction synergically catalyzed by (*S*)-proline (**1**) and thiourea **48** (Scheme 2.14). The assumed transition state **A** includes a noncovalent interaction by the thiourea moiety of the bipyridine derivative **48** and the carboxylic function of the (*S*)-proline (**1**).

Equally, chiral Brønsted acids were adopted in the asymmetric Mannich reaction. Gong's group has relied on the ability of phosphoric acid **52** to catalyze the reaction among heterocyclic ketones **50**, variously substituted aldehydes **9**, and aniline derivatives **51** (Scheme 2.15a, up to 98% ee) [37]. The assumed mechanistic pathway indicates that after the imine *in situ* generation, the carbonyl compound, in the presence of the chiral acid, enolizes and gives a nucleophilic attack on the imine intermediate, affording *anti*-diastereoselectivity in the expected product **53** [6, 38].

Likewise, a study reported by Zhu described an efficient synthesis of *anti*-1,3-diamines **56** employing the Brønsted acid **55** as catalyst (Scheme 2.15b, up to 99% ee) [39]. In the mechanistic proposal, the chiral phosphoric acid **55** should activate both the enecarbamate **54** and the *in situ* formed imine, followed by a *si*-face nucleophilic attack that affords the acylimine **B**, which is in turn trapped by added ethanol. At this stage, the initial product **C** is reduced *in situ* by $NaBH_3CN$ to generate the corresponding *anti*-1,3-diamine **56** (Scheme 2.16).

Similarly, Schneider and coworkers developed a vinylogous Mannich reaction via a nucleophilic attack of the silyldienolate **57** on the *in situ* generated imine between various aldehydes **9** and the *p*-anisidine (**3**). After the hydrolysis, they have obtained the vinylogous Mannich product **59** with good to excellent ee (Scheme 2.17, up to 97%) [40].

Furthermore, chiral bisphosphorylimides have been investigated within Brønsted acid catalysis. Specifically, the presence of two BINOL-derived phosphorylimide moieties in the same catalyst **62** introduces more steric hindrance in spite of a mono-axially chiral BINOL catalyst (Scheme 2.18). This particular dimer BINOL imine were synthetized by Zheng's group and employed to catalyze the reaction among cyclohexanone **60**, several aldehydes **9**, and the simplest aniline **61** in toluene at room temperature [41]. They attained extremely high ee (up to >99% ee) and yields (up to >99%), as well as remarkable diastereomeric ratio (up to >99:1). In order to investigate the scalability of the reaction, the gram-scale procedure were performed employing 1 mol% of catalyst loading and achieving the titled *syn*-product **63** with comparable yield, ee and dr.

In addition to the just depicted direct Mannich reaction, in recent times, the nitro-Mannich has been widely investigated,

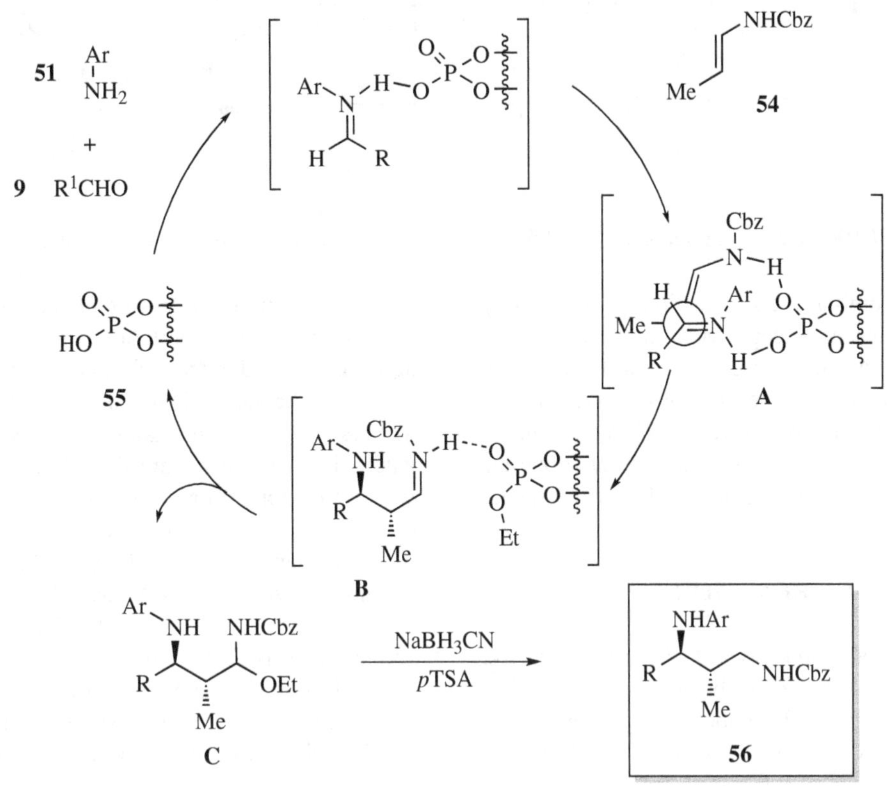

SCHEME 2.16 Proposed mechanistic pathway for a Brønsted acid-catalyzed synthesis of 1,3-diamine.

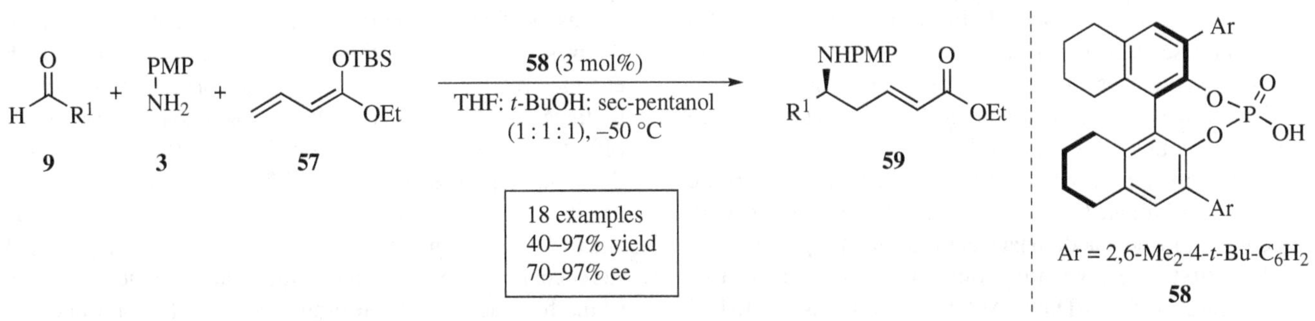

SCHEME 2.17 Axially chiral phosphoric acid catalysis of vinylogous Mannich reaction.

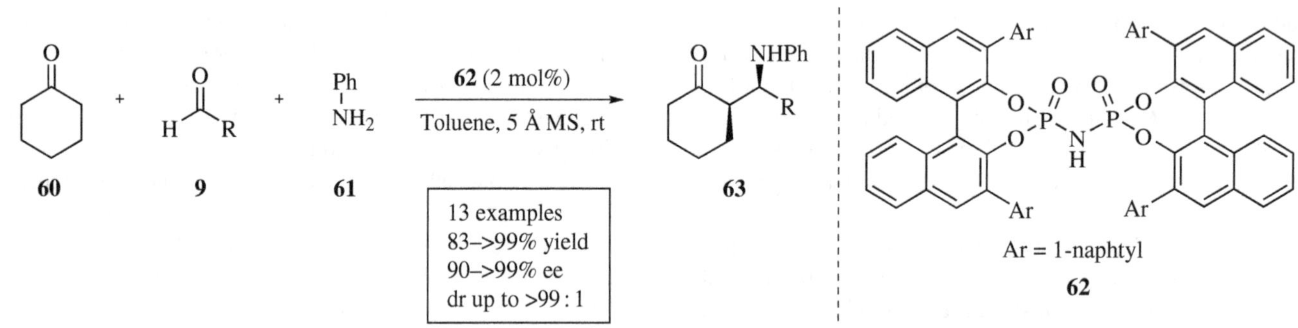

SCHEME 2.18 Biphosphorylimides as novel Brønsted acid catalysts.

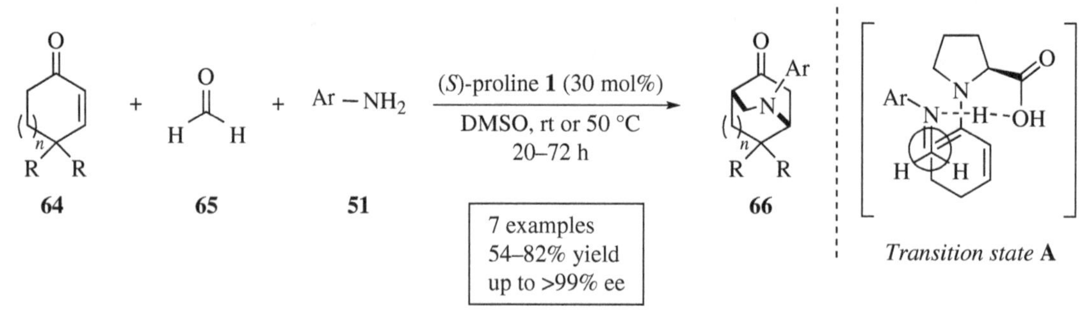

SCHEME 2.19 Proline-mediated three-component enantioselective aza-Diels–Alder reaction.

mostly included in domino/cascade reaction processes [42]. As an example of this useful strategy, Dixon et al. have worked on the synthesis of several heterocycles by a multicomponent symmetric DABCO-catalyzed nitro-Mannich reaction/lactamization [43]. Even though the reaction is not asymmetrically catalyzed, they have started from a chiral substrate, achieving the desired products in outstandingly dr and ee.

2.3 CYCLOADDITION REACTION

In the vast field covered by organocatalysis, enantioselective cycloaddition reactions have found broad success affording the desired products in high selectivities and excellent yields, combined with outstanding enantioselectivities [44]. Among the different classes of organocatalysts fruitfully used on this kind of reactions, we can cite several cinchona alkaloid, bifunctional thiourea, diarylprolinolsilylether salt, and phosphoric acid derivatives. Highly enantioselective Diels–Alder, inverse electron demand hetero-Diels–Alder, aza- and oxa-Diels–Alder reactions, and three-component 1,3-dipolar cycloadditions can be cited among the most currently used transformations in this area. In this chapter section, selected examples are going to be illustrated. However, the readers are invited to go through the "Development of Drug Intermediates" (Section 2.5) where additional examples are also given.

The first one-pot three-component enantioselective organocatalyzed aza-Diels–Alder reaction was published in 2005 by Córdova and coworkers [45]. The reaction proceeds through (*S*)-proline (**1**) catalysis in the presence of α,β-unsaturated cyclohexenones and heptenones **64**, aqueous formaldehyde (**65**), and aniline or *p*-anisidine (**3**). Based on the stereochemical outcome of the reaction, the authors suggest a mechanism via the transition state **A**, where the chiral iminium salt undergoes a selective 6-*endo-trig* cyclization followed by hydrolysis giving rise to the expected azabicycles **66** (Scheme 2.19).

An elegant example of cycloaddition reaction within organocatalyzed processes was published in 2011 by Chen, Jørgensen, and coworkers [46]. By associating the trienamine catalysis (e.g., activation of 2,4-dienals via HOMO raising through condensation with amines) with the Diels–Alder reaction, they devised the first tandem reaction upon cycle-specific trienamine and enamine catalysis (Scheme 2.20a). Thus, condensation of (*E*,*E*)-2,4-hexadienal (**67**) with catalyst (*S*)-**70** affords the trienamine intermediate, which upon cycloaddition reaction with the 3-olefinic oxindole **68** (which acts as the dienophile) affords the Diels–Alder adduct. Then, the process evolves, in the presence of catalyst **70**, through enamine catalysis to give the expected compounds **71** and **72** by their trapping with ethyl 2-(diethoxyphosphoryl)acrylate (**69**) (the electrophilic partner) (Scheme 2.20b). Remarkably, this tandem process gave **71** and **72** in 89% of overall yield and in a diastereomeric ratio of 85:15 at the α-position of the aldehyde. Concerning the second uncontrolled newly created stereocenter at the γ-position of the produced aldehydes (due to the high acidity of the proton between the phosphonate and ester moieties), an additional sequential transformation

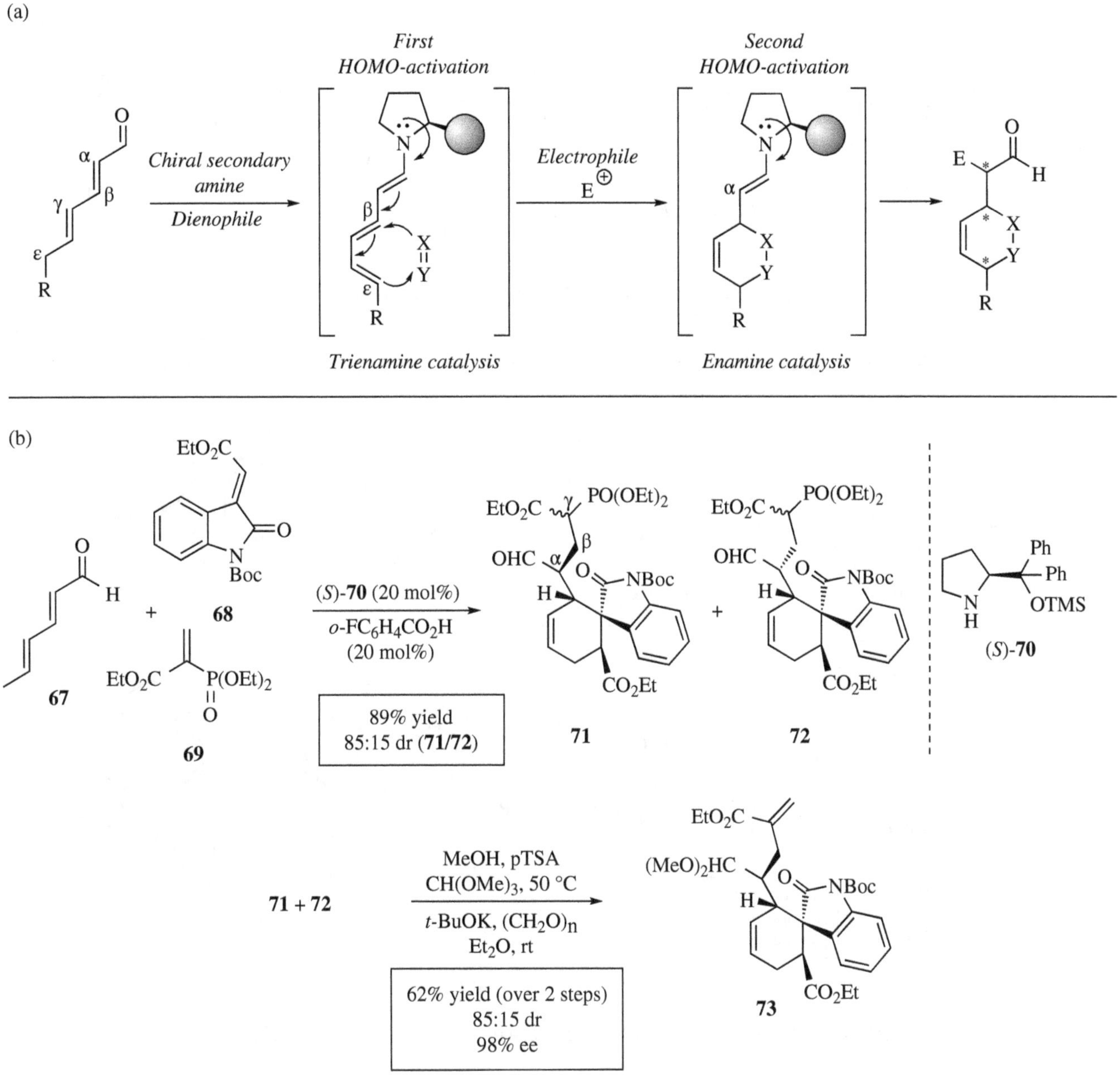

SCHEME 2.20 Organocatalytic trienamine–enamine multicomponent tandem reaction.

(e.g. protection/Horner–Wadsworth–Emmons reaction) afforded compound **73** via an elimination process.

The scope of organocatalyzed cycloaddition reactions was also extended to different syntheses of pyrrolidine derivatives. In this direction, Córdova and coworkers explored the combination of diphenylprolinol **76** and simple oxime cocatalyst **77** to promote the synthesis of functionalized proline derivatives bearing four contiguous stereocenters where one is an all-carbon quaternary one (Scheme 2.21) [47]. The success of the transformation relies on a cooperative combination of iminium and H bond activation to afford chemo- and stereoselective [3+2] cycloaddition under kinetic control. The reaction proceeds by grouping aromatic aldehydes **4**, protected α-cyanoglycine esters **74**, and enals **75**. It might be noted that the choice of the good catalytic system combination was crucial to insure a chemoselective process (dipolar cycloaddition vs. Michael addition) as well as to accelerate both imine formation and cycloaddition reaction.

Another example to illustrate the synthesis of proline derivatives **78** was published by Vicario and coworkers in 2013 [48]. Starting from aromatic aldehydes, diethyl malonate, and α,β-unsaturated aldehydes upon catalysis of unprotected diphenylprolinol, the synthesis of polysubstituted pyrrolidines was achieved in a one-pot condensation and [3+2] cycloaddition reaction in aqueous medium.

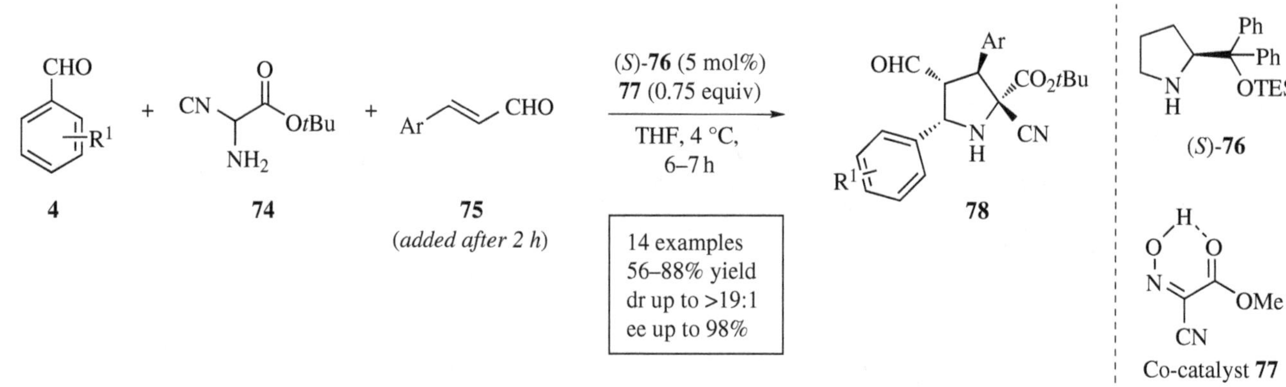

SCHEME 2.21 Organocatalytic one-pot three-component synthesis of polysubstituted pyrrolidines **78**.

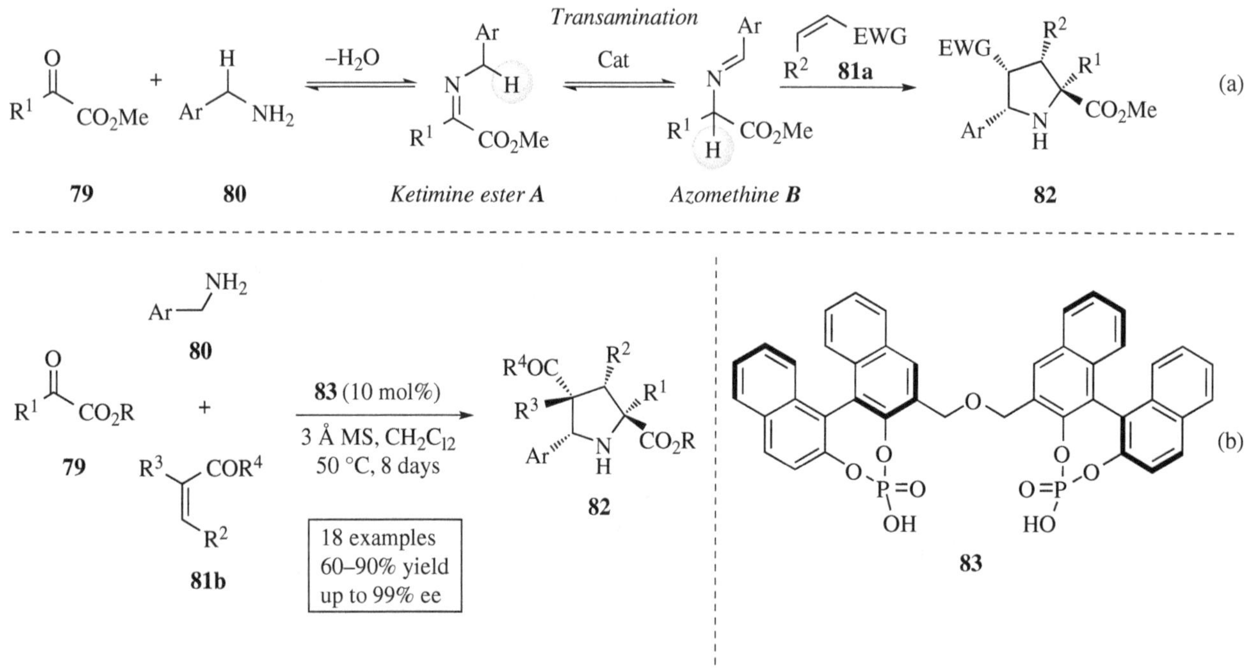

SCHEME 2.22 Biomimetic asymmetric three-component 1,3-dipolar cycloaddition.

The group of Gong and coworkers explored a biomimetic 1,3-dipolar cycloaddition between α-ketoester **79** and benzylamine derivatives **80** with electron-deficient olefins **81a,b** to devise a straightforward route to proline derivatives **82** in high yields and enantioselectivities [49]. The proposed biomimetic three-component 1,3-dipolar cycloaddition proceeds as illustrated in Scheme 2.22a. The azomethine ylide **B** is formed, via a transamination from ketimine ester **A**, which is in turn prepared from α-ketoesters **79** and benzylamine derivatives **80**; then, the 1,3-dipolar cycloaddition with electron-poor olefins **81a** takes place. For this purpose, the bisphosphoric acid **83** was found to be the catalyst of choice to promote such transformation (Scheme 2.22b). Replacing dimethyl maleate (previously used as deficient olefins) by methyleneindolinones, the same approach could be extended to spirooxindoles synthesis in high yields and enantioselectivities, also mediated by means of phosphoric acid catalysis.

The Feng group designed a three-component synthesis of spiro[5,5]undecane-1,5,9-trione analogues **87** via Knoevenagel/Diels–Alder reaction of enones **84**, aldehydes **9**, and Meldrum's acid (**85**) by means of 9-amino-9-deoxy-epi-quinine (**86**) (Scheme 2.23) [50]. The carbonyl group of enones and the alkylidene moieties are both supposed to be activated by the cinchona alkaloid-based primary amine; the former with the primary amine and the latter with the protonated quinuclidine framework (see intermediate **A**). Meanwhile, this conformation brings the two substrates into vicinal chiral environment and, consequently, facilitates the [4+2] cycloaddition reaction.

Chiral Brønsted acid catalyst **89** was used by Gong and coworkers in an asymmetric synthesis of 3-amino-δ-lactams **90** via a formal [4+2] cycloaddition reaction of

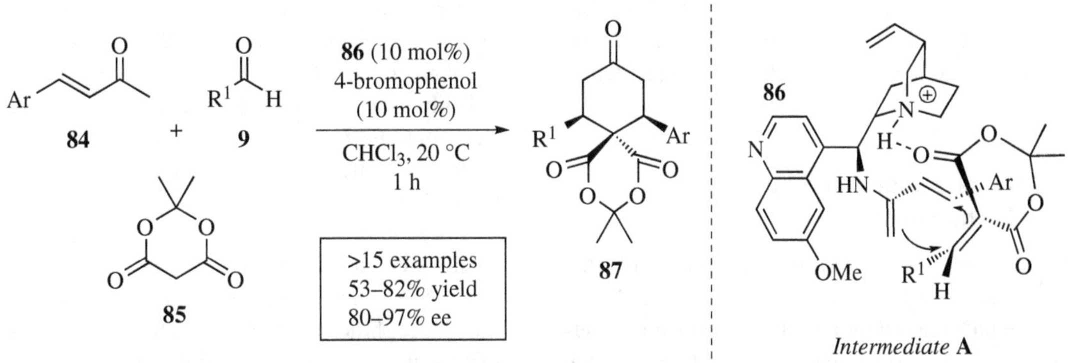

SCHEME 2.23 Asymmetric synthesis of spiro[5,5]undecane-1,5,9-trione derivatives **87**.

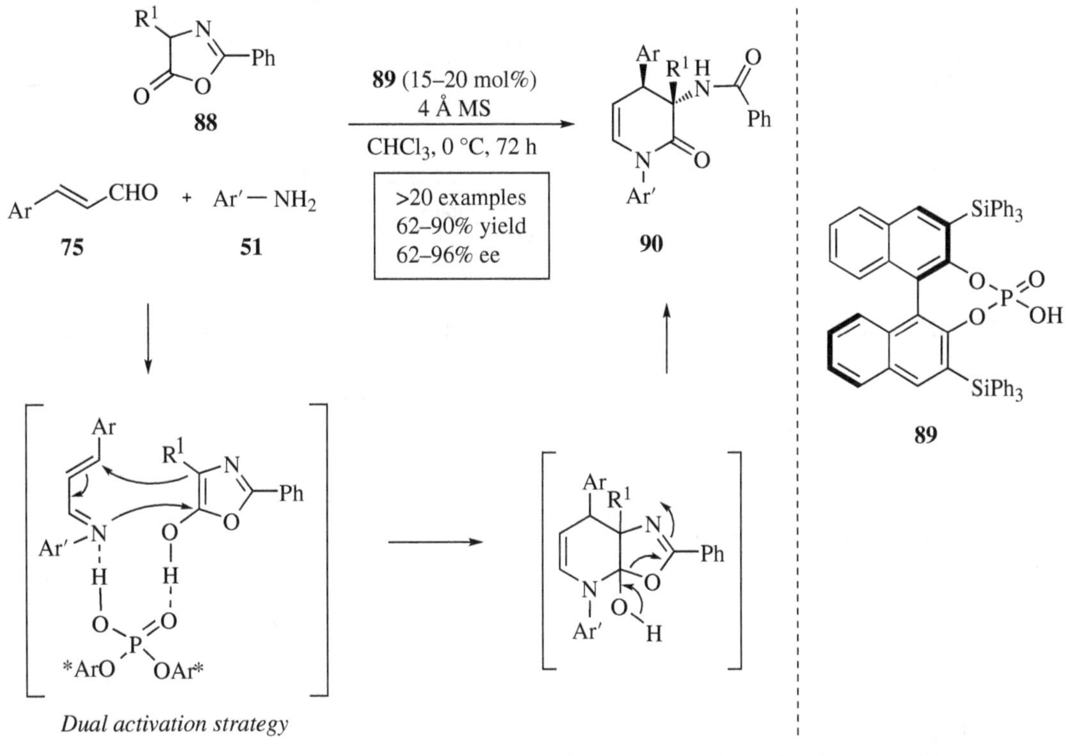

SCHEME 2.24 Three-component synthesis of 3-amino-δ-lactams via a formal [4+2] cycloaddition reaction.

azalactones **88**, aldehydes **75**, and aniline derivatives **51** (Scheme 2.24) [51]. By using substituted aryl ethylamines instead of aniline derivatives, benzo[*a*]quinolizidines can be isolated in moderate-to-good yields and high enantioselectivities.

2.4 ORGANOCATALYTIC MULTICOMPONENT DOMINO ASYMMETRIC REACTIONS

Prompted by the groundbreaking work of Enders [52], who outstandingly disclosed the potential of organocatalysis toward the straightforward construction of complex molecules in just one-pot procedure, a wider and wider segment of organic chemists have been involved in the improvement and development of new multicomponent tandem or domino-type reactions [2f, 53]. In the first subsection (Section 2.4.1), a particular attention is reserved to those procedures that involve, at least, a Michael addition, one of the most exploited and fruitful reactions in this field. Indeed, the Michael system could be regarded as bifunctional electrophile, which upon the addition of a suitable nucleophile could generate a new reactive moiety, which easily undergoes a subsequent electrophilic step [54]. The herein disclosed processes is organized with respect to the number of involved reactions and the activation mode.

The second subsection (Section 2.4.2) deals with selected examples of other multicomponent tandem or domino-type reactions, which should be mentioned for their synthetic interest. Moreover, since the high value of several developed procedures, we take into account not only the processes that resemble the initial definition of MCRs but also few procedures in which one or more reagents were subsequently added.

2.4.1 Michael-Type Multicomponent Process: Cyclic Carbon Frameworks

2.4.1.1 Double Cascade Reactions Starting with the double cascade reactions, Gong et al. reported a straightforward access to enantiomerically enriched dihydropyridines by means of an asymmetric three-component cyclization of cinnamaldehydes **75**, an aromatic primary amine **51** and 1,3-dicarbony compounds **91** [55]. The catalytic action of the chiral Brønsted acid **92** furnished the titled 1,4-dihydropyridines **93** with excellent enantioselectivities (up to 98% ee) and moderate good yields (up to 97%) tolerating important modification in both electron and steric demand (Scheme 2.25).

The author suggested that under acidic condition, α,β-unsaturated aldehyde **75** condenses with aromatic primary amine **51** giving the corresponding imine **94**, which undergoes the nucleophilic attack of the acetoacetates **91**. The formation of an H bond between the chiral Brønsted acid [BH*(**92**)] and

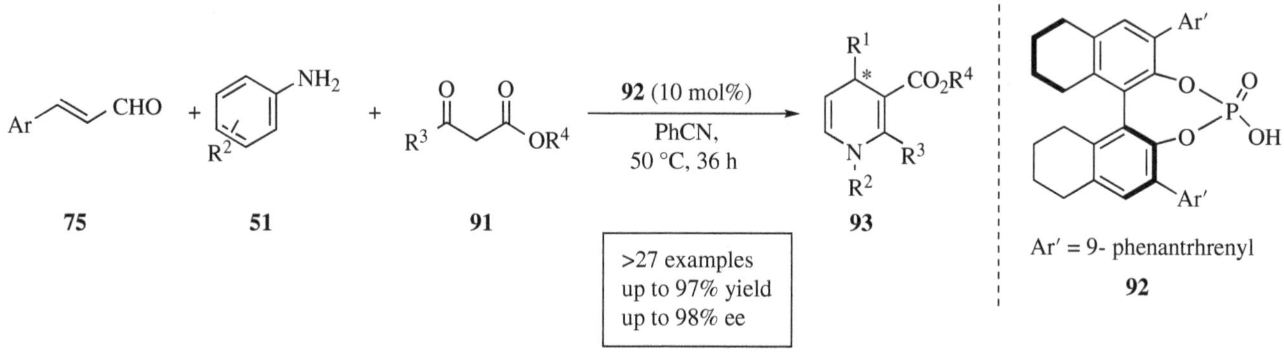

SCHEME 2.25 Cyclization reaction of cinnamaldehydes **75** and anilines **51** with 1,3-dicarbonyl compounds **91**.

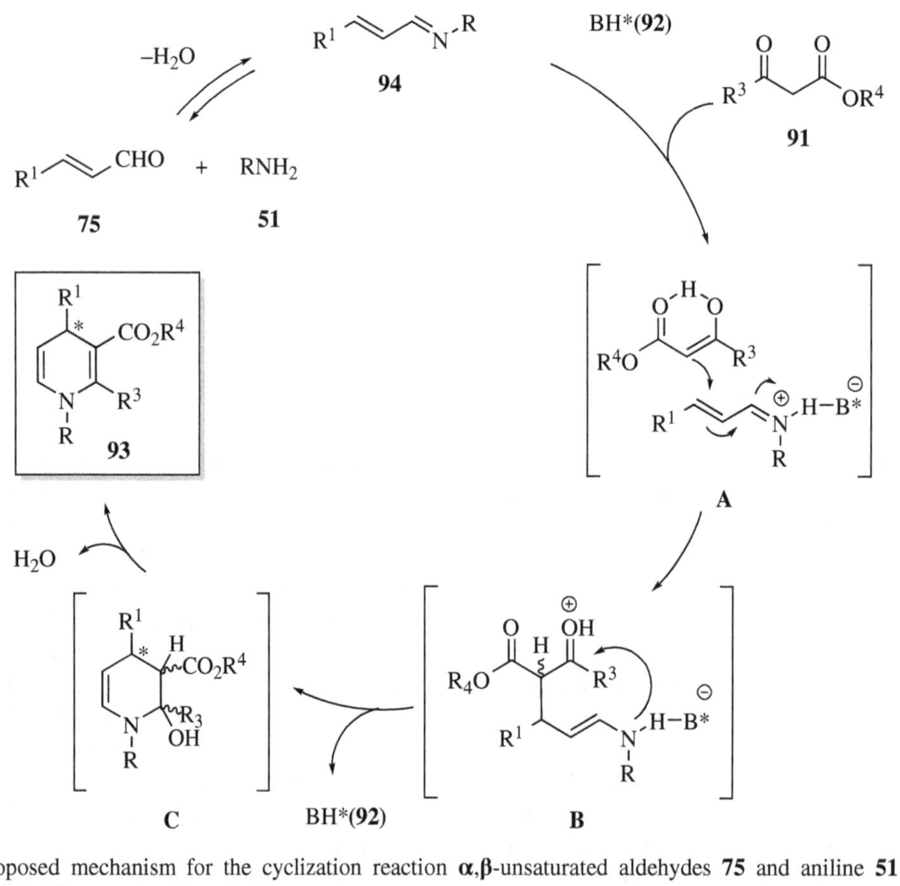

SCHEME 2.26 Proposed mechanism for the cyclization reaction **α,β**-unsaturated aldehydes **75** and aniline **51** with 1,3-dicarbonyl compounds **91** under chiral Brønsted acid catalysis.

imines **94** makes the latter more electron deficient and thereby promotes the addition of **91 to 94** (**A**, Scheme 2.26).

The resulting intermediate **B** cyclizes to furnish the intermediate **C**, which upon dehydration affords the optically active aromatic 1,4-dihydropyridines **93**.

A prominent segment of asymmetric organocatalytic MCRs takes advantage of both enamine and iminium ion activation mode involving amine catalysis. Indeed, as often highlighted, the catalyst structure could be easily adapted to the electronic and steric demands of various and different reactions. Within this context, pursuing the MacMillan proline-catalyzed cyclopropanation reaction [56] and their own research concerning the epoxidation of electron-poor olefins [57], Jørgensen et al. firstly presented a multicomponent domino conjugated addition/amination reaction of α,β-unsaturated aldehydes **95** catalyzed by (*S*)-α,α-diarylprolinol **97** (Scheme 2.27) [58].

The initial study was performed using various thiols (R^2SH) as nucleophiles, which afforded 1,2-aminothiol derivatives **98** bearing two adjacent stereogenic centers in one-pot good yield process (up to 72%) with marvelous stereoselectivity (dr up to 96:4, >99% ee) [58a]. Likewise, the subsequent employment of an aza-nucleophile gave access to *syn*-diaminated aldehydes **99** and **100**, easily converted in 1,2,3-triaminated products or 1,3-diamino alcohols [58b]. In both cases, the mechanistic pathway involves the first soft nucleophile attack to the iminium intermediate **A**, followed by the addition of a suitable electrophile **96** to the enamine intermediate **B**, giving the nearly enantiopure products **98–100** (Scheme 2.28).

Analogously, Melchiorre et al. developed an iminium–enamine tandem sequence, which relies on the chiral primary amine catalyst **101**, directly derived from natural cinchona alkaloids (Scheme 2.29) [59].

The proposed Michael addition/amination pathway furnished a straightforward access to compounds **102** and **103**, high valuable precursor of α-amino acid bearing two adjacent stereogenic centers with an impressive high enantiomeric purity (up to 99% ee, with dr from 3:1 to 20:1).

The employment of bifunctional urea–tertiary amine **104** or monothiourea **105** catalysts selectively promotes either the Michael addition or cycloaddition process, respectively, [60]. As depicted in Scheme 2.30, the tertiary amine group would activate the *in situ* formed α-amino esters to produce azomethine ylides **A** (or enolates), whereas the nitroalkene counterparts would be activated by the thiourea (urea) moiety through a double H bonding interaction (**B**).

Therefore, the synergic communication through bifunctional catalysis ensures a facile access to diverse chiral nitrogen-containing compounds of type **106** and **107** in a quite good yield (up to 93% yield) and stereoselectivity (94–96% ee).

Among the most remarkable double cascade reactions involving a Michael addition, MacMillan reported a noteworthy example of orthogonal mode activation in terms of LUMO-lowering (iminium) and HOMO-raising (enamine) catalysis. Taking advantage from their previous research [56, 61], the American group supposed that the exposure of a suitable aldehyde to imidazolidinone catalyst **108** should

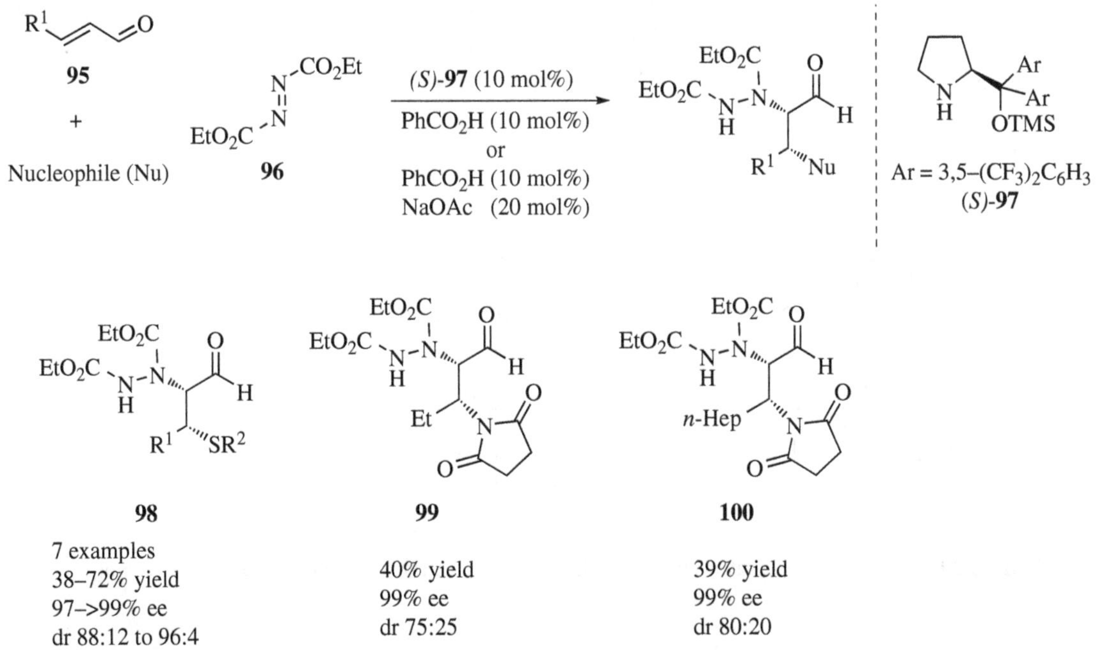

SCHEME 2.27 Organocatalyzed conjugated addition/amination reaction of **α,β**-unsaturated aldehydes.

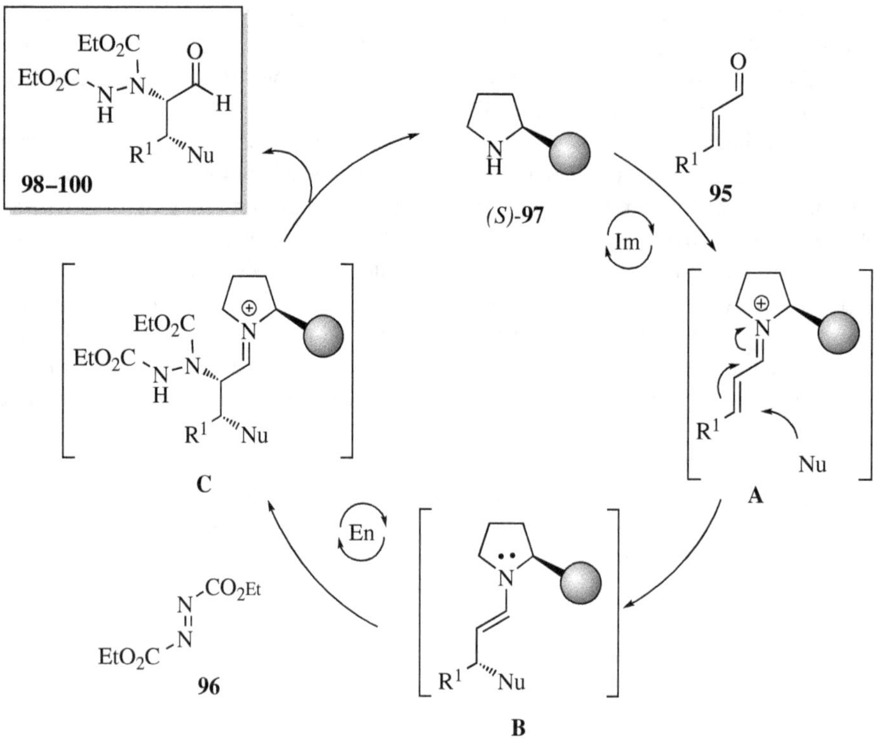

SCHEME 2.28 Combination of enamine and iminium ion activation for the enantioselective conjugated addition/amination reaction of α,β-unsaturated aldehydes.

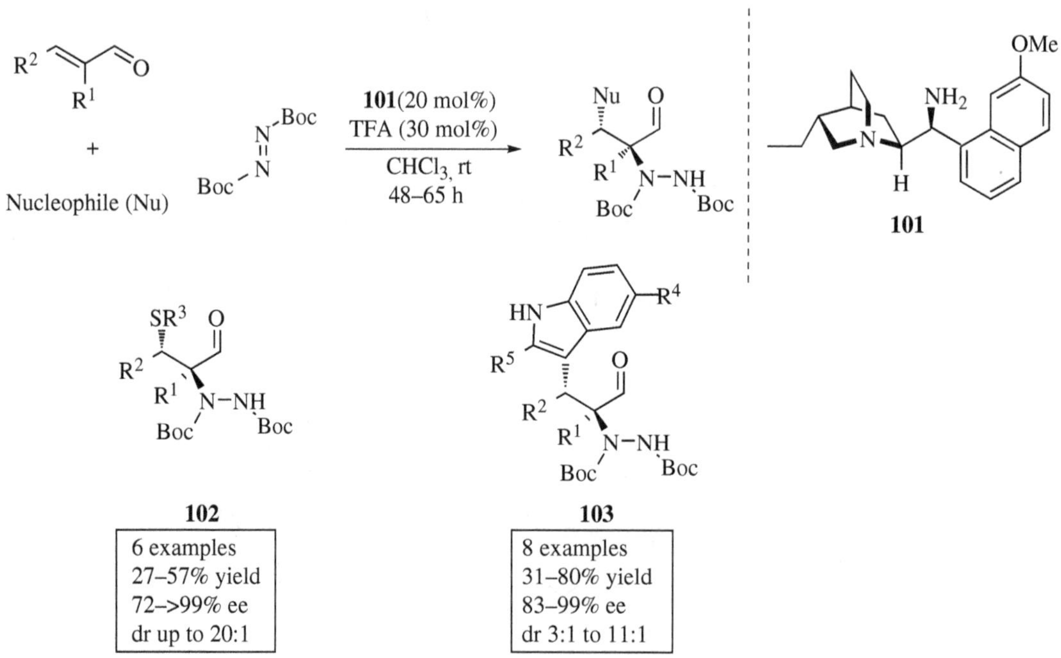

SCHEME 2.29 Organocascade catalysis with combination of azodicarboxylate and either indole or thiol derivatives.

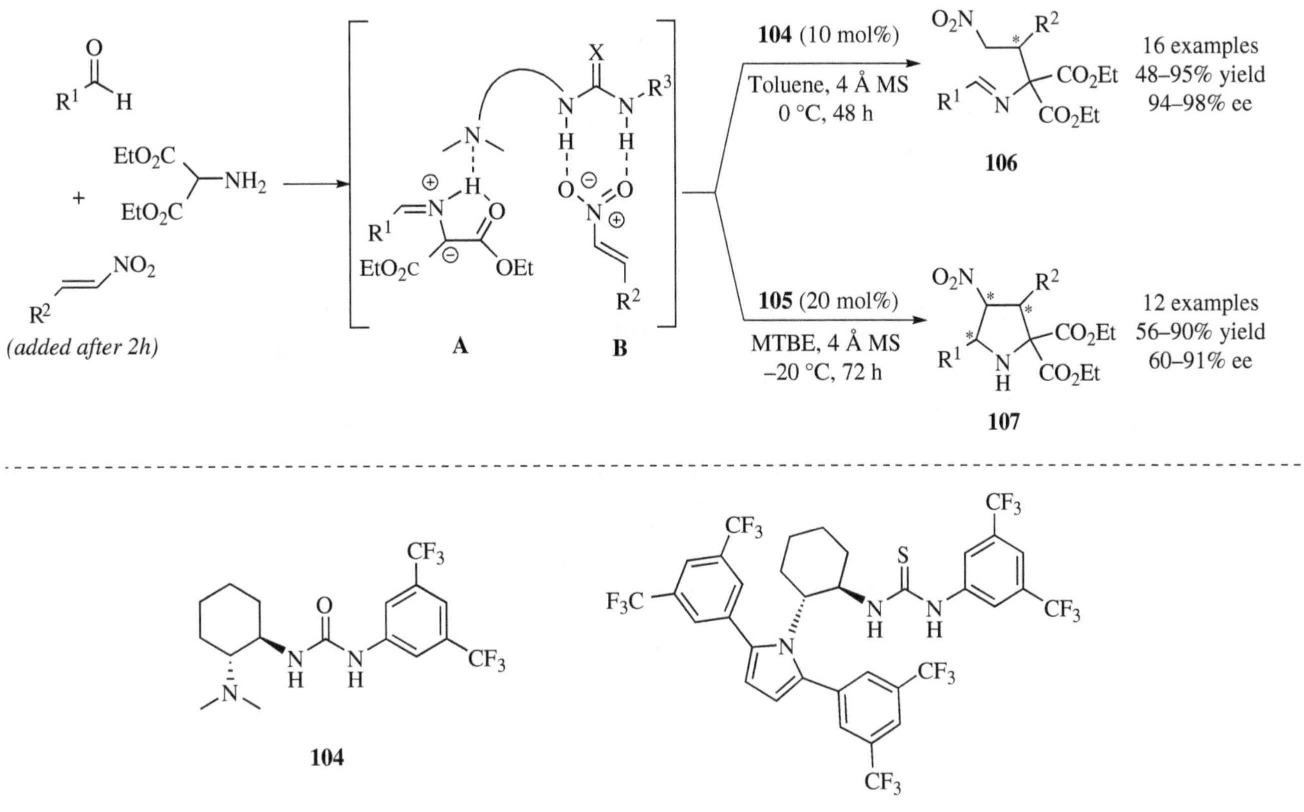

SCHEME 2.30 Reaction control in the organocatalytic asymmetric one-pot, three-component reaction of aldehyde, diethyl α-aminomanolate, and nitroalkenes.

furnish the activated iminium ion **A**, which subsequently would undergo the attack of a generic π- or hydrido-nucleophile (Nu) [62]. The rapid hydrolysis of **C** should generate the intermediate **109** that enters in the second catalytic cycle wherein the enamine activation should allow the final addition of a suitable electrophile (Scheme 2.31).

The validation of such hypothesis employing **111** as electrophile and furans, thiophenes, butenolides, and tertiary amino lactone equivalent as nucleophiles enabled the synthesis of all desired products with high level of stereocontrol (99% ee) (Scheme 2.32).

The concept of multicomponent processes associated with organocascade reactions allows the installation of new functionalities, which easily give access to quite complex structures. Recently, Jørgensen and coworkers reported a remarkable application of such definition toward the high stereocontrolled synthesis of optically active pyrroloisoquinolines **115** (dr up to >95, up to 99% ee) (Scheme 2.33) [63].

Actually, the formal [3+2]-cycloaddition reaction proceeds through a two-step sequence involving (i) a Michael addition of the *in situ* generated ylide **A** to the γ-carbon atom of the iminium ion **B** and (ii) a subsequent cyclization of the resulting intermediate **C** that affords the optically active cation **D** that was trapped by the triphenylcarbethoxymethylenephosphorane to furnish the final conjugated ester **115**.

A quite elegant example of one-pot three-component reaction via a domino Michael/cyclization sequence was reported by Yuan et al. to provide a wide range of optically active spiro[4*H*-pyran-3,3′-oxindoles] **117** [64]. Among all the tested chiral organocatalyst, cupreine (CPN) **116** resulted to be the most effective in terms of yield (up to 99%), enantioselectivity (up to 97% ee), and tolerance to the substrate modification (Scheme 2.34).

Truly, the overall process involves a noncatalyzed fast Knoevenagel condensation of isatin **118** with malononitrile **119** to afford the oxindole **120**, which enters in the catalytic cycle reported below. Indeed, as a Michael acceptor, the intermediate **120** undergoes the nucleophilic attack of **121** in the presence of catalytic amount of CPN (**117**, 10 mol%) (Scheme 2.35).

The reaction that initially implicates a keto–enol tautomerism equilibrium between **B** and **C** progresses with an intramolecular cycloaddition involving the CN group activated by the phenolic OH, followed by a tautomerization. The formation of the desired spiro derivatives **117** releases the catalyst **116**, which begins another catalytic cycle.

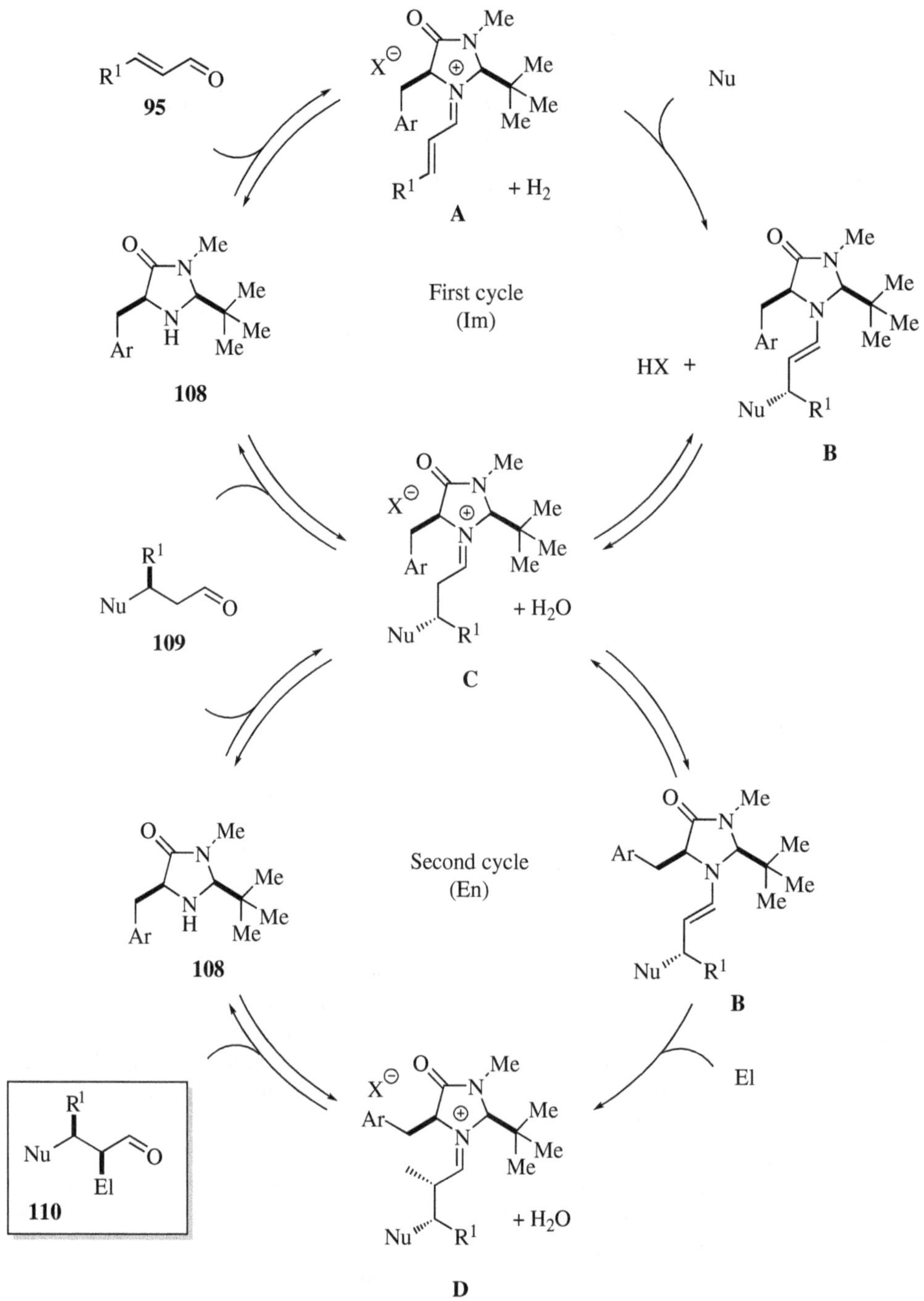

SCHEME 2.31 Proposed cascade iminium–enamine activation.

As initially stated, the definition of multicomponent reaction could be extended also to those reactions that implicate the addition of further reagents only after a suitable time from the very beginning of the procedure. Among others, few Chinese groups reported two notable examples of one-pot Michael addition/Pictet–Spengler cyclization sequence. In particular, in 2011, Zhao et al. developed a high enantioselective organocatalytic synthesis of several indoloquinolizidine derivatives **122** exploiting the alkyl propriolate ability to produce rich reaction site intermediates [65]. The organocatalyst-controlled conjugate addition of *in situ* generated β-enaminoesters and optically active α,β-enals, and the subsequent substrate-controlled Pictet–Spengler cyclization furnished the sought products **122** in good yields (up to 88%) and quite high enantioselectivity (up to 96% ee). The overall stereocontrol results to be a

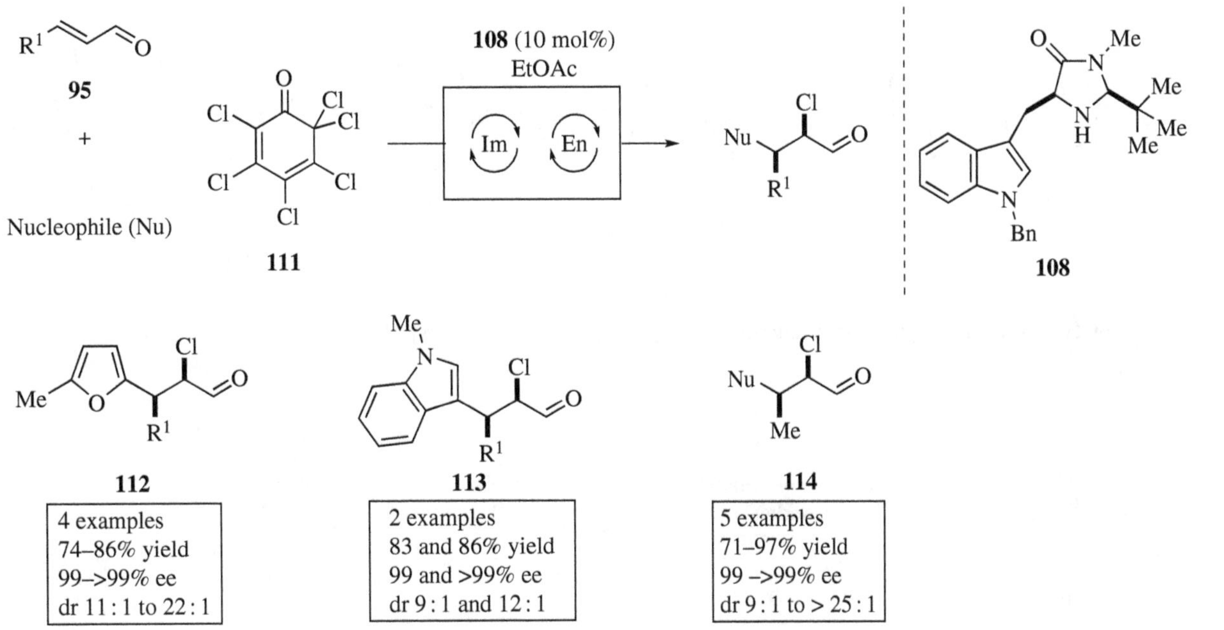

SCHEME 2.32 Examples of enantioselective organocascade catalysis.

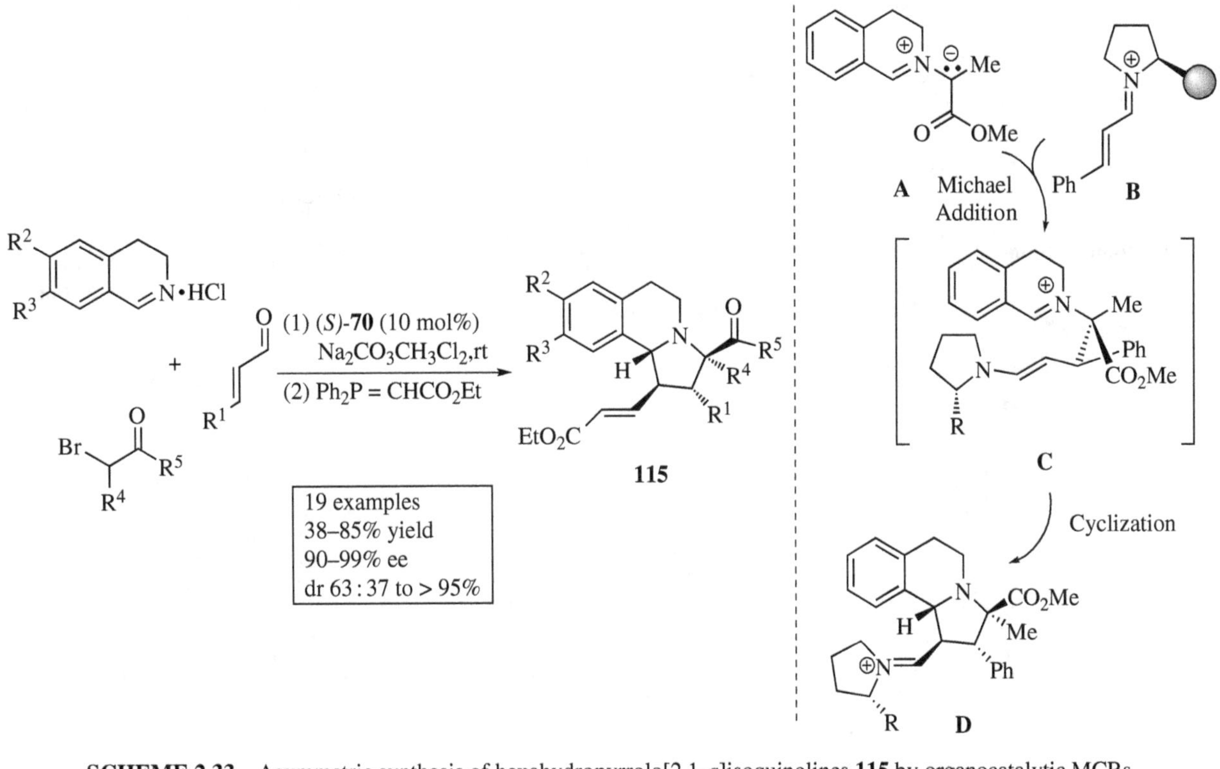

SCHEME 2.33 Asymmetric synthesis of hexahydropyrrolo[2,1-*a*]isoquinolines **115** by organocatalytic MCRs.

consequence of the initial attack of enaminoesters to the less hindered *re*-face of the α,β-unsaturated iminium intermediate (**A**, Scheme 2.36), which sets up the chiral center at the C2 position of the final indoloquinolizidines **122**.

In 2012, Xu suggested a similar approach to butyrolactam-fused indoloquinolizidines that were achieved in interesting yields and elevated stereocontrol (dr up to >20:1, 90–97% ee), using as key reagent the α-oxo-γ-butyrolactam **123**, a readily available *N*-containing pronucleophile (Scheme 2.37) [66]. On the basis of iminium ion activation strategy, the author assumed that the prolinol derivative (*R*)-**124** activates the enal **95**, shielding the *si*-face from the stereoselective addition

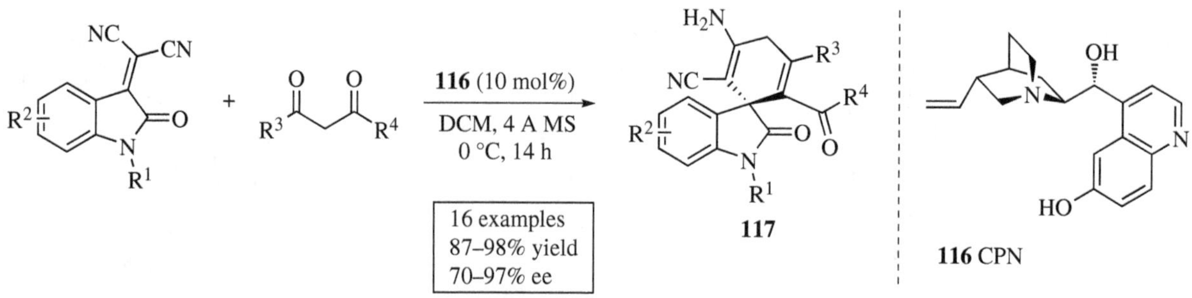

SCHEME 2.34 Asymmetric synthesis of spiro[4*H*-pyran-3,3′-oxindoles] **117** via three-component reactions.

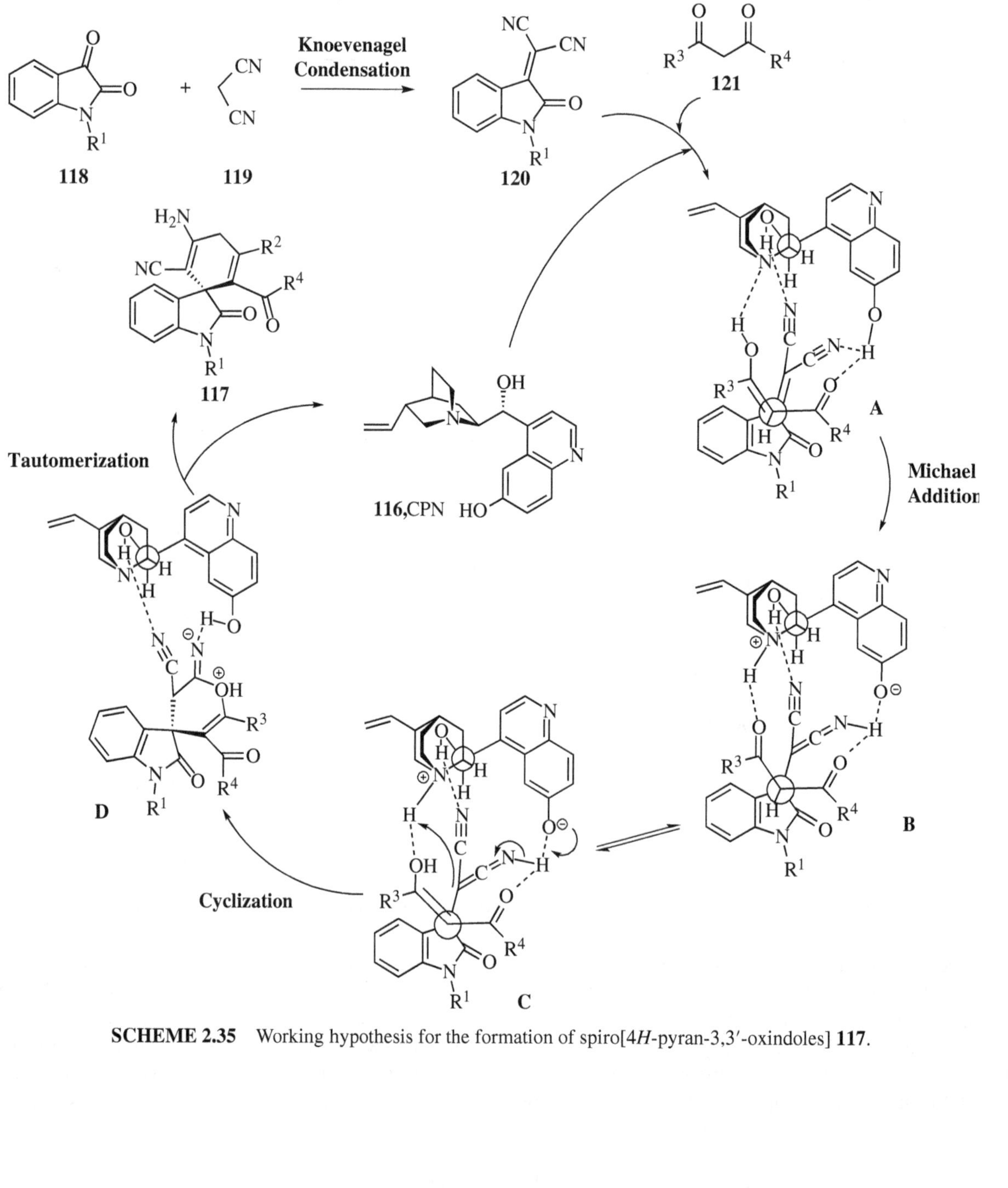

SCHEME 2.35 Working hypothesis for the formation of spiro[4*H*-pyran-3,3′-oxindoles] **117**.

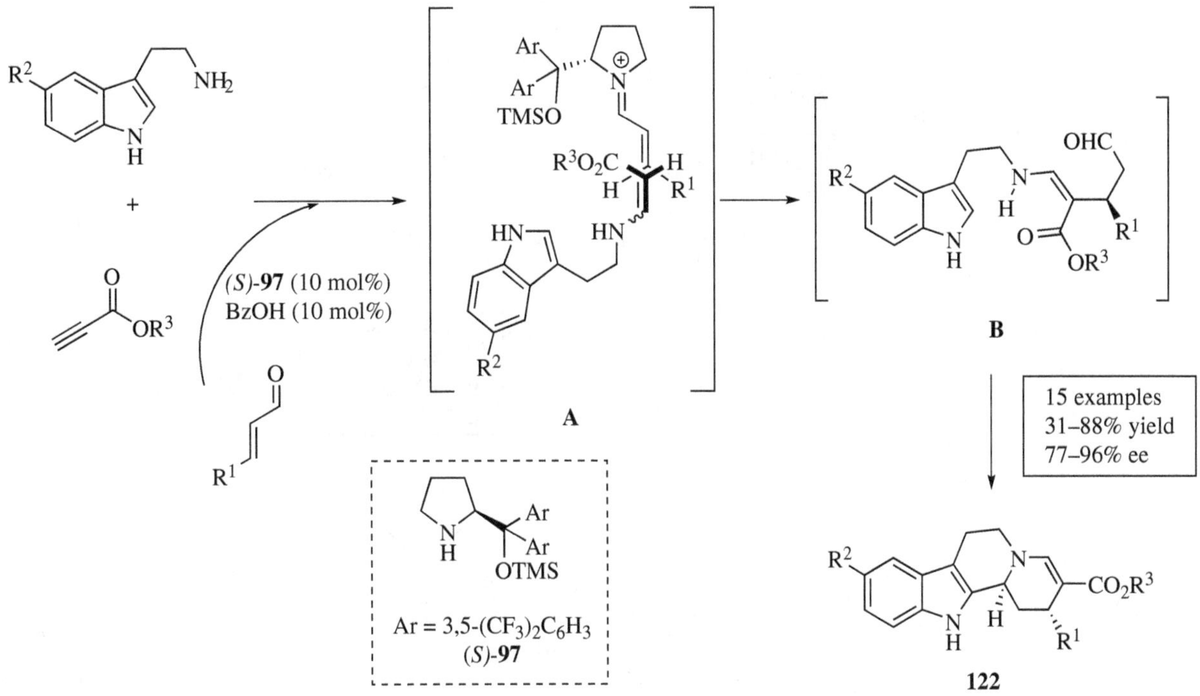

SCHEME 2.36 One-pot three-component synthesis of indoloquinolizidines **122**.

of the α-oxo-γ-butyrolactam **123**. The just generated optically active intermediate **B** undergoes to a straightforward tautomerization and hydrolysis, wherein the masked 1,5-dicarbonyl compounds **C** efficiently condense with tryptamine **125**. The resulting iminium ion **D** performs diastereoselective Pictet–Spengler cyclization to afford the butyrolactam-fused indoloquinolizidine **126** owing to the steric hindrance of the R^1 group.

The ability of MCRs involving a domino process in order to create in one-pot sequence molecular complexity and diversity was notably verified by Zhao and coworkers pursuing the synthesis of 6-amino-5-cyanodihydropyrano[2,3-*c*]pyrazoles **128** (Scheme 2.38) [67].

Once identified the cinchona alkaloid **127** as the most effective catalyst for their tandem Michael addition/Thorpe–Ziegler reaction, the American group obtained the desired product **128** via the planned enantioselective procedure (up to 96% ee, up to 89% yield) (Scheme 2.38).

Additionally, among the double cascade reactions, a remarkable example was very recently reported by Bugaut and Constantieux toward the synthesis of 2,6-diazabicyclo[2.2.2]octanones (Scheme 2.39) [68]. Ensuing their studies to access highly elaborated polycyclic molecules [69], the French group developed an enantioselective multicomponent reaction, which admirably combines β-ketoamide **129**, acrolein **131**, and aminophenol **130** upon a Michael addition/double iminium trapping sequence catalyzed by thiourea **132**. The resulting 2,6-DABCO core **133–138** due to the formation of five new bonds and bearing three stereocenters was assembled in high yields (up to 92%) and impressive stereoselectivities (dr up to >20:1, up to 96% ee).

2.4.1.2 Triple Cascade Reactions The segment of triple cascade reaction has been undoubtedly dominated by the innovative research of Enders group, whose Michael/Michael/aldol condensation sequence represents a milestone in the three-component domino reaction field [52]. In particular, the German group developed a very elegant strategy, which, using enamine activation of the first substrate to start a triple cascade, allowed the formation of four stereogenic centers with high diastereoselectivity and complete enantioselectivity (Scheme 2.40).

The simple chiral secondary amine (*S*)-**70** activates the linear aldehyde **139** by enamine formation, which selectively attacks the nitroalkenes **140** in a Michael-type reaction without any interference by the α,β-unsaturated aldehyde **95**. Indeed, the latter prefers to be activated as iminium ion by the catalyst (*S*)-**70** and to undergo the subsequent conjugate addition to form the Michael adduct **B** (Scheme 2.41).

The subsequent ring closure and hydrolysis allow the regeneration of the catalyst (*S*)-**70** and the formation of the wanted α,β-unsaturated aldehyde **142** too sterically hindered for a further Michael addition compared to **95**. The just disclosed protocol furnishes a simple and flexible access to polyfunctional cyclohexane derivatives as key building blocks toward the synthesis of more complex molecules. This hypothesis was further validated by the same German group that, one year later, offered a straightforward

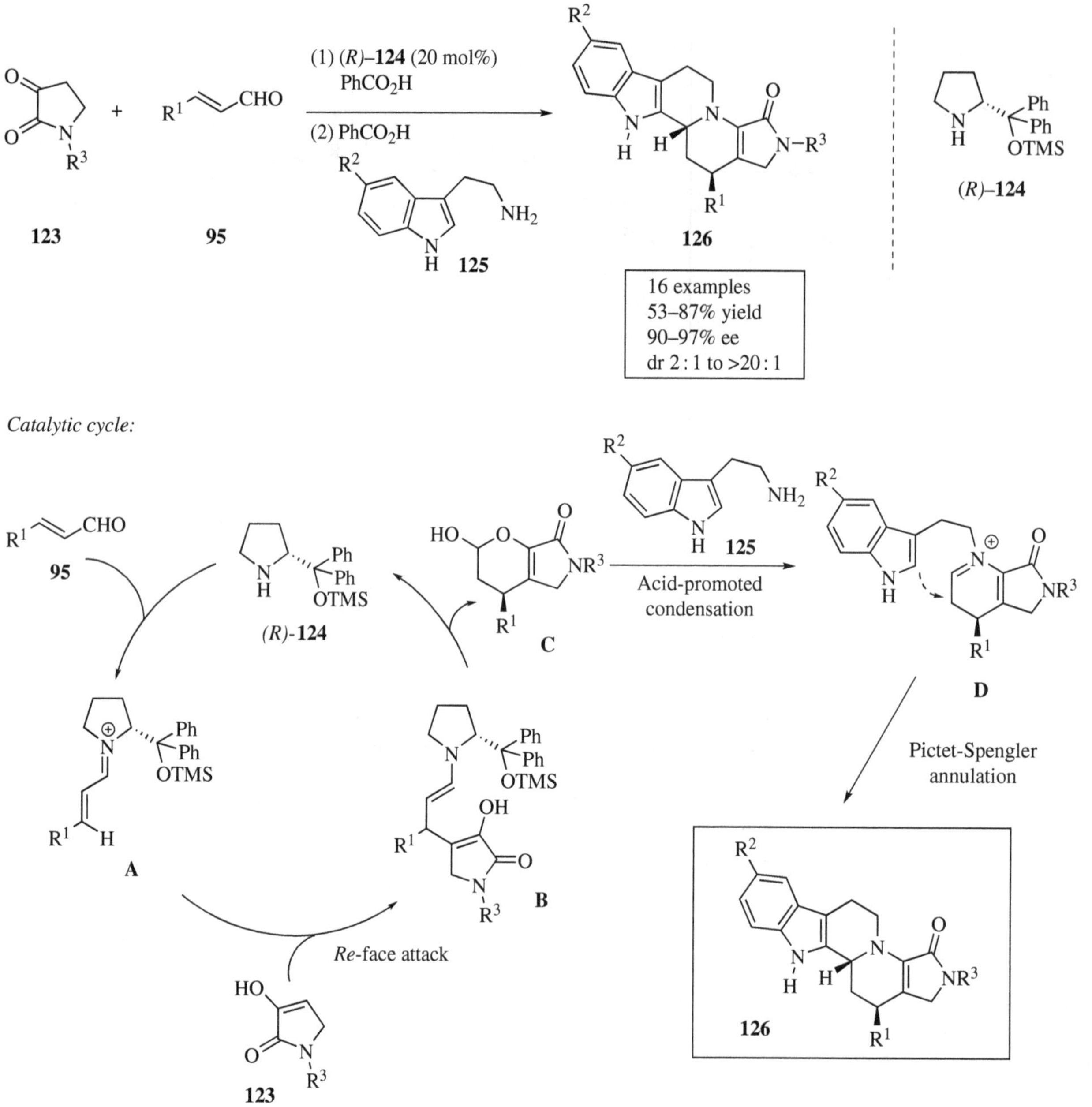

SCHEME 2.37 One-pot Michael/Pictet–Spengler sequence.

access to a functionalized tricyclic carbon framework exploiting a triple cascade sequence followed by a Diels–Alder reaction [70]. Actually, the overall process relies on the selection of a linear aldehyde, which should contain a diene moiety suitable to promote the final intramolecular Diels–Alder reaction (Scheme 2.42).

In such a way, the formation of five new C–C bonds allowed the synthesis of the desired decahydroacenaphthylene and decahydrophenalene cores bearing eight novel stereogenic centers with high optical purity (97 to >99% ee). Likewise, employing a linear aldehyde bearing a thioester moiety, the same enamine–iminium–enamine activation sequence furnished to Enders et al. a facile access to functionalized, highly substituted *cis*-thiadecalins [71].

On the wave of such impressive results, other notable research groups addressed their efforts to the development of alternative and complementary triple cascade procedures. For instance, amazing contributes have been published, over the years, by Melchiorre et al. that successfully suggested a practical synthesis of several spirocyclic oxindoles and their analogues [72]. Exploiting the enamine–iminium–enamine activation mode of aminocatalysis, the organocatalytic triple cascade process disclosed by the Spanish group provided the stereoselective construction of all-carbon quaternary

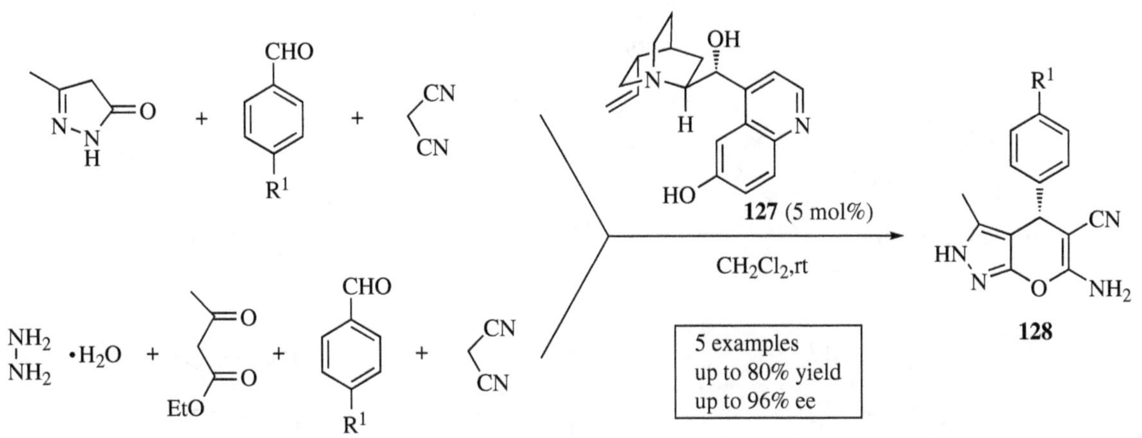

SCHEME 2.38 Enantioselective three- and four-component reactions for the synthesis of pyranopyrazoles **128**.

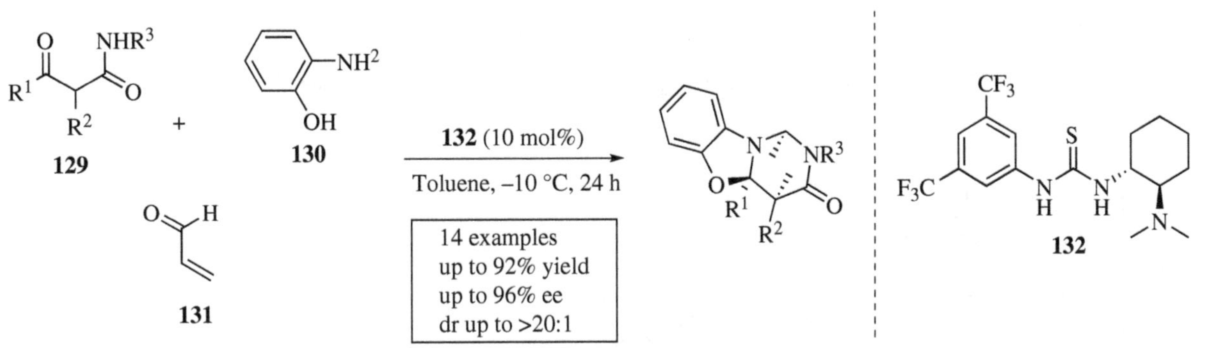

Few representative examples:

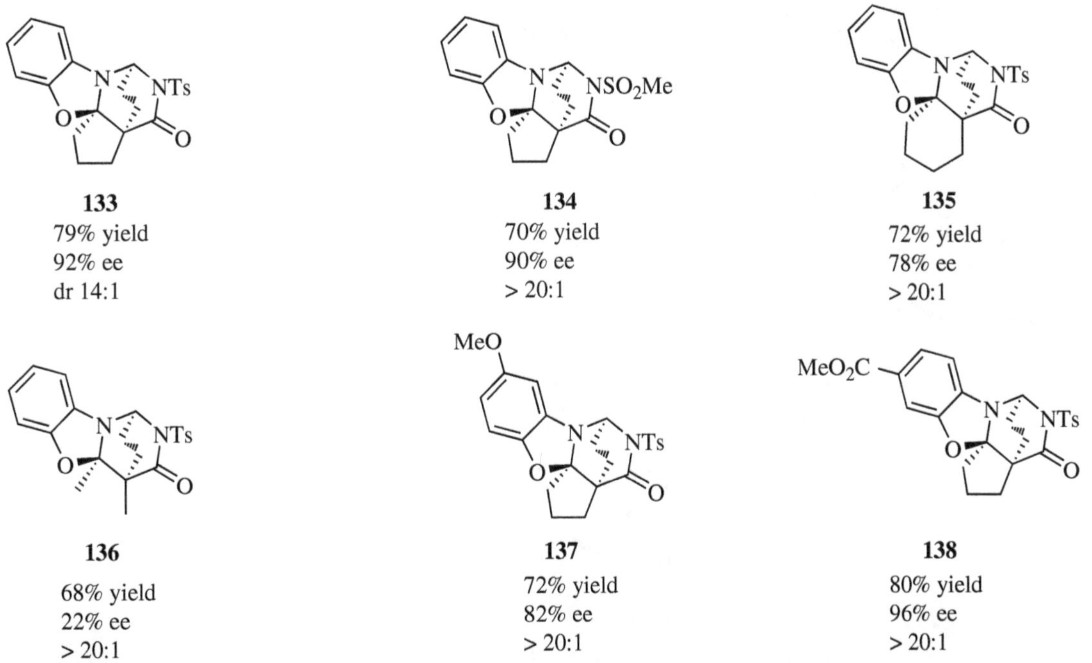

SCHEME 2.39 Enantioselective organocatalytic multicomponent synthesis of 2,6-diazabicyclo[2.2.2]octanones.

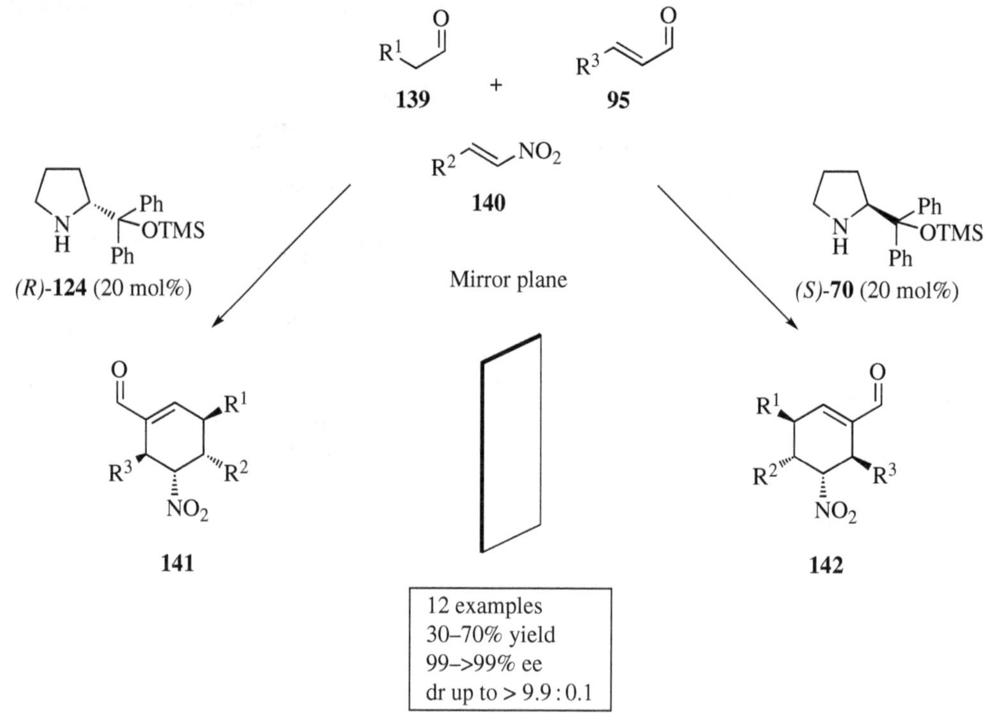

SCHEME 2.40 Asymmetric organocatalytic three-component cascade reaction.

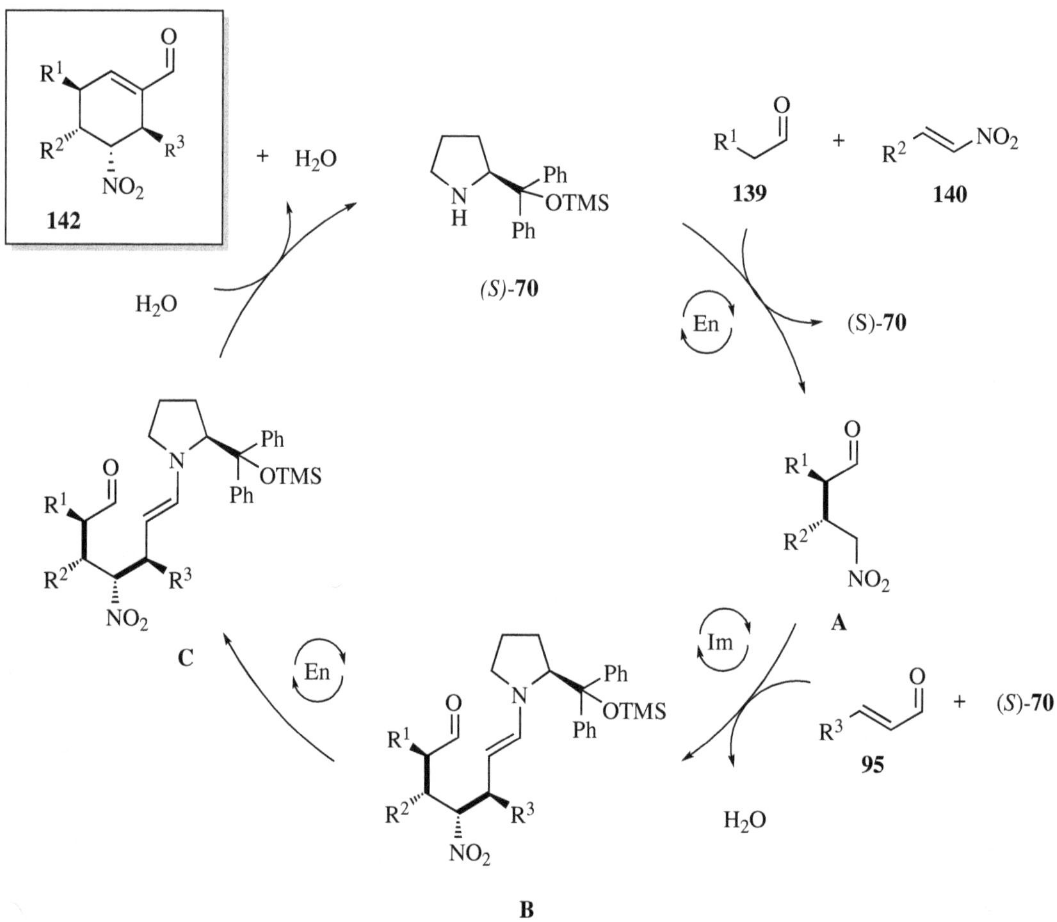

SCHEME 2.41 Proposed catalytic cycle of the triple cascade with the stereocontrolled formation of four new stereogenic centers.

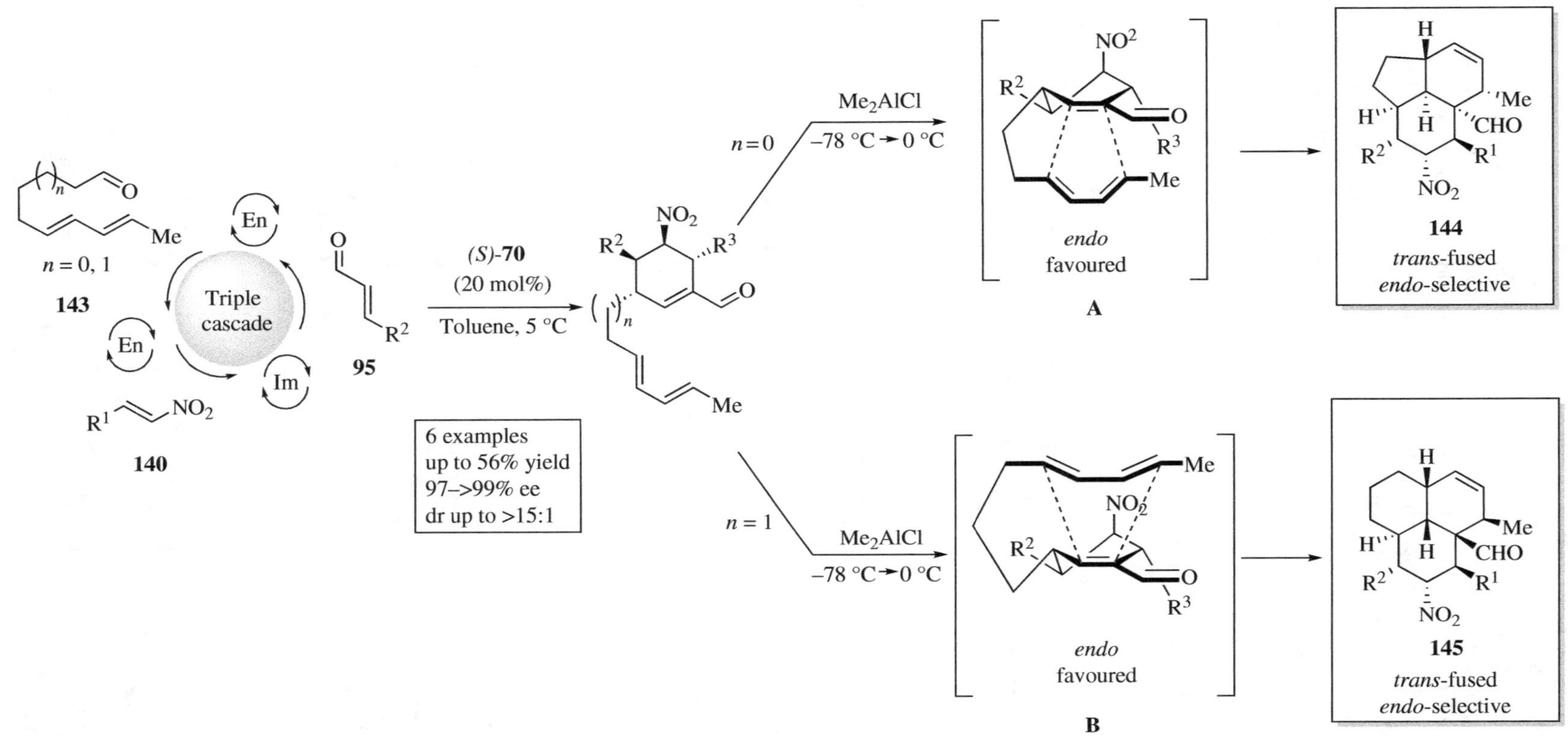

SCHEME 2.42 One-pot procedure for the synthesis of the tricyclic carbaldehyde **144** and **145**.

stereogenic centers in complex organic molecules. In particular, the isolated cyclohexane carbaldehyde bearing three or four stereogenic carbon atoms was prepared in high diastereomeric and enantiomeric control (Scheme 2.43).

The extension of such a protocol to more electron-poor olefins such as α-alkylideneoxindoles and benzofuranones derivatives outstandingly afforded the corresponding spiro compounds **147** and **148** [73] with very high outcomes in terms of yields (up to 74%) and diastereo- and enantioselectivity (dr up to 19:1, up to >99% ee).

Likewise, the application of a similar protocol to unsaturated pyrazolones **149** allowed Rios and coworkers to accomplish the synthesis of several spiropyrazolones **150** bearing four chiral centers (Scheme 2.44) [74]. The proposed Michael/Michael/aldol reaction resulted to be quite efficient in terms of both yields (up to 65%) and stereoselectivity (dr up to >25:1, ee up to >99%) when the (*S*)-α,α-diphenylprolinol **70** was employed as catalyst together with benzoic acid as additive.

Conversely, in 2007, Jørgensen disclosed a sequential iminium–iminium–enamine catalysis as an alternative to

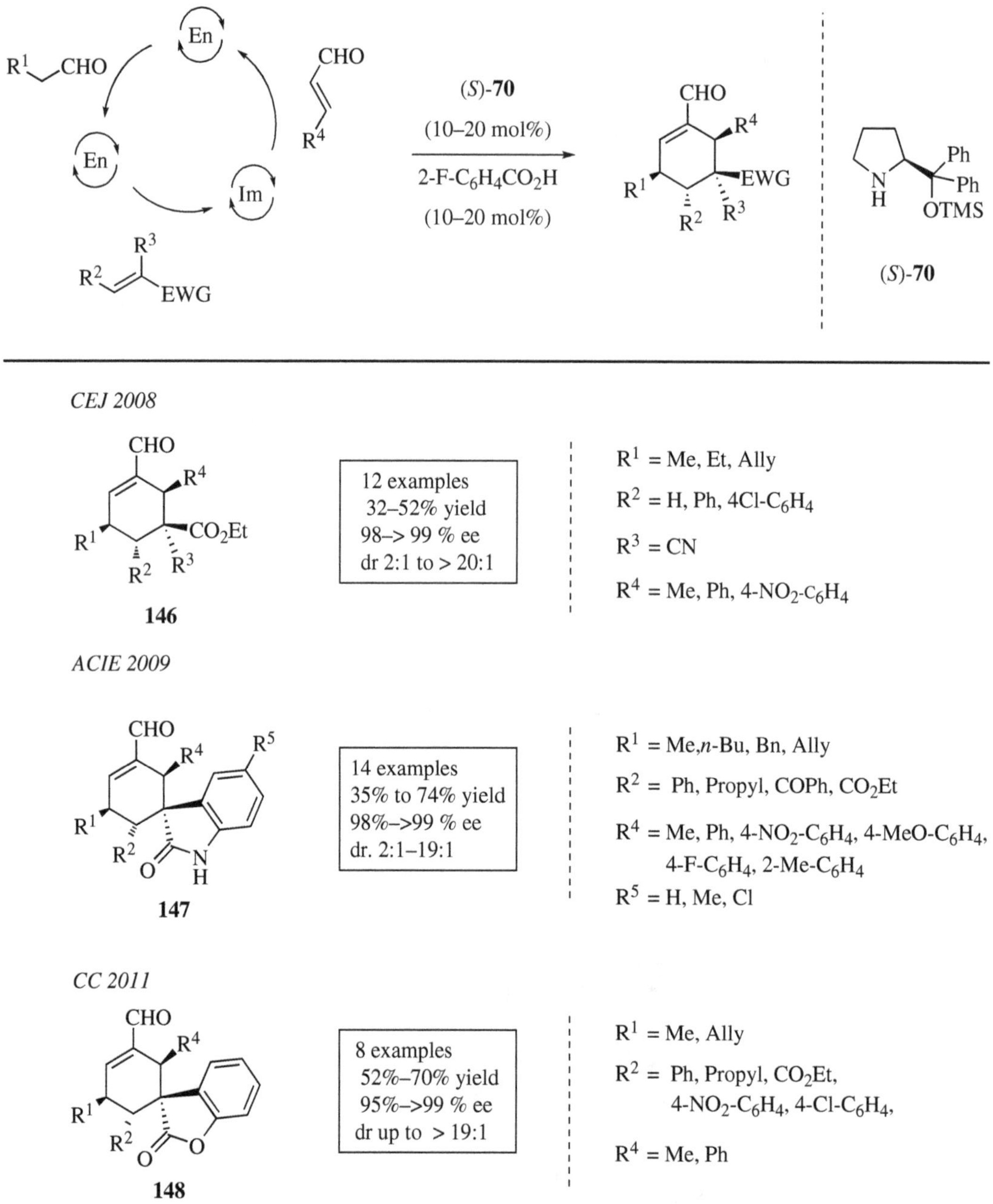

SCHEME 2.43 Melchiorre's approach to complex cyclohexyl carbaldehydes.

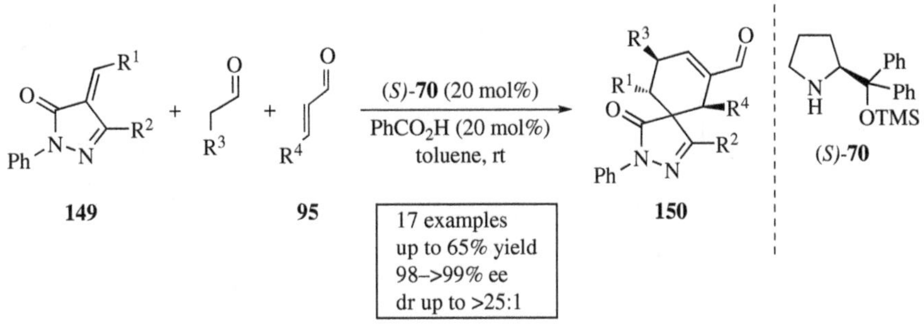

SCHEME 2.44 Asymmetric synthesis of pyrazolones **150**.

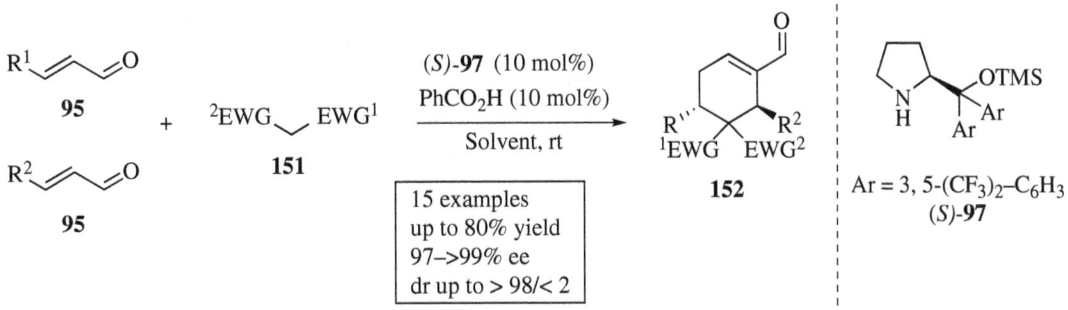

SCHEME 2.45 Multicomponent domino organocatalytic reactions toward cyclohexane derivatives **152**.

Enders triple cascade reaction for the formation of three new C—C bonds and providing enantiopure cyclohexane derivatives **152** through reaction of α,β-unsaturated aldehydes **95** with activated methylene compounds **151** using his famous Jørgensen–Hayashi catalyst (*S*)-**97** (Scheme 2.45) [75].

The observed high stereoselectivities was explained introducing the two catalytic cycles reported in Scheme 2.46. In particular, the aldehyde **95** is initially activated by catalyst (*S*)-**97** as iminium ion **A** toward the nucleophilic attack of a methylene compound **151** to give the enamine **B**, which immediately undergoes a fast hydrolysis producing the intermediate **C**. The just regenerated catalyst reenters in the second cycle to form a new iminium ion **D**, which reacts with **C** to generate an additional stereocenter in the resulting intermediate **E**. Such peculiar enamine mediates the final ring-closure reaction by an intramolecular aldol condensation affording the desired cyclic products **152** with very high enantiopurity (97 to >99% ee).

Likewise, the use of the simpler nitromethane (**153**) as nucleophile and the less hindered diphenylprolinol TMS-ether as catalyst allowed Enders and coworkers the achievement of 5-nitrocyclohexane carbaldehydes **154** in acceptable yields as pure stereoisomers bearing three contiguous stereocenters (Scheme 2.47) [76].

Once again, Melchiorre et al. took advantage of such concurrent organocatalytic MCRs to demonstrate the possibility of producing oxygenated spirocyclic derivatives **156** by using the simplest benzofuranone **155** as nucleophile and a Michael/Michael/intramolecular aldol condensation sequence as aminocatalyzed cascade reaction (Scheme 2.48) [73].

All the reported examples are characterized by high level of diastereo- and enantiocontrol (dr up to 19:1, >99% ee) even though the overall yields were not always satisfying.

Among all the possible approach to polysubstituted cyclohexane derivatives, Dixon et al. proposed a further and very elegant alternative approach wherein the asymmetric organocatalytic cascade reaction relies on the synergic action of a bifunctional base/Brønsted acid catalyst **158** and a cyclic secondary amine (*S*)-**76** (Scheme 2.49) [77].

The detailed mechanistic explanation pictures the initial weak interaction among the bifunctional thiourea catalyst **158**, the malonic ester **157** as nucleophile, and the nitroalkane **140** as electrophile that should promote the first chemo- and stereoselective Michael addition. The resulting adduct **A** would be poised to participate directly in the second catalytic cycle by acting as donor in a regioselective nitro-Michael reaction with the α,β-unsaturated aldehyde **95**, here activated as iminium ion by the secondary amine catalyst (*S*)-**76**. The new intermediate **B**, with its aldehyde and malonate moiety suitably spatial disposed, would undergo a base-promoted aldol cyclization to efficiently generate the planned cyclohexanol **159** in moderate yield (up to 87%) and marvelous enantioselectivity (up to >99% ee).

Among the triple cascade reaction that involves at least a Michael addition, Córdova et al. developed a tandem asymmetric cyclopropanation/esterification combining amine

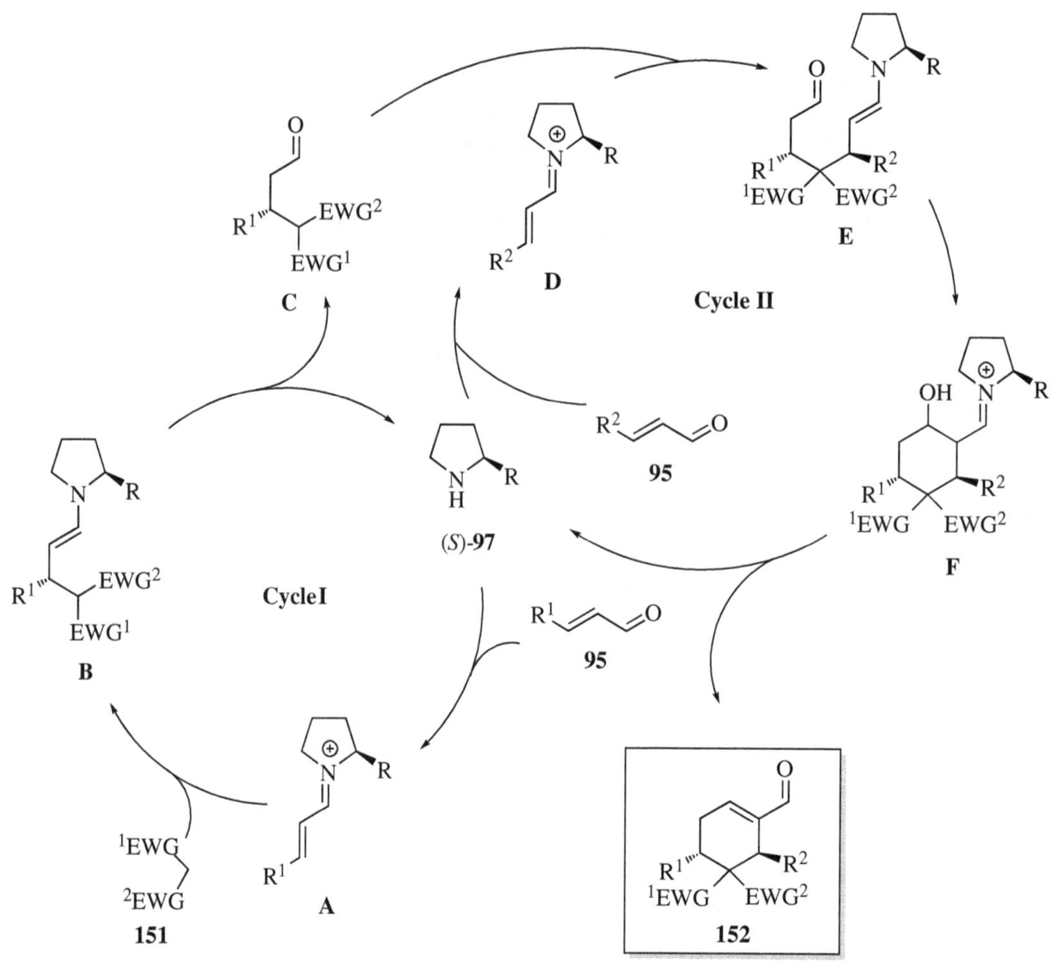

SCHEME 2.46 Proposed iminium–iminium–enamine sequential activation of α,β-unsaturated aldehyde **95**.

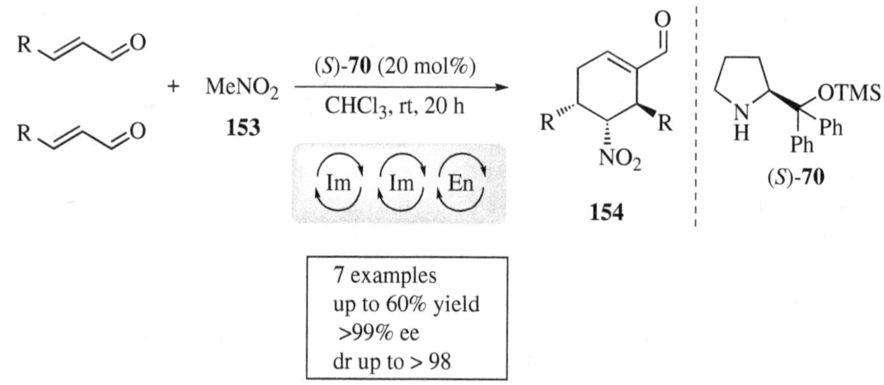

SCHEME 2.47 Organocatalytic asymmetric triple domino reaction of nitromethane (**153**).

and heterocyclic carbene catalysis (AHCC) for the direct conversion of enals **95** to valuable β-malonate esters **162** with high optical purity (94–97% ee) (Scheme 2.50) [78].

The postulated mechanism starts with the diastereo- and enantioselective domino iminium–enamine activation of enals **95** in the presence of malonate **160** to form the corresponding 2-formylcyclopopane intermediate **163**. Next, the base-generated heterocyclic carbine catalyst promotes the *in situ* C—C bond cleavage/ring-opening reaction followed by a concomitant oxidation of the resulting aldehyde and a subsequent esterification, furnishing the sought optically active β-malonate esters **162** (Scheme 2.51).

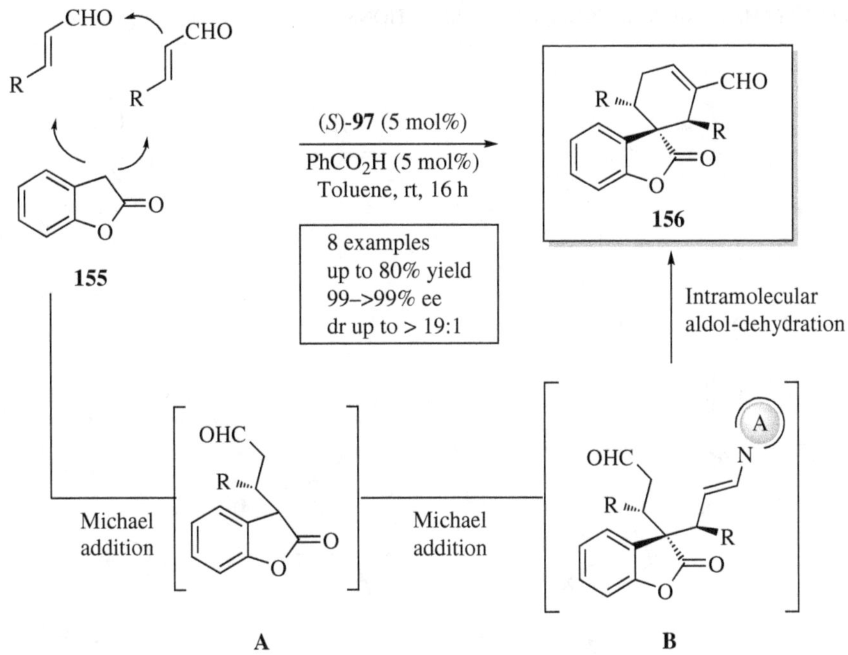

SCHEME 2.48 Preparation of spirocyclic benzofuranones **156** by aminocatalytic cascade reaction with enals.

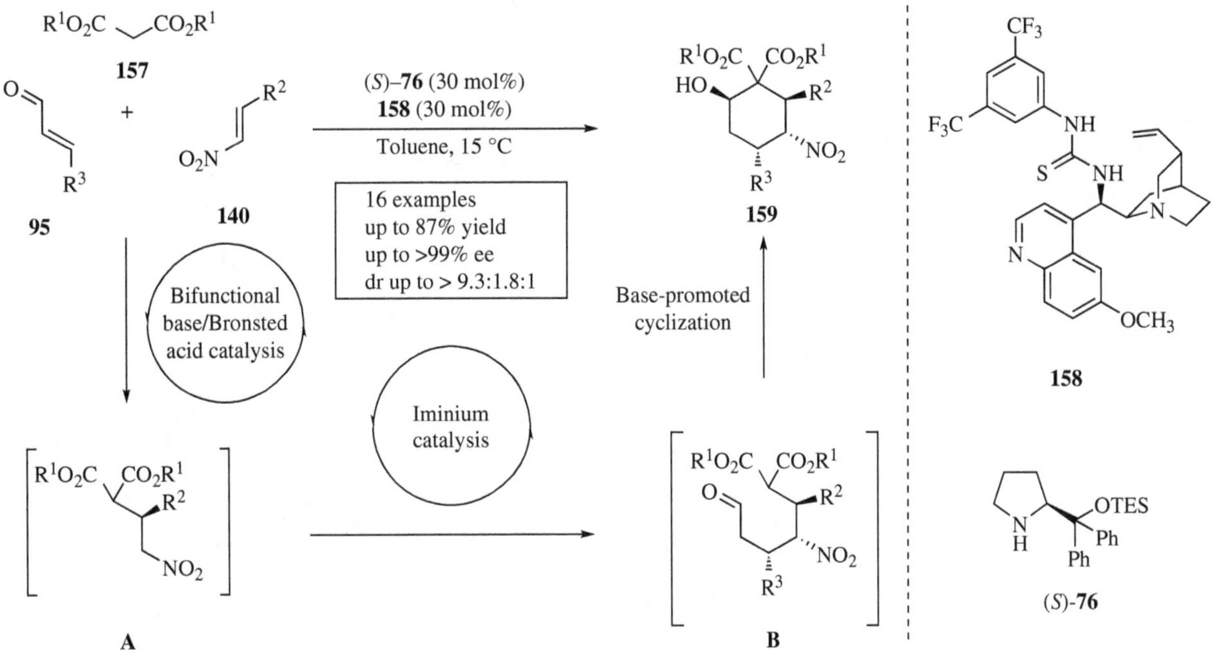

SCHEME 2.49 Asymmetric organocatalytic relay cascade (AORC) approach to functionalized cyclohexanes **159**.

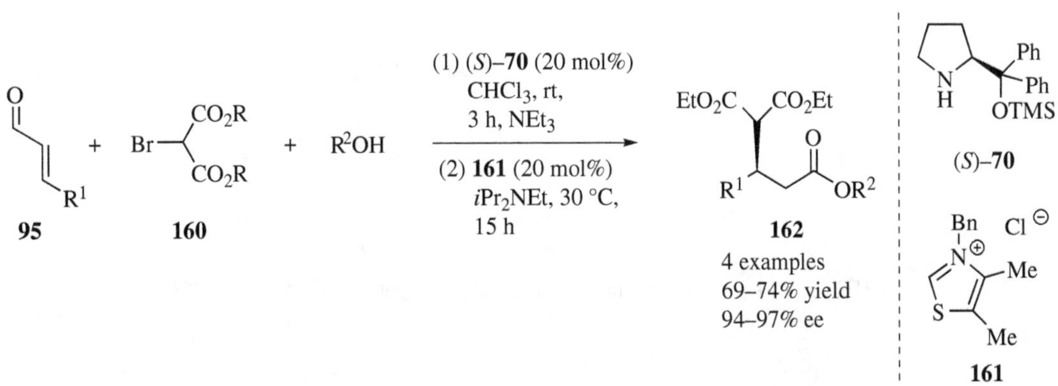

SCHEME 2.50 Organocatalytic domino Michael/α-alkylation reaction.

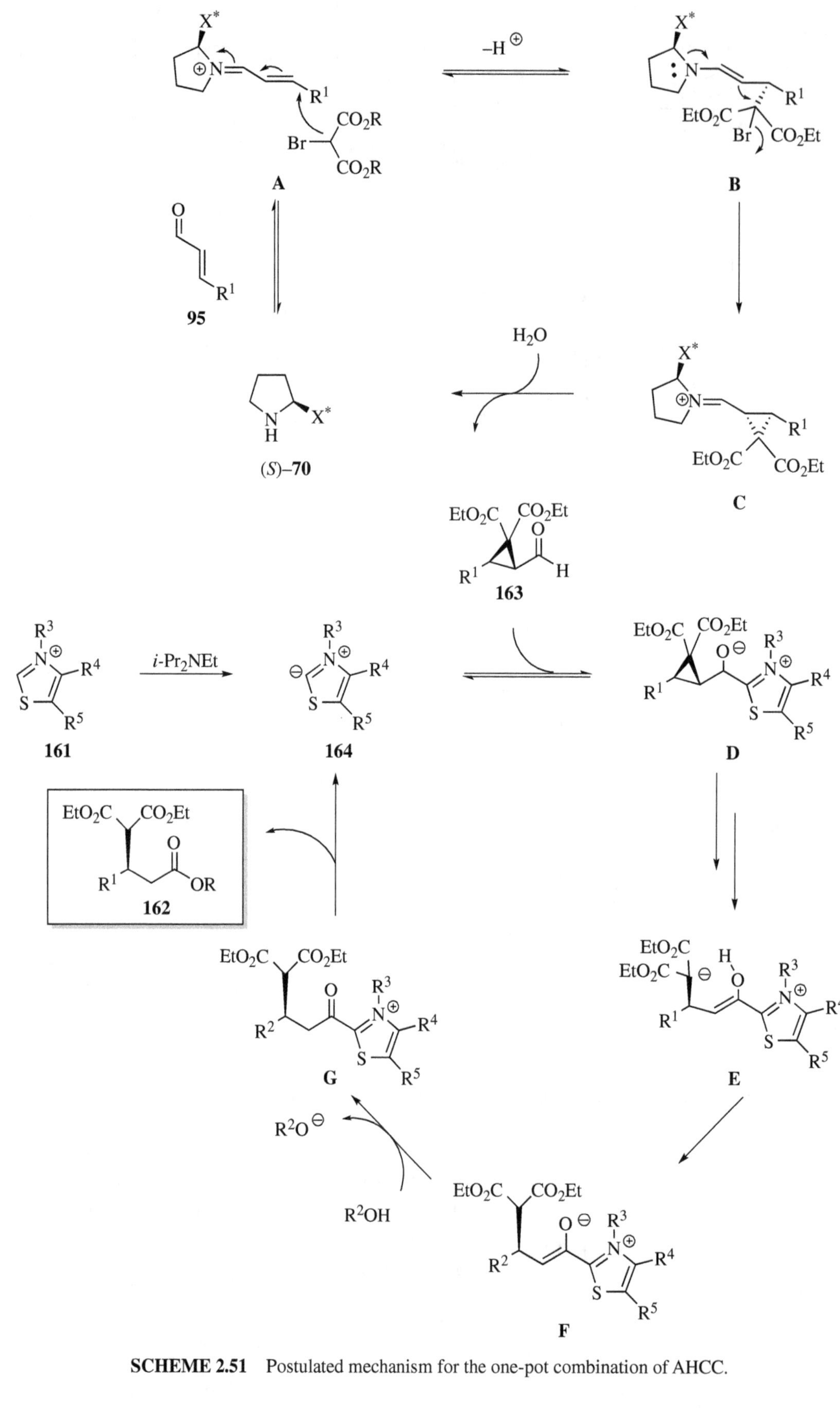

SCHEME 2.51 Postulated mechanism for the one-pot combination of AHCC.

A combination of Michael addition, Mannich reaction, and intramolecular condensation allowed Xu and coworkers to get a quite facile access to tetrahydropyridines **165** with C3 all-carbon quaternary stereocenters in moderate yields and good optical purity (up to 74% ee) [79]. The developed organocatalytic enantioselective multicomponent cascade reaction relies on the catalytic ability of the simple (*S*)-proline (**1**) that quickly reacts with the intermediate **A**, generated in turn via a Knoevenagel reaction between the β-ketoester **91** and formaldehyde **65**. The resulting iminium ion **B** undergoes the nucleophilic attack of a second moiety of β-ketoester **91** producing the Michael adduct **D**. Such intermediate enamine is then involved in the Mannich reaction with the imine **E** (due to the *in situ* condensation between primary amine **51** and formaldehyde **65**) to furnish the advanced intermediate **F**, which after an intramolecular condensation releases the (*S*)-proline (**1**), and the desired product **165** (Scheme 2.52).

2.4.1.3 Quadruple Cascade Reactions The impressive progress reached in the field of double and triple cascade reactions spurs chemists to develop more and more complex higher-order MCRs, mimicking as well as possible Nature's performance. With this aim, various quadruple cascade processes have been developed, allowing the preparation of remarkable complexity molecules. The first example was elegantly described in 2009 by Hong and coworkers, who concocted an efficient procedure for the asymmetric synthesis of tetrahydro-6*H*-benzo[*c*]chromenes **168** [80]. The three-component cascade reaction constructs four new bonds and five stereocenters on the basis of a domino oxa-Michael/Michael/Michael/aldol condensation reaction reaching high level of optical purity (up to 74% ee). Indeed, of the possible 32 different stereoisomers, only one enantiomer was observed and isolated. Such impressive stereoselectivity was ascribed, by the authors, to the first oxa-Michael addition, whereas the resulting intermediate influences the stereocontrol of the subsequent reaction sequence (Scheme 2.53).

Remaining in the field of hetero-Michael reaction, Gong et al. disclosed a four-component quadruple cascade reaction activation initiated by oxa-Michael addition of alcohol to acroleins providing an easy and direct access to highly functionalized chiral trisubstituted cyclohexene derivatives **170** (Scheme 2.54) [81].

The proposed sequence begins with the condensation between the catalyst (*S*)-**70** and the acrolein **131** to give the iminium intermediate **A**, which selectively reacts with the hard oxygen nucleophile **169**, to afford the enamine intermediate **B**. The preferred Michael addition to nitroalkene **140** furnishes **C**, which in the third step reacts with a second equivalent of **A** to generate the unstable enamine **D**, which under a fast intramolecular aldol condensation yields the wanted product **170**.

Relying, as in the previous studies, on the capability of secondary amine catalysts to activate both nucleophile and electrophile, Enders assembled a quadruple domino Friedel–Crafts-type/Michael/Michael/aldol condensation sequence, affording an efficient and direct asymmetric synthesis of polyfunctionalized 3-(cyclohexenylmethyl)-indoles **172** with moderate to high yields (Scheme 2.55) [82].

The observed excellent stereoselectivities (dr = 91:9 to >95:5, 94 to >99% ee) could be ascribed to the steric hindrance created by the employed catalyst in each step of the catalytic cycle reported below (Scheme 2.56). Once the chiral amine (*S*)-**70** activates the acrolein **131** as electrophile by generating the vinylogous iminium ion **A**, the indole **171** performs an intermolecular Friedel–Crafts-type reaction. The resulting enamine **B** acts as nucleophile in the Michael addition of the nitroalkene **140** leading to the iminium ion **D**, which upon hydrolysis liberates the catalyst and yields the intermediate aldehyde **173**. The latter compound enters in the second cycle by reacting with the iminium ion **A**, previously formed by the free catalyst. The subsequent intramolecular enamine-mediated aldol reaction of **E** completes the ring closure generating the intermediate **F**, which after dehydration and hydrolysis is transformed in the desired indole **172**.

More recently, Enders et al. disclosed a facile access to tetracyclic double annulated indole derivatives **175**, which basically relies on the chemistry of the acidic 2-substituted indole and its nitrogen nucleophilicity. Indeed, the employed quadruple cascade is initiated by the asymmetric aza-Michael-type *N*-alkylation of indole-2-methylene malononitrile derivative **174** to α,β-unsaturated aldehydes **95** under iminium activation (Scheme 2.57). The next well-known enamine–iminium–enamine sequence, which practically is realized with an intramolecular Michael addition followed by a further intermolecular Michael and aldol reactions, gives access to the titled tetracyclic indole scaffold **175** with *N*-fused 5-membered rings annulated to cyclohexanes in both diastereo- and enantioselectivity [83].

As often demonstrated by the amazing Melchiorre's work (see Section 2.4.1.2), spirooxindole motif incorporating a six-membered cyclic moiety is a common feature in a number of natural products as well as bioactive important compounds, whose synthesis is still a challenging goal. Taking advantage from Melchiorre's studies and their own efforts in three-component cascade reaction, Chen et al. successfully developed a Michael/Michael/Michael/aldol process achieving a number of densely substituted hydrindane derivatives incorporating a spirooxindole core **177** (Scheme 2.58) [84]. The formation of six contiguous stereocenters should be easily rationalized invoking quadruple iminium–enamine–iminium–enamine catalysis, which involves the α,α-diphenylprolinol (*S*)-**70** as catalyst, two molecules of α,β-unsaturated aldehydes **95**, and the

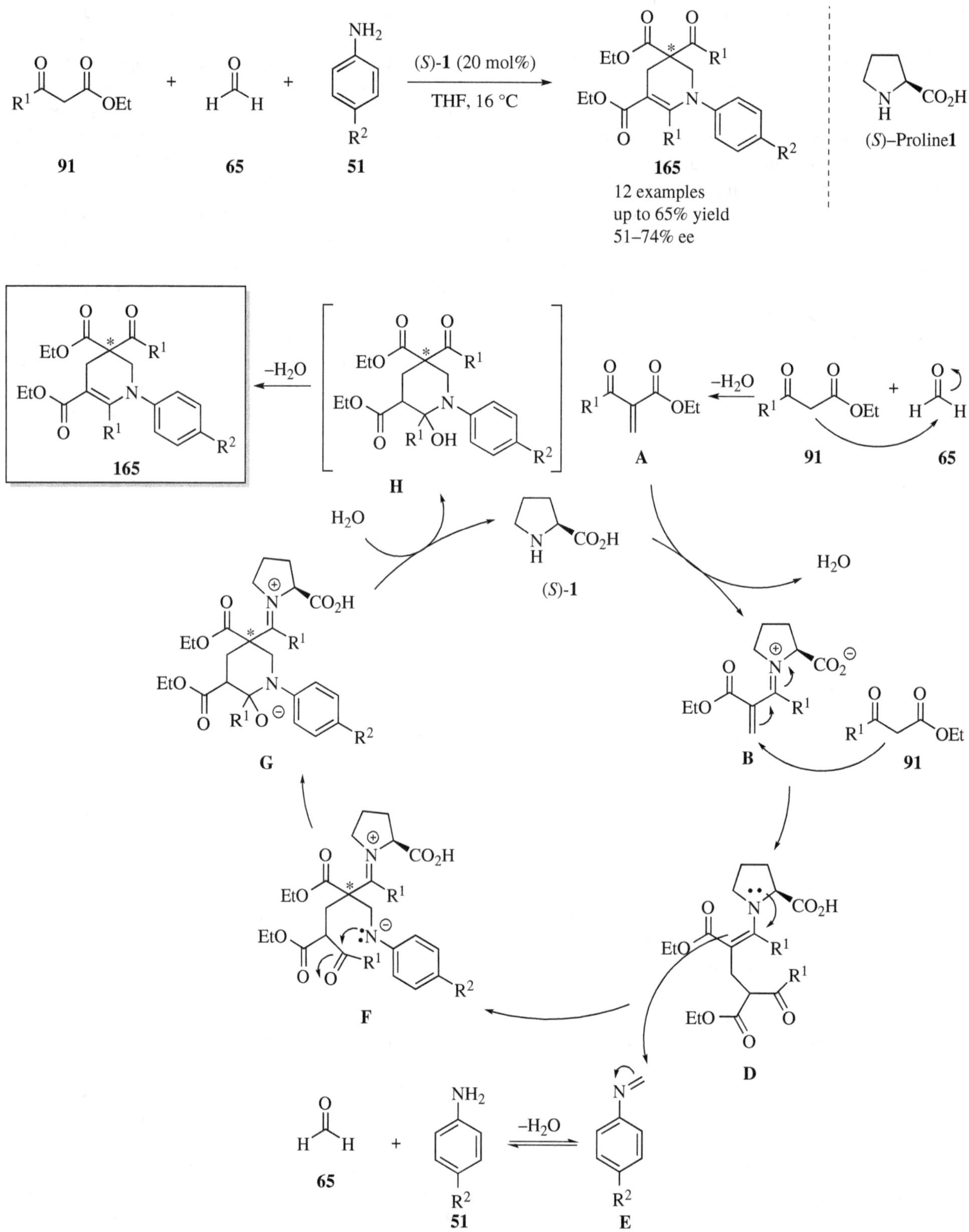

SCHEME 2.52 Synthesis of tetrahydropyridines via Michael/Mannich/cyclization MCR.

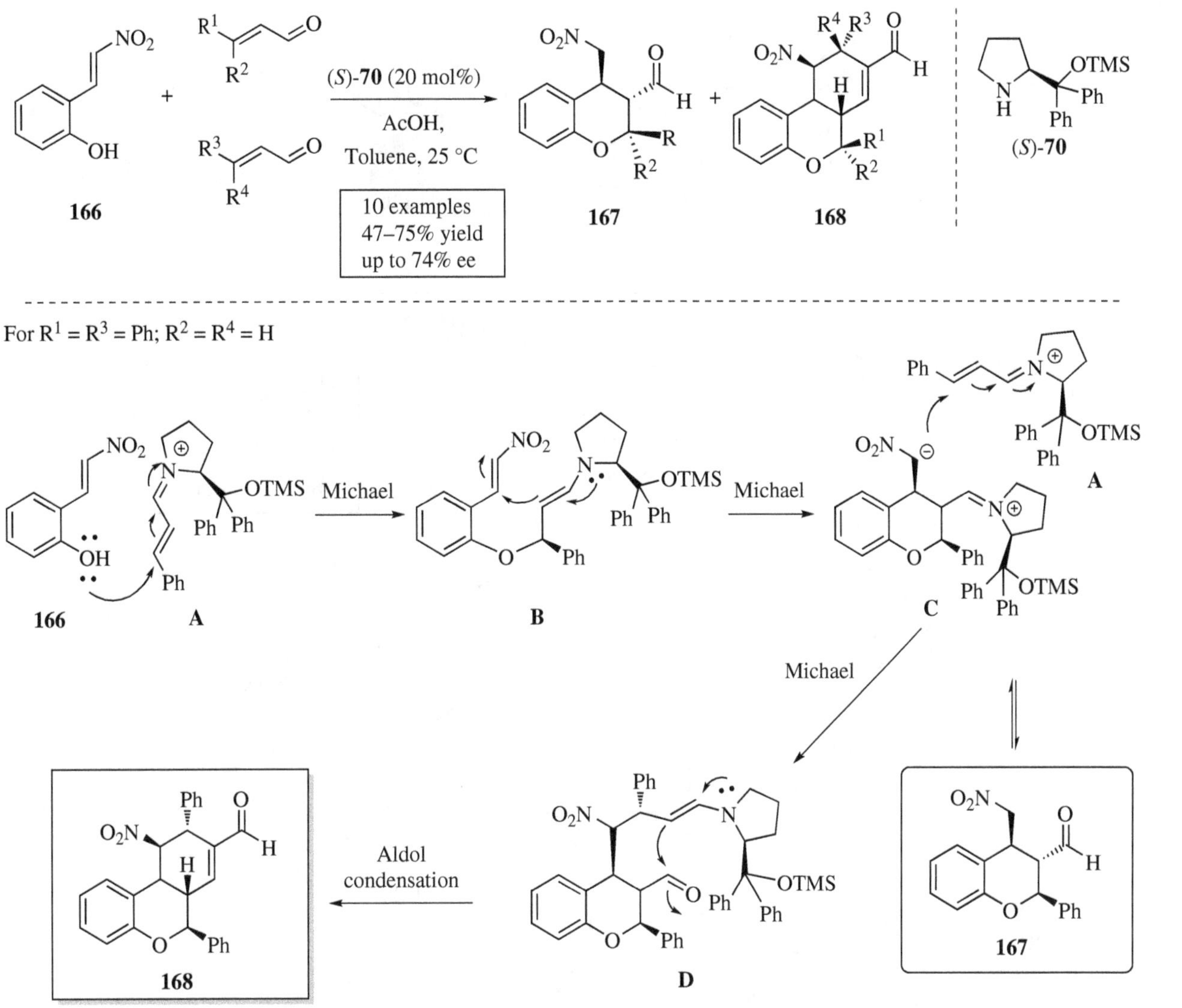

SCHEME 2.53 Synthesis of tetrahydro-6*H*-benzo[*c*]chromenes **168** via quadruple cascade reaction.

3-oxo-4-(2-oxoindolin-3-ylidene)butanoate **176** with its multiple reactive sites.

2.4.2 Miscellaneous Domino Reactions

In addition to the just disclosed MCRs, which involve at least a Michael addition, several other procedures have been developed for the construction of quite complex chiral molecular architectures from simple materials. Among them, Barbas et al. reported the first highly diastereoselective organocatalytic direct heterodomino Knoevenagel/Diels–Alder/epimerization [85]. These reactions gave highly substituted prochiralspiro[cyclohexane-1,2′-indan]-1′,3′,4-trione **179** in excellent yields (up to 99%) and a highly diastereoselective fashion [dr (*cis/trans*) up to 99:1] but with low to moderate enantioselectivity, employing easily available starting materials, that is, 4-substituted-3-buten-ones **84**, aldehydes **9**, and 1,3-indandione **178**. This three-component heterodomino Knoevenagel/Diels–Alder/epimerization catalyzed by (*S*)-proline (**1**) results to be significantly solvent and temperature dependent in terms of reaction rates, yields, and drs (Scheme 2.59).

On the basis of the proposed catalytic cycle, aldehyde **9** reacts with 1,3-indandione **178** by amine-catalyzed Knoevenagel condensation to provide the benzylidene-indandione **B** via iminium ion **A**, which is an excellent organic Lewis acid. This intermediate thus undergoes a Diels–Alder or a double Michael reaction with the soft nucleophile 2-amino-1,3-butadiene **C** generated *in situ* from enone **84** and the amine catalyst to produce the

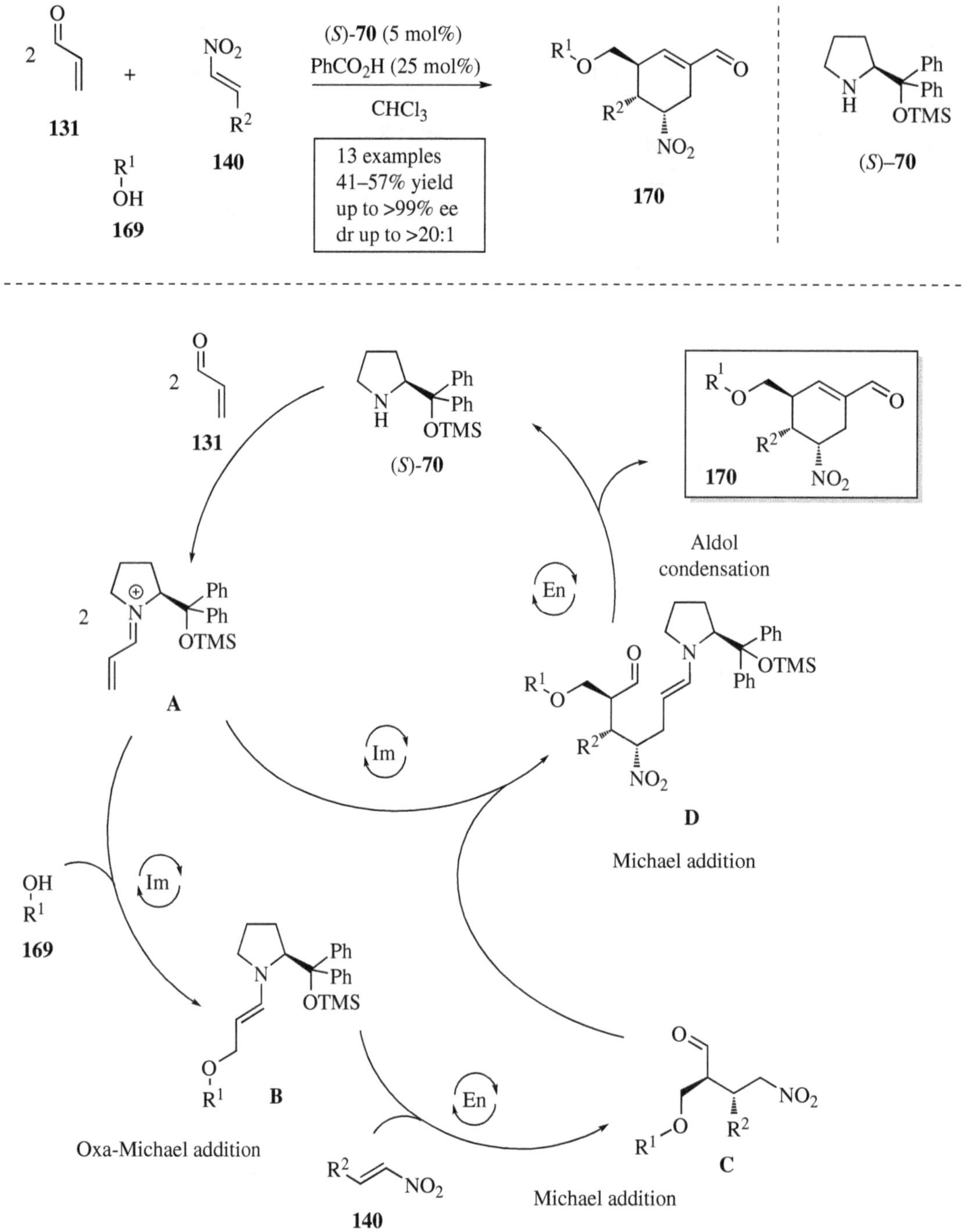

SCHEME 2.54 Trisubstituted cyclohexane derivatives **170** via quadruple cascade reaction.

diastereoisomers **D**. Afterward, the minor kinetic isomer *trans*-spiranes was epimerized to the thermodynamically stable *cis*-spiranes **179** via deprotonation/reprotonation or retro-Michael/Michael reactions catalyzed by (*S*)-proline (**1**) too (Scheme 2.59).

The same group reported an organocatalytic diastereospecific and enantioselective domino Knoevenagel/Diels–Alder procedure to obtain highly substituted spiro[5,5] undecane-1,5,9-triones **181** (attractive intermediates in the synthesis of natural products and in medicinal chemistry) from commercially available 4-substitued-3-buten-2-ones **84**, aldehydes **9**, and 2,2-dimethyl-1,3-dioxane-4,6-dione **85** (Meldrum's acid) [86]. In this asymmetric three-component sequence catalyzed directly by proline derivative **180**, only the solvent had a significant effect on the rates, yields, and ee values (Scheme 2.60).

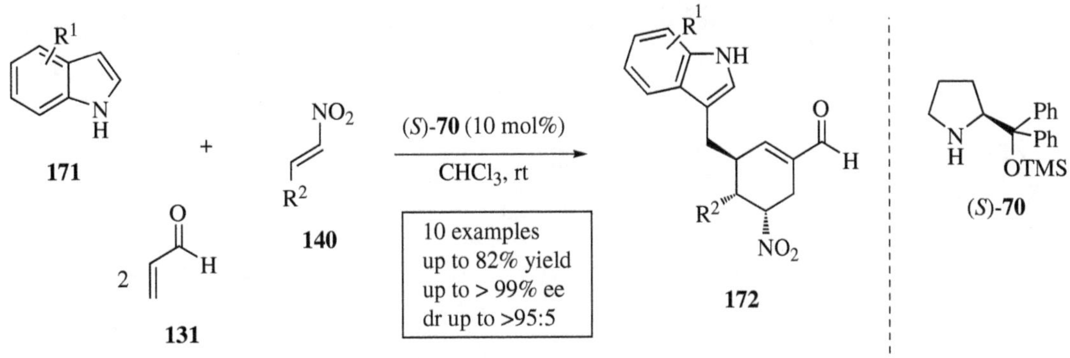

SCHEME 2.55 Enders's Friedel–Crafts-type/Michael/Michael/Aldol condensation domino sequence.

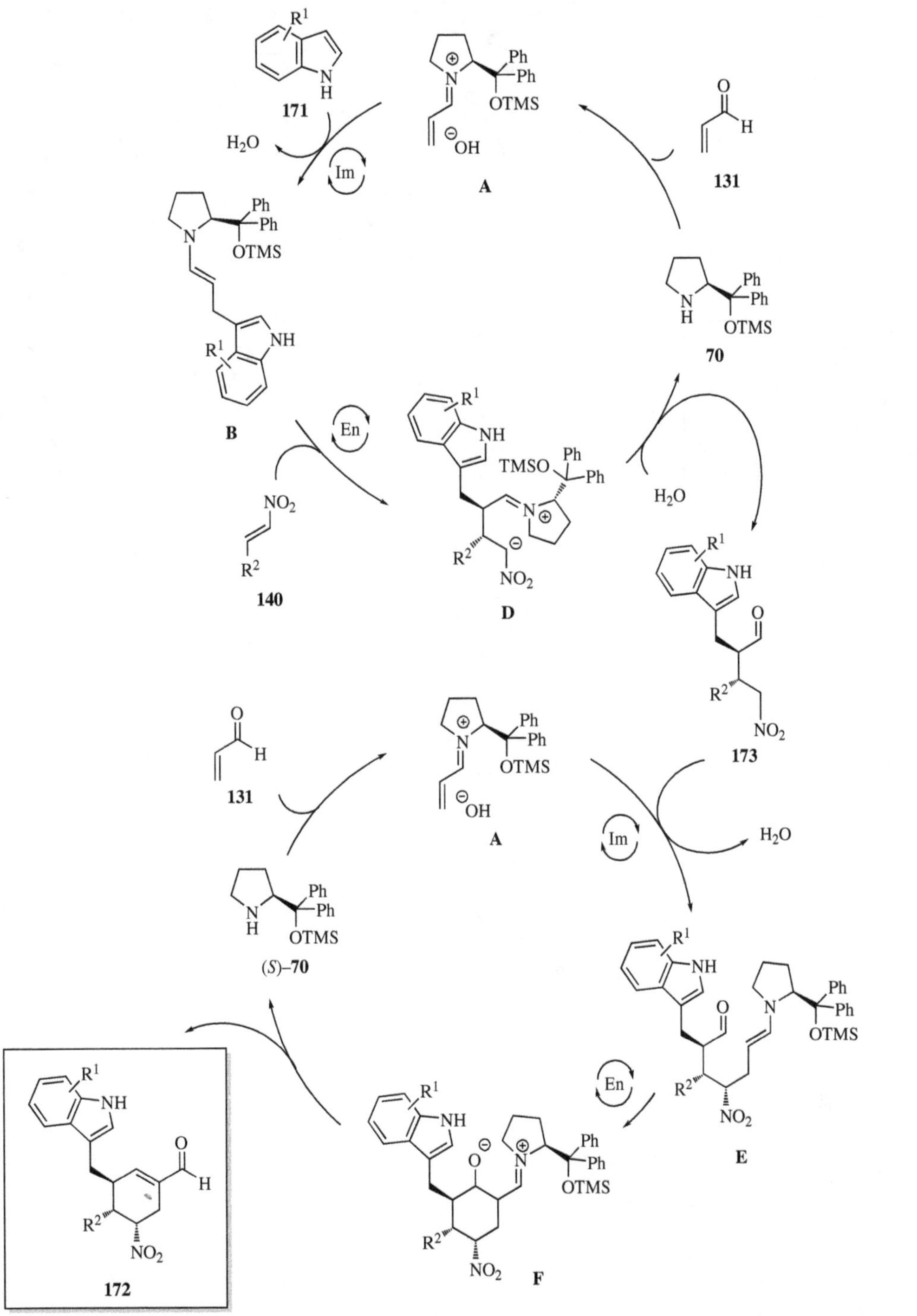

SCHEME 2.56 Postulated mechanism for the Friedel–Crafts-type/Michael/Michael/Aldol domino sequence.

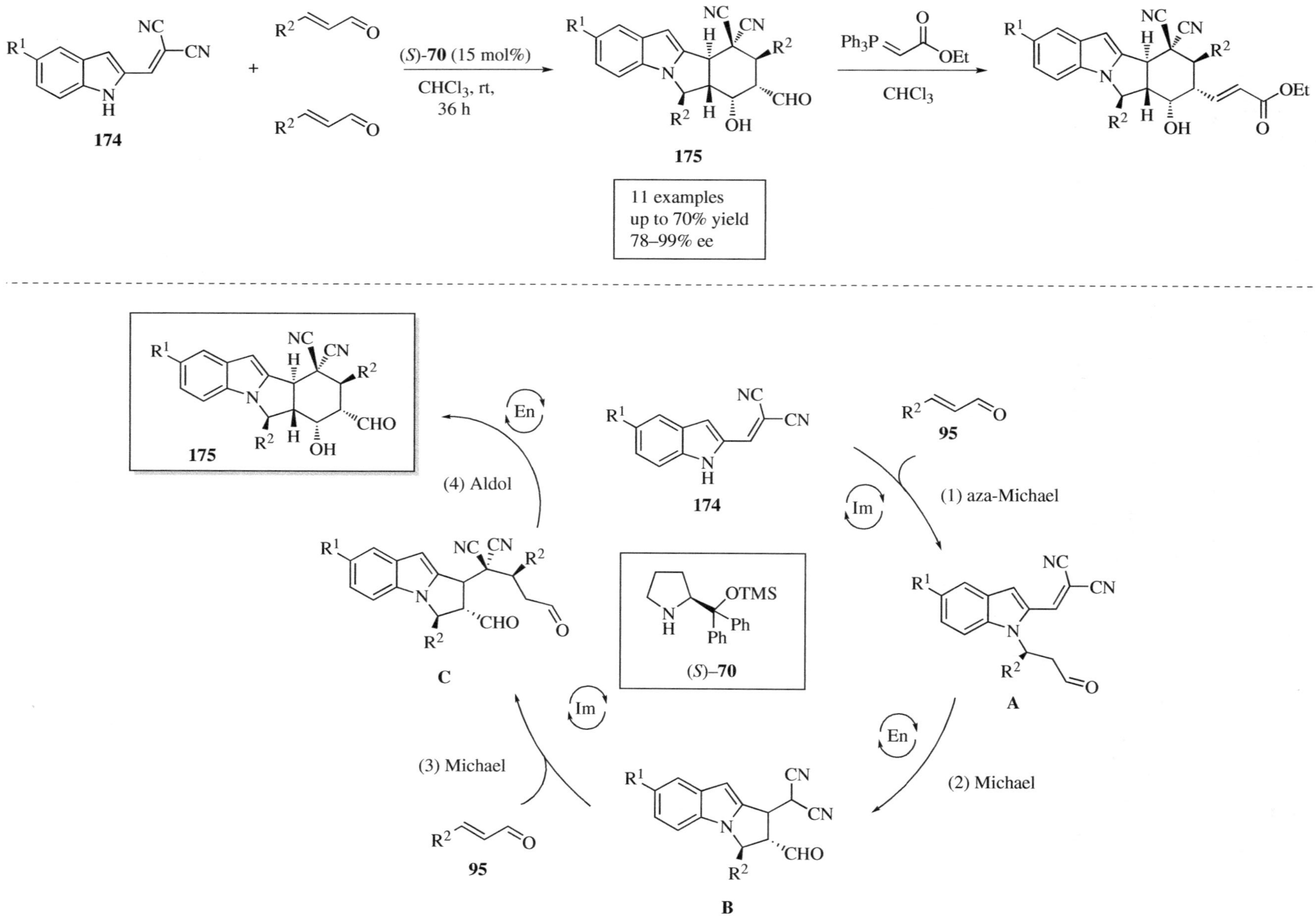

SCHEME 2.57 Asymmetric aza-Michael/Michael/Michael/Aldol sequence toward tetracyclic indole structures.

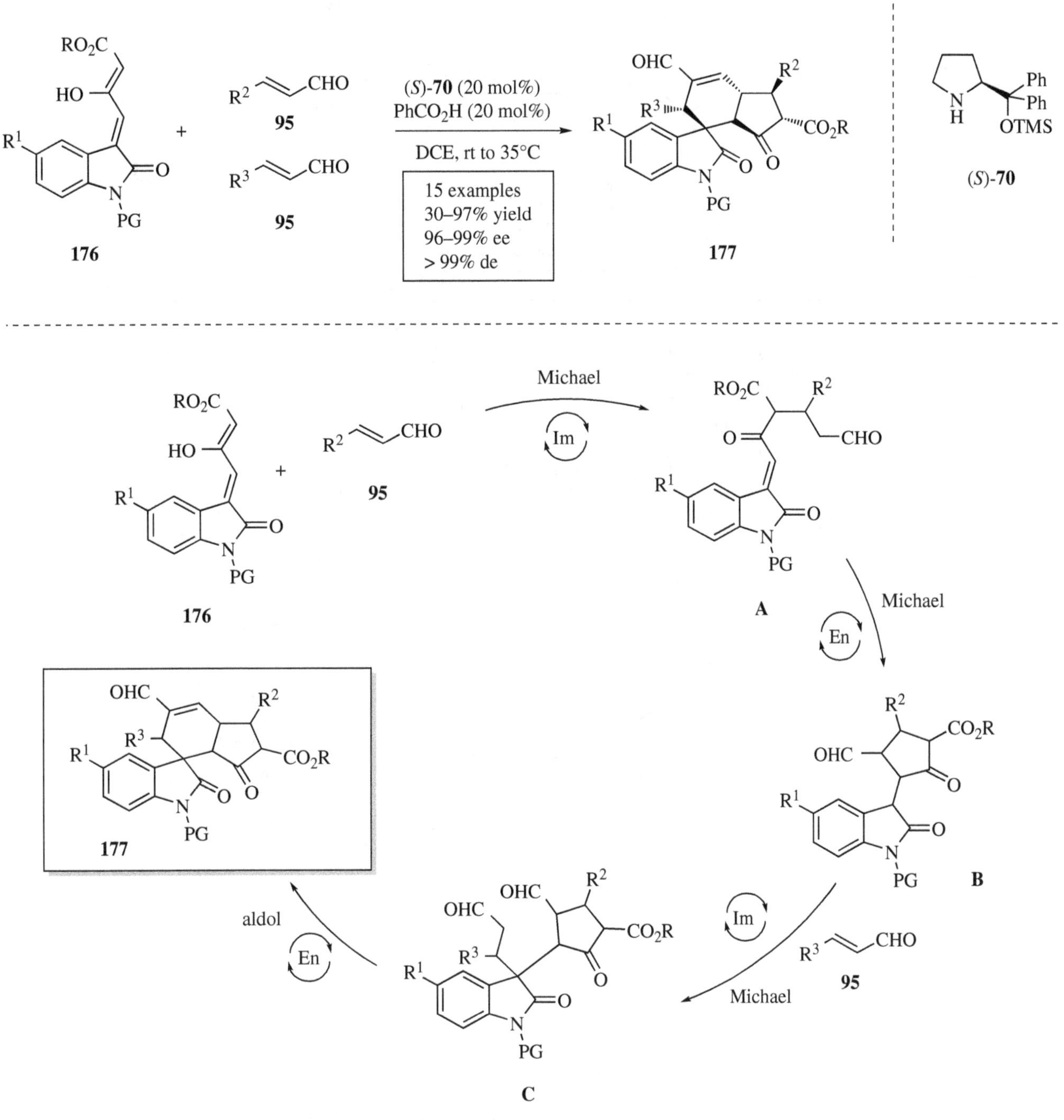

SCHEME 2.58 Quadruple iminium–enamine–iminium–enamine catalysis toward spirooxindole derivatives **177**.

The combination of the three-component domino Knoevenagel/Diels–Alder/epimerization allowed the construction of structures that with suitable modification gave the natural product-like cores as reported in a work of 2007 [87]. Hence, 1,3-dicarbonyl compounds **182** with benzaldehydes **4** and (*E*)-4-phenyl-3-buten-2-one **183** in the presence of a catalytic amount of (*S*)-5,5-dimethyl thiazolidinium-4-carboxylate **184** produced corresponding spiro compounds **185** in good yields (50–80%) and high diastereoselectivity [dr (*cis/trans*) >99:1] (Scheme 2.61).

In 2007, Ramachary et al. reported an asymmetric Knoevenagel/hydrogenation/Robinson annulation sequence to obtain Wieland–Miescher ketone **189** [88] (Scheme 2.62). The reaction of 5 equiv of aldehyde **9** with the 1,3-dicarbonyl compounds **186** (with CH acid) and Hantzsch ester **187** under proline catalysis furnished the expected cyclohexane-1,3-dione **B** in good yields. Once the solvent was removed by vacuum pump, the crude reaction mixture was diluted with DMF and treated with methyl vinyl ketone **188** in the presence of (*S*)-proline (**1**) furnishing the expected

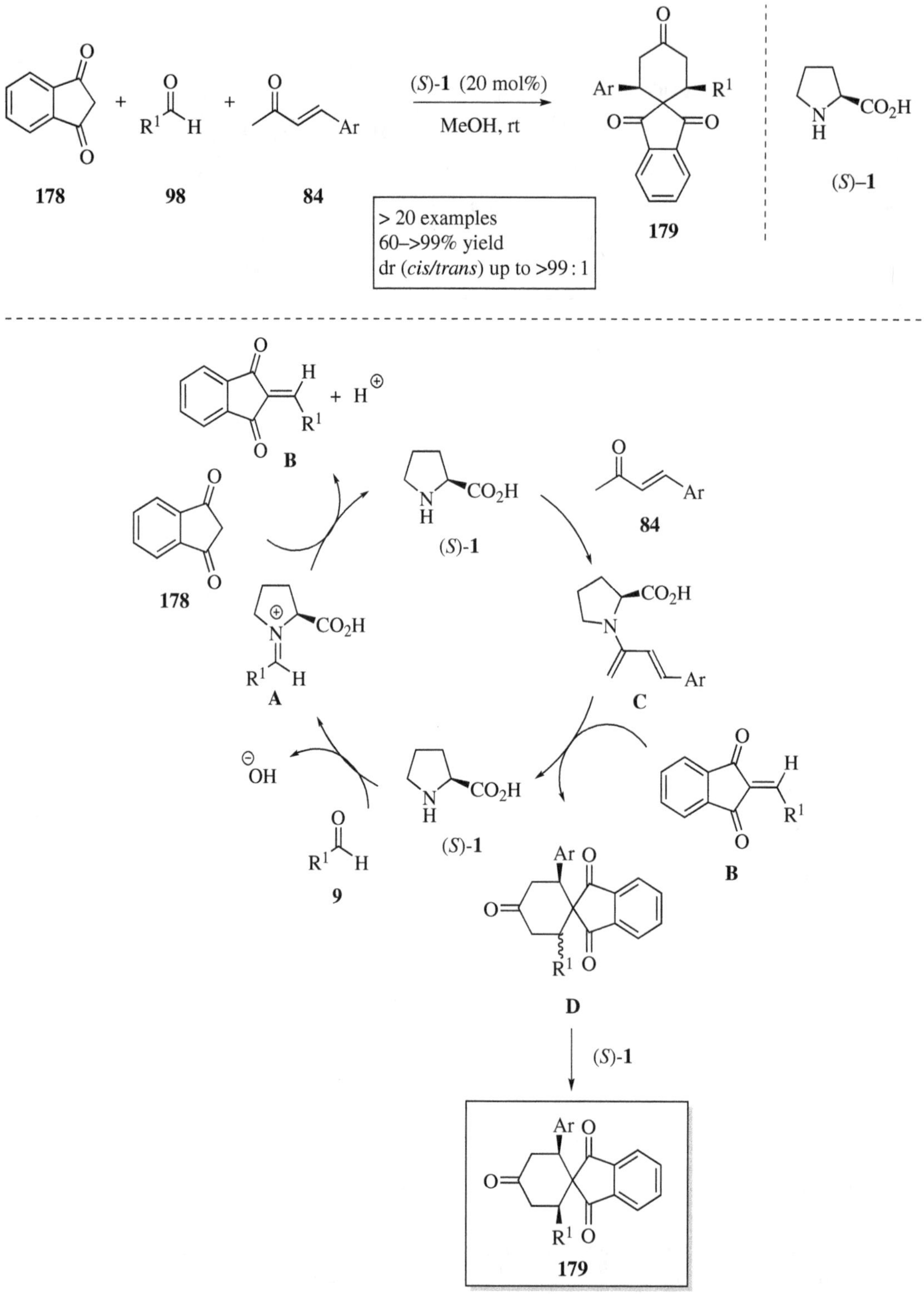

SCHEME 2.59 Heterodomino Knoevenagel/Diels–Alder/epimerization and its postulated mechanism.

Wieland–Miescher ketone **189** that is formed in quite moderate yields (up to 50%) with interesting optical purity (74% ee).

According to the proposed mechanism, in the first step, the (*S*)-proline (**1**) activates aldehyde **9** most likely by iminium ion formation, which then selectively adds to the 1,3-dicarbonyl compound **186** via Mannich and retro-Mannich-type reaction to generate active olefin **A**. Such intermediate is immediately reduced by the Hantzsch ester **187** to produce **B** through self-catalysis. In the third step, the Michael addition of **B** to

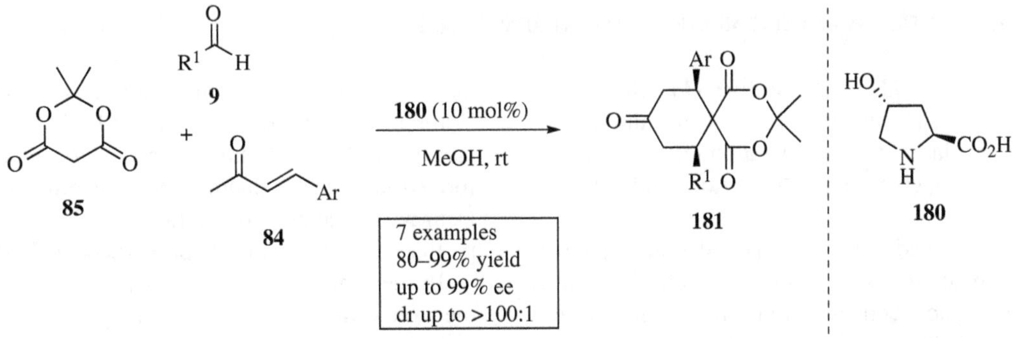

SCHEME 2.60 Synthesis of spiro[5,5]undecane-1,5,9-triones **181** via domino Knoevenagel/Diels–Alder.

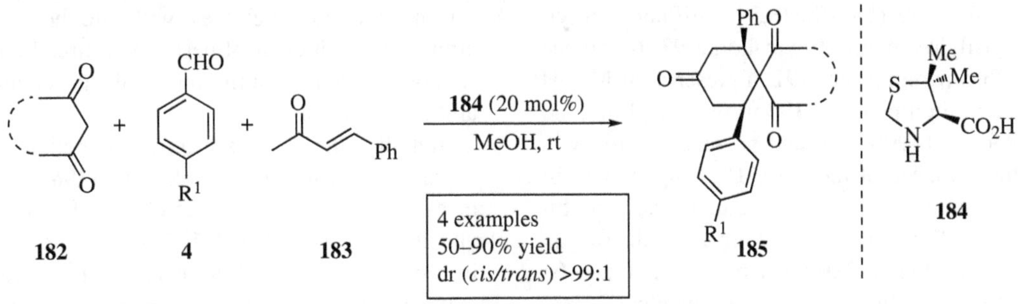

SCHEME 2.61 Three-component domino Knoevenagel/Diels–Alder/epimerization.

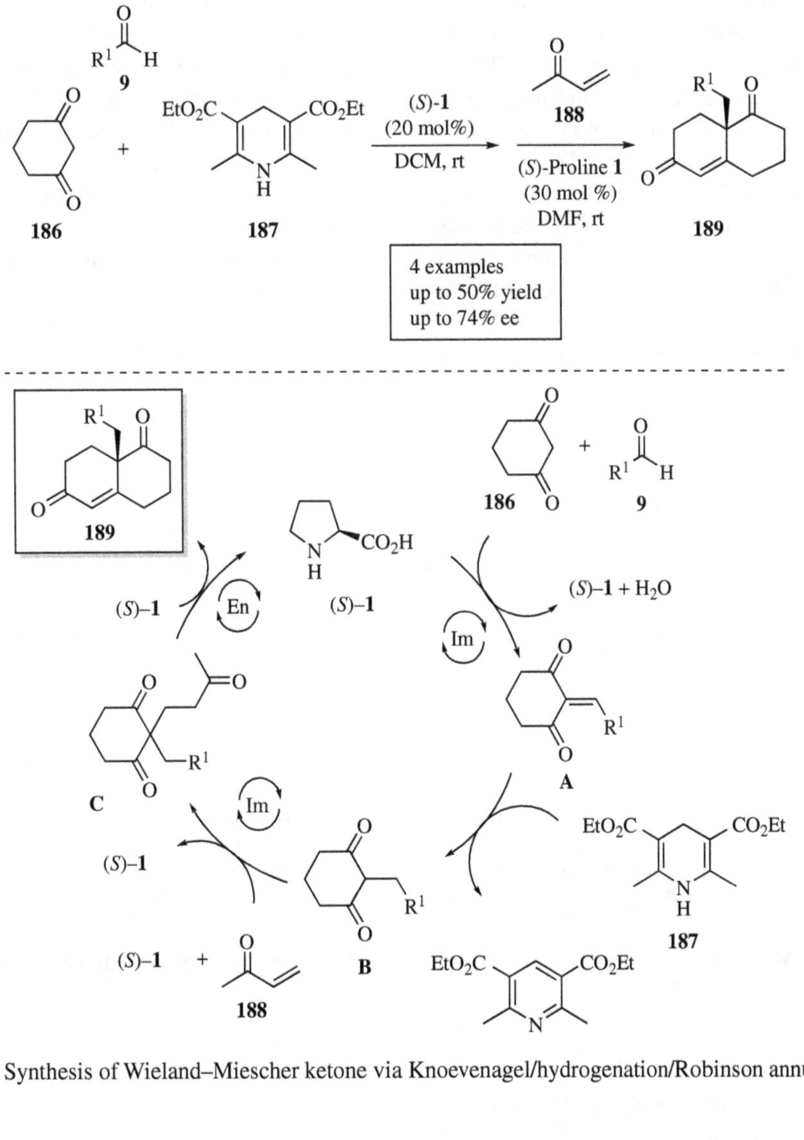

SCHEME 2.62 Synthesis of Wieland–Miescher ketone via Knoevenagel/hydrogenation/Robinson annulation sequence.

methyl vinyl ketone **188** by (*S*)-proline (**1**) affords the adduct **C**. Moreover, (*S*)-proline (**1**) catalyzed the asymmetric intramolecular aldol condensation of **C**, and the subsequent hydrolysis gives the desired ketone **189** regenerating the catalyst.

In 2009, Jørgensen and coworkers reported a simple, direct asymmetric catalytic one-pot approach to optically active propargylic and allylic fluorides starting from commercially available compounds (Scheme 2.63) [89]. In particular, chiral active propargylic fluorides **192** can be obtained by reacting aldehydes **9** with *N*-fluoro-dibenzene-sulfonimide (**190**, NFSI) in the presence of (*S*)-2-[bis(3,5-bis-trifluoromethyl-phenyl)trimethylsilyloxymethyl]-pyrrolidine **97** as catalyst and the Ohira–Bestmann reagent **191** in presence of MeOH and K_2CO_3. The desired products **192** were formed in good yields (up to 69%, 91–99% ee), and no racemization was observed for the presence of the base. The employment of quite different substrates, such as linear saturated and unsaturated aldehydes of different lengths, did not affect the overall process, wherein the Ohira–Bestmann reagent **191** can be generated directly *in situ* by the addition of dimethyl 2-oxo-propylphosphonate and 4-acetamidobenzenesulfonyl azide to the reaction mixture. Additionally, the preparation of chiral allylic fluorides was accomplished performing the reaction in the same conditions but introducing the commercially available methyl (triphenylphosphoranylidene) acetate (**193**).

Noteworthy are the quite complex mechanisms proposed by the authors for the alternative formation of propargylic and allylic fluorides as depicted in Scheme 2.64. The initial condensation of the catalyst (*S*)-**97** and aldehyde **9** leads to the formation of a reactive enamine species **B**. The intermediate product **D** obtained from iminium C, which is formed via electrophilic fluorination of B with 190, can be trapped following two alternative pathways. In the first case, methyl (triphenylphosphoranylidene) acetate (**193**) forms the oxaphosphetane intermediate **E**, and the allylic fluorides **194** are obtained via a Wittig mechanism by elimination of Ph_3PO. Otherwise, in the presence of Ohira–Bestmann reagent **191**, either preformed or formed *in situ*, the fluoroaldehyde intermediate **D** gives **G**, from which the propargylic fluoride **192** is formed by a 1,2-proton shift and elimination of N_2 (Scheme 2.64).

In the range of the study of new processes to improve the atom efficiency, the discovering versatility of a catalyst as multifunctional catalyst in a transformation is important. On the strength of this concept, Jiao and coworkers reported an enantioselective reduction and alkylation reaction of α,β-unsaturated aldehydes with alcohols, in which the ammonium salt catalyst performed three kinds of catalytic functions, namely, iminium catalysis, enamine catalysis, and acid catalysis [90].

Initially, the enantioselective reduction and alkylation reaction of *trans*-cinnamaldehyde (**196**, $R^1=R^2=H$) was investigated, with ethyl Hantzsch ester **187** and bis(4-dimethylamino-phenyl)-methanol (**195**, Ar=4-dimethylamino-phenyl) in the presence of different catalysts using toluene as a solvent. It was notable that the combination of catalysts **197** and **198** gave the formation of desired product (**199**, Ar=4-dimethylamino-phenyl) with the best result of 95% yield and 87% ee (Scheme 2.65).

This tandem enantioselective reduction and alkylation reaction was expanded to a variety of enals and benzylic alcohols without any considerable loss in terms of yield and stereoselectivity. In particular, cinnamaldehyde derivatives **196** bearing both electron-donating groups and electron-withdrawing groups smoothly underwent this transformations generating the corresponding products **199** in excellent yields (86–95%) with quite good ee values (81–90% ee). The proposed three catalytic cycles (iminium–enamine–acid catalysis) are in detail depicted in Scheme 2.66. Initially, α,β-unsaturated aldehyde **196** reacts with ammonium salt

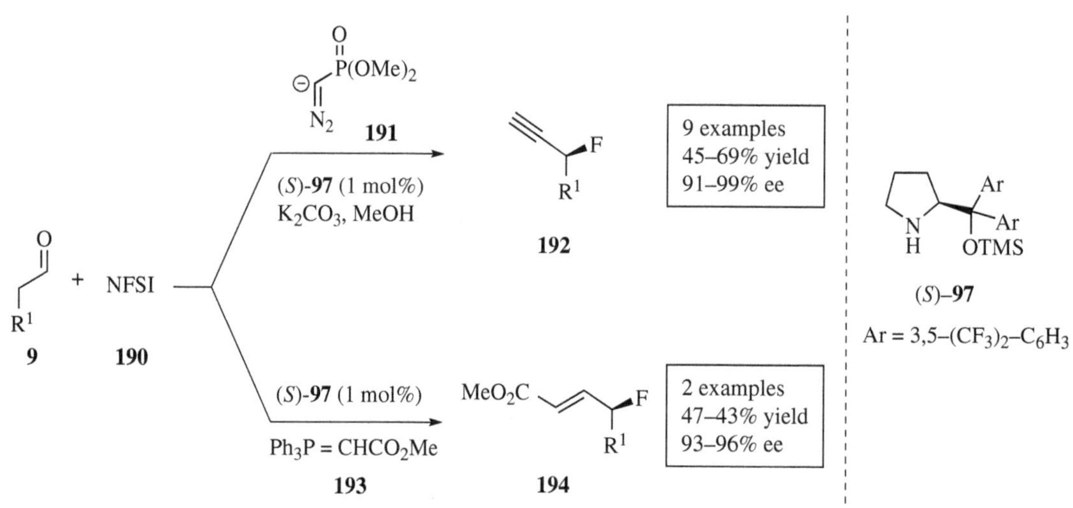

SCHEME 2.63 Jørgensen's approach to optically active propargylic **192** and allylic **194** fluorides.

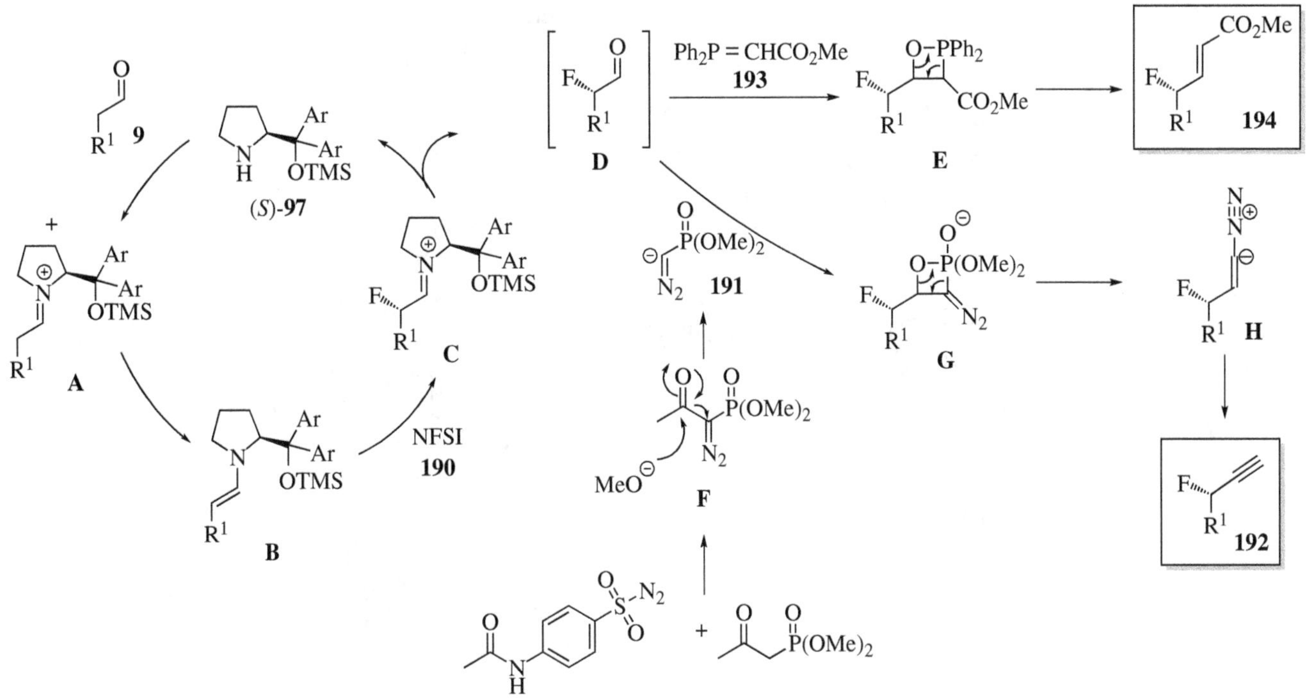

SCHEME 2.64 Proposed reaction pathway toward the preparation of propargylic and allylic fluorides.

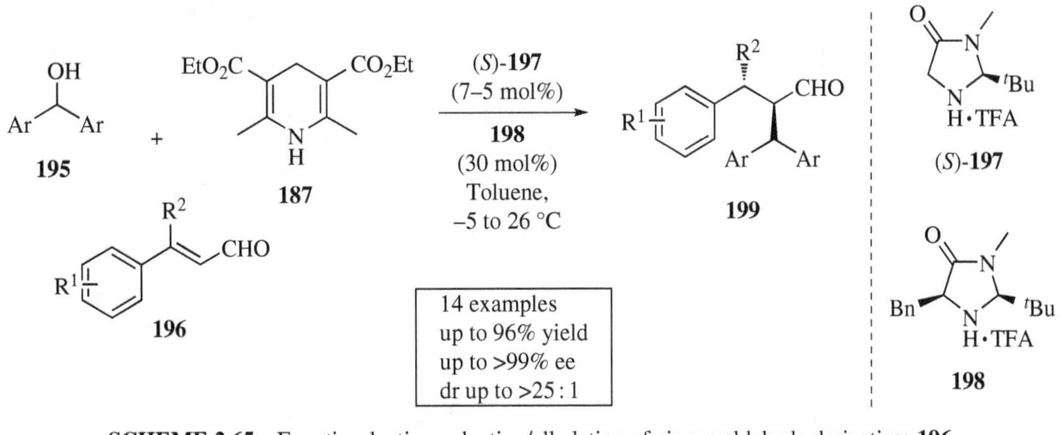

SCHEME 2.65 Enantioselective reduction/alkylation of cinnamaldehyde derivatives **196**.

catalysts to give activated iminium species **A**, which is attacked by hydrido-nucleophile ethyl Hantzsch ester **187** to generate the saturated aldehyde **200**, which subsequently undergoes the second cycle. Either the ammonium salt catalysts or the free acid (HX) could promote the formation of the carbocation **F** from the alcohol in the acid catalytic cycle. Subsequently, the activated enamine species **D** reacts with carbocation **F** to form the iminium intermediates **E**. The hydrolysis of **E** furnishes the desired product **199**.

Recently, in 2013, Lin and coworkers reported a one-pot multicomponent reaction of aromatic aldehydes **4**, anilines **51**, and β-ketoesters **91** catalyzed by the spirocyclic phosphoric acid **201** to give functionalized enantioenriched tetrahydropyridines **202** with high levels of stereocontrol [91]. The reaction proceeded using a 10 mol% chiral phosphoric acid catalyst in the presence of powdered 4 Å molecular sieves to afford the desired optically active tetrahydropyridine. The substituent on the aromatic ring of the aldehyde had a dramatic effect on the activity and stereoselectivity, and variation of the alkyl substituent of β-ketoesters could be well tolerated to provide the desired products **202** (Scheme 2.67).

Finally, the sequential iminium–enamine asymmetric catalysis was notably exploited by Roy and Chen in the kinetic resolution of nitroallylic acetates **203** (Scheme 2.68) [92].

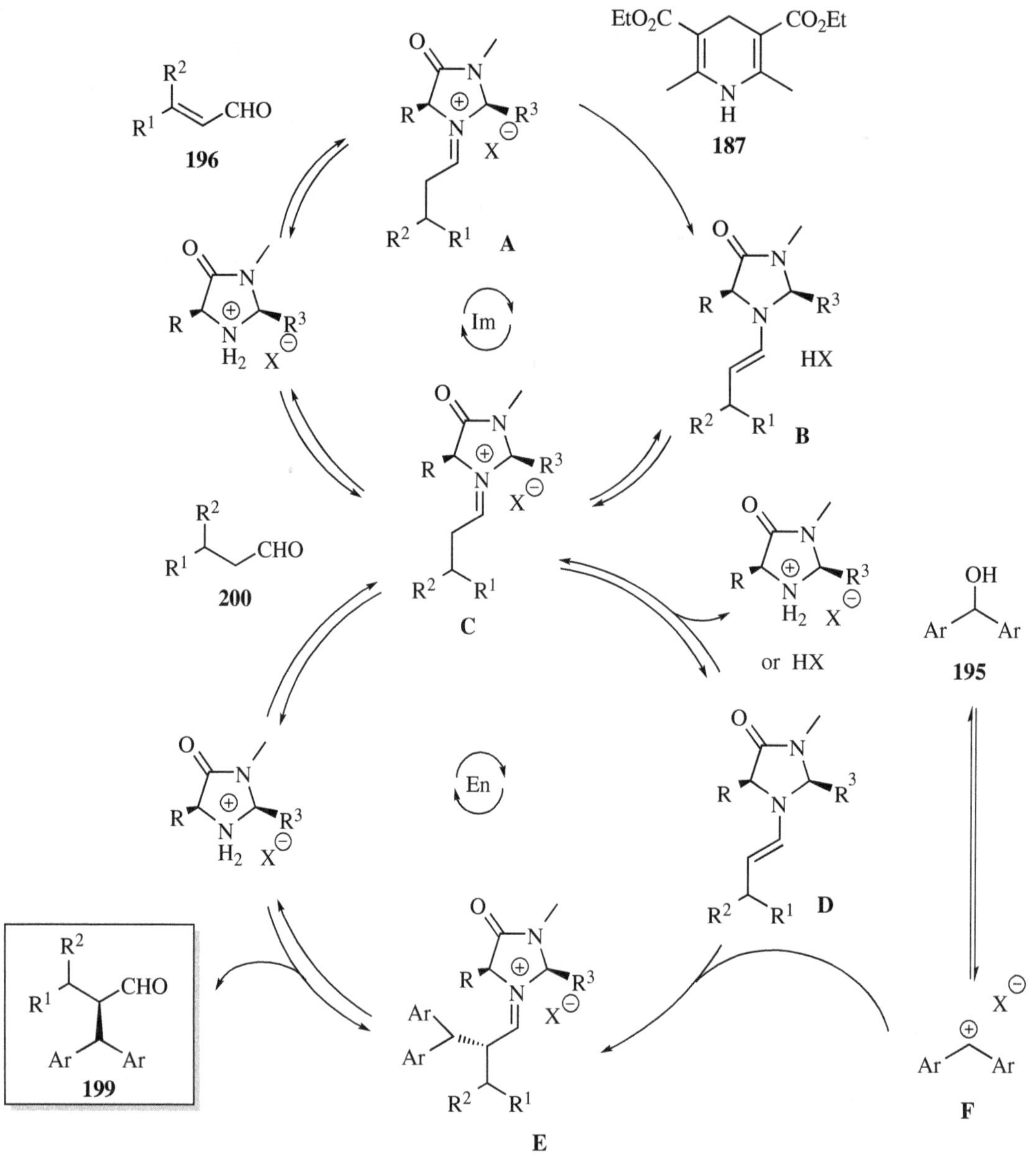

SCHEME 2.66 Proposed mechanistic pathway for the reduction/alkylation sequence.

The enantioenriched 3-alkylated indoles **204** were isolated in high-to-excellent stereoselectivity upon a Friedel–Crafts/S_N2' addition–elimination sequence, which also allowed recovering the less reactive starting material (*R*)-**203** with high optical purity.

2.5 DEVELOPMENT OF DRUG INTERMEDIATES

As illustrated previously, the organocatalyzed asymmetric MCRs are powerful processes that prompt the conversion of at least three starting materials to highly functionalized products displaying profuse molecular diversity and outstanding selectivities in one pot. In addition, this concept well corroborates with the idea of a sustainable chemistry, where several steps from simple starting materials can be engaged in a highly atom economy manner by reducing the number of reaction steps and waste. Therefore, it is not surprising that these kinds of transformations, mediated by organocatalysts, have become a prevailing tool to achieve diastereo- and enantioselective synthesis of active pharmaceutical ingredients (APIs) for drug development [93].

Among the selected examples, we can cite the remarkable Hayashi's syntheses of (-)-oseltamivir [94] and ABT-341 [95] both catalyzed by diphenylprolinol silyl ether (*R*)-**124** (Scheme 2.69). In both cases, a highly substituted cyclohexane moiety **B** (obtained via an asymmetric Michael addition and a domino Michael reaction/Horner–Wadsworth–Emmons

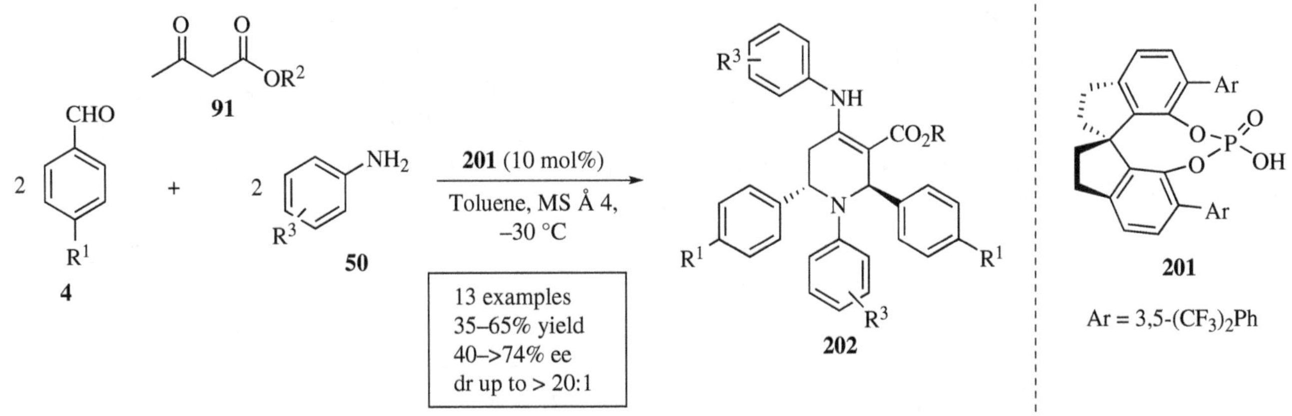

SCHEME 2.67 Preparation of enantioenriched tetrahydropyridines **202** via organocatalytic MCRs.

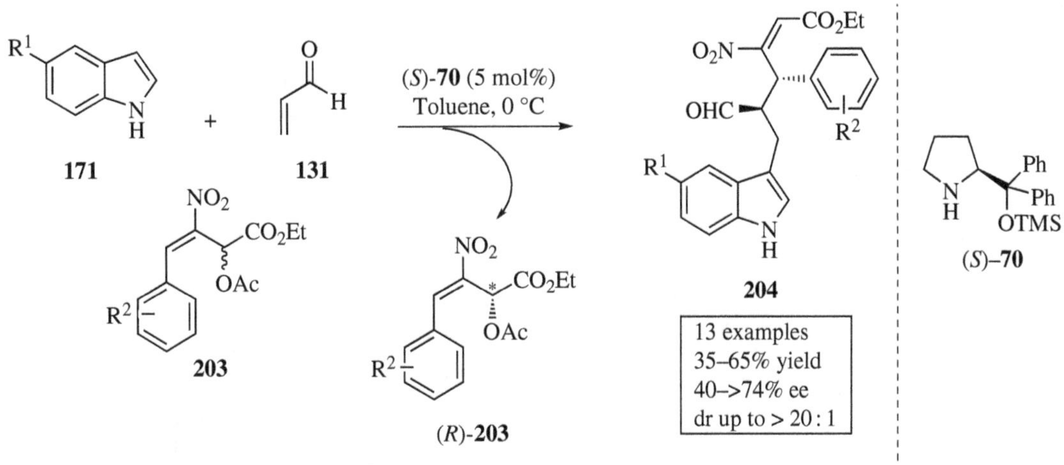

SCHEME 2.68 Kinetic resolution of nitroallylic acetates **203** with stereoselective formation of 3-alkylated indoles **204**.

reaction in a one-pot operation) was used as a key intermediate starting from readily available aldehydes, nitroalkenes, and dialkylvinylphosphonate derivatives.

The potential of one-pot three-component coupling reaction was further explored by the same group for the synthesis of substituted tetrahydropyranols **207** Ishikawa [96]. As for the two previous examples, the isolation of the Michael products **A** in suitable yields is only allowed when the addition of the different starting materials proceeds in a well-established order, thus avoiding undesired side reactions. Furthermore, this three-component coupling process could be successfully combined with an additional nucleophilic addition step to allow an asymmetric one-pot four-component coupling reaction to give rise to tetrahydropyrans **208** that are present in many natural products (Scheme 2.70).

Octahydroacridines are tetrahydroquinoline core structures with potentially interesting biological properties. An elegant example for their synthesis in a one-pot strategy was published by Jørgensen and coworkers [97]. The transformation goes through an enantioselective Michael addition between α,β-unsaturated aldehydes **95** and malononitrile **209** in the presence of (*S*)-**97** as organocatalyst. Then, the obtained intermediate evolves through condensation/Povarov cascade, in the presence of anilines **51**, under acidic conditions to give the desired octahydroacridines **210** in high diastereoselectivities (Scheme 2.71).

The combination of α,β-unsaturated aldehydes **95** and dimethyl 3-oxapentanedioate (**211**, involved twice on the reaction) in a cascade process successfully afforded optically active bicyclo[3.3.1]non-2-ene compounds **212** [98]. The reaction allows selective creation of 4 new C—C bonds, 6 stereocenters, and, quite impressive, 1 out of 64 possible stereoisomers with high selectivities in a one-pot approach (Scheme 2.72).

The construction of optically active imidazoles, oxazoles, and thiazoles **214** was accomplished in a highly effective one-pot reaction cascade that consists of an organocatalytic either aziridination or epoxidation of α,β-unsaturated aldehydes **95** combined with a [3+2]-annulation with suitable

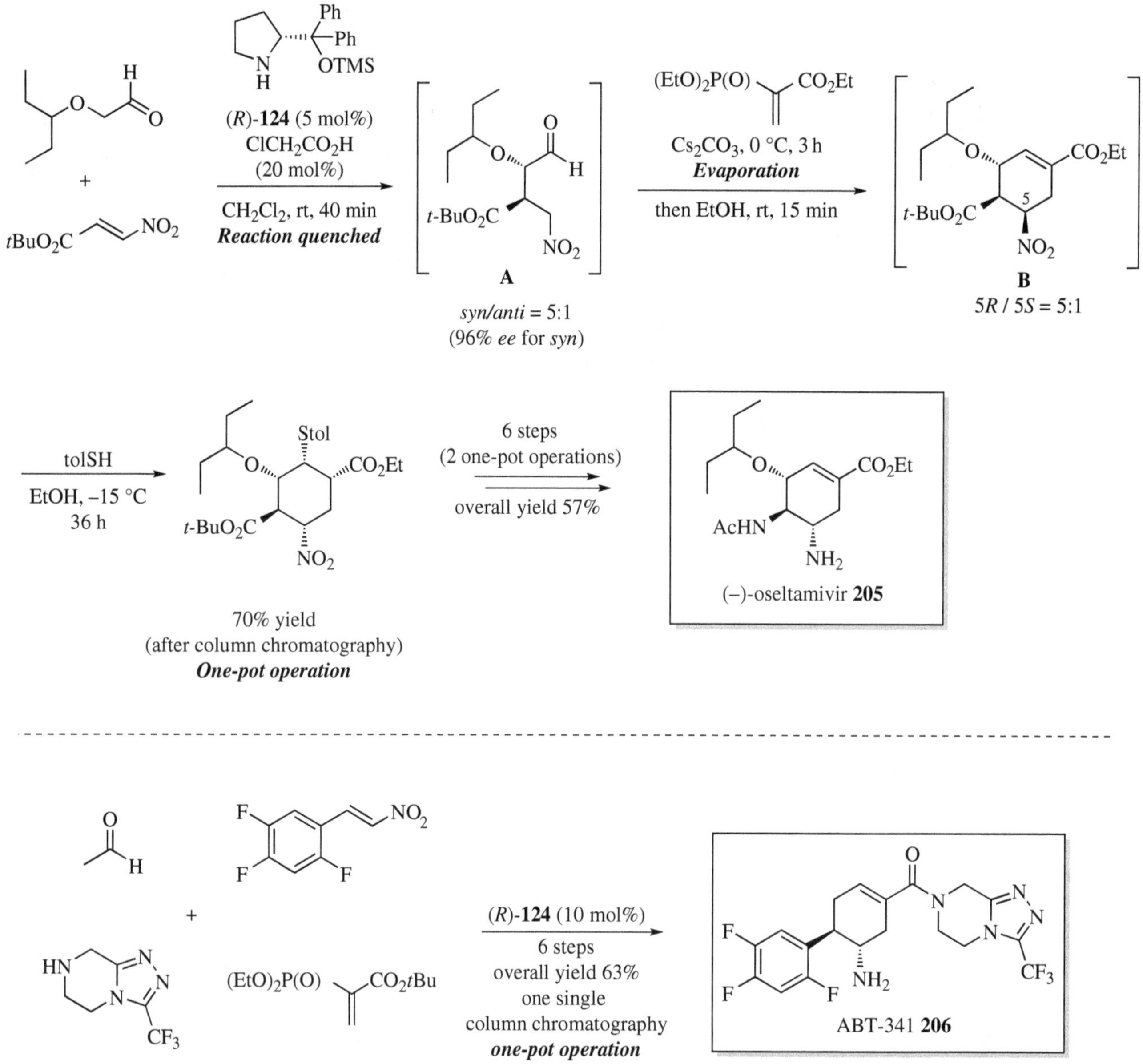

SCHEME 2.69 Asymmetric synthesis of (–)-oseltamivir **205** and ABT-341 **206**.

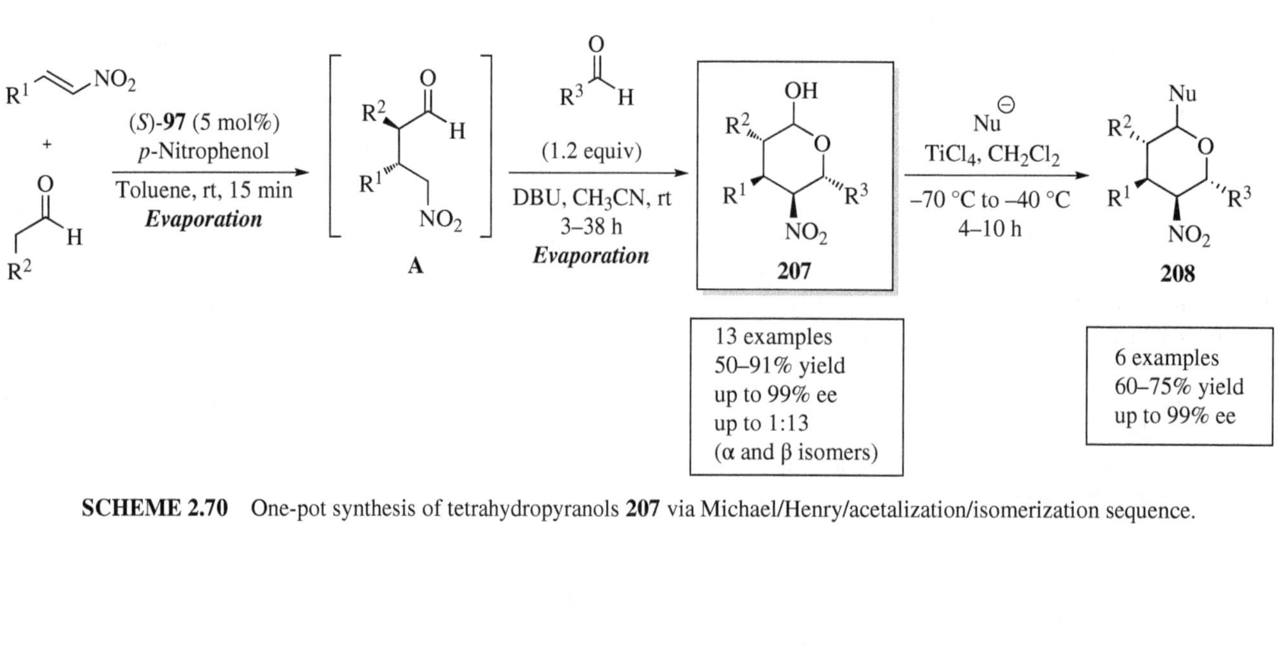

SCHEME 2.70 One-pot synthesis of tetrahydropyranols **207** via Michael/Henry/acetalization/isomerization sequence.

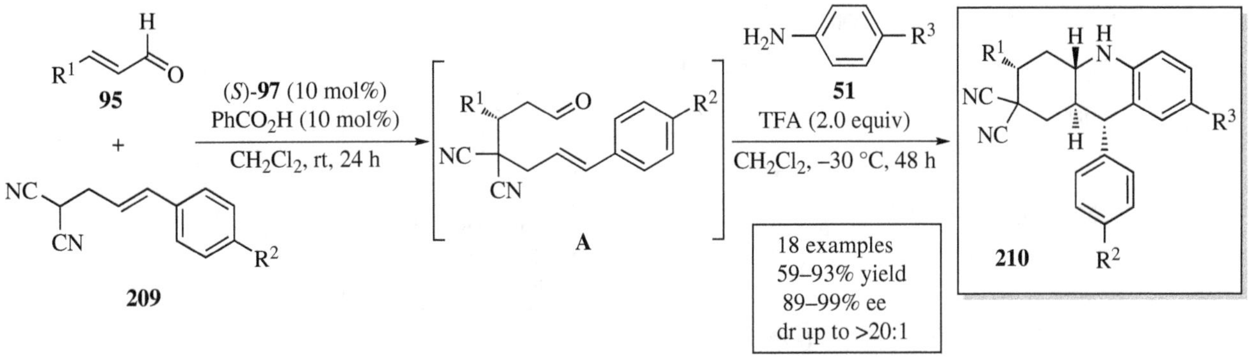

SCHEME 2.71 Organocatalytic one-pot strategy to octahydroacridines **210**.

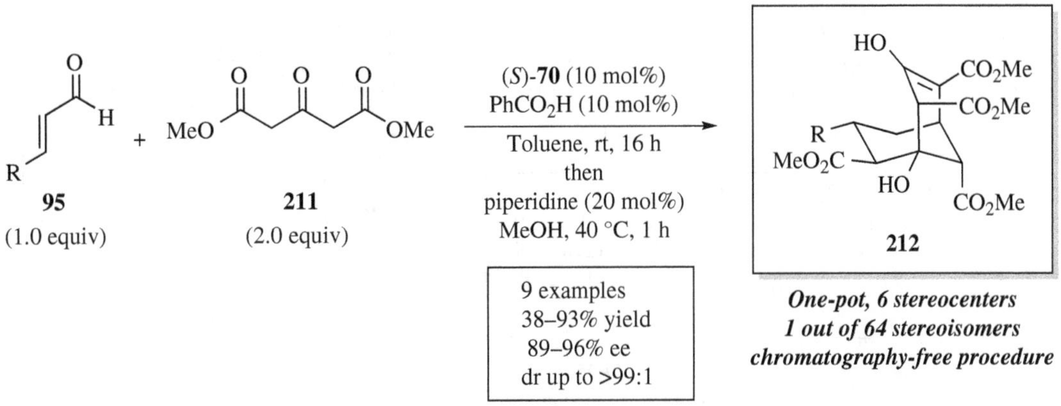

SCHEME 2.72 Organocatalytic one-pot strategy to bicyclo[3.3.1]non-2-ene compounds **212**.

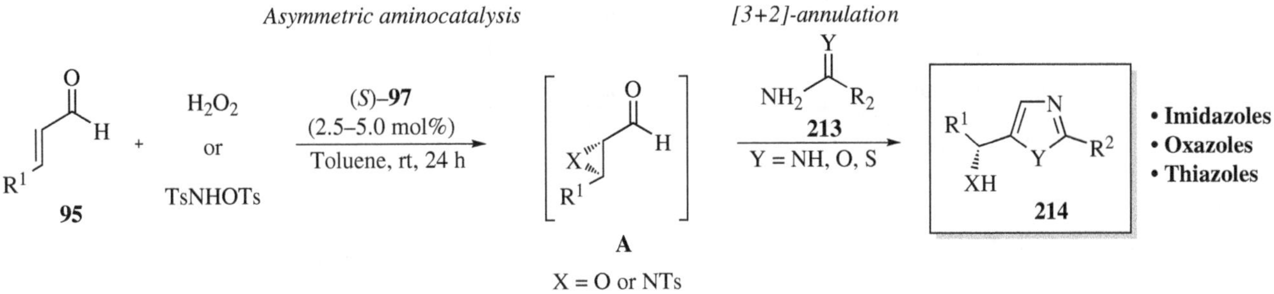

SCHEME 2.73 Asymmetric one-pot strategy for the synthesis of imidazoles, oxazoles, and thiazoles.

1,3-dinucleophilic species **213** (e.g., amidines, ureas, or thioureas) (Scheme 2.73) [99].

An elegant example of MCRs was devised in 2008 with the development of a strategy based on a double cascade proline-catalyzed five-component olefination/Diels–Alder/epimerization/olefination/hydrogenation (O–DA–E–O–H) reaction of enones **84**, aromatic aldehydes **4**, cyanoacetic acid alkyl esters **215**, and Hantzsch ester (**187**) in one pot. This atom-economic transformation gave rise to a versatile methodology that allows the preparation of functionalized cyclohexanes **216** with interesting potential for preparing cardiovascular and hypnotic active agents (Scheme 2.74) [100]. An original synthesis of functionalized tetrahydro-isobenzofuran-1,5-diones, an important class of heterocycles with broad biological activities, was disclosed by means of proline asymmetric catalysis [101]. The process goes through sequential cascade three-component reductive alkylation (TCRA) coupled with Michael/aldol reactions. Likewise, proline was used to catalyze diastereoselective three-component synthesis of optically pure functionalized spiro[5,5]undecane-1,5,9-triones exhibiting potential activity as HIV-1 inhibitors [102].

Spirooxindole derivatives are of considerable interest because of their broad biological properties. They are also structural motifs found in many classes of alkaloids and medicinal privileged compounds. In 2011, Macaev and coworkers demonstrated that (−)-(*S*)-brevicolline (a β-carboline alkaloid) can be used, in a cycloaddition process, to catalyze one-pot multicomponent synthesis of spiro[oxoindole-3,4′-4′

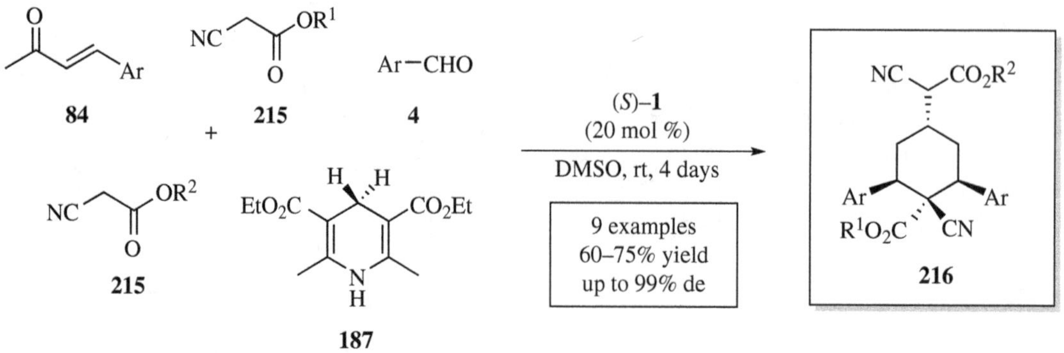

SCHEME 2.74 Double cascade proline-catalyzed five-component O–DA–E–O–H reaction.

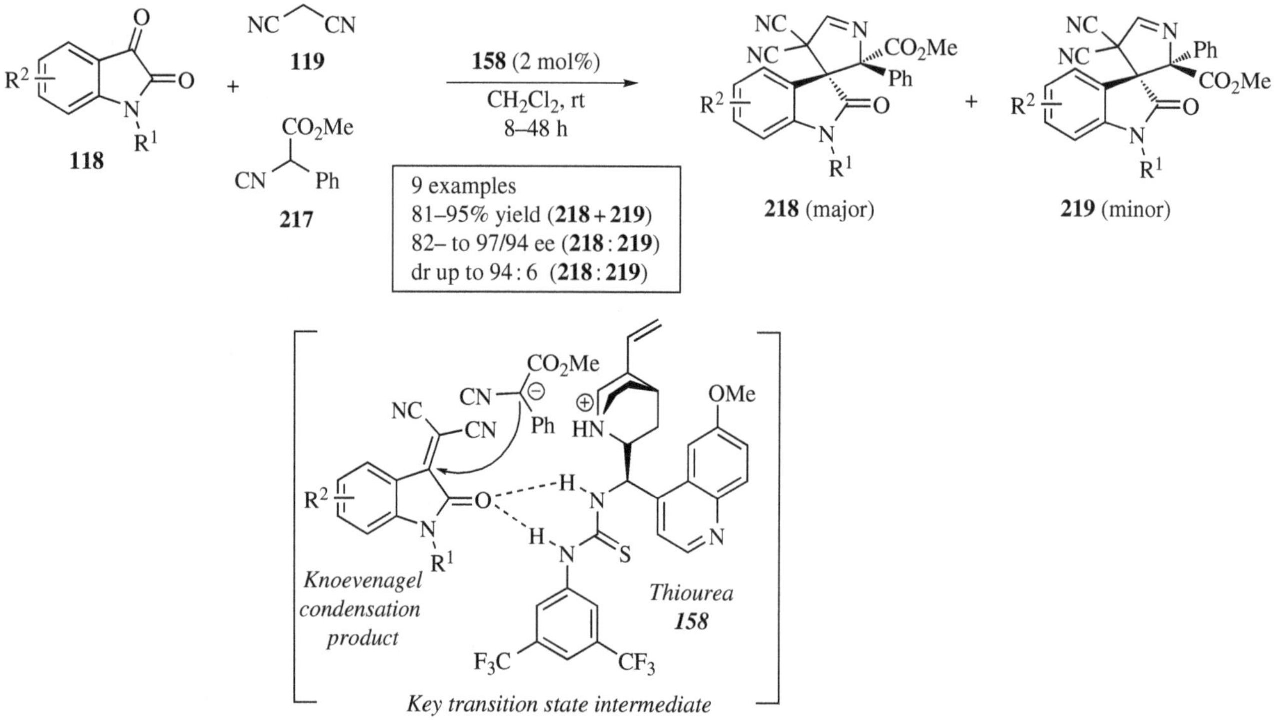

SCHEME 2.75 Thiourea-catalyzed three-component reaction for preparing dihydropyrryl-spirooxindoles.

H-pyrane] derivatives in the presence of isatins, malononitrile, and acetylacetone/ethyl 3-oxobutanoate [103]. Yan and coworkers showed in 2012 that chiral tertiary amine–thiourea (**158**) derived from quinine can catalyze a three-component reaction between isatins **118**, malononitrile (**119**), and α-phenyl-isocyanoacetate (**217**) (Scheme 2.75) [104]. The process affords dihydropyrryl-spirooxindoles **218** and involves an initial Knoevenagel condensation of **118** and **119** followed by the nucleophilic anion attack of **217** (see the key transition state intermediate on Scheme 2.75). Final intramolecular cycloaddition affords the expected compounds where H bond interactions are supposed to direct the attack of isocyanate anion and, consequently, control the enantioselectivity. One year later, Xu's group used a bifunctional cinchona-based squaramide to catalyze multicomponent cascade reaction to synthesize spiro[pyrrolidin-3,2′-oxindoles] via 1,3-proton shift and [3+2] cycloaddition. They have employed isatin, benzylamine, and aromatic nitroalkene derivatives as starting materials [105].

In 2011, a three-component 1,3-dipolar cycloaddition of prenyl aldehyde **220**, diethyl 2-aminomalonate (**221**), and methyl-2-(4-methoxy-2-nitrophenyl)acrylate (**222**) catalyzed by the bisphosphoric acid **83** was the key step on the synthesis of diastereoisomers of alkaloid spirotryprostatin A (Scheme 2.76) [106]. This 1,3-dipolar cycloaddition strategy using the same catalyst was further transposed to isatins as azomethine precursors allowing a straightforward way to biologically important spiro[pyrrolidin-3,2′-oxindole] frameworks with various stereogenic centers including one of two quaternary ones [107].

Chiral isoindoline derivatives **226** were successfully prepared by a Brønsted acid-catalyzed 1,3-dipolar addition followed by organic base-induced isomerization [108].

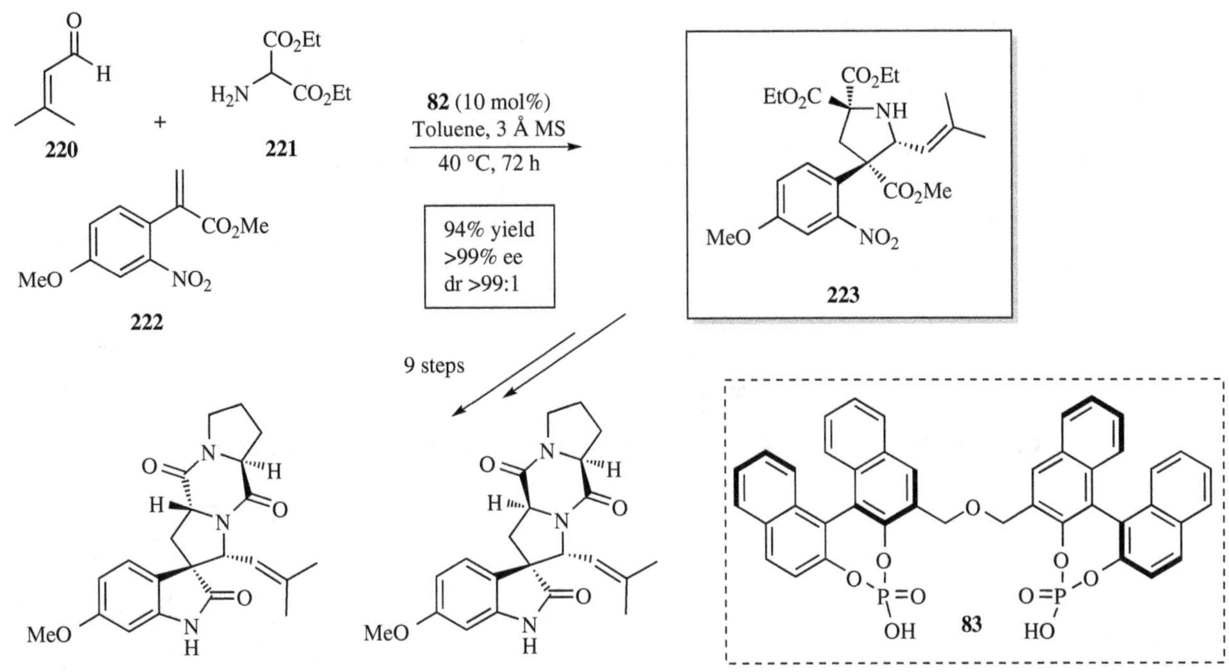

9,18-bis-epispirotryprostatin A 18-epispirotryprostatin A

SCHEME 2.76 Application of a three-component 1,3-dipolar cycloaddition reaction.

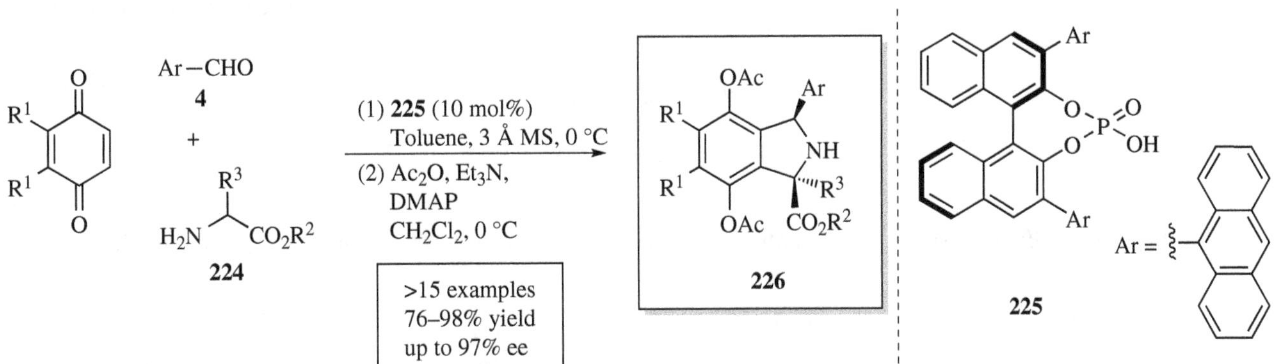

9-Anthracenyl

SCHEME 2.77 Formal double arylation of azomethines for the synthesis of isoindolines.

The reaction is supposed to proceed via a formal double arylation of azomethines (obtained from **4** and **224**) (Scheme 2.77).

The first catalytic asymmetric formal [3+3] cycloaddition of isatin-3-indolylmethanol derivatives and *in situ* generated azomethine ylides was recently developed to synthesize structurally diverse spiro[indoline-3,4′-pyridoindoles] **228** bearing one all-carbon quaternary stereogenic center [109]. The reaction takes place with a range of *N*-substituted isatin-derived 3-indolylmethanols, aromatic aldehydes **4**, and diethyl 2-aminomalonate **221** and is catalyzed with chiral phosphoric acid **227** with the bulky 9-phenanthrenyl group at the 3,3′-positions of the BINOL moiety (Scheme 2.78).

The hexahydro-6*H*-benzo[*c*]chromene framework, present in a broad number of natural products and APIs, displays a large panel of biological activities. In 2010, Hong and coworkers have used an organocatalyzed approach based on a domino oxa-Michael/Michael/Michael/aldol condensation to devise the first enantioselective total synthesis of the marine meroterpene (+)-conicol (**232**) [110]. This strategy affords a straightforward entry to functionalized hexahydro-6*H*-benzo[*c*] chromene skeleton **231** in high enantioselectivity from three simple starting materials (e.g., (*E*)-2-(2-nitrovinyl)-benzene-1,4-diol (**229**), 3-methylbut-2-enal (**220**), and 4,4-dimethoxy-but-2-enal (**230**)) (Scheme 2.79).

In addition to the aforementioned examples, we can also cite, in a nonexhaustive manner, the synthesis of tetrahydrobenzo[*b*]pyran derivatives by means of (*S*)-proline (**1**) in aqueous media via a one-pot three-component tandem

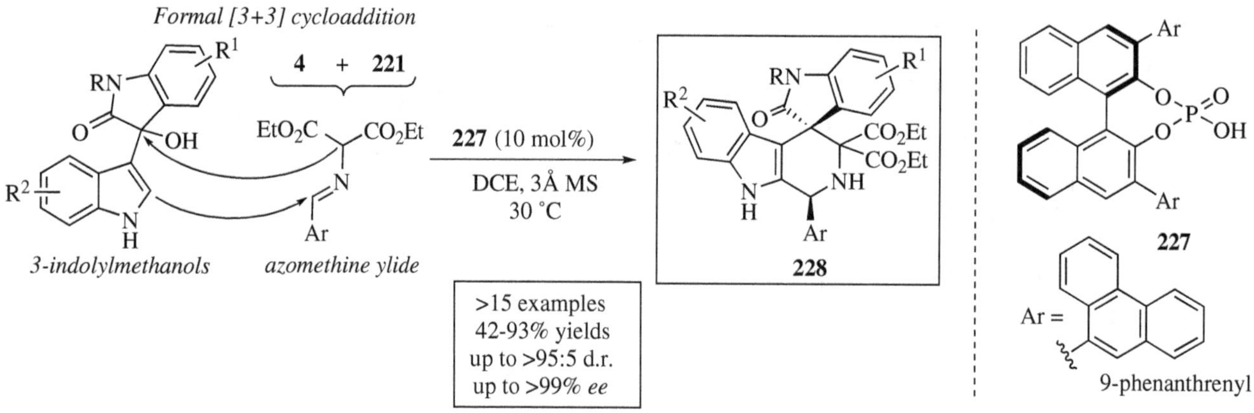

SCHEME 2.78 Formal [3 + 3] cycloadditions for the synthesis of piperidine frameworks **228**.

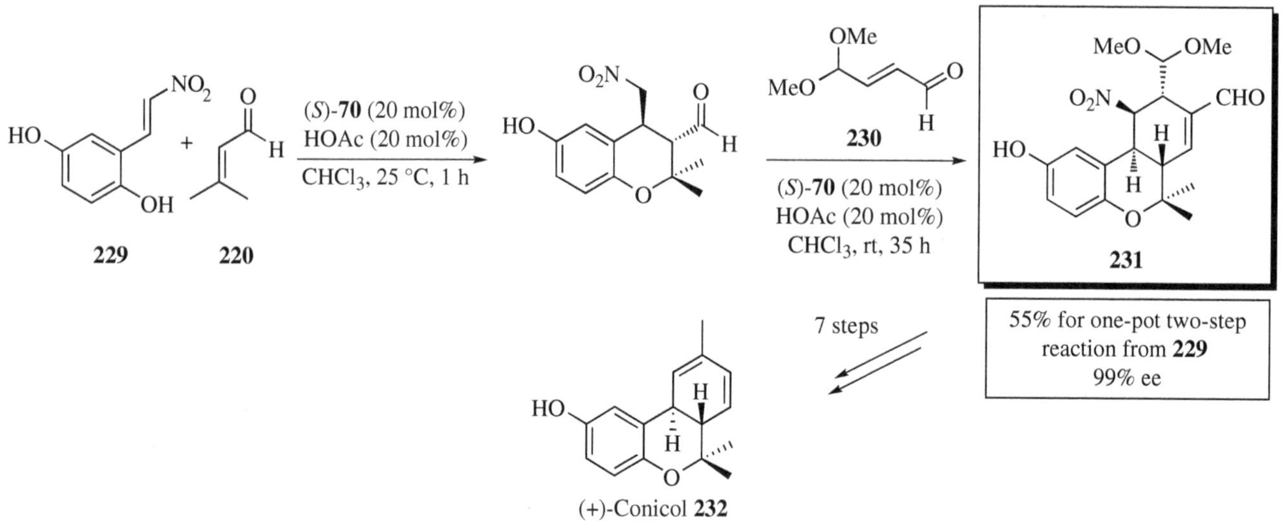

SCHEME 2.79 Oxa-Michael/Michael/Michael/aldol condensation on the first total synthesis of (+)-conicol **232**.

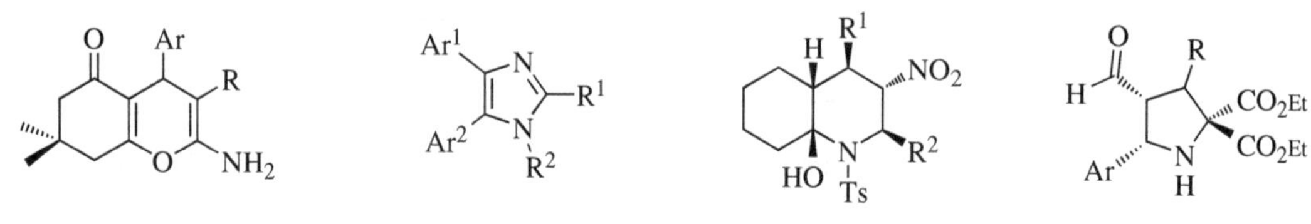

233	**234**	**235**	**236**
Tetrahydrobenzo[*b*]pyrans	Substituted imidazoles	Decahydroquinolines	Pyrrolidines
>20 examples	>20 examples	16 examples	16 examples
67–98% yield	42–92% yield	69–94% yield	50–63% yield
		up to 98% ee	up to 98% ee
			dr up to >10:1

FIGURE 2.1 Selected organocatalyzed MCRs for structure–diversity on APIs.

Knoevenagel/cyclocondensation reaction [111]; the synthesis of tri- and tetrasubstituted imidazoles by three-component cyclocondensation of 1,2-dicarbonyl compounds, aldehydes, and ammonium acetate using (*S*)-proline (**1**) [112]; the preparation of functionalized *cis*-decahydroquinolines (framework widespread in biologically active alkaloids) by means of a Michael/aza-Henry/hemiaminalization reaction sequence catalyzed by (*S*)-**70** [113]; and the synthesis of functionalized pyrrolidine derivatives based on an asymmetric multicomponent 1,3-dipolar cycloaddition with (*S*)-**70** [114]. All of these structures are important building blocks for APIs found in several bioactive molecules (Fig. 2.1).

2.6 MISCELLANEOUS REACTION

Finally, few MCRs, which proceed in an organocatalytic manner but cannot be ascribed to the just disclosed categories, deserve to be mentioned. Outstanding results have recently been reported by Armas and Garcia-Tellado, whose efforts have been devoted to the development and implementation of H bond-based organocatalysis performed in the presence of water conditions [115]. In support of this concept, the Spanish group has described a stereoselective multicomponent aza-Henry reaction catalyzed by the synergic action of a chiral H bond donor (thioureas **237** and **238** or squaramide **239**) and a Lewis base (*N*,*N*-dimethylcyclohexylamine) in the presence of water. As depicted in the Scheme 2.80, when aromatic aldehyde are employed, Jacobsen's catalyst **237** is the best one in terms of chemical efficiency, chiral induction, and preparation, whereas the presence of aliphatic aldehyde required the employment of **238** as catalyst in order to keep the same high level of yield and stereoselectivity. However, in both cases, squaramide **239** resulted to be the most efficient and effective catalyst in order to obtain the corresponding enantiomers of the all titled products **240** in very high optical purity (up to 99% ee).

Optically active phosphonates, in particular α-aminophosphonates, are widely studied in medicinal chemistry due to their broad spectrum of biological activities including peptidomimetics, antibiotics, herbicides, etc. Due to such features, various methodologies have been reported for the preparation of such entities. However, the literature is quite bare concerning the chiral α-aminophosphonates. Among the very few examples, notable are the procedures independently reported by Ma [116] and Bhusare [117]. In 2010, the Chinese group suggested a direct three-component hydrophosphonylation reaction of a series of aldehydes (either **4** or **75**) with 4-methoxyaniline **3**, diisopropyl phosphate **241** (as source of phosphate moiety), and the acid **242** as organocatalyst (Scheme 2.81). The reaction smoothly proceeded

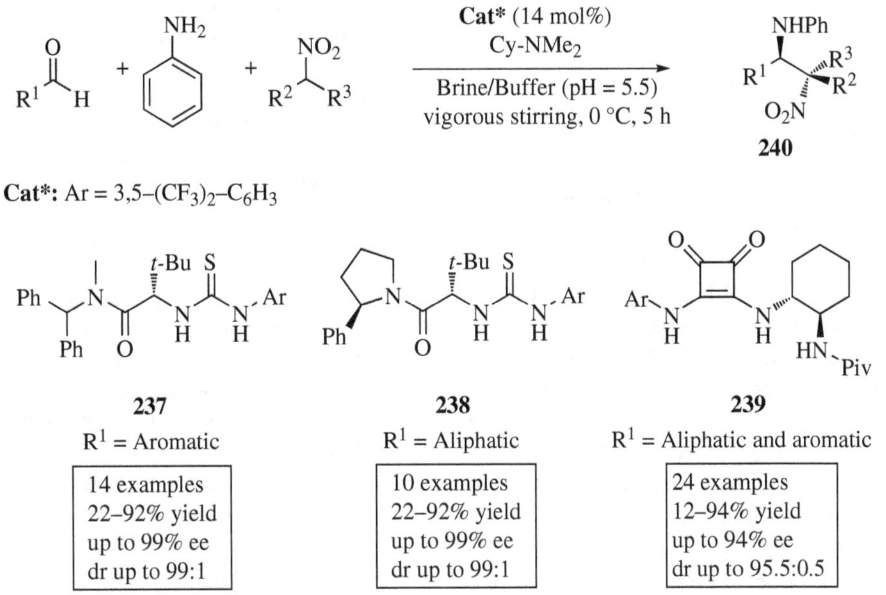

SCHEME 2.80 Multicomponent aza-Henry reaction via water-compatible H bond activation.

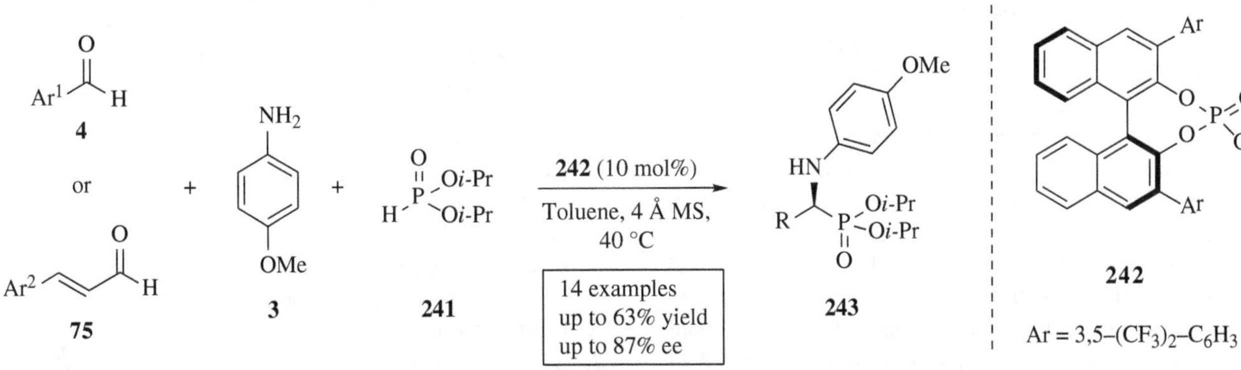

SCHEME 2.81 Ma's approach to **α**-aminophosphonates **243** via organocatalytic MCRs.

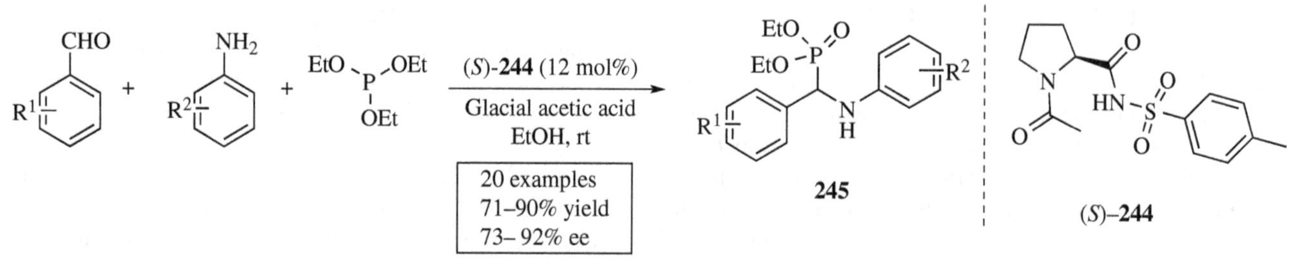

SCHEME 2.82 Bhusare's approach to **α**-aminophosphonates **245** via organocatalytic MCRs.

to the title enantioenriched compound **243**, which were isolated in good yields (up to 63%) and interesting optical purity (up to 87% ee).

More recently, in 2013, Bhusare and coworkers developed a direct one-pot procedure to analogues α-aminophosphonates **245**, which were successfully achieved by employing pyrrolidine derivative (*S*)-**244** as catalyst. The presented reaction resulted to be quite efficient in terms of both yield (71–90%) and stereoselectivity (73–92% ee) (Scheme 2.82) [117].

2.7 CONCLUSIONS

With the aim of taking advantage from Nature and particularly by mimicking the catalytic efficiency, selectivity, and high turnover of enzymes, organocatalysts have appeared as a powerful tool to mediate highly complex transformations, thanks to their diversified modes of activation. In the last 14 years, the constant growth of this field has attracted many research groups to investigate their utilization in order to pave the way to very impressive and outstanding combination of reactions in a one-pot process. Consequently, due to the high flexibility of activation modes of organocatalysts, as well as their combination, the construction of highly functionalized chiral complex molecules was made possible, many of these products being highly suitable for diastereo- and enantioselective synthesis of APIs for drug development. In this book chapter, rather than an exhaustive detailed approach on organocatalyzed asymmetric MCRs, we gave an overall overview on selected remarkable examples. To date, an impressive number of organocatalyzed asymmetric MCRs have appeared, and due to the suitable combination of transformations, very challenging scaffolds were isolated in a single operation by combining very simple, almost always, commercially available, starting materials. Nevertheless, it remains a very challenging field in where each step envisaged might be carefully analyzed in order to well establish compatibility between all substrates and synthetic intermediates. We hope that our selection of examples could demonstrate the potential of organocatalysis for such a purpose.

REFERENCES

1 D. W. C. MacMillan, *Nature* 2008, 455, 304–308. The advent and development of organocatalysis.

2 (a) P. I. Dalko, L. Moisan, *Angew. Chem. Int. Ed.* 2001, 40, 3726–3748. Enantioselective organocatalysis. (b) M. Benaglia, A. Puglisi, F. Cozzi, *Chem. Rev.* 2003, 103, 3401–3429. Polymer-supported organic catalysts. (c) K. N. Houk, B. List, *Acc. Chem. Res.* 2004, 37, 487–487. Asymmetric organocatalysis. (d) B. List, *Acc. Chem. Res.* 2004, 37, 548–557. Enamine catalysis is a powerful strategy for the catalytic generation and use of carbanion equivalents. (e) B. List, *Adv. Synth. Catal.* 2004, 346, 1021–1021. Organocatalysis: a complementary catalysis strategy advances organic synthesis. (f) C. M. R. Volla, L. Atodiresei, M. Rueping, *Chem. Rev.* 2014, 114, 2390–2431. Catalytic C-C bond-forming multi-component cascade or domino reactions: pushing the boundaries of complexity in asymmetric organocatalysis. (g) A. Dondoni, A. Massi, *Angew. Chem. Int. Ed.* 2008, 47, 4638–4660. Asymmetric organocatalysis: from infancy to adolescence. (h) L. Albrecht, H. Jiang, K. A. Jørgensen, *Angew. Chem. Int. Ed.* 2011, 50, 8492–8509. A simple recipe for sophisticated cocktails: organocatalytic one-pot reactions-concept, nomenclature, and future perspectives. (i) A. Berkessel, H. Gröger, *Asymmetric organocatalysis*. Wiley-VCH Verlag GmbH, Weinheim, Germany, 2005.

3 B. List, R. A. Lerner, C. F. Barbas, *J. Am. Chem. Soc.* 2000, 122, 2395–2396. Proline-catalyzed direct asymmetric aldol reactions.

4 (a) B. List, Synlett 2001, 1675–1686. Asymmetric aminocatalysis. (b) M. Bella, T. Gasperi, Synthesis 2009, 1583. Organocatalytic formation of quaternary stereocenters. (c) R. M. de Figueiredo, A. Mazziotta, D. P. de Sant'Ana, C. Palumbo, T. Gasperi, *Curr. Org. Chem.* 2012, 16, 2231–2289. Active methylene compounds in asymmetric organocatalytic synthesis of natural products and pharmaceutical scaffolds.

5 (a) G. Guillena, D. J. Ramón, M. Yus, *Tetrahedron: Asymmetry* 2007, 18, 693–700. Organocatalytic enantioselective multicomponent reactions (OEMCRs). (b) C. de Graaff, E. Ruijter, R. V. A. Orru, *Chem. Soc. Rev.* 2012, 41, 3969–4009. Recent developments in asymmetric multicomponent reactions.

6 C. Mannich, W. Krösche, *Arch. Pharm.* 1912, 250, 647–667. Ueber ein Kondensationsprodukt aus Formaldehyd, Ammoniak und Antipyrin.

7 (a) E. M. Ferstl, H. Venkatesan, D. C. Liotta, 2000, ORGN-006. Asymmetric synthesis of β-amino carbonyl compounds. Book

of Abstracts, 219th ACS National Meeting, San Francisco, CA, March 26–30, 2000. (b) E. Juaristi, V.A. Soloshonok, *Enantioselective synthesis of β-amino acids*. Wiley-VCH Verlag GmbH, Weinheim, Germany, 2005. (c) X.-H. Cai, G. Hui, X. Bing, *Eur. J. Chem.* 2012, 3, 258–266. Recent progress in the asymmetric Mannich reaction. (d) A. Ricci, *Amino group chemistry*. Wiley-VCH Verlag GmbH, Weinheim, Germany, 2008.

8 (a) W. Notz, F. Tanaka, C. F. Barbas, *Acc. Chem. Res.* 2004, 37, 580–591. Enamine-based organocatalysis with proline and diamines: the development of direct catalytic asymmetric aldol, Mannich, Michael, and Diels–Alder reactions. (b) A. Ting, S. E. Schaus, Eur. J. Org. Chem. 2007, 5797–5815. Organocatalytic asymmetric Mannich reactions: new methodology, catalyst design, and synthetic applications.

9 B. List, *J. Am. Chem. Soc.* 2000, 122, 9336–9337. The direct catalytic asymmetric three-component Mannich reaction.

10 B. List, P. Pojarliev, W. T. Biller, H. J. Martin, *J. Am. Chem. Soc.* 2002, 124, 827–833. The proline-catalyzed direct asymmetric three-component Mannich reaction: scope, optimization, and application to the highly enantioselective synthesis of 1,2-amino alcohols.

11 S. Bahmanyar, K. N. Houk, *Org. Lett.* 2003, 5, 1249–1251. Origins of opposite absolute stereoselectivities in proline-catalyzed direct Mannich and aldol reactions.

12 A. Córdova, Synlett 2003, 1651–1654. One-pot organocatalytic direct asymmetric synthesis of γ-amino alcohol derivatives.

13 Y. Hayashi, W. Tsuboi, I. Ashimine, T. Urushima, M. Shoji, K. Sakai, *Angew. Chem. Int. Ed.* 2003, 42, 3677–3680. The direct and enantioselective, one-pot, three-component, cross-Mannich reaction of aldehydes.

14 A. Córdova, *Chem.–Eur. J.* 2004, 10, 1987–1997. The direct catalytic asymmetric cross-Mannich reaction: a highly enantioselective route to 3-amino alcohols and α-amino acid derivatives.

15 W. Notz, S.-I. Watanabe, N. S. Chowdari, G. Zhong, J. M. Betancort, F. Tanaka, C. F. Barbas, *Adv. Synth. Catal.* 2004, 346, 1131–1140. The scope of the direct proline-catalyzed asymmetric addition of ketones to imines.

16 D. Enders, C. Grondal, M. Vrettou, G. Raabe, *Angew. Chem. Int. Ed.* 2005, 44, 4079–4083. Asymmetric synthesis of selectively protected amino sugars and derivatives by a direct organo-catalytic Mannich reaction.

17 I. Ibrahem, W. Zou, J. Casas, H. Sunden, A. Córdova, *Tetrahedron* 2006, 62, 357–364. Direct organocatalytic enantioselective α-aminomethylation of ketones.

18 B. Rodriguez, C. Bolm, *J. Org. Chem.* 2006, 71, 2888–2891. Thermal effects in the organocatalytic asymmetric Mannich reaction.

19 Y. Hayashi, W. Tsuboi, M. Shoji, N. Suzuki, *J. Am. Chem. Soc.* 2003, 125, 11208–11209. Application of high pressure induced by water-freezing to the direct catalytic asymmetric three-component List–Barbas–Mannich reaction.

20 G. Zhong, M. Lu, Y. Lu, P. Tan, Q. Lau, Synlett 2011, 477–480. Highly enantioselective synthesis of fluorinated β-amino ketones via asymmetric organocatalytic Mannich reactions: a case study of unusual reversal of regioselectivity.

21 (a) Q. Gu, L.-X. Jiang, K. Yuan, L. Zhang, X.-Y. Wu, *Synth. Commun.* 2008, 38, 4198–4206. Organocatalytic, asymmetric, one-pot, three-component Mannich reaction of hydroxyacetone. (b) Q. Gu, J.-J. Gong, J. Feng, X.-Y. Wu, Q.-L. Zhou, *Chin. J. Chem.* 2008, 26, 1902–1906. Organocatalytic direct asymmetric Mannich reactions of hydroxyacetone by (2*S*,5*S*)-pyrrolidine-2,5-dicarboxylic acid.

22 H. Yang, R. G. Carter, *J. Org. Chem.* 2009, 74, 2246–2249. Enantioselective Mannich reactions with the practical proline mimetic N-(p-dodecylphenyl-sulfonyl)-2-pyrrolidinecarboxamide.

23 Y. Hayashi, T. Urushima, S. Aratake, T. Okano, K. Obi, *Org. Lett.* 2008, 10, 21–24. Organic solvent-free, enantio- and diastereoselective, direct Mannich reaction in the presence of water.

24 A. Sasaoka, M. I. Uddin, A. Shimomoto, Y. Ichikawa, M. Shiro, H. Kotsuki, *Tetrahedron: Asymmetry* 2006, 17, 2963–2969. A novel design of roof-shaped anthracene-fused chiral prolines as organocatalysts for asymmetric Mannich reactions.

25 Y. An, Q. Qin, C. Wang, J. Tao, *Chin. J. Chem.* 2011, 29, 1511–1517. Isosteviol-amino acid conjugates as highly efficient organocatalysts for the asymmetric one-pot three-component Mannich reactions.

26 Y.-J. An, C.-C. Wang, Z.-P. Liu, J.-C. Tao, *Helv. Chim. Acta* 2012, 95, 43–51. Isosteviol-proline conjugates as highly efficient amphiphilic organocatalysts for asymmetric three-component Mannich reactions in the presence of water.

27 S. S. V. Ramasastry, H. Zhang, F. Tanaka, C. F. Barbas, III, *J. Am. Chem. Soc.* 2007, 129, 288–289. Direct catalytic asymmetric synthesis of anti-1,2-amino alcohols and syn-1,2-diols through organocatalytic anti-Mannich and syn-aldol reactions.

28 C. Wu, X. Fu, X. Ma, S. Li, C. Li, *Tetrahedron Lett.* 2010, 51, 5775–5777. Threonine-surfactant organocatalysts for the highly diastereo- and enantioselective direct anti-Mannich reactions of hydroxyacetone.

29 C. Wu, X. Fu, S. Li, *Tetrahedron: Asymmetry* 2011, 22, 1063–1073. Simple and inexpensive threonine-based organocatalysts for the highly diastereo- and enantioselective direct large-scale syn-aldol and anti-Mannich reactions of α-hydroxyacetone.

30 Y.-C. Teo, J.-J. Lau, M.-C. Wu, *Tetrahedron: Asymmetry* 2008, 19, 186–190. Direct asymmetric three-component Mannich reactions catalyzed by a siloxy serine organocatalyst in water.

31 F.-F. Yong, Y.-C. Teo, *Synth. Commun.* 2011, 41, 1293–1300. Recyclable siloxy serine organocatalyst for the direct asymmetric Mannich reactions in ionic liquids.

32 N. S. Chowdari, D. B. Ramachary, C. F. Barbas, III, Synlett 2003, 1906–1909. Organocatalysis in ionic liquids: highly efficient L-proline-catalyzed direct asymmetric Mannich reactions involving ketone and aldehyde nucleophiles.

33 X. Zheng, Y.-B. Qian, Y. Wang, *Eur. J. Org. Chem.* 2010, 2010, 515–522. 2-Pyrrolidinecarboxylic acid ionic liquid as a highly efficient organocatalyst for the asymmetric one-pot Mannich reaction.

34 I. Ibrahem, W. Zou, M. Engqvist, Y. Xu, A. Córdova, *Chem.–Eur. J.* 2005, 11, 7024–7029. Acyclic chiral amines and amino

acids as inexpensive and readily tunable catalysts for the direct asymmetric three-component Mannich reaction.

35 Q. Guo, J. C.-G. Zhao, *Org. Lett.* 2013, 15, 508–511. Highly enantioselective three-component direct Mannich reactions of unfunctionalized ketones catalyzed by bifunctional organocatalysts.

36 A. S. Demir, S. Basceken, *Tetrahedron: Asymmetry* 2013, 24, 515–525. Study of asymmetric aldol and Mannich reactions catalyzed by proline-thiourea host-guest complexes in nonpolar solvents.

37 Q.-X. Guo, H. Liu, C. Guo, S.-W. Luo, Y. Gu, L.-Z. Gong, *J. Am. Chem. Soc.* 2007, 129, 3790–3791. Chiral Brønsted acid-catalyzed direct asymmetric Mannich reaction.

38 T. Akiyama, K. Matsuda, K. Fuchibe, Synlett 2005, 322–324. HCl-Catalyzed stereoselective Mannich reaction in H_2O-SDS system.

39 G. Dagousset, F. Drouet, G. Masson, J. Zhu, *Org. Lett.* 2009, 11, 5546–5549. Chiral Brønsted acid-catalyzed enantioselective multicomponent Mannich reaction: synthesis of *anti*-1, 3-diamines using enecarbamates as nucleophiles.

40 M. Sickert, F. Abels, M. Lang, J. Sieler, C. Birkemeyer, C. Schneider, *Chem.–Eur. J.* 2010, 16, 2806–2818. The Brønsted acid catalyzed, enantioselective vinylogous Mannich reaction.

41 Y.-Y. Chen, Y.-J. Jiang, Y.-S. Fan, D. Sha, Q. Wang, G. Zhang, L. Zheng, S. Zhang, *Tetrahedron: Asymmetry* 2012, 23, 904–909. Double axially chiral bisphosphorylimides as novel Brønsted acids in asymmetric three-component Mannich reaction.

42 A. Noble, J. C. Anderson, *Chem. Rev.* 2013, 113, 2887–2939. Nitro-Mannich reaction.

43 P. Jakubec, D. M. Cockfield, M. Helliwell, J. Raftery, D. J. Dixon, *Beilstein J. Org. Chem.* 2012, 8, 567–578. Stereoselective, nitro-Mannich/lactamisation cascades for the direct synthesis of heavily decorated 5-nitropiperidin-2-ones and related heterocycles.

44 H. Pellissier, *Recent developments in asymmetric organocatalysis*. RSC Publishing, Cambridge, UK, 2010.

45 H. Sundén, I. Ibrahem, L. Eriksson, A. Córdova, *Angew. Chem. Int. Ed.* 2005, 44, 4877–4880. Direct catalytic enantioselective aza-Diels-Alder reactions.

46 Z. J. Jia, H. Jiang, J. L. Li, B. Gschwend, Q. Z. Li, X. Yin, J. Grouleff, Y. C. Chen, K. A. Jørgensen, *J. Am. Chem. Soc.* 2011, 133, 5053–5061. Trienamines in asymmetric organocatalysis: Diels-Alder and tandem reactions.

47 S. Lin, L. Deiana, G. Zhao, A. Córdova, J. Sun, *Angew. Chem. Int. Ed.* 2011, 50, 7624–7630. Dynamic one-pot three-component catalytic asymmetric transformation by combination of hydrogen-bond-donating and amine catalysts.

48 S. Reboredo, J. L. Vicario, L. Carrillo, E. Reyes, U. Uria, *Synthesis* 2013, 45, 2669–2678. A simple synthesis of polysubstituted pyrrolidines by an organocatalytic three-component approach featuring a one-pot condensation and [3+2]-cycloaddition reaction in aqueous medium.

49 J. S. Chang Guo, L.-Z. Gong, *Org. Lett.* 2013, 15, 2676–2679. Biomimetic asymmetric 1,3-dioplar cycloaddition: amino acid precursors in biosynthesis serve as latent azomethine ylides.

50 J. Shi, Y.-B. Liu, M. Wang, L.-L. Lin, X.-H. Liu, X.-M. Feng, *Tetrahedron* 2011, 67, 1781–1787. Asymmetric synthesis of multi-substituted spiro[5,5]undecane-1,5,9-triones *via* organocatalytic three-component reaction.

51 J. Jiang, J. Qing, L. Z. Gong, *Chem.–Eur. J.* 2009, 15, 7031–7034. Asymmetric synthesis of 3-amino-δ-lactams and benzo[a]quinolizidines by catalytic cyclization reactions involving azlactones.

52 D. Enders, M. R. Hüttl, C. Grondal, G. Raabe, *Nature* 2006, 441, 861–863. Control of four stereocentres in a triple cascade organocatalytic reaction.

53 (a) H. Pellissier, *Adv. Synth. Catal.* 2012, 354, 237–294. Recent developments in asymmetric organocatalytic domino reactions. (b) B. Westermann, M. Ayaz, S. Van Berkel, *Angew. Chem. Int. Ed.* 2010, 49, 846. Enantiodivergent organocascade reactions.

54 (a) S. B. Tsogoeva, Eur. J. Org. Chem. 2007, 2007, 1701–1716. Recent advances in asymmetric organocatalytic 1,4-conjugate additions. (b) J. Vicario, D. Badía, L. Carrillo, Synthesis 2007, 2065–2092. Organocatalytic enantioselective Michael and hetero-Michael reactions.

55 J. Jiang, J. Yu, X.-X. Sun, Q.-Q. Rao, L.-Z. Gong, *Angew. Chem. Int. Ed.* 2008, 47, 2458–2462. Organocatalytic asymmetric three-component cyclization of cinnamaldehydes and primary amines with 1,3-dicarbonyl compounds: straightforward access to enantiomerically enriched dihydropyridines.

56 R. K. Kunz, D. W. C. MacMillan, *J. Am. Chem. Soc.* 2005, 127, 3240–3241. Enantioselective organocatalytic cyclopropanations. The identification of a new class of iminium catalyst based upon directed electrostatic activation.

57 M. Marigo, J. Franzen, T. B. Poulsen, W. Zhuang, K. A. Jørgensen, *J. Am. Chem. Soc.* 2005, 127, 6964–6965. Asymmetric organocatalytic epoxidation of α,β-unsaturated aldehydes with hydrogen peroxide.

58 (a) M. Marigo, T. Schulte, J. Franzen, K. A. Jørgensen, *J. Am. Chem. Soc.* 2005, 127, 15710–15711. Asymmetric multicomponent domino reactions and highly enantioselective conjugated addition of thiols to α,β-unsaturated aldehydes. (b) H. Jiang, J. B. Nielsen, M. Nielsen, K. A. Jørgensen, *Chem. Eur. J.* 2007, 13, 9068–9075. Organocatalyzed asymmetric β-amination and multicomponent syn-selective diamination of α,β-unsaturated aldehydes.

59 P. Galzerano, F. Pesciaioli, A. Mazzanti, G. Bartoli, P. Melchiorre, *Angew. Chem. Int. Ed.* 2009, 48, 7892–7894. Asymmetric organocatalytic cascade reactions with α-substituted α,β-unsaturated aldehydes.

60 Y.-K. Liu, H. Liu, W. Du, L. Yue, Y.-C. Chen, *Chem.–Eur. J.* 2008, 14, 9873–9877. Reaction control in the organocatalytic asymmetric one-pot, three-component reaction of aldehydes, diethyl α-aminomalonate and nitroalkenes: toward diversity-oriented synthesis.

61 (a) S. G. Ouellet, J. B. Tuttle, D. W. C. MacMillan, *J. Am. Chem. Soc.* 2005, 127, 32–33. Enantioselective organocatalytic hydride reduction. (b) T. D. Beeson, D. W. C. MacMillan, *J. Am. Chem. Soc.* 2005, 127, 8826–8828. Enantioselective organocatalytic α-fluorination of aldehydes.

62 Y. Huang, A. M. Walji, C. H. Larsen, D. W. MacMillan, *J. Am. Chem. Soc.* 2005, 127, 15051–15053. Enantioselective organo-cascade catalysis.

63 A. Fraile, D. M. Scarpino Schietroma, A. Albrecht, R. L. Davis, K. A. Jørgensen, *Chem.–Eur. J.* 2012, 18, 2773–2776. Asymmetric synthesis of hexahydropyrrolo-isoquinolines by an organocatalytic three-component reaction.

64 W.-B. Chen, Z.-J. Wu, Q.-L. Pei, L.-F. Cun, X.-M. Zhang, W.-C. Yuan, *Org. Lett.* 2010, 12, 3132–3135. Highly enantioselective construction of spiro[4*H*-pyran-3,3′-oxindoles] through a domino Knoevenagel/Michael/cyclization sequence catalyzed by cupreine.

65 X.-Y. Wu, X.-Y. Dai, H.-H. Fang, L.-L. Nie, J. Chen, W.-G. Cao, G. Zhao, *Chem.–Eur. J.* 2011, 17, 10510–10514. One-pot three-component syntheses of indoloquinolizidine derivatives using an organocatalytic Michael addition and subsequent Pictet-Spengler cyclization.

66 H.-L. Zhu, J.-B. Ling, P.-F. Xu, *J. Org. Chem.* 2012, 77, 7737–7743. α-Oxo-γ-butyrolactam, *N*-containing pronucleophile in organocatalytic one-pot assembly of butyrolactam-fused indoloquinolizidines.

67 S. Gogoi, C.-G. Zhao, *Tetrahedron Lett.* 2009, 50, 2252–2255. Organocatalyzed enantioselective synthesis of 6-amino-5-cyanodihydropyrano[2,3-c]pyrazoles.

68 D. M. d. M. Sanchez, O. Basle, Y. Genisson, J.-C. Plaquevent, X. Bugaut, T. Constantieux, J. Rodriguez, *Angew. Chem. Int. Ed.* 2013, 52, 14143–14146. Enantioselective organocatalytic multicomponent synthesis of 2,6-diazabicyclo[2.2.2] octanones.

69 (a) C. Simon, T. Constantieux, J. Rodriguez, *Eur. J. Org. Chem.* 2004, 4957–4980. Utilisation of 1,3-dicarbonyl derivatives in multicomponent reactions. (b) D. Bonne, Y. Coquerel, T. Constantieux, J. Rodriguez, *Tetrahedron-Asymmetry* 2010, 21, 1085–1109. 1,3-Dicarbonyl compounds in stereoselective domino and multicomponent reactions. (c) F. Lieby-Muller, C. Simon, T. Constantieux, J. Rodriguez, *Qsar & Combinatorial Science* 2006, 25, 432–438. Current developments in Michael addition-based multicomponent domino reactions involving 1,3-dicarbonyls and derivatives. (d) X. Bugaut, D. Bonne, Y. Coquerel, J. Rodriguez, T. Constantieux, *Curr. Org. Chem.* 2013, 17, 1920–1928. Michael addition-initiated sequential reactions from 1,3-dicarbonyls for the synthesis of polycyclic heterocycles.

70 (a) D. Enders, M. R. M. Hüttl, J. Runsink, G. Raabe, B. Wendt, *Angew. Chem. Int. Ed.* 2007, 46, 467–469. Organocatalytic one-pot asymmetric synthesis of functionalized tricyclic carbon frameworks from a triple-cascade/Diels-Alder sequence. (b) D. Enders, M. R. M. Hüttl, G. Raabe, J. W. Bats, *Adv. Synth. Catal.* 2008, 350, 267–279. Asymmetric synthesis of polyfunctionalized mono-, bi-, and tricyclic carbon frameworks *via* organocatalytic domino reactions.

71 D. Enders, B. Schmid, N. Erdmann, G. Raabe, *Synthesis* 2010, 13, 2271–2277. Asymmetric synthesis of thiadecalins *via* an organocatalytic triple cascade/sulfa-Michael sequence.

72 (a) O. Penon, A. Carlone, A. Mazzanti, M. Locatelli, L. Sambri, G. Bartoli, P. Melchiorre, *Chem. Eur. J.* 2008, 14, 4788–4791. Quaternary stereogenic carbon atoms in complex molecules by an asymmetric, organocatalytic, triple-cascade reaction. (b) G. Bencivenni, L. Y. Wu, A. Mazzanti, B. Giannichi, F. Pesciaioli, M. P. Song, G. Bartoli, P. Melchiorre, *Angew. Chem. Int. Ed.* 2009, 48, 7200–7203. Targeting structural and stereochemical complexity by organocascade catalysis: construction of spirocyclic oxindoles having multiple stereocenters.

73 C. Cassani, X. Tian, E. C. Escudero-Adan, P. Melchiorre, *Chem. Commun.* 2011, 47, 233–235. Multiple approaches to enantiopure spirocyclic benzofuranones using organocatalytic cascade reactions.

74 A. Zea, A.-N. R. Alba, A. Mazzanti, A. Moyano, R. Rios, *Org. Biomol. Chem.* 2011, 9, 6519–6523. Highly enantioselective cascade synthesis of spiropyrazolones.

75 A. Carlone, S. Cabrera, M. Marigo, K. A. Jørgensen, *Angew. Chem. Int. Ed.* 2007, 46, 1101–1104. A new approach for an organocatalytic multicomponent domino asymmetric reaction.

76 D. Enders, M. Jeanty, J. Bats, Synlett 2009, 3175–3178. Organocatalytic asymmetric triple domino reactions of nitromethane with α,β-unsaturated aldehydes.

77 Y. Wang, R. G. Han, Y. L. Zhao, S. Yang, P. F. Xu, D. J. Dixon, *Angew. Chem. Int. Ed.* 2009, 48, 9834–9838. Asymmetric organocatalytic relay cascades: catalyst-controlled stereoisomer selection in the synthesis of functionalized cyclohexanes.

78 I. Ibrahem, G. L. Zhao, R. Rios, J. Vesely, H. Sunden, P. Dziedzic, A. Córdova, *Chem.–Eur. J.* 2008, 14, 7867–7879. One-pot organocatalytic domino Michael/α-alkylation reactions: direct catalytic enantioselective cyclopropanation and cyclopentanation reactions.

79 D.-F. Yu, Y. Wang, P.-F. Xu, *Tetrahedron* 2011, 67, 3273–3277. Organocatalytic enantioselective multicomponent cascade reaction: facile access to tetrahydropyridines with C3 all-carbon quaternary stereocenters.

80 P. Kotame, B.-C. Hong, J.-H. Liao, *Tetrahedron Lett.* 2009, 50, 704–707. Enantioselective synthesis of the tetrahydro-6*H*-benzo[*c*]chromenes *via* domino Michael-aldol condensation: control of five stereocenters in a quadruple-cascade organocatalytic multi-component reaction.

81 F.-L. Zhang, A.-W. Xu, Y.-F. Gong, M.-H. Wei, X.-L. Yang, *Chem.–Eur. J.* 2009, 15, 6815–6818. Asymmetric organocatalytic four-component quadruple domino reaction initiated by oxa-Michael addition of alcohols to acrolein.

82 D. Enders, C. Wang, M. Mukanova, A. Greb, *Chem. Commun.* 2010, 46, 2447–2449. Organocatalytic asymmetric synthesis of polyfunctionalized 3-(cyclohexenylmethyl)-indoles *via* a quadruple domino Friedel-Crafts-type/Michael/Michael/aldol condensation reaction.

83 D. Enders, A. Greb, K. Deckers, P. Selig, C. Merkens, *Chem.–Eur. J.* 2012, 18, 10226–10229. Quadruple domino organocatalysis: an asymmetric aza-Michael/Michael/Michael/aldol reaction sequence leading to tetracyclic indole structures with six stereocenters.

84 K. Jiang, Z.-J. Jia, X. Yin, L. Wu, Y.-C. Chen, *Org. Lett.* 2010, 12, 2766–2769. Asymmetric quadruple aminocatalytic domino reactions to fused carbocycles incorporating a spirooxindole motif.

85 (a) C. F. Barbas III, D. B. Ramachary, N. S. Chowdari, Synlett 2003, 1910–1914. The first organocatalytic hetero-domino Knoevenagel-Diels-Alder-epimerization reactions: diastereoselective synthesis of highly substituted spiro[cyclohexane-1,2′-indan]-1′,3′,4-triones. (b) D. B. Ramachary, K. Anebouselvy, N. S. Chowdari, C. F. Barbas III, J. Org. Chem. 2004, 5838–5849. Direct organocatalytic asymmetric heterodomino reactions: the Knoevenagel/Diels-Alder/epimerization sequence for the highly diastereoselective synthesis of symmetrical and nonsymmetrical synthons of benzoannelated centropolyquinanes.

86 D. B. Ramachary, N. S. Chowdari, C. F. Barbas III, *Angew. Chem. Int. Ed. Engl.* 2003, 42, 4233–4237. Organocatalytic asymmetric domino Knoevenagel/Diels-Alder reactions: a bioorganic approach to the diastereospecific and enantioselective construction of highly substituted spiro[5,5] undecane-1,5,9-triones.

87 D. Pizzirani, M. Roberti, M. Recanatini, *Tetrahedron Lett.* 2007, 48, 7120–7124. Domino Knoevenagel/Diels-Alder sequence coupled to Suzuki reaction: a valuable synthetic platform for chemical biology.

88 D. B. Ramachary, M. Kishor, *J. Org. Chem.* 2007, 72, 5056–5068. Organocatalytic sequential one-pot double cascade asymmetric synthesis of Wieland-Miescher ketone analogues from a Knoevenagel/hydrogenation/Robinson annulation sequence: scope and applications of organocatalytic biomimetic reductions.

89 H. Jiang, A. Falcicchio, K. L. Jensen, M. W. Paixão, S. Bertelsen, K. A. Jørgensen, *J. Am. Chem. Soc.* 2009, 131, 7153–7157. Target-directed organocatalysis: a direct asymmetric catalytic approach to chiral propargylic and allylic fluorides.

90 S. K. Xiang, B. Zhang, L. H. Zhang, Y. Cui, N. Jiao, *Chem. Commun.* 2011, 47, 5007–5009. The versatile roles of ammonium salt catalysts in enantioselective reduction and alkylation of α,β-unsaturated aldehydes: iminium catalysis, enamine catalysis and acid catalysis.

91 X. Li, Y. Zhao, H. Qu, Z. Mao, X. Lin, *Chem. Commun.* 2013, 49, 1401–1403. Organocatalytic asymmetric multicomponent reactions of aromatic aldehydes and anilines with β-ketoesters: facile and atom-economical access to chiral tetrahydropyridines.

92 S. Roy, K. Chen, *Org. Lett.* 2012, 14, 2496–2499. Three-component organocascade kinetic resolution of racemic nitroallylic acetates *via* sequential iminium/enamine asymmetric catalysis.

93 C. Hulme, V. Gore, *Curr. Med. Chem.* 2003, 10, 51–80. Multicomponent reactions: emerging chemistry in drug discovery.

94 (a) H. Ishikawa, T. Suzuki, Y. Hayashi, *Angew. Chem. Int. Ed.* 2009, 48, 1304–1307. High-yielding synthesis of the anti-influenza neuramidase inhibitor (–)-oseltamivir by three "one-pot" operations. (b) H. Ishikawa, T. Suzuki, H. Orita, T. Uchimaru, Y. Hayashi, *Chem. Eur. J.* 2010, 16, 12616–12626. High-yielding synthesis of the anti-influenza neuraminidase inhibitor (–)-oseltamivir by two "one-pot" sequences. (c) K. Patora-Komisarska, M. Benohoud, H. Ishikawa, D. Seebach, Y. Hayashi, *Helv. Chim. Acta* 2011, 94, 719–745. Organocatalyzed Michael addition of aldehydes to nitro alkenes—generally accepted mechanism revisited and revised.

95 H. Ishikawa, M. Honma, Y. Hayashi, *Angew. Chem. Int. Ed.* 2011, 50, 2824–2827. One-pot high-yielding synthesis of the DPP4-selective inhibitor ABT-341 by a four-component coupling mediated by a diphenylprolinol silyl ether.

96 H. Ishikawa, S. Sawano, Y. Yasui, Y. Shibata, Y. Hayashi, *Angew. Chem. Int. Ed.* 2011, 50, 3774–3779. Asymmetric one-pot four-component coupling reaction: synthesis of substituted tetrahydropyrans catalyzed by diphenylprolinol silyl ether.

97 G. Dickmeiss, K. L. Jensen, D. Worgull, P. T. Franke, K. A. Jørgensen, *Angew. Chem. Int. Ed.* 2011, 50, 1580–1583. An asymmetric organocatalytic one-pot strategy to octahydroacridines.

98 S. Bertelsen, R. L. Johansen, K. A. Jørgensen, Chem. Commun. 2008, 3016–3018. Controlling the formation of 1 out of 64 stereoisomers using organocatalysis.

99 L. Albrecht, L. K. Ransborg, A. Albrecht, L. Lykke, K. A. Jørgensen, *Chem.–Eur. J.* 2011, 17, 13240–13246. Organocatalytic strategies for the construction of optically active imidazoles, oxazoles, and thiazoles.

100 D. B. Ramachary, Y. V. Reddy, B. V. Prakash, *Org. Biomol. Chem.* 2008, 6, 719–726. Double cascade reactions based on the Barbas dienamine platform: highly stereoselective synthesis of functionalized cyclohexanes for cardiovascular agents.

101 D. B. Ramachary, M. Kishor, *Org. Biomol. Chem.* 2010, 8, 2859–2867. Direct catalytic asymmetric synthesis of highly functionalized tetronic acids/tetrahydro-isobenzofuran-1, 5-diones *via* combination of cascade three-component reductive alkylations and Michael-aldol reactions.

102 D. B. Ramachary, Y. V. Reddy, A. Banerjee, S. Banerjee, *Org. Biomol. Chem.* 2011, 9, 7282–7286. Design, synthesis and biological evaluation of optically pure functionalized spiro[5,5]undecane-1,5,9-triones as HIV-1 inhibitors.

103 F. Macaev, N. Sucman, F. Shepeli, M. Zveaghintseva, V. Pogrebnoi, *Symmetry* 2011, 3, 165–170. Facile and convenient one-pot process for the synthesis of spirooxindole derivatives in high optical purity using (–)-(*S*)-brevicolline as an organocatalyst.

104 W.-T. Wei, C.-X. Chen, R.-J. Lu, J.-J. Wang, X.-J. Zhang, M. Yan, *Org. Biomol. Chem.* 2012, 10, 5245–5252. Enantioselective synthesis of 3,3′-dihydropyrryl-spirooxindoles *via* an organocatalytic three-component reaction.

105 L. Tian, X.-Q. Hu, Y.-H. Li, P.-F. Xu, *Chem. Commun.* 2013, 49, 7213–7215. Organocatalytic asymmetric multicomponent cascade reaction via 1,3-proton shift and [3+2] cycloaddition: an efficient strategy for the synthesis of oxindole derivatives.

106 M. N. Cheng, H. Wang, L.-Z. Gong, *Org. Lett.* 2011, 13, 2418–2421. Asymmetric organocatalytic 1,3-dipolar cycloaddition of azomethine ylide to methyl 2-(2-nitrophenyl)acrylate for the synthesis of diastereoisomers of spirotryprostatin A.

107 F. Shi, S. Tu, Z. Tao, S. Luo, L. Gong, *Chem.–Eur. J.* 2012, 18, 6885–6894. Scaffold-inspired enantioselective synthesis of biologically important spiro[pyrrolidin-3,2′-oxindoles] with structural diversity through catalytic isatin-derived 1,3-dipolar cycloadditions.

108 C. Wang, X. H. Chen, S. M. Zhou, L. Z. Gong, *Chem. Commun.* 2010, 46, 1275–1277. Asymmetric organocatalytic formal double-arylation of azomethines for the synthesis of highly enantiomerically enriched isoindolines.

109 F. Shi, R.-Y. Zhu, W. Dai, C.-S. Wang, S.-J. Tu, *Chem.–Eur. J.* 2014, 20, 2597–2604. Catalytic asymmetric formal [3+3] cycloaddition of an azomethine ylide with 3-indolylmethanol: enantioselective construction of a six-membered piperidine framework.

110 B.-C. Hong, P. Kotame, C.-W. Tsai, J.-H. Liao, *Org. Lett.* 2010, 12, 776–779. Enantioselective total synthesis of (+)-conicol *via* cascade three-component organocatalysis.

111 S. Balalaie, M. Bararjanian, A. M. Amani, B. Movassagh, Synlett 2006, 263–266. (*S*)-Proline as a neutral and efficient catalyst for the one-pot synthesis of tetrahydrobenzo[*b*]pyran derivatives in aqueous media.

112 S. Samai, G. C. Nandi, P. Singh, M. S. Singh, *Tetrahedron* 2009, 65, 10155–10161. L-Proline: an efficient catalyst for the one-pot synthesis of 2,4,5-trisubstituted and 1,2,4,5-tetrasubstituted imidazoles.

113 L. Yadav, A. Rai, A. Singh, S. Singh, Synlett 2011, 335–340. Chiral amine-triggered triple cascade reactions: a new approach to functionalized decahydroquinolines.

114 I. Ibrahem, R. Rios, J. Vesely, A. Córdova, *Tetrahedron Lett.* 2007, 48, 6252–6257. Organocatalytic asymmetric multicomponent [C+NC+CC] synthesis of highly functionalized pyrrolidine derivatives.

115 F. Cruz-Acosta, P. de Armas, F. Garcia-Tellado, *Chem.–Eur. J.* 2013, 19, 16550–16554. Water-compatible hydrogen-bond activation: a scalable and organocatalytic model for the stereoselective multicomponent aza-Henry reaction.

116 L. Wang, S. M. Cui, W. Meng, G. W. Zhang, J. Nie, J. A. Ma, *Chin. Sci. Bull.* 2010, 55, 1729–1731. Asymmetric synthesis of α-aminophosphonates by means of direct organocatalytic three-component hydrophosphonylation.

117 P. B. Thorat, S. V. Goswami, R. L. Magar, B. R. Patil, S. R. Bhusare, *Eur. J. Org. Chem.* 2013, 2013, 5509–5516. An efficient organocatalysis: a one-pot highly enantioselective synthesis of α-aminophosphonates.

3

METAL-CATALYZED MULTICOMPONENT REACTIONS

ABEL ROS AND INMACULADA MATAS

Instituto de Investigaciones Químicas (Consejo Superior de Investigaciones Científicas—Universidad de Sevilla), Sevilla, Spain

3.1 INTRODUCTION

One of the challenging goals in synthetic chemistry is the discovery of efficient new routes for the single-step elaboration of an array of relevant products from simple and readily available building blocks. The growing understanding of transition metal catalysis has provided the opportunity to design such reactions, which rely upon the countless bond-forming processes that can be accomplished with metal catalysts. In these catalytic reactions, several synthetic operations occur in sequence, with the reactivity of transition metal complexes activating basic building blocks toward reaction. Overall, this allows the synthesis of target products in one step, in high yield, with minimal waste, and with high molecular diversity. Hence, the development of transition metal-mediated multicomponent reactions (MCRs) has been the subject of intense research over the past several years.

MCRs are defined as reactions that occur in one reaction vessel and involve more than two starting reagents that form a single product that contains the essential parts of the starting materials. Due to the vast number of MCRs based on metal catalysis, the present work will only focus on the "ideal" MCRs, which involve the simultaneous addition of reactants, reagents, and catalyst at the beginning of the reaction, and will exclude those where the addition of the components is carried out sequentially or that require adjustment of the reaction conditions throughout the process.

Although most of the established metal-catalyzed MCRs are based on palladium or copper catalysts, the search for new MCR products has resulted in the development of new catalytic systems in the last decade. This chapter has been divided into sections dealing with different metals, that is, palladium, nickel, group 11 metals, rhodium, iron, and ruthenium, which correspond to the most fruitful examples reported since 2005.

3.2 PALLADIUM-CATALYZED MCRs

3.2.1 Palladium-Catalyzed Carbonylation Reactions

Coupling reactions based on palladium-catalyzed carbonylations are an attractive class of multicomponent transformations that benefit from the use of the readily available and most important C_1 building block, carbon monoxide, which is incorporated into a substrate to produce carbonyl-containing compounds. Several carbonylation processes have been developed in organic syntheses and are applied in the pharmaceutical industry. Among those, the carbonylative coupling reactions of aryl halides and imines have been intensively explored in the last decade.

3.2.1.1 Carbonylative Coupling of Aryl Halides Since the pioneering work by Heck et al. in 1974 that described the alkoxycarbonylation of aryl and vinyl halides [1], research in this area has made an impressive progress in the last years in terms of reaction conditions and catalyst productivity, significantly broadening the scope of the method [2]. Thus, Buchwald reported in 2007 the first general method for the aminocarbonylation of aryl chlorides, with reactions proceeding at atmospheric CO pressure and moderate temperatures. A wide range of aryl chlorides and amines were all successfully transformed into the corresponding amides **1**

Multicomponent Reactions: Concepts and Applications for Design and Synthesis, First Edition. Edited by Raquel P. Herrera and Eugenia Marqués-López.

in palladium-catalyzed one-pot reactions, using an electron-rich bulky bisphosphine as the ligand (Scheme 3.1). The strategy relied on the use of sodium phenoxide as the basic additive, which also facilitates acyl transfer through the formation of phenyl esters **2** as intermediates and catalyzes the conversion of the phenyl ester into the final amide product **1** [3].

The extensive work carried out by Beller in recent years has contributed to the field with a series of interesting examples, most of them based on the system palladium/di-1-adamantyl-*n*-butylphosphine ($BuPAd_2$). Thus, they reported the use of ammonia as the nitrogen source for the palladium-catalyzed aminocarbonylation of aryl halides, providing a straightforward method to prepare primary amides with excellent applicability and functional group tolerance [4]. In the same group, the first carbonylative C—H activation reactions of heteroarenes to form diaryl ketones **4** were developed [5]. A series of nonpreactivated heteroarenes **3** (oxazoles, thiazoles, and imidazoles) were successfully carbonylative coupled to a wide range of aryl iodides (Scheme 3.2). To avoid the formation of the noncarbonylative coupled product, 40 atm CO and a bidentate chelating phosphine as the ligand were used in the palladium/copper system.

The scope of the carbonylative Sonogashira coupling for the synthesis of alkynones from aryl halides and terminal alkynes has also been broadened, allowing the use of inexpensive aryl bromides as starting materials instead of the more common aryl iodides or iodoalkenes [6]. Beller et al. found that the palladium/$BuPAd_2$ system in the presence of K_2CO_3 as the base afforded the corresponding alkynones **5** at low CO pressures from a series of aryl bromides and terminal alkynes in good yields (Scheme 3.3). The optimum reaction temperature was set at 100 °C, since decarbonylation is favored at higher temperatures and only the noncarbonylative coupled product 1,2-diphenylethyne is formed. The same group has developed a general palladium-catalyzed process that allows Heck carbonylation of aryl halides with various important classes of olefins such styrenes, acrylates, and enol ethers, providing an extension of the carbonylative Suzuki and Sonogashira reactions [7].

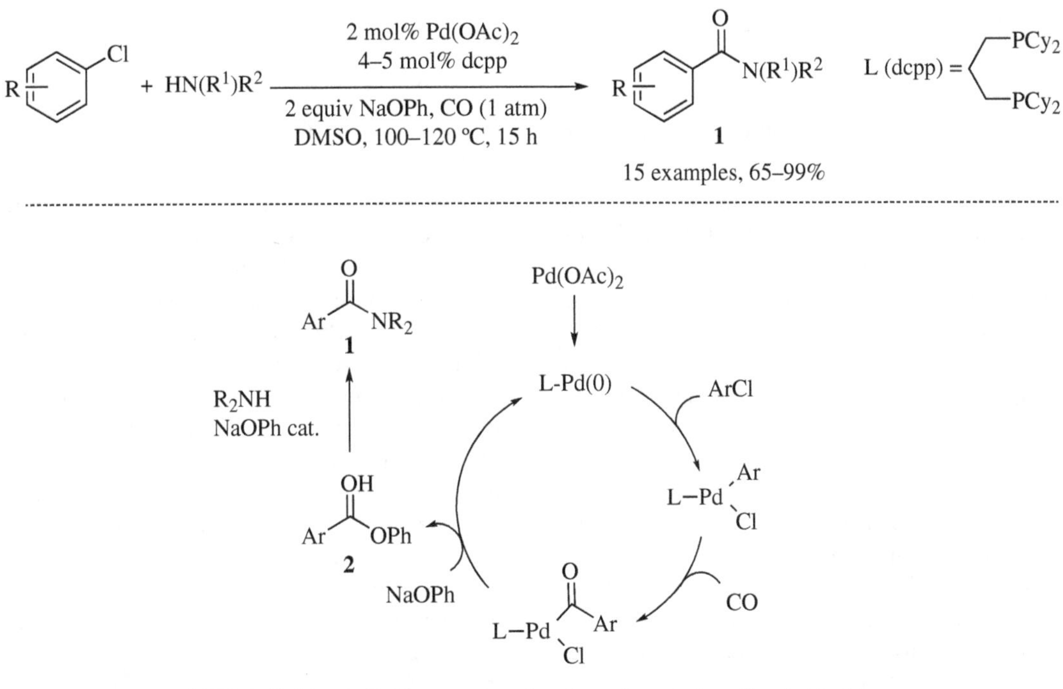

SCHEME 3.1 Palladium-catalyzed aminocarbonylation of aryl chlorides.

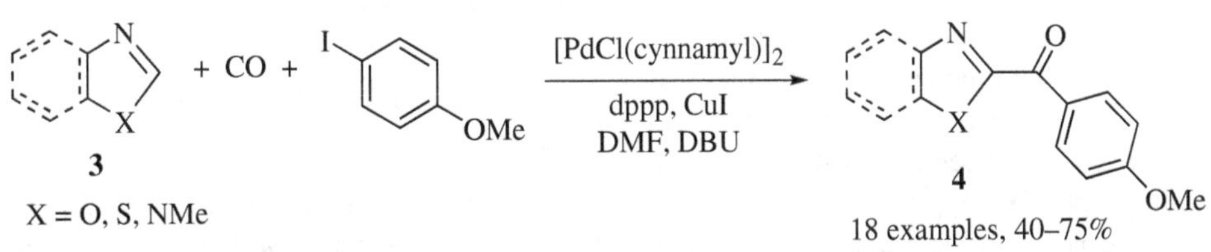

SCHEME 3.2 Carbonylative coupling of heteroarenes and aryl iodides *via* C—H activation.

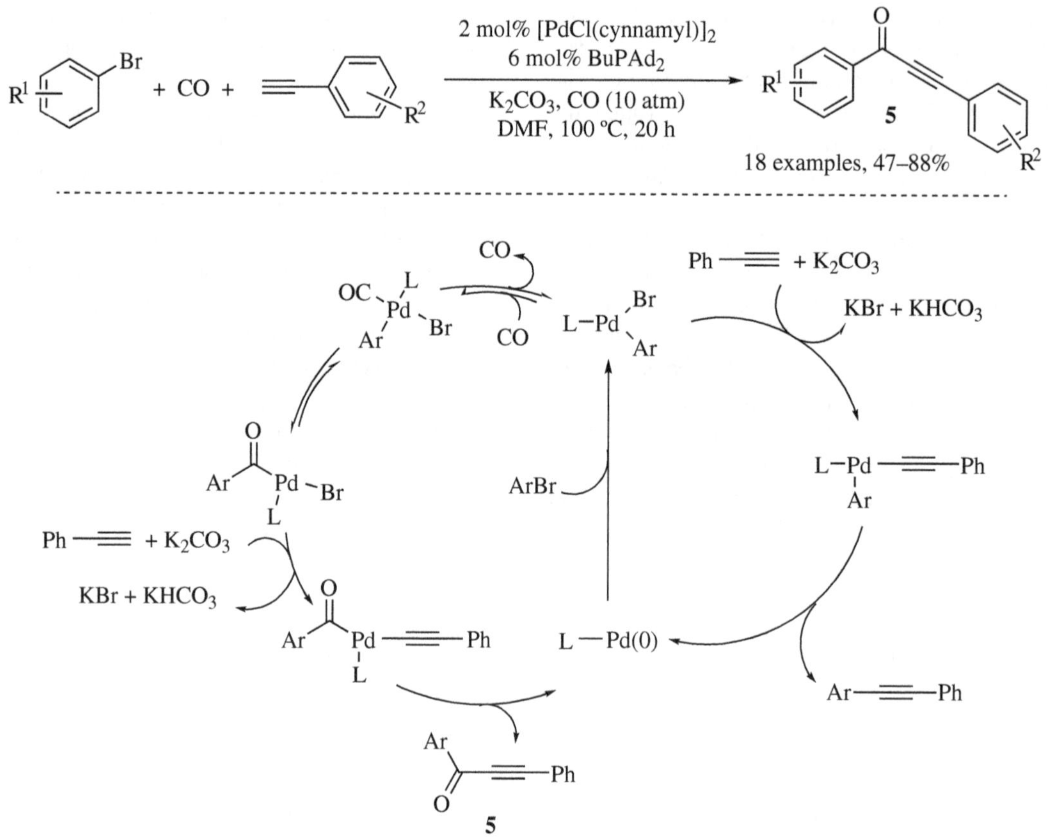

SCHEME 3.3 Palladium-catalyzed carbonylative coupling of aryl bromides and alkynes.

The formylation of aryl halides reported by Beller in 2006 constituted an improvement of the methodologies available to date for the synthesis of aromatic and heteroaromatic aldehydes. The palladium/BuPAd$_2$ system allows using synthesis gas (CO/H_2) as the CO and hydrogen source under mild conditions (5 atm), thus avoiding limited functional group tolerance, high catalyst loadings, environment unfriendly reagents and conditions, and side reactions such as reductive dehalogenation [8].

3.2.1.2 Coupling Reactions with Imines and Carbon Monoxide The versatile insertion chemistry of CO has led to the development of efficient synthetic methodologies to heterocycles based on multicomponent imine/CO couplings. Arndtsen exploited the reactivity of the mesoionic compound **6a** that is obtained upon activation of imines by acid chlorides in the presence of CO, commonly referred to as a Münchnone, which undergoes facile 1,3-dipolar cycloaddition reactions with unsaturated substrates to generate nitrogen heterocycles (Scheme 3.4). A series of multicomponent synthetic methodologies emerged from the latter reaction, the success of which relies on the high selectivity of the sequential steps that take place in the multicomponent mixture [9].

Thus, substituted pyrroles can be obtained with high selectivities and yields when an alkyne is added to the multicomponent system [10]. When the formation of the Münchnones occurs in the presence of electron-poor tosylimines, imidazoles are obtained [11]. Amide-substituted β-lactams are the product of the coupling of a second molecule of imine, which undergoes a formal [2+2] cycloaddition with the ketene isomer of the Münchnone **6b** [12]. The addition of imines in the presence of an acid catalyst leads to the formation of imidazolium salts, which yield imidazolines upon decarboxylation [13]. Imidazolium salts can also be prepared by using aryl halides as a greener alternative to acid chlorides, since the palladium-catalyzed carbonylation of aryl halides generates analogous intermediates than those obtained with the acid chlorides [14]. Very recently, this group reported an interesting application of this palladium-catalyzed multicomponent procedure for the synthesis of π-conjugated oligomers, which is based on the simultaneous assembly of dialdehydes, di(acyl chlorides), and imines in the presence of CO to provide a well-defined repeat unit (Scheme 3.5) [15].

3.2.2 Palladium-Catalyzed MCRs Involving Isocyanides

Whereas the use of CO as C_1 building block in palladium-catalyzed cascade reactions is well established, the application of isoelectronic isocyanides in similar processes is

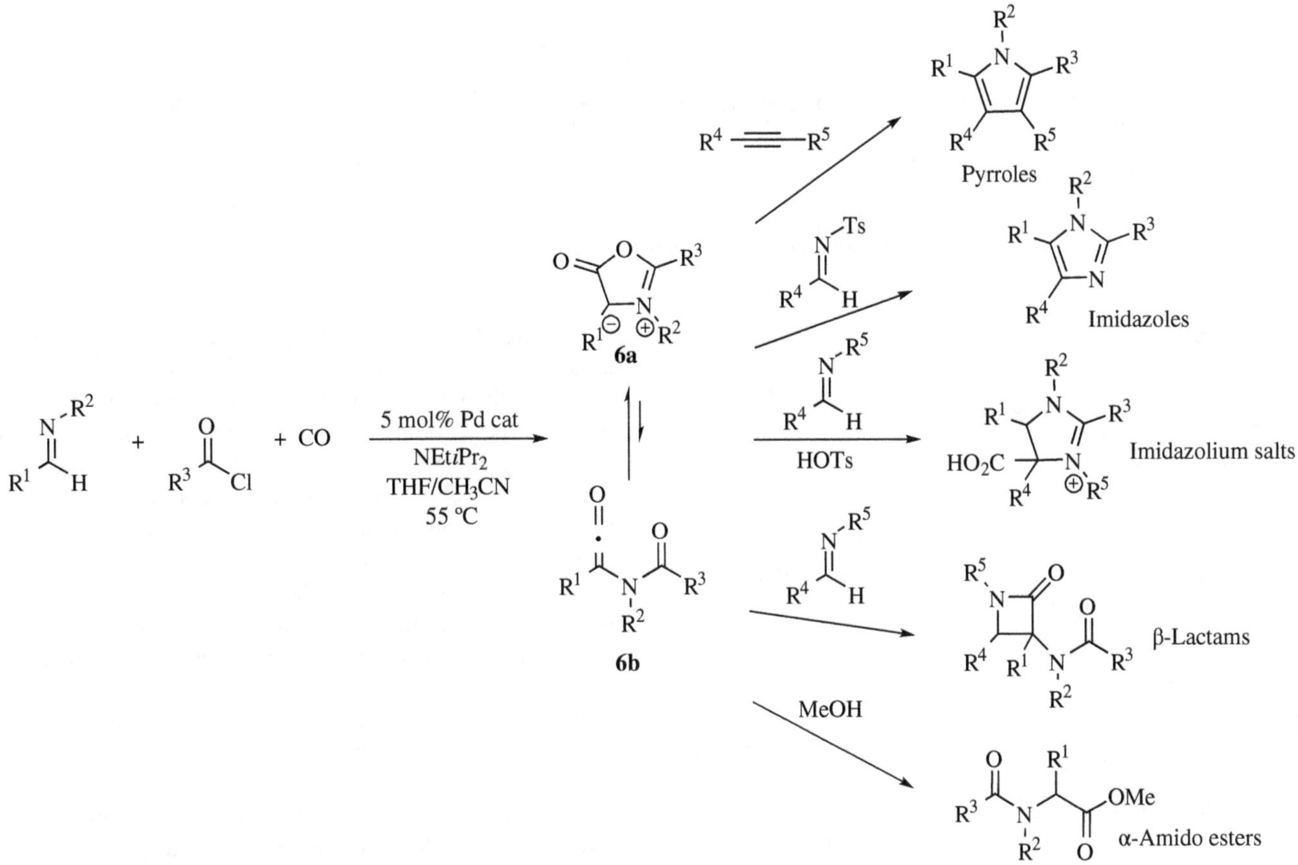

SCHEME 3.4 Synthesis of nitrogenated heterocycles *via* Münchnone intermediate.

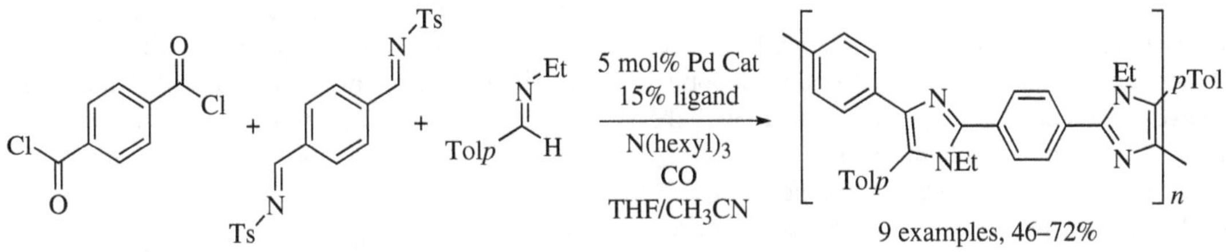

SCHEME 3.5 Synthesis of π-conjugated oligomers *via* carbonylative coupling of dialdehydes, di(acyl chloride)s, and imines.

much less explored, despite their widespread use in other types of MCRs [16]. This is somewhat surprising, considering the similar reactivity toward palladium of both types of molecules and the advantages that isocyanides offer over CO, these including their easier handling, the homogeneous character of their reactions, and the presence of one diversity point in the molecule that increases their versatility. It was in 2000 with the report by Whitby et al. concerning the preparation of amidines **7** by the palladium-catalyzed coupling of an aryl halide, isocyanide, and amine that palladium-catalyzed isocyanide reactions caught the attention of researchers (Scheme 3.6) [17]. Since then, MCRs involving palladium-catalyzed isocyanide insertions have experienced a significant surge of interest [18], being successfully applied in the synthesis of a wide range of nitrogen heterocycles.

In general, these reactions proceed *via* a catalytic cycle similar to that of typical carbonylative catalysis. As depicted in Scheme 3.7, it involves an initial oxidative addition of aryl halide to the Pd(0) species to give a Pd(II) intermediate **I**, which can then undergo insertion of an isocyanide molecule into the palladium–carbon bond to give **II** and subsequent reaction with a nucleophile to eventually afford the amidine **7** by reductive elimination from **III**.

The original limited scope with regard to the isocyanide was soon overcome. Lang et al. adapted this methodology for the synthesis of benzoxazoles **8** and oxazolines **9** by using

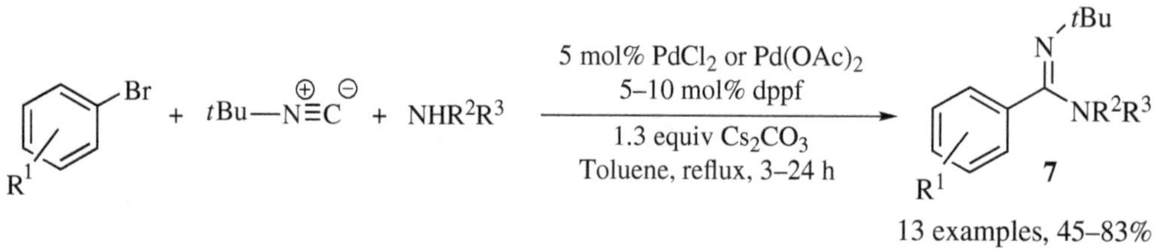

SCHEME 3.6 Palladium-catalyzed coupling of aryl halides, isocyanide, and amines.

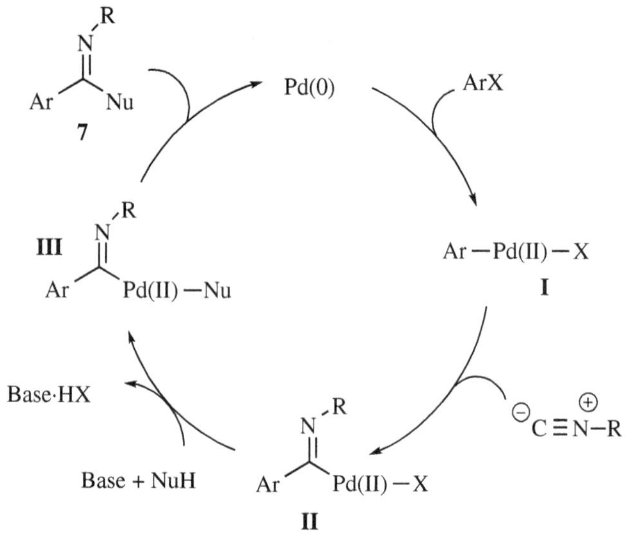

SCHEME 3.7 Proposed mechanism for the palladium-catalyzed isocyanide reactions.

2-aminophenols and ethanolamine as the nucleophile, respectively. Since these amines are linked to a second nucleophile, an additional cyclization takes place, which ends up with the elimination of *tert*-butylamine (Scheme 3.8) [19].

In the quest for an efficient route for the generation of quinazolin-4(3*H*)-imines **10** (Scheme 3.9), Pu and Wu developed a three-component methodology based on a palladium-catalyzed isocyanide insertion involving a *N*-(2-bromoaryl) carbodiimide, isocyanide, and a nucleophile (amine or alcohol) [20]. When phosphites are used as nucleophiles, 4-imino-3,4-dihydroquinazolin-2-ylphosphonates are obtained [21]. The same authors extended the procedure to include a second cyclization by employing symmetric carbodiimides containing two aryl iodide moieties [22].

Orru and Ruijter developed an efficient palladium-catalyzed multicomponent method toward 4-aminophthalazin-1 (2*H*)-ones **11** (Scheme 3.10), combining isocyanides, hydrazines, and 2-halobenzoates in an extremely fast and regioselective reaction that proceeds *via* isocyanide insertion followed by lactamization. Although the scope with regard to the isocyanide and the hydrazine is limited, a simple dealkylation procedure affords the unprotected products, which can be further derivatized [23].

A straightforward procedure for the synthesis of a new class of substituted 4-amine-benzo[*b*][1,4]oxazepines **12** *via* palladium-catalyzed three-component reaction of bromoalkynes, *o*-aminophenols, and isocyanides was reported by Jiang et al. The seven-membered heterocycles were obtained smoothly and in high yields from a variety of bromoalkynes and *o*-aminophenols. Several control experiments were carried out to gain more insight into the mechanism of the reaction. The authors propose that the process involves an initial nucleophilic addition of *o*-aminophenols to bromoalkynes that affords intermediate **IV**, which enters the catalytic cycle undergoing oxidative addition to Pd(0) to generate a vinyl palladium species **V**. Subsequent migratory insertion of isocyanide and reaction with the base result in the formation of the eight-membered azapalladacyclic intermediate **VI**. Finally, reductive elimination affords the unstable product **VII**, which isomerizes to give the final product **12** (Scheme 3.11) [24]. Another domino reaction for the synthesis of 1,4-benzoxazepines has been very recently developed by Cai et al., which consists in the palladium-catalyzed three-component reaction of *N*-tosylaziridines, 2-iodophenols, and isocyanides [25]. This process involved aziridine ring-opening reaction with 2-iodophenol, followed by palladium-catalyzed isocyanide insertion, and affords the desired products in moderate to high yields.

3.2.3 Carbopalladation of Unsaturated C–C π-Components

The palladium-catalyzed three-component coupling involving the oxidative addition of an organic halide to palladium followed by insertion of a carbon–carbon multiple bond and termination by a nucleophile is a well-known and efficient method for the construction of π-conjugated molecules. Alkynes, alkenes, conjugated dienes, and allenes have been used as unsaturated π-components, while common terminators are terminal alkynes, alkenes, and organometallic reagents [26]. Despite being extensively studied, these reactions still constitute an active area of research that has given rise to a number of interesting contributions in the last decade, as illustrated by the following examples.

Following their report on the stereo- and regioselective synthesis of tetrasubstituted olefins by the palladium-catalyzed cross-coupling of aryl halides, internal alkynes, and

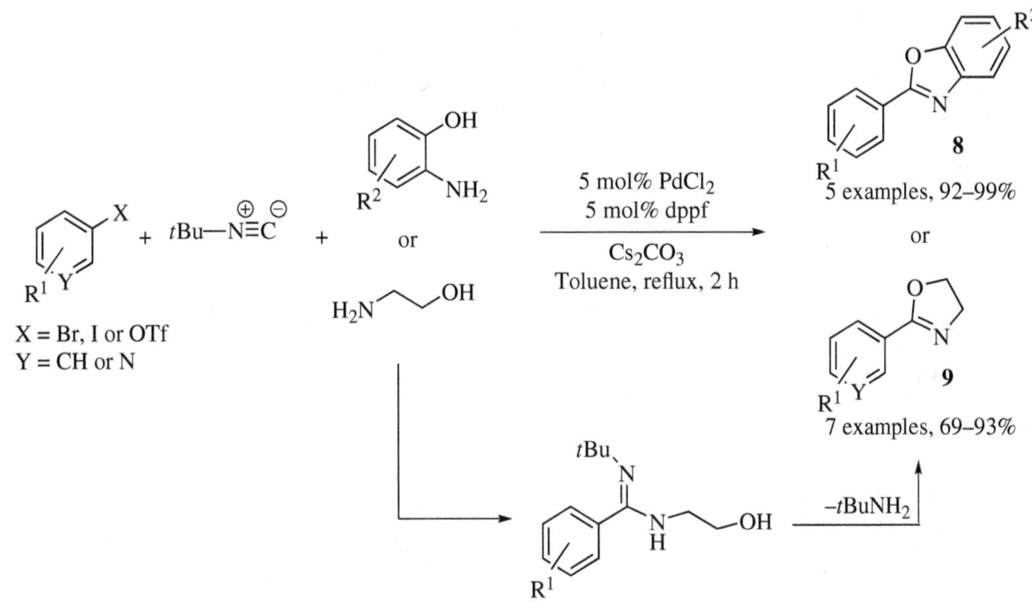

SCHEME 3.8 Synthesis of benzoxazoles and oxazolines by palladium-catalyzed isocyanide reactions.

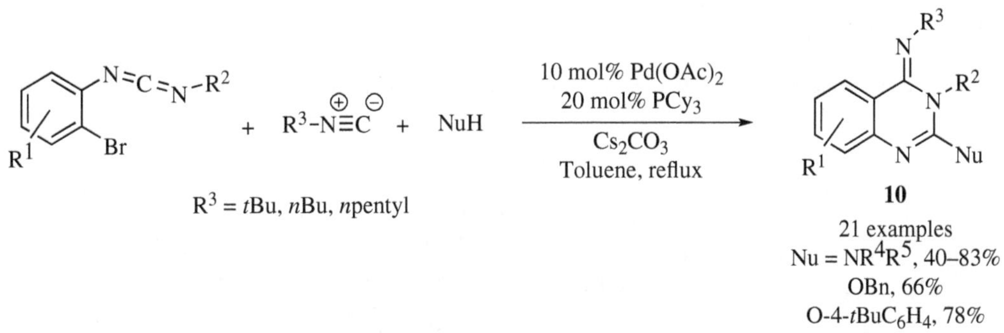

SCHEME 3.9 Synthesis of quinazolin-4(3*H*)-imines by palladium-catalyzed isocyanide reactions.

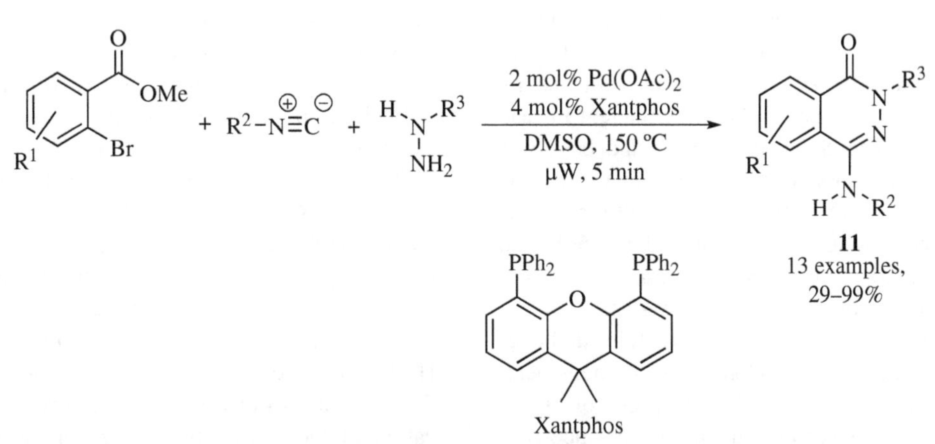

SCHEME 3.10 Synthesis of 4-aminophthalazin-1(2*H*)-ones by palladium-catalyzed isocyanide reactions.

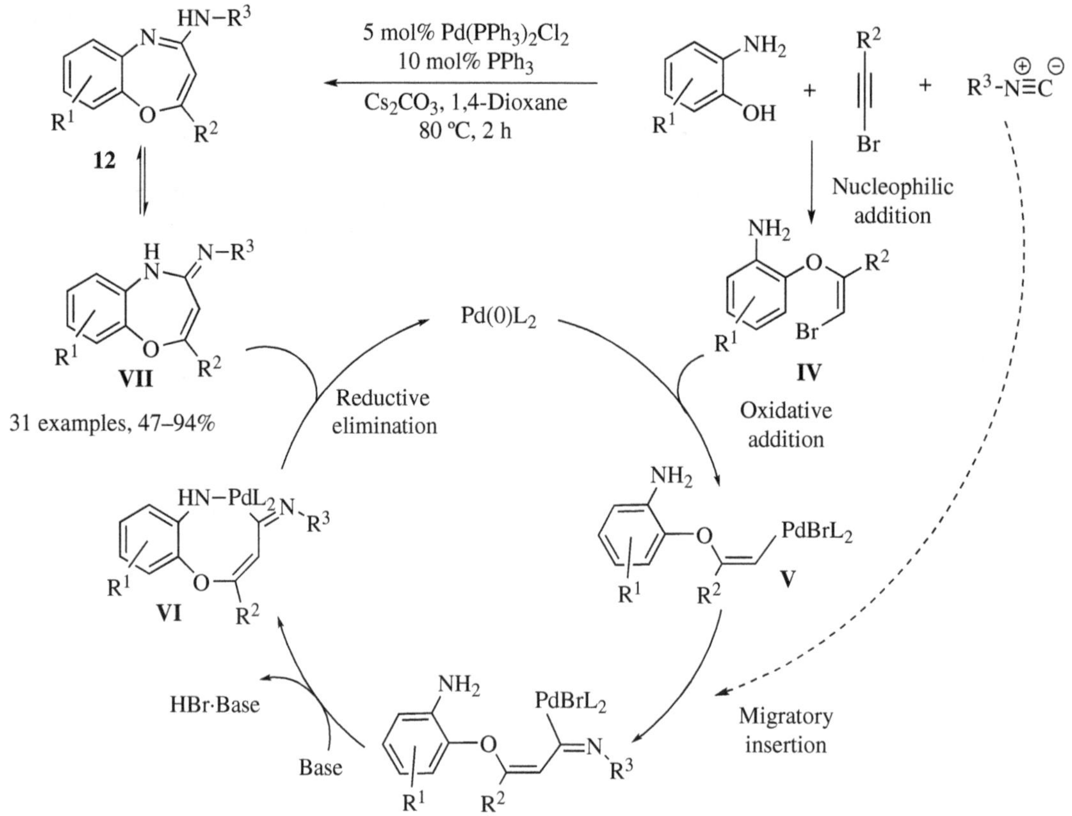

SCHEME 3.11 Proposed mechanism for the synthesis of 4-amine-benzo[*b*][1,4]oxazepines by palladium-catalyzed isocyanide reactions.

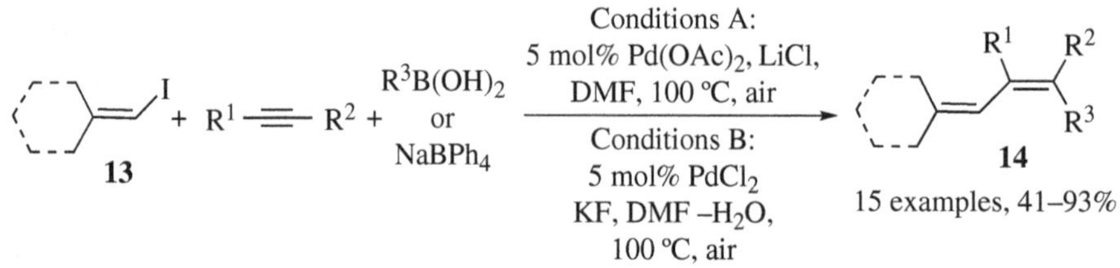

SCHEME 3.12 Synthesis of 1,3-dienes and trienes by palladium-catalyzed cross-coupling of aryl halides, internal alkynes, and arylboronic acids.

arylboronic acids [27], Larock et al. presented the one-pot intermolecular palladium-catalyzed synthesis of highly substituted 1,3-dienes and trienes **14**. This was achieved by using vinylic iodides **13** as the electrophilic reagent instead of aryl halides [28]. The choice of solvent and base was crucial to avoid direct cross-coupling between the vinylic halide and the organoboron compound or multiple alkyne insertions. Two condition sets were employed depending on the nature of the organoboron compound (tetraarylborates or arylboronic and vinylboronic acids; Scheme 3.12).

Another strategy to highly substituted 1,3-butadienes consists of the 1:2:1 coupling of iodobenzene, diphenylacetylene, and arylboronic acids reported by the group of Miura. In this reaction, which is carried out in the presence of a silver salt as a base and an oxidant, the insertion of two molecules of alkyne into the aryl palladium species gives a dienylpalladium intermediate **VIII** that undergoes a transmetalation with the boronic acid to afford **IX**. A subsequent reductive elimination eventually furnishes the 1,4-diarylated 1,3-diene **15** (Scheme 3.13) [29].

The scope of the synthetic procedure to substituted 1,3-butadienes **16** was later broadened by the same group, with the report of 18 examples of the 1:1:1 coupling of iodoarenes, diarylacetylenes, and monosubstituted alkenes. The mixture $Pd(OAc)_2$, LiCl, and $NaHCO_3$ led to the desired products in moderate yields, although minor amounts of the Mizoroki–Heck-type product (ArCH=CHR) were also formed [30]. For the synthesis of 1,3,5-hexatriene derivatives **17**, a modification of the former procedure consisting of using silver acetate as a base leads to the 1:2:1 coupling of iodoarenes,

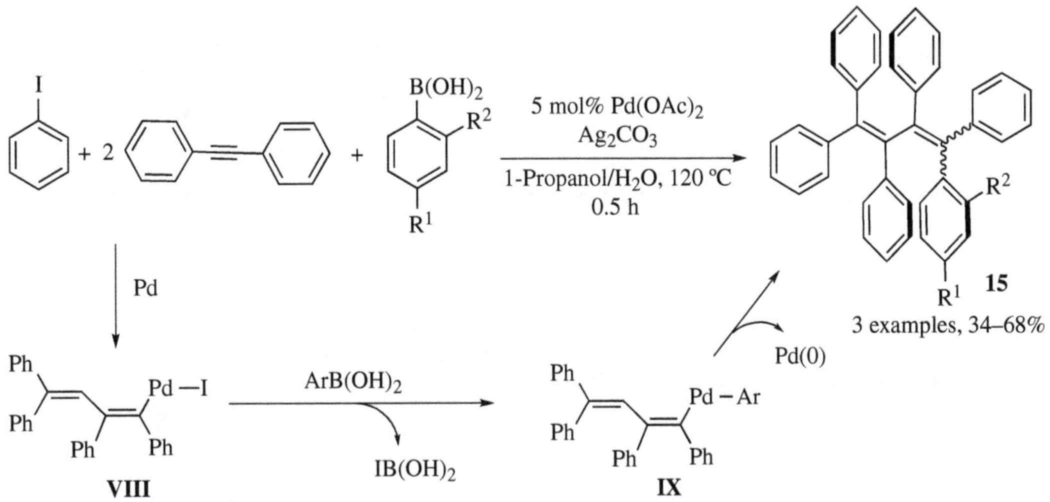

SCHEME 3.13 Synthesis of 1,3-butadienes by palladium-catalyzed cross-coupling of iodobenzene, diphenylacetylene, and arylboronic acids.

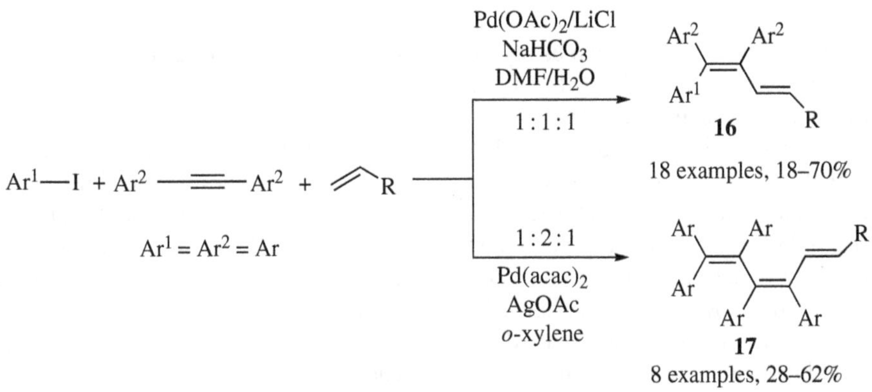

SCHEME 3.14 Synthesis of 1,3-butadienes and 1,3,5-hexatrienes by palladium-catalyzed cross-coupling of aryl iodides, diarylacetylenes, and monosubstituted alkenes.

diarylacetylenes, and monosubstituted alkenes (Scheme 3.14) [31]. The use of AgOAc as a base, in contrast to $NaHCO_3$, which gave the 1:1:1 coupling, seems to promote the double alkyne insertion by eliminating the iodide ligand on the iodopalladium intermediate. Minor amounts of 1:1:1 and 1:2:0 by-products are obtained in this case.

The synthesis of 1,3,5-hexatrienes **18** by the palladium-catalyzed 1:1:1 coupling of β-bromostyrenes with internal alkynes and monosubstituted alkenes was also achieved (Scheme 3.15) [30b].

Very recently, Jiang et al. developed a palladium-catalyzed three-component coupling reaction for the synthesis of 1,3-butadienes that involves the sequential formation of a $C(sp^3)$—$C(sp^2)$ bond and a $C(sp^2)$—$C(sp^2)$ bond from a benzyl chloride, a terminal alkyne, and a monosubstituted alkene [32].

As illustrated in Scheme 3.16, Sakai et al. have recently reported the synthesis of 1,1,2-trisubstituted enynes **19** based on the palladium-catalyzed arylalkynylation of aryl iodides, internal alkynes, and alkynylsilanes. Relatively mild conditions consisting of a methanol solution containing K_2CO_3 must be used to obtain preferentially the enyne derivative over the Sonogashira product (diarylacetylene). The formation of the latter compound is favored when using an activating reagent such as potassium fluoride, which permits facile cleavage of the C—Si bond, therefore favoring transmetalation of an acetylide ion over insertion of an internal alkyne on the Pd center [33].

The use of benzyne derivatives in intermolecular carbopalladation processes has stood out as a powerful tool for the preparation of 1,2-functionalized arenes in the last decade. Since the pioneering work by Yamamoto in 2000 [34], which introduced arynes as reactive partners in palladium-catalyzed reactions, several groups have exploited the carbopalladation of benzyne in three-component couplings. In general, 2-(trimethylsilyl)phenyl triflate derivatives, in combination with a fluoride salt such as CsF, are used as benzyne precursors.

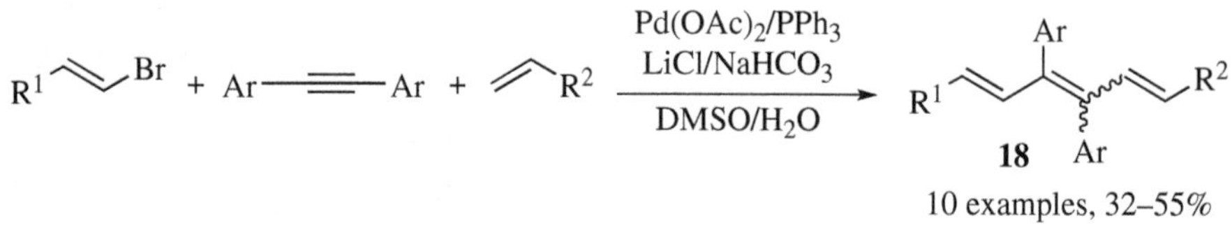

SCHEME 3.15 Synthesis of 1,3,5-hexatrienes by palladium-catalyzed cross-coupling of β-bromostyrenes, diarylacetylenes, and monosubstituted alkenes.

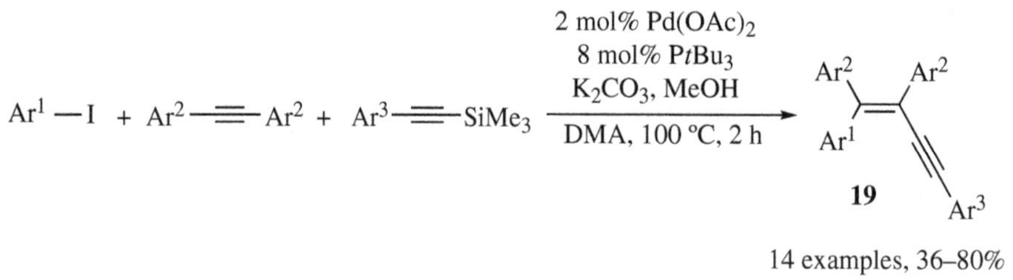

SCHEME 3.16 Palladium-catalyzed arylalkynylation of aryl iodides, internal alkynes, and alkynylsilanes. Synthesis of 1,1,2-trisubstituted enynes.

Cheng developed a three-component palladium-catalyzed assembling reaction of aryne precursors **20** with allylic chlorides **21** and alkynylstannanes **22** that affords 1-allyl-2-alkynylbenzenes **23** in excellent yields (Scheme 3.17) [35]. Based on the known palladium-catalyzed chemistry, two mechanisms are conceivable. After the initial oxidative addition of allyl chloride to Pd(0) to give a π-allyl palladium complex (route A), insertion of the benzyne to afford an aryl palladium intermediate would follow. Transmetalation with the alkynylstannane would give an alkynyl aryl palladium intermediate, which would undergo reductive elimination to afford the final product and regenerate the Pd(0) catalyst. The second possibility (route B) involves transmetalation of the alkynylstannane with the π-allyl palladium complex to give a σ-alkynyl π-allyl palladium intermediate and subsequent transfer of the alkynyl and allyl groups to the benzyne moiety. Although the authors favor the first route, the second option cannot be ruled out. A modification of this procedure that involves the use of aryl boronic acids **24** as the nucleophilic reagents instead of the alkynylstannanes allowed for the synthesis of *o*-allylbiaryls **25** in good yields (Scheme 3.17) [36].

Another interesting example reported by this group consists of the palladium-catalyzed carbocyclization of aryl iodides, bicyclic alkenes, and arynes. Annulated 9,10-dihydrophenanthrene derivatives **26** are obtained through a mechanism that involves C—H activation promoted by the fluoride anion as the key step [37]. On the basis of the regiochemistry of the products obtained and supported by the isolation of complexes similar to those involved in the process, the authors propose the catalytic cycle depicted in Scheme 3.18.

A related methodology to substituted phenanthrenes **27** was developed by Larock, which involves the palladium-catalyzed cross-coupling of aryl halides, acetylenes, and arynes (Scheme 3.19). The authors propose a mechanism that comprises carbopalladation of an internal alkyne and subsequent carbopalladation of the aryne [38].

Greaney et al. introduced the Heck reaction to the benzyne three-component coupling to prepare 1,2-functionalized arenes **28** and **29**. Alternatives to allyl chlorides as the initial carbopalladation electrophile were presented, that is, methyl bromoacetate and benzyl bromides, thus avoiding the formation of benzyne/benzyne/allyl by-products and therefore improving the efficiency of the reaction [39]. When using aryl iodides as the electrophilic partners, highly functionalized biaryl compounds **30** are obtained (Scheme 3.20) [40].

3.2.4 Amines as Building Blocks

3.2.4.1 Aminosulfonylations A number of reports on the palladium-catalyzed three-component aminosulfonylation have recently appeared as a result of the interest for developing catalytic methods to introduce sulfur dioxide into simple organic molecules. Willis et al. have shown that aryl *N*-aminosulfonamides may be prepared by coupling of organic iodides, hydrazines, and DABCO·$(SO_2)_2$ as a stable, easy-to-handle source of sulfur dioxide. Efficient aminosulfonylation reactions between a range of aryl, heteroaryl, and alkenyl iodides and *N*,*N*-dialkyl hydrazines were carried out providing aryl *N*-aminosulfonamides in good to excellent yields (Scheme 3.21) [41].

Wu adapted this reaction for the palladium-catalyzed three-component coupling of arylboronic acids, DABCO·$(SO_2)_2$, and hydrazines (Scheme3.22). The reactions proceeded smoothly in the presence of oxygen to afford the aryl *N*-aminosulfonamides in good to excellent yields. The authors proposed

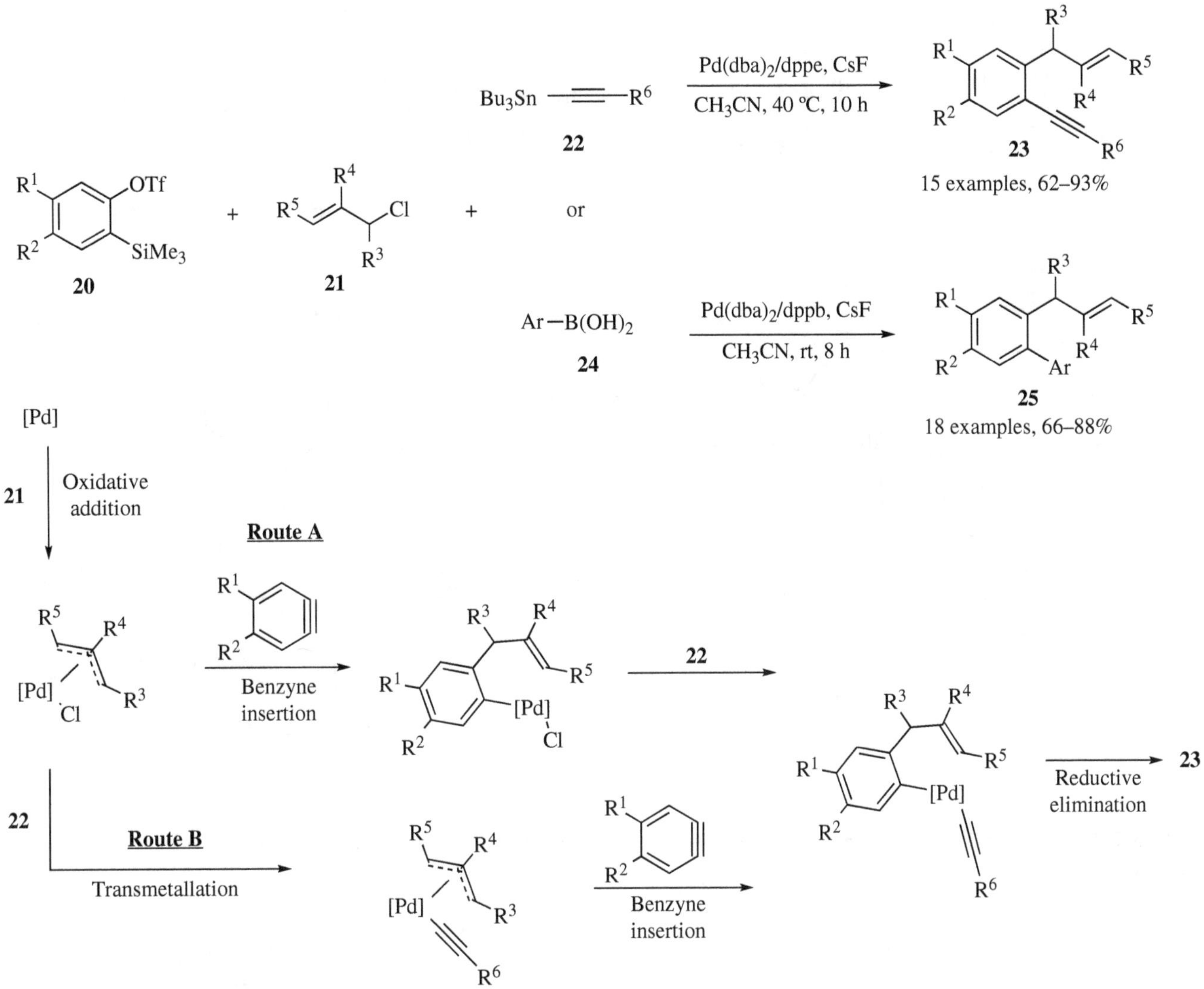

SCHEME 3.17 Palladium-catalyzed synthesis of 1-allyl-2-alkynylbenzenes and *o*-allylbiaryls.

the mechanism depicted in Scheme 3.22, which involves initial transmetalation of Pd(II) with arylboronic acid **31** to generate a Pd(II) species **X**. The subsequent coordination and insertion of sulfur dioxide would give rise to intermediate **XI**. Then a nucleophilic attack of hydrazine **32** would take place to produce the coupling product **33** and Pd(0). The latter would be oxidized in the presence of dioxygen to afford Pd(II), which would reenter the catalytic cycle [42].

The same group reported shortly after the use of potassium bisulfite as a convenient equivalent of sulfur dioxide for the coupling with aryl iodides or bromides and hydrazines. A wide range of aryl *N*-aminosulfonamides were prepared in good yields using $Pd(OAc)_2/PtBu_3$ as the catalyst [43].

3.2.4.2 Amines in Reactions Based on Buchwald–Hartwig Amination A number of strategies of indole derivatives based on palladium-catalyzed MCRs that proceed through Buchwald–Hartwig amination processes have recently been reported [44]. Barluenga et al. presented the cascade synthesis of indoles from primary amines, bromoalkenes, and dihalobenzenes [45]. Three different reactions take place by the action of the same Pd catalyst with excellent chemoselectivity, which ensures the success of the methodology: formation of the imine **XII** by alkenyl amination, σ-arylation of the imine to give **XIII**, and intramolecular N-arylation to afford **34** (Scheme 3.23).

Based on the aforementioned amination process, Xi developed a one-pot synthesis of 2-alkynylindoles **36** involving the multicomponent coupling of *o*-bromo-(2,2-dibromovinyl)benzenes **35**, arylamines, and terminal alkynes. The reaction is proposed to proceed through a Pd-catalyzed Sonogashira coupling leading to a monoalkynylated product **XIV**, followed by an intermolecular Buchwald–Hartwig amination to **XV** and a

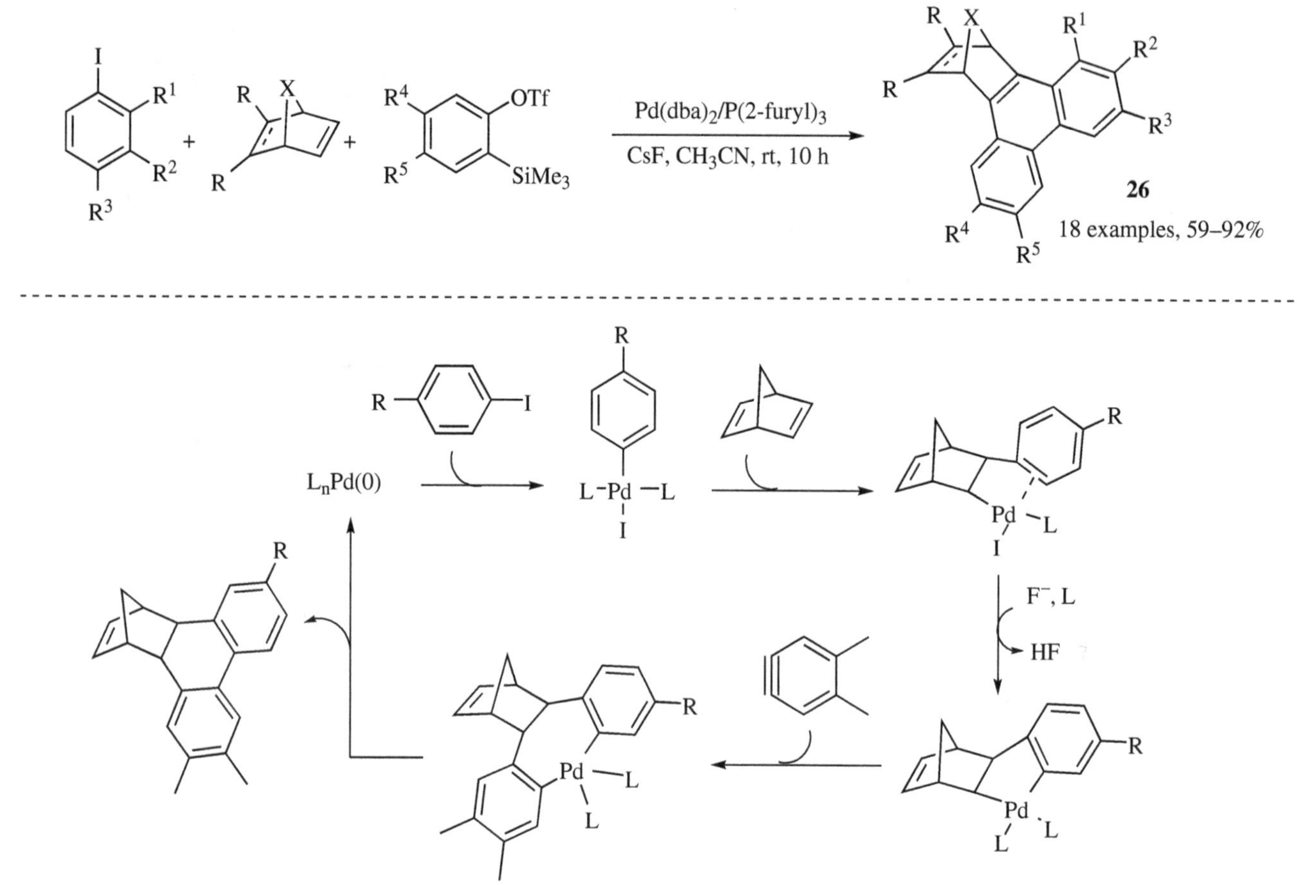

SCHEME 3.18 Palladium-catalyzed carbocyclization of aryl iodides, bicyclic alkenes, and arynes. Synthesis of annulated 9,10-dihydrophenanthrenes.

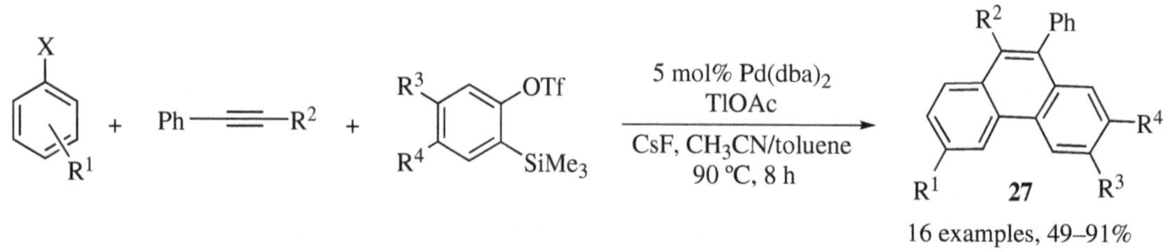

SCHEME 3.19 Synthesis of phenanthrenes *via* palladium-catalyzed cross-coupling of aryl halides, acetylenes, and arynes.

subsequent intramolecular amination to afford the α-alkynyl indoles **36** in moderate to good yields (Scheme 3.24) [46].

An efficient palladium-catalyzed multicomponent system that combines two N-arylations with an S-arylation was developed for the synthesis of phenothiazine derivatives [47]. The one-pot reaction of primary amines, 2-bromothiophenol, and substituted 1-bromo-2-iodobenzenes, catalyzed by $Pd_2(dba)_3$ and ferrocene ligands, affords the phenothiazines **37** in good yields (Scheme 3.25).

3.2.4.3 Miscellaneous Reactions Involving Amines as Coupling Partners A Pd(II)-catalyzed three-component cascade reaction starting from an alkynol, an aldehyde, and a primary amine was reported by Barluenga et al. for the synthesis of spiroacetals **38** [48]. Scheme 3.26 displays the proposed mechanism for this reaction. Coordination of the triple bond to the Pd(II) cationic complex promotes the attack of the hydroxyl group to give intermediate **XVI**. A protodemetalation follows, which affords the methylidenefuran **XVII**. This reacts with the imine activated by the Pd(II) species through a Mannich-type process. Addition of the phenol to the oxonium in **XVIII** leads to spiroacetal **38**. To avoid the formation of an equimolar amount of two diastereomers,

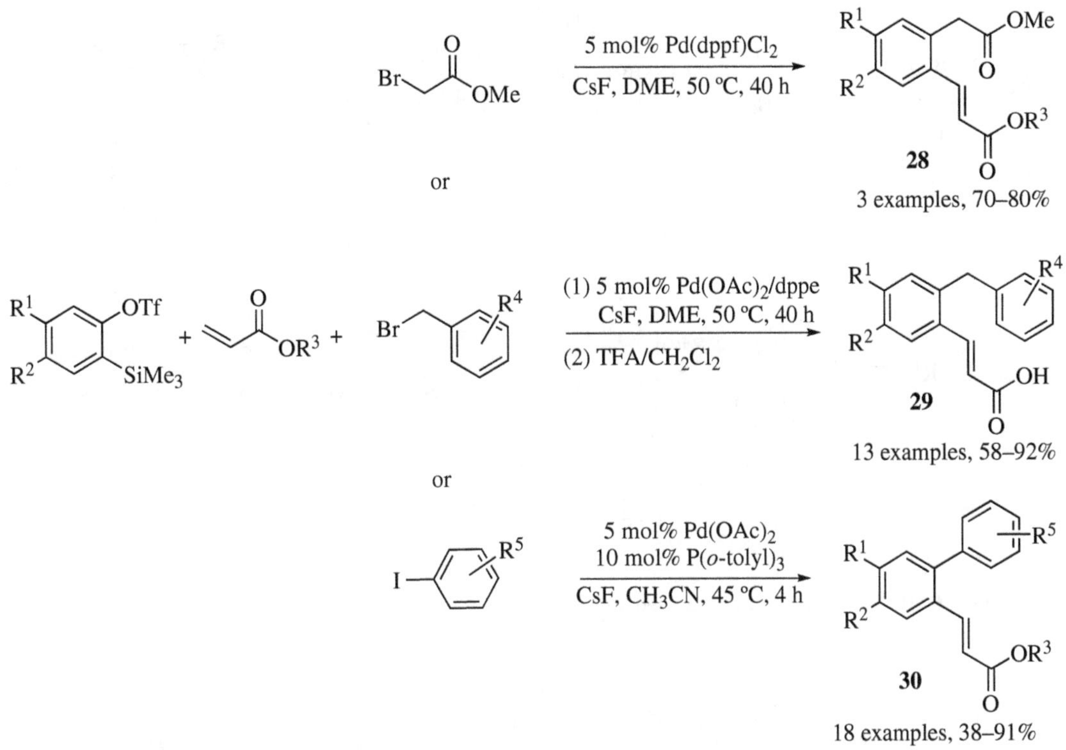

SCHEME 3.20 Palladium-catalyzed Heck-type benzyne three-component coupling.

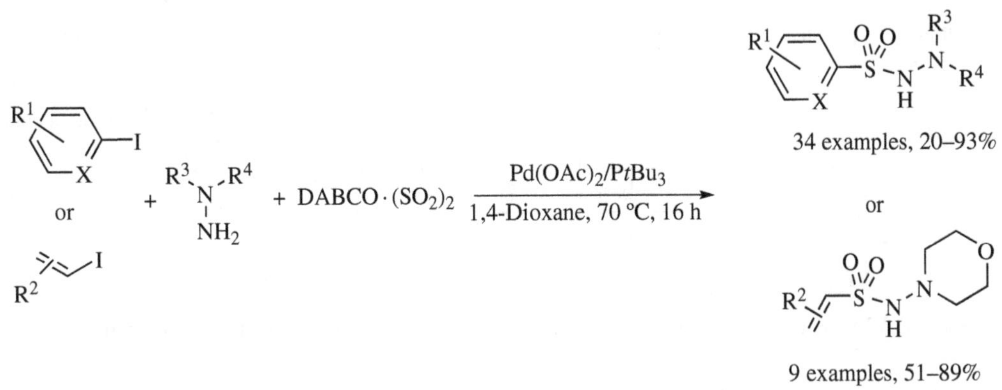

SCHEME 3.21 Palladium-catalyzed aminosulfonylation of aryl and vinyl iodides.

further treatment of the crude mixture with $Mg(ClO_4)_2$ and $HClO_4$ in CH_2Cl_2/CH_3CN at room temperature is required (Scheme 3.26).

Duchêne and Parrain developed a one-pot allylic amination/palladium-catalyzed Sonogashira cross-coupling and heterocyclization process for the preparation of 1,2,4-trisubstituted and 1,3-disubstituted pyrroles starting from diiodobutenoic acid, a primary amine, and a terminal alkyne [49]. Scheme 3.27 shows a plausible mechanism for this transformation. The initial C—N allylic amination, followed by a Sonogashira cross-coupling and an intramolecular hydroamination, affords a dihydroexoalkylidene pyrrole **XIX**, which rearranges into pyrrole **39**. The reaction is influenced by the nature of the nitrogen nucleophile, failing with tosylamine and benzylcarbamate, while being successful with aryl-, alkyl-, and benzylamines.

3.3 NICKEL-CATALYZED MCRs

3.3.1 Nickel-Catalyzed Cross-Trimerization of Alkynes

Based on previous results for the 1:2 cross-trimerization of alkynes [50], Ogata, Fukuzawa et al. [51] developed the first 1:1:1 cross-trimerization of alkynes employing $Ni(cod)_2/2PPh_3$ (10 mol%) as the catalyst. A high

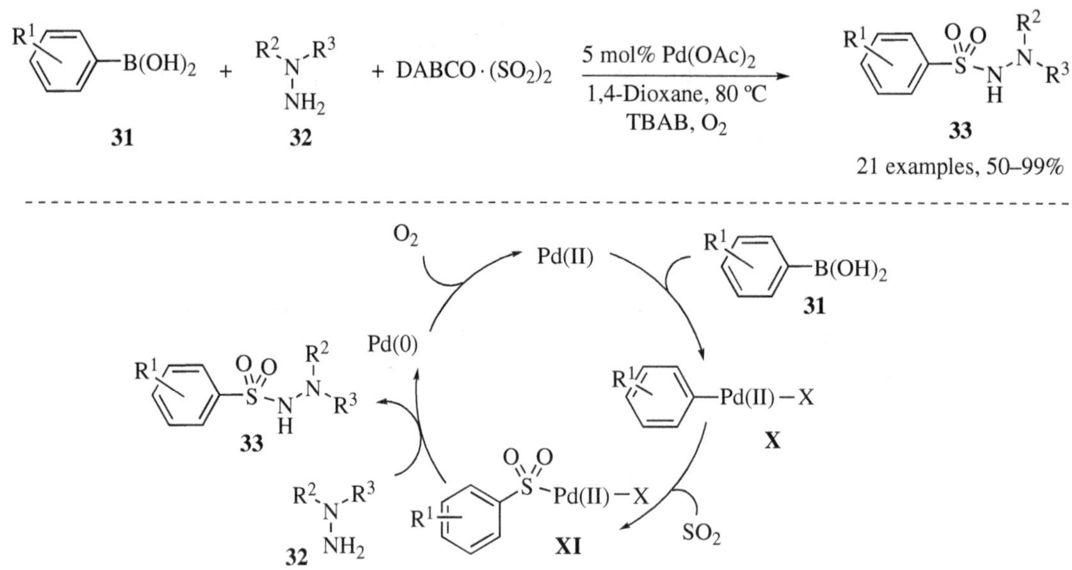

SCHEME 3.22 Palladium-catalyzed aminosulfonylation of boronic acids. Proposed mechanism.

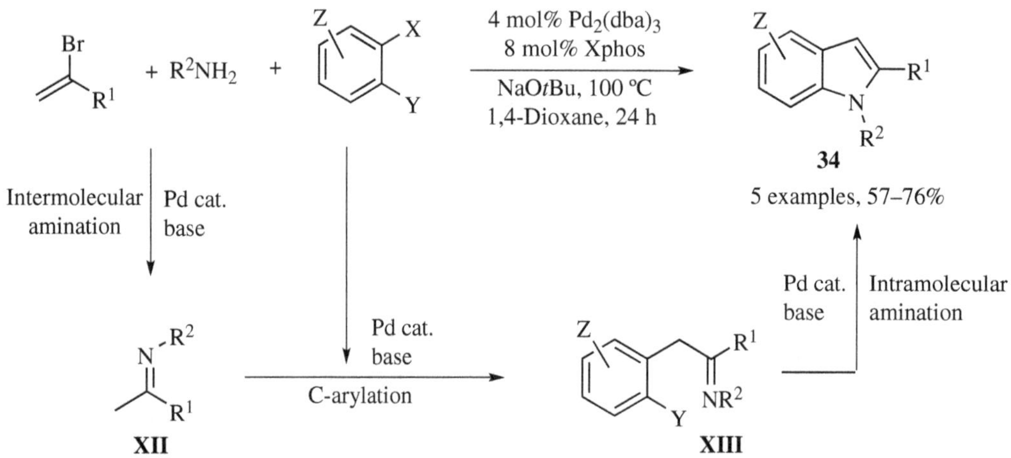

SCHEME 3.23 Palladium-catalyzed multicomponent reaction *via* Buchwald–Hartwig amination process.

chemo-, regio-, and stereoselective control was only observed when triisopropylsilylacetylene **40**, an ether-functionalized unsymmetrical internal alkyne **41**, and a symmetrical internal alkyl alkyne **42** were used (Scheme 3.28). The reaction was successful with different alkynes **41** presenting electron-donating groups in the aryl moiety, with good to high yields and regioselectivities. However, a drop in the regioselectivity was observed with substrates bearing electron-withdrawing groups. The methodology was also applicable to other alkynes **42** to afford the desired 1,3-dien-5-ynes **43** in good to excellent yields and high regioselectivities.

To explain the high chemo- and regioselectivities observed and the need of an oxygenated group in **41**, the authors proposed the mechanism depicted in Scheme 3.29. The catalytic cycle starts with the oxidative addition of **40** to the Ni(0) center to form the Ni-hydride **XX**, followed by the chemoselective insertion of alkyne **41** into the Ni—H bond to generate intermediate **XXI**, which is most likely driven by the coordination of the oxygenated group to the nickel center. The regioselectivity in the insertion step can be explained in terms of steric effects, with alkyne **41** avoiding the steric hindrance between the aryl fragment and the metal center. Next, insertion of symmetric alkyne **42** into the Ni—C bond of the sterically hindered intermediate **XXI** and subsequent reductive elimination afford product **43**.

This methodology was also applied to the 1:1:1 cross-trimerization between an internal alkyne **44** and two distinct terminal alkynes **40** and **45** [52], which is normally hampered by competition of the terminal alkynes for the oxidative addition to the metal center and for insertion into the

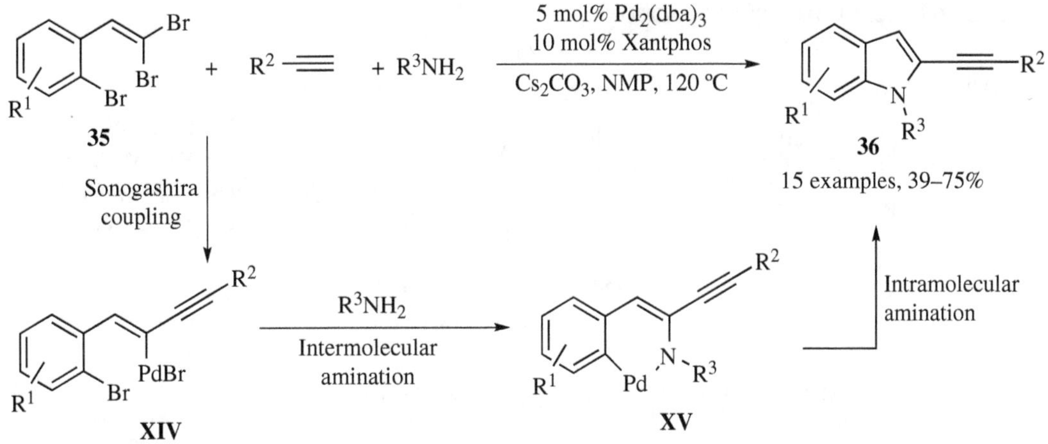

SCHEME 3.24 Sonogashira/Buchwald–Hartwig amination processes in Pd-catalyzed MCR.

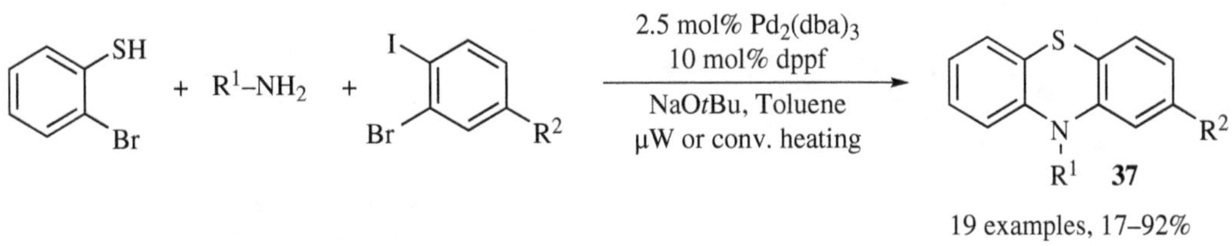

SCHEME 3.25 Synthesis of phenothiazines *via* Pd-catalyzed Buchwald–Hartwig amination and thiolation processes.

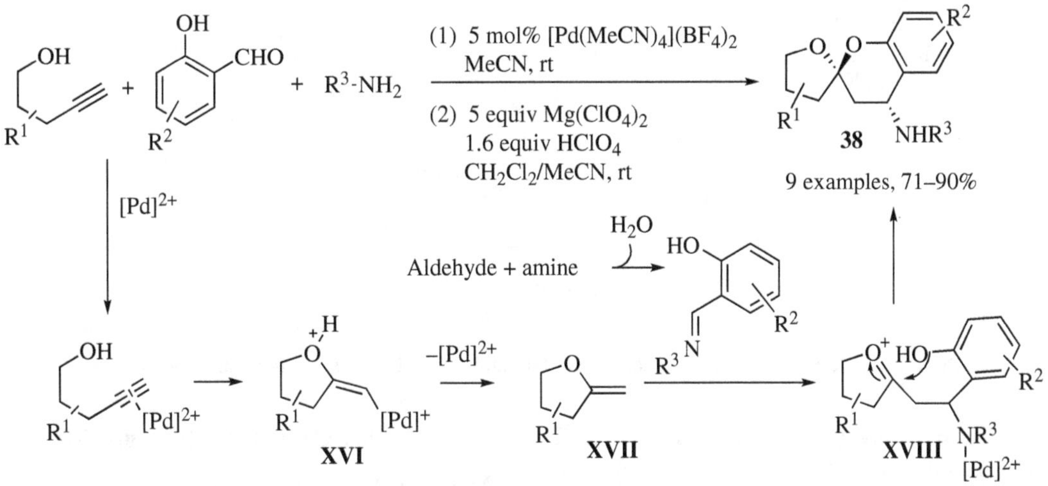

SCHEME 3.26 Synthesis of spiroacetals *via* palladium-catalyzed alkynol cyclization/Mannich-type processes.

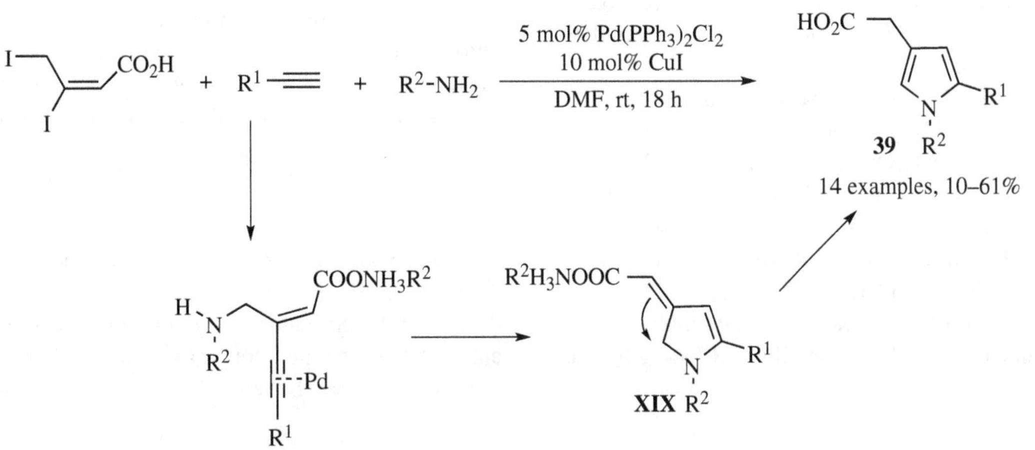

SCHEME 3.27 Palladium-catalyzed synthesis of 1,2,4-trisubstituted and 1,3-disubstituted pyrroles.

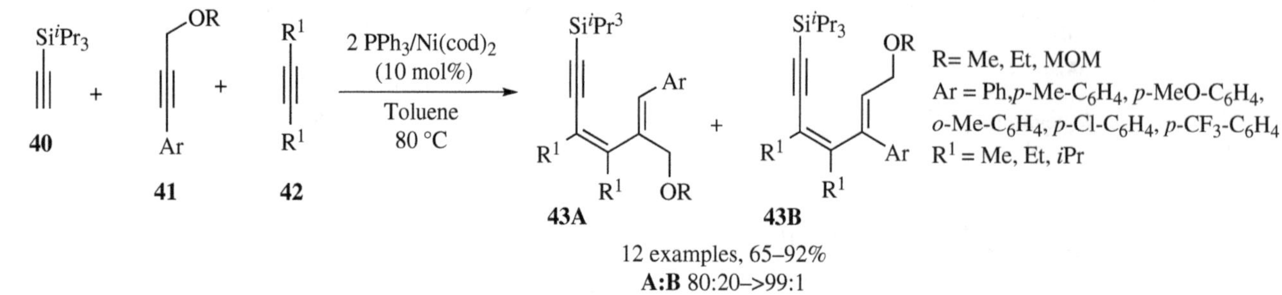

SCHEME 3.28 Ni-catalyzed 1:1:1 cross-trimerization of alkynes.

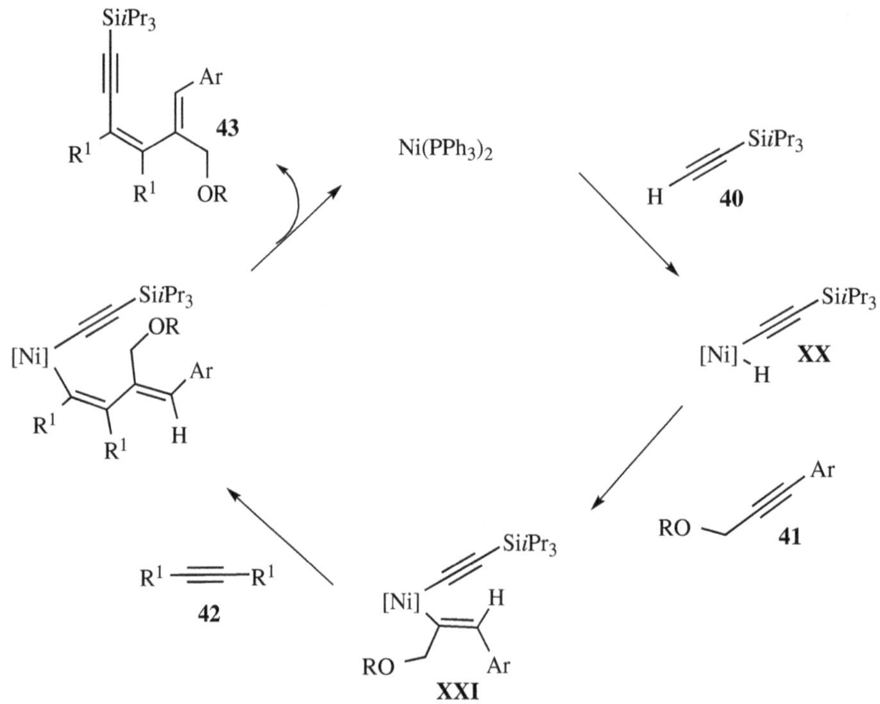

SCHEME 3.29 Proposed mechanism for Ni-catalyzed 1:1:1 cross-trimerization of alkynes.

Ni—C bond of the sterically hindered intermediate **XXI** (Scheme 3.29). Using the catalyst 2P(p-CF$_3$-C$_6$H$_4$)$_3$/Ni(cod)$_2$ (10 mol%), the reactions were carried out at room temperature with a broad variety of terminal alkynes **45** and internal alkynes **44** to give the desired 1,3-dien-5-ynes **46** in good to excellent yields and complete chemo- and regioselectivities (Scheme 3.30, Eq. 1).

The synthesis of 1,5-enynes **48** by replacing the symmetrical internal alkynes **42** by norbornenes **47** has also been described [53]. The catalytic system 2PMePh$_2$/Ni(cod)$_2$ (10 mol%) afforded the desired 1,5-enynes **48** in good to excellent yields, with high control of the chemoselectivity between alkene **47** and two alkynes **40–41** and high levels of regioselectivity (Scheme 3.30, Eq. 2). Another family of 1,5-enynes **48** can be prepared by reaction of two molecules of **40** and one molecule of **47** using 2PEt$_3$/Ni(cod)$_2$ as the catalyst.

3.3.2 Nickel-Catalyzed π-Systems Couplings

Metal-catalyzed sequential coupling of an organometallic reagent or metal hydride with two different π-components is a practical method for the synthesis of complex molecules by one-pot assembly of three different stable components. During the last years, several groups have been focused in the development of new three-component coupling methodologies employing different π-components such as benzynes, alkynes, electron-poor alkenes or aldehydes and boronic acids, boranes, or silanes as terminating agents.

3.3.2.1 Nickel-Catalyzed Couplings between Alkynes, Alkenes, and Boron Derivatives Cheng et al. [54a] described the Ni-catalyzed coupling between *in situ* generated arynes, enones, or α,β-unsaturated esters **49** and boronic acids **50** (Scheme 3.31, Eq. 1). Three-component

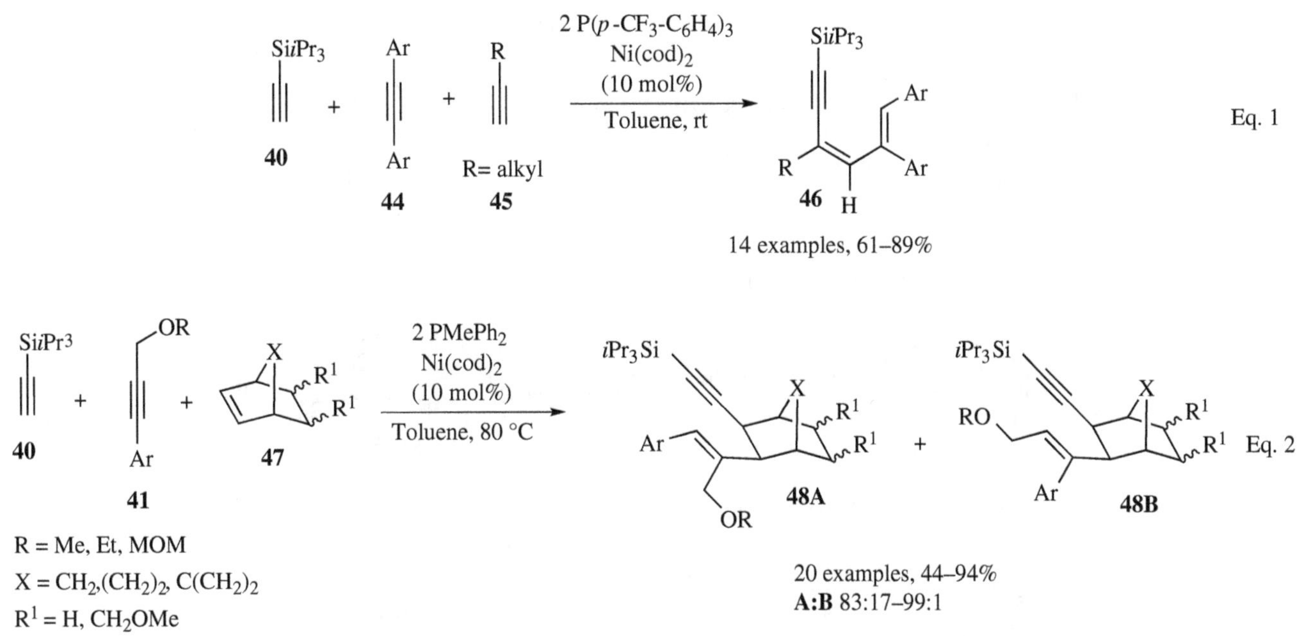

SCHEME 3.30 Ni-catalyzed synthesis of 1,3-dien-5-ynes and 1,5-enynes.

coupling products **51** were obtained in 60–95% yields employing $2PPh_3$/$Ni(cod)_2$ as the catalyst and CsF as the base under mild reaction conditions (40 °C). This methodology was also applied replacing arynes for alkynes [54b] and enones for oxabicyclic alkenes [54c]. In 2009 [54d], this group also reported the preparation of synthetically useful alkenyl boronic esters **52** by a three-component coupling of alkynes, alkenes, and bis(pinacolato)diboron (Scheme 3.31, Eq. 2).

3.3.2.2 Nickel-Catalyzed Reductive Couplings between Two π-Components Jamison et al. [55a] reported the synthesis of O-silylated allylic alcohols by a Ni-catalyzed reductive coupling between two π-components such as alkenes and aldehydes. They found that the desired O-silylated allylic alcohols **55** and **57** were obtained in 44–95% yield when the catalytic system $2P(o\text{-anisyl})_3$/$Ni(cod)_2$ (20 mol%), silyl triflate **54**, and an amine as the base were used. The reaction is successful with a range of alkenes **56** and aromatic aldehydes **53** under mild conditions of temperature and pressure (Scheme 3.32, Eq. 1). When using alpha olefins, the formation of homoallylic byproducts **58** was observed. The synthetic value of homoallylic products prompted the authors to use PPh_3 or $EtOPPh_2$ as the ligands in order to increase the carbonyl-ene-type reaction [55b]. These reactions take place under mild conditions with a broad variety of alkenes **56** and aromatic aldehydes **53**, as well as *t*BuCHO, to give the O-silylated homoallylic alcohols **58** in good to excellent yields and 70:30 to >95:5 *E:Z* diastereoselectivities (Scheme 3.32, Eq. 2). The synthesis of the homoallylic alcohol derivatives **58** have been solved with homoallylic/allylic ratios of >95:5 in most cases, although yields lower than 50% for the allylic derivatives **57** are obtained [55a]. To overcome this, Jamison et al. [55c] decided to modify the electronic properties of the catalyst combining a strong σ-donor ligand (N-heterocyclic carbene (NHC)) and a strong π-acceptor ($P(OPh)_3$), with the idea of accelerating the reductive elimination step. In these new conditions, allylic alcohol derivatives **57** were obtained in excellent yields with a broad substrate scope, and only traces of **58** were observed in some cases (Scheme 3.32, Eq. 3). The proposed mechanism explaining the role of the $P(OPh)_3$ ligand is detailed in Scheme 3.32.

Later, other strategies for the synthesis of allylic alcohol derivatives **61** using silane reducing agents have been described. Based on the previous work for the synthesis of *Z*-enol silanes by reductive coupling between alkynes and enals [56a], Montgomery et al. described a nickel-catalyzed reductive coupling of alkynes **60** or aldehydes **59** employing NHC as the ligands and trialkylsilanes as reducing agents [56b]. They found that the regioselectivity in the alkyne insertion step is controlled by steric hindrance of the carbene ligand. The formation of isomers **61B** as major products is observed when sterically hindered carbenes are used, while less hindered carbenes promote the formation of isomers **61A** (Scheme 3.33). To demonstrate the synthetic utility of this methodology, the authors developed the stereoselective synthesis of glycosylated allylic alcohols, which can undergo an intramolecular stereoselective glycosylation process [56c], as well as the total synthesis of 12-membered macrocycles [56d].

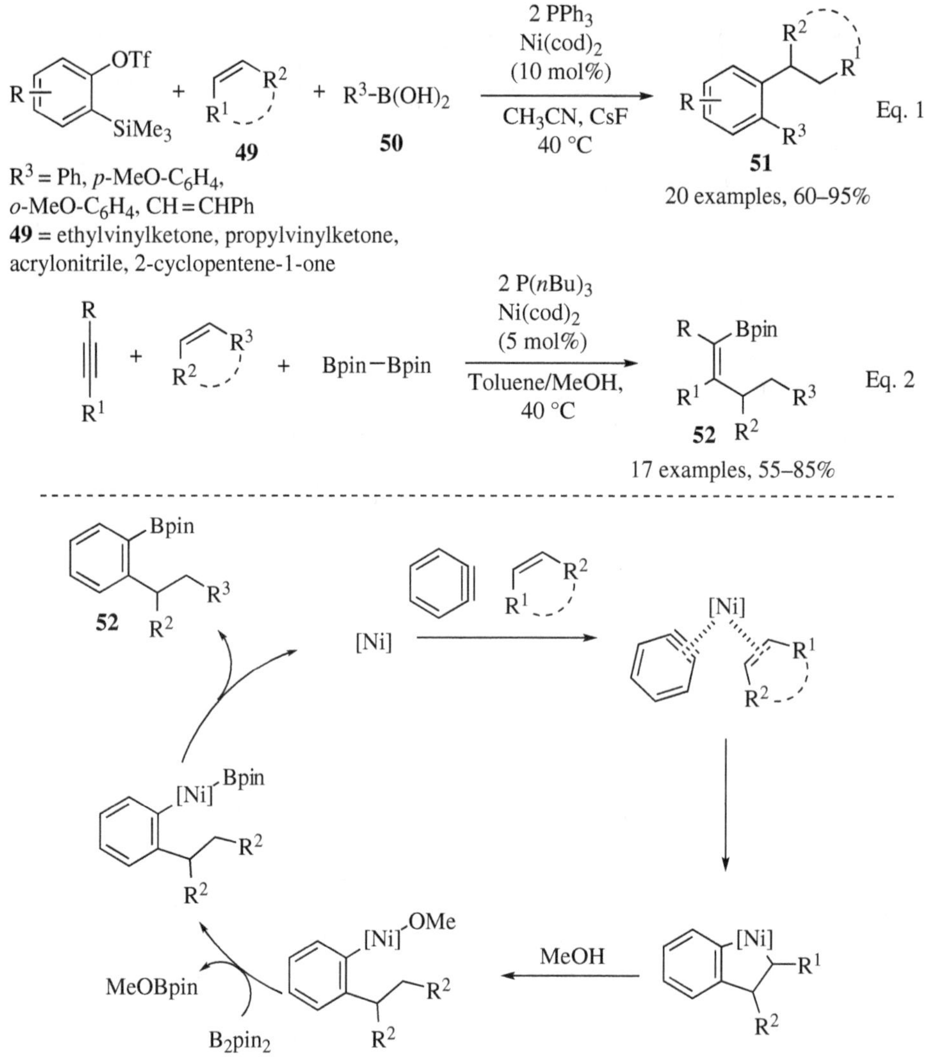

SCHEME 3.31 Ni-catalyzed coupling between arynes, enones, or **α,β**-unsaturated esters and boronic acids.

Other NHC/Ni-catalyzed methodologies for the synthesis of allylic or homoallylic alcohol derivatives are the reductive coupling between ynamides, aldehydes, and silanes to afford γ-silyloxyenamide derivatives in a highly regio- and stereoselective manner [57]; the ring-opening three-component coupling of methylenecyclopropanes, aldehydes, and silanes [58]; and the asymmetric synthesis of homoallylic alcohol derivatives by reductive coupling of 1,3-dienes, aldehydes, and silanes using chiral NHC [59].

3.3.2.3 Nickel-Catalyzed Couplings via Internal Redox Reactions In contrast with the reductive couplings discussed earlier, where two π-systems are joined *via* C–C bond formation and a net two-electron reduction takes place by the action of a reducing agent, Montgomery et al. observed that an alkyne and an enal can be coupled *via* an internal redox process where the aldehyde is oxidized and the alkyne is reduced [60]. A three-component reaction involving enone **62**, alkyne **63**, and aldehyde **64**, employing bulky NHC/Ni(cod)$_2$ or PCy$_3$/Ni(cod)$_2$ as the catalyst, was carried out to give 1,3-diketones **65** in good yields with a high degree of chemoselectivity. Regioselectivity for the insertion of alkyne **63** can be controlled choosing the appropriate ligand (Scheme 3.34). Thus, bulky carbene ligands promote the insertion of the alkyne through the carbon adjacent to the large group (R^L), whereas the opposite regioselectivity is observed when using PCy$_3$.

3.3.3 Ni-Catalyzed Reductive Conjugate Addition

Silyl enol ethers are very useful synthetic intermediates, but unfortunately, their synthesis, normally by conjugate addition–silyl enol ether formation, requires the use of air and moisture unstable organometallic reagents with the consequent problems of functional group incompatibilities, and

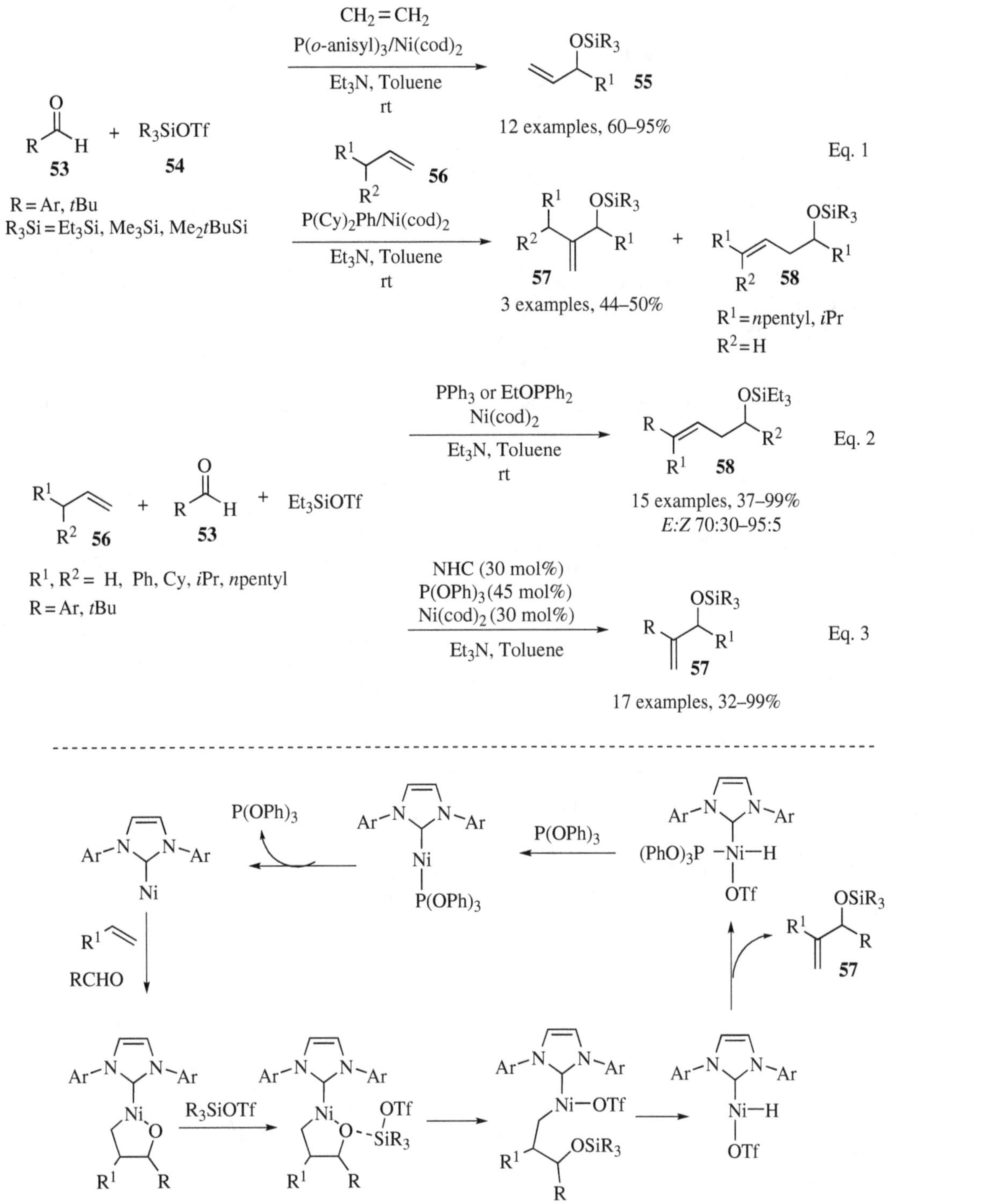

SCHEME 3.32 Ni-catalyzed synthesis of allylic and homoallylic *O*-silyl ethers by reductive coupling between alkenes and aldehydes.

even cryogenic temperatures are often required. To avoid the problems associated with the use of organometallic reagents, Weix et al. decided to develop a reductive approach [61]. The combination of nitrogenated ligands, such as 2,2′:6,2′tri-*tert*-butyl-terpyridine or 2,9-dimethyl-1,10-phenanthroline, and Ni(acac)$_2$ proved to be a very efficient catalyst for the reductive coupling of aryl/alkyl halides, enones, and trialkylchlorosilane reagents (Scheme 3.35). The presence of a stoichiometric amount of Mn powder is required. The reaction is successful with a broad family of unactivated hindered primary, secondary, and tertiary alkyl bromides **66** [61a], as well as vinyl and aryl bromide/iodides **67** [61b], to

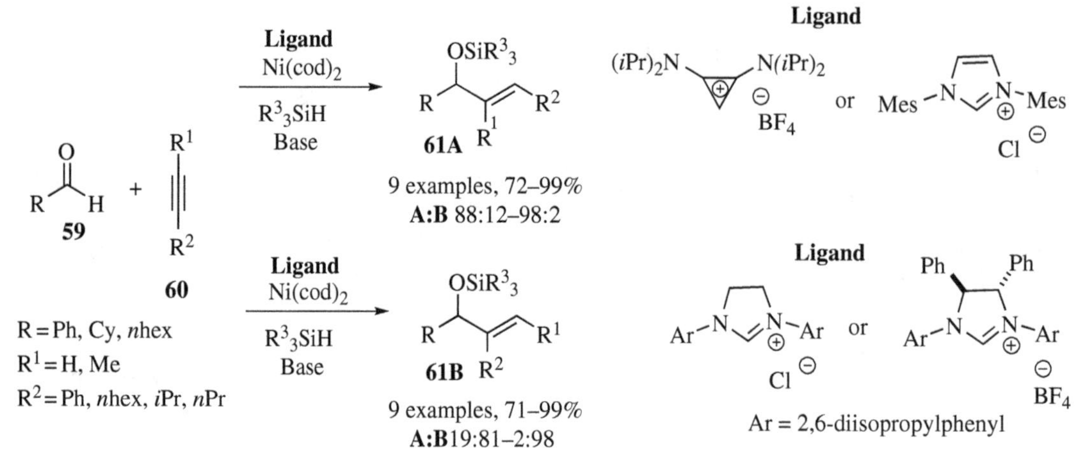

SCHEME 3.33 Ni-catalyzed synthesis of allylic *O*-silyl ethers by reductive coupling between alkynes and aldehydes.

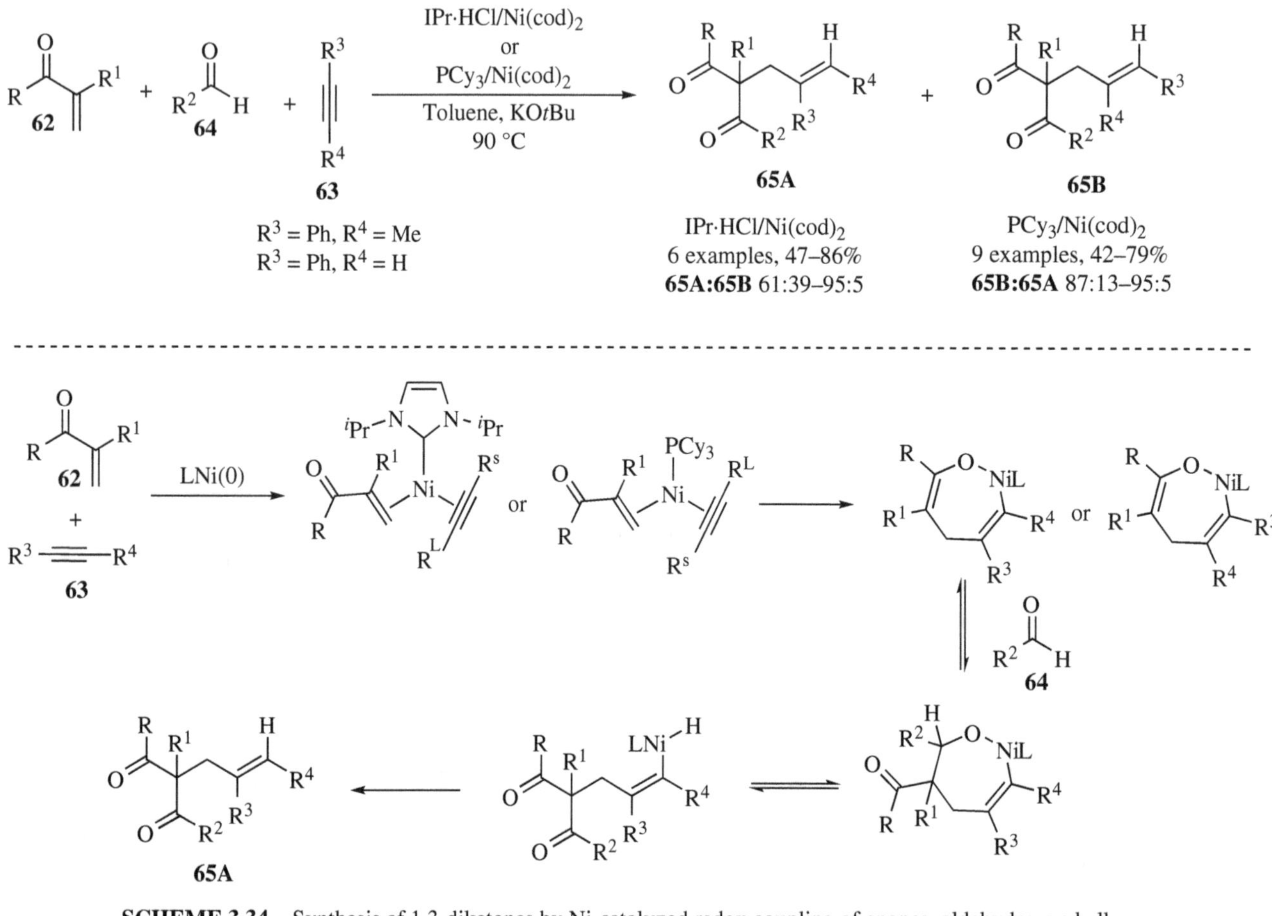

SCHEME 3.34 Synthesis of 1,3-diketones by Ni-catalyzed redox coupling of enones, aldehydes, and alkynes.

afford the enol silyl ethers **68–69** in moderate to good yields. The authors suggested that the rapid formation of an allylnickel intermediate **XXII** compared to that of arylnickel or alkylnickel species, and its subsequent reaction with aryl/alkyl halides **66–67**, may account for the observed selectivity.

The reductive aldol reaction employing trialkylboranes as reducing agents has been also developed by Montgomery et al. [62], who observed that the addition of aryl iodides is required to initiate the reaction. The corresponding aldol products were obtained in 60–91% yields and dr 86:14–96:4.

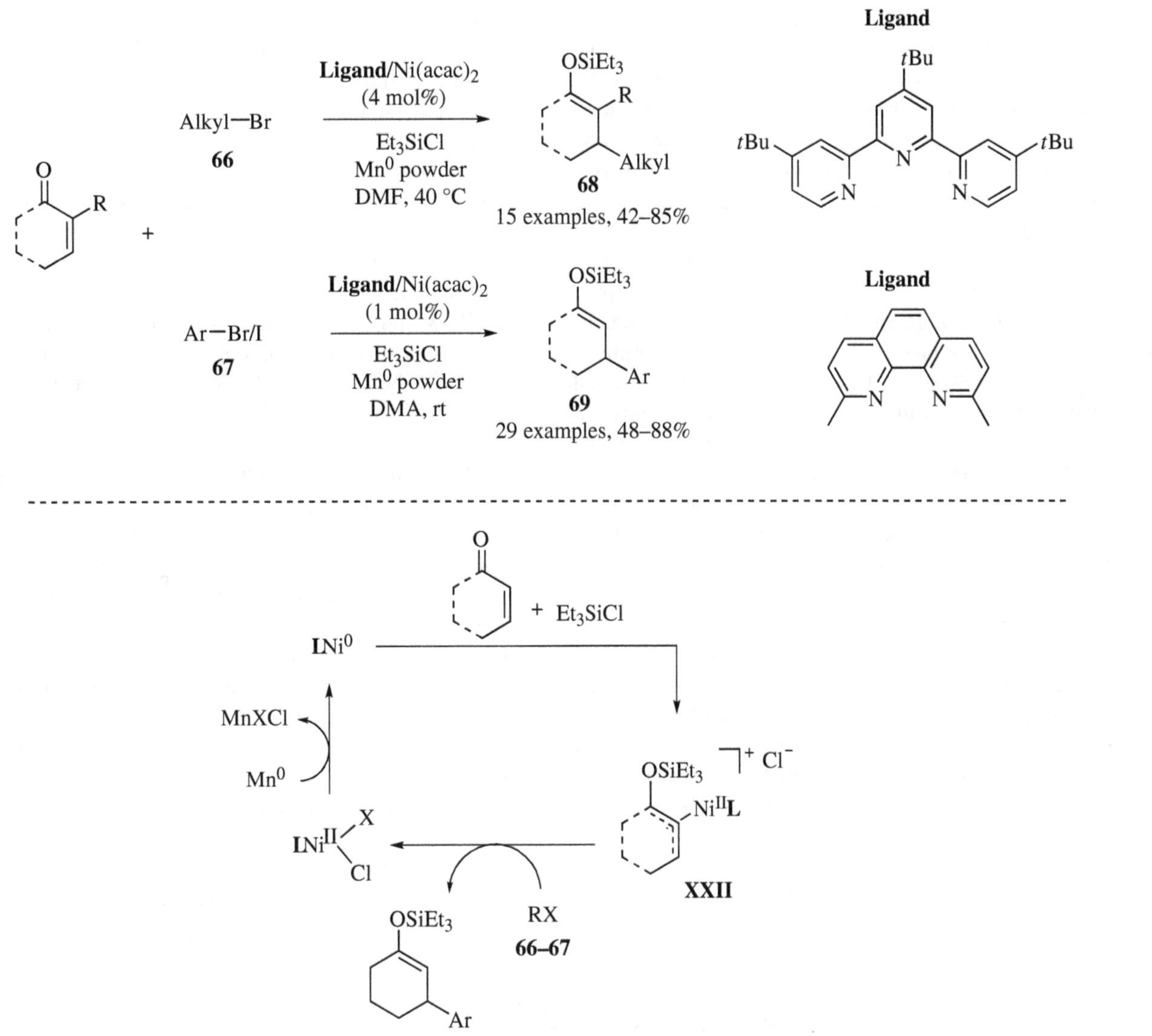

SCHEME 3.35 Synthesis of enol silyl ethers by Ni-catalyzed reductive coupling of aryl/alkyl halides, enones, and trialkylchlorosilane.

3.4 GROUP 11 METAL-CATALYZED MCRs

3.4.1 Copper-Catalyzed Azide–Alkyne Cycloaddition

Since the pioneering work by Sharpless et al. in 2001 [63], the copper-catalyzed azide–alkyne cycloaddition (CuAAC) reaction has been widely recognized as a highly reliable and straightforward method for the construction of new covalent bonds between building blocks containing different functional groups [64]. Different copper sources are commonly employed for these reactions, mainly copper(I) salts (halide and acetate) and coordination complexes such as $[Cu(CH_3CN)_4]PF_6$ and $[Cu(CH_3CN)_4]OTf$ or copper(II) salts (sulfate or acetate) in conjunction with ascorbate as a reductant, the latter system being widely accepted as the method of choice for the synthesis of 1,2,3-triazoles. More recently, heterogeneous systems based on Cu nanoparticles or copper species immobilized onto diverse supports have been used to this aim. The versatility of this reaction has given rise to a variety of synthetic methodologies, being of special interest those involving multicomponent processes, which are reviewed in the following text.

3.4.1.1 Synthesis of Triazoles A number of MCRs that use an epoxide–azide–alkyne Cu(I)-catalyzed cycloaddition sequence to provide substituted β-hydroxy-1,2,3-triazoles **70** (Scheme 3.36) have been recently developed. The azidation of an epoxide with sodium azide to give 2-azidoalcohols and subsequent reaction with a terminal alkyne have been carried out under mild conditions in environmentally friendly solvents, such as water, using the catalytic system $CuSO_4$ sodium ascorbate [65] (see the example depicted in Scheme 3.36) [65a] or a porphyrinatocopper(II) complex

[66], or PEG 400 as a reaction medium with CuI as the catalyst [67]. In the last case, *N*-tosylaziridines are also used instead of epoxides, providing *N*-tosylamino 1,2,3-triazoles. In addition, the latter contribution describes the application of this methodology to enantiomerically pure chiral epoxides and aziridines, yielding the corresponding β-hydroxy- and *N*-tosylamino 1,2,3-triazoles with high yields and excellent enantiomeric excess.

Moreover, heterogeneous versions of this reaction have been reported, as a result of the interest for developing recoverable and reusable catalysts. Sharghi et al. demonstrated that their [*meso*-tetrakis(*o*-chlorophenyl)porphyrinato]copper(II) catalyst can be impregnated onto activated multiwalled carbon nanotubes with an increase in yields and a decrease in reaction times while maintaining the same degree of regioselectivity in comparison with the homogeneous system [66]. Recently, copper nanoparticles have been supported on activated carbon [68] or zeolites [69] to catalyze the epoxide–azide cycloaddition in water, providing a green route to the multicomponent synthesis of β-hydroxy-1,2,3-triazoles.

Other copper-based heterogeneous methodologies recently reported afford 1,2,3-triazoles from organic halides, aromatic terminal alkynes, and sodium azide [70]. As illustrated in Scheme 3.37, Wan and Cai described the efficient and green synthesis of 1,4-substituted 1,2,3-triazoles **71** catalyzed by silica-immobilized *N*-heterocyclic carbene-Cu(I) **72** (Scheme 3.37).

These reactions have also been performed in ionic liquids, as reported by Wang and Yan [71]. The system CuI/amino acid ionic liquid (AAIL) was proved to catalyze the synthesis of 1,4-substituted 1,2,3-triazoles in 1-butyl-3-methylimidazolium tetrafluoroborate ([BMIM][BF_4]). The authors proposed the mechanism depicted in Scheme 3.38, with AAIL acting as a ligand. The reaction would start with the coordination of proline to Cu(I), and the subsequent generation of copper(I) acetylide **XXIII** by reaction with the alkyne. The concerted cycloaddition would be then performed to generate intermediate **XXIV**, in which the 1,4-isomer is favored due to steric hindrance. Protonation of **XXIV** would give the final 1,4-disubstituted 1,2,3-triazoles **73**. Noteworthy, the components CuI, AAIL, and [BMIM][BF_4] can be recovered for six consecutive trials without significant loss of activity.

3.4.1.2 CuAAC Reactions Involving N-Sulfonyl Azides In 2005, Chang et al. demonstrated that replacing the sodium azide by *N*-sulfonyl azide entails a significant change in reactivity, affording amidines instead of triazoles when *p*-toluenesulfonyl azide and phenylacetylene were treated with diisopropylamine [72]. The authors attribute this striking difference to the dissimilar reactivity between sulfonyl and alkyl(aryl) azides toward Cu acetylides. This unexpected finding led to the development of a highly efficient catalytic method for the synthesis of *N*-sulfonylamidines **74**, which consists in the copper-catalyzed three-component coupling of alkyne, sulfonyl azide, and amine (Scheme 3.39). The reaction showed a very broad scope with respect to the three components, being successful with tosyl, pyridine, and methanesulfonyl azide, a series of

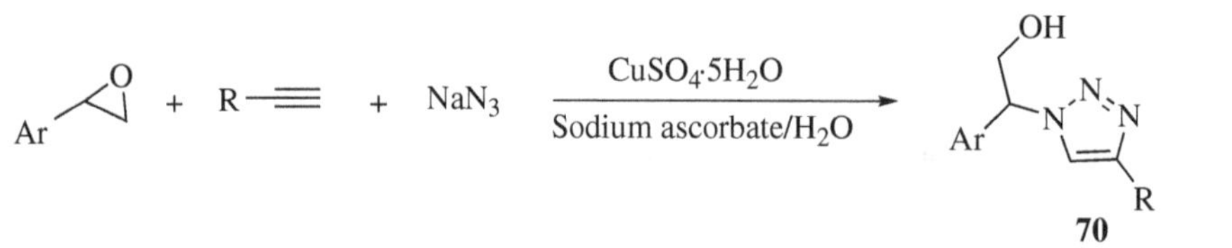

70

14 examples, 75–93%

SCHEME 3.36 β-Hydroxi-1,2,3-triazoles synthesis *via* CuAAC.

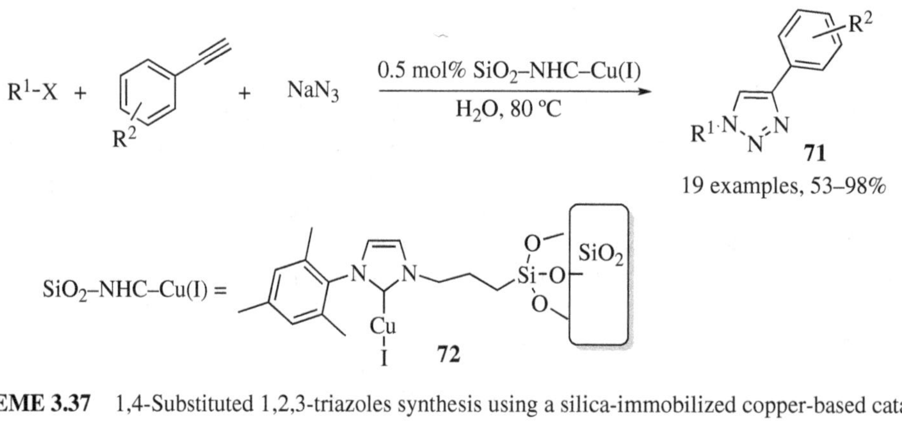

71

19 examples, 53–98%

72

SCHEME 3.37 1,4-Substituted 1,2,3-triazoles synthesis using a silica-immobilized copper-based catalyst.

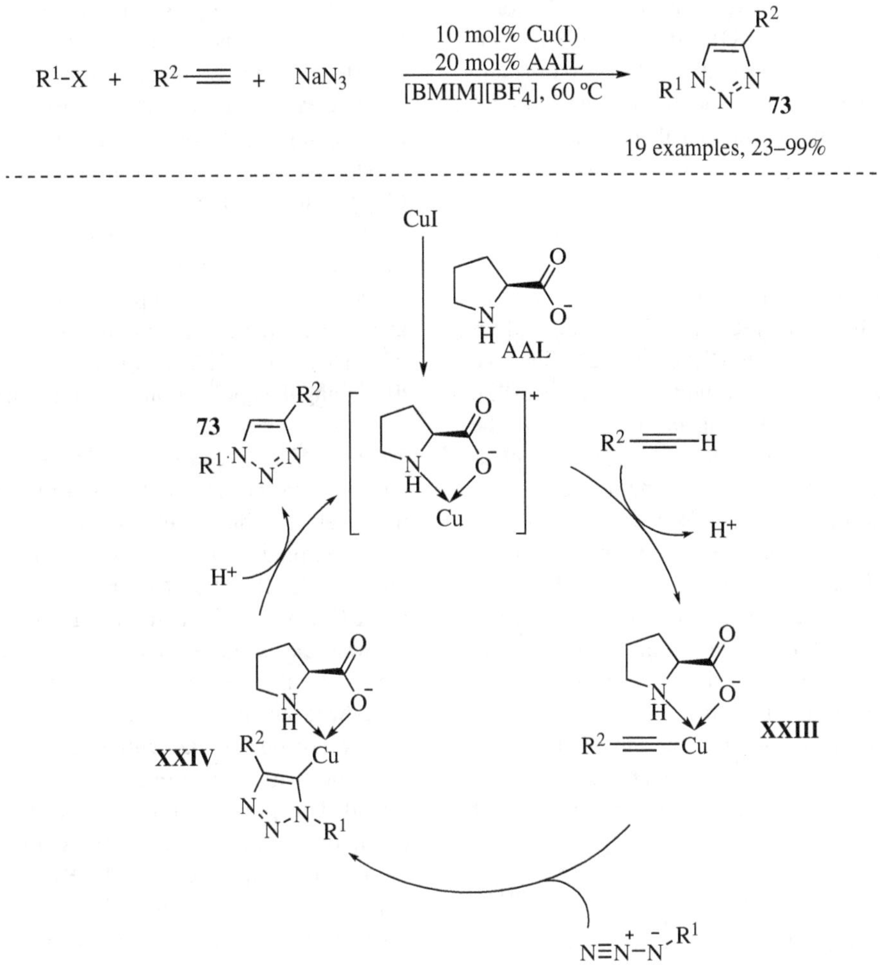

SCHEME 3.38 Synthesis of 1,4-substituted 1,2,3-triazoles by CuAAC in ionic liquid media.

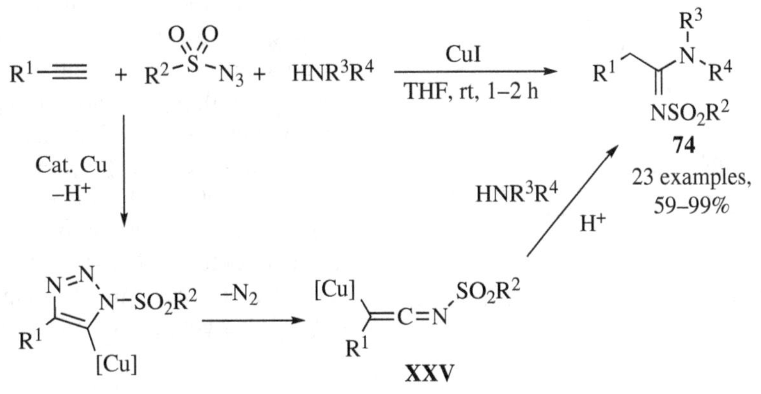

SCHEME 3.39 Synthesis of *N*-sulfonylamidines by CuAAC reactions.

alkynes bearing different functional groups and an array of primary, secondary, aliphatic, aryl, acyclic, or cyclic amines. The following studies demonstrated that this reaction proceeds *via* a ketenimine intermediate **XXV**, generated *in situ* by the Cu-catalyzed coupling of 1-alkynes and sulfonyl azides upon release of N_2 (Scheme 3.39) [73]. This methodology was subsequently applied to the synthesis of related amidine compounds [74], including phosphorus amidines [75], benzoxazoline-amidines [76], and benzimidazoles [77].

These Cu-catalyzed MCRs were further extended with the incorporation of different nucleophiles such as water [78], alcohol [79], imines [80], salicylaldehyde and acetophenones [81], aziridines [82], β-ketoesters [83], indoles [84], hydrazide [85], or pyrrole [86], which provided efficient routes to amides, amidates, azetidinimines and pyrimidines, iminocoumarins, pyrrolines, furans, functionalized indoles, isoquinolines, or functionalized pyrroles, respectively.

3.4.2 A³-Coupling

The reaction of an aldehyde, an alkyne, and an amine (A³-coupling) represents the most direct and efficient method for the preparation of propargylamines (Scheme 3.40). Since the first copper-catalyzed example reported in 1998 [87], significant progress has been made leading to the development of multicomponent processes, mainly catalyzed by copper, gold, and silver. The generalized mechanism for these reactions is depicted in Scheme 3.40 [88].

In general, the A³-coupling is relatively easier and can be performed under milder conditions with secondary than with primary amines, mainly due to the fact that iminium ions, derived from secondary amines, are much more electrophilic than their imine counterparts derived from primary amines. Therefore, it is not surprising that this reaction has mainly been optimized for secondary amines and anilines to afford tertiary alkylpropargylamines and *N*-aryl propargylamines, respectively. There are few examples of the synthesis of

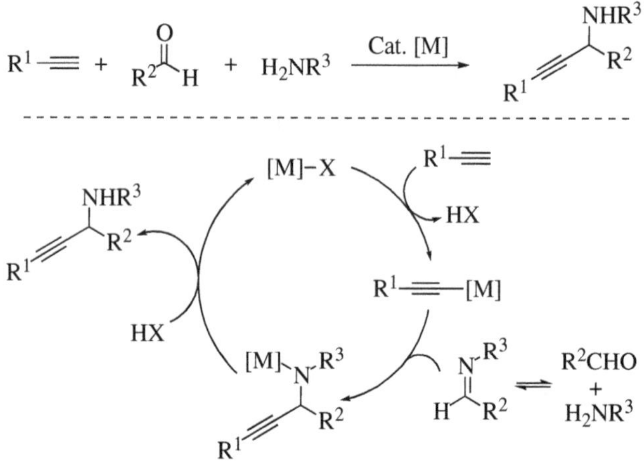

SCHEME 3.40 Propargylamine synthesis by metal-catalyzed A³-coupling reactions.

secondary alkylpropargylamines by A³-coupling processes [89], and it was not until 2010 when Van der Eycken et al. reported the first systematic study of such reaction [90]. The authors presented a fast microwave-assisted CuBr-catalyzed A³-coupling protocol for the preparation of secondary alkylpropargylamines starting from primary aliphatic amines (26 examples, 41–89% yield). These reactions were also successful with sterically hindered amines and aldehydes. Shortly after, the same group extended the scope of this A³-coupling with the copper-catalyzed coupling of a ketone, an alkyne, and a primary amine (Scheme 3.41). This reaction, which they called KA²-coupling, constituted the first example describing the application of ketones in the A³-coupling [91].

3.4.2.1 Heterogeneous A³-Coupling Another important challenge in the A³-coupling field involves the recycling of the catalyst, which seeks to avoid the loss of the catalyst after completion of the reaction. In 2005, Park and Alper reported the first study on the recyclability of the catalyst using CuCN in an ionic liquid, [BMIM][PF_6] [92]. The product was extracted from the reaction mixture with diethyl ether, and the recovered catalytic system was reused five times without any significant loss of activity.

The last decade has witnessed an enormous progress in the field with the application of heterogeneous systems intended to preserve the catalyst [93]. In 2006, Lingaiah et al. described the A³-coupling using the silver salt of 12-tungstophosphoric acid ($Ag_3PW_{12}O_{40}$) as a solid catalyst, which can be easily recovered by filtration and reused several times [94]. On the other hand, the first A³-coupling catalyzed by recyclable gold nanoparticles was reported by Kidwai et al. in 2007 [95]. Since then, a number of papers have appeared describing the use of Cu-, Ag-, or Au-supported catalysts in the multicomponent synthesis of propargylamines. These heterogeneous catalysts have been prepared by immobilization of metallic particles in materials such as silica gel [96], polymers such as polystyrene (Scheme 3.42) [97] or polyalanine [98], or impregnation in inorganic materials such as zeolites [99], nanocrystalline CeO_2 and ZrO_2 [100] or MgO [101], molecular sieves [102], magnetite [103], or mesoporous carbon nitride [104]. The microwave-assisted continuous-flow organic synthesis (MACOS) has also been applied in the A³-coupling process by irradiation of gold and copper thin films, as described by Organ, Li,

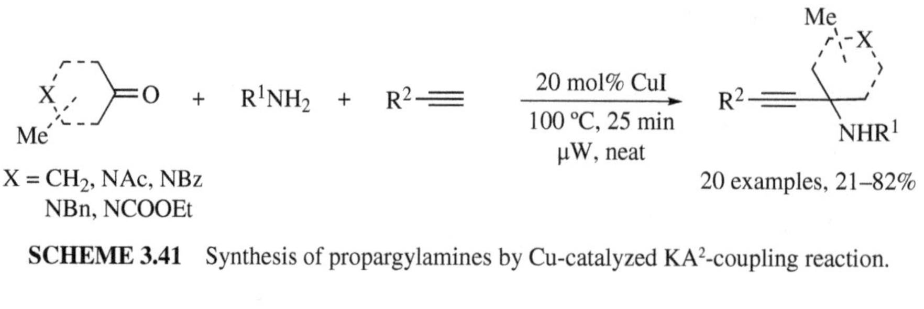

SCHEME 3.41 Synthesis of propargylamines by Cu-catalyzed KA²-coupling reaction.

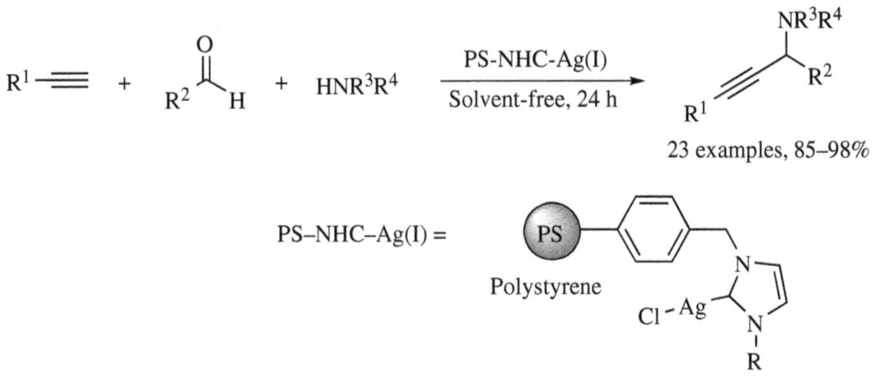

SCHEME 3.42 A^3-coupling synthesis of propargylamines using a supported silver-based catalyst.

et al. [105]. In addition, Sun et al. recently reported that the silver macromolecule Ag[1,4-bis(4,5-dihydro-2-oxazolyl) benzene](NO_3) [106], which forms a three-dimensional supramolecular structure by linkage of its one-dimensional chains by hydrogen bonding, can catalyze the A^3-coupling synthesis of propargylamines under mild conditions without the need for any additional cocatalyst or activator. The same group developed a protocol for preparing propargylamines at room temperature in air using silver oxide nanoparticles as the catalyst [107], which can be recovered readily and reused directly up to six times without purification.

3.4.2.2 Synthesis of Heterocycles through Propargylamines The synthetic utility of the A^3-coupling lies in the fact that propargylamines are key intermediates for the construction of a variety of heterocycles. Either as the first step of a series of consecutive reactions or by its involvement in a cascade process, the A^3-coupling provides an efficient route to more complex structures using propargylamines as the synthetic platform.

In 2005, Yamamoto et al. reported the synthesis of polycyclic pyrrole-2-carboxylates **76** *via* a $CuBr_2$-catalyzed three-component coupling of *N*-benzylallylamine, ethyl glyoxalate, and terminal alkynes, and subsequent transformation of the glycine-tethered 1,6-enynes **75** thus obtained through a cycloisomerization/Diels–Alder cycloaddition/dehydrogenation sequence under iridium-catalyzed conditions (Scheme 3.43) [108].

Recently, Van der Eycken et al. described the two-step synthesis of 3-benzazepines by coupling a Cu(I)-catalyzed A^3-coupling with a regio- and stereoselective Pd-catalyzed intramolecular acetylene hydroarylation [109]. The same group had previously reported on the preparation of triazolodibenzo[1,5-*a*]azocines **78**, which involves the cyclization of functionalized propargylamine intermediates **77** previously prepared *via* A^3-coupling (Scheme 3.44) [110]. In both examples, the key steps of the reactions benefit from focused microwave irradiation, affording the desired products with high yields and purity.

Propargylamines **79** derived from Cu-catalyzed A^3-coupling have been used as precursors for the synthesis of a series of imidazole derivatives **81–83** (Scheme 3.45). As described by Looper in 2009, von Braun cyanation of tertiary propargylamines **79** followed by lanthanide-mediated hydroamination/isomerization of the resultant propargylcyanamides **80** provide an efficient route to polysubstituted 2-dialkylamino-imidazoles **81** [111]. The same group extended this methodology to generate 2-thio- and 2-oxoimidazoles (**82** and **83**) by reacting propargylcyanamides **80** with thiol and alcohol nucleophiles [112]. Shortly after, Van der Eycken et al. applied the A^3-coupling to the preparation of tertiary *N*-methylallylpropargylamines, which were used as precursors in the synthesis of polysubstituted 2-aminoimidazole alkaloids [113].

Functionalized propargylamines can undergo spontaneous cyclization, providing one-pot cascade routes for the synthesis of heterocycles through A^3-coupling/cyclization sequences. The presence of an additional nucleophilic functionality in one of the precursors (aldehyde, alkyne, or amine) provides a reactive center toward cyclization, which is prompted by activation of the triple bond by the metal catalyst in the propargylamine intermediate. In 2007, Liu et al. presented the synthesis of aminoindolizines **84** by means of a gold(III)-catalyzed A^3-coupling/cycloisomerization reaction starting from pyridine-2-carboxaldehyde, amines, and alkynes under solvent-free conditions or in water [114]. The authors proposed the mechanism depicted in Scheme 3.46, which involves initial formation of NR^{1}R^2-substituted propargylic pyridine **XXVI**. Coordination of the triple bond in **XXVI** to the gold catalyst enhances the electrophilicity of the alkyne and the subsequent nucleophilic attack of the nitrogen lone pair to produce the cation **XXVII**, which undergoes deprotonation followed by demetalation to afford indolizines **84**. A similar reaction catalyzed by a silver salt has recently been reported [115].

A number of protocols based on the same strategy have recently appeared. Chernyak and Gevorgyan developed a methodology for the synthesis of 3-aminoindolines and

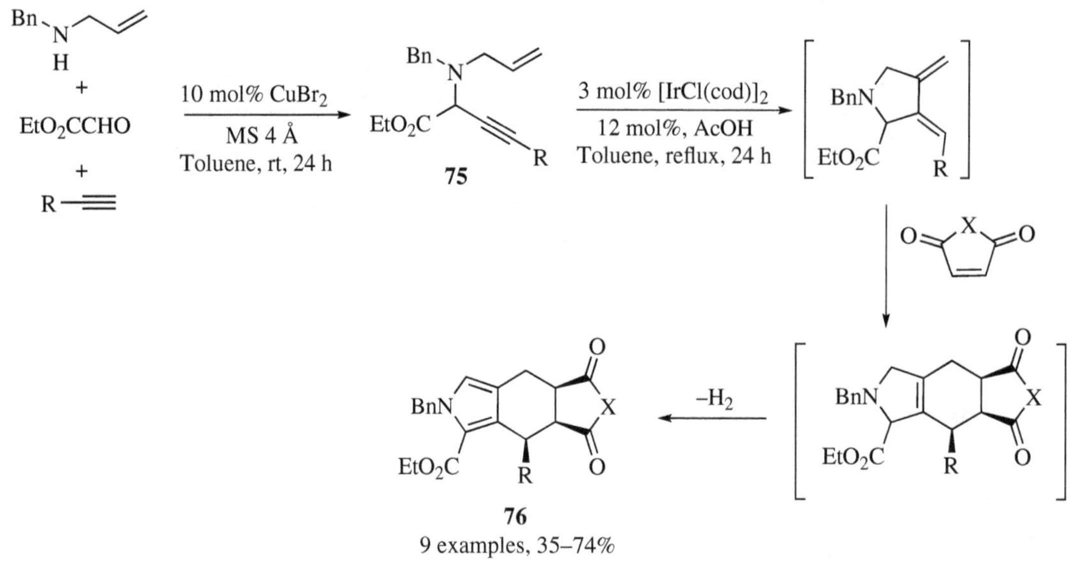

SCHEME 3.43 Synthesis of polycyclic pyrrole-2-carboxylates *via* CuAAC/cycloisomerization/Diels–Alder cycloaddition/dehydrogenation sequence.

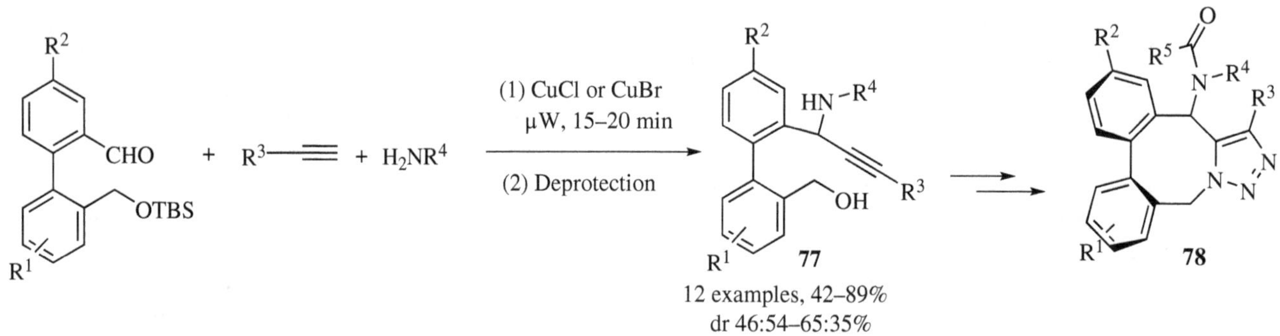

SCHEME 3.44 CuAAC in the synthesis of triazolodibenzo[1,5-*a*]azocines.

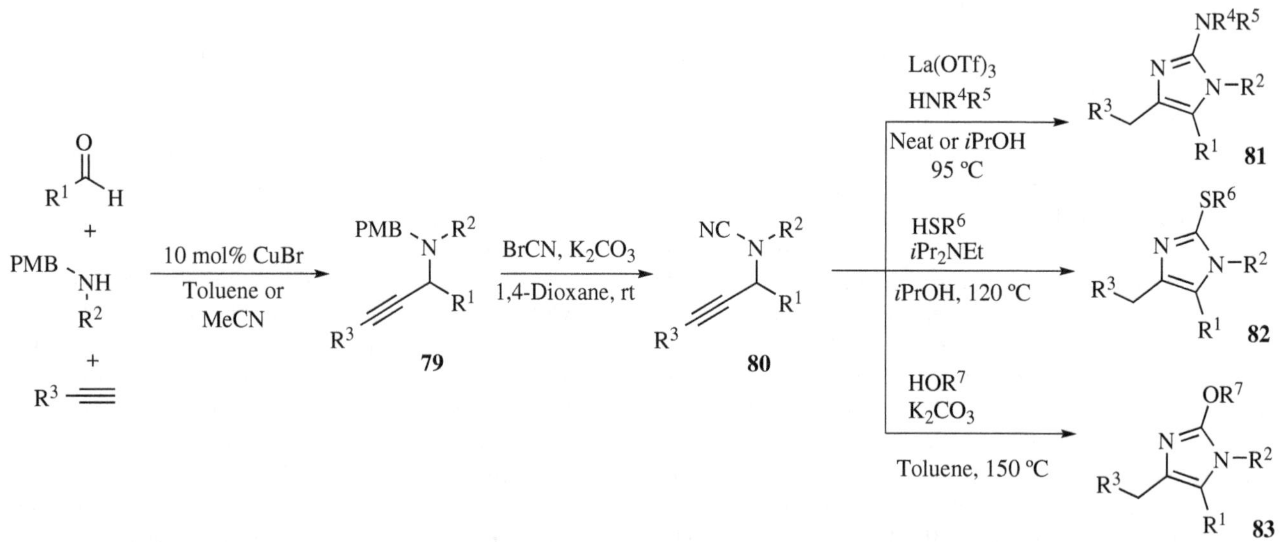

SCHEME 3.45 Synthesis of 2-dialkylaminoimidazoles and 2-thio- and 2-oxoimidazoles from propargylcyanamides.

3-aminoindoles, which consists of the copper-catalyzed A^3-coupling/cyclization involving 2-aminobenzaldehyde, an alkyne, and a secondary amine [116]. On the other hand, imidazoheterocycles are obtained using 2-aminopyridine, an aldehyde, and an alkyne as substrates [117]. Ohno and Fujii reported the copper-catalyzed synthesis of 2-(aminomethyl) indoles **86** (Scheme 3.47) [118]. In this case, the nucleophilic attack on the triple bond that causes the intramolecular cyclization of propargylamine intermediate **XXVIII** is accomplished by the amino group of ethynylaniline derivatives **85**.

1,2-Dihydroquinolines are obtained *via* Cu- or Ag-catalyzed A^3-coupling/cyclization when using 2-alkynylbenzaldehydes, amines, and ketones, which combines metal and enamine catalysis, as demonstrated by Wu et al. [119]. 2-Alkynylbenzaldehyde derivatives have also been used as the additional nucleophile carrier in the silver-catalyzed synthesis of naphthyridines and thienopyridines recently reported by Verma et al. [120]. A variety of naphthyridines and thienopyridines were prepared *via* dual activation using L-proline as the organocatalyst, whereas for other naphthyridines, an enolate chemistry approach without any organocatalyst was used.

In 2010, Ji et al. described the three-component Au(III)-catalyzed coupling of glyoxylic acid **87**, an amine, and an alkyne to afford two different classes of butenolides **89** and **90**, with one of the OH groups of **87** playing the role of the additional nucleophile [121]. In this case, the A^3-coupling/cyclization sequence gives an intermediate **XXIX**, which is trapped by an electrophile. The resultant butenolide depends on the nature of the alkyne and the electrophile. Thus, intermediate **XXIX** is trapped by a proton in reactions with aliphatic alkynes, affording butenolides of type **89** (Scheme 3.48), whereas complex butenolides of type **90** containing an α-amino acid moiety at the γ position of the butenolide core are obtained when the intermediate **XXIX** derived from aromatic alkynes is trapped by the iminium ion **88**.

Several examples of A^3-coupling processes that involve more than three components have been reported. In 2008, Li et al. described the preparation of tetrasubstituted isoindolines from three alkyne units, two aldehyde units, and a

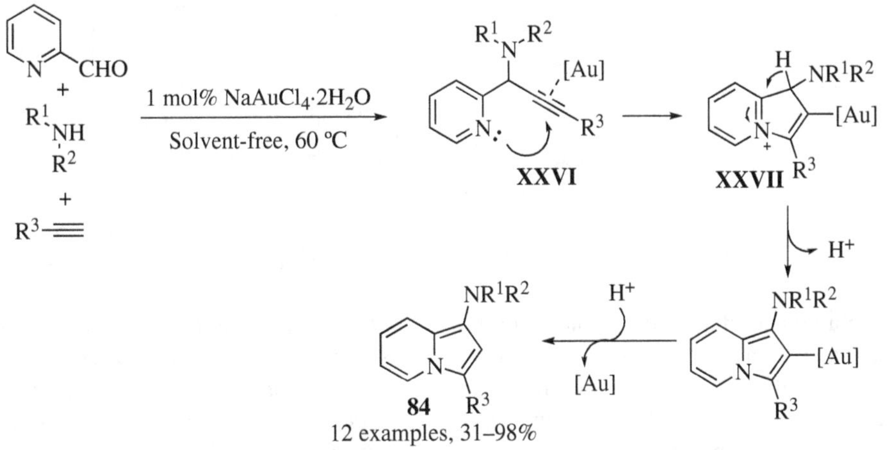

SCHEME 3.46 Proposed mechanism for the gold(III)-catalyzed A^3-coupling/cycloisomerization reaction.

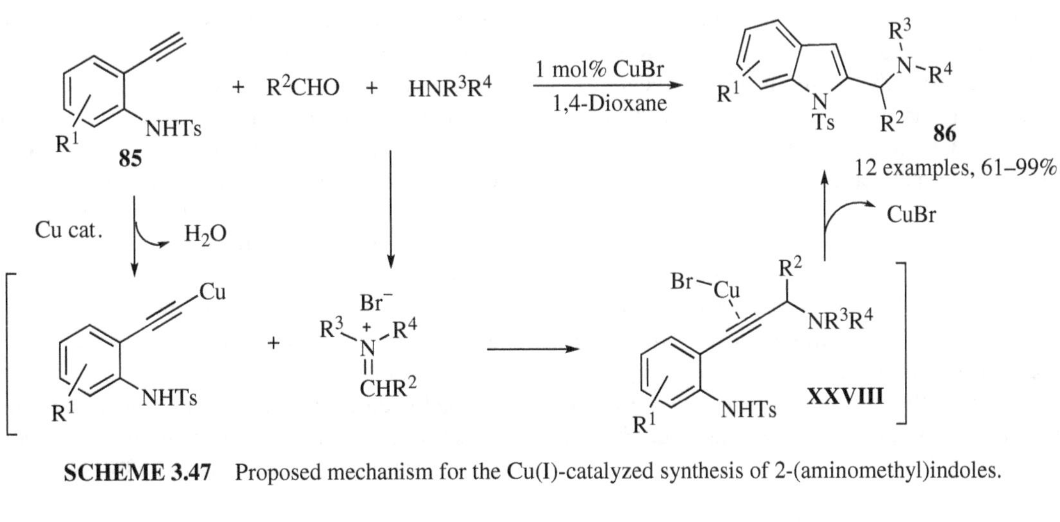

SCHEME 3.47 Proposed mechanism for the Cu(I)-catalyzed synthesis of 2-(aminomethyl)indoles.

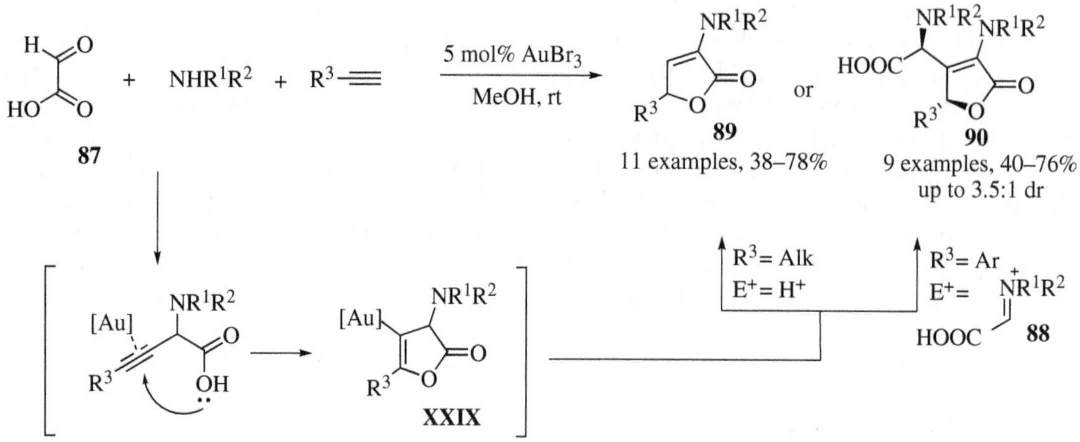

SCHEME 3.48 Synthesis of butenolides *via* gold(III)-catalyzed A^3-coupling/cyclization sequence.

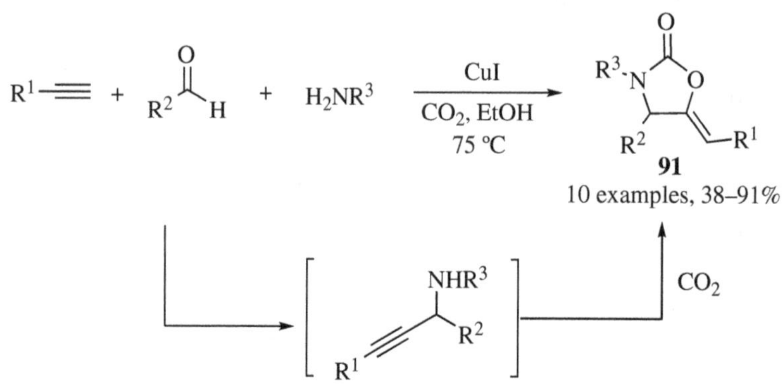

SCHEME 3.49 Cu-catalyzed A^3-coupling synthesis of oxazolidinones.

primary amine through a process that involves three CuBr-catalyzed A^3-couplings and a final [2+2+2] cycloaddition catalyzed by Wilkinson's catalyst [122]. Soon after that, the same group reported on a copper-catalyzed A^3-coupling performed under a carbon dioxide atmosphere, which affords oxazolidinones bearing exocyclic alkenes **91** (Scheme 3.49) [123]. In this four-component reaction, the carbon dioxide is proposed to play a dual role, both as a promoter and substrate for the tandem reaction. A similar reaction was later reported by Jiang et al. for the synthesis of 5-alkylene oxazolidinones, in which the aldehyde unit is replaced by a ketone [124]. In this Cu/Sn-catalyzed A^3-coupling/cycloaddition, the propargylamine intermediate is also subsequently trapped by CO_2.

3.4.2.3 ***Asymmetric A^3-Coupling*** After the first enantioselective version of an A^3-coupling published by Wei and Li in 2002 [125], a number of reports on the asymmetric multicomponent copper-catalyzed synthesis of propargylamines have subsequently appeared. The most significant examples reported from 2005 on are examined in the following text.

In general, chiral tridentate N-ligands of the pyridine bis(oxazoline) (pybox) type seem to be the best choice for the performance of asymmetric A^3-couplings with anilines. In 2006, Bisai reported the use of a chiral catalyst prepared from $CuPF_6$ and a C_2-symmetric pybox ligand **92** (Scheme 3.50), which afforded a wide variety of propargylamines **93** in good to excellent yields with excellent enantioselectivities (up to 99%) [126]. The reaction was restricted to aromatic alkynes, since the use of aliphatic alkynes resulted in diminished ee. This was overcome by Shao and Chan with the report of a protocol for the synthesis of chiral β,γ-alkynyl α-amino acid ethyl esters in good yields and enantiomeric excesses, applicable to both aromatic and aliphatic alkynes [127]. In 2010, Nakamura et al. reported a modification of these procedures in which the pybox ligand was replaced by a bis(imidazoline) pyridine (pybim) ligand [128]. The CuOTf/pybim system was optimized for aliphatic alkynes and provided good yields and enantiomeric excesses (up to 98%).

On the other hand, chiral bidentate P,N-ligands have been successfully applied to the asymmetric A^3-coupling using secondary amines as substrates [88b]. Knochel et al. have

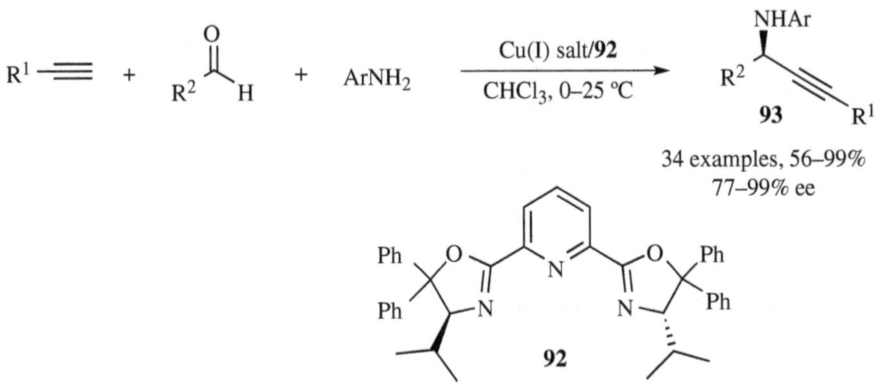

SCHEME 3.50 Asymmetric synthesis of propargylamines by Cu-catalyzed A³-couplings.

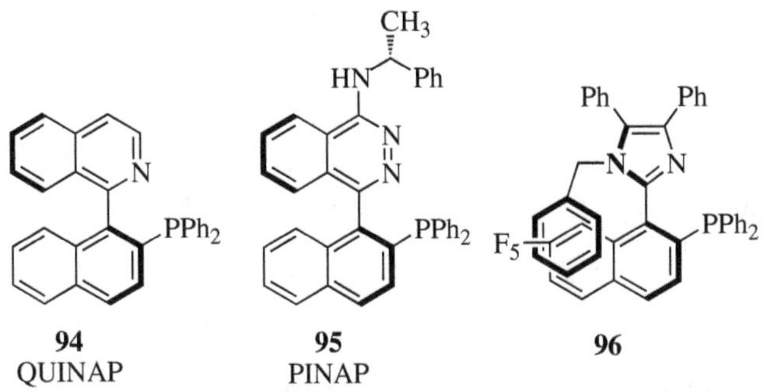

FIGURE 3.1 Chiral P,N-ligands used in asymmetric A³-coupling.

extensively studied the use of the Quinap ligand **94** (Fig. 3.1) [88c], and some of their latest results involve the application of bis(phenallyl)amine as amine component in the CuBr/Quinap-catalyzed asymmetric A³-coupling, which furnished bis(2-phenallyl)-protected propargylamines [129]. 2-Phenallyl protecting group can be easily removed with a Pd(0)-catalyzed allylic substitution. The authors also provided an extensive mechanistic study of their Cu/Quinap-catalyzed asymmetric A³-coupling procedures [130].

Related Pinap ligand **95** [131] (Fig. 3.1) was first introduced by Carreira et al. as an alternative to the expensive and difficult to prepare Quinap [132], resulting successful in the asymmetric A³-coupling. They found that CuBr/Pinap-catalyzed reactions could be performed using 4-piperidone hydrochloride hydrate **97** as amine component, the subsequent deprotection of propargylamines **98** with a supported scavenger amine affording primary propargylamines as their hydrochloric acid salts **99** (Scheme 3.51) [133]. Very recently, Pinap ligand has been used in the CuI-catalyzed highly enantioselective synthesis of chiral tetrahydroisoquinolines through the α-alkynylation of 1,2-unsubstituted tetrahydroisoquinolines with aldehydes and terminal alkynes [134].

Very recently, an imidazole-based chiral biaryl P,N-ligand **96** (Fig. 3.1) has been developed and applied in the enantioselective A³-coupling reaction [135]. The incorporation of an aromatic heterocycle in the biaryl motif was achieved by stabilizing the chiral ground-state conformation by π-stacking interactions, which leads to an increase in the barrier to rotation in biaryls.

3.4.2.4 *Modifications of A³-Coupling*

Related coupling reactions in which one of the A³-coupling substrates (aldehyde, amine, or alkyne) is replaced by other molecule affording mechanistically similar processes have recently appeared. Thus, as mentioned earlier (Scheme 3.41), the aldehyde component can be replaced by a ketone to give propargylamines bearing a quaternary carbon center [91, 136].

In 2008, Li et al. reported a copper-catalyzed amine–alkyne–alkyne addition reaction as an efficient method for the synthesis of γ,δ-alkynyl-β-amino acid derivatives **102** (Scheme 3.52) [137]. In this case, the first step of the reaction is proposed to be the hydroamination of the electron-deficient alkyne **100**, which plays the role of the aldehyde component. Subsequent reaction of the resultant intermediate **XXX** with alkyne **101** would afford intermediate **XXXI**, which would be then protonated to give an iminium intermediate **XXXII**. Finally, an intramolecular transfer of the alkyne moiety to the iminium ion would yield the γ,δ-alkynyl-β-amino ester **102** and regenerate the catalyst. The reaction was later extended using chiral prolinol derivatives as the amine component, which afforded the corresponding γ,δ-alkynyl-β-amino acid derivatives with excellent diastereoselectivities (up to >99:1) [138].

The aldehyde component has also been replaced by diethyl formylphosphonate hydrate in the silver-catalyzed synthesis of *N*-PMP-protected α-aminopropargylphosphonates [139] and by 2-oxoacetic acid [140] or glyoxylic acid [141] in copper-catalyzed microwave-assisted decarboxylative three-component couplings. In addition, glyoxylic acid has been coupled with amines and alkynes in the gold-catalyzed synthesis of butenolides (see Scheme 3.48) [121].

An interesting variation of the A³-coupling consists in the replacement of the aldehyde by a haloalkane. These

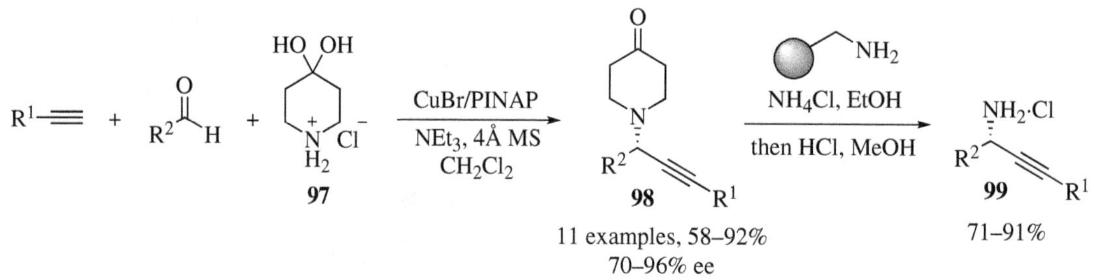

SCHEME 3.51 PINAP ligand in the Cu-catalyzed asymmetric A^3-coupling reaction.

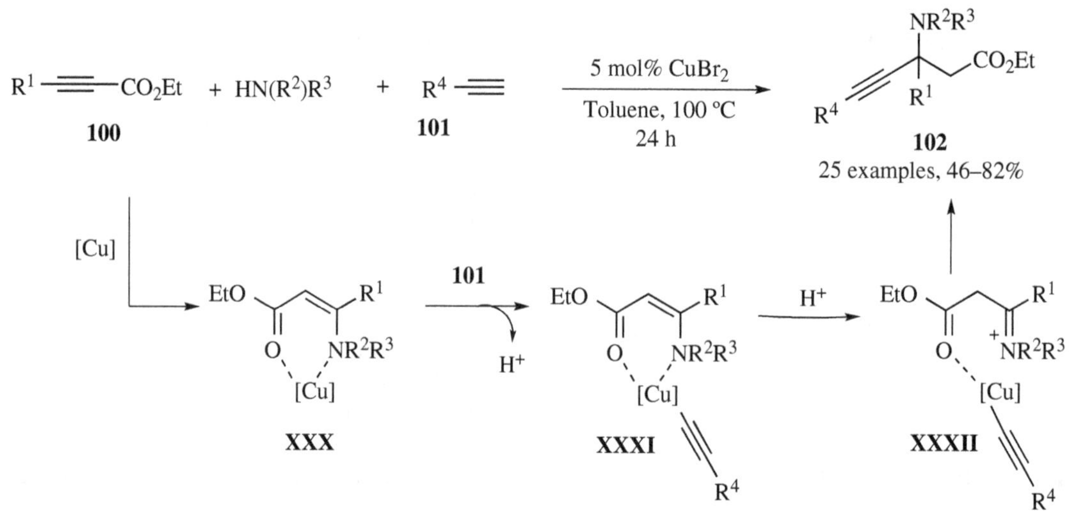

SCHEME 3.52 Cu-catalyzed amine–alkyne–alkyne coupling reaction.

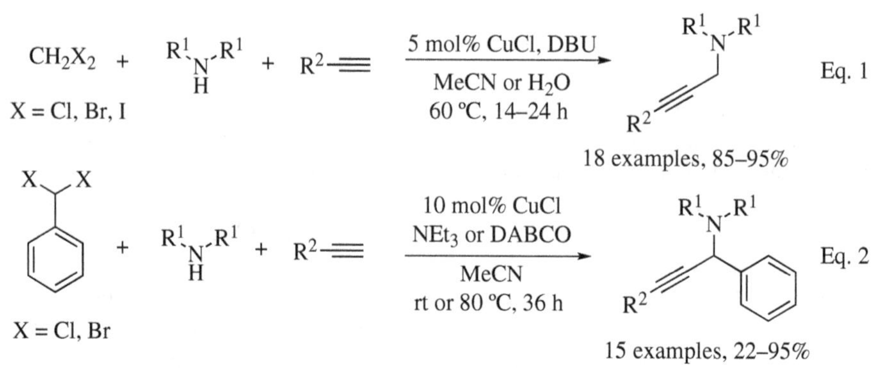

SCHEME 3.53 The use of dihaloalkanes in the Cu-catalyzed A^3-coupling reaction.

reactions, which involve C—H and C—halogen bond activation, have been catalyzed by copper (Scheme 3.53) [142] and gold [143]. The substrate scope for both processes is limited to secondary aliphatic amines and dihalomethanes (Eq. 1, Scheme 3.53) or benzal halides (Eq. 2, Scheme 3.53).

The amine component was successfully replaced by carbamates **103** in the CuOTf-catalyzed synthesis of propargylcarbamates **104** (Scheme 3.54), as reported by He, Li et al. Best results were obtained with *n*-butyl carbamate and aromatic alkynes and aldehydes [144].

Alkyne replacements have also been reported. In 2008, Sakai et al. described the use of alkynylsilanes instead of terminal alkynes in the Cu-catalyzed synthesis of propargylamines, which was applicable to both secondary and primary aliphatic amines, although the latter afforded the corresponding products in low yield [145]. Very recently, Van der Eycken et al. applied the A^3-coupling to the C—H alkylation of azoles through a copper-catalyzed heteroarene–amine–aldehyde/ketone coupling [146]. This reaction is proposed to proceed through the initial condensation of

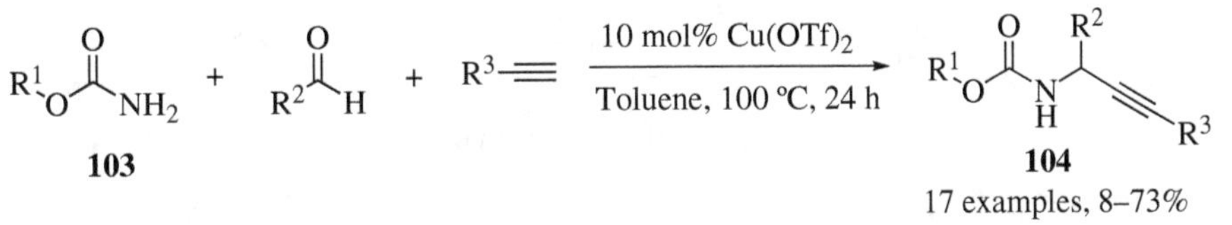

SCHEME 3.54 The use of carbamates in the Cu-catalyzed A³-coupling reaction.

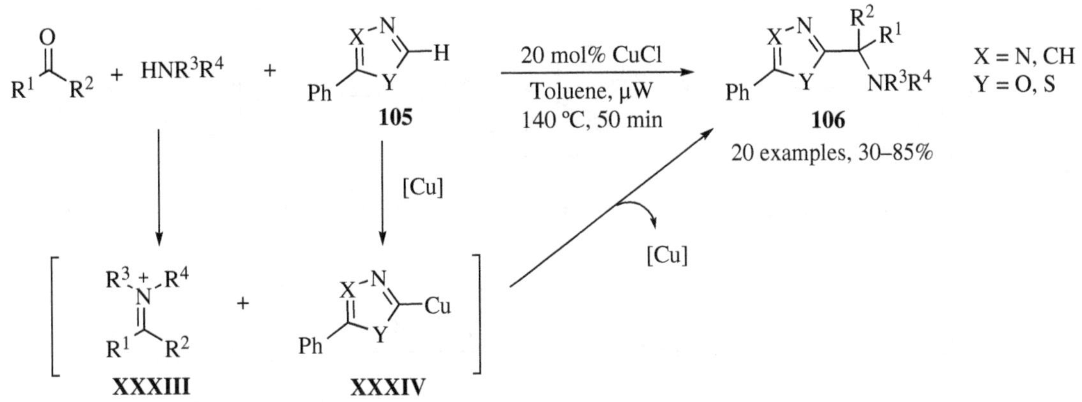

SCHEME 3.55 C—H alkylation of azoles *via* Cu-catalyzed A³-coupling reaction.

the aldehyde or ketone with the amine, leading to the corresponding iminium ion **XXXIII**, which is subsequently attacked by the azole–Cu complex **XXXIV** formed *in situ*, therefore installing the alkyl chain in the azole **105** affording **106** (Scheme 3.55).

3.4.3 Miscellaneous

Other examples of group 11 metal-catalyzed MCRs besides those mentioned in previous sections are listed in the following text.

Recent approaches to the copper-catalyzed three-component coupling of organometallic reagents with imines and acid chlorides employ organoindium compounds [147] or organostannanes [148]. In the first case, α-substituted amides or *N*-protected amines are obtained, whereas the second process provides hexahydro-1*H*-isoindolones *via* a tandem Cu-catalyzed three-component coupling/intramolecular Diels–Alder reaction.

Recently, Ma et al. developed a three-component reaction for the synthesis of 2-*N*-substituted benzothiazoles **107** *via* the copper-catalyzed coupling of 2-iodoanilines, carbon disulfide, and secondary amines (Scheme 3.56) [149]. In these reactions, the condensation of CS_2 with an amine in the presence of a base generates dithiocarbamate salts **XXXV**, which undergo coupling with 2-iodoanilines to give **XXXVI** and subsequent intramolecular condensation and elimination to afford the substituted benzothiazoles **107**.

Recently reported silver-catalyzed three-component reactions include the asymmetric coupling of an aldehyde, an alkene, and Oppolzer's chiral glycyl sultam as the amine component, which affords highly functionalized pyrrolidines [150]; the synthesis of arylnaphthalene lactones *via* the three-component coupling reaction between phenylacetylene, carbon dioxide, and 3-bromo-1-phenyl-1-propene [151]; and the three-component [2+2+1] cycloaddition of allenoates, dual activated olefins, and isocyanides to give functionalized cyclopentanone derivatives [152].

Regarding gold catalysts, recent contributions are the oxidative oxyarylation of alkenes, which consists of the coupling reaction of alkenes, arylboronic acids, and several types of oxygen nucleophiles, including alcohols, carboxylic acids, and water (Scheme 3.57) [153]; the enantioselective synthesis of cyclic carbamimidates *via* a three-component reaction of imines, terminal alkynes, and *p*-toluenesulfonyl isocyanate [154]; and the coupling of an aldehyde, an alkyne, and an alcohol to generate β-alkoxyketones [155].

3.5 RHODIUM-CATALYZED MCRs

3.5.1 Rhodium-Catalyzed MCRs *via* Onium Ylide Intermediates

It is well established that the rhodium-carbene species generated upon activation of diazo compounds by rhodium complexes can undergo insertion into a X—H bond (X= C, Si, O, N) to form a new C—X bond under mild conditions (Scheme 3.58). This reaction involves formation of an onium ylide intermediate and subsequent proton transfer [156]. The high reactivity of these onium ylide species, which can be trapped by an electrophile prior to the proton transfer leading to a second bond formation,

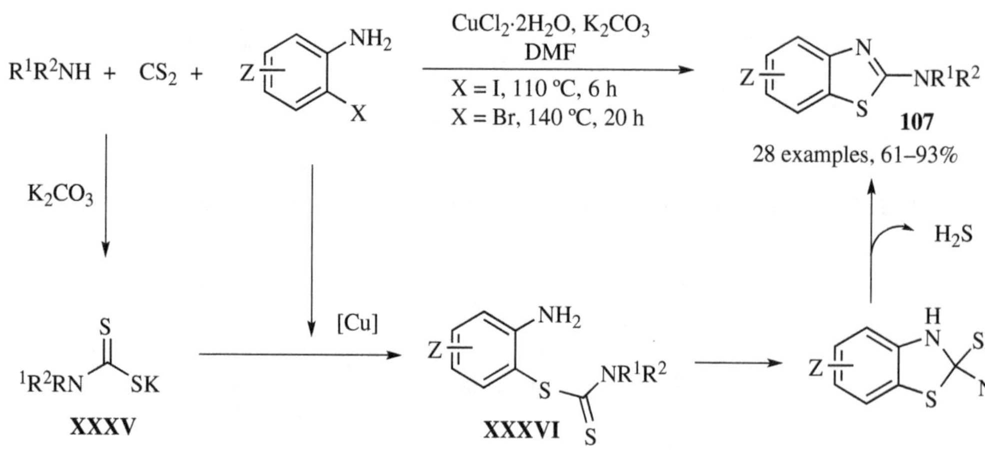

SCHEME 3.56 Synthesis of benzothiazoles *via* Cu-catalyzed coupling of 2-iodoanilines, carbon disulfide, and secondary amines.

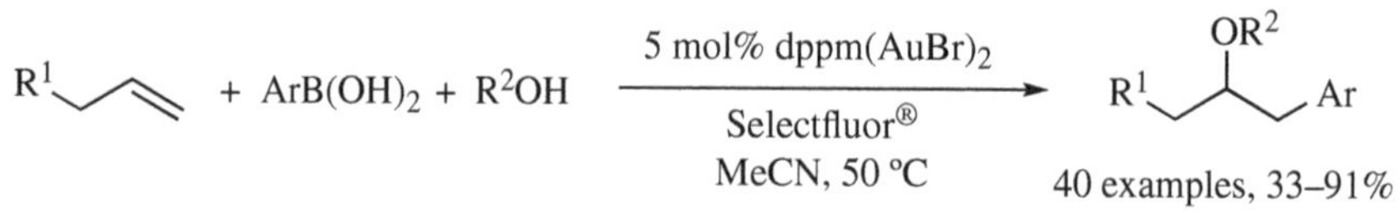

SCHEME 3.57 Au(I)-catalyzed oxidative oxyarylation of alkenes.

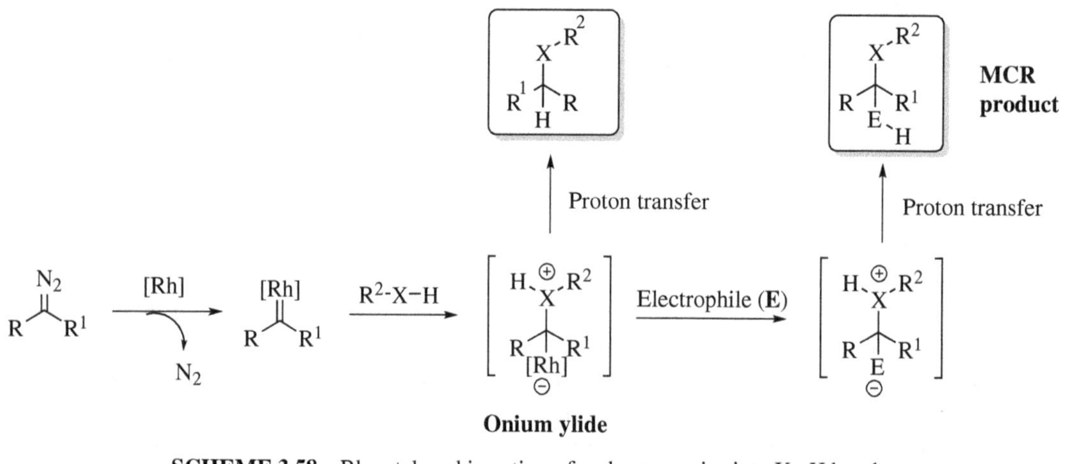

SCHEME 3.58 Rh-catalyzed insertion of carbene species into X—H bonds.

has prompted the development of new Rh-catalyzed MCRs [157] as illustrated in the following text.

3.5.1.1 ***Rhodium-Catalyzed Reactions of Oxonium Ylide Intermediates*** Hu et al. [158] reported the preparation of α,β-dihydroxyl and α-alkoxy-β-alkylamino acid derivatives through a rhodium-catalyzed three-component reaction starting from a diazo compound **108**, an alcohol **109**, and different electrophiles such as aldehydes **110** or imines **111**. The reaction proceeds through formation of a Rh-oxonium ylide species, generated from the diazo compound **108** and the alcohol **109**, which is subsequently trapped by the electrophile. Employing different Rh(II) catalysts, *erythro/threo* mixtures of α,β-dihydroxyl and α-alkoxy-β-alkylamino esters **112** and **113** could be obtained in moderate to good yields from a range of aryldiazoacetates, alcohols, and aldehydes or imines (Scheme 3.59, Eq. 1). The moderate yields may be explained as a result of the formation of proton transfer by-products **114**, which could be minimized by employing catalysts with more electron-withdrawing ligands or, in the case of imines, using 2-aminophenol derivatives where activation by the phenol moiety through intramolecular hydrogen bonding enhances the electrophilicity of the imine [159]. Unfortunately, this three-component methodology was limited to aryldiazoacetates, and racemic products were obtained when chiral catalysts Rh_2(S-DOSP)$_4$ and Rh_2(S-nttl)$_4$ were used. A single *erythro* (*R,R,R*)-isomer could be obtained employing (*R*)-*N*-(*tert*-butylsulfinyl)amine as chiral auxiliary [159]. Shortly

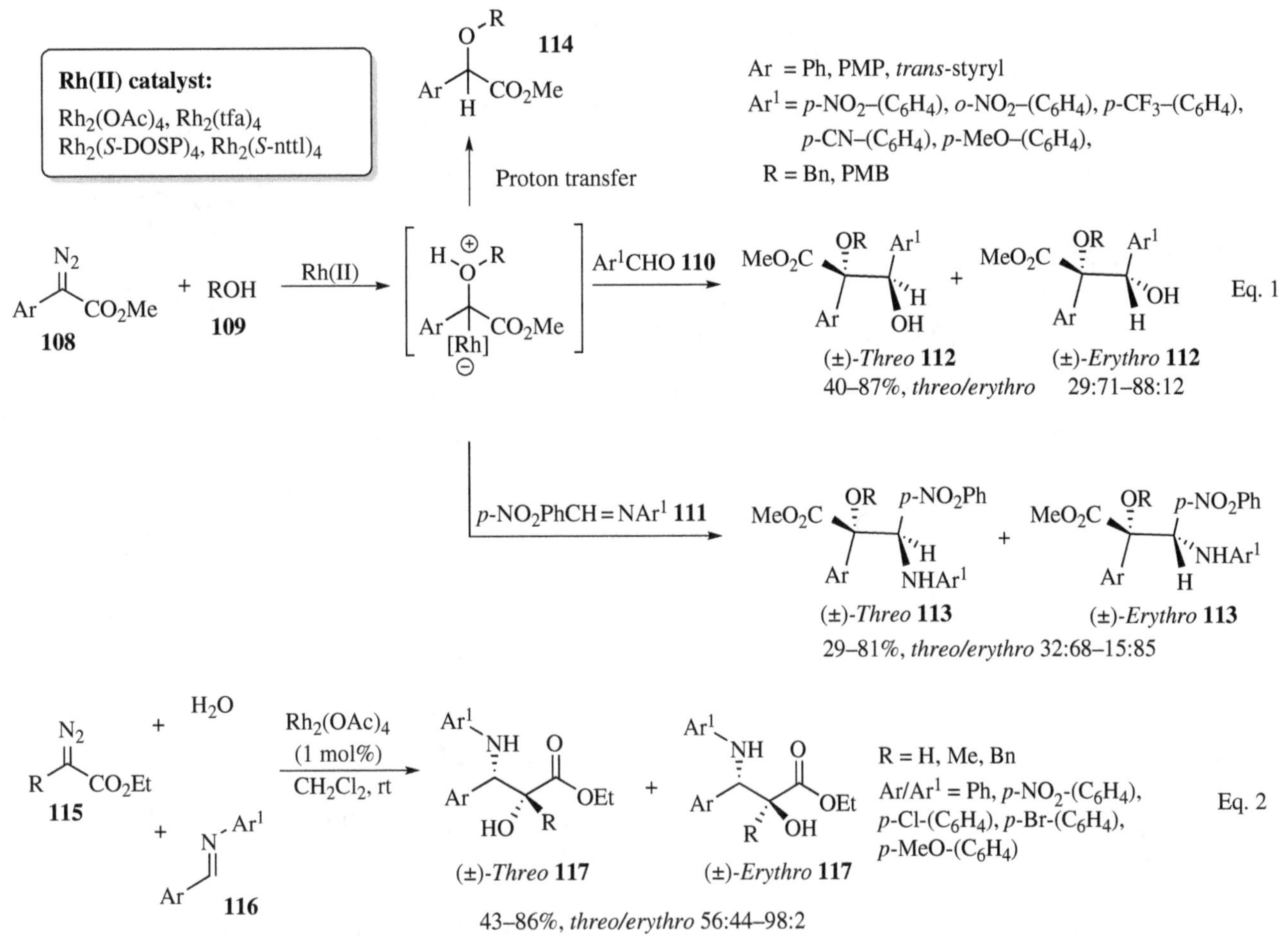

SCHEME 3.59 Rh-catalyzed reaction of oxonium ylides with aldehydes and imines.

after, this methodology was also extended to 2-alkyl-1-diazoacetates **115**, and a highly diastereoselective one-step synthesis of β-aryl isoserine esters was developed [160]. The oxonium ylides generated from the reaction of alkyl diazo compounds **115** with H_2O in the presence of $Rh_2(OAc)_4$ reacted with a broad family of aryl imines **116** to afford racemic β-aryl isoserine esters **117** in good yields and high diastereoselectivities (Scheme 3.59, Eq. 2).

The catalytic asymmetric versions of these reactions were also developed by Hu et al. employing Rh(II)-catalysts to generate the oxonium ylide intermediate and a chiral Lewis acid or chiral Brønsted acid to activate the aldehyde [161] or the imine [162], respectively (Scheme 3.60). Different chiral Lewis acids *in situ* generated using chiral bisoxazolines as the ligands and different metallic salts such as $Cu(OTf)_2$, $Yb(OTf)_3$, $Mg(ClO_4)_2$, or $Sn(OTf)_2$ were tested for the activation of aldehydes. Only the zirconium–BINOL system afforded the desired α,β-dihydroxyl acid derivatives **118** in good yields, good *erythro/threo* ratio, and high enantioselectivities [161]. Although this three-component reaction works well with a broad family of aromatic aldehydes bearing electron-donating and electron-withdrawing groups and high diastereoselectivities (*erythro/threo* 70:30–92:8) and enantioselectivities (>90% ee) were observed in most cases, it has only been described with benzylic alcohol, and it fails when using aliphatic aldehydes and aliphatic diazo compounds.

When using imines as the trapping electrophile, chiral phosphoric acids derived from BINOL proved to be very efficient catalyst for the 1,2-addition step, giving the desired β-amino-α-hydroxyl acid derivatives **119** in high yields. Reaction conditions were optimized for a bulky alcohol, such as anthracen-9-yl-methanol, and a 9-phenanthryl-derived catalyst (see Scheme 3.60), obtaining complete *syn*-diastereoselectivity and enantioselectivities higher than 90% when using $Rh_2(OAc)_4$ (2 mol%) and a chiral phosphoric acid (2 mol%) as the catalysts in CH_2Cl_2 at −20 °C [162a]. The reaction is successful with a broad family of aryl diazo compounds and aromatic aldimines, although with the aldimine derived from cyclohexylcarbaldehyde, a poor yield (34%) and 49% ee were observed.

To overcome the lack of reactivity of the chiral phosphoric acid toward aldehydes, Hu et al. [162b] decided to develop a four-component reaction for the synthesis of chiral β-amino-α-hydroxyesters **120** (Scheme 3.61, Eq. 1), where a Brønsted acid acts as catalyst for the formation

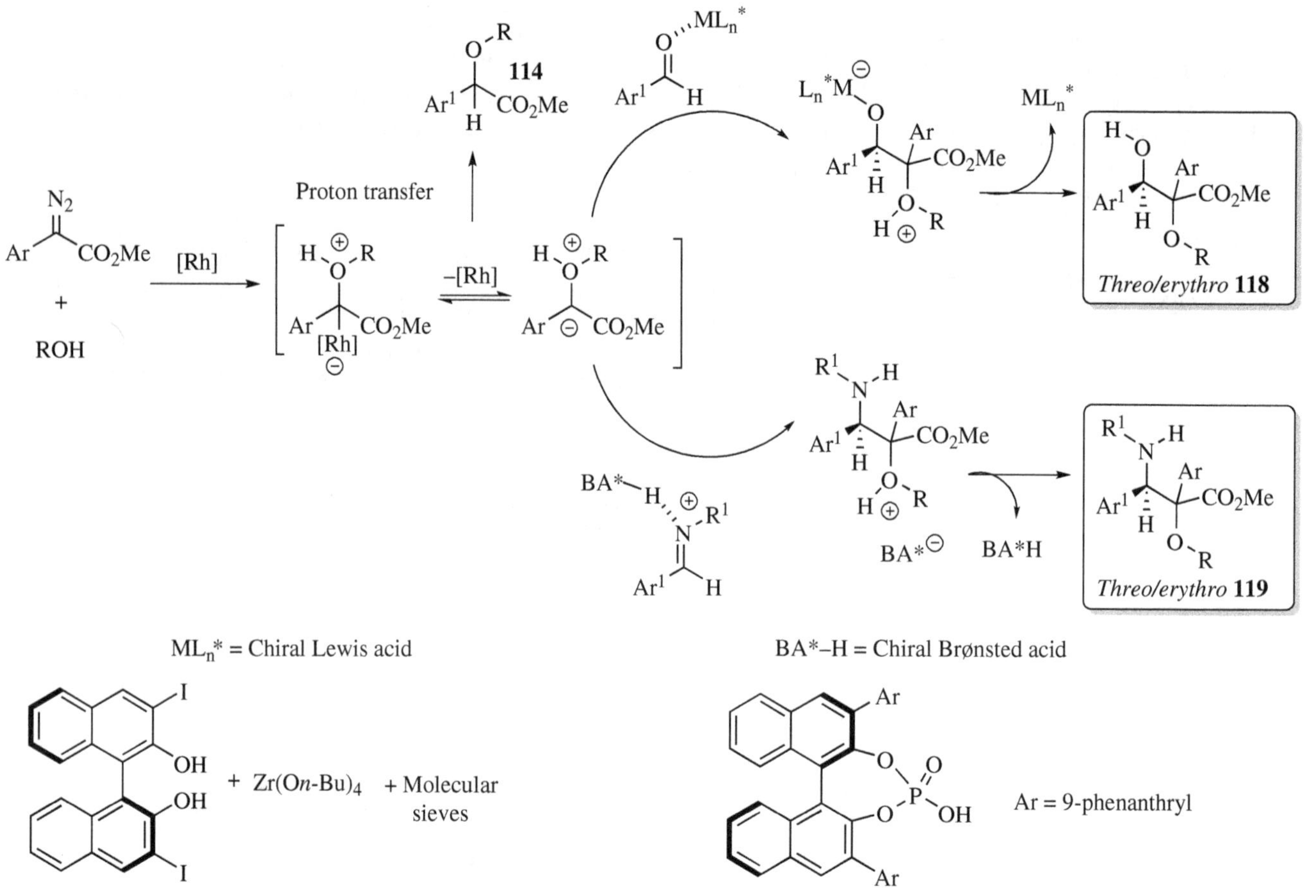

SCHEME 3.60 Asymmetric version of the reaction of oxonium ylides with aldehydes and imines.

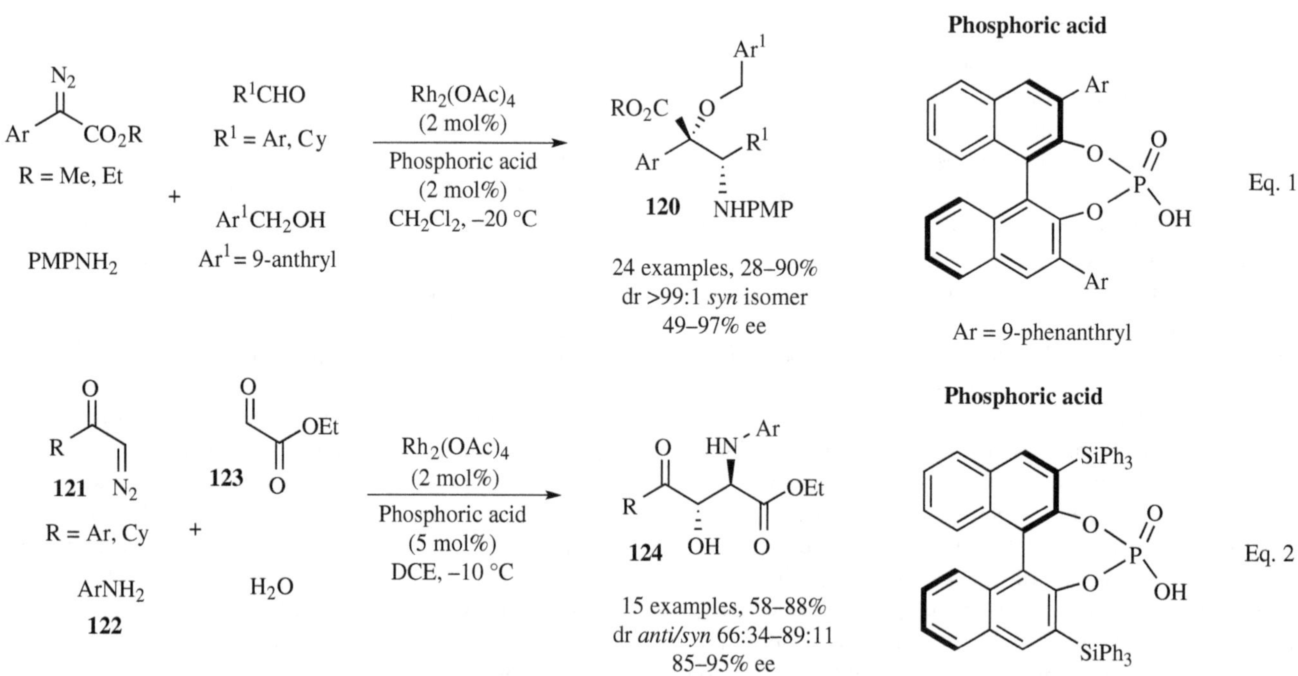

SCHEME 3.61 Rh-catalyzed four-component reactions of oxonium ylides with imines.

of the imine and its activator. When H_2O was used to generate the oxonium ylide, lower diastereoselectivities were observed, probably due to the higher reactivity of the oxonium ylide derived from H_2O [162c]. This methodology may be also applied to the four-component Mannich-type reaction between diazoketones **121**, H_2O, amines **122**, and ether glyoxylate **123** for the asymmetric synthesis of β-hydroxy-α-amino acid derivatives **124** [162d] (Scheme 3.61, Eq. 2).

Other similar Rh-catalyzed 1,2-addition of oxonium ylides to aldehydes or imines have been described [163]. Moreover, this methodology has been applied to the 1,2-addition of oxonium ylides to isatins [164a], β,γ-unsaturated α-ketoesters [164b], and azetidine-2,3-diones [164c].

In order to expand the synthetic applicability of the methodology, Hu et al. decided to employ Michael acceptors as other type of electrophiles. They developed a three-component reaction between an aryl diazo compound, H_2O, and an enone, for the generation of γ-hydroxyketones **126** bearing a quaternary carbon stereocenter, catalyzed by the system $Rh_2(OAc)_4$ (2 mol%)/chiral Lewis acid (30 mol%) [165a]. Different Lewis acids were tested as cocatalysts, but only the Sc-, Yb-, and Zn-based catalysts proved to be active, the *t*Bu-bisoxazoline/$Zn(OTf)_2$ system being the most active and selective. The catalyst combination $Rh_2(OAc)_4$ (2 mol%), (*S*)-*t*Bu-box–$Zn(OTf)_2$ (30 mol%), and TsOH (40 mol%) was applied to a broad family of acyl imidazoles **125** affording the desired products in good to excellent yields, almost complete *syn*-diastereoselectivities and 85–97% ee, when reactions were carried out in dichloromethane at −8 °C (Scheme 3.62, Eq. 1). When Ar = Ph and Ar^1 = *p*-Br-C_6H_4, product **126** can be further transformed into the corresponding methyl ester **127** in 63% yield.

Oxonium ylides generated from alcohols and diazo compounds using a Rh(II) catalyst were also trapped with other α,β-unsaturated electrophiles [165b], such as benzylidene Meldrum's acid **128** and 4-oxo-enoates **129** (Scheme 3.62, Eq. 2). Other Michael acceptors such as α,β-unsaturated methyl esters, α,β-unsaturated sulfones, and nitroalkene resulted unreactive. These reactions were successful with a range of aliphatic alcohols **130** and aromatic and aliphatic diazo compounds **131**, affording the desired products **132**–**133** with almost complete diastereoselectivity and high yields. To demonstrate the synthetic utility of the obtained products, **133** was selectively reduced using K-selectride, and subsequent quenching with methanol afforded the *anti*-γ-butyrolactone **134** in 62% yield.

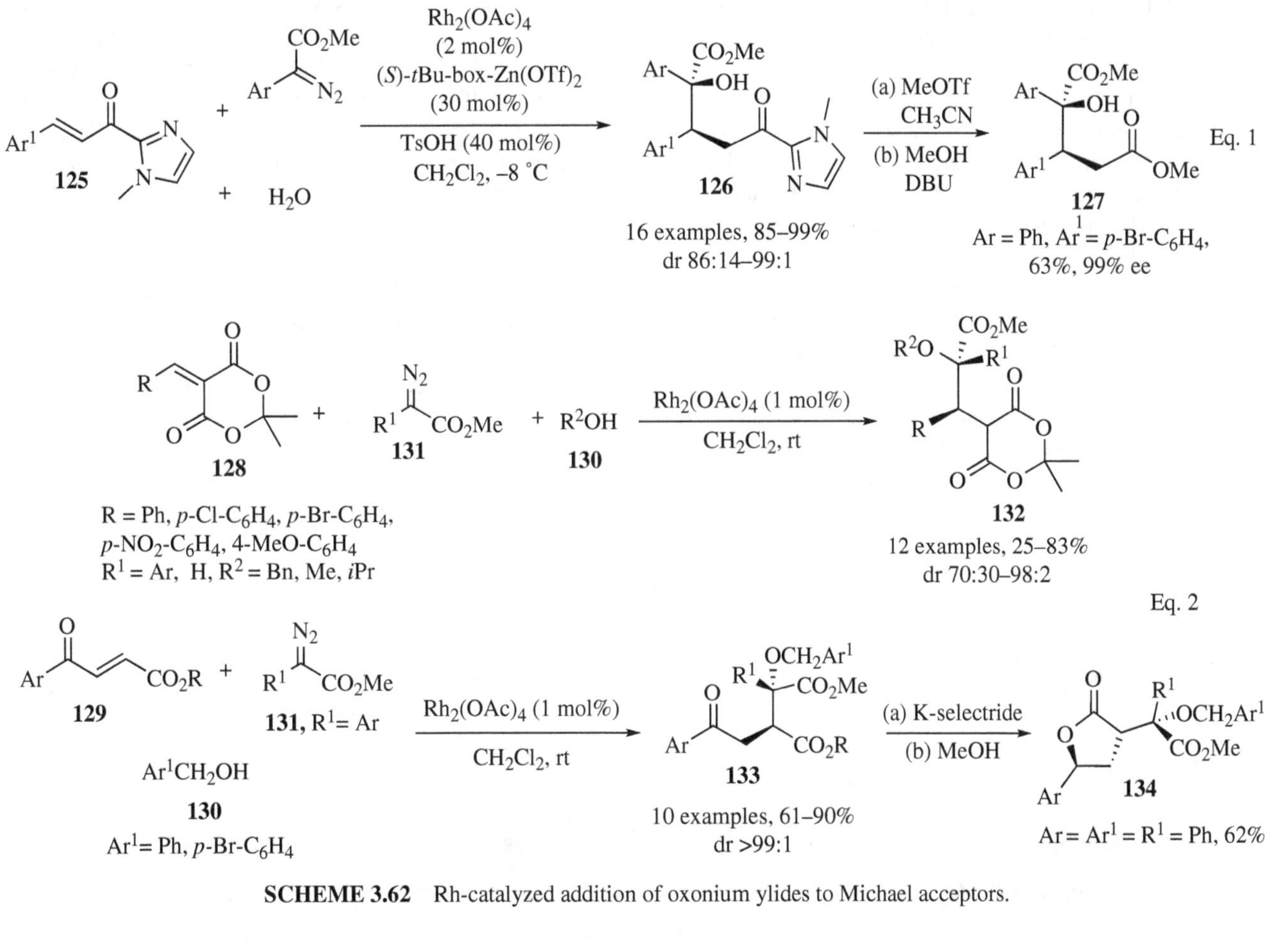

SCHEME 3.62 Rh-catalyzed addition of oxonium ylides to Michael acceptors.

With this new methodology in hands, Hu et al. [166] explored the trapping of the 1,4-addition intermediate with a different electrophile for the development of a new MCR. $Rh_2(OAc)_4$ was again the most active catalyst in the 1,4-addition/aldol-type intramolecular cascade reaction. Under the optimized reaction conditions, this three-component reaction worked well with a broad family of bifunctional substrates **135** bearing different substituents on the aryl group next to the enone moiety and a variety of alcohols **136** (Scheme 3.63). In all cases, 1-indanols **137** were obtained in 60–83% yield and with complete diastereoselectivity. Enantiopure 1-indanol was obtained employing a L-menthol-derived diazo compound. The intermolecular four-component version was also attempted, but the formation of the desired product was not observed.

3.5.1.2 Rhodium-Catalyzed Reactions via Ammonium Ylide Intermediates The chemistry of ammonium ylides, which are generated from a diazo compound and an amine in presence of a rhodium catalyst, has been less explored than that of the oxonium analogues. This is probably due to the lower stability of ammonium ylides, which are prone to undergo 1,2-proton transfer. Since Hu et al. [167] discovered that rhodium(II) acetate can promote the formation of an ammonium ylide, the subsequent trapping of which with benzaldehydes affords β-aryl-β-hydroxy α-amino acids, only a limited number of related reactions have been described.

Ammonium ylides **138**, generated from anilines and aryl-diazoacetates in the presence of a Rh(II) catalyst, can react with β,γ-unsaturated α-ketoesters to give the Michael addition/cyclization products **139** as mixtures of 4 diastereoisomers [168]. The reaction mixtures can be simplified by treatment with 20 mol% of citric acid monohydrate in toluene, affording *cis/trans* mixtures of the 2,3-dihydropyrrole derivatives **140** in moderate yields (Scheme 3.64). When diazoacetophenones **141** were used instead of aryldiazoacetates **131**, highly functionalized pyrrolidines **142** were obtained in good yields and high diastereoselectivities [168c]. Less hindered diazoacetates **143** react at the carbonyl center giving the 1,2-addition products **144** instead of the 1,4-addition [168b].

To increase the synthetic applicability of this methodology, Hu et al. decided to replace the arylamines by *N*-boc-protected amines **145** in order to develop the asymmetric version of the reaction [169]. Noteworthy, they found that the *anti*- or *syn*-diastereoselectivities can be controlled by using a different chiral phosphoric acid **146** or **147** as the cocatalyst, respectively. In both cases, *anti*-**148** and *syn*-**149** products were obtained in high yields and high enantioselectivities (Scheme 3.65). Addition of 20 mol% L-tartaric acid allowed decreasing the cocatalyst loading to 5 mol% by releasing the phosphoric acid sequestered by the diamine product.

3.5.1.3 Rhodium-Catalyzed Reactions via Zwitterionic Intermediates Diazo compounds are also known to undergo insertion into C—H bonds by action of a rhodium-based catalyst, giving rise to a zwitterionic species characterized by a similar reactivity to that of onium ylide species [170]. Recently, Hu et al. [171] have described that zwitterionic intermediates **150**, obtained by carbene insertion into a C—H bond in indoles, can be trapped by imine **151** activated by a chiral Brønsted acid. After optimization of conditions, three-component reactions carried out at −10 °C in toluene afforded the desired products **152** in high yields, >20:1 diastereoselectivities for the *anti*-isomer, and 84–99% ee (Scheme 3.66).

This methodology was also applied to other diazo compounds such as 3-diazooxindoles [172], which in the presence of $Rh_2(OAc)_4$ can be inserted into the C—H bond in indoles to afford a less reactive zwitterionic intermediates. Due to this lower reactivity, a highly activated electrophile such as ethyl glyoxylate was chosen as the trapping agent, affording the desired products in high yields and complete diastereoselectivities.

3.5.1.4 Rhodium-Catalyzed Cycloaddition Reactions of Carbonyl and Azomethine Ylides Aldehydes or imines can also react with rhodium-carbene species to generate carbonyl or azomethine ylides **153**, which can be subsequently trapped by an electrophile through a 1,3-cycloaddition reaction (Scheme 3.67).

Somfai et al. [173] described a three-component reaction for the synthesis of α-hydroxy-β-amino esters, employing

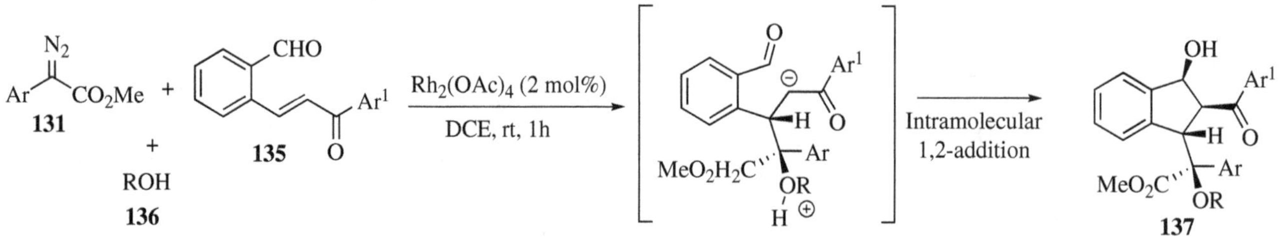

R = Bn, Me, Et, *i*Pr

17 examples, 60–83%

SCHEME 3.63 Synthesis of indanols by 1,4-addition/aldol-type intramolecular cascade reaction.

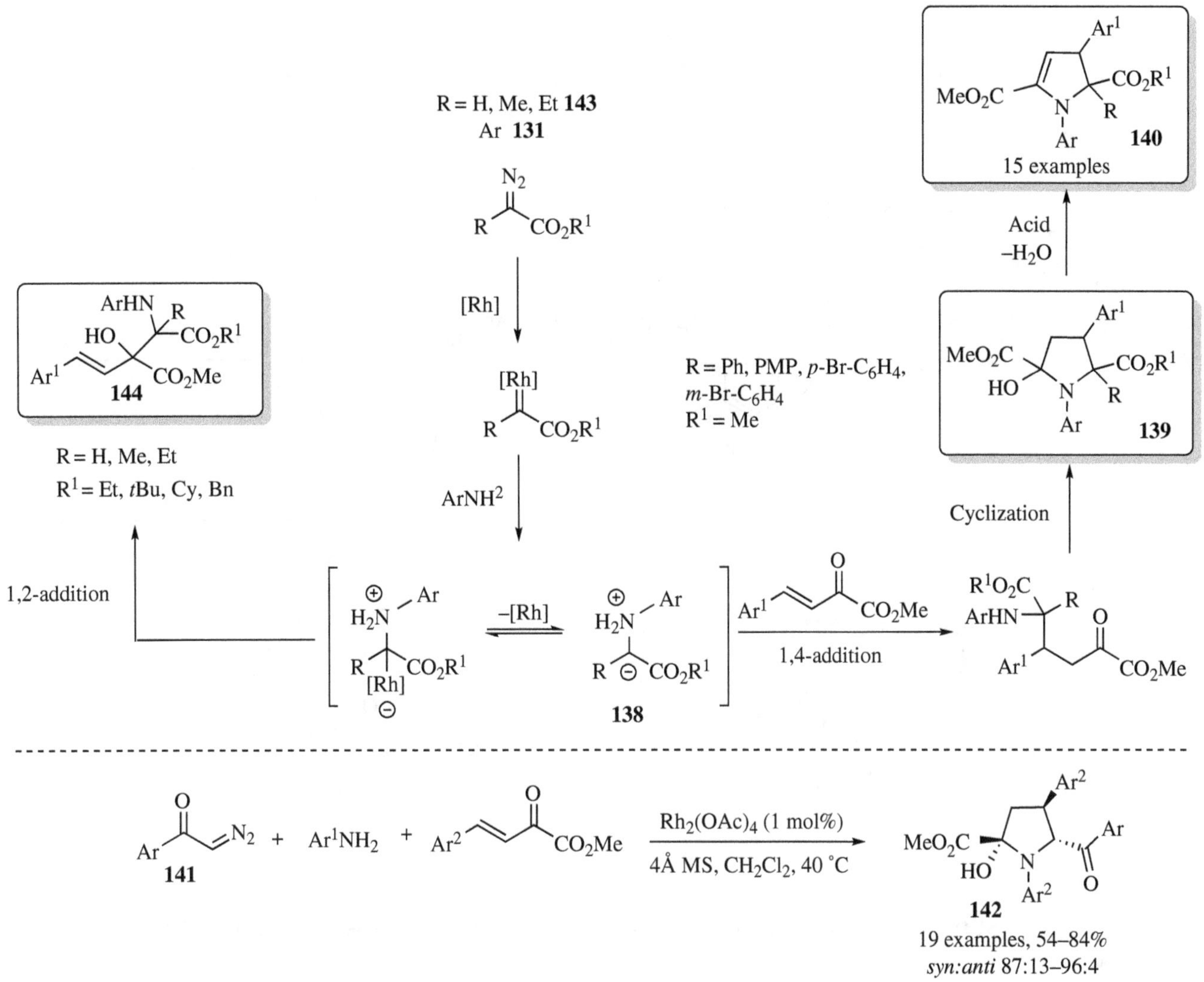

SCHEME 3.64 Rh-catalyzed 1,2- and 1,4-addition of ammonium ylides.

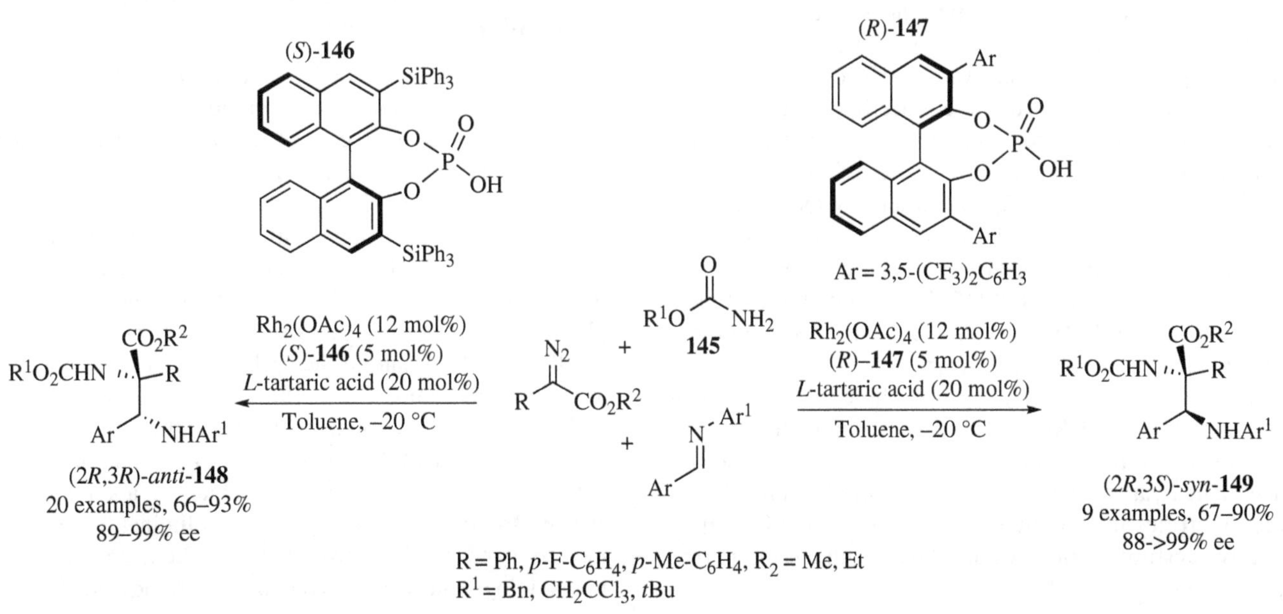

SCHEME 3.65 Asymmetric addition of ammonium ylides to imines.

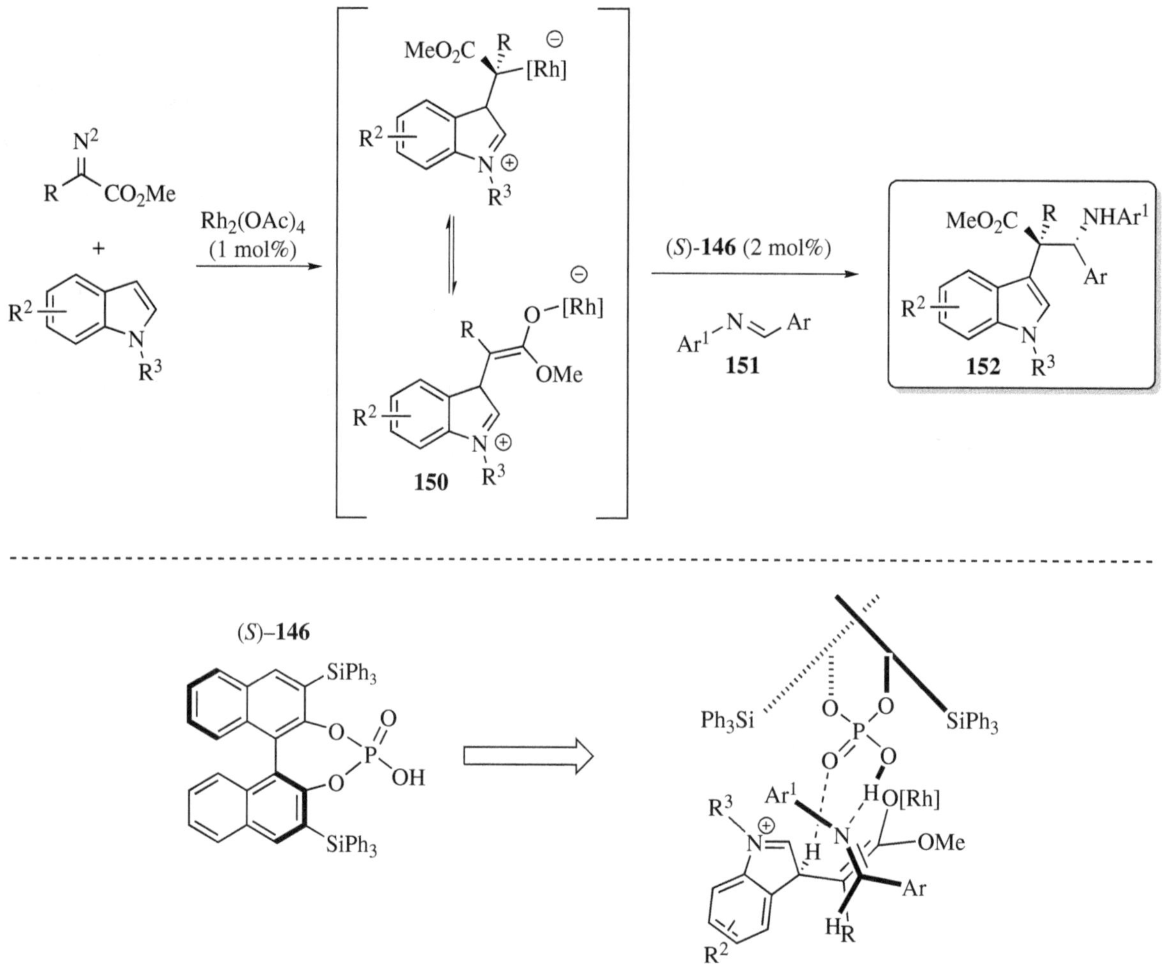

SCHEME 3.66 Rh-catalyzed asymmetric addition of zwitterionic intermediates to imines.

benzaldehyde instead of alcohol or water, *via* the 1,3-dipolar cycloaddition between carbonyl ylides and imines **154** (Scheme 3.67, Eq. 1). After acid treatment, *syn*-α-hydroxy-β-amino esters **155** were obtained in moderate to good yields and excellent diastereoselectivities. Asymmetric versions were developed using chiral auxiliaries or chiral Rh(II) catalysts, but modest enantioselectivities (er up to 82:18) were observed [173b]. Carbonyl ylide intermediates can also be trapped by other electrophiles such as enones **156**, α,β-unsaturated esters **157**, alkynes **158**, and α,β-unsaturated amides **159** [174] to afford chiral tetrahydrofuran rings **160–162** in high yields and diastereoselectivities (Scheme 3.67, Eq. 2).

Although the chemistry of azomethine ylides has been less explored in MCRs, the trapping of azomethine ylides generated from an imine and a diazo compound by acetylenedicarboxylates has been reported to afford 1,2-diarylpyrroles in a single multicomponent step [175a] or highly substituted spirooxindolyl oxazolidines and pyrrolines in good to excellent yields [175b].

3.5.2 Rhodium-Catalyzed Three-Component Cross-Addition Reactions

Aryl-rhodium species readily undergo the 1,4-addition to enones under mild conditions, although elevated temperatures are required in the case of alkynes. However, in 2006, Shinokubo et al. [176] observed that an emulsion of phenylboronic acid, 5-decyne, and methyl acrylate, in the presence of $[Rh(OH)(cod)]_2$ (4 mol%), afforded the three-component reaction product 4-butyl-5-phenylnona-2,4-dienoic acid methyl ester in 81% yield. The reaction, which only takes place as an emulsion in H_2O, can be explained by coordination of alkyne and acrylate to the aryl-Rh species **163** obtained after transmetalation of arylboronic acid, followed by addition of the aryl fragment to the carbon–carbon triple bond of alkyne and the double bond of acrylate in either a concerted or stepwise step. The obtained alkyl rhodium species **164** can undergo a β-hydride elimination to afford the observed product **167** (Scheme 3.68, Eq. 1). This methodology can be extended to other boronic acids **165** and

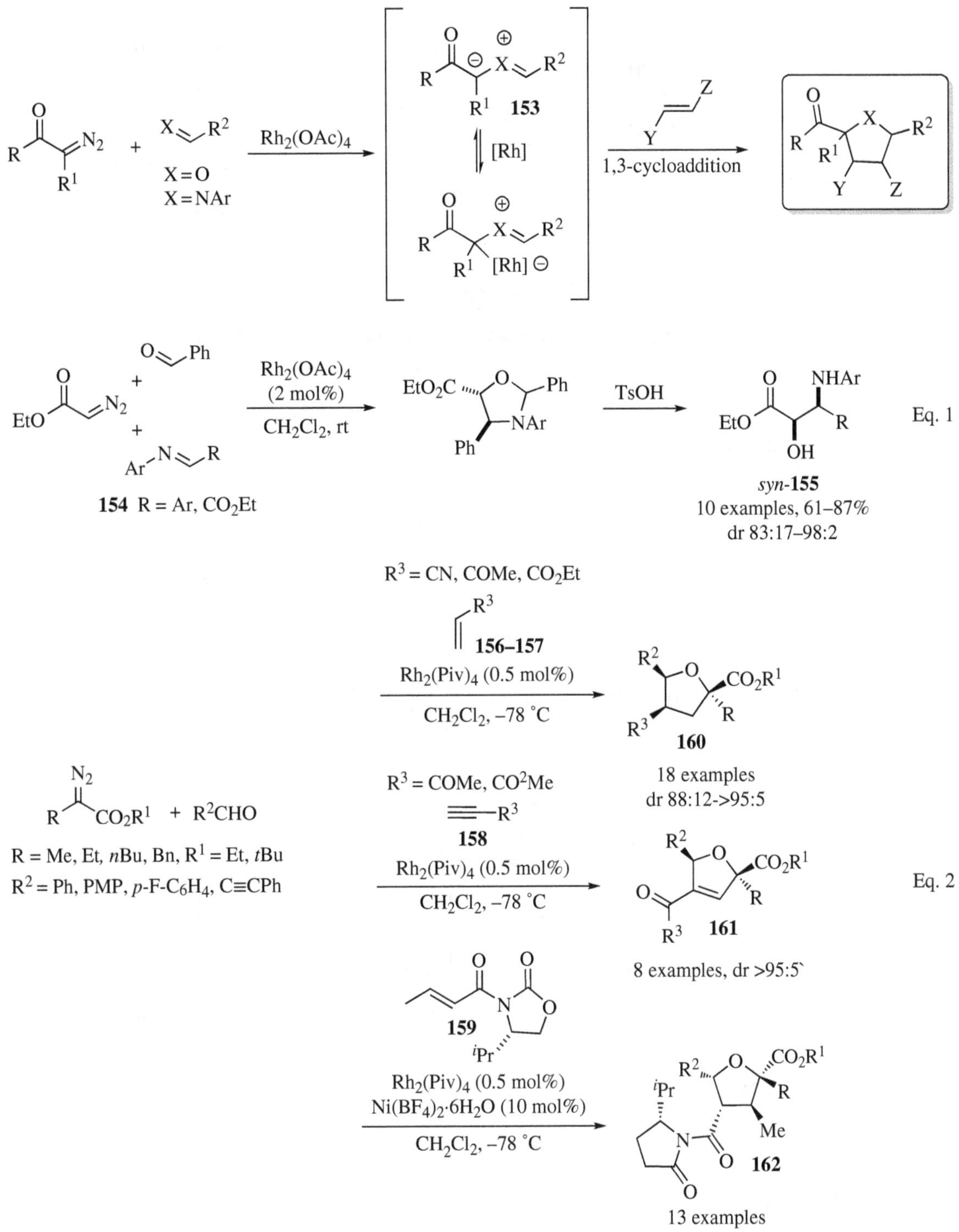

SCHEME 3.67 Rh-catalyzed 1,3-cycloaddition reactions of carbonyl and azomethine ylides.

alkynes **166** to give the three-component products **167** in moderate yields together with 1,4-addition and Heck-type compounds as by-products.

This methodology, typically catalyzed by nickel complexes (see Section 3.3.2.1), has recently been modified by Tanaka et al. [177], replacing the organic boronic acids **165** by silylacetylenes **168**. The (alkenyl)rhodium species obtained by reaction of the rhodium acetylide with alkynyl ester **169** can react with activated alkenes **170** or isocyanides **171** to give an alkyl rhodium intermediate. After suffering a protonation, the latter species affords the 1,3-enynes **172–173** in good yields and with high chemo- and regioselective

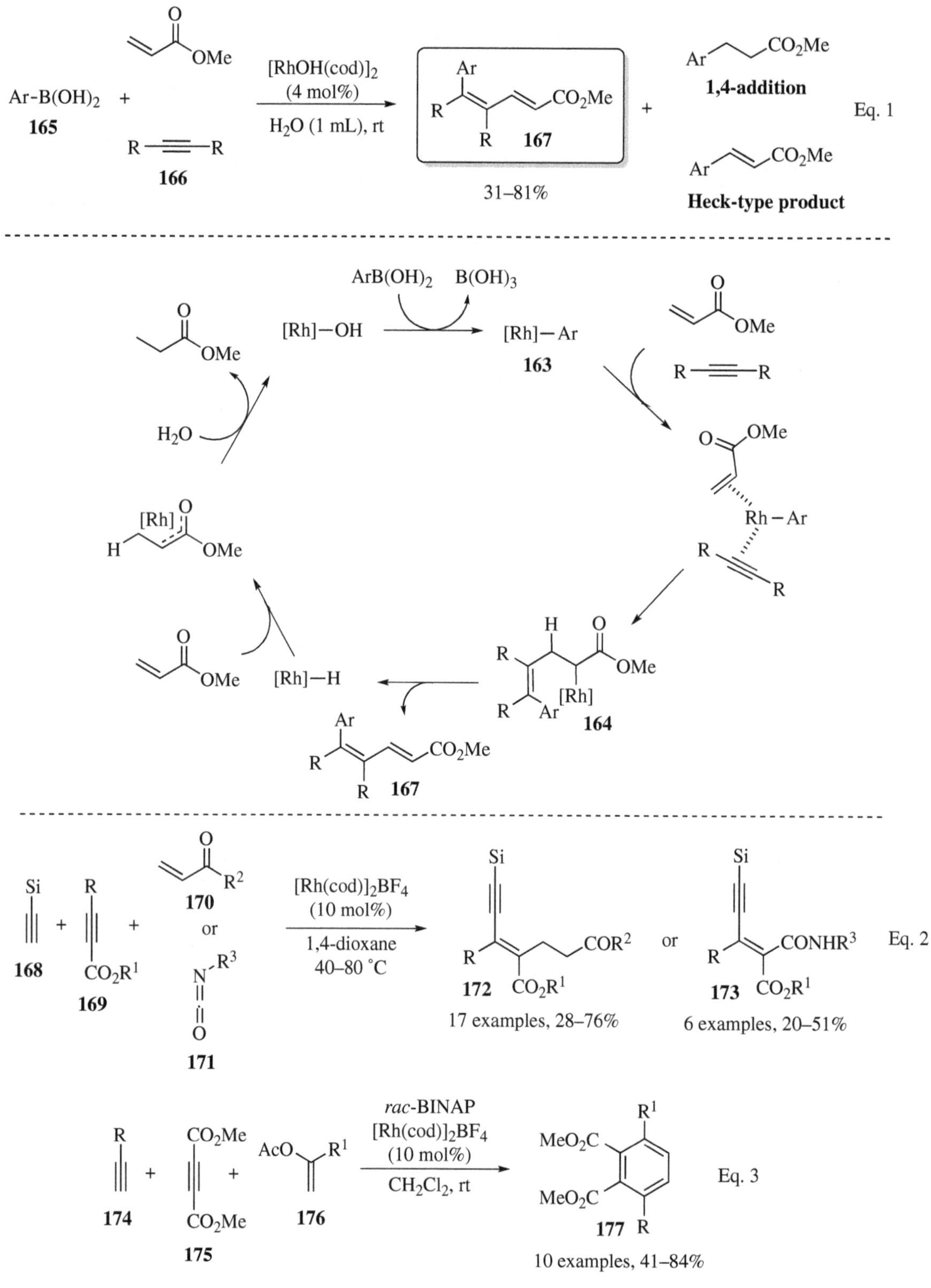

SCHEME 3.68 Rh-catalyzed three-component cross-addition reactions.

control (Scheme 3.68, Eq. 2). Following this methodology, a wide variety of synthetically useful 1,3-enynes can be obtained. This reaction was also extended to the cyclotrimerization of two different alkynes **174–175** with an alkene **176** to make substituted benzenes **177** in moderate yields and as single regioisomers (Scheme 3.68, Eq. 3) [178a] and to the cross-trimerization of two different alkenes with an alkyne, together with its asymmetric version [178b].

Very recently, Jun et al. [179] have described the synthesis of isoquinolines **181** and pyridines **182** by the $[Cp^*RhCl_2]_2$-catalyzed *N*-annulation reaction between aryl or α,β-unsaturated ketones **179–180**, alkynes **178**, and NH_4OAc, employing stoichiometric amounts of $Cu(OAc)_2$ as the oxidant and microwave irradiation. The reaction is successful with a wide family of arylketones **179** and enones **180** to afford the desired isoquinolines **181** and pyridines **182** in good to excellent yields (Scheme 3.69). A four-component reaction could be developed by *in situ* generation of enones by aldol reaction between an enolizable ketone and formaldehyde, affording the desired products in moderate yields. The proposed mechanism involves the Rh^{III}-promoted consecutive N—H and β-C—H bond activation of imine **XXXVII**, generated from ketone/enone and NH_3, followed by coordination of alkyne **178** to afford a rhodacyclic complex **XXXVIII**. Carbometalation of the alkyne in **XXXVIII** gives rise to seven-membered rhodacycle **XXXIX**. Reductive elimination gives the desired isoquinoline/pyridine **181–182** and generates a Rh^{I} species that is oxidized to the catalytically active Rh^{III} complex by $Cu(OA)_2$.

3.6 GROUP 8 METAL-CATALYZED MCRs

3.6.1 Iron-Catalyzed MCRs

Due to the increased interest for the utilization of cheap and environmentally friendly catalysts, during the last years, new variants of previously described metal-catalyzed MCRs have been developed employing iron-based catalysts.

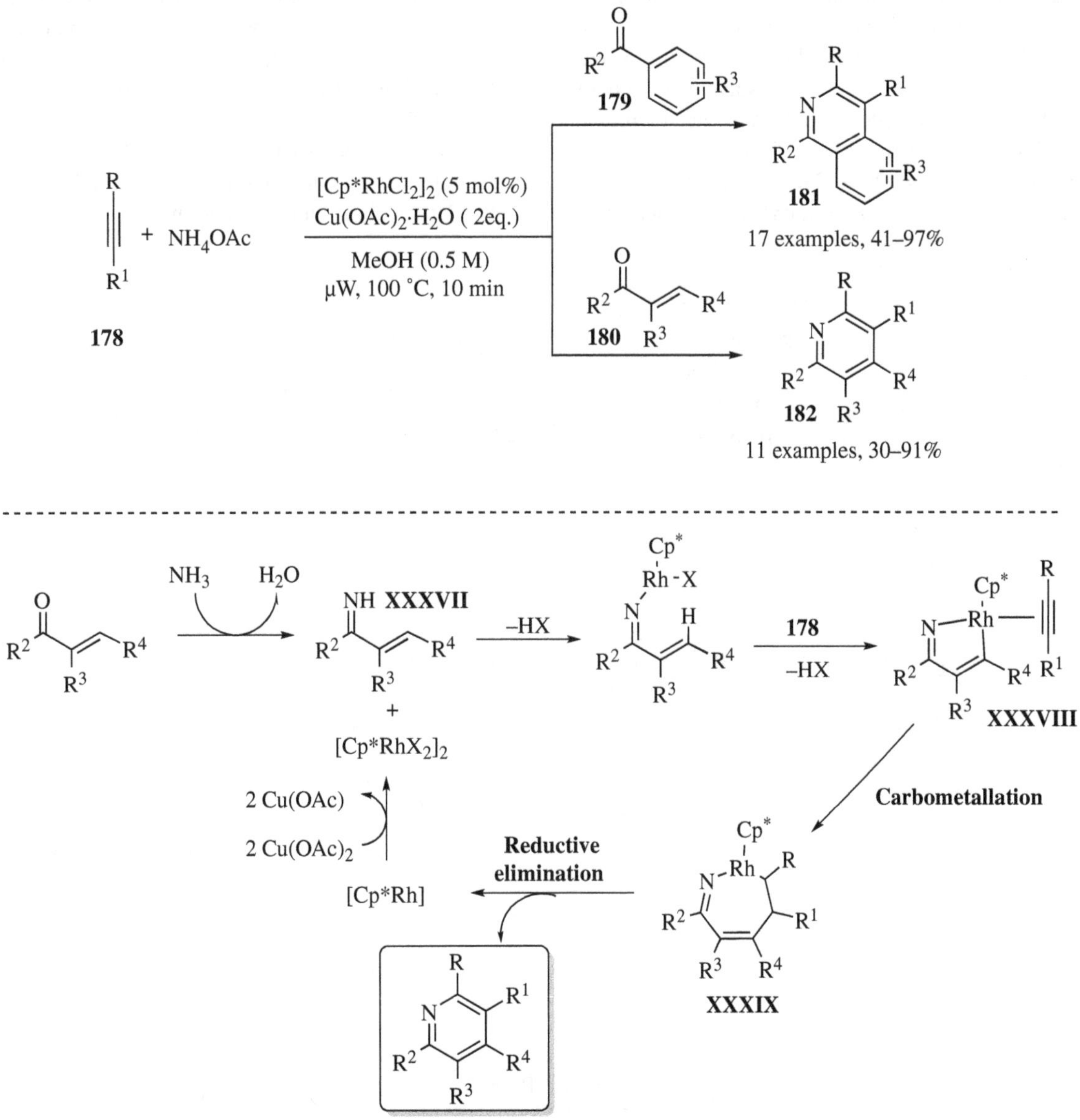

SCHEME 3.69 Synthesis of isoquinolines and pyridines by Rh-catalyzed annulation reactions.

3.6.1.1 A³-Coupling and Related Reactions Three-component coupling of aldehydes, alkynes, and amines (A^3-coupling) is a powerful methodology for the synthesis of propargylamines [88], which has been widely explored using Cu-, Ag-, or Au-based catalysts (see Section 3.4.2).

In order to develop a new version of the A^3-coupling employing a cheap and nontoxic metal, Li [180], Wang [181], and Tu [182] independently described the $FeCl_3$-catalyzed A^3-coupling reaction between aldehydes **183**, alkynes **184**, and aliphatic amines **185** to afford the corresponding propargylamines **186** in good to excellent yields (Scheme 3.70). When aromatic amines **187** are used [182], intermediate **XL** can suffer an intramolecular hydroarylation of the alkyne to give a dihydroquinoline intermediate **XLI**, which can be subsequently oxidized by air to the desired quinolines **188** in moderate to excellent yields. In a similar manner, aminoindolizine derivatives are obtained from 2-formylpyridine and 2-quinolinecarboxaldehyde [182b]. The synthesis of quinolines **188** was improved by using $Fe(OTf)_3$ as the catalyst under solvent-free conditions [182c]. Although reaction yields were not increased, catalyst loading could be reduced to 5 mol%, and reaction times were shortened from 24 to 3 h. The catalyst could be recycled at least three times while maintaining its activity.

This intramolecular hydroarylation of alkynes was also applied to the synthesis of imidazo[1,2-*a*]pyridines by A^3-coupling of 2-aminopyridine, alkynes, and aldehydes [183]. Singh et al. reported that this reaction is successfully catalyzed by a heterogeneous catalyst composed by nano-Fe_3O_4 and $KHSO_4{\cdot}SiO_2$, affording the desired imidazo[1,2-*a*]pyridines in 55–89% yields. The catalyst could be recovered by applying an external magnetic field and reused six times without significant loss of activity.

Quinoline-2,4-dicarboxylates **191** can also be synthesized by replacing alkynes **184** by dicarboxylic derivatives with acidic α-protons **190** and using more activated aldehydes such as glyoxylic esters **189** (Scheme 3.71, Eq. 1). Wang, You, et al. [184] described that $FeCl_3$ (5 mol%) can catalyze this reaction in acetonitrile and on air to afford the quinoline-2,4-dicarboxylates **191** in 31–88% yields after 30–36 h of reaction time. Quinazoline derivatives **194** can also be prepared by a solvent-free three-component methodology (Scheme 3.71, Eq. 2), starting from 2-benzoylanilines **192** and aldehydes **193** and using [BMIM][$FeCl_4$] as the catalyst [185].

It is known that pyrroles can be synthesized by the Michael reaction of β-enamino ketones or esters and nitroalkenes followed by cyclization [186], but this methodology requires the

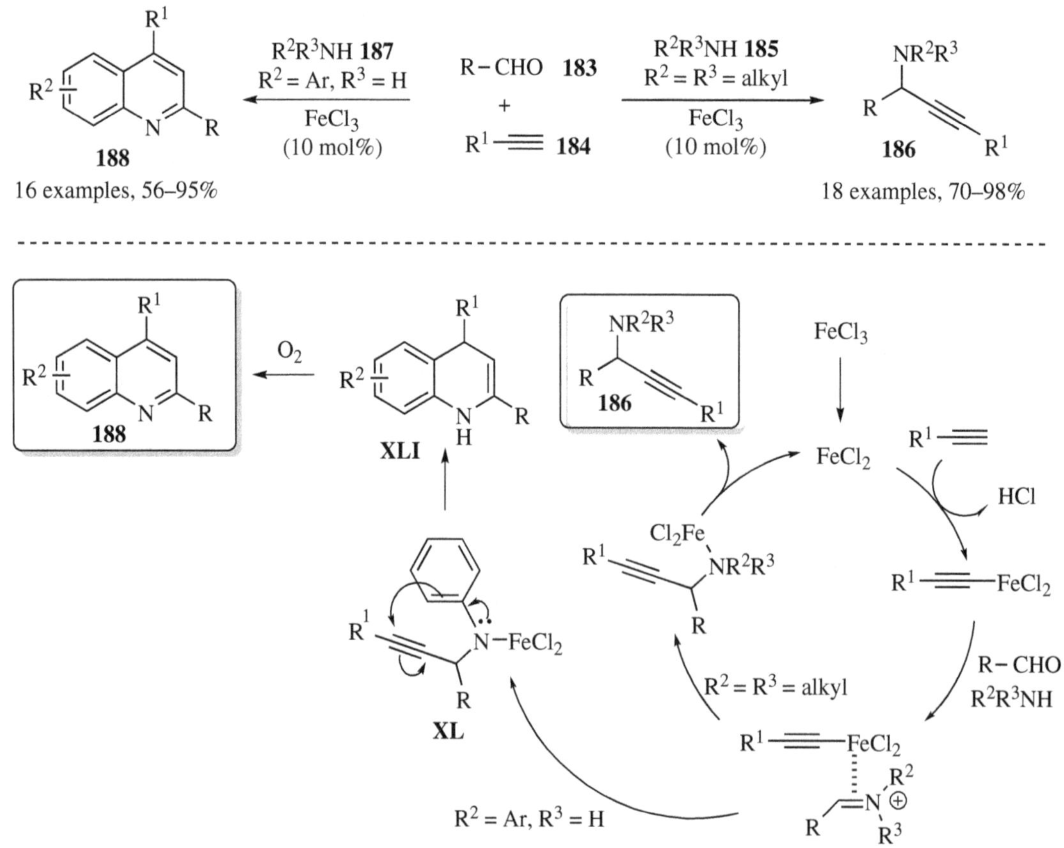

SCHEME 3.70 Iron(III)-catalyzed A^3-coupling reactions.

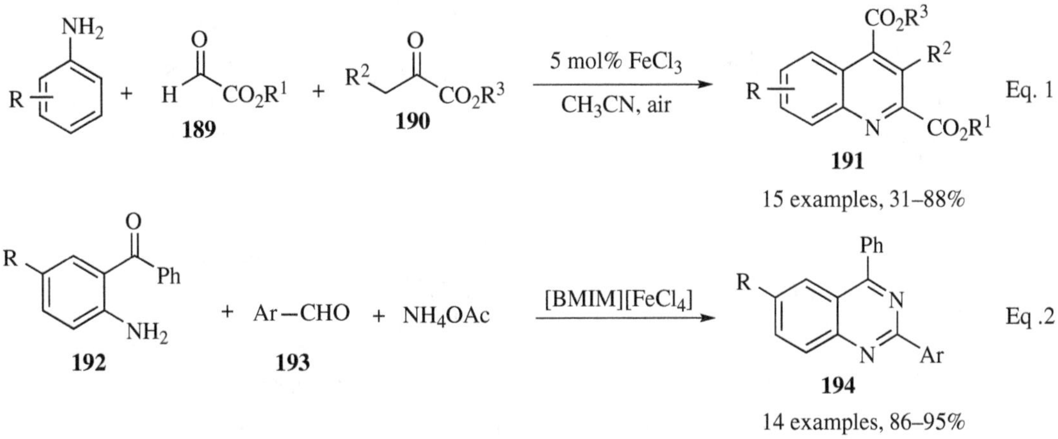

SCHEME 3.71 Fe(III)-catalyzed synthesis of quinoline-2,4-dicarboxylates and quinazolines *via* A^3-type coupling reactions.

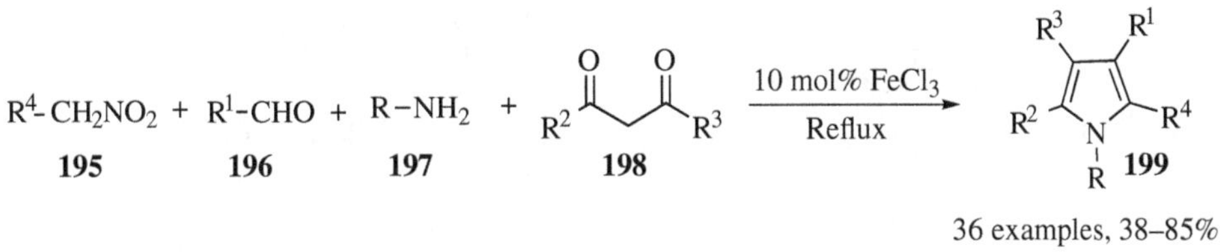

SCHEME 3.72 Four-component iron(III)-catalyzed synthesis of pyrroles.

previous synthesis of nitroalkenes and β-enaminocarbonyl derivatives. Based on this idea, Jana et al. [187] developed a four-component iron-catalyzed synthesis of pyrroles **199** from aldehydes **196**, amines **197**, 1,3-dicarbonyl compounds **198**, and nitroalkanes **195** (Scheme 3.72). The reaction is successful with a wide variety of aldehydes presenting electron-donating and electron-withdrawing groups, aliphatic and aromatic amines, and a range of 1,3-diketones and β-ketoesters. In addition, heteroaromatic motifs such as thiophene and furan were compatible with the reaction conditions.

3.6.1.2 Synthesis of Amines One of the most common strategies for the synthesis of branched amines is the 1,2-addition of a nucleophile to the previously synthesized aldimines or ketimines. Avoiding the imine synthesis, though, is highly desirable. MCR-based methodologies toward this end have been developed.

Tian et al. [188a] described the synthesis of very useful protected homoallylic amines by a four-component reaction catalyzed by the inexpensive and environmentally friendly $FeSO_4·7H_2O$. The reaction between carbonyl compounds (aldehydes, ketones, ketals, acetals) **200**, CbzCl, HMDS, and allyltrimethylsilane afforded the corresponding homoallylic amines **201** in 43–87% yields (Scheme 3.73, Eq. 1) under mild conditions. This methodology was also applied to the synthesis of *N*-protected primary and branched amines **202** by replacing the allyltrimethylsilane for Et_3SiH [188b], which gave rise to the desired products in 40–95% yields (Scheme 3.73, Eq. 2).

3.6.1.3 MCRs via Ammonium Ylide Intermediates MCRs based on the trapping of ammonium ylide intermediates with electrophiles, typically catalyzed by Rh(II) catalysts (see Section 3.5.1.2), are limited to aromatic amines with low nucleophilicity or *N*-boc-protected amines to avoid the catalyst poisoning and secondary side reactions. With the aim of solving this problem and at the same time developing cheaper catalysts, Hu, Xing, et al. [189] developed an iron-catalyzed three-component reaction between ethyldiazoacetate, aliphatic amines **203**, and β,γ-unsaturated α-ketoesters **204** employing a porphyrin–iron complex as the catalyst (Scheme 3.74). A wide range of β-hydroxy-α-amino esters **205** were obtained in good yields, with complete 1,2- versus 1,4-addition regioselectivity and moderate *syn-/anti*-stereoselectivity.

3.6.1.4 Miscellaneous Other interesting Fe-catalyzed MCRs that have been described in the last years are the alkoxy allylation of activated double bonds [190], one-pot synthesis of β-acetamido carbonyl compounds [191], and Ritter and Strecker reactions [192].

3.6.2 Ruthenium-Catalyzed MCRs

3.6.2.1 Intermolecular Alkyne Cyclotrimerization The cyclotrimerization of three different alkynes to give linear products have been described mainly using Ni-based catalysts (see Section 3.3.1) [51, 52]. Nevertheless, Yamamoto

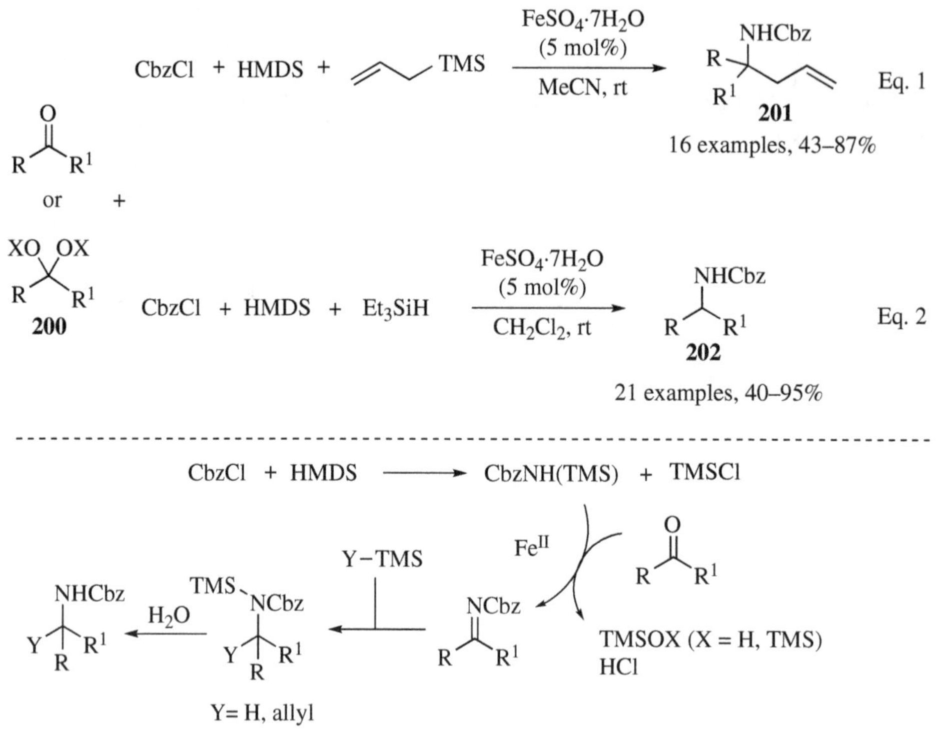

SCHEME 3.73 Four-component Fe(II)-catalyzed synthesis of branched amines.

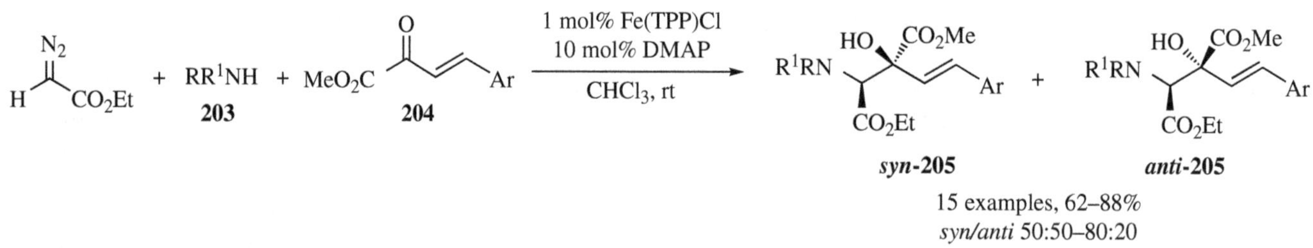

SCHEME 3.74 Iron-catalyzed 1,2-addition of ammonium ylides to β,γ-unsaturated α-ketoesters.

et al. had previously reported the Pd-catalyzed [2+2+2] regio- and chemoselective intermolecular trimerization of alkynes [193] and the Ru-catalyzed intermolecular cyclotrimerization of three unsymmetrical alkynes [194]. The latter reaction takes place with different alkynylboronates **206**, propargyl alcohols **207**, and terminal alkynes **208** to give cyclic arylboronates **209**, which can then undergo a rhodium-catalyzed Michael addition to afford **210**, be oxidized to the corresponding phenol derivative **211**, or be transformed into substituted phthalides **212** by palladium(II)-catalyzed carbonylation in moderate to good yields (Scheme 3.75).

The proposed mechanism (Scheme 3.76) starts with the oxidative cyclization of diyne **XLII** on the Cp*RuCl fragment, followed by coordination of **208** and formal [2+2] coupling. Subsequent Ru—C cleavage and reductive elimination afford the desired product **209** and regenerate the ruthenium active species. The complete regioselectivity observed in the insertion of alkyne **208** can be explained in terms of steric influence of the substituent R^2 and electronic directing effect of the electron-deficient boron center.

3.6.2.2 Synthesis of Pyrroles Due to the interest of pyrrole derivatives in the preparation of pharmaceuticals, agrochemicals, or dyes and based on *hydrogen autotransfer* reactions for the more benign amination and alkylation of alcohols [195], Beller et al. [196] have recently developed a Ru-catalyzed three-component synthesis of pyrroles starting from easily available substrates such as ketones **213**, amines **214**, and diols **215–216**. When the reactions were carried out in *t*-amyl alcohol at 130 °C, using Xantphos/[$Ru_3(CO)_{12}$] (3:1) as the catalyst and a catalytic amount of K_2CO_3, the desired 1,3-, 1,2,3-, and 1,2,3,5-substituted pyrroles **217–219** were obtained in moderate to excellent yields (Scheme 3.77). It is noteworthy that this methodology achieves complete regioselectivity when unsymmetrical diols **216** are employed (Scheme 3.77, Eq. 2). Moreover, ammonia can be used instead of amines **214**

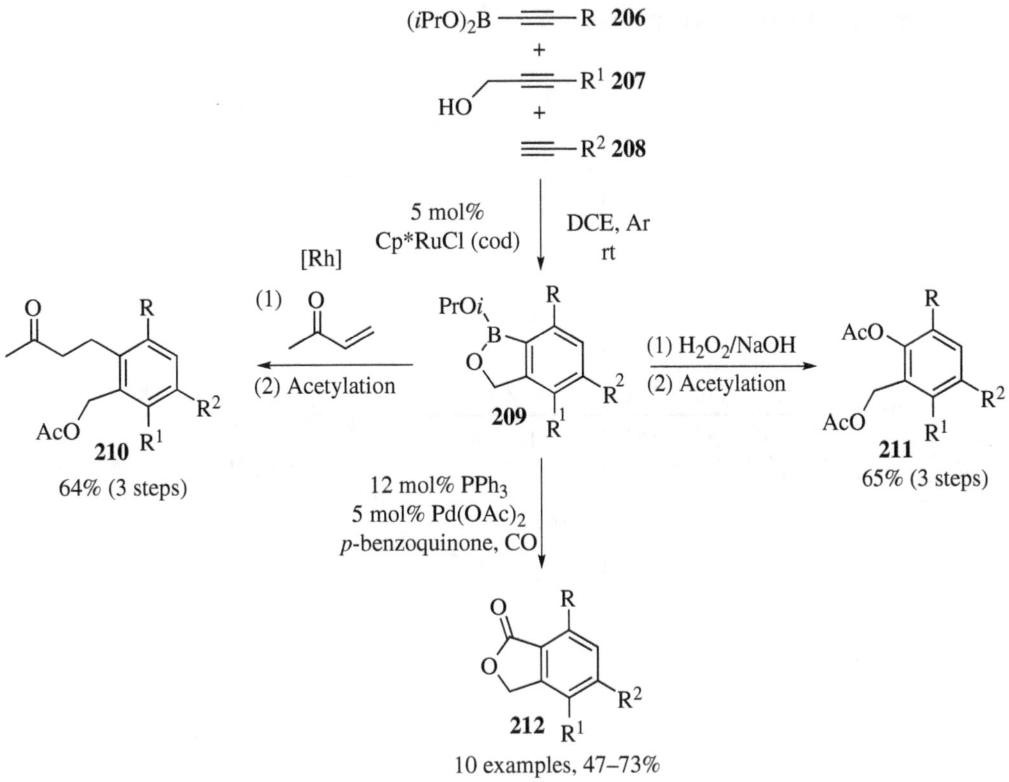

SCHEME 3.75 Ru(II)-catalyzed intermolecular cyclotrimerization of alkynes.

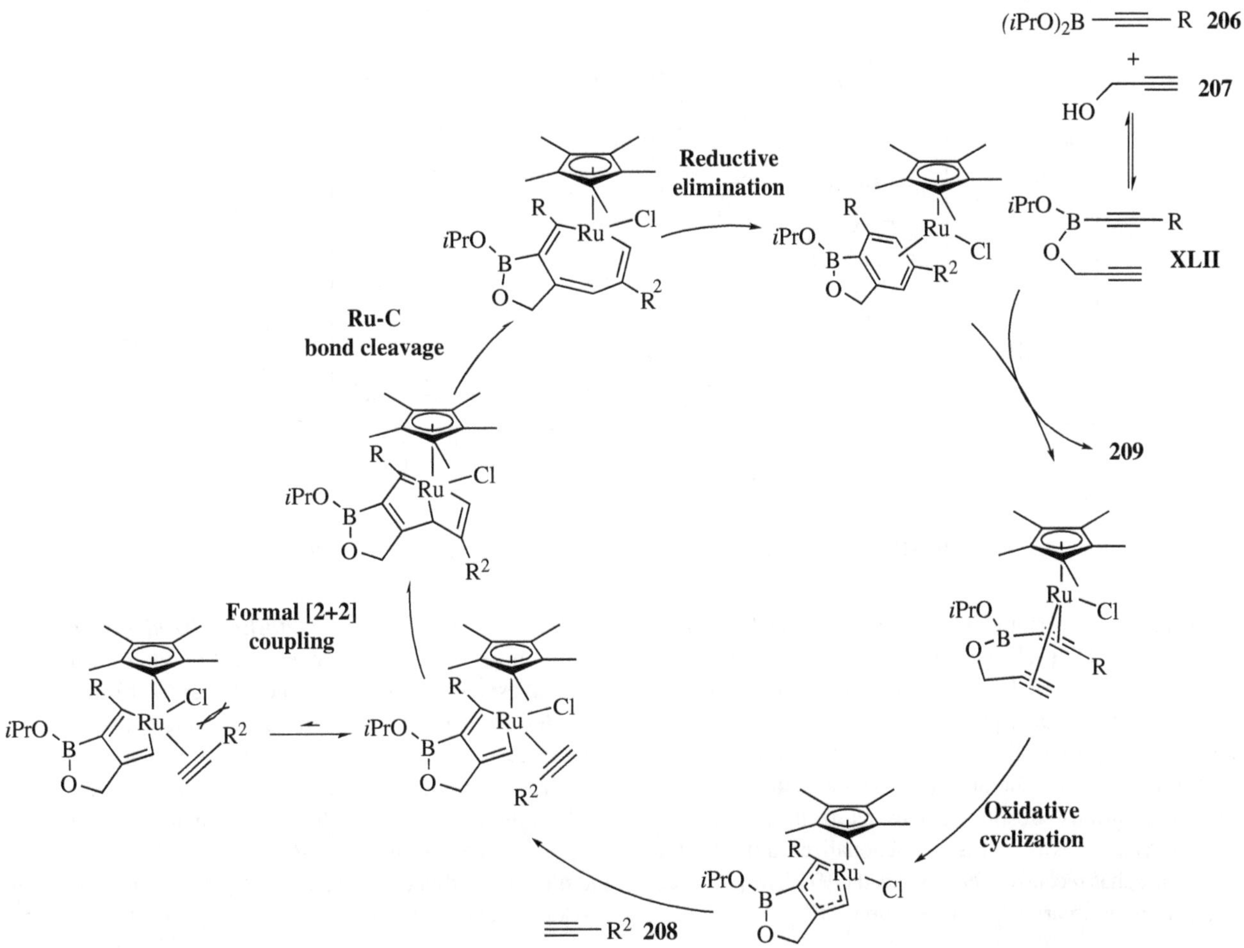

SCHEME 3.76 Proposed mechanism for the Ru(II)-catalyzed intermolecular cyclotrimerization of alkynes.

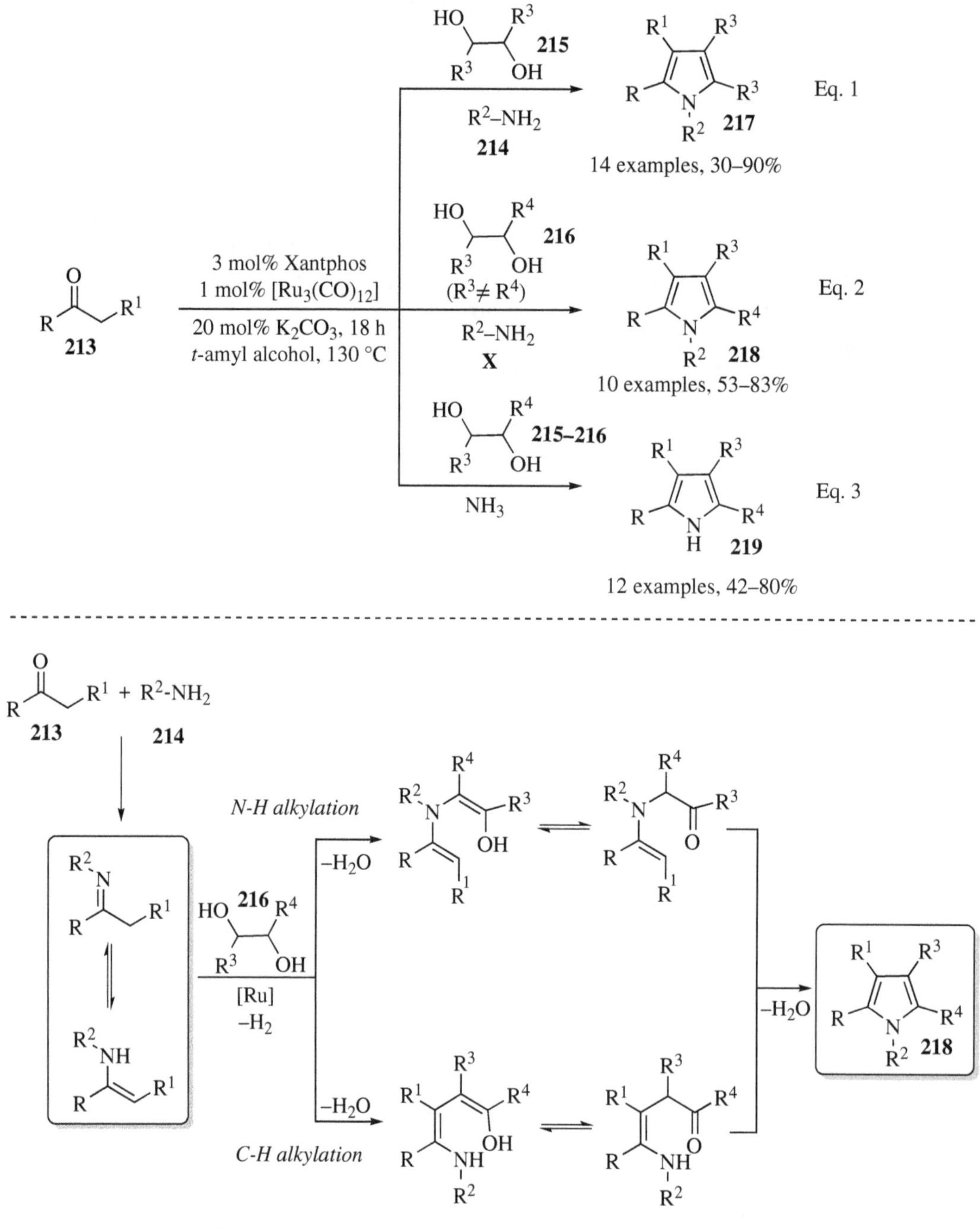

SCHEME 3.77 Ru-catalyzed three-component synthesis of pyrroles.

for the synthesis of versatile intermediates such as *N*-nonsubstituted pyrroles **219** (Scheme 3.77, Eq. 3). This methodology is restricted to benzylic ketones and sterically less hindered vicinal diols, but a subsequent modification of the reaction conditions ($[Ru(p\text{-cymene})Cl_2]_2$/Xantphos/*t*BuOK) overcame these limitations. A general and versatile method for the synthesis of various substituted pyrroles, from less reactive aryl and alkyl ketones as well as α-functionalized and activated benzylic ones, has recently been described [197]. The proposed mechanism is summarized in the Scheme 3.77.

3.6.2.3 Cross-Enyne Metathesis/Diels–Alder MCRs

Ruthenium-based complexes are known to catalyze the metathesis between alkynes and alkenes to afford 1,3-dienes. Fustero, del Pozo et al. further exploited this cross-enyne metathesis (CEYM) by trapping the 1,3-diene with a dienophile *via* an intermolecular Diels–Alder reaction [198]. Thus, the Ru-catalyzed MCR between alkynes **220–221** and 1,7-octadiene as an *in situ* source of ethylene by RCM generates a 1,3-diene that can subsequently undergo a Diels–Alder reaction with a wide variety of dienophiles **222**

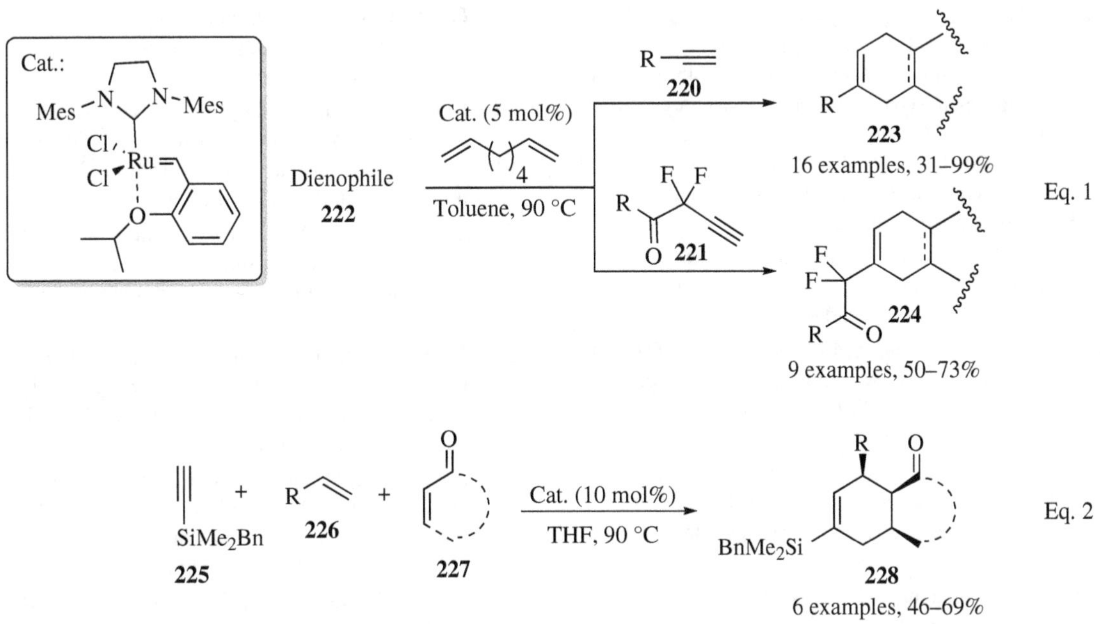

SCHEME 3.78 Ru-catalyzed cross-enyne metathesis/Diels–Alder multicomponent reactions.

(Scheme 3.78, Eq. 1). Reactions were carried out at 90 °C in toluene employing the Hoveyda–Grubbs second-generation catalyst (5 mol%) to afford the three-component cycloadducts **223** and difluorinated derivatives **224** in 31–99% and 50–73% yields, respectively. At the same time, Welker and Junker [199] reported a similar reaction employing 4-alkyl- and 4-aryl-2-silyl-1,3 dienes, generated *via* cross-enyne metathesis between silylated alkyne **225** and alkenes **226–227**, which can be *in situ* trapped in a highly stereoselective Diels–Alder reaction to afford the corresponding cycloadducts **228** in moderate yields (Scheme 3.78, Eq. 2).

3.7 CONCLUSIONS

The chemistry reviewed in this chapter illustrates how metal-catalyzed processes have clearly emerged as outstanding tools for the development of a wide variety of MCRs. The rich chemistry of transition metal complexes and their ability to catalyze multistep processes constitute inestimable platforms for reaction design, allowing to devise new methodologies toward complex scaffolds achieved in a programmed fashion. Mainly catalyzed by palladium and copper but also by nickel, iron, or ruthenium, these processes have experienced a considerable progress in the last decade, with advances ranging from the development of asymmetric versions of already known MCRs to the application of heterogeneous systems intended to preserve the catalysts.

Undoubtedly, there are still many metal-mediated multicomponent transformations waiting to be discovered in this challenging and appealing area of research.

REFERENCES

1 (a) A. Schoenberg, I. Bartoletti, R. F. Heck, *J. Org. Chem.* 1974, 39, 3318–3326. Palladium-catalyzed carboalkoxylation of aryl, benzyl, and vinylic halides. (b) A. Schoenberg, R. F. Heck, *J. Org. Chem.* 1974, 39, 3327–3331. Palladium-catalyzed amidation of aryl, heterocyclic, and vinylic halides. (c) A. Schoenberg, R. F. Heck, *J. Am. Chem. Soc.* 1974, 96, 7761–7764. Palladium-catalyzed formylation of aryl, heterocyclic, and vinylic halides.

2 A. Brennführer, H. Neumann, M. Beller, *Angew. Chem. Int. Ed.* 2009, 48, 4114–4133. Palladium-catalyzed carbonylation reactions of aryl halides and related compounds.

3 J. R. Martinelli, T. P. Clark, D. A. Watson, R. H. Munday, S. L. Buchwald, *Angew. Chem. Int. Ed.* 2007, 46, 8460–8463. Palladium-catalyzed aminocarbonylation of aryl chlorides at atmospheric pressure: the dual role of sodium phenoxide.

4 X.-F. Wu, H. Neumann, M. Beller, *Chem.–Eur. J.* 2010, 16, 9750–9753. Selective palladium-catalyzed aminocarbonylation of aryl halides with CO and ammonia.

5 X.-F. Wu, P. Anbarasan, H. Neumann, M. Beller, *Angew. Chem. Int. Ed.* 2010, 49, 7316–7319. Palladium-catalyzed carbonylative C—H activation of heteroarenes.

6 X.-F. Wu, H. Neumann, M. Beller, *Chem.–Eur. J.* 2010, 16, 12104–12107. A general and convenient palladium-catalyzed carbonylative sonogashira coupling of aryl bromides.

7 (a) X.-F. Wu, H. Neumann, A. Spannenberg, T. Schulz, H. Jiao, M. Beller, *J. Am. Chem. Soc.* 2010, 132, 14596–14602. Development of a general palladium-catalyzed carbonylative Heck reaction of aryl halides. (b) X.-F. Wu, H. Jiao, H. Neumann, M. Beller, *ChemCatChem* 2011, 3, 726–733. Progress in carbonylative-Heck reactions of aryl bromides: catalysis and DFT studies.

8 (a) S. Klaus, H. Neumann, A. Zapf, D. Strübing, S. Hübner, J. Almena, T. Riermeier, P. Groß, M. Sarich, W.-R. Krahnert, K. Rossen, M. Beller, *Angew. Chem. Int. Ed.* 2006, 45, 154–158. A general and efficient method for the formylation of aryl and heteroaryl bromides. (b) A. G. Sergeev, A. Spannenberg, M. Beller, *J. Am. Chem. Soc.* 2008, 130, 15549–15563. Palladium-catalyzed formylation of aryl bromides: elucidation of the catalytic cycle of an industrially applied coupling reaction.

9 J. S. Quesnel, B. A. Arndtsen, *Pure Appl. Chem.* 2013, 85, 377–384. Transition-metal-catalyzed multicomponent coupling reactions with imines and carbon monoxide.

10 R. Dhawan, B. A. Arndtsen, *J. Am. Chem. Soc.* 2004, 126, 468–469. Palladium-catalyzed multicomponent coupling of alkynes, imines, and acid chlorides: a direct and modular approach to pyrrole synthesis.

11 A. R. Siamaki, B. A. Arndtsen, *J. Am. Chem. Soc.* 2006, 128, 6050–6051. A direct, one step synthesis of imidazoles from imines and acid chlorides: a palladium catalyzed multicomponent coupling approach.

12 R. Dhawan, R. D. Dghaym, D. J. St. Cyr, B. A. Arndtsen, *Org. Lett.* 2006, 8, 3927–3930. Direct, palladium-catalyzed, multicomponent synthesis of β-lactams from imines, acid chloride, and carbon monoxide.

13 K. Worrall, B. Xu, S. Bontemps, B. A. Arndtsen, *J. Org. Chem.* 2011, 76, 170–180. A palladium-catalyzed multicomponent synthesis of imidazolinium salts and imidazolines from imines, acid chlorides, and carbon monoxide.

14 S. Bontemps, J. S. Quesnel, K. Worrall, B. A. Arndtsen, *Angew. Chem. Int. Ed.* 2011, 50, 8948–8951. Palladium-catalyzed aryl iodide carbonylation as a route to imidazoline synthesis: design of a five-component coupling reaction.

15 A. R. Siamaki, M. Sakalauskas, B. A. Arndtsen, *Angew. Chem. Int. Ed.* 2011, 50, 6552–6556. A palladium-catalyzed multicomponent coupling approach to π-conjugated oligomers: assembling imidazole-based materials from imines and acyl chlorides.

16 (a) A. Dömling, I. Ugi, *Angew. Chem. Int. Ed.* 2000, 39, 3168–3210. Multicomponent reactions with isocyanides. (b) J. Zhu, *Eur. J. Org. Chem.* 2003, 1133–1144. Recent developments in the isonitrile-based multicomponent synthesis of heterocycles. (c) H. Eckert, *Molecules* 2012, 17, 1074–1102. Diversity oriented syntheses of conventional heterocycles by smart multi component reactions (MCRs) of the last decade.

17 C. G. Saluste, R. J. Whitby, M. Furber, *Angew. Chem. Int. Ed.* 2000, 39, 4156–4158. A palladium-catalyzed synthesis of amidines from aryl halides.

18 (a) S. Lang, *Chem. Soc. Rev.* 2013, 42, 4867–4880. Unravelling the labyrinth of palladium-catalysed reactions involving isocyanides. (b) T. Vlaar, E. Ruijter, B. U. W. Maes, R. V. A. Orru, *Angew. Chem. Int. Ed.* 2013, 52, 7084–7097. Palladium-catalyzed migratory insertion of isocyanides: an emerging platform in cross-coupling chemistry.

19 P. J. Boissarie, Z. E. Hamilton, S. Lang, J. A. Murphy, C. J. Suckling, *Org. Lett.* 2011, 13, 6256–6259. A powerful palladium-catalyzed multicomponent process for the preparation of oxazolines and benzoxazoles.

20 G. Qiu, G. Liu, S. Pu, J. Wu, *Chem. Commun.* 2012, 48, 2903–2905. A palladium-catalyzed three-component reaction for the preparation of quinazolin-4(3*H*)-imines.

21 G. Qiu, Y. Lu, J. Wu, *Org. Biomol. Chem.* 2013, 11, 798–802. A concise synthesis of 4-imino-3,4-dihydroquinazolin-2-ylphosphonates *via* a palladium-catalyzed reaction of carbodiimide, isocyanide, and phosphite.

22 G. Qiu, Y. He, J. Wu, *Chem. Commun.* 2012, 48, 3836–3838. Preparation of quinazolino[3,2-*a*]quinazolines *via* a palladium-catalyzed three-component reaction of carbodiimide, isocyanide, and amine.

23 T. Vlaar, E. Ruijter, A. Znabet, E. Janssen, F. J. J. de Kanter, B. U. W. Maes, R. V. A. Orru, *Org. Lett.* 2011, 13, 6496–6499. Palladium-catalyzed synthesis of 4-aminophthalazin-1(2*H*)-ones by isocyanide insertion.

24 B. Liu, Y. Li, M. Yin, W. Wu, H. Jiang, *Chem. Commun.* 2012, 48, 11446–11448. Palladium-catalyzed tandem reaction of *o*-aminophenols, bromoalkynes and isocyanides to give 4-amine-benzo[*b*][1,4]oxazepines.

25 F. Ji, M.-F. Lv, W.-B. Yi, C. Cai, *Adv. Synth. Catal.* 2013, 355, 3401–3406. Synthesis of 1,4-benzoxazepine derivatives *via* a novel domino aziridine ring-opening and isocyanide-insertion reaction.

26 (a) M. Jeganmohan, C.-H. Cheng, *Chem. Commun.* 2008, 3101–3117. Transition metal-catalyzed three-component coupling of allenes and the related allylation reactions. (b) G. Balme, D. Bouyssi, N. Monteiro, *Pure Appl. Chem.* 2006, 78, 231–239. Palladium-mediated cascade or multicomponent reactions: a new route to carbo- and heterocyclic compounds. (c) T. Vlaar, E. Ruijter, R. V. A. Orru, *Adv. Synth. Catal.* 2011, 353, 809–841. Recent advances in palladium-catalyzed cascade cyclizations. (d) G. Balme, D. Bouyssi, N. Monteiro in *Multicomponent reactions* (J. Zhu, H. Bienaymé, Eds.). Wiley-VCH Verlag GmbH, Weinheim, Germany, 2005, pp. 224–276. Metal-catalyzed multicomponent reactions.

27 (a) C. Zhou, D. E. Emrich, R. C. Larock, *Org. Lett.* 2003, 5, 1579–1582. An efficient, regio- and stereoselective palladium-catalyzed route to tetrasubstituted olefins. (b) C. Zhou, R. C. Larock, *J. Org. Chem.* 2005, 70, 3765–3777. Regio- and stereoselective route to tetrasubstituted olefins by the palladium-catalyzed three-component coupling of aryl iodides, internal alkynes, and arylboronic acids.

28 X. Zhang, R. C. Larock, *Org. Lett.* 2003, 5, 2993–2996. Synthesis of highly substituted 1,3-dienes and 1,3,5-trienes by the palladium-catalyzed coupling of vinylic halides, internal alkynes, and organoboranes.

29 T. Satoh, S. Ogino, M. Miura, M. Nomura, *Angew. Chem. Int. Ed.* 2004, 43, 5063–5065. Synthesis of highly substituted 1,3-butadienes by palladium-catalyzed arylation of internal alkynes.

30 (a) K. Shibata, T. Satoh, M. Miura, *Org. Lett.* 2005, 7, 1781–1783. Palladium-catalyzed intermolecular three component coupling of aryl iodides, alkynes, and alkenes to produce 1,3-butadiene derivatives. (b) K. Shibata, T. Satoh, M. Miura, *Adv. Synth. Catal.* 2007, 349, 2317–2325. Palladium-catalyzed intermolecular three-component coupling of organic halides with alkynes and alkenes: efficient synthesis of oligoene compounds.

31 H. Horiguchi, K. Hirano, T. Satoh, M. Miura, *Adv. Synth. Catal.* 2009, 351, 1431–1436. Palladium-catalyzed three-component 1:2:1 coupling of aryl iodides, alkynes, and alkenes to produce 1,3,5-hexatriene derivatives.

32 Y. Wen, L. Huang, H. Jiang, *J. Org. Chem.* 2012, 77, 5418–5422. Access to $C(sp^3)$–$C(sp^2)$ and $C(sp^2)$–$C(sp^2)$ bond formation *via* sequential intermolecular carbopalladation of multiple carbon–carbon bonds.

33 N. Sakai, R. Komatsu, N. Uchida, R. Ikeda, T. Konakahara, *Org. Lett.* 2010, 12, 1300–1303. A single-step synthesis of enynes: Pd-catalyzed arylalkynylation of aryl iodides, internal alkynes, and alkynylsilanes.

34 (a) E. Yoshikawa, K. V. Radhakrishnan, Y. Yamamoto, *J. Am. Chem. Soc.* 2000, 122, 7280–7286. Palladium-catalyzed controlled carbopalladation of benzyne. (b) E. Yoshikawa, Y. Yamamoto, *Angew. Chem. Int. Ed.* 2000, 39, 173–175. Palladium-catalyzed intermolecular controlled insertion of benzyne-benzyne-alkene and benzyne-alkyne-alkene-synthesis of phenanthrene and naphthalene derivatives.

35 M. Jeganmohan, C.-H. Cheng, *Org. Lett.* 2004, 6, 2821–2824. Palladium-catalyzed allylalkynylation of benzynes: a highly efficient route to substituted 1-allyl-2-alkynylbenzenes.

36 T. T. Jayanth, M. Jeganmohan, C.-H. Cheng, *Org. Lett.* 2005, 7, 2921–2924. Highly efficient route to *o*-allylbiaryls *via* palladium-catalyzed three-component coupling of benzynes, allylic halides, and aryl organometallic reagents.

37 S. Bhuvaneswari, M. Jeganmohan, C.-H. Cheng, *Org. Lett.* 2006, 8, 5581–5584. Carbocyclization of aromatic iodides, bicyclic alkenes, and benzynes involving a palladium-catalyzed C–H bond activation as a key step.

38 Z. Liu, R. C. Larock, *Angew. Chem. Int. Ed.* 2007, 46, 2535–2538. Palladium-catalyzed, sequential, three-component cross-coupling of aryl halides, alkynes, and arynes.

39 J. L. Henderson, A. S. Edwards, M. F. Greaney, *J. Am. Chem. Soc.* 2006, 128, 7426–7427. Three-component coupling of benzyne: domino intermolecular carbopalladation.

40 J. L. Henderson, A. S. Edwards, M. F. Greaney, *Org. Lett.* 2007, 9, 5589–5592. Biaryl synthesis *via* palladium-catalyzed aryne multicomponent coupling.

41 (a) B. Nguyen, E. J. Emmett, M. C. Willis, *J. Am. Chem. Soc.* 2010, 132, 16372–16373. Palladium-catalyzed aminosulfonylation of aryl halides. (b) E. J. Emmett, C. S. Richards-Taylor, B. Nguyen, A. Garcia-Rubia, B. R. Hayter, M. C. Willis, *Org. Biomol. Chem.* 2012, 10, 4007–4014. Palladium-catalysed aminosulfonylation of aryl-, alkenyl- and heteroaryl halides: scope of the three-component synthesis of *N*-aminosulfonamides.

42 S. Ye, J. Wu, *Chem. Commun.* 2012, 48, 7753–7755. A palladium-catalyzed three-component coupling of arylboronic acids, sulfur dioxide and hydrazines.

43 S. Ye, J. Wu, *Chem. Commun.* 2012, 48, 10037–10039. A palladium-catalyzed reaction of aryl halides, potassium metabisulfite, and hydrazines.

44 D. Bouyssi, N. Monteiro, G. Balme, *Beilstein J. Org. Chem.* 2011, 7, 1387–1406. Amines as key building blocks in Pd-assisted multicomponent processes.

45 J. Barluenga, A. Jiménez-Aquino, C. Valdés, F. Aznar, *Angew. Chem. Int. Ed.* 2007, 46, 1529–1532. The azaallylic anion as a synthon for Pd-catalyzed synthesis of heterocycles: domino two- and three-component synthesis of indoles.

46 Y. Liang, T. Meng, H.-J. Zhang, Z. Xi, *Synlett* 2011, 911–914. Palladium-catalyzed, one-pot, three-component approach to α-alkynyl indoles from *o*-bromo-(2,2-dibromovinyl)benzenes, terminal alkynes and arylamines.

47 T. Dahl, C. W. Tornøe, B. Bang-Andersen, P. Nielsen, M. Jørgensen, *Angew. Chem. Int. Ed.* 2008, 47, 1726–1728. Palladium-catalyzed three-component approach to promazine with formation of one carbon–sulfur and two carbon–nitrogen bonds.

48 J. Barluenga, A. Mendoza, F. Rodríguez, F. J. Fañanás, *Angew. Chem. Int. Ed.* 2009, 48, 1644–1647. A palladium(II)-catalyzed synthesis of spiroacetals through a one-pot multicomponent cascade reaction.

49 S. Lamande-Langle, M. Abarbri, J. Thibonnet, A. Duchêne, J.-L. Parrain, *Chem. Commun.* 2010, 46, 5157–5159. Domino allylic amination/Sonogashira/heterocyclisation reactions: palladium-catalysed three-component synthesis of pyrroles.

50 K. Ogata, H. Murayama, J. Sugasawa, N. Suzuki, S.-I. Fukuzawa, *J. Am. Chem. Soc.* 2009, 131, 3176–3177. Nickel-catalyzed highly regio- and stereoselective cross-trimerization between triisopropylsilylacetylene and internal alkynes leading to 1,3-diene-5-ynes.

51 K. Ogata, J. Sugasawa, S.-I. Fukuzawa, *Angew. Chem. Int. Ed.* 2009, 48, 6078–6080. Highly chemoselective nickel-catalyzed three-component cross-trimerization of three distinct alkynes leading to 1,3-dien-5-ynes.

52 K. Ogata, Y. Atsuumi, S.-I. Fukuzawa, *Org. Lett.* 2011, 13, 122–125. Highly chemoselective nickel-catalyzed three-component cross trimerization between two distinct terminal alkynes and an internal alkyne.

53 K. Ogata, J. Sugasawa, Y. Atsuumi, S.-I. Fukuzawa, *Org. Lett.* 2010, 12, 148–151. Nickel-catalyzed highly chemo-, regio- and stereoselective three component reaction of norbornene with two alkynes.

54 (a) T. T. Jayanth, C.-H. Cheng, *Angew. Chem. Int. Ed.* 2007, 46, 5921–5924. Nickel-catalyzed coupling of arynes, alkenes and boronic acids: dual role of the boronic acid. (b) C.-M. Yang, M. Jeganmohan, K. Parthasarathy, C.-H. Cheng, *Org. Lett.* 2010, 12, 3610–3613. Highly selective nickel-catalyzed three-component coupling of alkynes with enones and alkenyl boronic acids: a novel route to substituted 1,3-dienes. (c) S. Mannathan, C.-H. Cheng, *Chem. Commun.* 2013, 49, 1557–1559. Nickel-catalyzed regio- and diastereoselective intermolecular three-component coupling of oxabicyclic alkenes with alkynes and organoboronic acids. (d)S. Mannathan, M. Jeganmohan, C.-H. Cheng, *Angew. Chem. Int. Ed.* 2009, 48, 2192–2195. Nickel-catalyzed borylative coupling of alkynes, enones, and bis(pinacolato)diboron as a route to substituted alkenyl boronates.

55 (a) S.-S. Ng, T. F. Jamison, *J. Am. Chem. Soc.* 2005, 127, 14194–14195. Simple alkenes as substitutes for organometallic reagents: nickel-catalyzed, intermolecular coupling of aldehydes, silyl triflates and alpha olefins. (b) C.-Y. Ho, S.-S.

Ng, T. F. Jamison, *J. Am. Chem. Soc.* 2006, 128, 5362–5363. Nickel-catalyzed, carbonyl-ene-type reactions: selective for alpha olefins and more efficient with electron-rich aldehydes. (c) C.-Y. Ho, T. F. Jamison, *Angew. Chem. Int. Ed.* 2007, 46, 782–785. Highly selective coupling of alkenes and aldehydes catalyzed by [Ni(NHC){$P(OPh)_3$}]: synergy between a strong σ donor and a strong π acceptor.

56 (a) A. Herath, J. Montgomery, *J. Am. Chem. Soc.* 2008, 130, 8132–8133. Highly chemoselective and stereoselective synthesis of *Z*-enol silanes. (b) H. A. Malik, G. J. Sormunen, J. Montgomery, *J. Am. Chem. Soc.* 2010, 132, 6304–6305. A general strategy for regiocontrol in nickel-catalyzed reductive couplings of aldehydes and alkynes. (c) K. M. Partridge, S. J. Bader, Z. A. Buchan, C. E. Taylor, J. Montgomery, *Angew. Chem. Int. Ed.* 2013, 52, 13647–13650. A streamlined strategy for aglycone assembly and glycosylation. (d) A.-R. Shareef, D. H. Sherman, J. Montgomery, *Chem. Sci.* 2012, 3, 892–895. Nickel-catalyzed regiodivergent approach to macrolide motifs.

57 N. Saito, T. Katayama, Y. Sato, *Org. Lett.* 2008, 10, 3829–3832. Nickel-catalyzed highly regioselective multicomponent coupling of ynamides, aldehydes and silane: a new access to functionalized enamides.

58 K. Ogata, Y. Atsuumi, S.-I. Fukuzawa, *Org. Lett.* 2010, 12, 4536–4539. Nickel-catalyzed ring-opening three-component coupling of methylenecyclopropane with aldehydes and silanes.

59 Y. Sato, Y. Hinata, R. Seki, Y. Oonishi, N. Saito, *Org. Lett.* 2007, 9, 5597–5599. Nickel-catalyzed enantio- and diastereoselective three-component coupling of 1,3-dienes, aldehydes, and silanes using chiral N-heterocyclic carbenes as ligands.

60 A. Herath, W. Li, J. Montgomery, *J. Am. Chem. Soc.* 2008, 130, 469–471. Fully intermolecular nickel-catalyzed three-component couplings *via* internal redox.

61 (a) R. Shrestha, D. J. Weix, *Org. Lett.* 2011, 13, 2766–2769. Reductive conjugate addition of haloalkanes to enones to form silyl enol ethers. (b) R. Shrestha, S. C. M. Dorn, D. J. Weix, *J. Am. Chem. Soc.* 2013, 135, 751–762. Nickel-catalyzed reductive conjugate addition to enones *via* allylnickel intermediates.

62 C. C. Chrovian, J. Montgomery, *Org. Lett.* 2007, 9, 537–540. Surprising role of aryl halides in nickel-catalyzed reductive aldol reactions.

63 H. C. Kolb, M. G. Finn, K. B. Sharpless, *Angew. Chem. Int. Ed.* 2001, 40, 2004–2021. Click chemistry: diverse chemical function from a few good reactions.

64 (a) V. V. Rostovtsev, L. G. Green, V. V. Fokin, K. B. Sharpless, *Angew. Chem. Int. Ed.* 2002, 41, 2596–2599. A stepwise huisgen cycloaddition process: copper(I)-catalyzed regioselective "ligation" of azides and terminal alkynes. (b) C. W. Tornøe, C. Christensen, M. Meldal, *J. Org.Chem.* 2002, 67, 3057–3064. Peptidotriazoles on solid phase: [1,2,3]-triazoles by regiospecific copper(I)-catalyzed 1,3-dipolar cycloadditions of terminal alkynes to azides. (c) J. E. Hein, V. V. Fokin, *Chem. Soc. Rev.* 2010, 39, 1302–1315. Copper-catalyzed azide-alkyne cycloaddition (CuAAC) and beyond: new reactivity of copper(I) acetylides.

65 (a) J. S. Yadav, B. V. S. Reddy, G. M. Reddy, D. N. Chary, *Tetrahedron Lett.* 2007, 48, 8773–8776. Three component, regioselective, one-pot synthesis of β-hydroxytriazoles from epoxides *via* "click reactions". (b) H. Sharghi, M. Hosseini-Sarvari, F. Moeini, R. Khalifeh, A. S. Beni, *Helv. Chim. Acta* 2010, 93, 435–449. One-pot, three-component synthesis of 1-(2-hydroxyethyl)-1*H*-1,2,3-triazole derivatives by copper-catalyzed 1,3-dipolar cycloaddition of 2-azido alcohols and terminal alkynes under mild conditions in water.

66 H. Sharghi, M. H. Beyzavi, A. Safavi, M. M. Doroodmand, R. Khalifeh, *Adv. Synth. Catal.* 2009, 351, 2391–2410. Immobilization of porphyrinatocopper nanoparticles onto activated multi-walled carbon nanotubes and a study of its catalytic activity as an efficient heterogeneous catalyst for a click approach to the three-component synthesis of 1,2,3-triazoles in water.

67 G. Kumaraswamy, K. Ankamma, A. Pitchaiah, *J. Org. Chem.* 2007, 72, 9822–9825. Tandem epoxide or aziridine ring opening by azide/copper catalyzed [3+2] cycloaddition: efficient synthesis of 1,2,3-triazolo β-hydroxy or β-tosylamino functionality motif.

68 F. Alonso, Y. Moglie, G. Radivoy, M. Yus, *J. Org. Chem.* 2011, 76, 8394–8405. Multicomponent click synthesis of 1,2,3-triazoles from epoxides in water catalyzed by copper nanoparticles on activated carbon.

69 T. Boningari, A. Olmos, B. M. Reddy, J. Sommer, P. Pale, *Eur. J. Org. Chem.* 2010, 2010, 6338–6347. Zeo-click chemistry: copper(I)–zeolite-catalyzed cascade reaction; one-pot epoxide ring-opening and cycloaddition.

70 (a) L. Wan, C. Cai, *Catal. Lett.* 2012, 142, 1134–1140. Multicomponent synthesis of 1,2,3-triazoles in water catalyzed by silica-immobilized NHC–Cu(I). (b) K. Chanda, S. Rej, M. H. Huang, *Chem. Eur. J.* 2013, 16036–16043. Facet-dependent catalytic activity of Cu_2O nanocrystals in the one-pot synthesis of 1,2,3-triazoles by multicomponent click reactions.

71 J. Yan, L. Wang, *Synthesis* 2010, 447–452. Synthesis of 1,4-disubstituted 1,2,3-triazoles by use of copper(I) and amino acids ionic liquid catalytic system.

72 I. Bae, H. Han, S. Chang, *J. Am. Chem. Soc.* 2005, 127, 2038–2039. Highly efficient one-pot synthesis of N-sulfonylamidines by Cu-catalyzed three-component coupling of sulfonyl azide, alkyne, and amine.

73 E. J. Yoo, M. Ahlquist, I. Bae, K. B. Sharpless, V. V. Fokin, S. Chang, *J. Org. Chem.* 2008, 73, 5520–5528. Mechanistic studies on the Cu-catalyzed three-component reactions of sulfonyl azides, 1-alkynes and amines, alcohols, or water: dichotomy *via* a common pathway.

74 (a) J. Kim, S. Y. Lee, J. Lee, Y. Do, S. Chang, *J. Org. Chem.* 2008, 73, 9454–9457. Synthetic utility of ammonium salts in a Cu-catalyzed three-component reaction as a facile coupling partner. (b) J. Y. Kim, S. H. Kim, S. Chang, *Tetrahedron Lett.* 2008, 49, 1745–1749. Highly efficient synthesis of α-amino amidines from ynamides by the Cu-catalyzed three-component coupling reactions. (c) I. Yavari, S. Ahmadian, M. Ghazanfarpur-Darjani, Y. Solgi, *Tetrahedron Lett.* 2011, 52, 668–670. Formation of *N*-sulfonylamidines by copper-catalyzed

coupling of sulfonyl azides, terminal alkynes, and trialkylamines. (d) M. Xu, C. Kuang, Z. Wang, Q. Yang, *Synlett* 2010, 2664–2666. Efficient synthesis of *N*-sulfonylacetamidines from propiolic acid by copper-catalyzed three-component coupling reactions.

75 S.-L. Cui, J. Wang, Y.-G. Wang, *Org. Lett.* 2008, 10, 1267–1269. Copper-catalyzed multicomponent reaction: facile access to novel phosphorus amidines.

76 Y. Shang, X. He, J. Hu, J. Wu, M. Zhang, S. Yu, Q. Zhang, *Adv. Synth. Catal.* 2009, 351, 2709–2713. Copper-catalyzed efficient multicomponent reaction: synthesis of benzoxazoline-amidine derivatives.

77 (a) H. Jin, X. Xu, J. Gao, J. Zhong, Y. Wang, *Adv. Synth. Catal.* 2010, 352, 347–350. Copper-catalyzed one-pot synthesis of substituted benzimidazoles. (b) Y. Kim, M. R. Kumar, N. Park, Y. Heo, S. Lee, *J. Org. Chem.* 2011, 76, 9577–9583. Copper-catalyzed, one-pot, three-component synthesis of benzimidazoles by condensation and C–N bond formation.

78 (a) S. H. Cho, E. J. Yoo, I. Bae, S. Chang, *J. Am. Chem. Soc.* 2005, 127, 16046–16047. Copper-catalyzed hydrative amide synthesis with terminal alkyne, sulfonyl azide, and water. (b) M. P. Cassidy, J. Raushel, V. V. Fokin, *Angew. Chem. Int. Ed.* 2006, 4, 3154–3157. Practical synthesis of amides from *in situ* generated copper(I) acetylides and sulfonyl azides. (c) S. H. Cho, S. Chang, *Angew. Chem. Int. Ed.* 2007, 46, 1897–1900. Rate-accelerated nonconventional amide synthesis in water: a practical catalytic aldol-surrogate reaction.

79 (a) E. J. Yoo, I. Bae, S. H. Cho, H. Han, S. Chang, *Org. Lett.* 2006, 8, 1347–1350. A facile access to *N*-sulfonylimidates and their synthetic utility for the transformation to amidines and amides. (b) E. J. Yoo, S. H. Park, S. H. Lee, S. Chang, *Org. Lett.* 2009, 11, 1155–1158. A new entry of copper-catalyzed four-component reaction: facile access to α-aryl β-hydroxy imidates. (c) R. Husmann, Y. S. Na, C. Bolm, S. Chang, *Chem. Commun.* 2010, 46, 5494–5496. Copper-catalyzed one-pot synthesis of a-functionalized imidates. (d) W. Song, W. Lu, J. Wang, P. Lu, Y. Wang, *J. Org. Chem.* 2010, 75, 3481–3483. A facile route to γ-nitro imidates *via* four-component reaction of alkynes with sulfonyl azides, alcohols, and nitroolefins. (e) G. Murugavel, T. Punniyamurthy, *Org. Lett.* 2013, 15, 3828–3831. Novel copper-catalyzed multicomponent cascade synthesis of iminocoumarin aryl methyl ethers.

80 (a) M. Whiting, V. V. Fokin, *Angew. Chem. Int. Ed.* 2006, 45, 3157–3161. Copper-catalyzed reaction cascade: direct conversion of alkynes into *N*-sulfonylazetidin-2-imines. (b) W. Lu, W. Song, D. Hong, P. Lu, Y. Wang, *Adv. Synth. Catal.* 2009, 351, 1768–1772. Copper-catalyzed one-pot synthesis of 2-alkylidene-1,2,3,4-tetrahydropyrimidines.

81 (a) S.-L. Cui, X.-F. Lin, Y.-G. Wang, *Org. Lett.* 2006, 8, 4517–4520. Novel and efficient synthesis of iminocoumarins *via* copper-catalyzed multicomponent reaction. (b) Y. Shen, S. Cui, J. Wang, X. Chen, P. Lu, Y. Wang, *Adv. Synth. Catal.* 2010, 352, 1139–1144. Copper-catalyzed three-component synthesis of 2-iminodihydrocoumarins and 2-iminocoumarins. (c) S.-L. Cui, J. Wang, Y.-G. Wang, *Tetrahedron* 2008, 64, 487–492. Efficient synthesis of 2-imino-1,2-dihydroquinolines and 2-imino-thiochromenes *via* copper-catalyzed domino reaction.

82 S.-L. Cui, J. Wang, Y.-G. Wang. *Org. Lett.* 2007, 9, 5023–5025. Copper-catalyzed multicomponent reaction: facile access to functionalized 5-arylidene-2-imino-3-pyrrolines.

83 Y. Shang, K. Ju, X. He, J. Hu, S. Yu, M. Zhang, K. Liao, L. Wang, P. Zhang, *J. Org. Chem.* 2010, 75, 5743–5745. Copper-catalyzed multicomponent reaction: synthesis of 4-arylsulfonylimino-4,5-dihydrofuran derivatives.

84 J. Wang, J. Wang, Y. Zhu, P. Lu, Y. Wang, *Chem. Commun.* 2011, 47, 3275–3277. Copper-cascade catalysis: synthesis of 3-functionalized indoles.

85 S. Li, Y. Luo, J. Wu, *Org. Lett.* 2011, 13, 4312–4315. Three-component reaction of N′-(2-alkynylbenzylidene)hydrazide, alkyne, with sulfonyl azide *via* a multicatalytic process: a novel and concise approach to 2-amino-H-pyrazolo[5,1-*a*]isoquinolines.

86 S. H. Cho, S. Chang, *Angew. Chem. Int. Ed.* 2008, 47, 2836–2839. Room temperature copper-catalyzed 2-functionalization of pyrrole rings by a three-component coupling reaction.

87 A. B. Dyatkin, R. A. Rivero, *Tetrahedron Lett.* 1998, 39, 3647–3650. The solid phase synthesis of complex propargylamines using the combination of Sonogashira and Mannich reactions.

88 (a) W.-J. Yoo, L. Zhao, C.-J. Li, *Aldrichimica Acta* 2011, 44, 43–51. The A^3-coupling (aldehyde-alkyne-amine) reaction: a versatile method for the preparation of propargyl amines. (b) V. A. Peshkov, O. P. Pereshivko, E. V. Van der Eycken, *Chem. Soc. Rev.* 2012, 41, 3790–3807. A walk around the A^3-coupling. (c) C. Koradin, K. Polborn, P. Knochel, *Angew. Chem. Int. Ed.* 2002, 41, 2535–2538. Enantioselective synthesis of propargylamines by copper-catalyzed addition of alkynes to enamines.

89 (a) J. S. Yadav, B. V. Subba Reddy, V. Naveenkumar, R. Srinivasa Rao, K. Nagaiah, *New J. Chem.* 2004, 28, 335–337. [bmim]PF_6/CuBr: a novel and recyclable catalytic system for the synthesis of propargyl amines. (b) W. S. Bremner, M. G. Organ, *J. Comb. Chem.* 2008, 10, 142–147. Formation of substituted pyrroles *via* an imine condensation/aza-Claisen rearrangement/imine–allene cyclization process by MAOS.

90 J. B. Bariwal, D. S. Ermolat'ev, E. V. Van der Eycken, *Chem.–Eur. J.* 2010, 16, 3281–3284. Efficient microwave-assisted synthesis of secondary alkylpropargylamines by using A^3-coupling with primary aliphatic amines.

91 O. P. Pereshivko, V. A. Peshkov, E. V. Van der Eycken, *Org. Lett.* 2010, 12, 2638–2641. Unprecedented Cu(I)-catalyzed microwave-assisted three-component coupling of a ketone, an alkyne, and a primary amine.

92 S. B. Park, H. Alper, Chem. Commun. 2005, 1315–1317. An efficient synthesis of propargylamines *via* C-H activation catalyzed by copper(I) in ionic liquids.

93 M. J. Climent, A. Corma, S. Iborra, *RSC Adv.* 2012, 2, 16–58. Homogeneous and heterogeneous catalysts for multicomponent reactions.

94 K. Mohan Reddy, N. Seshu Babu, I. Suryanarayana, P. S. Sai Prasad, N. Lingaiah, *Tetrahedron Lett.* 2006, 47, 7563–7566. The silver salt of 12-tungstophosphoric acid: an efficient catalyst for the three-component coupling of an aldehyde, an amine and an alkyne.

95 M. Kidwai, V. Bansal, A. Kumar, S. Mozumdar, *Green Chem.* 2007, 9, 742–745. The first Au-nanoparticles catalyzed green synthesis of propargylamines *via* a three-component coupling reaction of aldehyde, alkyne and amine.

96 (a) P. Li, L. Wang, *Tetrahedron* 2007, 63, 5455–5459. A highly efficient three-component coupling of aldehyde, terminal alkyne, and amine *via* C–H activation catalyzed by reusable immobilized copper in organic–inorganic hybrid materials under solvent-free reaction conditions. (b) M. Wang, P. Li, L. Wang, *Eur. J. Org. Chem.* 2008, 2255–2261. Silica-immobilized NHC–CuI complex: an efficient and reusable catalyst for A^3-coupling (aldehyde–alkyne–amine) under solventless reaction conditions. (c) B. Karimi, M. Gholinejad, M. Khorasani, *Chem. Commun.* 2012, 48, 8961–8963. Highly efficient three-component coupling reaction catalyzed by gold nanoparticles supported on periodic mesoporous organosilica with ionic liquid framework.

97 P. Li, L. Wang, Y. Zhang, M. Wang, *Tetrahedron Lett.* 2008, 49, 6650–6654. Highly efficient three-component (aldehyde–alkyne–amine) coupling reactions catalyzed by a reusable PS-supported NHC–Ag(I) under solvent-free reaction conditions.

98 S. Palaniappan, B. Rajender, *Adv. Synth. Catal.* 2010, 352, 2507–2514. A novel polyaniline-silver nitrate-*p*-toluenesulfonic acid salt as recyclable catalyst in the stereoselective synthesis of β-amino ketones: "one-pot" synthesis in water medium.

99 (a) M. K. Patil, M. Keller, B. M. Reddy, P. Pale, J. Sommer, *Eur. J. Org. Chem.* 2008, 4440–4445. Copper zeolites as green catalysts for multicomponent reactions of aldehydes, terminal alkynes and amines: an efficient and green synthesis of propargylamines. (b) H. Naeimi, M. Moradian, *Appl. Catal. A Gen.* 2013, 467, 400–406. Encapsulation of copper(I)-Schiff base complex in NaY nanoporosity: an efficient and reusable catalyst in the synthesis of propargylamines *via* A^3-coupling (aldehyde-amine-alkyne) reactions.

100 X. Zhang, A. Corma, *Angew. Chem. Int. Ed.* 2008, 47, 4358–4361. Supported gold(III) catalysts for highly efficient three-component coupling reactions.

101 K. Layek, R. Chakravarti, M. Lakshmi Kantam, H. Maheswaran, A. Vinu, *Green Chem.* 2011, 13, 2878–2887. Nanocrystalline magnesium oxide stabilized gold nanoparticles: an advanced nanotechnology based recyclable heterogeneous catalyst platform for the one-pot synthesis of propargylamines.

102 A. Fodor, A. Kiss, N. Debreczeni, Z. Hell, I. Gresits, *Org. Biomol. Chem.* 2010, 8, 4575–4581. A simple method for the preparation of propargylamines using molecular sieve modified with copper(II).

103 M. J. Aliaga, D. J. Ramon, M. Yus, *Org. Biomol. Chem.* 2010, 8, 43. Impregnated copper on magnetite: an efficient and green catalyst for the multicomponent preparation of propargylamines under solvent free conditions.

104 K. K. R. Datta, B. V. S. Reddy, K. Ariga, A. Vinu, *Angew. Chem. Int. Ed.* 2010, 49, 5961–5965. Gold nanoparticles embedded in a mesoporous carbon nitride stabilizer for highly efficient three-component coupling reaction.

105 G. Shore, W.-J. Yoo, C.-J. Li, M. G. Organ, *Chem.–Eur. J.* 2010, 16, 126–133. Propargyl amine synthesis catalysed by gold and copper thin films by using microwave-assisted continuous-flow organic synthesis (MACOS).

106 Y. Zhao, X. Zhou, T.-A. Okamura, M. Chen, Y. Lu, W.-Y. Sun, J.-Q. Yu, *Dalton Trans.* 2012, 41, 5889–5896. Silver supramolecule catalyzed multicomponent reactions under mild conditions.

107 X. Zhou, Y. Lu, L.-L. Zhai, Y. Zhao, Q. Liu, W.-Y. Sun, *RSC Adv.* 2013, 3, 1732–1734. Propargylamines formed from three-component coupling reactions catalyzed by silver oxide nanoparticles.

108 Y. Yamamoto, H. Hayashi, T. Saigoku, H. Nishiyama, *J. Am. Chem. Soc.* 2005, 127, 10804–10805. Domino coupling relay approach to polycyclic pyrrole-2-carboxylates.

109 V. A. Peshkov, O. P. Pereshivko, P. A. Donets, V. P. Mehta, E. V. Van der Eycken, *Eur. J. Org. Chem.* 2010, 2010, 4861–4867. Diversity-oriented microwave-assisted synthesis of the 3-benzazepine framework.

110 N. Mont, V. Pravinchandra Mehta, P. Appukkuttan, T. Beryozkina, S. Toppet, K. Van Hecke, L. Van Meervelt, A. Voet, M. DeMaeyer, E. Van der Eycken, *J. Org. Chem.* 2008, 73, 7509–7516. Diversity oriented microwave-assisted synthesis of (–)-steganacin aza-analogues.

111 R. L. Giles, J. D. Sullivan, A. M. Steiner, R. E. Looper, *Angew. Chem. Int. Ed.* 2009, 48, 3116–3120. Addition–hydroamination reactions of propargyl cyanamides: rapid access to highly substituted 2-aminoimidazoles.

112 R. L. Giles, R. A. Nkansah, R. E. Looper, *J. Org. Chem.* 2010, 75, 261–264. Synthesis of 2-thio- and 2-oxoimidazoles *via* cascade addition–cycloisomerization reactions of propargylcyanamides.

113 D. S. Ermolat'ev, J. B. Bariwal, H. P. L. Steenackers, S. C. J. De Keersmaecker, E. V. Van der Eycken, *Angew. Chem. Int. Ed.* 2010, 49, 9465–9468. Concise and diversity-oriented route toward polysubstituted 2-aminoimidazole alkaloids and their analogues.

114 B. Yan, Y. Liu, *Org. Lett.* 2007, 9, 4323–4326. Gold-catalyzed multicomponent synthesis of aminoindolizines from aldehydes, amines, and alkynes under solvent-free conditions or in water.

115 Y. Bai, J. Zeng, J. Ma, B. K. Gorityala, X.-W. Liu, *J. Comb. Chem.* 2010, 12, 696–699. Quick access to druglike heterocycles: facile silver-catalyzed one-pot multicomponent synthesis of aminoindolizines.

116 D. Chernyak, N. Chernyak, V. Gevorgyan, *Adv. Synth. Catal.* 2010, 352, 961–966. Efficient and general synthesis of 3-aminoindolines and 3-aminoindoles *via* copper-catalyzed three-component coupling reaction.

117 N. Chernyak, V. Gevorgyan, *Angew. Chem. Int. Ed.* 2010, 49, 2743–2746. General and efficient copper-catalyzed three-component coupling reaction towards imidazoheterocycles: one-pot synthesis of Alpidem and Zolpidem.

118 (a) H. Ohno, Y. Ohta, S. Oishi, N. Fujii, *Angew. Chem. Int. Ed.* 2007, 46, 2295–2298. Direct synthesis of 2-(aminomethyl) indoles through copper(I)-catalyzed domino three-omponent

coupling and cyclization reactions. (b) Y. Ohta, H. Chiba, S. Oishi, N. Fujii, H. Ohno, *J. Org. Chem.* 2009, 74, 7052–7058. Construction of nitrogen heterocycles bearing an aminomethyl group by copper-catalyzed domino three-component coupling–cyclization.

119 (a) Y. Ye, Q. Ding, J. Wu, *Tetrahedron* 2008, 64, 1378–1382. Three-component reaction of 2-alkynylbenzaldehyde, amine, and nucleophile using Lewis acid-surfactant combined catalyst in water. (b) Q. Ding, J. Wu, *Org. Lett.* 2007, 9, 4959–4962. Lewis acid- and organocatalyst-cocatalyzed multicomponent reactions of 2-alkynylbenzaldehydes, amines, and ketones. (c) S. Ye, J. Wu, *Tetrahedron Lett.* 2009, 50, 6273–6275. Silver triflate and triphenylphosphine co-catalyzed reactions of 2-alkynylbenzaldehyde, amine, and α,β-unsaturated ketone.

120 A. K. Verma, S. K. R. Kotla, D. Choudhary, M. Patel, R. K. Tiwari, *J. Org. Chem.* 2013, 78, 4386–4401. Silver-catalyzed tandem synthesis of naphthyridines and thienopyridines *via* three-component reaction.

121 Q. Zhang, M. Cheng, X. Hu, B.-G. Li, J.-X. Ji, *J. Am. Chem. Soc.* 2010, 132, 7256–7257. Gold-catalyzed three-component tandem process: an efficient and facile assembly of complex butenolides from alkynes, amines, and glyoxylic acid.

122 E. R. Bonfield, C.-J. Li, *Adv. Synth. Catal.* 2008, 350, 370–374. Efficient preparation of the isoindoline framework *via* a six component, tandem double A^3-coupling and [2+2+2] cycloaddition reaction.

123 W.-J. Yoo, C.-J. Li, *Adv. Synth. Catal.* 2008, 350, 1503–1506. Copper-catalyzed four-component coupling between aldehydes, amines, alkynes, and carbon dioxide.

124 J. Zhao, H. Huang, C. Qi, H. Jiang, *Eur. J. Org. Chem.* 2012, 2012, 5665–5667. CuI/$SnCl_2$ Co-catalyzed four-component reaction of ketones, amines, alkynes, and carbon dioxide.

125 C. Wei, C.-J. Li, *J. Am. Chem. Soc.* 2002, 124, 5638–5639. Enantioselective direct-addition of terminal alkynes to imines catalyzed by copper(I)pybox complex in water and in toluene.

126 A. Bisai, V. K. Singh, *Org. Lett.* 2006, 8, 2405–2408. Enantioselective one-pot three-component synthesis of propargylamines.

127 Z. Shao, X. Pu, X. Li, B. Fan, A. S. C. Chan, *Tetrahedron: Asymmetry* 2009, 20, 225–229. Enantioselective, copper(I)-catalyzed three-component reaction for the synthesis of β,γ-alkynyl α-amino acid derivatives.

128 S. Nakamura, M. Ohara, Y. Nakamura, N. Shibata, T. Toru, *Chem.–Eur. J.* 2010, 16, 2360–2362. Copper-catalyzed enantioselective three-component synthesis of optically active propargylamines from aldehydes, amines, and aliphatic alkynes.

129 N. Gommermann, P. Knochel, *Chem. Commun.* 2005, 4175–4177. 2-Phenallyl as a versatile protecting group for the asymmetric one-pot three-component synthesis of propargylamines.

130 N. Gommermann, P. Knochel, *Chem.–Eur. J.* 2006, 12, 4380–4392. Practical highly enantioselective synthesis of propargylamines through a copper-catalyzed one-pot three-component condensation reaction.

131 T. F. Knöpfel, P. Aschwanden, T. Ichikawa, T. Watanabe, E. M. Carreira, *Angew. Chem. Int. Ed.* 2004, 43, 5971–5973. Readily available biaryl P,N ligands for asymmetric catalysis.

132 N. W. Alcock, J. M. Brown, D. I. Hulmes, *Tetrahedron: Asymmetry* 1993, 4, 743–756. Synthesis and resolution of 1-(2-diphenylphosphino-1-naphthyl)isoquinoline; a P-N chelating ligand for asymmetric catalysis.

133 P. Aschwanden, C. R. J. Stephenson, E. M. Carreira, *Org. Lett.* 2006, 8, 2437–2440. Highly enantioselective access to primary propargylamines: 4-piperidinone as a convenient protecting group.

134 W. Lin, T. Cao, W. Fan, Y. Han, J. Kuang, H. Luo, B. Miao, X. Tang, Q. Yu, W. Yuan, J. Zhang, C. Zhu, S. Ma, *Angew. Chem. Int. Ed.* 2014, 53, 277–281. Enantioselective double manipulation of tetrahydroisoquinolines with terminal alkynes and aldehydes under copper(I) catalysis.

135 F. S. P. Cardoso, K. A. Abboud, A. Aponick, *J. Am. Chem. Soc.* 2013, 135, 14548–14551. Design, preparation, and implementation of an imidazole-based chiral biaryl P,N-ligand for asymmetric catalysis.

136 (a) X. Tang, J. Kuang, S. Ma, *Chem. Commun.* 2013, 49, 8976–8978. CuBr for KA^2 reaction: en route to propargylic amines bearing a quaternary carbon center. (b) M. Cheng, Q. Zhang, X.-Y. Hu, B.-G. Li, J.-X. Ji, A. S. C. Chan, *Adv. Synth. Catal.* 2011, 353, 1274–1278. Gold-catalyzed direct intermolecular coupling of ketones, secondary amines, and alkynes: a facile and versatile access to propargylic amines containing a quaternary carbon center. (c) Y. Suzuki, S. Naoe, S. Oishi, N. Fujii, H. Ohno, *Org. Lett.* 2011, 14, 326–329. Gold-catalyzed three-component annulation: efficient synthesis of highly functionalized dihydropyrazoles from alkynes, hydrazines, and aldehydes or ketones.

137 L. Zhou, H.-F. Jiang, C.-J. Li, *Adv. Synth. Catal.* 2008, 350, 2226–2230. Efficient synthesis of γ,δ-alkynyl-β-amino acid derivatives by a new copper-catalyzed amine-alkyne-alkyne addition reaction.

138 L. Zhou, Q. Shuai, H.-F. Jiang, C.-J. Li, *Chem.–Eur. J.* 2009, 15, 11668–11674. Copper-catalyzed amine–alkyne–alkyne addition reaction: an efficient method for the synthesis of γ,δ-alkynyl-β-amino acid derivatives.

139 R. Dodda, C.-G. Zhao, *Org. Lett.* 2007, 9, 165–167. Silver(I) triflate-catalyzed direct synthesis of *N*-PMP protected α-aminopropargylphosphonates from terminal alkynes.

140 H. Feng, D. S. Ermolat'ev, G. Song, E. V. Van der Eycken, *J. Org. Chem.* 2011, 76, 7608–7613. Microwave-assisted decarboxylative three-component coupling of a 2-oxoacetic acid, an amine, and an alkyne.

141 H. Feng, D. S. Ermolat'ev, G. Song, E. V. Van der Eycken, *Org. Lett.* 2012, 14, 1942–1945. Regioselective Cu(I)-catalyzed tandem A^3-coupling/decarboxylative coupling to 3-amino-1,4-enynes.

142 (a) Z. Lin, D. Yu, Y. Zhang, *Tetrahedron Lett.* 2011, 52, 4967–4970. Propargylic amines constructed *via* copper-catalyzed three-component coupling of terminal alkynes, benzal halides and amines. (b) D. Yu, Y. Zhang, *Adv. Synth. Catal.*

2011, 353, 163–169. Copper-catalyzed three-component coupling of terminal alkyne, dihalomethane and amine to propargylic amines.

143 D. Aguilar, M. Contel, E. P. Urriolabeitia, *Chem.–Eur. J.* 2010, 16, 9287–9296. Mechanistic insights into the one-pot synthesis of propargylamines from terminal alkynes and amines in chlorinated solvents catalyzed by gold compounds and nanoparticles.

144 X.-Y. Dou, Q. Shuai, L.-N. He, C.-J. Li, *Adv. Synth. Catal.* 2010, 352, 2437–2440. Copper(II) triflate-catalyzed three-component coupling of aldehydes, alkynes and carbamates.

145 N. Sakai, N. Uchida, T. Konakahara, *Synlett* 2008, 1515–1519. Facile and selective synthesis of propargylic amines and 1,6-diynes: one-pot three-component coupling reactions of alkynylsilanes, aldehydes and amines by a cooperative catalytic system comprised of CuCl and $Cu(OTf)_2$.

146 D. D. Vachhani, A. Sharma, E. Van der Eycken, *Angew. Chem. Int. Ed.* 2013, 52, 2547–2550. Copper-catalyzed direct secondary and tertiary C-H alkylation of azoles through a heteroarene–amine–aldehyde/ketone coupling reaction.

147 (a) D. A. Black, B. A. Arndtsen, *Org. Lett.* 2006, 8, 1991–1993. General approach to the coupling of organoindium reagents with imines *via* copper catalysis. (b) R. E. Beveridge, D. A. Black, B. A. Arndtsen, *Eur. J. Org. Chem.* 2010, 3650–3656. Copper-catalyzed multicomponent coupling of organoindium reagents with nitrogen-containing aromatic heterocycles.

148 L. Zhang, H. C. Malinakova, *J. Org. Chem.* 2007, 72, 1484–1487. Copper-catalyzed multicomponent cascade process for the synthesis of hexahydro-1H-isoindolones.

149 D. Ma, X. Lu, L. Shi, H. Zhang, Y. Jiang, X. Liu, *Angew. Chem. Int. Ed.* 2011, 50, 1118–1121. Domino condensation/S-arylation/heterocyclization reactions: copper-catalyzed three-component synthesis of 2-*N*-substituted benzothiazoles.

150 P. Garner, H. Ü. Kaniskan, J. Hu, W. J. Youngs, M. Panzner, *Org. Lett.* 2006, 8, 3647–3650. Asymmetric multicomponent [C+NC+CC] synthesis of highly functionalized pyrrolidines catalyzed by silver(I).

151 N. Eghbali, J. Eddy, P. T. Anastas, *J. Org. Chem.* 2008, 73, 6932–6935. Silver-catalyzed one-pot synthesis of arylnaphthalene lactones.

152 J. Li, Y. Liu, C. Li, X. Jia, *Adv. Synth. Catal.* 2011, 353, 913–917. Silver hexafluoroantimonate-catalyzed three-component [2+2+1] cycloadditions of allenoates, dual activated olefins, and isocyanides.

153 A. D. Melhado, W. E. Brenzovich, A. D. Lackner, F. D. Toste, *J. Am. Chem. Soc.* 2010, 132, 8885–8887. Gold-catalyzed three-component coupling: oxidative oxyarylation of alkenes.

154 M. J. Campbell, F. D. Toste, *Chem. Sci.* 2011, 2, 1369–1378. Enantioselective synthesis of cyclic carbamimidates *via* a three-component reaction of imines, terminal alkynes, and p-toluenesulfonylisocyanate using a monophosphine gold(I) catalyst.

155 D. Schultz, N. R. Schultz, J. Babij, D. Wolfe, *Adv. Synth. Catal.* 2012, 354, 3451–3455. Intermolecular gold(I)-catalyzed alkyne carboalkoxylation reactions for the multicomponent assembly of β-alkoxy ketones.

156 (a) M. P. Doyle, M. A. McKervey, T. Ye (Eds.) *Modern catalytic methods for organic synthesis with diazo compounds*. John Wiley & Sons, Inc., New York, 1998, Chapter 8. (b) H. M. L. Davies, J. R. Manning, *Nature* 2008, 451, 417–424. Catalytic C–H functionalization by metal carbenoid and nitrenoid insertion.

157 X. Guo, W. Hu, *Acc. Chem. Res.* 2013, 46, 2427–2440. Novel multicomponent reactions *via* trapping of protic onium ylides with electrophiles.

158 C.-D. Lu, H. Liu, Z.-Y. Chen, W.-H. Hu, A.-Q. Mi, *Org. Lett.* 2005, 7, 83–86. Three-component reaction of aryl diazoacetates, alcohols, and aldehydes (or imines): evidence of alcoholic oxonium ylide intermediates.

159 H. Huang, X. Guo, W. Hu, *Angew. Chem. Int. Ed.* 2007, 46, 1337–1339. Efficient trapping of oxonium ylides with imines: a highly diastereoselective three-component reaction for the synthesis of β-amino-α-hydroxyesters with quaternary stereocenters.

160 Z. Guo, T. Shi, J. Jiang, L. Yang, W. Hu, *Org. Biomol. Chem.* 2009, 7, 5028–5033. Component match in rhodium catalyzed three-component reactions of ethyl diazoacetate, H_2O and aryl imines: a highly diastereoselective one-step synthesis of β-aryl isoserine derivatives.

161 X. Zhang, H. Huang, X. Guo, X. Guan, L. Yang, W. Hu, *Angew. Chem. Int. Ed.* 2008, 47, 6647–6649. Catalytic enantioselective trapping of an alcoholic oxonium ylide with aldehydes: Rh^{II}/Zr^{IV}-cocatalyzed three-component reactions of aryl diazoacetates, benzyl alcohol and aldehydes.

162 (a) W. Hu, X. Xu, J. Zhou, W.-J. Liu, H. Huang, J. Hu, L. Yang, L.-Z. Gong, *J. Am. Chem. Soc.* 2008, 130, 7782–7783. Cooperative catalysis with chiral Brønsted acid $Rh_2(OAc)4$: highly enantioselective three-component reactions of diazo compounds with alcohols and imines. (b) X. Xu, J. Zhou, L. Yang, W. Hu, *Chem. Commun.* 2008, 6564–6566. Selectivity control in enantioselective four-component reactions of aryl diazoacetates with alcohols, aldehydes and amines: an efficient approach to synthesizing chiral β-amino-α-hydroxyesters. (c) Y. Qian, C. Jing, T. Shi, J. Ji, M. Tang, J. Zhou, C. Zhai, W. Hu, *ChemCatChem* 2011, 3, 653–656. Dual catalysis in highly enantioselective multicomponent reaction with water: an efficient approach to chiral β-amino-α-hydroxy acid derivatives. (d)Y. Qian, C. Jing, S. Liu, W. Hu, *Chem. Commun.* 2013, 49, 2700–2702. A highly enantioselective four-component reaction for the efficient construction of chiral β-hydroxy-α-amino acid derivatives.

163 (a) Y. Qian, X. Xu, L. Jiang, D. Prajapati, W. Hu, *J. Org. Chem.* 2010, 75, 7483–7486. A strategy to synthesize taxol side chain and (–)-epi cytoxazone *via* chiral brønsted acid-$Rh_2(OAc)4$ cocatalyzed enantioselective three-component reactions. (b) X. Xu, Y. Qian, L. Yang, W. Hu, *Chem. Commun.* 2011, 47, 797–799. Cooperative catalysis in highly enantioselective Mannich-type three-component reaction of a diazoacetophenone with an alcohol and an imine.

164 (a) X. Guo, H. Huang, L. Yang, W. Hu, *Org. Lett.* 2007, 9, 4721–4723. Trapping of oxonium ylide with isatins: efficient and stereoselective construction of adjacent quaternary carbon centers. (b) J. Ji, X. Zhang, Y. Zhu, Y. Qian, J. Zhou, L. Yang, W. Hu, *J. Org. Chem.* 2011, 76, 5821–5824. Diastereoselectivity switch in cooperatively catalyzed three-component reactions of an aryldiazoacetate, an alcohol, and a β,γ-unsaturated α-keto ester. (c) B. Alcaide, P. Almendros, C. Aragoncillo, R. Callejo, M. P. Ruiz, M. R. Torres, *Eur. J. Org. Chem.* 2012, 2359–2366. Diastereoselective synthesis of β-lactam–oxindole hybrids through a three-component reaction of azetidine-2,3-diones, α-diazo-oxindoles and alcohols catalyzed by $[Rh_2(OAc)_4]$.

165 (a) X.-Y. Guan, L.-P. Yang, W. Hu, *Angew. Chem. Int. Ed.* 2010, 49, 2190–2192. Cooperative catalysis in multi-component reactions: highly enantioselective synthesis of γ-hydroxyketones with a quaternary carbon stereocenter. (b) X. Han, M. Gan, H. Qiu, J. Ji, X. Zhang, L. Jiang, W. Hu, *Synlett* 2011, 1717–1722. Trapping of oxonium ylides with Michael acceptors: highly diastereoselective three-component reactions of diazo compounds with alcohols and benzylidene Meldrum's acids/4-oxo-enoates.

166 J. Jiang, X. Guan, S. Liu, B. Ren, X. Ma, X. Guo, F. Lv, X. Wu, W. Hu, *Angew. Chem. Int. Ed.* 2013, 52, 1539–1542. Highly diastereoselective multicomponent cascade reactions: efficient synthesis of functionalized 1-indanols.

167 Y.-H. Wang, Z.-Y. Chen, A.-Q. Mi, W. Hu, *Chem. Commun.* 2004, 2486–2487. Novel C–C bond formation through addition of ammonium ylides to arylaldehydes: a facile approach to β-aryl-β-hydroxy α-amino acid frameworks.

168 (a) Y. Zhu, C. Zhai, Y. Yue, L. Yang, W. Hu, *Chem. Commun.* 2009, 1362–1364. One-pot three-component tandem reaction of diazo compounds with anilines and unsaturated ketoesters: a novel synthesis of 2,3-dihydropyrrole derivatives. (b) Y. Zhu, C. Zhai, L. Yang, W. Hu, *Eur. J. Org. Chem.* 2011, 1113–1124. Highly regioselective, three-component reactions of diazoacetates with anilines and β,γ-unsaturated α-keto esters: 1,2-addition *versus* 1,4-addition. (c)X. Zhang, J. Ji, Y. Zhu, C. Jing, M. Li, W. Hu, *Org. Biomol. Chem.* 2012, 10, 2133–2138. A highly diastereoselective three-component tandem 1,4-conjugated addition–cyclization reaction to multisubstituted pyrrolidines.

169 J. Jiang, H.-D. Xu, J.-B. Xi, B.-Y. Ren, F.-P. Lv, X. Guo, L.-Q. Jiang, Z.-Y. Zhang, W. Hu, *J. Am. Chem. Soc.* 2011, 133, 8428–8431. Diastereoselectively switchable enantioselective trapping of carbamate ammonium ylides with imines.

170 M. P. Doyle, R. Duffy, M. Ratnikov, L. Zhou, *Chem. Rev.* 2010, 110, 704–724. Catalytic carbene insertion into C–H bonds.

171 H. Qiu, M. Li, L.-Q. Jiang, F.-P. Lv, L. Zan, C.-W. Zhai, M. P. Doyle, W.-H. Hu, *Nat. Chem.* 2012, 4, 733–738. Highly enantioselective trapping of zwitterionic intermediates by imines.

172 D. Xing, C. Jing, X. Li, H. Qiu, W. Hu, *Org. Lett.* 2013, 15, 3578-3581. Highly efficient synthesis of mixed 3,3-bisindoles *via* Rh(II)-catalyzed three-component reaction of 3-diazooxindoles with indoles and ethyl glyoxylate.

173 (a) S. Torssell, M. Kienle, P. Somfai, *Angew. Chem. Int. Ed.* 2005, 44, 3096–3099. 1,3-Dipolar cycloadditions of carbonyl ylides to aldimines: a three-component approach to *syn*-α-hydroxy-β-amino esters. (b) S. Torssell, P. Somfai, *Adv. Synth. Catal.* 2006, 348, 2421–2430. 1,3-Dipolar cycloadditions of carbonyl ylides to aldimines: scope, limitations and asymmetric cycloadditions.

174 (a) A. DeAngelis, M. T. Taylor, J. M. Fox, *J. Am. Chem. Soc.* 2009, 131, 1101–1105. Unusually reactive and selective carbonyl ylides for three-component cycloaddition reactions. (b) Y. Hashimoto, K. Itoh, A. Kakehi, M. Shiro, H. Suga, *J. Org. Chem.* 2013, 78, 6182–6195. Diastereoselective synthesis of tetrahydrofurans by Lewis acid catalyzed intermolecular carbenoid-carbonyl reaction cycloaddition sequences: unusual diastereoselectivity of Lewis acid catalyzed cycloadditions.

175 (a) C. V. Galliford, K. A. Scheidt, *J. Org. Chem.* 2007, 72, 1811–1813. Catalytic multicomponent reactions for the synthesis of *N*-aryl trisubstituted pyrroles. (b) T. Rajasekaran, G. Karthik, B. Sridhar, B. V. S. Reddy, *Org. Lett.* 2013, 15, 1512–1515. Dual behavior of isatin-based cyclic ketimines with dicarbomethoxy carbene: expedient synthesis of highly functionalized spirooxindolyl oxazolidines and pyrrolines.

176 T. Kurahashi, H. Shinokubo, A. Osuka, *Angew. Chem. Int. Ed.* 2006, 45, 6336–6338. Intermolecular rhodium-catalyzed carbometalation/Heck-type reaction in water.

177 Y. Hoshino, Y. Shibata, K. Tanaka, *Angew. Chem. Int. Ed.* 2012, 51, 9407–9411. Rhodium-catalyzed three-component cross-addition of silylacetylenes, alkynyl esters and electron-deficient alkenes or isocyanates.

178 (a) H. Hara, M. Hirano, K. Tanaka, *Org. Lett.* 2008, 10, 2537–2540. Liquid enol ethers and acetates as gaseous alkyne equivalents in Rh-catalyzed chemo- and regioselective formal cross-alkyne cyclotrimerization. (b) M. Kobayashi, K. Tanaka, *Chem.–Eur. J.* 2012, 18, 9225–9229. Rhodium-catalyzed linear cross-trimerization of two different alkynes with an alkene and two different alkenes with an alkyne.

179 H. Lee, Y.-K. Sim, J.-W. Park, C.-H. Jun, *Chem.–Eur. J.* 2014, 20, 323–333. Microwave-assisted, rhodium(III)-catalyzed N-annulation reactions of aryl and α,β-unsaturated ketones with alkynes.

180 W.-W. Chen, R. V. Nguyen, C.-J. Li, *Tetrahedron Lett.* 2009, 50, 2895–2898. Iron-catalyzed three-component coupling of aldehyde, alkyne, and amine under neat conditions in air.

181 P. Li, Y. Zhang, L. Wang, *Chem.–Eur. J.* 2009, 15, 2045–2049. Iron-catalyzed ligand-free three-component coupling reactions of aldehydes, terminal alkynes and amines.

182 (a) K. Cao, F.-M. Zhang, Y.-Q. Tu, X.-T. Zhuo, C.-A. Fan, *Chem. Eur. J.* 2009, 15, 6332–6334. Iron (III)-catalyzed and air-mediated tandem reaction of aldehydes, alkynes and amines: an efficient approach to substituted quinolines. (b) Patil, S.; Patil, S. V.; Bobade, V. D. *Synlett* 2011, 16, 2379–2383. Synthesis of aminoindolizine and quinoline derivatives *via* $Fe(acac)_3$/TBA-OH-catalyzed sequential cross-coupling–cycloisomerization reactions. (c) C. Yao, B. Qin, H. Zhang, J. Lu, D. Wang, S. Tu, *RSC Adv.* 2012, 2, 3759–3764. One-pot solvent-free synthesis of quinolines by C–H activation/C–C bond formation catalyzed by recyclable iron(III) triflate.

183 T. Guntreddi, B. K. Allam, K. N. Singh, *Synlett* 2012, 2635–2638. Simple and efficient one-pot synthesis of imidazo[1,2-a] pyridines catalyzed by magnetic nano-Fe_3O_4–$KHSO_4$·SiO_2.

184 W. Wei, J. Wen, D. Yang, X. Sun, J. You, Y. Suo, H. Wang, *Tetrahedron* 2013, 69, 10747–10751. Iron-catalyzed three-component tandem process: a novel and convenient synthetic route to quinoline-2,4-dicarboxylates from arylamines, glyoxylic esters and α-ketoesters.

185 S. K. Panja, S. Saha, *RSC Adv.* 2013, 3, 14495–14500. Recyclable, magnetic ionic liquid bmim[$FeCl_4$]-catalyzed, multicomponent, solvent-free, green synthesis of quinazolines.

186 C. A. Grob, K. Camenisch, *Helv. Chim. Acta* 1953, 36, 49–58. Eine neue pyrrolring-synthese.

187 S. Maiti, S. Biswas, U. Jana, *J. Org. Chem.* 2010, 75, 1674–1683. Iron(III)-catalyzed four-component coupling reaction of 1,3-dicarbonyl compounds, amines, aldehydes and nitroalkanes: a simple and direct synthesis of functionalized pyrroles.

188 (a) Q.-Y. Song, B.-L. Yang, S.-K. Tian, *J. Org. Chem.* 2007, 72, 5407–5410. $FeSO_4$·$7H_2O$-catalyzed four-component synthesis of protected homoallylic amines. (b) B.-L. Yang, S.-K. Tian, *Eur. J. Org. Chem.* 2007, 4646–4650. Iron-catalyzed four-component reaction for the synthesis of protected primary amines.

189 C. Ma, D. Xing, C. Zhai, J. Che, S. Liu, J. Wang, W. Hu, *Org. Lett.* 2013, 15, 6140–6143. Iron porphyrin-catalyzed three-component reaction of ethyl diazoacetate with aliphatic amines and β,γ-unsaturated α-keto esters.

190 A. P. Dieskau, M. S. Holzwarth, B. Plietker, *Chem.–Eur. J.* 2012, 18, 2423–2429. Fe-catalyzed multicomponent reactions: the regioselective alkoxy allylation of activated olefins and its application in sequential Fe catalysis.

191 A. T. Khan, T. Parvin, L. H. Choudhury, *Tetrahedron* 2007, 63, 5593–5601. Iron(III) chloride-catalyzed convenient one-pot synthesis of β-acetamido carbonyl compounds.

192 M. B. Gawande, A. K. Rathi, I. D. Nogueira, R. S. Varma, P. S. Branco, *Green Chem.* 2013, 15, 1895–1899. Magnetite-supported sulfonic acid: a retrievable nanocatalyst for the Ritter reaction and multicomponent reactions.

193 V. Gevorgyan, U. Radhakrishnan, A. Takeda, M. Rubina, M. Rubin, Y. Yamamoto, *J. Org. Chem.* 2001, 66, 2835–2841. Palladium-catalyzed highly chemo- and regioselective formal [2+2+2] sequential cycloaddition of alkynes: a renaissance of the well known trimerization reaction?

194 Y. Yamamoto, J.-I. Ishii, H. Nishiyama, K. Itoh, *J. Am. Chem. Soc.* 2005, 127, 9625–9631. Cp*RuCl-catalyzed formal intermolecular cyclotrimerization of three unsymmetrical alkynes through a boron temporary tether: regioselective four-component coupling synthesis of phthalides.

195 G. Guillena, D. J. Ramón, M. Yus, *Chem. Rev.* 2010, 110, 1611–1641. Hydrogen autotransfer in the *N*-alkylation of amines and related compounds using alcohols and amines as electrophiles.

196 M. Zhang, H. Neumann, M. Beller, *Angew. Chem. Int. Ed.* 2013, 52, 597–601. Selective ruthenium-catalyzed three-component synthesis of pyrroles.

197 M. Zhang, X. Fang, H. Neumann, M. Beller, *J. Am. Chem. Soc.* 2013, 135, 11384–11388. General and regioselective synthesis of pyrroles *via* ruthenium-catalyzed multicomponent reactions.

198 S. Fustero, P. Bello, J. Miró, A. Simón, C. del Pozo, *Chem.–Eur. J.* 2012, 18, 10991–10997. 1,7-Octadiene-assisted tandem multicomponent cross-enyne metathesis (CEYM)-Diels-Alder reactions: a useful alternative to Mori's condition.

199 C. S. Junker, M. E. Welker, *Tetrahedron* 2012, 68, 5341–5345. Ruthenium carbenes as catalysts in stereoselective ene-yne metathesis/Diels-Alder and ene-yne metathesis/Diels-Alder/cross coupling multicomponent reactions.

4

MULTICOMPONENT REACTIONS WITH ORGANOBORON COMPOUNDS

Eugenia Marqués-López and Raquel P. Herrera

Departamento de Química Orgánica, Instituto de Síntesis Química y Catálisis Homogénea (ISQCH), CSIC-Universidad de Zaragoza, Zaragoza, Spain

4.1 INTRODUCTION

Organoboron reagents, especially boronic acids and boronates, have been widely explored as suitable and powerful reagents for the formation of new C—C bonds in organic synthesis, medicinal chemistry, and material science [1]. This kind of compounds has received the attention of the scientific community not only because of their wide commercial availability but also, and more important, because of their stability toward air and water, as well as the low toxicity and tolerance to several functional groups. Since these species have been the center of numerous studies giving rise to a complete area of investigation, we have decided to dedicate a chapter to briefly cover some remarkable aspects of this chemistry. In this context, we want to show here those multicomponent reactions (MCRs) in which the participation of organoboron compounds has been crucial for the success of the process. Herein, we will cover examples where the boron atom remains in the final structure of the product and other cases in which the boron atom is released through the course of the process.

4.2 CATALYTIC MCRS WITH ORGANOBORON COMPOUNDS

The preparation of more complex architectures and the concern for the sustainable chemistry have attracted the interest of many researchers in the development of new catalytic procedures. In this context, many efforts have been devoted in new catalytic multicomponent transformations, and we disclose in this chapter some pivotal advances in the field of organoboron compounds [2].

4.2.1 Cobalt-Catalyzed MCRs Containing Organoboron Compounds

In this field, interesting examples have been disclosed by Hilt and coworkers, who, based on their pioneering work of cobalt(I)-catalyzed Diels–Alder reaction employing alkynyl boron derivatives and 1,3-dienes for the generation of dihydroaromatic building blocks [3], have recently explored more complex procedures with the presence of four components (Schemes 4.1 [4] and 4.2 [5]).

In this first example, a cobalt(I)-catalyzed Diels–Alder reaction employing alkynyl boron derivatives **2** and 1,3-diene **1** for preparation of dihydroaromatic building blocks **4** is disclosed. Ensuing, dihydroaromatic boronic ester intermediates **3** were used as useful synthetic scaffolds in an unusual thermal Diels–Alder reaction with different dienophiles, generating highly substituted and functionalized products in acceptable to very good yields by recrystallization after four synthetic steps. Five new C—C bonds from four simple components were formed in this process.

More recently, the same group has developed a one-pot sequence cobalt-catalyzed 1,4-hydrovinylation/allylboration. The scope of the reaction was explored for a wide range of aliphatic, aromatic, heteroaromatic, and vinylic aldehydes affording interesting building blocks, for the synthesis of tetrasubstituted tetrahydropyrans in subsequent synthetic steps, with excellent yields and *anti*-diastereoselectivity as represented in Scheme 4.2 [5a].

Multicomponent Reactions: Concepts and Applications for Design and Synthesis, First Edition. Edited by Raquel P. Herrera and Eugenia Marqués-López.

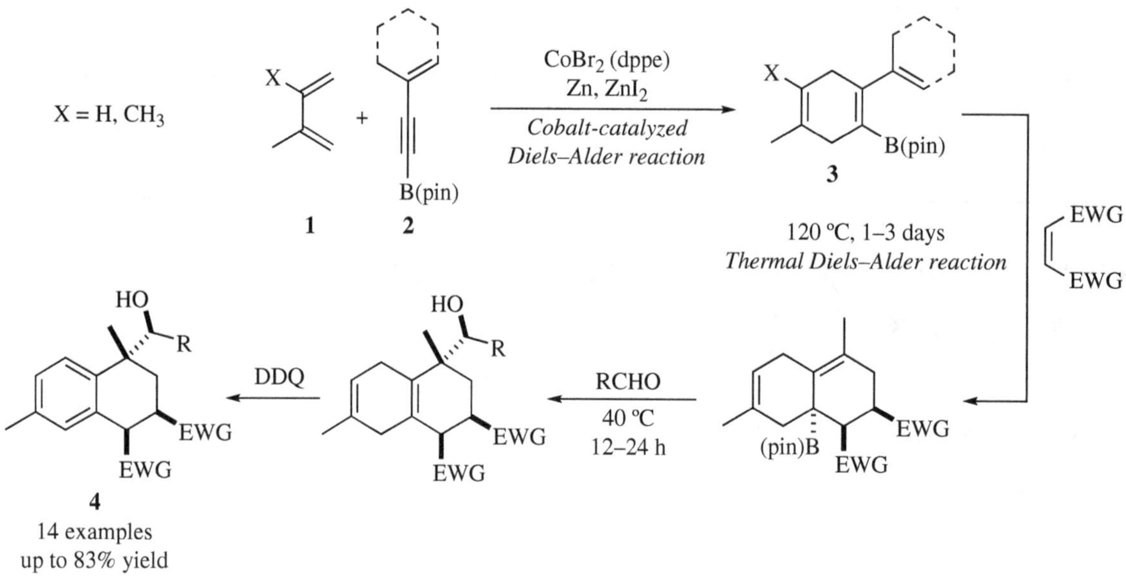

SCHEME 4.1 Cobalt(I)-catalyzed Diels–Alder reaction for the synthesis of polycyclic multifunctionalized products.

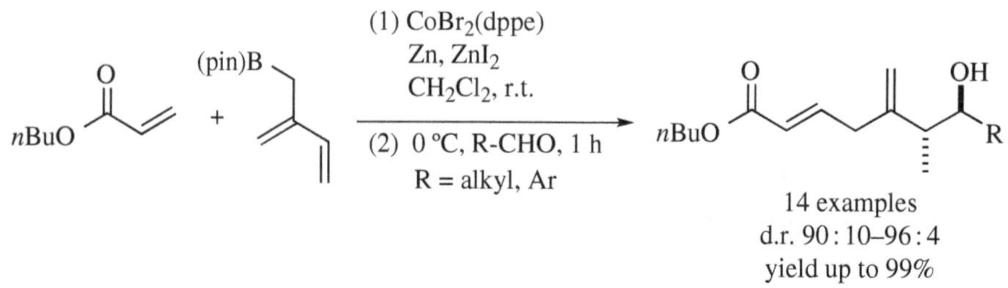

SCHEME 4.2 Multicomponent cobalt-catalyzed 1,4-hydrovinylation/allylboration. dppe, 1,2-bis(diphenylphosphino)ethane.

As extension of this work, two examples of four-component one-pot reaction were also reported using 1,3-dienes **5a,b** at the end of the sequence (Scheme 4.3).

A more complex example of a triple-cobalt-catalyzed sequence with five components for the synthesis of increasingly complex products has been later reported by the same group, as shown in Scheme 4.4 [5b].

In this work, the possibility of carrying out triple-cobalt catalysis utilizing different types of cobalt-catalyzed reactions was demonstrated, and it was also proved that they are capable of producing complex molecules from rather simple starting materials with a variety of functional groups.

4.2.2 Palladium-Catalyzed MCRs Containing Organoboron Compounds

Carboni and coworkers developed the application of a one-pot palladium-catalyzed cycloisomerization reaction between enynes **6**/Diels–Alder cycloaddition/allylboration sequence to efficiently generate tricyclic structures **7** with complete control of the four new stereogenic centers formed in this process (Scheme 4.5) [6].

4.3 MULTICOMPONENT ASSEMBLY OF ORGANOBORON COMPOUNDS: EFFICIENT APPROACH TO SUPRAMOLECULAR CHEMISTRY

In this section, interesting synthetic strategies for the construction of boron-based macrocycles and dendrimers [7] are described, as appealing building blocks for applications as sensors [8], catalysts, nanoarchitectures [9], separation, and delivery [10].

In this field, great achievements have been reported by the group of Severin and coworkers who have focused part of their research on the development of this new area, and they have reached interesting results that will be summarized here.

Therefore, for the construction of boron-based pentameric macrocycles **9a–f**, they performed a condensation of aryl- and alkylboronic acids with 3,4-dihydroxypyridine **8** in which 5 boronate esters are connected by dative B—N bonds (Scheme 4.6) [11]. This procedure was inspired in their previous work [12]. The boron atoms of these complex structures represent chiral centers, and the assembly process is highly diastereoselective.

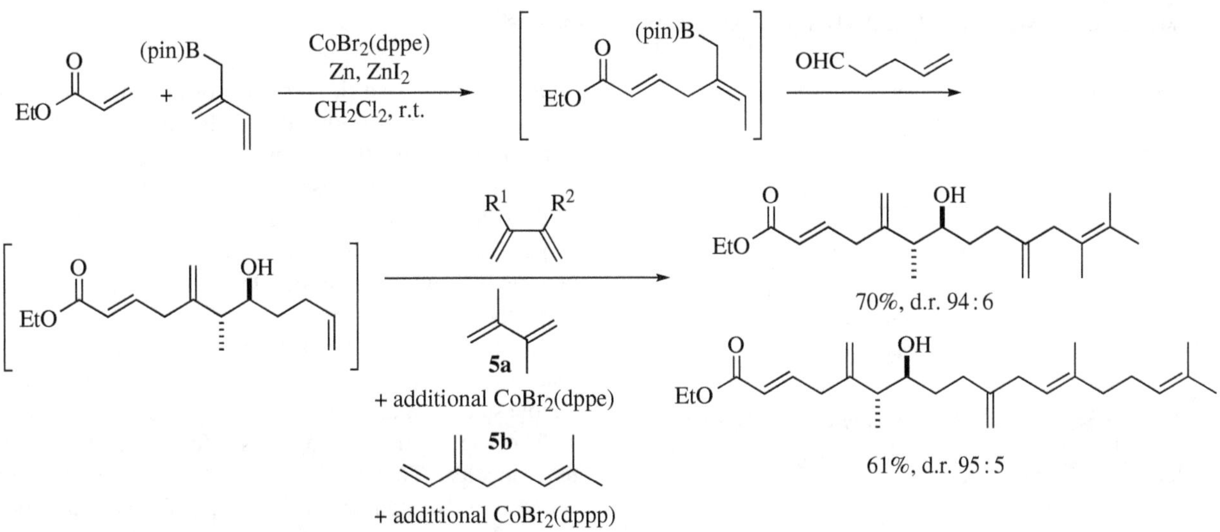

SCHEME 4.3 Four-component one-pot reaction *via* two cobalt-catalyzed 1,4-hydrovinylations and allylboration. dppp, 1,2-bis(diphenylphosphino)propane.

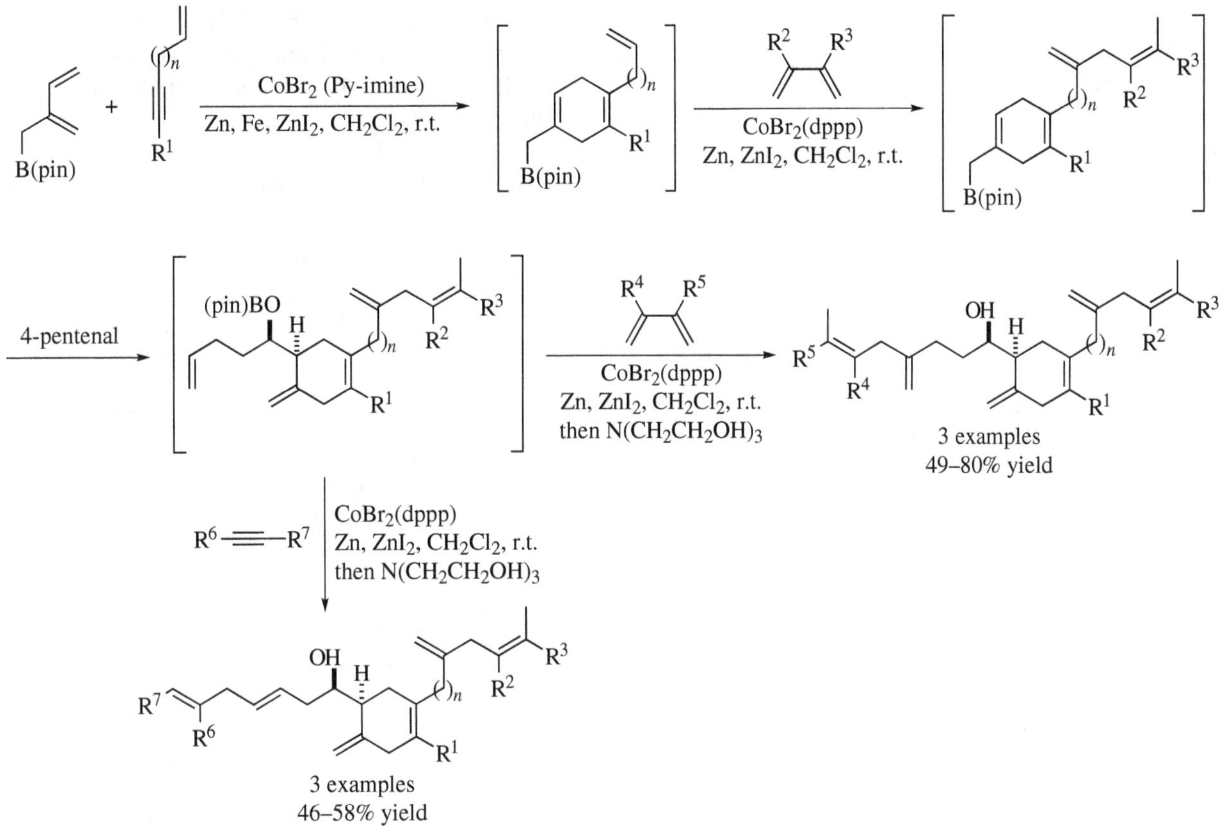

SCHEME 4.4 Five-component one-pot reaction *via* triple-cobalt-catalyzed reaction sequences. py-imine, 2,4,6-trimethylphenyl-*N*-(pyridin-2-ylmethylene)aniline.

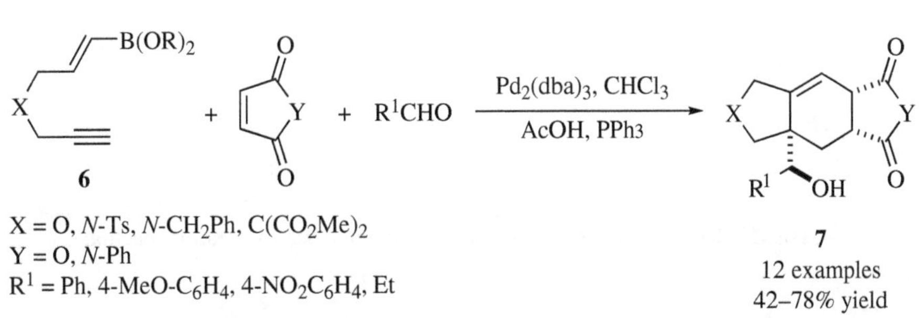

SCHEME 4.5 Scope of the cycloisomerization/Diels–Alder cycloaddition/allylboration multicomponent process.

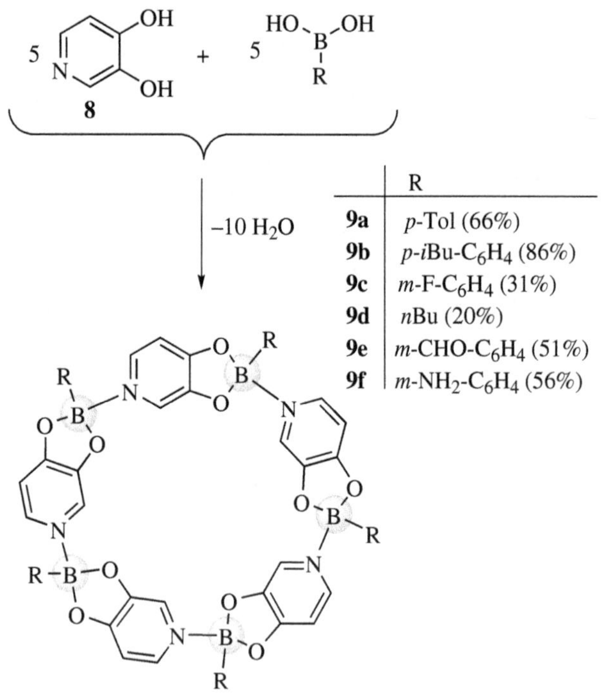

	R
9a	*p*-Tol (66%)
9b	*p*-*i*Bu-C_6H_4 (86%)
9c	*m*-F-C_6H_4 (31%)
9d	*n*Bu (20%)
9e	*m*-CHO-C_6H_4 (51%)
9f	*m*-NH_2-C_6H_4 (56%)

SCHEME 4.6 Synthesis of macrocyclic boron containing compounds.

It was found that the presence of an amino or aldehyde groups in the *meta*-position of the arylboronic acid did not interfere in the macrocyclization reaction. This allowed the effective assembly of functionalized boronic acids, dihydroxypyridine ligands, and amines or aldehydes, respectively, in a multicomponent procedure affording pentameric boronic acid macrocycles (Scheme 4.7) and tetramer derivatives (Scheme 4.8).

Later, the same group extended their methodology to other macrocycles with a cage of unprecedented size from very simple starting materials as described in Scheme 4.9 [13].

The use of different amines afforded nanometer-sized macrocycles. This is the case when 4-formylphenylboronic acid **10**, pentaerythritol **11**, and tris(2-aminoethyl) amine **12** are employed, which gave macrobicycle **9l** in 82% through the formation of 18 new covalent bonds (Scheme 4.10) [13].

To further enlarge the scope of this synthetic approach, they simultaneously combined three types of reversible interactions, 3-chloro-4-formylpyridine **14** and 3-aminophenylboronic acid **13** through a condensation as well as metal–ligand interactions using $ReBr(CO)_5$ closing the cycle [13]. The presence of metal could assist the macrocycle formation, in which 12 building blocks are

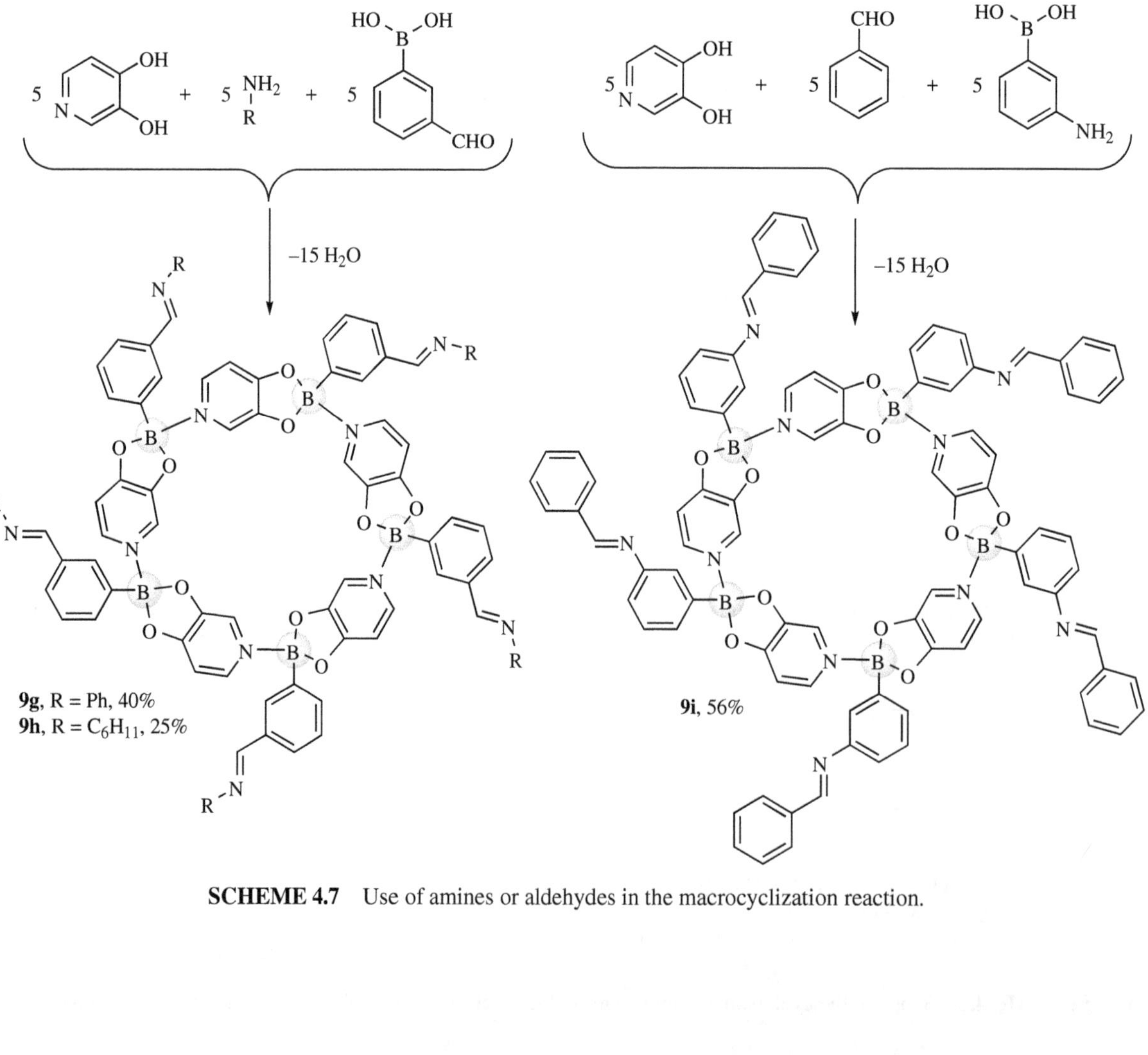

SCHEME 4.7 Use of amines or aldehydes in the macrocyclization reaction.

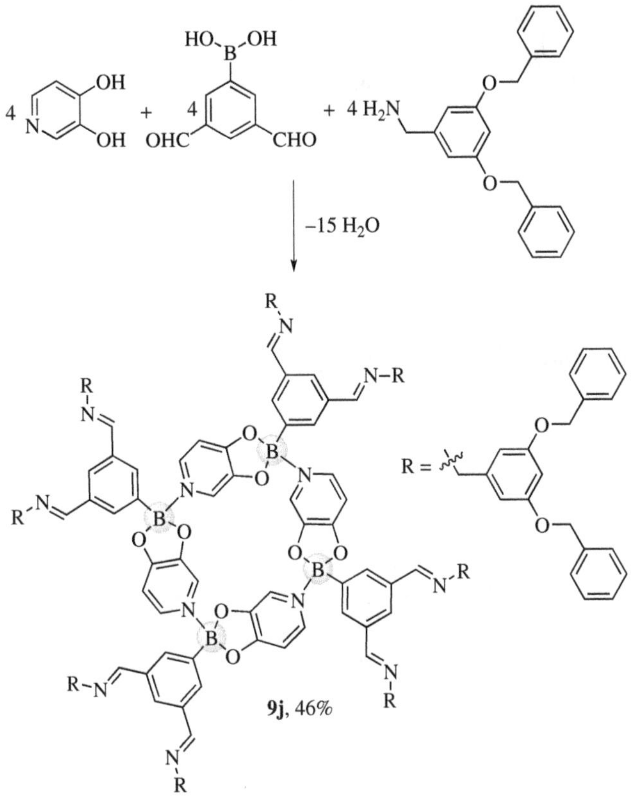

SCHEME 4.8 Formation of dendritic nanostructures based on [4+4+4]-type self-assembly.

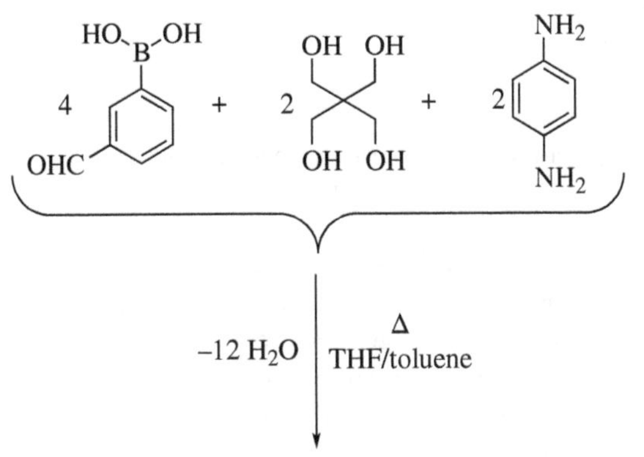

SCHEME 4.9 Synthesis of macrocycles through a [4+2+2] condensation reaction.

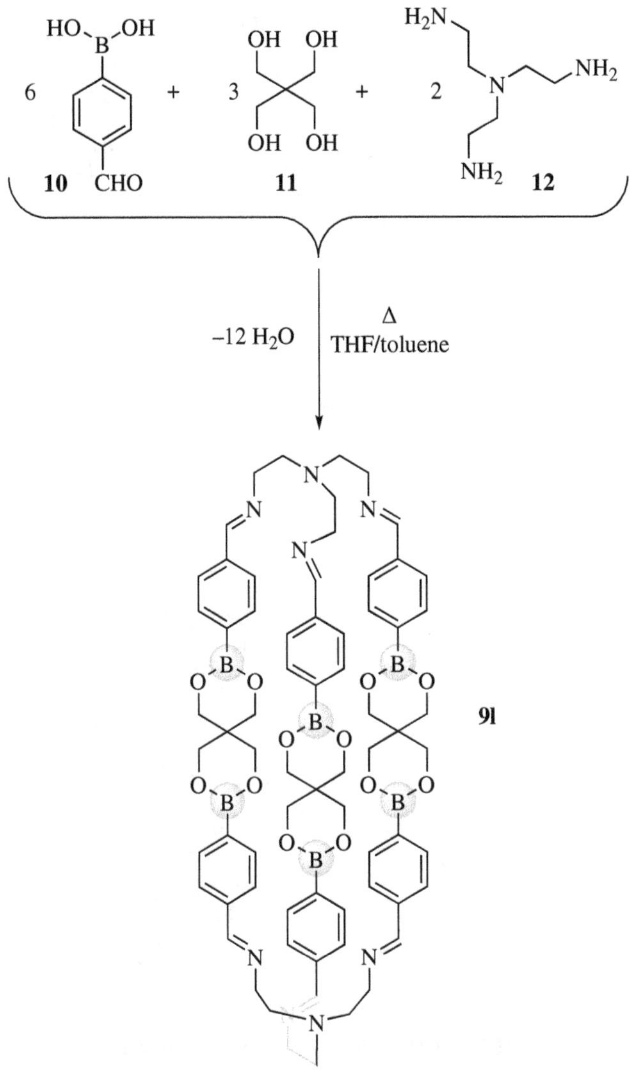

SCHEME 4.10 Multicomponent synthesis of macrobicycle **9l**.

impressively cyclized in a multicomponent procedure (Scheme 4.11).

Severin's group also prepared the first examples of boron-based rotaxanes utilizing boronic esters as stoppers. The MCR of 1,2-di(4-pyridyl) ethylene **16**, catechol **17**, 3,5-bis(trifluoromethyl)phenylboronic acid **18**, and 1,5-dinaphtho-38-crown-10 **19** resulted in the formation of rotaxane **20** (Scheme 4.12) [14].

The boronic ester exhibits a dual role: (i) this acts as a Lewis acid, and (ii) this plays as a stopper in the cyclization. However, these compounds have shown a disadvantage for potential applications in molecular devices because the B—N bond is highly dynamic and it cannot easily be fixed (e.g., by changing the temperature or solvent).

More recently, a MCR between 3 diboronic acids **22** with 6 catechols **23** and 2 tripyridyl linkers **21** allowed the formation of trigonal prismatic cages **24** (Scheme 4.13) [15]. These cages feature six dative boron–nitrogen bonds as

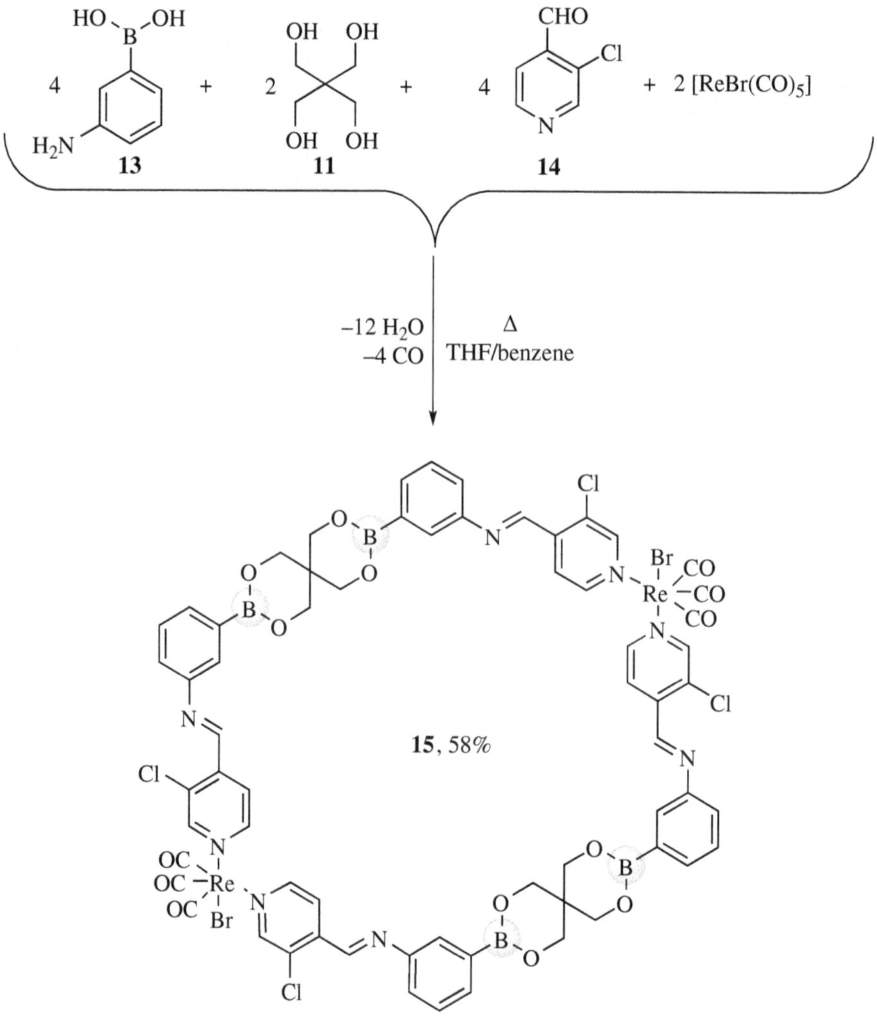

SCHEME 4.11 Metal–ligand coordination assisted boronic acid-based macrocycle **15**, from 12 molecular building blocks.

structure-directing elements. The size of the cages was varied by changing the diboronic acid building block. The authors used these complex structures to encapsulate polyaromatic molecules such as triphenylene or coronene.

The achievements in this field are very numerous and other pivotal strategies have been also developed for the preparation of boron-based molecular cages following a multicomponent protocol [16].

4.4 MULTICOMPONENT PETASIS-BORONO–MANNICH REACTION

Organoboron compounds have gained special attention and importance in the last years due to their use in different processes, but one of the most important and efficient protocols is the multicomponent Petasis-borono–Mannich (PBM) reaction [17–19]. In this approach, an amine, an aldehyde, and a boronic acid react to give access to a new amine derived from a C—C bond-forming reaction (Scheme 4.14). The limiting step of this strategy is the migration of the boronic acid substituent to the carbon, which bears an iminium ion intermediate **25**, and it could be promoted by using high temperatures. However, the presence of an OH group in either the aldehyde or amine component can activate the boronic acid by formation of a boron tetrahedral intermediate as appears in the intermediate **25**.

After the pioneering work reported by Petasis and coworkers in 1993 for the preparation of tertiary allylamines and applied to the synthesis of natifine [17a], the PBM MCR has been intensively explored for the obtainment of several organic molecules bearing distinct functional groups such as α-amino acids [20], α-hydroxyl amines [21], 2-hydroxymorpholines [22], alkylaminophenols, and 2*H*-chromenes [23, 24].

Although interesting revisions have been published covering this area [18, 24], here, we would like to show some key works developed during the second part of the last decade.

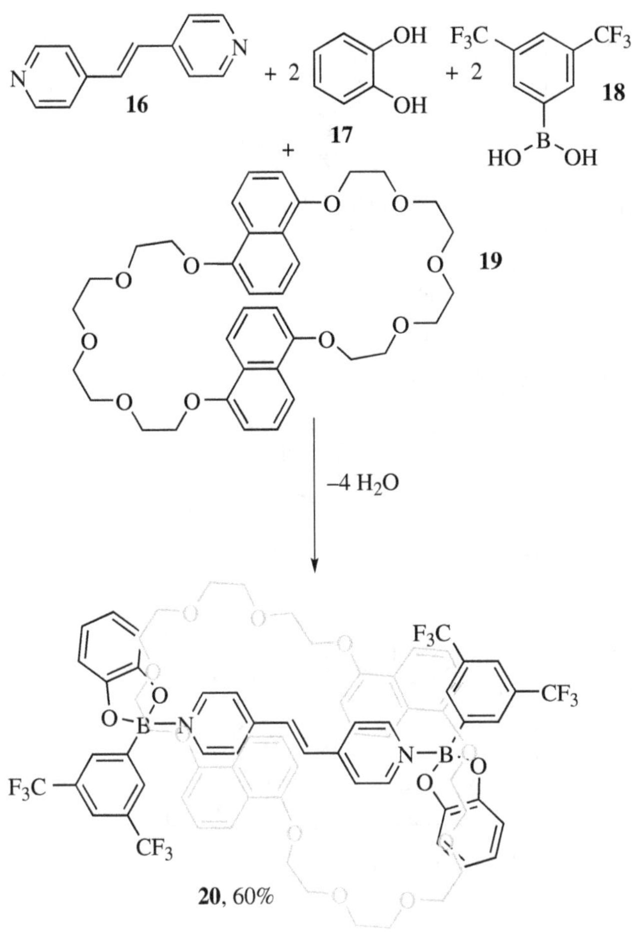

SCHEME 4.12 Synthesis of rotaxane **20** by multicomponent self-assembly.

4.4.1 Organocatalytic Enantioselective Petasis-Type Reaction

Some remarkable results have been achieved on asymmetric induction by using chiral α-hydroxy aldehydes (Scheme 4.14) [21a, 25], chiral amines [17b, c], or chiral boronates [26], affording high diastereoselection. However, the first catalytic enantioselective example of PBM has been only recently reported by Takemoto and coworkers in the field of asymmetric organocatalysis using thiourea **28** as chiral catalyst (Scheme 4.15 and Table 4.1) [27].

In this pivotal example, the authors have demonstrated the ability of bifunctional thiourea organocatalyst **28** to provide the sufficient activation of organoboronic acids in the Petasis reaction using quinolones **26**, as a powerful method for the enantio- and regioselective synthesis of 1,2-adducts **29** against the possibility of generation of 1,4-adducts (Table 4.1) [28].

Based on the formation of a reactive "ate" complex **25** (Scheme 4.14) in the Petasis reaction, as a key species playing an important role in the reactivity and diastereoselectivity of the process, the authors designed several bifunctional thiourea catalysts having a chelating functionality and envisioning that these could activate the boronic acids and direct the stereochemical outcome, as represented in Figure 4.1. In this sense, the thiourea would play a dual role activating at the same time the electrophile and the nucleophile of this process.

Later, Schaus' group reported a new organocatalytic enantioselective example of Petasis reaction employing chiral biphenol-derived diol **30** as effective catalyst to promote

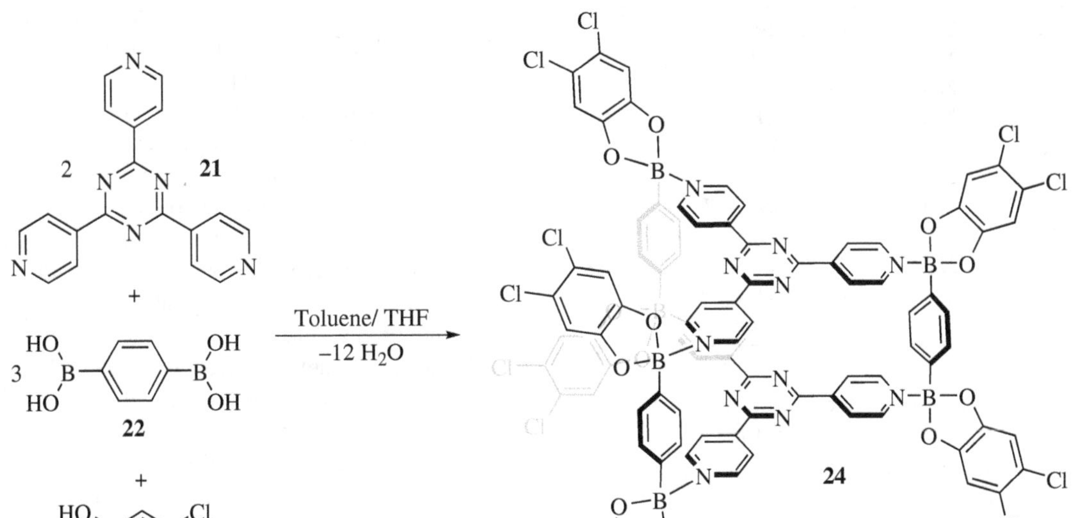

SCHEME 4.13 Multicomponent synthesis of trigonal prismatic cages **24**.

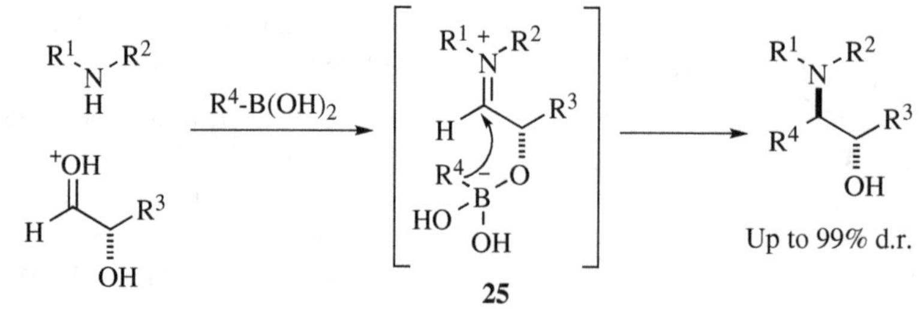

SCHEME 4.14 Petasis reaction using α-hydroxy aldehydes.

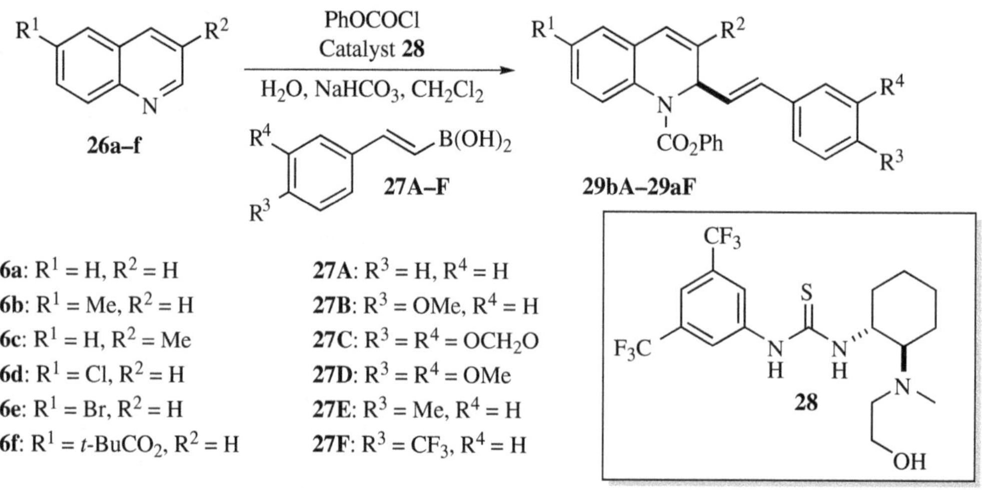

SCHEME 4.15 Thiourea organocatalyzed Petasis reaction.

TABLE 4.1 Thiourea organocatalyzed Petasis reaction

Entry	Substrate	Boronic acid	T (°C)	Product (yield)	ee (%)
1	**26a**	**27A**	−65	**29aA** (65%)	94
2	**26b**	**27A**	−65	**29bA** (75%)	95
3	**26c**	**27A**	−78	**29cA** (70%)	96
4	**26d**	**27A**	−65	**29dA** (63%)	94
5	**26e**	**27A**	−65	**29eA** (78%)	95
6	**26f**	**27A**	−65	**29fA** (61%)	96
7	**26a**	**27B**	−78	**29aB** (70%)	97
8	**26a**	**27C**	−78	**29aC** (59%)	82
9	**26a**	**27D**	−78	**29aD** (60%)	89
10	**26a**	**27E**	−65	**29aE** (60%)	91
11	**26a**	**27F**	−40	**29aF** (28%)	95

asymmetric induction between alkenyl boronates, secondary amines, and glyoxylate (Scheme 4.16) [29]. The use of glyoxylate in the Petasis reaction renders the methodology useful for the construction of chiral α-amino acids [30] with high yields and excellent enantioselectivities.

4.4.2 Metal-Catalyzed Four-Component PBM Reaction

Although the aforementioned method proceeds with high regio- and stereocontrol, the synthetic scope is limited by the scarce availability of obtaining functionalized allyl boronates. Focused on this drawback, Szabó and coworkers developed the first four-component PBM involving *in situ* generation of the allyl boronate reagent **35** of the process, starting from allyl alcohols **31** in order to extend the scope of this process (Scheme 4.17, Table 4.2) [31].

The method was efficient for primary, secondary, and tertiary alcohols and tolerated benzyloxy and carbethoxy/methoxy groups (Table 4.2). The four-component coupling reactions proceeded very cleanly, with formation of a single product with a very high selectivity, affording final α-amino acid derivatives **37a–k** as single regio- and stereoisomers from a variety of available allyl alcohols **31**.

Although the mechanism of the reaction is not fully understood, the authors strongly suggest that the catalytic transfer of the boronate group from diboronic acid **32a** or its derivatives proceeds as depicted in Scheme 4.18.

The boronate group would transfer from diboronate **38** to catalyst **33c**, affording η[1]-boronato complex **33d**. Subsequently, the boronato group substitutes the activated hydroxy group of **40** to give allyl boronic acid **35** and boric acid. Using functionalized allyl alcohols, the boronation reaction takes place with an excellent regioselectivity [32], which is very important for the subsequent allylation processes [33]. The boronation reaction shows a clear

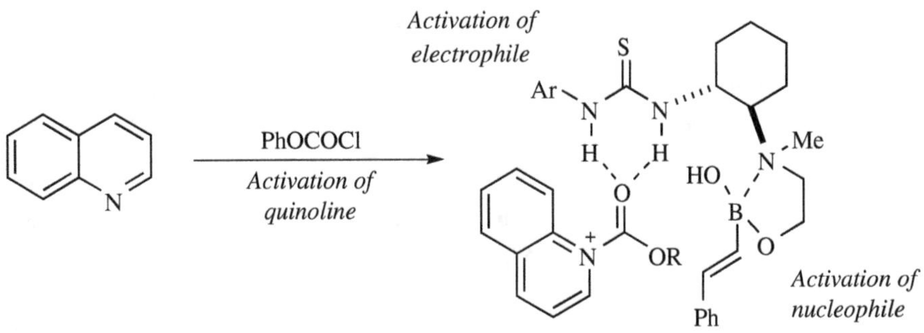

FIGURE 4.1 Mode of dual activation in the Petasis reaction.

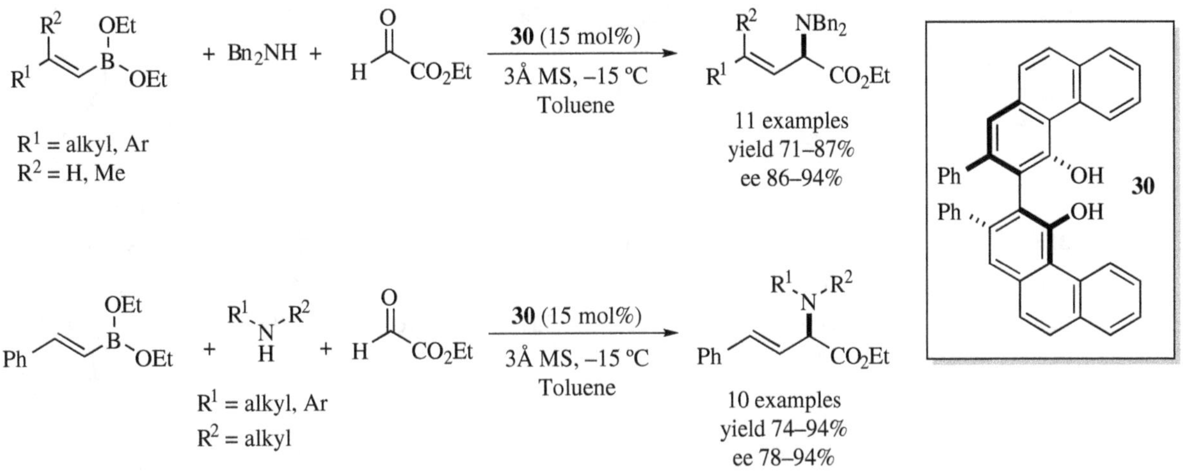

SCHEME 4.16 Organocatalyzed Petasis multicomponent reaction.

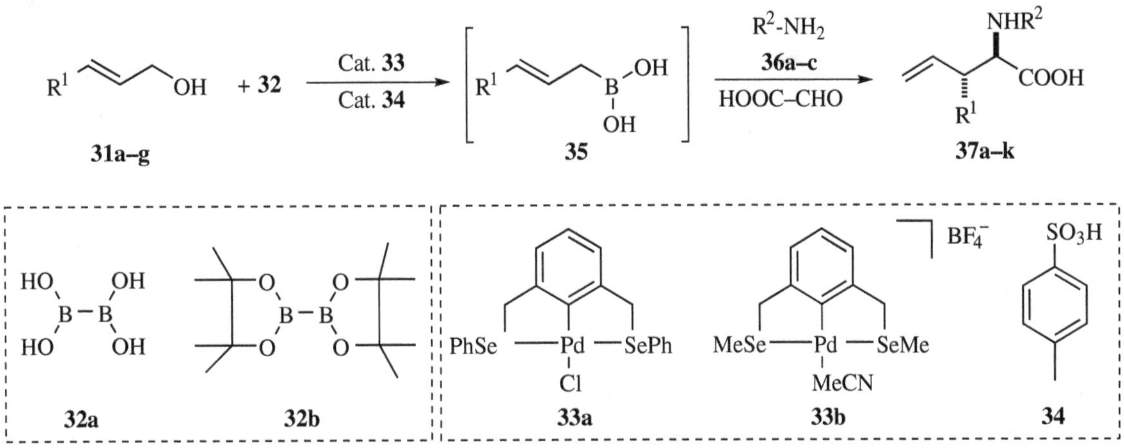

SCHEME 4.17 One-pot synthesis of **α**-amino acids *via* catalytic generation of allyl boronates.

preference for the formation of the linear allyl boronates, as both primary and secondary allyl alcohols give the corresponding terminally borylated products [32b]. In the subsequent step, the allylation of *in situ* generated imine, which proceeds with an excellent stereo- and regioselectivity, the very high stereoselectivity is probably explained by the fact that in the transition state the generated imine and the boron atom of the allyl boronate **35** come to a very close proximity as proceeds with aldehydes [34].

4.4.3 Synthetic Applications of PBM

The PBM methodology has been used as the key step in the synthesis of interesting molecules such as the pioneering

TABLE 4.2 Four-component synthesis of α-amino acids from allyl alcohols *via* PBM

Entry	Alcohol	Amine	Diboron	Cat.	Cond. A[a]	Cond. B[b]	Product **37**
1	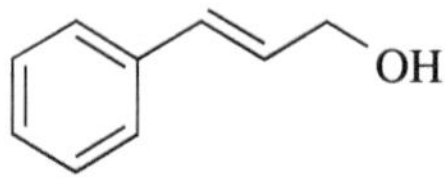 **31a**	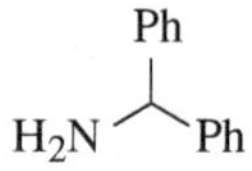 **36a**	**32a**	**33a**	40/16	25/16	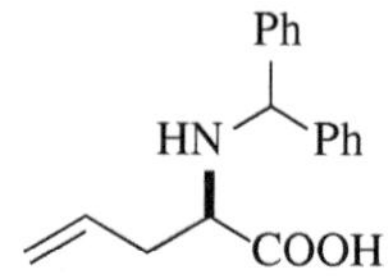 **37a** (76%)
2	**31a**	**36a**	**32b**	**33a**	50/16	25/16	**37a** (67%)
3	**31a**	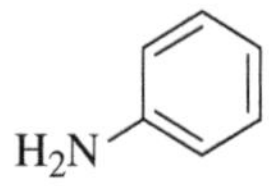 **36b**	**32a**	**33a**	40/16	25/16	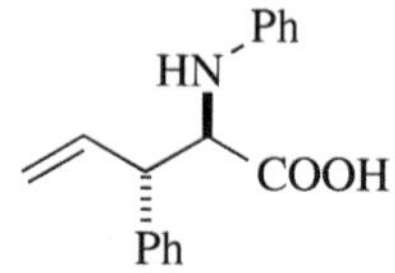 **37b** (78%)
4	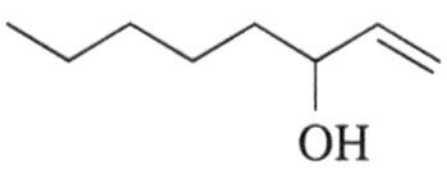 **31b**	**36a**	**32a**	**33a**	50/16	25/8	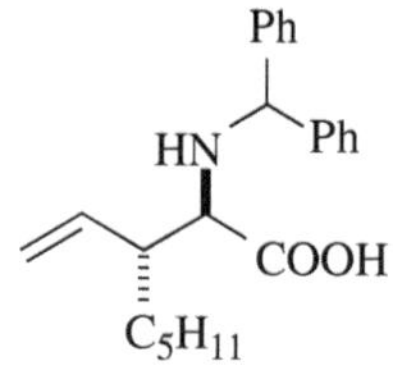 **37c** (75%)
5	**31b**	**36a**	**32a**	**33b**	50/4	25/16	**37c** (83%)
6	**31b**	**36a**	**32b**	**33b**	50/4	25/16	**37c** (77%)
7	**31b**	**36b**	**32a**	**33a**	50/16	25/16	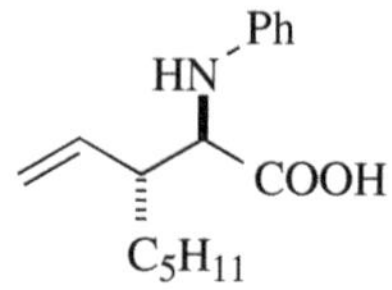 **37d** (77%)
8	**31b**	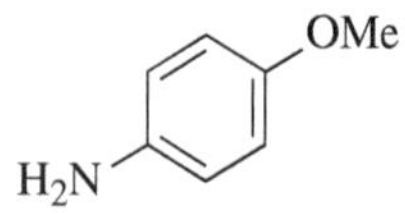 **36c**	**32a**	**33a**	50/16	25/8	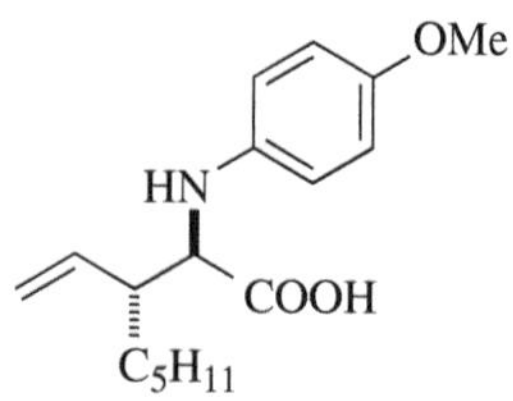 **37e** (52%)
9	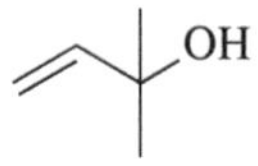 **31c**	**36a**	**32a**	**33a**	50/16	25/24	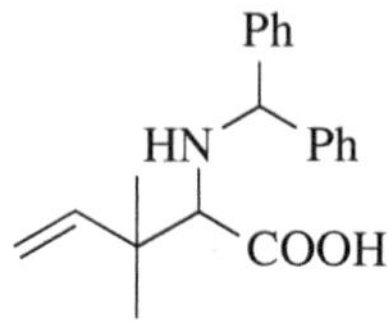 **37f** (60%)
10	 **31d**	**36a**	**32a**	**33a**	40/8	25/16	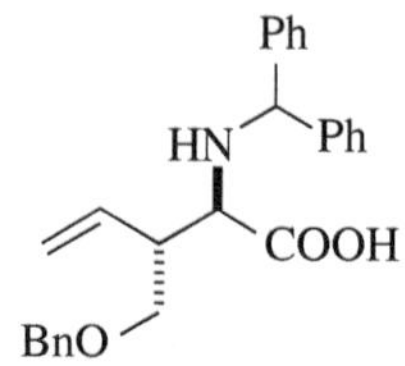 **37g** (75%)
11	**31d**	**36b**	**32a**	**33a**	40/8	25/16	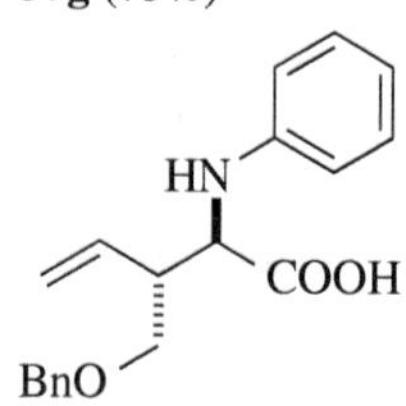 **37h** (77%)

TABLE 4.2 ***(Continued)***

Entry	Alcohol	Amine	Diboron	Cat.	Cond. A[a]	Cond. B[b]	Product **37**
12	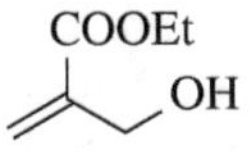 **31e**	**36a**	**32a**	**33a**	50/16	25/16	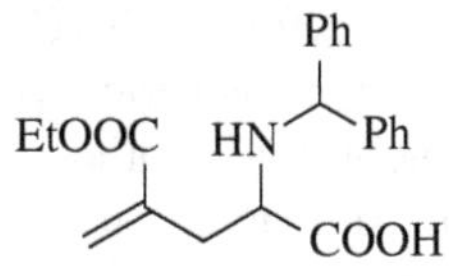 **37i** (68%)
13	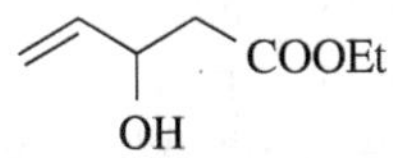 **31f**	**36a**	**32a**	**33a**	50/16	25/16	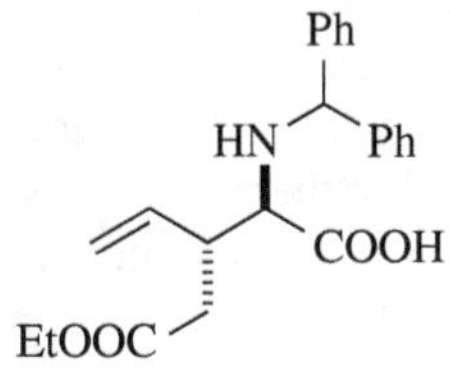 **37j** (78%)
14	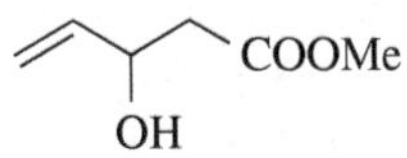 **31g**	**36a**	**32a**	**33a**	50/16	25/16	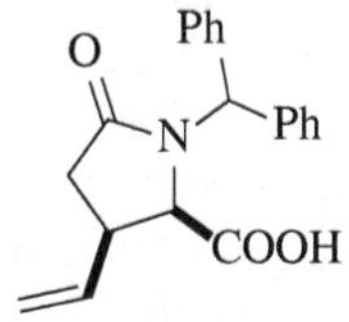 **37k** (80%)

[a]Conditions A: temperature/time (°C/h) for the boronation reaction.
[b]Conditions B: temperature/time (°C/h) for the allylation reaction.

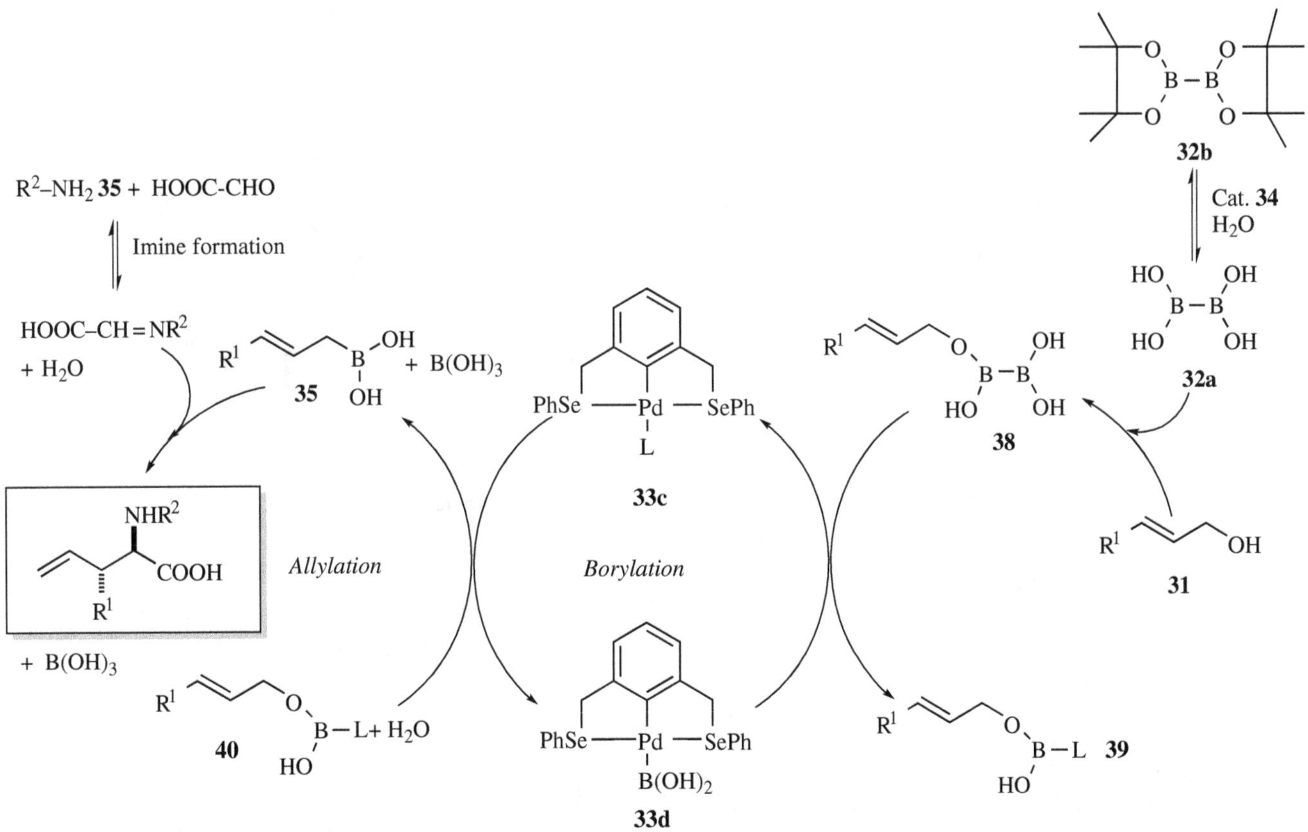

SCHEME 4.18 Plausible mechanism for 4-component PBM reaction.

example reported by Hansen and coworkers for the preparation of the hormone secretagogue NN703 **41** (Scheme 4.19) [35].

Kalinski and coworkers explored the potential of different powerful MCRs (Ugi, Petasis, and Passerini) in the production of generic drugs such as the antiplatelet agent clopidogrel **42** synthesized in the racemic form (Scheme 4.20) [36]. Although the use of Ugi 3-component reaction afforded better yields in this synthesis, the PBM reaction allowed the preparation of the desired product **42** in 44% overall yield.

Schaus and coworkers envisioned the preparation of CCR5 antagonist maraviroc **43**, a new class of compounds for HIV therapy and responsible for recognition of the virus, through an asymmetric allylboration of benzoyl imines. It is remarkable that this is not a multicomponent protocol since the imine is previously formed (Scheme 4.21) [37]. (*S*)-3,3′-Ph_2-BINOL was the catalyst of choice to promote this process. It was found that polar noncoordinating solvents gave faster rates and higher selectivities and the addition of molecular sieves to the reaction medium was crucial avoiding the decomposition of unstable acyl imine. An additional mechanistic study of this process carried out by the same authors indicated that a dissymmetrical boronate complex with the catalyst, without the formation of the corresponding cyclic boronate, was the active species for the chiral induction of this transformation.

An interesting example starting from L-xylose as the aldehyde component was addressed by Pyne and coworkers for the synthesis of uniflorine A **45** using the PBM reaction and ring-closing metathesis as the key steps (Scheme 4.22) [38]. After the initial PMB step affording intermediate **44**, an efficient 8-step synthesis finally rendered uniflorine A **45** [39]. Boronate intermediate **46** was proposed to explain the stereochemical outcome of this reaction, in which the reactive conformation is one that minimized 1,3-allylic strain as depicted in Scheme 4.22.

It must be mentioned that the authors corrected the structure of this compound in a later article Figure 4.2 [40]. The same group applied this strategy as a key step for the synthesis of other alkaloids such as polyhydroxyindolizidine castanospermine **47** [41], the corrected structure of (+)-uniflorine A **48** [40], and the australine derivatives **49** and **50** [42]. All these structures are synthesized from a common initial tetrahydroxy amine with the desired configurations.

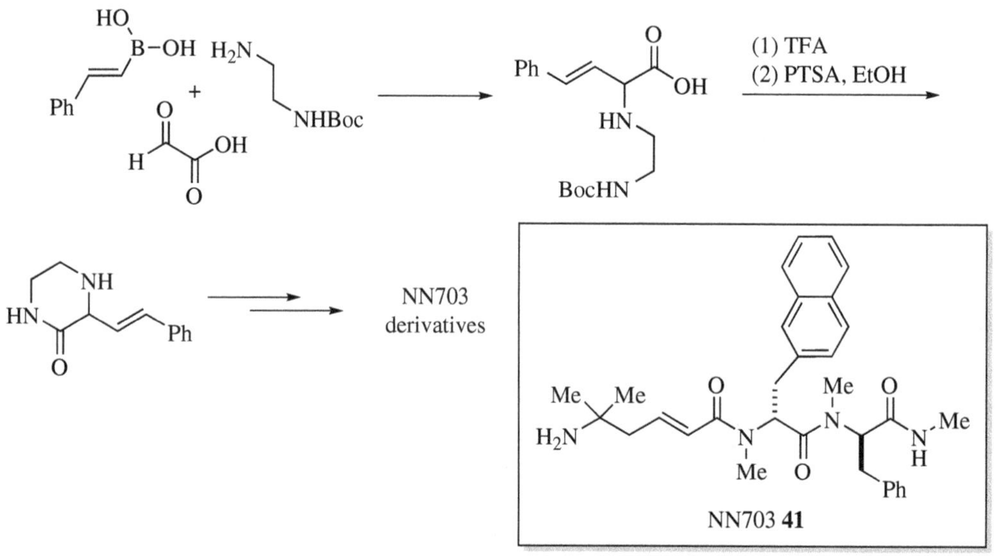

SCHEME 4.19 Synthesis of active NN703 derivatives through PMB protocol.

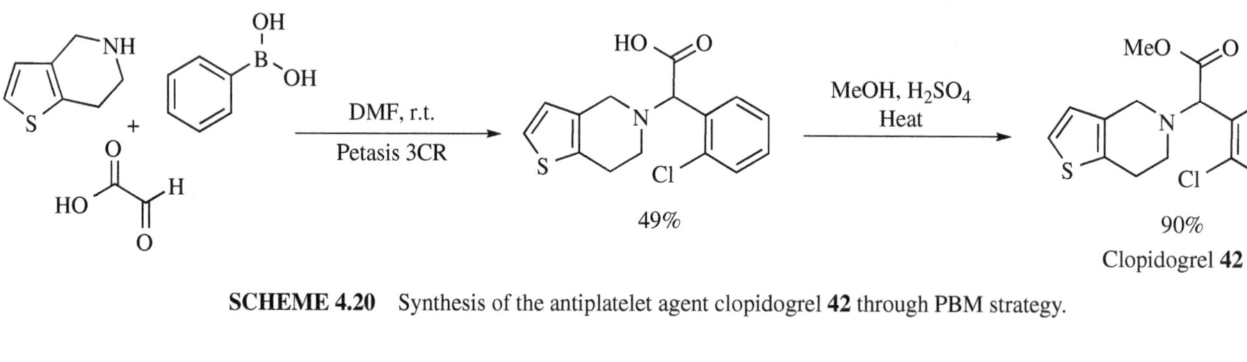

SCHEME 4.20 Synthesis of the antiplatelet agent clopidogrel **42** through PBM strategy.

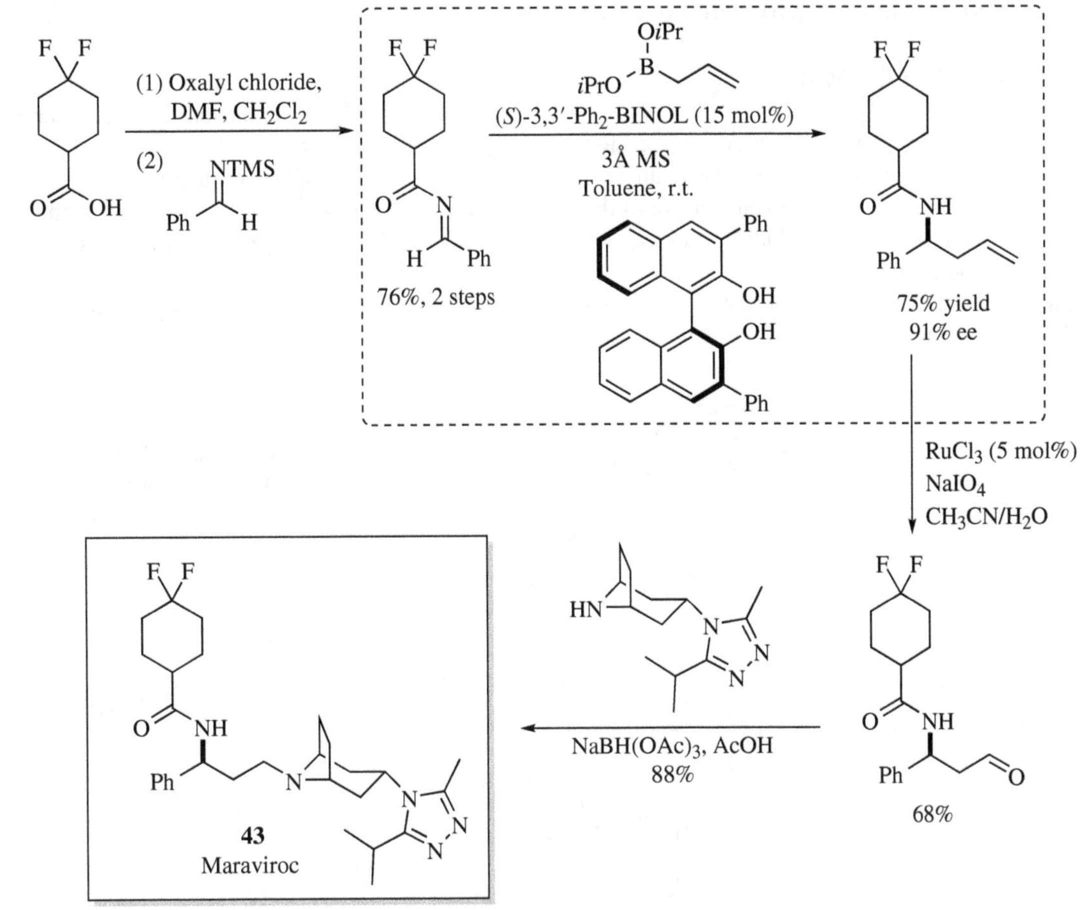

SCHEME 4.21 Synthesis of maraviroc **43** *via* PMB protocol.

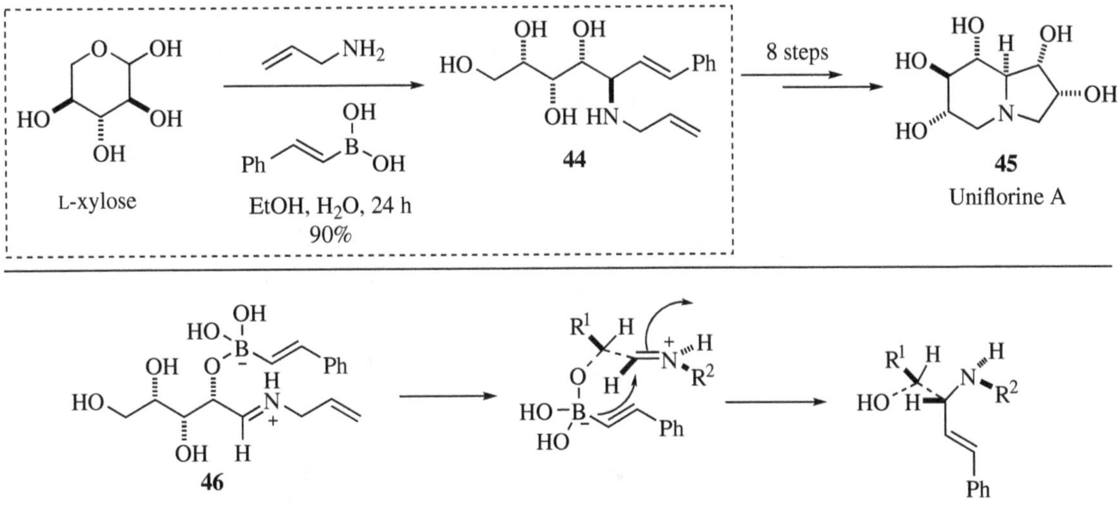

SCHEME 4.22 Synthesis of uniflorine A **45**.

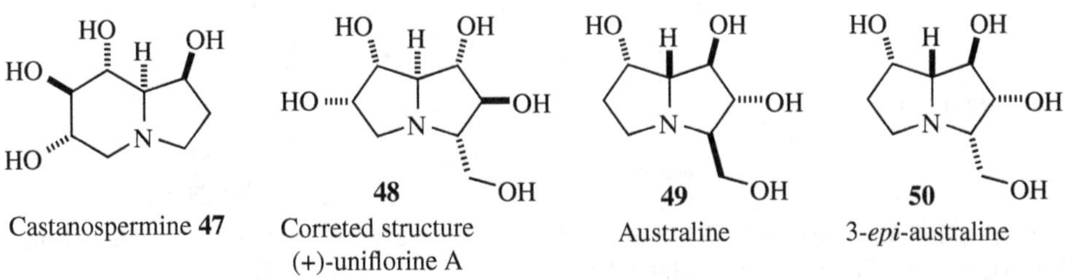

FIGURE 4.2 Target compounds synthesized through a PMB protocol.

4.5 ALLENYLBORATES IN MCRs

Yoshida MCR is a palladium-catalyzed coupling reaction between allenylboronates, secondary amines, and an aryl iodide to give allylic amines **51** with complete control of the olefin geometry, which could be isolated or used for subsequent reactions (Scheme 4.23) [43].

Intermediate **51**, containing a vinyl boronate, could then be functionalized and transformed in different interesting scaffolds depending on the amine group through several steps as shown in Scheme 4.23 as a proof of fact of the diversity and usefulness of this protocol.

The applicability of this efficient methodology was demonstrated by the synthesis of the anti-inflammatory agent and antidepressant rolipram **52** in four steps with a 48% overall yield and starting from a Yoshida MCR (Scheme 4.24) [43].

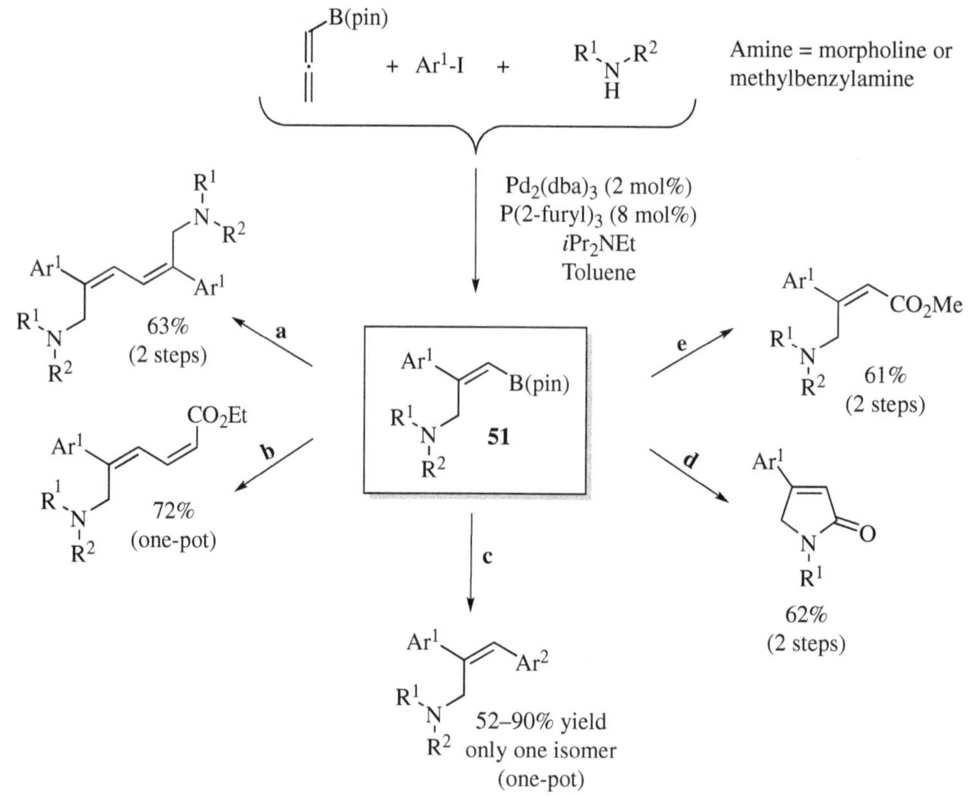

SCHEME 4.23 Yoshida multicomponent reaction: (a) (Ar^1=*p*-tolyl) $Pd(OAc)_2$, PPh_3, Cs_2CO_3, H_2O, toluene, 90 °C, under air. (b) Ethyl *cis*-3-iodoacrylate, Cs_2CO_3, H_2O, 90 °C. (c) Ar^2I, Cs_2CO_3, H_2O. (d) $Pd(OAc)_2$, PPh_3, *p*-benzoquinone, CO (1 atm), MeOH, 50 °C (amine is benzylamine). (e) $Pd(OAc)_2$, PPh_3, *p*-benzoquinone, CO (1 atm), MeOH, 50 °C (amine is morpholine).

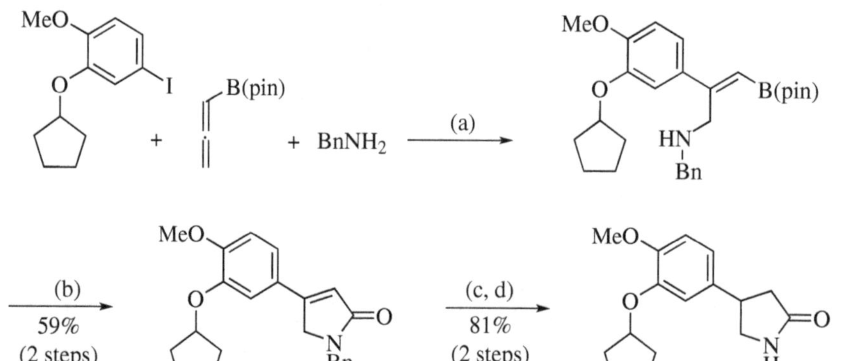

52
Rolipram

SCHEME 4.24 Synthesis of rolipram: (a) $Pd_2(dba)_3$, P(2-furyl)$_3$, iPr_2NEt, toluene, 80 °C; (b) $Pd(OAc)_2$, PPh_3, *p*-benzoquinone, CO (1 atm), MeOH, 50 °C; (c) Pd/C, H_2 (60 atm), EtOH, 60 °C; (d) Li, liq. NH_3, −40 °C.

4.6 MULTICOMPONENT HETERO-DIELS–ALDER/ALLYLBORATION

4.6.1 Chiral Catalyzed One-Pot [4 + 2] Cycloaddition/Allylboration

Although a multicomponent [4 + 2] cycloaddition/allylboration reaction is not a new protocol [44] and this sequence has been explored to trap the apparently unstable intermediate with an aldehyde to afford the corresponding alcohols directly after hydrolysis, the pioneering chiral catalytic example was independently reported by Hall's [45] and Carreaux's [46] groups in 2003, and afterward, Hall's research group has extended the methodology (Scheme 4.25) [47].

The authors developed the first chiral catalyzed inverse electron demand hetero-[4 + 2] cycloaddition between 3-boronoacrolein esters **53** and alkyl enol ether **54** giving allylborate product **56** with high conversion. Subsequent addition of alkyl and aryl aldehydes, after completion of the first cyclization step, afforded α-hydroxyalkyl dihydropyrans **57** under an allylboration step with very high yield and ee. It is noteworthy that the reactions were carried out under solventless conditions, which are ideal from both an economical and environmental point of view. The conversion of the carbon–boron bond into a carbon–oxygen bond through H_2O_2 oxidation was also alternatively demonstrated, showing the versatility of this methodology sequence and the versatility of vinyl borates in general.

4.6.1.1 Application of Hall's [4 + 2] Cycloaddition/Allylboration To demonstrate the applicability of the aforementioned methodology, Hall's group envisioned that the pyran core of a potent marine antibiotic [48] could be synthesized applying an *endo*-selective three-component hetero-[4 + 2] cycloaddition/allylboration approach as a key step in the catalytic asymmetric synthesis of thiomarinol derivative, as shortly depicted in Scheme 4.26 [49].

Following this route, the authors achieved the synthesis of 10 analogues of compound **58**, in a high global yield, where the crucial step of this total synthesis was the efficient catalytic enantio-, regio-, *E*/*Z*-, and diastereoselective three-component inverse electron demand hetero-[4 + 2] cycloaddition/allylboration sequence. This key process provides a rare example of an enantioselective hetero-Diels–Alder reaction involving acyclic 2-substituted enol ethers. Additionally, these compounds were evaluated for antimicrobial activity, and two of them showed more activity than the original thiomarinol H.

More recently, the same authors have also synthesized the pyran core of psymberin **59**, a natural product isolated from marine sponge [50], following their own developed procedure as shown in Scheme 4.27 [51].

The direct synthesis of the crucial intermediate **60** for the preparation of psymberin **59** is depicted in Scheme 4.28.

An interesting overview of the use of boron 1,3-diene or heterodienes in the synthesis of appealing structures through a multicomponent strategy has been recently compiled by Carboni and coworkers [52].

4.6.2 Polymer-Supported MCRs

The chemistry of the boron compounds has been also developed using solid supports. Hall's group performed a

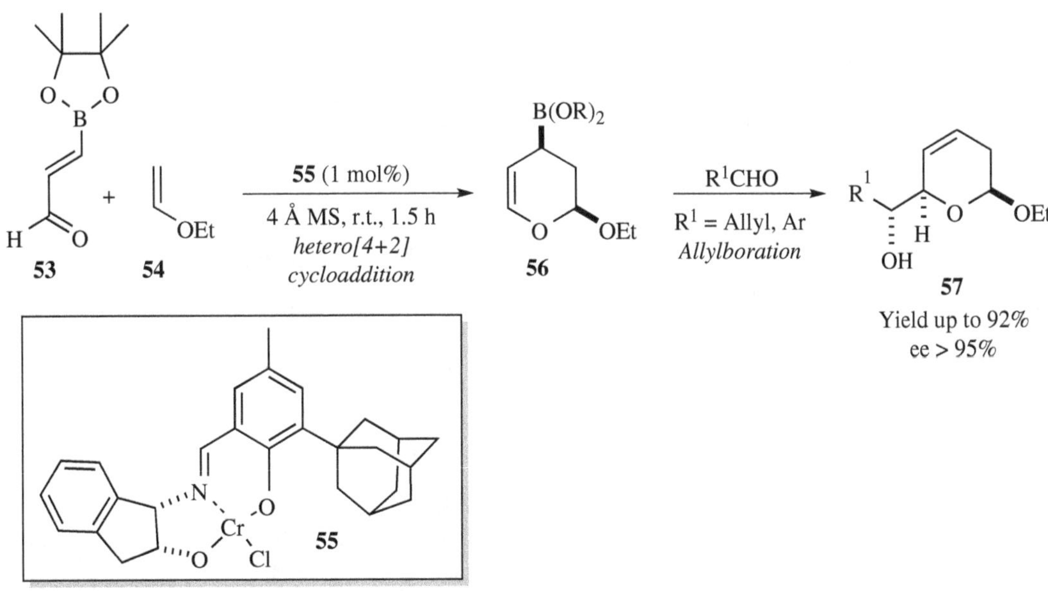

SCHEME 4.25 Chiral cycloaddition [4 + 2]/allylboration using aldehydes.

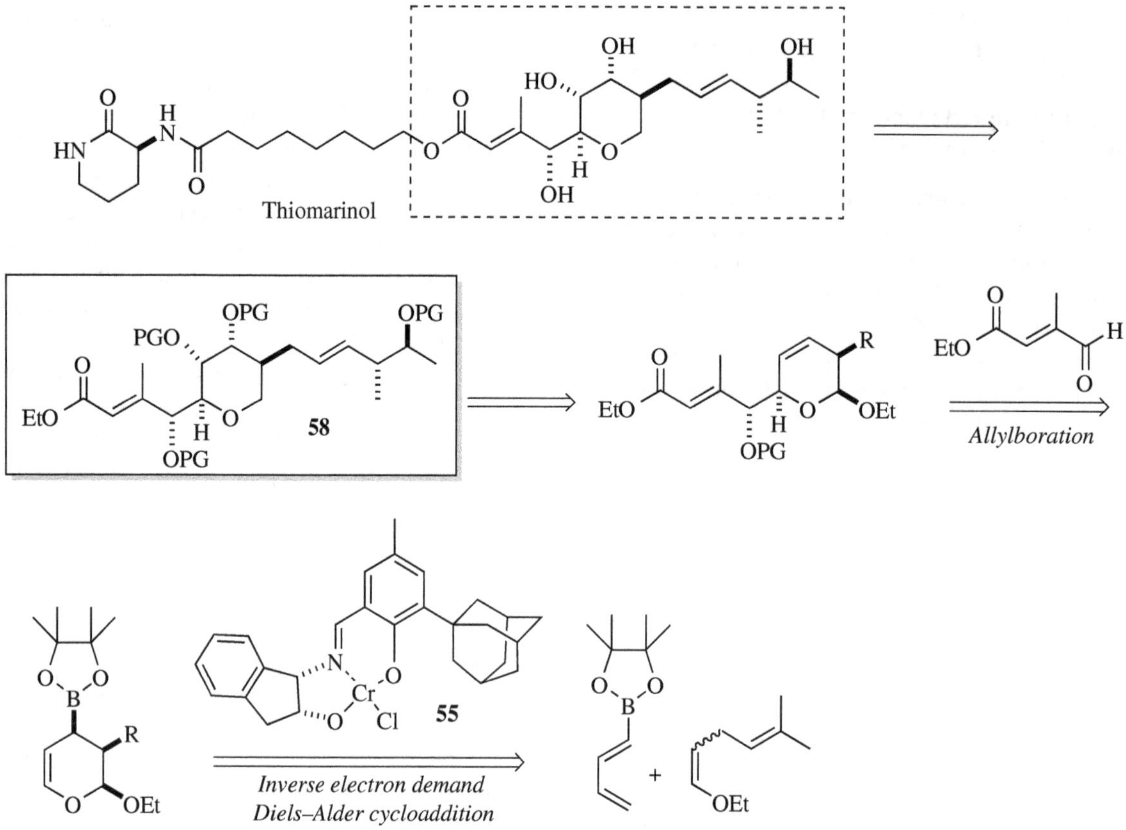

SCHEME 4.26 Retrosynthetic analysis for the preparation of compound **58**.

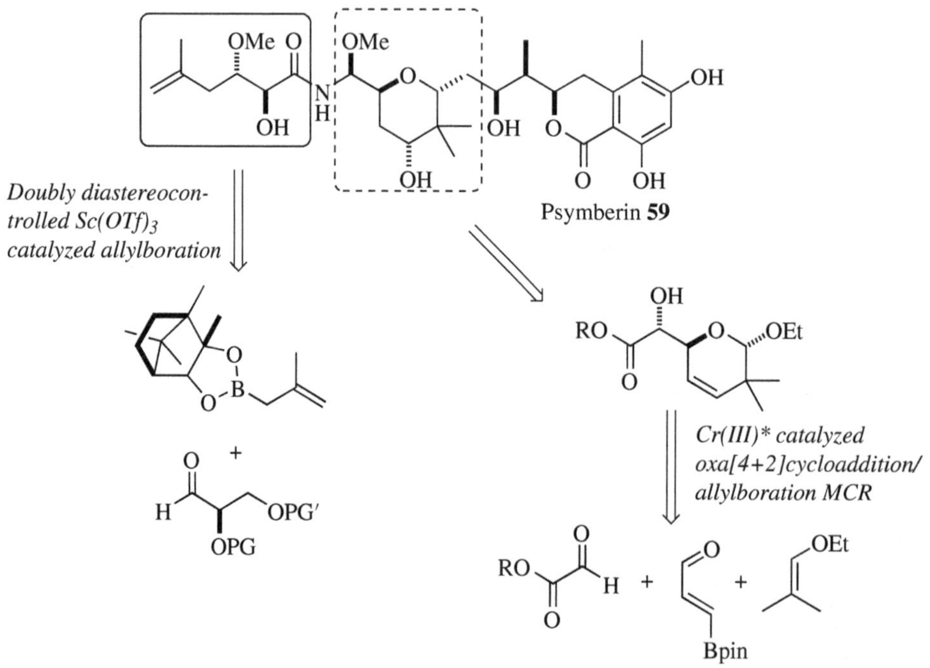

SCHEME 4.27 Retrosynthetic analysis for the preparation of compound **59**.

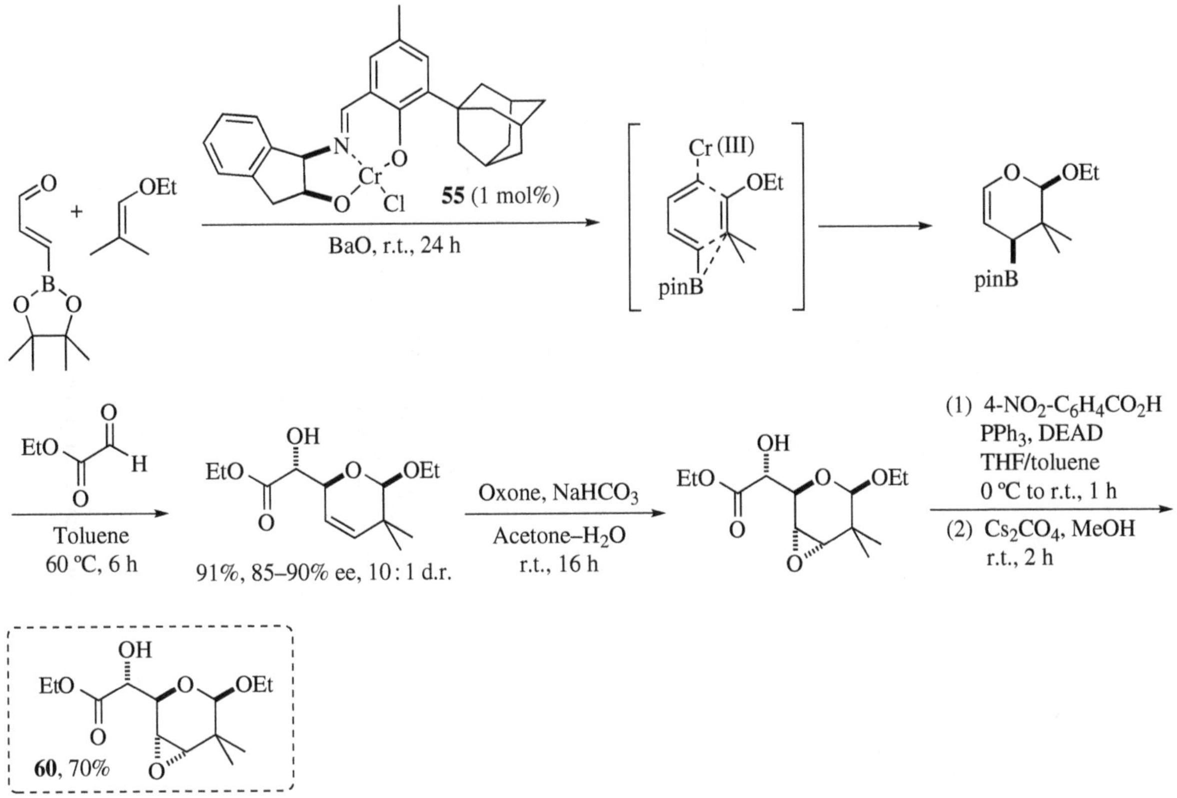

SCHEME 4.28 Synthesis of intermediate **60**.

multicomponent aza-[4+2] cycloaddition/allylboration reaction between 1-aza-4-boronobutadienes, maleimides, and aldehydes on a solid support (Scheme 4.29) [53]. The process was performed either with the immobilization of maleimide on a Sasrin resin or the anchoring of *m*-hydroxybenzaldehyde *via* a silyloxy linker as depicted in Scheme 4.29.

Interestingly, the same strategy of a tandem aza-[4+2] cycloaddition/allylboration MCR, although in solution-phase, was employed to obtain a library of 944 polysubstituted piperidines in a high degree of purity suitable for biological screening for a panel of phosphatase enzymes, demonstrating the utility of the method.

4.7 PALLADIUM-CATALYZED ASYMMETRIC ALLENE DIBORATION/α-AMINOALLYLATION

Morken and coworkers extended their previously developed palladium-catalyzed asymmetric diboration of prochiral allenes [54] to the subsequent α-aminoallylation of the *in situ* generated allyl and vinylboronate **62**, as versatile intermediate for allylation reactions (Scheme 4.30) [55].

This methodology produces extremely useful chiral β-ketone amines **63** with very high enantioselectivities, generated from simple starting materials. In addition, the allylation intermediate could be protonated with propionic acid or coupling with iodobenzene under Suzuki–Miyaura cross-coupling conditions, to give allylic amines **64** with high yield and without loss of enantioselectivity (Scheme 4.31).

4.8 SYNTHETIC APPLICATIONS OF BORON-BASED MCRs

She and coworkers successfully applied the one-pot double allylboration reaction previously reported by Roush's group [56] for the enantio- and diastereoselective preparation of a (*E*)-1,5-*anti*-ene-diol subunit, a key fragment in the total synthesis of (+)-strictifolione **65** [57], a 6-susbtituted α-pyrone with antifungal activity (Scheme 4.32) [58].

The synthesis of (+)-strictifolione **65** is accomplished in 5 synthetic steps with an overall yield of 23%. This synthesis demonstrates the utility of organoboron compounds to introduce the desired stereogenic center through a MCRs.

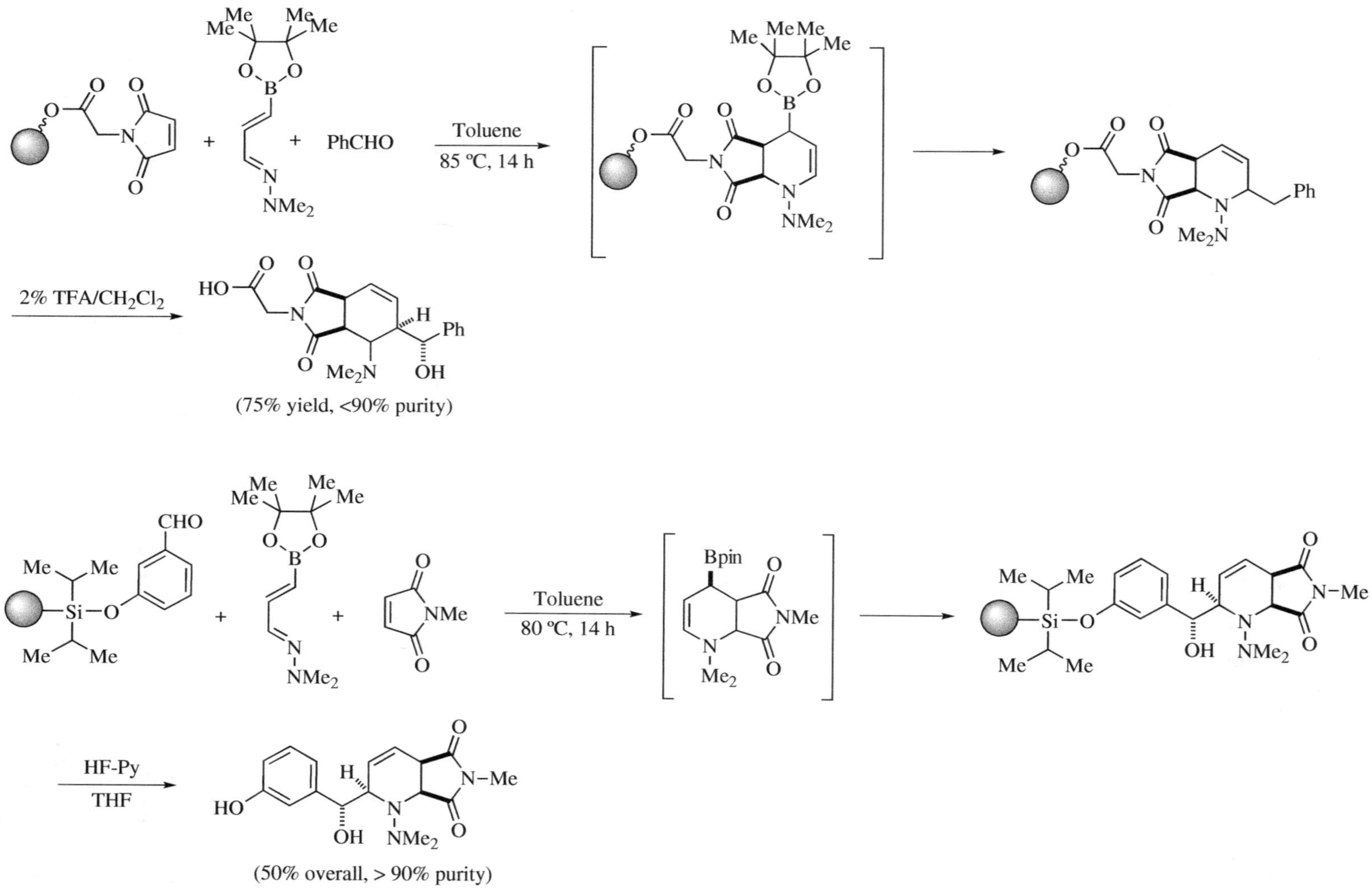

SCHEME 4.29 Solid-phase synthesis approach with a supported aldehyde or the maleimide component.

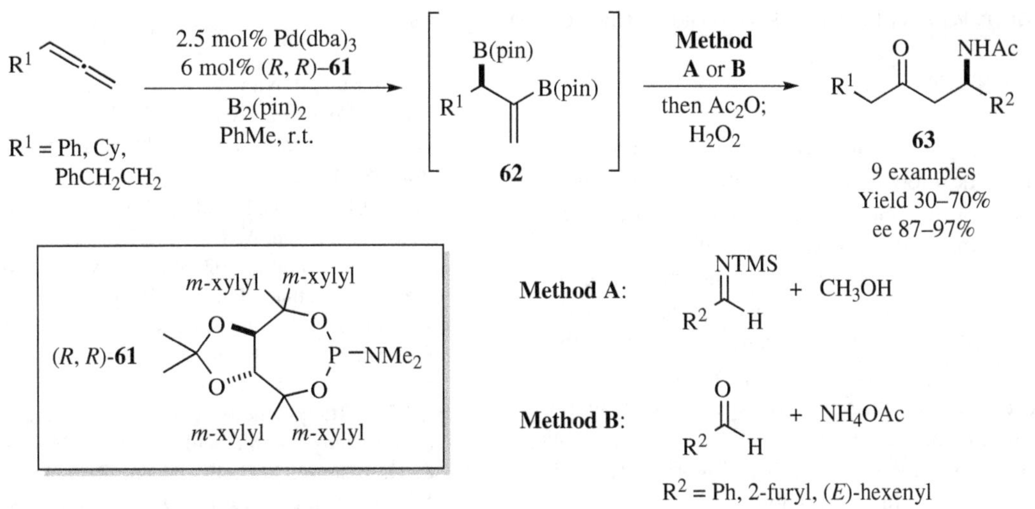

SCHEME 4.30 Palladium-catalyzed enantioselective diboration/allylation.

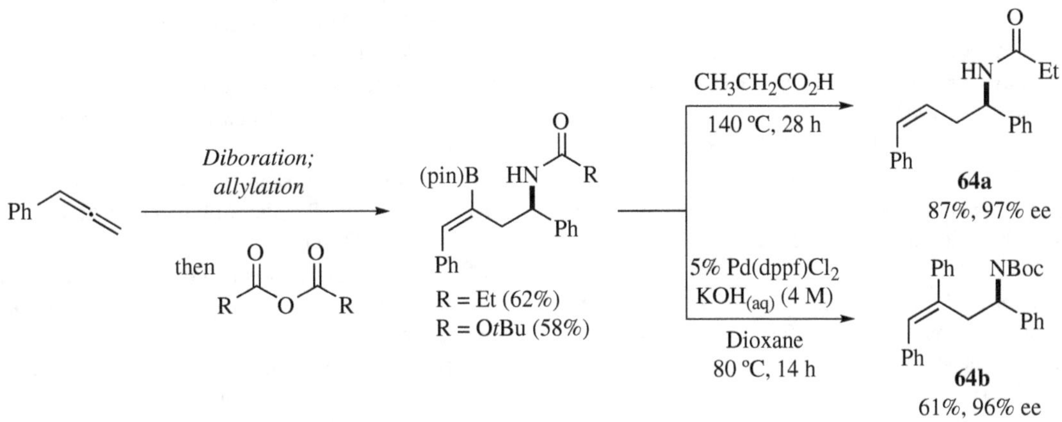

SCHEME 4.31 Synthesis of allylic amines **64**.

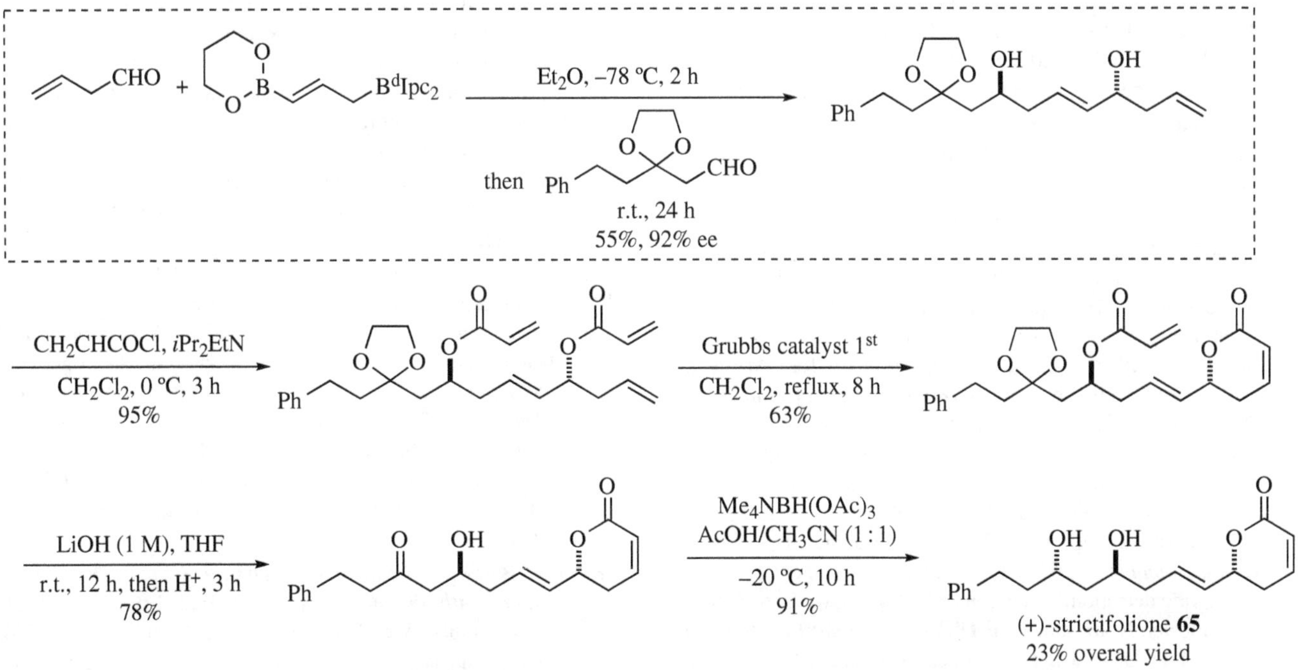

SCHEME 4.32 Total synthesis of (+)-strictifolione **65**.

4.9 CONCLUSION

In this chapter, we have given a brief overview of the interesting chemistry developed with organoboron compounds related with their participation in multicomponent processes. We have given to the reader a broad vision of the interesting examples reported in this field. Although the importance of the organoboron compounds has been reflected with the examples disclosed in this chapter, many other works also formed part of this huge area of research, and they have been crucial for the advance of this field. We expect more research in the asymmetric catalytic version that in the last decade has been the center of an impressive number of works due to the continuous concern with the sustainability.

REFERENCES

1 D. Hall (Ed.), *Boronic acids: preparation and applications in organic synthesis and medicine*, Wiley-VCH Verlag GmbH, Weinheim, Germany, 2005.

2 A. Suzuki, *Heterocycles* 2010, 80, 15–43. Organoboranes in organic synthesis including Suzuki coupling reaction.

3 G. Hilt, K. I. Smolko, *Angew. Chem. Int. Ed.* 2003, 42, 2795–2797. Alkynylboronic esters as efficient dienophiles in cobalt-catalyzed Diels-Alder reactions.

4 G. Hilt, S. Lüers, K. I. Smolko, *Org. Lett.* 2005, 7, 251–253. A two-step reaction sequence for the syntheses of tetrahydronaphthalenes.

5 (a) F. Erver, G. Hilt, *Org. Lett.* 2011, 13, 5700–5703. Multicomponent regio- and diastereoselective cobalt-catalyzed hydrovinylation/allylboration reaction sequence. (b) F. Erver, G. Hilt, *Org. Lett.* 2012, 14, 1884–1887. Double- and triple-cobalt catalysis in multicomponent reactions.

6 A. Hercouet, F. Berrée, C. H. Lin, L. Toupet, B. Carboni, *Org. Lett.* 2007, 9, 1717–1720. Boronated enynes as versatile sources of stereodefined and skeletally diverse molecules.

7 R. Nishiyabu, Y. Kubo, T. D. James, J. S. Fossey, *Chem. Commun.* 2011, 47, 1124–1150. Boronic acid building blocks: tools for self assembly.

8 For some selected examples, see: (a) A. M. Kelly, Y. Pérez-Fuertes, J. S. Fossey, S. L. Yeste, S. D. Bull, T. D. James, *Nat. Protoc.* 2008, 3, 215–219. Simple protocols for NMR analysis of the enantiomeric purity of chiral diols. (b) Y. Pérez-Fuertes, A. M. Kelly, J. S. Fossey, M. E. Powell, S. D. Bull, T. D. James, *Nat. Protoc.* 2008, 3, 210–214. Simple protocol for NMR analysis of the enantiomeric purity of chiral primary amines. (c) D. K. Scrafton, J. E. Taylor, M. F. Mahon, J. S. Fossey, T. D. James, *J. Org. Chem.* 2008, 73, 2871–2874. "Click-fluors": modular fluorescent saccharide sensors based on a 1,2,3-triazole ring. (d) W. M. J. Ma, M. P. P. Morais, F. D'Hooge, J. M. H. van den Elsen, J. P. L. Cox, T. D. James, J. S. Fossey, *Chem. Commun.* 2009, 532–534. Dye displacement assay for saccharide detection with boronate hydrogels. (e) S. A. Elfeky, S. E. Flower, N. Masumoto, F. D'Hooge, L. Labarthe, W. Chen, C. Len, T. D. James, J. S. Fossey, *Chem. Asian J.* 2010, 5, 581–588. Diol appended quenchers for fluorescein boronic acid.

9 S. Tamesue, M. Numata, K. Kaneko, T. D. James, S. Shinkai, *Chem. Commun.* 2008, 4478–4480. Hierarchical carbon nanotube assemblies created by sugar–boric or boronic acid interactions.

10 (a) F. D'Hooge, D. Rogalle, M. J. Thatcher, S. P. Perera, J. M. H. van den Elsen, A. T. A. Jenkins, T. D. James, J. S. Fossey, *Polymer* 2008, 49, 3362–3365. Polymerisation resistant synthesis of methacrylamido phenylboronic acids. (b) T. R. Jackson, J. S. Springall, D. Rogalle, N. Masumoto, H. C. Li, F. D'Hooge, S. P. Perera, A. T. A. Jenkins, T. D. James, J. S. Fossey, J. M. H. van den Elsen, *Electrophoresis* 2008, 29, 4185–4191. Polymerisation resistant synthesis of methacrylamido phenylboronic acids. (c) M. P. Morais, J. D. Mackay, S. K. Bhamra, J. G. Buchanan, T. D. James, J. S. Fossey, J. M. van den Elsen, *Proteomics* 2010, 10, 48–58. Analysis of protein glycation using phenylboronate acrylamide gel electrophoresis.

11 N. Christinat, R. Scopelliti, K. Severin *J. Org. Chem.* 2007, 72, 2192–2200. Multicomponent assembly of boron-based dendritic nanostructures.

12 N. Christinat, R. Scopelliti, K. Severin, *Chem. Commun.* 2004, 1158–1159. A new method for the synthesis of boronate macrocycles.

13 N. Christinat, R. Scopelliti, K. Severin, *Angew. Chem. Int. Ed.* 2008, 47, 1848–1852. Multicomponent assembly of boronic acid based macrocycles and cages.

14 N. Christinat, R. Scopelliti, K. Severin, *Chem. Commun.* 2008, 3660–3662. Boron-based rotaxanes by multicomponent self-assembly.

15 B. Icli, E. Sheepwash, T. Riis-Johannessen, K. Schenk, Y. Filinchuk, R. Scopelliti, K. Severin, *Chem. Sci.* 2011, 2, 1719–1721. Dative boron–nitrogen bonds in structural supramolecular chemistry: multicomponent assembly of prismatic organic cages.

16 For interesting examples, see: (a) N. Christinat, E. Croisier, R. Scopelliti, M. Cascella, U. Röthlisberger, K. Severin, *Eur. J. Inorg. Chem.* 2007, 5177–5181. Formation of boronate ester polymers with efficient intrastrand charge-transfer transitions by three-component reactions. (b) M. Hutin, G. Bernardinelli, J. R. Nitschke, *Chem. Eur. J.* 2008, 14, 4585–4593. An iminoboronate construction set for subcomponent self-assembly. (c) E. Sheepwash, V. Krampl, R. Scopelliti, O. Sereda, A. Neels, K. Severin, *Angew. Chem. Int. Ed.* 2011, 50, 3034–3037. Molecular networks based on dative boron–nitrogen bonds.

17 For pioneering examples, see: (a) N. A. Petasis, I. Akritopoulou, *Tetrahedron Lett.* 1993, 34, 583–586. The boronic acid Mannich reaction: a new method for the synthesis of geometrically pure allylamines. (b) N. A. Petasis, I. A. Zavialov, *J. Am. Chem. Soc.* 1997, 119, 445–446. A new and practical synthesis of α-amino acids from alkenyl boronic acids. (c) N. A. Petasis, A. Goodman, I. A. Zavialov, *Tetrahedron* 1997, 53, 16463–16470. A new synthesis of α-arylglycines from aryl boronic acids.

18 R. A. Batey in *Boronic acids: preparation and applications in organic synthesis and medicine* (D. Hall, Ed.). Wiley-VCH Verlag GmbH, Weinheim, Germany, 2005, pp. 279–304. Nucleophilic addition reactions of aryl and alkenylboronic acids and their derivatives to imines and iminium ions.

19 For the use of NH_3, see: (a) M. Sugiura, K. Hirano, S. Kobayashi, *J. Am. Chem. Soc.* 2004, 126, 7182–7183. α-Aminoallylation of aldehydes with ammonia: stereoselective synthesis of homoallylic primary amines. (b) S. Kobayashi, K. Hirano, M. Sugiura, *Chem. Commun.* 2005, 104–106. α-Aminoallylation of aldehydes in aqueous ammonia. (c) B. Dhudshia, J. Tiburcio, A. N. Thadani, *Chem. Commun.* 2005, 5551–5553. Diastereoselective allylation and crotylation of *N*-unsubstituted imines derived from ketones.

20 P. F. Kaiser, Q. I. Churches, C. A. Hutton, *Aust. J. Chem.* 2007, 60, 799–810. Anomalies in the stereoselectivity of the Petasis reaction using styrenyl boronic acids.

21 (a) G. K. S. Prakash, M. Mandal, S. Schweizer, N. A. Petasis, G. A. Olah, *Org. Lett.* 2000, 2, 3173–3176. A facile stereocontrolled synthesis of *anti*-α-(trifluoromethyl)-β-amino alcohols. (b) C. W. G. Au, S. G. Pyne, *J. Org. Chem.* 2006, 71, 7097–7099. Asymmetric synthesis of *anti*-1,2-amino alcohols *via* the borono-Mannich reaction: a formal synthesis of (–)-swainsonine.

22 F. Berrée, A. M. Debache, Y. Marsac, B. Collet, P. Girard-Le Bleiz, B. Carboni, *Tetrahedron* 2006, 62, 4027–4037. Stereoselective synthesis of 2-hydroxymorpholines and aminodiols *via* a three-component boro-Mannich reaction.

23 (a) Q. Wang, M. G. Finn, *Org. Lett.* 2000, 2, 4063–4065. 2*H*-Chromenes from salicylaldehydes by a catalytic Petasis reaction. (b) N. A. Petasis, A. N. Butkevich, *J. Organomet. Chem.* 2009, 694, 1747–1753. Synthesis of 2*H*-chromenes and 1,2-dihydroquinolines from aryl aldehydes, amines, and alkenylboron compounds.

24 (a) N. A. *Petasis in Multicomponent reactions* (J. Zhu, H. Bienaymé, Eds.). Wiley-VCH Verlag GmbH, Weinheim, Germany, 2005, pp. 199–223. Multicomponent reactions with organoboron compounds. (b) N. R. Candeias, F. Montalbano, P. M. S. D. Cal, P. M. P. Gois, *Chem. Rev.* 2010, 110, 6169–6193. Boronic acids and esters in the Petasis-borono Mannich multicomponent reaction.

25 (a) N. A. Petasis, I. A. Zavialov, *J. Am. Chem. Soc.* 1998, 120, 11798–11799. Highly stereocontrolled one-step synthesis of *anti*-β-amino alcohols from organoboronic acids, amines, and α-hydroxy aldehydes. (b) G. K. S. Prakash, M. Mandal, S. Schweizer, N. A. Petasis, G. O. Olah, *J. Org. Chem.* 2002, 67, 3718–3723. Stereoselective synthesis of *anti*-α-(difluoromethyl)-β-amino alcohols by boronic acid based three-component condensation. Stereoselective preparation of (2*S*,3*R*)-difluorothreonine.

26 (a) L. M. Harwood, G. S. Currie, M. G. B. Drew, R. W. A. Luke, *Chem. Commun.* 1996, 1953–1954. Asymmetry in the boronic acid Mannich reaction: diastereocontrolled addition to chiral iminium species derived from aldehydes and (*S*)-5-phenylmorpholin-2-ones. (b) T. Koolmeister, M. Södergren, M. Scobie, *Tetrahedron Lett.* 2002, 43, 5969–5970. The first example of chiral induction using homochiral boronic esters in the Petasis reaction. (c) K. K. Nanda, B. W. Trotter, *Tetrahedron Lett.* 2005, 46, 2025–2028. Diastereoselective Petasis Mannich reactions accelerated by hexafluoroisopropanol: a pyrrolidine-derived arylglycine synthesis. (d) T. J. Southwood, M. C. Curry, C. A. Hutton, *Tetrahedron* 2006, 62, 236–242. Factors affecting the efficiency and stereoselectivity of α-amino acid synthesis by the Petasis reaction.

27 Y. Yamaoka, H. Miyabe, Y. Takemoto, *J. Am. Chem. Soc.* 2007, 129, 6686–6687. Catalytic enantioselective Petasis-type reaction of quinolines catalyzed by a newly designed thiourea catalyst.

28 (a) A. Alexakis, F. Amiot, *Tetrahedron: Asymmetry* 2002, 13, 2117–2122. Enantioselective addition of organolithium reagents on isoquinoline. (b) F. Amiot, L. Cointeaux, E. J. Silve, A. Alexakis, *Tetrahedron* 2004, 60, 8221–8231. Enantioselective nucleophilic addition of organometallic reagents to quinoline: regio-, stereo- and enantioselectivity.

29 S. Lou, S. E. Schaus, *J. Am. Chem. Soc.* 2008, 130, 6922–6923. Asymmetric Petasis reactions catalyzed by chiral biphenols.

30 (a) K. Maruoka, T. Ooi, *Chem. Rev.* 2003, 103, 3013–3028. Enantioselective amino acid synthesis by chiral phase-transfer catalysis. (b) J.-A. Ma, *Angew. Chem. Int. Ed.* 2003, 42, 4290–4299. Recent developments in the catalytic asymmetric synthesis of α- and β-amino acids. (c) C. Najera, J. M. Sansano, *Chem. Rev.* 2007, 107, 4584–4671. Catalytic asymmetric synthesis of α-amino acids.

31 N. Selander, A. Kipke, S. Sebelius, K. J. Szabó, *J. Am. Chem. Soc.* 2007, 129, 13723–13731. Petasis borono-Mannich reaction and allylation of carbonyl compounds *via* transient allyl boronates generated by palladium-catalyzed substitution of allyl alcohols. An efficient one-pot route to stereodefined α-amino acids and homoallyl alcohols.

32 (a) S. Sebelius, V. J. Olsson, K. J. Szabó, *J. Am. Chem. Soc.* 2005, 127, 10478–10479. Palladium pincer complex catalyzed substitution of vinyl cyclopropanes, vinyl aziridines, and allyl acetates with tetrahydroxydiboron. An efficient route to functionalized allylboronic acids and potassium trifluoro(allyl) borates. (b) V. J. Olsson, S. Sebelius, N. Selander, K. J. Szabó, *J. Am. Chem. Soc.* 2006, 128, 4588–4589. Direct boronation of allyl alcohols with diboronic acid using palladium pincer-complex catalysis. A remarkably facile allylic displacement of the hydroxy group under mild reaction conditions.

33 (a) S. Sebelius, V. J. Olsson, O. A. Wallner, K. J. Szabó, *J. Am. Chem. Soc.* 2006, 128, 8150–8151. Palladium-catalyzed coupling of allylboronic acids with iodobenzenes. Selective formation of the branched allylic product in the absence of directing groups. (b) N. Selander, S. Sebelius, C. Estay, K. Szabó, *Eur. J. Org. Chem.* 2006, 4085–4087. Highly selective and robust palladium-catalyzed carbon–carbon coupling between allyl alcohols and aldehydes *via* transient allylboronic acids.

34 (a) R. W. Hoffmann, *Angew. Chem. Int. Ed. Engl.* 1982, 21, 555–566. Diastereogenic addition of crotylmetal compounds to aldehydes. (b) B. W. Gung, X. Xue, W. R. Roush, *J. Am. Chem. Soc.* 2002, 124, 10692–10697. The origin of diastereofacial control in allylboration reactions using tartrate ester derived allylboronates: attractive interactions between the Lewis acid coordinated aldehyde carbonyl group and an ester carbonyl oxygen.

35 For a pioneering example, see: T. K. Hansen, N. Schlienger, B. S. Hansen, P. H. Andersen, M. R. Bryce, *Tetrahedron Lett.* 1999, 40, 3651–3654. Synthesis of piperazinones and their application in constrained mimetics of the growth hormone secretagogue NN703.

36 C. Kalinski, H. Lemoine, J. Schmidt, C. Burdack, J. Kolb, M. Umkehrer, G. Ross, *Synthesis* 2008, 4007–4011. Multicomponent reactions as a powerful tool for generic drug synthesis.

37 S. Lou, P. N. Moquist, S. E. Schaus, *J. Am. Chem. Soc.* 2007, 129, 15398–15404. Asymmetric allylboration of acyl imines catalyzed by chiral diols.

38 S. G. Pyne, C. W. G. Au, A. S. Davis, I. R. Morgan, T. Ritthiwigrom, A. Yazici, *Pure Appl. Chem.* 2008, 80, 751–762. Exploiting the borono-Mannich reaction in bioactive alkaloid synthesis.

39 A. S. Davis, S. G. Pyne, B. W. Skelton, A. H. White, *J. Org. Chem.* 2004, 69, 3139–3143. Synthesis of putative uniflorine A.

40 T. Ritthiwigrom, S. G. Pyne, *Org. Lett.* 2008, 10, 2769–2771. Synthesis of (+)-uniflorine A: a structural reassignment and a configurational assignment.

41 T. Machan, A. S. Davis, B. Liawruangrath, S. G. Pyne, *Tetrahedron* 2008, 64, 2725–2732. Synthesis of castanospermine.

42 T. Ritthiwigrom, A. C. Willis, S. G. Pyne, *J. Org. Chem.* 2010, 75, 815–824. Total synthesis of uniflorine A, casuarine, australine, 3-*epi*-australine, and 3,7-di-*epi*-australine from a common precursor.

43 K. Tonogaki, K. Itami, J.-I. Yoshida, *J. Am. Chem. Soc.* 2006, 128, 1464–1465. Catalytic four-component assembly based on allenylboronate platform: new access to privileged allylic amine structures.

44 For selected works, see: (a) P. Y. Renard, J. Y. Lallemand, *Tetrahedron: Asymmetry* 1996, 7, 2523–2524. 1,3-Dienylboronates in Diels-Alder reactions: part II. (b) A. Zhang, Y. Kan, B. Jiang, *Tetrahedron* 2001, 57, 2305–2309. Asymmetric hetero-Diels-Alder reaction of chiral pinanediol 1,3-dienylboronates with azo-compounds. (c) J. Y. Lallemand, Y. Six, L. Ricard, *Eur. J. Org. Chem.* 2002, 503–513. A concise synthesis of an advanced clerodin intermediate through a Vaultier tandem reaction. (d) X. Gao, D. G. Hall, *Tetrahedron Lett.* 2003, 44, 2231–2235. New electronically enriched boronobutadienes for the synthesis of hydroxylated cyclohexenes *via* tandem [4+2]/allylboration. (e) J. Taylor, D. G. Hall, *Org. Lett.* 2000, 2, 3715–3718. Tandem aza[4+2]/allylboration: a novel multicomponent reaction for the stereocontrolled synthesis of α-hydroxyalkyl piperidine derivatives. (f) B. B. Touré, H. R. Hoveyda, J. Taylor, A. Ulaczyk-Lesanko, D. G. Hall, *Chem. Eur. J.* 2003, 9, 466–474. A three-component reaction for diversity-oriented synthesis of polysubstituted piperidines: solution and solid-phase optimization of the first tandem aza[4+2]/allylboration. (g) M. Deligny, F. Carreaux, B. Carboni, L. Toupet, G. Dujardin, *Chem. Commun.* 2003, 276–277. A novel diastereoselective route to α-hydroxyalkyl dihydropyrans using an hetero Diels-Alder/allylboration sequence. (h) B. B. Touré, D. G. Hall, *Angew. Chem. Int. Ed.* 2004, 43, 2001–2004. Three-component sequential aza[4+2] cycloaddition/allylboration/retro-sulfinyl-ene reaction: a new stereocontrolled entry to palustrine alkaloids and other 2,6-disubstituted piperidines. (i) B. B. Touré, D. G. Hall, *J. Org. Chem.* 2004, 69, 8429–8436. Design of a nonreductive method for chemoselective cleavage of hydrazines in the presence of unsaturations: application to a stereoconvergent three-component synthesis of (–)-methyl palustramate.

45 X. Gao, D. G. Hall, *J. Am. Chem. Soc.* 2003, 125, 9308–9309. 3-Boronoacrolein as an exceptional heterodiene in the highly enantio- and diastereoselective Cr(III)-catalyzed three-component [4+2]/allylboration.

46 M. Deligny, F. Carreaux, L. Toupet, B. Carboni, *Adv. Synth. Catal.* 2003, 345, 1215–1219. Efficient asymmetric synthesis of 2,6-disubstituted 2H-dihydropyrans *via* a catalytic hetero-Diels-Alder/allylboration sequence.

47 X. Gao, D. G. Hall, M. Deligny, A. Favre, F. Carreaux, B. Carboni, *Chem.–Eur. J.* 2006, 12, 3132–3142. Catalytic enantioselective three-component hetero-[4+2] cycloaddition/allylboration approach to α-hydroxyalkyl pyrans: scope, limitations, and mechanistic proposal.

48 D. B. Stierle, A. A. Stierle, *Experientia* 1992, 48, 1165–1169. Pseudomonic acid derivatives from a marine bacterium.

49 O. Marion, X. Gao, S. Marcus, D. G. Hall, *Bioorg. Med. Chem.* 2009, 17, 1006–1017. Synthesis and preliminary antibacterial evaluation of simplified thiomarinol analogs.

50 (a) R. H. Cichewicz, F. A. Valeriote, P. Crews, *Org. Lett.* 2004, 6, 1951–1954. Psymberin, a potent sponge-derived cytotoxin from *Psammocinia* distantly related to the pederin family. (b) G. R. Pettit, J. P. Xu, J.-C. Chapuis, R. K. Pettit, L. P. Tackett, D. L. Doubek, J. N. A. Hooper, J. M. Schmidt, *J. Med. Chem.* 2004, 47, 1149–1152. Antineoplastic agents. 520. Isolation and structure of irciniastatins A and B from the indo-pacific marine sponge *Ircinia ramose*.

51 H. Lachance, O. Marion, D. G. Hall, *Tetrahedron Lett.* 2008, 49, 6061–6064. Synthetic studies toward the pyran core and the amide side chain of psymberin.

52 L. Eberlin, F. Tripoteau, F. Carreaux, A. Whiting, B. Carboni, *Beilstein J. Org. Chem.* 2014, 10, 237–250. Boron-substituted 1,3-dienes and heterodienes as key elements in multicomponent processes.

53 A. Ulaczyk-Lesanko, E. Pelletier, M. Lee, H. Prinz, H. Waldmann, D. G. Hall, *J. Comb. Chem.* 2007, 9, 695–703. Optimization of three- and four-component reactions for polysubstituted piperidines: application to the synthesis and preliminary biological screening of a prototype library.

54 (a) N. F. Pelz, A. R. Woodward, H. E. Burks, J. D. Sieber, J. P. Morken, *J. Am. Chem. Soc.* 2004, 126, 16328–16329. Palladium-catalyzed enantioselective diboration of prochiral allenes. (b) A. R. Woodward, H. E. Burks, L. M. Chan, J. P. Morken, *Org. Lett.* 2005, 7, 5505–5507. Concatenated catalytic asymmetric allene diboration/allylation/functionalization.

55 J. D. Sieber, J. P. Morken, *J. Am. Chem. Soc.* 2006, 128, 74–75. Sequential Pd-catalyzed asymmetric allene diboration/α-aminoallylation.

56 (a) E. M. Flamme, W. R. Roush, *J. Am. Chem. Soc.* 2002, 124, 13644–13645. Enantioselective synthesis of 1,5-*anti*- and 1,5-*syn*-diols using a highly diastereoselective one-pot double allylboration reaction sequence. (b) E. M. Flamme, W. R. Roush, *Org. Lett.* 2005, 7, 1411–1414. Synthesis of the C(1)-C(25) fragment of amphidinol 3: application of the double-allylboration reaction for synthesis of 1,5-diols. (c) R. M. Owen, W. R. Roush, *Org. Lett.* 2005, 7, 3941–3944. Stereoselective synthesis of the C(1)-C(11) fragment of peloruside A.

57 L. D. Juliawaty, M. Kitajima, H. Takayama, S. A. Achmad, N. Aimi, *Phytochemistry* 2000, 54, 989–993. A 6-substituted-5,6-dihydro-2-pyrone from *Cryptocarya strictifolia*.

58 S. Tang, X. Xie, X. Wang, L. He, K. Xu, X. She, *J. Org. Chem.* 2010, 75, 8234–8240. Concise total syntheses of (+)-strictifolione and (6*R*)-6-[(4*R*,6*R*)-4,6-dihydroxy-10-phenyldec-1-enyl]-5,6-dihydro-2*H*-pyran-2-one.

5

CARBENE-PROMOTED MULTICOMPONENT REACTIONS

DAVID MONGE[1], JAVIER IGLESIAS-SIGÜENZA[1] AND ELENA MAS-MARZÁ[2]

[1] *Departamento de Química Orgánica, Universidad de Sevilla, Sevilla, Spain*

[2] *Departamento de Física, Área Física Aplicada, Universidad Jaume I, Grupo de Dispositivos Fotovoltaicos y Optoelectrónicos, Castellón, Spain*

5.1 INTRODUCTION

Multicomponent reactions (MCRs), defined as one-pot reactions in which three or more starting materials are involved in sequential reactions, have been progressively acquiring importance in all areas of organic chemistry. In such reactions, the generation of reactive intermediates, employment of special reagents, and activation of functional groups by robust catalysts are normally the *war horses* for developing new technologies. In this context, carbenes appear as privileged entities able to define different sceneries depending on their roles in the processes, which have been used to classify in this chapter (Scheme 5.1):

1. **Nucleophilic carbenes**, such as dimethoxycarbenes (Section 5.2.1) or N-heterocyclic carbenes (NHCs) (Section 5.2.2), are used **as reagents in the synthesis of heterocycles**, becoming part of the final molecule. These MCRs involve zwitterionic species generated by the addition of nucleophilic carbenes to multiple bonds and subsequent reactions with electrophilic compounds.
2. **Fischer carbene complexes (FCCs)**, organometallic compounds, are used as reagents in MCRs. These species transfer the organic carbene moiety from the metal center to the final target (Section 5.2.3).
3. **Carbenes** are employed as catalysts to drive the MCRs (Section 5.3). This section illustrates the formidable performance of NHCs as organocatalysts (Section 5.3.1) and ligands for organometallic catalysts (Section 5.3.2) in MCRs.

This chapter is not meant to be an exhaustive review of the literature but rather illustrates the main roles of carbenes in MCRs.

5.2 MCRs INVOLVING CARBENES AS KEY COMPONENTS

Carbenes are neutral species containing a carbon atom with only six valence electrons. In the case of carbenes containing electron-rich substituents (–OR, –NR_2, –SR), the heteroatom lone pair donation into the formally vacant p-orbital of the carbene carbon produces stabilization of the singlet state of the carbene [1], enhancing its nucleophilicity (Fig. 5.1).

In this section, MCRs driven by nucleophilicity of dimethoxycarbenes and NHCs where the structure of the final product contains the carbene are analyzed (Scheme 5.2).

5.2.1 MCRs of Dimethoxycarbenes

Dialkoxycarbenes, such as dimethoxycarbene **I**, constitute a very reactive class of compounds that were initially difficult to handle [2, 3]. Fortunately, a convenient, efficient, and safe procedure was developed by Warkentin and coworkers [4]. The protocol generates dimethoxycarbene, together with acetone and nitrogen, by thermolysis of the corresponding 2,2-dimethoxy-5,5-dimethyl-2,5-dihydro-1,3,4-oxadiazole **1** (Scheme 5.3).

Consequently, dimethoxycarbene **I** has gained increasing interest as key component to promote MCRs, as it allows rapid access to highly functionalized molecules. As presented

Multicomponent Reactions: Concepts and Applications for Design and Synthesis, First Edition. Edited by Raquel P. Herrera and Eugenia Marqués-López.

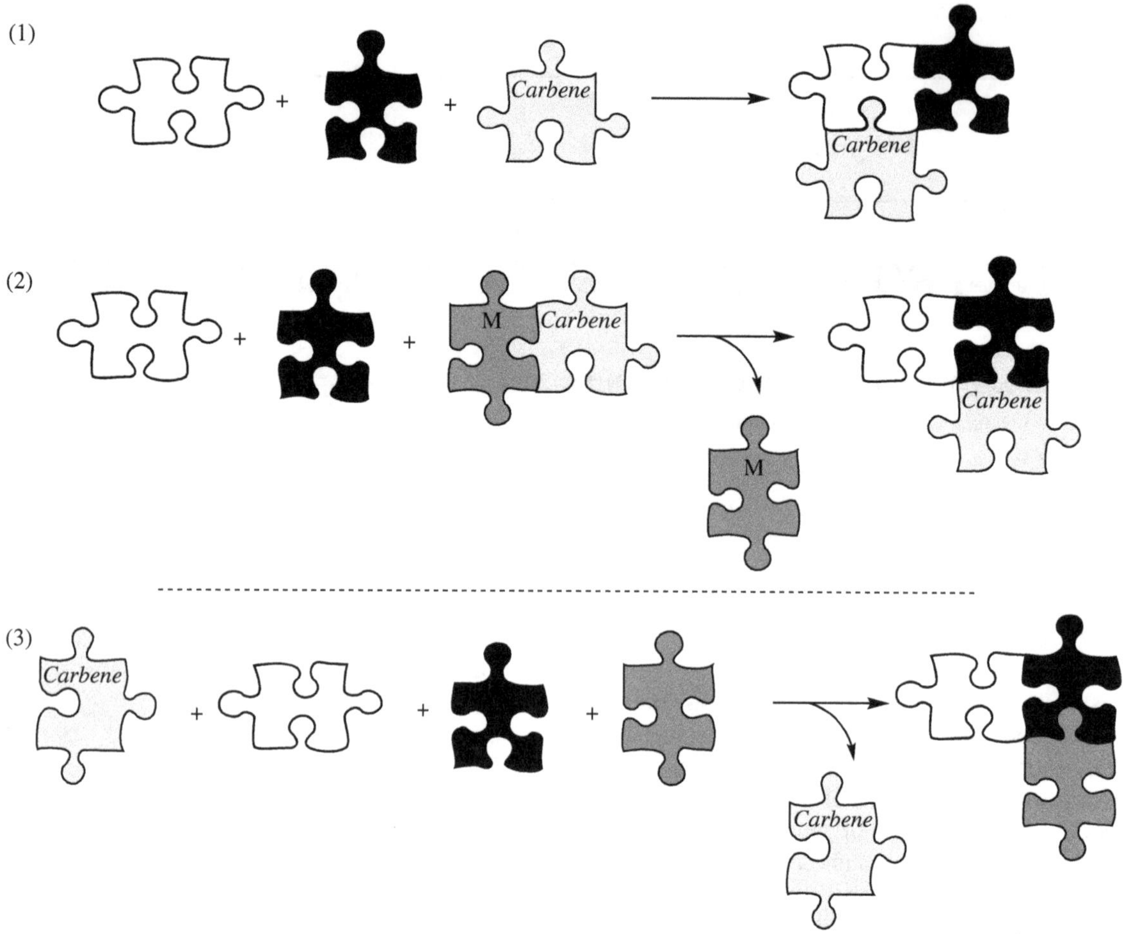

SCHEME 5.1 Different carbene-promoted multicomponent reactions.

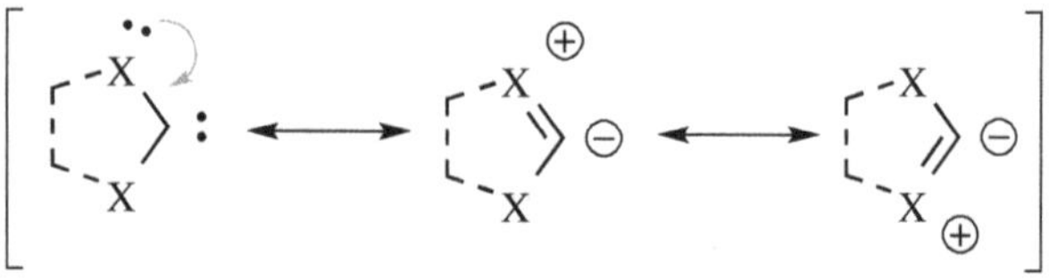

FIGURE 5.1 Nucleophilic carbenes as 1,1-dipole equivalents.

in Scheme 5.4, the common mechanism of these reactions involves formation of zwitterion **I1** [from dimethoxycarbene **I** and dimethylacetylenedicarboxylate (DMAD)]. After that, interception of dipolar **I1** with a third component (electrophilic carbonyl compound) constitutes a facile synthesis of dihydrofuran derivatives [5].

Following the general strategy depicted in Scheme 5.4, Nair and coworkers have described an easy route for the synthesis of highly substituted dihydrofuran derivatives **2–5** by the one-pot reaction of dimethoxycarbene **I**, generated *in situ* from oxadiazole **1**, DMAD, and different carbonyl compounds, such as aldehydes (***a***) [5, 6], ketones (***b***) [7], α,β-unsaturated ketones (***c***) [5, 8], or quinones (***d***) [5, 6] (Scheme 5.5). In the specific case of enones (***c***), the addition of the zwitterion **I1** took place in the carbonyl group regioselectively.

The same reaction can be extended to the synthesis of spirodihydrofurans **6–8** or cyclopentene acetals **9**, employing diverse electrophilic partners such as anhydrides (***a***) [7], *N*-alkyl isatines (***b***) [7], 1,2-dicarbonyl compounds (***c***) [7], or benzylidenemalononitriles (***d***) [9] (Scheme 5.6).

Employing isocyanates instead of DMAD, the reaction with dimethoxycarbene in toluene produces hydantoins **10**. However, the outcome of this MCR can be modified depending on the reaction media as performing the reaction in chlorobenzene allows to obtain β-lactams **11** [10] (Scheme 5.7).

All the compounds presented in this section play an important function in medicinal chemistry. Some of them, like spirodihydrofurans **6–8**, are connected with the treatment/prevention of atherosclerosis, and others like β-lactams **11** are the prevalent substructures of the most widely used family of antibiotics.

5.2.2 MCRs of NHCs

Since the isolation by Arduengo and coworkers, in the early 1990s, of the first crystalline and stable NHC [11], the interest and curiosity aroused by stable carbenes have been

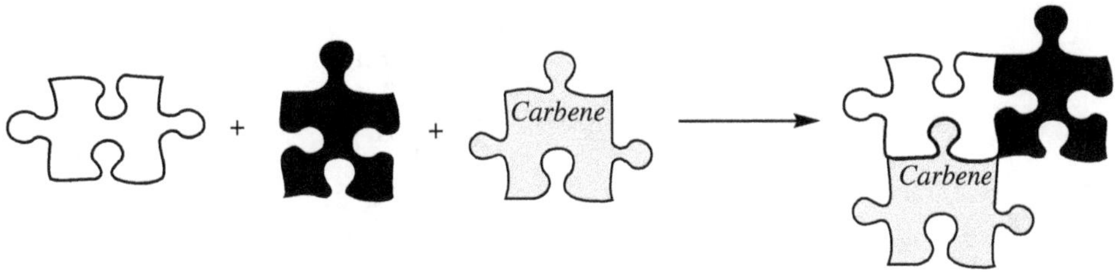

SCHEME 5.2 MCRs: carbenes as components.

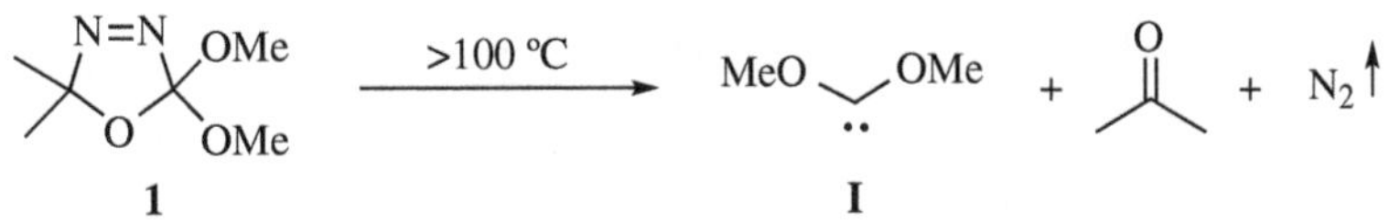

SCHEME 5.3 Oxadiazoline **1** as precursor of dimethoxycarbene **I**.

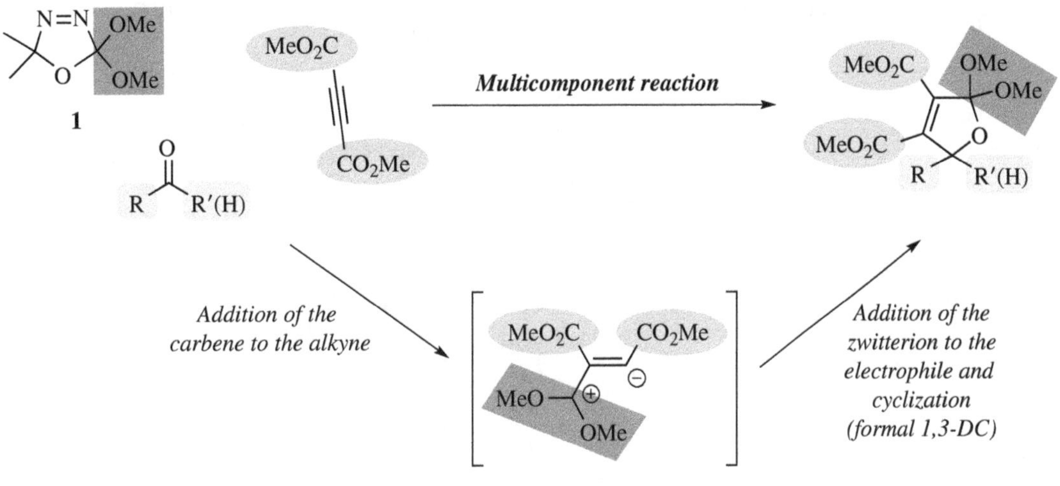

SCHEME 5.4 General mechanism of MCR to obtain dihydrofuran derivatives starting from **1**.

enormous due to their extraordinary properties (relative stability, moderate nucleophilicity, or strong basicity) [12], which have motivated a growing effort of the scientific community to obtain NHCs with the greater possible structural variety [13]. Therefore, these great synthetic advances have been exploited in MCRs where NHCs are the key components for the synthesis of furan derivatives containing bis- and fused-heterocycles (motifs with important pharmacological properties: furan-fused 1,4-thiazepine derivatives (Section 5.2.2.1), fused-ring azepine derivatives (Section 5.2.2.2), and triazole-containing compounds (Section 5.2.2.3).

The general strategy for carrying out these MCRs starts with the generation of the NHC, which involves the deprotonation of the corresponding azolium salt by an appropriate base. Next, this carbene reacts with several electrophiles to produce reactive zwitterionic intermediates (typically from reaction with carbonyl compound or activated alkynes), followed by different rearrangements, rendering the final product. In this section, only examples based on thiazol-2-ylidenes **II**, imidazol-2-ylidenes **III** (or imidazolin-2-ylidenes **IV**), and triazol-5-ylidenes **V** are covered.

5.2.2.1 MCRs Involving Thiazol-2-ylidenes as Key Components In a similar procedure to the one described in Section 5.2.1 for the synthesis of dihydrofuran derivatives **2–9**, highly substituted 3-aminofuran derivatives **13** can be obtained employing thiazolium salts **12** as key precursors [14] (Scheme 5.8). The reaction starts with a nucleophilic addition of thiazol-2-ylidene **II** to the aldehyde followed by the addition to DMAD to give spirocycle **I2**. Then, the ring opening of this spirocycle, releasing H_2, followed by hydrolysis yields 3-aminofurans **13**.

Furthermore, highly substituted furan-fused 1,4-thiazepine derivatives **14** can be prepared by a three-component reaction (3CR) involving thiazolium salts **12**, DMAD, and ketenes or acid derivatives [15] (Scheme 5.9).

A plausible mechanism for this reaction is depicted in Scheme 5.10. Firstly, a nucleophilic attack of thiazol-2-ylidene

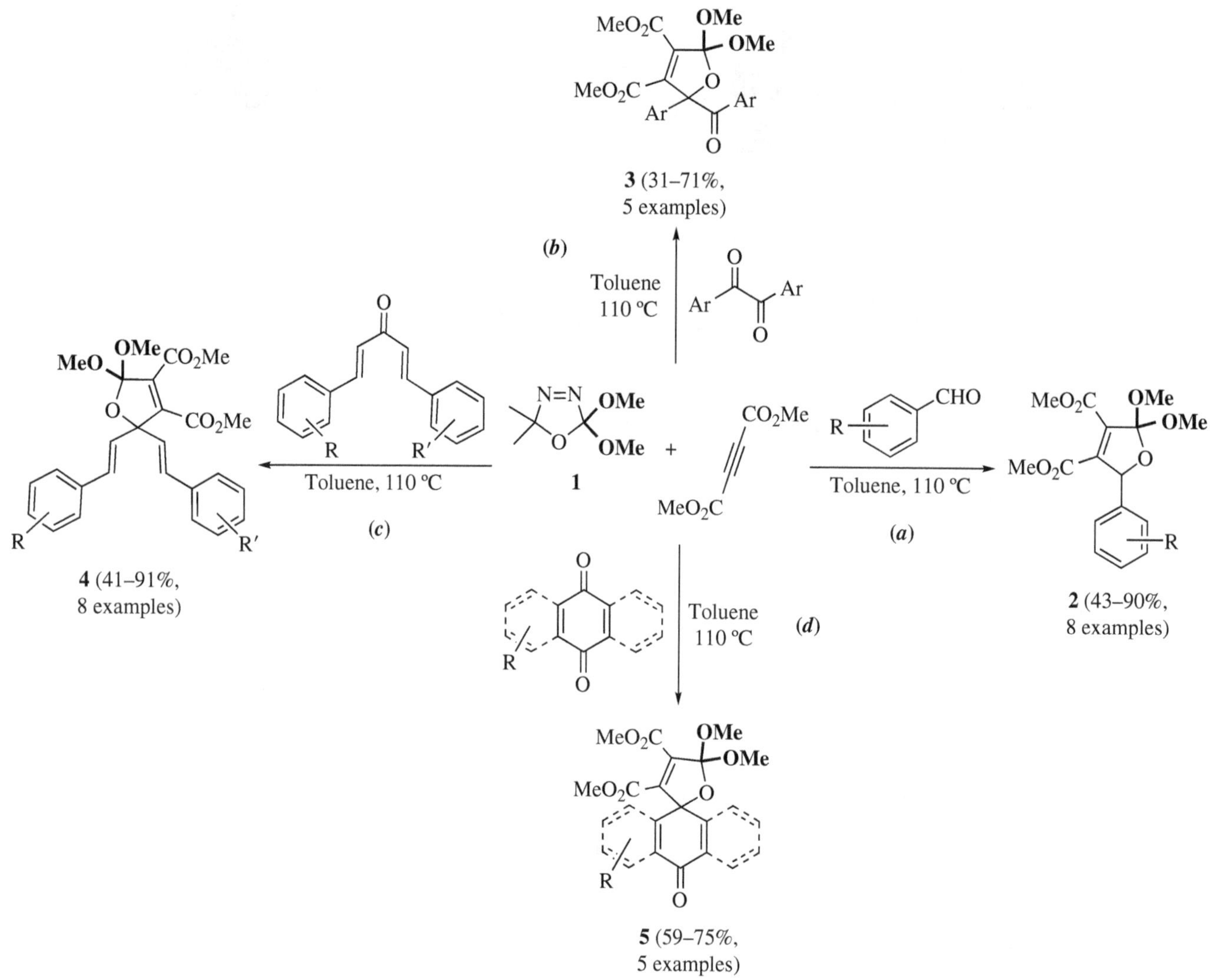

SCHEME 5.5 Synthesis of substituted dihydrofuran derivatives by MCR of dimethoxycarbene **I**.

II to the ketene (carbonyl group) gives zwitterion **I3**, which subsequently adds to the activated alkyne forming **I4**. Successive ring expansion provides furothiazepine **14** through a [1, 3]-sigmatropic sulfur shift.

5.2.2.2 MCRs Involving Imidazol-2-ylidenes or Imidazolin-2-ylidenes as Key Components Imidazol-2-ylidenes **III** can also achieve similar reactivity to thiazol-2-ylidenes **II**. In 2003, Nair and coworkers described the first-ever MCR of this family of NHCs, wherein two related carbenes reacted differently. As shown in Scheme 5.11, reaction of imidazol-2-ylidenes **III** (R = aryl) with DMAD and aldehydes afforded furanones **15**. However, when more nucleophilic imidazolin-2-ylidenes **IV** (R = aryl) were used, under the same reaction conditions, 2-oxy-maleate derivatives **16** were obtained [16].

A reasonable explanation for these two different pathways is given in Scheme 5.12. With imidazol-2-ylidenes **III** (featuring aromatic stabilization), the reaction begins with the formation of zwitterionic species **I5** by the addition of the carbene to the activated alkyne, which later adds to the aldehyde to form zwitterion **I6**. This species undergoes an intramolecular nucleophilic acyl substitution between alkoxide anion and an ester carbonyl group, giving **I7**. Subsequent elimination of methanol yields the corresponding furanone derivatives **15**. On the other hand, in the case of the imidazolin-2-ylidenes **IV** (no aromatic stabilization), the carbene adds first to the aldehyde to give the electron-rich enaminol intermediate **I8**, which endures a conjugated addition on the alkyne through the harder oxygen atom, followed by proton abstraction to yield 2-oxy-maleate derivatives **16**.

Another interesting feature of the reactivity of imidazol/imidazolin-2-ylidenes in these MCRs is the influence of the N-substituents. The same authors observed that employing *N*-alkyl imidazol-2-ylidenes, the reaction took an entirely

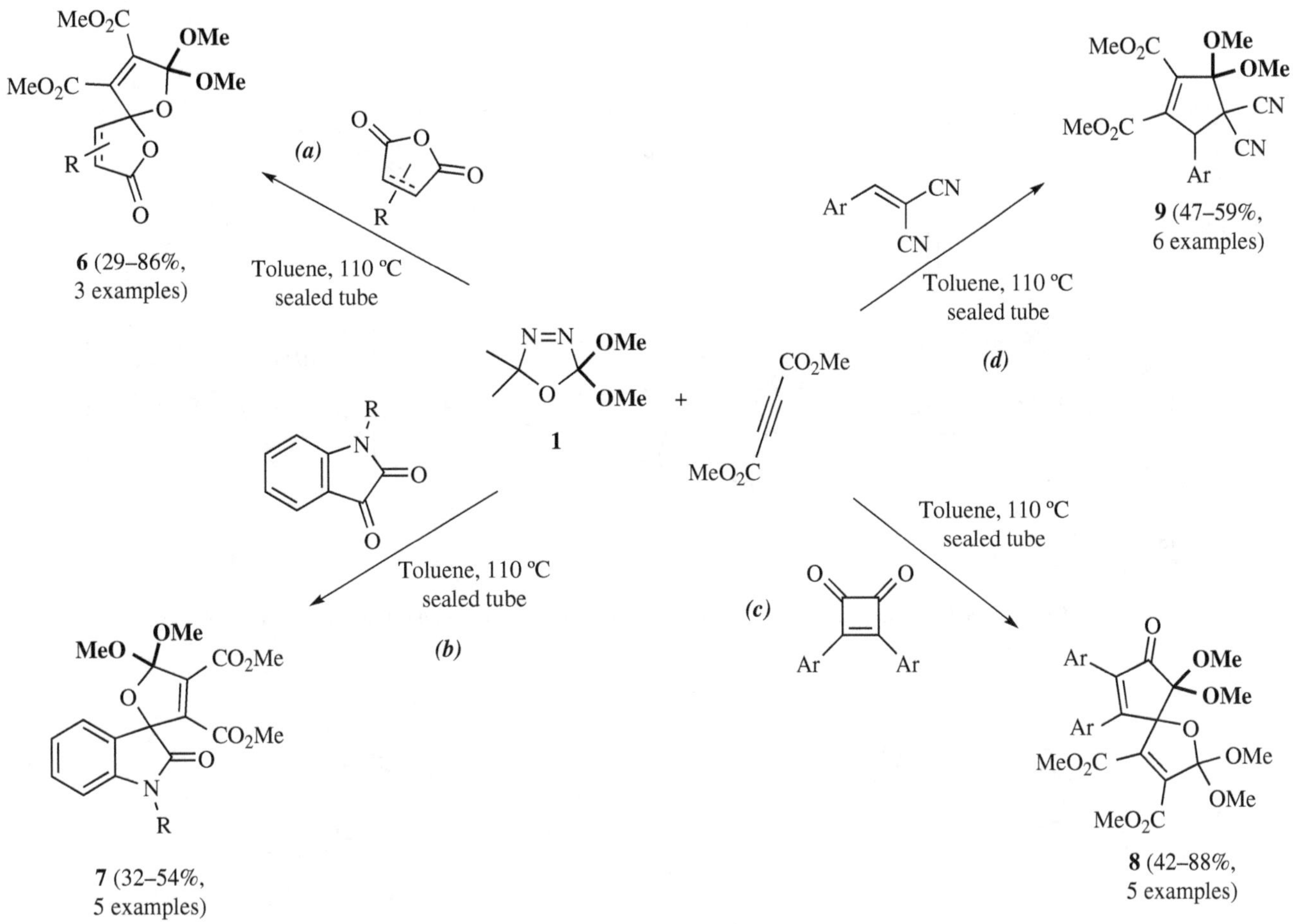

SCHEME 5.6 Synthesis of spirodihydrofurans **6–8** and cyclopentene acetals **9** by MCR.

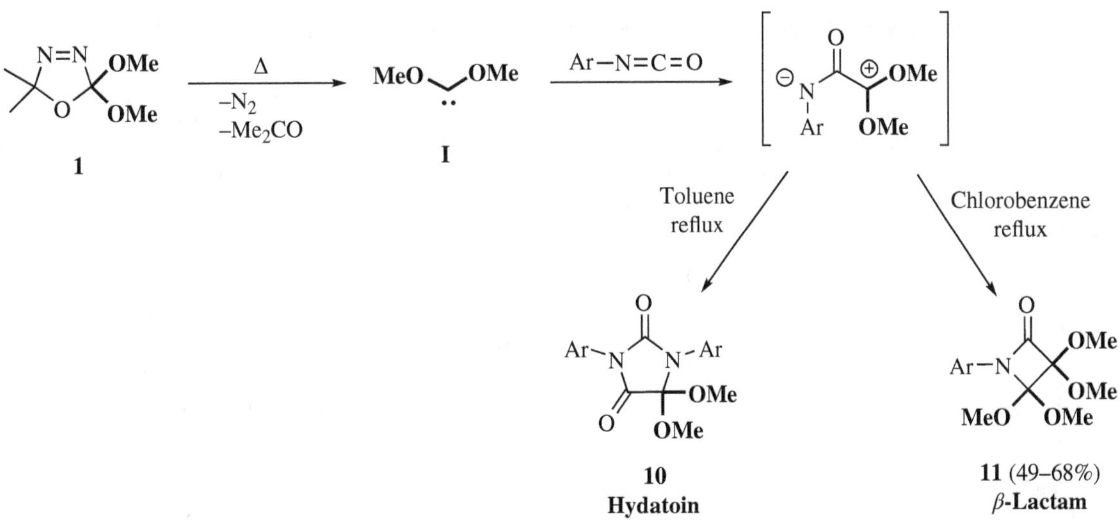

SCHEME 5.7 Solvent effect on dimethoxycarbene addition to aryl isocyanates.

different pathway and a novel 4CR afforded 2-oxy-maleates **17** [17] (Scheme 5.13). Noteworthy, when the same reaction was carried out with *N*-alkyl imidazolin-2-ylidenes, aminofuran derivatives **18** were obtained instead of 2-oxy-maleates **16**.

In 2010, Cheng et al. published the first MCR of imidazo[1,5-*a*]pyridinium salts **19** with aromatic aldehydes and allenoates to provide fully substituted furans **20–21** [18] (Scheme 5.14).

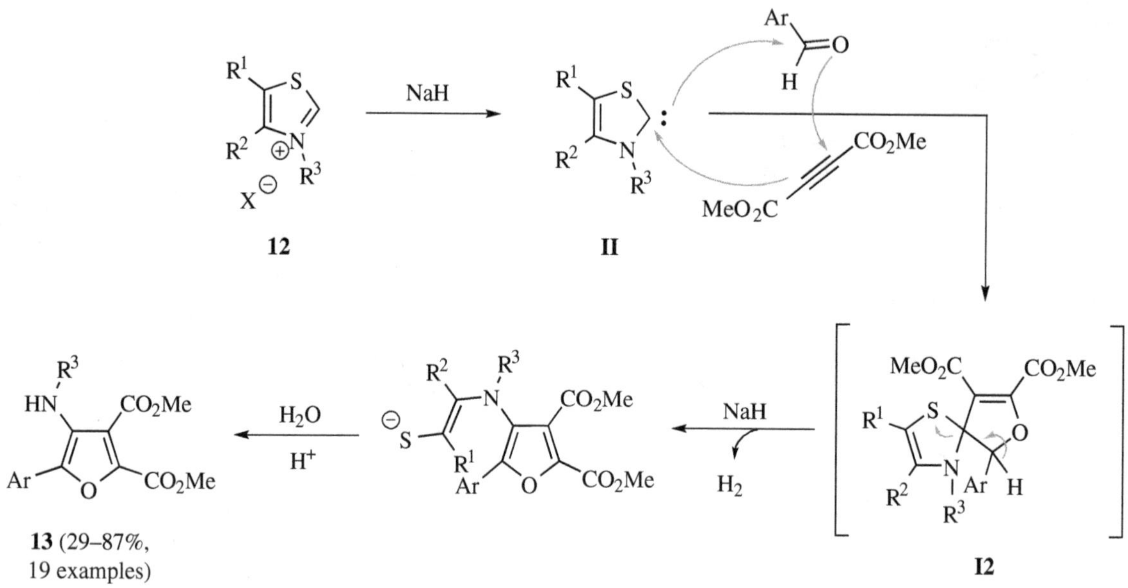

SCHEME 5.8 Thiazolium-mediated MCR for the synthesis of 3-aminofuran derivatives **13**.

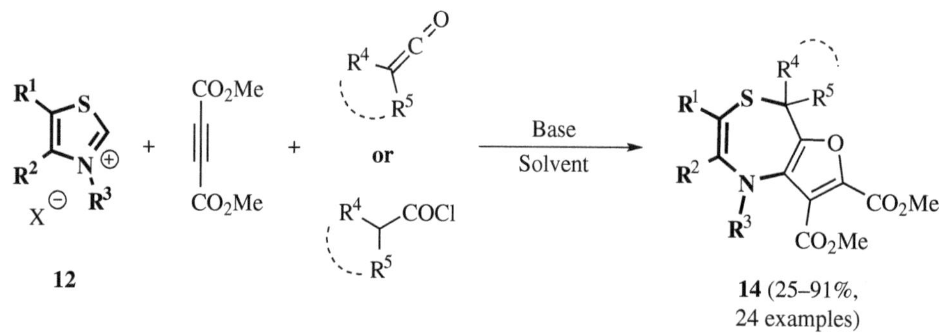

SCHEME 5.9 Synthesis of polysubstituted furan-fused 1,4-thiazepine derivatives **14**.

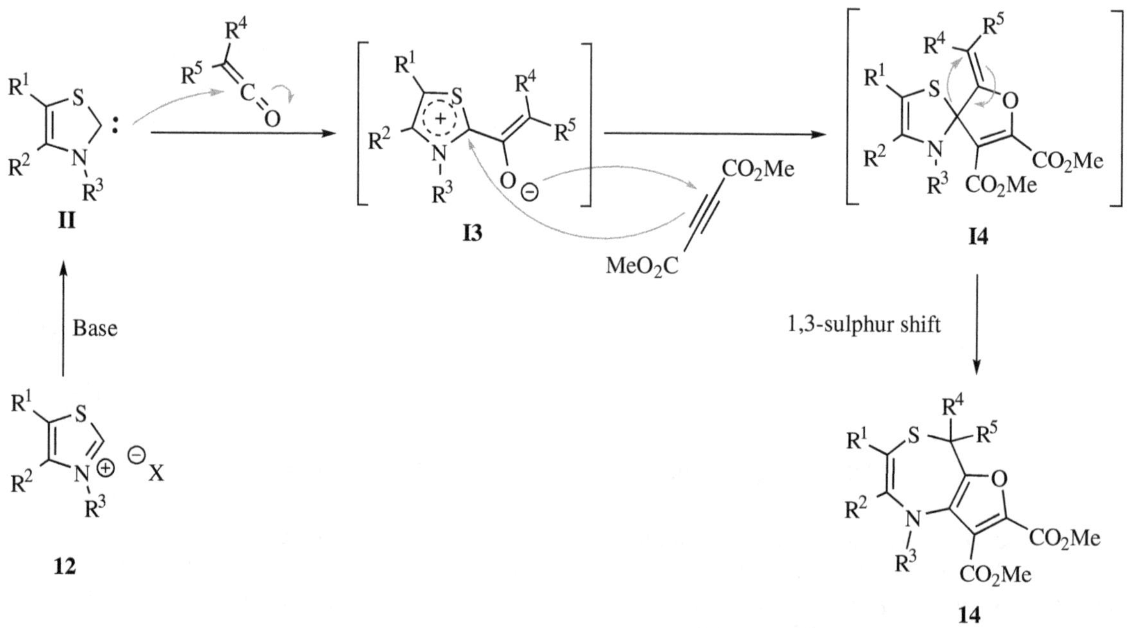

SCHEME 5.10 Plausible mechanism of the synthesis of 1,4-thiazepines.

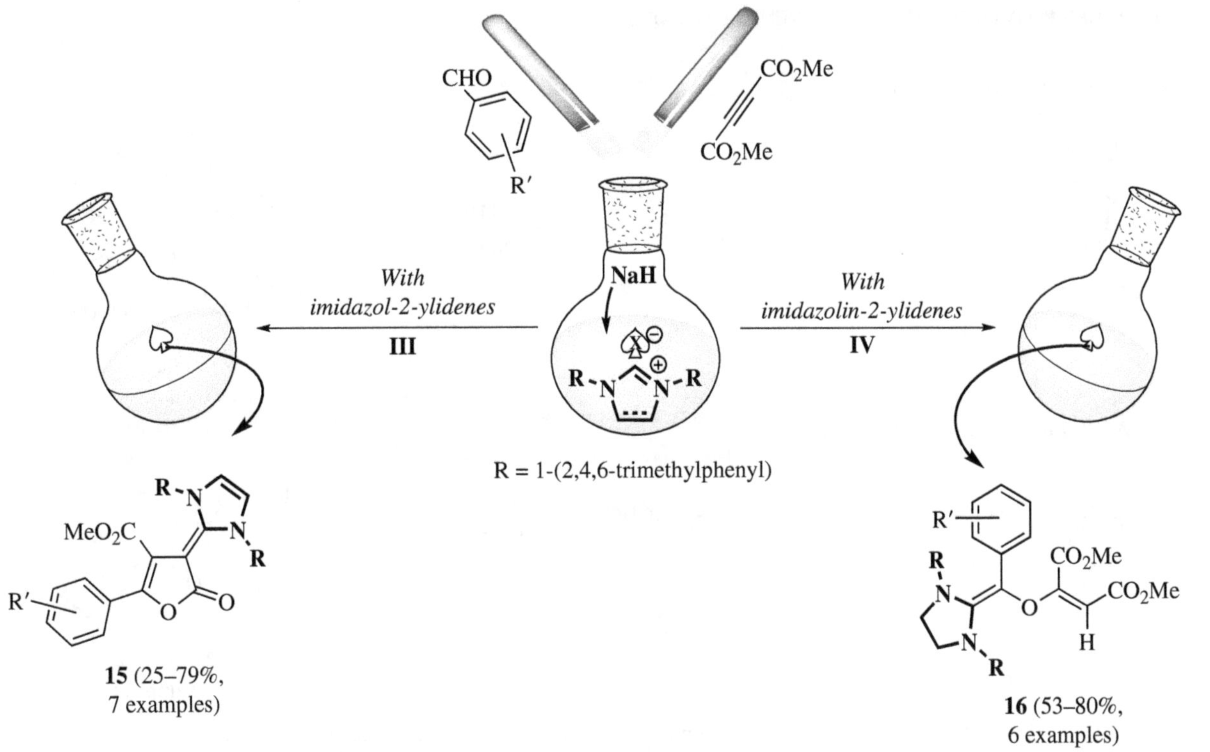

SCHEME 5.11 MCRs involving imidazol/imidazolin-2-ylidenes.

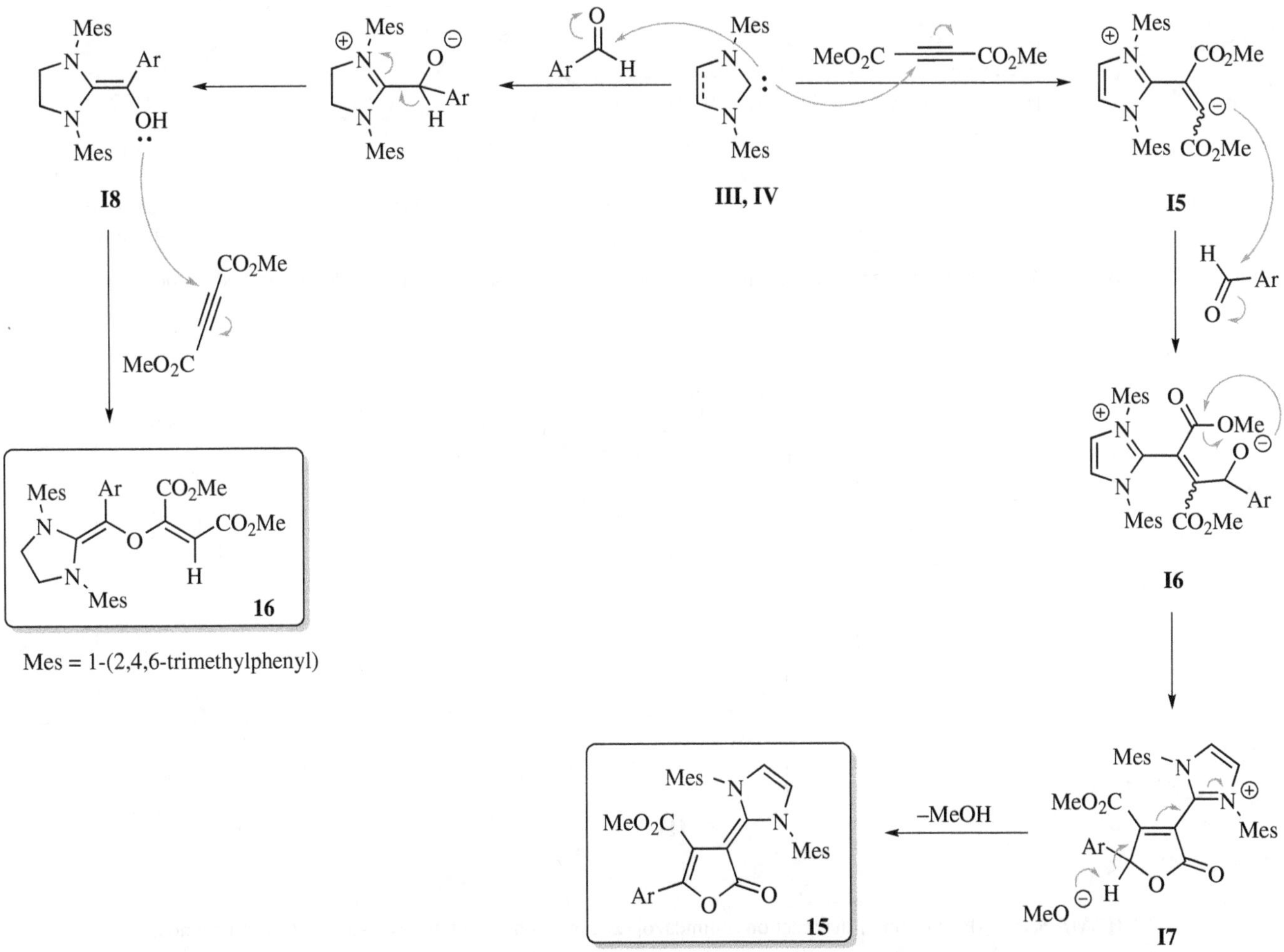

SCHEME 5.12 Plausible mechanism for MCR of imidazol-2-ylidenes **III** and imidazolin-2-ylidenes **IV**.

With N-alkyl Imidazol-2-ylidenes:

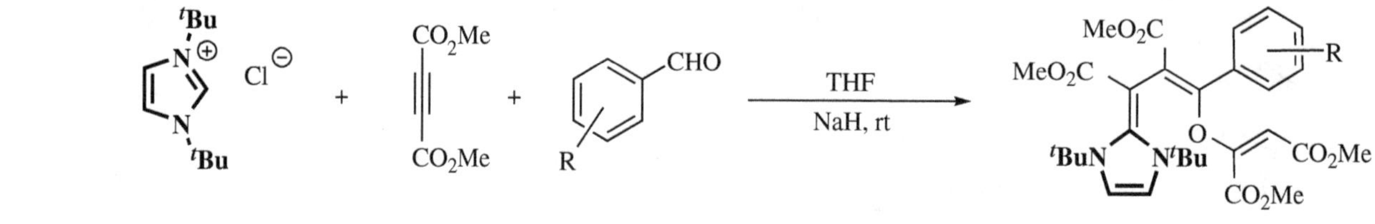

17 (20–65%, 8 examples)

With N-alkyl Imidazolin-2-ylidenes:

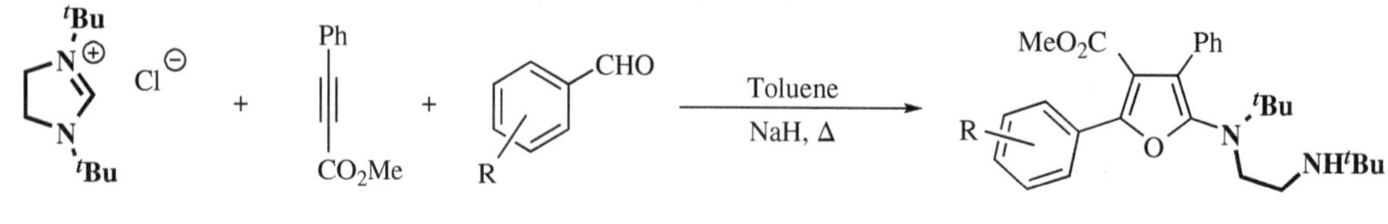

18 (40–65%, 7 examples)

SCHEME 5.13 MCR involving *N,N*-di-*tert*-butyl imidazol/imidazolin-2-ylidenes.

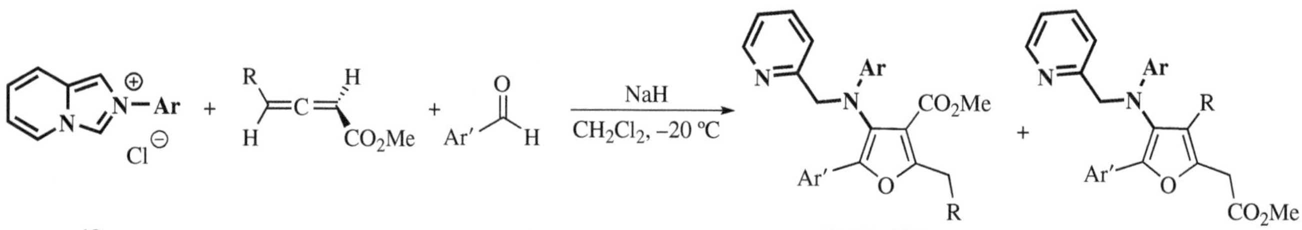

19 **20** (40–53%, 9 examples) **21** (<10%, 9 examples)

SCHEME 5.14 Multicomponent reaction involving imidazo[1,5-*a*]pyridin-3-ylidenes, aldehydes, and allenoates.

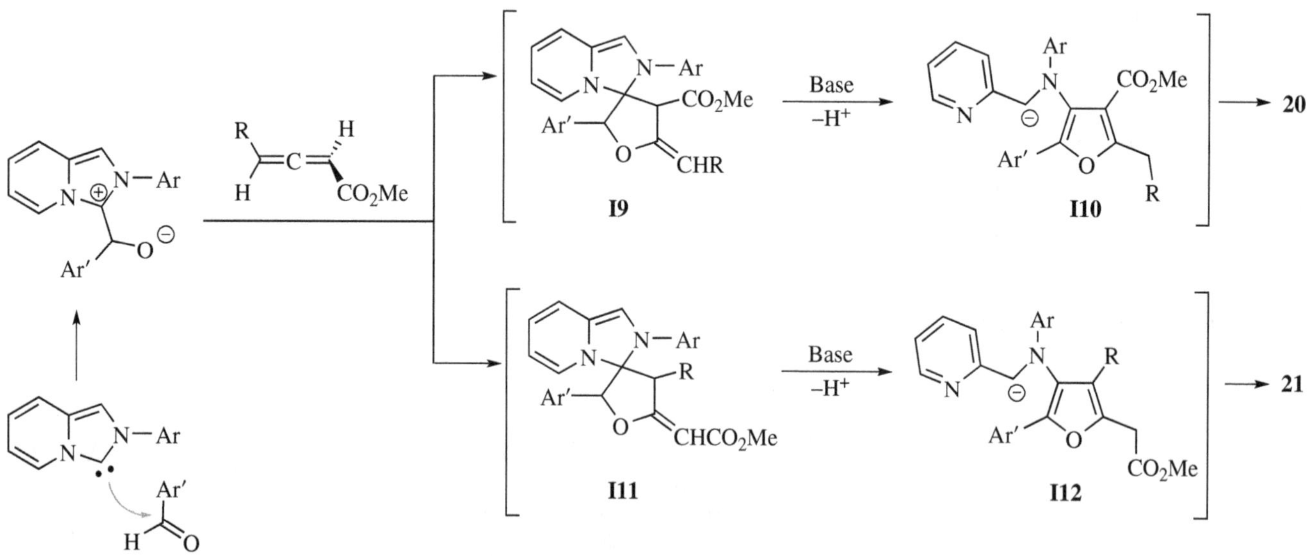

SCHEME 5.15 Mechanism of the reaction of imidazo[1,5-*a*]pyridin-3-ylidenes with aldehydes and allenoates.

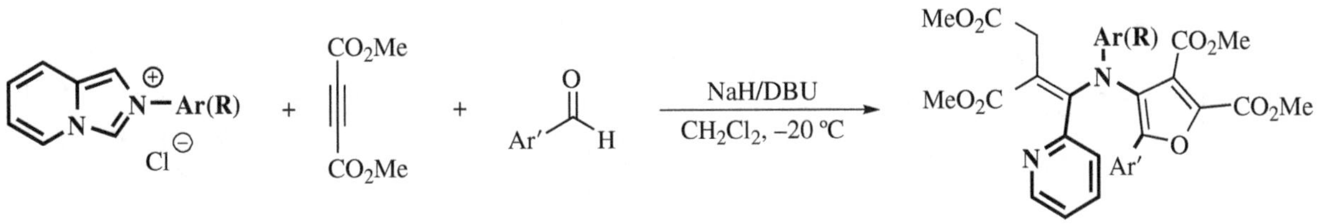

SCHEME 5.16 Multicomponent reaction involving imidazo[1,5-*a*]pyridin-3-ylidenes, aldehydes, and DMAD.

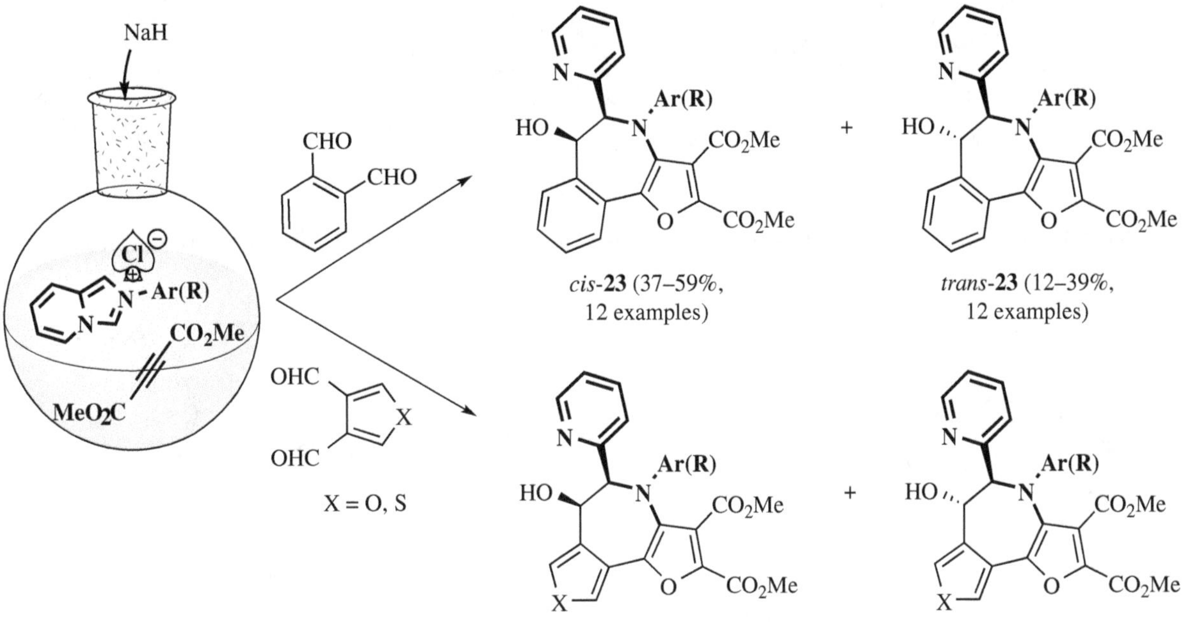

SCHEME 5.17 MCR of imidazo[1,5-*a*]pyridin-3-ylidenes with *ortho*-dialdehydes and DMAD.

The reaction starts with the nucleophilic addition of the carbene to the aldehyde followed by [3+2] cycloaddition with the allenoate to furnish spirodihydrofuran intermediates **I9/I11**. In the presence of a base, opening of the imidazole ring and rearrangement of the exocyclic to endocyclic carbon–carbon double bond afford carbanions **I10/I12**. Finally, protonation of these intermediates yields furans **20** and **21** (Scheme 5.15).

When more reactive and less stereohindered DMAD is used, a different reactivity pattern was observed. In this particular case, similar anionic intermediate to **I10/I12** is added nucleophilically to another molecule of DMAD affording furans **22** (Scheme 5.16).

The reactivity of this class of carbenes has also been explored in order to prepare novel tricyclic fused-ring systems **23–24** (azepine derivatives) by a multicomponent procedure employing *ortho*-dialdehydes, such as phthalaldehydes [19] or furan(thiophene)-3,4-dicarbaldehydes [20] (Scheme 5.17).

The reaction is *cis*-diastereoselective, yielding preferentially the most thermodynamically stable *cis*-isomer favored by an intramolecular hydrogen bond between the hydroxyl group and the pyridine nitrogen atom. The mechanism for the synthesis of these compounds, similar to mechanism in Scheme 5.15, is shown in Scheme 5.18 (represented for furan(thiophene)-3,4-dicarbaldehyde).

5.2.2.3 MCRs Involving Triazol-5-ylidenes as Key Components MCRs involving triazol-5-ylidenes **V** are more difficult as a result of the diminished nucleophilicity of these carbenes compared with previously described thiazol-2-ylidenes **II** and imidazol-2-ylidenes **III**. However, the importance of triazole moieties in relevant organic compounds has encouraged the use of triazole carbenes as key components in MCRs. To this aim, the reaction of triazol-5-ylidene **V**, DMAD, and different aromatic aldehydes provided fully substituted furanone derivatives **25** [21]

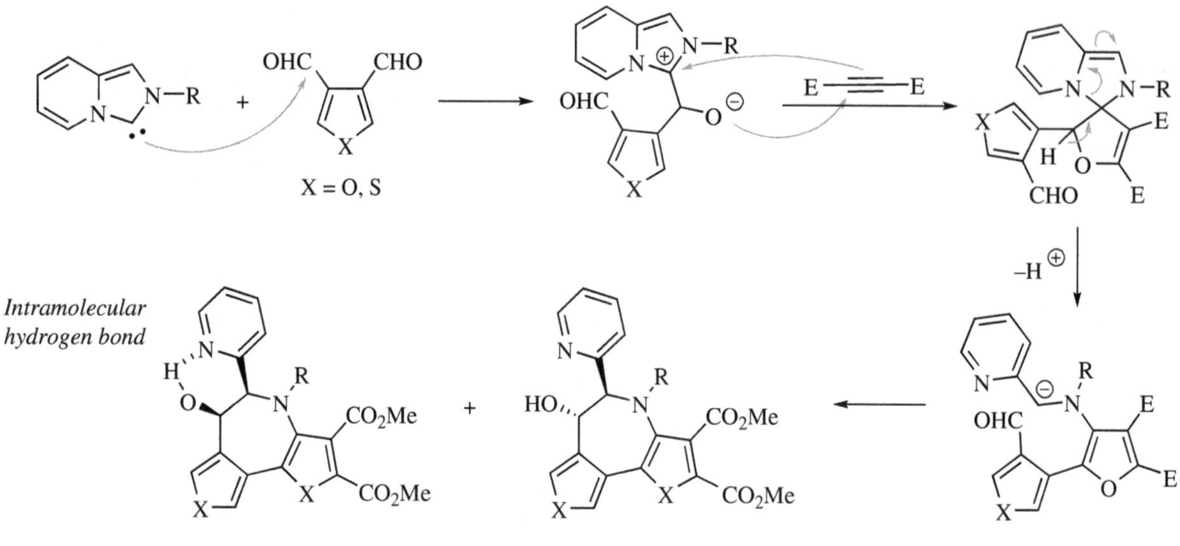

cis-**24** *trans*-**24**

SCHEME 5.18 Mechanism of MCR of imidazo[1,5-*a*]pyridin-3-ylidenes with *ortho*-dialdehydes and activated alkynes.

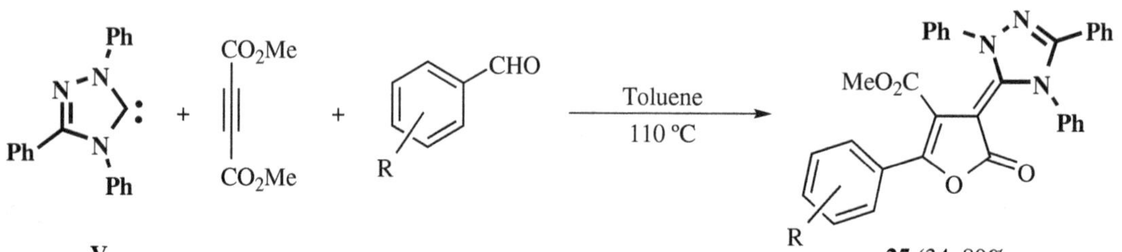

V **25** (34–80%, 9 examples)

SCHEME 5.19 MCR involving triazol-5-ylidene **V**, DMAD, and aldehydes.

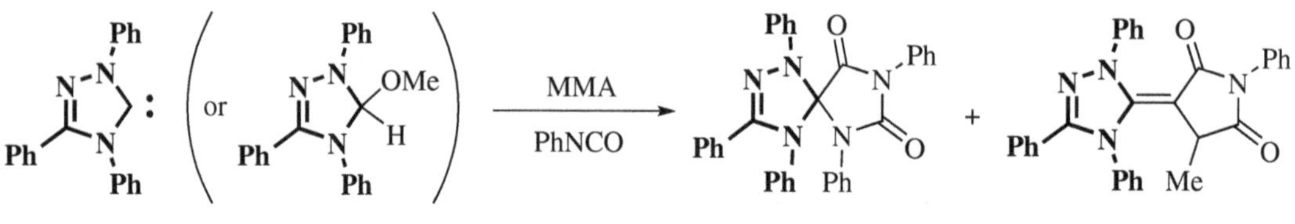

V **26** **27**

SCHEME 5.20 MCR of triazol-5-ylidene **V**, MMA, and PhNCO.

(Scheme 5.19). Formation of these furanone derivatives **25** can be rationalized following the mechanism described for imidazol-2-ylidenes showed in Scheme 5.12.

Recently, triazol-5-ylidene **V** has also been described in a novel MCR involving esters [methyl methacrylate (MMA)] and isocyanates generating a mixture of two different compounds, **26** and **27** (Scheme 5.20), where the ratio depends on the equivalents of the reagents used [22].

NHC **V** adds to the isocyanate to quantitatively produce zwitterion **I13** (Scheme 5.21). Depending on the amount of isocyanate, the reaction progresses in two different ways. In the presence of two equivalents, zwitterion **I13** reacts with the second isocyanate to afford **26**. If there is only one equivalent of isocyanate, zwitterion **I13** reverts partially to **V**, which reacts with MMA and PhNCO, through a tandem umpolung/annulation sequence via deoxy-Breslow intermediate **I14**, to generate product **27**.

5.2.3 FCCs as Reagents: Approach to Highly Substituted Carbo- and Heterocycles

Heteroatom-stabilized FCCs, firstly reported in 1964 [23], are organometallic species that consist of a transition metal in low oxidation state, having π-acceptor ligands (CO) and a

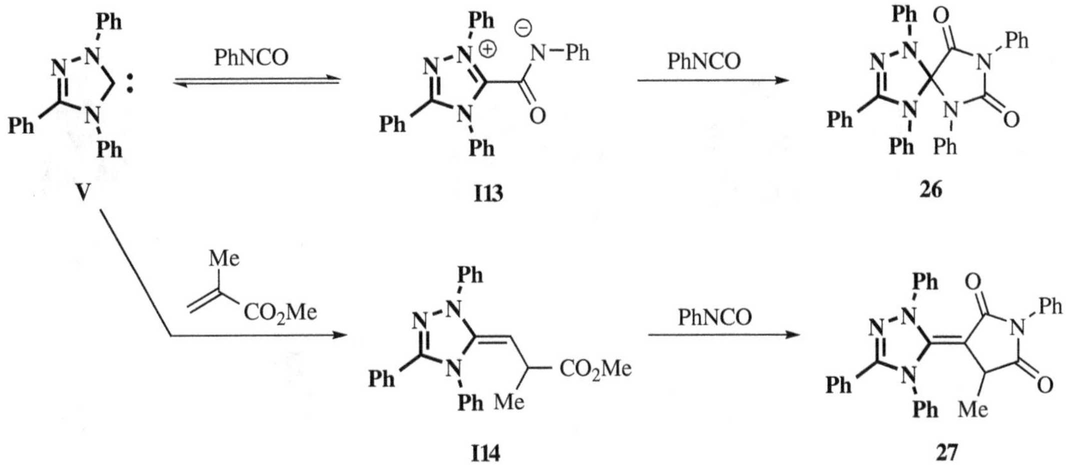

SCHEME 5.21 MCR of **V**, MMA, and PhNCO across the equilibrium between **V** and **I13**.

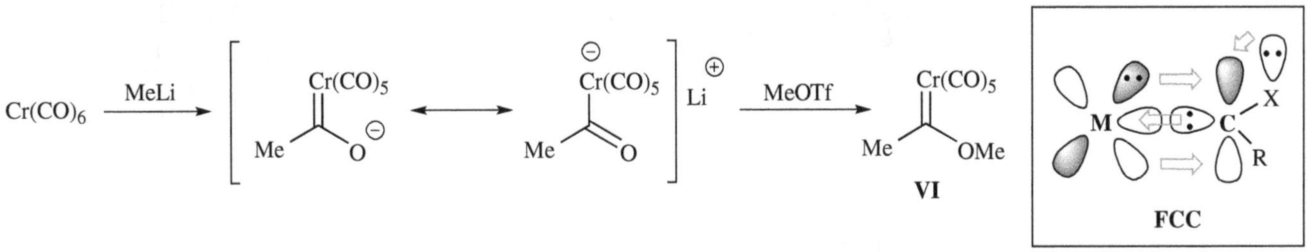

SCHEME 5.22 Synthesis of Fischer carbene complex **VI**.

carbene ligand bearing π-donor substituents (OMe or NMe_2). The chemistry of group VI (tungsten and chromium) FCCs is well documented, and many different synthetic methods have been reported for their preparation with different substituents at the carbene carbon. Scheme 5.22 shows the synthetic route to prepare Fischer carbene **VI**. The addition of a strong nucleophile, such as methyllithium, to $Cr(CO)_6$ and the subsequent reaction with an alkyl electrophile (methyl triflate) achieve compound **VI** with an alkoxy substituent at the carbene carbon atom.

The interaction between the carbene and the metal is dominated by the σ-donor component, which leads to a polarization with the positive charge centered in the carbon atom. Consequently, this carbenic carbon atom has an electrophilic character.

FCCs are considered important building blocks in synthetic organic chemistry, mainly due to their high versatility in C—C and C–heteroatom bond formation, as well as their often surprising behavior, which allows the incorporation of a carbonyl moiety from the metal fragment in the final product. For that reason, the use of FCCs in MCRs with the aim to prepare highly functionalized structures has emerged as an efficient strategy in modern organic chemistry [24].

The ability of heteroatom-stabilized FCCs to participate in MCRs, principally those based on chromium, was remarkably reviewed in 2005 by Barluenga [25] and later updated in 2010 by Aguilar [26]. Therefore, this section has been conceived to give an overview of the more recent results developed since then. With this idea, we have divided this section into two parts (Scheme 5.23): (i) *FCCs in MCRs without CO insertion* (Section 5.2.3.1) and (ii) *FCCs in MCRs involving CO insertion* (CO might be considered as an additional component coming from the FCC) (Section 5.2.3.2). Furthermore, in the second part, a diastereoselective MCR involving FCCs is also presented.

5.2.3.1 FCCs in MCRs without CO Insertion

Many research groups have explored the use of FCCs for the development of new synthetic routes to obtain highly functionalized heterocyclic molecules with important biological properties.

Following 3CR involving a cyclization/Diels–Alder sequence, Ghorai and coworkers reported the synthesis of a wide set of N-containing heterocyclic ring systems (quinoxaline, phenazine, and phenanthridine derivatives). These processes implied the reaction of FCC **VI**, an *o*-alkynyl carbonyl heteroaryl **28–30** and a dienophile [27, 28]. All these transformations proceeded through the common mechanistic pathway outlined in Scheme 5.24, which consecutively involves the following: (i) loss of one CO ligand from FCC and insertion of the alkynyl into the C═Cr bond lead to organometallic intermediate **I15**, where

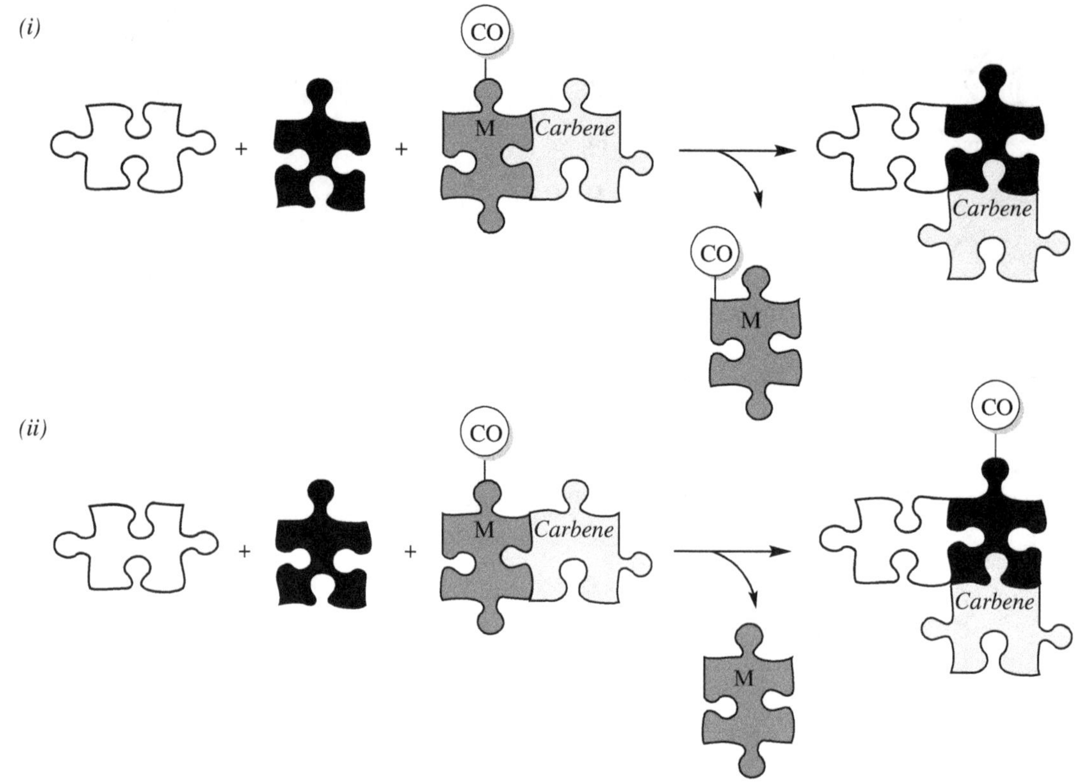

SCHEME 5.23 Fischer carbene complexes in MCRs.

a new C=Cr bond is formed; (ii) subsequent loss of a CO ligand and ring closure between the carbonyl moiety and the carbene carbon atom in **I15** afford furo-intermediate **I16**; and (iii) finally, species **I16** is trapped by a dienophile in a Diels–Alder cycloaddition affording [4+2] oxa-bridged adduct **31** and/or the corresponding aromatic heterocyclic product **32** depending on the dienophile, as shown in Table 5.1. Additionally, in some cases, oxa adduct **31** could be converted into the corresponding species **32** by extrusion of water, favored by the formation of an aromatic ring.

The same authors extended this procedure to α,β-unsaturated FCC **VII** (Scheme 5.25). In these reactions, the corresponding furo-intermediate **I17a/b** formed (analogue to **I16** in Scheme 5.24) contained a tetraene carbon framework, which conditioned the reactivity observed. For example, in the reaction of carbene **VIIa**, alkynyl heteroaryl ketone **33**, and DMAD, benzannulation product **34a** was isolated as the only product. This process involves a twofold Diels–Alder reaction of furo-intermediate **I17a** (with a rigid tetraene acting as two independent diene fragments) and DMAD, followed by ring opening and extrusion of water (favored by the formation of the aromatic system in **34a**). On the contrary, when the reaction was performed starting with carbene **VIIb**, intermediate **I17b** reacted with DMAD in a [8+2] cycloaddition, achieving oxabicyclo derivatives **34b**. This different reactivity can be attributed to the flexible disposition of the tetraene moiety in **I17b**, which can behave as an 8π-electron fragment [29].

*5.2.3.2 **FCCs in MCRs Involving CO Insertion*** As previously mentioned, in some reactions, FCCs allow the incorporation of carbonyl ligands along the reaction pathway. In view of this particular reactivity, in 2011, Ghorai reported a new general route for the synthesis of furoquinoline/furoisoquinoline derivatives **37** and **38**, in moderate yield, by simultaneous construction of furan and benzene rings [30] (Scheme 5.26). All the reactions were performed in the presence of PPh_3 as a ligand additive to avoid a possible competing ligation of the sp^2 nitrogen. As depicted in Scheme 5.26, coupling of complex **VI** with pyridine-bridged enyne derivatives **35** led to the formation of intermediate **I20**, upon insertion of the alkynyl moiety in the carbene and incorporation of a carbonyl fragment from the carbene complex. Later, cyclization of **I20** gave phenol derivative **36**, which was converted into the final products **37** and **38** by treatment with acid. The scope and generality of this reaction were tested using different enynes. In some cases, especially when the reaction was performed with a pyridine enyne with the alkene in *cis*-disposition, quinoline/isoquinoline species **39** and **40** were isolated.

The same authors have recently observed that coupling of γ,δ-unsaturated FCCs **VIII** with carbonyl derivative **41a,b** formed intermediates **I21**, which proceeded following

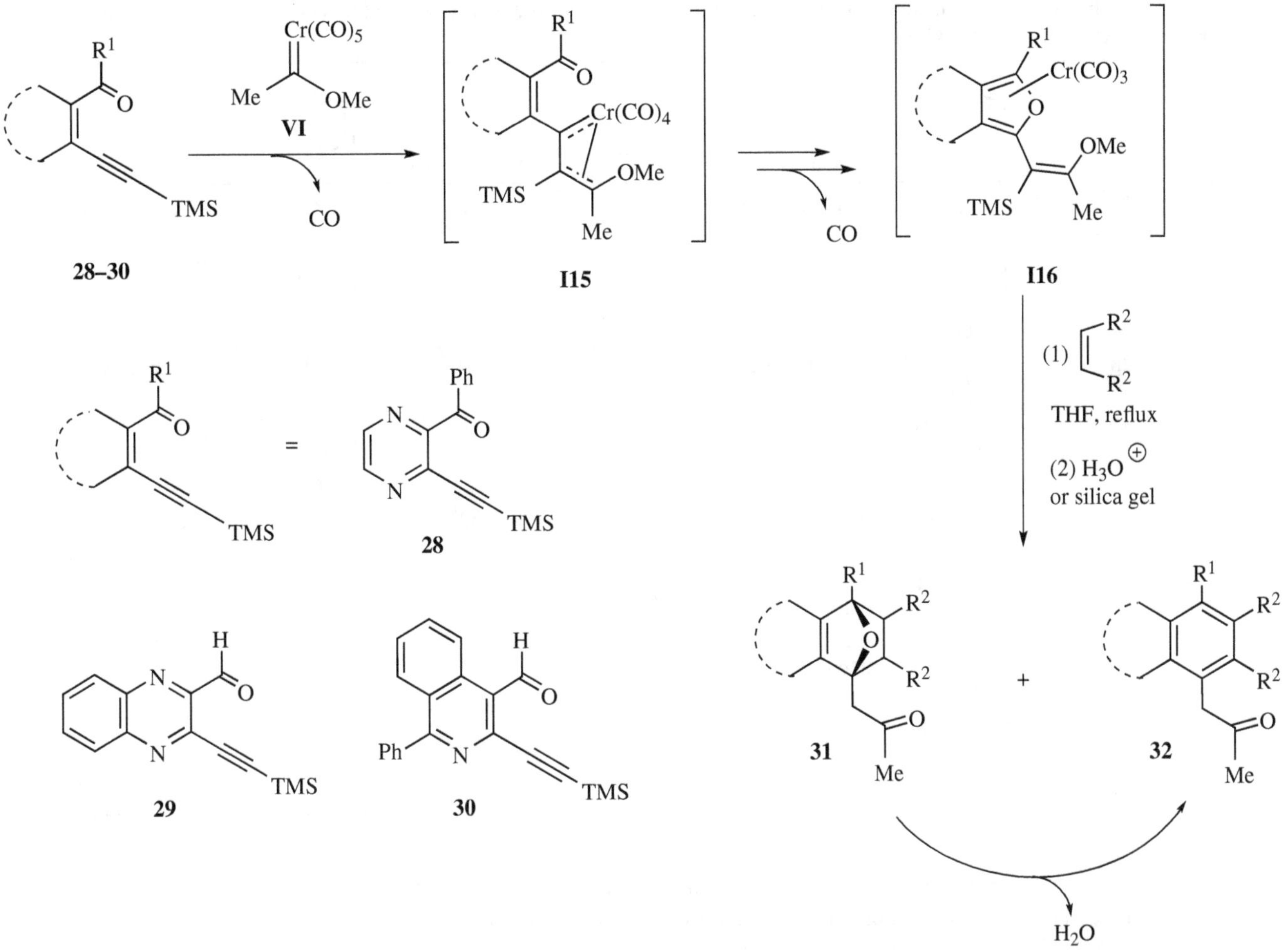

SCHEME 5.24 Synthesis of *N*-containing heterocyclic ring systems by coupling of FCC **VI** with an *o*-alkynyl carbonyl heteroaryl **28–30** and a dienophile.

different mechanistic pathways depending on the carbonyl moiety [31] (Scheme 5.27). When the reaction was performed with aldehyde **41a** (R^1 = H), CO insertion occurred achieving pyrone **I23**. Subsequent intramolecular Diels–Alder reaction led to a mixture of enol ether **42** and lactone **43**. If the same reaction was performed using ketone **41b** (R^1 = aryl), no CO insertion took place, and compounds **44** were isolated as the sole products. This different reactivity behavior might be attributed to the reduced nucleophilicity of the carbonyl oxygen in aldehydes, which would allow CO insertion from the metal fragment in **I21**, whereas in the case of ketones, the carbonyl oxygen could attack the carbene carbon atom in **I21** forming furo-intermediate **I24**.

Wulff has described a new method to successfully prepare chromenes **46** from α,β-unsaturated carbene complexes **IX** and alkoxyenynes **45** [32]. This reaction was efficient for a reasonable number of substituent patterns with different sizes either in the carbene or the enynes. Scheme 5.28 shows the mechanism that explains this transformation. First, benzannulation and migration of a CO moiety from **IX** form phenol intermediate **I26**, and then elimination of HOTMS generates the species **I27**, which after an electrocyclization process gives chromenes **46**. A similar synthetic strategy has recently been applied to prepare [m,n]cyclophanes [33].

The great potential of FCCs is also reflected in their use as precursors in MCRs for diastereoselective and enantioselective synthesis. Following a diastereoselective MCR involving CO insertion, Barluenga and coworkers have described the enantioselective synthesis of 4-allenyl-4-hydroxy-2-cyclohexenones **49** [34] and fused 6-5 bicyclic 2-butenolides [35] using chiral nonracemic enolates, FCCs, and organometallic reagents. For instance, as shown in Scheme 5.29, enantioenriched or enantiomerically pure compound **49** was obtained (in moderate yield) from chiral lithium enolate **47** (prepared from (*S*)-3-acetyl-4-benzyl-2-oxazolidinone and LDA), aryl/heteroaryl FCC **X**, and alkynyl magnesium bromide **48**.

The authors proposed the mechanism depicted in Scheme 5.30a to explain the stereoselective synthesis of these 4-hydroxy-2-cyclohexenones **49**. This mechanistic pathway

TABLE 5.1 Synthesis of oxa-bridged adduct 31 and aromatized *N*-containing heterocyclic 32 by coupling of FCC VI with an *o*-alkynyl carbonyl heteroaryl 28–30 and a dienophile

Entry	Carbonyl reagent	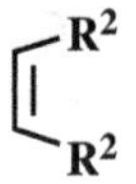	Products (yield)	Entry	Carbonyl reagent	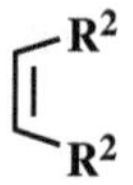	Products (yield)
1	**28**	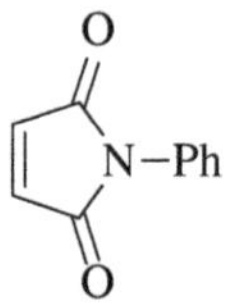	**31a** (42%), **32a** (30%)	5	**29**	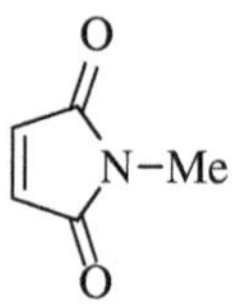	—, **32e** (55%)
2	**28**	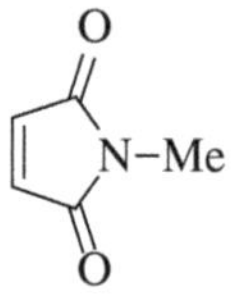	—, **32b** (55%)	6	**30**	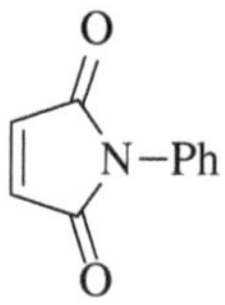	—, **32f** (39%)
3	**28**	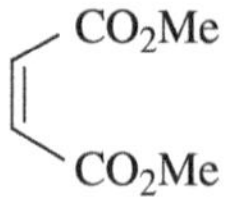	**31c** (40%), —	7	**30**	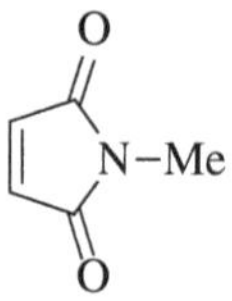	**31g** (30%), **32g** (32%)
4	**29**	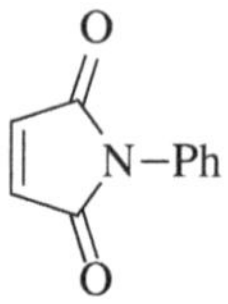	**31d** (10%), **32d** (52%)	8	**30**	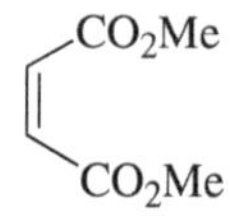	—, **32h** (44%)

involves the following: (i) the initial addition of chiral lithium enolate **47** to the carbene carbon atom of carbene **X** affords lithium intermediate **I28**; (ii) the successive regioselective double addition of organomagnesium **48** to the exocyclic carbonyl group of *N*-acyl-2-oxazolidinone unit leads to the removal of this chiral auxiliary group, incorporating two allenyl moieties and forming adduct **I29**; (iii) the insertion of a CO moiety from chromium fragment into the C(sp^3)—Cr bond of **I29** reaches lithium acetyl tetracarbonylchromate species **I30**; (iv) the chelation of lithium to both the sp^3-hybridized oxygen atoms in **I30** assists the subsequent regioselective intramolecular carbometalation of the terminal C—C bond of one allene group, forming allylchromate **I31**; and (v) the final protonation of **I31** gives cyclohexenone **49**.

The absolute stereoinduction of this process is detailed in Scheme 5.30b. As illustrated in model **I32** (and corresponding Newman projection **I32***), the chelated enolate approaches the carbene complex from the less hindered face. This approximation in an *anti*-orientation of donor and acceptor π systems places the bulky substituent of the enolate away from the $Cr(CO)_5$ fragment, avoiding steric interactions. Approximation from the opposite prochiral face of carbene complex **X** would provoke severe steric interactions with the oxazolidinone group as indicated in model **I33** [34].

5.3 MCRs INVOLVING CARBENES AS CATALYSTS

MCRs driven by catalytic species possess all the credentials to be considered an ideal tool for organic synthesis in terms of atom economy and efficiency. In this section, an overview of such processes involving NHCs is discussed from two perspectives (Scheme 5.31): (i) NHCs acting as organocatalysts and (ii) NHCs as ligands in transition metal-based catalysts.

MCRs involving the catalytic generation of metallocarbenes (e.g., from diazo compounds and Rh(II) catalysts) have been also reported. These types of reactions are not considered herein since recent examples are discussed in Chapter 3.

5.3.1 NHCs as Organocatalysts in MCRs

Reactions employing NHCs as organocatalysts have emerged as an exceptionally prolific research area in organic chemistry in recent years [36]. The phenomenal success of NHCs in organocatalysis can be attributed to their electronic properties, which, as illustrated in Scheme 5.32, allow different types of substrate activation. For instance, the pronounced nucleophilicity of NHCs allows the addition to electrophiles such as aldehydes leading to the formation of acyl anion

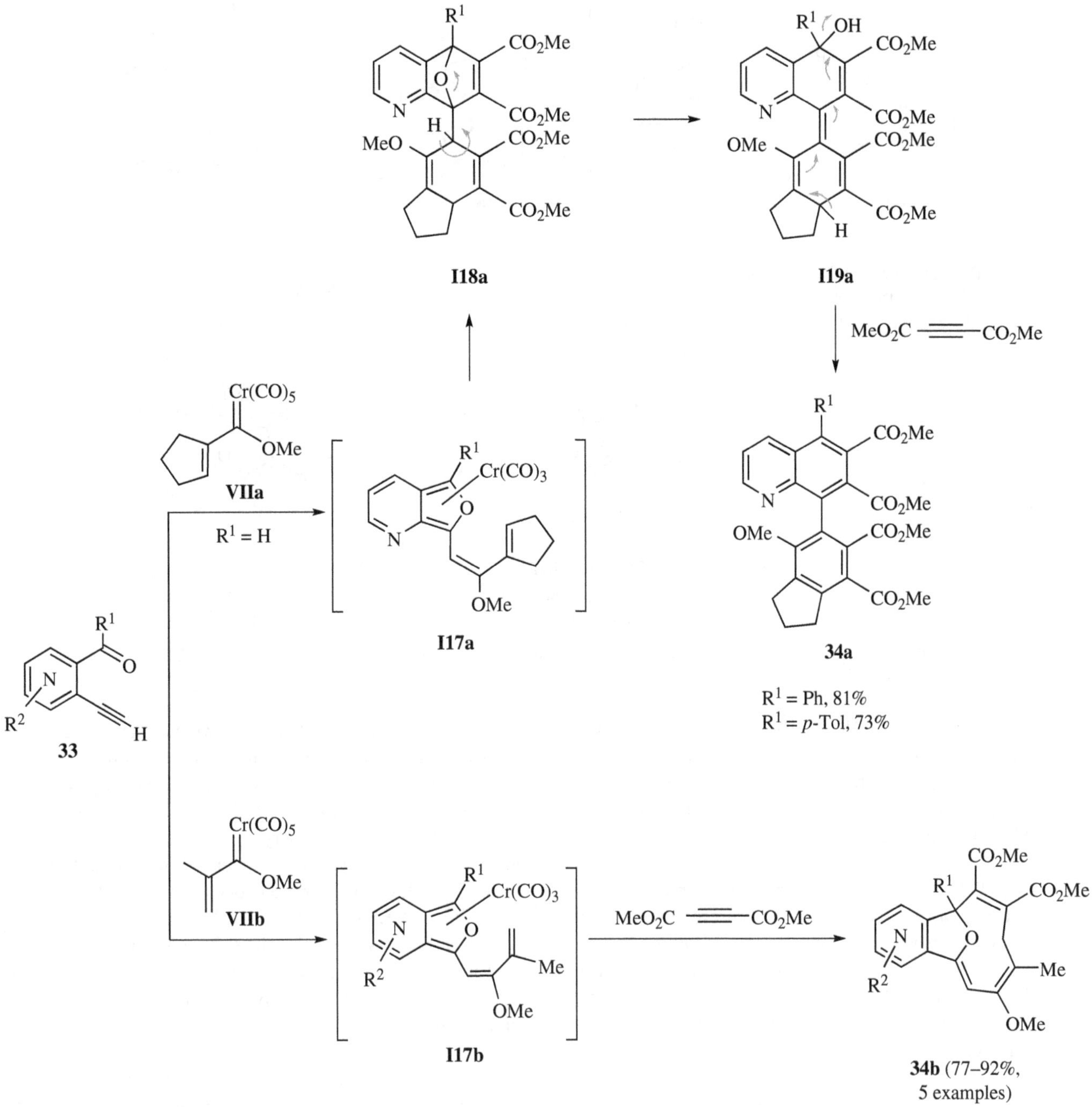

SCHEME 5.25 Coupling of **α,β**-unsaturated carbene **VIIa/b** with alkynyl heteroaryl ketone **33** and DMAD.

equivalents, referred to as "Breslow intermediates" (Scheme 5.32, Eq. 1). Key examples for this reaction pathway are benzoin condensation [37], wherein an aryl aldehyde acts as the second reaction component, and Stetter reaction [38], where a Michael acceptor takes the role. It is worth pointing out that employing functionalized aldehydes (enals, α-hydroxyaldehydes, ynals, etc.) as donor substrate opens new mechanistic pathways and gives rise to new synthetic strategies, especially in MCRs. With esters as substrates, an activated carboxylate is generated (Scheme 5.32, Eq. 2) allowing transesterifications, while three-membered ring epoxides or aziridines are also prone to ring opening by NHC catalysis (Scheme 5.32, Eq. 3).

In contrast, the high Lewis basicity of NHCs also explains their capability to activate silylated nucleophiles by coordination to the silicon atom (Scheme 5.32, Eq. 4), while their Brønsted basicity allows for activation of alcohols or amines via hydrogen bonding.

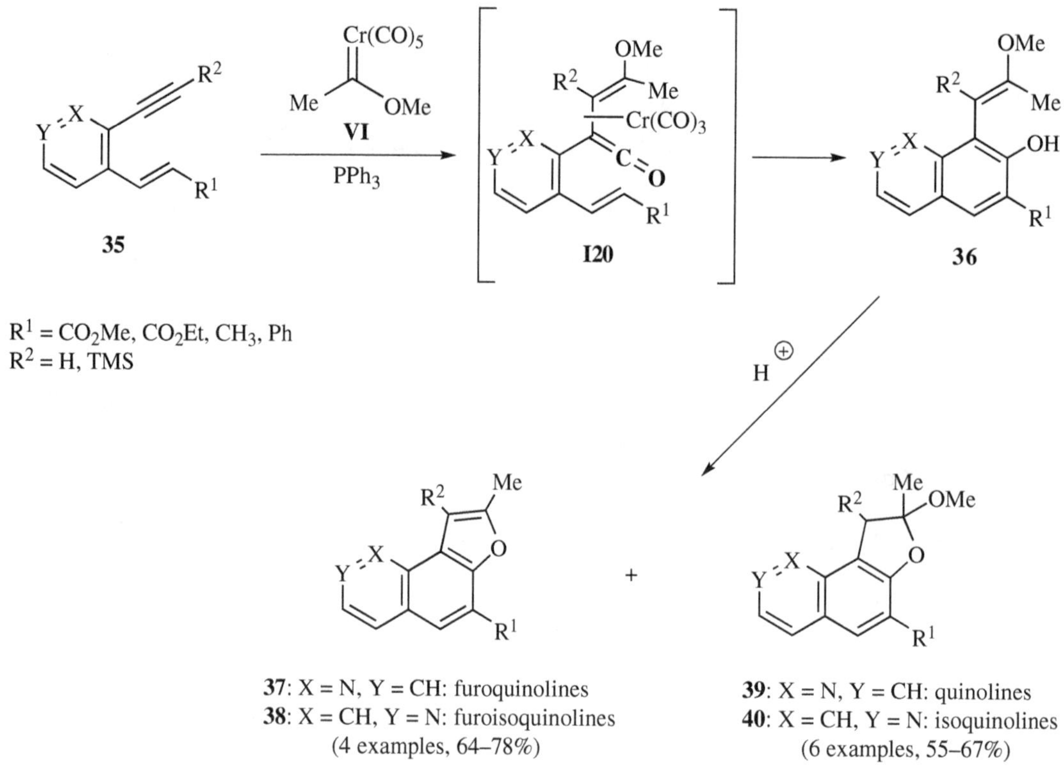

SCHEME 5.26 Synthesis of furoquinoline/furoisoquinoline derivatives **37** and **38** from FCC **VI**.

Although numerous NHC-catalyzed reactions are well documented in the literature and MCRs incorporating them as components have also been explored (Section 5.2.2), the application of some of the aforementioned activation modes on NHC-catalyzed MCRs still remains underexploited.

5.3.1.1 MCRs Involving Amidation of Carbonyls with Nitroso Compounds NHC-catalyzed direct amidation of carbonyls with nitroso compounds is a powerful method for the synthesis of *N*-arylhydroxamic acids **50** and their derivatives [39] (Scheme 5.33). The strategy exploits the reactivity of nitroso compounds as electrophiles, which allows reaction with intermediate **I34**, forming the corresponding C—N bonds.

Employing this C—N bond-forming reaction, Cheng and coworkers have described 3CR of α,β-unsaturated aldehydes (enals), nitrosoarenes, and α,β-unsaturated ketones to synthesize *N,O*-functionalized *N*-hydroxyacrylamides **52** in moderate to good yields (Scheme 5.34) [40].

Under presented conditions (triazolium bromide **XI** and DBU in CH_2Cl_2), the reaction proceeds by a cascade reaction: (i) first, amidation (aza-benzoin condensation between enal and nitrosoarene, as outlined in Scheme 5.33) produces *N*-hydroxyacrylamide **51**, and (ii) second, in the presence of base such as DBU or triazole carbene [41], oxo-Michael addition of **51** to enones would afford products **52** (Scheme 5.35).

In the particular case of acrolein, which can be used both as a donor and acceptor component (3 equiv to nitrosobenzene), a novel domino amidation–redox amination took place under similar conditions (triazolium **XII**/DBU, CH_2Cl_2, rt), yielding *N*-propionyloxy-*N*-phenylacrylamide **54** as the major product (75%) [42]. In this reaction, as depicted in Scheme 5.36, two catalytic cycles seem to operate, one forming the *N*-phenylhydroxamic acid **53** by reaction of nucleophilic intermediate **I35** (as d^1) with nitrosobenzene. The second one operates through β-protonation of **I35** (as d^3) forming the activated ester **I36**, which then reacts with the hydroxamate of **53** yielding **54** along with NHC catalyst.

A similar NHC-catalyzed reaction between benzaldehyde, nitrosobenzene, and acrolyl derivatives **55** employing triazolium salt **XII** and NaH for the synthesis of *N*-arylaziridines **56** has been reported (Scheme 5.37) [42]. The mechanism starts with the formation of hydroxamic acid anion **I37**. Next, isomerization toward *O*-benzoyl derivative **I39** occurs, through oxaziridine **I38**. Although attempts to capture or detect spectroscopically such intermediates have not been successful, in principle, any of them may act transferring the nitrogen-containing component to **55** (**I39** represented as the nucleophilic nitrogen transfer agent in Scheme 5.37).

5.3.1.2 MCRs through Homoenolates Since the pioneering works by Bode [43] and Glorius and Burstein [44] describing the first NHC-catalyzed generation of reactive

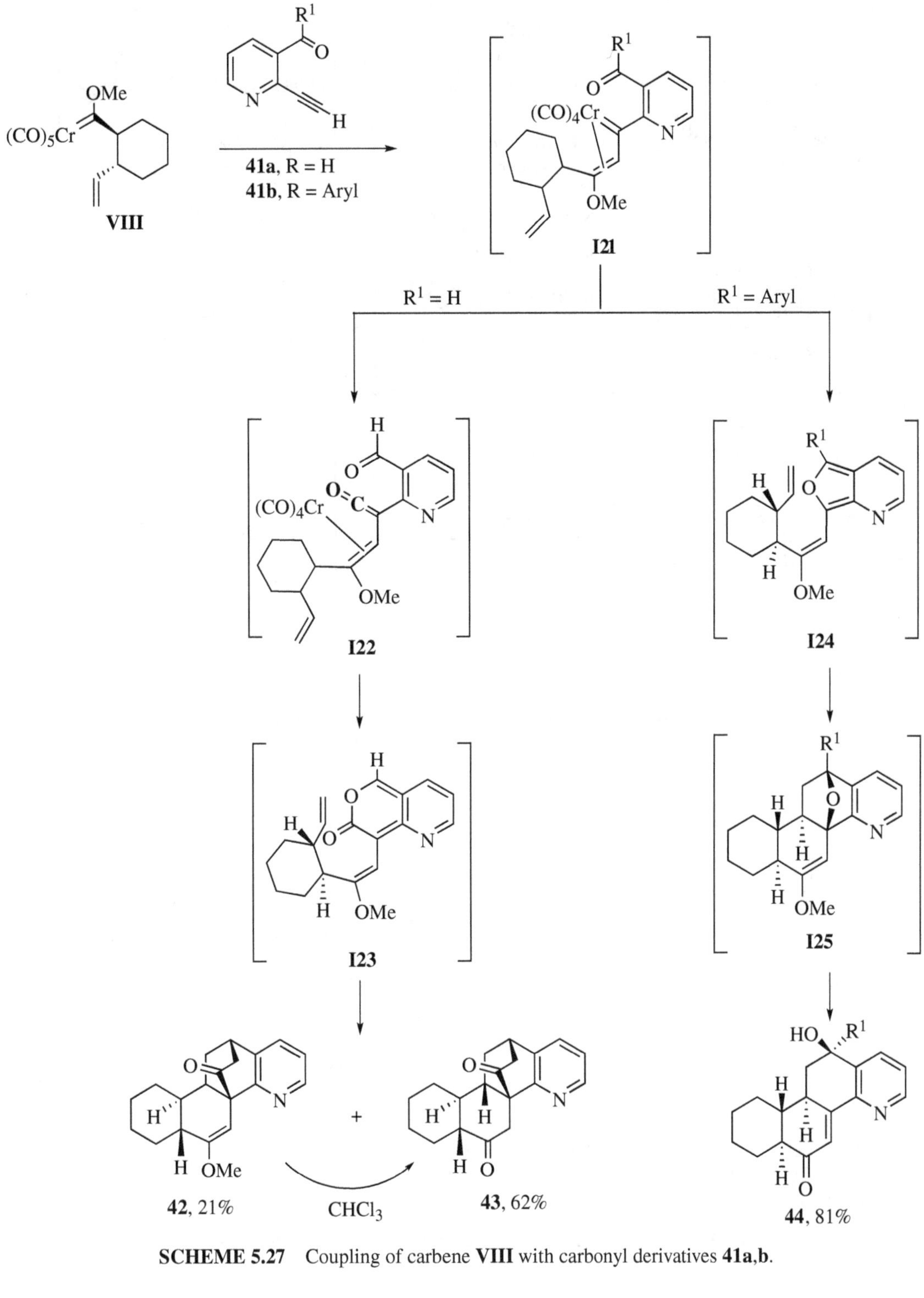

SCHEME 5.27 Coupling of carbene **VIII** with carbonyl derivatives **41a,b**.

homoenolate intermediates **I40** from enals (Scheme 5.38), the involvement of these intermediates (acting as d^3-nucleophiles) in reactions with a variety of Michael acceptor has gradually been recognized as an important synthetic tool. Altogether, this process converts α-reducible aldehydes into their corresponding esters/amides **57** (Nu=OR or NRR') [45].

The general catalytic cycle presented in Scheme 5.38 appears a priori as one of the fundamental strategies for designing MCRs (from 3CR; enal + electrophile + nucleophile), and relevant contributions have been already described.

In this context, Nair and coworkers reported three-component NHC-catalyzed domino reactions with enals,

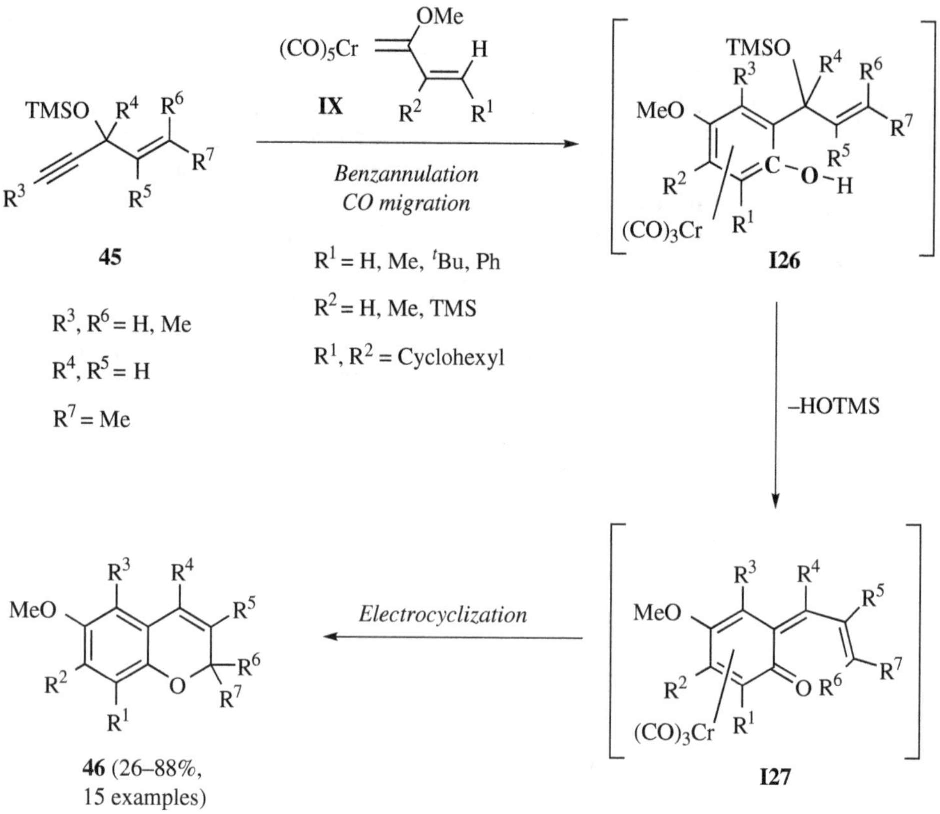

SCHEME 5.28 Synthesis of chromenes **46** from FCC **IX**.

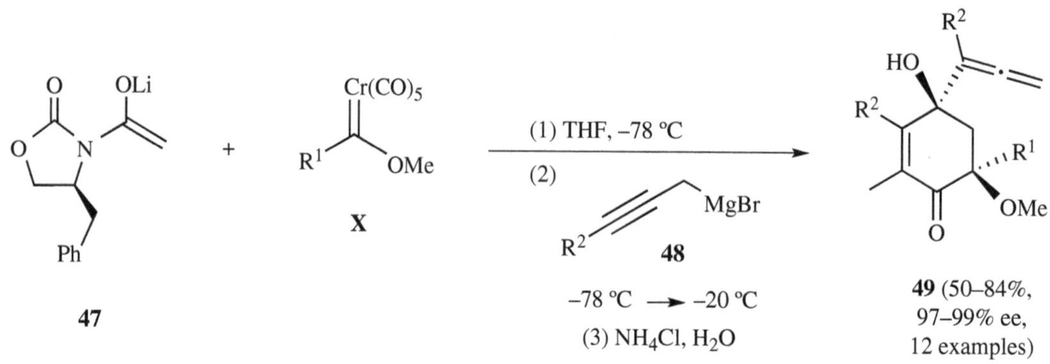

R^1 = Ph, 2-naphthyl, p-$MeOC_6H_4$, p-$TMSOC_6H_4$, p-ClC_6H_4, p-BrC_6H_4, 5-TMS-2-furyl, 5-TMS-2-thienyl
R^2 = Me, Et, Bu, Ph

SCHEME 5.29 Synthesis of 4-allenyl-4-hydroxy-2-cyclohexenones **49**.

chalcones, and methanol to constitute functionalized cyclopentanes **58** as major products and *trans*-acyclic δ-ketoesters **59** as minor ones (Scheme 5.39) [46]. Among the various catalysts screened, benzimidazole carbene (from **XIII**) and imidazole carbene (from **IPr-Cl**) showed low activity, while triazole carbenes (from **XII** and **XIV**) afforded the redox product **60** (from the enal). However, employment of imidazolium salt **IMes-Cl** and DBU in methanol led to methyl β-hydroxycyclopentanecarboxylates **58** in good yields with variable amounts of acyclic δ-ketoesters **59**. It is noteworthy that only one diastereomer (four contiguous chiral centers in **58**) was formed in each case.

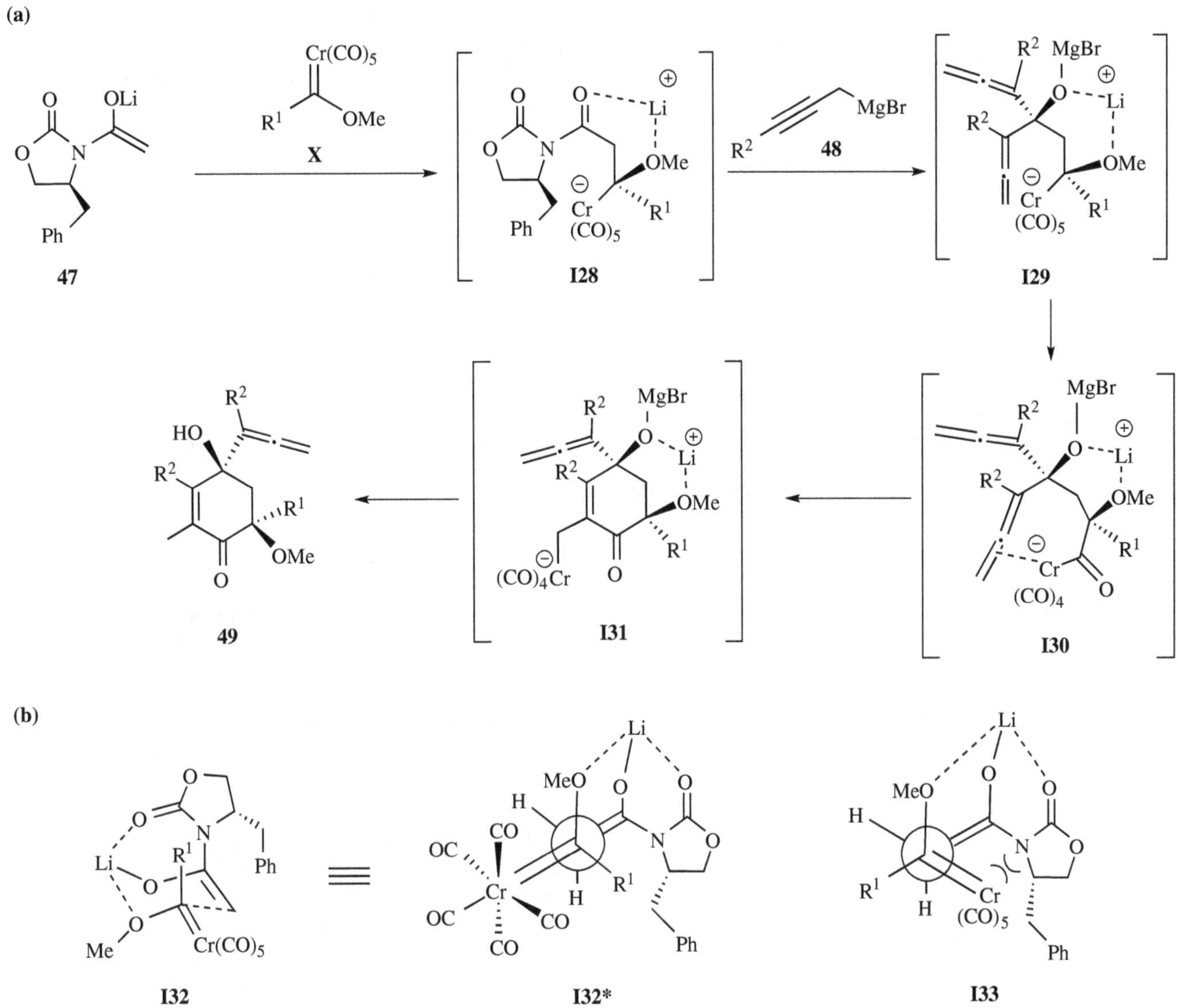

SCHEME 5.30 (a) Mechanistic pathway for the synthesis of cyclohexenone **49**. (b) Stereochemical models.

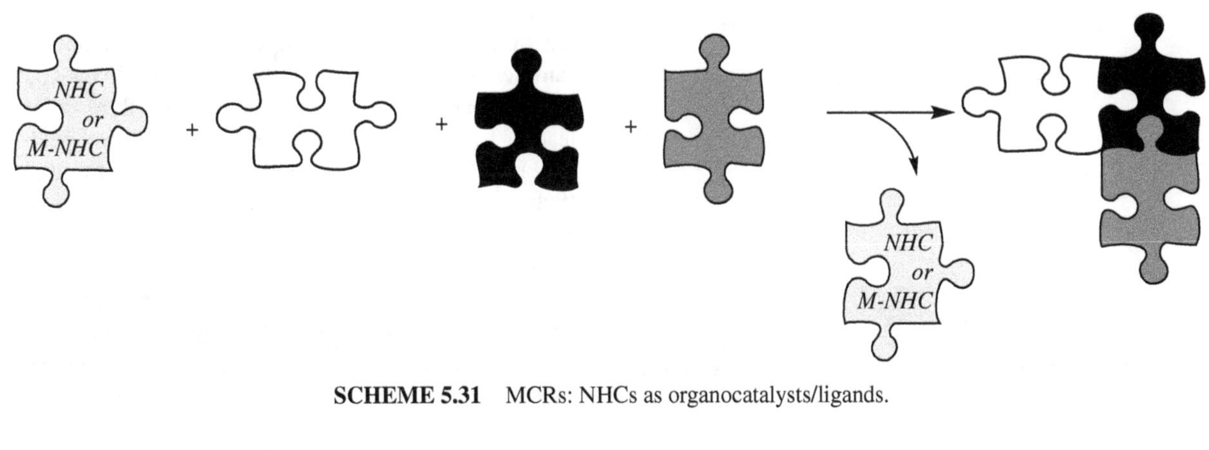

SCHEME 5.31 MCRs: NHCs as organocatalysts/ligands.

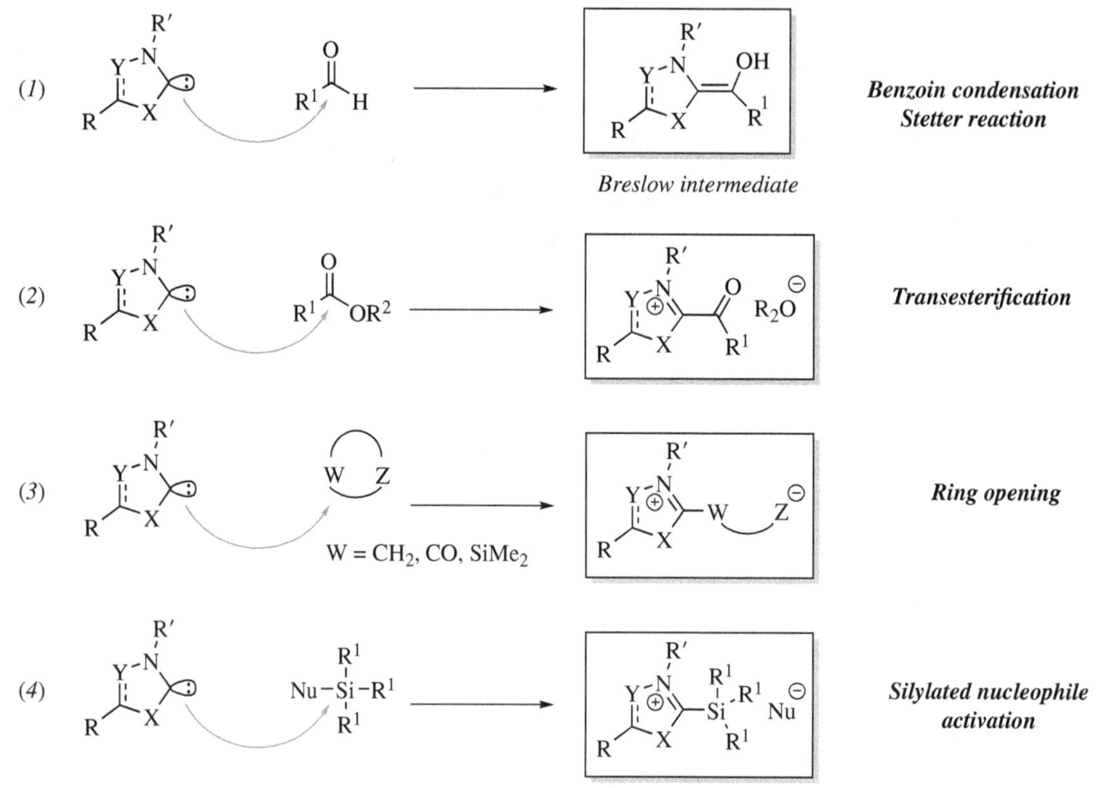

SCHEME 5.32 Basic modes of substrate activation by NHCs in organocatalysis.

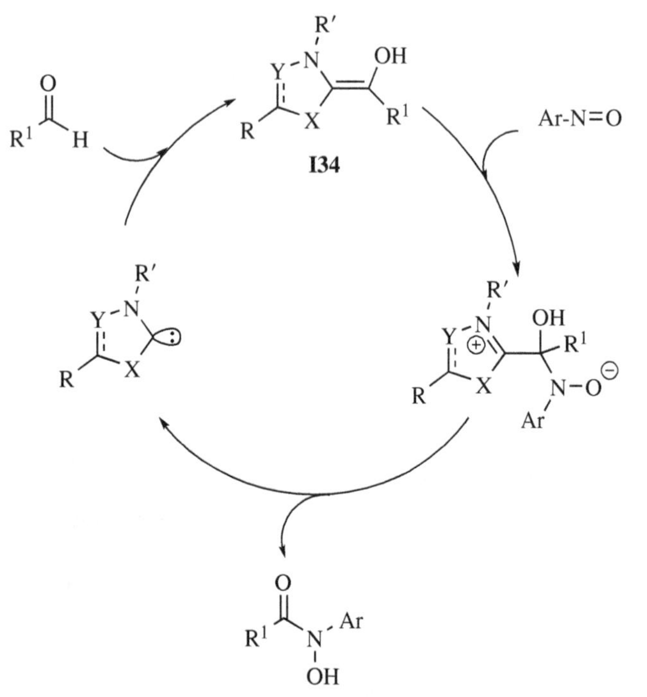

SCHEME 5.33 Proposed amidation of carbonyls using nitrosoarenes.

The postulated catalytic cycle for the formation of **58** is shown in Scheme 5.40. First, addition of homoenolate to chalcone affords 1,5-dicarbonyl species **I41**. Then, the corresponding NHC-bound enol form **I42** undergoes intramolecular cyclization to the ketocarbonyl followed by methanolysis (of **I43**) to give the final product **58**, releasing **IMes**.

In the course of a mechanistic investigation by Bode and coworkers, the same substrates were subjected to a mixed solvent of ethanol and 1,2-dichloroethane with triazolium salt **XV** as precatalyst. In this case, the MCR was not observed, yielding cyclopentene **61** as the major product [47] (Scheme 5.41, *left*).

Later, Chen and coworkers showed the third possible reaction pathway with identical substrates employing the same catalytic system described by Nair (**IMes-Cl**, DBU) with propargylic alcohol as nucleophilic third component, leading to acyclic δ-ketoesters **62** with excellent yields (up to 98%) and high *cis*-stereoselectivities (up to 20:1) [48] (Scheme 5.41, *right*). The relative stereochemistry might be explained by a cascade crossed-benzoin/oxy-Cope rearrangement as enlightened by Bode's mechanism [47]. In Scheme 5.42, a plausible catalytic cycle (different from the one presented as general catalytic cycle in Scheme 5.38) is depicted. A crossed-benzoin reaction occurs between the catalyst-bound Breslow intermediate **I44** (acting as d^1) and

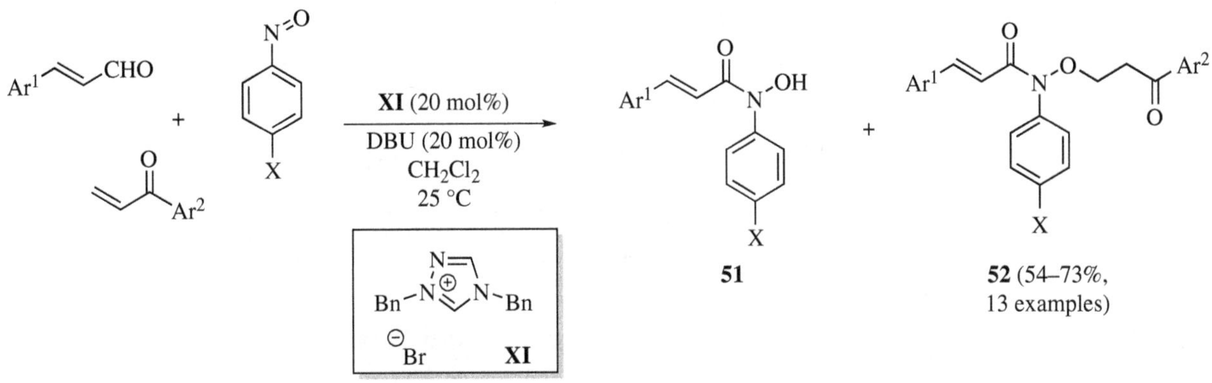

SCHEME 5.34 NHC-catalyzed 3CR of enals, nitrosoarenes, and enones.

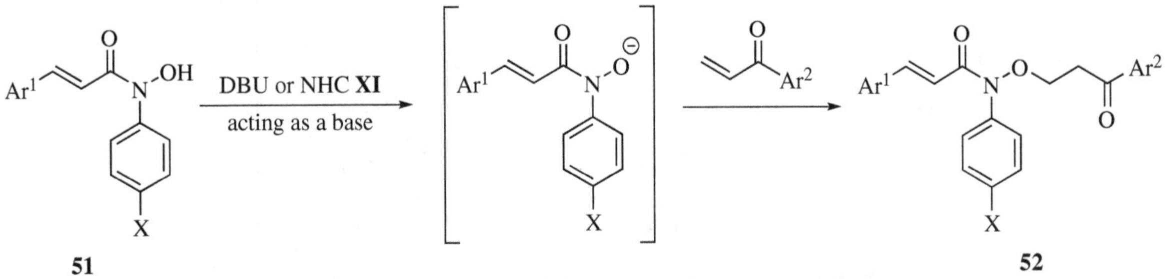

SCHEME 5.35 Base promoted oxo-Michael addition of **51** to enones.

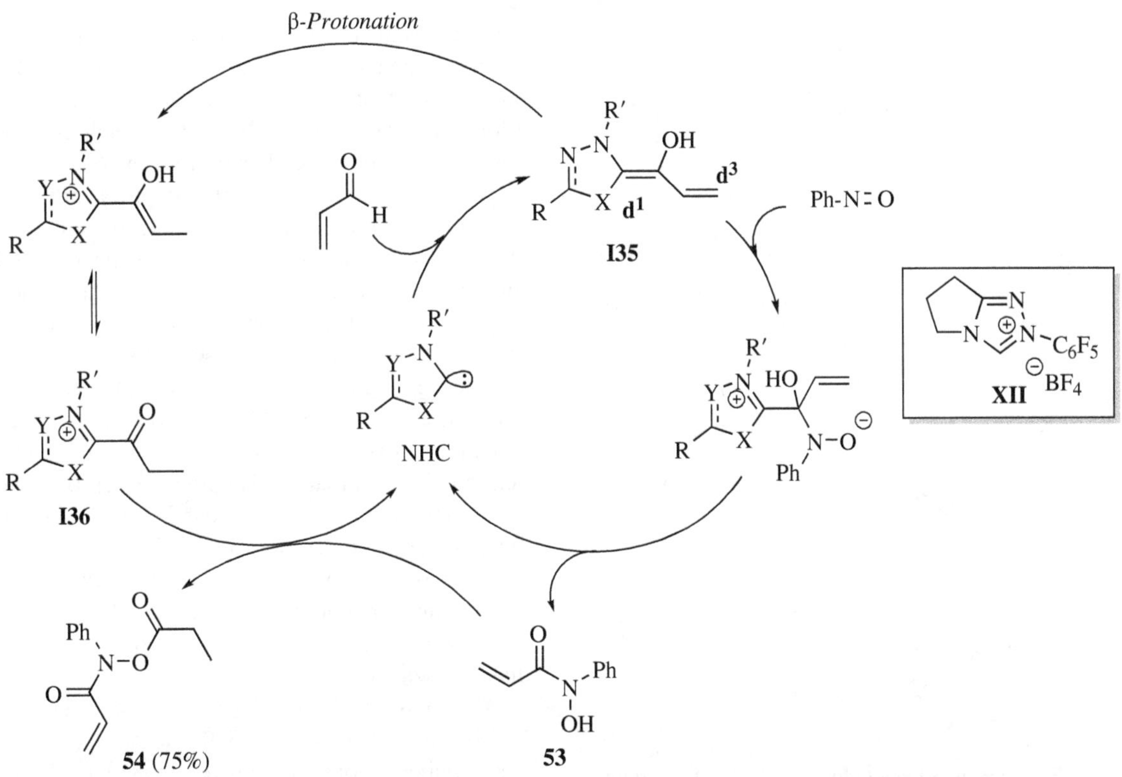

SCHEME 5.36 Proposed domino amidation–redox amination of acrolein with nitrosobenzene.

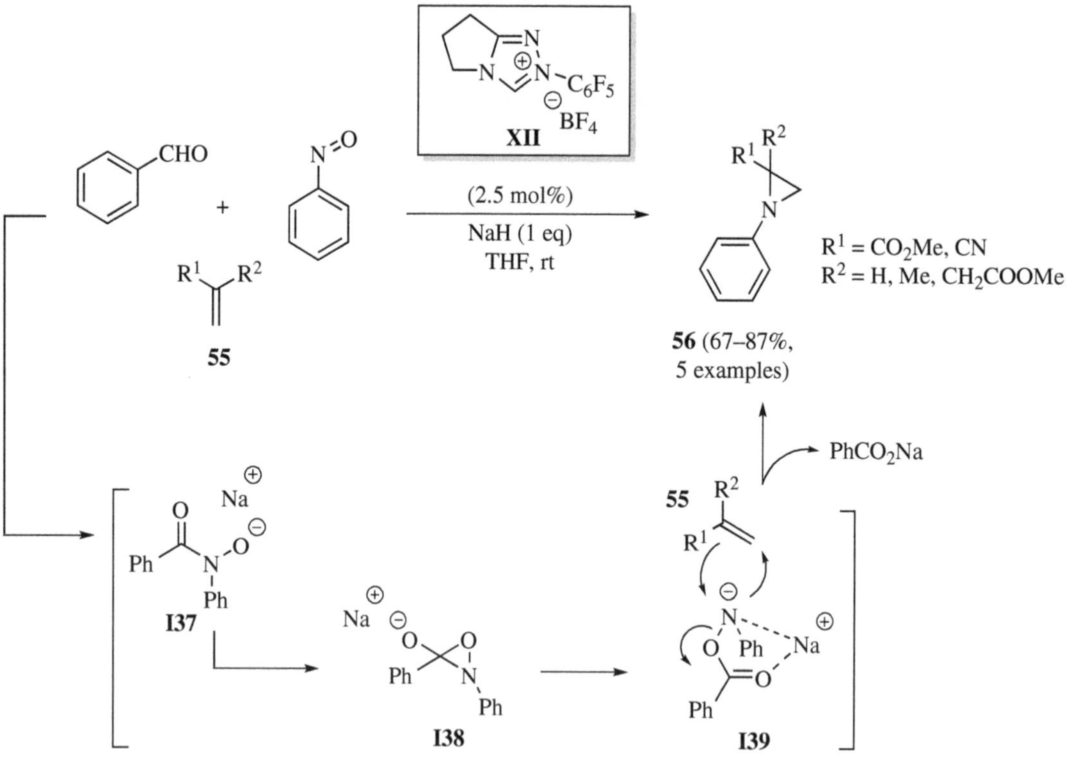

SCHEME 5.37 3CR for the synthesis of *N*-arylaziridines **56**.

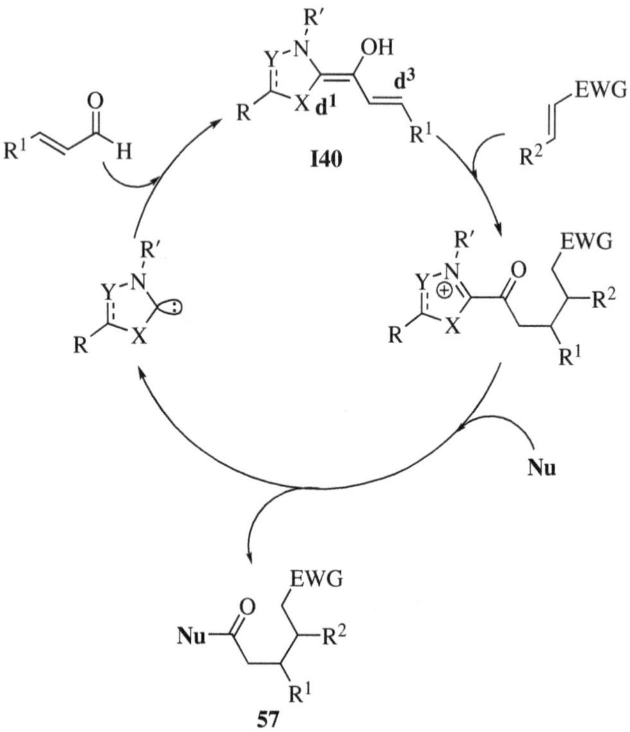

SCHEME 5.38 General strategy for MCR design through homoenolates.

the chalcone, leading to intermediate **I45**. This intermediate undergoes an oxy-Cope rearrangement to give **I46** with *cis*-relative configuration. Finally, tautomerization/protonation affords activated carboxylate **I47**, which is subsequently trapped by propargylic alcohol, forming **62**.

Among different Michael acceptors, nitroalkenes are probably the most synthetically useful electrophiles. In 2009, Nair and coworkers reported the use of imidazolinium-based carbenes (from **SIMes-Cl**) as catalysts in the addition of α,β-unsaturated aromatic aldehydes, via homoenolate intermediates (**I40**, Scheme 5.38), to β-nitrostyrenes, giving racemic δ-nitroester **63** [49] (Scheme 5.43, *left*). Three years later, Liu and coworkers developed the enantioselective version of this reaction employing indanol-derived triazolium salt **XVI** as precatalyst [50] (Scheme 5.43, *right*). The scope was also expanded, and a broad range of enals in conjunction with a variety of nitroalkenes, such as nitrodienes, nitroenynes, and nitrostyrenes, could be transformed into enantiomerically enriched δ-nitroester **63** in good yields, high enantioselectivities, and acceptable diastereoselectivities.

As previously discussed, based on the general strategy for MCRs presented in Scheme 5.38, various types of reactive a^3 electrophiles (enones and nitroalkenes) have been included to prepare synthetically valuable molecular scaffolds. In

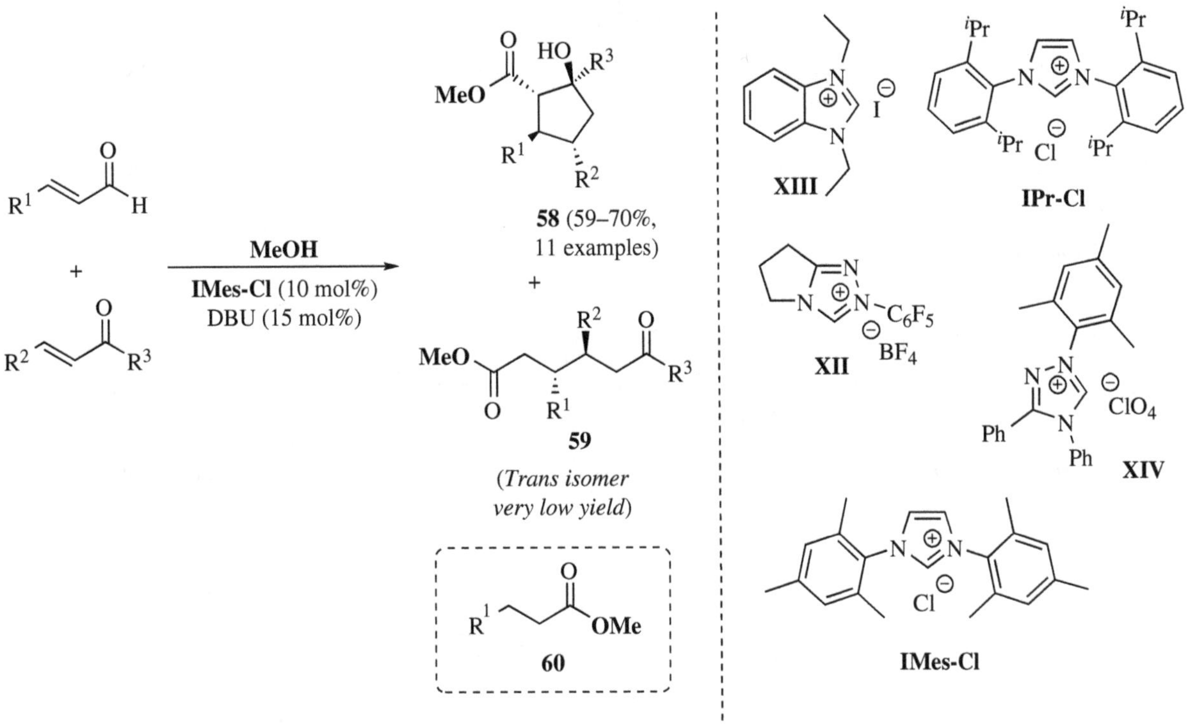

SCHEME 5.39 NHC-catalyzed 3CR of enals, chalcones, and methanol for the synthesis of **58**.

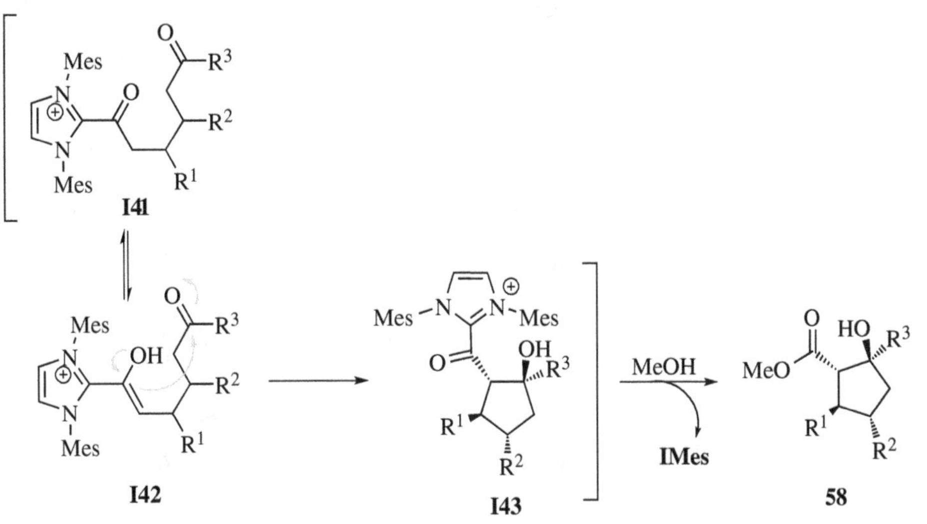

SCHEME 5.40 Proposed annulation of **I41** to access cyclopentanes **58**.

other interesting examples, Nair and coworkers employed sulfonimines as a[1] electrophiles to develop a novel *pseudo*-4CR (Scheme 5.44, *left*) and 3CR (Scheme 5.44, *right*) for the synthesis of γ-aminobutyric acid derivatives **64** and **65**, respectively [51].

Subjecting *p*-tolylsulfonamide and enal (4 equiv.) to the catalytic system **IMes-Cl**/DBU in methanol, **64** was obtained stereoselectively in high yields. As postulated in Scheme 5.45, homoenolate **I48** undergoes facile 1,2-addition to a sulfonimine generated *in situ* from the enal and the sulfonamide, and subsequent methanolysis of intermediate **I50** affords **64**. Complementary, in the absence of methanol, intermediate **I50** intramolecularly cyclizes onto the carboxy surrogate to afford γ-lactam derivative **65** with good diastereomeric ratio.

The examples of this section illustrate how slightly modifications of reaction conditions exert remarkable influence on the MCR outcomes.

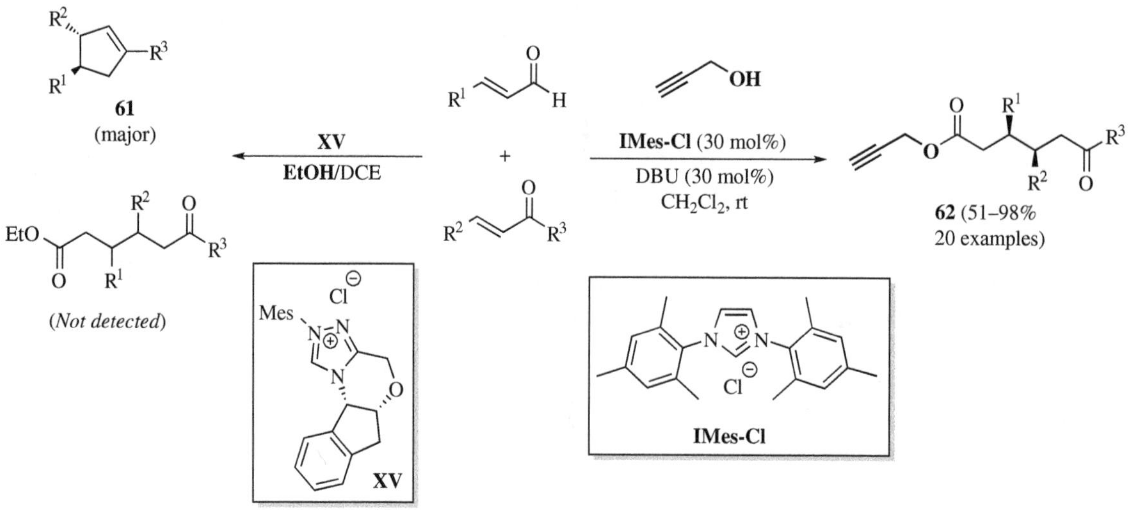

SCHEME 5.41 Different reaction pathways in NHC-catalyzed 3CR of enals, chalcones, and alcohols.

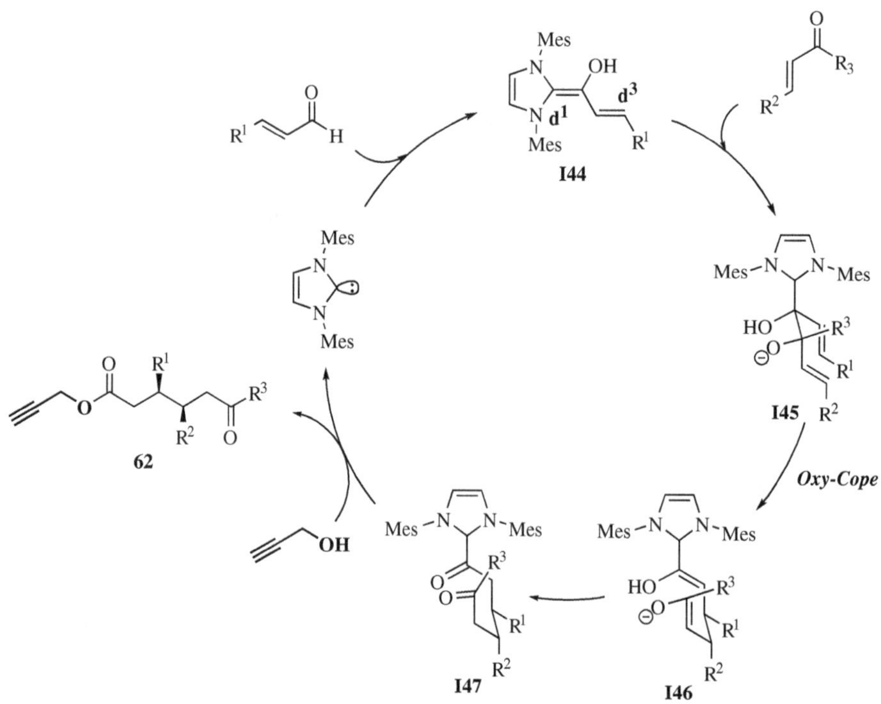

SCHEME 5.42 Plausible mechanism for the formation of **62**.

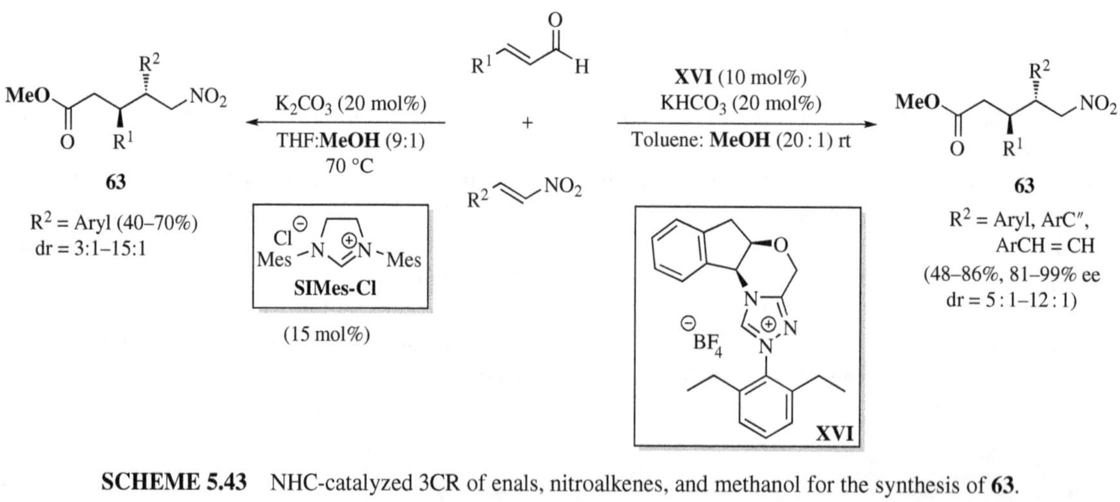

SCHEME 5.43 NHC-catalyzed 3CR of enals, nitroalkenes, and methanol for the synthesis of **63**.

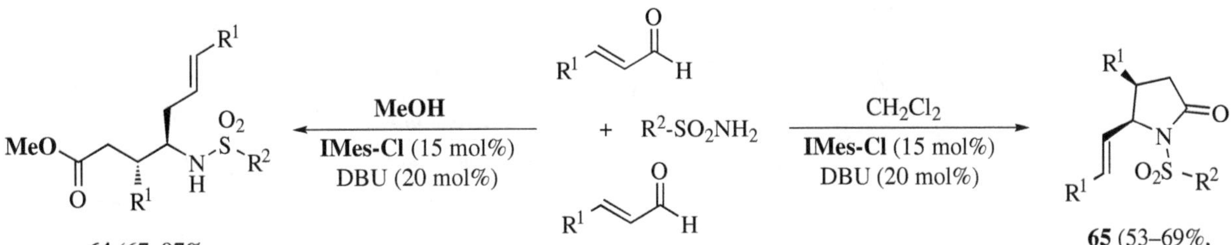

64 (67–97%, 12 examples)

65 (53–69%, dr up to 10:1, 9 examples)

SCHEME 5.44 NHC-catalyzed 4CR or 3CR of enals and *p*-tolylsulfonamide for the synthesis of **64** and **65**.

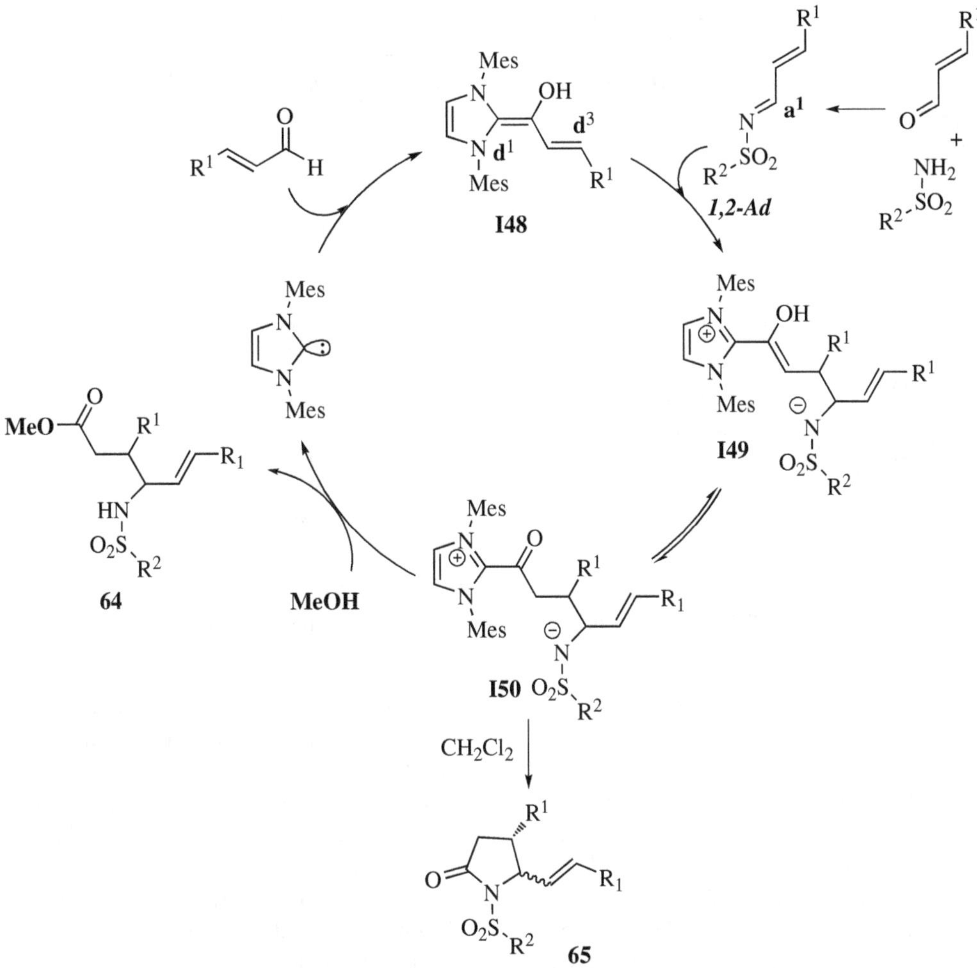

SCHEME 5.45 Postulated catalytic cycle.

5.3.1.3 MCRs from Ynals Activation of α,β-unsaturated ynals (alkynyl aldehydes) by NHCs has been recently recognized as a fascinating tool to catalytically generate α,β-unsaturated acyl azolium **I51**, as double electrophilic reactive species [52] (Scheme 5.46). Therefore, a rapidly growing number of applications have been reported employing this strategy in cascade reactions [53].

In 2012, Du and Lu exploited this activation mode in a three-component (two alkynyl aldehydes + one oxindole) domino reaction for the synthesis of spirooxindole 4*H*-pyran-2-one derivatives **66** (Scheme 5.47) [54]. Extensive optimizations led to identify carbene precursor **IPr-Cl** (30 mol%) and ^{t}BuOK (30 mol%) in THF at 65 °C as the best reaction conditions to avoid Knoevenagel product **67**, affording **66** in reasonable

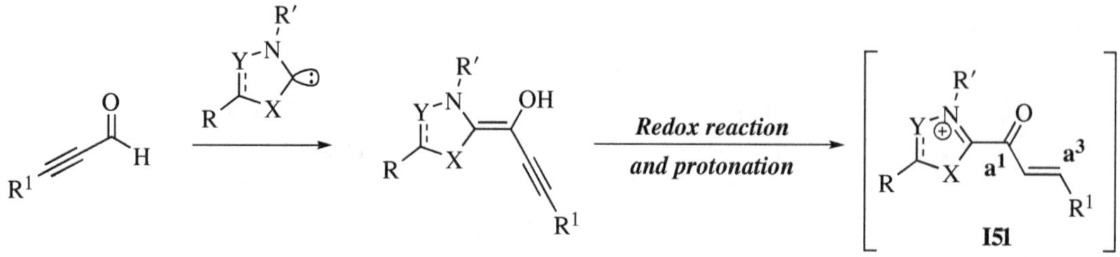

SCHEME 5.46 Catalytic generation of α,β-unsaturated acyl azolium **I51**.

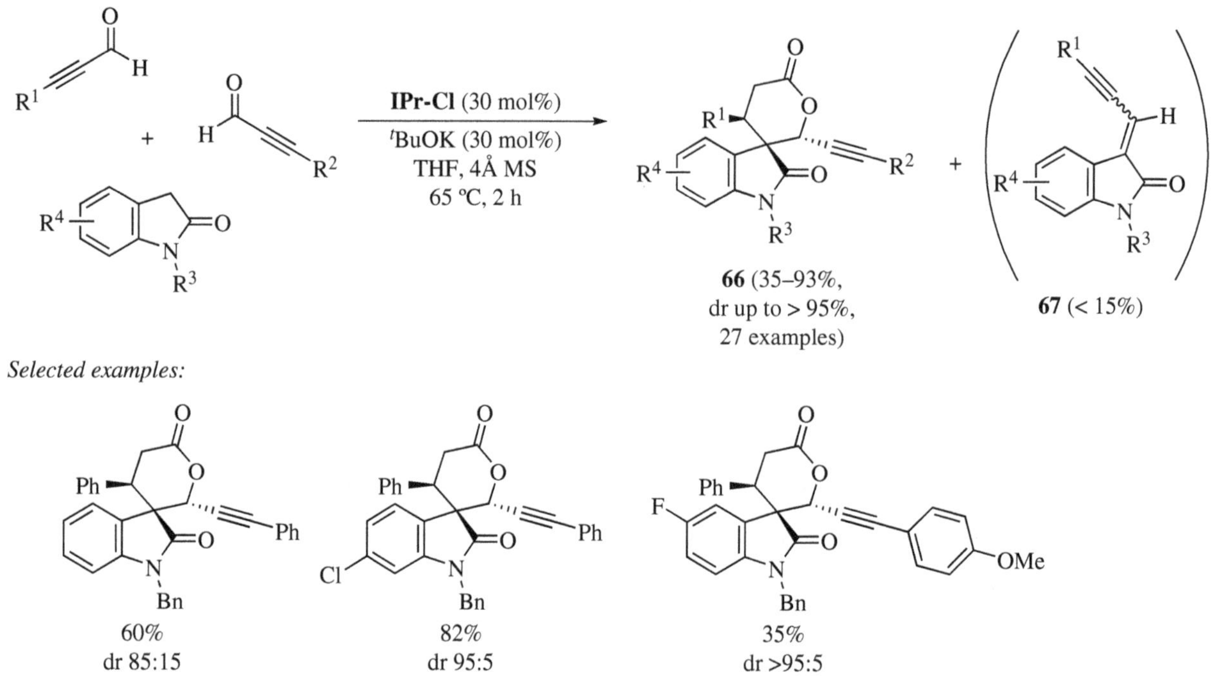

SCHEME 5.47 NHC-catalyzed 3CR of alkynyl aldehydes with oxindoles.

yields (35–93%) with good to high diastereoselectivities (*dr* 85:15–95:5).

A mechanistic rationalization for this reaction is presented in Scheme 5.48. Domino process is initiated with generation of key intermediate **I52**, which undergoes a 1,4-addition of a deprotonated oxindole to form **I53**. After H-migration and following aldol reaction with another molecule of alkynyl aldehyde, intermediate **I55** is formed. Subsequent lactonization delivers product **66** and regenerates NHC for the next catalytic cycle. The stereochemistry of the products may be determined by the most sterically favored chair transition state in **I55**.

5.3.1.4 NHC-Catalyzed MCRs in Heterocyclic Chemistry In 2001, Müller and coworkers developed an elegant example where chalcones could be transformed in a one-pot sequence to the corresponding pyrrole derivatives [55], stating the potential of NHC-catalyzed umpolung strategies in heterocyclic chemistry. Scheidt and coworkers described a catalytic multicomponent synthesis of pyrroles via sila-Stetter/Paal–Knorr strategy in a one-pot fashion [56]. Later, Yao and coworkers reported an efficient NHC-catalyzed 3CR for the synthesis of 2,3-diarylated fully substituted furans **68** from inexpensive starting materials (aromatic aldehydes and malononitrile or ethyl cyanoacetate) [57] (Scheme 5.49).

A plausible reaction mechanism is illustrated in Scheme 5.50. Initially, Knoevenagel condensation between an aryl aldehyde and malononitrile, under basic conditions, gives Michael acceptor **69**, which reacts with intermediate **I56** via intermolecular Stetter reaction, leading to adduct **70**. Then, upon effect of the base, an intramolecular cyclization occurs to give furan **68**.

5.3.2 Metal-Catalyzed MCRs Involving NHCs as Ligands

It is possible to say without any doubt that the popularity of NHC ligands in the design of homogeneous catalysis is

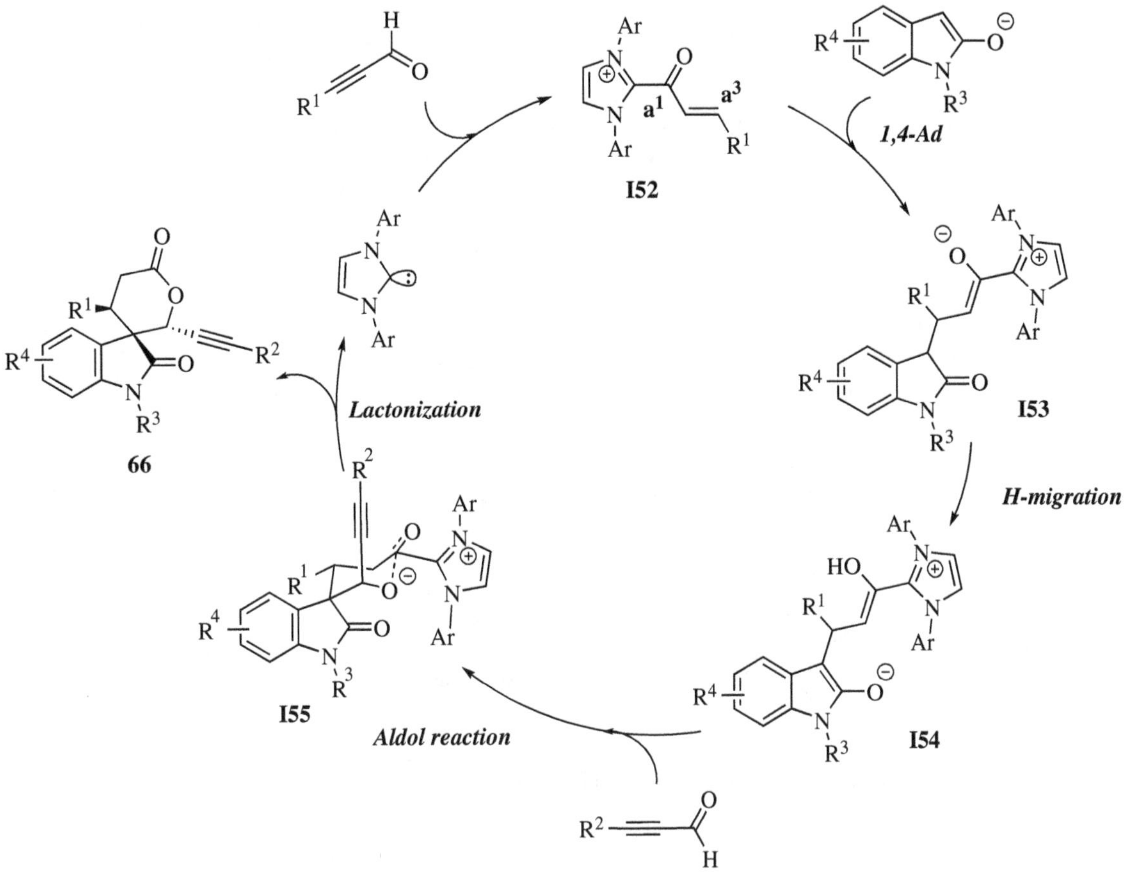

SCHEME 5.48 Proposed mechanism for NHC-catalyzed 3CR of alkynyl aldehydes with oxindoles.

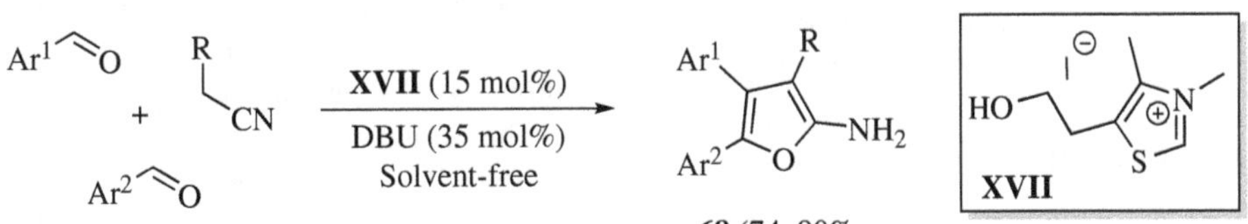

68 (74–90%, 15 examples)

SCHEME 5.49 NHC-catalyzed 3CR for the synthesis of furans **68**.

constantly increasing. Since the first application of an M-NHC complex in a homogeneous catalysis by Herrmann [58], a great number of NHC-based catalysts have been employed in a wide variety of reactions. The importance of NHC as ligands is reflected in the number of reviews covering aspects such as preparation [59], stability [1], stereoelectronic properties [60], and coordination strategies [59c, 61]. Excellent reviews have also appeared dealing with the design of chiral NHC systems and their applications in asymmetric catalysis [62]. In addition, monographs [63] and special issues [64] have been utterly devoted to this research area.

NHC-based catalysts feature robust carbon–metal bonds that provide high thermal stability, low dissociation rates, and, consequently, better resistance against oxidation or leaching phenomena, making the use of ligand excess unnecessary (Fig. 5.2).

The remarkable catalytic performances shown by M-NHC complexes (compatibility with many reagents, reaction conditions, other catalysts, and activation modes) have allowed their increasing application in MCRs. In this section, the most representative and recent outcomes developed in this field are presented. With the aim of facilitating the reading, this section has been divided according to the different transition metals used and the specific MCR described.

5.3.2.1 MCRs Involving Pd–NHC Catalysts Palladium is, along with ruthenium, the most widely used transition metal with NHC ligands in catalytic transformations, and

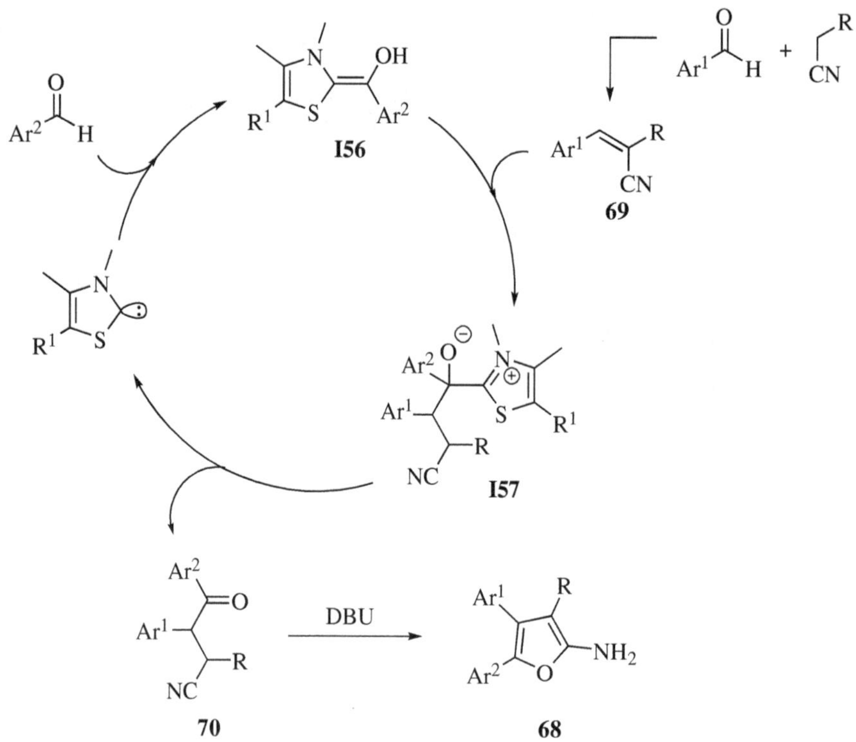

SCHEME 5.50 A plausible mechanism for NHC-catalyzed 3CR for the synthesis of **68**.

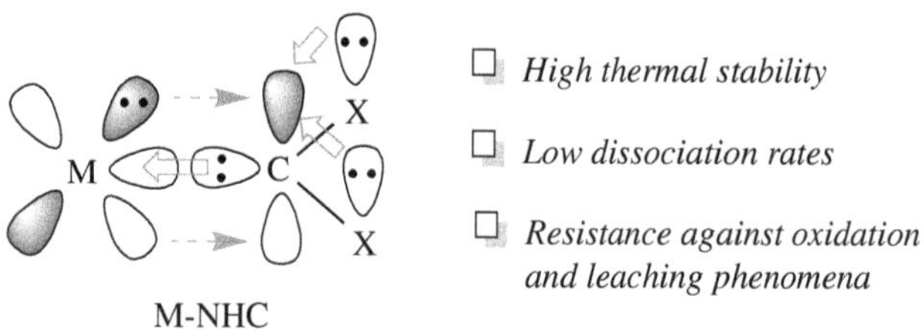

FIGURE 5.2 Features of M-NHC in organometallic catalysis.

thus, their applications in homogeneous catalysis have been comprehensibly reviewed [65]. The significant performances shown by Pd–NHC complexes have allowed the development of new catalytic methodologies; therefore, examples of MCRs where Pd–NHC complexes are used as catalysts have recently appeared in the literature. In this section, some recent examples are presented.

The Strecker Reaction The Strecker reaction [66] is one of the most general methods for synthesis of α-aminonitriles **71** (Scheme 5.51), which are intermediates of amino acids and pharmacologically valuable compounds. This procedure implies MCR of aldehydes or ketones, amines, and a cyanide source. However, the majority of the studies are restricted to aldehydes, mainly due to the harsh reaction conditions and difficult manipulations required when ketones are used [67]. Recently, both Pd and NHC systems shown in Scheme 5.51 have been found to be active catalysts in one-pot Strecker reactions using ketones, aniline or benzylamine, and trimethylsilyl cyanide (TMSCN).

Compound **XVIII** was reported by Jung in 2009 and is the first example of a Pd–NHC used in a Strecker procedure. This complex acts as a Lewis acid to favor the addition of cyanide to the imine generated *in situ* (by condensation of amine and carbonyl reagent) [68].

The organometallic hollow sphere (OHS) bearing {Pd(NHC)} species, compound **XIX**, was described by Son in 2010. This compound acts as an efficient heterogeneous catalyst, and it could be recovered from the reaction media and reused without observing any loss of catalytic activity for three consecutive runs [69].

Carbonylative Suzuki Cross-Coupling Reactions Carbonylative Suzuki cross-coupling is a useful 3CR (boronic acid + aryl halide + CO) to prepare asymmetric substituted ketones [70]. One of the main drawbacks of this methodology is the direct coupling reaction, which forms biaryl products without carbon monoxide insertion, particularly with electron-deficient aryl halides.

In 2007, Castanet reported an efficient carbonylative Suzuki reaction of pyridyl halides with different boronic acids using 50 bar of carbon monoxide. The catalyst used is a combination of $Pd(OAc)_2$ and imidazolium salt precursors NHC-Cl. In all cases, carbonylated species **72** were obtained as the major products (Scheme 5.52). The best results were found with ligand precursor **IMes-Cl**, being also active in

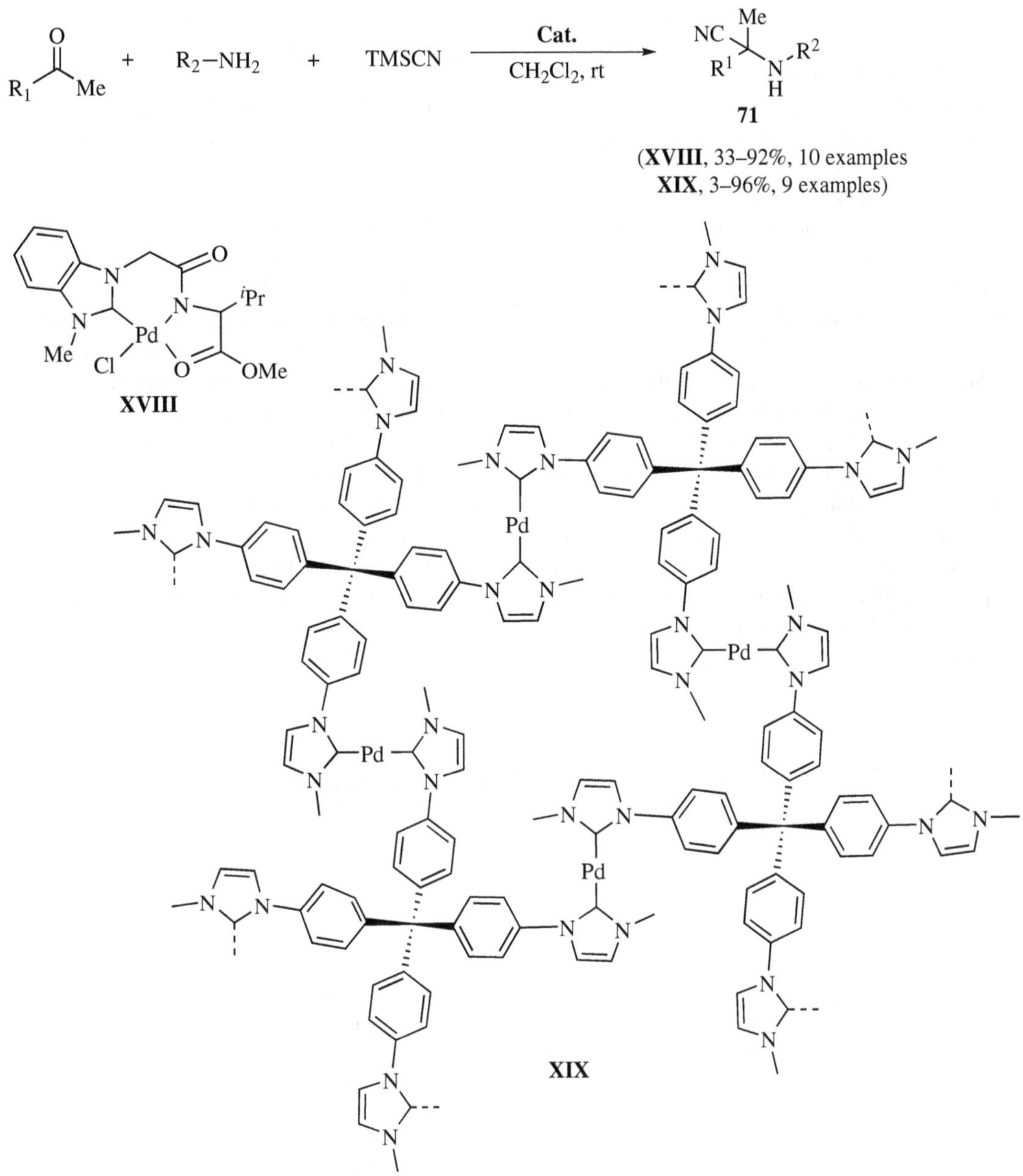

SCHEME 5.51 One-pot Strecker reaction using Pd–NHC catalysts **XVIII** and **XIX**.

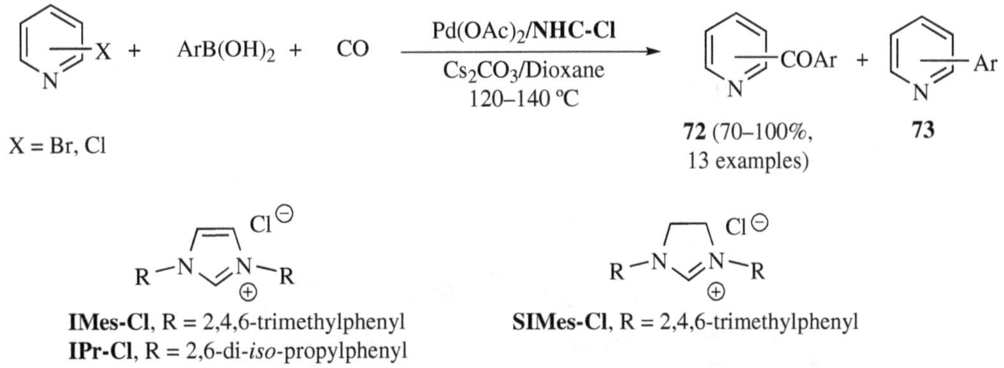

SCHEME 5.52 Pd–NHC-catalyzed carbonylative Suzuki reaction.

the activation of economically attractive but less reactive chloropyridines [71].

The same year, Xia described the carbonylative coupling of sodium tetraphenylborate or phenylboronic acid with aryl iodides using carbon monoxide at atmospheric pressure. In this case, the employed catalysts were the preformed Pd–NHC complexes **XX–XXII**, yielding in all cases carbonylated species **74** as the major products (Scheme 5.53). Catalyst **XX** was also active with electron-deficient aryl iodides achieving the corresponding aryl ketone in good selectivities and excellent yields [72].

Carbonylative Amination Among cross-coupling reactions, palladium-catalyzed aminocarbonylation is an interesting regioselective procedure to prepare amides [73]. The methodology consists of three-component coupling of aryl or alkynyl halides, primary/secondary amines, and carbon monoxide.

In 2012, Orellana showed that Pd–NHC complex **XXIII** catalyzed the coupling of pyrrole with aryl and heteroaryl iodides in the presence of carbon monoxide at atmospheric pressure (Scheme 5.54). This reaction could be performed with different functional groups on the aryl ring. In most cases, the corresponding acyl pyrroles **76** were isolated in good to excellent yields [74].

Simultaneously to Orellana's results, Bhanage and coworkers described a "*green* catalytic procedure" for the aminocarbonylation of aryl iodides with amines. The reaction was carried out using water as solvent and a polymer-supported Pd–NHC catalyst (PS-Pd–NHC, **XXIV**; Scheme 5.55), which could be easily recycled up to four consecutive runs without loss of activity or selectivity. This protocol allowed the aminocarbonylation of different aryl iodides with both aryl and alkyl primary/secondary amines using 100 psi of carbon monoxide. In all studied examples, the corresponding amides **77** were obtained in good to excellent yields, except for aromatic secondary amines [75].

The robust allylic Pd–NHC complex **XXV** (Fig. 5.3) has also been described to catalyze aminocarbonylation of various hetero(aryl) iodides with primary/secondary amines under atmospheric carbon monoxide pressure. This catalyst tolerated a broad range of functional groups, electronic properties, as well as bulkiness on hetero(aryl) iodides and amines, affording in all cases the corresponding amides **77** in good to excellent yields [76].

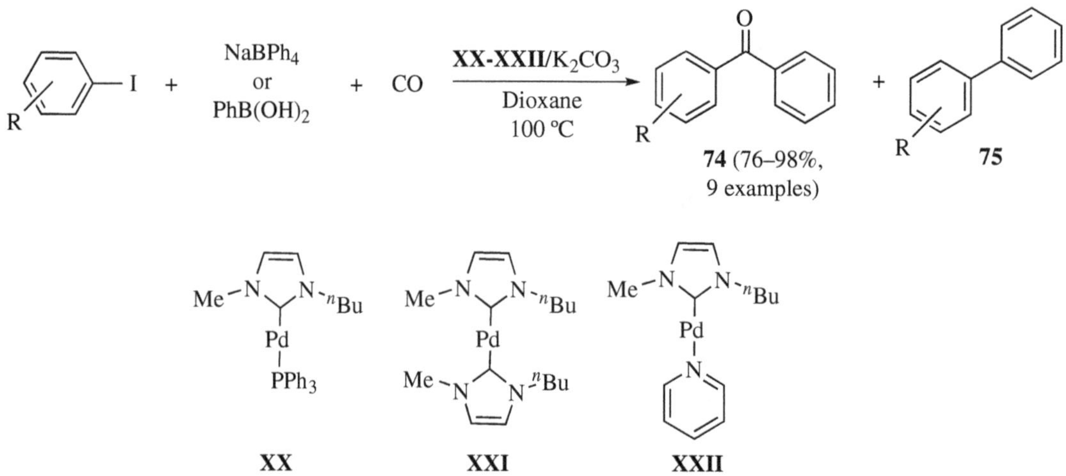

SCHEME 5.53 Pd–NHC-catalyzed carbonylative Suzuki reaction using **XX–XXII** as catalyst.

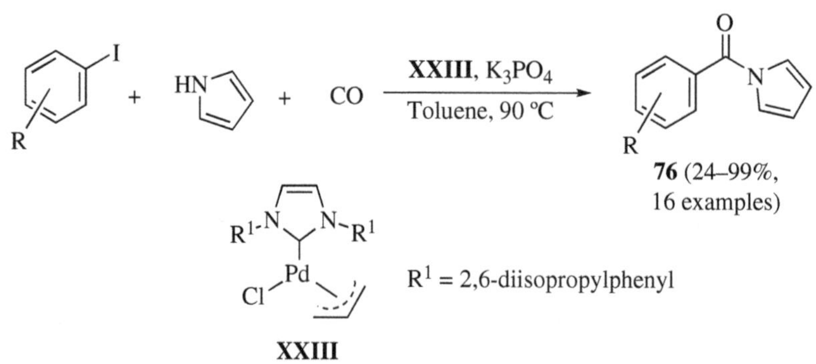

SCHEME 5.54 Synthesis of acyl pyrroles via Pd–NHC-catalyzed carbonylative amination.

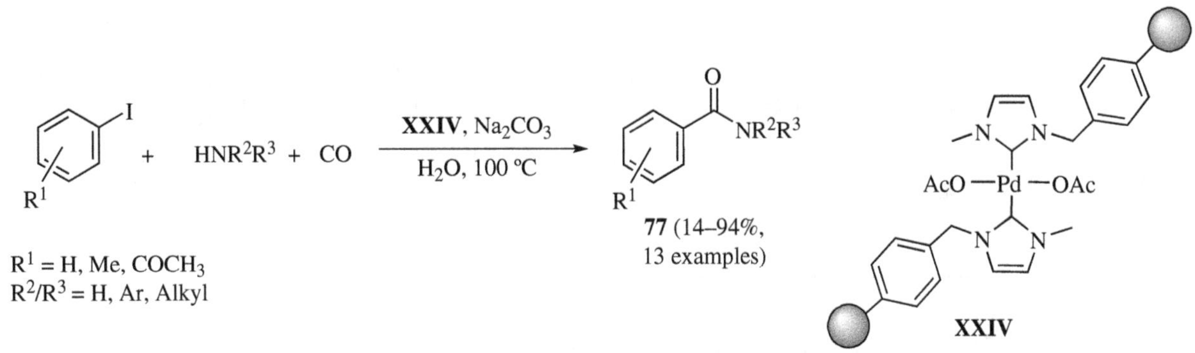

SCHEME 5.55 Aminocarbonylation of aryl iodides with primary/secondary amines using PS-Pd–NHC **XXIV** (PS denotes poly(imidazoliummethyl styrene)-surface grafted polystyrene resin).

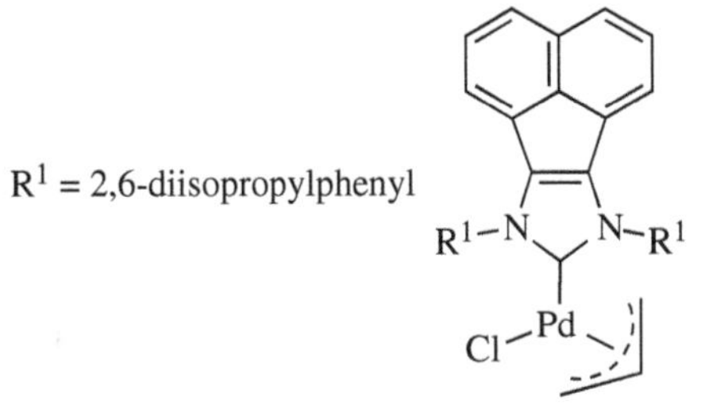

XXV

FIGURE 5.3 Robust allylic Pd–NHC catalysts **XXV**.

"Supported ionic liquid phase" (SILP) methodology combines the advantage of ionic liquids and those of heterogeneous supports [77]. Recently, various SILP–Pd catalysts **XXVI**, based on imidazolium ionic liquids coordinated to Pd and supported on silica gel, have been prepared. These species were active in the double carbonylation of iodobenzene in the presence of secondary amines, yielding α-ketoamides **78** (Scheme 5.56). In general, the SILP–Pd catalysts **XXVI** showed selectivity toward double carbonylation, except for the case of using bulky amines, where amides **79** were obtained as the main products. Moreover, under the optimal conditions, these catalysts could be recycled up to six times [78].

5.3.2.2 MCRs Involving Ru–NHC Catalysts Ruthenium-based Grubbs catalysts **XXVII–XXIX** (Fig. 5.4) are indispensable tools for the formation of carbon–carbon double bonds through olefin metathesis [79]. Importantly, second-generation catalysts **XXVIII** and **XXIX**, which consists of ruthenium alkylidene complexes stabilized by a saturated NHC, provide higher stability and reactivity and broad functional group tolerance (attractive features for designing MCRs).

In 2009, Botta and coworkers described a MCR of different alkynes with ethyl vinyl ether and ethyl glyoxalate, in the presence of **XXVIII**, to afford 2,3-dihydropyrans **80** as a mixture of 2:1 *trans-/cis*-diastereoisomers, which could be subsequent equilibrated in the presence of a Lewis acid, such as $ZnCl_2$ [80] (Scheme 5.57).

In a similar reaction, Welker et al. combined different alkenes and α,β-unsaturated carbonylic compounds with benzyldimethylsilyl alkyne, in the presence of **XXIX**, to yield selectively cycloadducts **81** in moderate to good yields [81] (Scheme 5.58).

In both cases, the mechanism of the reactions can be described like an ene–yne metathesis catalyzed by the Ru–NHC complex [82], followed by a Diels–Alder reaction. The mechanism of these reactions (exemplified for the synthesis of **81**) is depicted in Scheme 5.59. Initial [2+2] cycloaddition between the alkyne and the olefin stemming from the metallic carbene achieves metallacyclobutene **I58**. Subsequently, cycloelimination and [2+2] cycloaddition with the alkene reagent give **I59**. A second cycloelimination generates a diene (by-product) and the active catalytic species **I60**, which initiates the catalytic cycle following successive cycloadditions and cycloeliminations up to afford diene **82**. Finally, a Diels–Alder reaction between diene **82** and α,β-unsaturated carbonyls (dienophile) produces exclusively the *syn-(endo-)*product **81**.

5.3.2.3 MCRs Involving Ni–NHC Catalysts Nickel-catalyzed reductive couplings and cyclizations allow access to important organic structures through MCRs that involve two or more π-components, like alkenes and carbonyl groups [83]. In this section, these types of catalytic processes based on Ni–NHC are discussed.

Reductive Couplings of 1,3-Dienes, Aldehydes, and Silanes Mori, Sato, and coworkers reported the application of Ni complexes as catalysts in reductive three-component coupling of 1,3-dienes, aldehydes, and silanes [84], in which regioselectivity, stereoselectivity, and reaction pathways are controlled by the properties of the ancillary ligand (Scheme 5.60). For instance, employing PPh_3 as ligand coupling product (*E*)-**83** was obtained via π-allylnickel intermediate **I61** (generated by the reaction of 1,3-diene and H–Ni(II)–SiR_3 species), whereas NHC ligands afforded (*Z*)-**83** exclusively. This product is obtained via σ-bond

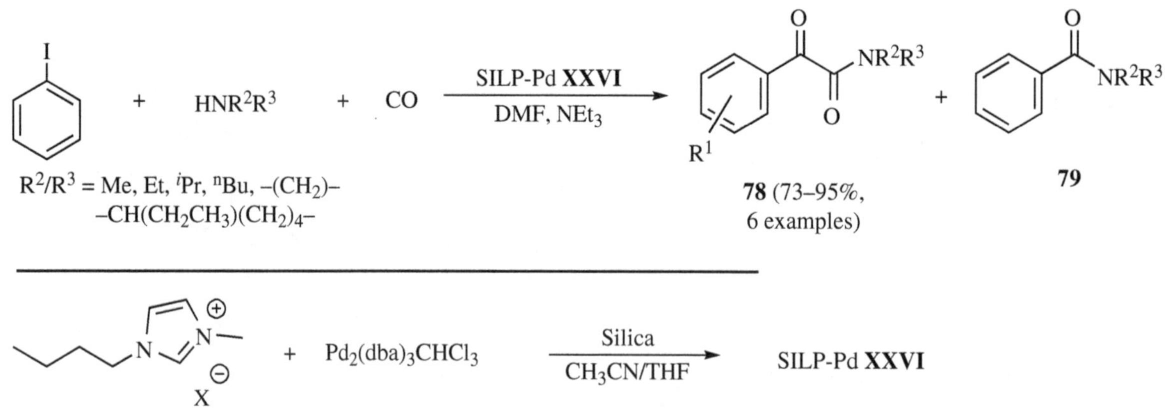

R^2/R^3 = Me, Et, ^{i}Pr, ^{n}Bu, $-(CH_2)-$ $-CH(CH_2CH_3)(CH_2)_4-$

78 (73–95%, 6 examples) **79**

$X = PF_6, BF_4$

SCHEME 5.56 Double carbonylation using SILP–Pd catalysts **XXVI**.

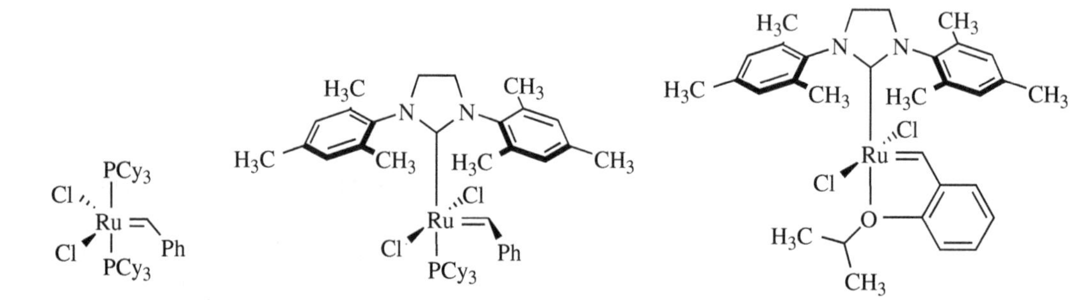

XXVII, 1st generation **XXVIII**, 2nd generation **XXIX**, Hoveyda-Grubbs

FIGURE 5.4 Grubbs catalysts.

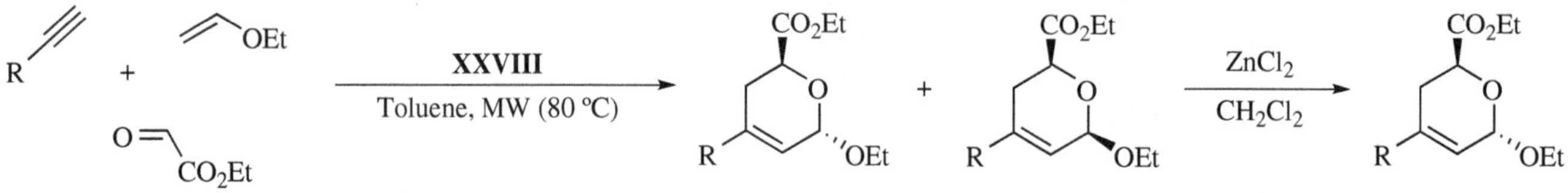

R = TMS, $BocNHCH_2$, $PMBOCH_2$, Ph, $BrCH_2$

trans-**80** *cis*-**80** *trans*-**80**, 49–75% (8 examples)

SCHEME 5.57 One-pot multicomponent synthesis of 2,3-dihydropyrans **80**.

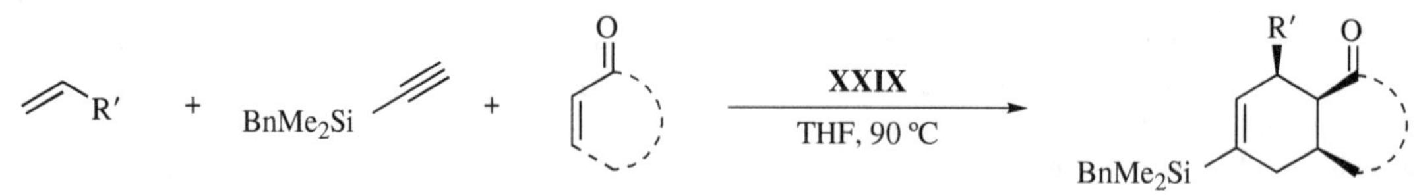

R′ = Ph, *p*-MeOPh, *p*-ClPh, o-CF_3Ph

NPM, ethyl acrylate, diethyl maleate

81 (46–49%, 7 examples)

SCHEME 5.58 Multicomponent synthesis of **81**.

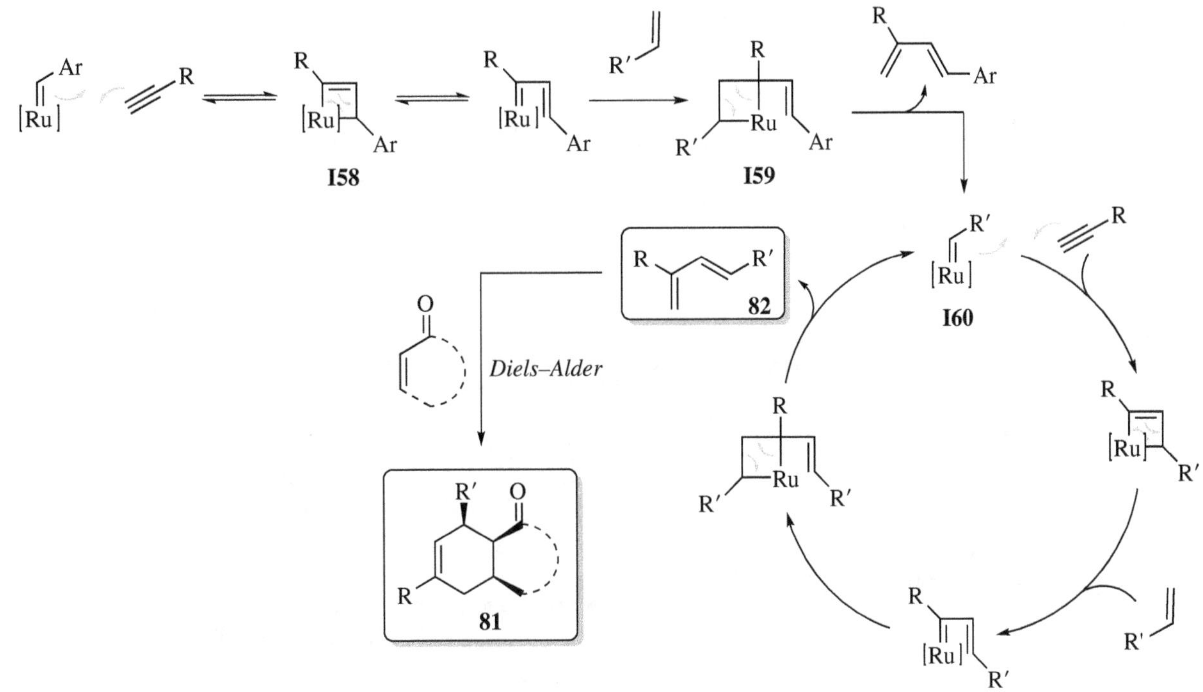

SCHEME 5.59 Mechanism of ene–yne metathesis/Diels–Alder multicomponent reaction.

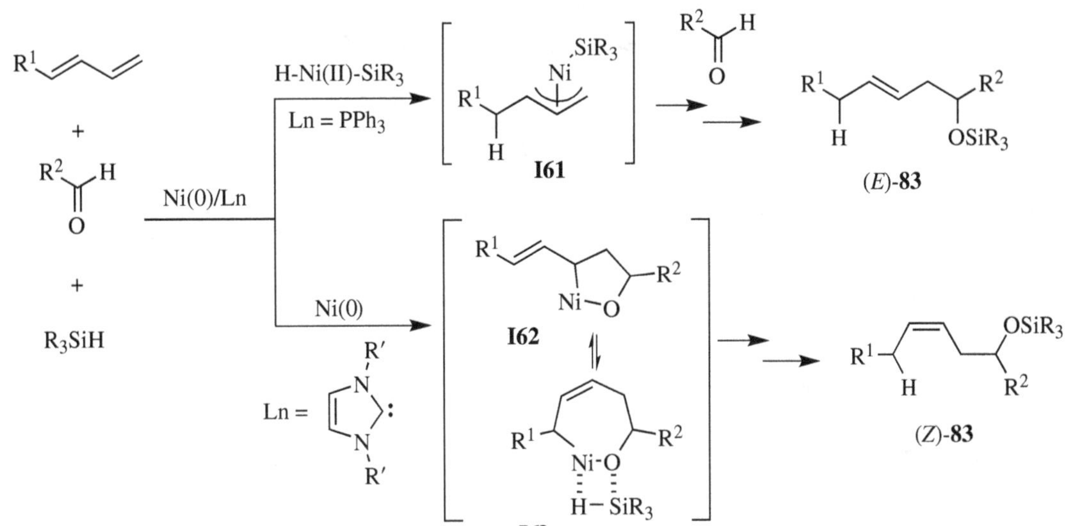

SCHEME 5.60 Reaction pathways in reductive couplings of 1,3-dienes, aldehydes, and silanes.

metathesis between nickelacycle **I63** (in equilibrium with **I62**) and R_3SiH.

Later, the same group described an asymmetric version of this 3CR employing a chiral NHC [85]. Coupling of (*E*)-1-(but-1,3-dien-1-yl)-4-[(methoxymethoxy)methyl]benzene, Et_3SiH, and anisaldehyde using $Ni(COD)_2$ (COD = 1,5-cyclooctadiene) and chiral imidazolium salt **NHC-Cl** (as NHC precursor) gave coupling product (*Z*)-**83a** stereoselectively (Scheme 5.61). Optimization of the ligand structure led to identify imidazolium salt **XXXIII** as the best NHC ligand precursor, in terms of yield and stereoselectivity.

Under optimized conditions, this coupling reaction was extended to various 1,3-dienes and aldehydes providing coupling products (*Z*)-**83** in good yields, with high regio-, diastereo- (anti selective), and enantioselectivities (up to 97% ee) (Scheme 5.62).

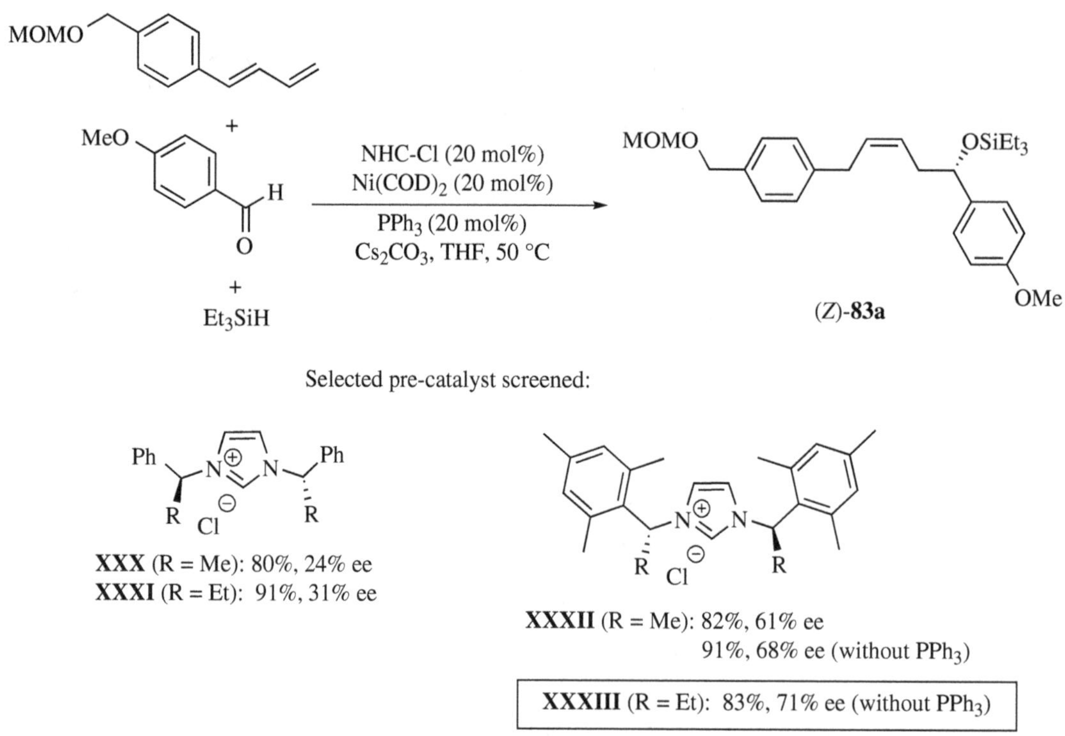

SCHEME 5.61 Asymmetric Ni–NHC-catalyzed reductive coupling of 1,3-dienes, aldehydes, and silanes.

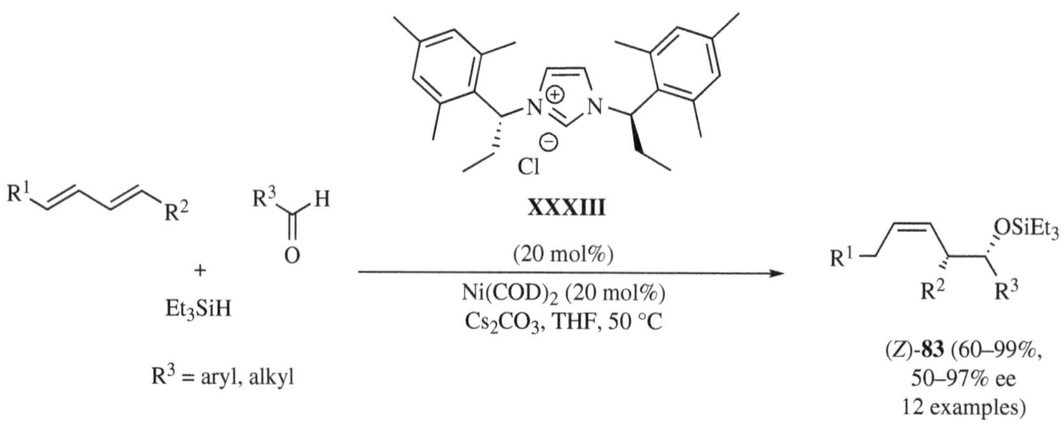

SCHEME 5.62 Asymmetric Ni–**XXXIII**-catalyzed reductive coupling of 1,3-dienes, aldehydes, and silanes.

Reductive Couplings of Alkynes, Aldehydes, and Silanes In 2004, Montgomery and coworkers reported the reductive coupling of alkynes and aldehydes using Ni(COD)–**IMes** as catalyst [86] (Scheme 5.63). The scope of this procedure was relatively broad, obtaining in all cases the product **84** in good to excellent yields with regioselectivities higher than 96%.

When the same reaction was performed with alkynes containing functionalities (working as regiocontrol elements) adjacent to the C≡C bond (ynamides, propargyl alcohols, and ynoates), 1,3- (Eq. 1 and 2, Scheme 5.64) and 1,4-difunctional compounds were obtained (Eq. 3, Scheme 5.64) [87]. In the last MCR (Scheme 5.64, Eq. 3), both PPh_3 and a saturated carbene **SIPr** were required as ligands.

The reductive coupling of aldehydes and alkynes catalyzed by Ni–NHC has provided a direct entry to allylic alcohols [88]. In 2010, continuous efforts by Montgomery's group in this topic led to discover a more general methodology in terms of regiocontrol [89]. In Table 5.2, a selection of experiments employing unbiased and biased alkynes is presented. Interestingly, a careful choice of NHC ligand and optimization of reaction conditions (**A–D**) afforded a general strategy based on steric control by the ligand structure.

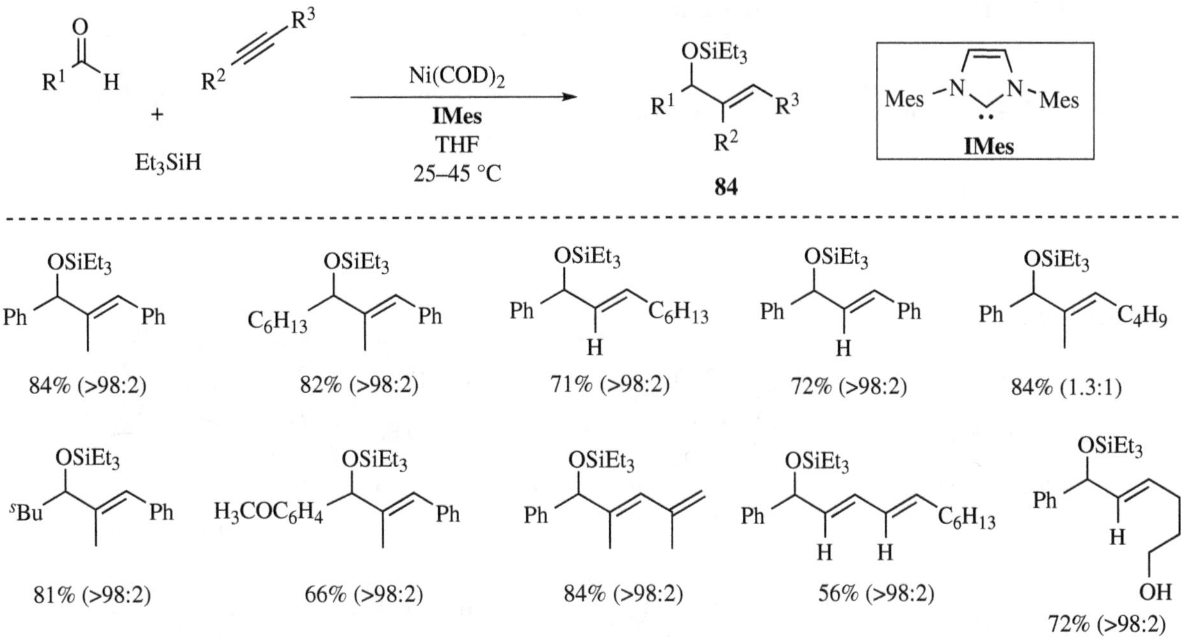

SCHEME 5.63 Ni(COD)–**IMes**-catalyzed reductive coupling of alkynes, aldehydes, and triethylsilanes.

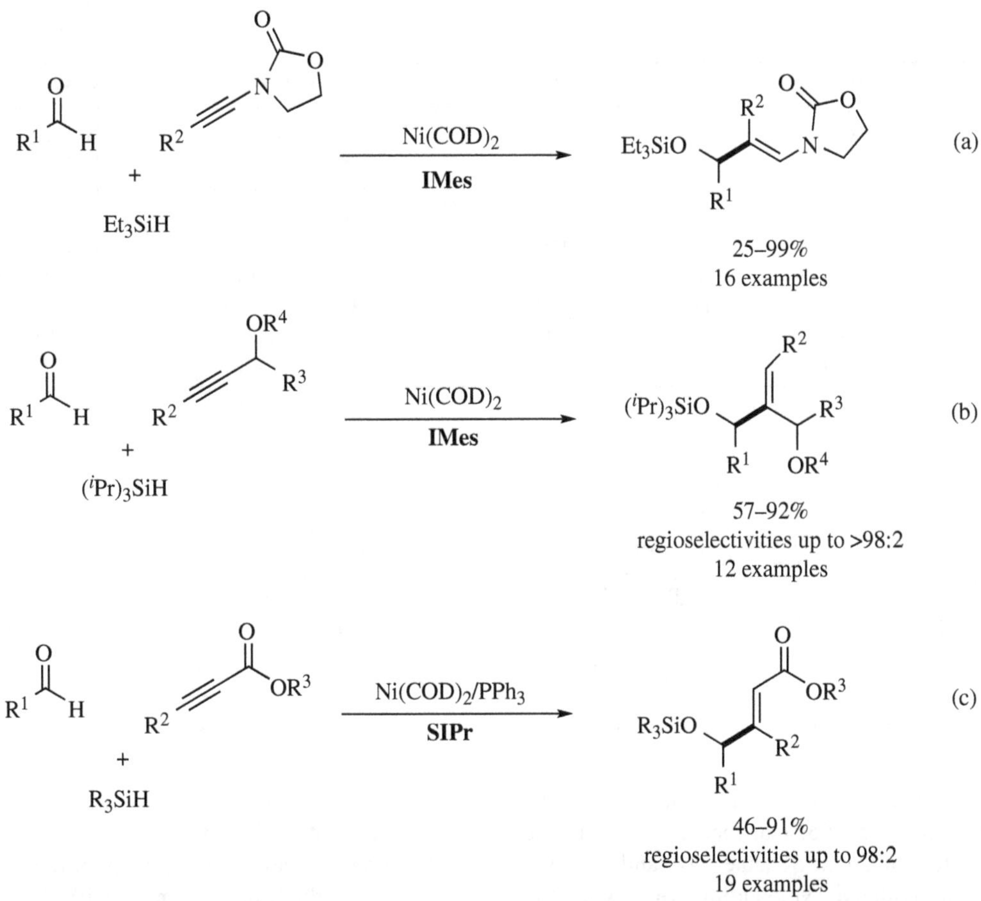

SCHEME 5.64 Ni(COD)–**NHC**-catalyzed reductive coupling of alkynes, aldehydes, and silanes.

TABLE 5.2 Ligand-controlled regioselectivity reversal

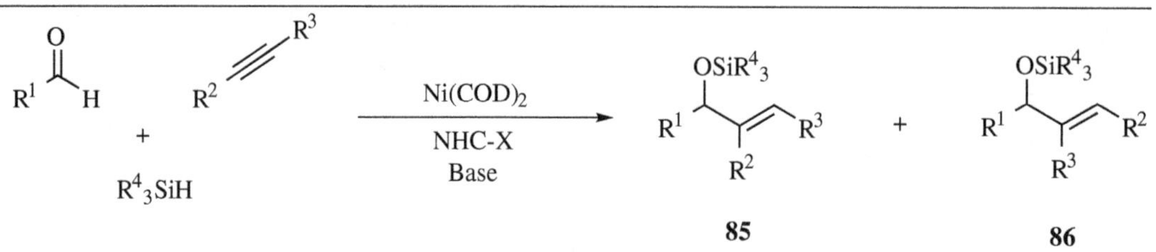

Conditions:

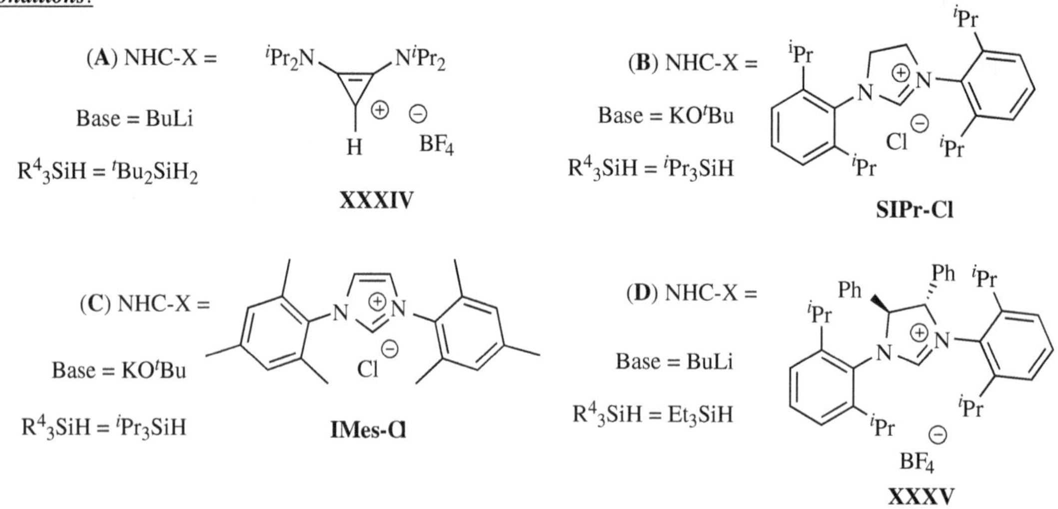

Entry	R^1	R^2	R^3	85:86 (% yield)	85:86 (% yield)
1	ⁿHex	Me	ⁿPr	**A**, 88:12 (78%)	**B**, 7:93 (85%)
2	ᶜHex	Me	ⁿPr	**A**, 82:18 (75%)	**B**, 5:95 (91%)
3	Ph	Me	ⁿPr	**A**, 84:16 (72%)	**B**, 2:98 (86%)
4	Ph	Me	ⁱPr	**A**, 97:3 (85%)	**B**, 10:90 (89%)
5	Ph	Me	Ph	**C**, 98:2 (84%)	**B**, 19:81 (99%)
6	ⁿHex	Me	ᶜHexenyl	**C**, 97:3 (99%)	**B**, 9:91 (77%)
7	Ph	H	CH_2OTBS	**C**, 93:7 (88%)	**D**, 15:85 (86%)
8	Ph	H	ⁿHex	**C**, 97:3 (82%)	**D**, 12:88 (71%)
9	ⁿHex	H	ⁱPr	**C**, 98:2 (74%)	**D**, 5:95 (76%)

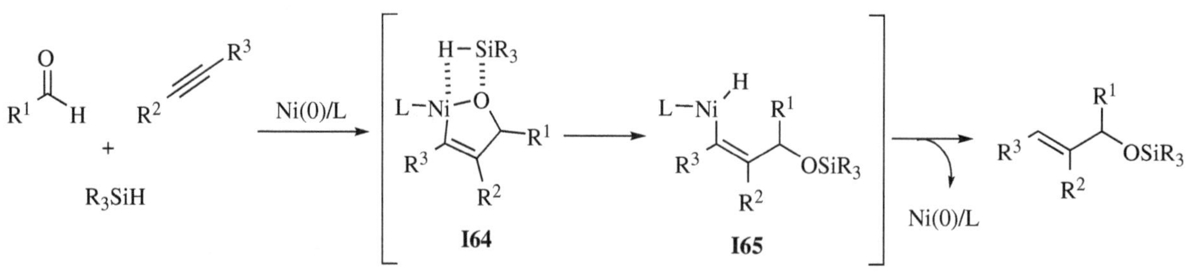

SCHEME 5.65 Proposed mechanism in reductive couplings of alkynes, aldehydes, and silanes.

A proposed mechanistic pathway based on experimental data is depicted in Scheme 5.65. This mechanism is related to that proposed for dienes in Scheme 5.60 and involves the following: first, oxidative cycloaddition of alkynes and aldehydes to a nickel(0) complex affords oxanickelacycle **I64**; then, cleavage of the Ni—O bond assisted by σ-bond metathesis of nickelacycle **I64** with silane leads to hydride–Ni intermediate **I65**; and finally, reductive elimination from **I65** would achieve the corresponding three-component coupling product, in agreement with the regio- and stereoselective C—C bond formation at the oxanickelacycle **I64** formation step.

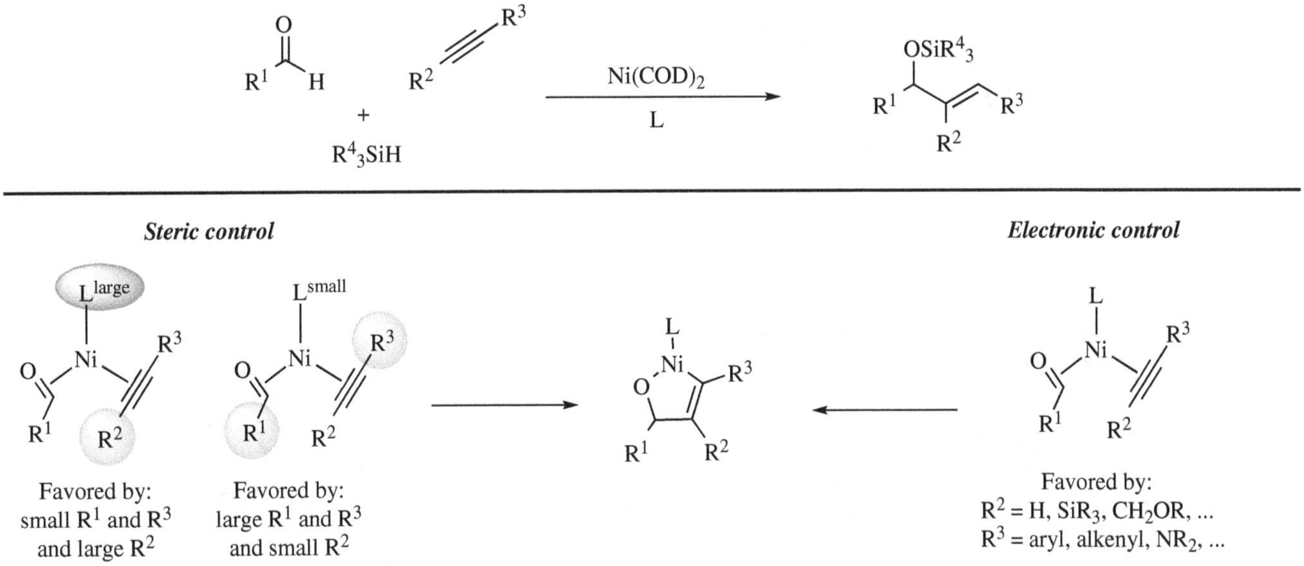

SCHEME 5.66 Predictive model for regiocontrol.

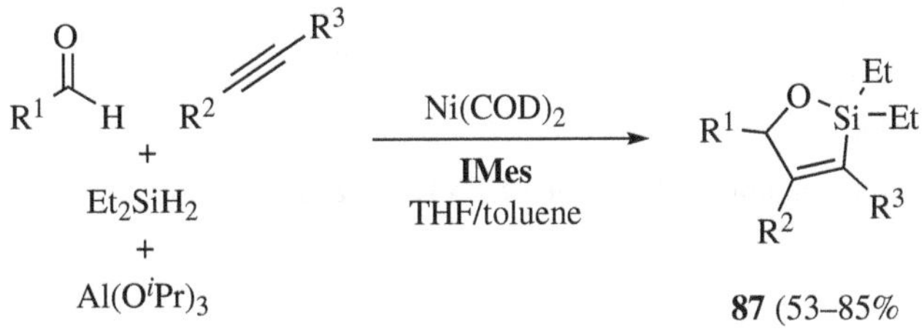

SCHEME 5.67 Synthesis of oxasilacyclopentanes **87**.

Interestingly, Montgomery formulated a simple model to explain the regioselectivities observed in these processes (Scheme 5.66).

This transformation has also been successfully applied to the construction of macrocycles [90] (in its intramolecular version) and oxasilacyclopentanes **87** from aldehydes, alkynes, and dihydrosilanes (Scheme 5.67) [91].

Additionally, the employment of enantiopure α-silyloxy aldehydes in these reactions provided a general entry to *anti*-1, 2-diols **88** in high diastereoselectivities [92] (Scheme 5.68).

An asymmetric version of reductive coupling of aldehydes and alkynes has also been reported employing a specially designed imidazolium salt **XXXVI** as carbene precursor (Scheme 5.69) [93]. This methodology provided an efficient strategy for the synthesis of enantiomerically enriched protected allylic alcohols **89** in good yields, moderate to good regioselectivities, and reasonable enantioselectivities.

Couplings of Other Systems Allenes or methylenecyclopropanes are also suitable components for reductive couplings (Scheme 5.70). In the presence of Ni(COD)$_2$/**IPr**, the reactions with allenes proceeded at low temperature with total *Z* selectivity and transfer of axial chirality (Scheme 5.70, Eq. 1) [94]. On the other hand, Ni(COD)$_2$/**IMes** afforded silylated allylic alcohols containing an alkyl group at α-position installed through the ring opening of the proximal C—C bond of cyclopropane ring [95] (Scheme 5.70, Eq. 2).

Fukuzawa and coworkers also reported a diastereoselective three-component coupling between aryl aldehydes, norbornenes, and silanes, leading to indanols **90** through an interesting C—H bond activation of aryl aldehydes at the *ortho*-position (Scheme 5.71) [96]. Optimization of NHC structure identified saturated carbene **XXXIX** as the most efficient ligand from the selection showed in Scheme 5.71. Remarkably, substitution pattern of aromatic rings strongly influences the reaction outcome as employing ligands **XL** and **XLI** resulted in low yields.

Sato and coworkers have expanded the scope of Ni(0)/**IMes** catalyst in alkylative three-component couplings of 1,3-dienes and aldehydes with tetraorganosilicon reagent **91** [97]. As depicted in Scheme 5.72, the use of **91** provided 1,3-*anti*-products **92** diastereoselectively, presumably through a nucleophilic backside attack of a pentacoordinate silicate (preformed from **91** and Cs$_2$CO$_3$) to allylnickel complexes **I67** (from **I66**).

5.3.2.4 MCRs Involving Ag–NHC Catalysts Since the publication of the first well-characterized Ag(I)–NHC complex by Arduengo's group in 1993 [98], numerous developments and applications of Ag(I)–NHC complexes in catalysis and medicinal chemistry have been reported [99]. In this section, recent advances on the Ag(I)–NHC-catalyzed three-component coupling of aldehydes, alkynes, and amines, referred to as A^3-coupling reaction [100], are covered (Scheme 5.73). Remarkably, the employment of Ag(I)–PR$_3$ complexes [101] produces the two-component aldehyde–alkyne product instead of the A^3-coupling afforded by Ag(I)–NHC systems.

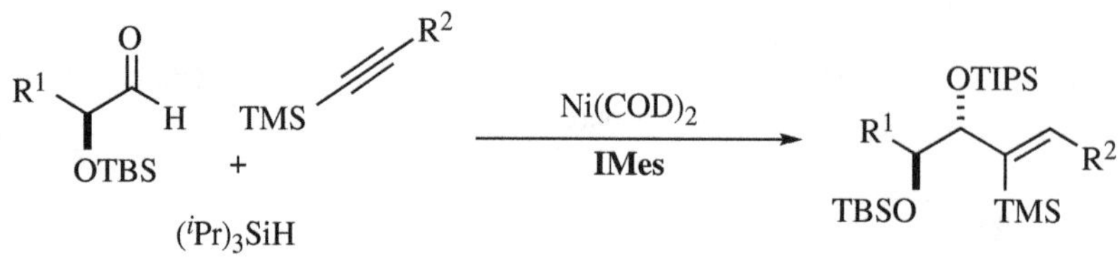

88
>80%
dr > 98 : 2
9 examples

SCHEME 5.68 Synthesis of *anti*-1,2-diols **88**.

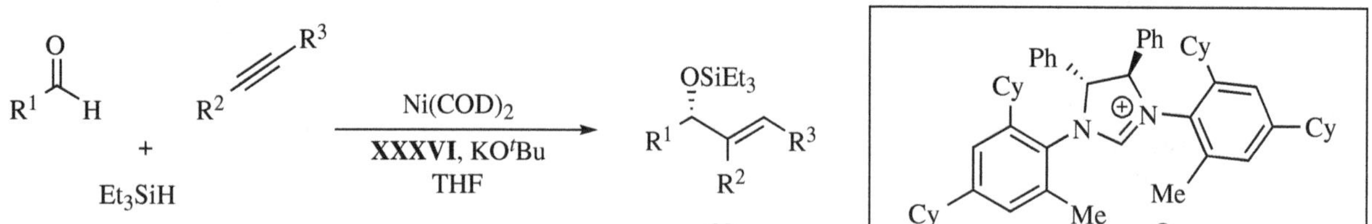

89
47–98%
65–85% ee
regioselectivity:
3 : 1–>19 : 1
12 examples

SCHEME 5.69 Asymmetric 3-C reductive coupling employing chiral imidazolium salt **XXXVI** as ligand precursor.

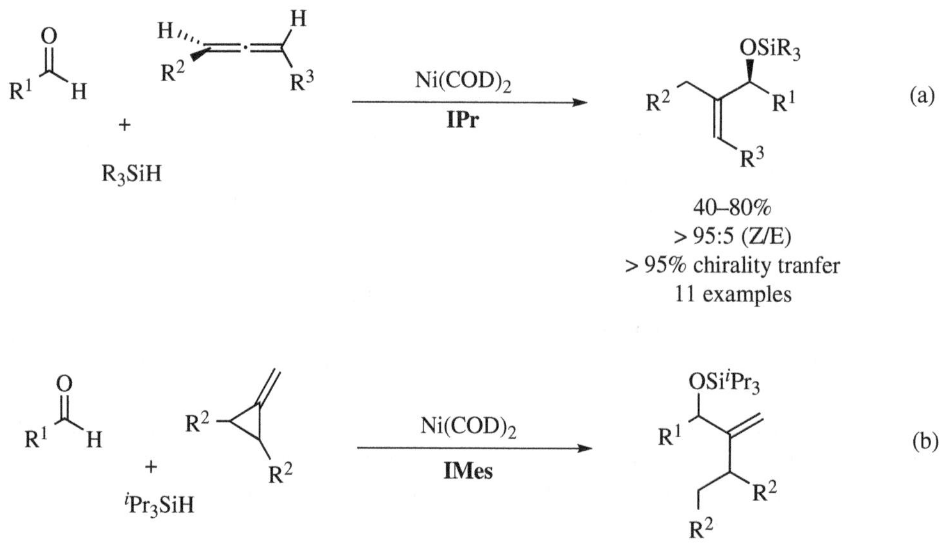

SCHEME 5.70 Ni–NHC-catalyzed 3-C couplings.

In 2003, Li and coworkers reported the first example of Ag(I)-catalyzed A^3 reaction [102], overpassing some limitations of the systems based on Cu(I) and Au(I) related to the undesired trimerization of the aldehyde component, particularly with aliphatic aldehydes.

A tentative mechanism for the A^3-coupling is depicted in Scheme 5.74. Firstly, the acetylene reacts with the Ag catalyst, forming π-metal–alkyne intermediate **I68**, which leads to the formation of species **I69**. This process is still poorly understood and presumably is the key step of the reaction,

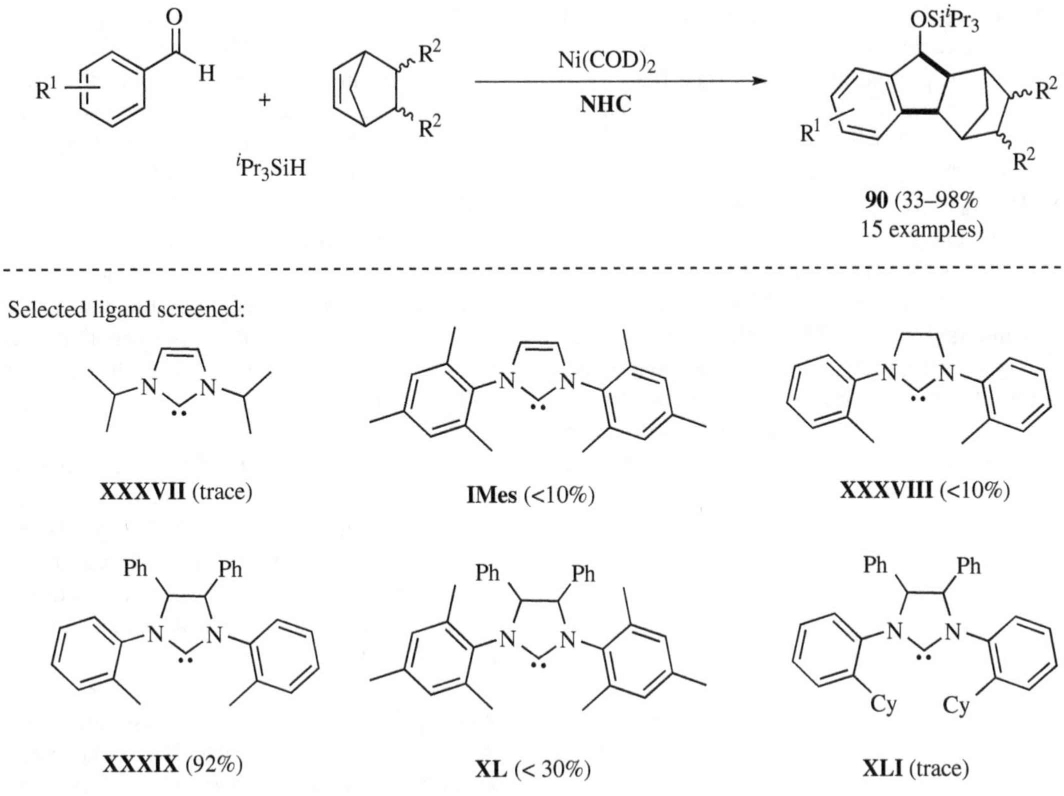

(Model reaction: benzaldehyde, norbornene and tri-*iso*-propylsilane)

SCHEME 5.71 Ni–NHC-catalyzed 3-C coupling between aryl aldehydes, norbornenes, and silanes.

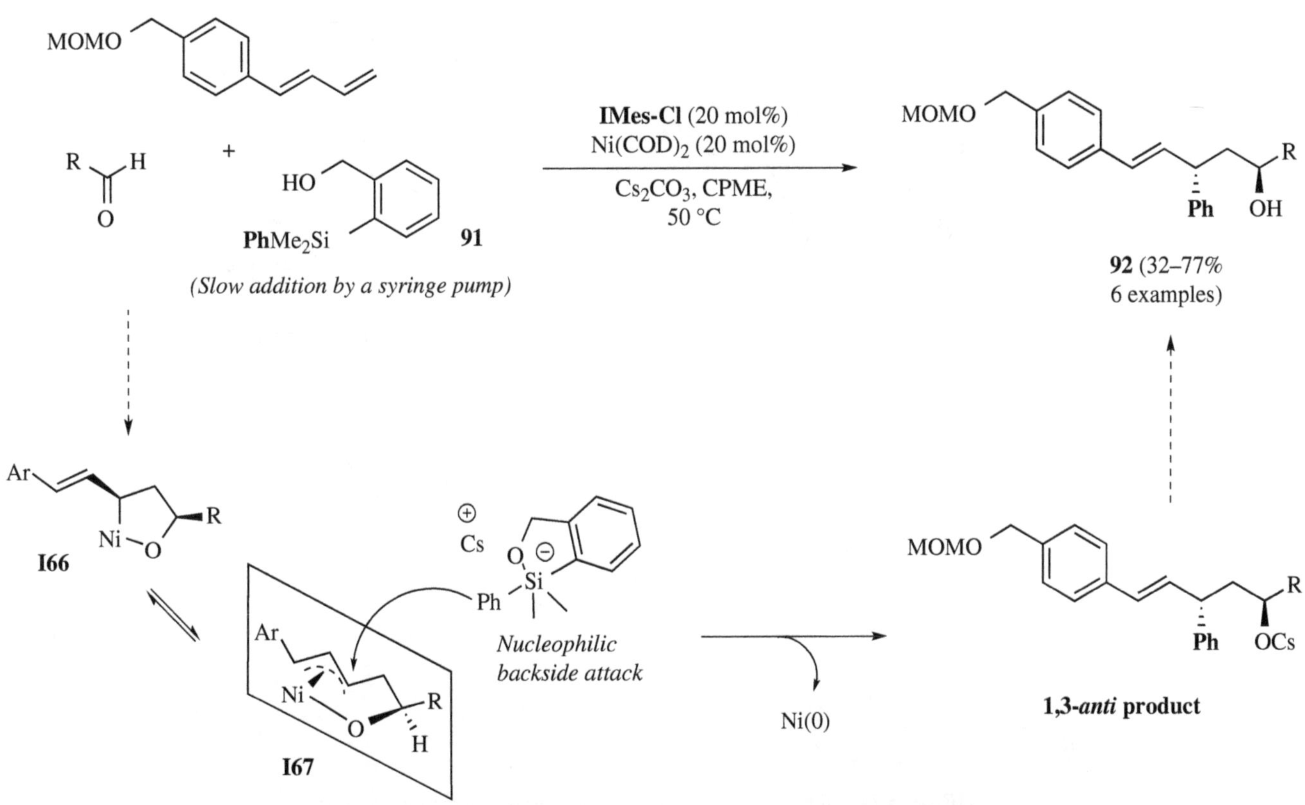

SCHEME 5.72 Ni–**IMes**-catalyzed alkylative 3-C couplings of 1,3-dienes, aldehydes, and tetraorganosilicon reagent **91**.

where different catalytic systems might have an impact modulating both electronic and steric properties in **I68** (e.g., in the acidity of alkyne proton) and intermediate **I69**. Then, **I69** reacts with iminium ion intermediate **I70**, yielding propargylamine **93** after releasing the metal catalyst.

In 2011, Zou and coworkers synthesized Ag(I)–NHC complexes **XLII** (Fig. 5.5). These compounds were efficient catalysts for the A^3-coupling of 3-phenylpropionaldehyde, piperidine, and phenylacetylene in air at high temperatures [103]. A catalyst design in complexes **XLII** led to tune the structures from dimers $[(NHC)AgX]_2$ **XLIIa,b** (silver–silver interaction in crystal structure) to monomers (NHC)AgX (**XLIIc,d**) by introducing more sterically demanding groups on the imidazole scaffold. In this way, catalysts **XLIIc,d** containing a more accessible silver center afforded better catalytic activities.

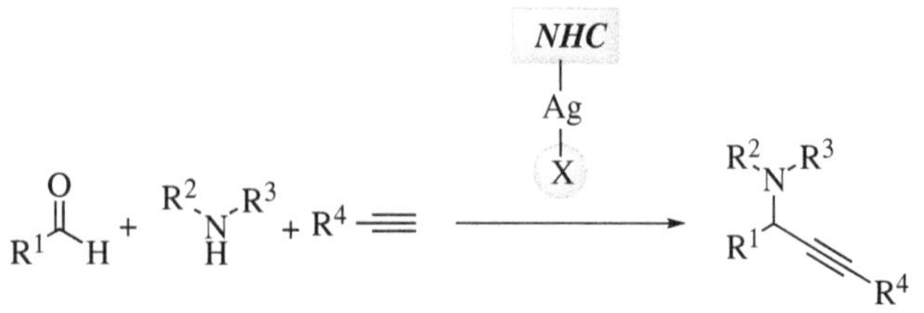

93

SCHEME 5.73 Ag(I)–NHC-catalyzed A^3-coupling reaction.

Recyclable polymer-supported Ag(I)–NHC complexes **XLIII** [104] and **XLIV** [105] have also been reported allowing to perform the reaction at room temperature (with or without solvent), albeit with long reaction times.

Recently, Navarro and coworkers described that well-defined complexes **XLV** and **XLVI** (Fig. 5.6) with commercially available NHC ligands allowed A^3-coupling reactions to take place at room temperature, in open air, and in much shorter reaction times [106]. The activities of **XLV** and **XLVI** were notably affected by the nature of the counteranion X (OAc > Cl > Br), suggesting that metal acetylide formation (**I69**; Scheme 5.74) might be favored with less bulky anions.

5.3.2.5 MCRs Involving Cu–NHC Catalysts Cu–NHC complexes act as catalysts in a growing variety of reactions [107]. Even more, the use of NHC ligands serve to protect Cu(I) from oxidation and disproportionation, thus enhancing its catalytic performances. In this section, some recent examples of Cu(I)–NHC species involved in MCRs are presented.

A^3-Coupling reaction As previously seen in Section 5.3.2.4, A^3-coupling is a very useful MCR to obtain propargylamines. In 2008, a silica-immobilized Cu(I)–NHC complex **XLVII** (Fig. 5.7) was successfully reported to catalyze such transformation with a broad range of aromatic and aliphatic

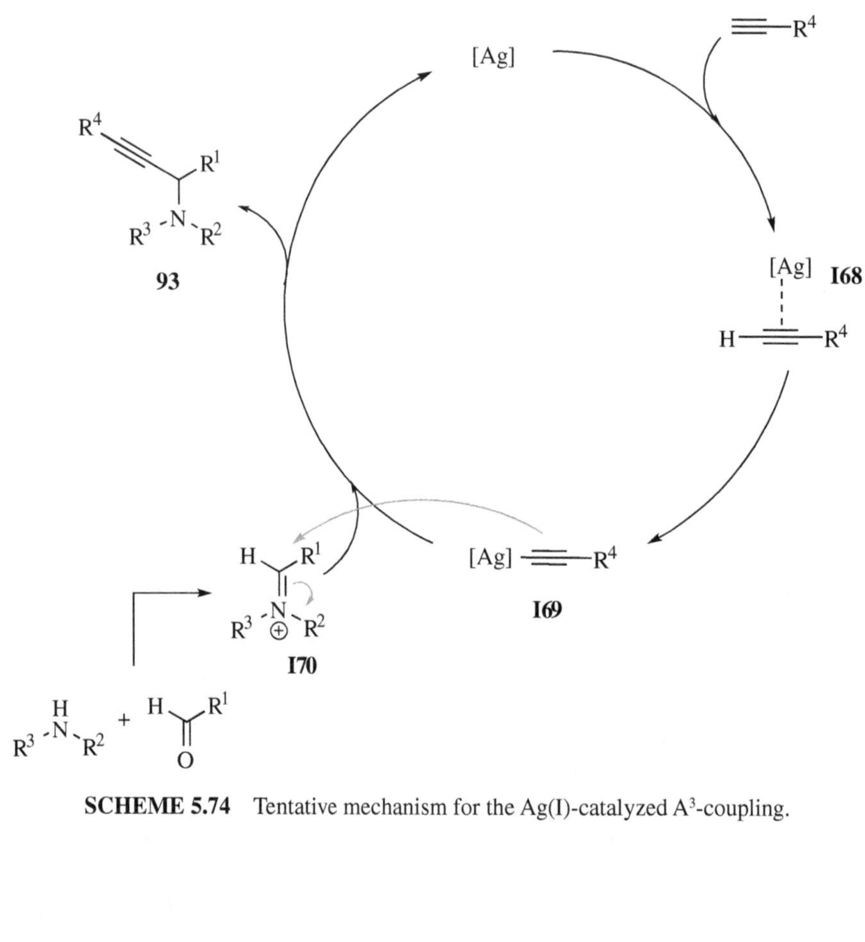

SCHEME 5.74 Tentative mechanism for the Ag(I)-catalyzed A^3-coupling.

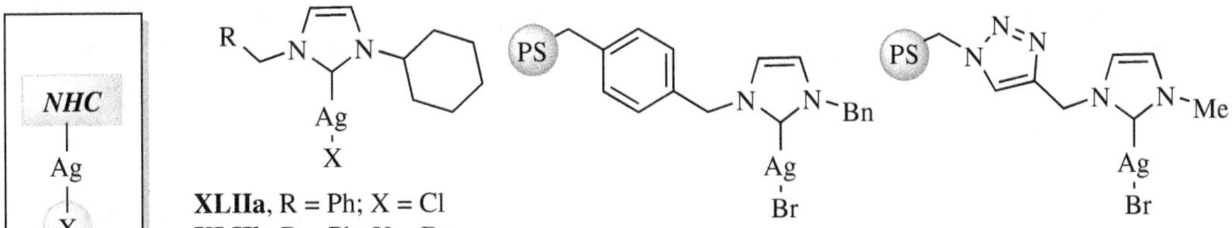

XLIIa, R = Ph; X = Cl
XLIIb, R = Ph; X = Br
XLIIc, R = naphthyl; X = Cl
XLIId, R = naphthyl; X = Br

XLIII

XLIV

FIGURE 5.5 Ag(I)–NHC systems for A^3-coupling.

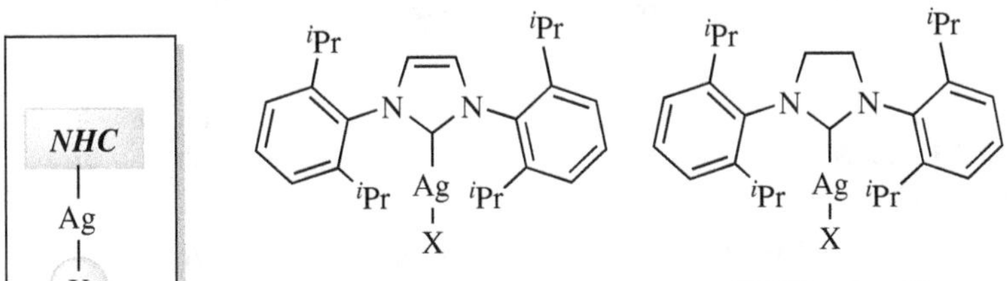

XLVa, X = Cl
XLVb, X = Br
XLVc, X = OAc

XLVIa, X = Cl
XLVIb, X = OAc

FIGURE 5.6 Ag(I)–NHC catalysts for A^3-coupling.

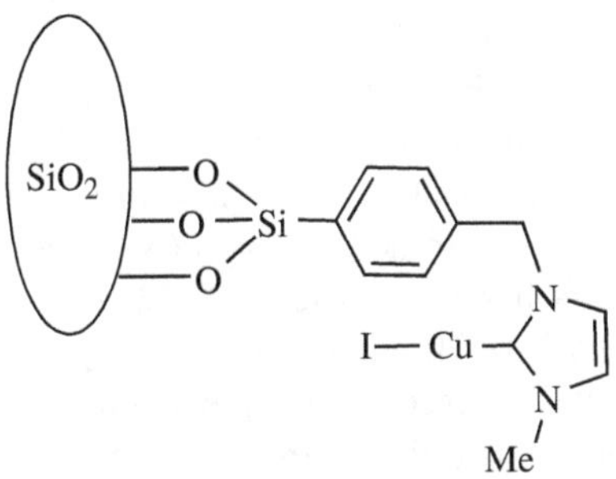

XLVII

FIGURE 5.7 Silica-immobilized Cu(I)–NHC complex **XLVII**.

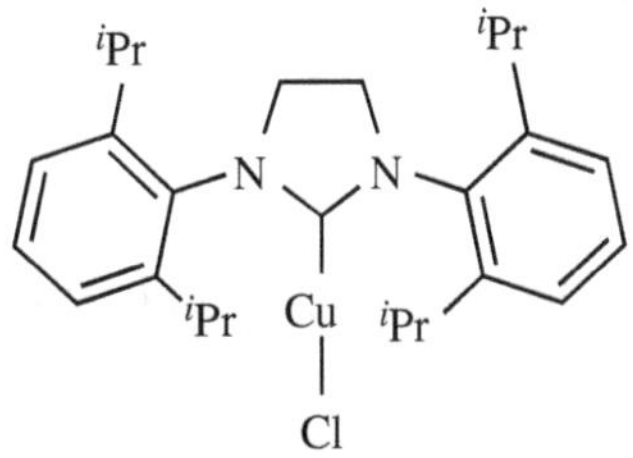

XLVIII

FIGURE 5.8 Neutral Cu(I)–NHC complex **XLVIII**.

substrates. In addition, **XLVII** allowed to run the experiments under solvent-free conditions and offered the possibility of recyclability [108].

Cu(I)-Catalyzed Azide–Alkyne Cycloaddition Cu(I)-catalyzed azide–alkyne cycloaddition (CuAAC) is one of the most popular click reactions, essentially due to its mild conditions and high efficiency [109]. Traditionally, this process was based on the reaction of an organic azide and an alkyne. With the aim to avoid the use of hazard organic azides, many efforts have been made over the last years in order to develop CuAAC-based MCRs.

In 2008, Nolan found that neutral Cu(I)–NHC complex **XLVIII** (Fig. 5.8) could catalyze the two-component click reaction of benzyl azide and phenylacetylene in water but only giving 26% conversion after one week [110].

In order to improve these preliminary results, Li and coworkers prepared the ionic functionalized **XLIXa–c** and **La–c** complexes introducing an ammonium moiety in the NHC to increase their solubility in water (Scheme 5.75). These complexes showed highly efficient reactivity toward 3-C CuAAC of a wide variety of benzyl bromides and alkynes with sodium azide at room temperature, affording the corresponding 1,2,3-triazoles **94** in good yields. Furthermore, compound **XLIXa** could be conveniently recycled up to three times without significant loss of its activity [111].

The silica-immobilized Cu(I)–NHC heterogeneous catalyst **LI** (Fig. 5.9) was also found to be active in this reaction. In this case, after the reaction, the catalyst was separated by simple filtration and could be reused directly without further purification up to six repetitive cycles [112].

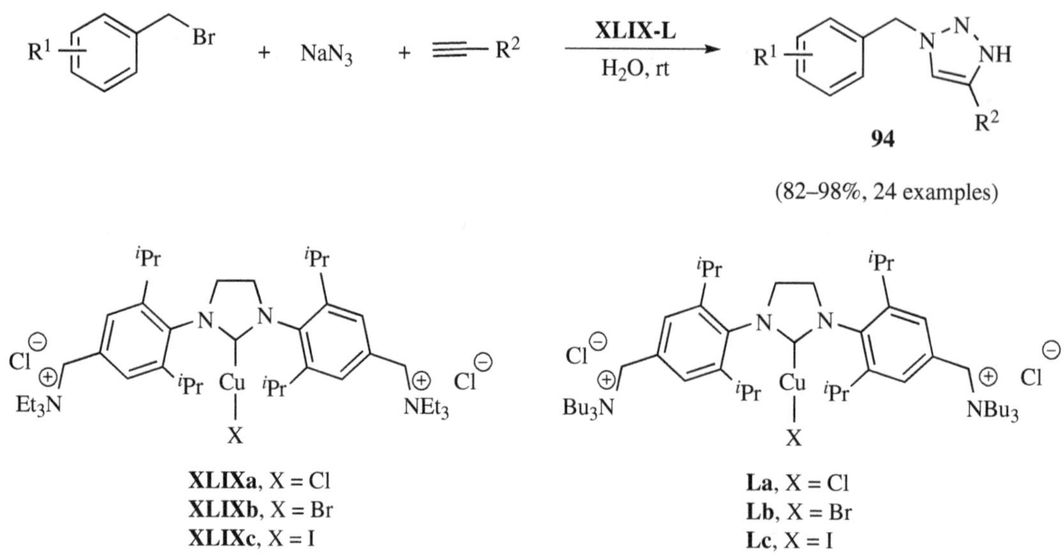

SCHEME 5.75 3-C CuAAC catalyzed by **XLIX/L**.

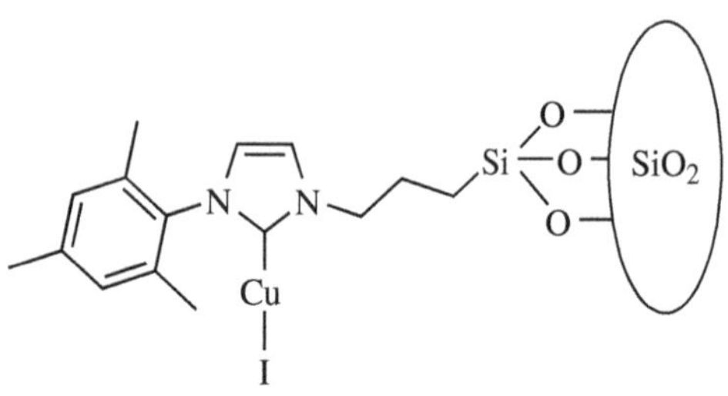

LI

FIGURE 5.9 Silica-immobilized Cu(I)–NHC heterogeneous catalyst **LI**.

5.4 SYNTHETIC UTILITY

In this chapter, the chemistry of carbenes in MCRs has been discussed. Among the reaction described, useful methodologies to prepare molecules with interesting architectures have been presented. Some of these procedures also accomplished the synthesis of molecules with added value in pharmacological industry, as well as organic molecules with structures resembling that of natural products. In this section, some specific examples for the synthesis of biologically active compounds, whose preparation involves MCRs using carbenes as components or catalysts, are presented.

5.4.1 Carbenes as Components

As seen in Section 5.2.1, the easy access to dimethoxycarbene **I** by simple thermolysis of oxadiazole **1** has been applied by Nair and coworkers in a multicomponent procedure for the synthesis of highly substituted dihydrofurans (Section 5.2.1, Scheme 5.5). This protocol offers great potential for the rapid access to a variety of important molecules. For example, as illustrated in Scheme 5.76, exposure of dihydrofuran derivatives **4** (presented in Scheme 5.5c) to Lewis acid (BF_3) underwent an interrupted Nazarov reaction yielding bicyclic lactones **95** [8], which structure appears in some biologically active monoterpenoids such as artemisia lactone, vulgaris lactone, or ilexlactone [113].

Section 5.2.3 shows that the use of FCC as building blocks in MCRs has emerged as a very useful strategy to prepared highly functionalized structures. In particular, the efficient synthesis of chromenes using FCCs reported by Wulff and coworkers has been described [32] (see Scheme 5.28 and Section 5.2.3). Highlighting the utility of this methodology, the authors presented the preparation of a mixture of eight stereoisomers (all-*rac*)-α-tocopherol **100**, which is the major commercial form of vitamin E. As shown in Scheme 5.77, enyne **97** was successfully prepared in 86% yield from Swern oxidation of commercially available (all-*rac*)-pythol **96**, followed by reaction with propynyl Grignard and protection of the alcohol as TBS ether. The reaction of FCC **LII** (analogue to **IX**, where R^1, R^2=Me; Scheme 5.28, Section 5.2.3) with enyne **97** led to chromene **98** in 85% yield. Quantitative hydrogenation of a double bond in **98** and successive cleavage of the methyl ether gave **100**, in 73% overall yield from carbene **LII**.

5.4.2 NHCs as Catalysts/Ligand

As mentioned in Section 5.3.1, enantioenriched δ-nitroester **63** can be prepared in MCR employing indanol-derived triazolium salt **XVI** as organocatalyst (Section 5.3.1, Scheme 5.43). This multicomponent procedure possesses great potential as initial step for the preparation of chiral substituted δ-lactams. As exemplified in Scheme 5.78, once enantiomerically enriched **63a** (78% ee) is obtained, it can

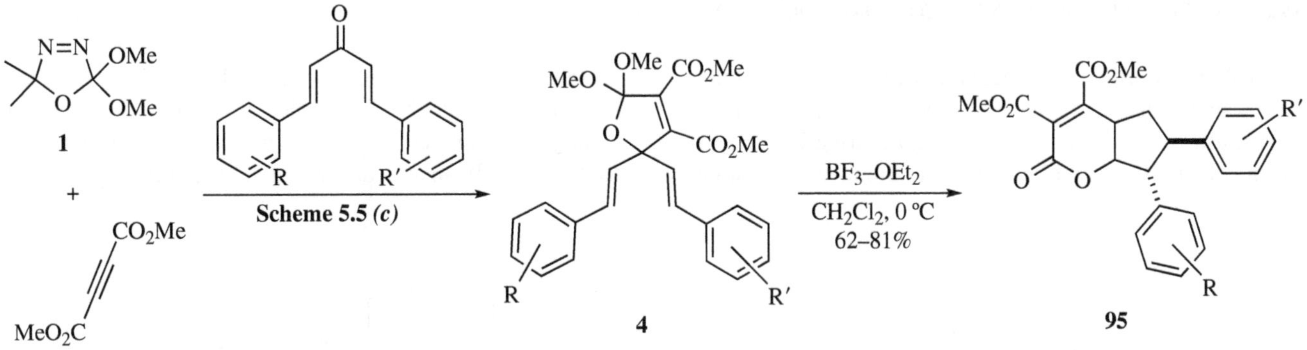

SCHEME 5.76 Synthesis of bicyclic lactones **95** starting from a MCR.

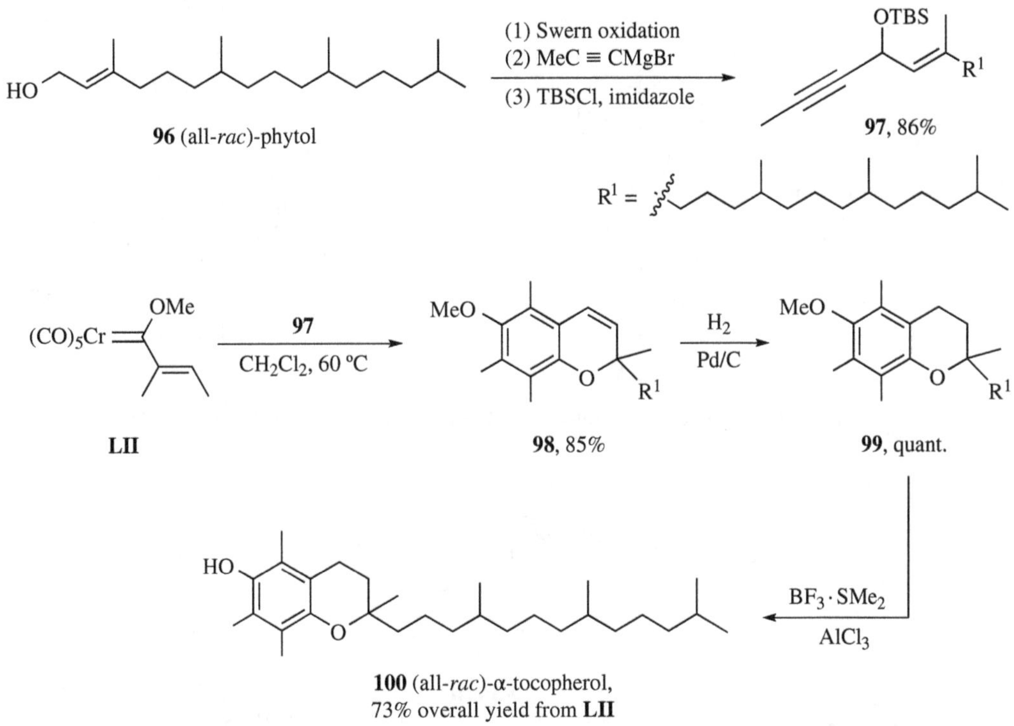

SCHEME 5.77 Synthesis of vitamin E following a MCR with FCC **LII**.

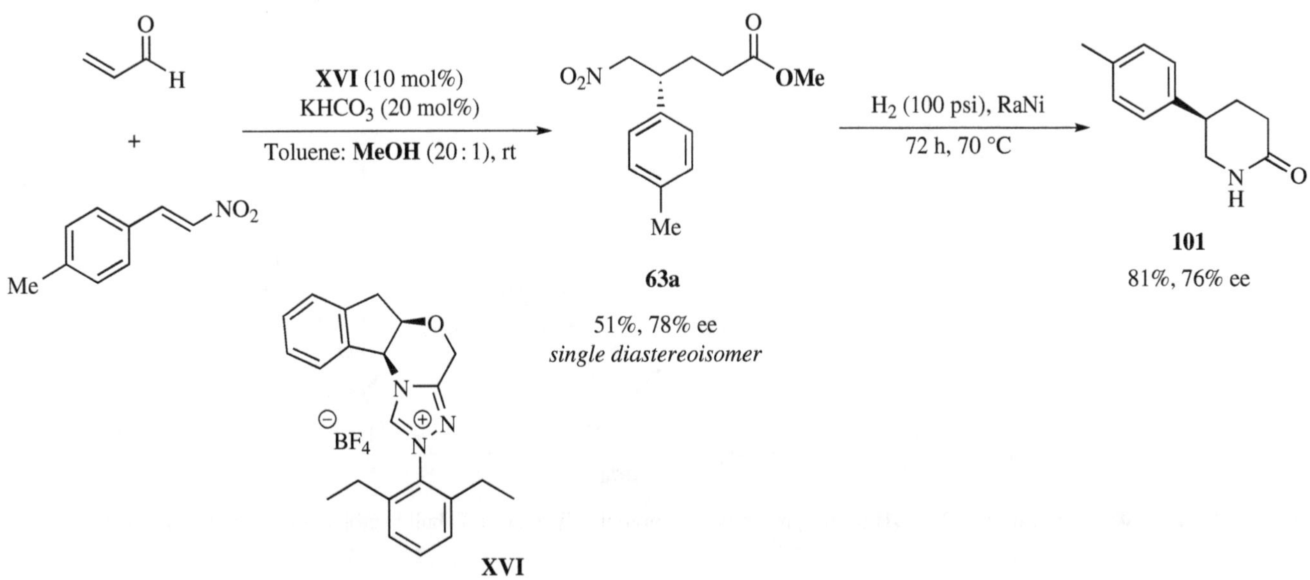

SCHEME 5.78 Transformation of δ-nitroester **63a** into δ-lactam **101**.

be easily transformed into δ-lactam **101** by simple treatment with Raney nickel under an atmosphere of hydrogen [50].

The use of Pd–NHC complexes has allowed the development of metal-catalyzed multicomponent aminocarbonylations (Section 5.3.2.1). Among the Pd–NHC described in that section, robust Pd–NHC complex **XXV** (Fig. 5.3) has efficiently been applied for the synthesis of tamibarotene **105** (Scheme 5.79), an anticancer drug that is normally obtained

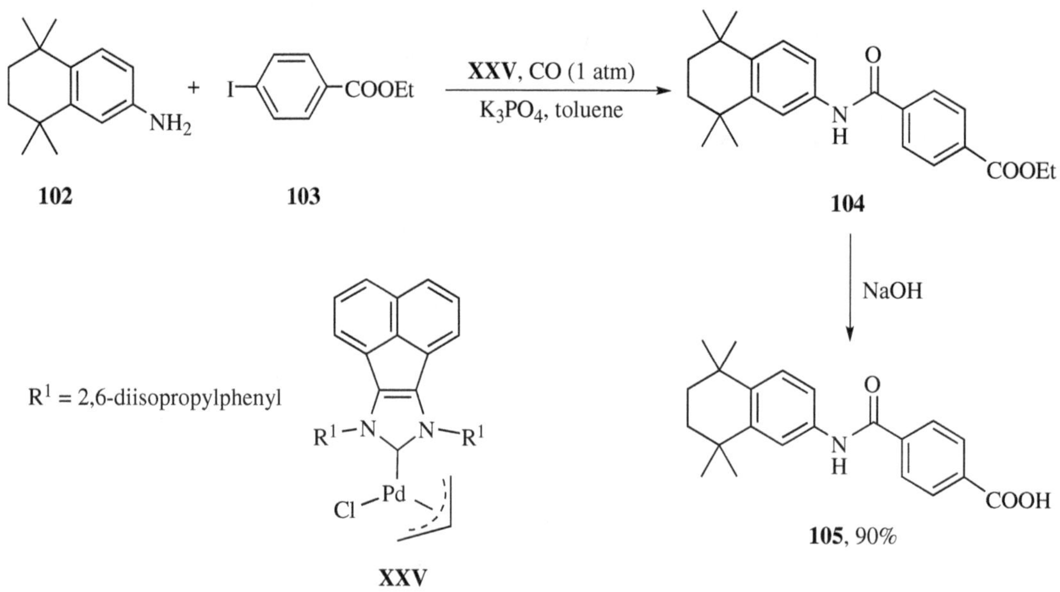

SCHEME 5.79 Synthesis of tamibarotene **105** using robust Pd–NHC complex **XXV**.

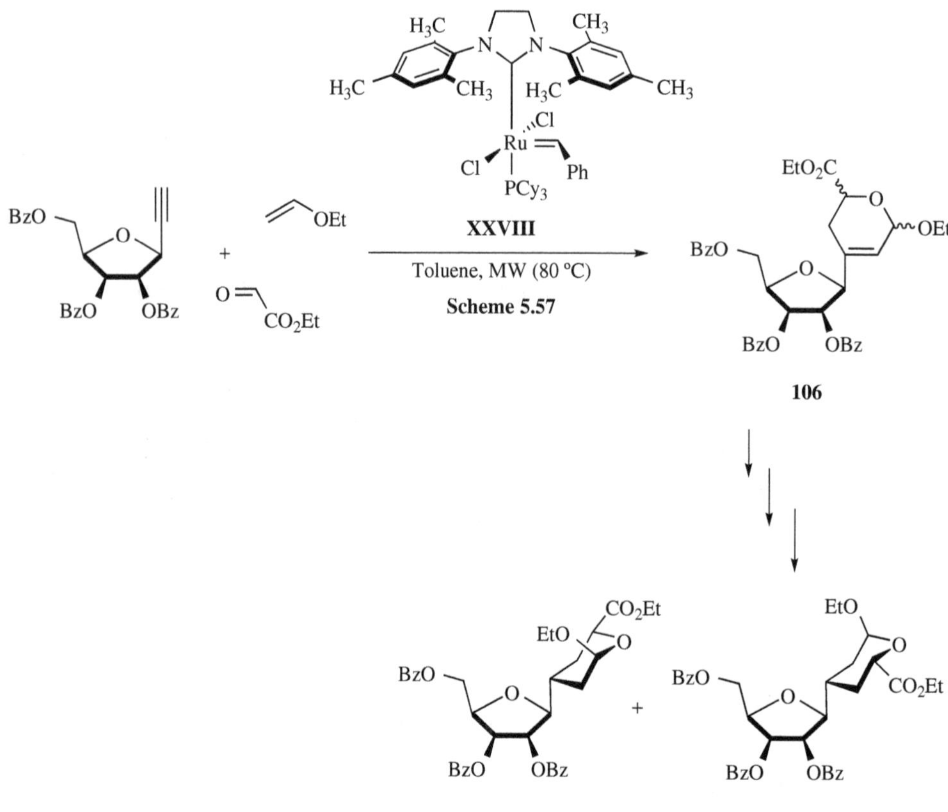

SCHEME 5.80 Application of **XXVIII** for the preparation of biologically interesting building blocks in carbohydrate chemistry.

following expensive and complicated synthetic procedures [114]. Coupling of amine **102** (prepared from aniline), ethyl 4-iodobenzoate **103**, and CO in the presence of **XXV** afforded amide **104**, which after hydrolysis gave tamibarotene **105** in 90% yield. The main advantages of this method are the use of inexpensive, available, and commercial reagents and the possibility to scale the procedure up to 5 mmol scale [76].

Grubbs catalyst second-generation **XXVIII** was used in 2009 by Botta and coworkers in a multicomponent procedure for the preparation of 2,3-dihydropyrans (Section 5.3.2.2, Scheme 5.57). As an application of this MCR, furanose–pyranose C—C-linked disaccharide **106** was obtained as a mixture of four diastereoisomers, which after equilibration with $ZnCl_2$ and hydrogenation afford **107a,b** (Scheme 5.80). This reaction constitutes a versatile and effective approach for the preparation of biologically interesting building blocks in carbohydrate chemistry [80].

5.5 CONCLUSION

As has been shown during this chapter, carbenes, initially considered laboratory curiosities, are privileged species for development of MCRs. As key components, their ability to generate reactive zwitterionic intermediates has been exploited in MCRs involving dimethoxycarbenes and NHCs for the synthesis of biological active molecules. Moreover, FCCs offer high versatility as reagents in C—C and C–heteroatom bond formation, which allow the incorporation of the organic fragment in the final product following different reaction pathways.

The contribution of NHCs in organocatalysis is especially relevant as has expanded the "normal" reactivity boundaries of carbonyl compounds offering powerful tools for designing new MCRs. The extraordinary properties of NHCs have also been applied in the design of transition metal-based homogeneous catalysis, where these carbenes act as supporting ligands, enhancing the catalytic activity of their metal complexes. The exceptional catalytic performances shown by M-NHC systems have allowed otherwise difficult transformations, reflected in the increasing number of applications. The incorporation of such complexes in MCRs opens new opportunities to obtain complex molecules from readily available compounds under mild conditions.

REFERENCES

1 D. Bourissou, O. Guerret, F. P. Gabbaï, G. Bertrand, *Chem. Rev.* 2000, 39, 39–91. Stable carbenes.

2 R. W. Hoffmann, *Angew. Chem. Int. Ed.* 1971, 10, 529–590. Generation of carbenes by thermal cycloelimination.

3 R. A. Moss, M. Wlostowski, J. Terpinski, G. Kmiecik-Lawrynowicz, K. Krogh-Jespersen, *J. Am. Chem. Soc.* 1987, 109, 3811. A diazirine precursor for a dioxacarbene: generation and reactions of methoxyphenoxycarbene.

4 P. Couture, J. K. Terlouw, J. Warkentin, *J. Am. Chem. Soc.* 1996, 118, 4214–4215. 2-Alkoxy-2-amino-Δ^3-1,3,4-oxadiazolines as novel sources of alkoxyaminocarbenes.

5 V. Nair, S. Bindu, V. Sreekumar, L. Balagopal, Synthesis 2003, 1446–1456. Novel dipolar cycloaddition reactions of zwitterionic species generated from dimethoxycarbene and dimethyl acetylenedicarboxylate with carbonyl compounds: facile synthesis of dihydrofuran derivatives.

6 V. Nair, S. Bindu, L. Balagopal, *Tetrahedron Lett.* 2001, 42, 2043–2044. A novel multicomponent reaction of dimethoxycarbene and DMAD with aldehydes and quinines: facile synthesis of dihydrofuran derivatives.

7 V. Nair, A. Deepthi, M. Poonoth, B. Santhamma, S. Vellalath, B. Pattoorpady, R. Mohan, E. Suresh, *J. Org. Chem.* 2006, 71, 2313–2319. Reaction of dimethoxycarbene-DMAD zwitterions with 1,2-diones and anhydrides: a novel synthesis of highly substituted dihydrofurans and spirodihydrofurans.

8 V. Nair, S. Bindu, V. Sreekumar, A. Chiaroni, *Org. Lett.* 2002, 4, 2821–2823. A novel approach to the synthesis of bicyclic lactones *via* an interrupted Nazarov reaction of *gem*-divinyl dihydrofurans.

9 V. Nair, P. Beneesh, V. Sreekumar, S. Bindu, R. Menon, A. Deepthi, *Tetrahedron Lett.* 2005, 46, 201–203. The multicomponent reaction of dimethoxycarbene, dimethyl butynedioate and electrophilic styrenes: an unprecedented synthesis of highly substituted cyclopentenone acetals.

10 (a) J. H. Rigby, J.-C. Brouet, P. J. Burke, S. Rohach, S. Sidique, M. J. Heeg, *Org. Lett.* 2006, 8, 3121–3123. New entry into β-lactams *via* reaction of dimethoxycarbene with isocyanates. (b) G. Mloston, H. Heimgartner, *Helv. Chim. Acta* 2007, 90, 1758–1764. The first reaction of dimethoxycarbene with imine moiety.

11 A. J. Arduengo, R. L. Harlow, M. Kline, *J. Am. Chem. Soc.* 1991, 113, 361–363. A stable crystalline carbene.

12 B. Maji, M. Breugst, H. Mayr, *Angew. Chem. Int. Ed.* 2011, 50, 6915–6919. N-Heterocyclic carbenes: organocatalysts with moderate nucleophilicity but extraordinarily high Lewis basicity.

13 T. Döge, F. Glorius, *Angew. Chem. Int. Ed.* 2010, 49, 6940–6952. The measure of all rings-N-heterocyclic carbenes.

14 (a) C. Ma, Y. Yang, *Org. Lett.* 2005, 7, 1343–1345. Thiazolium-mediated multicomponent reactions: a facile synthesis of 3-aminofuran derivatives. (b) C. Ma, H. Ding, G. Wu, Y. Yang, *J. Org. Chem.* 2005, 70, 8919–8923. Facile synthesis of highly substituted 3-aminofurans from thiazolium salts, aldehydes and dimethyl acetylenedicarboxylate.

15 (a) C. Ma, H. Ding, Y. Zhang, M. Bian, *Angew. Chem. Int. Ed.* 2006, 45, 7793–7797. A ring-expansion methodology involving multicomponent reactions: highly efficient access to polysubstituted furan-fused 1,4-thiazepine derivatives. (b) H. Ding, Y. Zhang, M. Bian, W. Yao, C. Ma, *J. Org. Chem.* 2008, 73, 578–584. Concise assembly of highly substituted furan-fused 1,4-thiazepines and their Diels-Alder reactions with benzynes.

16 V. Nair, S. Bindu, V. Sreekumar, N. Rath, *Org. Lett.* 2003, 5, 665–667. Unprecedented reactivity of N-heterocyclic carbenes toward DMAD and aldehydes leading to novel multicomponent reactions.

17 V. Nair, V. Sreekumar, S. Bindu, E. Suresh, *Org. Lett.* 2005, 7, 2297–2300. Two unprecedented multicomponent reactions involving N-heterocyclic carbenes, activated acetylenes and aldehydes.

18 H.-R. Pan, Y.-J. Li, C.-X. Yan, J. Xing, Y. Cheng, *J. Org. Chem.* 2010, 75, 6644–6652. Multicomponent reaction of imidazo[1,5-*a*]pyridine carbenes with aldehydes and dimethyl acetylenedicarboxylate or allenoates: a straightforward approach to fully substituted furans.

19 H.-R. Pan, X.-R. Wang, C.-X. Yan, Z.-X. Sun, Y. Cheng, *Org. Biomol. Chem.* 2011, 9, 2166–2174. The multicomponent reaction of imidazo[1,5-*a*]pyridine carbenes with phthalaldehydes and dimethyl acetylenedicarboxylate: a facile construction of benzo[*d*]furo[3,2-*b*]azepines.

20 C.-X. Yan, Z.-X. Sun, Y. Cheng, Synthesis 2012, 865–874. Construction of novel tricyclic fused-ring systems by the multicomponent reactions of imidazo[1,5-*a*]pyridine carbenes with aromatic *ortho*-dialdehydes and electron-deficient alkynes.

21 V. Nair, S. C. Mathew, S. Vellalath, A. N. Pillai, E. Suresh, Synthesis 2008, 551–554. An efficient three-component reaction involving triazolylidene carbenes, DMAD and aldehydes for the synthesis of furanone derivatives.

22 S. Matsuoka, Y. Tochigi, K. Takagi, M. Suzuki, *Tetrahedron* 2012, 68, 9836–9841. Sequential one-pot and three-component reactions of an N-heterocyclic carbene to form 4-(1,2,4-triazol-5-ylidene)pyrrolidine-2,5-diones: a tandem umpolung/annulation sequence *via* deoxy-Breslow intermediates.

23 E. O. Fischer, A. Maasböl, *Angew. Chem. Int. Ed.* 1964, 3, 580–581. On the existence of a tungsten carbonyl carbene complex.

24 For recent reviews, see: (a) J. Santamaría, *Curr. Org. Chem.* 2009, 13, 31–46. Fischer-type group 6 carbene complexes in the synthesis of optically active molecules. (b) K. H. Dötz, J. Stendel, *Chem. Rev.* 2009, 109, 3227–3274. Fischer carbene complexes in organic synthesis: metal-assisted and metal-templated reactions. (c) J. W. Herndon, *Coord. Chem. Rev.* 2010, 254, 103–194. The chemistry of the carbon-transition metal double and triple bond: annual survey covering the year 2008.

25 J. Barluenga, M. A. Fernández-Rodríguez, E. Aguilar, *J. Organomet. Chem.* 2005, 690, 539–587. Group 6 Fischer carbene complexes: "chemical multitalents" for multi-component reactions.

26 M. A. Fernández-Rodríguez, P. García-García, E. Aguilar, *Chem. Commun.* 2010, 46, 7670–7687. Multicomponent reactions involving group 6 Fischer carbene complexes: a source of inspiration for future catalytic transformations.

27 (a) P. Roy, B. K. Ghorai, *Beilstein J. Org. Chem.* 2010, 6, 52. One-pot three-component synthesis of quinoxaline and phenazine ring systems using Fischer carbene complexes. (b) S. Mukherjee, P. Roy, B. K. Ghorai, Synthesis 2011, 1419–1426. One-pot three-component synthesis of quinoxaline, quinazoline and phenazine ring systems using Fischer carbene complexes.

28 G. P. Jana, S. Mukherjee, B. K. Ghorai, Synthesis 2010, 3179–3187. Multicomponent approach for the synthesis of phenanthridine and acridine ring systems *via* the coupling of Fischer carbene complexes with heteroaromatic *o*-alkynyl carbonyl derivatives.

29 P. Roy, B. K. Ghorai, *Tetrahedron Lett.* 2011, 52, 5668–5671. Annulation of furan-bridged 10-membered rings on *N*-heterocycles through [8+2] cycloaddition of dienylazaisobenzofurans and dimethyl acetylenedicarboxylate.

30 P. Roy, B. K. Ghorai, *Tetrahedron Lett.* 2011, 52, 251–253. One-pot synthesis of furo[2,3-*h*]quinoline and furo[2,3-h]isoquinoline derivatives using Fischer carbene complex.

31 P. Roy, P. Mitra, B. K. Ghorai, *Tetrahedron Lett.* 2013, 54, 1440–1443. Synthesis of azahomosteroid ring system through intramolecular [4+2] cycloaddition of in situ generated azaisobenzofuran intermediates.

32 N. Majumdar, K. A. Korthals, W. D. Wulff, *J. Am. Chem. Soc.* 2012, 134, 1357–1362. Simultaneous synthesis of both rings of chromenes *via* a benzannulation/*O*-quinone methide formation/electrocyclization cascade.

33 H. Wang, A. V. Predeus, W. D. Wulff, *Chem.–Eur. J.* 2013, 19, 8261–8267. Synthesis of [*m,n*]cyclophanes: regiochemistry transfer from vinyl halides to cyclophane *via* Fischer carbene complexes.

34 J. Barluenga, M. G. Suero, R. De la Campa, J. Flórez, *Angew. Chem. Int. Ed.* 2010, 49, 9720–9724. Enantioselective synthesis of 4-hydroxy-2-cyclohexenones through a multicomponent cyclization.

35 M. G. Suero, R. De la Campa, L. Torre-Fernández, S. García-García, J. Flórez, *Chem.–Eur. J.* 2012, 18, 7287–7295. Enantioselective multicomponent synthesis of fused 6-5 bicyclic 2-butenolides by a cascade heterobicyclisation process.

36 For reviews on NHC as organocatalyst, see: (a) D. Enders, T. Balensiefer, *Acc. Chem. Res.* 2004, 37, 534–541. Nucleophilic carbenes in asymmetric organocatalysis. (b) N. Marion, S. Díez-González, S. P. Nolan, *Angew. Chem. Int. Ed.* 2007, 46, 2988–3000. N-Heterocyclic carbenes as organocatalysts. (c) D. Enders, O. Niemeier, A. Henseler, *Chem. Rev.* 2007, 107, 5606–5655. Organocatalysis by N-heterocyclic carbenes.

37 (a) D. Enders, U. Kallfass, *Angew. Chem. Int. Ed.* 2002, 41, 1743–1745. An efficient nucleophilic carbene catalyst for the asymmetric benzoin condensation. (b) J. S. Johnson, *Angew. Chem. Int. Ed.* 2004, 43, 1326–1328. Catalyzed reactions of acyl anion equivalents.

38 M. S. Kerr, J. R. de Alaniz, T. Rovis, *J. Am. Chem. Soc.* 2000, 124, 10298–10299. A highly enantioselective catalytic intramolecular Stetter reaction.

39 J. Seayad, P. K. Patra, Y. Zhang, J. Y. Ying, *Org. Lett.* 2008, 10, 953–956. Organocatalytic synthesis of *N*-phenylisoxazolidin-5-ones and a one-pot synthesis of β-amino acid esters.

40 S. Z.-X. Sun, Y. Cheng, Eur. J. Org. Chem. 2012, 4982–4987. One-step formation both of C-N and of C-O bonds of *N*-alkoxyamides through NHC-catalyzed three component reactions of enals, nitrosoarenes and enones.

41 NHC have been reported to catalyze conjugate additions of alcohols to enones: E. M. Phillips, M. Riedrich, K. A. Scheidt,

J. Am. Chem. Soc. 2010, 132, 13179–13181. N-heterocyclic carbene-catalyzed conjugate additions of alcohols.

42 F. T. Wong, P. K. Patra, J. Seayad, Y. Zhang, J. Y. Ying, *Org. Lett.* 2008, 10, 2333–2336. N-heterocyclic carbene (NHC)-catalyzed direct amidation of aldehydes with nitroso compounds.

43 (a) K. Y. Chow, J. W. Bode, *J. Am. Chem. Soc.* 2004, 126, 8126–8127. Catalytic generation of activated carboxylates: direct stereoselective synthesis of β-hydroxyesters from epoxyaldehydes. (b) S. S. Sohn, E. L. Rosen, J. W. Bode, *J. Am. Chem. Soc.* 2004, 126, 14370–14371. N-heterocyclic carbene-catalyzed generation of homoenolates: γ-butyrolactones by direct annulations of enals and aldehydes.

44 C. Burstein, F. Glorius, *Angew. Chem. Int. Ed.* 2004, 43, 6205–6208. Organocatalyzed conjugate umpolung of α,β-unsaturated aldehydes for the synthesis of γ-butyrolactones.

45 S. De Sarkar, A. Biswas, R. C. Samanta, A. Studer, *Chem.–Eur. J.* 2013, 19, 4664–4678. Catalysis with N-heterocyclic carbenes under oxidative conditions.

46 V. Nair, B. P. Babu, S. Vellalath, V. Varghese, A. E. Raveendran, E. Suresh, *Org. Lett.* 2009, 11, 2507–2510. Nucleophilic heterocyclic carbene catalyzed annulation of enals to chalcones in methanol: a stereoselective synthesis of highly functionalized cyclopentanes.

47 P. C. Chiang, J. Kaeobamrung, J. W. Bode, *J. Am. Chem. Soc.* 2007, 129, 3520–3521. Enantioselective cyclopentene-forming annulations *via* NHC-catalyzed benzoin-oxy-Cope reactions.

48 J. Ma, Y. Huang, R. Chen, *Org. Biomol. Chem.* 2011, 9, 1791–1798. N-heterocyclic carbene-catalyzed (NHC) three-component domino reactions: highly stereoselective synthesis of functionalized acyclic ε-ketoesters.

49 V. Nair, C. R. Sinu, B. P. Babu, V. Varghese, A. Jose, E. Suresh, *Org. Lett.* 2009, 11, 5570–5573. Novel nucleophilic heterocyclic carbene mediated stereoselective conjugate addition of enals to nitrostyrenes *via* homoenolate.

50 B. Maji, L. Ji, S. Wang, S. Vedachalam, R. Ganguly, X.-W. Liu, *Angew. Chem. Int. Ed.* 2012, 51, 8276–8280. N-heterocyclic carbene catalyzed homoenolate-addition reaction of enals and nitroalkenes: asymmetric synthesis of 5-carbon-synthon δ-nitroesters.

51 V. Nair, V. Varghese, B. P. Babu, C. R. Sinu, E. Suresh, *Org. Biomol. Chem.* 2010, 8, 761–764. A novel *pseudo* four component reaction involving homoenolate for the synthesis of γ-aminobutyric acid (GABA) derivatives.

52 Selected examples: (a) J. Kaeobamrung, J. Mahatthananchai, P. Zheng, J. W. Bode, *J. Am. Chem. Soc.* 2010, 132, 8810–8812. An enantioselective Claisen rearrangement catalyzed by N-heterocyclic carbenes. (b) Z. Q. Zhu, J. C. Xiao, *Adv. Synth. Catal.* 2010, 352, 2455–2458. N-heterocyclic carbene-catalyzed reaction of alkynyl aldehydes with 1,3-ketoesters or 1,3-diketones.

53 (a) Z. Q. Zhu, X.-L. Zheng, N.-F. Jiang, X. Wan, J.-C. Xiao, *Chem. Commun.* 2011, 47, 8670–8672. Chiral N-heterocyclic carbene catalyzed annulation of α,β-unsaturated aldehydes with 1,3-dicarbonyls. (b) B. Zhou, Z. Luo, Y. Li, *Chem. Eur. J.* 2013, 19, 4428–4431. Assembly of spirooxindole derivatives containing four consecutive stereocenters by using cascade reactions catalyzed by an N-heterocyclic carbene.

54 D. Du, Z. Hu, J. Jin, Y. Lu, W. Tang, B. Wang, T. Lu, *Org. Lett.* 2012, 14, 1274–1277. N-heterocyclic carbene-catalyzed three-component domino reaction of alkynyl aldehydes with oxindoles.

55 R. U. Braun, K. Zeitler, T. J. Müller, *Org. Lett.* 2001, 3, 3297–3300. A novel one-pot pyrrole synthesis *via* a coupling-isomerization-Stetter-Paal-Knorr sequence.

56 A. R. Bharadwaj, K. A. Scheidt, *Org. Lett.* 2001, 3, 2465–2468. Catalytic multicomponent synthesis of highly substituted pyrroles utilizing a one-pot sila-Stetter/Paal-Knorr strategy.

57 C. Yu, J. Lu, T. Li, D. Wang, B. Qin, H. Zhang, C. Yao, Synlett 2011, 2420–2424. A NHC-involved, cascade, metal-free and three-component synthesis of 2,3-diarylated fully substituted furans under solvent-free conditions.

58 W. A. Herrmann, M. Elison, J. Fischer, C. Kocher, G. R. J. Artus, *Angew. Chem. Int. Ed.* 1995, 34, 2371–2374. Metal complexes of N-heterocyclic carbenes—a new structural principle for catalysts in homogeneous catalysis.

59 (a) W. A. Herrmann, C. Kocher, *Angew. Chem. Int. Ed.* 1997, 36, 2162–2187. N-heterocyclic carbenes. (b) F. E. Hahn, *Angew. Chem. Int. Ed.* 2006, 45, 1348–1352. Heterocyclic carbenes. (c) F. E. Hahn, M. C. Jahnke, *Angew. Chem. Int. Ed.* 2008, 47, 3122–3172. Heterocyclic carbenes: synthesis and coordination chemistry.

60 (a) L. Cavallo, A. Correa, C. Costabile, H. Jacobsen, *J. Organomet. Chem.* 2005, 690, 5407–5413. Steric and electronic effects in the bonding of N-heterocyclic ligands to transition metals. (b) R. Dorta, E. D. Stevens, N. M. Scott, C. Costabile, L. Cavallo, C. D. Hoff, S. P. Nolan, *J. Am. Chem. Soc.* 2005, 127, 2485–2495. Steric and electronic properties of N-heterocyclic carbenes (NHC): a detailed study on their interaction with $Ni(CO)_4$. (c) S. Díez-González, S. P. Nolan, *Coord. Chem. Rev.* 2007, 251, 874–883. Stereoelectronic parameters associated with N-heterocyclic carbene (NHC) ligands: a quest for understanding.

61 E. Peris, *Top. Organomet. Chem.* 2007, 41, 83–116. Routes to N-heterocyclic carbene complexes.

62 (a) L. H. Gade, S. Bellemin-Laponnaz, *Top. Organomet. Chem.* 2007, 41, 117–157. Chiral N-heterocyclic carbenes as stereodirecting ligands in asymmetric catalysis. (b) L. H. Gade, S. Bellemin-Laponnaz, *Coord. Chem. Rev.* 2007, 251, 718–725. Mixed oxazoline-carbenes as stereodirecting ligands for asymmetric catalysis.

63 (a) S. P. Nolan (Ed.), *N-heterocyclic carbenes in synthesis*. Wiley-VCH Verlag GmbH, Weinheim, Germany, 2006. (b) F. Glorius, *Topics in organometallic chemistry*. Springer-Verlag, Berlin, 2007. N-heterocyclic carbenes in transition metal catalysis.

64 (a) G. Bertrand, *J. Organomet. Chem.* 2005, 690, 5397–6252. Carbene chemistry. (b) R. H. Crabtree, *Coord. Chem. Rev.* 2007, 251, 595–896. Recent developments in the organometallic chemistry of N-heterocyclic carbenes.

65 (a) E. A. B. Kantchev, C. J. O'Brien, M. G. Organ, *Angew. Chem. Int. Ed.* 2007, 46, 2768–2813. Palladium complexes of N-heterocyclic carbenes as catalysts for cross-coupling reactions-a synthetic chemist's perspective. (b) S. Díez-González, S. P. Nolan, *Top. Organomet. Chem.* 2007, 21, 47–82.

Palladium-catalyzed reactions using NHC ligands. (c) J. A. Mata, M. Poyatos, *Curr. Org. Chem.* 2011, 15, 3309–3324. Recent developments in the applications of palladium complexes bearing *N*-heterocyclic carbene ligands.

66 A. Strecker, *Ann. Chem. Pharm.* 1850, 75, 27–45. Ueber die künstliche Bildung der Milchsäure und einen neuen, dem Glycocoll homologen Körper.

67 C. Najera, J. M. Sansano, *Chem. Rev.* 2007, 107, 4584–4671. Catalytic asymmetric synthesis of α-amino acids.

68 J. Jarusiewiez, Y. Choe, K. S. Yoo, C. P. Park, K. W. Jung, *J. Org. Chem.* 2009, 74, 2873–2876. Efficient three-component Strecker reaction of aldehydes/ketones *via* NHC-amidate palladium(II) complex catalysis.

69 J. Choi, H. Y. Yang, H. J. Kim, S. U. Son, *Angew. Chem. Int. Ed.* 2010, 49, 7718–7722. Organometallic hollow spheres bearing bis(N-heterocyclic carbene)-palladium species: catalytic application in three-component Strecker reactions.

70 X.-F. Wu, H. Neumann, M. Beller, *Chem. Soc. Rev.* 2011, 40, 4986–5009. Palladium-catalyzed carbonylative coupling reactions between Ar-X and carbon nucleophiles.

71 E. Maerten, M. Sauthier, A. Mortreux, Y. Castanet, *Tetrahedron*, 2007, 63, 682–689. Palladium-N-heterocyclic carbene an efficient catalytic system for the carbonylative cross-coupling of pyridine halides with boronic acids.

72 S. Z. Zheng, L. W. Xu, C. G. Xia, *Appl. Organomet. Chem.* 2007, 21, 772–776. Highly efficient *N*-heterocyclic carbene–palladium complex-catalyzed multicomponent carbonylative Suzuki reaction: novel practical synthesis of unsymmetric aryl ketones.

73 C. L. Allena, J. M. J. Williams, *Chem. Soc. Rev.* 2011, 40, 3405–3415. Metal-catalyzed approaches to amide bond formation.

74 S. Ho, G. Bondarenko, D. Rosa, B. Dragisic, A. Orellana, *J. Org. Chem.* 2012, 77, 2008–2012. Synthesis of acyl pyrroles *via* palladium-catalyzed carbonylative amination of aryl and alkenyl iodides.

75 Z. S. Qureshi, S. A. Revankar, M. V. Khedkar, B. M. Bhanage, *Catal. Today* 2012, 198, 148–153. Aminocarbonylation of aryl iodides with primary and secondary amines in aqueous medium using polymer supported palladium N-heterocyclic carbene complex as an efficient and heterogeneous recyclable catalyst.

76 W. Fang, Q. Deng, M. Xu, T. Tu, *Org. Lett.* 2013, 15, 3678–3681. Highly efficient aminocarbonylation of iodoarenes at atmospheric pressure catalyzed by a robust acenaphthoimidazolyidene allylic palladium complex.

77 (a) Y. Gu, G. Li, *Adv. Synth. Catal.* 2009, 351, 817–847. Ionic liquids-based catalysis with solids: state of the art. (b) Q.H. Zhang, S. G. Zhang, Y. Q. Deng, *Green Chem.* 2011, 13, 2619–2637. Recent advances in ionic liquid catalysis.

78 M. Papp, R. Skoda-Földes, *J. Mol. Catal. A: Chem.* 2013, 378, 193–199. Phosphine-free double carbonylation of iodobenzene in the presence of reusable supported palladium catalysts.

79 (a) F. Hamad, T. Sun, S. Xiao, F. Verpoort, *Coord. Chem. Rev.* 2013, 257, 2274–2292. Olefin metathesis ruthenium catalysts bearing unsymmetrical heterocyclic carbenes. (b) S. Kress, S. Blechert, *Chem. Soc. Rev.* 2012, 41, 4389–4408. Asymmetric catalysts for stereocontrolled olefin metathesis reactions. (c) T. J. Donohoe, J. F. Bower, L. Chan, *Org. Biomol. Chem.* 2012, 10, 1322–1328. Olefin cross-metathesis for the synthesis of heteroaromatic compounds.

80 D. Castagnolo, L. Botta, M. Botta, *Tetrahedron Lett.* 2009, 50, 1526–1528. One pot multicomponent synthesis of 2,3-dihydropyrans: new access to furanose-pyranose 1,3-C-C-linked-disaccharides.

81 C. S. Junker, M. E. Welker, *Tetrahedron* 2012, 68, 5341–5345. Ruthenium carbenes as catalysts in stereoselective ene-yne metathesis/Diels-Alder and ene-yne metathesis/Diels-Alder/cross coupling multicomponent reaction.

82 C. Fischmeister, C. Bruneau, *Beilstein J. Org. Chem.* 2011, 7, 156–166. Ene-yne cross metathesis with ruthenium carbene catalysts.

83 J. Montgomery, *Angew. Chem. Int. Ed.* 2004, 43, 3890–3908. Nickel-catalyzed reductive cyclizations and couplings.

84 (a) M. Takimoto, Y. Hiraga, Y. Sato, M. Mori, *Tetrahedron Lett.* 1998, 39, 4543–4546. Nickel-catalyzed regio- and stereo-selective synthesis of homoallylic alcohol derivatives from dienes and aldehydes. (b) Y. Sato, R. Sawaki, M. Mori, *Organometallics* 2001, 20, 5510–5512. Dramatic influence on the olefinic geometry in the nickel(0)-catalyzed coupling reaction of 1,3-dienes and aldehydes using N-heterocyclic carbene as a ligand. (c) R. Sawaki, Y. Sato, M. Mori, *Org. Lett.* 2004, 6, 1131–1133. Ligand-controlled highly stereoselective syntheses of *E*- and *Z*-allylsilanes from dienes and aldehydes using nickel complex.

85 Y. Sato, Y. Hinata, R. Seki, Y. Oonishi, N. Saito, *Org. Lett.* 2007, 9, 5597–5599. Nickel-catalyzed enantio- and diastereoselective three-component coupling of 1,3-dienes, aldehydes, and silanes using chiral N-heterocyclic carbenes as ligands.

86 G. M. Mahandru, G. Liu, J. Montgomery, *J. Am. Chem. Soc.* 2004, 126, 3698–3699. Ligand-dependent scope and divergent mechanistic behavior in nickel-catalyzed reductive couplings of aldehydes and alkynes.

87 Ynamides: (a) N. Saito, T. Katayama, Y. Sato, *Org. Lett.* 2008, 10, 3829–3832. Nickel-catalyzed highly regioselective multicomponent coupling of ynamides, aldehydes and silane: a new access to functionalized enamides. Propargyl alcohols: (b) H. A. Malik, M. R. Chaulagain, J. Montgomery, *Org. Lett.* 2009, 11, 5734–5737. Cooperativity of regiochemistry control strategies in reductive couplings of propargyl alcohols and aldehydes. Ynoates: (c) S. Rodrigo, H. Guan, *J. Org. Chem.* 2012, 77, 8303–8309. Quick installation of a 1,4-difunctionality *via* regioselective nickel-catalyzed reductive coupling of ynoates and aldehydes.

88 A. Lumbroso, M. L. Cooke, B. Breit, *Angew. Chem. Int. Ed.* 2013, 52, 1890–1932. Catalytic asymmetric synthesis of allylic alcohols and derivatives and their applications in organic synthesis.

89 H. A. Malik, G. J. Sormunen, J. Montgomery, *J. Am. Chem. Soc.* 2010, 132, 6304–6305. A general strategy for regiocontrol in nickel-catalyzed reductive couplings of aldehydes and alkynes.

90 B. Knapp-Reed, G. M. Mahandru, J. Montgomery, *J. Am. Chem. Soc.* 2005, 127, 13156–13157. Access to macrocyclic endocyclic and exocyclic allylic alcohols by nickel-catalyzed reductive cyclization of ynals.

91 R. D. Baxter, J. Montgomery, *J. Am. Chem. Soc.* 2008, 130, 9662–9663. Dehydrogenative cyclocondensation of aldehydes, alkynes and dialkylsilanes.

92 K. Sa-ei, J. Montgomery, *Org. Lett.* 2006, 8, 4441–4443. Highly diastereoselective preparation of anti-1,2-diols by catalytic addition of alkynylsilanes to α-silyloxyaldehydes.

93 M. R. Chaulagain, G. J. Sormunene, J. Montgomery, *J. Am. Chem. Soc.* 2007, 129, 9568–9569. New N-heterocyclic carbene ligand and its application in asymmetric nickel-catalyzed aldehyde/alkyne reductive couplings.

94 S.-S. Ng, T. F. Jamison, *J. Am. Chem. Soc.* 2005, 127, 7320–7321. Highly enantioselective and regioselective nickel-catalyzed coupling of allenes, aldehydes and silanes.

95 K. Ogata, Y. Atsuumi, S.-I. Fukuzawa, *Org. Lett.* 2010, 12, 4536–4539. Nickel-catalyzed ring-opening three-component coupling of methylenecyclopropane with aldehydes and silanes.

96 K. Ogata, Y. Atsuumi, D. Shimada, S.-I. Fukuzawa, *Angew. Chem. Int. Ed.* 2011, 50, 5896–5899. Nickel-catalyzed three-component coupling between aryl aldehydes, norbornenes and silanes leading to indanols through aromatic C-H bond activation of aryl aldehydes.

97 N. Saito, T. Yamazaki, Y. Sato, *Tetrahedron Lett.* 2008, 49, 6747–6750. Nickel(0)-catalyzed diastereoselective three-component coupling of 1,3-dienes, aldehydes and organometallic reagents: influence of organometallic reagents on diastereoselectivity.

98 A. J. Arduengo, H. V. R. Dias, J. C. Calabrese, F. Davidson, *Organometallics* 1993, 12, 3405–3409. Homoleptic carbene-silver(I) and carbene-copper(I) complexes.

99 (a) J. C. Garrison, W. J. Youngs, *Chem. Rev.* 2005, 105, 3978–4008. Ag(I) N-heterocyclic carbene complexes: synthesis, structure, and application. (b) I. J. B. Lin, C. S. Vasam, *Coord. Chem. Rev.* 2007, 251, 642–670. Preparation and application of N-heterocyclic carbene complexes of Ag(I). (c) A. Kascatan-Nebioglu, M. J. Panzner, C. A. Tessier, C. L. Cannon, W. J. Youngs, *Coord. Chem. Rev.* 2007, 251, 884–895. N-heterocyclic carbene-silver complexes: a new class of antibiotics.

100 V. A. Peshkov, O. P. Pereshivko, E. V. Van der Eycken, *Chem. Soc. Rev.* 2012, 41, 3790–3807. A walk around the A^3-coupling.

101 X. Yao, C.-J. Li, *Org. Lett.* 2005, 7, 4395–4398. Phosphine-triggered complete chemo-switch: from efficient aldehyde-alkyne-amine coupling to efficient aldehyde-alkyne coupling in water.

102 C. Wei, Z. Li, C.-J. Li, *Org. Lett.* 2003, 5, 4473–4475. The first silver-catalyzed three-component coupling of aldehyde, alkyne and amine.

103 Y. Li, X. Chen, Y. Song, L. Fang, G. Zou, *Dalton Trans.* 2011, 40, 2046–2052. Well-defined *N*-heterocyclic carbene silver halides of 1-cyclohexyl-3-arylmethylimidazolylidenes: synthesis, structure and catalysis in A^3-reaction of aldehydes, amines and alkynes.

104 P. Li, L. Wang, Y. Zhang, M. Wang, *Tetrahedron Lett.* 2008, 49, 6650–6654. Highly efficient three-component (aldehyde-alkyne-amine) coupling reactions catalyzed by a reusable PS-supported NHC-Ag(I) under solvent-free reaction conditions.

105 Y. He, M.-F. Lv, C. Cai, *Dalton Trans.* 2012, 41, 12428–12433. A simple procedure for polymer-supported N-heterocyclic carbene silver complex *via* click chemistry: an efficient and recyclable catalyst for the one-pot synthesis of propargylamines.

106 M.-T. Chen, B. Landers, O. Navarro, *Org. Biomol. Chem.* 2012, 10, 2206–2208. Well-defined (N-heterocyclic carbene)-Ag(I) complexes as catalysts for A^3 reactions.

107 J. D. Egbert, C. S. J. Cazin, S. P. Nolan, *Catal. Sci. Technol.* 2013, 3, 912–926. Copper N-heterocyclic carbene complexes in catalysis.

108 M. Wang, P. Li, L. Wang, Eur. J. Org. Chem. 2008, 2255–2261. Silica-immobilized NHC-Cu(I) complex: an efficient and reusable catalyst for A3-coupling (aldehydes-alkyne-amine) under solventless reaction conditions.

109 (a) J. E. Moses. A. D. Moorhouse, *Chem. Soc. Rev.* 2007, 36, 1249–1262. The growing applications of click chemistry. (b) M. Meldal. C. W. Tornøe, *Chem. Rev.* 2008, 108, 2952–3015. Cu-catalyzed azide-alkyne cycloaddition.

110 S. Díez-González, E. D. Stevens, S. P. Nolan, Chem. Commun. 2008, 4747–4749. A [(NHC)CuCl] complex as a latent click catalyst.

111 W. Wang, J. Wu, C. Xia, F. Li, *Green Chem.* 2011, 13, 3440–3445. Reusable ammonium salt-tagged NHC-Cu(I) complexes: preparation and catalytic application in the three component click reaction.

112 L. Wan, C. Cai, *Catal. Lett.* 2012, 142, 1134–1140. Multicomponent synthesis of 1,2,3-triazoles in water catalyzed by silica-immobilized NHC-Cu(I).

113 (a) S. Ravi, A. Laxman, *Indian J. Chem. B* 2001, 40, 443–446. Phytoconstituents of *Artemisia vulgaris*. (b) T. Hartmut, B. Herbert, *Phytochemistry* 1980, 19, 1866–1868. Ilex-lacton ein bisnormonoterpen neuartiger struktur aus *Ilex aquifolium*.

114 (a) J. Xiao, L. Luo, S. He, *Chin. J. Med. Chem.* 2009, 4, 268–269. Improved synthesis of anti-leukemia drug tamibarotene. (b) S. D. Friis, T. L. Andersen, T. Skrydstrup, *Org. Lett.* 2013, 15, 1378–1381. Palladium-catalyzed synthesis of aromatic carboxylic acids with silacarboxylic acids.

6

MULTICOMPONENT REACTIONS IN THE SYNTHESIS OF TARGET MOLECULES

Lea Radtke[1], Eugenia Marqués-López[2] and Raquel P. Herrera[2]

[1] *Institute of Pharmaceutical Sciences, Department of Chemistry and Applied Biosciences, Swiss Federal Institute of Technology (ETH) Zürich, Zurich, Switzerland*

[2] *Departamento de Química Orgánica, Instituto de Síntesis Química y Catálisis Homogénea (ISQCH), CSIC-Universidad de Zaragoza, Zaragoza, Spain*

6.1 INTRODUCTION

Multicomponent reactions (MCRs) are a powerful tool in the synthesis of complex target molecules because multiple bonds are effectively built up in one step from relatively simple building blocks. Several interesting reviews covering this topic until 2009 have been published [1]; in this sense, we will cover the most important examples published after this date. Heterocycles are structural motifs that are often found in bioactive molecules or drugs [2]; therefore, the high-yielding and fewer-step synthesis of these compounds is of great interest for medicinal chemistry [3]. MCRs often give access to heterocycles like in the classical Hantzsch or Biginelli reaction, which will then be part of the final molecule (Fig. 6.1). Heterocycles can also be starting materials in the used MCR, taking advantage of the rich heterocyclic reactivity. Here, we will discuss some multicomponent strategies to synthesize heterocycles like indoles, pyrroles, spiro-type polyheterocycles, fused polyheterocycles, dihydropyrimidines (DHPMs), thiazines, as well as amino acids.

6.2 MCRs IN DRUG DISCOVERY AND FOR THE SYNTHESIS OF BIOLOGICALLY IMPORTANT MOLECULES

Efficiency in total synthesis means the minimization of the number of steps along with a maximum input of structural complexity. In this context, the excellent properties showed by multicomponent methodologies, such as the facility of application, shorter reaction times, the versatility and diversity of final scaffolds, and the very large chemical spectrum of molecules, have enabled an easy access to biologically active compounds. Many research groups and pharmaceutical companies have focused their efforts on the design and application of new multicomponent strategies for the discovery of new synthetic methods giving access to these active products and many of them with industrial interest. Although the biological activities of MCR-derived structures have been revised in previous works [6], an interesting review has recently appeared covering a great number of important examples of bioactive compounds synthesized by MCR chemistry [7]. For this reason, we only selected a short number of these cases to discuss here. In this area, the 3- and 4-multicomponent Ugi reaction has represented a key and powerful tool for the development of this field (see Chapter 7) (Fig. 6.2).

MCR chemistry was also used in the synthesis of the blood coagulation factor VIIa (FVIIa) inhibitor **23** (Scheme 6.1). The synthesis of this inhibitor is achieved through the BF_3-catalyzed addition of the diaroyl Schiff base to the isocyanide (benzyl or morpholinoethyl) to afford intermediate **22**, which is hydrolyzed to carboxylic acid **23** [15].

Compound **23** has good potency and selectivity and is orally active as a double prodrug in guinea pigs [16]. Kobayashi and coworkers demonstrated the efficiency of an intermolecular Ugi four-component reaction in the synthesis of the naturally derived *20S* proteasome inhibitor, omuralide **28**

Multicomponent Reactions: Concepts and Applications for Design and Synthesis, First Edition. Edited by Raquel P. Herrera and Eugenia Marqués-López.

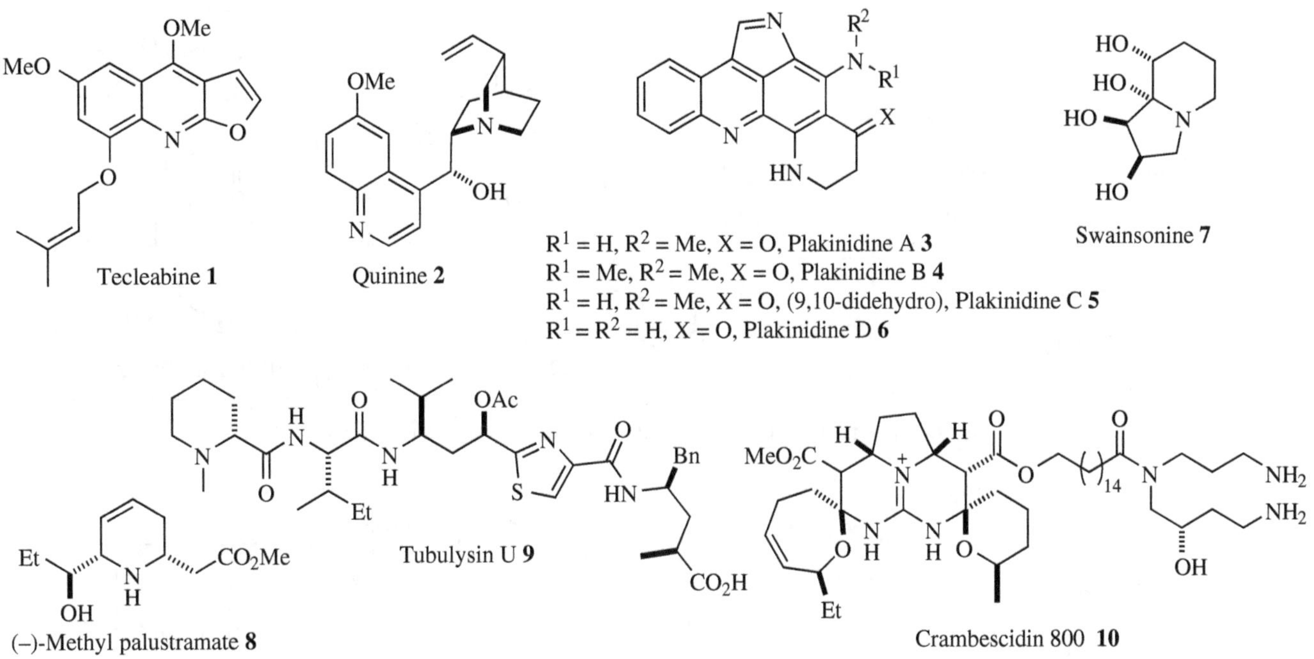

FIGURE 6.1 Examples of heterocycles and complex natural products synthesized by MCR chemistry [4, 5].

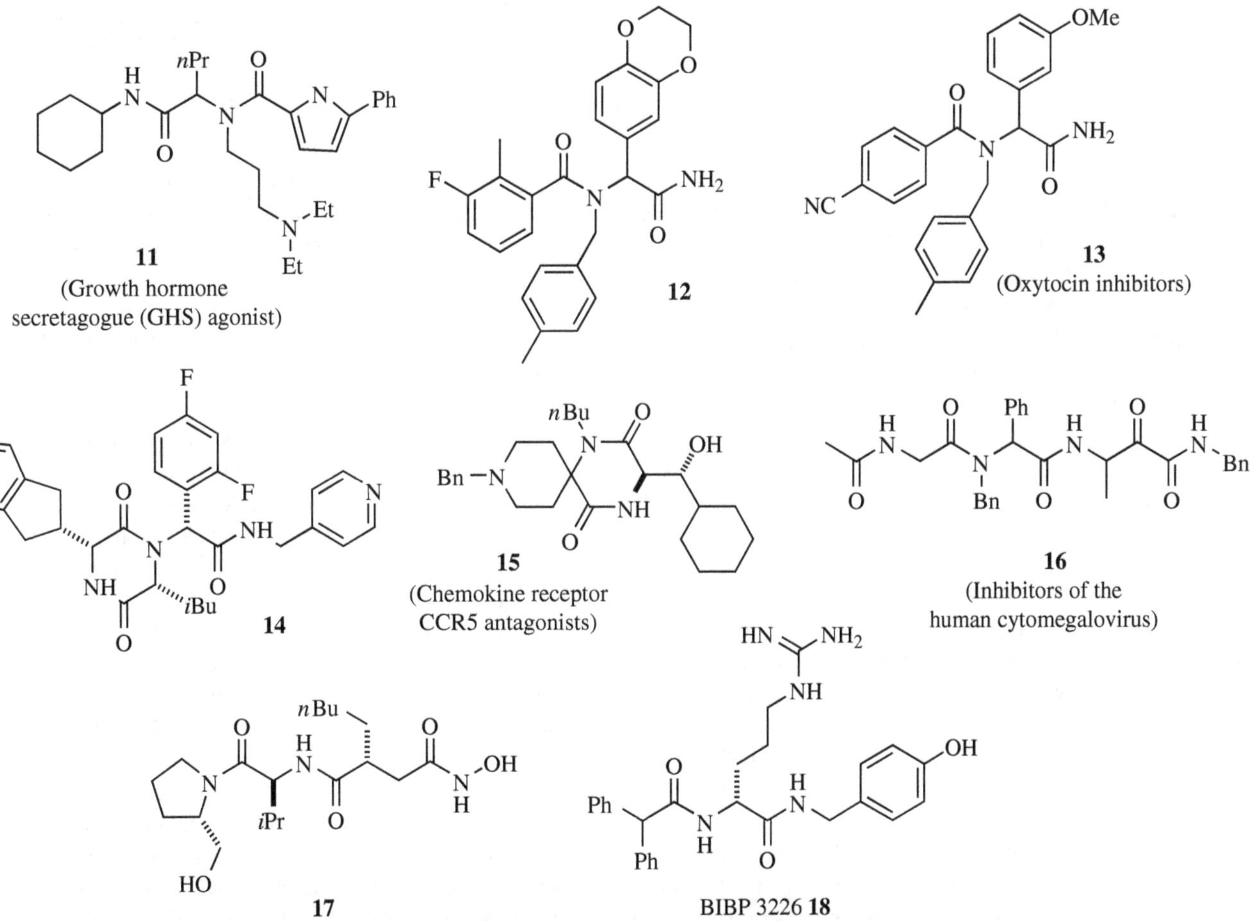

FIGURE 6.2 Biologically active products synthesized by a multicomponent Ugi approach [8–14].

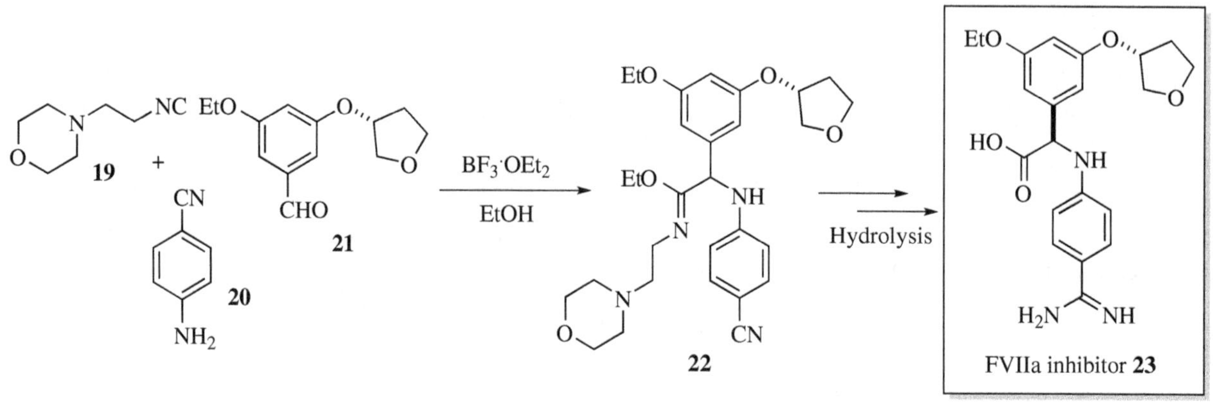

SCHEME 6.1 U-3CR reaction in the synthesis of FVIIa inhibitor **23**.

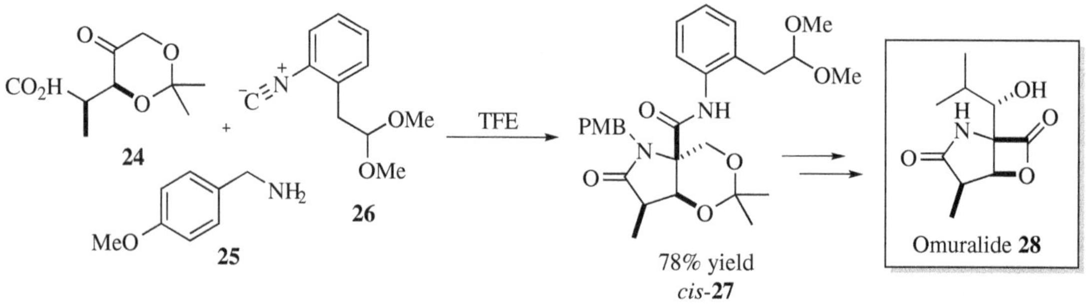

SCHEME 6.2 Stereoselective intermolecular U-4CR in the synthesis of omuralide.

(Scheme 6.2) [17]. Herein, acid **24** reacts with a novel isocyanide **26**, which was introduced by two different groups independently [18], and benzylamine **25** to stereoselectively yield only the *cis*-diastereoisomer **27**. The anilide can be easily hydrolyzed to give a free pyroglutamic acid. This strategy can be used in the synthesis of molecules bearing a pyroglutamic acid residue like as omuralide **28** or salinosporamide A.

Wessjohann and coworkers combined three different MCRs to efficiently create analogues of tubulysin, called tubugis [19]. Compound **36** was synthesized according to an Ugi-Nenajdenko MCR in 67% yield and a separable diastereomeric ratio of 1:1, while the thiazole **37** was conducted *via* a Passerini–Dömling MCR (Scheme 6.3) [20]. Compounds **36** and **37** were further connected by an Ugi four-component reaction. The so obtained analogues **38** showed similar subnanomolar cytotoxicities as tubulysin A.

Ruijter and coworkers elegantly combined stereoselective MCR chemistry with biocatalysis to obtain telaprevir **47**, which is a pharmaceutical drug (Incivek® or Incivo®) for the treatment of hepatitis C (Scheme 6.4) [21]. In the first step, a Passerini three-component reaction gives access to form **42**, which is, upon treatment with NMM (*N*-methylmorpholine) and triphosgene, converted to isocyanide **43**. The latter further reacts with the pyrroline derivative **45** (arose from the desymmetrization of the 3,4-disubstituted *meso*-pyrrolidine with monoamine oxidase-N (MAO-N, derived from *Aspergillus niger*)) and the acid **44** to the Ugi compound **46** in 76% yield. Only few steps are then necessary to achieve the synthesis of the target molecule, telaprevir **47**.

6.3 SYNTHESIS OF NATURAL PRODUCTS IN AN EFFICIENT MANNER

MCRs are convergent reactions in which three or more components react to form a single product. They therefore offer means to construct complex and structurally diverse compounds in an almost simple and elegant manner (Fig. 6.3).

Among MCRs, the Ugi reaction is a powerful three- or four-component coupling reaction with isocyanide as one of the reactants to construct multiple bonds in complex molecules. Applications of Ugi reactions in total synthesis are rather rare. One of the most advanced examples is the use in the total synthesis of ecteinascidin 743 **48** by Fukuyama and coworkers, as depicted in Scheme 6.5 [22].

Ecteinascidin 743 is a marine tetrahydroisoquinoline alkaloid that was isolated from *Ecteinascidia turbinata*, a mangrove tunicate, and is now approved by the FDA as an orphan drug under the trade name Yondelis® against soft tissue sarcoma and ovarian cancer [27]. Amine **53**, isocyanide **54**, acetaldehyde **55**, and carboxylic acid **56** were heated in MeOH at 70°C to build up the dipeptide **60** in 90% yield (Scheme 6.5).

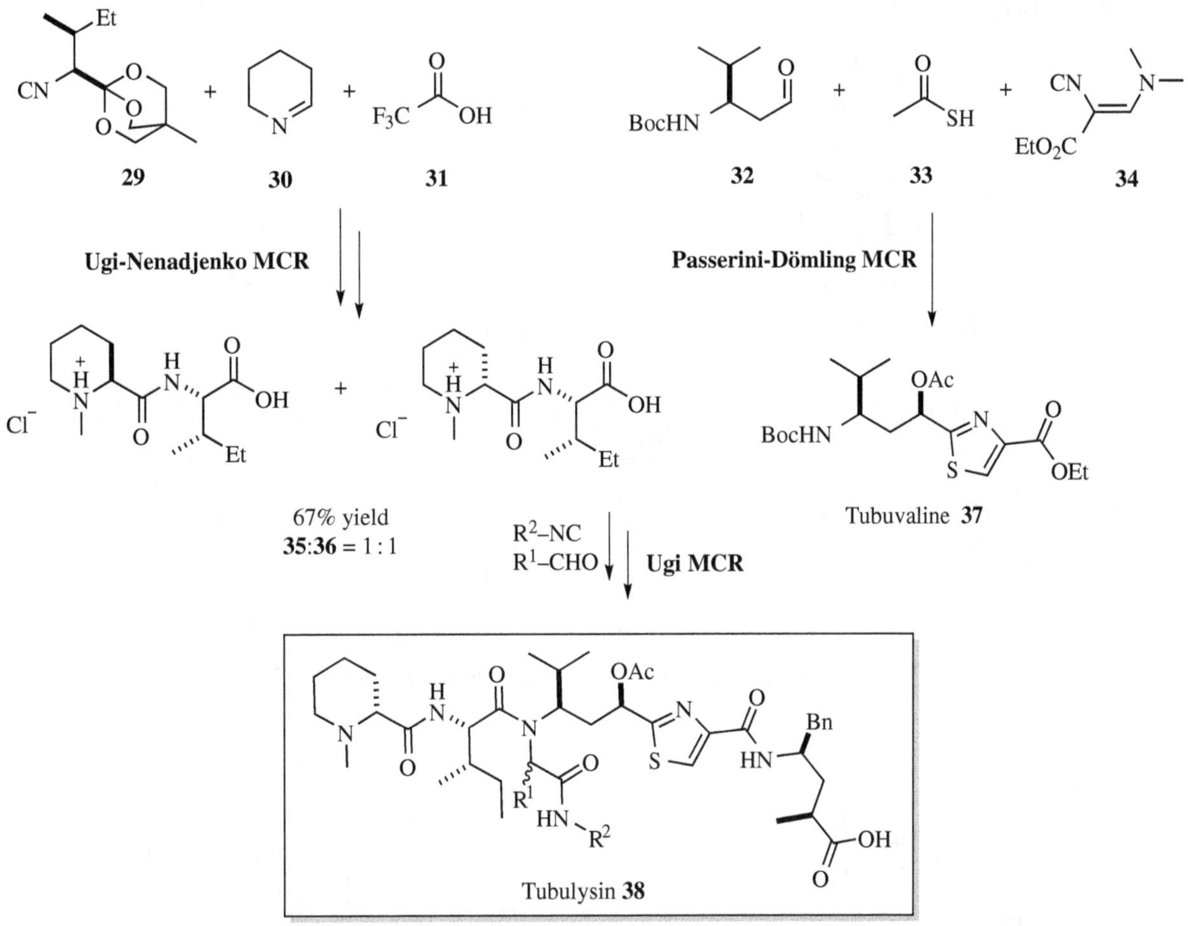

SCHEME 6.3 Combined MCRs in the synthesis of tubulysin analogues **38**.

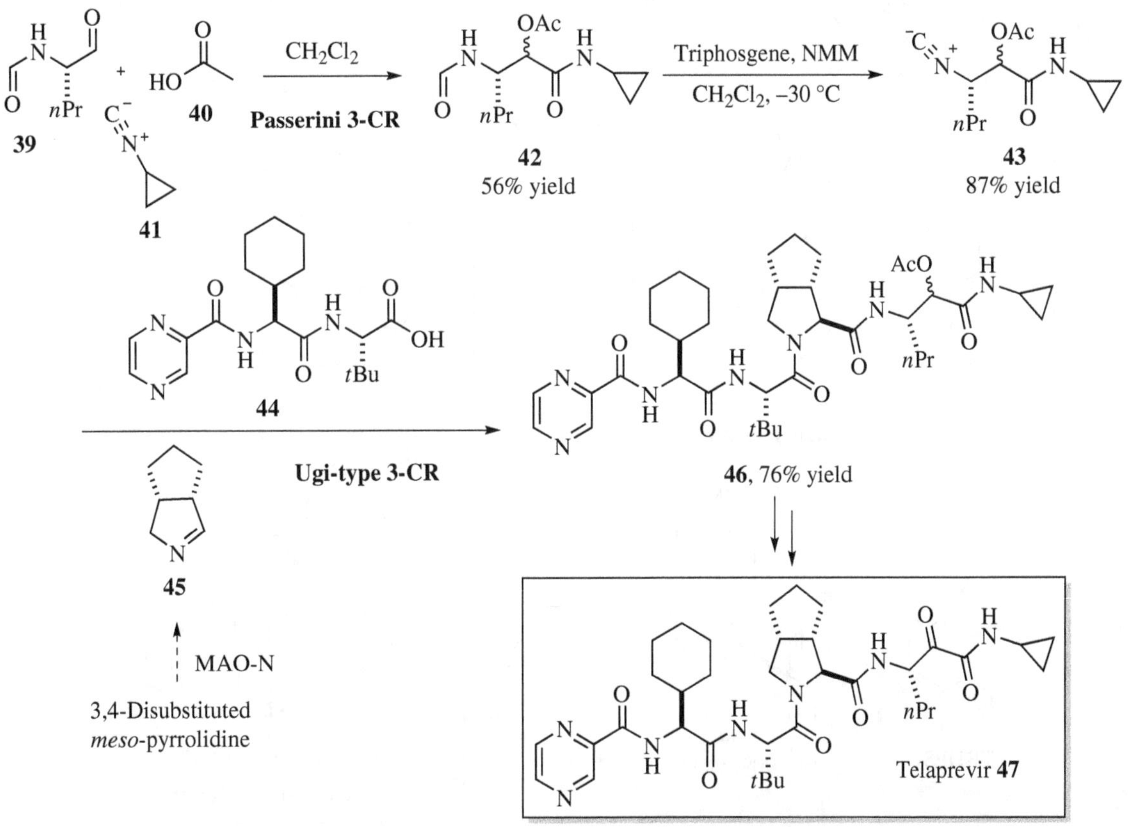

SCHEME 6.4 MCR chemistry combined with biocatalysis in the synthesis of telaprevir **47**.

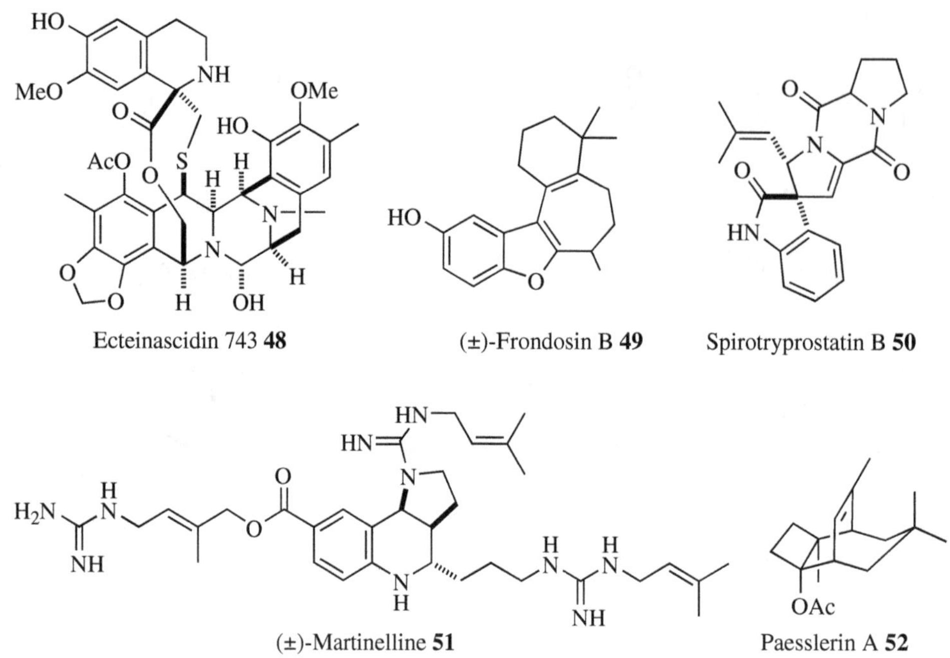

FIGURE 6.3 MCRs as powerful tool for natural products synthesis [22–26].

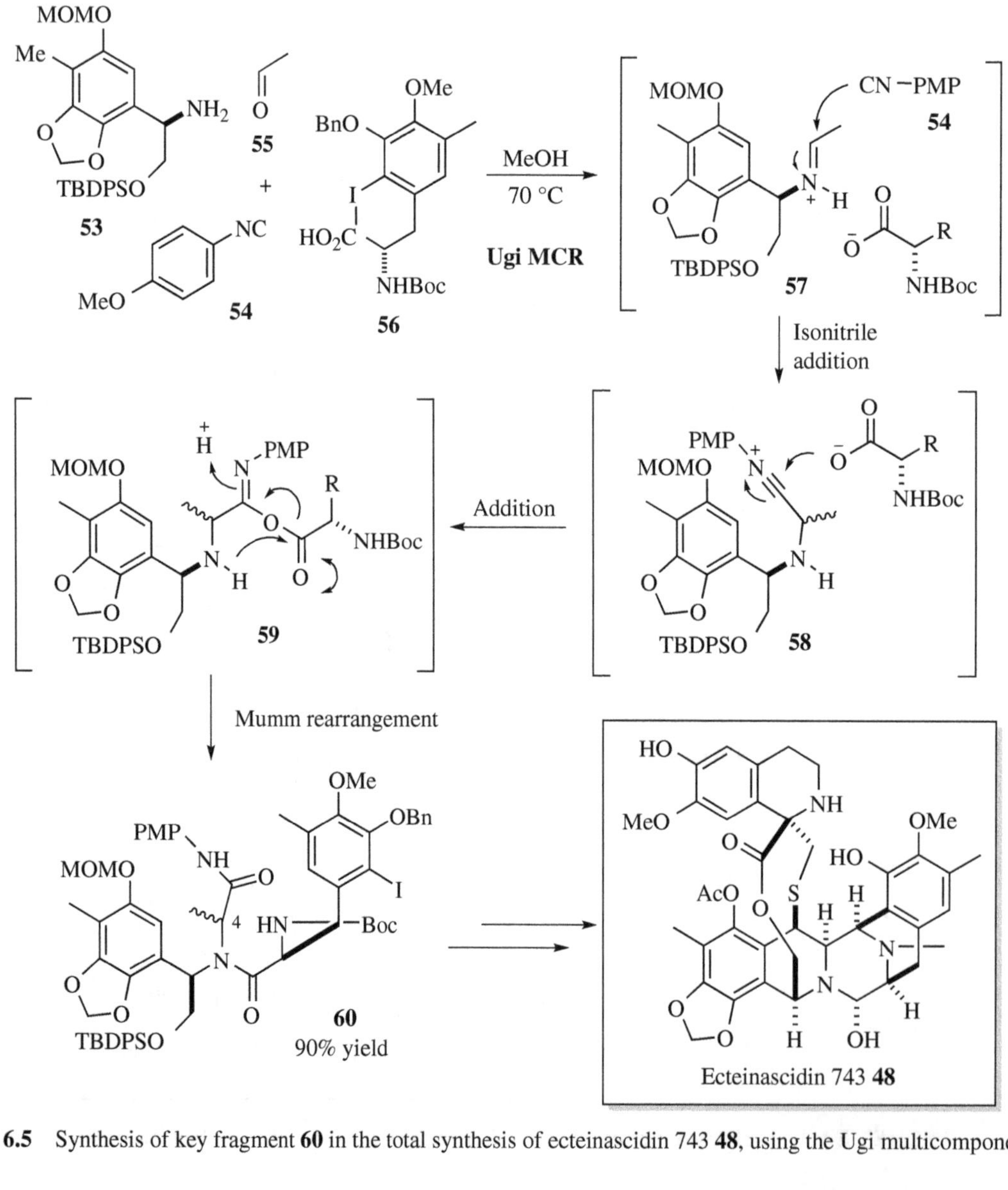

SCHEME 6.5 Synthesis of key fragment **60** in the total synthesis of ecteinascidin 743 **48**, using the Ugi multicomponent reaction.

Acid-catalyzed iminium ion **57** formation is followed by isocyanide **54** addition and capture of the remaining nitrilium ion **58** by the carboxylate of **56** to generate compound **59**. A rapid O→N acyl transfer takes then place in order to obtain dipeptide **60** as a mixture of epimers at C4. Although the control of the stereochemistry of the newly formed stereocenter still remains a problem of the Ugi reaction, it nicely demonstrates the force of the Ugi MCR, as all of the carbon atoms needed are successively assembled in a single step [22].

Mannich-type 4CR are often used for the synthesis of natural products. One example is the synthesis of the alkaloid (±)-roelactamine **65** (Scheme 6.6). The reaction of piperonal **61** with methylamine, Grignard reagent **62**, and acid chloride **63** results in the formation of amide **64**, which then undergoes acid-catalyzed double cyclization to give the latter compound **65** [28].

Williams has reported a MCR in the synthesis of spirotryprostatin B **50**, a spirooxindole-substituted diketopiperazine with a prenylsubstituent at C-18 (Scheme 6.7). In the key step, diphenylmorpholinone **66** reacts with isovaleraldehyde **67** to form azomethine ylide **69**, which is attacked by the oxindole **68** to build the latter **70** in a [3+2]-dipolar cycloaddition [24].

Ihara, Takasu, and coworker described a three-component [4+2]/[2+2] cycloaddition, which was applied to the total synthesis of paesslerin A (**52**). The intermediate **73**, in the synthesis of the desired product **52**, was obtained in 92% yield using $EtAlCl_2$ as catalyst (Scheme 6.8) [26].

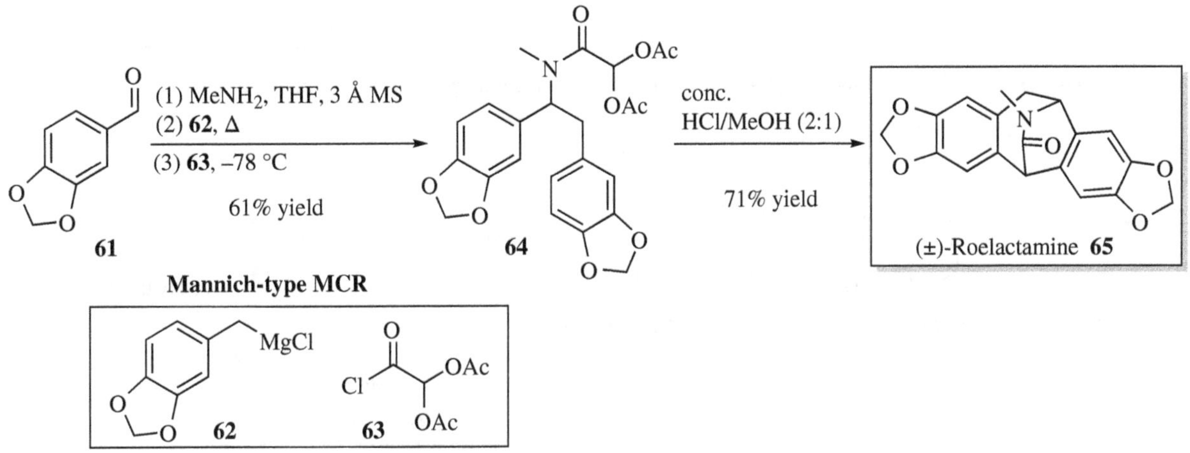

SCHEME 6.6 Mannich-type 4CR in the synthesis of (±)-roelactamine **65**.

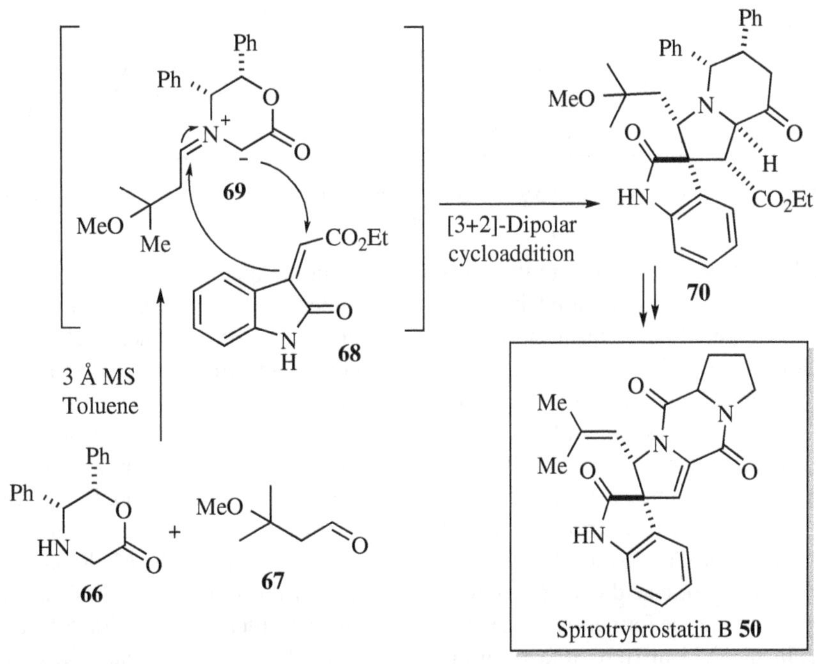

SCHEME 6.7 [3+2]-Dipolar cycloaddition in the synthesis of spirotryprostatin B **50**.

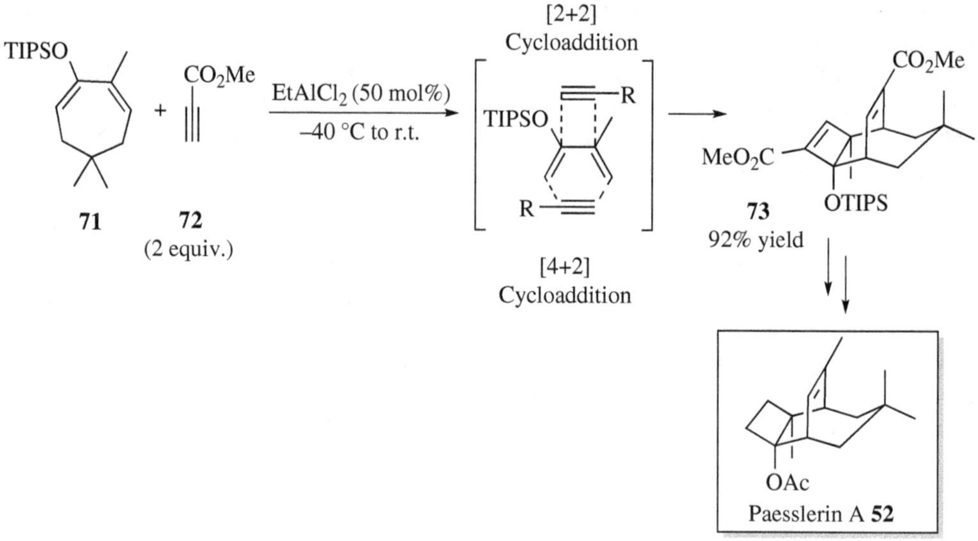

SCHEME 6.8 Three-component [4 + 2]/[2 + 2] cycloaddition cascade in the total synthesis of paesslerin A **52**.

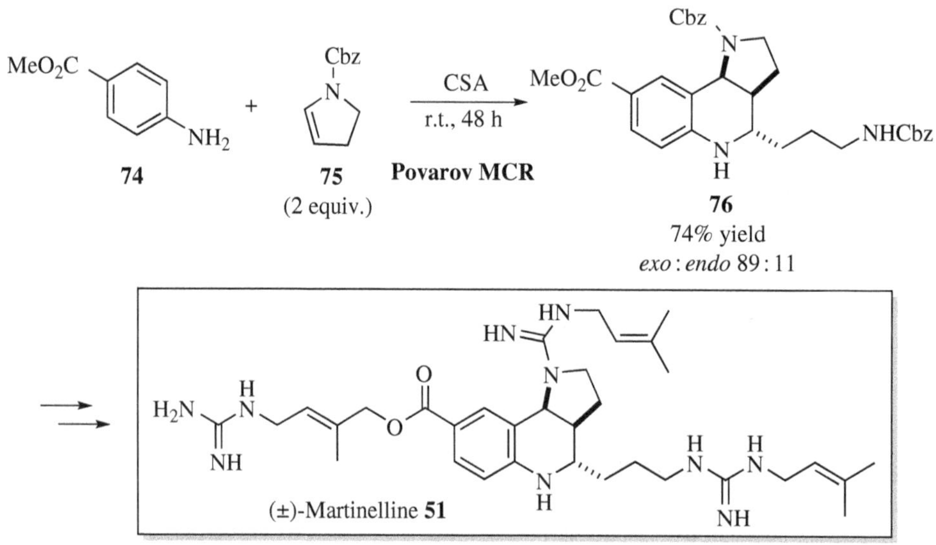

SCHEME 6.9 Povarov multicomponent reaction applied in the total synthesis of (±)-martinelline **51**.

The alkaloid martinelline **51** was isolated from the roots of *Martinella iquitosensis* vine, which was used in traditional medicine by indigenous people in Amazonas lowland rainforest to treat inflammations of the eyes. The latter displays an interesting pharmacological profile as it is the first naturally occurring nonpeptidic bradykinin B2 receptor antagonist (Scheme 6.9) [29]. Batey and Powell were convinced that martinelline **51** could be accessible through a biomimetic Povarov reaction [25]. Therefore, methyl 4-aminobenzoate **74** was treated with Cbz-protected 2-pyrroline **75** and catalytic amounts of the protic acid CSA (camphorsulfonic acid; 5 mol%), resulting in the formation of the core structure of martinelline, in an 89:11 mixture in favor of the desired *exo*-product **76**.

Multicomponent domino reactions are often used in the synthesis of complex natural products. One very nice example was recently demonstrated by the group of Bodwell. They combined six steps in the synthesis of cannabinol **82**, using a Knoevenagel reaction, *trans*-esterification, enamine formation, inverse electron demand Diels–Alder (IEDDA) reaction, 1,2-elimination, and transfer hydrogenation. Both the diene and the dienophile for the IEDDA are generated *in situ* by secondary amine-mediated reactions (Scheme 6.10) [30]. The yields of the MCRs are in most cases higher than those of the stepwise ones. Thus, aldehyde **77** reacted well with the other reagents (**78–80**), affording intermediate **81** (48% yield) on a 1.2 g scale. This product was converted into cannabinol **82** by two different four-step pathways.

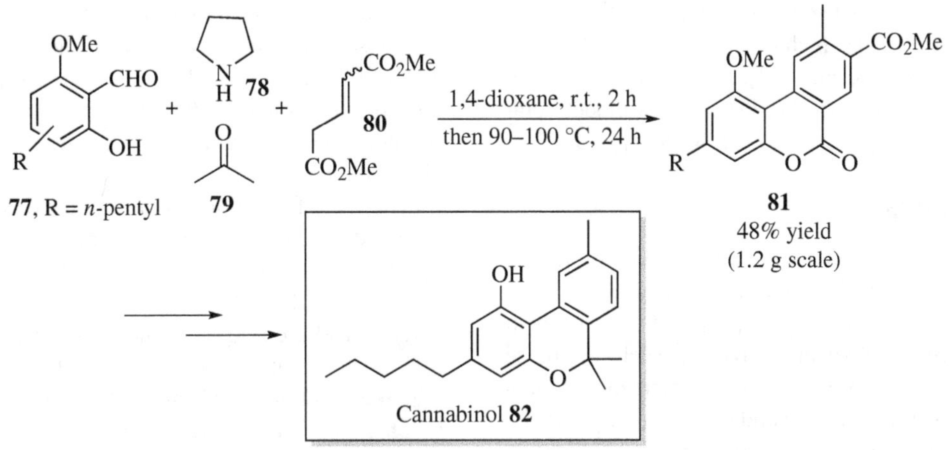

SCHEME 6.10 Synthesis of cannabinol **83** in a multicomponent domino reaction.

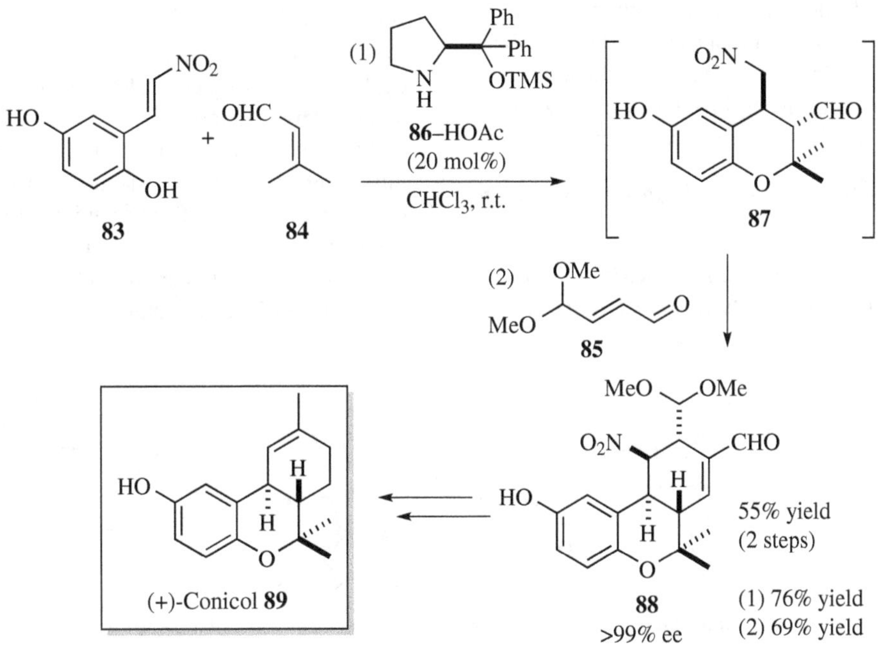

SCHEME 6.11 Organocatalytic oxa-Michael–Michael–Michael-aldol as the key step in the synthesis of (+)-conicol **89**.

Hong and coworkers have demonstrated the three-component organocatalytic oxa-Michael–Michael–Michael–aldol condensation of α,β-unsaturated aldehyde **84** with 2-((*E*)-2-nitrovinyl)benzene-1,4-diol (**83**), as the key step in the total synthesis of (+)-conicol **89** (Scheme 6.11) [31].

The reaction started with the tandem oxa-Michael–Michael reaction of **83** and **84** to give **87** in 76% yield and greater than 99% ee. The following Michael–aldol condensation with 4,4-dimethoxy-but-2-enal **85** works nicely in 69% yield and also excellent enantioselectivity. The obtained hexahydro-6*H*-benzo[*c*]chromene **88** is a highly functionalized intermediate in the total synthesis of (+)-conicol **89**. The two-step reaction could be achieved in one pot from **83** and **84**, without isolation of the intermediate **87**, with a 55% overall isolated yield of **88**.

6.4 HETEROCYCLES AS KEY SUBSTRATES IN MCRs

Heterocycles constitute the core of a large number of natural bioactive compounds and many drugs, being recognized as privileged structural motifs in nature [2]. With the growing interest for the synthesis of compounds with molecular skeletons found in natural products or active compounds, the development of new synthetic methods for the generation of heterocyclic compounds with molecular diversity and structural complexity has become one of the main goals in organic and medicinal chemistry [32]. Besides the well-established synthetic methods for the preparation of heterocycles, efforts have been done in the development of new MCRs leading to highly functionalized heterocycles, taking profit of their interesting advantages such as high flexibility, chemoselectivity, convergency, and atom efficiency [33]. In this chapter, selected key heterocyclic structures will be displayed as model compounds such as indoles, spirocycles, DHPMs, and pyrroles.

6.4.1 Synthesis of Indoles

The indole ring system is among the most important heterocyclic systems present in a large number of natural products, bioactive compounds, and medicines [34], and many of them display remarkable biological activities, such as those depicted in Figure 6.4 [35]. Therefore, interesting works have been developed for over one and a half century with the aim of synthesizing [36] and functionalizing [37] indole structural scaffolds. For this purpose, MCRs have been efficiently used in the construction and functionalization of indoles, and some of these important examples will be described here [38]. Although numerous works have been reported to synthesize indole derivatives, the efficient tool of asymmetric catalysis has been employed only in a few examples, and we want to collect these remarkable cases here.

6.4.1.1 Synthesis of Interesting Indole Structures through Asymmetric Catalysis

Organocatalysis (See Chapter 2) MacMillan's group applied multiple asymmetric catalytic strategies to develop an effective organocatalytic synthetic approach for the highly enantioselective synthesis of chlorinated indole derivatives **101** [39]. The amine **100**-mediated cascade reaction uses an α,β-unsaturated carbonyl compound **98**, indoles **97**, as the nucleophile, and *in situ* generated Cl^+ (from **99**) as the electrophile (Scheme 6.12).

The role of the imidazolidinone catalyst **100** employed is depicted in Scheme 6.13. In the proposed cycle, the nucleophile (Nu) is added over the first formed α,β-unsaturated iminium ion **102**. After a fast hydrolysis, adduct **104** and catalyst **100** are released. The adduct **104** would then enter in the second catalytic cycle to form the active enamine **105**, which could trap the electrophile (E) present in the reaction in a highly diastereoselective addition.

As a remarkable fact, the induction in the enamine addition step (second cycle) is believed to arise from catalyst control that would provide high levels of diastereoselectivity for the overall process. As represented, a single imidazolidinone catalyst could enable both activation cycles, modulating control of the enforced sense of enantio- and diastereoinduction.

Fréchet's group reported a similar cascade reaction but using two immobilized amino catalyst derivatives (Scheme 6.14) [40]. They found that the combination of two catalysts, the salt of imidazolidinone with polystyrene sulfonate **114** and polymer-supported diphenylprolinol methyl ether **115**, efficiently catalyzed a MCR of two consecutive

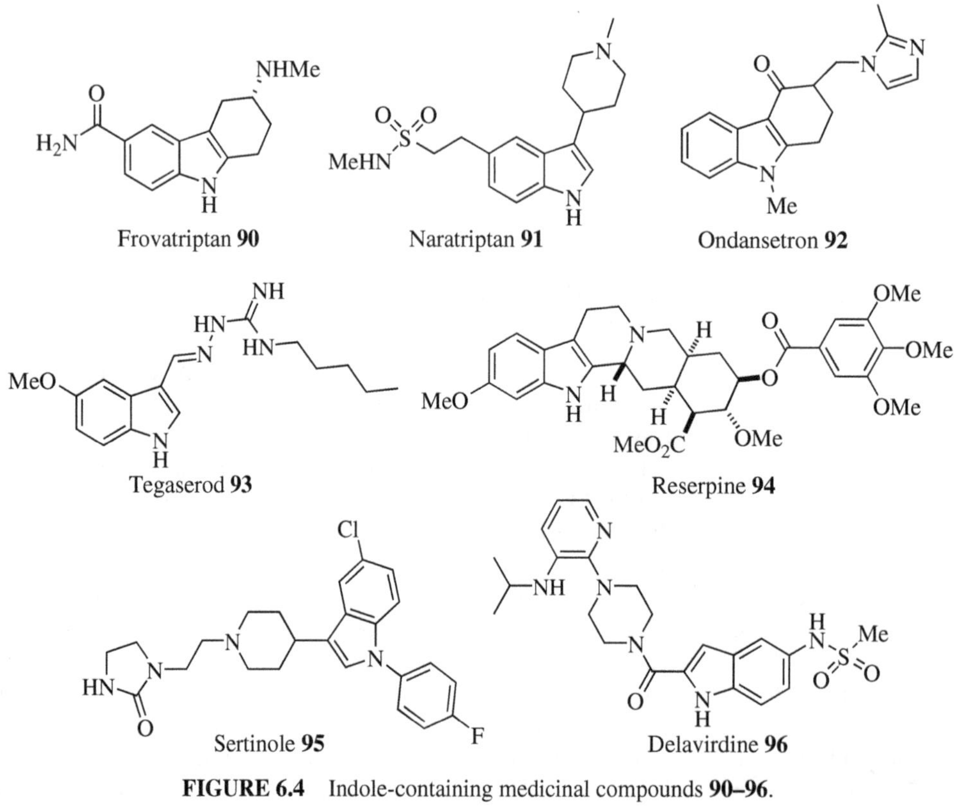

FIGURE 6.4 Indole-containing medicinal compounds **90–96**.

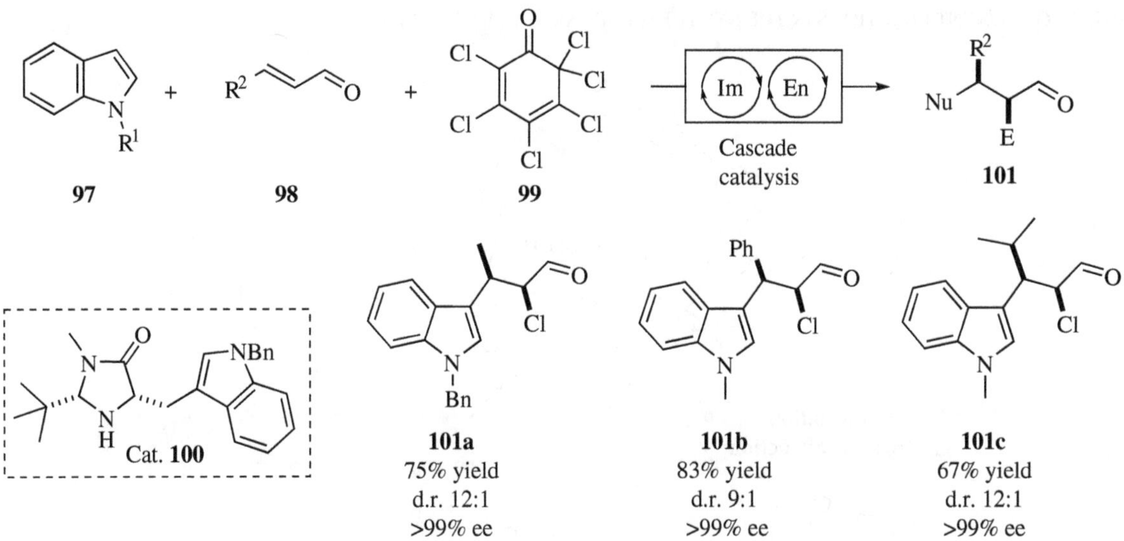

SCHEME 6.12 Cascade reaction through iminium–enamine catalysis.

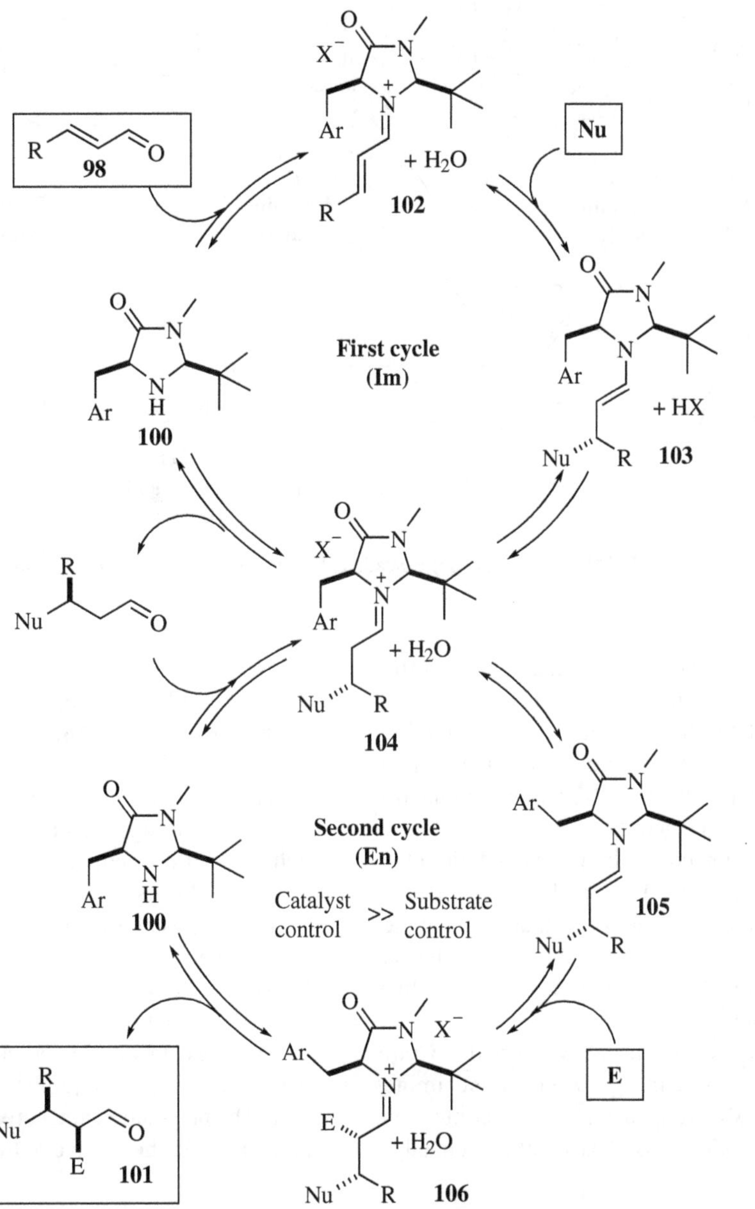

SCHEME 6.13 Catalytic cascade cycle.

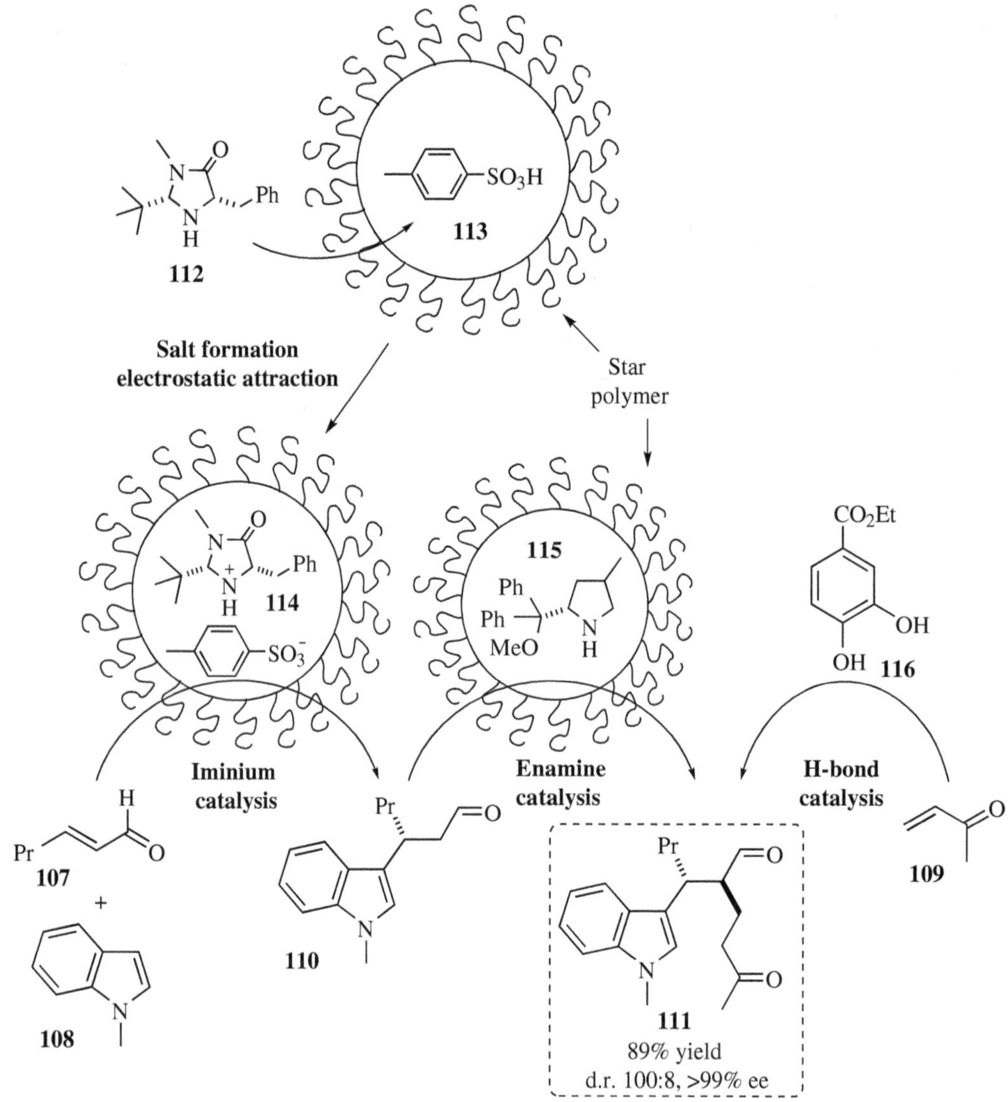

SCHEME 6.14 One-pot cascade catalysis with star polymer catalysts.

Michael addition reactions between *N*-methylindole (**108**) and hexenal (**107**), and the resulting adduct **110** with methyl vinyl ketone (MVK, **109**). The addition of a H bonding catalyst **116** to activate the MVK improved the yield of final indole products **111**, from 33% to 89%, while the enantiomeric excess remained unchanged (>99%).

The force of this reaction comes from the catalytic integrity of the active star polymers **113** and **115**, as they cannot penetrate each other's core. In contrast to that, small molecules and catalysts can freely diffuse to the core of the star polymers **113** to form the desired salt **114**, which will then react as optimal iminium catalyst.

A highly enantioselective organocatalytic Friedel–Crafts aminoalkylation of indoles with *in situ* generated imines was reported by Zhang's group using a multicomponent approach. Chiral phosphoric acid **121** as an efficient catalyst is used to promote the synthesis of fluorinated indole derivatives **122** and **123**, starting from trifluoroacetaldehyde methyl hemiacetal **118**, 3,4,5-trimethoxyaniline **120**, and indoles **117**, in high yields (80–99%) and with excellent enantioselectivities (79 to >99% ee) (Scheme 6.15) [41, 42]. This methodology was further extended to difluoroacetaldehyde methyl hemiacetal **119**, which afforded the final products **123**, also with excellent results.

The proposed transition state **TS-1** to explain the role of the catalyst is depicted in Figure 6.5. The first reaction step between trifluoroacetaldehyde methyl hemiacetal **118** and the aniline **120** provides the corresponding *N,O*-acetal in the presence of catalyst **121** and molecular sieves. The *N,O*-acetal is expected to be in equilibrium with the imine **124**, which would be protonated and activated by the chiral phosphoric acid to shield the *si*-face of the imine. Contemporaneously,

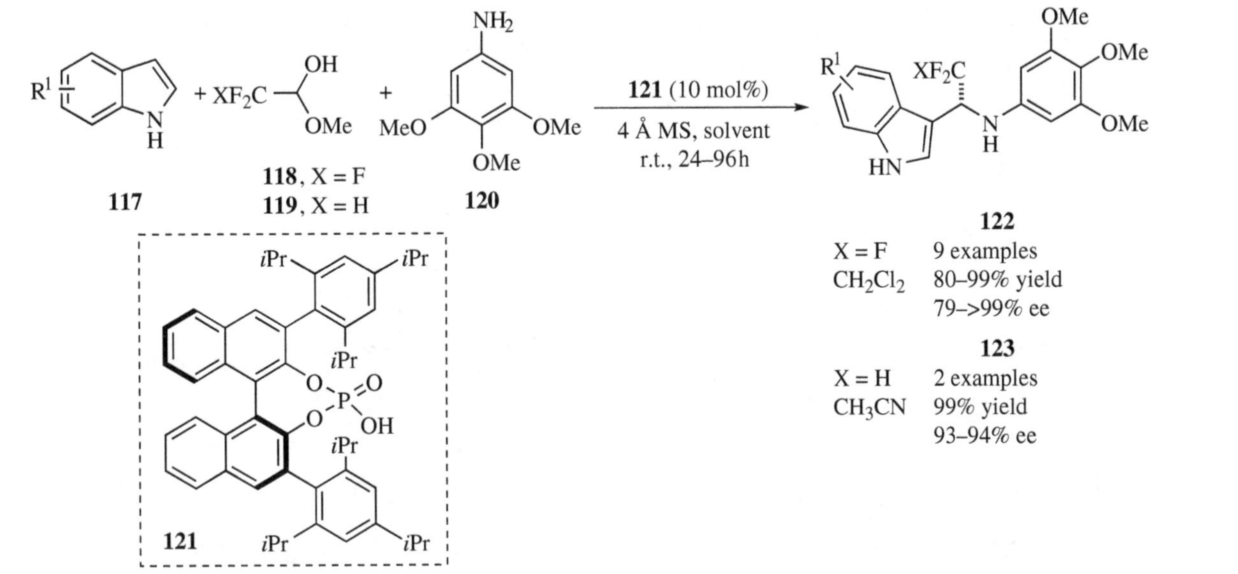

SCHEME 6.15 Phosphoric acid **121**-catalyzed three-component reaction.

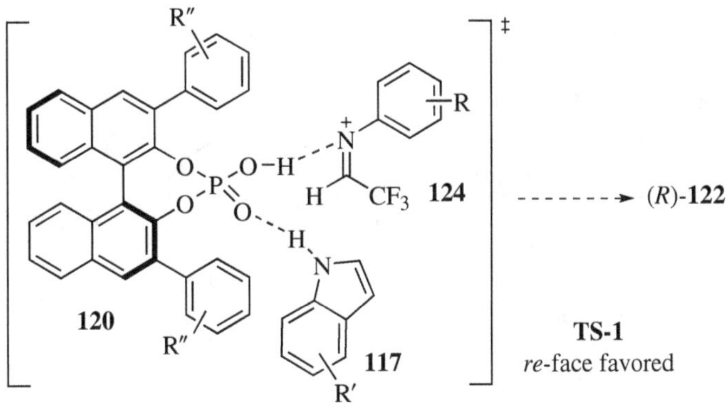

FIGURE 6.5 Activated transition state **TS-1**.

the indole **117** would attack mainly from the *re*-face of the imine to give the corresponding final products **122** with an *R* configuration.

An interesting example of aminocatalysis where a chiral primary amine activates α-substituted α,β-unsaturated aldehydes in a cascade sequence for the addition of indoles and azocarboxylate was reported by Melchiorre and coworkers (Scheme 6.16) [43]. Primary amine **128** directly derived from natural cinchona alkaloid was the catalyst of choice, which promoted this cascade reaction providing final adducts **129**, as valuable precursors for α-amino acids having two adjacent stereogenic centers, one of which is quaternary, with very high enantioselectivities (83–99%). The yields are moderate to good (31–80%), and different substituents on the indole core (Cl, OMe, Me) are well tolerated.

Zhao and coworkers have reported an enantioselective three-component Michael addition/Pictet–Spengler reaction sequence to synthesize highly substituted indoloquinolizidines **135** (Scheme 6.17), in moderate to excellent yields (51–95%) and good to excellent enantioselectivities (80–96%) [44, 45].

The stereochemical outcome of this reaction is explained through the formation of transition state **TS-2**, where less steric hindrance between the equatorial β-proton and the indole moiety leads to the exclusive formation of product **135** (Scheme 6.18).

Recently, Zhu, Sun, and coworkers nicely showed the first catalytic asymmetric three-component aza-Diels–Alder reaction to create highly substituted indoles **140** in moderate to excellent yields (45–93%) and the same for the enantioselectivity values (50–92% ee) (Scheme 6.19) [46].

The rigid oxetane next to the aldehyde is essential for the excellent diastereoselectivities, as well as the enantioselectivities. It is thought that it serves as a hydrogen bond acceptor. Randomly, three of the synthesized indoles **140k-m**

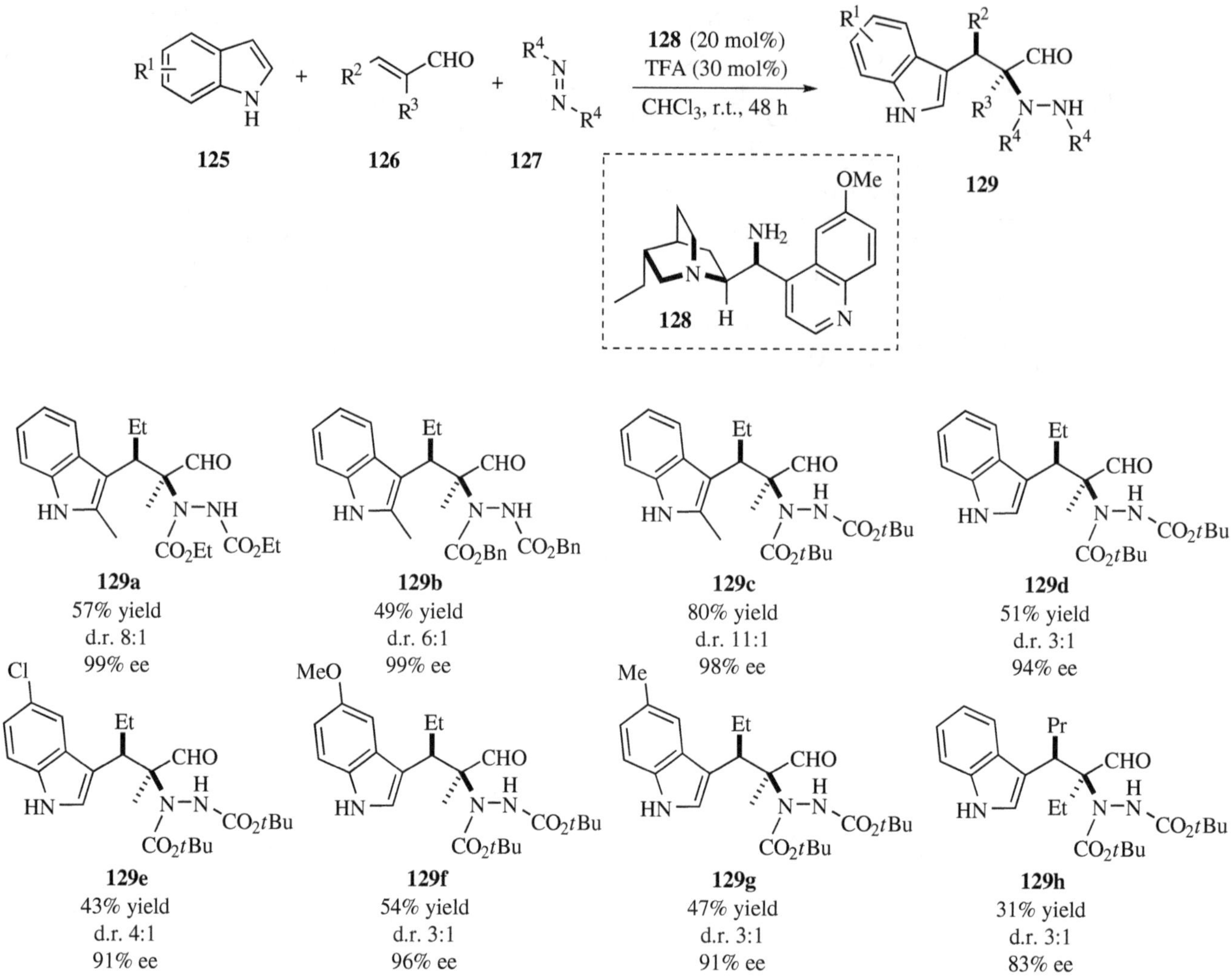

SCHEME 6.16 Primary amine **128**-catalyzed multicomponent reaction, to give chiral indole derivatives **129** [43].

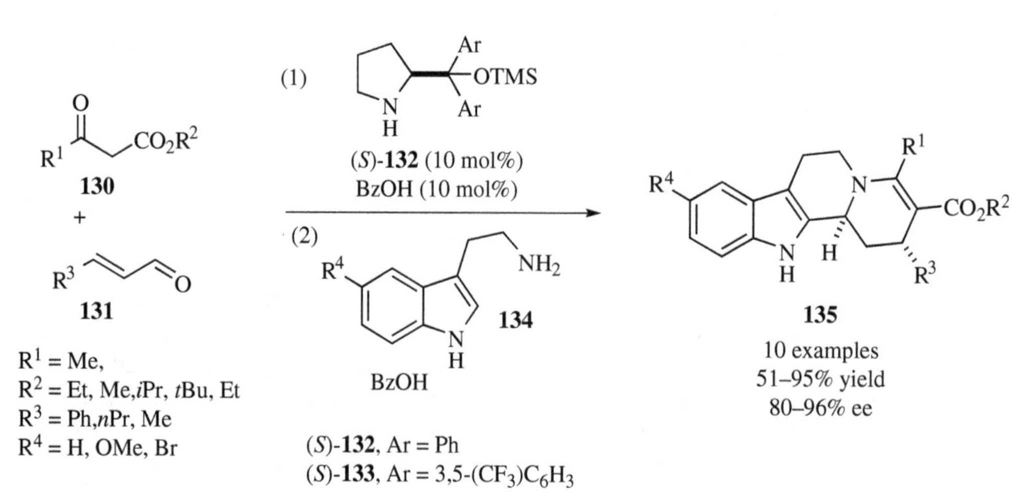

SCHEME 6.17 Enantioselective three-component Michael addition/Pictet–Spengler reaction sequence to synthesize highly substituted indoloquinolizidines **135**.

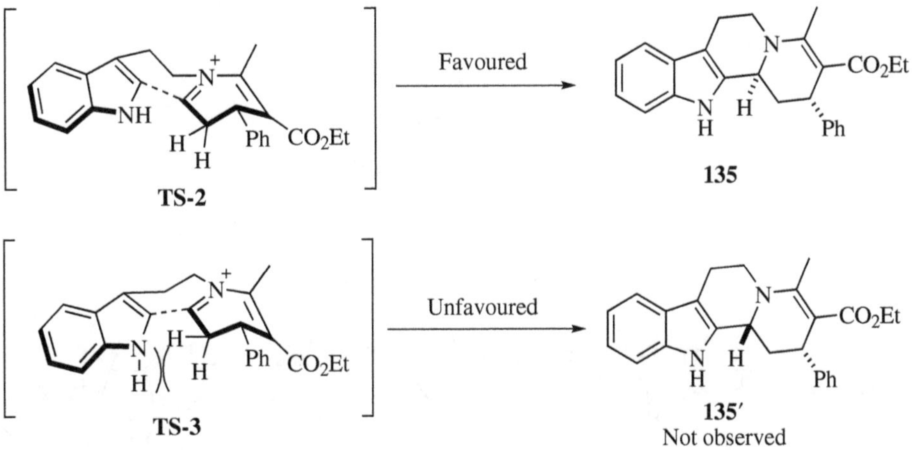

SCHEME 6.18 Transition state (**TS-2** and **TS-3**) of the Pictet–Spengler cyclization.

were tested in two cell lines (A549, human lung carcinoma; HeLa, human cervical carcinoma) because they show a similar molecular structure as some biological interesting compounds such as (+)-melonine **141** (Fig. 6.6).

And indeed, the three compounds showed cytotoxicity in the low micromolar range. This example nicely demonstrates the force of asymmetric MCRs in the diversity-oriented synthesis and drug discovery.

Metal Catalysis Arai and Yokoyama have developed a chiral imidazoline–aminophenol–CuOTf complex-catalyzed three-component tandem reaction to create acyclic products **146** with three contiguous stereocenters (Scheme 6.20) [47]. Product **146** is mainly formed along with **147**, as the minor product. The yield can be enhanced by the addition of HFIP, as well as the stereoselectivity of the major product. Interesting fully substituted tetrahydro-β-carbolines (THBCs) **148** can be obtained after reduction of the nitro group of **146** and subsequent Pictet–Spengler cyclization [48].

6.4.2 Synthesis of Fused Polyheterocycles

Fused polyheterocycles are also often present in active natural products like those depicted in Figure 6.7. In the previous section, few examples of efficient synthesis of this kind of molecules, containing indole in their structure, have been shown. Some further instances of spiroindoline preparation will be compiled in the section of dedicated to spiro-type polyheterocyclic compounds (Section 6.4.3).

Flynn and coworkers have used a palladium-catalyzed MCR to construct the benzo[*b*]furan core of frondosin B **49** (Scheme 6.21) [23], a strategy that they have already established for the synthesis of some analogues of the anticancer compound combretastatin A-4 [49]. Bromide **157** and alkyne **158** react with MeMgBr to afford enyne **161** *via* a Kumada coupling. After oxidative addition of the palladium catalyst to compound **162**, and alkyne coordination, the π-complex **163** was formed, which released magnesium bromide to build the benzofuran **164**. Reductive elimination of the catalyst gives then the desired product **165** in 48% yield. Frondosin B **49** can be obtained in only few more steps.

Chromene derivatives are interesting structural motifs as they are present in many natural products. Despite the already existing strategies for the synthesis of chromenes, there is need for improvement to shorten reaction times and reduce the loading of the catalyst and the number of steps. In this context, Perumal and coworkers have reported the efficient regioselective and chemoselective one-pot synthesis of highly functionalized chromene-5-ones **171** and **172**. (Scheme 6.22) [50]. To achieve good results, they used NMSM ((*E*)-*N*-methyl-1-(methylthio)-2-nitro-ethenamine) **170** because it contains an electrophilic, as well as a nucleophilic site. Furthermore, there are four active sites in the molecule with three functional groups on the ethane core. The nitroethylene substructure is a good Michael acceptor, the thiomethyl group is a good leaving group, and the methylamino group is a good Michael donor. Perumal and coworkers have used a three-component reaction of NMSM **170**, aromatic aldehydes **169**, and dimedone **167** or 4-hydroxycumarin **168**. The desired chromene-5-ones were obtained in high yield and short reaction times (2–5 h). No electronic effects were observed regarding different residues at the aromatic aldehyde. The products obtained are valuable compounds for further functionalizations because they contain free methylamine and nitro substituents in 2- and 3-positions, respectively.

Nuevamine **179** is an alkaloid that was isolated from *Berberis darwinii*, and it is the first tetrahydroisoindolo[1,2-*a*]isoquinoline amide found in natural sources [51]. An Ugi

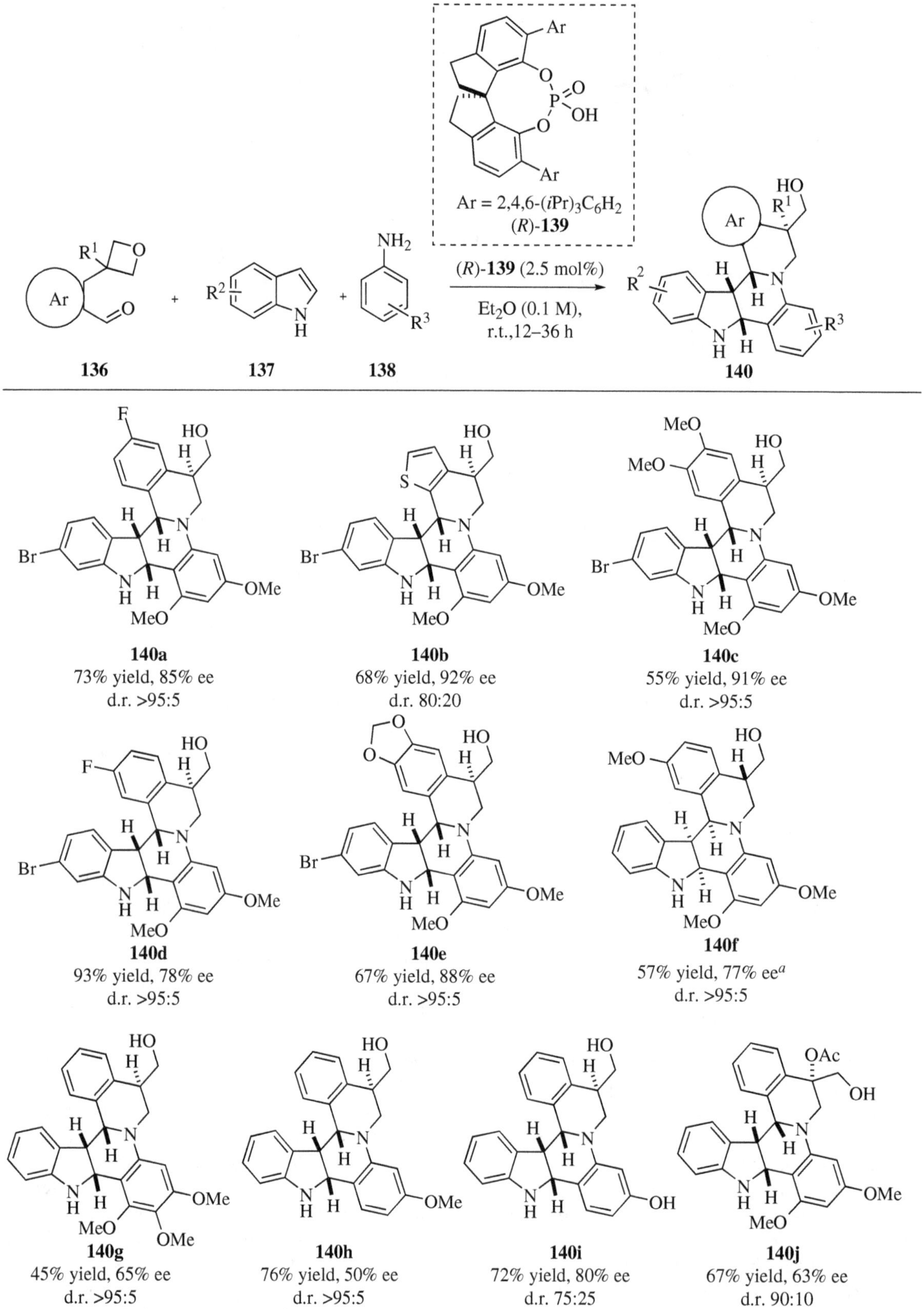

SCHEME 6.19 Three-component aza-Diels–Alder reaction to create indoles derivatives **140**. [a]Catalyst (*S*)-**139** was used.

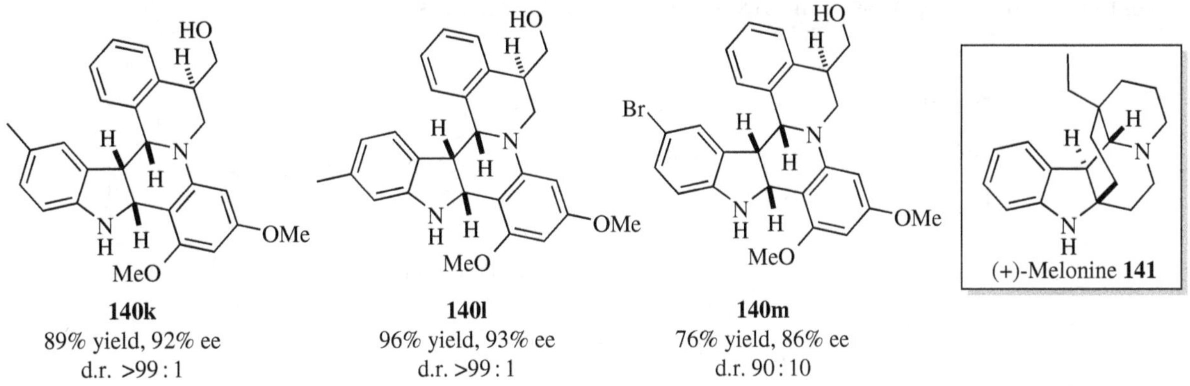

FIGURE 6.6 Indoles **140k-m**, which were tested against different cell lines.

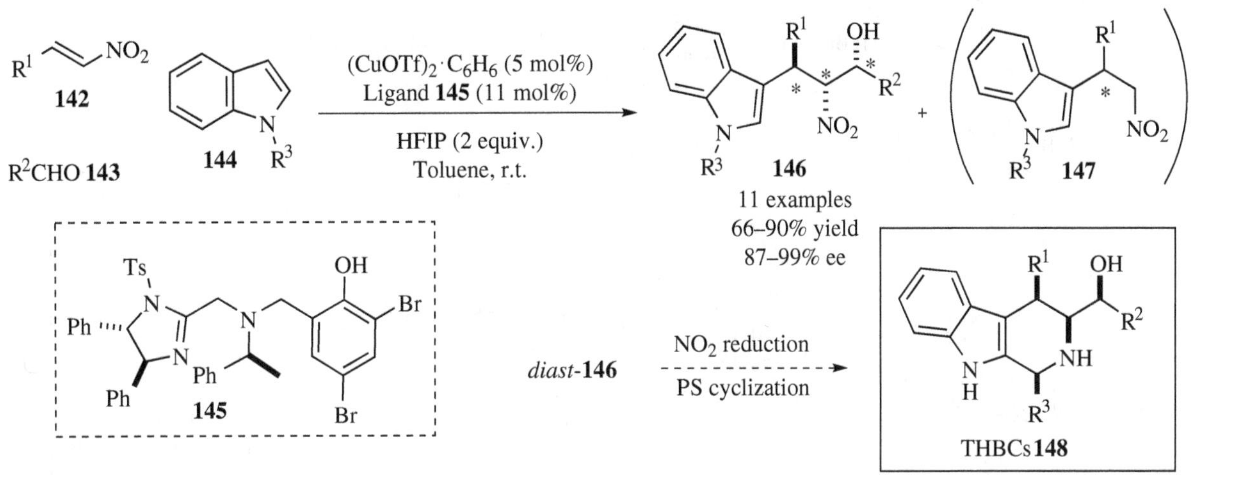

SCHEME 6.20 Three-component tandem reaction to synthesize acyclic products—indole derivatives **146**—with three contiguous stereocenters.

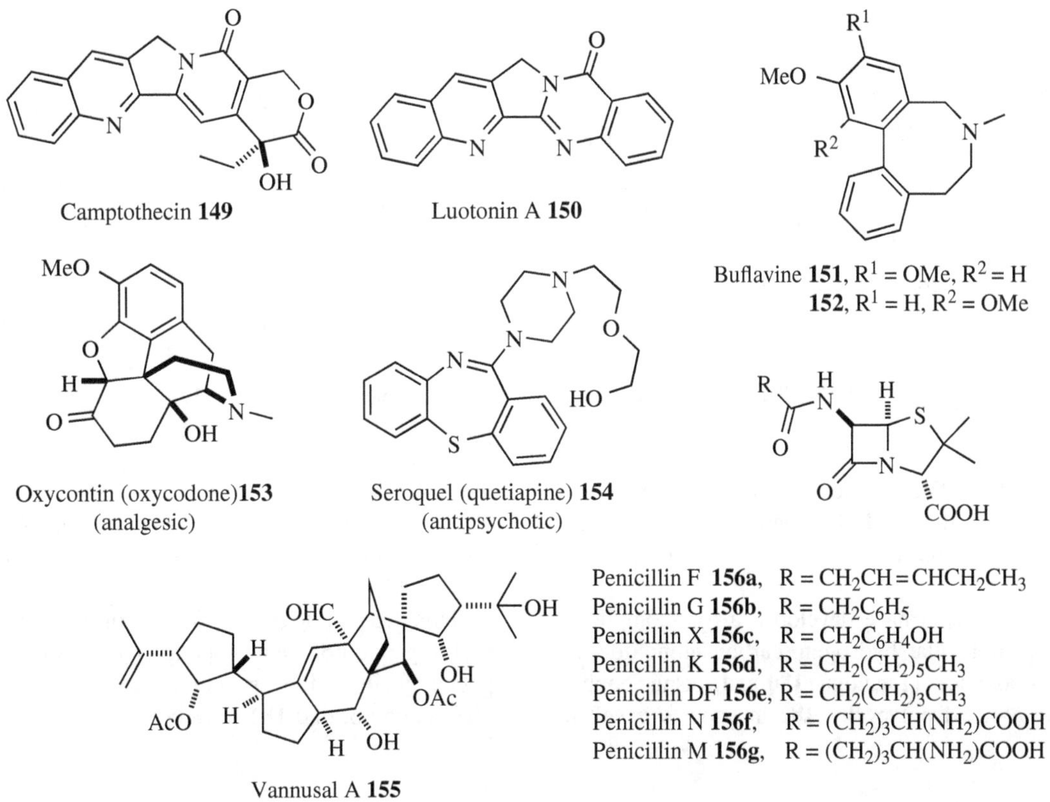

FIGURE 6.7 Natural products **149–156**, containing fused heterocycles.

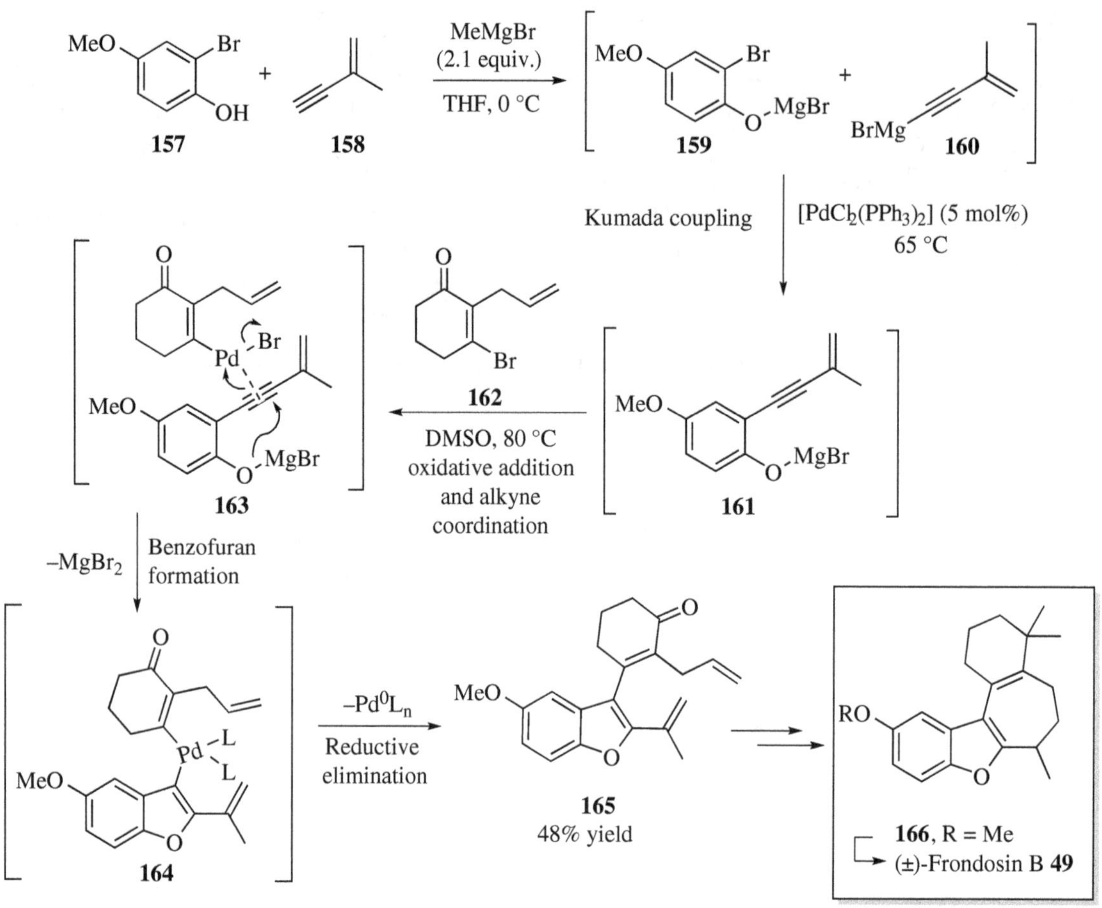

SCHEME 6.21 Pd-catalyzed multicomponent reaction in the synthesis of (±)-frondosin B **49** [23].

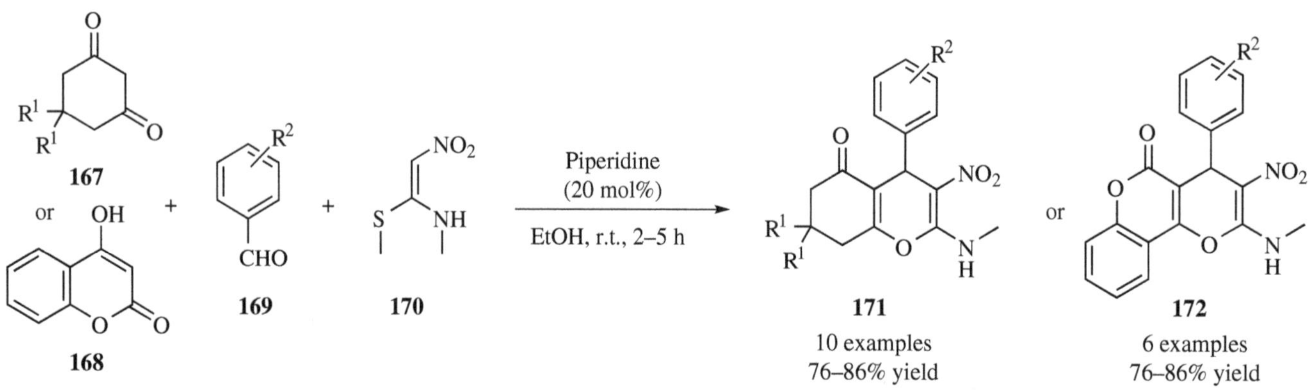

SCHEME 6.22 Efficient one-pot synthesis of chromene-5-ones **171** or **172**.

three-component/aza-Diels–Alder reaction discloses access to nuevamine aza-analogues **178** (Scheme 6.23) [52]. Those were synthesized according to the previously established procedure [53].

Mann and coworkers have developed a four-component reaction using molecular hydrogen, carbon monoxide, an unsaturated carboxylic acid analogue **180**, and a bisnucleophile **181** to create *N*-fused polycycles **182** in good to excellent yields (54–98%) (Scheme 6.24) [54]. Not only amino alcohols but also amino acids could be used as the binucleophile in this sequence leading to the desired products in good yields.

This methodology was further implemented to a high-yielding one-pot Pictet–Spengler reaction using 3-MeO-phenethylamine **183**, tryptamine **184**, or *O*-Me-tryptophan **185**, as compound **181** in Scheme 6.24 (Scheme 6.25).

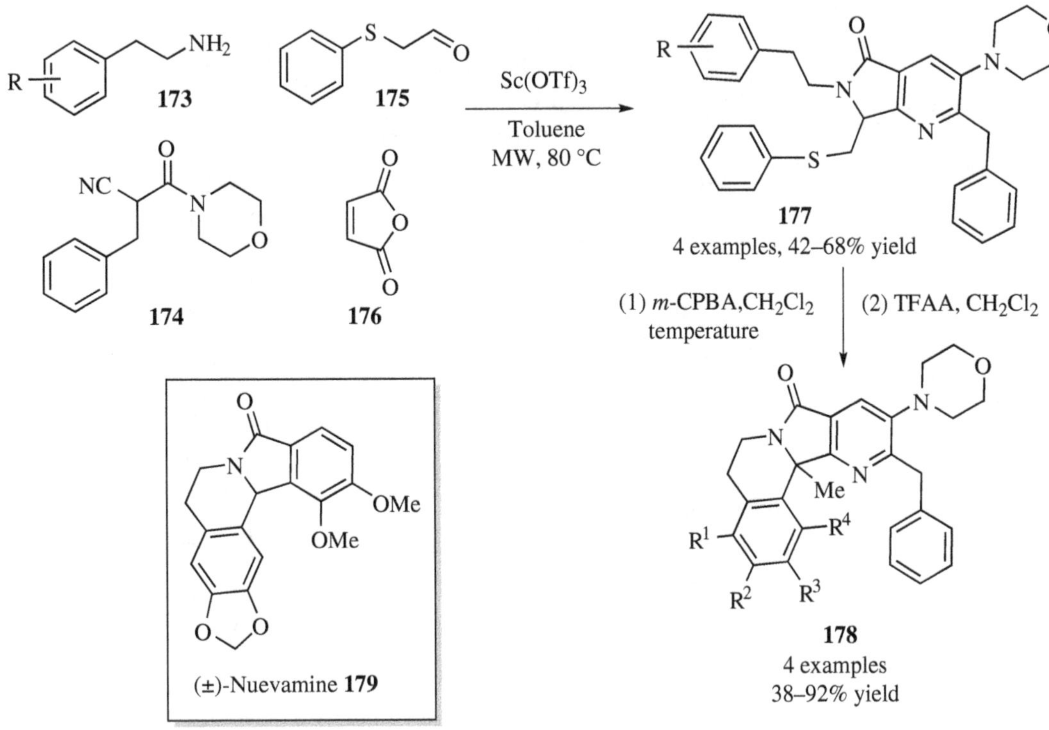

SCHEME 6.23 Synthesis of nuevamine aza-analogues **178**.

An asymmetric total synthesis of the alkaloid dendrobatid 251 F **192** using a Noyori-type three-component reaction was described by Aubé and coworkers (Scheme 6.26) [55]. 1,4-Cuprate addition to enone **189** is achieved from the *exo*-face, and subsequent aldol condensation results in the formation of only one diastereoisomer **191** in 65% yield. Dendrobatid 251 F **192** is achieved in only six further steps.

Shibasaki and coworkers have used the Noyori-type three-component reaction in the synthesis of the fused polyheterocycle garsubellin A **195** (Scheme 6.27) [56]. The chiral cyclohexenone **193** reacts with methylmagnesium bromide, followed by trapping of the enolate with isopropanaldehyde. The key intermediate **194** is isolated in 61% yield and diastereomeric ratio of 4:1.

Feng and coworkers have reported an asymmetric three-component inverse electron demand aza-Diels–Alder (IEDDA) reaction catalyzed by *N,N'*-dioxide-Sc(OTf)$_3$ to give access to ring-fused tetrahydroquinolines [57]. A variety of aldehydes was tested leading to the desired products **200** in good to high yields (up to 99%) and excellent stereoselectivities (up to *cis:trans* >99:1 and 99% ee) (Scheme 6.28). Three stereogenic centers are formed during this one-pot process, and they can also do this process in large scale although with a slight decrease of the yield (74%) and enantioselectivity (93% ee).

A highly enantioselective Povarov reaction using enethioureas **203**, catalyzed by chiral phosphoric acid **121** to get hexahydropyrroloquinolines **204**, was described by Zhu and coworkers in 2012 (Scheme 6.29) [58]. The desired *endo*-products **204** were obtained in good to excellent yields (57–93%) and high diastereo- and enantioselectivities.

The stereochemical outcome of this reaction can be explained through mechanistic rationale shown in Figure 6.8.

The chiral phosphoric acid catalyst **121** forms hydrogen bonds with enethiourea **203** and the imine, formed *in situ* from amine **201** and aldehyde **202**. The enethiourea **203** attacks then that imine in a pseudointramolecular manner from the *si*-face to form iminium ion depicted in **TS-5**. After intramolecular aza-Friedel–Crafts reaction, hexahydropyrroloquinolines **204** were obtained.

Li and Tu have described an innovative three-component domino reaction for the synthesis of multifunctionalized tetracyclic indeno[1,2-*b*]indole derivatives (Scheme 6.30) [59]. Ninhydrin **205** was used to test the allylic esterification, but the desired product **208** was not observed during this process. Indeed, a novel polysubstituted tetracyclic indeno[1,2-*b*]indole derivative **207** was obtained. Li and Tu tested then the best conditions and found out that the reaction proceeds nicely in acetic anhydride under microwave conditions in 20–32 min with 63–87% yield (Scheme 6.30a). Electron-withdrawing substituents like fluoro, chloro, and bromo in the *para*-position of the benzene ring as well as electron-donating substituents (Me, OMe) are well tolerated in this reaction. Bulky residues do not cause any problems.

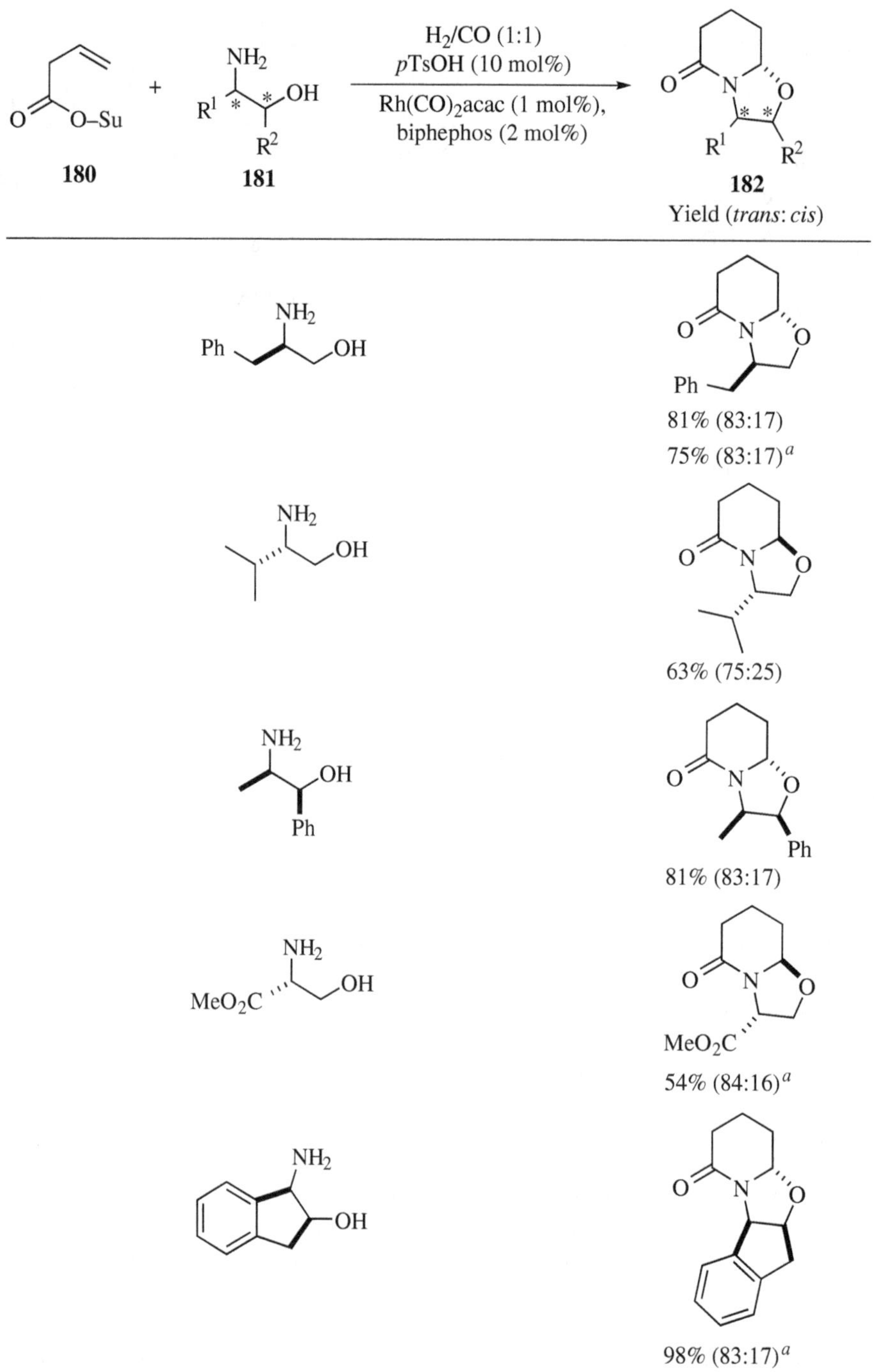

SCHEME 6.24 Four-component reaction to create *N*-fused polycycles **182**. [a]PPTS (5 mol%) and MW were used.

The allylic amination was then tested, and it was thought that it might be easy with aromatic amines because of their stronger nucleophilicity (Scheme 6.30b), but the desired product **210** could not be isolated. The unexpected formation of compounds **209** was observed in 65–86% yield. The best results were obtained in acetic acid. Again, electron-withdrawing as well as electron-donating substituents are well tolerated. Water is a by-product during the synthesis and makes the work-up very convenient. The reactions are finished in 15–32 min and showed excellent stereo- and regioselectivitiy as only the *cis*-isomer is formed. The mechanism of this original strategy is depicted in Scheme 6.31. The initial nucleophilic substitution (**205** → **211**) is followed by intramolecular cyclization (**211** → **212**). Water is then released (**212** → **213**), and isomerization of the carbocation (**213** → **214**) and methyl migration (**214** → **215**) give access to a new carbocation **215**, which leads, after deprotonation

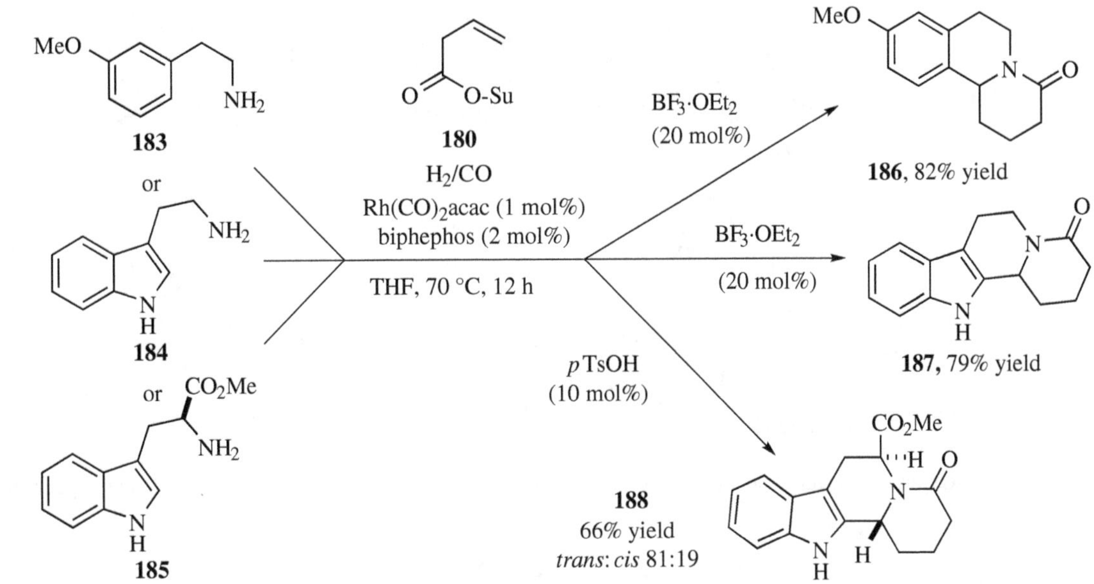

SCHEME 6.25 Multicomponent Pictet–Spengler cyclizations to obtain fused polycycles **186–188**.

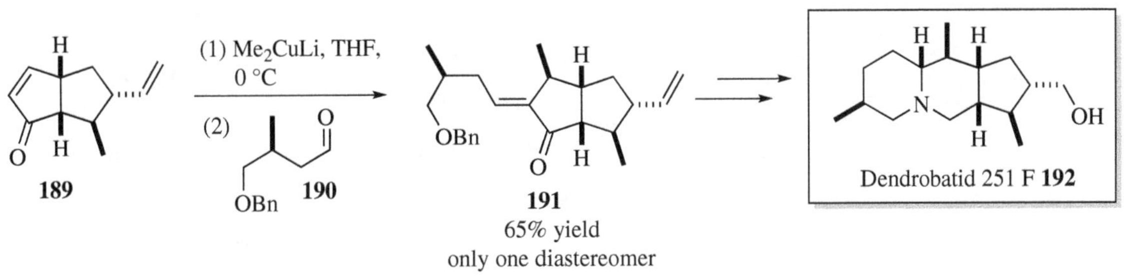

SCHEME 6.26 Noyori-type three-component reaction in the total synthesis of dendrobatid 251 F **192**.

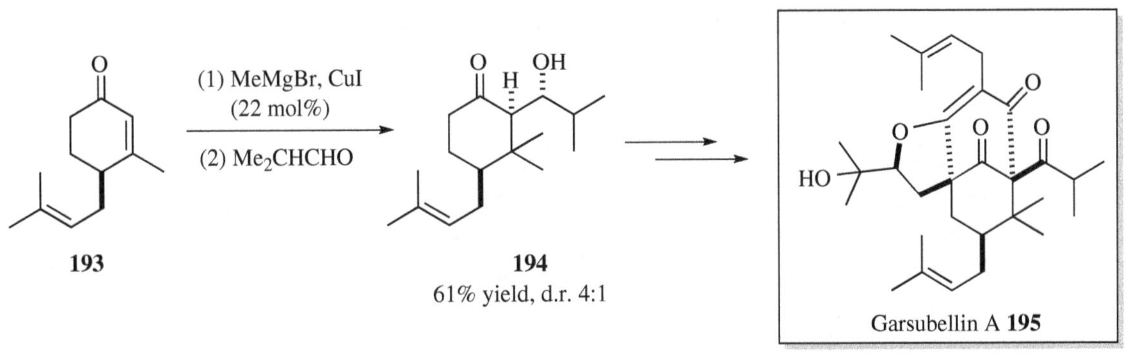

SCHEME 6.27 Noyori-type three-component reaction in the total synthesis of garsubellin A **195**.

(**215**→**216**), final esterification, and aromatization (**215**→**216**), to compound **207**. The easiness of the methyl migration is explained through the stability of the resulting carbocation **215**. In analogy to this, compound **209** is formed during the protonation and release of water (**217**→**218**) to generate a carbocation **218**, which undergoes S_N1 type reaction with aromatic amines. The stereochemistry is explained through the intramolecular hydrogen bonds (**217**→**218**).

6.4.3 Synthesis of Spiro-Type Polyheterocyclic Compounds

Spirocycles have received considerable interest in the last decade as a privileged structural motif. The conformational restriction associated with the structural rigidity and the chirality of their quaternary center influences considerably their biological activities (antimicrobial, antitumor, antibiotic)

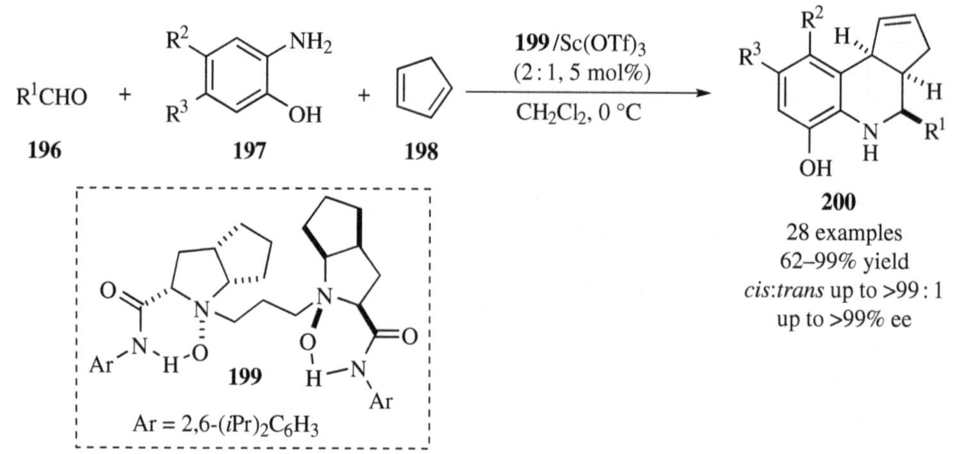

SCHEME 6.28 Asymmetric inverse electron demand aza-Diels–Alder (IEDDA) reaction to yield fused heterocycles **200**.

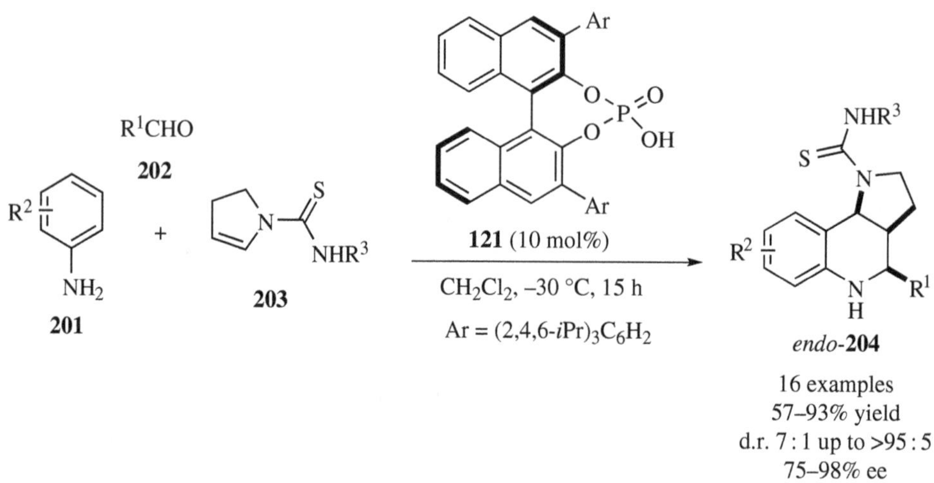

SCHEME 6.29 Phosphoric acid-catalyzed Povarov reaction.

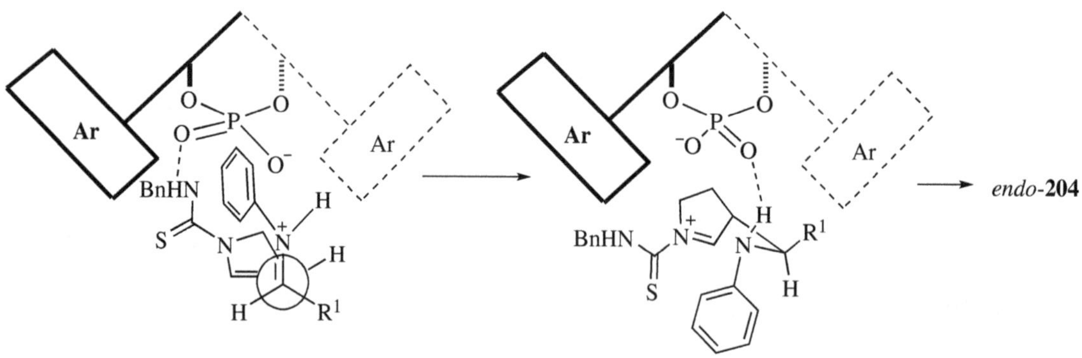

TS-4 **TS-5**

FIGURE 6.8 Rationale for the *endo*-selectivity of reaction depicted in Scheme 6.29.

and inhibitors of human NK-1 receptor [60]. Between them, spirooxindoles have attracted the attention of several research groups for their challenging structural complexity and because they exhibit interesting biological activities suitable for drugs candidates [61]. The spirooxindole structural core represents an important unit present in many synthetic pharmaceuticals and in a number of bioactive natural products such as those depicted in Figure 6.9 [62].

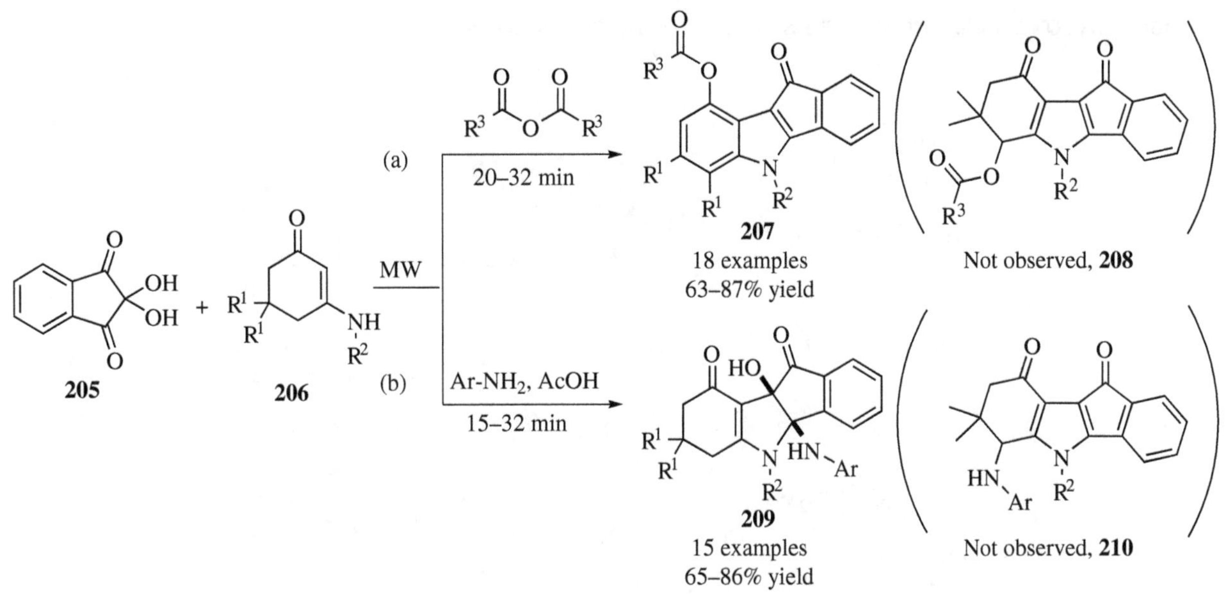

SCHEME 6.30 Synthesis of multifunctionalized tetracyclic indeno[1,2-*b*]indole derivatives **207** and **209**.

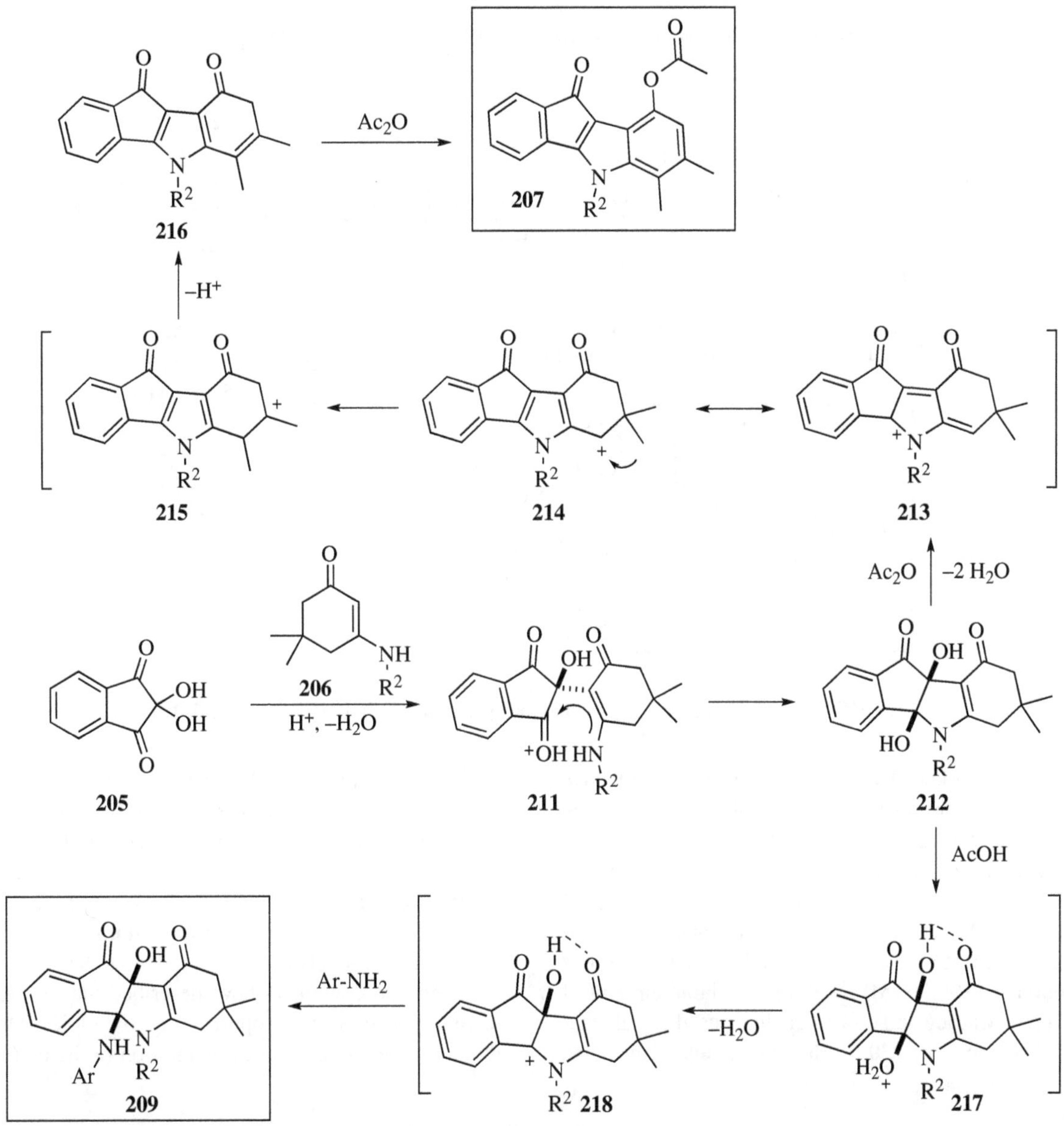

SCHEME 6.31 Proposed mechanism for the synthesis of the multifunctionalized tetracyclic indeno[1,2-*b*]indole derivatives **207** and **209**.

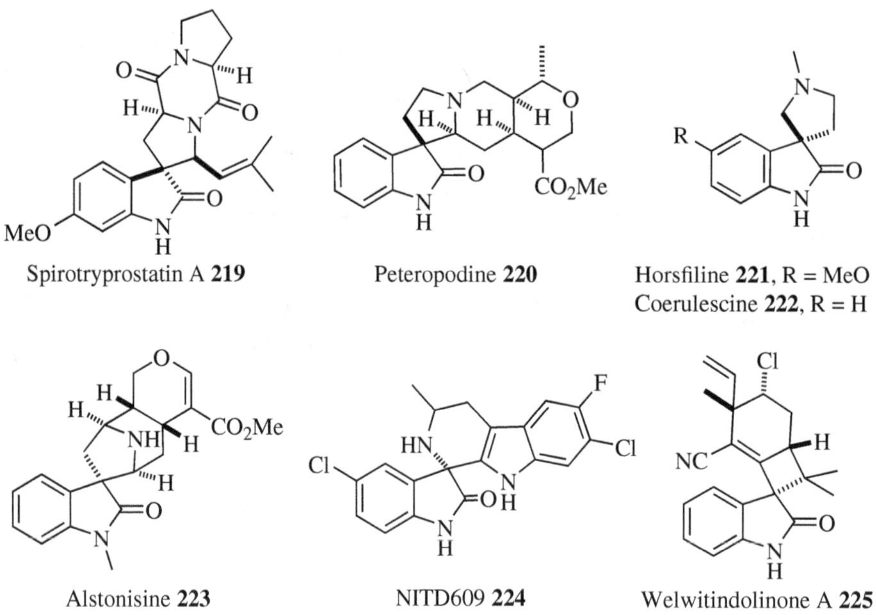

FIGURE 6.9 Representative spirocycles **219–225** with biological properties.

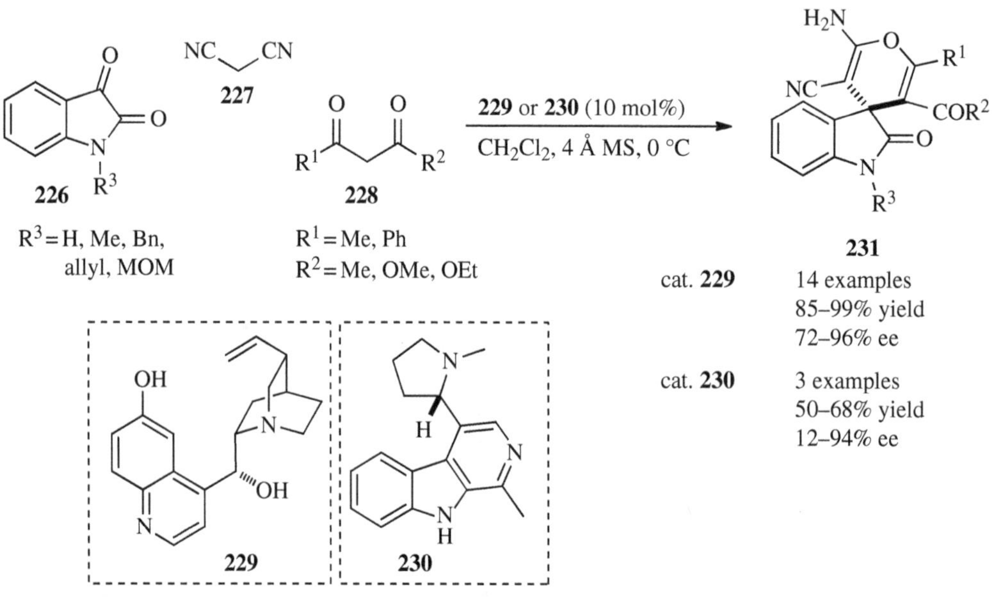

SCHEME 6.32 MCR strategy used in the synthesis of spirooxindole pyrans **231**.

Yuan and coworkers have reported an enantioselective three-component domino Knoevenagel/Michael/cyclization sequence using **226**, a 1,3-dicarbonyl compound **228**, and malononitrile **227** catalyzed by cinchona alkaloid cupreine **229** (Scheme 6.32). The spirocyclic oxindole pyrans **231** were obtained in high yields (85–99%) and enantioselectivities (up to 96% ee) [63]. Inconveniently, the reaction requires high dilutions (0.1–0.005 M) resulting in a large amount of used solvents. Macaev and coworkers have used chiral catalyst (−)-(*S*)-brevicolline **230** for the obtainment of the same spirooxindole pyrans **231** in 50–68% yield and 12–94% ee (Scheme 6.32) [64]. Several racemic examples of this reaction have also been reported previously [65].

Shi has reported a three-component reaction in water to build spirooxindoles **235** (Scheme 6.33) [66]. The reaction is catalyzed with 10 mol% of cerammonium nitrate (CAN) and leads to the products in high yields (up to 91%).

Gong and coworkers have described the first enantioselective synthesis of spiropyrrolidine oxindoles **240** using a three-component dipolar cycloaddition (Scheme 6.34) [67].

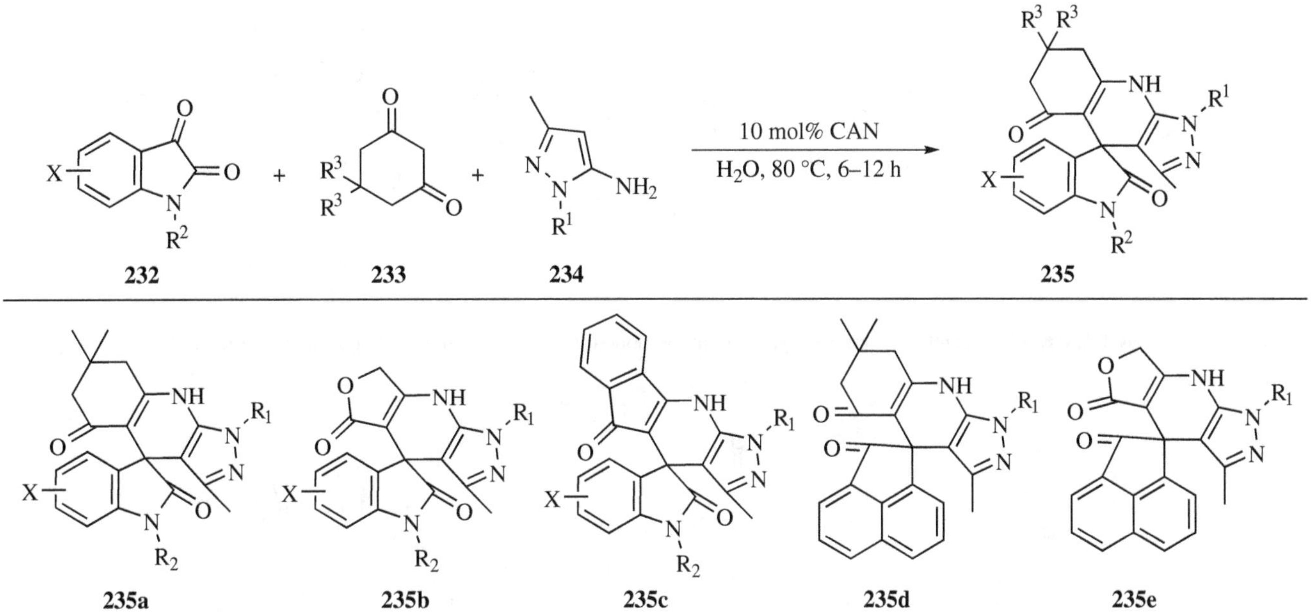

24 examples, yields up to 91%

SCHEME 6.33 CAN-catalyzed synthesis of spirooxindole derivatives **235** *via* three-component reaction in aqueous media.

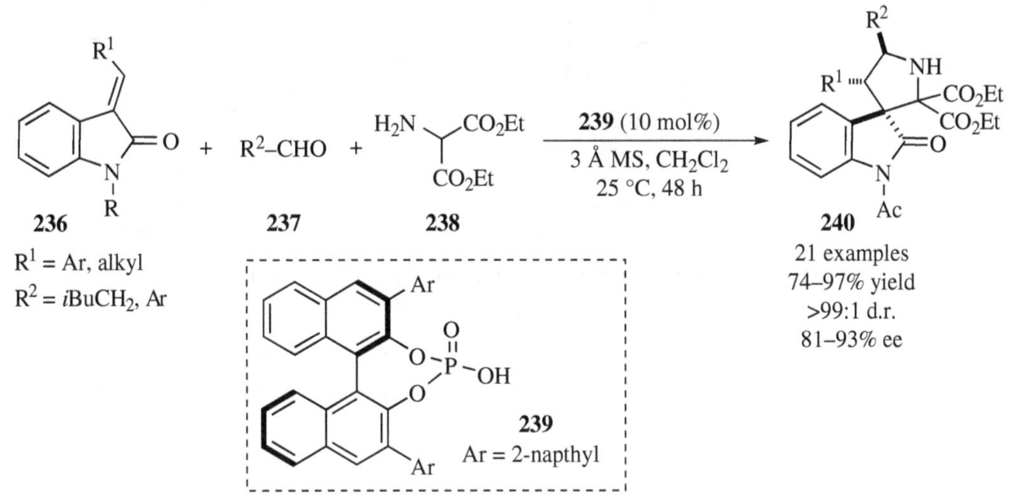

SCHEME 6.34 First enantioselective synthesis of spiropyrrolidine oxindoles **240** using BINOL-derived phosphoric acid **239**.

This reaction is catalyzed by a BINOL-derived phosphoric acid **239** and proceeds with high yields (74–97%) and high enantioselectivities (81–93% ee).

The good enantio- and regioselectivity can be explained through transition state **TS-6** where the azomethine ylide and the methyleneindolinone build hydrogen bonds with the chiral BINOL-phosphoric acid catalyst **239** (Fig. 6.10).

A triethylamine-catalyzed MCR with isatin **241** to synthesize a spiroheterocycle **243** in excellent yield was reported by Kumar and Arya (Scheme 6.35) [68]. Eighteen different spiroheterocycles were synthesized using this method with variation of the diketone **241** and the 2-aminobenzothiazole **242**.

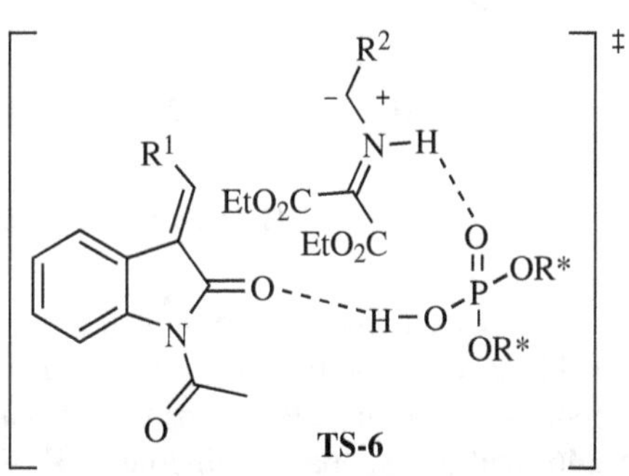

FIGURE 6.10 Proposed transition state **TS-6** for the reaction depicted in Scheme 6.34.

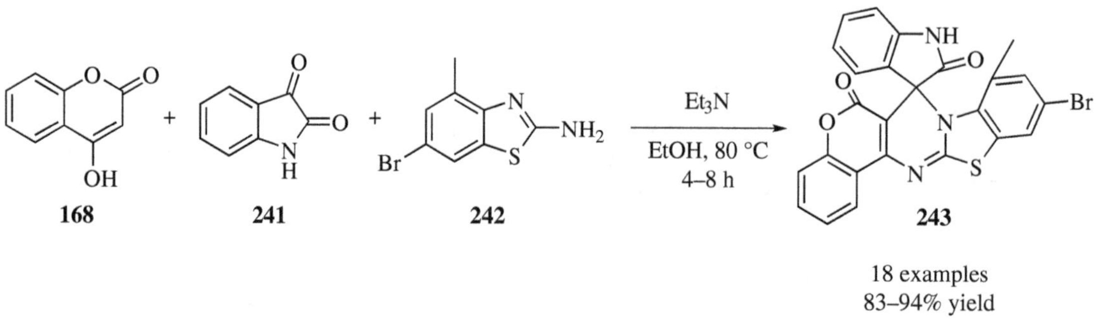

SCHEME 6.35 Triethylamine-catalyzed multicomponent reaction to synthesize spiroheterocycles **243**.

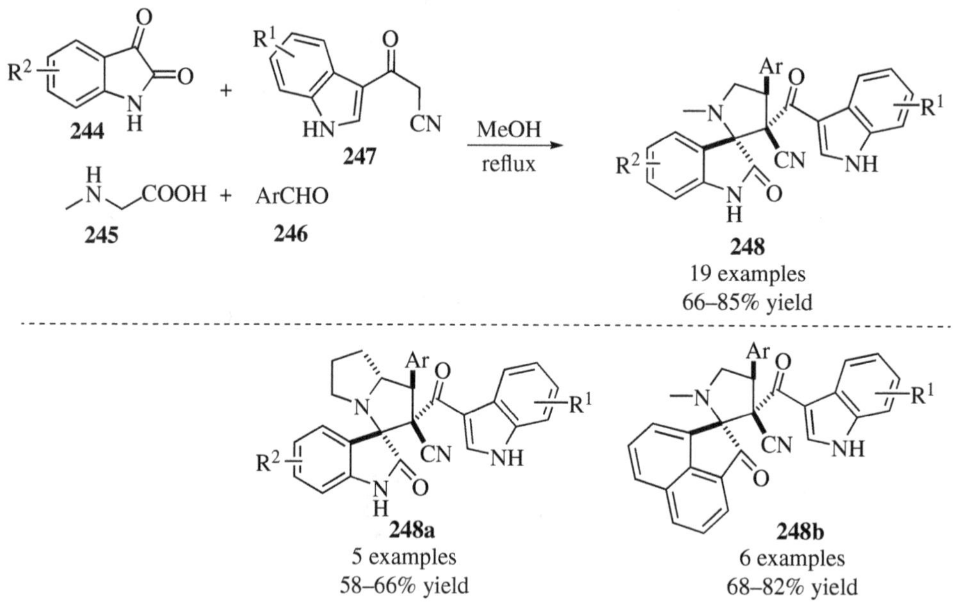

SCHEME 6.36 One-pot four-component [3+2]-cycloaddition.

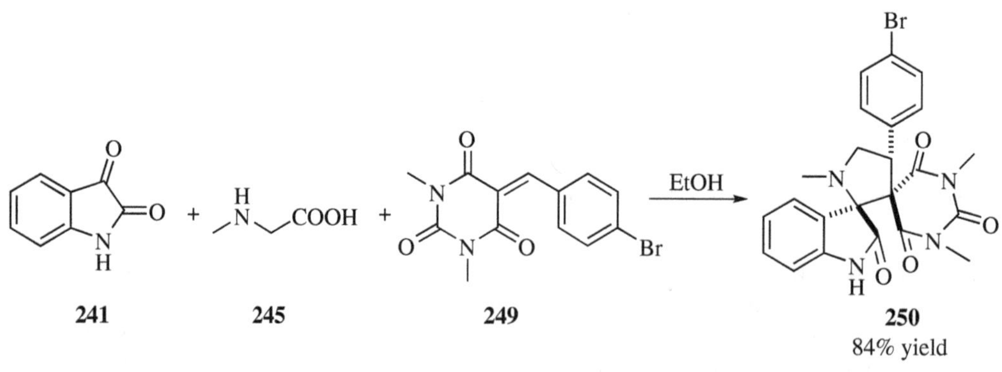

SCHEME 6.37 Synthesis of dispirooxindoles *via* multicomponent reaction.

Xu, Ji, and coworkers have demonstrated a one-pot four-component [3+2]-cycloaddition to obtain functionalized spiro compounds [69]. Isatin **244**, sarcosine **245**, an aromatic aldehyde **246**, and 3-(cyanoacetyl)indoles **247** were used to synthesize 3-spiropyrrolidine oxindoles **248** in good to excellent yields (66–85%) (Scheme 6.36). Pyrrolizidine oxindoles **248a** could be obtained in good yields when L-proline was used (instead of **245**), and acenaphthenones **248b** when acenapthylene-1,2-dione was used (instead of **244**).

Wang and Shi have used a three-component reaction between isatin **241**, sarcosine **245**, and dimethylpyrimidinone **249** to synthesize dispirooxindoles **250** (Scheme 6.37) [70].

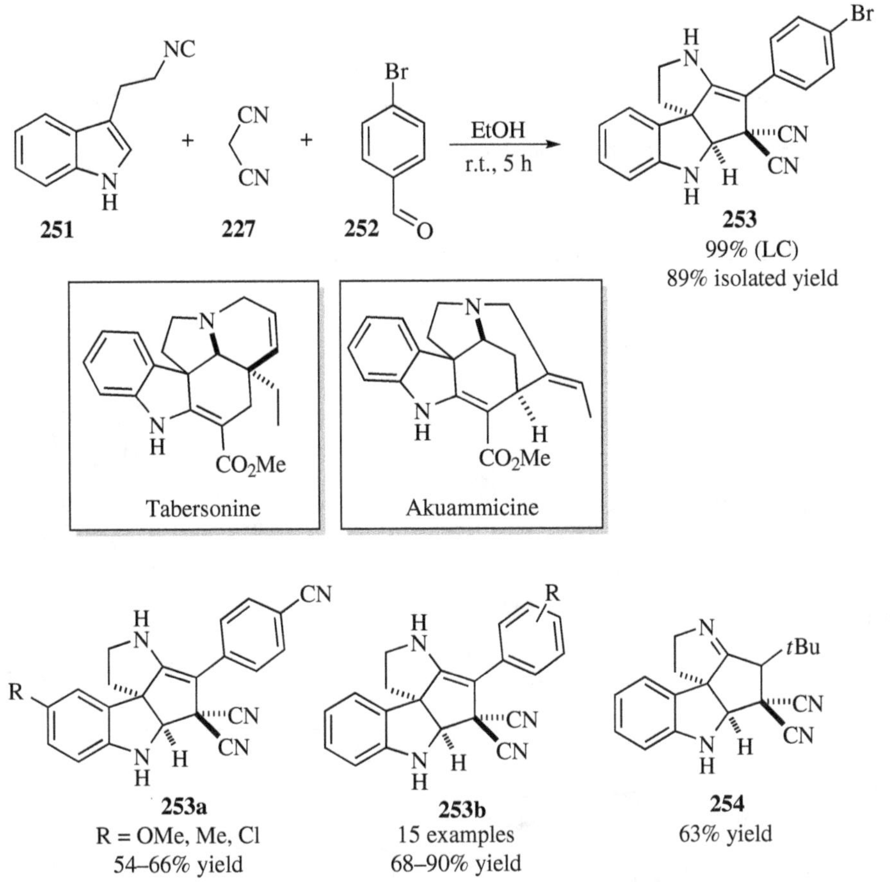

SCHEME 6.38 Isocyanide-based multicomponent reaction to build polycyclic spiroindolines **253** and natural products tabersonine and akuammicine.

L-thioproline and acenaphthenequinone could also be used and provide the desired products in high yields.

Ji and Wang have recently published a catalyst-free isocyanide-based MCR to construct polycyclic spiroindolines **253** stereoselectively (Scheme 6.38) [71, 72]. This method attractively offers an entry to this structural motif that is often present in biologically active natural products, as well as in pharmaceutical compounds like tabersonine and akuammicine.

In this reaction, isocyanide **251** reacts with malononitrile **227** and 4-bromobenzaldehyde **252** to form only one diastereoisomer. Ji and Wang have tested different solvents and have obtained the best results in EtOH. They further tested the scope of this reaction by variation of the readily available starting materials. The desired products were obtained in good yield with high diastereoselectivities, which reflects the force of this reaction. Electron-donating as well as electron-deficient functional groups are well tolerated. One drawback of this reaction is the requirement of aromatic aldehydes. If aliphatic aldehyde, for example, pivalaldehyde, is used, unrearranged product **254** was obtained.

Morpholines are interesting scaffolds as they are often present in natural compounds or biologically active molecules

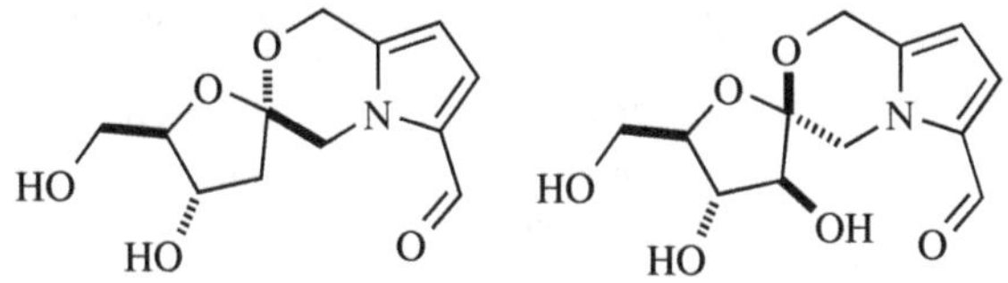

FIGURE 6.11 Naturally derived bioactive spiro-morpholines **255** and **256**.

(Fig. 6.11) [73]. They display for example antitumor [74], antidepressant [75], antioxidant [76], or antibacterial [77] properties.

Yeung and coworkers have developed an electrophilic multicomponent approach using an olefin, nosyl amide, epichlorohydrin, and *N*-bromosuccinimide (NBS), followed by basic cyclization, to obtain halogenated 2,2,6-trisubstituted morpholines in high yield and regioselectivity [78]. If enantiopure epichlorohydrin **259** is used, the product **260** is obtained in 99% enantiomeric excess (Scheme 6.39).

Several ferrocene-grafted pyrrolidine spiroheterocycles **263** are accessible through a four-component [3+2]-cycloaddition [79]. This method is operationally simple,

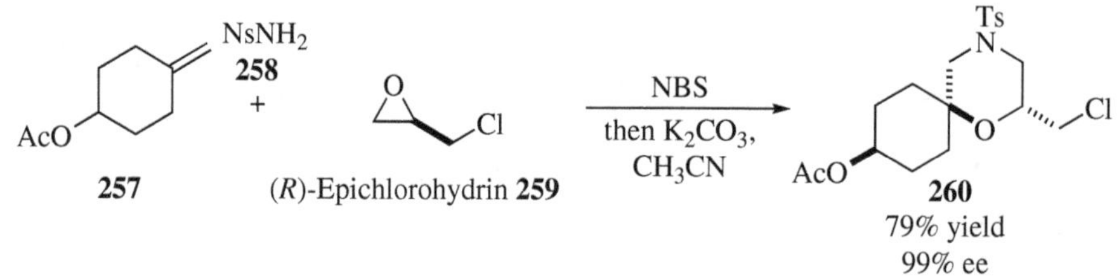

SCHEME 6.39 Electrophilic multicomponent approach for the synthesis of spiro-morpholine **260**.

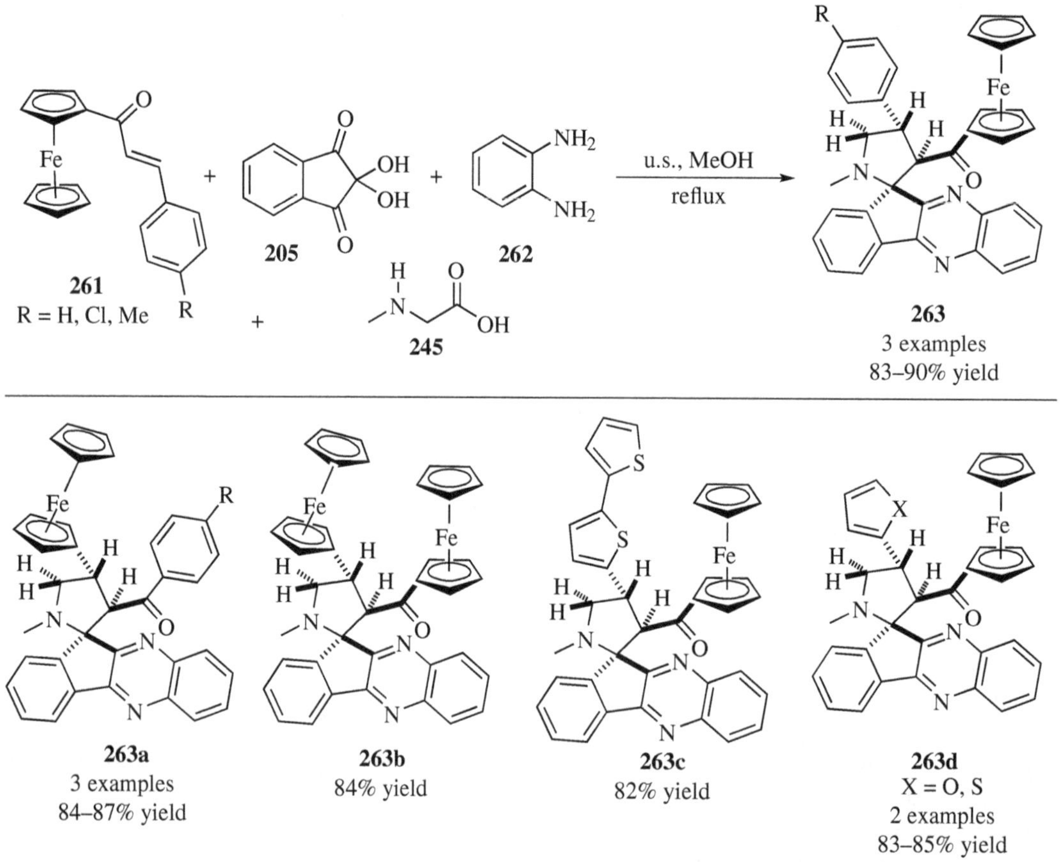

SCHEME 6.40 Synthesis of ferrocene-grafted pyrrolidine spiroheterocycles.

requires mild reaction conditions, and leads to the formation of the products very quickly from easily accessible starting materials (Scheme 6.40).

Perumal and coworkers have described the synthesis of highly functionalized novel dispiropyrrolidines **267–271** in moderate yields (Scheme 6.41) [80].

The synthesized compounds were tested for their *in vitro* activity against *Mycobacterium tuberculosis H37Rv* (MTB). Compound **268a** showed the most promising result with an MIC (minimum inhibitory concentration) value of 1.98 μM against MTB. It was also up to 26 times more potent than ethambutol and pyrazinamide, the standard first line TB drugs.

Santra and Andreana have used an Ugi/Michael/aza-Michael (UMAM) cascade reaction to synthesize fused azaspiro polycyclic alkaloids **276–278** in excellent yields and high diastereomeric ratios (Scheme 6.42) [81]. This concept nicely demonstrates the potency of MCRs. Six contiguous bonds are formed in one step, a quaternary center and up to four stereogenic centers are built during the reaction, and a huge variety of different substituents are tolerated.

Tu, Li, and coworkers have developed a microwave-assisted method to synthesize pentacyclic azaspiro **279** and oxa-azabridged heterocycles **280** (Fig. 6.12) [82].

6.4.4 Synthesis of DHPMs and Thiazines

DHPMs are interesting heterocycles as they are present in some pharmacologically active natural products as

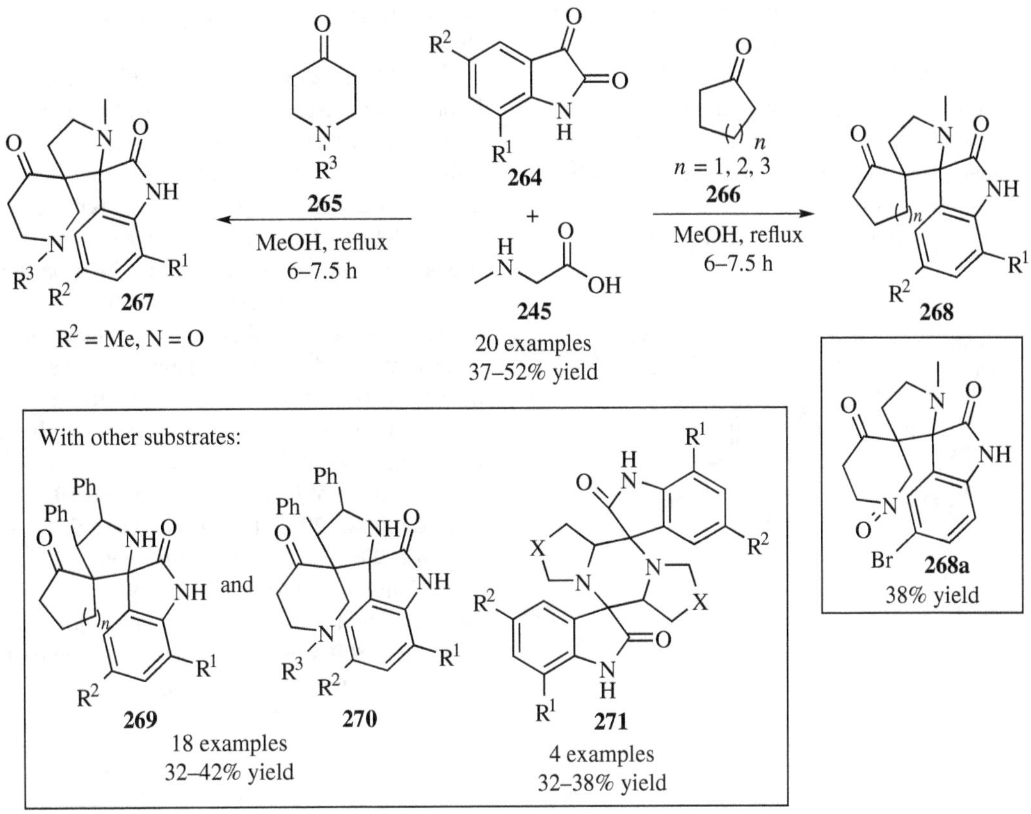

SCHEME 6.41 Synthesis of highly functionalized novel dispiropyrrolidines **267–271**.

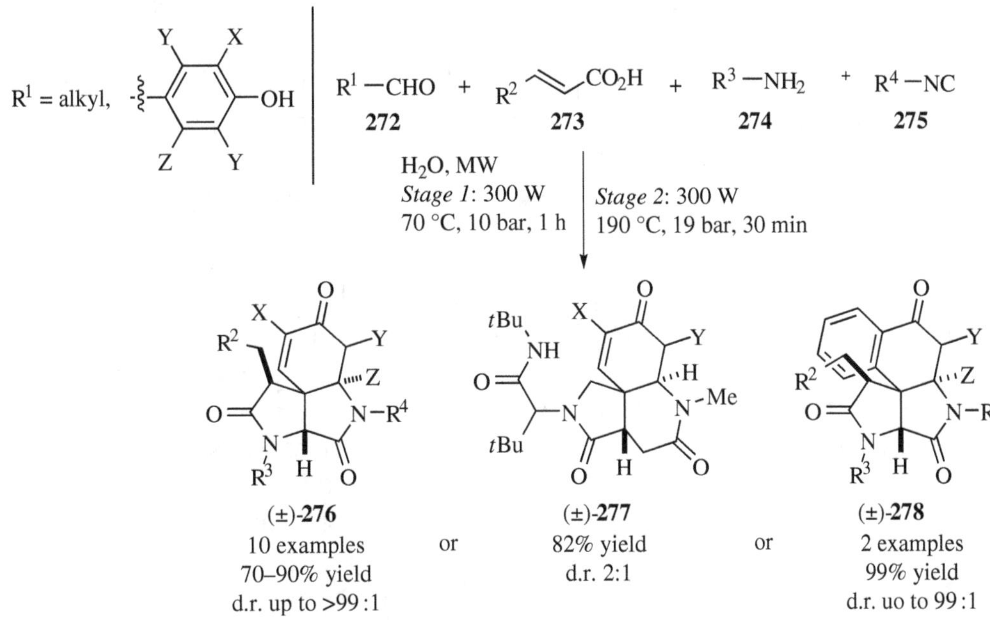

SCHEME 6.42 Synthesis of polycyclic alkaloids using an Ugi/Michael/aza-Michael cascade reaction.

depicted in Figure 6.13. They exhibit, for example, anti-inflammatory, antiviral, antitumor, and antibacterial activities. Some of them are used as antihypertensive agents or as calcium channel modulators. The biological activity of these scaffolds depends on the absolute configuration at C4. It is thought that the orientation of the substituent acts as a molecular switch between agonist and antagonist activities [86]. Thiazines are the 2-thio analogues of DHPM and are used for dyes (methylene blue), tranquilizers, and insecticides.

Classically, they are built through a Biginelli reaction between urea, aldehydes, and β-ketoesters in protic solvents under strongly acidic conditions [87]. The yield of this original reaction is rather low and side reactions occur very often [88]. In this sense, great effort has been devoted in the last years, focused on the research of this MCR (see Chapter 9). However, the discovery of new reactions for the preparation of this kind of compounds is of great interest [89]. Herein, some examples of interest are shown.

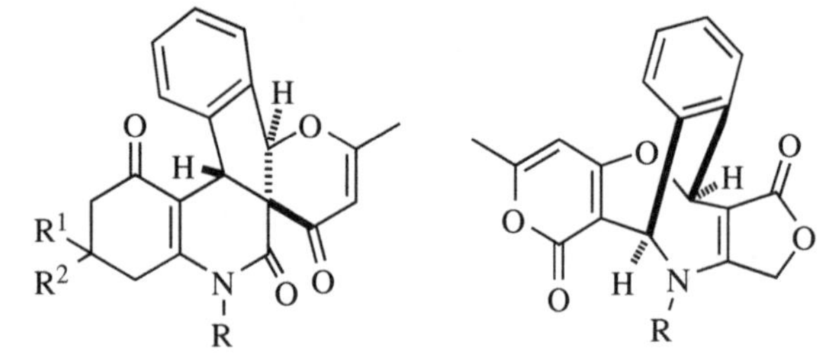

279
16 examples
44–68% yield

280
7 examples
71–84% yield

FIGURE 6.12 Synthesis of pentacyclic indeno[2,1-*c*]quinolones **279** and pyrano[4,3-*b*]oxepines **280**.

SNAP-7941 **293** is a chiral DHPM, which is currently used in preclinical tests as anorectic, antidepressant, and anxiolytic agent. It is also studied because of its role as MCH1 receptor. This compound can be synthesized through two different enantioselective organocatalytic syntheses, as reported by Schaus and Goss in 2008 (Scheme 6.43) [90]. One of them in an asymmetric Biginelli reaction catalyzed by the chiral phosphoric acid **291** (10mol%) leading to DHPM **292** in 96% yield and 89% ee. Four further steps led to the final target molecule **293**.

Overman and coworkers have showed an elegant example of a three-component Biginelli reaction in the synthesis of the complex natural product crambescidin 800 **10** [91]. In Scheme 6.44, the preparation of the key intermediate **302** in the obtainment of final target compound **10** is shown. Pyrazole carboxamidine **294**, ethyl acetoacetate **295**, and the masked aldehyde **296** react in a base-mediated Biginelli reaction to afford 2-iminoDHPM **299** in 62% yield. After further transformations including guanidine protection and aminolysis, compound **300** was obtained in 88% yield. Acid-catalyzed cleavage of the carbamate and acetalization, followed by

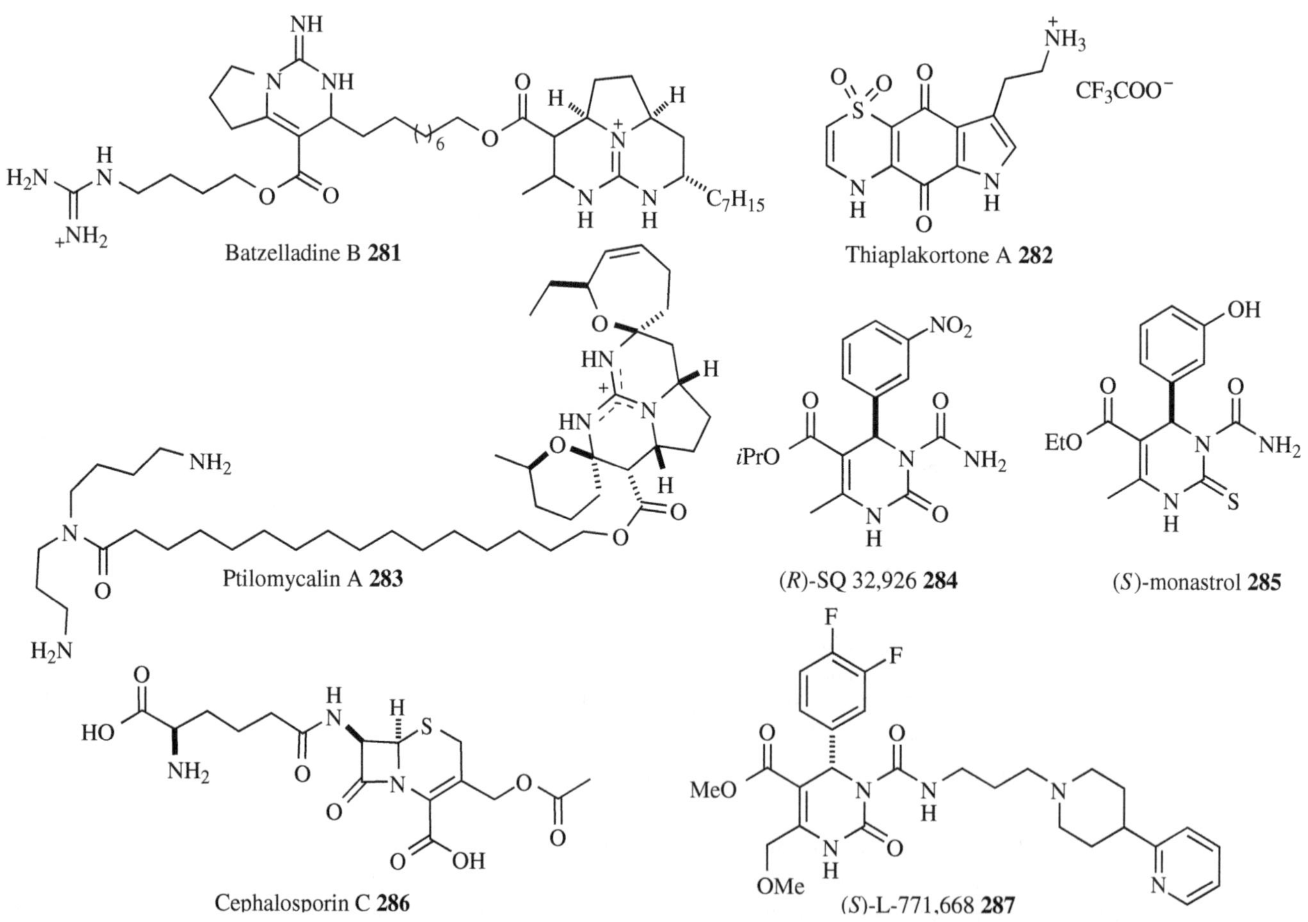

FIGURE 6.13 Interesting natural compounds containing dihydropyrimidines or thiazines [83–85].

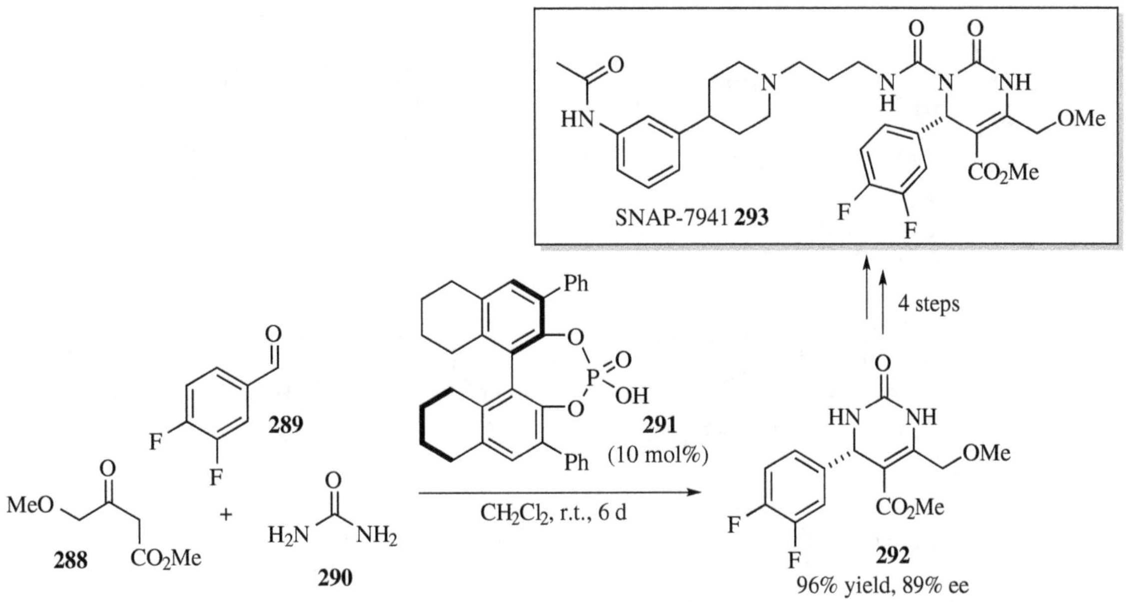

SCHEME 6.43 Two asymmetric routes to the synthesis of MCH inhibitor SNAP-7941 **293** [90].

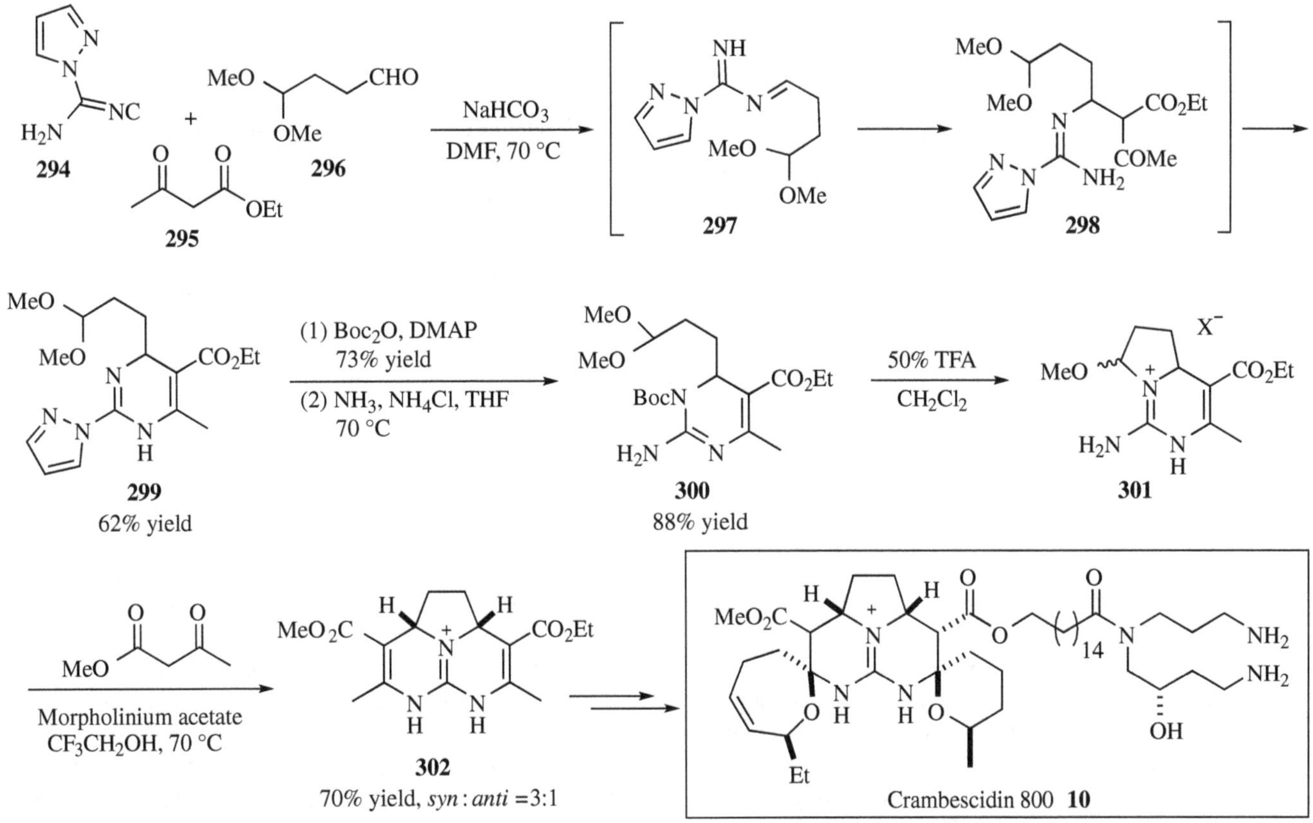

SCHEME 6.44 Multicomponent approach to the core of crambescidin 800 **10**.

treatment of methyl acetoacetate, furnished tricyclic guanidine compound **302** in 70% yield and diastereomeric ratio of 3:1 [92].

DHPMs **306** of biological interest have been synthesized by Kumar and coworkers, using *p*-toluene sulfonic acid as catalyst (Scheme 6.45) [93]. The 33 synthesized compounds were tested for their *in vitro* antioxidant activity as well as for their anticancer activity against human cancer breast cell lines (MCF-7). Among them, compounds **306a** and **306b**

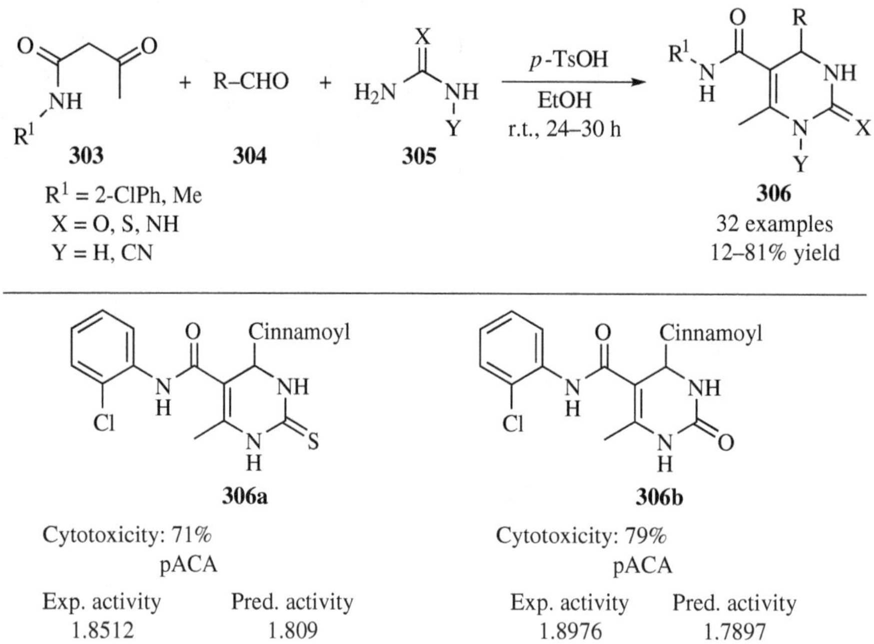

SCHEME 6.45 Synthesis of biologically interesting dihydropyrimidines **306**.

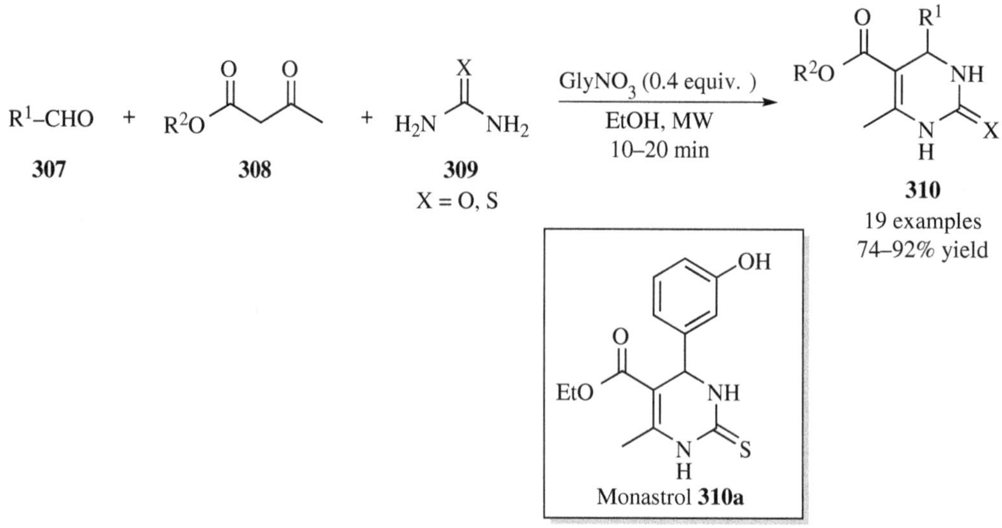

SCHEME 6.46 Synthesis of dihydropyrimidinones **310** by the use of ionic liquid. Gram-scale synthesis of monastrol **310a**.

have demonstrated significant cytotoxic activity. They have been further elucidated in a comparative molecular similarity indices analysis (CoMSIA) toward their anticancer activity.

Sinha and coworkers have demonstrated the use of the ionic liquid glycine nitrate in the synthesis of dihydropyrimidinones **310** (Scheme 6.46) [94]. The reaction is microwave assisted and the desired products **310** were formed after 10–20 min in high yields (74–92%). Glycine nitrate can be recycled without compromising the yield (up to 10 cycles with yields to 85%) and is biodegradable. This process has been applied in gram scale to the synthesis of monastrol **310a** (72% yield), a potent mitotic kinesin Eg5 inhibitor.

Finally, two examples of preparation of thiazines, the 2-thio analogues of DHPM, are depicted in Scheme 6.47. On the one hand, thiazine-2-thiones **313** have been synthesized in a one-pot procedure by Orru and coworkers *via* a Horner–Wadsworth–Emmons approach (Scheme 6.47a) [95]. The compounds were obtained in low to good yields (30–72%). The authors point out that the products are not very stable and slowly decompose over time in solutions, but purified products can be stored under inert conditions. And on the other hand, Shang and coworkers reported the DABCO-catalyzed synthesis of 3,4-dihydro-2*H*-1,4-benzo[*b*]thiazine derivatives **317** in good yields (62–88%) (Scheme 6.47b) [96].

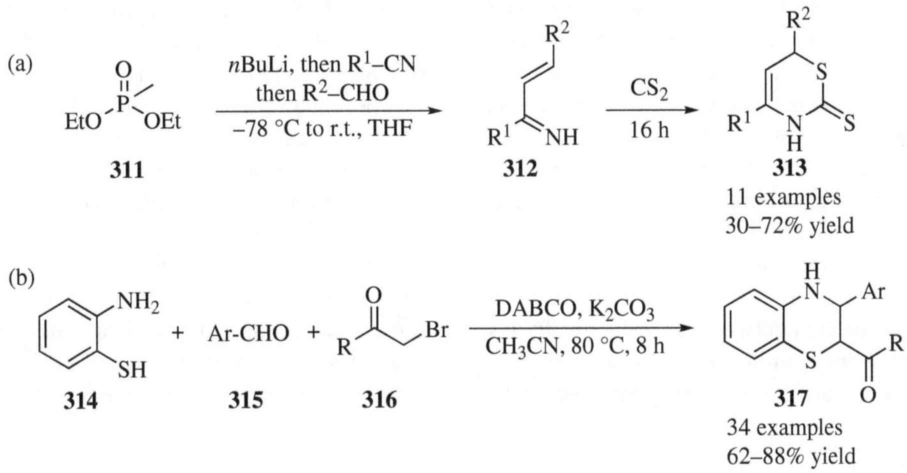

SCHEME 6.47 Synthesis of 3,6-dihydro-2*H*-1,3-thiazine-2-thiones **313** and 3,4-dihydro-2*H*-1,4-benzo[*b*]thiazine derivatives **317**.

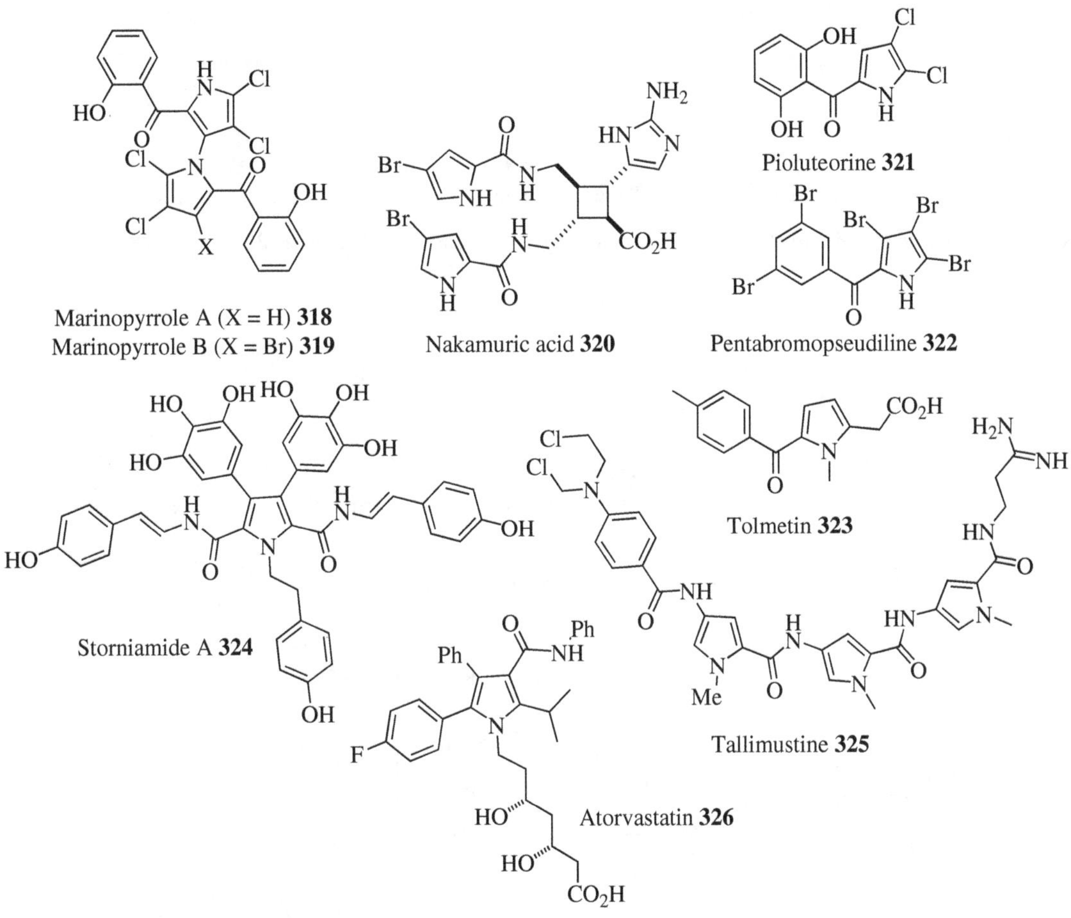

FIGURE 6.14 Interesting pyrrole-derived natural products **318–326** [101].

6.4.5 Synthesis of Pyrroles

Pyrroles are one of the most important classes of heterocyclic compounds that exhibit interesting biological activities [97], and their skeleton is related with several bioactive natural products [98], pharmaceutical agents [99], and material science [100]. Some interesting pyrroles are depicted in Figure 6.14.

Due to the well-recognized value of pyrroles as versatile synthetic intermediates in organic chemistry, a wide range of synthetic procedures for the construction of this structural fragment has been reported [102, 103]. These developed strategies generally involve multistep synthetic procedures; thus, in the last decade, more efforts have been dedicated to the design of new multicomponent approaches giving access to highly functionalized pyrroles (Fig. 6.15) [107].

Maiti and coworkers have recently reported an excellent work about the synthesis of pentasubstituted pyrroles starting from three, four, and seven components of common precursors under very mild reaction conditions using $CeCl_3{\cdot}7H_2O$ as activator (Scheme 6.48) [108].

A four-component reaction between aromatic aldehydes **335**, β-ketoesters **336**, benzylamines **337**, and nitromethane **338** described by Khan and coworkers gives access to tetrasubstituted *N*-benzylpyrrole derivatives **339**, catalyzed by bromodimethylsulfonium bromide (BDMS) in good yields (60–82%) (Scheme 6.49) [109].

The same group has reported the synthesis of tetrasubstituted pyrroles catalyzed by nickel(II)chloride hexahydrate in good yields. Some of them seemed to be good inhibitors of phosphodiesterase 4B (PDE4B) enzymes [110].

Zhang and coworkers used gluconic acid aqueous solution (GAAS) in the strategy to obtain tetrasubstituted pyrroles, which were obtained in high yields (up to 93%) [111]. Functional groups are well tolerated in this process (40 examples), the green media can be reused, and it is completely inexpensive and environmentally friendly.

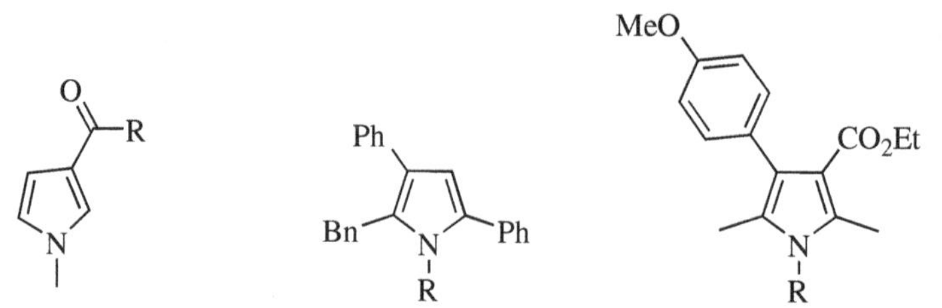

327 *N*-Methyl-3-acylpyrroles **328** Tetrasubstituted pyrroles **329** Pentasubstituted pyrroles

FIGURE 6.15 Selected pyrrole structures synthesized through a multicomponent reaction: **327** [104], **328** [105], and **329** [106].

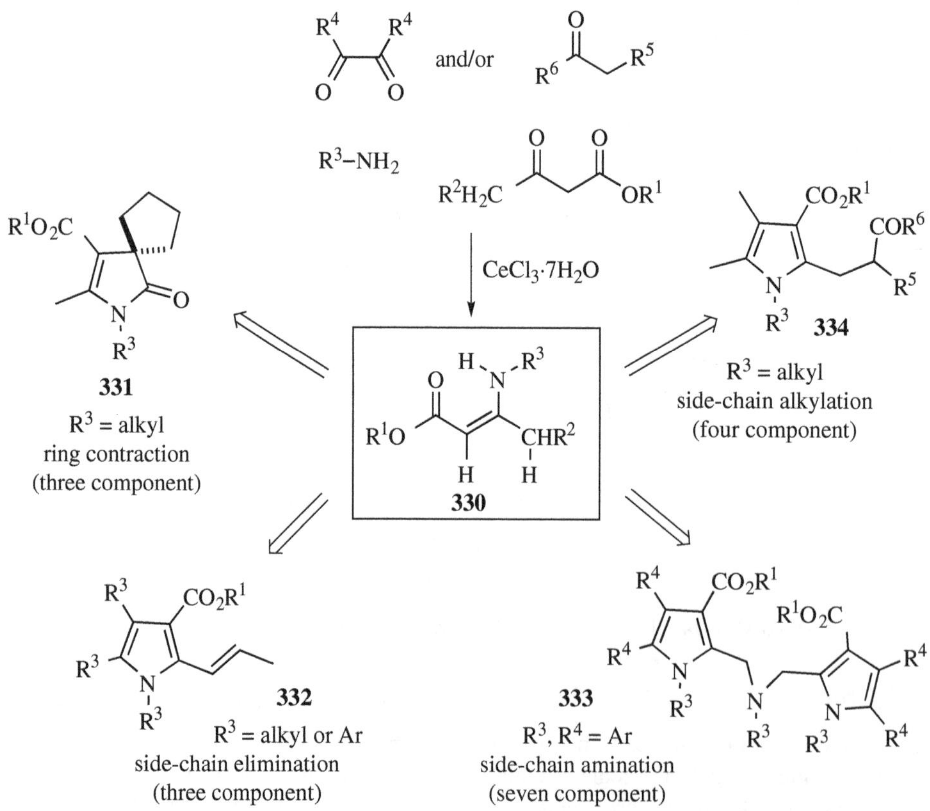

SCHEME 6.48 Synthesizing pentasubstituted pyrroles **331–334**.

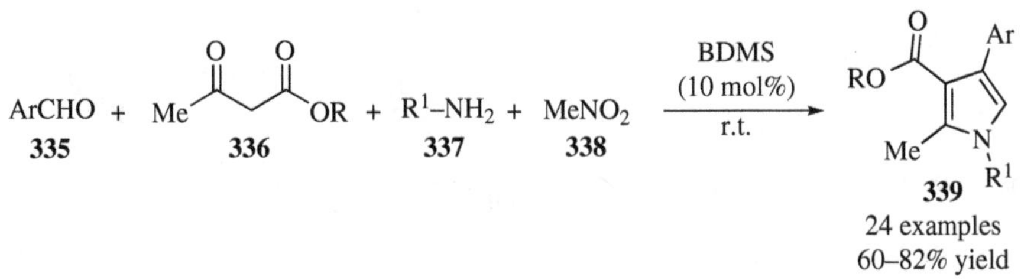

SCHEME 6.49 BDMS-catalyzed MCR for the obtainment of pyrroles **339**.

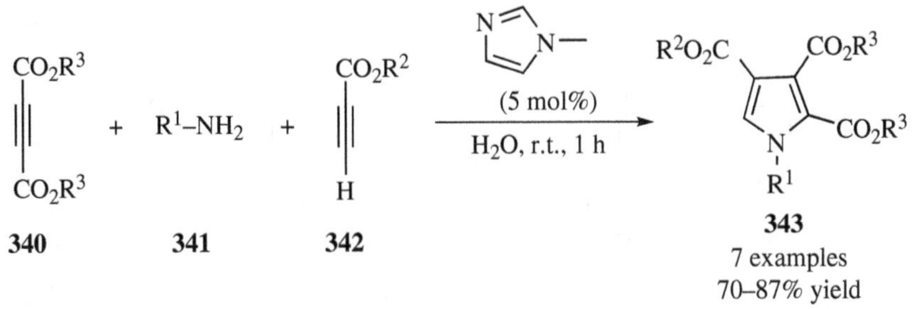

SCHEME 6.50 *N*-Methylimidazole-catalyzed synthesis of tetrasubstituted pyrroles **343** in water.

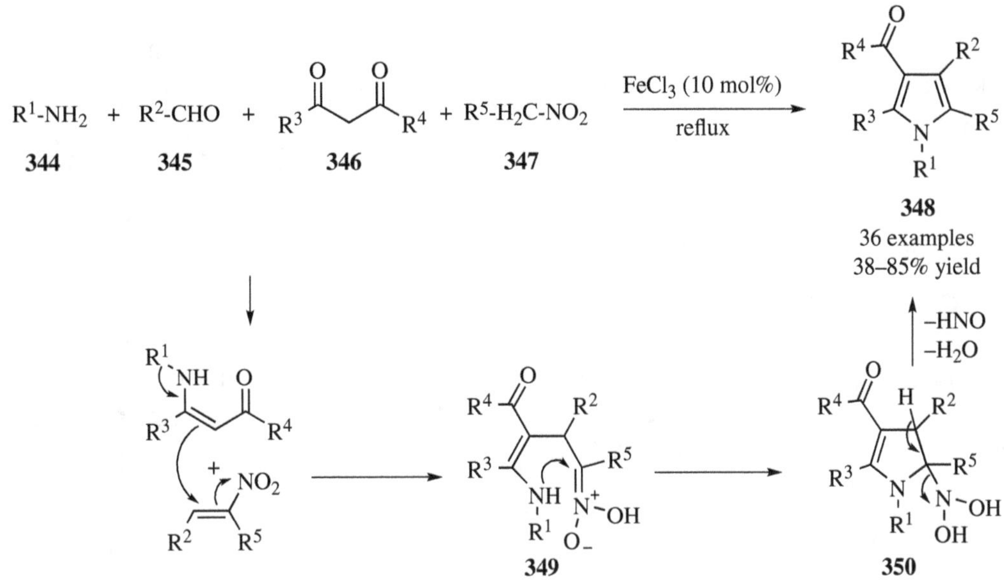

SCHEME 6.51 $FeCl_3$-catalyzed four-component coupling synthesis of functionalized pyrroles **348**.

The group of Sayyed-Alangi has demonstrated an *N*-methylimidazole-mediated pyrrole synthesis in water [112]. The desired tetrasubstituted pyrroles **343** were obtained in high yields (70–87%) (Scheme 6.50).

Jana's group has reported a simple and interesting iron(III)-catalyzed multicomponent coupling method for the synthesis of pentasubstituted pyrroles **348** (Scheme 6.51) [113]. This four-component coupling strategy of amines **344**, aldehydes **345**, 1,3-dicarbonyl compounds **346**, and nitroalkanes **347** provides an alternative approach for easy access of a wide range of highly functionalized pyrroles in moderate to very good yields (38–85%). This reaction proceeds *via* amination/Michael reaction and cycloisomerization as shown in Scheme 6.51.

Jeong and Atar were the first who recently developed the synthesis of tetrasubstituted pyrroles **356** sustained by silica gel-supported tungstic acid (STA) **355** [114]. The reactions proceed in high yields (60–88%), and the catalyst could be reused in several reactions without losing its catalytic activity (Scheme 6.52).

Shi, Huang, and coworkers have developed a catalyst-free MCR of an aniline **357**, arylglyoxal monohydrate **358**, dialkyl

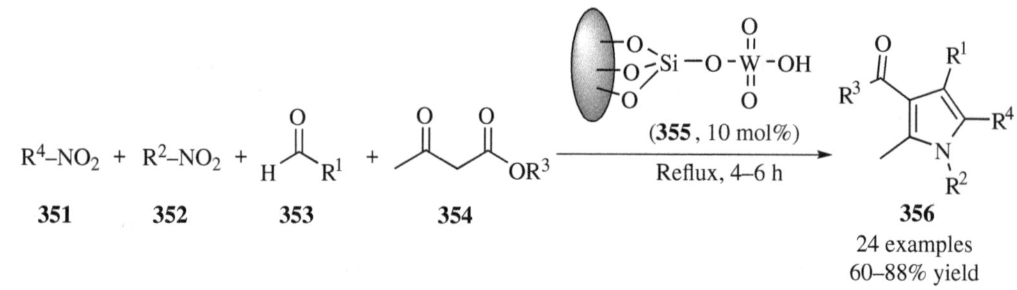

SCHEME 6.52 Silica gel-supported tungstic acid as catalyst for the synthesis of pyrroles **356**.

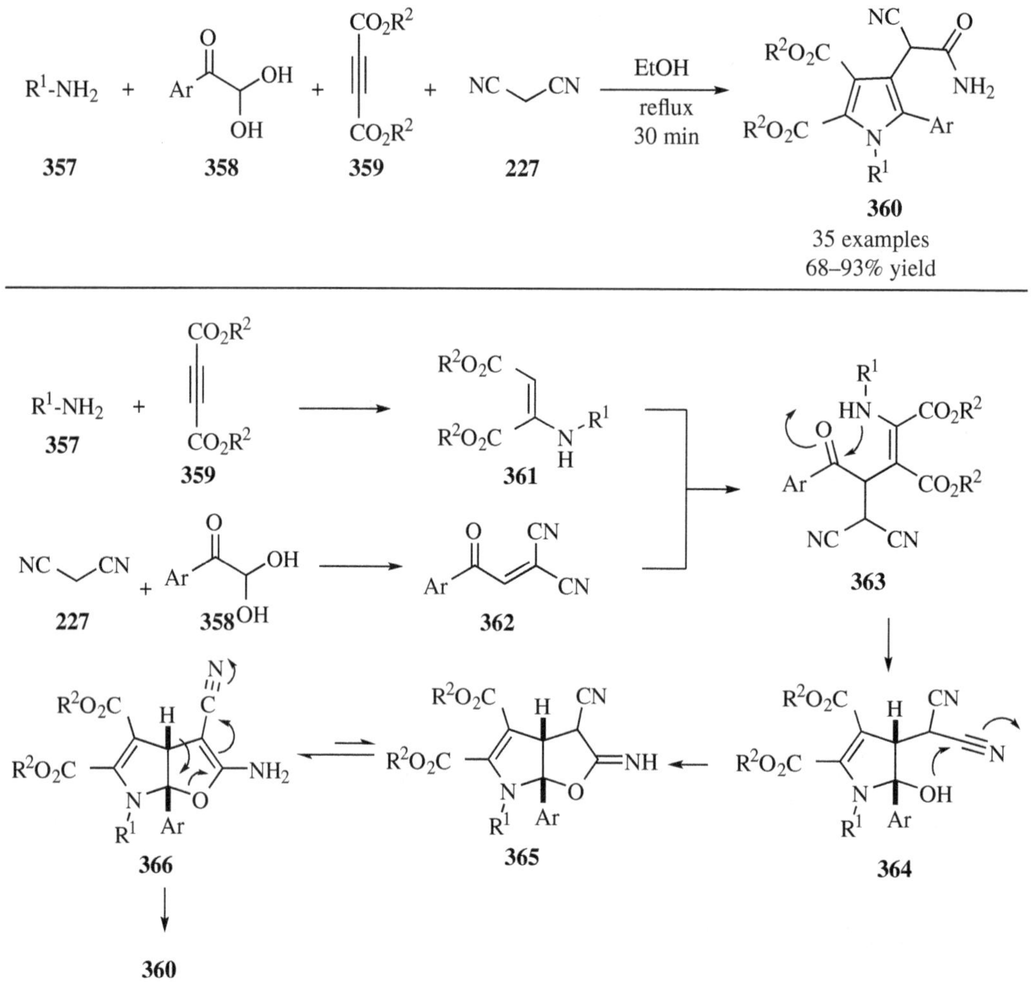

SCHEME 6.53 Catalyst-free synthesis of polysubstituted pyrroles **360** and proposed mechanism.

acetylenedicarboxylates or alkyl acetoacetate **359**, and malononitrile **227** in refluxing ethanol to obtain polysubstituted pyrroles **360** in good yields (68–93%) (Scheme 6.53) [115–117]. The mechanism is believed to proceed through the formation of 6,6a-dihydrofuro[2,3-*b*]pyrrole **366**, which is opened to the desired product [118].

Tu, Jiang, and coworkers have developed the asymmetric acid-catalyzed three-component reaction in water to obtain polysubstituted *syn*-4,5-dihydropyrroles **370** in good yields (53–74%) (Scheme 6.54) [119].

As shown in Scheme 6.54, the mechanism of this process starts with the formation of imine **371**, from compounds **368** and **369**. Then, imine **371** is attacked by enamine **367**. After proton transfer and cyclization, product **370** is obtained. The pure *syn*-stereoselectivity is assumed over the hydrogen bond of intermediate **373**.

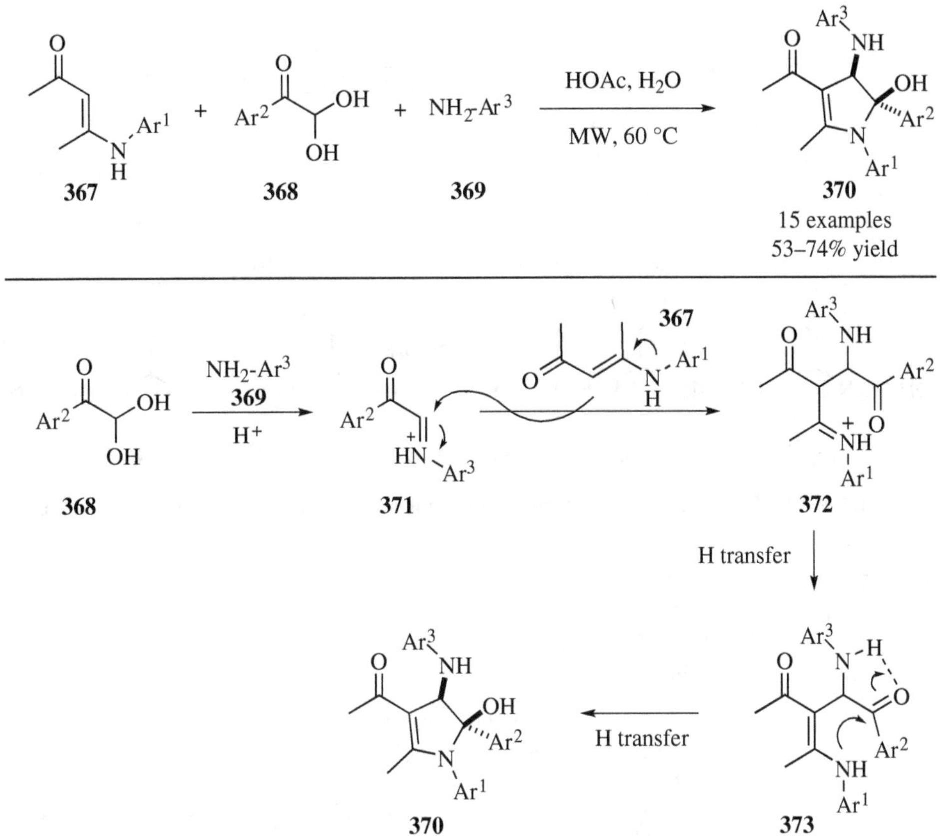

SCHEME 6.54 Asymmetric synthesis of *syn*-4,5-dihydropyrroles **370** in water and proposed mechanism.

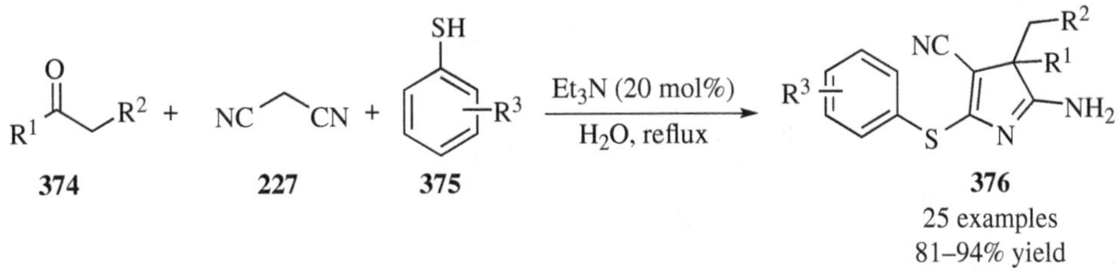

SCHEME 6.55 Synthesis of 3*H*-pyrroles **376**.

Recently, a huge variety of substituted 3*H*-pyrroles **376** have been synthesized through an organocatalyzed sequence using ketones **374**, malononitrile **227**, and thioles **375** in water as depicted in Scheme 6.55 [120]. Not only electron-withdrawing and electron-donating residues are well tolerated but also cycles like thiophene or 9*H*-fluorene affording the desired pyrroles in excellent yields (81–94%). The remaining cyano group in the molecules can be further used for the synthesis of heterocycles.

6.5 AMINO ACID DERIVATIVES BY MCRs

There is no doubt about the importance of the development of efficient method for the synthesis of amino acids, a kind of molecules of great value. In this sense, MCRs provide access to (un)natural monosubstituted (*R*)-amino acids, and herein, few examples are described.

tert-Butyl isocyanide **377**, ammonium formate **378**, and aldehydes or ketones **379** react in a large-scale Ugi reaction followed by acidic hydrolysis to the unnatural α-substituted amino acids **380** (Scheme 6.56) [121].

Kazmaier and coworkers have observed the formation of eight-membered rings or linear side products when ammonia is used as amine partner in the aforementioned reaction [122]. The desired amino acids are obtained as minor product or only in traces. These side reactions can be avoided when nonnucleophilic alcohols are used like trifluoroethanol; sterically demanding aldehydes, like pivalaldehyde; and sterically demanding carboxylic acids.

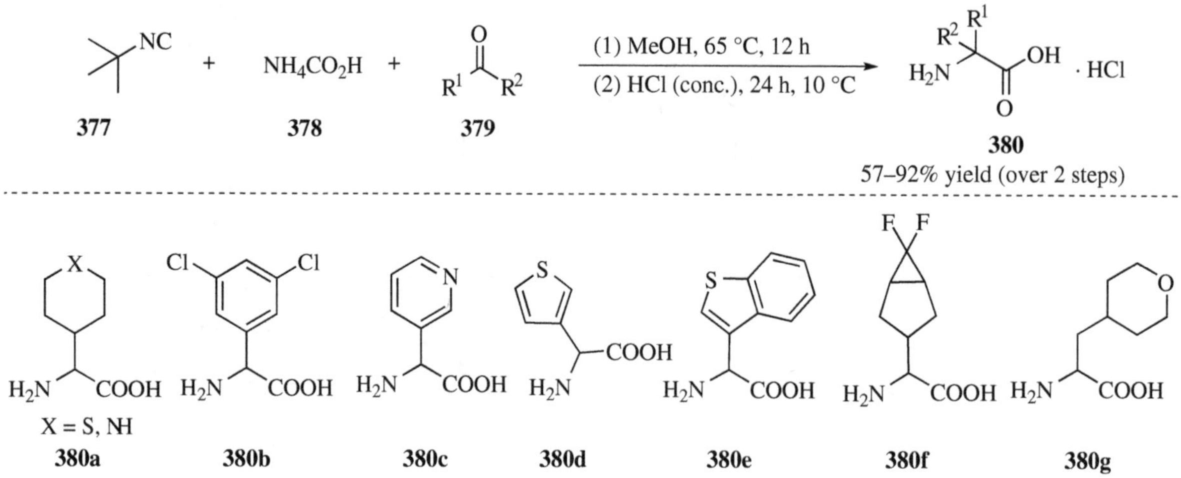

SCHEME 6.56 Large-scale synthesis of unnatural amino acids **380** using an Ugi reaction.

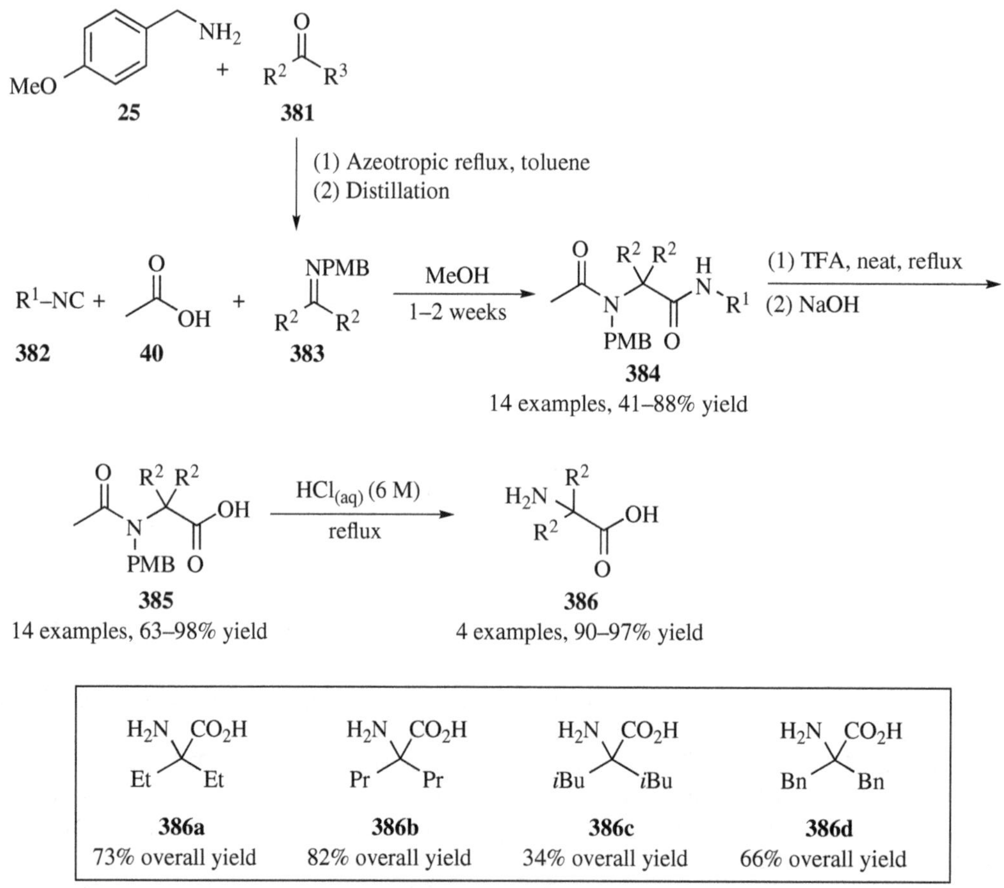

SCHEME 6.57 Ugi reaction in the synthesis of α,α-disubstituted amino acids **386**.

Costa and coworkers have shown the application of an Ugi reaction for the synthesis of α,α-disubstituted amino acids **386** (Scheme 6.57) [123]. The imine **383** required for the reaction is first synthesized through azeotropic reflux and subsequent distillation of the ketone **381** and *p*-methoxybenzylamine **25**. This imine is then treated with acetic acid **40** and isocyanide **382** to obtain peptide **384**, which is further converted into α,α-disubstituted PMB-protected amino acid **385**, followed by acidic hydrolysis to obtain the final products **386**. The authors suggest that this is the best route to the synthesis of bulky α,α-dialkyl glycines, as it involves simple procedures in clean reactions leading to high overall yields, usually above 60%.

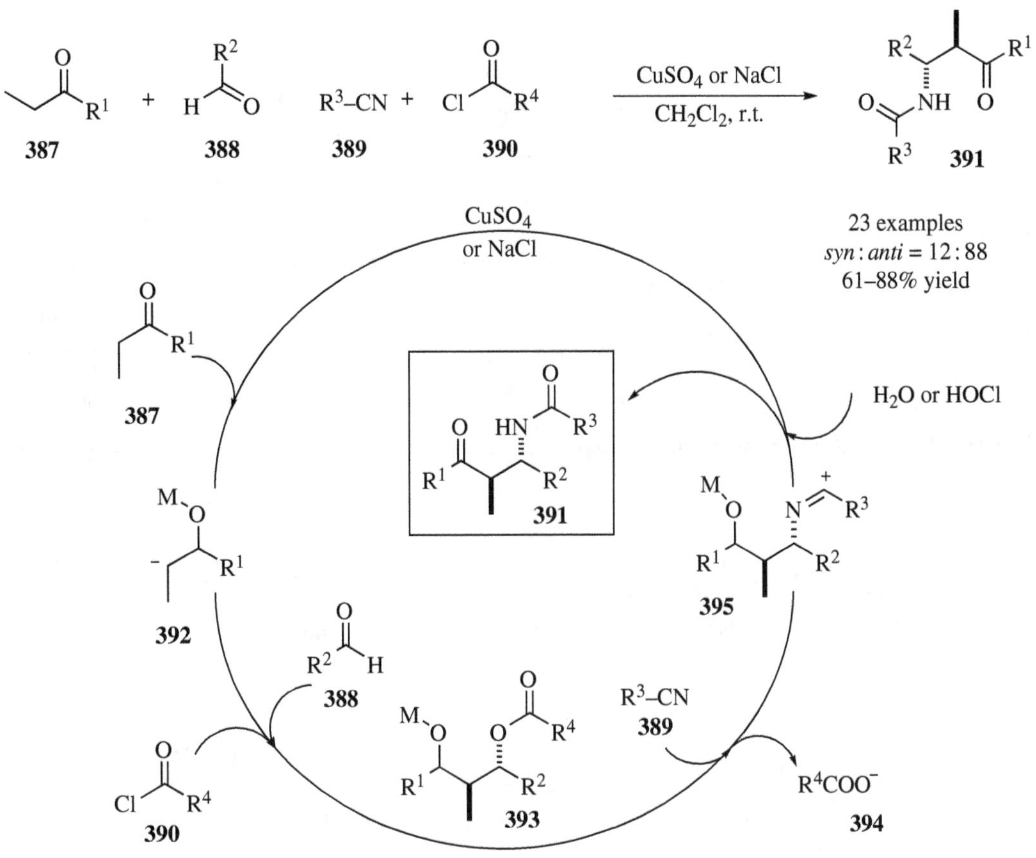

M = Cu or Na

SCHEME 6.58 *Anti*-Selective synthesis of β-amino acid derivatives **391** and catalytic cycle.

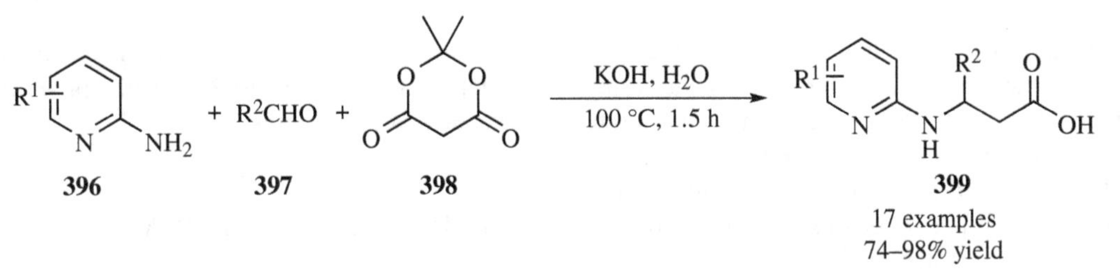

SCHEME 6.59 Three-component reaction to synthesize β-amino acids **399**.

The *anti*-selective synthesis of β-amino acid derivatives **391** was achieved through a copper sulfate- or sodium chloride-mediated process between a ketone **387**, an aldehyde **388**, a nitrile **389**, and an acyl chloride **390**, yielding the products **391** in good yields (61–88%) and high diastereoselectivities (Scheme 6.58) [124].

The reaction starts with the complexation of the carbonyl group of ketene **387** to form enolate **392** (Scheme 6.58). This sterically hindered enolate attacks acyl chloride **390** and aldehyde **388** to form adduct **393**. The acyloxy group is then replaced by nitrile **389** to form the stable cation **395**. Addition of water or hypochlorous acid leads then to the formation of the desired β-amino acid derivative **391**. The high *anti*-selectivity can be explained on the basis of the described catalytic cycle. The steric interaction of the acyloxy group of the aldehyde carbon and the more hindered α-substituted enolate forces the addition to take place from the less hindered site to form the *anti*-product.

Another process where β-amino acids are synthesized has been developed by Adib and coworkers. In a three-component reaction using 2-aminopyridine **396**, aldehydes **397**, and Meldrum's acid **398**, the desired products **399** were obtained in 74–98% yields (Scheme 6.59) [125].

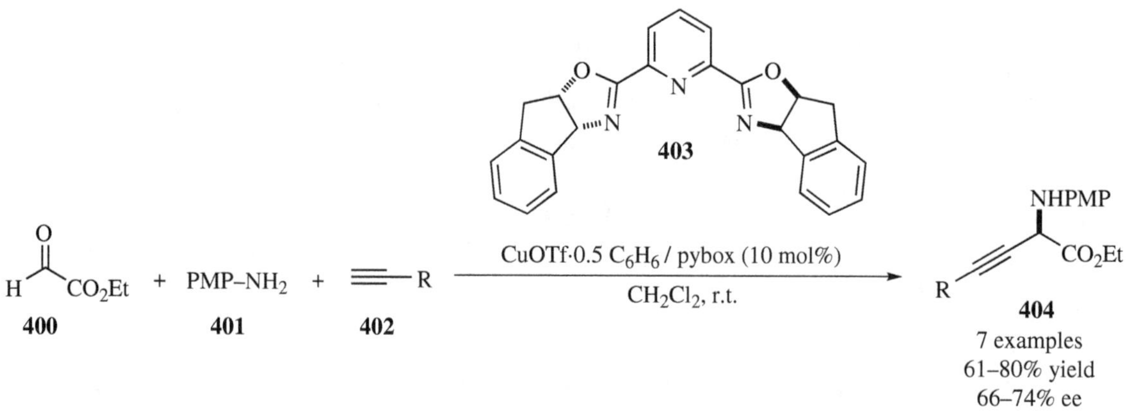

SCHEME 6.60 Asymmetric multicomponent reaction of chiral β,γ-alkynyl α-amino acid derivatives **404**.

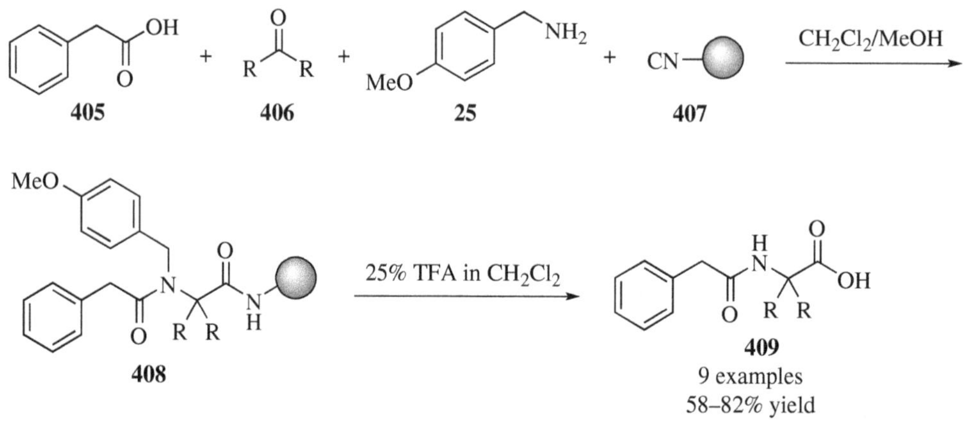

SCHEME 6.61 Solid-phase Ugi reaction for the synthesis of *N*-acylated α,α-dialkylglycines **409**.

Shao, Chan, and coworkers have developed the first catalytic asymmetric synthesis of chiral β,γ-alkynyl α-amino acid derivatives **404** in 61–80% yields and moderate enantioselectivities (66–74%), using ethyl glyoxylate **400**, *p*-anisidine **401**, and aliphatic or aromatic alkynes **402** (Scheme 6.60) [126]. This process is catalyzed by a catalyst system between Cu(I) triflate benzene complex and 10 mol% of pybox catalyst **403**.

A solid-phase Ugi synthesis gives access to *N*-acylated α,α-dialkylglycines **409** in good yields (58–92%), employing phenyl acetic acid **405**, ketones **406**, *p*-methoxybenzylamine **25**, and cyanides **407**, which were attached to the resin [127]. After acidic cleavage of the resin, the desired *N*-acylated α,α-dialkylglycines **408** are obtained (Scheme 6.61).

γ,δ-Alkynyl β-amino acid derivatives **413** can be synthesized through a copper-catalyzed amine–alkyne–alkyne MCR in high yields (Scheme 6.62) [128]. When chiral prolinol derivatives **410b** are used, the products **413b** could be obtained in excellent diastereoselectivities (up to >99:1).

Dömling has reported the synthesis of seleno amino acid derivatives **418** *via* Ugi reaction (Scheme 6.63) [129]. Methylselenocysteine **414**, ketones **415**, and isocyanides **416** are used to generate the latter in 48–95% yield. Selenomethionine is used as well, providing the desired products in good yields.

6.6 INDUSTRIAL APPLICATIONS

Cheap and efficient methods in the synthesis of drugs in big scale are of great value for industrial applications. Simple, low cost, and commercially available starting materials are very important as cost of goods play more and more a central role in industry. MCRs fulfill these conditions as complex structures can be assembled in only one step [130].

More than 200 million people are worldwide affected by schistosomiasis, which is a neglected tropical sickness. Praziquantel **424** is currently the only drug approved against this disease. Dömling and Khoury have reported a very short Ugi four-component reaction followed by Pictet–Spengler cyclization to obtain praziquantel **424** in greater than 75% yield (Scheme 6.64) [131]. Ninety percent of infected people live in sub-Saharan Africa where cheap preparations of drugs

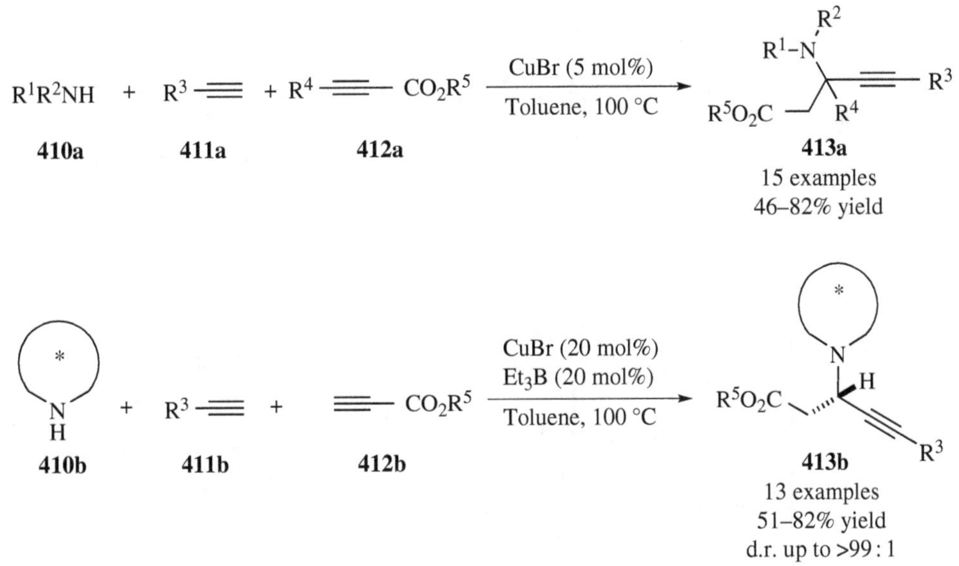

SCHEME 6.62 Synthesis of γ,δ-alkynyl β-amino acid derivatives.

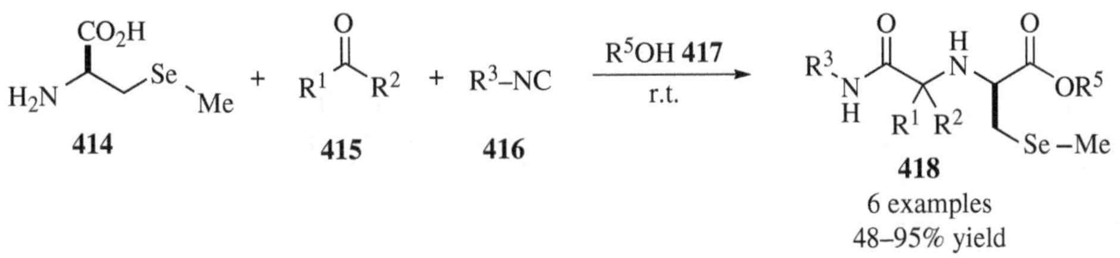

SCHEME 6.63 Ugi four-component reaction allows the synthesis of seleno amino acid derivatives **418**.

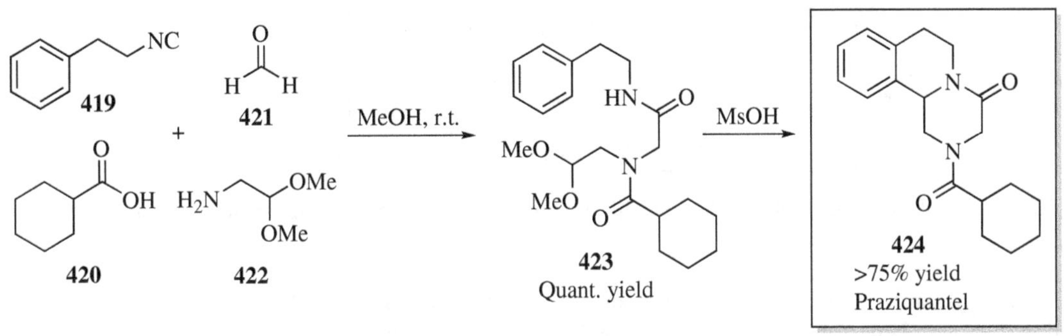

SCHEME 6.64 Four-component Ugi reaction followed by Pictet–Spengler cyclization to obtain praziquantel **424**.

are of great importance. In this reaction, all goods are very cheap except for phenethyl isocyanate **419** that can be easily prepared starting from phenylethylamine. This Ugi approach is more efficient than the industrial synthesis, which needs five steps.

Another example is the synthesis of *rac*-clopidogrel (Plavix®) **430** that is used as antiplatelet agent to inhibit the formation of blood clots in coronary artery disease (Scheme 6.65) [132]. Ross and coworkers have used a three-component Ugi reaction from commercially available starting materials (**425–427**) to construct intermediate **428** [133]. After acidic treatment of the enamine and esterification of the resulting acid **429**, they obtain *rac*-clopidogrel **430** in 73% yield (over three steps). The racemic compound is obtained in only three steps; however, the drug on the market is the (*R*)-enantiomer.

HIV protease inhibitors are often quite large and therefore, the synthesis needs up to 20 steps. The development of much shorter reactions will thus be of great interest. The piperazine key intermediate **435** in the synthesis of HIV protease inhibitor Crixivan (Indinavir®) **436** can be synthesized with an Ugi four-multicomponent reaction (Scheme 6.66) [134].

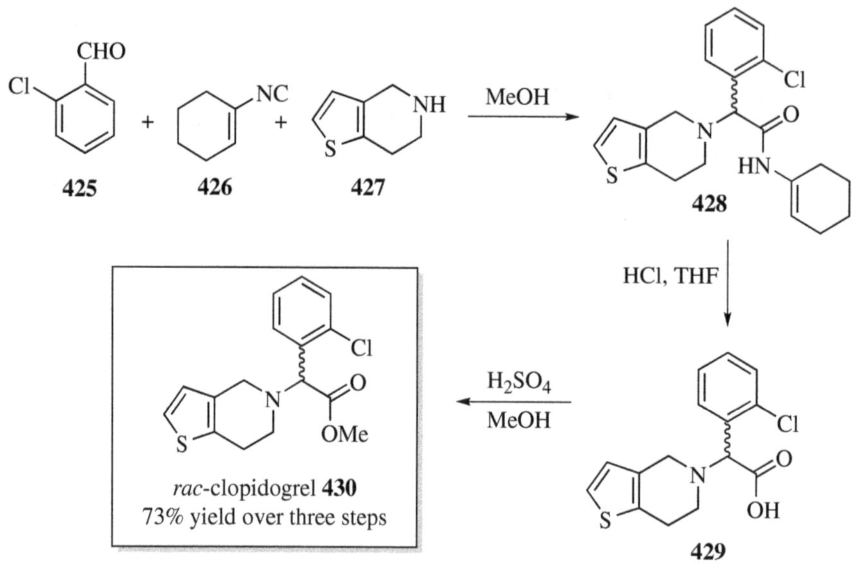

SCHEME 6.65 Multicomponent reaction of *rac*-clopidogrel **430**.

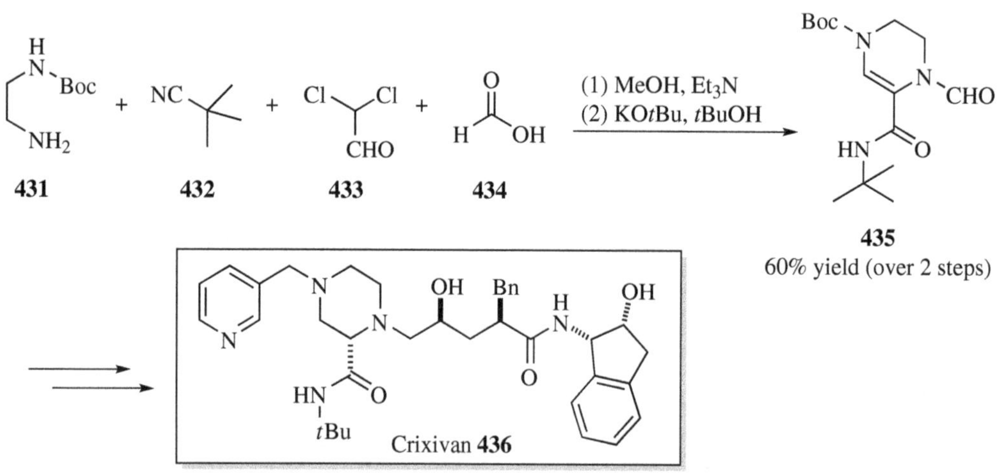

SCHEME 6.66 Four-component Ugi reaction in the synthesis of key intermediate **435** in the synthesis of Crixivan **436**.

FTY 720 **440** is synthesized by a three-multicomponent Petasis reaction involving benzylamine **437**, glycerine **438**, and boronic acid **439** (Scheme 6.67). The desired product is obtained in 50% yield, which is further hydrogenated to FTY 720 (Gilenya®) **440** (45% yield over 2 steps) [135]. The latter is a clinically investigated immunosuppressive and shows also promising clinical results in the treatment of multiple sclerosis.

Mandipropamid **444** is highly active against the economically most important phytopathogens *Phytophthora infestans* (potato and tomato late blight) and *Plasmopara viticola* (grape downy mildew), which caused complete devastation of the Irish potato crop between 1845 and 1849, when one million of Irish people died of starvation and two millions emigrated to North America [136]. The compound is rapidly absorbed to the wax layer of the plant surface, and since 2007, it is on the agrochemical market under the trade names Revus© and Pergado©. There are several synthetic routes for the obtainment of the fungicide; herein, a synthesis based on a Passerini-type reaction will be shown. Thus, Zeun and coworkers from Syngenta Crop Protection have used Seebach's variation of the Passerini reaction taking advantage of the directly formed α-hydroxycarboxamide **443** mediated by titanium(IV) chloride (Scheme 6.68) [137]. Using Seebach's variation, the application of Lewis acids, like $TiCl_4$, instead of carboxylic acids as acid component, delivers not the common α-acyloxycarboxamides, but directly α-hydroxycarboxamides. Then, the intermediate **443** is further alkylated with propargylbromide to yield the marketed product mandipropamid **444** in 79% yield.

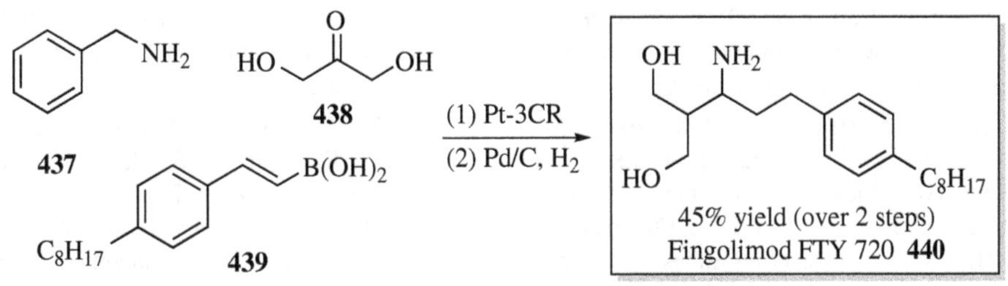

SCHEME 6.67 Multicomponent reaction to synthesize fingolimod FTY 720 **440**.

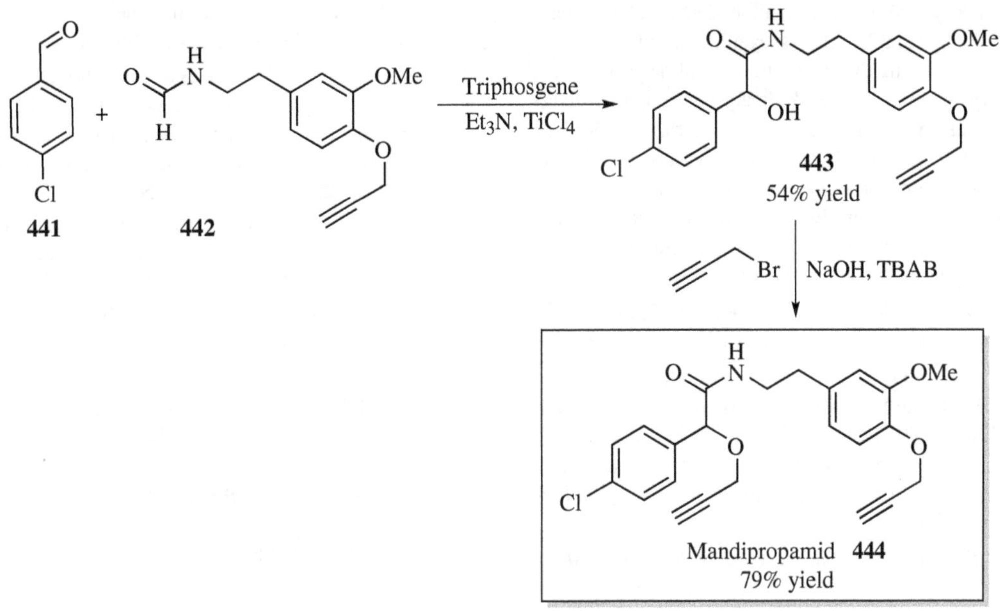

SCHEME 6.68 Seebach's variation of the Passerini reaction to synthesize fungicide mandipropamid **444**.

6.7 CONCLUSION

MCRs are powerful means for the construction of multiple bonds in functionalized molecules. They introduce molecular diversity in only one step and do so in a highly efficient and convergent manner. More than one and a half century after their discovery, MCRs have proven to be a very efficient manner to reach complex organic molecules with relevant biological activities, from polyheterocyclic compounds to peptide derivatives.

The simplicity and the large scope of those reactions have made them a powerful tool for combinatorial chemistry. Furthermore, by enabling the formation of many bonds in only one step and with usually few by-products, MCRs deal perfectly with the recent environmental issues and the concepts of "green chemistry" [138] and "atom economy" [139] and suit therefore perfectly to industrial applications.

The examples shown here demonstrate the force of MCRs in the formation of indoles, DHPMs, pyrroles, and amino acids. There is still place for improvement, for example, for the synthesis of thiazines. More often, several MCRs are combined in one reaction to get access to complicate natural products very efficiently.

REFERENCES

1 (a) N. Isambert, R. Lavilla, *Chem.–Eur. J.* 2008, 14, 8444–8454. Heterocycles as key substrates in multicomponent reactions: the fast lane towards molecular complexity. (b) B. B. Touré, D. G. Hall, *Chem. Rev.* 2009, 109, 4439–4486. Natural product synthesis using multicomponent reaction strategies.

2 (a) C. Merlot, D. Domine, C. Cleva, D. J. Church, *Drug Discov. Today* 2003, 8, 594–602. Chemical substructures in drug discovery. (b) L. Constantino, D. Barlocco, *Curr. Med. Chem.* 2006, 13, 65–85. Privileged structures as leads in medicinal chemistry.

3 (a) D. A. Horton, G. T. Bourne, M. L. Smythe, *Chem. Rev.* 2003, 103, 893–930. The combinatorial synthesis of bicyclic privileged structures or privileged substructures. For an overview of the field, see: (b) G. Wess, M. Urmann, B. Sickenberger, *Angew. Chem. Int. Ed.* 2001, 40, 3341–3350. Medicinal chemistry: challenges and opportunities.

4 A. Ulaczyk-Lesanko, D. G. Hall, *Curr. Opin. Chem. Biol.* 2005, 9, 266–276. Wanted: new multicomponent reactions for generating libraries of polycyclic natural products.

5 (a) B. C. O'Reilly, K. S. Atwal, *Heterocycles* 1987, 26, 1185–1188. Synthesis of substituted 1,2,3,4-tetrahydro-6-methyl-2-oxo-5-pyrimidinecarboxylic acid esters: the Biginelli condensation revisited. (b) B. L. Nilsson, L. E. Overman, *J. Org. Chem.* 2006, 71, 7706–7714. Concise synthesis of guanidine-containing heterocycles using the Biginelli reaction.

6 (a) A. Dömling, *Chem. Rev.* 2006, 106, 17–89. Recent developments in isocyanide based multicomponent reactions in applied chemistry. (b) I. Akritopoulou-Zanze, *Curr. Opin. Chem. Biol.* 2008, 12, 324–331. Isocyanide-based multicomponent reactions in drug discovery. (c) C. Hulme, J. Dietrich, *Mol. Diversity* 2009, 13, 195–207. Emerging molecular diversity from the intra-molecular Ugi reaction: iterative efficiency in medicinal chemistry. (d) I. Akritopoulou-Zanze, S. Djuric in *Synthesis of heterocycles via multicomponent reactions II* (R. V. A. Orru, E. Ruijter, Eds.). Springer-Verlag, Berlin, Germany, 2010, Vol. 25, pp. 231–289. Applications of MCR-derived heterocycles in drug discovery.

7 A. Dömling, W. Wang, K. Wang, *Chem. Rev.* 2012, 112, 3083–3135. Chemistry and biology of multicomponent reactions.

8 M. Shoda, T. Harada, Y. Kogami, R. Tsujita, H. Akashi, H. Kouji, F. L. Stahura, L. Xue, J. Bajorath, *J. Med. Chem.* 2004, 47, 4286–4290. Identification of structurally diverse growth hormone secretagogue agonists by virtual screening and structure-activity relationship analysis of 2-formylamino-acetamide derivatives.

9 S. Guery, M. Schmitt, J.-J. Bourguignon, *Synlett* 2002, 2003–2006. An expeditive method for the synthesis of arginine-based chemical libraries.

10 D. R. Armour, A. S. Bell, P. J. Edwards, D. Ellis, D. Hepworth, M. L. Lewis, C. R. Smith, 2004, WO 2004020414. Preparation of heterocyclylcarboxamides as oxytocin inhibitors.

11 A. D. Borthwick, R. J. Hatley, D. M. B. Hickey, J. Liddle, D. G. H. Livermore, A. M. Mason, N. D. Miller, F. Nerozzi, S. L. Sollis, A. K. Szardenings, P. G. Wyatt, 2003, WO 2003053443. Preparation of substituted diketopiperazines as oxytocin antagonists.

12 R. Nishizawa, Y. Takaoka, S. Shibayama, 2004, WO 2004080966. Preparation of nitrogen-containing heterocyclic derivatives as chemokine receptor CCR5 antagonists and drugs containing the same as the active ingredient.

13 X. P. Lin, L. Wenwei, Z. Xiaomin, *Synthesis* 2002, 1017–1026. Synthesis of a peptidomimetic HCMV protease inhibitor library.

14 M. Thormann, M. Almstetter, 2004, WO 2004099124. Method of preparation of bioisosteres of actinonin of interest as metalloproteinase inhibitors.

15 (a) K. G. Zbinden, U. Obst-Sander, K. Hilpert, H. Kühne, D. W. Banner, H.-J. Böhm, M. Stahl, J. Ackermann, L. Alig, L. Weber, H. P. Wessel, M. A. Riederer, T. B. Tschopp, T. Lavé, *Bioorg. Med. Chem. Lett.* 2005, 15, 5344–5352. Selective and orally bioavailable phenylglycine tissue factor/factor VIIa inhibitors. (b) K. G. Zbinden, D. W. Banner, J. Ackermann, A. D'arcy, D. Kirchhofer, Y.-H. Ji, T. B. Tschopp, S. Wallbaum, L. Weber, *Bioorg. Med. Chem. Lett.* 2005, 15, 817–822. Design of selective phenylglycine amide tissue factor/factor VIIa inhibitors.

16 K. G. Zbinden, D. W. Banner, K. Hilpert, J. Himber, T. Lavé, M. A. Riederer, M. Stahl, T. B. Tschopp, U. Obst-Sander, *Bioorg. Med. Chem.* 2006, 14, 5357–5369. Dose-dependent antithrombotic activity of an orally active tissue factor/factor VIIa inhibitor without concomitant enhancement of bleeding propensity.

17 C. B. Gilley, M. J. Buller, Y. Kobayashi, *Org. Lett.* 2007, 9, 3631–3634. New entry to convertible isocyanides for the Ugi reaction and its application to the stereocontrolled formal total synthesis of the proteasome inhibitor omuralide.

18 (a) S. Lage, I. Villaluenga, N. Sotomayor, E. Lete, *Synlett* 2008, 3188–3192. Synthesis of pyrrolo[1,2-*b*]isoquinolines through mesityllithium-mediated intramolecular carbolithiation. (b) O. Kreye, B. Westermann, L. A. Wessjohann, *Synlett* 2007, 3188–3192. A stable, convertible isonitrile as a formic acid carbanion [$^{-}$COOH] equivalent and its application in multicomponent reactions. (c) J. Isaacson, C. B. Gilley, Y. Kobayashi, *J. Org. Chem.* 2007, 72, 3913–3916. Expeditious access to unprotected racemic pyroglutamic acids.

19 O. Pando, S. Stark, A. Denkert, A. Porzel, R. Preusentanz, L. A. Wessjohann, *J. Am. Chem. Soc.* 2011, 133, 7692–7695. The multiple multicomponent approach to natural product mimics: tubugis, N-substituted anticancer peptides with picomolar activity.

20 A. Dömling, B. Beck, U. Eichelberger, S. Sakamuri, S. Menon, Q.-Z. Chen, Y. Lu, L. A. Wessjohann, *Angew. Chem. Int. Ed.* 2007, 46, 2347–2348. Total synthesis of tubulysin U and V.

21 A. Znabet, M. M. Polak, E. Janssen, F. J. J. de Kanter, N. J. Turner, R. V. A. Orru, E. Ruijter, *Chem. Commun.* 2010, 46, 7918–7920. A highly efficient synthesis of telaprevir by strategic use of biocatalysis and multicomponent reactions.

22 A. Endo, A. Yanagisawa, M. Abe, S. Tohma, T. Kan, T. Fukuyama, *J. Am. Chem. Soc.* 2002, 124, 6552–6554. Total synthesis of ecteinascidin 743.

23 D. J. Kerr, A. C. Willis, B. L. Flynn, *Org. Lett.* 2004, 6, 457–460. Multicomponent coupling approach to (±)-frondosin B and a ring-expanded analogue.

24 (a) P. R. Sebahar, R. M. Williams, *J. Am. Chem. Soc.* 2000, 122, 5666–5667. The asymmetric total synthesis of (+)- and (−)-spirotryprostatin B. (b) A. Ulaczyk-Lesanko, D. G. Hall, *Curr. Opin. Chem. Biol.* 2005, 9, 266–276. Wanted: new multicomponent reactions for generating libraries of polycyclic natural products.

25 D. A. Powell, R. A. Batey, *Org. Lett.* 2002, 4, 2913–2916. Total synthesis of the alkaloids martinelline and martinellic acid *via* a hetero Diels-Alder multicomponent coupling reaction.

26 K. Inanaga, K. Takasu, M. Ihara, *J. Am. Chem. Soc.* 2004, 126, 1352–1353. Rapid assembly of polycyclic substances by a multicomponent cascade [4+2]–[2+2] cycloadditions: total synthesis of the proposed structure of paesslerin A.

27 (a) K. L. Rinehart, T. G. Holt, N. L. Fregeau, J. G. Stroh, P. A. Keifer, F. Sun, L. H. Li, D. G. Martin, *J. Org. Chem.* 1990, 55,

4512–4515. Ecteinascidins 729, 743, 745, 759A, 759B, and 770: potent antitumor agents from the Caribbean tunicate *Ecteinascidia turbinata*. (b) R. Sakai, K. L. Rinehart, Y. Guan, A. H. Wang, *Proc. Natl. Acad. Sci. U.S.A.* 1992, 89, 11456–11460. Additional antitumor ecteinascidins from a Caribbean tunicate: crystal structures and activities *in vivo*. For a review on antitumor compounds from tunicates, see: (c) K. L. Rinehart, *Med. Res. Rev.* 2000, 20, 1–27. Antitumor compounds from tunicates.

28 J. D. Sunderhaus, C. Dockendorff, S. F. Martin, *Org. Lett.* 2007, 9, 4223–4226. Applications of multicomponent reactions for the synthesis of diverse heterocyclic scaffolds.

29 K. M. Witherup, R. W. Ransom, A. C. Graham, A. M. Bernard, M. J. Salvatore, W. C. Lumma, P. S. Anderson, S. M. Pitzenberger, S. L. Varga, *J. Am. Chem. Soc.* 1995, 117, 6682–6685. Martinelline and martinellic acid, novel G-protein linked receptor antagonists from the tropical plant *Martinella iquitosensis* (bignoniaceae).

30 (a) P. R. Nandaluru, G. J. Bodwell, *Org. Lett.* 2012, 14, 310–313. Multicomponent synthesis of 6*H*-dibenzo[*b*,*d*]pyran-6-ones and a total synthesis of cannabinol. (b) J. W. Huffman, S. A. Hepburn, N. Lyutenko, A. L. S. Thompson, J. L. Wiley, D. E. Selley, B. R. Martin, *Bioorg. Med. Chem.* 2010, 18, 7809–7815. 1-Bromo-3-(1′,1′-dimethylalkyl)-1-deoxy-Δ^8-tetrahydrocannabinols: new selective ligands for the cannabinoid CB2 receptor.

31 (a) P. Kotame, B.-C. Hong, J.-H. Liao, *Tetrahedron Lett.* 2009, 50, 704–707. Enantioselective synthesis of the tetrahydro-6*H*-benzo[*c*]chromenes *via* domino Michael-Aldol condensation: control of five stereocenters in a quadruple-cascade organocatalytic multi-component reaction. (b) B.-C. Hong, P. Kotame, C.-W. Tsai, J.-H. Liao, *Org. Lett.* 2010, 12, 776–779. Enantioselective total synthesis of (+)-conicol *via* cascade three-component organocatalysis.

32 (a) D. A. Horton, G. T. Bourne, M. L. Smythe, *Chem. Rev.* 2003, 103, 893–930. The combinatorial synthesis of bicyclic privileged structures or privileged substructures. (b) For an overview of the field, see: G. Wess, M. Urmann, B. Sickenberger, *Angew. Chem. Int. Ed.* 2001, 40, 3341–3350. Medicinal chemistry: challenges and opportunities. For a recent review, see: (c) R. Dua, S. Shrivastava, S. K. Sonwane, S. K. Srivastava, *Adv. Biol. Res.* 2011, 5, 120–144. Pharmacological significance of synthetic heterocycles scaffold: a review.

33 (a) J. Sapi, J.-Y. Laronze, *Arkivoc* 2004, (vii), 208–222. Indole based multicomponent reactions towards functionalized heterocycles. (b) D. M. D'Souza, T. J. J. Müller, *Chem. Soc. Rev.* 2007, 36, 1095–1108. Multi-component syntheses of heterocycles by transition-metal catalysis. (c) N. Isambert, R. Lavilla, *Chem.–Eur. J.* 2008, 14, 8444–8454. Heterocycles as key substrates in multicomponent reactions: the fast lane towards molecular complexity. (d) J. D. Sunderhaus, S. F. Martin, *Chem.–Eur. J.* 2009, 15, 1300–1308. Applications of multicomponent reactions to the synthesis of diverse heterocyclic scaffolds.

34 K. L. Vine, L. Matesic, J. M. Locke, M. Ranson, D. Skropeta, *Anticancer Agents Med. Chem.* 2009, 9, 397–414. Cytotoxic and anticancer activities of isatin and its derivatives: a comprehensive review from 2000–2008.

35 G. W. Gribble (Ed.) in *Top. Heterocycl. Chem.* Springer-Verlag, Berlin, Germany, 2010, Vol. 26.

36 (a) G. Battistuzzi, S. Cacchi, G. Fabrizi, *Eur. J. Org. Chem.* 2002, 2671–2681. The aminopalladation/reductive elimination domino reaction in the construction of functionalized indole rings. (b) S. Cacchi, G. Fabrizi, *Chem. Rev.* 2005, 105, 2873–2920. Synthesis and functionalization of indoles through palladium-catalyzed reactions. (c) G. R. Humphrey, J. T. Kuethe, *Chem. Rev.* 2006, 106, 2875–2911. Practical methodologies for the synthesis of indoles. (d) J. Campo, M. García-Valverde, S. Marcaccini, M. J. Rojo, T. Torroba, *Org. Biomol. Chem.* 2006, 4, 757–765. Synthesis of indole derivatives *via* isocyanides. (e) J. Barluenga, F. Rodríguez, F. J. Fañanás, *Chem. Asian J.* 2009, 4, 1036–1048. From simple starting materials to functionalized indoles and quinolones.

37 (a) M. Bandini, A. Eichholzer, *Angew. Chem. Int. Ed.* 2009, 48, 9608–9644. Catalytic functionalization of indoles in a new dimension. (b) E. Marqués-López, A. Diez-Martinez, P. Merino, R. P. Herrera, *Curr. Org. Chem.* 2009, 13, 1585–1609. The role of the indole in important organocatalytic enantioselective Friedel-Crafts alkylation reactions. (c) M. Shiri, M. A. Zolfigol, H. G. Kruger, Z. Tanbakouchian, *Chem. Rev.* 2010, 110, 2250–2293. Multicomponent reactions of isocyanides in the synthesis of heterocycles. (d) G. Bartoli, G. Benciveni, R. Dalpozzo, *Chem. Soc. Rev.* 2010, 39, 4449–4465. Multicomponent reactions of isocyanides in the synthesis of heterocycles.

38 (a) J. Sapi, J.-Y. Laronze, *Arkivoc* 2004, (vii), 208–222. Indole based multicomponent reactions towards functionalized heterocycles. (b) S. Sadjadi, M. M. Heravi, *Tetrahedron* 2011, 67, 2707–2752. Recent application of isocyanides in synthesis of heterocycles. (c) A. V. Ivachtchenko, Y. A. Ivanenkov, V. M. Kysil, M. Y. Krasavin, A. P. Ilyin, *Russ. Chem. Rev.* 2010, 79, 787–817. Multicomponent reactions of isocyanides in the synthesis of heterocycles. (d) M. Shiri, *Chem. Rev.* 2012, 112, 3508–3549. Indoles in multicomponent processes (MCPs).

39 Y. Huang, A. M. Walji, C. H. Larsen, D. W. C. MacMillan, *J. Am. Chem. Soc.* 2005, 127, 15051–15053. Enantioselective organo-cascade catalysis.

40 Y. Chi, S. T. Scroggins, J. M. J. Fréchet, *J. Am. Chem. Soc.* 2008, 130, 6322–6323. One-pot multi-component asymmetric cascade reactions catalyzed by soluble star polymers with highly branched non-interpenetrating catalytic cores.

41 G.-W. Zhang, L. Wang, J. Nie, J.-A. Ma, *Adv. Synth. Catal.* 2008, 350, 1457–1463. Chiral Brønsted acid-mediated enantioselective organocatalytic three-component reaction for the construction of trifluoromethyl-containing molecules.

42 For other examples of use of chiral phosphoric acid as catalyst in the synthesis of indole derivatives, see: (a) Q. Kang, Z.-A. Zhao, S.-L. You, *Tetrahedron* 2009, 65, 1603–1607. Enantioselective synthesis of (3-indolyl)glycine derivatives *via* asymmetric Friedel-Crafts reaction between indoles and glyoxylate imines. (b) M. Terada, K. Machioka, K. Sorimachi, *Angew. Chem. Int. Ed.* 2009, 48, 2553–2556. Activation of hemiaminal ethers by chiral Brønsted acids for facile access to enantioselective two-carbon homologation using enecarbamates.

43 P. Galzerano, F. Pesciaioli, A. Mazzanti, G. Bartoli, P. Melchiorre, *Angew. Chem. Int. Ed.* 2009, 48, 7892–7894. Asymmetric organocatalytic cascade reactions with α-substituted α,β-unsaturated aldehydes.

44 X. Wu, X. Dai, L. Nie, H. Fang, J. Chen, Z. Ren, W. Cao, G. Zhao, *Chem. Commun.* 2010, 46, 2733–2735. Organocatalyzed enantioselective one-pot three-component access to indoloquinolizidines by a Michael addition-Pictet-Spengler sequence.

45 For other example of the use of same type of catalyst, in the obtainment of indole derivatives, see: D. Enders, C. Wang, M. Mukanova, A. Greb, *Chem. Commun.* 2010, 46, 2447–2449. Organocatalytic asymmetric synthesis of polyfunctionalized 3-(cyclohexenylmethyl)-indoles *via* a quadruple domino Friedel-Crafts-type/Michael/Michael/aldol condensation reaction.

46 Z. Chen, B. Wang, Z. Wang, G. Zhu, J. Sun, *Angew. Chem. Int. Ed.* 2013, 52, 2027–2031. Complex bioactive alkaloid-type polycycles through efficient catalytic asymmetric multicomponent aza-Diels-Alder reaction of indoles with oxetane as directing group.

47 T. Arai, N. Yokoyama, *Angew. Chem. Int. Ed.* 2008, 47, 4989–4992. Tandem catalytic asymmetric Friedel-Crafts/Henry reaction: control of three contiguous acyclic stereocenters.

48 T. Arai, M. Wasai, N. Yokoyama, *J. Org. Chem.* 2011, 76, 2909–2912. Easy access to fully functionalized chiral tetrahydro-β-carboline alkaloids.

49 (a) B. L. Flynn, G. P. Flynn, E. Hamel, M. K. Jung, *Bioorg. Med. Chem. Lett.* 2001, 11, 2341–2343. The synthesis and tubulin binding activity of thiophene-based analogues of combretastatin A-4. (b) D. J. Kerr, E. Hamel, M. K. Jung, B. L. Flynn, *Bioorg. Med. Chem.* 2007, 15, 3290–3298. The concise synthesis of chalcone, indanone and indenone analogues of combretastatin A4.

50 J. Kamalraja, D. Muralidharan, P. T. Perumal, *Synlett* 2012, 23, 2894–2898. An efficient, one-pot regioselective synthesis of highly functionalized chromene-5-ones and pyrano[3,2-*c*] chromene-5-ones *via* a tandem Knoevenagel-Michael-cyclization sequence.

51 (a) E. Valencia, A. J. Freyer, M. Shamma, V. Fajardo, *Tetrahedron Lett.* 1984, 25, 599–602. (±)-Nuevamine, an isoindoloisoquinoline alkaloid, and (±)-lennoxamine, an isoindolobenzazepine. (b) R. Alonso, L. Castedo, D. Dominguez, *Tetrahedron Lett.* 1985, 26, 2925–2928. Synthesis of isoindoloisoquinoline alkaloids. A revision of the structure of (±)-nuevamine.

52 A. Islas-Jácome, L. E. Cárdenas-Galindo, A. V. Jerezano, J. Tamariz, E. González-Zamora, R. Gámez-Montaño, *Synlett* 2012, 23, 2951–2956. Synthesis of nuevamine aza-analogues by a sequence: I-MCR-aza-Diels-Alder-Pictet-Spengler.

53 (a) A. Zamudio-Medina, M. C. García-Gonzalez, J. Padilla, E. González-Zamora, *Tetrahedron Lett.* 2010, 51, 4837–4839. Synthesis of a tetracyclic lactam system of nuevamine by four-component reaction and free radical cyclization. (b) A. Islas-Jácome, E. González-Zamora, R. Gámez-Montaño, *Tetrahedron Lett.* 2011, 52, 5245–5248. A short microwave-assisted synthesis of tetrahydroisoquinolin-pyrrolopyridinones by a triple process: Ugi-3CR-aza Diels-Alder/S-oxidation/Pummerer.

54 E. Airiau, N. Girard, A. Mann, J. Salvadori, M. Taddei, *Org. Lett.* 2009, 11, 5314–5317. Four-component reactions toward fused heterocyclic rings.

55 A. Wrobleski, K. Sahasrabudhe, J. Aubé, *J. Am. Chem. Soc.* 2002, 124, 9974–9975. Asymmetric total synthesis of dendrobatid alkaloid 251 F.

56 A. Kuramochi, H. Usuda, K. Yamatsugu, M. Kanai, M. Shibasaki, *J. Am. Chem. Soc.* 2005, 127, 14200–14201. Total synthesis of (±)-garsubellin A.

57 M. Xie, X. Chen, Y. Zhu, B. Gao, L. Lin, X. Liu, X. Feng, *Angew. Chem. Int. Ed.* 2010, 49, 3799–3802. Asymmetric three-component inverse electron-demand aza-Diels-Alder reaction: efficient synthesis of ring-fused tetrahydroquinolines.

58 G. Dagousset, P. Retailleau, G. Masson, J. Zhu, *Chem.–Eur. J.* 2012, 18, 5869–5873. Chiral phosphoric acid-catalyzed enantioselective three-component Povarov reaction using cyclic enethioureas as dienophiles: stereocontrolled access to enantio-enriched hexahydropyrroloquinolines.

59 B. Jiang, Q.-Y. Li, S.-J. Tu, G. Li, *Org. Lett.* 2012, 14, 5210–5213. Three-component domino reactions for selective formation of indeno[1,2-*b*]indole derivatives.

60 R. R. Kumar, S. Perumal, P. Senthilkumar, P. Yogeeswari, D. Sriram, *Eur. J. Med. Chem.* 2009, 44, 3821–3829. Facile synthesis and antimycobacterial evaluation of novel spiro-pyrido-pyrrolizines and pyrrolidines.

61 (a) A. H. Abdel-Rahman, E. M. Keshk, M. A. Hanna, S. M. El-Bady, *Bioorg. Med. Chem.* 2004, 12, 2483–2488. Synthesis and evaluation of some new spiro indoline-based heterocycles as potentially active antimicrobial agents. (b) A. P. Antonchick, C. Gerding-Reimers, M. Catarinella, M. Schürmann, H. Preut, S. Ziegler, D. Rauh, H. Waldmann, *Nat. Chem.* 2010, 2, 735–740. Highly enantioselective synthesis and cellular evaluation of spirooxindoles inspired by natural products. (c) S. M. Rajesh, S. Perumal, J. C. Menendez, P. Yogeeswari, D. Sriram, *Med. Chem. Commun.* 2011, 2, 626–630. Antimycobacterial activity of spirooxindolo-pyrrolidine, pyrrolizine and pyrrolothiazole hybrids obtained by a three-component regio- and stereoselective 1,3-dipolar cycloaddition. (d) V. V. Vintonyak, K. Warburg, H. Kruse, S. Grimme, K. Hübel, D. Rauh, H. Waldmann, *Angew. Chem. Int. Ed.* 2010, 49, 5902–5905. Identification of thiazolidinones spiro-fused to indolin-2-ones as potent and selective inhibitors of the mycobacterium tuberculosis protein tyrosine phosphatase B.

62 (a) C. V. Galliford, K. A. Scheidt, *Angew. Chem. Int. Ed.* 2007, 46, 8748–8758. Pyrrolidinyl-spirooxindole natural products as inspirations for the development of therapeutic agents. (b) B. M. Trost, M. K. Brennan, *Synthesis* 2009, 3003–3025. Asymmetric syntheses of oxindole and indole spirocyclic alkaloid natural products. (c) N. R. Ball-Jones, J. J. Badillo, A. K. Franz, *Org. Biomol. Chem.* 2012, 10, 5165–5181. Strategies for the enantioselective synthesis of spirooxindoles.

63 W.-B. Chen, Z.-J. Wu, Q.-L. Pei, L.-F. Cun, X.-M. Zhang, W.-C. Yuan, *Org. Lett.* 2010, 12, 3132–3135. Highly enantioselective construction of spiro[*4H*-pyran-3,3′-oxindoles] through a domino Knoevenagel/Michael/cyclization sequence catalyzed by cupreine.

64 N. Sucman, F. Shepeli, M. Zveaghintseva, V. Pogrebnoi, F. Macaev, *Symmetry* 2011, 3, 165–170. Facile and convenient

one-pot process for the synthesis of spirooxindole derivatives in high optical purity using (−)-(*S*)-brevicolline as an organocatalyst.

65 (a) G. Shanthi, G. Subbulakshmi, P. T. Perumal, *Tetrahedron* 2007, 63, 2057–2063. A new $InCl_3$-catalyzed, facile and efficient method for the synthesis of spirooxindoles under conventional and solvent-free microwave conditions. (b) S.-L. Zhu, S.-J. Ji, Y. Zhang, *Tetrahedron* 2007, 63, 9365–9372. A simple and clean procedure for three-component synthesis of spirooxindoles in aqueous medium.

66 H. Chen, D. Shi, *J. Comb. Chem.* 2010, 12, 571–576. Efficient one-pot synthesis of novel spirooxindole derivatives *via* three-component reaction in aqueous medium.

67 X.-H. Chen, Q. Wei, S.-W. Luo, H. Xiao, L.-Z. Gong, *J. Am. Chem. Soc.* 2009, 131, 13819–13825. Organocatalytic synthesis of spiro[pyrrolidin-3,3′-oxindoles] with high enantiopurity and structural diversity.

68 A. K. Arya, M. Kumar, *Mol. Divers.* 2011, 15, 781–789. Base catalyzed multicomponent synthesis of spiroheterocycles with fused heterosystems.

69 K. Zhao, S.-L. Zhu, D.-Q. Shi, X.-P. Xu, S.-J. Ji, *Synthesis* 2010, 1793–1803. Regioselective synthesis of highly functionalized 3-spiropyrrolidine/pyrrolizidine oxindoles and acenaphthenones *via* one-pot four-component [3+2] cycloaddition.

70 Z. Huang, Q. Zhao, G. Chen, H. Wang, W. Lin, L. Xu, H. Liu, J. Wang, D. Shi, Y. Wang, *Molecules* 2012, 17, 12704–12717. An efficient synthesis of novel dispirooxindole derivatives *via* one-pot three component 1,3-dipolar cycloaddition reactions.

71 X. Wang, S.-Y. Wang, S.-J. Ji, *Org. Lett.* 2013, 15, 1954–1957. Isocyanide-based multicomponent reactions: catalyst-free stereoselective construction of polycyclic spiroindolines.

72 For other example of a multicomponent preparation of spiroindolines, see: L. El Kaïm, L. Grimaud, X.-F. Le Goff, M. Menes-Arzate, L. D. Miranda, *Chem. Commun.* 2011, 47, 8145–8147. Straightforward four-component access to spiroindolines.

73 (a) X.-G. Tong, L.-L. Zhou, Y.-H. Wang, C. Xia, Y. Wang, M. Liang, F.-F. Hou, Y.-X. Cheng, *Org. Lett.* 2010, 12, 1844–1847. Acortatarins A and B, two novel antioxidative spiroalkaloids with a naturally unusual morpholine motif from *Acorus tatarinowii*. (b) G. Sudhakar, V. D. Kadam, S. Bayya, G. Pranitha, B. Jagadeesh, *Org. Lett.* 2011, 13, 5452-5455. Total synthesis and stereochemical revision of acortatarins A and B. (c) R. Wijtmans, M. K. S. Vink, H. E. Shoemaker, F. L. van Delft, R. H. Blaauw, F. P. J. T. Rutjes, *Synthesis* 2004, 641–662. Biological relevance and synthesis of *C*-substituted morpholine derivatives.

74 D. Kim, D. H. Ryu, J. Y. Lee, N. Lee, Y. Kim, J. Kim, K. Chang, G. Im, T. Kim, W. Choi, *J. Med. Chem.* 2001, 44, 1594–1602. Synthesis and biological evaluation of novel A-ring modified hexacyclic camptothecin analogues.

75 P. Melloni, A. Della Torre, E. Lazzari, G. Mazzini, M. Meroni, *Tetrahedron* 1985, 41, 1393–1399. Configurational studies on 2-[α-(2-ethoxyphenoxy)benzyl] morpholine fce 20124.

76 C. Guilloneau, Y. Charton, Y. Ginot, M. Fouquier-d'Hérouël, M. Bertrand, B. Lockhart, P. Lestage, S. Goldstein, *Eur. J. Med. Chem.* 2003, 38, 1–11. Synthesis and pharmacological evaluation of new 1,2-dithiolane based antioxidants.

77 K. Araki, T. Kuroda, S. Uemori, A. Moriguchi, Y. Ikeda, F. Hirayama, Y. Yokoyama, E. Iwao, T. Yakushiji, *J. Med. Chem.* 1993, 36, 1356–1363. Quinolone antimicrobial agents substituted with morpholines at the 7-position. Syntheses and structure-activity relationships.

78 J. Zhou, L. Zhou, Y.-Y. Yeung, *Org. Lett.* 2012, 14, 5250–5253. Multicomponent approach in the synthesis of 2,2,6-trisubstituted morpholine derivatives.

79 A. R. S. Babu, D. Gavaskar, R. Raghunathan, *J. Organomet. Chem.* 2013, 745–746, 409–416. An expedient ultrasonic assisted one-pot four component synthesis of novel ferrocene grafted pyrrolidine heterocycles *via* [3+2]-cycloaddition of azomethine ylides.

80 R. S. Kumar, S. M. Rajesh, D. Banerjee, P. Yogeeswari, D. Sriram, S. Perumal, *Eur. J. Med. Chem.* 2010, 45, 411–422. Novel three-component domino reactions of ketones, isatin and amino acids: synthesis and discovery of antimycobacterial activity of highly functionalised novel dispiropyrrolidines.

81 S. Santra, P. R. Andreana, *Angew. Chem. Int. Ed.* 2011, 50, 9418–9422. A bioinspired Ugi/Michael/aza-Michael cascade reaction in aqueous media: natural-product-like molecular diversity.

82 B. Jiang, B.-M. Feng, S.-L. Wang, S.-J. Tu, G. Li, *Chem.–Eur. J.* 2012, 18, 9823–9826. Domino constructions of pentacyclic indeno[2,1-*c*]quinolines and pyrano[4,3-*b*]oxepines by [4+1]/[3+2+1]/[5+1] and [4+3] multiple cyclizations.

83 M. A. Arnold, K. A. Day, S. G. Durón, D. Y. Gin, *J. Am. Chem. Soc.* 2006, 128, 13255–13260. Total synthesis of (+)-batzelladine A and (−)-batzelladine D *via* [4+2]-annulation of vinyl carbodiimides with N-alkyl imines.

84 R. A. Davis, S. Duffy, S. Fletcher, V. M. Avery, R. J. Quinn, *J. Org. Chem.* 2013, 78, 9608–9613. Thiaplakortones A-D: antimalarial thiazine alkaloids from the Australian marine sponge *Plakortis lita*.

85 E. P. Abraham, G. G. F. Newton, *Biochem. J.* 1961, 79, 377–393. The structure of cephalosporin C.

86 G. C. Rovnyak, S. D. Kimball, B. Beyer, G. Cucinotta, J. D. Dimarco, J. Gougoutas, A. Hedberg, B. C. Oreilly, *J. Med. Chem.* 1991, 34, 806–811. Dihydropyrimidine calcium channel blockers. 3. 3-Carbamoyl-4-aryl-1,2,3,4-tetrahydro-6-methyl-5-pyrimidinecarboxylic acid esters as orally effective antihypertensive agents.

87 P. Biginelli, *Gazz. Chim. Ital.* 1893, 360–413. Aldehyde-urea derivatives of aceto- and oxaloacetic acids.

88 C. O. Kappe, *QSAR Comb. Sci.* 2003, 22, 630–645. The generation of dihydropyrimidine libraries utilizing Biginelli multicomponent chemistry.

89 For synthesis of DHPMs *via* four component reactions between phosphonates, nitriles, aldehydes and isocyanates, see: (a) D. J. Vugts, M. M. Koningstein, R. F. Schmitz, F. J. J. de Kanter, M. B. Groen, R. V. A. Orru, *Chem.–Eur. J.* 2006, 12, 7178–7189. Multicomponent synthesis of dihydropyrimidines and thiazines. (b) D. J. Vugts, H. Jansen, R. F. Schmitz, F. J. J. de Kanter, R. V. A. Orru, *Chem. Commun.* 2003, 2594–2595.

A novel four-component reaction for the synthesis of functionalised dihydropyrimidines.

90 J. M. Goss, S. E. Schaus, *J. Org. Chem.* 2008, 73, 7651–7656. Enantioselective synthesis of SNAP-7941: chiral dihydropyrimidone inhibitor of MCH1-R.

91 (a) Z. F. Aron, L. E. Overman, *J. Am. Chem. Soc.* 2005, 127, 3380–3390. Total synthesis and properties of the crambescidin core zwitterionic acid and crambescidin 359. (b) Y. H. Rhee, L. E. Overman, *J. Am. Chem. Soc.* 2005, 127, 15652–15658. Total synthesis of (−)-crambidine and definition of the relative configuration of its unique tetracyclic guanidinium core. (c) F. Cohen, L. E. Overman, *J. Am. Chem. Soc.* 2006, 128, 2594–2604. Evolution of a strategy for the synthesis of structurally complex batzelladine alkaloids. Enantioselective total synthesis of the proposed structure of batzelladine F and structural revision.

92 B. A. Nilsson, L. E. Overman, *J. Org. Chem.* 2006, 71, 7706–7714. Concise synthesis of guanidine-containing heterocycles using the Biginelli reaction.

93 G. Sankar, R. B. N. Baig, S. Chandrashekaran, B. R. P. Kumar, *Eur. J. Med. Chem.* 2009, 44, 4192–4198. Novel Biginelli dihydropyrimidines with potential anticancer activity: a parallel synthesis and CoMSIA study.

94 N. Sharma, U. K. Sharma, R. Kumar, Richa, A. K. Sinha, *RSC Adv.* 2012, 2, 10648–10651. Green and recyclable glycine nitrate ($GlyNO_3$) ionic liquid triggered multicomponent Biginelli reaction for the efficient synthesis of dihydropyrimidinones.

95 A. Kruithof, M. L. Ploeger, E. Janssen, M. Helliwell, F. J. J. de Kanter, E. Ruijter, R. V. A. Orru, *Molecules* 2012, 17, 1675–1685. Multicomponent synthesis of 3,6-dihydro-2*H*-1,3-thiazine-2-thiones.

96 J. Wu, C. Wang, X. He, Z. Yan, M. Hu, F. Zhou, Y. Shang, *RSC Adv.* 2013, 3, 4643–4651. Synthesis of 3,4-dihydro-2*H*-1,4-benzo[*b*]thiazine derivatives *via* DABCO-catalyzed one-pot three-component condensation reactions.

97 (a) P. A. Jacobi, L. D. Coults, J. S. Guo, S. I. Leung, *J. Org. Chem.* 2000, 65, 205–213. New strategies for the synthesis of biologically important tetrapyrroles. The "B,C+D+A" approach to linear tetrapyrroles. (b) A. Fürstner, *Angew. Chem. Int. Ed.* 2003, 42, 3528–3603. Chemistry and biology of roseophilin and the prodigiosin alkaloids: a survey of the last 2500 years.

98 (a) D. O'Hagan, *Nat. Prod. Rep.* 2000, 17, 435–446. Pyrrole, pyrrolidine, pyridine, piperidine and tropane alkaloids. (b) C. T. Walsh, S. Garneau-Tsodikova, A. R. Howard-Jones, *Nat. Prod. Rep.* 2006, 23, 517–531. Biological formation of pyrroles: nature's logic and enzymatic machinery.

99 (a) M. Kidwai, R. Venkataramanan, R. Mohan, P. Sapra, *Curr. Med. Chem.* 2002, 9, 1209–1228. Cancer chemotherapy and heterocyclic compounds. (b) G. LaRegina, R. Silvestri, M. Artico, A. Lavechia, E. Novellino, O. Befani, P. Turini, E. Agostinelli, *J. Med. Chem.* 2007, 50, 922–931. New pyrrole inhibitors of monoamine oxidase: synthesis, biological evaluation, and structural determinants of MAO-A and MAO-B selectivity.

100 A. Loudet, K. Burgess, *Chem. Rev.* 2007, 107, 4891–4932. BODIPY dyes and their derivatives: syntheses and spectroscopic properties.

101 (a) F. Bellina, R. Rossi, *Tetrahedron* 2006, 62, 7213–7256. Synthesis and biological activity of pyrrole, pyrroline and pyrrolidine derivatives with two aryl groups on adjacent positions. (b) H. N. Fan, J. Peng, M. T. Hamann, J. F. Hu, *Chem. Rev.* 2008, 108, 264–287. Lamellarins and related pyrrole-derived alkaloids from marine organisms.

102 (a) V. F. Ferreira, M. C. B. V. de Souza, A. C. Cunha, L. O. R. Pereira, M. L. G. Ferreira, *Org. Prep. Proced. Int.* 2001, 33, 411–454. Recent advances in the synthesis of pyrroles. (b) M. Abid, S. M. Landge, B. Török, *Org. Prep. Proced. Int.* 2006, 38, 495–500. An efficient and rapid synthesis of *N*-substituted pyrroles by microwave assisted solid acid catalysis.

103 (a) O. V. Larionov, A. Meijere, *Angew. Chem. Int. Ed.* 2005, 44, 5664–5667. Versatile direct synthesis of oligosubstituted pyrroles by cycloaddition of α-metalated isocyanides to acetylenes. (b) G. Minetto, L. F. Raveglia, A. Sega, M. Taddei, *Eur. J. Org. Chem.* 2005, 24, 5277–5288. Microwave-assisted Paal-Knorr reaction—three-step regiocontrolled synthesis of polysubstituted furans, pyrroles and thiophenes. (c) V. Cadierno, P. Crochet, *Curr. Org. Synth.* 2008, 5, 343–364. Ruthenium-catalyzed furan- and pyrrole-ring formation.

104 H. Valizadeh, A. Fakhari, *Molecules* 2010, 15, 2972–2979. A mild and facile one-pot synthesis of *N*-methyl-3-acyl-pyrroles.

105 M. Lin, L. Hao, R.-D. Ma, Z.-P. Zhan, *Synlett* 2010, 2345–2351. A novel indium-catalyzed three-component reaction: general and efficient one-pot synthesis of substituted pyrroles.

106 V. Cadierno, J. Gimeno, N. Nebra, *Chem.–Eur. J.* 2007, 13, 9973–9981. One-pot three-component catalytic synthesis of fully substituted pyrroles from readily available propargylic alcohols, 1,3-dicarbonyl compounds and primary amines.

107 (a) G. Balme, *Angew. Chem. Int. Ed.* 2004, 43, 6238–6241. Pyrrole syntheses by multicomponent coupling reactions. (b) G. Balme, D. Bouyssi, N. Monteiro, *Heterocycles* 2007, 73, 87–124. Functionalized acetylenes as versatile building-blocks for the multicomponent assembling of polysubstituted furans and pyrroles. (c) V. Estévez, M. Villacampa, J. C. Menéndez, *Chem. Soc. Rev.* 2010, 39, 4402–4421. Multicomponent reactions for the synthesis of pyrroles. (d) S. Madabhushi, V. S. Vangipuram, K. K. R. Mallu, R. Jillella, S. Kurva, S. R. Pamulaparthi, *Tetrahedron Lett.* 2013, 54, 6737–6739. Ceric (IV) ammonium nitrate (CAN) mediated synthesis of pyrrole-2,3,4,5-tetracarboxylates by reaction of dimethyl acetylenedicarboxylate with an amine. (e) D. Kumar, S. Shamim, I. R. Siddiqui, *J. Heterocyclic Chem.* 2013, 50, E111–E115. Ionic liquid promoted multicomponent reaction: a good strategy for the eco-compatible synthesis of functionalized pyrroles. (f) B. M. Babu, G. S. Kumar, P. B. Thakur, V. M. Bangade, H. M. Meshram, *Tetrahedron Lett.* 2013, 54, 2296–2302. Catalyst-free four-component protocol for the synthesis of substituted pyrroles under reusable reaction media. (g) C. Martín-Santos, C. Jarava-Barrera, A. Parra, F. Esteban, C. Navarro-Ranninger, J. Alemán, *ChemCatChem*

2012, 4, 976–979. Modular three-component organocatalytic synthesis of 3,4-disubstituted pyrroles by a one-pot domino reaction. (g) T. B. Kumar, C. Sumanth, S. Vaishaly, M. S. Rao, K. B. C. Sekhar, C. L. T. Meda, A. Kandale, D. Rambabu, G. R. Krishna, C. M. Reddy, K. S. Kumar, K. V. L. Parsa, M. Pal, *Bioorg. Med. Chem. Lett.* 2012, 22, 5639–5647. Pd-mediated functionalization of polysubstituted pyrroles: their evaluation as potential inhibitors of PDE4. (i) V. Y. Korotaev, A. Y. Barkov, I. V. Kotovich, V. Y. Sosnovskikh, *J. Fluorine Chem.* 2012, 138, 42–47. Three-component synthesis of substituted β-(trifluoromethyl)pyrroles *via* Grob cyclization of 1,1,1-trifluoro-3-nitrobut-2-ene with 1,3-dicarbonylic compounds and ammonia or primary amines. (j) S. I. Bhat, D. R. Trivedi, *Tetrahedron Lett.* 2013, 54, 5577–5582. A catalyst- and solvent-free three-component reaction for the regioselective one-pot access to polyfunctionalized pyrroles. (k) M. Zhang, X. Fang, H. Neumann, M. Beller, *J. Am. Chem. Soc.* 2013, 135, 11384–11388. General and regioselective synthesis of pyrroles *via* ruthenium-catalyzed multicomponent reactions. (l) M. Zhang, H. Neumann, M. Beller, *Angew. Chem. Int. Ed.* 2013, 52, 597–601. Selective ruthenium-catalyzed three-component synthesis of pyrroles. (m) M. Farahi, F. Tamaddon, *Synlett* 2012, 23, 1379–1383. A new three-component reaction catalyzed by silica sulfuric acid: synthesis of tetrasubstituted pyrroles.

108 D. Dhara, K. S. Gayen, S. Khamarui, P. Pandit, S. Ghosh, D. K. Maiti, *J. Org. Chem.* 2012, 77, 10441–10449. $CeCl_3{\cdot}7H_2O$ catalyzed C—C and C—N bond-forming cascade cyclization with subsequent side-chain functionalization and rearrangement: a domino approach to pentasubstituted pyrrole analogues.

109 P. R. Bagdi, R. S. Basha, M. Lal, A. T. Khan, *Chem. Lett.* 2013, 42, 939–941. Bromodimethylsulfonium bromide (BDMS)-catalyzed synthesis of substituted pyrroles through a one-pot four-component reaction.

110 M. Lal, P. R. Bagdi, R. S. Basha, P. Saravanan, S. Patra, A. T. Khan, *Tetrahedron Lett.* 2012, 53, 4145–4150. Synthesis of tetra-substituted pyrroles, a potential phosphodiesterase 4B inhibitor, through nickel(II) chloride hexahydrate catalyzed one-pot four-component reaction.

111 B.-L. Li, P.-H. Li, X.-N. Fang, C.-X. Li, J.-L. Sun, L.-P. Mo, Z.-H. Zhang, *Tetrahedron* 2013, 69, 7011–7018. One-pot four-component synthesis of highly substituted pyrroles in gluconic acid aqueous solution.

112 Z. Hossaini, F. Rostami-Charati, S. Z. Sayyed-Alangi, *Chin. Chem. Lett.* 2012, 23, 1119–1121. Synthesis of highly functionalized pyrroles from primary amines and activated acetylenes in water.

113 S. Maiti, S. Biswas, U. Jana, *J. Org. Chem.* 2010, 75, 1674–1683. Iron(III)-catalyzed four-component coupling reaction of 1,3-dicarbonyl compounds, amines, aldehydes, and nitroalkanes: a simple and direct synthesis of functionalized pyrroles.

114 A. B. Atar, Y. T. Jeong, *Tetrahedron Lett.* 2013, 54, 5624–5628. Heterogenized tungsten complex: an efficient and high yielding catalyst for the synthesis of structurally diverse tetra substituted pyrrole derivatives *via* four-component assembly.

115 X. Feng, Q. Wang, W. Lin, G.-L. Dou, Z.-B. Huang, D.-Q. Shi, *Org. Lett.* 2013, 15, 2542–2545. Highly efficient synthesis of polysubstituted pyrroles *via* four-component domino reaction.

116 For another approach where dialkyl acetylenedicarboxylates were used in the synthesis of polysubstituted pyrroles, see: N. Bhunia, B. Das, *Synthesis* 2013, 45, 1045–1050. One-pot synthesis of pentasubstituted pyrroles from propargylic alcohols, amines, and dialkyl acetylenedicarboxylates; tandem amination, propargylation and cycloisomerization catalyzed by molecular iodine.

117 For other example of catalyst free multicomponent synthesis of pyrroles, see: X. Wang, X.-P. Xu, S.-Y. Wang, W. Zhou, S.-J. Ji, *Org. Lett.* 2013, 15, 4246–4249. Highly efficient chemoselective synthesis of polysubstituted pyrroles via isocyanide-based multicomponent domino reaction.

118 For a similar approach, see: W. M. Shu, Y. Yang, D. X. Zhang, L. M. Wu, Y. P. Zhu, G. D. Yin, A. X. Wu, *Org. Lett.* 2013, 14, 456–459. Mild reaction conditions for the terminal deuteration of alkynes.

119 G. Tong, H. Xu, W. Fan, B. Jiang, S. Wang, S. Tu, *Chin. J. Chem.* 2013, 31, 1039–1044. A highly diastereoselective three-component domino reaction in water yielding polysubstituted 4,5-dihydropyrroles.

120 P. Das, S. Ray, C. Mukhopadhyay, *Org. Lett.* 2013, 15, 5622–5625. Exploitation of dual character of CN moiety in the synthesis of uniquely decorated 3*H*-pyrroles: a rare observation.

121 S. Alla, S.-Y. Choi, C. Condeiu, D. Dhanoa, E. Difranco, G. Krokhina, K. Li, B. Thiagarajan, W.-C. Zhang. 2003, WO/2003/010130. Preparation of amino acid derivatives as pharmaceuticals.

122 U. Kazmaier, C. Hebach, *Synlett* 2003, 1591–1594. Peptide syntheses *via* Ugi reactions with ammonia.

123 S. P. Costa, H. L. Maia, S. M. Pereira-Lima, *Org. Biomol. Chem.* 2003, 1, 1475–1479. An improved approach for the synthesis of α,α-dialkyl glycine derivatives by the Ugi-Passerini reaction.

124 D. Bahulayan, V. S. Shinu, P. Pramitha, S. Arun, B. Sheeja, *Synth. Commun.* 2012, 42, 1162–1176. Highly *anti*-selective one-pot multicomponent synthesis of Mannich-type *N*-acylated β-amino acid derivatives by copper or sodium salt catalysis.

125 S. Fatemi, M. Nosrati, H. R. Bijanzadeh, M. Adib, *Synlett* 2008, 3177–3179. One-pot three-component reaction between 2-aminopyridines, aldehydes and Meldrum's acid in water: an efficient synthesis of β-amino acids.

126 (a) X. Pu, X. Li, B. Fan, A. S. C. Chan, Z. Shao, *Tetrahedron: Asymmetry* 2009, 20, 225–229. Enantioselective, copper(I)-catalyzed three-component reaction for the synthesis of β,γ-alkynyl α-amino acid derivatives. (b) J.-X. Ji, J. Wu, A. S. C. Chan, *Proc. Natl. Acad. Sci. U.S.A.* 2005, 102, 11196–11200. Catalytic asymmetric alkynylation of α-imino ester: a versatile approach to optically active unnatural α-amino acid derivatives.

127 N. R. Aguiam, V. I. Castro, A. I. F. Ribeiro, R. D. V. Fernandes, C. M. Carvalho, S. P. G. Costa, S. M. M. A. Pereira-Lima, *Tetrahedron* 2013, 69, 9161–9165. α,α-Dialkylglycines obtained by solid phase Ugi reaction performed over isocyanide functionalized resins.

128 L. Zhou, Q. Shuai, H. Jiang, C.-J. Li, *Chem.–Eur. J.* 2009, 15, 11668–11674. Copper-catalyzed amine-alkyne-alkyne addition reaction: an efficient method for the synthesis of γ,δ-alkynyl β-amino acid derivatives.

129 H. Liu, A. Dömling, *Chem. Biol. Drug Des.* 2009, 74, 302–308. One-pot synthesis of highly functionalized seleno amino acid derivatives.

130 (a) C. Kalinski, M. Umkehrer, L. Weber, J. Kolb, C. Burdack, G. Ross, *Mol. Divers.* 2010, 14, 513–522. On the industrial applications of MCRs: molecular diversity in drug discovery and generic drug synthesis. (b) P. Slobbe, E. Ruijter, R. V. A. Orru, *Med. Chem. Commun.* 2012, 3, 1189–1218. Recent applications of multicomponent reactions in medicinal chemistry.

131 A. Dömling, K. Khoury, *ChemMedChem* 2010, 5, 1420–1434. Praziquantel and schistosomiasis.

132 P. A. Gurbel, C. M. O'Connor, C. C. Cummings, V. L. Serebruany, *Pharmacol. Res.* 1999, 40, 107–111. Clopidogrel: the future choice for preventing platelet activation during coronary stenting?.

133 C. Kalinski, H. Lemoine, J. Schmidt, C. Burdack, J. Kolb, M. Umkehrer, G. Ross, *Synthesis* 2008, 4007–4011. Multicomponent reactions as a powerful tool for generic drug synthesis.

134 K. Rossen, P. J. Pye, L. M. Dimichele, R. P. Volante, P. J. Reider, *Tetrahedron Lett.* 1998, 39, 6823–6826. An efficient asymmetric hydrogenation approach to the synthesis of the Crixivan® piperazine intermediate.

135 S. Sugiyama, S. Arai, M. Kiriyama, K. Ishii, *Chem. Pharm. Bull.* 2005, 53, 100–102. A convenient synthesis of immunosuppressive agent FTY720 using the Petasis reaction.

136 T. E. Jordan, *J. R. Soc. Health* 1997, 117, 216–227. An awful visitation of providence: the Irish famine of 1845–49.

137 C. Lamberth, A. Jeanguenat, F. Cederbaum, A. de Mesmaeker, M. Zeller, H.-J. Kempf, R. Zeun, *Bioorg. Med. Chem.* 2008, 16, 1531–1545. Multicomponent reactions in fungicide research: the discovery of mandipropamid.

138 R. Noyori, *Chem. Commun.* 2005, 1807–1811. Pursuing practical elegance in chemical synthesis.

139 (a) B. M. Trost, *Science* 1991, 254, 1471–1477. The atom economy—a search for synthetic efficiency. (b) B. M. Trost, *Angew. Chem. Int. Ed.* 1995, 34, 259–281. Atom economy—a challenge for organic synthesis: homogeneous catalysis leads the way.

7

RECENT ADVANCES IN THE UGI MULTICOMPONENT REACTIONS

JOSÉ ALEMÁN[1], SILVIA CABRERA[2] AND CUAUHTÉMOC ALVARADO[3]

[1] *Departamento de Química Orgánica (módulo-1), Universidad Autónoma de Madrid, Cantoblanco, Madrid, Spain*

[2] *Departamento de Química Inorgánica (módulo-7), Universidad Autónoma de Madrid, Cantoblanco, Madrid, Spain*

[3] *División Académica de Ciencias Básicas, Universidad Juárez Autónoma de Tabasco, Cunduacán, Tabasco, Mexico*

7.1 INTRODUCTION

In 1959, Ugi reported for the first time the multicomponent reaction (MCR) between a carboxylic acid, an amine, a carbonyl compound, and an isocyanide, which is now named Ugi four-component reaction and also abbreviated as Ugi-4CR [1]. After this successful reaction, different versions of the original Ugi reaction have been developed. Thus, the elimination of one of the components (usually the carboxylic acid) provided the Ugi-3CR and different advances, for example, its asymmetric version, have been accomplished. Nowadays, the number of components can be more than four, achieving extraordinary complex examples with up to eight reagents. Therefore, the use of different types of starting materials such as nitrophenol derivatives (e.g., Ugi–Smiles coupling) has allowed the variation of the final product. In addition, the size of the chemical space of the Ugi scaffolds has exponentially increased due to the incorporation of postmodifications that allow the preparation of molecules with high complexity and diversity in short pathways, reducing time and tedious purifications.

The aim of this chapter is to highlight the literature of the last 8 years (2006–2014) concerning the use of the Ugi strategy for the preparation of different complex molecules.

This chapter has been organized in four main sections (Fig. 7.1). Thus, in the first part of this chapter, some representative advances in the Ugi-3CR and Ugi-4CR have been presented, followed by some extraordinary examples in the combination of the Ugi with other MCRs (up to eight components). Then, other successful variations, like the Ugi–Smiles coupling, have been described and the most representative examples in this area have been outlined. Finally, the combination of the Ugi reaction with other transformations (postmodifications) to increase the complexity of the obtained Ugi-adducts has also been described.

During the last decades, excellent revisions concerning the reactivity and use of the Ugi reaction [2], isocyanides [3–8], MCRs [9–12], and the Ugi–Smiles coupling [13] have been reported. All of them contain information related to the topic of this chapter, and they should be consulted for obtaining detailed information about the literature that appeared in the last years.

7.2 UGI THREE-COMPONENT REACTIONS

Modifications to the Ugi-3CR have been conducted in three main directions: (i) changes to the methodology, such as the use of different catalysts, the *in situ* generation of imines, and the use of convertible isonitriles; (ii) development of enantio- and diastereoselective Ugi reactions; and (iii) applications of the methodology for the synthesis of new compounds with important biological activity.

The carboxylic acid plays a dual role in the Ugi four-component reaction (see Scheme 7.14); they act as Brønsted acid activating the imine intermediate and they are also incorporated in the final Ugi product. However, some research groups have showed that acids can be replaced by other activators.

Multicomponent Reactions: Concepts and Applications for Design and Synthesis, First Edition. Edited by Raquel P. Herrera and Eugenia Marqués-López.

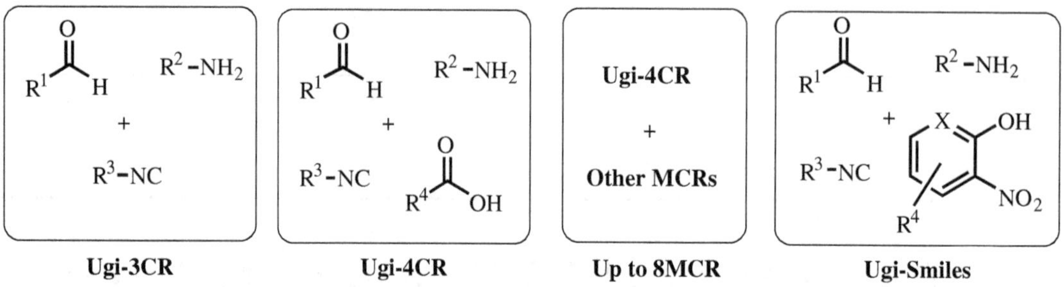

FIGURE 7.1 Different Ugi reactions and variations highlighted in this chapter.

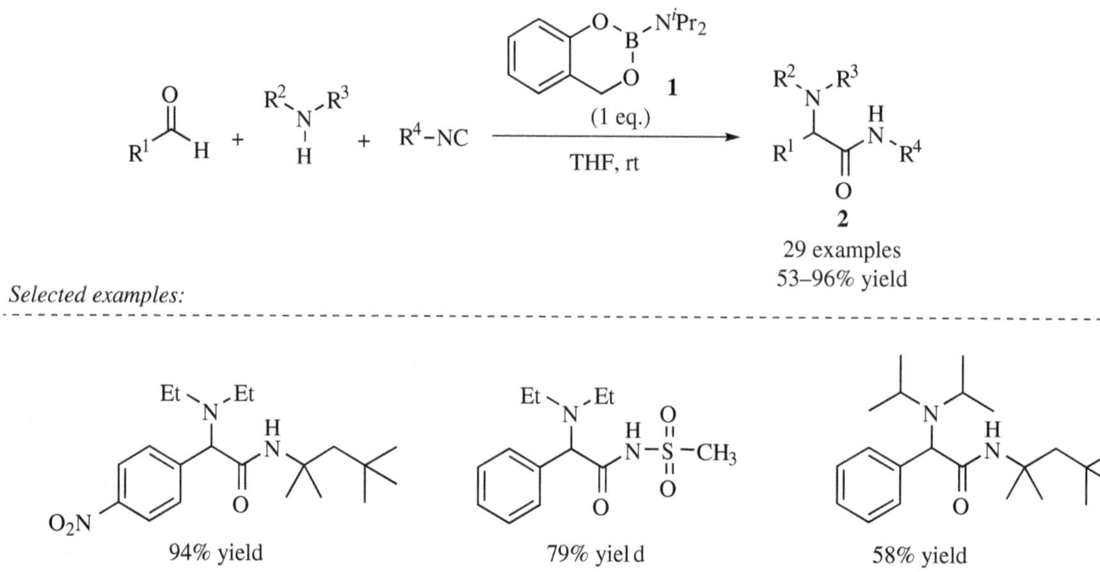

SCHEME 7.1 Synthesis of α-amino amides **2** using aminoborane **1** as catalyst.

In this sense, in 2007, Suginome and coworkers developed a pioneering example of a free acid Ugi-type reaction using the aminoborane **1** that yielded the final amides (Scheme 7.1) [14]. In this methodology, a number of aldehydes, secondary amines, and isocyanides were mixed in the presence of aminoborane **1** at room temperature, giving the corresponding α-amino amides **2** in good to excellent yields. A broad scope was demonstrated for this reaction, since up to 29 different compounds were prepared. The authors also showed that the use of 1-isocyano-4-methoxybenzene (R^4=PMP) allowed the deprotection of the obtained Ugi-amide with CAN to get the free amide.

The authors proposed the mechanism shown in Scheme 7.2. First, an amine exchange reaction in catalyst **1** afforded the aminoborane **3**. This borane **3** reacted with the aldehyde through nucleophilic attack of the amino group of **3** to the carbonyl group followed by B—O bond formation. Next, the intermediate **4** would evolve to the formation of the iminium ion intermediate, which in the presence of isocyanide gave the α-amino amides **2** and the by-product **5** following an Ugi-mechanism-type reaction.

In 2008, List's group reported for the first time the use of phenyl phosphinic acid (**6**) as catalyst in the Ugi-3CR [15]. In this example, water was used as a nucleophile instead of an organic acid. The yields of the obtained α-amino amides **2** ranged from moderate to good (Scheme 7.3). In order to explain the results, a mechanistic proposal was envisaged by the authors where the phenyl phosphinic acid (**6**) was proposed to have a dual role, as Brønsted acid by protonation of the imine intermediate and as a Lewis base by trapping the nitrilium ion intermediate formed between the aldehyde, the amine, and the isonitrile, forming intermediate **7**. In the last step, the water released from the imine formation reacts with intermediate **7** to give **8**, which was transformed into the compound **2** and catalyst **6**.

Other catalysts, such as $ZrCl_4$ [16] and cellulose sulfuric acid (CSA) [17], have been also reported to activate the Ugi-3CR. In this context, Khan et al. have described the use of bromodimethylsulfonium bromide (**9**) (BDMS) as catalyst for the synthesis of α-amino amidines **10** by the reaction of aromatic aldehydes, aromatic amines, and isocyanides (Scheme 7.4) [18].

The formation of the α-amino amidines **10** was explained as follows: initially, the aromatic aldehyde reacts with the aromatic amine to form an intermediate imine **11** and water

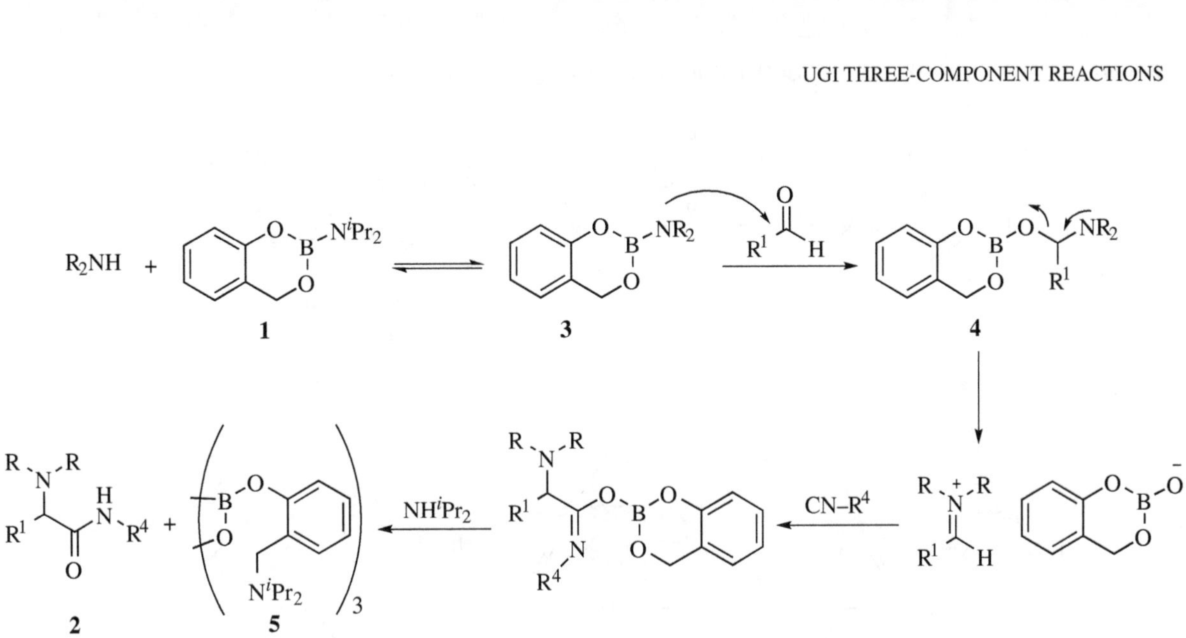

SCHEME 7.2 Mechanism for the Ugi-3CR catalyzed by aminoborane.

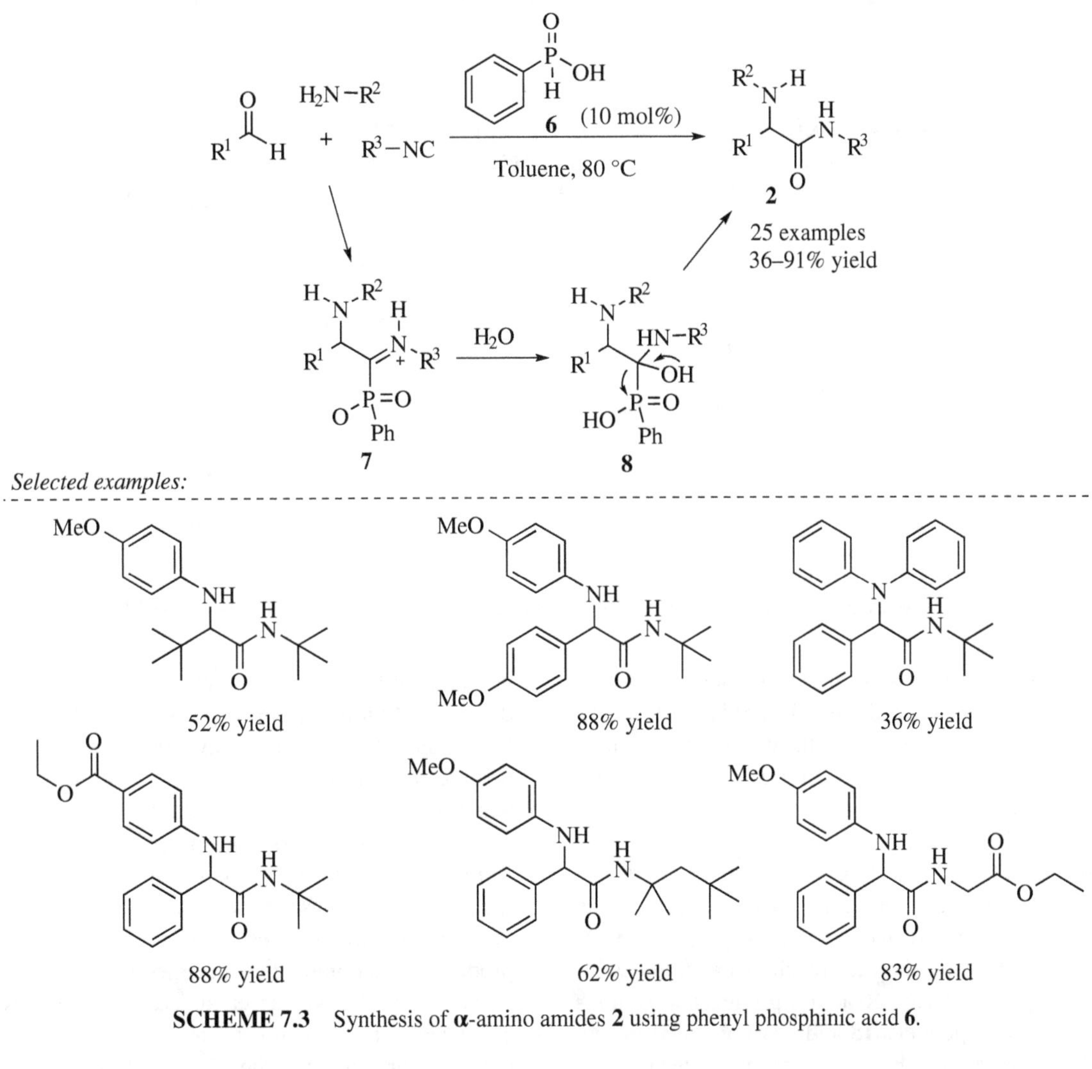

SCHEME 7.3 Synthesis of α-amino amides **2** using phenyl phosphinic acid **6**.

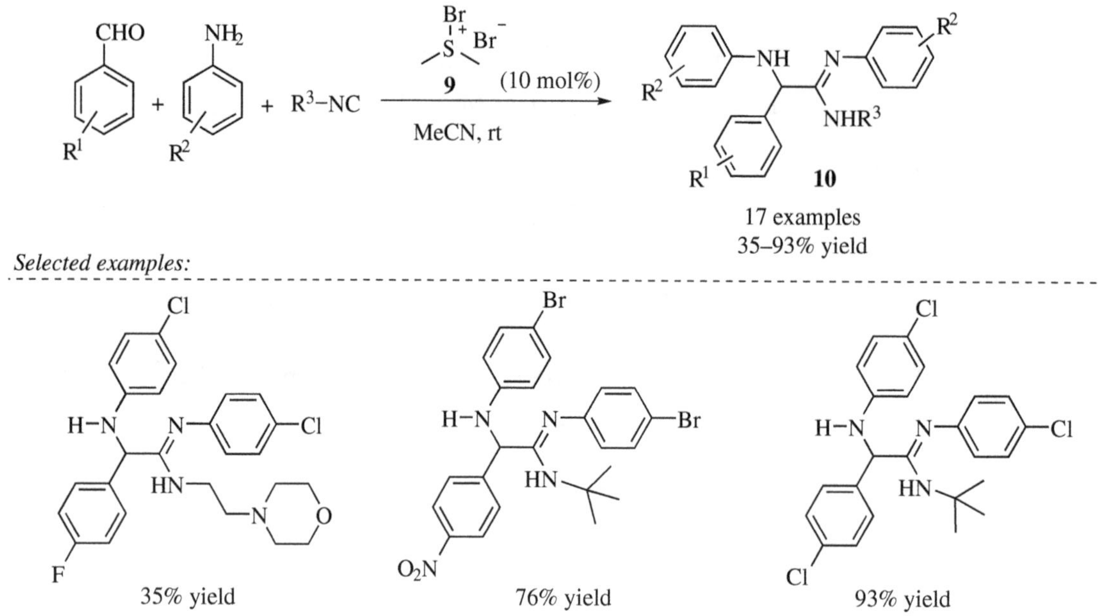

SCHEME 7.4 Synthesis of α-amino amidines **10** by using BDMS **9** as catalyst.

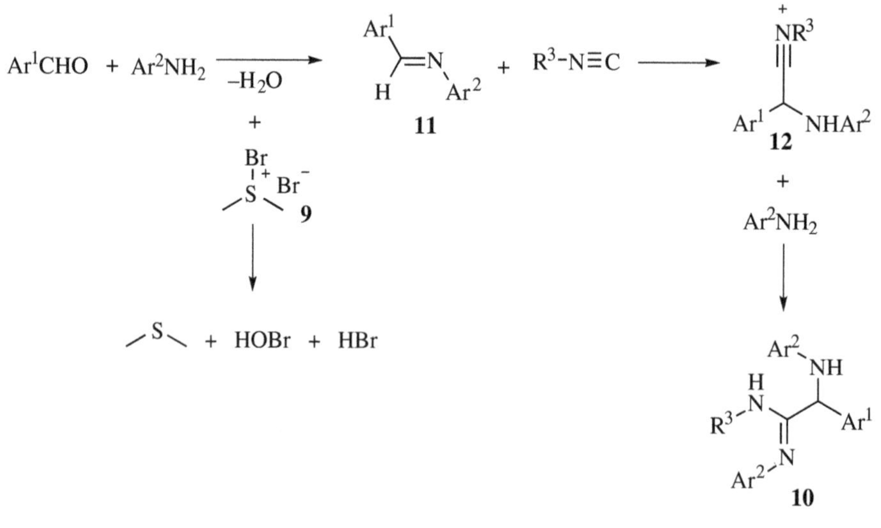

SCHEME 7.5 Reaction mechanism for the synthesis of α-amino amidines by using BDMS.

(Scheme 7.5). The released water molecule then reacts with BDMS (**9**) to generate HBr, dimethyl sulfide, and HOBr. Since there is no water in the system, the water can not attack the nitrilium ion intermediate **12**. Therefore, a second molecule of amine could react with the nitrilium ion, generating the final product **10**.

An enzyme-catalyzed Ugi-3CR condensation of an amine, aldehyde, and ethyl or benzyl isocyanoacetate was developed by Ostaszewski and coworkers [19]. Different enzymes were tested and the best results were found with Novozym 435 (Scheme 7.6). Reaction products were non-naturally occurring dipeptides **13** and were obtained in low to moderate yields. Although it is known that water is often not compatible with the Ugi reaction due to isocyanide decomposition in the presence of acids, this reaction could be conducted in organic solvents, water, or a mixture of organic solvent/water (5% v/v). Nevertheless, the use of water in the reaction was not totally understood, because in some cases, its addition improved the yield or enabled the formation of products, as long as in other cases, its addition just caused a decrease in the yield (Scheme 7.6). The authors observed that the reactions were not stereoselective and the products were obtained as racemic mixtures.

Very recently, Kumar et al. described the first Ugi-3CR catalyzed by nanoparticles of ZnO (ZnO-NP) wherein 2-arylamino-2-phenylacetimidamide derivatives were

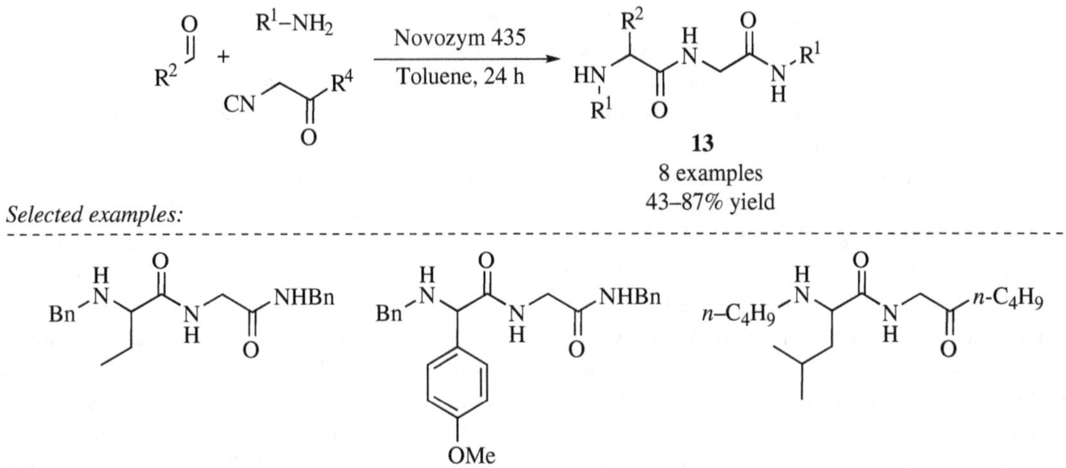

Yield (%) without H_2O	76	0	10
Yield (%) with H_2O (5% v/v)	87	43	0

SCHEME 7.6 Synthesis of nonnaturally occurring dipeptides by using Novozym 435.

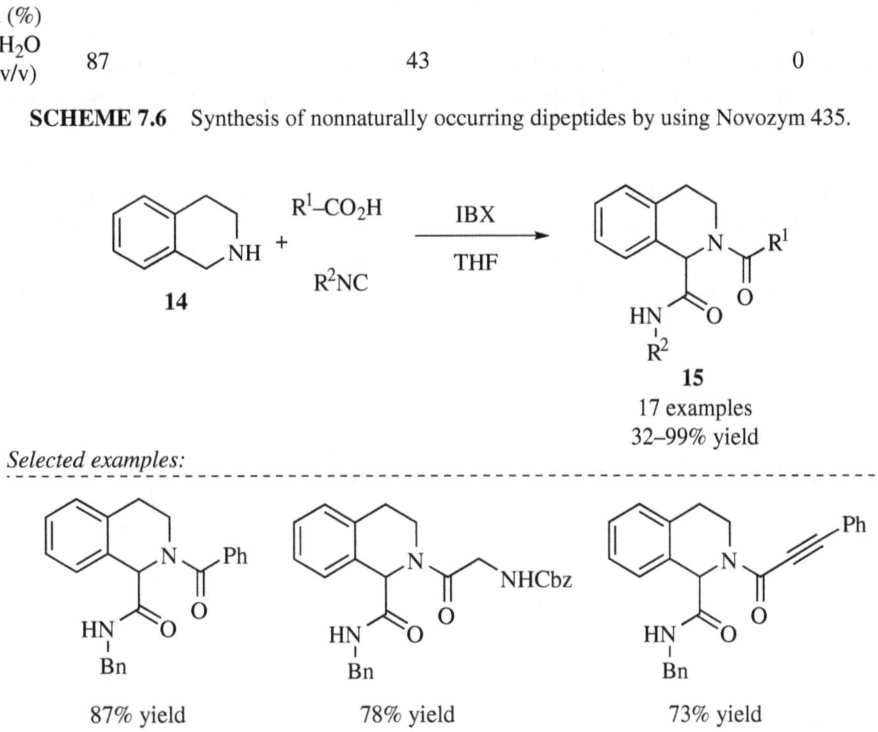

SCHEME 7.7 The first oxidative Ugi-3CR for the synthesis of tetrahydroisoquinolines **15**.

synthesized from aldehydes, anilines, and isocyanides [20]. The reaction was efficient and products were obtained in good yields. In addition, the process was carried out in water and had good atom economy (95.2%) and good atom efficiency (85.6%).

Other direction in the Ugi-3CR is the *in situ* generation of imines by oxidation. Thus, the development of the first oxidative Ugi-3CR was reported by Zhu and Ngouansavanh [21]. A dual acylation of the nitrogen and α-carbon centers of tetrahydroisoquinoline (**14**) by a carboxylic acid and an isocyanide was achieved under mild conditions in good to high yields (Scheme 7.7). In such methodology, the electrophilic iminium species were generated by *in situ* oxidation of the tetrahydroisoquinoline (**14**) with 2-iodoxybenzoic acid (IBX). Later on, Zhu and Masson published the oxidation of alcohols with IBX to aldehydes that were used *in situ* as starting materials for the Ugi four-component reaction [22].

Another oxidative Ugi-3CR was reported by Xie et al. for the synthesis of α-amino imides from tertiary amines by direct activation of sp^3 C—Hs adjacent to nitrogen using an oxidant (*tert*-butyl hydroperoxide (TBHP)) and a copper salt [23]. Notably, the reaction could be conducted under mild conditions and demonstrated tolerance to a broad range of functional groups. In addition, Che and coworkers developed a new protocol in which secondary amines were oxidized to imines by singlet oxygen in the presence of tetraphenylporphyrin as photosensitizer [24].

The increase in the number of applications of the Ugi reaction in pharmaceutical and agricultural chemistry had provoked the search for asymmetric variations in this field because of the possible different activities of the different enantiomers. Most stereoselective Ugi-type methodologies developed during the period 2006–2014 were related to the generation of diastereoisomeric compounds, as long as only few concerned to the enantioselective methods. In addition, most of the strategies applied to control the chirality of the newly formed stereocenter used chiral substrates or chiral auxiliaries [25]. In 2010, Orru et al. reported the highly stereoselective synthesis of substituted prolyl peptides **17** by an Ugi-3CR (Scheme 7.8) [26]. The first step of the approach was the preparation of imine **16**, obtained by oxidation with the monoaminooxidase N (MAO-N). Then, the imine **16** was reacted with a number of isonitriles and organic acids, yielding substituted prolyl peptides **17** in good yields and high diastereo- and enantioselectivity.

Up to 2009, the development of an enantioselective version of these reactions remained a significant challenge. In that year, Wang and Zhu reported the first Brønsted acid-catalyzed enantioselective Ugi-3CR [27], wherein a phosphoric acid **19** catalyzed the reaction of aldehydes, anilines, and α-isocyanoacetamides **18**, leading to 2-(1-aminoalkyl)-5-aminoxazoles **20** in excellent yields and moderate to good enantioselectivities (Scheme 7.9).

In order to explain the stereochemical outcome of the reaction, a mechanism was proposed by the authors. Thereby, the protonation of imine **21** by the organocatalyst would lead to the formation of the ionic pair **22** (Scheme 7.10). Then, the nucleophilic addition of isonitrile **18** onto the *Si* face of the iminium ion **22** would afford the nitrilium ion intermediate **23**, which after cyclization gave **24**. Ultimately, deprotonation on C4 of **24** would yield the 5-aminooxazole **20**.

In 2012, Maruoka and coworkers developed a new catalytic asymmetric Ugi-3CR through the use of an axially chiral dicarboxylic acid **27** [28]. In the approach, a variety of aldehydes, 2-benzoyloxyphenyl isocyanide (**26**), and an acyclic azomethine imine **25** were initially reacted, yielding heterocyclic compounds **28** in excellent yields and good enantiomeric excesses (Scheme 7.11). Upon acid hydrolysis of the heterocyclic compound **28a**, the corresponding α-hydrazino amide **29** was obtained without loss of enantioselectivity (bottom-left, Scheme 7.11).

Very recently, Wulff's group has published a chiral catalyst for the three-component Ugi reaction based on BOROX catalysts **30** (Scheme 7.12) [29]. This BOROX catalyst was formed *in situ* by combination of dibenzylamine, $BH_3 \cdot SMe_2$, an alcohol (or phenol), and the biaryl ligand. After a large

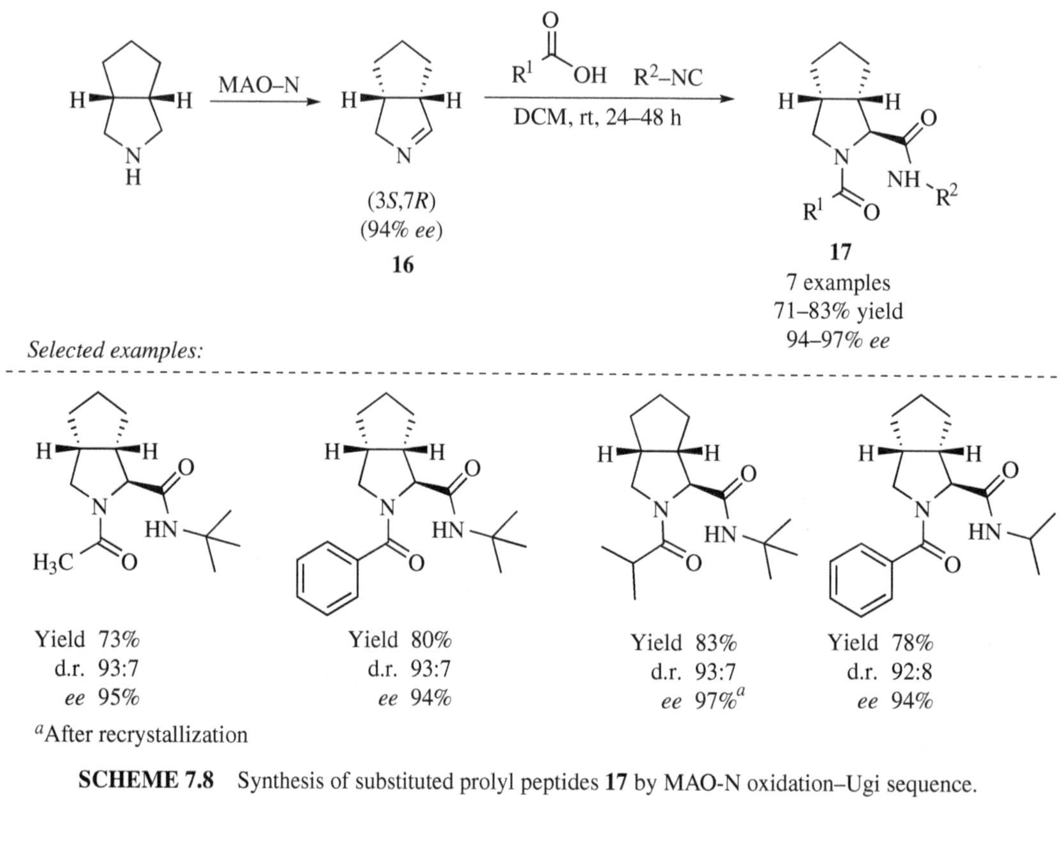

SCHEME 7.8 Synthesis of substituted prolyl peptides **17** by MAO-N oxidation–Ugi sequence.

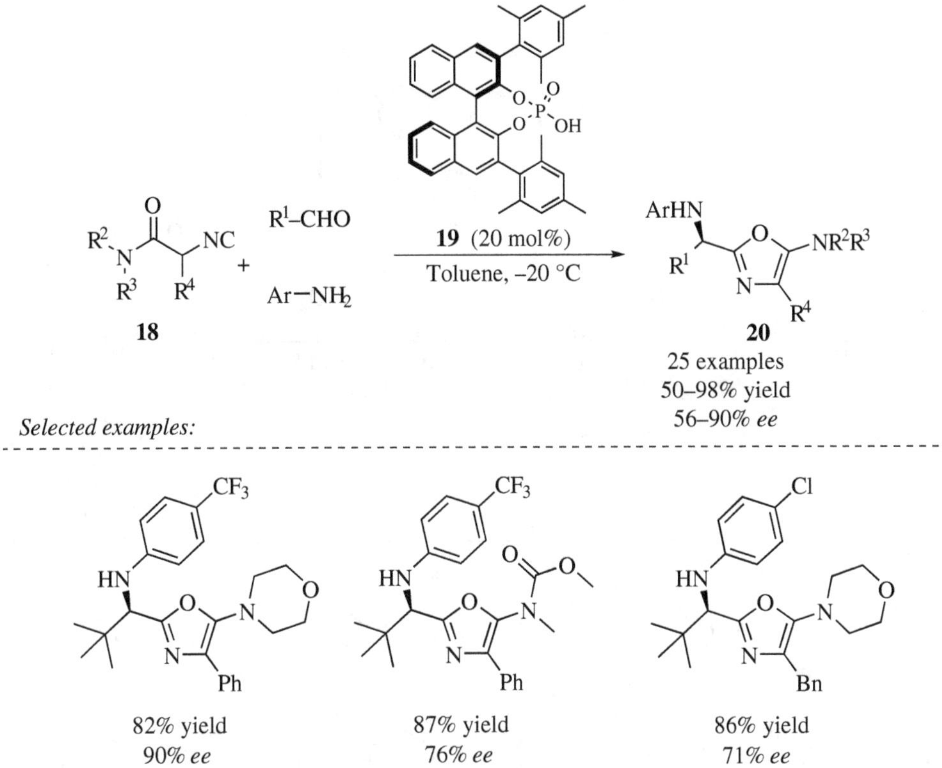

SCHEME 7.9 Enantioselective synthesis of 2-(1-aminoalkyl)-5-aminoxazoles **20**.

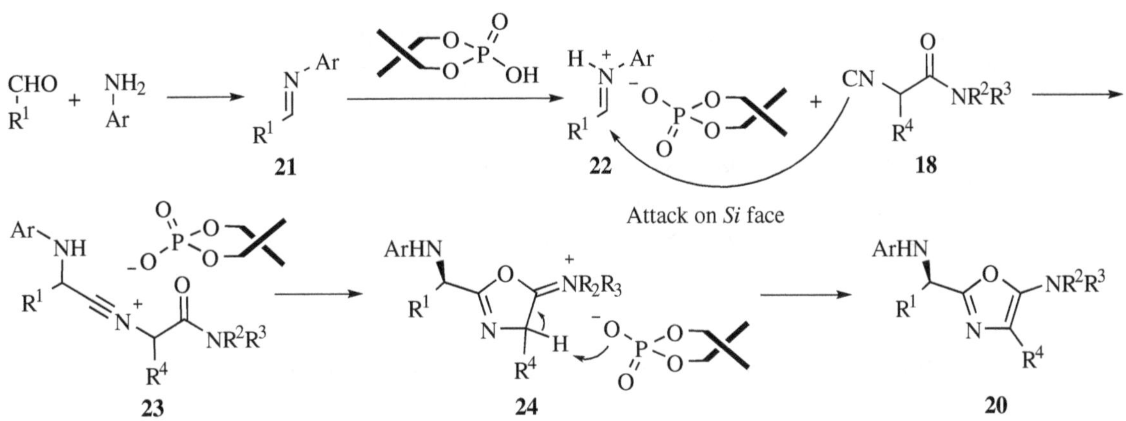

SCHEME 7.10 Mechanism proposal for the asymmetric synthesis of aminoxazoles **20**.

screening, the authors found an optimal catalyst system, providing Ugi products **31** with good enantioselectivity.

MCRs have been recognized as valuable tools in the preparation of diversity-oriented synthesis (DOS) and divergent total synthesis (DTS), for the construction of low molecular weight compound libraries through combinatorial strategies and parallel synthesis, and for rapid access to natural product and drug-like small molecules with complex and diverse molecular structures [12]. Thus, in 2011, Van der Eycken and coworkers reported a microwave-assisted diastereoselective Ugi-4C-3CR synthesis of dibenzo[*c*,*e*]azepinones **35** (Scheme 7.13) [30]. In the beginning of the synthesis, commercially available substituted *o*-bromo benzoates **32** were coupled with substituted 2-formylphenylboronic acids **33** under Suzuki–Miyaura coupling conditions to obtain biaryl-formyl ester scaffolds that were afterward hydrolyzed to give **34**. Secondly, an array of differently substituted biaryl compounds obtained were reacted with various amines and isocyanides to afford the corresponding dibenzo[*c*,*e*]azepinones **35** also in good to excellent yields (40–98%) as single diastereoisomer.

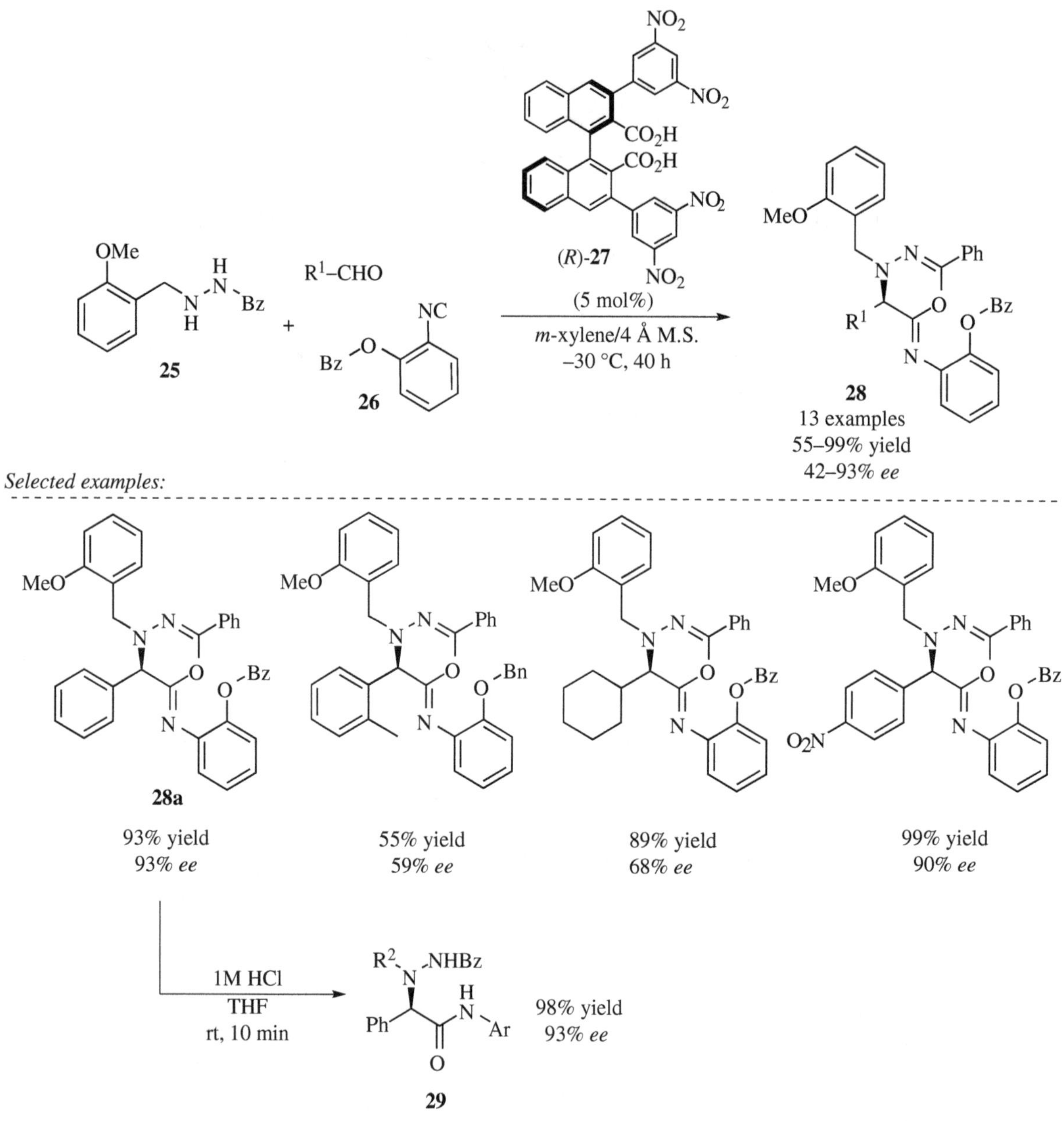

SCHEME 7.11 Synthesis of enantiomerically enriched heterocyclic compounds **28**.

7.3 UGI FOUR-COMPONENT REACTIONS

The classical Ugi four-component reaction (U-4CR) originally described by Ugi et al. [1] consists in the reaction of a primary amine, a carbonyl compound (aldehyde or ketone), an isocyanide, and a carboxylic acid to afford α-acylamino amides **36** (Scheme 7.14) [31, 32]. Rapidly, other components or adducts were found to give the Ugi-4CR—that is, ammonia, hydroxylamine, or hydrazine as amine component or thiosulfates [33], thiocarboxylic acids [34], isothiocyanic acids [35], hydrazoic acid [36], or carbonic acid monoesters [37] as acid component—giving access to structurally different skeletons.

The most accepted mechanism for the Ugi-4CR is still the one proposed originally by Ugi (Scheme 7.15) [38]. In the first step, the aldehyde (or ketone) and the amine react to form the corresponding imine. Then, the acid protonates the nitrogen atom of the imine to increase the electrophilicity of the C=N bond. As other reactions of imines, its electrophilicity can be also increased by the addition of Lewis acids such as $TiCl_4$, $Sc(OTf)_3$, or $Yb(OTf)_3$ [39]. Next, the addition of the isocyanate to the imine gives a nitrilium ion **37**, which

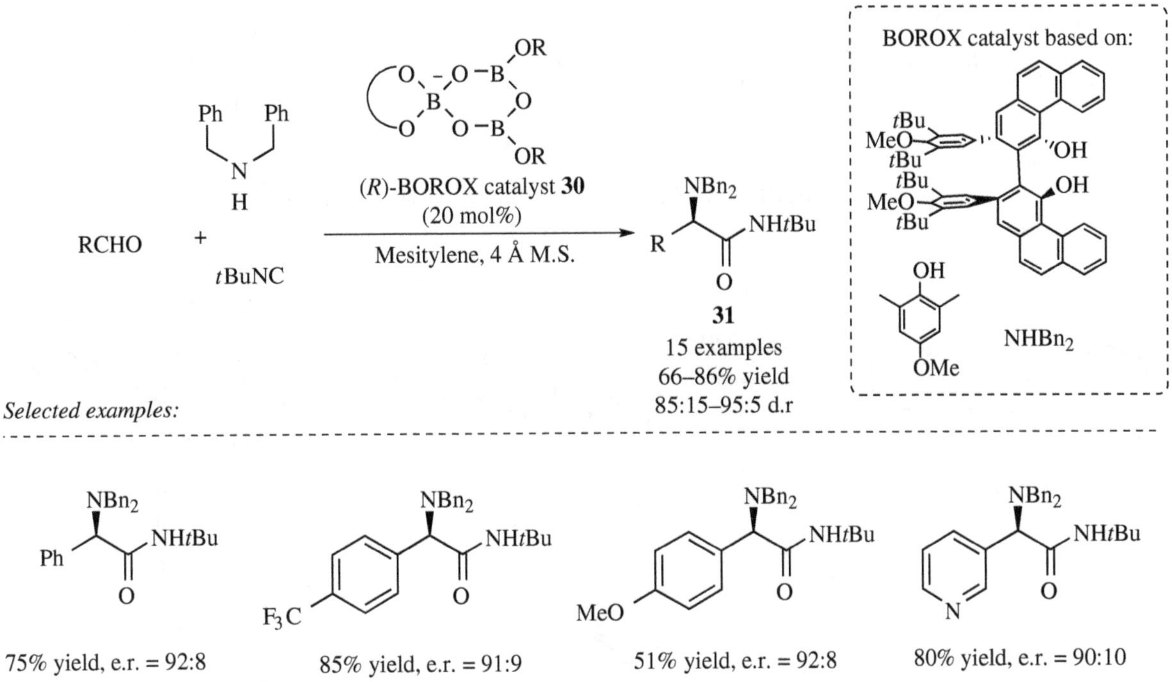

SCHEME 7.12 The three-component Ugi reaction based on BOROX catalyst **30**.

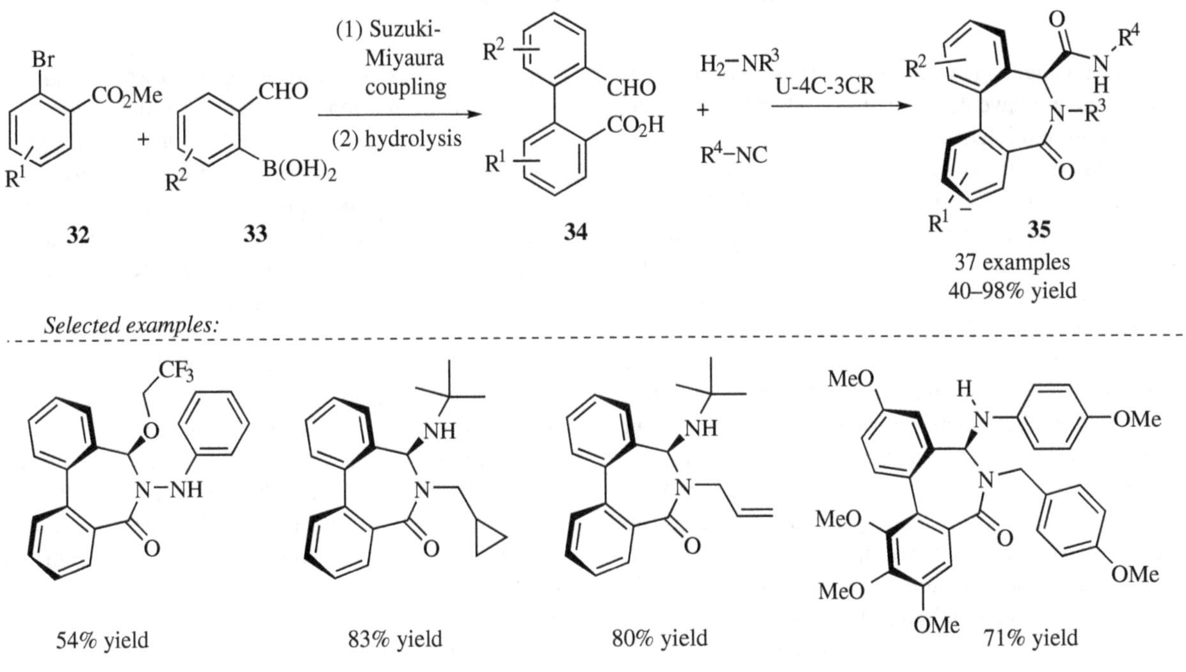

SCHEME 7.13 Synthesis of dibenzo[*c*,*e*]azepinones **35** by an Ugi-4C-3CR.

is transformed into an *O*-acylimidate **38** by the addition of the carboxylate anion. The final and nonreversible step—known as Mumm rearrangement—is the intramolecular transfer of the acyl moiety from the oxygen to the nitrogen atom that yields the Ugi products **36**. Recently, some evidences and studies have been carried out to support this mechanism [40], such as the theoretical studies performed by Fleurat-Lessard's group [41]. In 2014, Neto et al. have

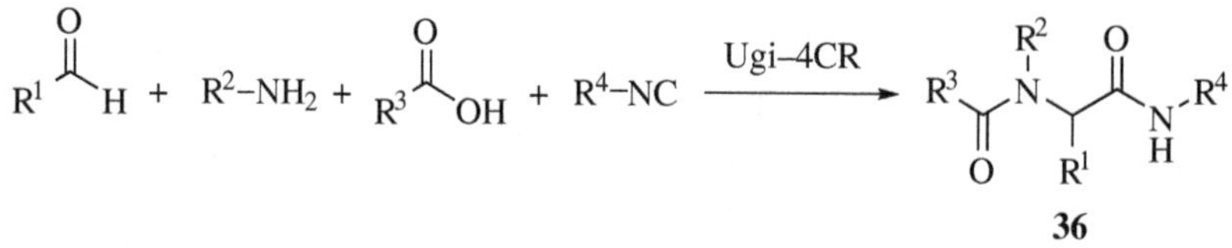

SCHEME 7.14 Classical Ugi four-component reaction.

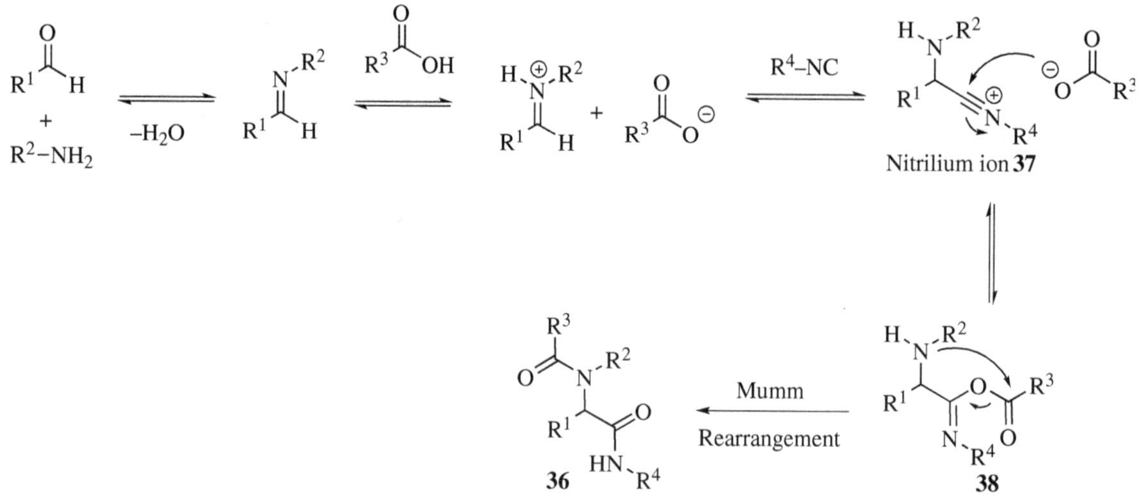

SCHEME 7.15 Proposed mechanism of the Ugi-4CR.

monitored the mechanism by ESI(+)-MS(/MS) using charge-tagged reagents and have been able to isolate and characterized the nitrilium ion intermediate [42].

In the last years, different advances have been achieved in the development of novel Ugi-4CRs. For example, Dömling et al. have reported a 5-center 4-component Ugi reaction using α-amino acids, an oxo component, isocyanide, and a primary or secondary amine instead of the previously used alcohol (Scheme 7.16) [43]. This reaction constitutes a challenge due to competing reactions such as the addition of the alcohol used as solvent to the six-membered intermediate **39** or the enamine formation between the oxo component and the amine (NHR^5R^6). However, mixing the four reagents, iminodicarboxamides **40** were obtained by the nucleophilic attack of the primary or secondary amine to the cyclic intermediate **39** formed from the α-amino acid, the oxo component, and isocyanide. The authors mentioned that a library of more than 400 compounds had been prepared in good yields using a broad variability of all components.

The Ugi-4CR is one of the most versatile tools for the construction of scaffolds with biological properties, such as diazepines [44] and peptoids or peptide–peptoid backbones [45]. In this context, the group of Andrade has prepared functionalized peptoids by combining methyl isocyanoacetate, paraformaldehyde, *N*-Cbz-glycine, and different primary amines (Scheme 7.17) [45c]. This procedure allowed the synthesis of several peptoids **41** in high yields (78–92%) and very short times (<5 min) using microwave irradiation and solvent-free conditions. In addition, after the ester hydrolysis of the peptoids **41** obtained, a consecutive Ugi reaction could be performed by the addition of paraformaldehyde, methyl isocyanoacetate, and a different amine, giving access to pentapeptoids **42** in short times (<5 min) and high yields.

Currently, the extension of the Ugi-4CR in the synthesis of molecules with various applications is one of the main directions in this field. Thus, due to the rapid assembly of building blocks into complex molecules, the Ugi-4CR has been the key step in the synthesis of the natural products massadine [46] and muraymycin analogues, among others [47], and also in the synthesis of macrocycles by multiple Ugi-4CR reactions [48]. In the medicinal field, the Ugi-4CR is a useful tool for the synthesis of libraries for combinatorial screening of drugs and to rapidly explore structure–activity relationships [49] and to the synthesis of drugs such as epelsiban [50].

One of the Ugi-4CR variants with high synthetic potential is the split-Ugi reaction, which was reported in 2006 by the group of Tron and Giovenzana (Scheme 7.18) [51]. In this reaction, after the formation of the imino-anhydride intermediate **43**, the Mumm rearrangement took place by the nucleophilic attack of the other amine present in the diamine reagent. In this manner, the backbone obtained in the general Ugi-4CR was split, and from the point of view of the diamine reagent, a desymmetrization process has occurred as one of

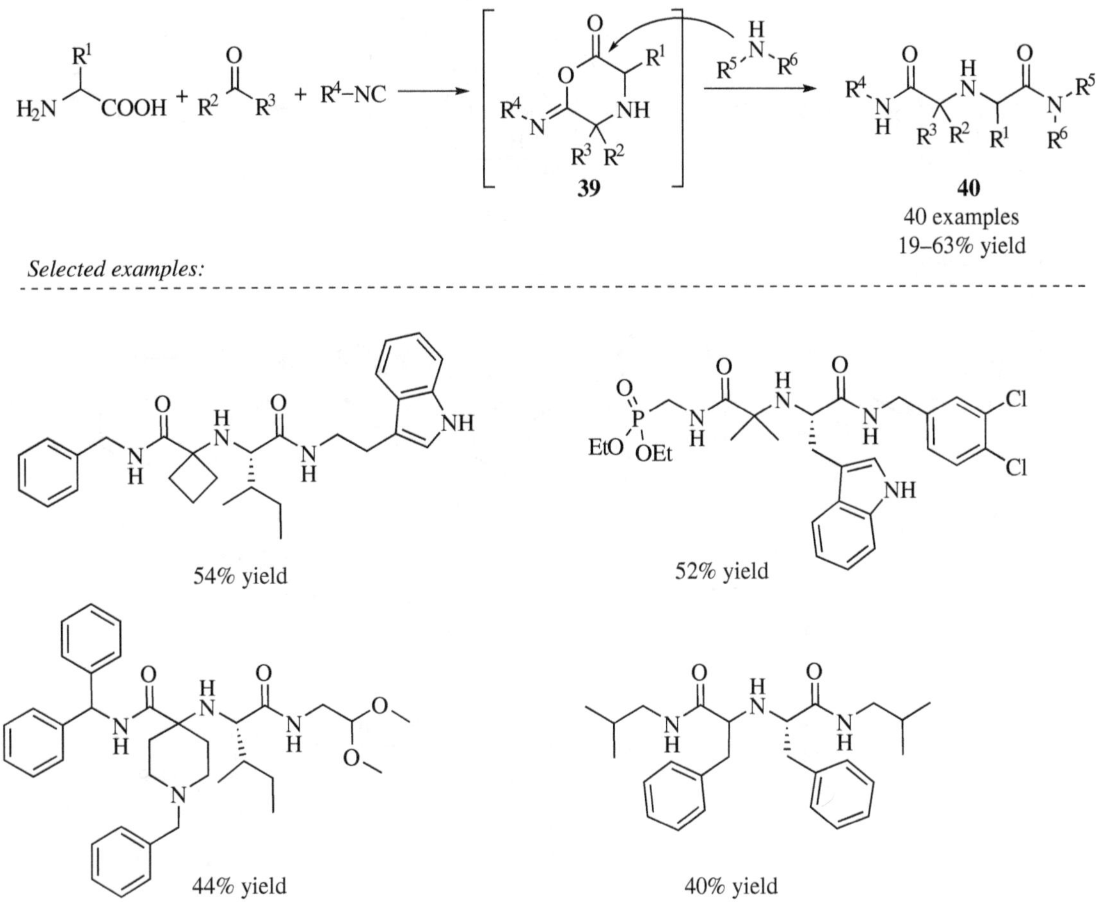

SCHEME 7.16 Synthesis of iminodicarboxamides **40** by an Ugi-5C-4CR.

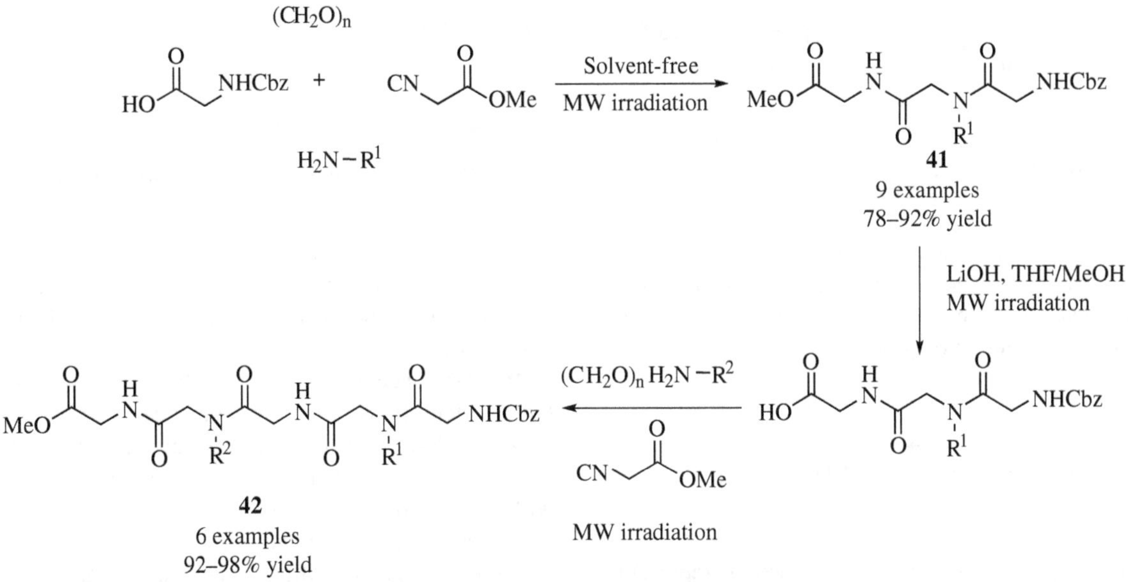

SCHEME 7.17 Synthesis of functionalized peptoids.

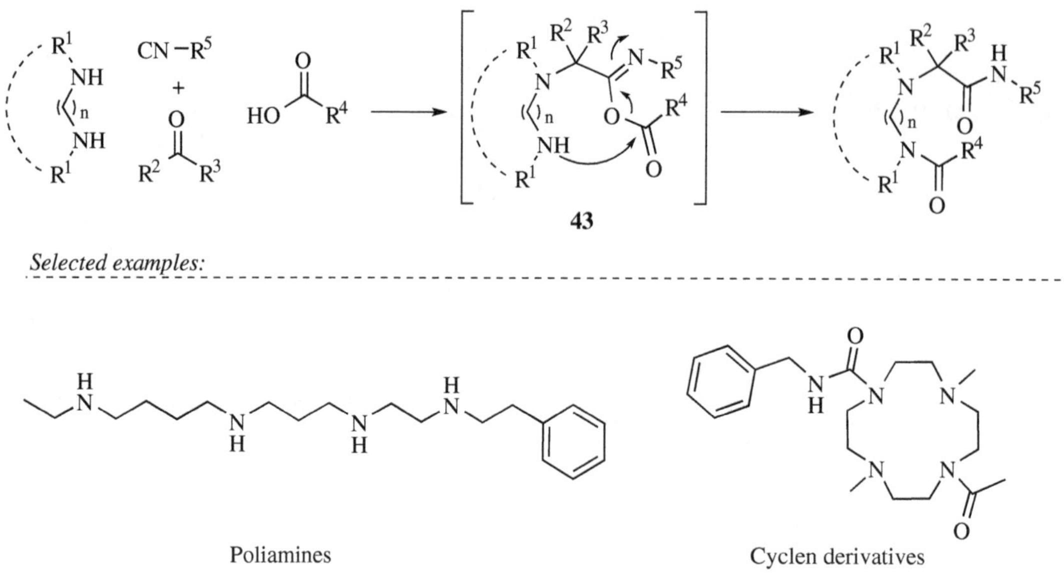

SCHEME 7.18 Split-Ugi reaction and selected examples.

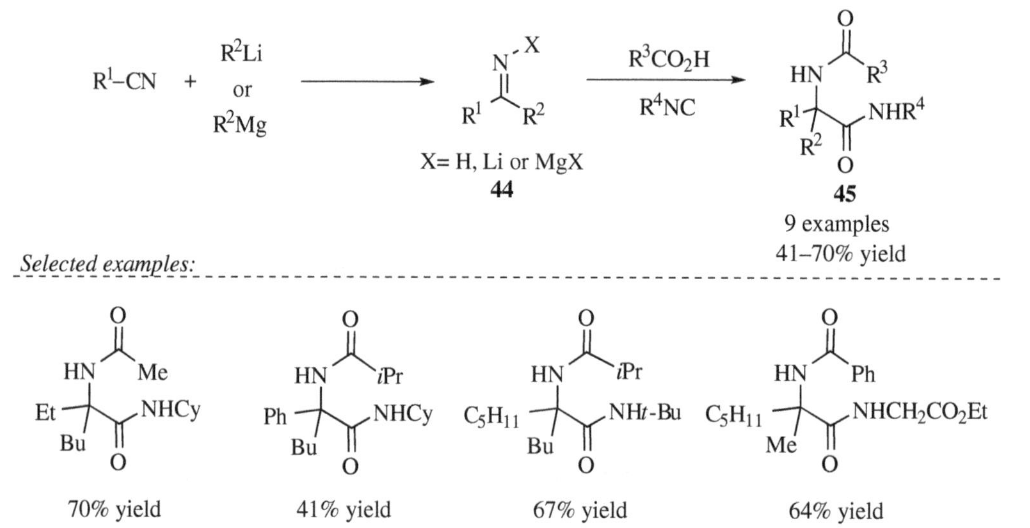

SCHEME 7.19 Sequential four-component synthesis of diamides **45** from organolithium and Grignard reagents.

the nitrogen atoms was acylated and the other was alkylated. The synthesis of several polyamines [52] and unsymmetrical derivatives of cyclen [53] have been possible without the use of protecting groups and in only one step.

7.4 FIVE-, SIX-, SEVEN-, AND EIGHT-COMPONENT REACTIONS BASED ON THE UGI REACTION

Although less explored, five-component reactions represent an interesting manner to obtain highly functionalized scaffolds, but only few examples have appeared after the pioneering work reported by Armstrong in 1998 [54]. The advantage of these methods is related to the high degree of diversity that can be attained by reacting, in a combinatorial way, five components in a one-pot or tandem Ugi/nucleophilic addition process.

In 2006, Ganem et al. described four- and five-component Ugi condensations based on the reaction of organometallic reagents (RLi and RMgX) with nitriles [55]. First, the authors studied the 4CR starting with the reaction of nitrile with organolithium or organomagnesium compounds, which generates the metalated ketimines **44** (Scheme 7.19). These ketimines are usually difficult to obtain by ammonia–ketone equilibrium. However, this methodology allowed an easy formation of ketimines **44** that reacted with the acid and isonitrile to form diamides **45** from moderate to good yields.

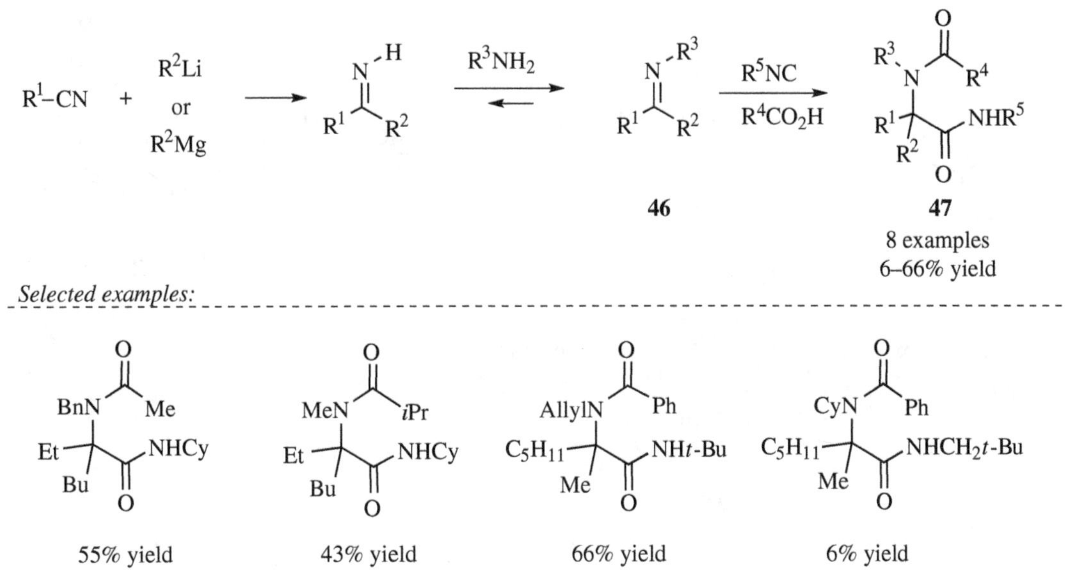

SCHEME 7.20 Sequential five-component synthesis of diamides **47** from organolithium or Grignard compounds.

In the same publication, the authors proposed the 5CR with an additional equivalent of the amine (R^3NH_2). Thus, the formation of the *N*-substituted ketimine **46** was expected to be favored by a delicate amine–imine interchange reaction (Scheme 7.20). Then, the ketimine **46** could be attacked by the isocyanide and the acid, giving the 5MC product **47** in good yield for most cases. However, the use of bulkier amines like cyclohexylamine provoked a decrease in the yield (6%) presumably due to the difficulty in the formation of the ketimine **46** with a secondary amine.

In 2009, Marcaccini et al. described a five-component synthesis of spiropyrrolidinochromanones **51**, which are natural products and have important biological applications [56]. First, the authors carried out an Ugi-4CR by reaction of a formylchromone **48** with an amine, isocyanide, and glyoxylic acid **49**, obtaining the Ugi product **50** (Scheme 7.21). Sequentially, an amine was added to the Ugi four-component product in a one-pot manner to obtain the final product **51**. The high electrophilic 2-position of the chromane intermediate **50** reacted with the amine (R^5NH_2), generating an enolate that intramolecularly reacted with the carbonyl group for giving the final five-component product **51**. Reactions proceeded in good yields and remarkable high diastereoselectivity (only one diastereoisomer was observed), forming four new stereogenic centers in the final product.

Riguet reported a methodology involving an enantioselective organocatalytic Friedel–Crafts alkylation followed by an Ugi 4-center 3-component reaction [57]. In the first reaction, the Friedel–Crafts reaction, 5-hydroxyfuran-2(5*H*)-one (**53**) was used as an electrophile to alkylate indoles **52** to give the intermediate **54**, using a diphenylprolinol silyl ether **56** as catalyst (Scheme 7.22). Then, adducts **54**—obtained from Friedel–Crafts reaction—were used in a one-pot manner as chiral scaffolds in the Ugi reaction to afford chiral five-membered lactams **55** in high yields and enantioselectivities. Unfortunately, Friedel–Crafts/Ugi products were obtained with low diastereoselectivities.

An efficient synthesis of polyfunctional *E*-enaminones [58] was reported by Balalaie in 2011 via one-pot five- and six-component sequential Ugi/nucleophilic addition reaction [59]. The method was highly chemo- and stereoselective, and the products **58** could be easily prepared through one-pot reaction from accessible starting materials (Scheme 7.23). This transformation started from propiolic acid (**57**), an aldehyde, an amine, and an isocyanate followed by the addition of different amines. Different examples were described with yields ranging from 72 to 96%.

A mechanism to explain the obtained products in this approach was proposed by the authors (Scheme 7.24). The Ugi product **59** contained an activated triple bond that could be attacked by a nucleophile (usually an amine) to give the intermediate **60**, which evolved to the final product **58**.

In the same manuscript, the authors described the synthesis of polyfunctionalized *Z*-dithiocarbamates **61**, which are important building blocks due to their numerous biological activities [59]. This 6-MCR proceeded through a one-pot and six-component Ugi/nucleophilic addition sequence in high yields (Scheme 7.25). The main difference with the aforementioned 5-MCR was the use of carbon disulfide and an amine for the nucleophilic addition.

The authors proposed a similar mechanism for the 6-MCR (Scheme 7.26). In both cases, the Ugi intermediate **59** contained an active triple bond, which was suitable for further nucleophilic addition reactions of the nucleophiles present in the medium. However, in this case, the nucleophile was formed by reaction of the amine to carbon disulfide.

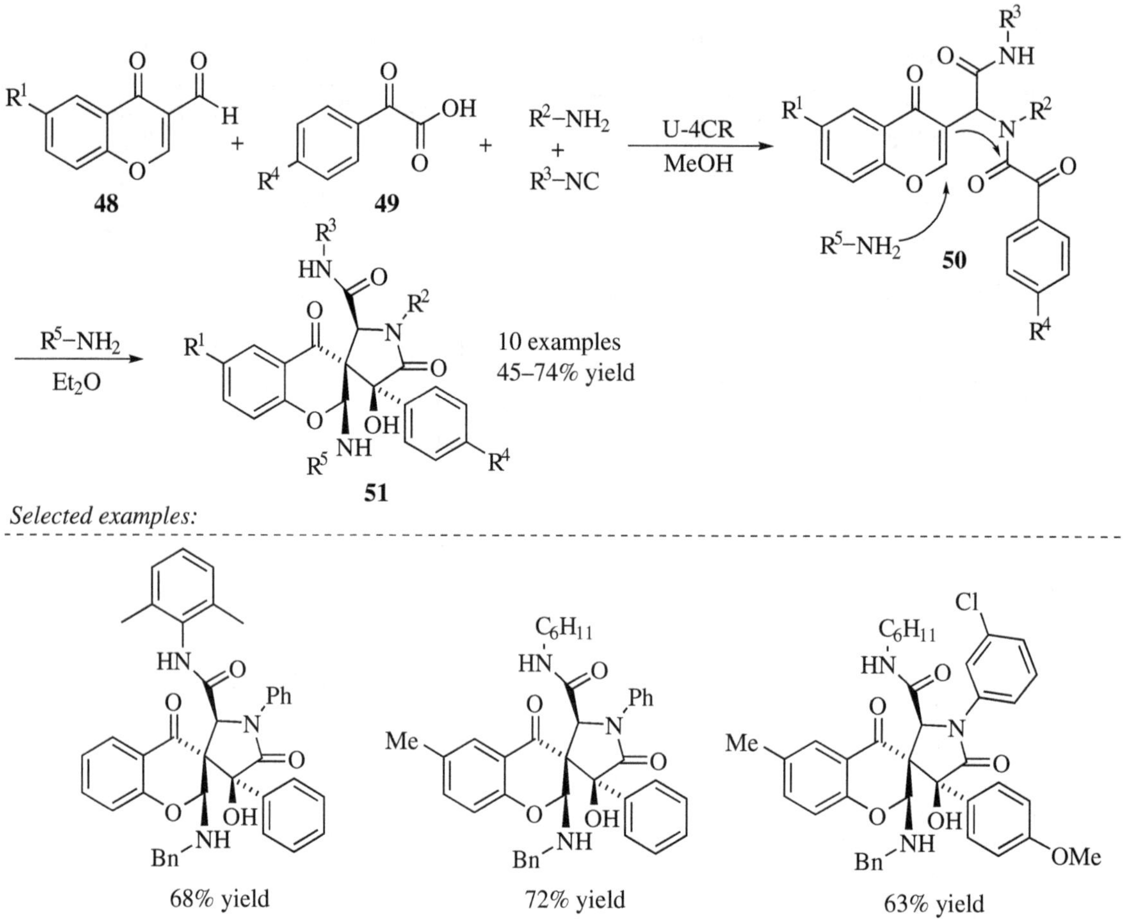

48 49 50 51

SCHEME 7.21 Sequential five-component synthesis of spiropyrrolidinochromanones **51**.

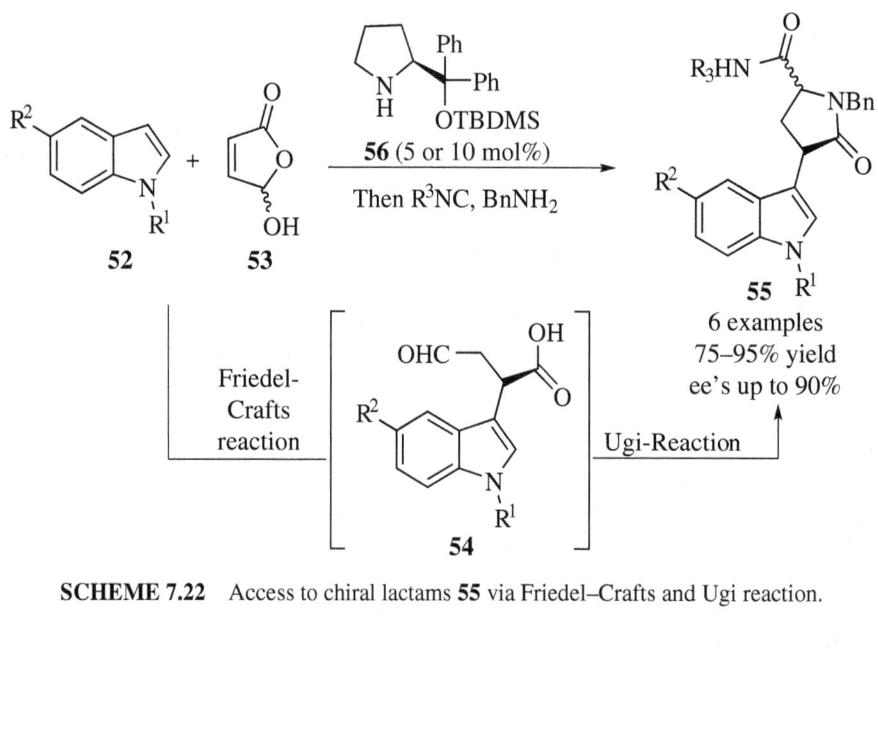

52 53 54 55

SCHEME 7.22 Access to chiral lactams **55** via Friedel–Crafts and Ugi reaction.

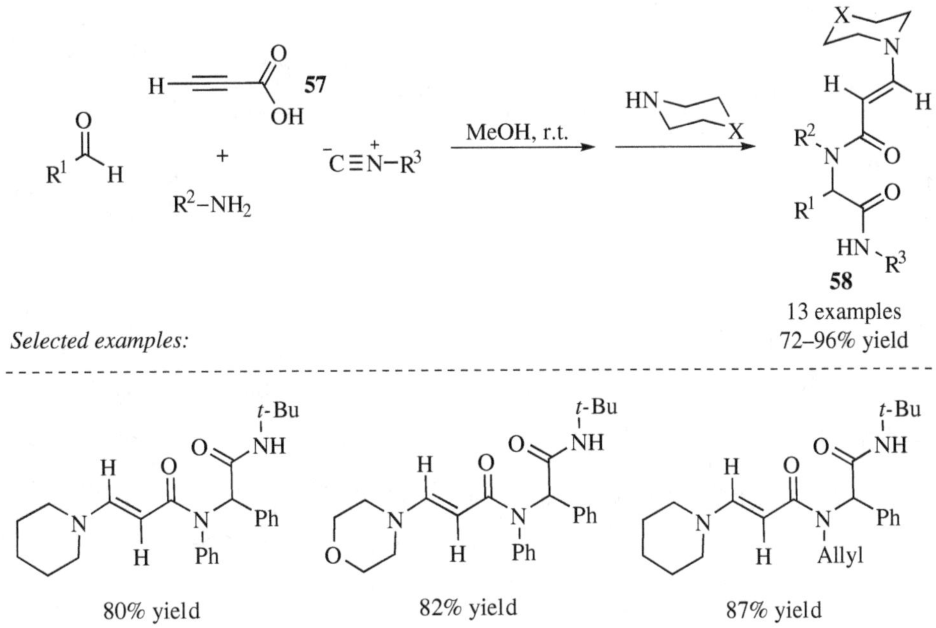

SCHEME 7.23 Multicomponent synthesis of *E*-enaminones **58** via 5-component reaction.

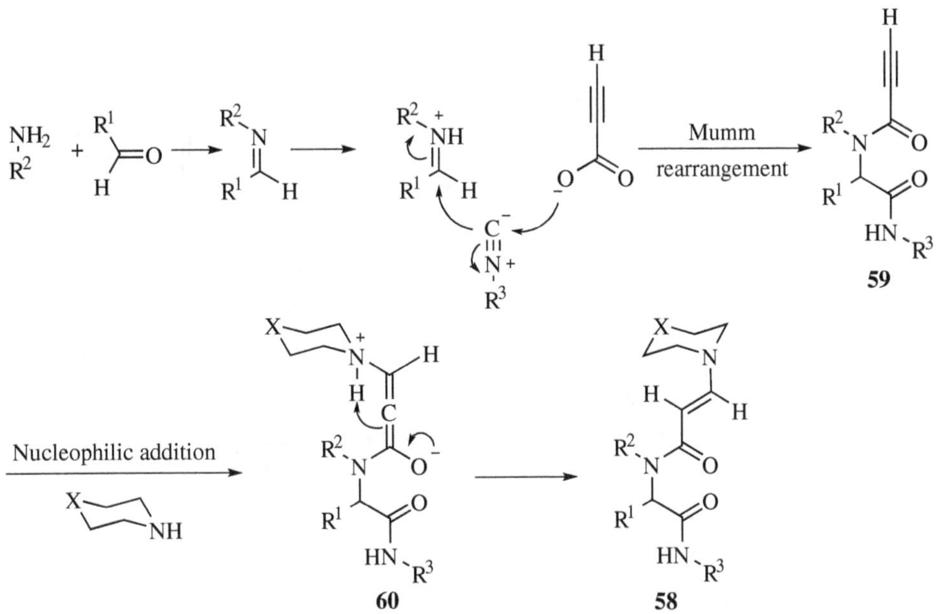

SCHEME 7.24 Mechanism proposal for the 5-component synthesis of *E*-enaminones **58**.

The same group has also described the synthesis of 1,4-disubstituted polyfunctional piperazines **65** through a one-pot Ugi-5-MCR (Scheme 7.27) [60]. In this case, the authors used piperazine (**62**), which attacked the intermediate **63**, and later, the formed intermediate **64** would attack to a second unit of the intermediate **63**, forming the final 1,4-disubstituted piperazine **65**. Thus, this reaction allowed the use of four points of diversity in the starting materials.

In 2010, Al-Tel's group described the synthesis of polyfunctional structures using a one-pot five- and six-component sequential process [61]. In this process, firstly the imine **66** was formed in the presence of $Sc(OTf)_3$ and then the solution could be directly used in the next [4+1]-cycloaddition (Groebke–Blackburn reaction). The cycloaddition product **67** contained an acid moiety, which allowed a subsequent Ugi-4CR. Thus, the final Ugi products **68** were obtained from isocyanides, aldehydes, and amines in good yields (Scheme 7.28).

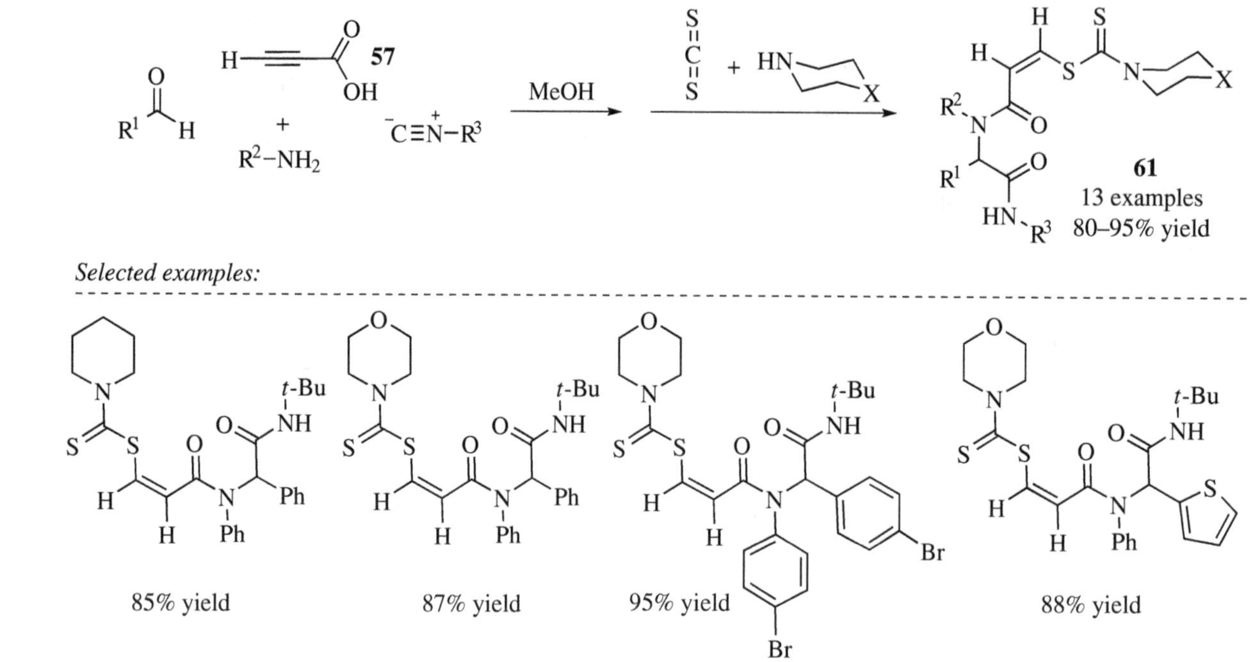

SCHEME 7.25 6-MCR for the synthesis of *Z*-dithiocarbamates **61**.

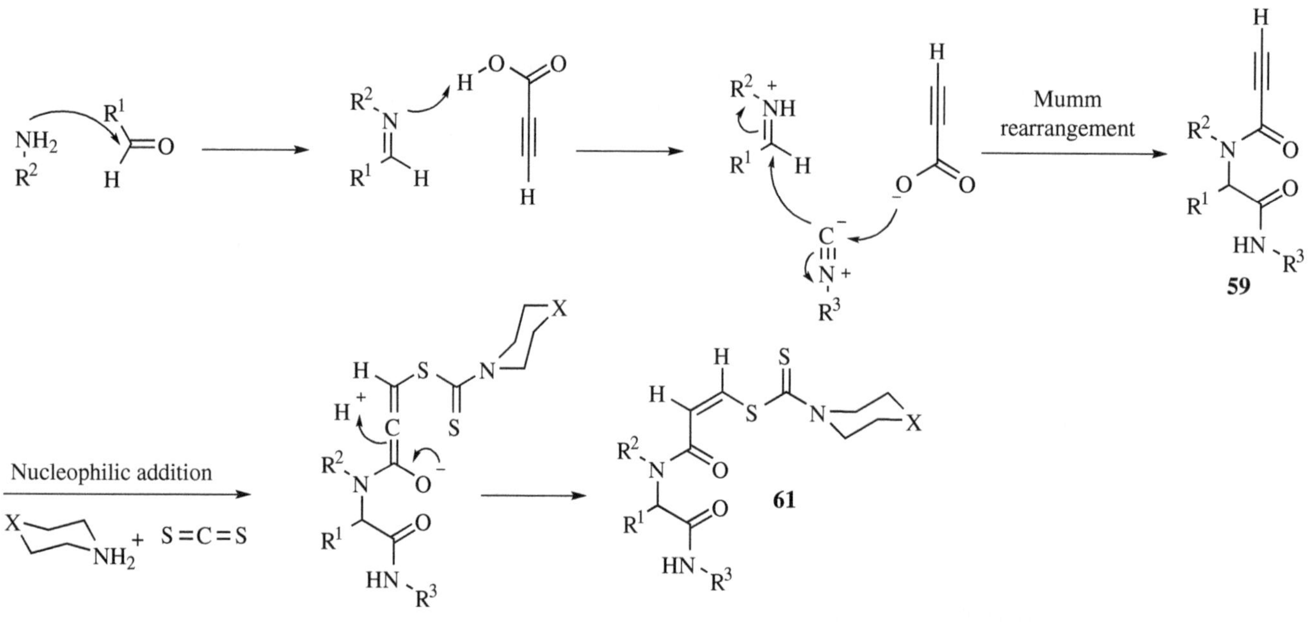

SCHEME 7.26 Mechanism proposal for the 6-MCR synthesis of *Z*-dithiocarbamates.

The same group also showed the use of other starting imines, which were formed in the same manner, and subsequently subjected to the Groebke–Blackburn/Ugi reaction (see starting imines in Fig. 7.2). In all cases, good yields were reported for the sequential process (55–73% yield).

In 2014, Shaabani's group described the synthesis of coumarin-3-carboxamides **73** based on a six-component reaction [62]. Final products **73** were synthesized in excellent yields following a one-pot six-component reaction: a tandem Knoevenagel/Ugi/click reaction sequence (Scheme 7.29). A plausible mechanism was indicated by the authors. The reaction would start by the initial formation of conjugated electrodeficient heterodiene **69** by standard Knoevenagel conditions. Then, an Ugi reaction starting from acid **70**, the isocyanate, and the *in situ* formed imine **71** led to the Ugi intermediate **72**, which was ready for a click reaction in the terminal triple bond.

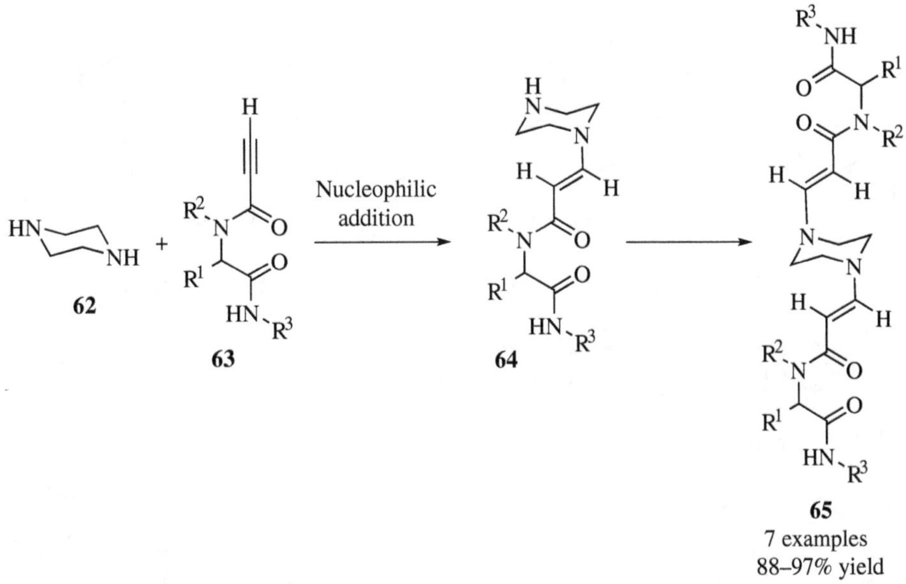

SCHEME 7.27 Synthesis of piperazine derivatives **65** by a 5-MCR.

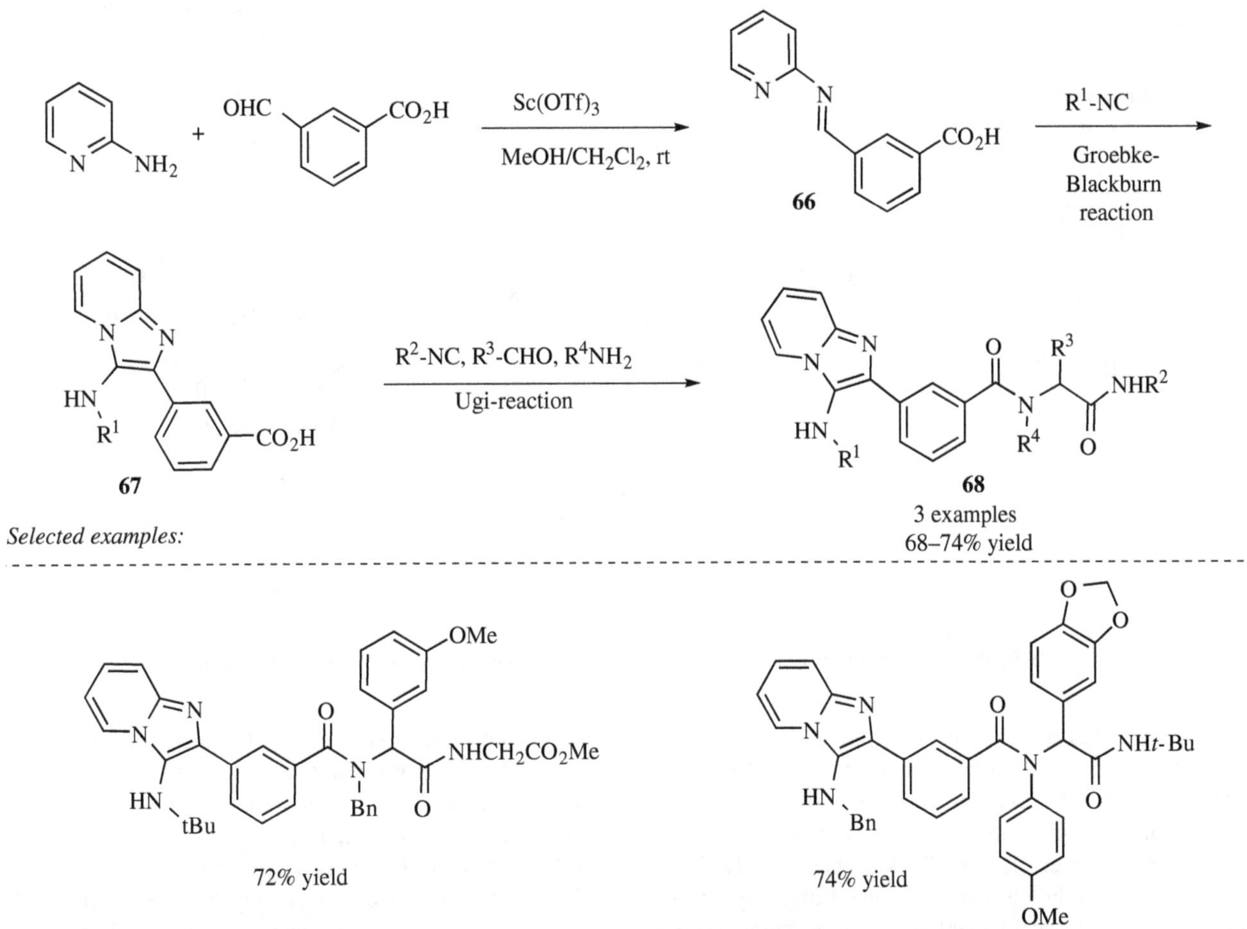

SCHEME 7.28 [4 + 1]-Cycloaddition (Groebke–Blackburn reaction) and Ugi reaction.

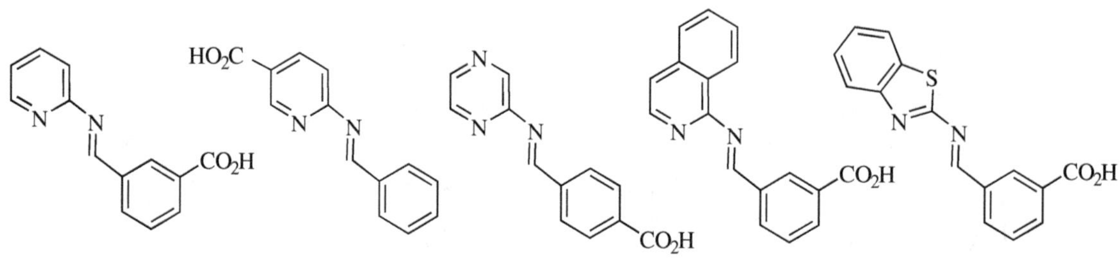

FIGURE 7.2 Different starting imines used in the Groebke–Blackburn/Ugi reaction.

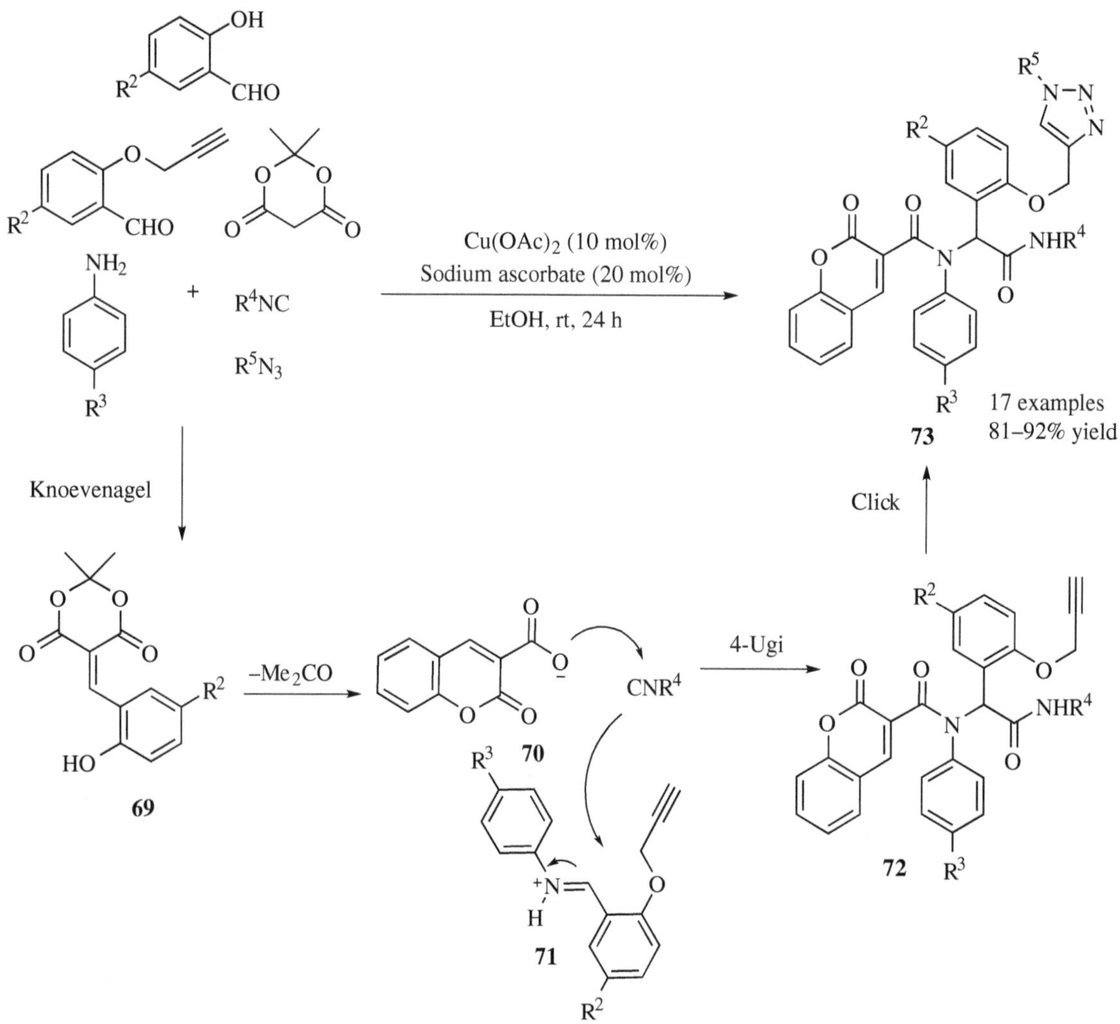

SCHEME 7.29 Synthesis of coumarin-3-carboxamides **73** based on a six-component reaction.

In 1993, Ugi described the seven-component reaction [63] and pointed out the possibility of carrying out 7-MCR based on this strategy. However, it was not until 2009 when an efficient one-pot reaction of eight-components was described by Orru et al. [64]. This process involved the formation of 9 new bonds and 11 points of diversity. In this study, a sequential development from six- to eight-component reaction was carried out. The MCR was initiated with the synthesis of intermediate **74**, which was formed from methyl 2-isocyanatopropanoate, acetone, and sodium glycinate (Scheme 7.30). Then, after the protonation of the intermediate **74** by HCl, an Ugi reaction was carried out by addition of *i*PrCHO, *n*-propylamine (or benzylamine), and *t*BuNC to give the corresponding 6-MCR products **75** in 38–62% as mixture of diastereoisomers (nearly to 50:50 in both cases).

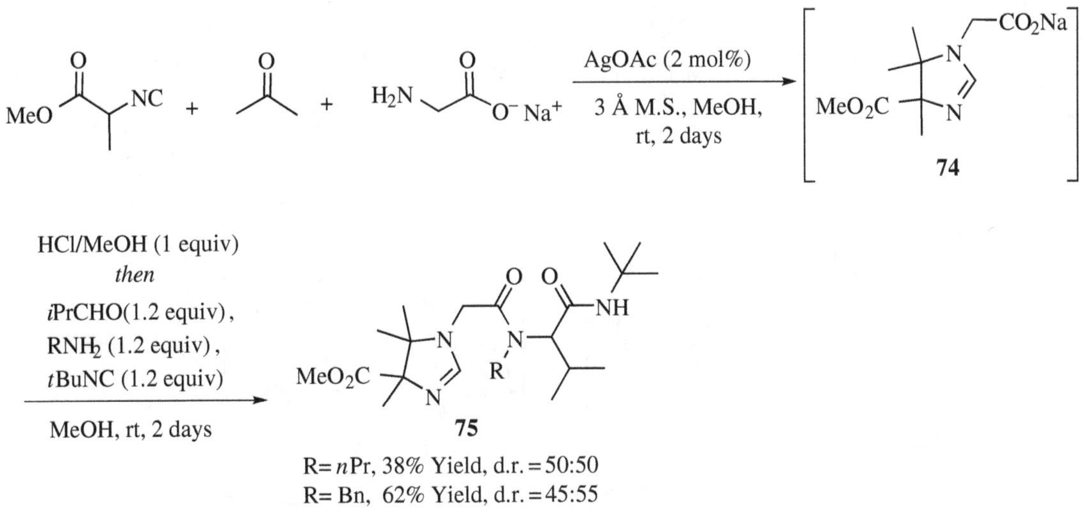

SCHEME 7.30 Six-multicomponent reaction involving an Ugi-4CR.

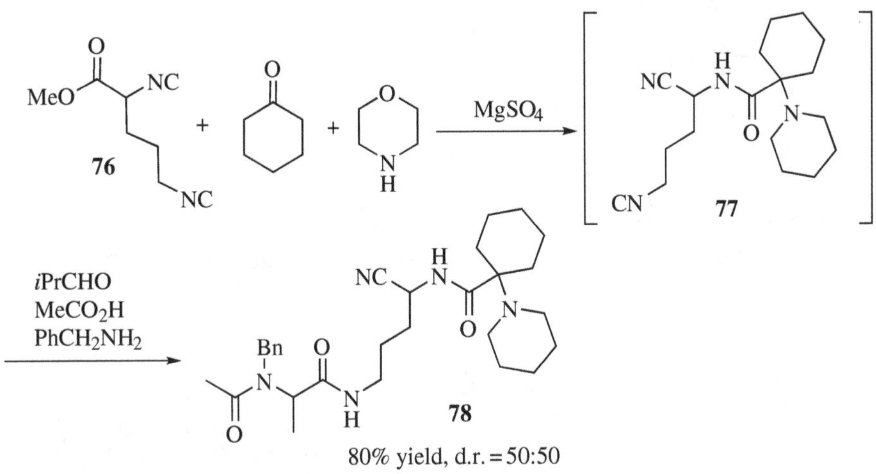

SCHEME 7.31 Use of diisocyanide and synthesis of intermediate **77**.

In the same publication, the authors studied the formation of another intermediate **77**, starting from diisocyanide **76**, that reacted under Ugi conditions to form **78** in 80% yield as mixture of diastereoisomers 50:50 (Scheme 7.31).

With these two intermediates on hand, the authors proposed the combination of the isonitrile intermediate **77** (from a U-3CR) and the acid intermediate **74** (from a U-3CR) to carry out the Ugi reaction by addition of the corresponding aldehyde and amine, obtaining in this manner the 8-MCR product **79** (Scheme 7.32).

7.5 UGI POSTMODIFICATION PROCESSES

One of the best strategies for the synthesis of valuable pharmaceutical compounds and also for the rapid generation of molecular diversity is the use of isocyanide-based multicomponent reactions (IMCRs) [3–8]. More specifically, the Ugi reaction has been used as platform in combination with other reactions (as postmodification process) to obtain dozens of new structures. Since 2006, there are several examples dealing with this strategy like Ugi/Diels–Alder [65], Ugi/Buchwald–Hartwig [66], Ugi/Heck [67], and Ugi/ring-closing metathesis [68], among others. In the following lines, we want to highlight some representative examples of this Ugi postmodification strategy since 2006.

Hulme et al. have described an Ugi/aldol sequence for the synthesis of different substituted nitrogen heterocycles [69] (Scheme 7.33). Thus, depending on the reaction conditions and also in the structure of the used reagents, the authors described three types of products: 5-hydroxypyridine-2-ones **80** and pyridinone-3-yl-benzimidazol-2-ones **81**, among others (Scheme 7.33).

Balalaie et al. described in 2010 a six-component reaction. The stereoselective synthesis of oxindole derivatives was carried out using a sequential one-pot Ugi/Heck carbocyclization/

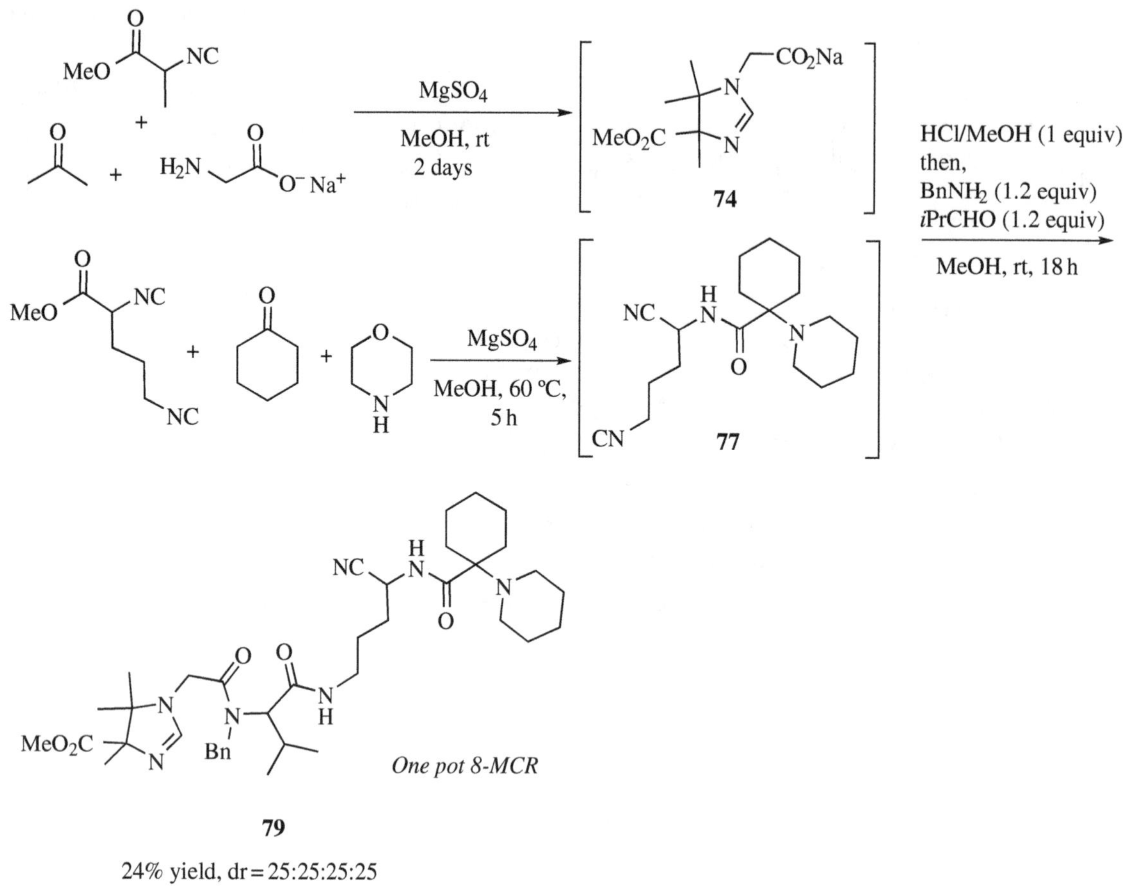

SCHEME 7.32 One-pot 8MCR based on the intermediates **74** and **77**.

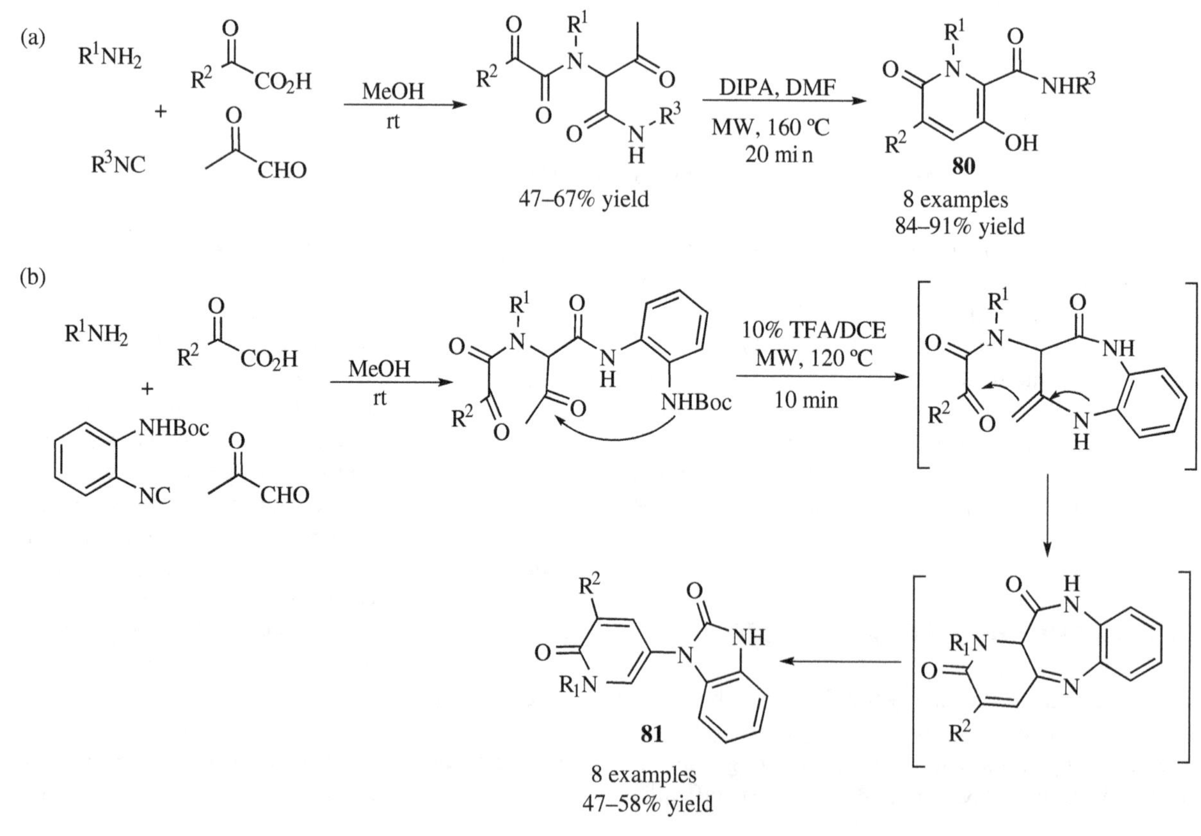

SCHEME 7.33 Synthesis of nitrogen heterocycles by Ugi/aldol sequence.

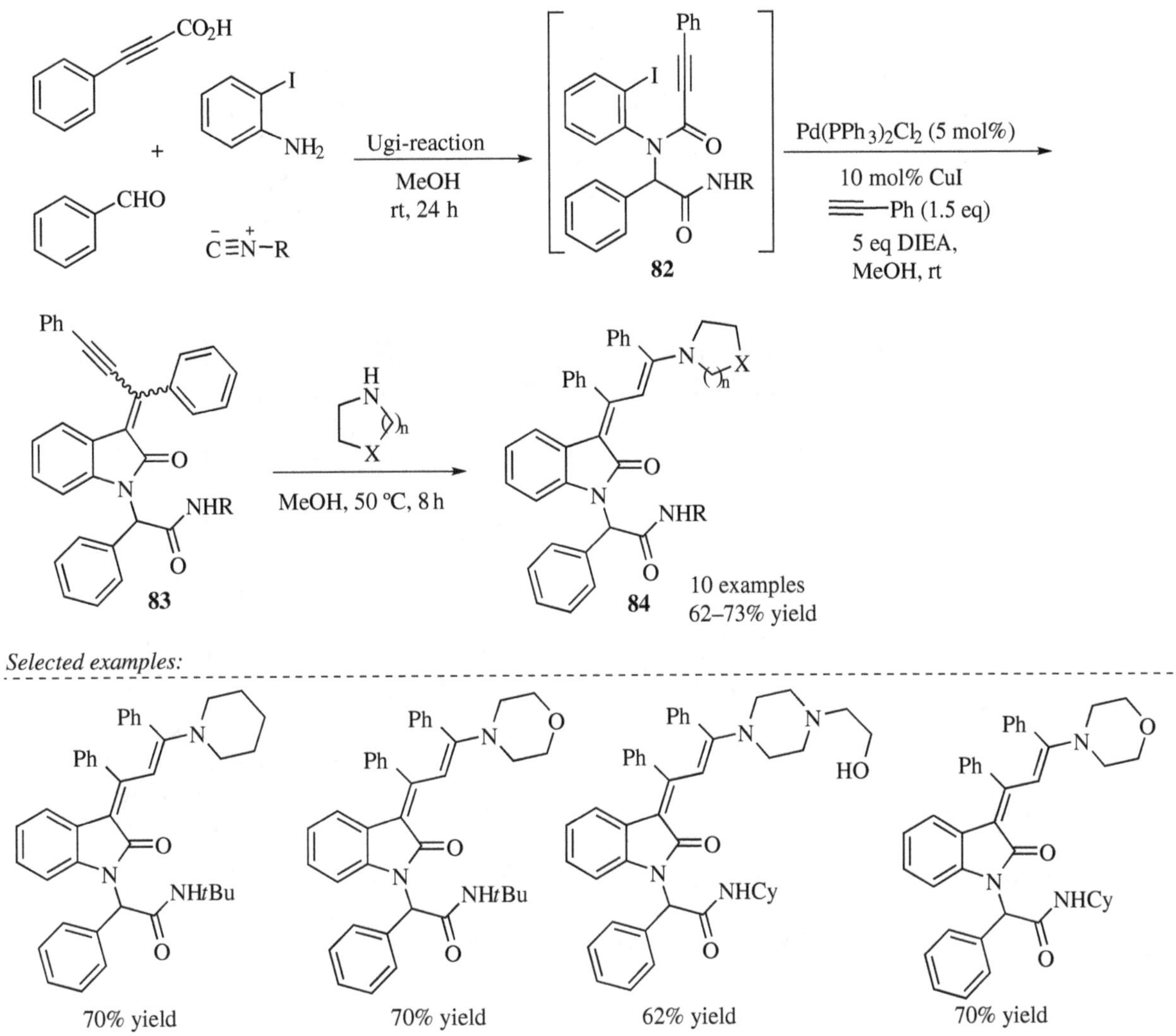

SCHEME 7.34 Synthesis of oxindoles **84** using a one-pot Ugi/Heck carbocyclization/Sonogashira/nucleophilic addition.

Sonogashira/nucleophilic addition (Scheme 7.34) [70]. The reaction started by the Ugi-4CR, giving the intermediate **82**, which was transformed into the enyne **83** by a Heck–Sonogashira tandem reaction. Finally, the activated enyne evolved an aza-Michael-type addition to give the final six-component product **84**.

In 2011, Andreana and Santra described an Ugi/Michael/aza-Michael cascade reaction in aqueous media as strategy for molecular diversity in natural products [71]. Fused aza-spiro tricycles and tetracycles were obtained by using microwave irradiation and water as solvent. In this remarkable work, up to four stereogenic centers and six contiguous bonds were created with high diastereoselectivity. The reaction started from an Ugi-4CR, which gave the intermediate **85**, which underwent a first Michael-type reaction that led intermediate **86**. Finally, an additional aza-Michael reaction on **86** yielded the final tricyclic product **87** (Scheme 7.35).

Very recently, the synthesis of benzo-1,4-diazepin-2,5-diones was reported, using the Ugi-4CR postmodification strategy [72]. A very similar strategy had already been used for the synthesis of benzodiazepine-3-ones **88** by Andreana's group [73]. Thus, the reaction involved the formation of the Ugi intermediate followed by the intramolecular lactamization by reduction of the nitro group (Scheme 7.36). The synthesis could be carried out in a one-pot manner and yields ranged from 35 to 80%. Miranda's group has also used the reductive methylation reaction protocol as a procedure for the synthesis of piperazinohydroisoquinoline ring system in Ugi postmodification processes [74].

In an elegant work, Kobayashi's group introduced the 1-isocyano-2-(2,2-dimethoxyethyl)benzene (indole-isonitrile) **90**, as a convertible isonitrile in the Ugi-4C-3CR, and demonstrated the selective cleavage of the resulting C-terminal amide bond [75]. To this end, a mixture of ammonium acetate and a γ- or δ-keto acid **89** was initially reacted (Scheme 7.37). Once the imine was formed, indole-isonitrile **90** was added so that the Ugi condensation products **91** were obtained (49–77%). Then, the Ugi product **91** was successfully con-

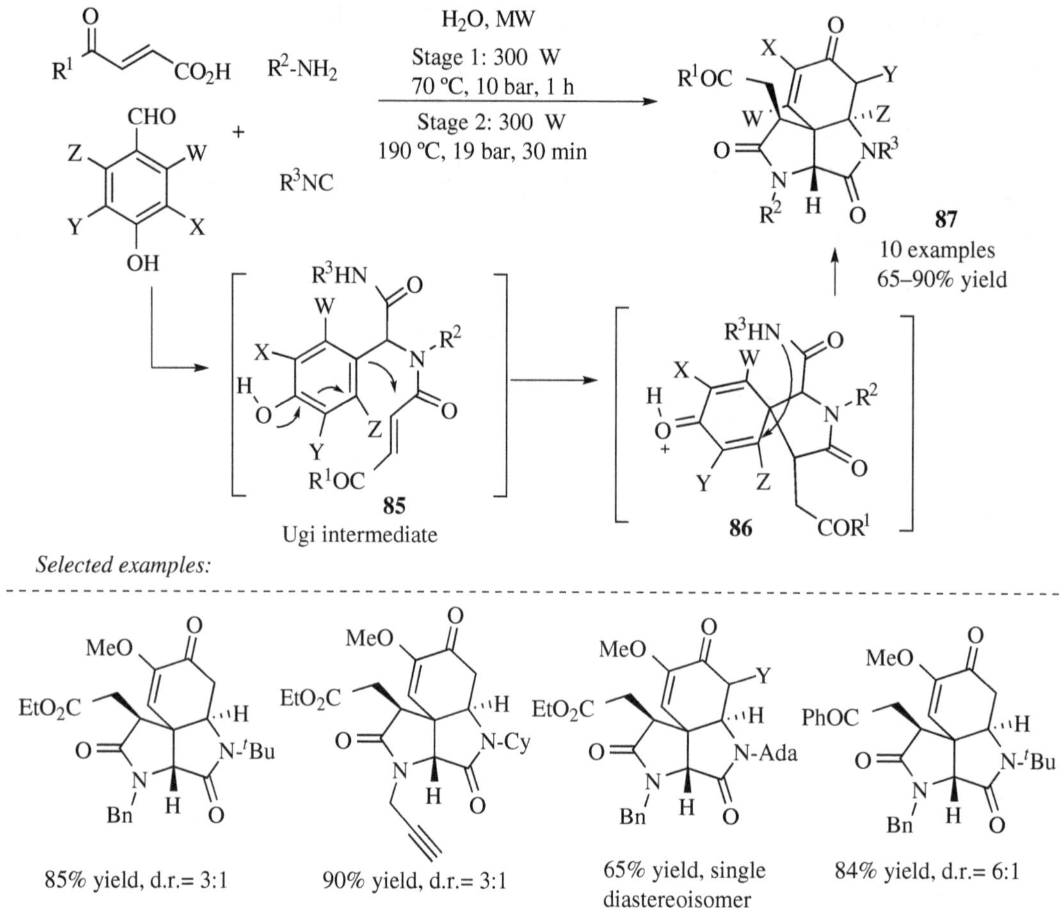

SCHEME 7.35 Synthesis of fused azaspiro tricycles **87**.

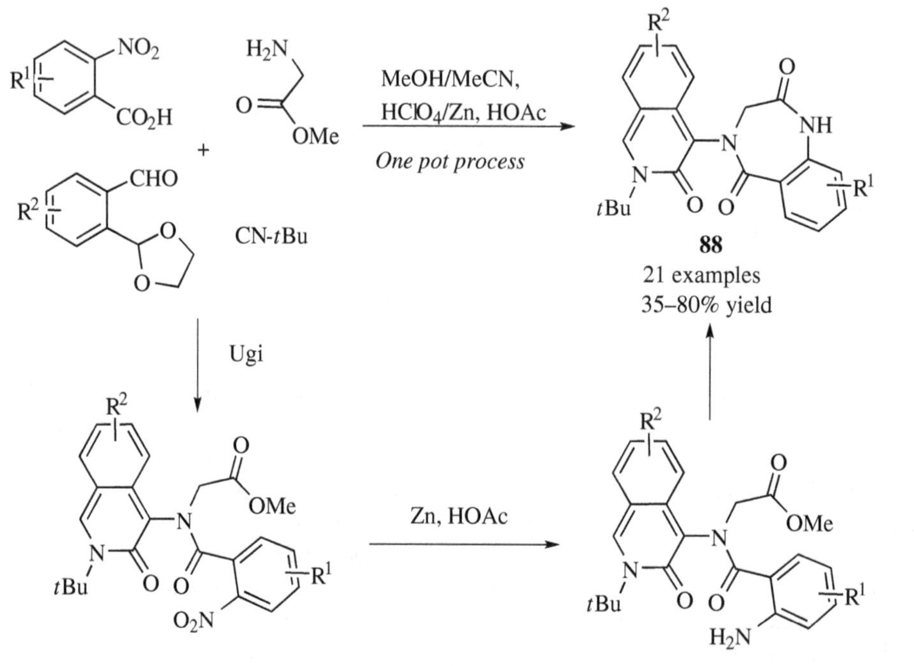

SCHEME 7.36 Synthesis of benzo-1,4-diazepin-2,5-diones **88** under Ugi and reductive conditions.

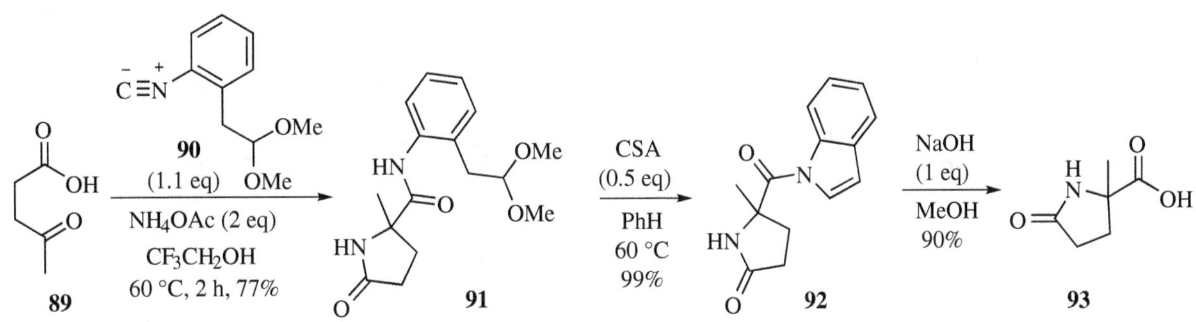

SCHEME 7.37 Synthesis of pyroglutamic acid derivatives by convertible indole-isonitrile methodology.

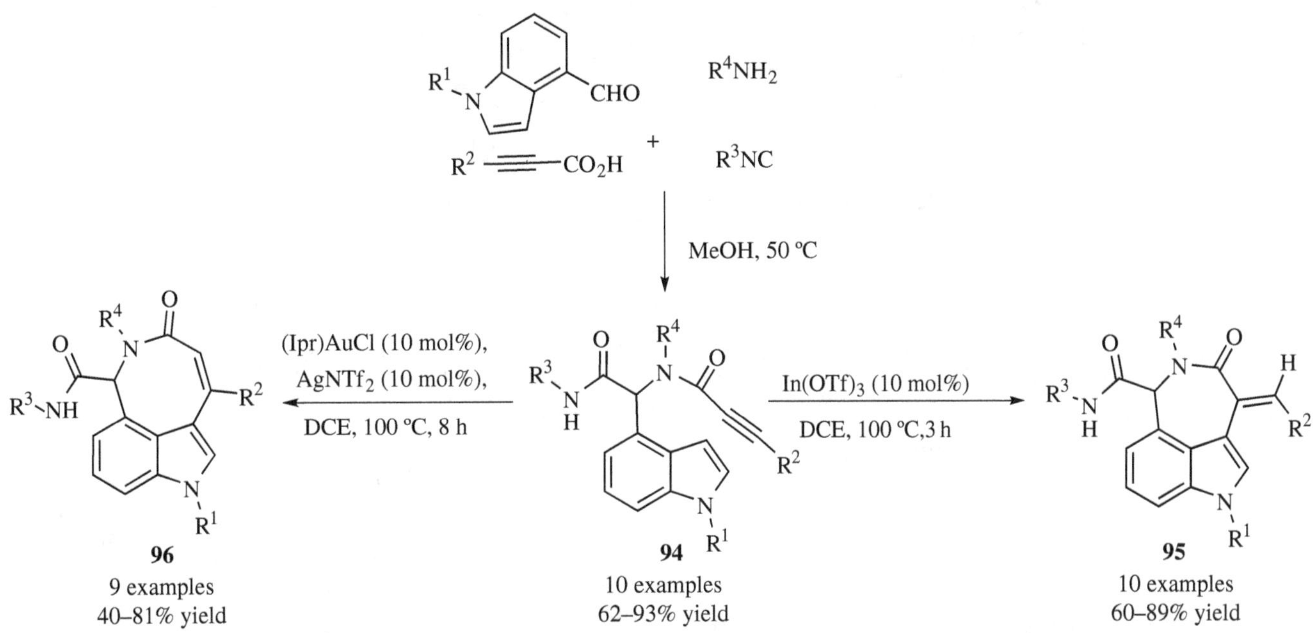

SCHEME 7.38 Synthesis of azepinoindolones **95** under In(I) catalysis (right) or azocinoindolones **96** under Au(I) catalysis (left).

verted to the *N*-acylindole scaffold **92** upon treatment with a catalytic amount of camphorsulfonic acid (CSA). Indole moiety was cleaved to give the free pyroglutamic acid derivative **93** by treatment with sodium hydroxide via the methyl ester intermediate.

Van der Eycken's group has published different papers employing the Ugi reaction followed by metal-catalyzed intramolecular reactions [76, 77]. Thus, the synthesis of the Ugi intermediate **94** was performed under standard conditions, and depending on the chosen metal, the cyclization proceeded to the synthesis of azepinoindolones **95** under In(I) catalysis or to azocinoindolones **96** under Au(I) catalysis (Scheme 7.38).

As aforementioned in the Ugi-3CR section, Zhu and Wang described the enantioselective organocatalytic synthesis of epoxy-tetrahydropyrrolo[3,4-*b*]pyridin-5-ones [27]. Later on, Zhu and Masson described the postmodification of the Ugi products **98** using the Diels–Alder reaction (Scheme 7.39) [78]. In this MCR, six bonds and five stereocenters including two quaternary ones were created in one pot using a BINOL-derived chiral phosphoric acid **97**, isolating the products **99** in high yields, and good to excellent diastereo- and enantioselectivities. The oxazole intermediate **98** could be isolated or *N*-acylated to undergo an intramolecular hetero-Diels–Alder reaction to give the oxatricyclic products **99**.

Later on, Hulme and coworkers reported a methodology based on an intramolecular U-4C-3CR postcondensation strategy that enabled production of nitrogen-enriched polycyclic scaffolds **103** and **104** endowed with benzimidazoles, lactams of various ring sizes, dihydroquinazolines, and other moieties [79] (Scheme 7.40). A range of tethered keto acids or aldehyde acids **100**, *tert*-butyl 2-isocyanophenylcarbamate (**101**), and amines or diamines **102** were successfully employed in intramolecular ring forming Ugi reactions, followed by either one or two consecutive aminocyclodehydrations. Products of a single aminocyclodehydration **103** and from two consecutive aminocyclodehydrations **104** are shown in Scheme 7.40.

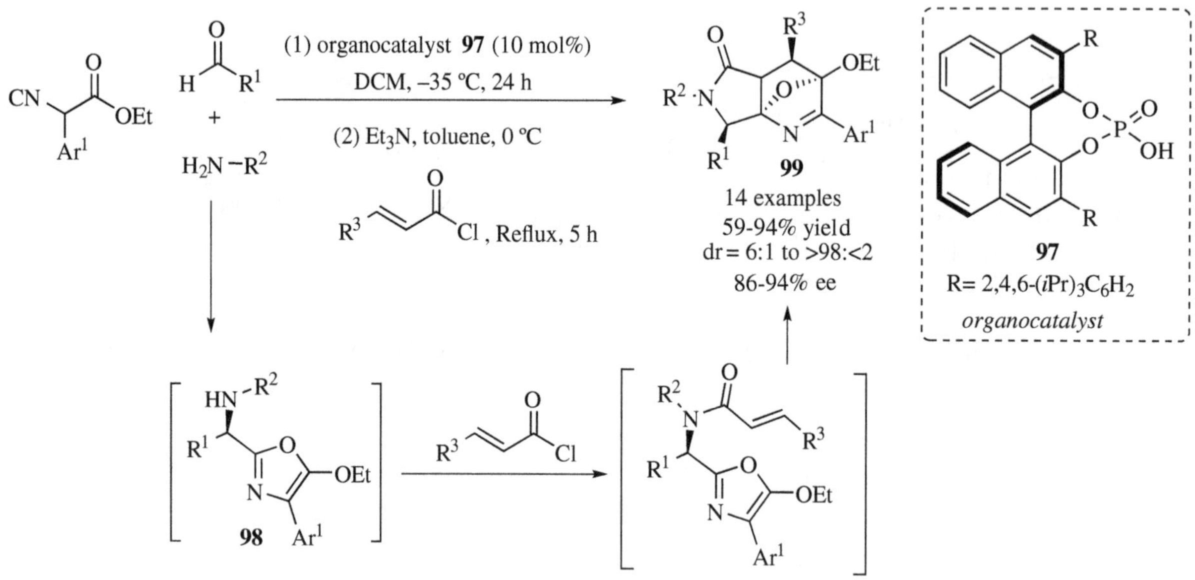

SCHEME 7.39 Postmodification of the Ugi products using Diels–Alder reaction.

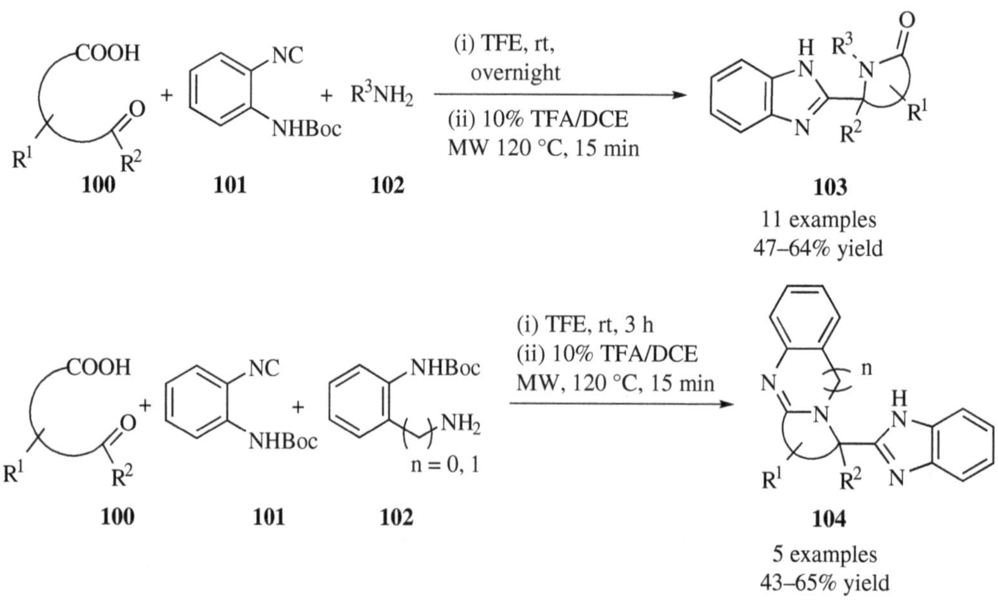

SCHEME 7.40 Synthesis of nitrogen-enriched polycyclic scaffolds by sequential postcondensation intramolecular U-4C-3CR strategy.

Other reactions that have been applied in the Ugi postmodification strategy are the aza-Wittig reaction [80–84] and copper-catalyzed processes [85, 86]. More recently, Dömling and coworkers have developed a new approach for the synthesis of polycyclic compounds via Ugi MCR followed by a Pictet–Spengler reaction [87–89]. In 2011, a protocol for the rapid access to quinolin-2-(1*H*)-one scaffold was reported by a sequential 4-component Ugi–Knoevenagel condensation [90]. Hulme et al. have introduced a one-pot, two-step protocol for large-scale production of libraries of novel peptidomimetic-like *bis*-pyrrolidinone tetrazoles via the Ugi-3CR–azide reaction of tethered keto ester methyl levulinate, primary amines, isocyanides, and azidotrimethylsilane, in the presence of trifluoroacetic acid [91].

7.6 UGI–SMILES APPROACH

One of the best strategies for discovering new reactions in the MCR field is the replacement of one of the components, which is also called single reactant replacement (SRR) [92]. This strategy requires a detailed analysis of the mechanism reaction. Following the accepted mechanism for the Ugi-4CR (see Scheme 7.15), the last step is the nonreversible Mumm-type

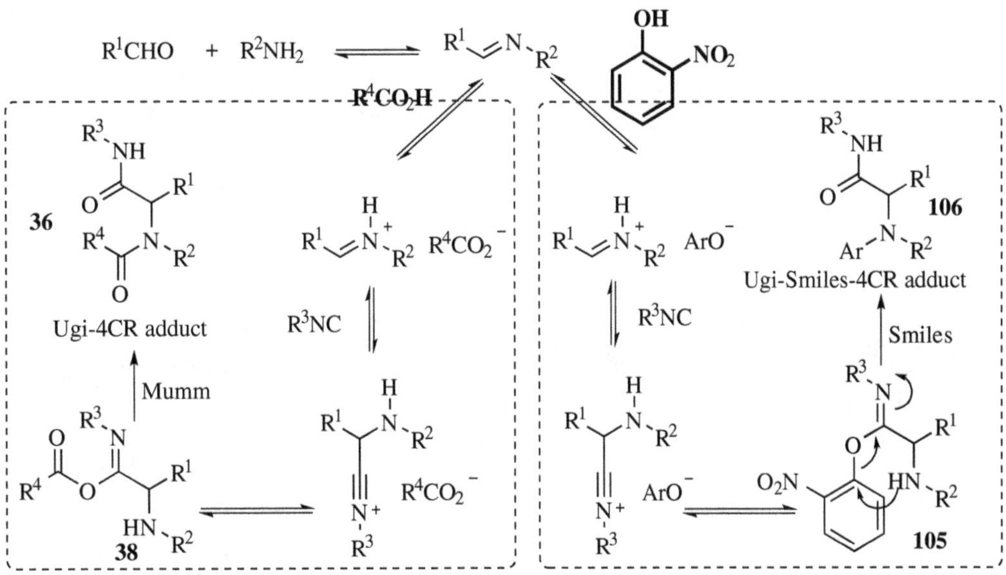

SCHEME 7.41 Comparison of the mechanism of Ugi-4CR and Ugi–Smiles-4CR.

rearrangement, which gives the final product (see left, Scheme 7.41). One of the keys in the high efficiency of the Ugi-4CR was related to the use of acid (commonly benzoic acid derivatives), which allowed the formation of the iminium, trapped the nitrilium intermediate and finally was also involved in the Mumm rearrangement. As aforementioned, the use of other acidic additives, like hydrogen thiosulfate, hydrazoic acid, carbonic acid monoesters, isocyanic and isothiocyanic acids, and more recently thiocarboxylic acids, has been also reported to undergo Ugi reaction.

The main difference between the Ugi-4CR and Ugi–Smiles-4CR is the use of phenol derivatives instead of acid derivatives (see right and left, Scheme 7.41). Some preliminary reactions reported by Passerini and Marcaccini showed that phenol derivatives could be involved as acidic compounds [93, 94]. However, after several decades, El Kaïm and Grimaud reported the first Ugi–Smiles coupling by heating stoichiometric amount of an *o*-nitrophenol, a primary amine, a carbonyl compound, and an isocyanide [95]. However, the use of phenol in the Ugi–Smiles-4CR implied a slightly different mechanism, which included the Smiles rearrangement (instead of the Mumm rearrangement; see intermediates **38** and **105**, Scheme 7.41). The Smiles rearrangement is an intramolecular migration of an aromatic ring between the two nucleophilic centers of the phenol imidate intermediate **105** [13]. This reaction allowed the preparation of different functionalized *N*-arylated products **106** with good yields under smooth conditions (Scheme 7.42). The reaction also worked with highly activated *p*-nitrophenols **107**, but with salicylic derivatives, lower yields were found [96]. The tropolone could be also used in the Ugi–Smiles coupling reaction, giving access to aminotropolone derivatives [97]. Unexpectedly, the polarity of the solvent had not a strong influence in the reactivity and the reaction could be carried out in toluene, in methanol, or even in water [98]. In addition, the effect of hydrogen bond in the nucleophilic aromatic substitution was studied using density functional theory [99].

El Kaïm et al. extended the Ugi–Smiles methodology to the preparation of new heterocyclic scaffolds **109** by coupling of 2-hydroxypyridines or pyrimidines (**108**) with carbonyl compounds, amines, and isocyanides [100]. The scope was further extended using heterocyclic thiols to form highly functionalized thioamides (Scheme 7.43) [101].

The same group also described the synthesis of new pyrazines **110** and quinoxaline derivatives using the Ugi–Smiles-4CR of isocyanides, aldehydes, primary amines, and pyrazone or quinoxalinones (Scheme 7.44) [102].

The reaction was also carried out with five-membered heterocycles, but only mercaptotriazoles **111** showed moderate reactivity for different substituents (Scheme 7.45a) [101]. In the same publication, the authors showed that mercaptobenzoxazoles and benzothiazoles (**112**) could also be used as starting materials for the Ugi–Smiles coupling, giving the final thioamides **113** in good yields with aliphatic aldehydes and ketones, and with lower yields in the case of aromatic aldehydes (Scheme 7.45b).

El Kaïm and Grimaud also used cyclic imines as starting materials for the Ugi–Smiles coupling [103]. Cyclic imines were obtained by oxidation with *N*-chlorosuccinimide (NCS) followed by elimination of HCl with a base (DBU). Then, the imine reacted with isocyanides and electrodeficient phenols to afford *N*-aryl piperidines **114** and pyrrolidines in good yields (Scheme 7.46).

Based on the previous work of El Kaïm and Grimaud and their own work [26], Orru's group combined MAO-N-catalyzed

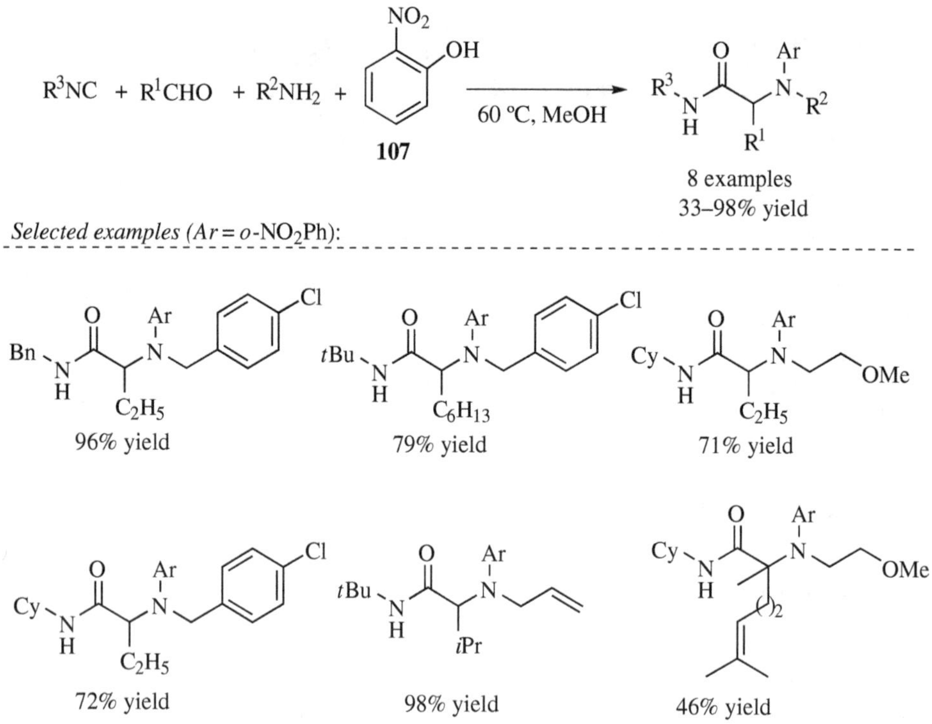

SCHEME 7.42 First Ugi–Smiles-4CR reaction described by El Kaïm and Grimaud.

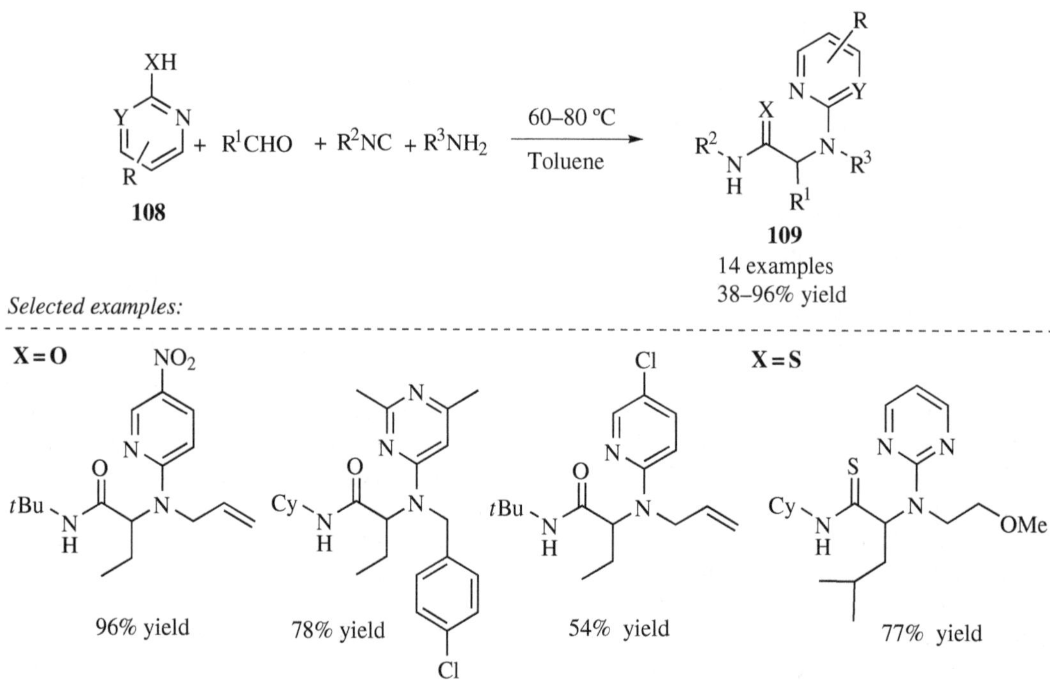

SCHEME 7.43 Heterocyclic products obtained using the Ugi–Smiles coupling.

desymmetrization of cyclic meso-amines with Ugi–Smiles coupling to afford optically pure *N*-aryl proline amides **115** (Scheme 7.47) [104]. This is the first publication related to the asymmetric Ugi–Smiles coupling.

More recently, Westermann et al. have described a seven-component reaction by a sequential Ugi–Mumm/Ugi–Smiles reaction [105]. In this work, the authors proposed the use of a different nitroaryl acid derivative (e.g., **116** or **118**;

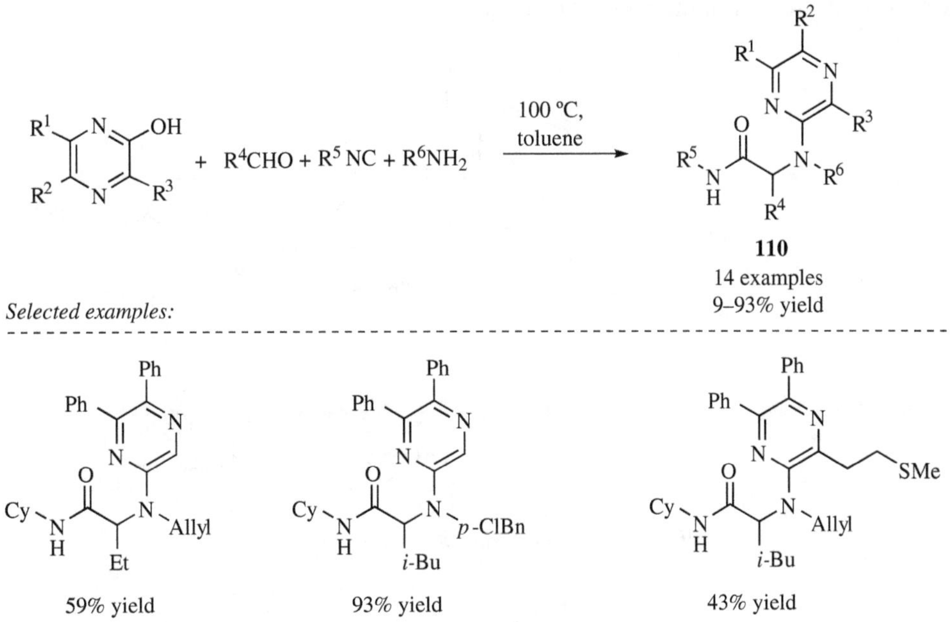

SCHEME 7.44 Pyrazine synthesis by Ugi–Smiles coupling.

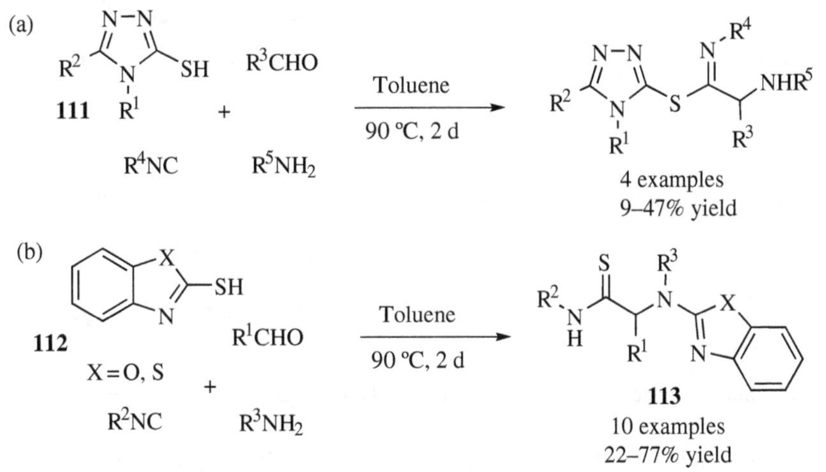

SCHEME 7.45 Mercapto-heterocycles in the Ugi–Smiles coupling.

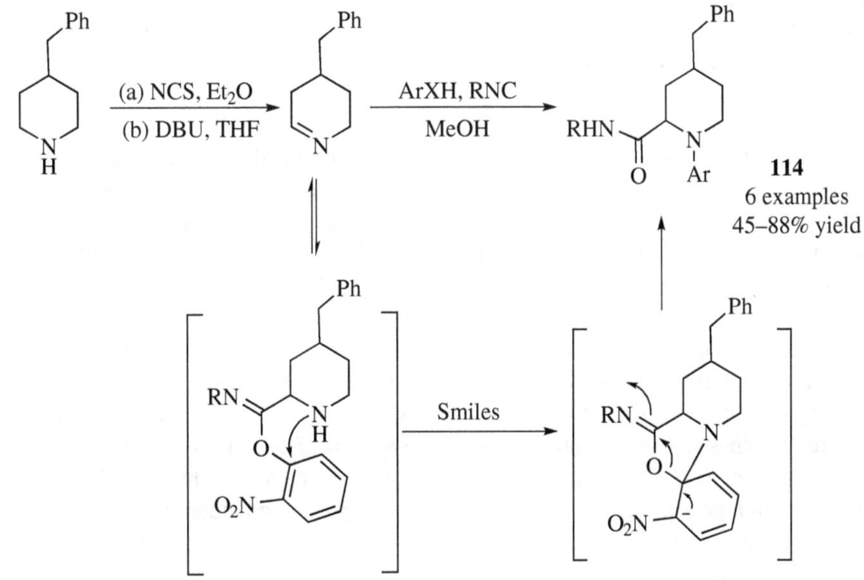

SCHEME 7.46 Piperidines and pyrrolidines via Ugi–Smiles coupling.

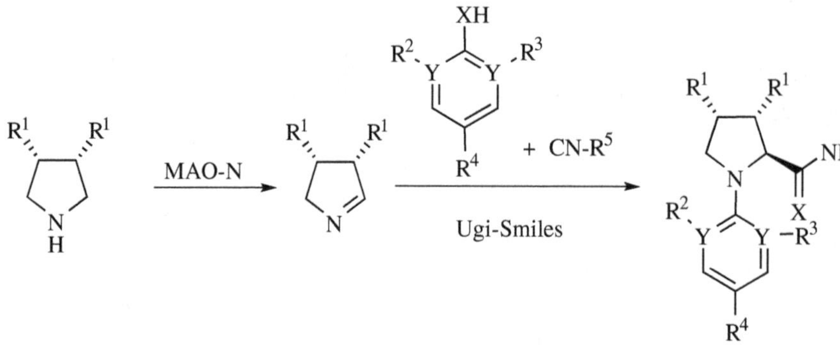

115
11 examples
37–72% yield

X = O, S
Y = C, N

SCHEME 7.47 First asymmetric Ugi–Smiles coupling.

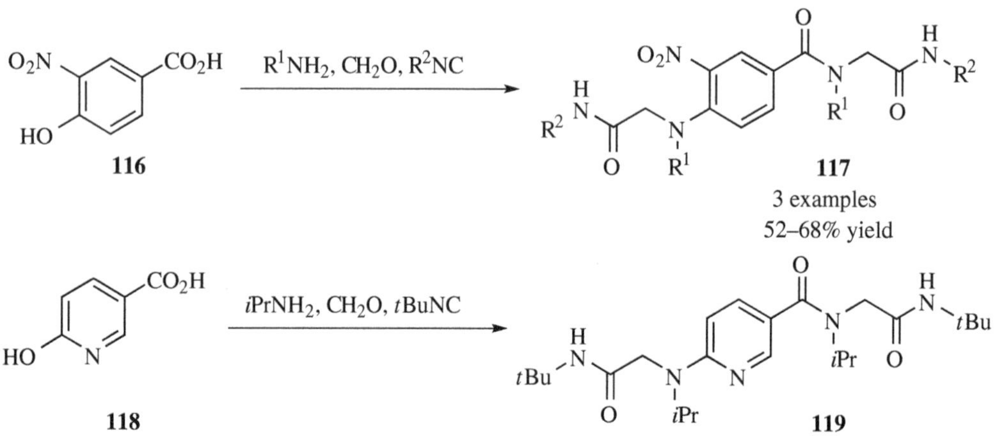

116

117
3 examples
52–68% yield

118

119
25% yield

SCHEME 7.48 Seven-component reactions by sequential chemoselective Ugi–Mumm/Ugi–Smiles reactions.

Scheme 7.48). The acid reacted following the Ugi–Mumm sequence, whereas the nitroaryl group reacted following an Ugi–Smiles coupling pathway. Thus, the sequential functionalization of different aryl acid derivatives was performed allowing the formation of eight bonds and the synthesis of highly functionalized amides **117** and **119** (Scheme 7.48).

7.7 UGI–SMILES POSTMODIFICATION PROCESSES

Once the Ugi–Smiles coupling was established, different works related to the derivatization of the products were carried out. Some of the examples were performed in a one-pot manner, but most of them required previous isolation of the Ugi–Smiles adducts. One of the first examples was developed by El Kaïm and Grimaud by combination of the Ugi–Smiles reaction and a ring-closure metathesis, affording interesting pyrimido azepine scaffolds **120** (Scheme 7.49) [106]. However, the one-pot procedure could not be carried out probably because of the degradation of the employed Hoveyda–Grubbs second-generation catalyst by the residual isocyanide.

Inspired in the combination of the Ugi–Smiles reaction with other transformations, the same group envisaged a new route for the construction of indole scaffolds **121** based on an extension of this novel Ugi–Smiles protocol with a subsequent Heck cyclization in an one-pot procedure (Scheme 7.50) [107].

Other metal-catalyzed reactions have been also combined with the Ugi–Smiles reaction. Thus, substituted pyrrolo[2,3-*d*]pyrimidines **122** were synthesized following an Ugi–Smiles/Sonogashira cascade followed by a base-catalyzed intramolecular cyclization (Scheme 7.51) [108].

Palladium-catalyzed ring-opening reactions have been also used for the synthesis of α,β-unsaturated aldehydes [109]. The synthesis started from the Ugi–Smiles coupling, giving the intermediate **123** under standard conditions (Scheme 7.52). Then, a fragmentation process of the furan ring **123** catalyzed by palladium gave direct access to the α,β-unsaturated aldehydes **124**.

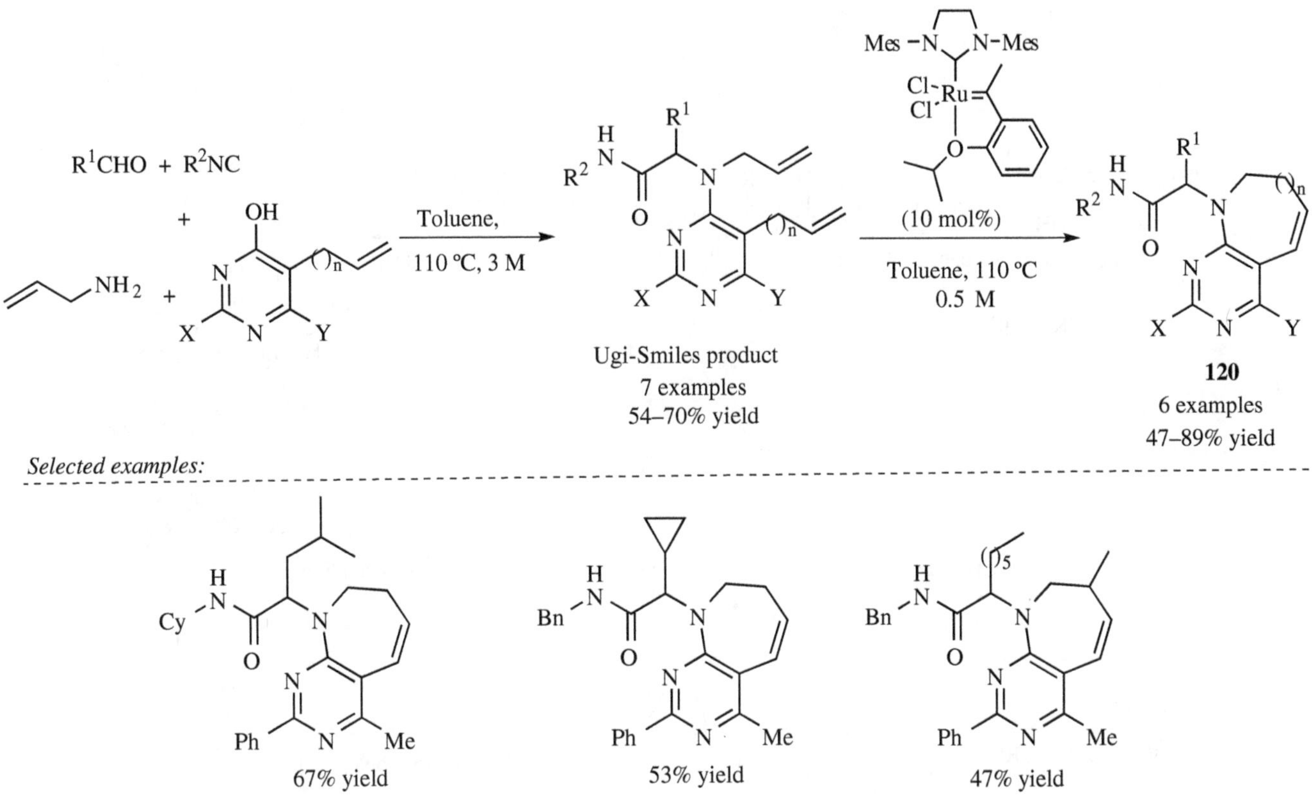

SCHEME 7.49 Ugi–Smiles combined with an olefin ring-closure metathesis.

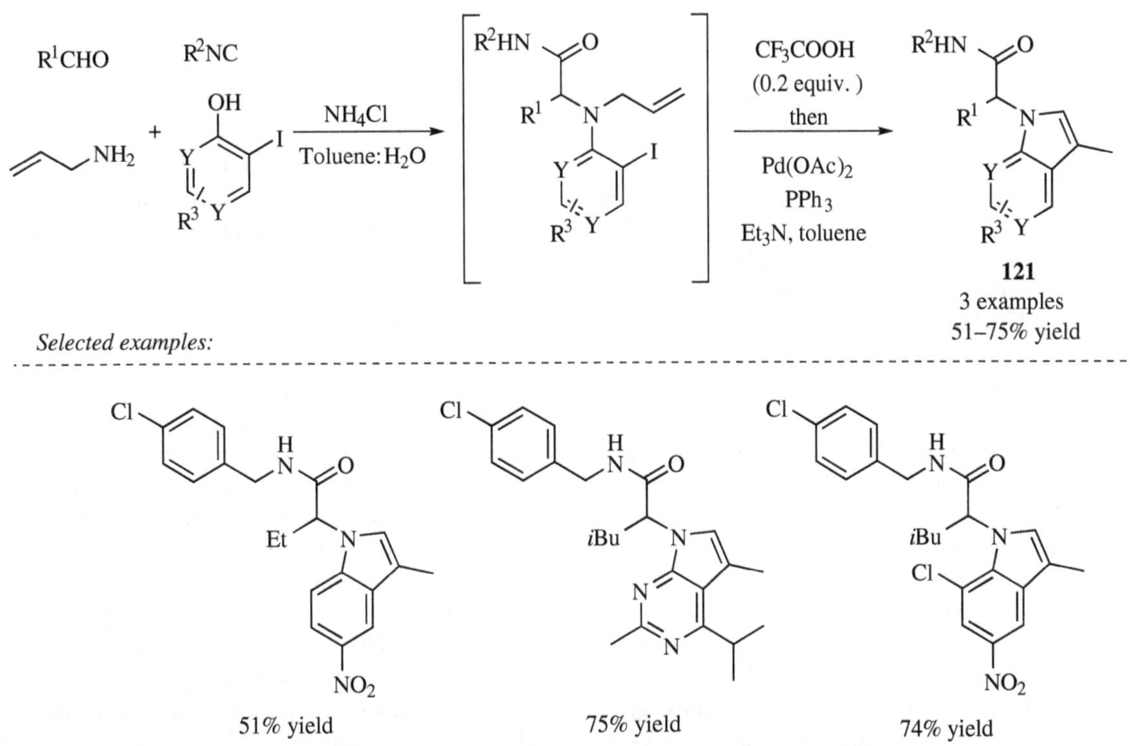

SCHEME 7.50 One-pot Ugi–Smiles reaction and Heck cyclization.

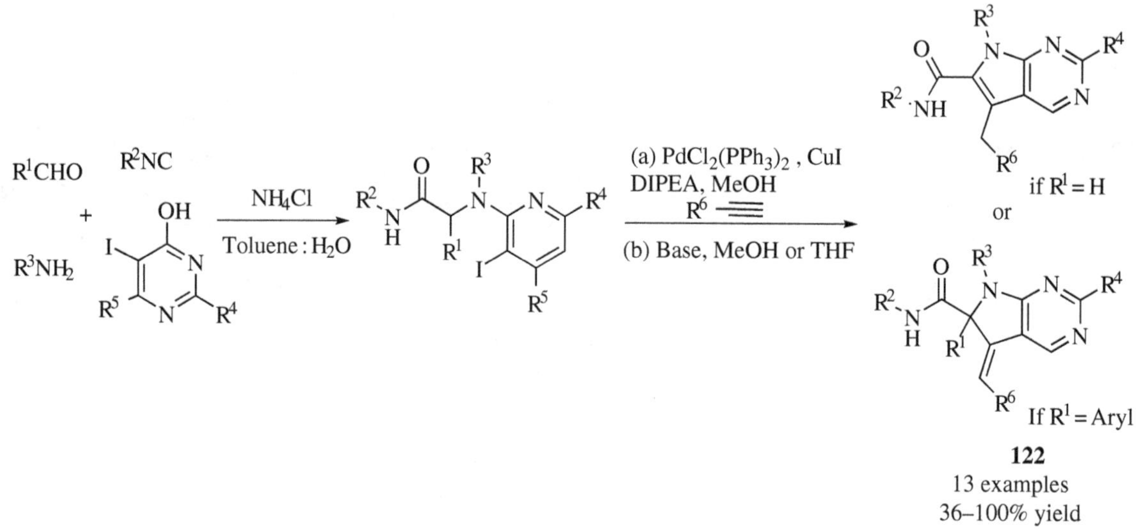

SCHEME 7.51 One-pot Ugi–Smiles reaction and Sonogashira cyclization.

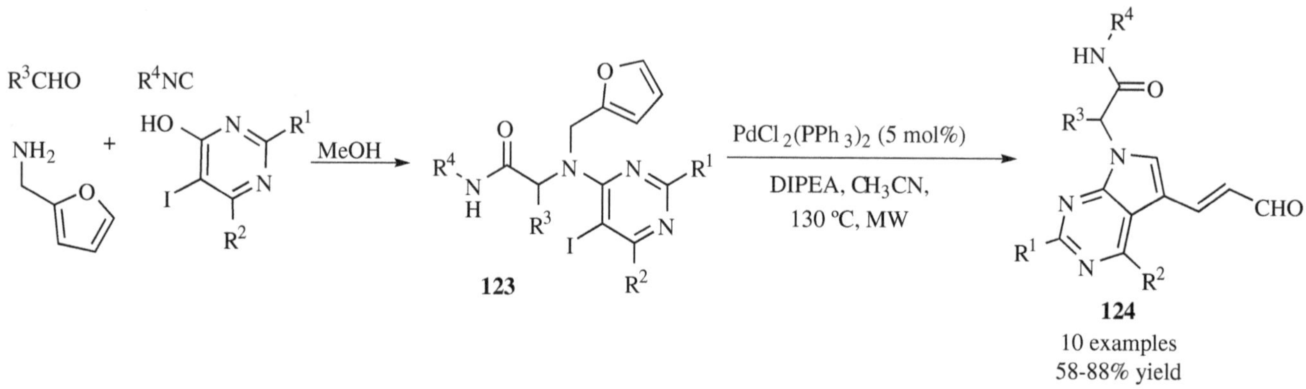

SCHEME 7.52 Ugi–Smiles reaction and palladium-catalyzed opening of furans.

The combination of the Ugi–Smiles coupling with the reduction of the nitro group of the aryl group has been used for the synthesis of benzimidazolopiperazines **128** [110]. Thus, the use of aminoacetaldehyde dimethyl acetal **125** with 2-nitrophenol **107** in the Ugi–Smiles reaction yielded the piperazine **127** (Scheme 7.53) through the opened intermediate **126**. Then, the hydrogenolysis of the piperazine **127** provided the fused system **128** at 60 °C in the presence of acetic acid to ensure a faster cyclization.

The same combination (Ugi–Smiles coupling and hydrogenolysis of the nitro group) offered a direct synthesis of benzotriazoles **132** (Scheme 7.54) [111]. The use of ammonia as nitrogen source in Ugi–Smiles couplings has provided the Ugi–Smiles products from moderate to good yields, but the use of microwave irradiation was required [112]. To overcome this issue, allylamine as surrogate of ammonia was used. Thus, the Ugi–Smiles coupling of allylamine (**129**), *o*-nitrophenol (**107**), aldehydes, and isocyanides gave adducts **130**, which were deallylated by a palladium-catalyzed process to give **131**. Then, the intermediate **131** was hydrogenated and treated with sodium nitrite to yield benzotriazol **132** (Scheme 7.54). In the same work, the authors proposed the use of other electrophiles, giving access to benzimidazoles.

After these pioneering examples, the Ugi–Smiles post-modification processes have been extended for a broad scope of interesting reactions. The synthesis of the compounds depicted in Figure 7.3 has been accomplished following a variation of the first discovered Ugi–Smiles procedure starting from different heterocycles. Thus, the reduction strategy has been also used for the synthesis of antimalarial analogues **133** starting from 4-hydroxy- and mercaptopyridines [113]. Isoquinolinones and isoindolinones (**134** and **135**) were obtained from nitromethyl salicylate after the Ugi–Smiles four-component coupling and a base-triggered cyclization [114]. Imidazolinium salts **136** were synthesized via a one-pot, two-step procedure involving an Ugi–Smiles coupling and copper-catalyzed reaction, starting from mercaptotriazine [115]. Diaminopyrimidines **137** were synthesized from thiouracil derivatives via Ugi–Smiles coupling and S_NAr [116].

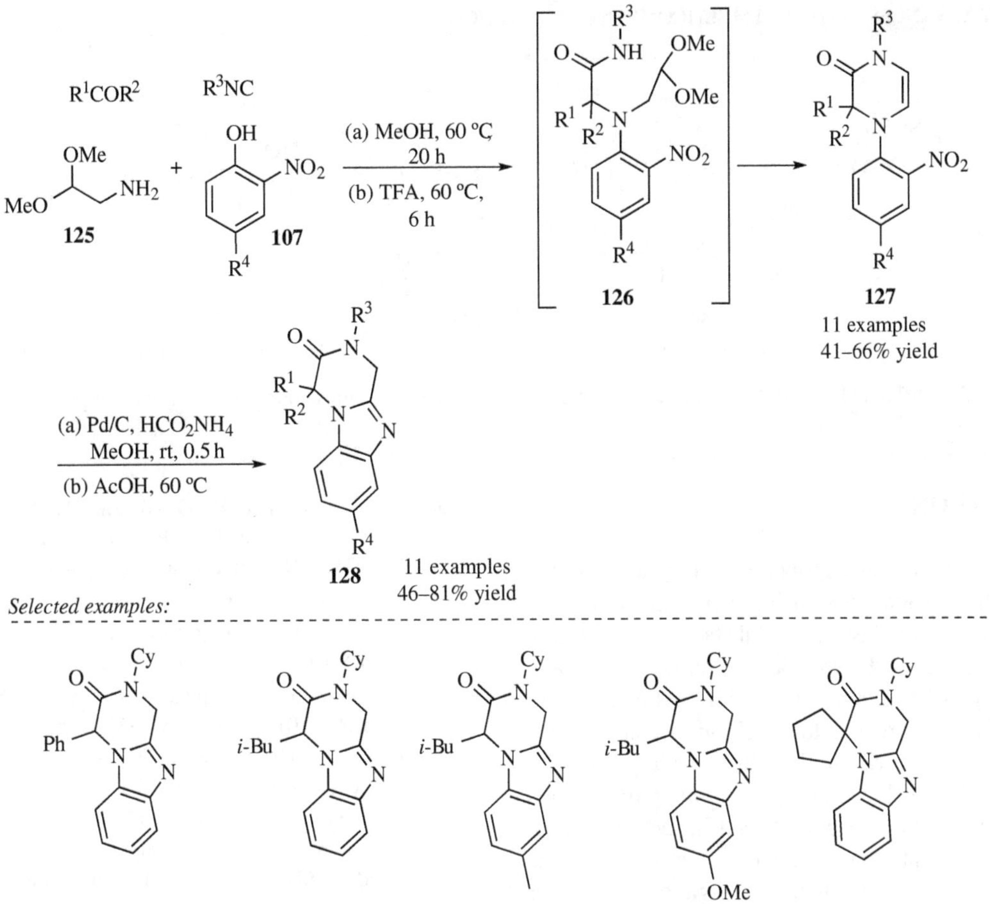

SCHEME 7.53 Combination of Ugi–Smiles coupling and a reduction process.

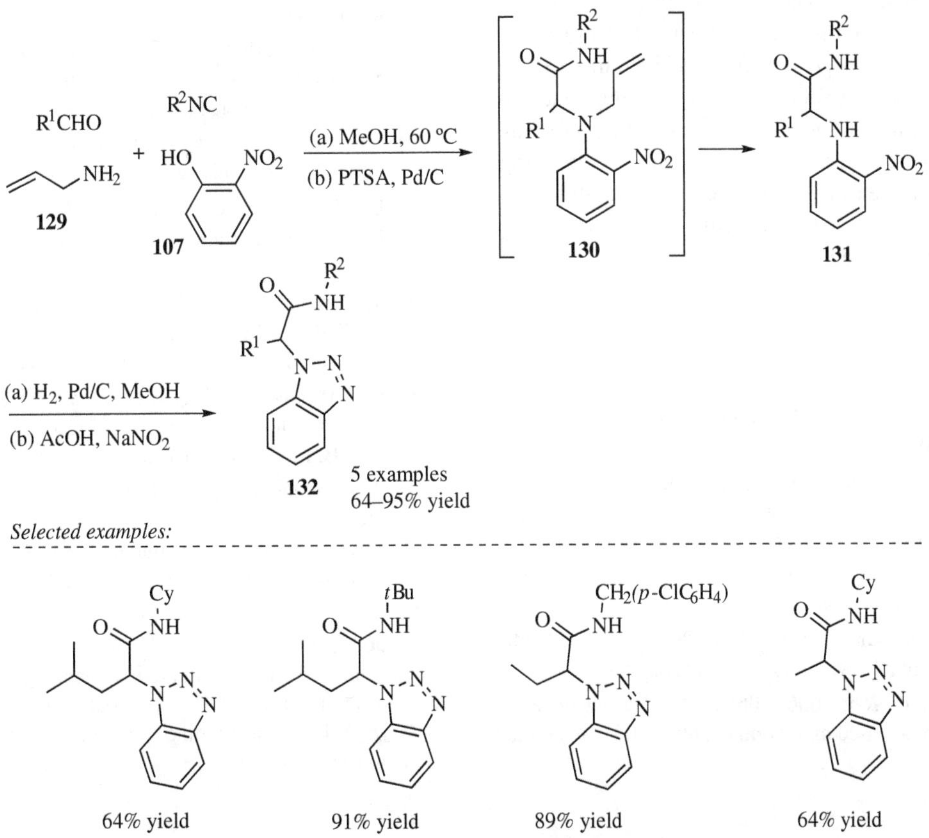

SCHEME 7.54 Combination of Ugi-Smiles coupling, reduction process, and nitration.

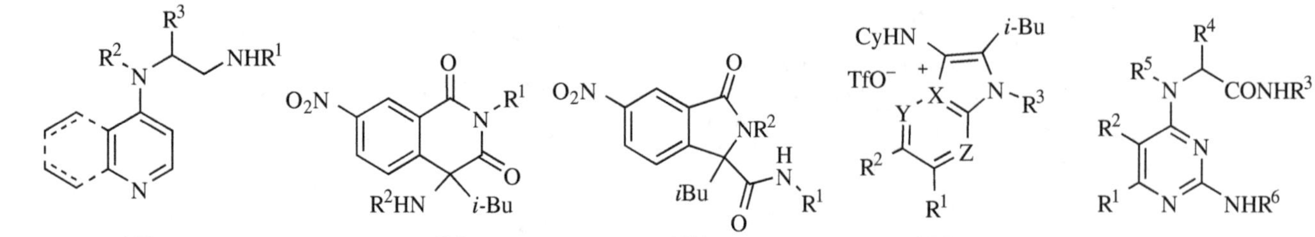

133 Chloroquine analogues | **134** Isoquinolinones | **135** Isoindolinones | **136** Imidazolinium salts | **137** Diaminopyrimidines

FIGURE 7.3 Different heterocycles obtained by combination of the Ugi–Smiles coupling and other one-pot transformations.

7.8 CONCLUSION

The field of the MCRs has exponentially increased by the presence of different reactions, being especially important the Ugi approach. In the last years, substantial advances in the Ugi-3CR and Ugi-4CR have been achieved, especially those concerning to the asymmetric versions. Although there are several publications related to the Ugi-diastereoselective version (e.g., using chiral auxiliaries), the asymmetric-catalyzed methods developed are rather limited and efforts in this direction should be accomplished in the next years. Moreover, developments concerning to the application of the Ugi reaction in the organometallic, organocatalytic, or biological fields will be discovered in the future. The chemical versatility of the Ugi reaction, starting point of many synthetically interesting transformations (postmodifications), will increase the synthetic usefulness of the Ugi research field, converting this area in a very fruitful tool for the creation of C—C bonds. After 35 years since the original Ugi reaction, the mere variation in one starting material, from an acid to a phenol derivative, has made possible the development of a new version: the Ugi–Smiles coupling. The replacement of one of the components, also called SRR, would be a good strategy for discovering new attractive Ugi variations in the next decades.

REFERENCES

1 I. Ugi, R. Meyr, U. Fetzer, C. Steinbrückner, *Angew. Chem.* 1959, 71, 386–386. Versuche mit Isonitrilen.

2 L. El Kaïm, L. Grimaud, *Tetrahedron* 2009, 65, 2153–2171. Beyond the Ugi reaction: less conventional interactions between isocyanides and iminium species.

3 D. Moderhack, *Tetrahedron* 2012, 68, 5949–5967. *N*-isocyanides-synthesis and reactions.

4 M. Tobisu, N. Chatani, *Chem. Lett.* 2011, 40, 330–340. Renaissance of organic synthesis using isocyanates.

5 A. Dömling, *Chem. Rev.* 2006, 106, 17–89. Recent developments in isocyanide based multicomponent reactions in applied chemistry.

6 S. S. van Berkel, B. G. M. Bögels, M. A. Wijdeven, B. Westermann, F. P. J. T. Rutjes, *Eur. J. Org. Chem.* 2012, 3543–3559. Recent advances in asymmetric isocyanide-based multicomponent reactions.

7 A. Dömling, I. Ugi, *Angew. Chem. Int. Ed.* 2000, 39, 3168–3210. Multicomponent reaction with isocyanides.

8 A. V. Gulevich, A. G. Zhdanko, R. V. A. Orru, V. G. Nenajdenko, *Chem. Rev.* 2010, 110, 5235–5331. Isocyanoacetate derivatives: synthesis, reactivity, and application.

9 F. De Moliner, L. Banfi, R. Riva, A. Basso, *Comb. Chem. High Throughput Screen.* 2011, 14, 782–810. Beyond Ugi and Passerini reactions. Multicomponent approaches based on isocyanides and alkynes as an efficient tool for diversity oriented synthesis.

10 A. V. Ivachtchenko, Y. A. Ivanenkov, V. M. Kysil, M. Y. Krasavin, A. P. Ilyin, *Russ. Chem. Rev.* 2010, 79, 787–817. Multicomponent reactions of isocyanides in the synthesis of heterocycles.

11 J. D. Sunderhaus, S. F. Martin, *Chem.–Eur. J.* 2009, 15, 1300–1308. Applications of multicomponent reactions to the synthesis of diverse heterocyclic scaffolds.

12 A. Dömling, W. Wang, K. Wang, *Chem. Rev.* 2012, 112, 3083–3135. Chemistry and biology of multicomponent reactions.

13 L. El Kaïm, L. Grimaud, *Mol. Divers.* 2010, 14, 855–867. Ugi-Smiles couplings: new entries to *N*-aryl carboxamide derivatives.

14 Y. Tanaka, T. Hasui, M. Suginome, *Org. Lett.* 2007, 9, 4407–4410. Acid-free, aminoborane-mediated Ugi-type reaction leading to general utilization of secondary amines.

15 S. C. Pan, B. List, *Angew. Chem. Int. Ed.* 2008, 47, 3622–3625. Catalytic three-component Ugi reaction. *Corrigendum*: S. C. Pan, B. List, *Angew. Chem. Int. Ed.* 2008, 47, 5490.

16 S. K. Guchait, C. Madaan, *Org. Biomol. Chem.* 2010, 8, 3631–3634. Towards molecular diversity: dealkylation of *tert*-butyl amine in Ugi-type multicomponent reaction product establishes *tert*-butyl isocyanide as a useful convertible isonitrile.

17 H. Mofakham, Z. Hezarkhani, A. Shaabani, *J. Mol. Cat.* 2012, 360, 26–34. Cellulose-SO_3H as a biodegradable solid acid catalyzed one-pot three-component Ugi reaction: synthesis of α-amino amide, 3,4-dihydroquinoxalin-2-amine, 4*H*-benzo [*b*][1,4]thiazin-2-amine and 1,6-dihydropyrazine-2,3-dicarbonitrile derivatives.

18 A. T. Khan, R. S. Basha, M. Lal, M. H. Mir, *RSC Adv.* 2012, 2, 5506–5509. Formation of unexpected α-amino amidine through three-component Ugi condensation reaction.

19 S. Kzossowski, B. Wiraszka, S. Berzozecki, R. Ostaszewski, *Org. Lett.* 2013, 15, 566–569. Model studies on the first enzyme-catalyzed Ugi reaction.

20 A. Kumar, D. Saxena, M. K. Gupta, *Green Chem.* 2013, 15, 2699–2703. Nanoparticle catalyzed reaction (NPCR): ZnO-NP catalyzed Ugi-reaction in aqueous medium.

21 T. Ngouansavanh, J. Zhu, *Angew. Chem. Int. Ed.* 2007, 46, 5775–5778. IBX-mediated oxidative Ugi-type multicomponent reactions: application to the N and C1 functionalization of tetrahydroisoquinoline.

22 F. Drouet, G. Masson, J. Zhu, *Org. Lett.* 2013, 15, 2854–2857. Ugi four-component reaction of alcohols: stoichiometric and catalytic oxidation/MCR sequences.

23 X. Ye, C. Xie, Y. Pan, L. Han, T. Xie, *Org. Lett.* 2010, 12, 4240–4243. Copper-catalyzed synthesis of α-amino imides from tertiary amines: Ugi-type three-component assemblies involving direct functionalization of sp^3 C-Hs adjacent to nitrogen atoms.

24 G. Jiang, J. Chen, J.-S. Huang, C.-M. Che, *Org. Lett.* 2009, 11, 4568–4571. Highly efficient oxidation of amines to imines by singlet oxygen and its application in Ugi-type reactions.

25 For a review on asymmetric MCRs, see:(a) D. J. Ramón, M. Yus, *Angew. Chem. Int. Ed.* 2005, 44, 1602–1634. Asymmetric multicomponent reactions (AMCRs): the new frontier. For a recent example, see:(b) D. Zhu, L. Xia, L. Pan, S. Li, R. Chen, Y. Mou, X. Chen, *J. Org. Chem.* 2012, 77, 1386–1395. An asymmetric Ugi three-component reaction induced by chiral cyclic imines: synthesis of morpholin- or piperazine-ketocarboxamide derivatives.

26 A. Znabet, E. Ruijter, F. J. J. de Kanter, V. Khler, M. Helliwell, N. J. Turner, R. V. A. Orru, *Angew. Chem. Int. Ed.* 2010, 49, 5289–5292. Highly stereoselective synthesis of substituted prolyl peptides using a combination of biocatalytic desymmetrization and multicomponent reactions.

27 T. Yue, M.-X. Wang, D.-X. Wang, G. Masson, J. Zhu, *Angew. Chem. Int. Ed.* 2009, 48, 6717–6721. Brønsted acid catalyzed enantioselective three-component reaction involving the α-addition of isocyanides to imines.

28 T. Hashimoto, H. Kimura, Y. Kawamata, K. Maruoka, *Angew. Chem. Int. Ed.* 2012, 51, 7279–7281. A catalytic asymmetric Ugi-type reaction with acyclic azomethine imines.

29 W. Zhao, L. Huang, Y. Guan, W. D. Wulff, *Angew. Chem. Int. Ed.* 2014, 53, 3436–3441. Three-component asymmetric catalytic Ugi reaction—concinnity from diversity by substrate-mediated catalyst assembly.

30 V. P. Mehta, S. G. Modha, E. Ruijter, K. Van Hecke, L. Van Meervelt, C. Pannecouque, B. Romano, R. V. A. Orru, E. Van der Eycken, *J. Org. Chem.* 2011, 76, 2828–2839. A microwave-assisted diastereoselective multicomponent reaction to access dibenzo[*c*,*e*]azepinones: synthesis and biological evaluation.

31 (a) A. Demharter, W. Hörl, E. Herdtweck, I. Ugi, *Angew. Chem. Int. Ed.* 1996, 35, 173–175. Synthesis of chiral 1,1′-iminodicarboxylic acid derivatives from α-amino acids, aldehydes, isocyanides, and alcohols by the diastereoselective five-center–four-component reaction. (b) I. Ugi, A. Demharter, W. Hörl, T. Schmid, *Tetrahedron* 1996, 52, 11657–11664. Ugi reactions with trifunctional α-amino acids, aldehydes, isocyanides, and alcohols.

32 S. Marcaccini, T. Torroba, *Nat. Protoc.* 2007, 2, 632–639. The use of the Ugi four-component condensation.

33 I. Ugi, C. Steinbruckner, *Angew. Chem.* 1960, 72, 267–268. Über ein neues Kondensations-Prinzip.

34 (a) S. Heck, A. Dömling, *Synlett* 2000, 424–426. A versatile multi-component one-pot thiazole synthesis. (b) J. Kolb, B. Beck, A. Dömling, *Tetrahedron Lett.* 2002, 43, 6897–6901. Simultaneous assembly of the *β*-lactam and thiazole moiety by a new multicomponent reaction. (c) J. Kolb, B. Beck, M. Almstetter, S. Heck, E. Herdtweck, A. Dömling, *Mol. Divers.* 2003, 6, 297–313. New MCRs: the first 4-component reaction leading to 2,4-disubstituted thiazoles. (d) B. Henkel, B. Westner, A. Dömling, *Synlett* 2003, 2410–2412. Polymer-bound 3-*N,N*-(dimethylamino)-2-isocyanoacrylate for the synthesis of thiazoles *via* a multicomponent reaction.

35 (a) I. Ugi, F. K. Rosendhal, F. Bodesheim, *Liebigs Ann. Chem.* 1963, 666, 54–61. Kondensation von primären Aminen und Ketonen mit Isonitrilen und Hodanwasserstoffsäure. (b) I. Ugi, K. Offerman, *Chem. Ber.* 1964, 97, 2276–2281. Isonitrile XVIII, hydantoin-imide-(4).

36 (a) I. Ugi, *Angew. Chem.* 1960, 72, 639. Neues über Isonitrile. (b) I. Ugi, C. Steinbrückner, *Chem. Ber.* 1961, 94, 734–742. Reaktion von Isonitrilen mit Carbonylverbindungen, Aminen und Stickstoffwasserstoffsäure.

37 I. Ugi, C. Steinbrückner, *Chem. Ber.* 1961, 94, 2802–2814. *α*-Addition von Immonium-ionen und Carbonsäure-anionen an Isonitrile.

38 I. Ugi, *Angew. Chem. Int. Ed.* 1962, 1, 8–21. The α-addition of immonium ions and anions to isonitriles accompanied by secondary reactions.

39 (a) T. Godet, Y. Bonvin, G. Vincent, D. Merle, A. Thozet, M. A. Ciufolini, *Org. Lett.* 2004, 6, 3281–3284. Titanium catalysis in the Ugi reaction of alpha-amino acids with aromatic aldehydes. (b) B. O. Okandeji, J. R. Gordon, J. K. Sello, *J. Org. Chem.* 2008, 73, 5595–5597. Catalysis of Ugi four component coupling reactions by rare earth metal triflates.

40 For experimental evidence of the formation of the imidate, see: C. Faggi, M. García-Valverde, S. Marcaccini, G. Menchi, *Org. Lett.* 2010, 12, 788–791. Isolation of Ugi four-component condensation primary adducts: a straightforward route to isocoumarins.

41 N. Chéron, R. Ramozzi, L. El Käim, L. Grimaud, P. Fleurat-Lessard, *J. Org. Chem.* 2012, 77, 1361–1366. Challenging 50 years of established views on Ugi reaction: a theoretical approach.

42 G. A. Medeiros, W. A. da Silva, G. A. Bataglion, D. A. C. Ferreira, H. C. B. de Oliveira, M. N. Eberlin, B. A. D. Neto, *Chem. Commun.* 2014, 50, 338–340. Probing the mechanism of the Ugi four-component reaction with charge-tagged reagents by ESI-MS(/MS).

43 (a) K. Khoury, M. K. Sinha, T. Nagashima, E. Herdtweck, A. Dömling, *Angew. Chem. Int. Ed.* 2012, 51, 10280–10283.

Efficient assembly of iminodicarboxamides by a "truly" four-component reaction. For the Ugi-4CR post-condensation reaction of this transformation, see:(b) M. K. Sinha, K. Khoury, E. Herdtweckb, A. Dömling, *Org. Biomol. Chem.* 2013, 11, 4792–4796. Various cyclization scaffolds by a truly Ugi 4-CR.

44 For recent examples, see:(a) N. Zohreh, A. Alizadeh, H. R. Bijanzadeh, L.-G. Zhu, *J. Comb. Chem.* 2010, 12, 497–502. Novel approach to 1,5-benzodiazepine-2-ones containing peptoid backbone *via* one-pot diketene-based Ugi-4CR. (b) R. S. Borisov, A. I. Polyakov, L. A. Medvedeva, V. N. Khrustalev, N. I. Guranova, L. G. Voskressensky, *Org. Lett.* 2010, 12, 3894–3897. Concise approach toward tetrazolo[1,5-*a*][1,4] benzodiazepines *via* a novel multicomponent isocyanide-based condensation. (c) R. Mossetti, D. Saggiorato, G. C. Tron, *J. Org. Chem.* 2011, 76, 10258–10262. Exploiting the acylating nature of the imide-Ugi intermediate: a straightforward synthesis of tetrahydro-1,4-benzodiazepin-2-ones, and references cited therein.

45 For recent examples of the synthesis of peptides and peptoids by the Ugi 4-CR, see:(a) O. E. Vercillo, C. K. Z. Andrade, L. A. Wessjohann, *Org. Lett.* 2008, 10, 205–208. Design and synthesis of cyclic RGD pentapeptoids by consecutive Ugi reactions. (b) O. Pando, S. Stark, A. Denkert, A. Porzel, R. Preusentanz, L. A. Wessjohann, *J. Am. Chem. Soc.* 2011, 133, 7692–7695. The multiple multicomponent approach to natural product mimics: tubugis, *N*-substituted anticancer peptides with picomolar activity. (c) A. Fátima, S. Barreto, O. E. Vercillo, M. A. Birkett, J. C. Caulfield, L. A. Wessjohann, C. K. Z. Andrade, *Org. Biomol. Chem.* 2011, 9, 5024–5027. Fast and efficient microwave-assisted synthesis of functionalized peptoids *via* Ugi reactions. (d) M. Abbas, L. A. Wessjohann, *Org. Biomol. Chem.* 2012, 10, 9330–9333. Direct synthesis of sensitive selenocysteine peptides by the Ugi reaction. For an application to peptide–peptoid podands by threefold Ugi-4CRs, see:(e) D. G. Rivera, F. León, O. Concepción, F. E. Morales, L. A. Wessjohann, *Chem.–Eur. J.* 2013, 19, 6417–6428. A multiple multicomponent approach to chimeric peptide–peptoid podands.

46 G. M. Chinigo, A. Breder, E. M. Carreira, *Org. Lett.* 2011, 13, 78–81. Ugi-4-component reaction enabling rapid access to the core fragment of Massadine.

47 (a) T. Tanino, S. Ichikawa, B. Al-Dabbagh, A. Bouhss, H. Oyama, A. Matsuda, *ACS Med. Chem. Lett.* 2010, 1, 258–262. Synthesis and biological evaluation of Muraymycin analogues active against anti-drug-resistant bacteria. (b) T. Tanino, B. Al-Dabbagh, D. Mengin-Lecreulx, A. Bouhss, H. Oyama, S. Ichikawa, A. Matsuda, *J. Med. Chem.* 2011, 54, 8421–8439. Mechanistic analysis of muraymycin analogues: a guide to the design of MraY inhibitors.

48 (a) D. G. Rivera, L. A. Wessjohann, *J. Am. Chem. Soc.* 2009, 131, 3721–3732. Architectural chemistry: synthesis of topologically diverse macromulticycles by sequential multiple multicomponent macrocyclizations. (b) L. A. Wessjohann, D. G. Rivera, O. E. Vercillo, *Chem. Rev.* 2009, 109, 796–814. Multiple multicomponent macrocyclizations (MiBs): a strategic development toward macrocycle diversity, and references cited therein.

49 J. Jacobs, V. Grum-Tokars, Y. Zhou, M. Turlington, S. A. Saldanha, P. Chase, A. Eggler, E. S. Dawson, Y. M. Baez-Santos, S. Tomar, A. M. Mielech, S. C. Baker, C. W. Lindsley, P. Hodder, A. Mesecar, S. R. Stauffer, *J. Med. Chem.* 2013, 56, 534–546. Discovery, synthesis, and structure-based optimization of a series of *N*-(*tert*-butyl)-2-(*N*-arylamido)-2-(pyridin-3-yl) acetamides (ML188) as potent noncovalent small molecule inhibitors of the severe acute respiratory syndrome coronavirus (SARS-CoV) 3CL protease.

50 A. D. Borthwick, J. Liddle, D. E. Davies, A. M. Exall, C. Hamlett, D. M. Hickey, A. M. Mason, I. E. D. Smith, F. Nerozzi, S. Peace, D. Pollard, S. L. Sollis, M. J. Allen, P. M. Woollard, M. A. Pullen, T. D. Westfall, D. J. Stanislaus, *J. Med. Chem.* 2012, 55, 783–796. Pyridyl-2,5-diketopiperazines as potent, selective, and orally bioavailable oxytocin antagonists: synthesis, pharmacokinetics, and *in vivo* potency.

51 (a) G. B. Giovenzana, G. C. Tron, S. Di Paola, I. G. Menegotto, T. Pirali, *Angew. Chem. Int. Ed.* 2006, 45, 1099–1102. A mimicry of primary amines by bis-secondary diamines as components in the Ugi four-component reaction. (b) G. C. Tron, Eur. *J. Med. Chem.* 2013, 1849–1859. Off the beaten track: the use of secondary amines in the Ugi reaction.

52 T. Pirali, G. Callipari, E. Ercolano, A. A. Genazzani, G. B. Giovenzana, G. C. Tron, *Org. Lett.* 2008, 10, 4199–4202. A concise entry into nonsymmetrical alkyl polyamines.

53 G. Piersanti, F. Remi, V. Fusi, M. Formica, L. Giorgi, G. Zappia, *Org. Lett.* 2009, 11, 417–420. Direct preparation of unsymmetrical difunctionalized cyclen derivatives by an Ugi multicomponent reaction.

54 T. A. Keating, R. W. Armstrong, *J. Org. Chem.* 1998, 63, 867–871. The Ugi five-component condensation using CO_2, CS_2, and COS as oxidized carbon sources.

55 C. A. Simoneau, E. A. George, B. Ganem, *Tetrahedron Lett.* 2006, 47, 1205–1207. A new approach to four- and five-component Ugi condensations starting from nitriles.

56 S. Marcaccini, A. G. Neo, C. F. Marcos, *J. Org. Chem.* 2009, 74, 6888–6890. Sequential five-component synthesis of spiropyrrolidinochromanones.

57 E. Riguet, *J. Org. Chem.* 2011, 76, 8143–8150. Enantioselective organocatalytic Friedel Crafts alkylation reaction of indoles with 5-hydroxyfuran-2(*5H*)-one: access to chiral γ-lactones and γ-lactams *via* a Ugi 4-center 3-component reaction.

58 A. A. Elassar, A. A. El-Khair, *Tetrahedron* 2003, 59, 8463–8480. Recent developments in the chemistry of enaminones.

59 M. Bararjanian, S. Balalaie, F. Rominger, B. Movassagh, H. R. Bijanzadeh, *Mol. Divers.* 2011, 15, 583–594. Novel and efficient one-pot five- and six-component reactions for the stereoselective synthesis of highly functionalized enaminones and dithiocarbamates.

60 M. Bararjanian, S. Balalaie, B. Movassagh, H. R. Bijanzadeh, *Tetrahedron Lett.* 2010, 51, 3277–3279. Efficient synthesis of 1,4-disubstituted polyfunctional piperazines *via* a sequential one-pot Ugi/nucleophilic addition five-component reaction.

61 T. H. Al-Tel, R. A. Al-Qawasmeh, W. Voelter, *Eur. J. Org. Chem.* 2010, 5586–5593. Rapid assembly of polyfunctional structure using a one-pot five- and six-component sequential Groebke-Blackburn/Ugi/Passerini process.

62 S. Shaabani, A. Shaabani, S. Weng Ng, *ACS Comb. Sci.* 2014, 16, 176–183. One-pot synthesis of coumarin-3-carboxamides containing a triazole ring *via* an isocyanide base six-component reaction.

63 A. Dömling, I. Ugi, *Angew. Chem. Int. Ed. Engl.* 1993, 32, 563–564. The seven component reaction.

64 N. Elders, D. van der Born, L. J. D. Hendrickx, B. J. J. Timmers, A. Krause, E. Janssen, F. J. J. de Kanter, E. Ruijter, R. V. A. Orru, *Angew. Chem. Int. Ed.* 2009, 48, 5856–5859. The efficient one-pot reaction of up to eight components by the union of multicomponent reactions.

65 (a) A. Basso, L. Banfi, R. Riva, *Eur. J. Org. Chem.* 2010, 1831–1841. A marriage of convenience: combining the power of isocyanide-based multicomponent reactions with the versatility of (hetero)norbornene chemistry. (b) A. Ilyin, V. Kysil, M. Krasavin, I. Kurashvili, A. V. Ivachtchenko, *J. Org. Chem.* 2006, 71, 9544–9547. Complexity-enhancing acid-promoted rearrangement of tricyclic products of tandem Ugi 4CC/intramolecular Diels–Alder reaction.

66 F. Bonnaterre, M. Bois-Choussy, J. Zhu, *Org. Lett.* 2006, 8, 4351–4354. Rapid access to oxindoles by the combined use of an Ugi four-component reaction and a microwave-assisted intramolecular Buchwald–Hartwig amidation reaction.

67 A. Salcedo, L. Neuville, C. Rondot, P. Retailleau, J. Zhu, *Org. Lett.* 2008, 10, 857–860. Palladium-catalyzed domino intramolecular *N*-arylation/intermolecular C—C bond formation for the synthesis of functionalized benzodiazepinediones.

68 T. P. Ribelin, A. S. Judd, I. Akritopoulou-Zanze, R. F. Henry, J. L. Cross, D. N. Whittern, S. W. Djuric, *Org. Lett.* 2007, 9, 5119–5122. Concise construction of novel bridged bicyclic lactams by sequenced Ugi/RCM/Heck reactions.

69 Z. Xu, F. de Molines, A. P. Cappelli, C. Hulme, *Angew. Chem. Int. Ed.* 2012, 51, 8037–8040. Ugi/Aldol sequence: expeditious entry to several families of densely substituted nitrogen heterocycles.

70 M. Bararjanian, S. Balalaie, F. Rominger, B. Movassagh, H. R. Bijanzadeh, *J. Org. Chem.* 2010, 75, 2806–2812. Six-component reactions for the stereoselective synthesis of 3-arylidene-2-oxindoles *via* sequential one-pot Ugi/Heck carbocyclization/Sonogashira/nucleophilic addition.

71 S. Santra, P. R. Andreana, *Angew. Chem. Int. Ed.* 2011, 50, 9418–9422. A bioinspired Ugi/Michael/Aza-Michael cascade reaction in aqueous media: natural-product-like molecular diversity.

72 C. Che, S. Li, Z. Yu, F. Li, S. Xin, L. Zhou, S. Lin, Z. Yang, *ACS Comb. Sci.* 2013, 15, 202–207. One-Pot syntheses of isoquinoline-3-ones and benzo-1,4-diazepin-2,5-diones utilizing Ugi-4CR post-transformation strategy.

73 R. A. De Silva, S. Santra, P. R. Andreana, *Org. Lett.* 2008, 10, 4541–4544. A tandem one-pot, microwave-assisted synthesis of regiochemically differentiated 1,2,4,5-tetrahydro-1,4-benzodiazepin-3-ones.

74 M. A. Cano-Herrera, L. D. Miranda, *Chem. Commun.* 2011, 47, 10770–10772. Expedient entry to the piperazinohydroisoquinoline ring system using a sequential Ugi/Pictet-Spengler/reductive methylation reaction protocol.

75 C. B. Gilley, M. J. Buller, Y. Kobayashi, *Org. Lett.* 2007, 9, 3631–3634. New entry to convertible isocyanides for the Ugi reaction and its application to the stereocontrolled formal total synthesis of the proteasome inhibitor Omuralide.

76 S. G. Modha, A. Kumar, D. D. Vachhani, S. K. Sharma, V. S. Parmar, E. V. Van der Eycken, *Chem. Commun.* 2012, 48, 10916–10918. Gold(I) and platinum(II) switch: a post-Ugi intramolecular hydroarylation to pyrrolopyridinones and pyrroloazepinones.

77 A. Kumar, Z. Li, S. K. Sharma, V. S. Parmar, E. V. Van der Eycken, *Chem. Commun.* 2013, 49, 6803–6805. Switching the regioselectivity *via* indium(III) and gold(I) catalysis: a post-Ugi intramolecular hydroarylation to azepino- and azocino-[*c*,*d*]indolones.

78 Y. Su, M. J. Bouma, L. Alcaraz, M. Stocks, M. Furber, G. Masson, J. Zhu, *Chem.–Eur. J.* 2012, 18, 12624–12627. Organocatalytic enantioselective one-pot four-component Ugi-type multicomponent reaction for the synthesis of epoxytetrahydropyrrolo[3,4-*b*]pyridin-5-ones.

79 Z. Xu, M. Ayaz, A. A. Cappelli, C. Hulme, *ACS Comb. Sci.* 2012, 14, 460–464. General one-pot, two-step protocol accessing a range of novel polycyclic heterocycles with high skeletal diversity.

80 M. Sañudo, M. García-Valverde, S. Marcaccini, J. J. Delgado, J. Rojo, T. Torroba, *J. Org. Chem.* 2009, 74, 2189–2192. Synthesis of benzodiazepine β-turn mimetics by an Ugi 4CC/Staudinger/aza-Wittig sequence. Solving the conformational behavior of the Ugi 4CC adducts.

81 P. He, J. Wu, Y.-B. Nie, M.-W. Ding, *Eur. J. Org. Chem.* 2010, 1088–1095. Unexpected synthesis of rearranged 3,4-dihydroquinazolines by a sequential Ugi 4CC/aza-Wittig/carbodiimide-mediated cyclization.

82 R. Mossetti, T. Pirali, D. Saggiorato, G. C. Tron, *Chem. Commun.* 2011, 47, 6966–6968. Imides: forgotten players in the Ugi reaction. One-pot multicomponent synthesis of quinazolines.

83 R. Mossetti, D. Saggiorato, G. C. Tron, *J. Org. Chem.* 2011, 76, 10258–10262. Exploiting the acylating nature of the imide-Ugi intermediate: a straightforward synthesis of tetrahydro-1,4-benzodiazepin-2-ones.

84 L. Moni, L. Banfi, A. Basso, A. Galatini, M. Spallarossa, R. Riva, *J. Org. Chem.* 2014, 79, 339–351. Enantio- and diastereoselective synthesis of highly substituted benzazepines by a multicomponent strategy coupled with organocatalytic and enzymatic procedures.

85 L. El Kaïm, L. Grimaud, X. F. Le Goff, M. Menes-Arzate, L. D. Miranda, *Chem. Commun.* 2011, 47, 8145–8147. Straightforward four-component access to spiroindolines.

86 L. Zhang, F. Zhao, M. Zheng, Y. Zhai, J. Wang, H. Liu, *Eur. J. Org. Chem.* 2013, 5710–5715. Selective synthesis of 5,6-dihydroindol[1,2-*a*]quinoxalines and 6,7-dihydroindolo[*2,3-c*]quinolines by orthogonal copper and palladium catalysis.

87 W. Wang, S. Ollio, E. Herdtweck, A. Dömling, *J. Org. Chem.* 2011, 76, 637–644. Polycyclic compound by Ugi-Pictet-Spengler sequence.

88 H. Liu, A. Dömling, *J. Org. Chem.* 2009, 74, 6895–6898. Efficient and diverse synthesis of indole derivatives.

89 M. K. Sinha, K. Khoury, E. Herdtweck, A. Dömling, *Chem.–Eur. J.* 2013, 19, 8048–8052. Tricycles by a new Ugi variation and Pictet-Spengler reaction in one-pot.

90 C. P. Gordon, K. A. Young, L. Hizartzidis, F. M. Deane, A. McCluskey, *Org. Biomol. Chem.* 2011, 9, 1419–1428. Investigation of the one-pot synthesis of quinolin-2-(*1H*)-ones and the discovery of a variation of the three-component Ugi reaction.

91 S. Gunawan, J. Petit, C. Hulme, *ACS Comb. Sci.* 2012, 14, 160–163. Concise one-pot preparation of unique bis-pyrrolidinone tetrazoles.

92 B. Ganem, *Acc. Chem. Res.* 2009, 42, 463–472. Strategies for innovation in multicomponent reaction design.

93 M. Passerini, *Gazz. Chim. Ital.* 1931, 61, 26–33. Sopra gli isonitrili (XVII). Reazioni con nitroso-derivati.

94 R. Bossio, S. Marcaccini, P. Paoli, R. Pepino, C. Polo, Synthesis 1991, 999–1000. Studies on isocyanides and related compounds. Synthesis of benzofuran derivatives.

95 (a) L. El Kaïm, L. Grimaud, J. Oble, *Angew. Chem. Int. Ed.* 2005, 44, 7961–7964. Phenol Ugi-Smiles systems: strategies for the multicomponent *N*-arylation of primary amines with isocyanides, aldehydes, and phenols. (b) L. El Kaïm, M. Gizolme, L. Grimaud, J. Oble, *J. Org. Chem.* 2007, 72, 4169–4180. Smiles rearrangements in Ugi- and Passerini-type couplings: new multicomponent access to *O*- and *N*-arylamides.

96 L. El Kaïm, M. Gizolme, L. Grimaud, J. Oble, *J. Org. Chem.* 2007, 72, 4169–4180. Smiles rearrangements in Ugi and Passerini-type couplings: new multicomponent accesses to *O*- and *N*-aryl amides.

97 A. Massoudi, I. Amini, A. Ramazani, F. Z. Nasrabadi, Y. Ahmadi, *Bull. Korean Chem. Soc.* 2012, 33, 39–42. Four-component synthesis of 2-(*N,N*-dialkylamino)-2,4-6-cycloheptatrien-1-one derivatives from tropolones, an isocyanide, a primary amine and an aldehyde *via* Ugi-Smiles coupling reaction.

98 L. El Kaïm, L. Grimaud, S. R. Purumandla, *Tetrahedron Lett.* 2010, 51, 4962–4964. Ugi-Smiles coupling in water.

99 N. Chéron, L. El Kaïm, L. Grimaud, P. Fleurat-Lessard, *Chem.–Eur. J.* 2011, 17, 14929–14934. Evidences for the key role of hydrogen bond in nucleophilic aromatic substitution reactions.

100 L. El Kaïm, M. Gizolme, L. Grimaud, J. Oble, *Org. Lett.* 2006, 8, 4019–4021. Direct access to heterocyclic scaffolds by new multicomponent Ugi-Smiles couplings.

101 A. Barthelon, L. El Kaïm, M. Gizolme, L. Grimaud, *Eur. J. Org. Chem.* 2008, 5974–5987. Thiols in Ugi- and Passerini-Smiles-type couplings.

102 For the synthesis of pyrazines, see: A. Barthelon, A. Dos Santos, L. El Kaïm, L. Grimaud, *Tetrahedron Lett.* 2008, 49, 3208–3211. Ugi/Smiles access to pyrazine scaffolds.

103 L. El Kaïm, L. Grimaud, J. Oble, S. Wagschal, *Tetrahedron Lett.* 2009, 50, 1741–1743. Three-component Ugi-Smiles couplings of cyclic imines.

104 A. Znabet, S. Blanken, E. Janssen, F. J. J. de Kanter, M. Helliwell, N. J. Turner, E. Ruijter, R. V. A. Orru, *Org. Biomol. Chem.* 2012, 10, 941–944. Stereoselective synthesis of *N*-aryl proline amides by biotransformation-Ugi-Smiles sequence.

105 S. Brauch, L. Gabriel, B. Westermann, Chem. Commun., 2010, 3387–3389. Seven-component reactions by sequential chemoselective Ugi-Mumm/Ugi-Smiles reactions.

106 L. El Kaïm, L. Grimaud, J. Oble, *J. Org. Chem.* 2007, 72, 5835–5838. New Ugi-Smiles-metathesis strategy toward the synthesis of pyrimido azepines.

107 L. El Kaïm, M. Gizzi, L. Grimaud, *Org. Lett.* 2008, 10, 3417–3419. New MCR-Heck-isomerization cascade toward indoles.

108 L. El Kaïm, L. Grimaud, S. Wagschal, *J. Org. Chem.* 2010, 75, 5343–5346. Toward pyrrolo[2,3-d]pyrimidine scaffolds.

109 L. El Kaïm, L. Grimaud, S. Wagschal, *Chem. Commun.* 2011, 47, 1887–1889. Palladium catalyzed ring opening of furans as a route to α,β-unsaturated aldehydes.

110 L. El Kaïm, L. Grimaud, S. R. Purumandla, *J. Org. Chem.* 2011, 76, 4728–4733. Multicomponent synthesis of fused benzimidazolopiperazines.

111 D. Coffinier, L. El Kaïm, L. Grimaud, *Org. Lett.* 2009, 11, 995–997. New benzotriazole and benzimidazole scaffolds from Ugi-Smiles couplings of isocyanides.

112 A. Barthelon, L. El Kaïm, M. Gizzi, L. Grimaud, *Synlett* 2010, 2784. Ammonia in Ugi-Smiles and Ugi couplings.

113 L. El Kaïm, L. Grimaud, P. Pravin, *Org. Lett.* 2012, 14, 476–478. Ugi-Smiles couplings of 4-substituted pyridine derivatives: a fast access to chloroquine analogues.

114 L. El Kaïm, L. Grimaud, X. F. Le Goff, A. Schiltz, *Org. Lett.* 2011, 13, 534–536. Smiles cascades toward heterocyclic scaffolds.

115 A. Barthelon, X.-F. Legoff, L. El Kaïm, L. Grimaud, *Synlett* 2010, 153–157. Four-component synthesis of imidazolinium-fused heterocycles from Ugi-Smiles couplings.

116 For the synthesis of 2,4-diaminopyrimidines, see: M. A. Sidhoum, L. El Kaïm, L. Grimaud, *Synlett* 2012, 632–636. Ugi-Smiles coupling of thiouracil derivatives towards 2,4-diamino pyrimidines.

8

PASSERINI MULTICOMPONENT REACTIONS

EUGENIA MARQUÉS-LÓPEZ AND RAQUEL P. HERRERA

Departamento de Química Orgánica, Instituto de Síntesis Química y Catálisis Homogénea (ISQCH), CSIC-Universidad de Zaragoza, Zaragoza, Spain

8.1 INTRODUCTION

If in Chapter 7 different aspects about Ugi reaction have been discussed, in this chapter, we are going to disclose to the reader a vision about the new contributions regarding other crucial isonitrile-based multicomponent reaction (MCR): the Passerini reaction (P-3CR) discovered in 1921 [1]. The traditional multicomponent Passerini reaction [2] is another isonitrile-based MCR that provides easy access to α-acyloxycarboxamides **4** in a one-pot synthesis involving an aldehyde **1**, a carboxylic acid **2**, and an isonitrile **3** (Scheme 8.1), which has been subject of intensive studies in the last decade [3]. The importance of using isocyanides lays in its dual role as nucleophile and electrophile, and moreover, if $R^1 \neq R^2$, a new stereocenter could be created under asymmetric conditions.

The resulting α-acyloxyamide motif **4** is present in the backbone of many bioactive products, such as drugs and active peptidomimetics. Additionally, Passerini MCR offers a cheap and rapid way to generate interesting and useful compound libraries.

An accepted mechanism for this important reaction is depicted in the following Scheme 8.2, where the reaction proceeds through intermediate **6** before the final Mumm rearrangement [4] to allow α-acyloxyamides **4**. However, the way how **6** is formed is still the focus of active debate [5].

During this century, variations on the Passerini reaction have been mainly based on one or several of the following possible points: (i) replacing the carbonyl component **1** by alcohols, (ii) replacing the starting isocyanide **3**, (iii) replacing the carboxylic acid **2**, and (iv) combining the Passerini reaction with further synthetic steps allowing the synthesis of more complex structures. Few of these interesting aspects will be covered in this chapter. Some of the Passerini adducts achieved during the last years are depicted in Figure 8.1.

We will describe in this chapter some important novel examples concerning some variations in Passerini reactions that have been reported since 2005 in order to give a broad overview of the recent trends in this crucial multicomponent protocol.

8.2 *O*-ALKYLATIVE AND SILYLATIVE PASSERINI THREE-COMPONENT REACTIONS

The mechanism of the Passerini reaction has been widely studied and the carboxylic acid **2** has been found to play a crucial role in the course of this process, where the activation of aldehyde **1** by the acid compound **2** would start the reaction as described in Scheme 8.2. Therefore, a carboxylic acid is normally required and it could represent a limitation for the applications of this process to the synthesis of more complex scaffolds. We disclose here some advances achieved in this field, in order to avoid the use of carboxylic acid and amplifying the scope of the process.

8.2.1 *O*-Arylative Passerini Three-Component Reactions

An interesting variation of the Passerini 3CR is the *O*-arylative Passerini reaction, where in such a process instead of the key carboxylic acid, an alkyl or aryl alcohol is employed

Multicomponent Reactions: Concepts and Applications for Design and Synthesis, First Edition. Edited by Raquel P. Herrera and Eugenia Marqués-López.

affording α-hydroxyamide derivatives after an irreversible Smiles rearrangement, as previously described in the Ugi reaction (Chapter 7) [13].

The pioneering example of this new protocol was also disclosed by El Kaim, Grimaud, and coworkers using phenol derivatives **14** as carboxylic acid surrogates in a three-component addition for rendering α-aryloxy amides **15** with moderate to good yields for aliphatic aldehydes. For some representative examples, see Scheme 8.3 [14].

The authors also extended the scope of the reaction to heterocyclic phenols such as hydroxypyrimidines and hydroxypyridines with moderate yields (Scheme 8.3).

Although the mechanism of this process was not studied, a plausible mechanistic hypothesis is invoked by the authors (Scheme 8.4). On the latter, the nitrophenol **16** is assumed to act as acid surrogate activating the aldehyde to promote the further isocyanide addition. The resulting nitrilium **17** is trapped by the nucleophilic phenoxide generating imidate intermediate **18**, which undergoes a final Smiles rearrangement [15, 16] to provide the more stable α-aryloxy amide **15** (Scheme 8.4).

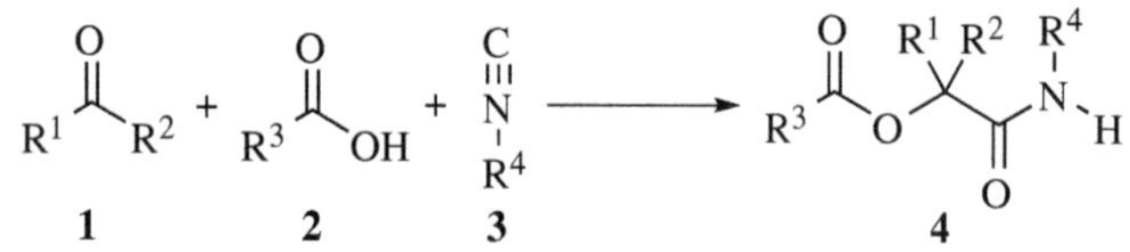

SCHEME 8.1 The classical Passerini three-component reaction.

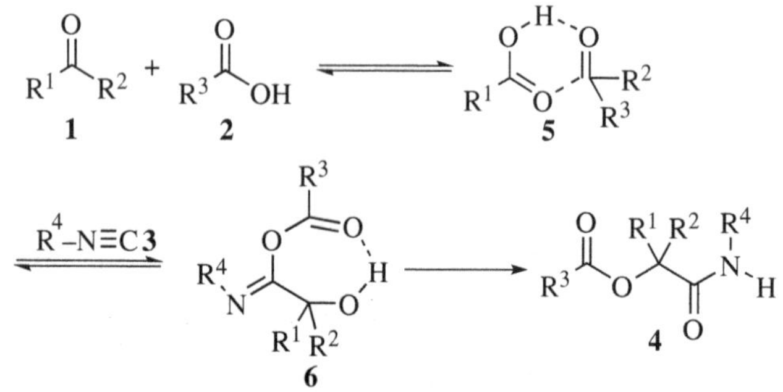

SCHEME 8.2 Generally accepted Passerini mechanism.

8.2.2 Metal-Catalyzed *O*-Alkylative Passerini Three-Component Reactions

Based on the previously reported methodology by El Kaim and Grimaud [14], Taguchi's group developed the first metal-catalyzed Passerini reaction using alcohols instead of the carboxylic acid component allowing the construction of a chemical library of α-alkoxy amide derivatives **19** (Scheme 8.5) [17].

A reaction pathway for the metal-catalyzed alkylative Passerini reaction is represented in Scheme 8.6.

$In(OTf)_3$ would initially promote the formation of oxocarbenium **21** by the reaction of aldehydes with alcohols. However, the authors believe that the addition of orthoester

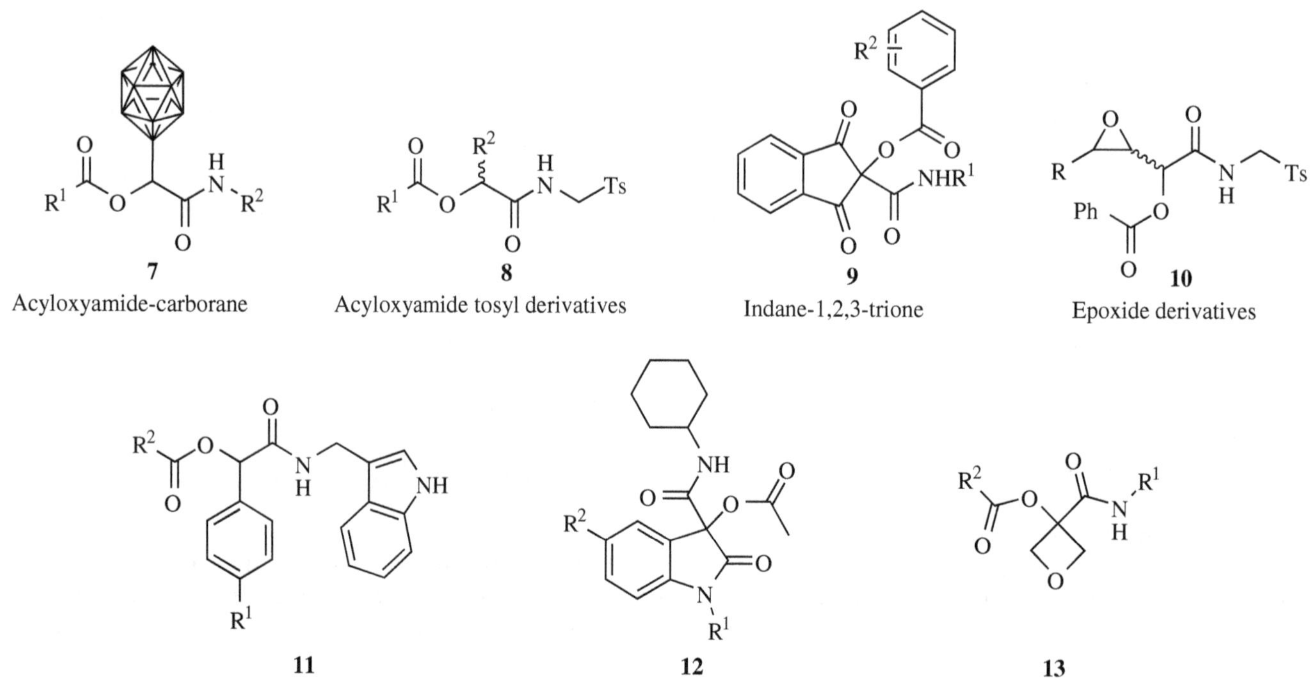

7 Acyloxyamide-carborane; **8** Acyloxyamide tosyl derivatives; **9** Indane-1,2,3-trione; **10** Epoxide derivatives; **11** 3-aminomethylindole dervative; **12** 3,3-disubstituted indol-2-ones; **13** 3,3-disubstituted oxetanes

FIGURE 8.1 Acyloxyamides prepared by a Passerini MCR: **7** [6], **8** [7], **9** [8], **10** [9], **11** [10], **12** [11], and **13** [12].

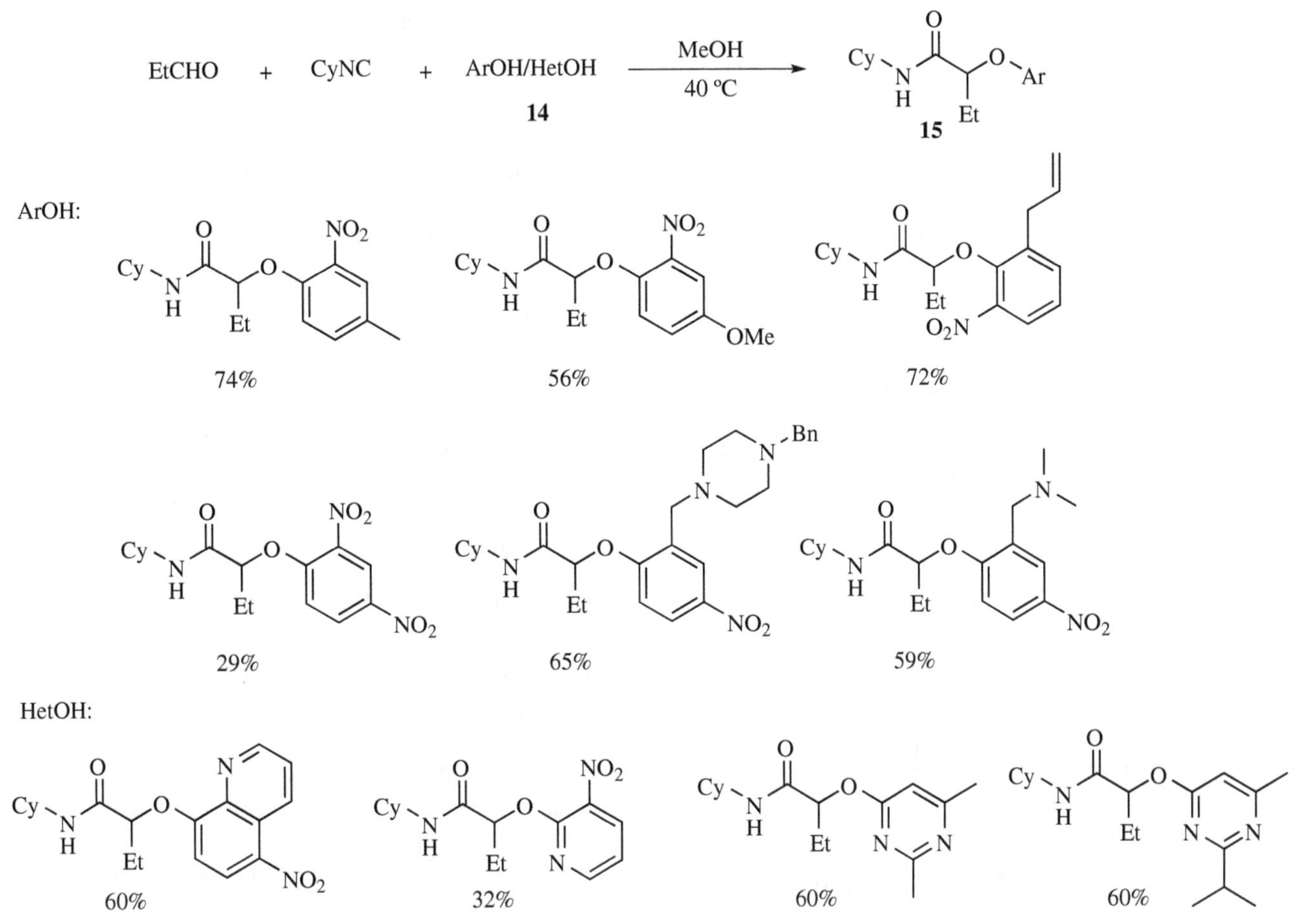

SCHEME 8.3 Modified nitrophenol (ArOH) and heteroaromatic phenol (HetOH) Passerini–Smiles coupling.

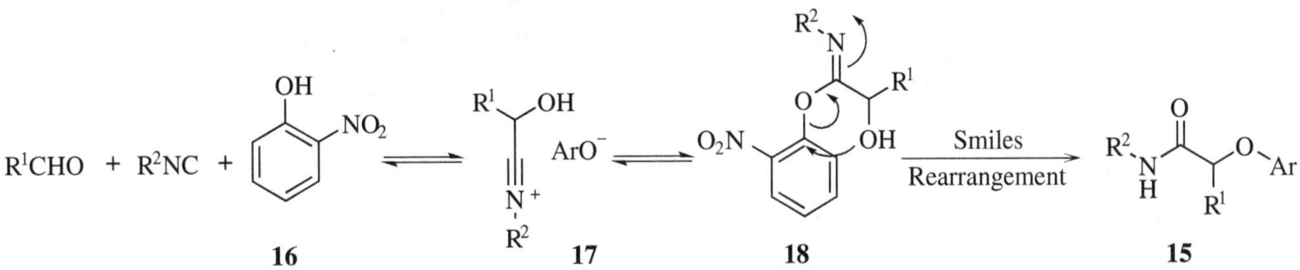

SCHEME 8.4 Mechanistic proposal for the *O*-arylative Passerini MCR.

20 probably could accelerate the formation of **21**. Furthermore, *tert*-butyl isocyanide would attack to oxocarbenium species **21** and the final hydrolysis of the resultant nitrilium intermediate **22** by H_2O would afford the alkylative Passerini products **19**. The mechanism is supported by additional experimental results performed by the same research group.

8.2.3 *O*-Silylative Passerini Three-Component Reactions

Recently, Soeta, Inomata, and coworkers successfully tried to supply the use of carboxylic acids by molecules possessing a similar acidic function, with an electrophilic and a nucleophilic moiety, such as the general form Z-OH [18]. For this purpose, the authors examined silanol derivatives **23** as suitable reagents for giving the corresponding α-siloxyamide derivatives **24** as versatile synthetic intermediates. A wide range of diverse aldehydes and isocyanides are applicable to this methodology affording final products in high yields (Scheme 8.7) [19].

On the basis of these results and additional control experiments, the authors proposed a plausible mechanism to explain the *O*-silylative Passerini reaction as depicted in Scheme 8.8.

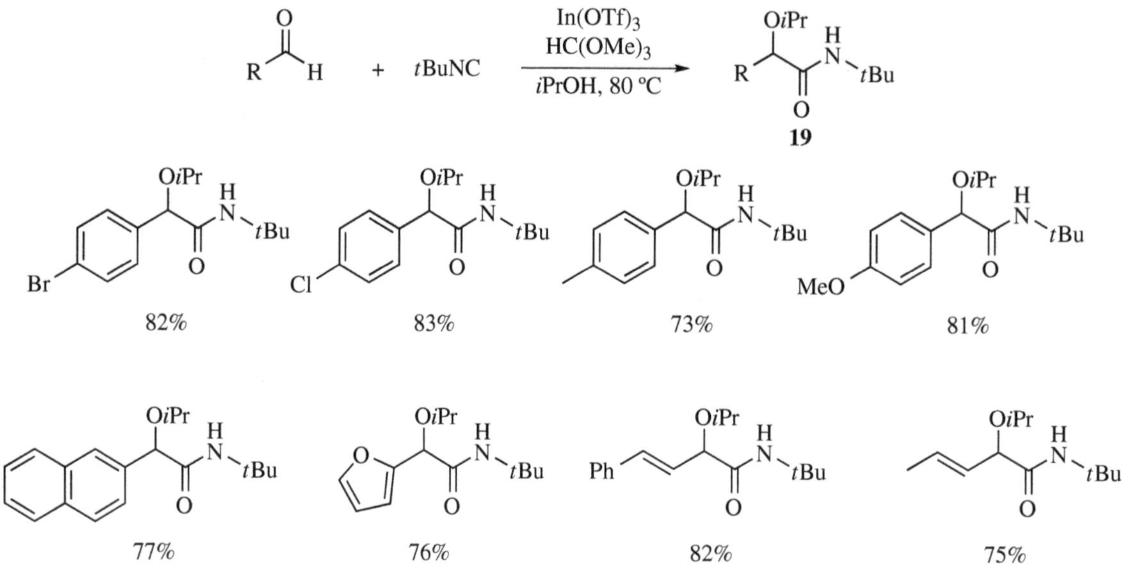

SCHEME 8.5 In(OTf)$_3$-catalyzed O-alkylative multicomponent P-3CR.

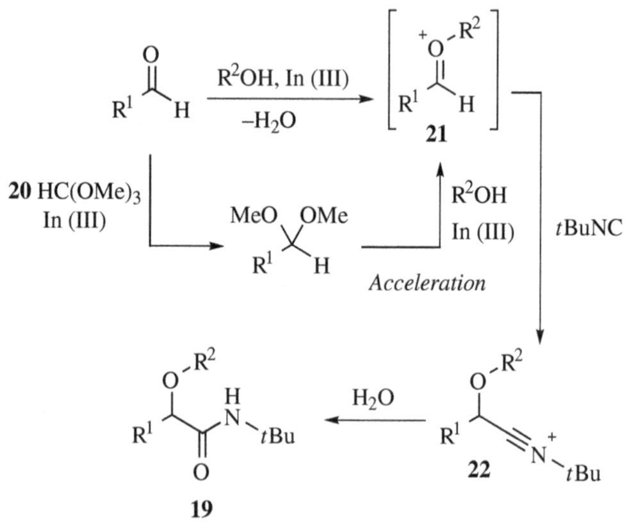

SCHEME 8.6 Proposed reaction mechanism.

In this mechanism, the silanol **23** would be coordinated to the carbonyl group of the aldehyde to promote the subsequent nucleophilic attack by the isocyanide giving a nitrilium intermediate **25**. Furthermore, the hydroxyl group on the silanol rearranges to the intermediate **26**, which finally tautomerizes to afford the corresponding α-siloxyamide **24**. More recently, and based on this idea, the same group has developed a similar method using phosphinic acid **27** instead of a carboxylic acid in the first one-pot *O*-phosphinative Passerini/Pudovik reaction (Scheme 8.9) [20]. The reaction was applied to a wide range of aldehydes providing the corresponding α-(phosphinyloxy)amides **28** in high yields (65–96%). The mechanism proposed to explain the final products **28** is similar to that described in Scheme 8.8.

8.3 PASSERINI 3CR UNDER OXIDATIVE CONDITIONS

The oxidation of alcohols providing useful carbonyl intermediates has been employed as an efficient synthetic tool in many processes [21].

In this context, Zhu and coworkers envisioned the great utility of alcohols in Passerini reaction instead of aldehydes, which are unstable and sometimes difficult to handle, since the carbonyl compounds can be obtained by direct oxidation of the corresponding alcohols. Thus, the authors generated *in situ* the aldehydes by oxidation of the alcohols **29** using IBX to obtain α-acyloxycarboxamide derivatives **30** in good yields (Scheme 8.10) [22].

This pioneering oxidation/P-3CR approach increases the versatility and the efficiency of the P-3CR processes and broadens the scope and versatility of this powerful methodology [23].

8.3.1 Metal-Catalyzed Oxidation Passerini 3CR

The necessity of using an excess of IBX and the subsequent removal at the end of the process prompted Zhu and coworkers to study the methodology under catalytic oxidative conditions using a combination of copper salt with TEMPO as catalyst and oxygen as terminal oxidant (Scheme 8.11) [24, 25]. Under the optimized reaction conditions, this oxidative strategy was applied in a great number of examples using representative primary alcohols **29**, isocyanides, and different carboxylic acids rendering final α-acyloxycarboxamide derivatives **30** in moderate to good yields (30–75%).

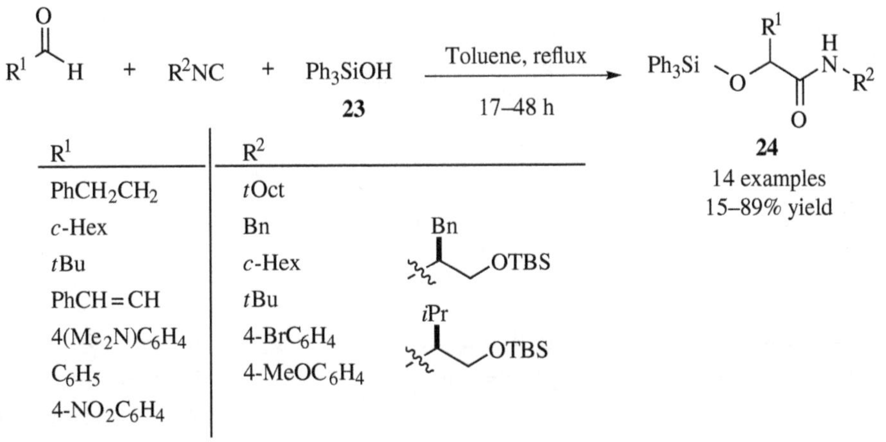

SCHEME 8.7 Scope of the O-silylative P-3CR.

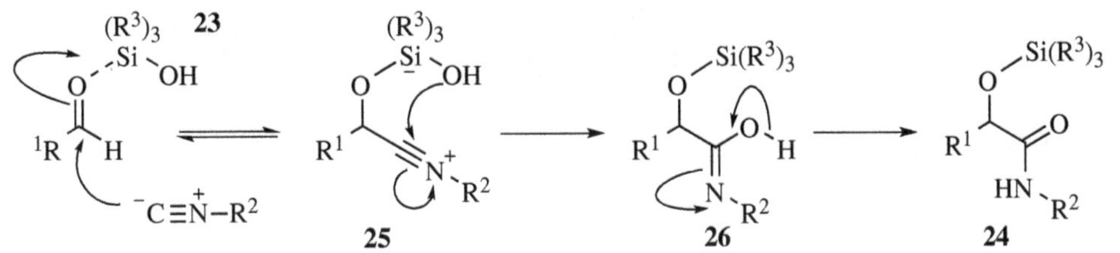

SCHEME 8.8 Mechanism for the *O*-silylative P-3CR.

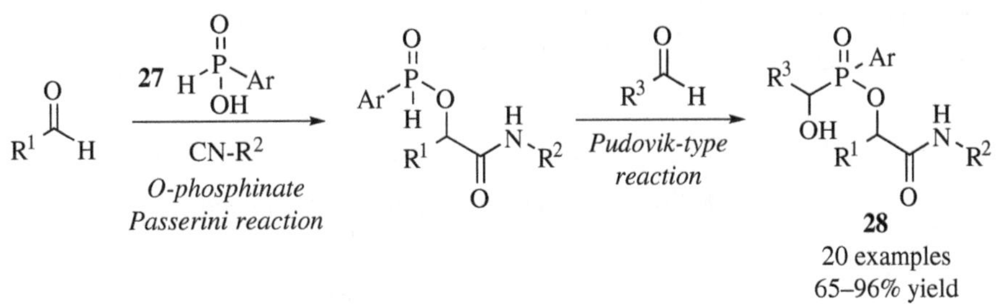

SCHEME 8.9 General reaction to give access to α-(phosphinyloxy)amides **28**.

A feasible reaction sequences for this aerobic oxidative P-3CR is proposed as follows (Scheme 8.12).

Formation of $CuCl_2$·TEMPO complex **31** followed by ligand exchange with the starting alcohol **29** would furnish the ternary complex **32**. From the latter, aldehyde is formed through an intramolecular hydrogen abstraction followed by fragmentation and with releasing of complex **33**. Subsequent multicomponent Passerini 3CR between the *in situ* formed aldehyde, the isocyanide, and the acidic component gave adducts **30**. Additionally, complex **33** would be oxidized by NO_2 regenerating the initial reactive complex copper(II)·TEMPO **31**. Although this proposal is based on the experimental results and in the fact that both $CuCl_2$ and $NaNO_2$ were important for the success of the reaction, the authors cannot discard other feasible pathways involving oxoammonium intermediate [26].

8.4 SYNTHESIS OF MACROCYCLES BY A PASSERINI REACTION

The interest by the synthesis of macrocycles has grown in the last years because these complex structures often display remarkable biological activities and many of them are used as drugs [27]. Their reduced number of accessible energetic conformations and the compromise between conformational preorganization and flexibility can influence their biological activities, which are responsible for their attractive behavior. The classic route giving access to macrocycles consists of a stepwise construction of a linear substrate followed by a ring closure [28]. This approach has been extensively employed in target-oriented synthesis of single compounds. However, this procedure presents limitations in diversity-oriented synthesis of macrocyclic libraries [29].

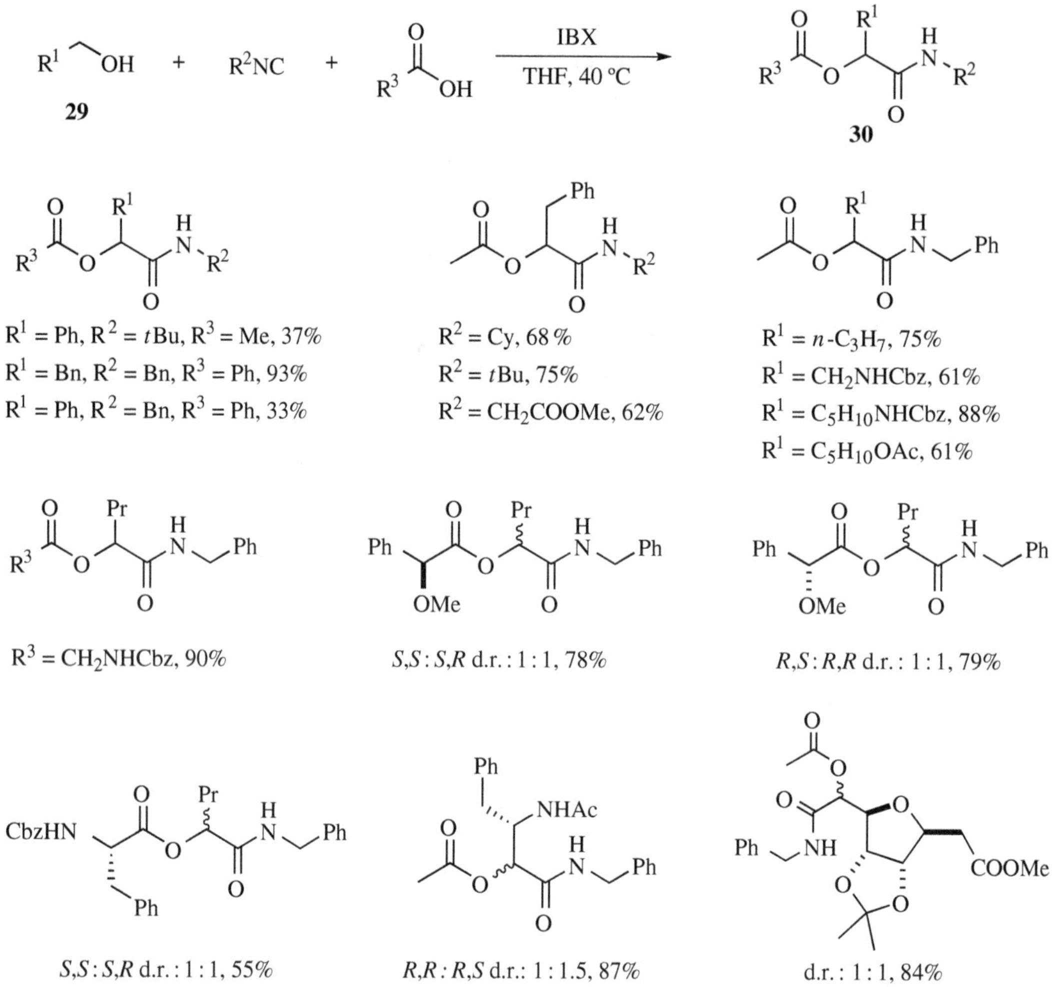

SCHEME 8.10 IBX-promoted oxidation/P-3CR reaction.

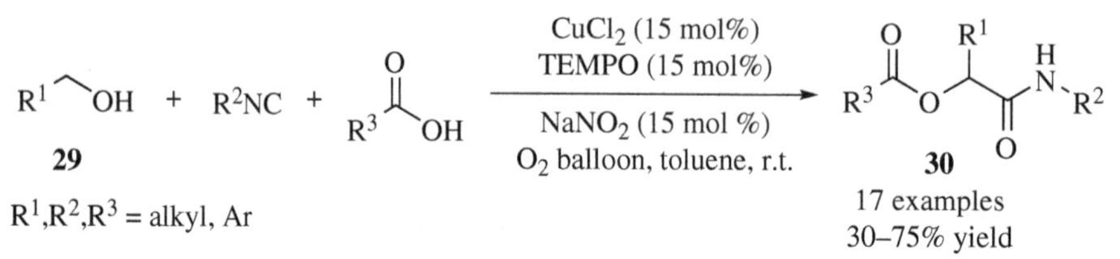

SCHEME 8.11 $CuCl_2$·TEMPO-catalyzed oxidative P-3CR.

Moreover, MCR procedures for the synthesis of macrocycles from readily accessible starting materials have been scarcely explored in the literature [30]. Although Passerini 3CR has been used previously in the synthesis of macrocycles, first forming the previous acyclic precursors that are later cyclized by, for instance, ring-closing metathesis [31], only recently, Wessjohann's group reported the first example where Passerini 3CR is employed as the ring-closing reaction itself for the synthesis of macrocycles, with low and moderate yields (Fig. 8.2) [32].

While the first two macromolecules **34** and **35** are obtained under very mild reaction conditions, heating of the multicomponent mixture was necessary for the last two products **36** and **37**, where alcohols were employed instead of unstable aldehydes, which are *in situ* oxidized with IBX (as described in Section 8.3). The use of the more easily accessible alcohols in the last approach under oxidative conditions increases the possibilities of this reaction giving easy access to difficult frameworks, which would not be available otherwise because of the lack of stability of the aldehydes employed.

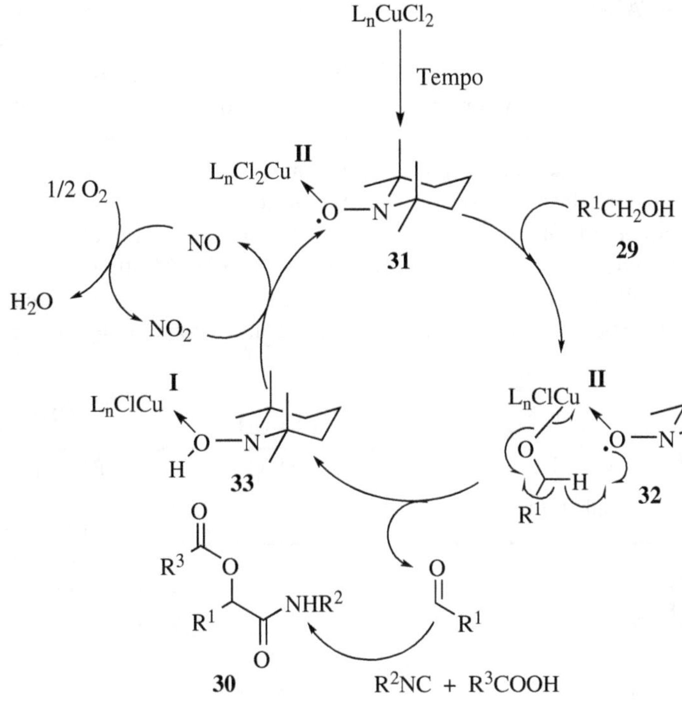

SCHEME 8.12 Mechanistic hypothesis.

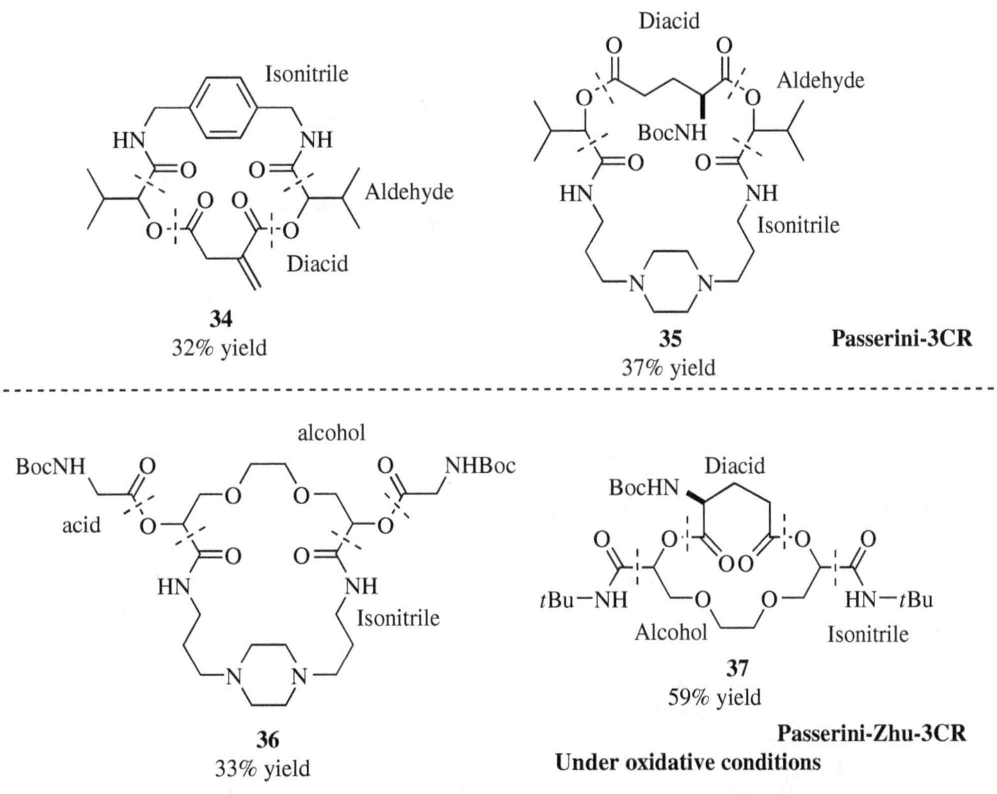

FIGURE 8.2 Depsipeptide-like macrocycles synthesized through a multicomponent P-3CR.

8.5 ENANTIOSELECTIVE METAL-CATALYZED PASSERINI REACTION

In sharp contrast to the importance acquired by asymmetric catalysis in the last decades, and although asymmetric induction has been achieved in Passerini reaction through diastereoselective approaches by using chiral substrates and/or chiral auxiliaries [33], the number of catalytic asymmetric procedures reported is still limited for this reaction and it remains as a significant challenge [34]. Several drawbacks exist that make this task particularly difficult to overcome: (i) the complexity of the reaction mechanism; (ii) the possible competitiveness of the uncatalyzed background reaction; (iii) the capacity of the three starting materials to coordinate the catalyst, as Lewis bases, and consequently deactivating this; and (iv) the possible inhibition of the catalyst by the formed product, decreasing the turnover.

After the pioneering examples reported by Denmark [35] (Scheme 8.13), Dömling [36] (Scheme 8.14), and Schreiber [37] (Scheme 8.15), only scarce examples have appeared in the literature concerning this appealing topic over the last

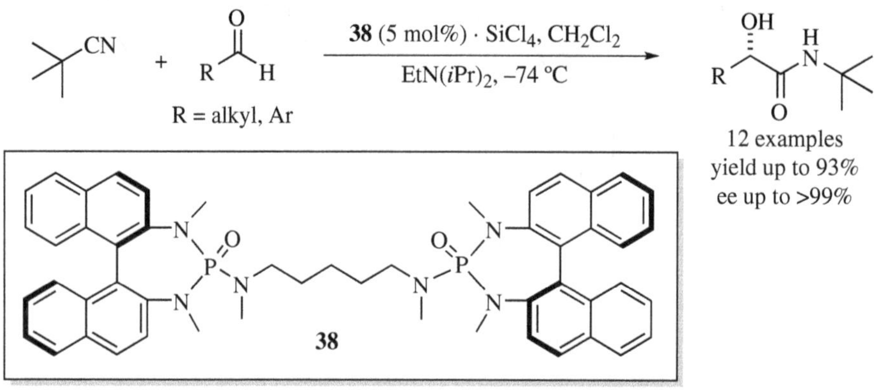

SCHEME 8.13 First catalytic enantioselective Passerini-type reaction.

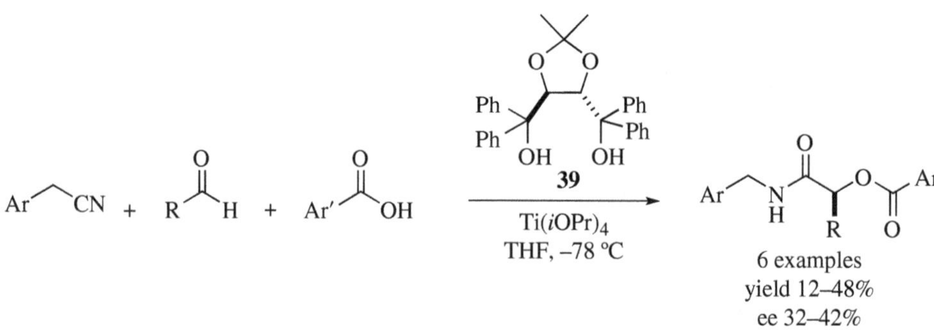

SCHEME 8.14 Ti-catalyzed asymmetric Passerini 3CR.

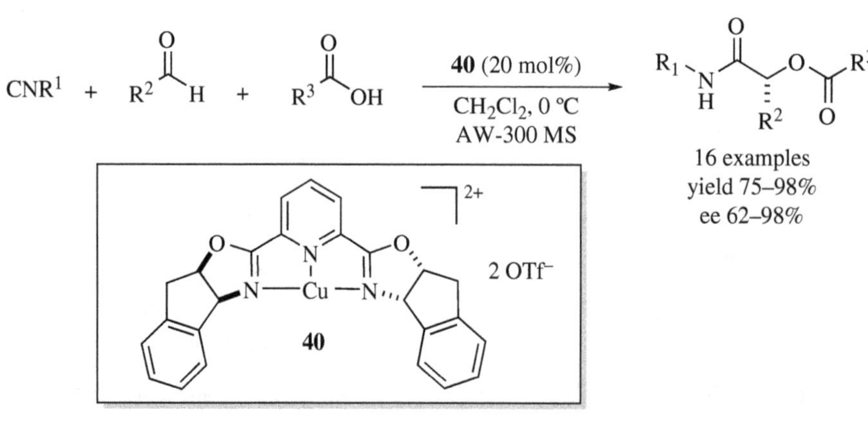

SCHEME 8.15 Cu-catalyzed asymmetric Passerini 3CR.

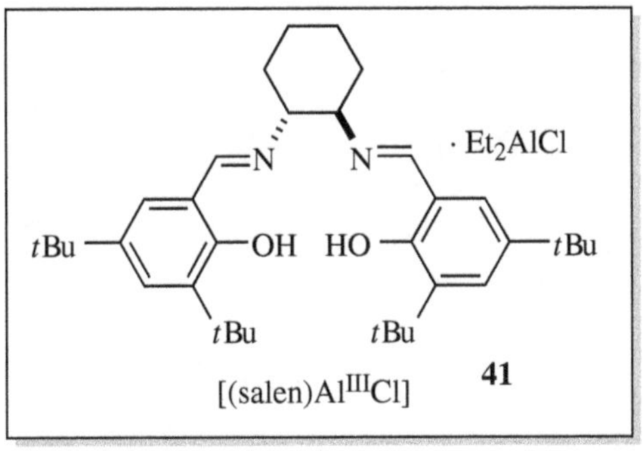

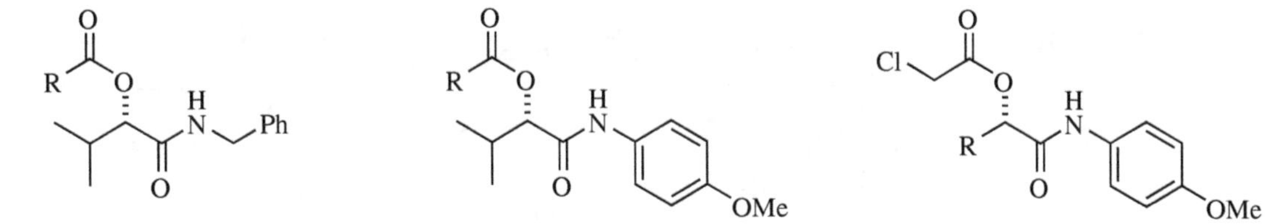

R = Ph 70% yield, 63% ee
R = Me 59% yield, 63% ee

R = Ph 63% yield, 84% ee
R = Me 60% yield, 84% ee
R = MeC = CH_2 62% yield, 80% ee
R = $SHCH_2$ 51% yield, 75% ee
R = $ClCH_2$ 64% yield, >99% ee

R = Me 66% yield, 87% ee
R = *n*-Hex 67% yield, 73% ee
R = CH_2CH_2Ph 68% yield, 71% ee

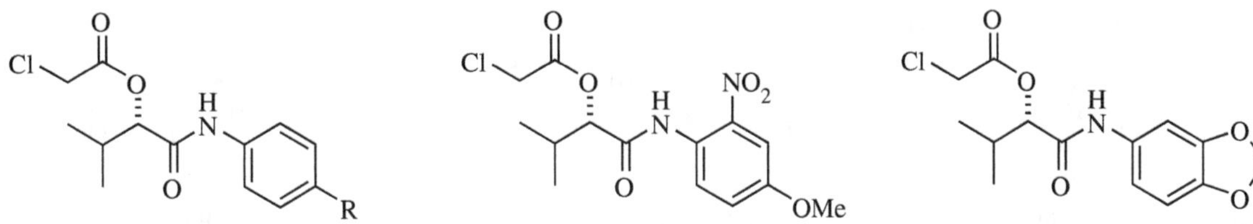

R = H 68% yield, 93% ee
R = Me 61% yield, 81% ee
R = Br 52% yield, 88% ee

66% yield, 75% ee

64% yield, 68% ee

FIGURE 8.3 Scope of the [(salen)Al^{III}Cl] **41**-catalyzed enantioselective Passerini reaction.

years, maybe due to some inherent drawbacks of this reaction in the moment of the asymmetric induction.

More recently, Zhu and coworkers reported an efficient chiral Lewis acid-catalyzed Passerini 3CR rendering final adducts in good to excellent enantioselectivities for a variety of nonchelating aldehydes, carboxylic acids, and isocyanides as represented in Figure 8.3 [38].

Almost simultaneously, the same research group disclosed a [(salen)Al^{III}Me] **43**-catalyzed enantioselective Passerini-type reaction of aldehydes, isocyanides, and hydrazoic acid **42**, affording a great scope of tetrazole derivatives **44** with good to excellent enantioselectivities (51–95%) (Scheme 8.16) [39].

A plausible mechanistic explanation to describe the formation of tetrazole derivatives is represented in Scheme 8.17.

The addition of the isocyanide to the active complex **A** gives the nitrilium ion **B**, which is further trapped by HN_3 **42** delivering tetrazole **44** as final product. Simultaneously, reaction of **B** with HCl could provide the chloroimidate **C**.

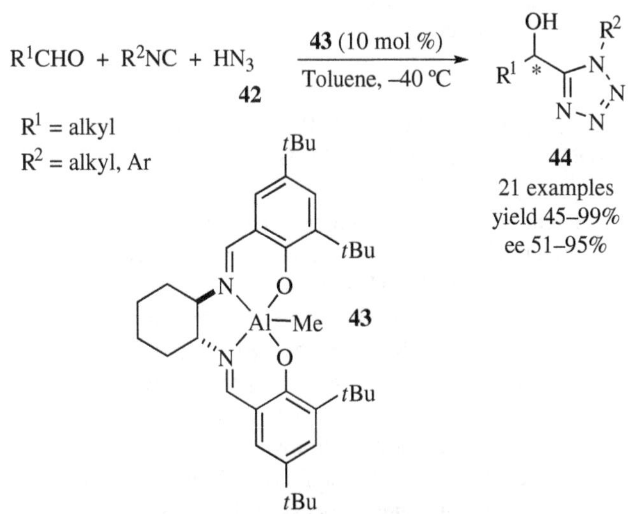

SCHEME 8.16 Enantioselective synthesis of tetrazole derivatives **44**.

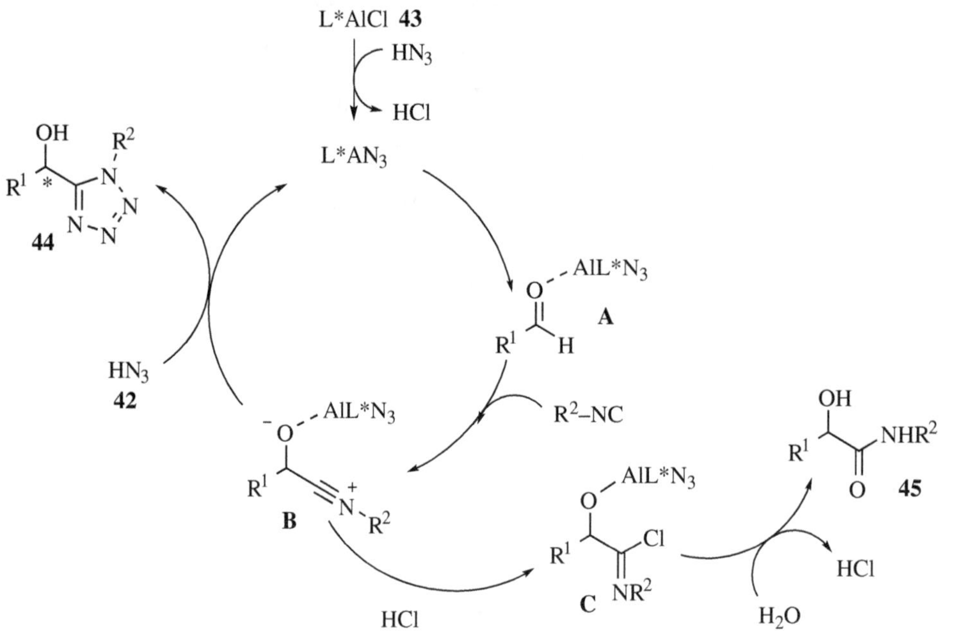

SCHEME 8.17 Mechanistic proposal for the formation of tetrazole derivatives **44**.

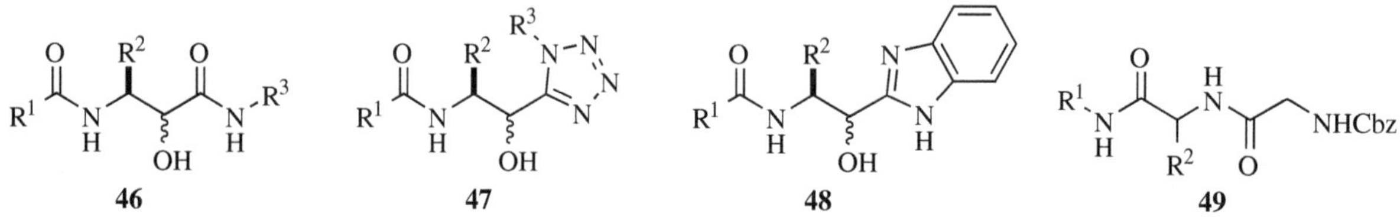

FIGURE 8.4 General structure of norstatines **46** [44], *cis*-constrained norstatine analogues **47** [40a], benzimidazole analogues **48** [45], and peptidic kinase inhibitors **49** [46].

This intermediate, in the presence of water, can be converted into the α-hydroxyamide **45**. It is worth noting that in this case the authors use HN_3 **42** as the acid instead of the usual carboxylic acid component.

8.6 SYNTHESIS OF PHARMACOLOGICALLY IMPORTANT PEPTIDOMIMETICS

A potentially useful synthetic extension in this area of research has been the Passerini reaction–amine deprotection–acyl migration (PADAM) sequence, first conceived as a tool for giving access to peptide-like structures [40] and mainly developed by Banfi, Riva, and coworkers [2a]. In this sense, PADAM sequence has been employed for the synthesis of interesting peptidomimetics since the resulting α-ketoamide scaffolds are useful for drug discovery [41] or in the synthesis of enzyme inhibitors [42], and this amplifies the scope of this valuable multicomponent process [43] (Fig. 8.4).

The general mechanism for a PADAM reaction involves a first condensation between N-Boc-protected α-aminoaldehydes **50** with various isocyanides and various carboxylic acids to give the corresponding P-3CR α-acyloxyamide product **51** (Scheme 8.18). Further one-pot Boc deprotection and acyl migration give rise to complex peptide-like substances **52** possessing a central α-hydroxy-β-amino acid unit. Furthermore, a possible subsequent oxidation would render oligopeptides **53** containing an α-oxo-β-amino acid unit, which is an even more attractive skeleton, due to its similarity with protease transition state [41].

In this context, and more recently, an interesting application of this methodology has been described by Faure, Aitken, and coworkers. Thus, they applied PADAM methodology (three-component Passerini reaction–amine deprotection–O,N-acyl migration) as key synthetic step of the linear pentapeptide intermediate in the formal total synthesis of cyclotheonamide C **54**, a potent inhibitor of serine protease such as thrombin belonging to a family of cyclic pentapeptides with potential biological activities (Scheme 8.19) [47].

The authors reasoned that the appropriate selection of the aldehyde **55**, isonitrile **56**, and carboxylic acid **57** fragments

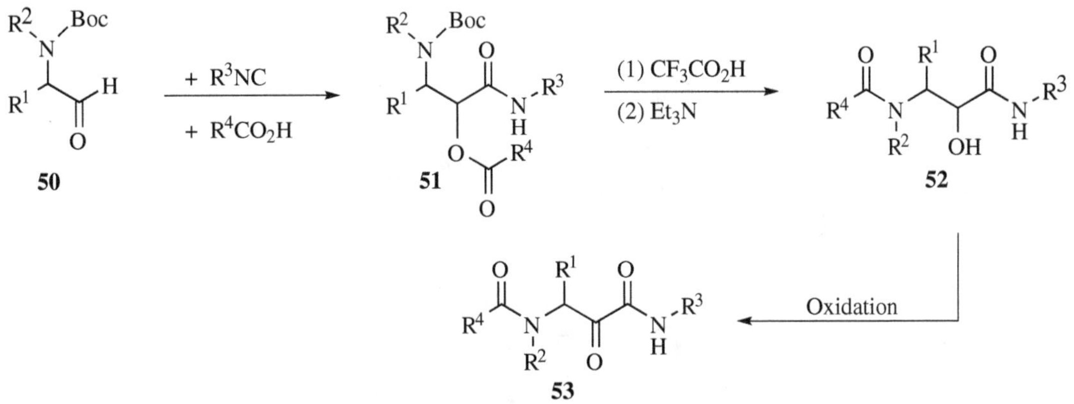

SCHEME 8.18 General mechanism for a PADAM strategy.

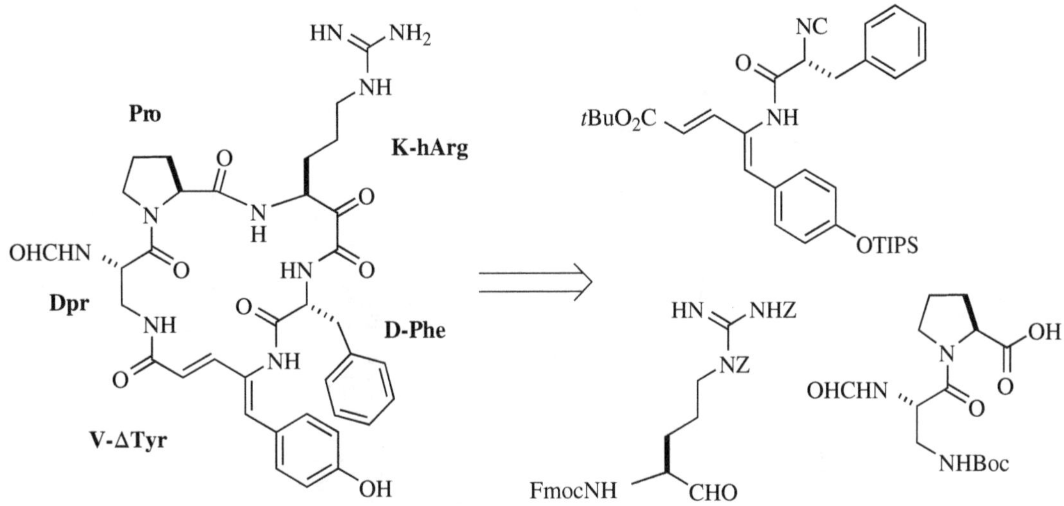

Cyclotheonamide C **54**

SCHEME 8.19 Retrosynthesis for formal synthesis of cyclotheonamide C (CtC) **54**.

through a useful PADAM sequence (Scheme 8.20) could provide the α-acyloxyamides **58**. Selective Fmoc deprotection of the pentadepsipeptide **58c** provided the key linear pentapeptide **59** in 88% yield, which is involved in the synthetic route of CtC **54**.

Since many types of tumor cells are associated with oxidative stress, Sasse, Jacob, and coworkers decided to use a Passerini MCR to elegantly synthesize a number of structures combining two, three, or even four redox centers with good yield, among them those containing multiple chalcogen, porphyrin metal binding, and quinone redox sites (Fig. 8.5) [48].

Additionally, the compounds were then examined against cancer cells showing a clear inhibition of the cell proliferation and inducing apoptotic cell death. Interestingly, some of these redox-active compounds exhibited quite low toxicity with normal cells.

8.7 MULTICOMPONENT PASSERINI APPROACH TO IMPORTANT TARGETS

Although maybe Passerini condensation has been less successful in its application for preparing biologically relevant molecules in comparison with the Ugi reaction, mainly because in the Passerini reaction there are only three variables that can be introduced, thus limiting the application of intramolecular versions to prepare heterocycles, we reflect here two attractive examples.

Liu, Lee, and coworkers applied the Passerini 3-MCR approach as the final key step for the synthesis of interesting podophyllotoxin derivatives **60** with good yields (36–85%) (Scheme 8.21) [49].

The *in vitro* cytotoxicity of final Passerini adducts was evaluated against four human tumor cell lines: A549 (non-small cell lung cancer), DU145 (prostate cancer cell line),

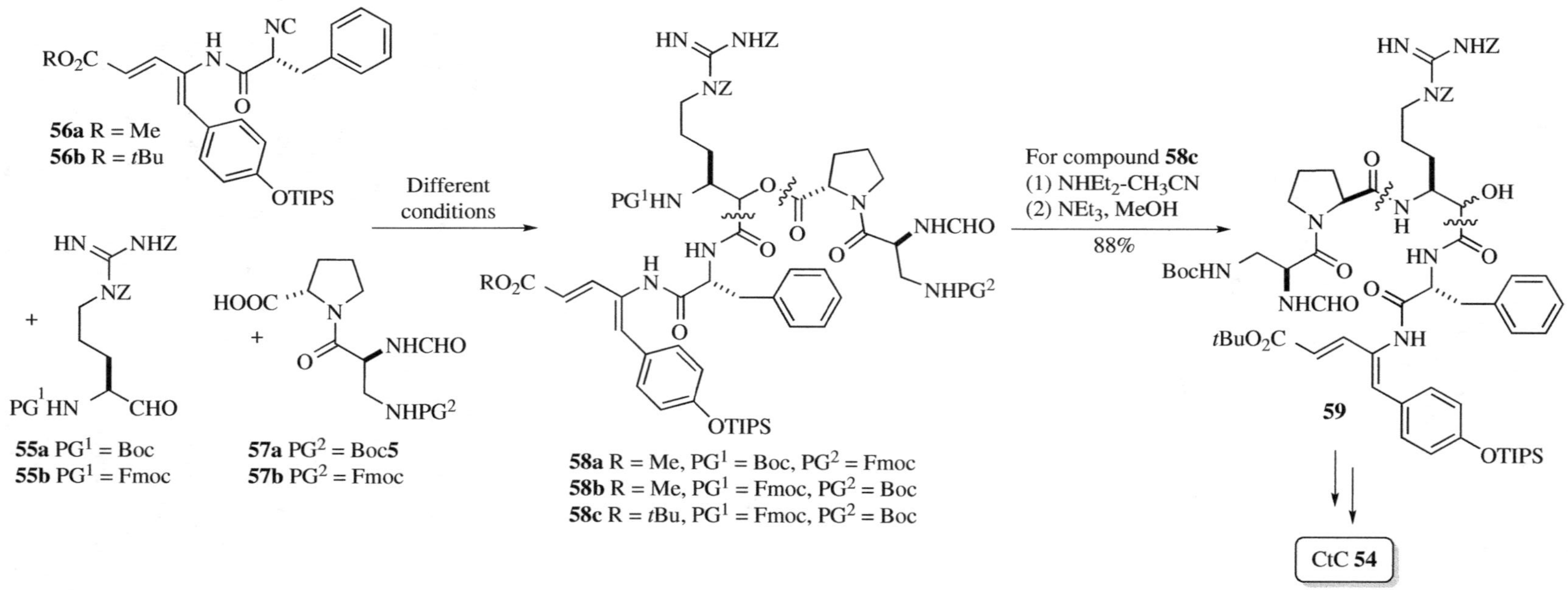

SCHEME 8.20 Synthesis of intermediate **59** in the synthesis of CtC **54**, using a PADAM sequence.

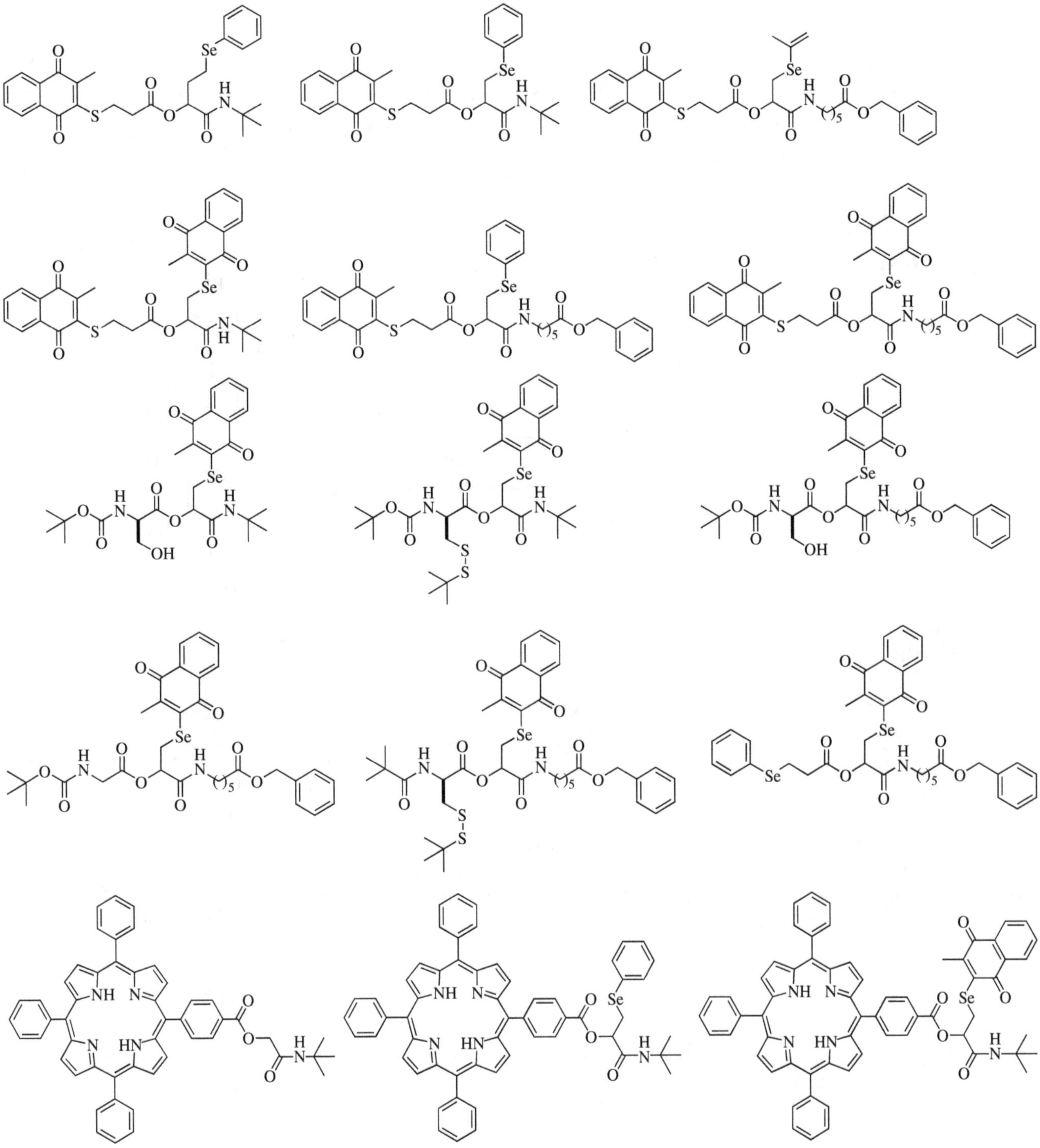

FIGURE 8.5 Synthesis of di-, tri-, and tetrafunctional redox agents containing multiple chalcogen and quinone redox sites.

KB (nasopharyngeal carcinoma), and KBvin [multidrug-resistant (MDR) KB subline selected using vincristine]. Interestingly, most of them exhibited high levels of cytotoxicity against the cells assayed and overcame acquired drug resistance in the latter cell line.

Ruijter and coworkers have revealed an elegant synthesis of the important drug candidate telaprevir **61**, a peptidic HCV NS3 protease inhibitor, as schematically represented in Scheme 8.22 [50]. The authors projected a convergent synthesis achieving a noteworthy increased both the atom and

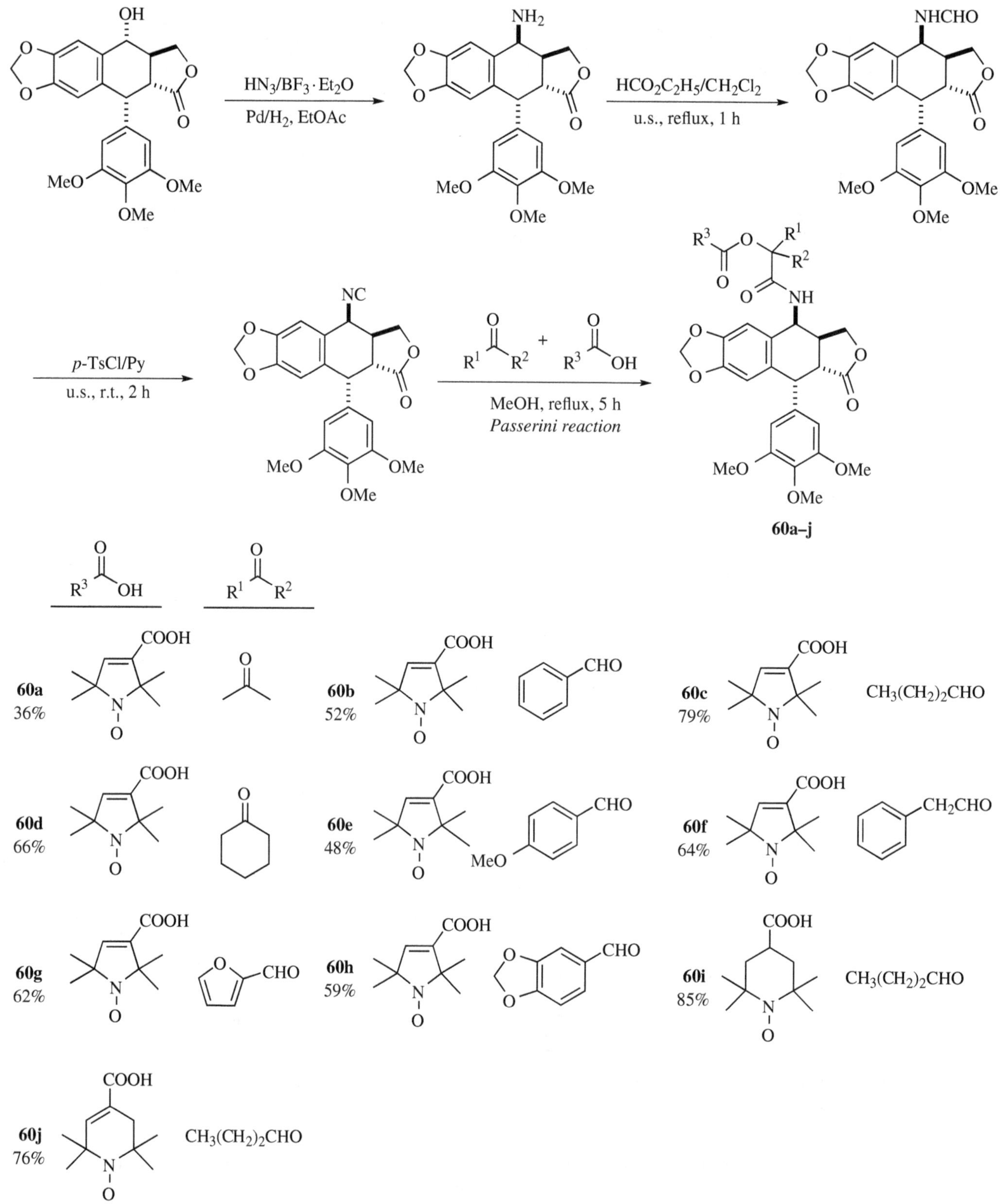

SCHEME 8.21 Synthesis of target compounds **60** *via* Passerine reaction.

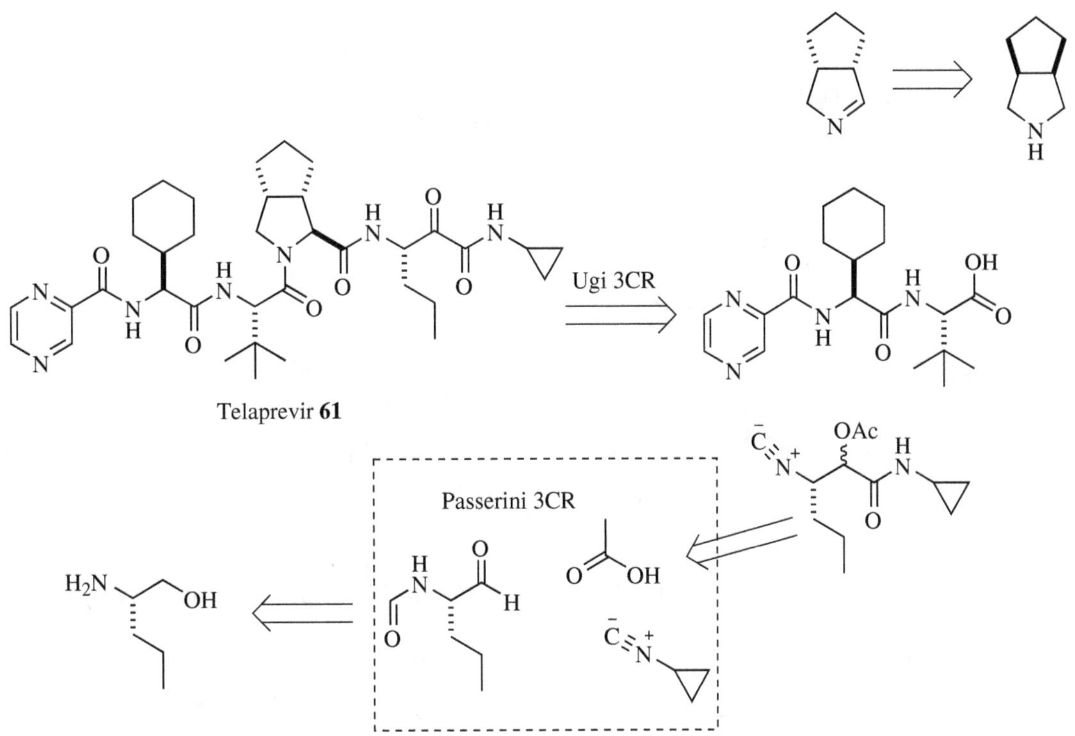

SCHEME 8.22 Retrosynthetic route of telaprevir **61**.

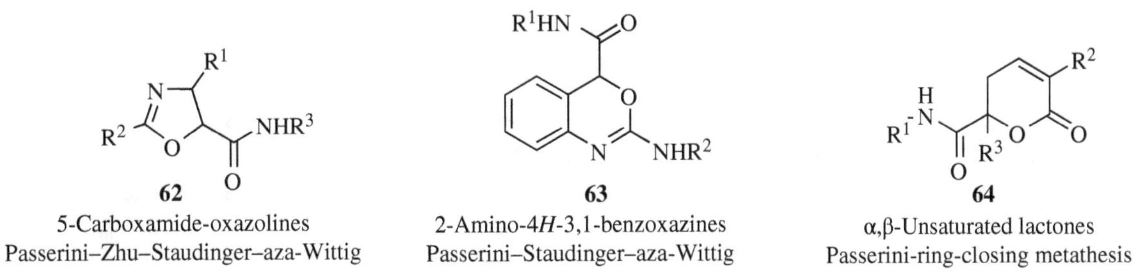

FIGURE 8.6 Preparation of complex molecules following a postcondensation transformation after an initial Passerini MCR: **62** [52], **63** [53], and **64** [54].

step economy and the overall yield compared with the previous procedures [51], using a biocatalytic desymmetrization and two MCRs (Passerini 3CR and Ugi-3CR) as the crucial synthetic steps.

8.8 α-HYDROXYCARBOXAMIDE, AN IMPORTANT INTERMEDIATE FOR CHEMICAL SYNTHESIS

The resulting α-hydroxycarboxamides have been interestingly employed as intermediate for the synthesis of more complex scaffolds. In fact, as in Ugi reaction (Chapter 7), the combination of these MCRs with a further transformation has gained much attention in the last years for the preparation of structurally diverse complex molecules. Many appealing structural cores have been prepared using different methodologies (Fig. 8.6).

Thus, Basso and coworkers reported the synthesis of a small library of novel class of oxazolines in two steps, following a Passerini-Zhu/Staudinger–aza-Wittig (SAW) protocol (Scheme 8.23) [55]. Interestingly, alcohols **65** are used instead of the corresponding aldehydes and they are oxidized in the medium following the aforementioned protocol reported by Zhu's group [22].

Since oxazolines are privileged structures in medicinal chemistry, the authors explored the stability of final oxazolines **62** and it was found that they were completely stable at pH 7.4, decomposition (10%) was observed under prolonged (24 h) exposure to pH 5 solutions, and complete hydrolysis was observed at pH 3. This is interesting since these molecules could exhibit biological applications.

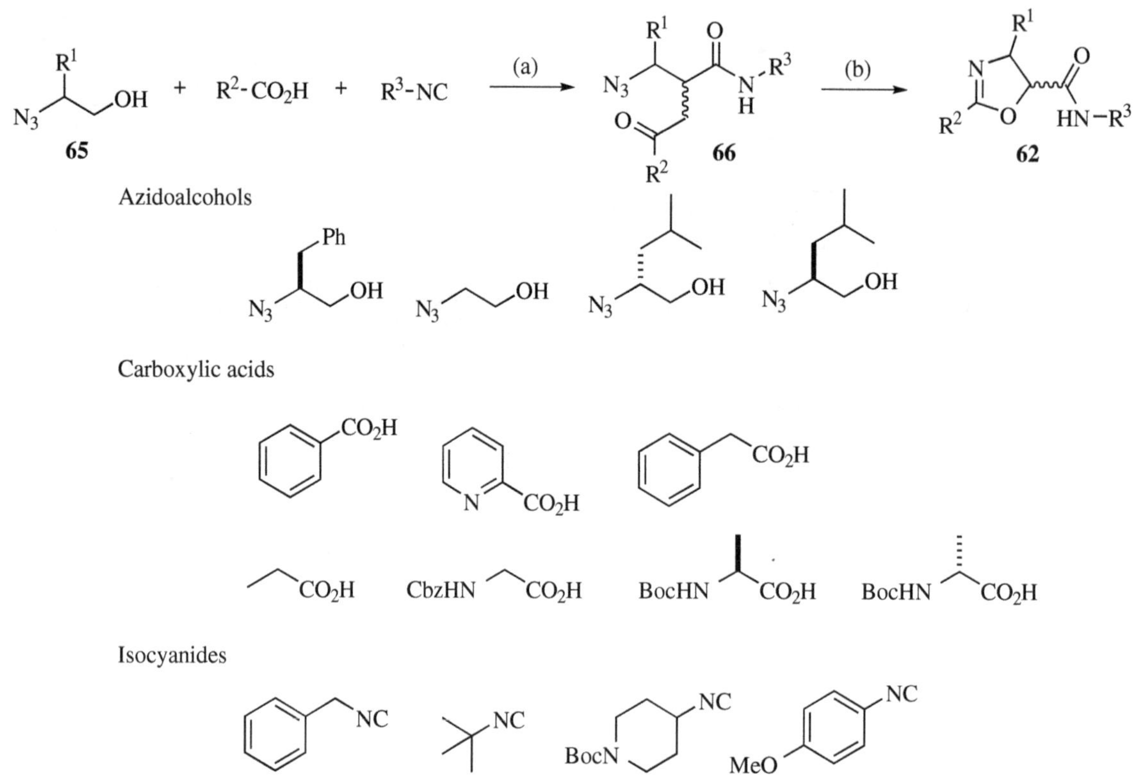

SCHEME 8.23 Passerini-Zhu/Staudinger–aza-Wittig protocol: (a) IBX, THF, MW (150 W), 100 °C; (b) polymer supported triphenylphosphine, CH_2Cl_2, MW (150 W), 100 °C.

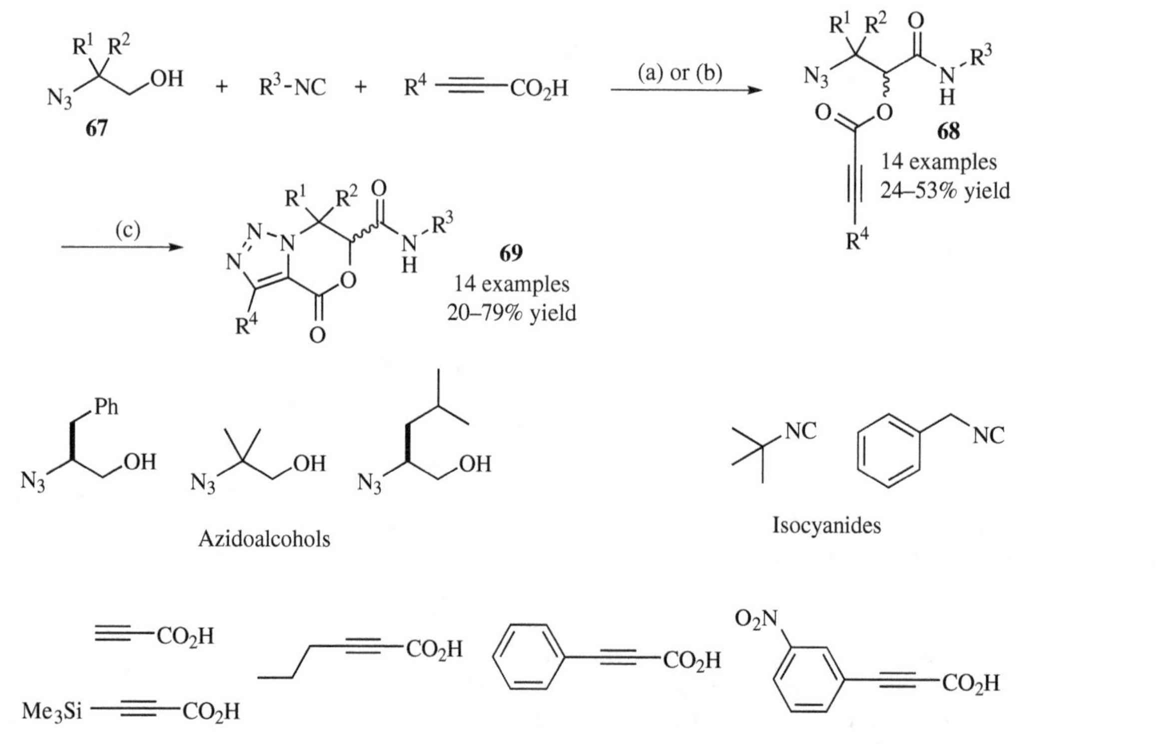

SCHEME 8.24 Passerini-Zhu/cycloaddition protocol: (a) IBX, THF, 100 °C (MW, 150 W), then R^3NC and $R^4C{\equiv}CCO_2H$, THF, r.t.; (b) R^3NC, $R^4C{\equiv}CCO_2H$, IBX, THF, 100 °C (MW, 150 W); (c) DMF, 150 °C (MW, 150 W).

The same group reported a variation of the valuable protocol using a Passerini-Zhu/cycloaddition strategy for the synthesis of triazolo-fused oxazinones starting from α-azido alcohols **67**, which are oxidized in the first step to generate the required aldehyde, and final adducts **69** are obtained in moderate to good yields (20–79%) (Scheme 8.24) [56].

Intermediate **68** undergoes a straightforward azide-alkyne dipolar cycloaddition to furnish triazolo-fused dihydrooxazinones **69**. These results, together with the previously reported synthesis of oxazolines **62** (Scheme 8.23), demonstrate the efficacy of novel α-azido aldehyde scaffolds and the versatility of the Passerini protocol to assemble valuable heterocyclic compounds.

8.9 PASSERINI 3CR UNDER ECO-FRIENDLY REACTION CONDITIONS

In the last decade, with the aim of improving several aspects of the Passerini reaction such as the yield, the cost of the process, the ecological impact, and the reaction time, a few and sustainable works have appeared achieving such optimizations under eco-friendly reaction conditions.

8.9.1 Aqueous Media

Among the different developed protocols, recent findings suggest that the Passerini reaction can be carried out in water instead of traditional organic solvents [57], such as methylene chloride or MeOH. In fact, it has been demonstrated that the rate of this reaction is accelerated when carried out in water. Here is disclosed the interesting finding reported by Vigalok's group carrying out the process in water and "on water" and observing a different regiochemistry depending on the conditions (Scheme 8.25) [58].

The addition of simple salts such as NaOTs or Na_2SO_4 to the aqueous phase provided an unusual variation of the selectivity of the Passerini-type MCR by shuttling the reactants between the two phases. This is an unusual fact since there are only few reports concerning the use of a salt showing a moderate effect on the ratio between isomeric products [59], but this study increases the possibilities for the development of new synthetic methodologies in aqueous media and using a Passerini methodology. The authors proposed a possible mechanism to explain the reactivity observed in each case and reflected in Scheme 8.25.

8.9.2 Ionic Liquids and PEG

Ionic liquids have become very popular and common as solvents in organic synthesis nowadays and constitute an environmentally friendly option instead of the use for volatile organic solvents [60].

In this context, the pioneering example performing contemporaneously the classical Passerini reaction using ionic liquids or polyethylene glycol (PEG 400) as green reaction media and as an alternative to volatile organic solvents was reported by Andrade and coworkers. The process was successfully carried out using both media affording final products with good yields under mild conditions. For some examples, see Figure 8.7 [61, 62].

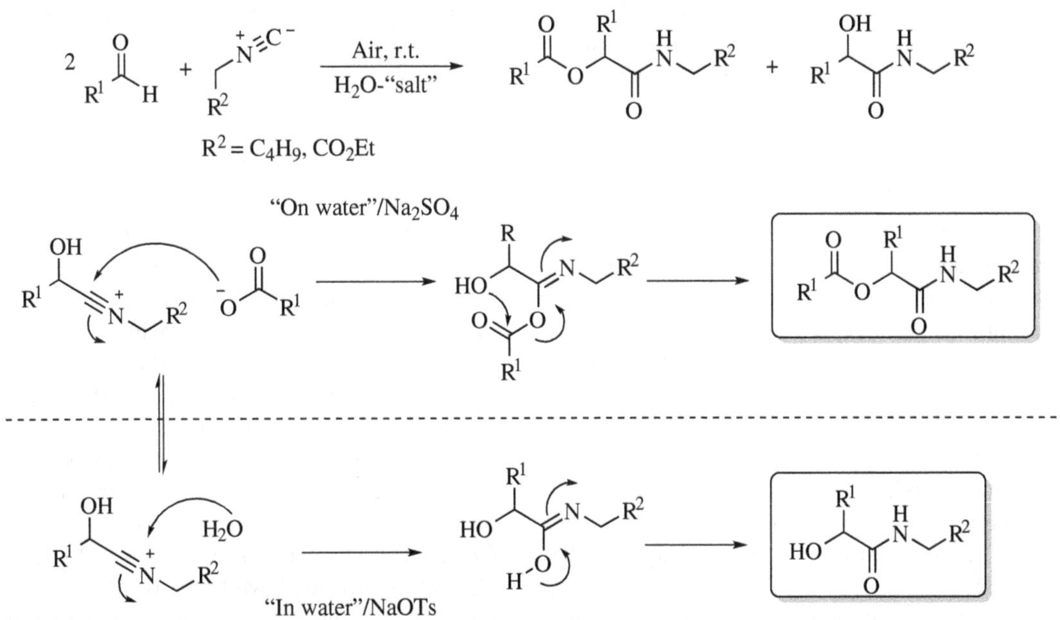

SCHEME 8.25 Competition mechanism between the reactions performed in water and "on water."

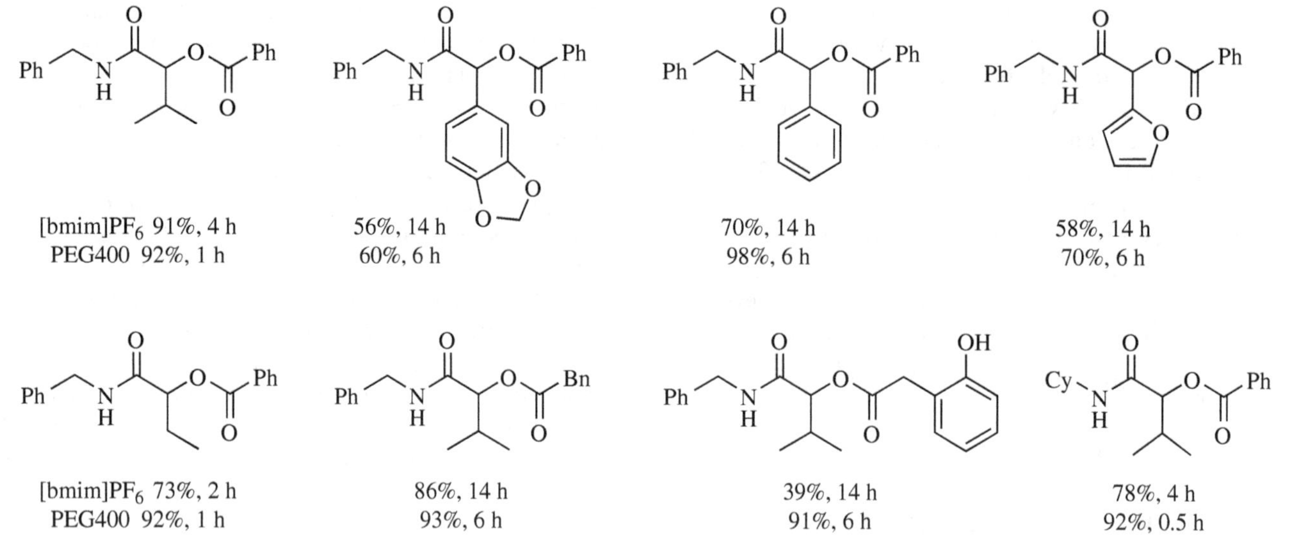

FIGURE 8.7 Passerini adducts under greener conditions.

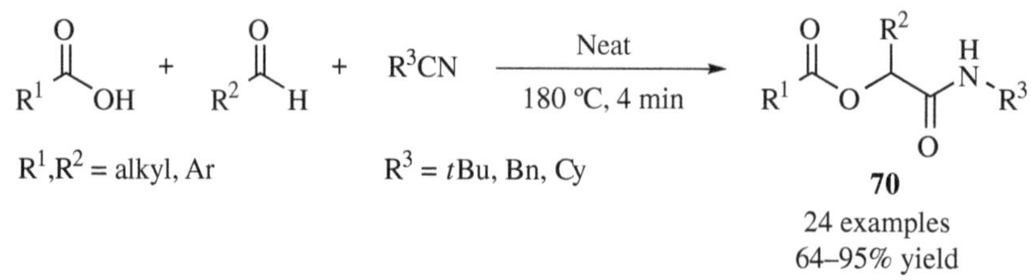

SCHEME 8.26 Passerini 3CR under solvent-free conditions.

The authors also proved the possibility of recycling and reuse these solvents, observing similar yields after five runs. For this propose, the products were extracted from the ionic liquids using diethyl ether as immiscible solvent. The remaining ionic liquids were reused after to dry it by heating under vacuum for 30 min. This strategy is a green example of Passerini protocol.

8.9.3 Solvent-Free Conditions

Performing the reactions under solvent-free conditions has been proved an alternative eco-friendly media [63] and in this sense recently has been reported a useful and efficient Passerini 3CR under solvent-free conditions at high temperature allowing the synthesis of a broad scope of α-acyloxyamides **70** with high yields (64–95%) in short reaction times (Scheme 8.26) [64], after the pioneering work appeared in 2008 [65].

Biju's group employed isatin derivatives **71** [66] as carbonyl component under solvent-free conditions for giving easy access to biologically important 3-acyloxy 3-carbamoyl indol-2-ones **72** from moderate to high yields (Scheme 8.27) [67].

This work is significant because is an efficient, atom-economic, and environmentally benign Passerini protocol and because a congestive ketone is used in this process. The use of bulky ketones is a normal hard work and other protocols failed in the attempt. Moreover, the presence in the final adducts of the isatin moiety provides to this work an additional interest from the potential biological properties point of view.

8.9.4 MW-Assisted Passerini Reaction

Contemporaneously, the use of microwave as promoter chemical reactions has become increasingly popular in organic synthesis [68]. This tool has been demonstrated to be efficient for increasing the rate of MCR reactions reducing the reaction times and improving the yields [69, 70]. However, the first example of MW-assisted Passerini reaction has been only recently disclosed (Scheme 8.28) [71].

The results obtained in this study in terms of reaction time (1 min or less at 120 °C) are remarkable as compared with our previous results on the Passerini reaction using, for instance, ionic liquids (2–14 h) [61], PEG (0.5–6 h) [61], CH_2Cl_2 (18 h) [57a], or water (3–3.5 h) [57a]. All these alternative solvents have been found to be efficient in this reaction, but in all cases, longer reaction times were required and lower yields were achieved.

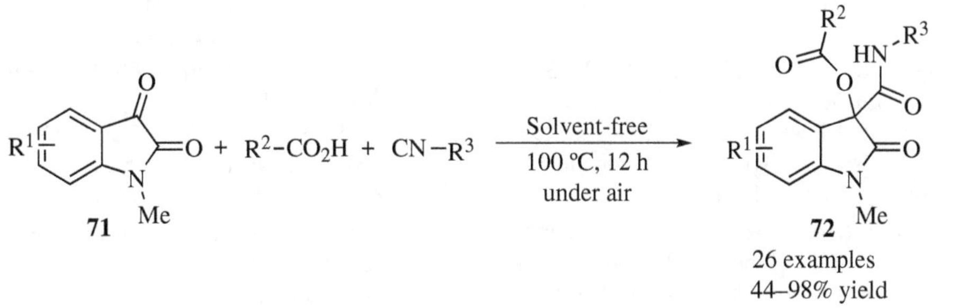

SCHEME 8.27 Isatin derivatives synthesized following a Passerini 3CR.

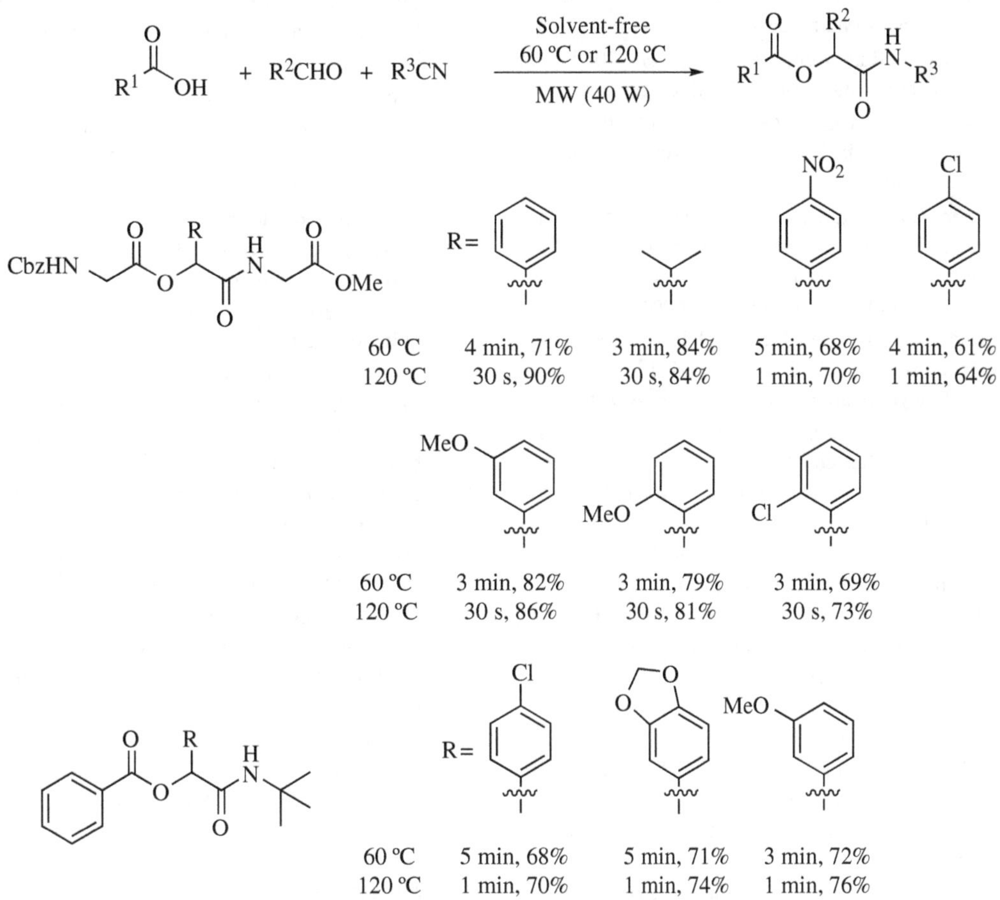

SCHEME 8.28 MW-promoted Passerini 3CR.

8.10 CONCLUSIONS

The Passerini reaction is a pivotal isocyanide-based MCR together with Ugi reaction—proof of that is the huge number of contributions concerning this key protocol for the obtainment of appealing and biologically active α-acyloxyamides. Here, we have remarked some pivotal examples covering different aspects of this reaction, but due to the high activity registered in this field, we are convinced that in the next years, more complex approaches involving this protocol will be discovered and others such as Ugi-Passerini strategies [72] or Passerini protocol in polymerization reactions [73] will become more usual tools. We have also disclosed some interesting targets synthesized following this MCR as a key synthetic step, which have been the focus of biological studies affording interesting results. Additionally, some interesting environmentally benign protocols have been remarked and more improvements in this respect are expected in the next years with the increasing concern about the sustainability of the processes.

REFERENCES

1 (a) M. Passerini, *Gazz. Chim. Ital.* 1921, 51, 126–129. Sopra gli isonitrili (I). Composto del *p*-isonitrili-azobenzole con acetone ed acido acetica. (b) M. Passerini, *Gazz. Chim. Ital.* 1921, 51, 181–188. Sopra gli isonitrili (II). Composti con aldeidi o con chetoni ed acidi organici monobasici.

2 (a) L. Banfi, R. Riva in *Organic reactions* (A. B. Charette, Ed.). John Wiley & Sons, Inc., Hoboken, NJ, 2005, Vol. 65, pp. 1–140. The Passerini reaction. (b) L. Banfi, R. Riva, A. Basso, *Synlett* 2010, 23–41. Coupling isocyanide-based multicomponent reactions with aliphatic or acyl nucleophilic substitution processes. (c) S. Sadjadi, M. M. Heravi, *Tetrahedron* 2011, 67, 2707–2752. Recent application of isocyanides in synthesis of heterocycles.

3 A. R. Kazemizadeh, A. Ramazani, *Curr. Org. Chem.* 2012, 16, 418–450. Synthetic applications of Passerini reaction.

4 O. Mumm, *Ber. Dstch. Chem. Ges.* 1910, 43, 886–893. Umsetzung von Säureimidchloriden mit Salzen organischer Säure und mit Cyankalium.

5 For a recent computational study, see: S. Maeda, S. Komagawa, M. Uchiyama, K. Morokuma, *Angew. Chem. Int. Ed.* 2011, 50, 644–649. Finding reaction pathways for multicomponent reactions: the Passerini reaction is a four-component reaction.

6 S. C. Jonnalagadda, J. S. Cruz, R. J. Connell, P. M. Scott, V. R. Mereddy, *Tetrahedron Lett.* 2009, 50, 4314–4317. Synthesis of α-carboranyl-α-acyloxy-amides as potential BNCT agents.

7 P. R. Krishna, G. Dayaker, P. V. N. Reddy, *Tetrahedron Lett.* 2006, 47, 5977–5980. Diastereoselective Passerini reactions using *p*-toluenesulfonylmethyl isocyanide (TosMIC) as the isonitrile component.

8 A. R. Kazemizadeh, A. Ramazani, *J. Braz. Chem. Soc.* 2009, 20, 309–312. Passerini multicomponent reaction of indane-1,2,3-trione: an efficient route for the one pot synthesis of sterically congested 2,2-disubstituted indane-1,3-dione derivatives.

9 P. R. Krishna, K. Lopinti, *Synlett* 2007, 83–86. Diastereoselective Passerini reaction of chiral 2,3-epoxy aldehydes with TosMIC.

10 C. G. Neochoritis, A. Dömling, *Org. Biomol. Chem.* 2014, 12, 1649–1651. Towards a facile and convenient synthesis of highly functionalized indole derivatives based on multi-component reactions.

11 A. A. Esmaeili, S. A. Ghalandarabad, S. Jannati, *Tetrahedron Lett.* 2013, 54, 406–408. A novel and efficient synthesis of 3,3-disubstituted indol-2-ones *via* Passerini three-component reactions in the presence of 4 Å molecular sieves.

12 B. O. Beasley, G. J. Clarkson, M. Shipman, *Tetrahedron Lett.* 2012, 53, 2951–2953. Passerini reactions for the efficient synthesis of 3,3-disubstituted oxetanes.

13 L. El Kaim, L. Grimaud, in *Isocyanide chemistry: applications in synthesis and material science* (V. Nenajdenko, Ed.). Wiley-VCH Verlag GmbH, Weinheim, Germany, 2012, pp. 159–194. Ugi and Passerini reactions with carboxylic acid surrogates.

14 (a) L. El Kaim, M. Gizolme, L. Grimaud, *Org. Lett.* 2006, 8, 5021–5023. *O*-arylative Passerine reactions. (b) L. El Kaim, M. Gizolme, L. Grimaud, J. Oble, *J. Org. Chem.* 2007, 72, 4169–4180. Smiles rearrangements in Ugi- and Passerini-type couplings: new multicomponent access to *O*- and *N*-arylamides.

15 (a) L. A. Warren, S. Smiles, *J. Chem. Soc.* 1930, 178, 956–963. *iso*-β-Naphthol sulphide. (b) L. A. Warren, S. Smiles, *J. Chem. Soc.* 1930, 1327–1331. Dehydro-2-naphtholsulphone.

16 For a double Smiles rearrangement of Passerini adducts, see: E. Martinand-Lurin, A. Dos Santos, L. El Kaim, L. Grimaud, P. Retailleau, *Chem. Commun.* 2014, 50, 2214–2217. Double Smiles rearrangement of Passerini adducts towards benzoxazinones.

17 H. Yanai, T. Oguchi, T. Taguchi, *J. Org. Chem.* 2009, 74, 3927–3929. Direct alkylative Passerini reaction of aldehydes, isocyanides, and free aliphatic alcohols catalyzed by indium(III) triflate.

18 T. Soeta, Y. Ukaji, *Chem. Rec.* 2014, 14, 101–116. Carboxylic acid free novel isocyanide-based reactions.

19 T. Soeta, Y. Kojima, Y. Ukaji, K. Inomata, *Org. Lett.* 2010, 12, 4341–4343. *O*-Silylative Passerini reaction: a new one-pot synthesis of α-siloxyamides.

20 T. Soeta, S. Matsuzaki, Y. Ukaji, *Chem.–Eur. J.* 2014, 20, 5007–5012. A one-pot *O*-phosphinative Passerini/Pudovik reaction: efficient synthesis of highly functionalized α-(phosphinyloxy)amide derivatives.

21 (a) R. J. K. Taylor, M. Reid, J. Foot, S. A. Raw, *Acc. Chem. Res.* 2005, 38, 851–869. Tandem oxidation processes using manganese dioxide: discovery, applications, and current studies. (b) L. F. Tietze, G. Brasche, K. M. Gericke (Eds.), *Domino reactions in organic synthesis*. Wiley-VCH Verlag GmbH, Weinheim, Germany, 2006. (c) K. Ekoue-Kovi, C. Wolf, *Chem.–Eur. J.* 2008, 14, 6302–6315. One-pot oxidative esterification and amidation of aldehydes.

22 T. Ngouansavanh, J. Zhu, *Angew. Chem. Int. Ed.* 2006, 45, 3495–3497. Alcohols in isonitrile-based multicomponent reaction: Passerini reaction of alcohols in the presence of *O*-iodoxybenzoic acid.

23 For more recent results, see also: F. de Moliner, M. Bigatti, L. Banfi, R. Riva, A. Basso, *Org. Lett.* 2014, 16, 2280–2283. OPHA (oxidation–Passerini–hydrolysis–alkylation) strategy: a four-step, one-pot improvement of the alkylative Passerini reaction.

24 J. Brioche, G. Masson, J. Zhu, *Org. Lett.* 2010, 12, 1432–1435. Passerini three-component reaction of alcohols under catalytic aerobic oxidative conditions.

25 For other metals, see also: B. Karimia, E. Farhangia, *Adv. Synth. Catal.* 2013, 355, 508–516. One-pot oxidative Passerini reaction of alcohols using a magnetically recyclable TEMPO under metal- and halogen-free conditions.

26 For a review concerning mechanistic discussion on the Cu-TEMPO-catalyzed aerobic oxidation, see: C. Galli, P. Gentili, O. Lanzalunga, *Angew. Chem. Int. Ed.* 2008, 47, 4790–4796. Hydrogen abstraction and electron transfer with aminoxyl radicals: synthetic and mechanistic issues.

27 (a) L. A. Wessjohann, E. Ruijter, D. Garcia-Rivera, W. Brandt, *Mol. Divers* 2005, 9, 171–186. What can a chemist learn from

nature's macrocycles?—a brief, conceptual view. (b) L. A. Wessjohann, C. K. Z. Andrade, O. E. Vercillo, D. G. Rivera in *Targets in heterocyclic systems: chemistry and properties* (O. A. Attanasi, D. Spinelli, Eds.). Societa Chimica Italiana, Italy, 2006, Vol. 10, pp. 24–53. Macrocyclic peptoids: N-alkylated cyclopeptides and depsipeptides. (c) F. von Nussbaum, M. Brands, B. Hinzen, S. Weingand, D. Häbich, *Angew. Chem. Int. Ed.* 2006, 45, 5072–5129. Antibacterial natural products in medicinal chemistry-exodus or revival? (d) E. M. Driggers, S. P. Hale, J. Lee, N. K. Terrett, *Nature* 2008, 7, 608–624. The exploration of macrocycles for drug discovery—an underexploited structural class.

28 (a) A. Parenty, X. Moreau, J.-M. Campagne, *Chem. Rev.* 2006, 106, 911–939. Macrolactonizations in the total synthesis of natural products. (b) S. E. Gibson, C. Lecci, *Angew. Chem. Int. Ed.* 2006, 45, 1364–1377. Amino acid derived macrocycles—an area driven by synthesis or application?

29 M. D. Burke, S. L. Schreiber, *Angew. Chem. Int. Ed.* 2004, 43, 46–58. A planning strategy for diversity-oriented synthesis.

30 For selected examples, see: (a) G. Zhao, X. Sun, H. Bienaymé, J. Zhu, *J. Am. Chem. Soc.* 2001, 123, 6700–6701. Activation of a terminal carboxylic acid by an internal oxazole: a novel synthesis of macrocyclodepsipeptide. (b) P. Janvier, M. Bois-Choussy, H. Bienaymé, J. Zhu, *Angew. Chem. Int. Ed.* 2003, 42, 811–814. A one-pot four-component (ABC_2) synthesis of macrocycles. (c) L. A. Wessjohann, E. Ruijter, *Mol. Divers* 2005, 9, 159–169. Macrocycles rapidly produced by multiple multicomponent reactions including bifunctional building blocks (MiBs). (d) C. Bughin, G. Zhao, H. Bienaymé, J. Zhu, *Chem.–Eur. J.* 2006, 12, 1174–1184. 5-Aminooxazole as an internal traceless activator of C-terminal carboxylic acid: rapid access to diversely functionalized cyclodepsipeptides. (e) T. Pirali, G. C. Tron, J. Zhu, *Org. Lett.* 2006, 8, 4145–4148. One-pot synthesis of macrocycles by a tandem three-component reaction and intramolecular [3 + 2] cycloaddition.

31 (a) T. D. Owens, G.-L. Araldi, R. F. Nutt, J. E. Semple, *Tetrahedron Lett.* 2001, 42, 6271–6274. Concise total synthesis of the prolyl endopeptidase inhibitor eurystatin A *via* a novel Passerini reaction–deprotection–acyl migration strategy. (b) B. Beck, G. Larbig, B. Mejat, M. Magnin-Lachaux, A. Picard, E. Herdtweck, A. Dömling, *Org. Lett.* 2003, 5, 1047–1050. Short and diverse route toward complex natural product-like macrocycles. (c) S. Marcaccini, T. Torroba, in *Multicomponent reactions* (J. Zhu, H. Bienaymé, Eds.), 1st ed. Wiley-VCH Verlag GmbH, Weinheim, Germany, 2005, pp. 33–75. Post-condensation modifications of the Passerini and Ugi reactions.

32 F. Leon, D. G. Rivera, L. A. Wessjohann, *J. Org. Chem.* 2008, 73, 1762–1767. Multiple multicomponent macrocyclizations including bifunctional building blocks (MiBs) based on Staudinger and Passerini three-component reactions.

33 (a) L. Banfi, A. Basso, G. Guanti, R. Riva in *Multicomponent reactions* (J. Zhu, H. Bienaymé, Eds.), 1st ed. Wiley-VCH Verlag GmbH: Weinheim, Germany, 2005, pp. 1–33. Asymmetric isocyanide-based MCRs. (b) B. Alcaide, P. Almendros, C. Aragoncillo, R. Callejo, M. P. Ruiz, *J. Org. Chem.* 2013, 78, 10154–10165. Organocatalyzed three-component Ugi and Passerini reactions of 4-oxoazetidine-2-carbaldehydes and azetidine-2,3-diones. Application to the synthesis of γ-lactams and γ-lactones.

34 (a) S. S. van Berkel, B. G. M. Bögels, M. A. Wijdeven, B. Westermann, F. P. J. T. Rutjes, *Eur. J. Org. Chem.* 2012, 3543–3559. Recent advances in asymmetric isocyanide-based multicomponent reactions. (b) Q. Wang, L. Zhu, M.-X. Wang in *Asymmetric synthesis II* (M. Christmann, S. Bräse, Eds.). Wiley-VCH Verlag GmbH, Weinheim, Germany, 2012, pp. 95–101. Enantioselective Passerini reaction.

35 (a) S. E. Denmark, Y. Fan, *J. Am. Chem. Soc.* 2003, 125, 7825–7827. The first catalytic, asymmetric α-additions of isocyanides Lewis-base-catalyzed, enantioselective Passerini-type reactions. (b) S. Denmark, Y. Fan, *J. Org. Chem.* 2005, 70, 9667–9676. Catalytic, enantioselective α-additions of isocyanides: Lewis base catalyzed Passerini-type reactions.

36 U. Kusebauch, B. Beck, K. Messer, E. Herdtweck, A. Dömling, *Org. Lett.* 2003, 5, 4021–4024. Massive parallel catalyst screening: towards asymmetric MCRs.

37 P. R. Andreana, C. C. Liu, S. L. Schreiber, *Org. Lett.* 2004, 6, 4231–4233. Stereochemical control of the Passerini reaction.

38 S.-X. Wang, M.-X. Wang, D.-X. Wang, J. Zhu, *Angew. Chem. Int. Ed.* 2008, 47, 388–391. Catalytic enantioselective Passerini three-component reaction.

39 T. Yue, M.-X. Wang, D.-X. Wang, J. Zhu, *Angew. Chem. Int. Ed.* 2008, 47, 9454–9457. Asymmetric synthesis of 5-(1-hydroxyalkyl)tetrazoles by catalytic enantioselective Passerini-type reactions.

40 For some previous examples, see: (a) L. Banfi, G. Guanti, R. Riva, *Chem. Commun.* 2000, 985–986. Passerini multicomponent reaction of protected α-aminoaldehydes as a tool for combinatorial synthesis of enzyme inhibitors. (b) T. D. Owens, J. E. Semple, *Org. Lett.* 2001, 3, 3301–3304. Atom-economical synthesis of the N(10)–C(17) fragment of cyclotheonamides *via* a novel Passerine reaction-deprotection-acyl migration strategy. (c) L. Banfi, G. Guanti, R. Riva, A. Basso, E. Calcagno, *Tetrahedron Lett.* 2002, 43, 4067–4069. Short synthesis of protease inhibitors *via* modified Passerini condensation of *N*-Boc-α-aminoaldehydes. Tetrahedron Lett. 2002, 43, 4067–4069. (d) A. Basso, L. Banfi, R. Riva, P. Piaggio, G. Guanti, *Tetrahedron Lett.* 2003, 44, 2367–2370. Solid-phase synthesis of modified oligopeptides *via* Passerini multicomponent reaction. (e) L. Banfi, A. Basso, G. Guanti, R. Riva, *Mol. Divers* 2003, 6, 227–235. Passerini reaction–amine deprotection–acyl migration (PADAM): a convenient strategy for the solid-phase preparation of peptidomimetic compounds. (f) F. Morana, A. Basso, R. Riva, V. Rocca, L. Banfi, *Chem.–Eur. J.* 2013, 19, 4563–4569. The homo-PADAM protocol: stereoselective and operationally simple synthesis of α-oxo- or α-hydroxy-γ-acylaminoamides and chromanes.

41 For selected examples: (a) Z. Li, A.-C. Ortega-Vilain, G. S. Patil, D.-L. Chu, J. E. Foreman, D. D. Eveleth, J. C. Powers, *J. Med. Chem.* 1996, 39, 4089–4098. Novel peptidyl α-keto amide inhibitors of calpains and other cysteine proteases. (b) J. Cacciola, R. S. Alexander, J. M. Fevig, P. F. W. Stouten, *Tetrahedron Lett.* 1997, 38, 5741–5744. The synthesis of

lysine α-ketoamide thrombin inhibitor *via* an epoxy amide ring opening. (c) S. L. Harbeson, S. M. Abelleira, A. Akiyama, R. Barrett III, R. M. Carroll, J. A. Straub, J. N. Tkacz, C. Wu, G. F. Musso, *J. Med. Chem.* 1994, 37, 2918–2929. Stereospecific synthesis of peptidyl α-keto amides as inhibitors of calpain. (d) O. Pando, S. Stark, A. Denkert, A. Porzel, R. Preusentanz, L. A. Wessjohann, *J. Am. Chem. Soc.* 2011, 133, 7692–7695. The multiple multicomponent approach to natural product mimics: tubugis, *N*-substituted anticancer peptides with picomolar activity. (e) D. Gravestock, A. L. Rousseau, A. C. U. Lourens, H. C. Hoppe, L. A. Nkabinde, M. L. Bode, *Tetrahedron Lett.* 2012, 53, 3225–3229. Novel branched isocyanides as useful building blocks in the Passerini-amine deprotection-acyl migration (PADAM) synthesis of potential HIV-1 protease inhibitors.

42 Selected examples: (a) B. E. Evans, K. E. Rittle, M. G. Bock, C. D. Bennett, R. M. DiPardo, J. Boger, M. Poe, E. H. Ulm, B. I. LaMont, E. H. Blaine, G. M. Fanelli, I. I. Stabilito, D. F. Veber, *J. Med. Chem.* 1985, 28, 1756–1759. *N*-(1,3,4,6,7,12b-hexahydro-2*H*-benzo[*b*]furo[2,3*a*]quinolizin-2-yl)-*N*-methyl-2-hydroxyethanesulfonamide: a potent and selective α_2-adrenoceptor antagonist. (b) T. Mimoto, J. Imai, S. Tanaka, N. Hattori, S. Kisanuki, K. Akaji, Y. Kiso, *Chem. Pharm. Bull.* 1991, 39, 3088–3090. KNI-102, a novel tripeptide HIV protease inhibitor containing allophenylnorstatine as a transition-state mimic.

43 L. A. Wessjohann, C. R. B. Rhoden, D. G. Rivera, O. E. Vercillo, *Top. Heterocycl. Chem.* 2010, 23, 199–226. Cyclic peptidomimetics and pseudopeptides from multicomponent reactions.

44 T. Nixey, C. Hulme, *Tetrahedron Lett.* 2002, 43, 6833–6835. Rapid generation of *cis*-constrained norstatine analogs using a $TMSN_3$-modified Passerini MCC/*N*-capping strategy.

45 A. Y. Shaw, F. Medda, C. Hulme, *Tetrahedron Lett.* 2012, 53, 1313–1315. Facile and rapid route for the synthesis of novel norstatine analogs *via* PADAM-cyclization methodology.

46 W. Szymanski, M. Zwolinska, S. Klossowski, I. Młynarczuk-Biały, Ł. Biały, T. Issat, J. Malejczyk, R. Ostaszewski, *Bioorg. Med. Chem.* 2014, 22, 1773–1781. Synthesis of novel, peptidic kinase inhibitors with cytostatic/cytotoxic activity.

47 S. Faure, T. Hjelmgaard, S. P. Roche, D. J. Aitken, *Org. Lett.* 2009, 11, 1167–1170. Passerini reaction-amine deprotection-acyl migration peptide assembly: efficient formal synthesis of cyclotheonamide C.

48 S. Shaaban, R. Diestel, B. Hinkelmann, Y. Muthukumar, R. P. Verma, F. Sasse, C. Jacob, *Eur. J. Med. Chem.* 2012, 58, 192–205. Novel peptidomimetic compounds containing redox active chalcogens and quinones as potential anticancer agents.

49 L. Kou, M.-J. Wang, L.-T. Wang, X.-B. Zhao, X. Nan, L. Yang, Y.-Q. Liu, S. L. Morris-Natschke, K.-H. Lee, *Eur. J. Med. Chem.* 2014, 75, 282–288. Toward synthesis of third-generation spin-labeled podophyllotoxin derivatives using isocyanide multicomponent reactions.

50 A. Znabet, M. M. Polak, E. Janssen, F. J. J. de Kanter, N. J. Turner, R. V. A. Orru, E. Ruijter, *Chem. Commun.* 2010, 46, 7918–7920. A highly efficient synthesis of telaprevir by strategic use of biocatalysis and multicomponent reactions.

51 (a) J. A. Monn, M. J. Valli, *J. Org. Chem.* 1994, 59, 2773–2778. A concise, stereocontrolled thiazolium ylide approach to kainic acid. (b) Y. Yip, F. Victor, J. Lamar, R. B. Johnson, Q. M. Wang, D. Barket, J. Glass, L. Jin, L. Liu, D. Venable, M. Wakulchik, C. Xie, B. Heinz, E. Villarreal, J. Colacino, N. Yumibe, M. Tebbe, J. Munroe, S.-H. Chen, *Bioorg. Med. Chem. Lett.* 2004, 14, 251–256. Discovery of a novel bicycloproline P2 bearing peptidyl α-ketoamide LY514962 as HCV protease inhibitor.

52 S. Wu, J.-C. Liu, L. Wang, M.-W. Ding, *Synlett* 2011, 2880–2882. Facile synthesis of 5-carboxamide-oxazolines *via* a Passerini 3CC–Staudinger–aza-Wittig sequence.

53 P. He, J. Wu, Y.-B. Nie, M.-W. Ding, *Tetrahedron* 2009, 65, 8563–8570. New efficient synthesis of 4-aminocarbonyl substituted 4*H*-3,1-benzoxazines by a Passerini 3CC/Staudinger/aza-Wittig sequence.

54 A. Schwäblein, J. Martens, *Eur. J. Org. Chem.* 2011, 4335–4344. First synthesis of α,β-unsaturated lactones with high diversity through the Passerini reaction and ring-closing metathesis (RCM).

55 F. de Moliner, S. Crosignani, L. Banfi, R. Riva, A. Basso, *J. Comb. Chem.* 2010, 12, 613–616. Synthesis of 5-carboxamide-oxazolines with a Passerini-Zhu/Staudinger-aza-Wittig two-step protocol.

56 F. de Moliner, S. Crosignani, A. Galatini, R. Riva, A. Basso, *ACS Comb. Sci.* 2011, 13, 453–457. Novel application of α-azido aldehydes in multicomponent reactions: synthesis of triazolo-fused dihydrooxazinones *via* a Passerini reaction-dipolar cycloaddition strategy.

57 For interesting methods developed in water, see: (a) M. C. Pirrung, K. D. Sarma, *J. Am. Chem. Soc.* 2004, 126, 444–445. Multicomponent reactions are accelerated in water. (b) M. C. Pirrung, K. D. Sarma, J. Wang, *J. Org. Chem.* 2008, 73, 8723–8730. Hydrophobicity and mixing effects on select heterogeneous, water-accelerated synthetic reactions. (c) N. Shapiro, A. Vigalok, *Angew. Chem. Int. Ed.* 2008, 47, 2849–2852. Highly efficient organic reactions "on water", "in water", and both. (d) M. M. Hooper, B. DeBoef, *J. Chem. Edu.* 2009, 86, 1077–1079. A green multicomponent reaction for the organic chemistry laboratory: the aqueous Passerini reaction. (e) T. Sela, A. Vigalok, *Org. Lett.* 2014, 16, 1964–1967. Organic synthesis "on water" *vs* "on liquids": a comparative analysis.

58 T. Sela, A. Vigalok, *Adv. Synth. Catal.* 2012, 354, 2407–2411. Salt-controlled selectivity in "on water" and "in water" Passerini-type multicomponent reactions.

59 (a) S. S. Pawar, U. Phalgune, A. Kumar, *J. Org. Chem.* 1999, 64, 7055–7060. Ionic effects on the stereoselectivities of Diels-Alder reactions: the cyclopentadiene–methyl acrylate reaction in aqueous LiCl, NaCl, NaBr, $CaCl_2$, $LiClO_4$, and guanidinium chloride. (b) D. Sarma, A. Kumar, *Org. Lett.* 2006, 8, 2199–2202. Hydrophobic effects are dominant over secondary orbital interactions for a simple Diels-Alder reaction in salt solutions.

60 (a) J. Dupont, R. F. de Souza, P. A. Z. Suarez, *Chem. Rev.* 2002, 102, 3667–3692. Ionic liquid (molten salt) phase organometallic catalysis. (b) H. Zhao, S. V. Malhotra, *Aldrichimica Acta* 2002, 35, 75–83. Applications of ionic liquids in organic

synthesis. (c) P. Wasserscheid, T. Welton (Eds.), *Ionic liquids in synthesis*. Wiley-VCH Verlag GmbH, Weinheim, Germany, 2003. (d) C. K. Z. Andrade, L. M. Alves, *Curr. Org. Chem.* 2005, 9, 195–218. Environmentally benign solvents in organic synthesis: current topics.

61 C. K. Z. Andrade, S. C. S. Takada, P. A. Z. Suarez, M. B. Alves, *Synlett* 2006, 1539–1542. Revisiting the Passerini reaction under eco-friendly reaction conditions.

62 For contemporaneous works, see also: (a) X. Fan, Y. Li, X. Zhang, G. Qu, J. Wang, *Can. J. Chem.* 2006, 84, 794–799. A novel and green version of the Passerini reaction in an ionic liquid ([bmim][BF_4]). (b) X. Y. Zhang, Y. Z. Li, X. S. Fan, G. R. Qu, J. J. Wang, X. Y. Hu, *Chin. Chem. Lett.* 2006, 17, 578–580. Ionic liquid [bmim][BF_4] as a green medium for Passerini reaction.

63 For some examples: (a) C. Cui, C. Zhu, X.-J. Du, Z.-P. Wang, Z.-M. Li, W.-G. Zhao, *Green Chem.* 2012, 14, 3157–3163. Ultrasound-promoted sterically congested Passerini reactions under solvent-free conditions. (b) K. Sato, T. Ozu, N. Takenaga, *Tetrahedron Lett.* 2013, 54, 661–664. Solvent-free synthesis of azulene derivatives *via* Passerini reaction by grinding.

64 T. Bousquet, M. Jida, M. Soueidan, R. Deprez-Poulain, F. Agbossou-Niedercorn, L. Pelinski, *Tetrahedron Lett.* 2012, 53, 306–308. Fast and efficient solvent-free Passerini reaction.

65 D. Koszelewski, W. Szymanski, J. Krysiak, R. Ostaszewski, *Synth. Commun.* 2008, 28, 1120–1127. Solvent-free Passerini reactions.

66 For important reviews about 3,3′-disubstituted oxindole structural motif, see: (a) F. Zhou, Y.-L. Liu, J. Zhou, *Adv. Synth. Catal.* 2010, 352, 1381–1407. Catalytic asymmetric synthesis of oxindoles bearing a tetrasubstituted stereocenter at the C-3 position. (b) J. J. Badillo, N. V. Hanhan, A. K. Franz, *Curr. Opin. Drug Discov. Dev.* 2010, 13, 758–776. Enantioselective synthesis of substituted oxindoles and spirooxindoles with applications in drug discovery. (c) A. Kumar, S. S. Chimni, *RSC Advances* 2012, 2, 9748–9762. Catalytic asymmetric synthesis of hydroxyoxindole: a potentially bioactive molecule. (d) S. Mohammadi, R. Heiran, R. P. Herrera, E. Marqués-López, *ChemCatChem* 2013, 5, 2131–2148. Isatin as a strategic motif for asymmetric catalysis.

67 T. Kaicharla, S. R. Yetra, T. Roy, A. T. Biju, *Green Chem.* 2013, 15, 1608–1614. Engaging isatins in solvent-free, sterically congested Passerini reaction.

68 C. O. Kappe, A. Stadler (Eds.), *Microwaves in organic and medicinal chemistry*. Wiley-VCH Verlag GmbH, Weinheim, Germany, 2005.

69 (a) C. O. Kappe, *Chem. Soc. Rev.* 2008, 37, 1127–1139. Microwave dielectric heating in synthesis organic chemistry. (b) B. Jiang, F. Shi, S.-J. Tu, *Curr. Org. Chem.* 2010, 14, 357–378. Microwave-assisted multicomponent reactions in the heterocyclic chemistry.

70 H. M. Hügel, *Molecules* 2009, 14, 4936–4972. Microwave multicomponent synthesis.

71 A. F. S. Barreto, O. E. Vercillo, C. K. Z. Andrade, *J. Braz. Chem. Soc.* 2011, 22, 462–467. Microwave-assisted Passerini reactions under solvent-free conditions.

72 (a) F. C. S. C. Pinto, S. M. M. A. Pereira-Lima, C. Ventura, L. Albuquerque, R. Gonçalves-Maia, H. L. S. Maia, *Tetrahedron* 2006, 62, 8184–8198. Synthesis and kinetic investigation of the selective acidolysis of *para*-substituted *N*-benzyl- or *N*-phenyl-*N*-phenylacetyla, α,α-dialkylglycine cyclohexylamides. (b) F. C. S. C. Pinto, S. M. M. A. Pereira-Lima, H. L. S. Maia, *Tetrahedron* 2009, 65, 9165–9179. Straightforward, racemization-free synthesis of peptides with fairly to very bulky di- and trisubstituted glycines. (c) R. Mossetti, T. Pirali, G. C. Tron, *J. Org. Chem.* 2009, 74, 4890–4892. Synthesis of Passerini-Ugi hybrids by a four-component reaction using the glycolaldehyde dimer.

73 (a) O. Kreye, T. Tóth, M. A. R. Meier, *J. Am. Chem. Soc.* 2011, 133, 1790–1792. Introducing multicomponent reactions to polymer science: Passerini reactions of renewable monomers. (b) L. Li, A. Lv, X.-X. Deng, F.-S. Du, Z.-C. Li, *Chem. Commun.* 2013, 49, 8549–8551. Facile synthesis of photo-cleavable polymers *via* Passerini reaction. (c) S. C. Solleder, M. A. R. Meier, *Angew. Chem. Int. Ed.* 2014, 53, 711–714. Sequence control in polymer chemistry through the Passerini three-component reaction.

9

BIGINELLI MULTICOMPONENT REACTIONS

Eugenia Marqués-López and Raquel P. Herrera

Departamento de Química Orgánica, Instituto de Síntesis Química y Catálisis Homogénea (ISQCH), CSIC-Universidad de Zaragoza, Zaragoza, Spain

9.1 INTRODUCTION

Biginelli reaction is one of the most remarkable examples of multicomponent reactions (MCRs) discovered in 1891 [1, 2], and proof of its importance is the high number of articles reported in the last decade covering this field. Very recently, an amazing and comprehensive review has appeared covering this reaction with more than 500 references [3]. In this chapter, we will just cover only some aspects overlooked in the incredible revision performed by Suresh and Sandhu and reported since 2006.

Although this approach has remained almost unexplored until the 1980s, considerable interest has been registered over the past decade. Biginelli reaction allows the synthesis of important building blocks and versatile synthons such as 3,4-dihydropyrimidin-2-(1*H*)-one (DHPM) derivatives [4], which are present motives in organic synthesis owing to their biological and pharmacological properties [5]. DHPMs can act as antihypertensive, antiviral, antibacterial, anti-inflammatory, or anticancer agents and potent calcium channel blockers, as examples of the wide range of biological activities that they can exhibit [6]. Moreover, some marine alkaloids containing a dihydropyrimidinone-5-carboxylate core (the batzelladine alkaloids) have been isolated, and these alkaloids were found to be potent HIV gp-120-CD4 inhibitors [7]. Furthermore, DHPMs could be obtained as chiral compounds, and the control of the stereochemistry at C(4) is crucial to determine their biological properties as in the structures disclosed in Figure 9.1, where the enantiomer (*R*)-SQ 32926 **9** exhibits an antihypertensive effect [16], or the different anticancer activities showed by (*S*)-monastrol **7** [17] against its (*R*)-enantiomer, or the properties revealed by (*R*)-mon97 **8** [14], among others.

Consequently, the discovery of milder and practical routes for the synthesis of these chiral dihydropyrimidin-2(1*H*)-ones is an active field of research, and it continues to attract the attention of a great number of research groups. Several improved procedures for the preparation of these DHPMs *via* a Biginelli protocol have been well documented, and these involve variations of the former reaction conditions, and we will briefly discuss in this chapter some of the most remarkable achievements reached over the last years in this field.

The original Biginelli condensation [18] involving the reaction of aldehydes **10**, urea **11**, and β-ketoesters **12** under strong acidic conditions to give 3,4-dihydropyrimidin-2-ones **13** often suffers from low yields when aliphatic and substituted aromatic aldehydes are employed. Several modifications and improvements concerning this aspect have resulted in the discovery of milder and more efficient procedures and active catalysts (Scheme 9.1).

9.2 MECHANISM

In spite of the importance of the Biginelli reaction, its mechanism is still under discussion. Thus, the original accepted mechanism initially proposed by Folkers and Johnson [19] was more recently supported by Kappe [20] (Scheme 9.2).

In this proposal, nucleophilic addition of urea **11** to benzaldehyde **14** affords intermediate *N*-(1-hydroxybenzyl)-urea **15**, which undergoes rapid dehydration in the presence of acid to lead to a highly reactive iminium ion **16**. In the

Multicomponent Reactions: Concepts and Applications for Design and Synthesis, First Edition. Edited by Raquel P. Herrera and Eugenia Marqués-López.

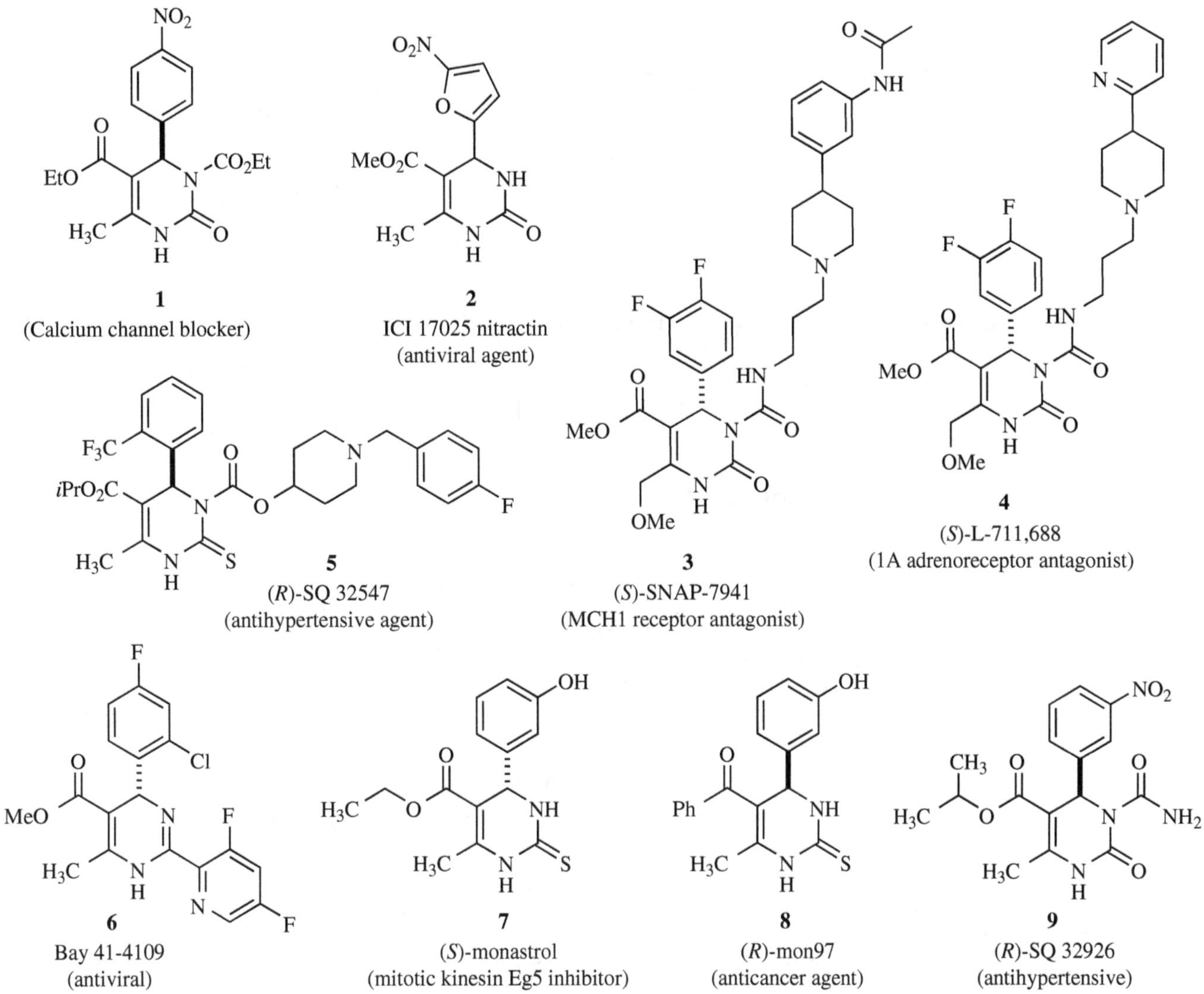

FIGURE 9.1 Biologically active DHPM structures: **1** [8], **2** [9], **3** [10], **4** [11], **5** [8], **6** [12], **7** [13], **8** [14], and **9** [15].

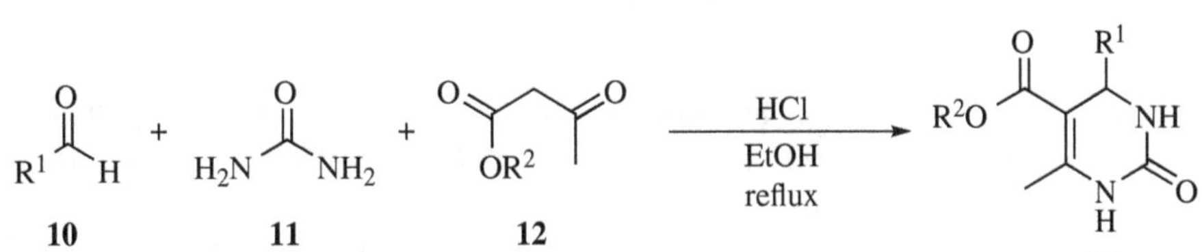

13

SCHEME 9.1 Original Biginelli reaction.

absence of the third component, 1,3-dicarbonyl compound, a second equivalent of urea, would be added to **16**, rendering bis-ureide **17**. However, if 1,3-dicarbonyl compound is present in the reaction medium, the electron-deficient iminium ion **16** is trapped by the corresponding enol tautomer to furnish intermediates **18**, which then would cyclize to the desired Biginelli adducts **19**. The first addition of urea to the aldehyde is assumed to be the rate-determining step, and the subsequent acid-catalyzed dehydration, the addition of a second equivalent of urea to the iminium ion, and the addition of 1,3-dicarbonyl compound to the iminium ion **16** are considered faster steps. This fact is in agreement with NMR experiments performed under typical Biginelli conditions. This mechanism has been further supported by other authors using mass spectrometry and density functional theory calculations [21].

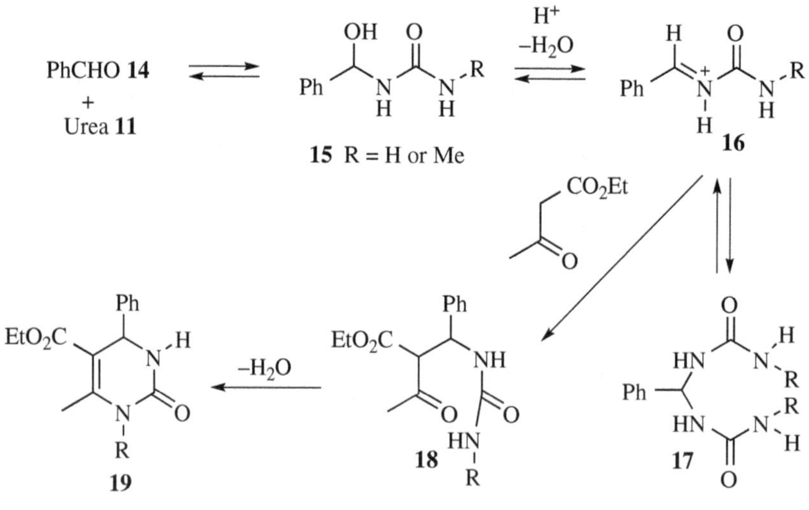

SCHEME 9.2 Currently accepted Biginelli mechanism.

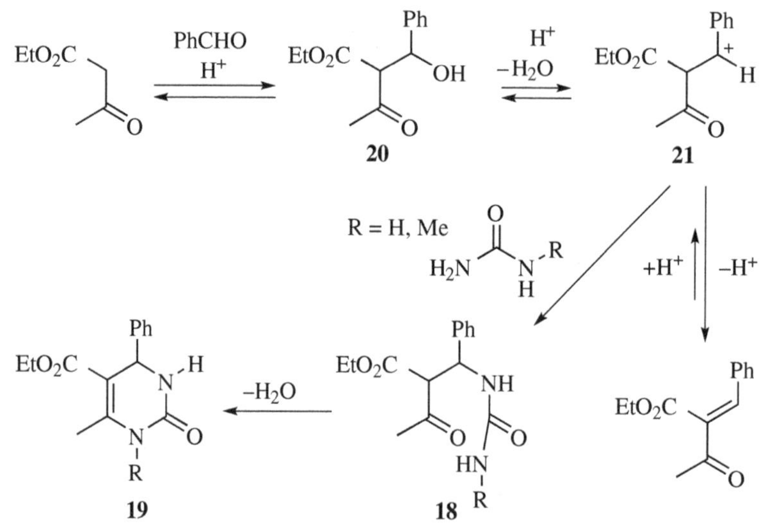

SCHEME 9.3 Sweet and Fissekis' Biginelli mechanism.

Kappe proposed the aforementioned mechanism (Scheme 9.2) [20] against a previous proposal reported by Sweet and Fissekis in 1973 [22] and disclosed in Scheme 9.3.

The authors suggested that the first rate-limiting step was an acid-catalyzed aldol condensation between the 1,3-dicarbonyl compound and the aldehyde affording aldol-type product **20**. The following acid-catalyzed dehydration would furnish stabilized carbocation **21**. Next, the urea would trap the carbocation **21** rendering the same intermediate **18**, previous to the final cyclization to furnish Biginelli products **19**.

Therefore, the discussions concerning the mechanism are still open, and in the future, more works supporting one of these mechanisms or even new ones are expected.

9.3 CHIRAL LEWIS- AND BRØNSTED ACID-CATALYZED BIGINELLI REACTIONS

The first acid-catalyzed (HCl/EtOH) one-pot Biginelli procedure suffered from some disadvantages such as long reaction times and low yields, mainly when aliphatic and substituted aromatic aldehydes were used. In this context, other methodologies have been further developed, especially in the last few years, with the aim of improving the results and overcoming these problems. For such a purpose, catalytic procedures using Lewis acids have been continuously explored in the last decade such as $LnCl_3{\cdot}7H_2O$ [23], $Cu(NH_2SO_3)_2$ [24], $Cu(BF_4)_2{\cdot}xH_2O$ [25], $Yb(OTf)_3$ [26], $TaBr_5$ [27], $RuCl_3$ [28], $Y(NO_3)_3{\cdot}6H_2O$ [29], $TiCl_4$-$MgCl_2$ [30], and $Cu(OAc)_2$ [31]. Other works have been devoted to the use of heterogeneous catalysts [32], ionic liquids [33], microwave irradiation [34], ultrasonic mediated methods [35], and other environmentally benign processes [36].

In spite of the importance of this reaction and the number of outcomes reached in this area, the chiral metal version of this process has been shortly reported. In this respect, the first highly enantioselective example was developed by Zhu and coworkers using a new chiral ytterbium **22** catalyst (Scheme 9.4) [37, 38].

This pioneering catalytic approach gives access to a wide range of optically active dihydropyrimidones with remarkable

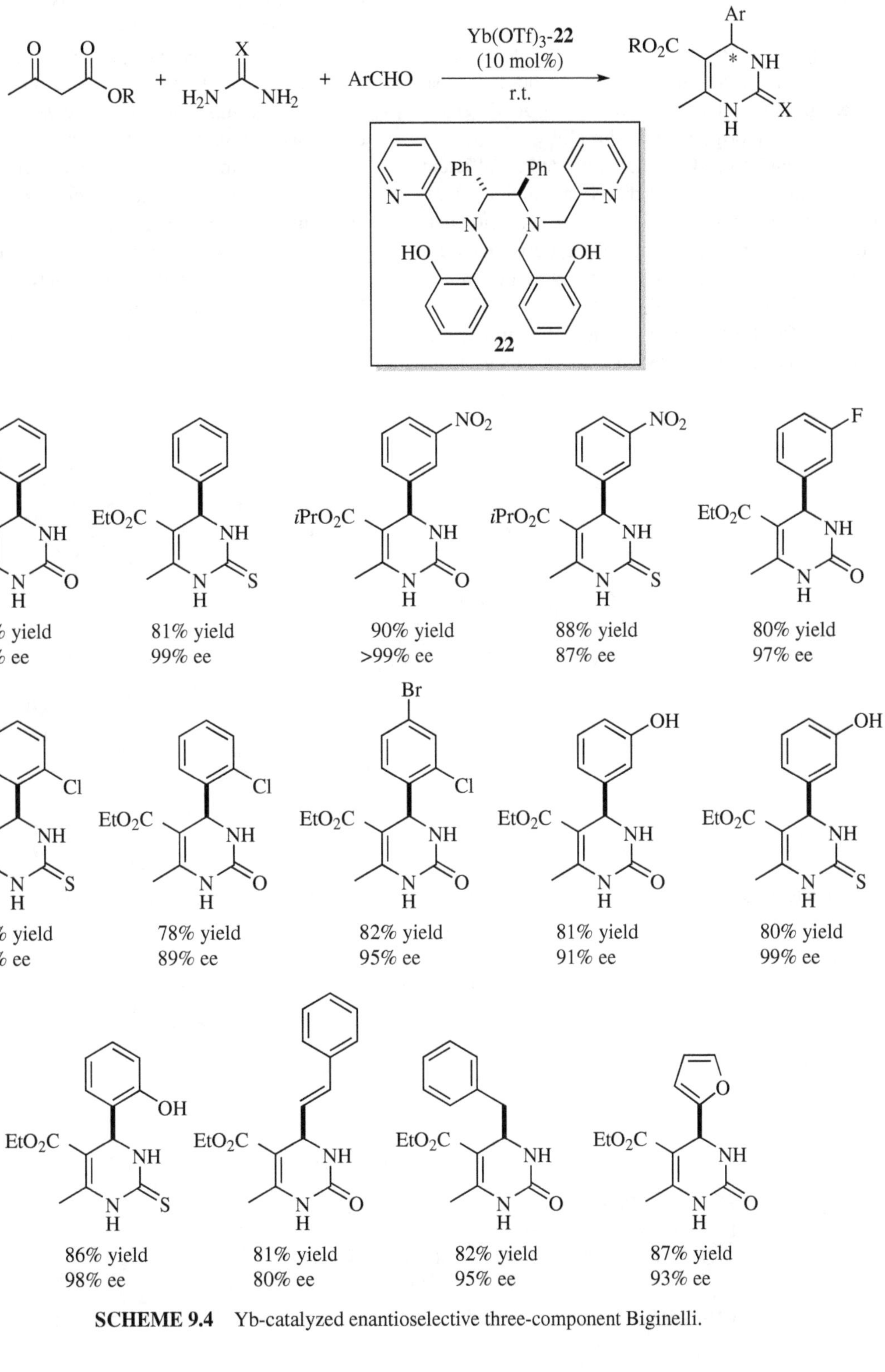

SCHEME 9.4 Yb-catalyzed enantioselective three-component Biginelli.

pharmacological interest in high yields and with good to excellent enantioselectivities under mild conditions.

9.4 BRØNSTED BASE-CATALYZED ONE-POT THREE-COMPONENT BIGINELLI-TYPE REACTIONS

Although the original Biginelli reaction is an acid-catalyzed process, more recently, base-catalyzed approaches have also gained interest because of their moderate to good results compared to that of the original procedure [39].

Remarkably, Ji and coworkers have reported an efficient one-pot synthesis of 4,5,6-triaryl-3,4-dihydropyrimidin-2(1*H*)-ones *via* a three-component Biginelli-type condensation between aldehydes **10**, 2-phenylacetophenone derivatives **23**, and urea **11**/thiourea **24** in the presence of a catalytic amount of *t*-BuOK (20 mol%) as described in Scheme 9.5 [39c].

The reactions afford the desired Biginelli dihydropyrimidin-2(1*H*)-one/thione products **25** and **26** in moderate to good yields at 70°C. Interestingly, the authors performed additional experiments in order to give a plausible mechanistic explanation. Moreover, two different pathways were proposed depending on the (thio)urea **11/24** used in the reaction medium (Schemes 9.6 and 9.7). The key step proceeds through enone **27** when thiourea **24** is used (Scheme 9.6), and bis-urea **28** was suggested as the key intermediate for reactions involving urea **11** (Scheme 9.7).

However, the authors prepared the bis-ureide **28** under acidic conditions to carry out the additional experiments, and subsequently, this isolated bis-ureide was reacted with different substrates under basic conditions to form the Biginelli product. This ambiguous proof with no details concerning the formation of bis-ureide **28** under basic conditions promoted an additional MS spectrometry study reported by Santhosh and coworkers [40] in order to clarify if other plausible pathways were able for this base-catalyzed Biginelli reaction. However, the authors were not able to find, under the same reaction conditions, the peak for the corresponding bis-ureide intermediate **28** proposed by Ji and coworkers [39c].

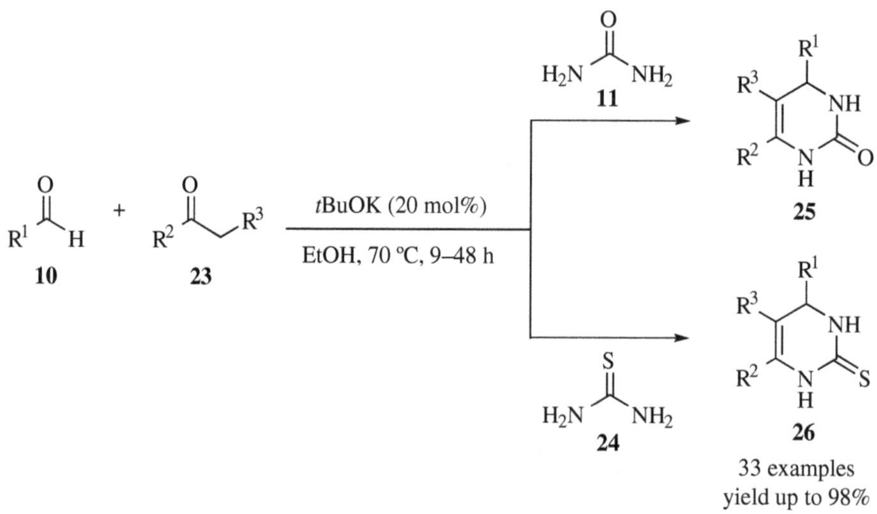

SCHEME 9.5 Base-catalyzed synthesis of 4,5,6-triaryl-3,4-dihydropyrimidin-2(1*H*)-ones/thiones **25** and **26**.

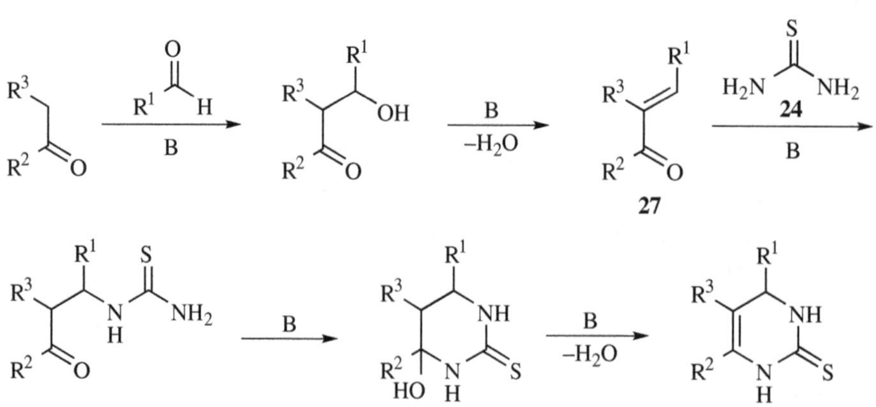

SCHEME 9.6 Proposed Biginelli reaction mechanism using thiourea **24** as substrate.

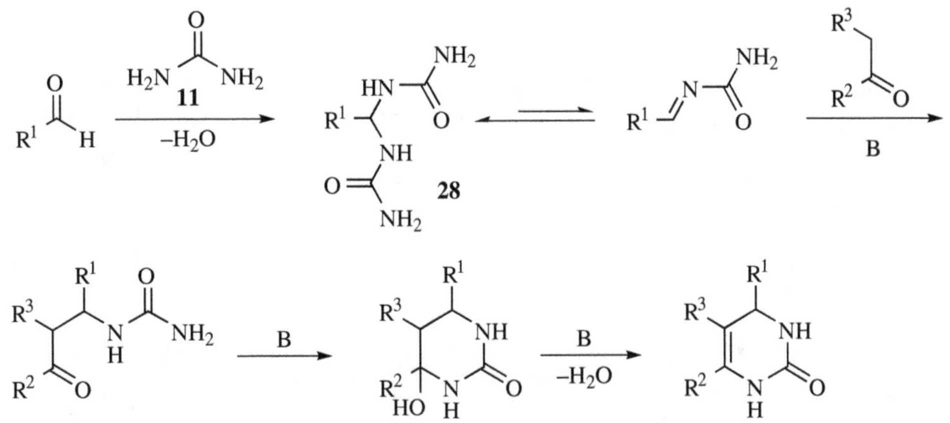

SCHEME 9.7 Proposed Biginelli reaction mechanism using urea **10** as substrate.

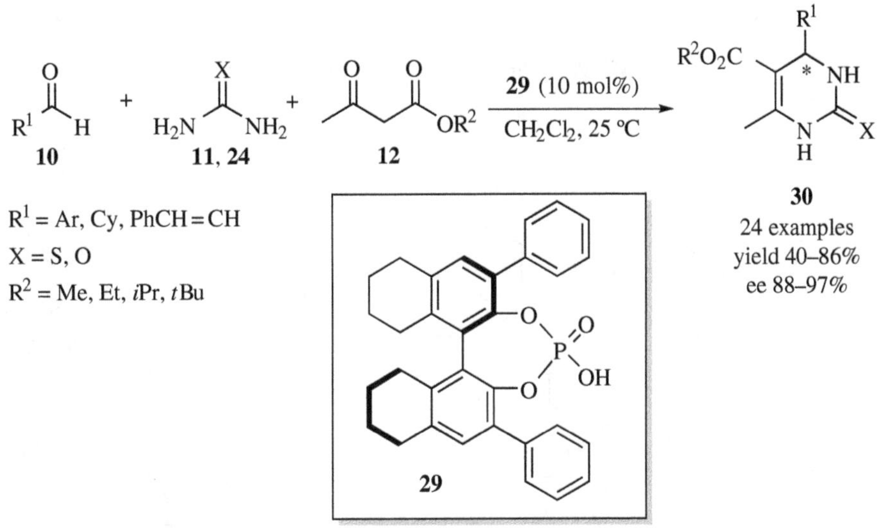

SCHEME 9.8 Phosphoric acid-catalyzed Biginelli reaction.

9.5 ORGANOCATALYTIC ENANTIOSELECTIVE BIGINELLI REACTIONS

The Biginelli reaction has also been the focus of an increased interest in the area of asymmetric organocatalysis especially for the development of more efficient procedures and rendering new chiral examples of this process [41]. Great improvements have been reached in this field [42, 43] compared with those obtained in the chiral metal version that has been little explored [37, 38].

9.5.1 Chiral Brønsted Acid-Organocatalyzed Biginelli Reactions

Gong and coworkers reported the first enantioselective organocatalytic Biginelli reaction (Scheme 9.8) [44]. Thus, chiral phosphoric acid **29** efficiently activated the process between an aldehyde **10**, a thiourea **24** or urea **11**, and acetoacetates **12**, furnishing highly functionalized 3,4-dihydropyrimidin-2-(1*H*)-ones or the respective thio-analogues **30** enantiomerically enriched.

Interestingly, the authors also demonstrated the synthetic application of this catalytic asymmetric Biginelli methodology by the preparation of the active pharmaceutical ingredient monastrol **7** in two steps and with high optical purity (91% *ee*) from TBS-protected 3-hydroxybenzaldehyde, thiourea, and ethyl acetoacetate.

The same authors envisioned that a phosphoric acid could effectively activate the formation of the chiral iminium ion pair intermediate **31** and induce the chirality in the subsequent Mannich-type addition of the acetoacetates **12** over the azomethine carbon of compound **31** to generate optically active adducts **32** (Scheme 9.9). This hypothesis was based on previous reported results since chiral phosphoric acids have been found to act as promoters for asymmetric additions of a variety of nucleophiles to

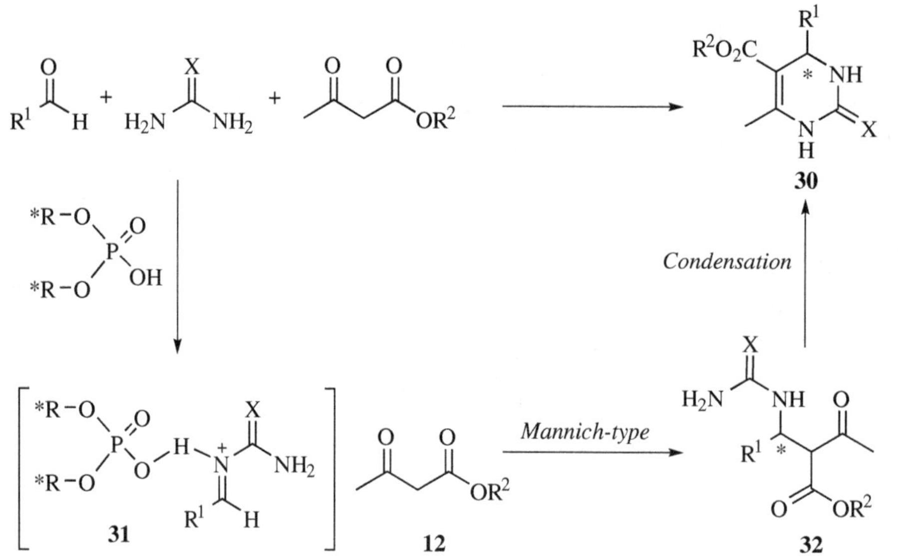

SCHEME 9.9 Proposed organocatalytic mechanism.

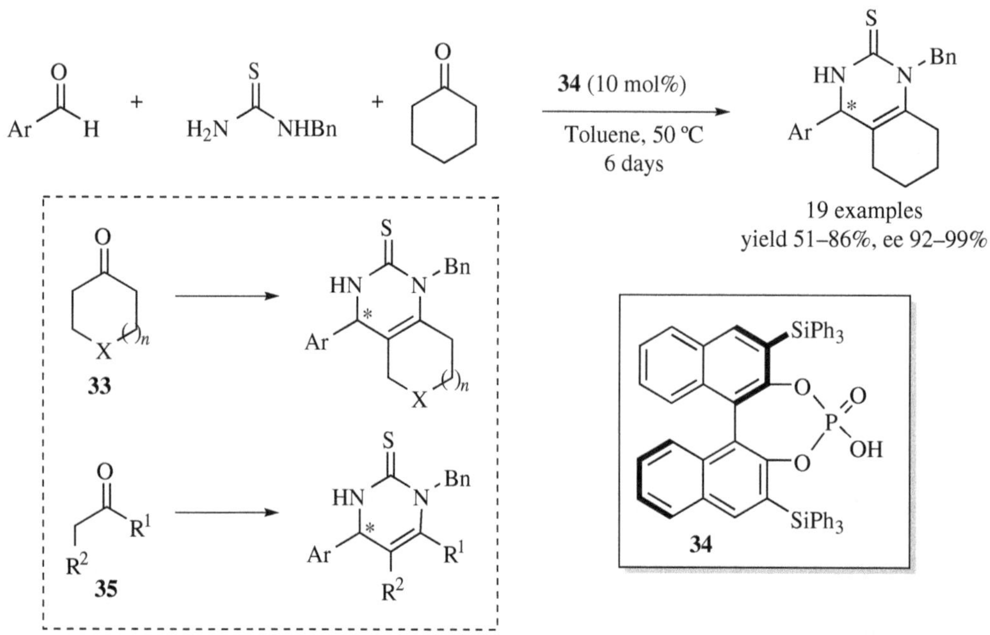

SCHEME 9.10 Use of cyclic ketones **33** in the enantioselective Biginelli-like reaction.

imines [45] and the original Biginelli reaction is a Brønsted acid-catalyzed process.

Gong's group reported the first catalytic enantioselective Biginelli-like reaction [46, 47] using cyclic and acyclic ketones (**33** and **35**, respectively), instead of β-ketoester and a BINOL-based phosphoric acid catalyst **34** (Scheme 9.10 and Fig. 9.2).

A plausible reaction mechanism was suggested by the authors on the basis of DFT calculations and in previous modes of activation for this kind of phosphoric catalysts (Scheme 9.11).

Thus, the reaction would start by the condensation between the aldehyde and the thiourea to give an imine **I** (Scheme 9.11, **A**), which would be promoted by Brønsted acid. The catalyst could be involved in the activation of such imine (**B**) for the subsequent attack of the β-ketoester (**C**), through formation of a chiral iminium species **II**, to undergo an enantioselective Mannich reaction, as suggested by DFT calculations. The ensuing cyclization reaction (**D**) and dehydration (**E**) afford the desired chiral Biginelli products and released the chiral phosphoric acid catalyst restarting the catalytic cycle.

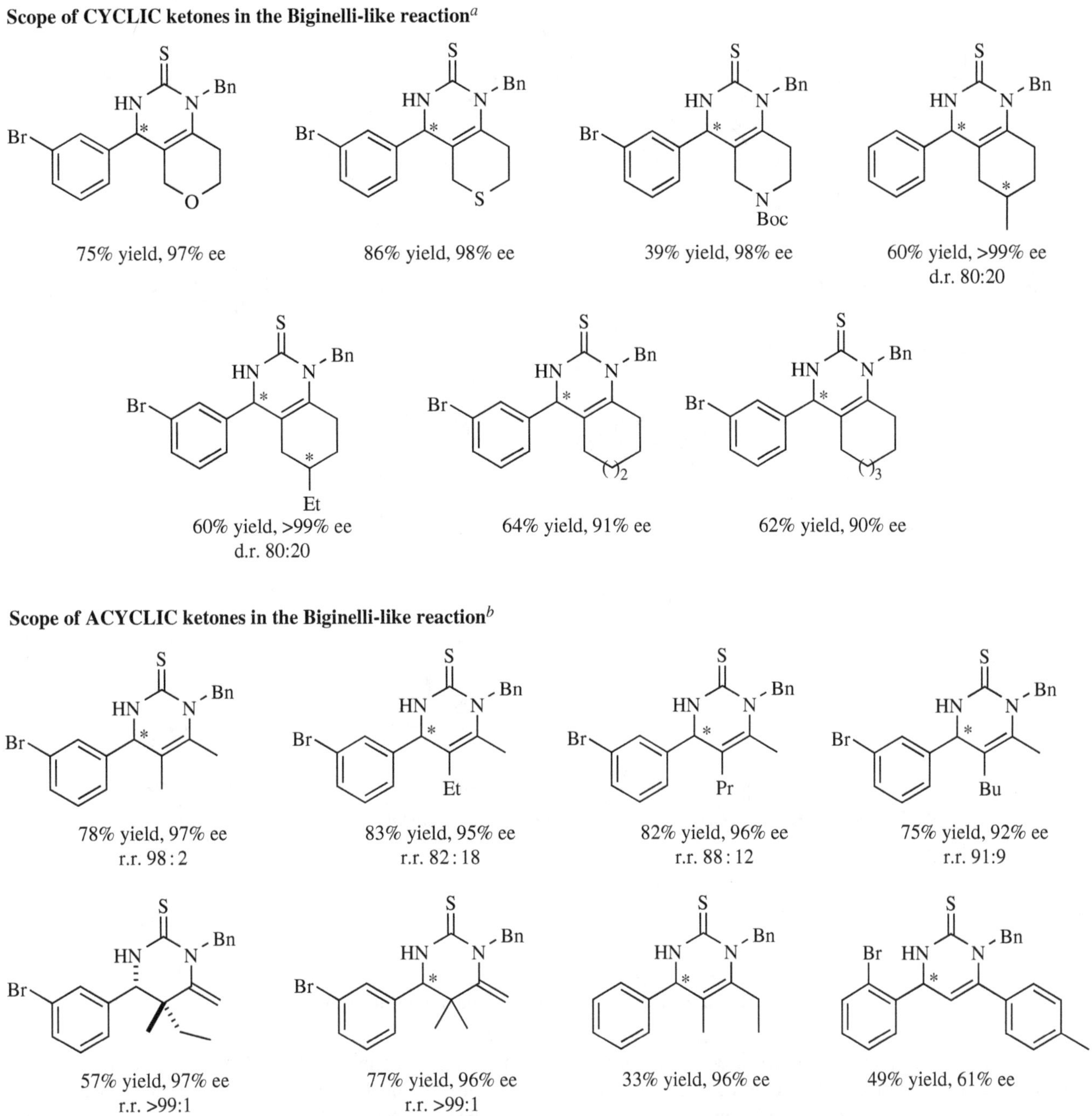

FIGURE 9.2 Use of cyclic **33** and acyclic ketones **35** in the enantioselective Biginelli-like reaction. [a]Same reaction conditions with those shown in Scheme 9.10. [b]Same reaction conditions with those shown in Scheme 9.10 but at 65°C instead of 50°C.

9.5.2 Aminocatalyzed Biginelli Reactions

Although L-proline and its corresponding hydrochloride derivative have been used as catalysts in the Biginelli reaction, the resulting DHPMs were found to be racemic [48]. In contrast, Juaristi and coworkers described the application of a chiral bicyclic diamine **36** as organocatalysts in the enantioselective Biginelli reaction, affording the expected DHPMs **37** in good yields but in moderate enantioselectivities (up to 46% ee) (Scheme 9.12) [49].

Based on previously reported mechanistic considerations concerning the aminocatalytic field [50], the authors proposed the catalytic cycle depicted in Scheme 9.13 to explain the stereoselection found in the final products **37**. In this proposal, the enamine intermediate **II** is the active species responsible of the enantioselectivity of this process.

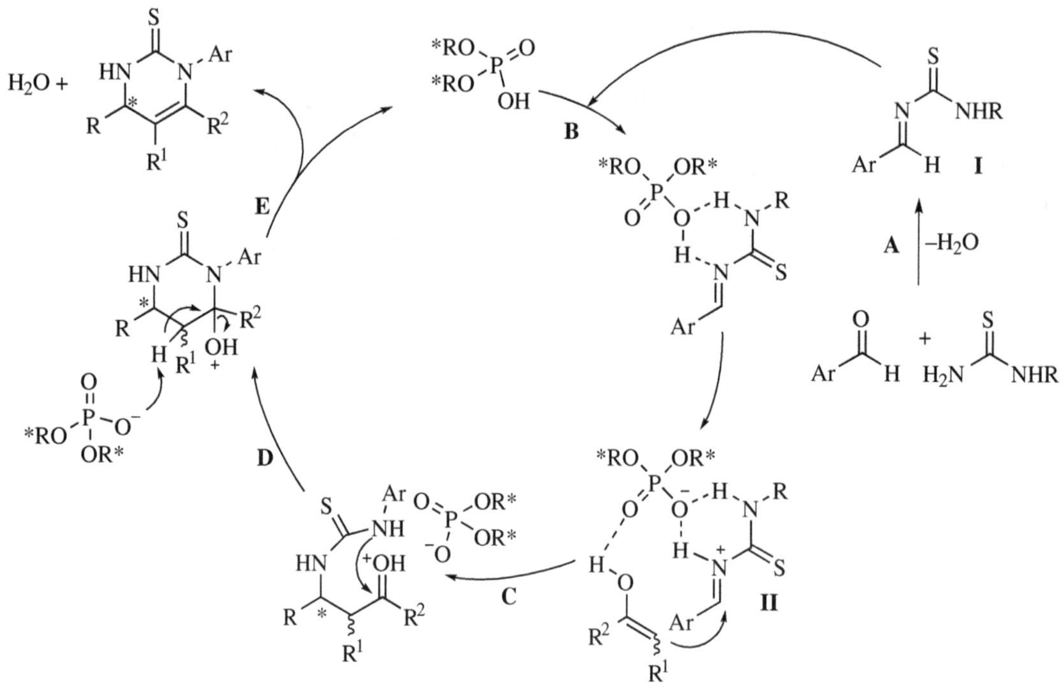

SCHEME 9.11 Mechanistic hypothesis.

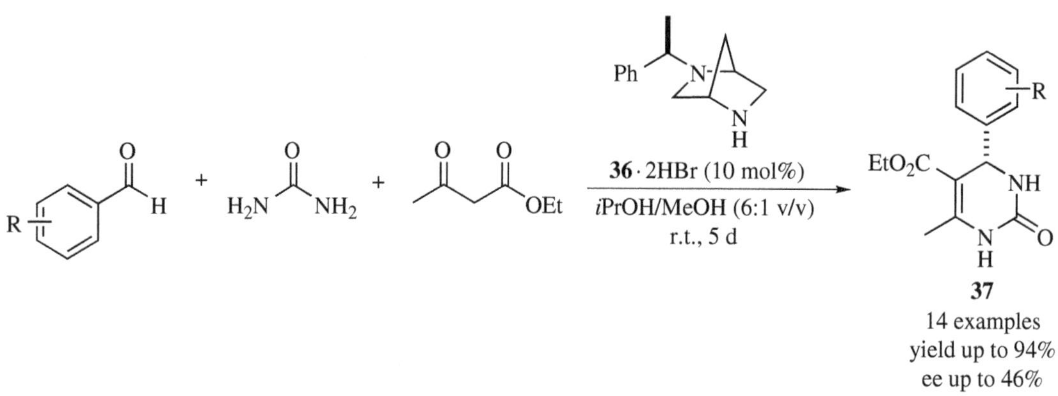

SCHEME 9.12 Biginelli reaction catalyzed by amine **36**.

Contemporaneously, Feng's group disclosed a *trans*-4-hydroxyprolinamide-derived **38** as chiral secondary amine catalyst in this enantioselective multicomponent Biginelli reaction [51]. The final products **39** were obtained with high yields and better enantioselectivities compared with the latter example (Scheme 9.14).

A plausible transition state (TS) to explain the origin of the activation and the asymmetric induction through a dual-activation mode by the catalyst has been envisioned by the authors (Fig. 9.3). The synergic effect provided by the involvement of both an achiral Brønsted acid and the chiral secondary amine **38** as the catalytic system in this multicomponent process was crucial for the success of this reaction since neither by themselves could catalyze the formation of DHPM efficiently. Thus, the acid additive could be involved in the formation of the *N*-acylimine intermediate **40** and in the generation of the enamine intermediate **41** between the chiral secondary amine and the β-ketoester, and it might also favor the activation of the *N*-acylimine **40** for the subsequent attack of the nucleophile participating in the asymmetric induction of the reaction. Based on the observed absolute configurations of the final products and the aforementioned hypothesis, the authors displayed two possible TS that require a dual activation in the asymmetric induction step (Fig. 9.3). Between both, **TS-1** would justify the stereoselection obtained in this process.

Further achievements have also been reached in the aminocatalysis field using the same class of chiral amine catalysts by other research groups [52]. In this sense, Zhao, Wang, and coworkers developed a new 5-(pyrrolidin-2-yl) tetrazole derivative **42** as an active catalyst in this MCR

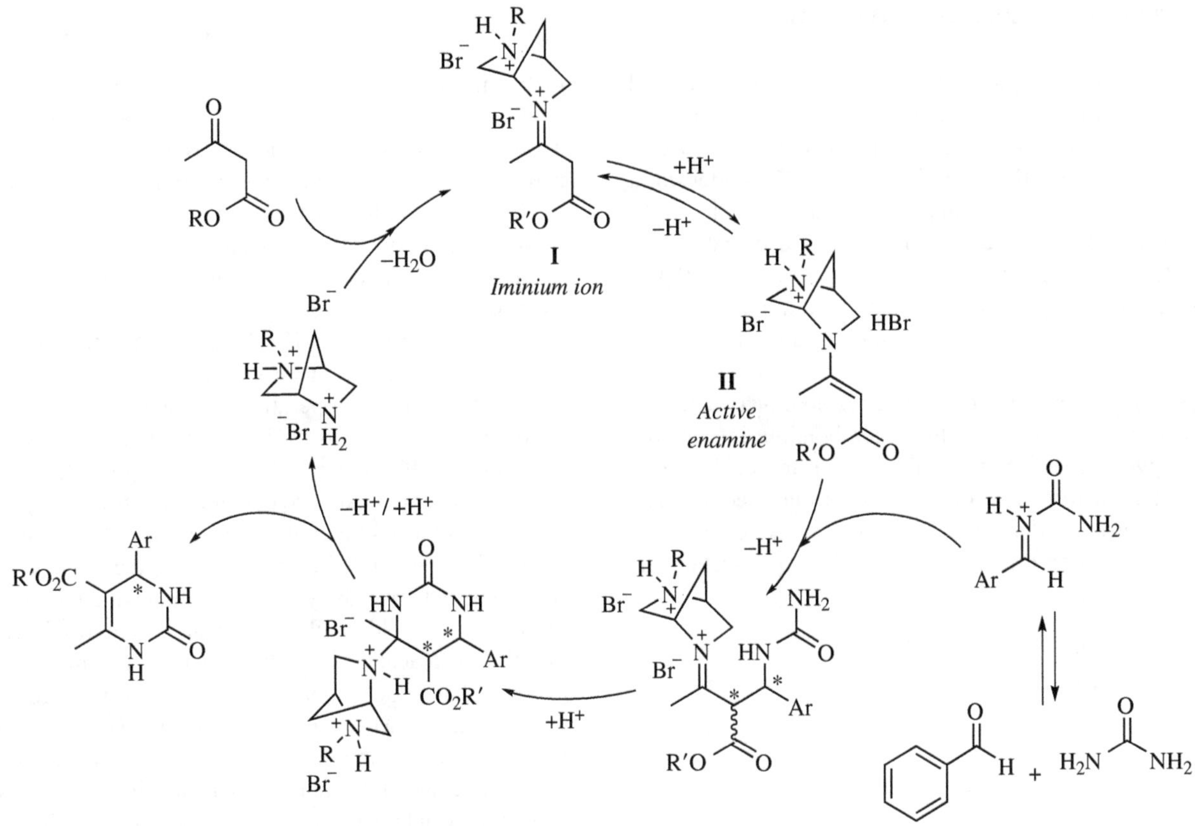

SCHEME 9.13 Proposed organocatalytic mechanism.

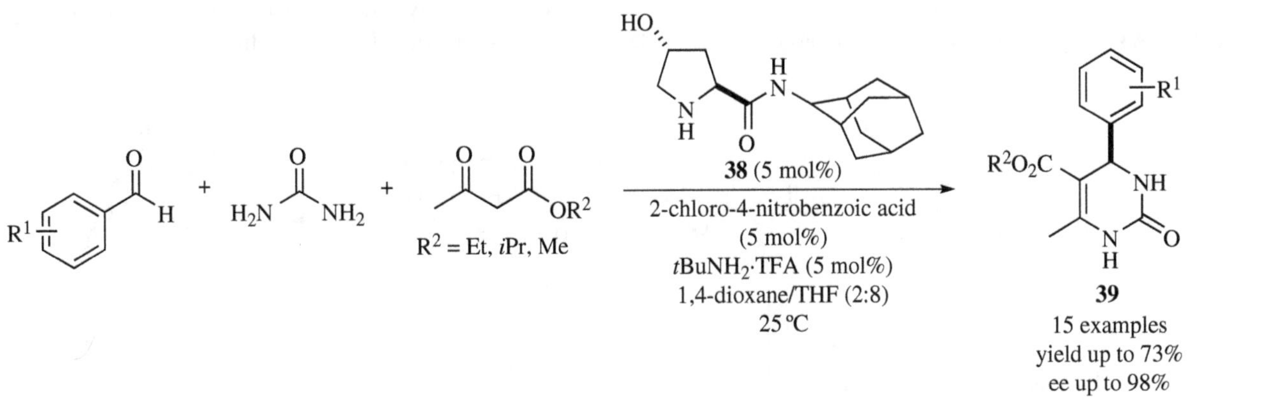

SCHEME 9.14 Biginelli reaction catalyzed by amine **38**.

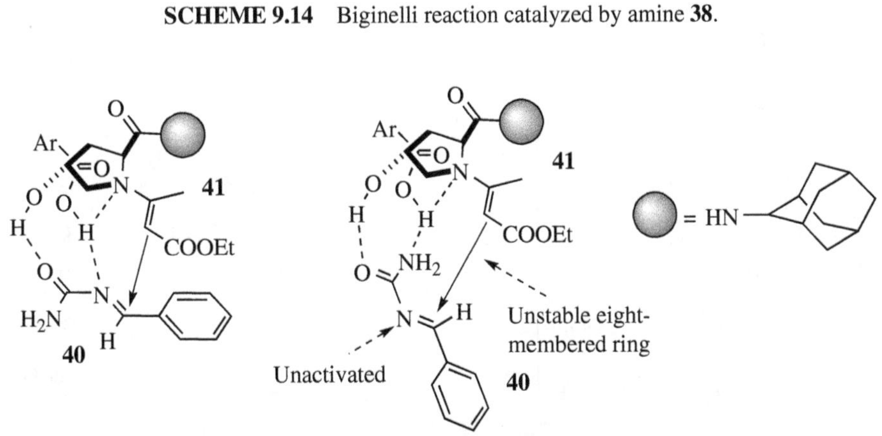

TS-1: *re* face favored
R-product generated

TS-2: *si* face disfavored
S-product generated

FIGURE 9.3 Proposed organocatalytic mechanism.

(Fig. 9.3) [53]. The optimization of catalyst structures and reaction conditions furnished the corresponding 3,4-dihydropyrimidin-2(1*H*)-one (DHPM) derivatives in good yields (63–88%) and enantioselectivities (68–81%) under mild reaction conditions. Based on previous studies of the mechanism of the Biginelli reaction for this kind of catalyst, a plausible TS with catalyst **42** is proposed in Figure 9.4 to explain the stereochemical results observed in this work.

A dual-activation way is invoked based on the similar widely accepted catalytic mode of activation for L-proline catalyst. In this context, the acidic tetrazole moiety activates the electrophilic imine by a hydrogen-bonding interaction, and the pyrrolidine moiety would activate the nucleophilic β-ketoester *via* formation of an enamine intermediate. The stereochemistry of the acidic tetrazole moiety seems to be crucial for the stereoselectivity in the final products. This fact is in agreement with two experimental observations: (i) the inversion of the absolute configuration of the product when catalysts with the opposite configuration were used and (ii) modification of the substituents at the 4-position of the catalysts that had no remarkable influence on the enantioselectivity of the products.

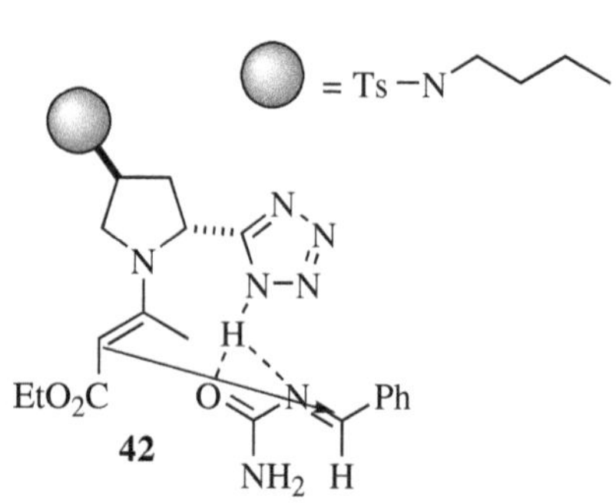

42

FIGURE 9.4 Proposed organocatalytic mechanism using catalyst **42**.

Miao, Chen, and coworkers have reported an efficient enantioselective Biginelli reaction catalyzed by a chiral bifunctional primary amine-thiourea derivative **43** with a carbohydrate in the catalyst skeleton and using an external Brønsted acid in a cooperative way, with *t*-$BuNH_2$·TFA as additive (Scheme 9.15) [54]. This protocol renders a wide range of optically active DHPMs **44** in high yields and with good to excellent enantioselectivities (up to 99% ee).

The possible TS for explaining the role of this bifunctional catalyst is shown in Scheme 9.15, where the thiourea moiety would interact through hydrogen bonding with the acyl group of the urea scaffold, while the primary amine would activate the ethyl acetoacetate through an enamine intermediate **I**. On the other, hand, the authors justify the (*S*) absolute configuration of final DHPMs **44**, reasoning that the *si* face of the imine was preferentially approached by the enamine intermediate in the TS. In contrast, the attack of the enamine to the *re* face of the imine would be restricted by the conformation of the cyclohexyl scaffold of the catalyst. In this sense, the thiourea moiety and the cyclohexyl scaffold of this bifunctional catalyst would play a dual crucial role in controlling the regio- and stereoselectivity of the Biginelli products.

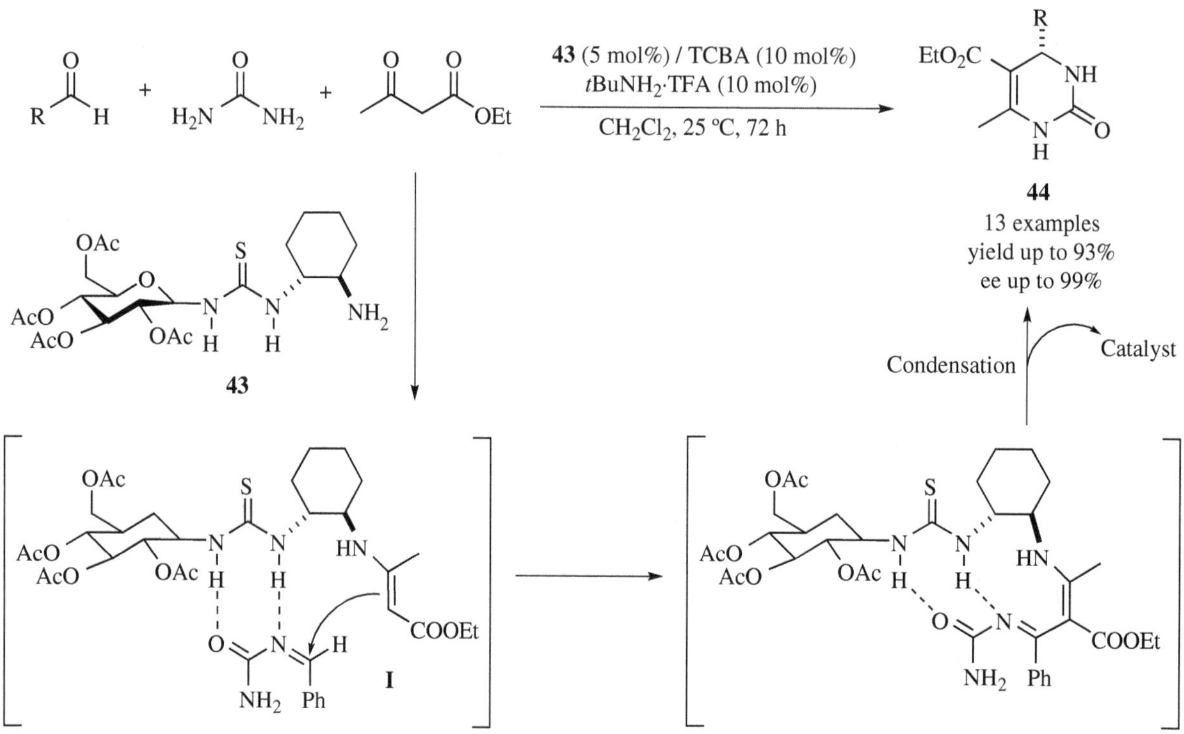

SCHEME 9.15 Biginelli reaction catalyzed by primary amine-thiourea **43**.

Zhao and Ding disclosed an interesting example using also a quinine-derived amine **45** as the best catalyst for promoting this MCR with moderate yields (14–81%) and good enantioselectivities (51–78%). The authors envisioned a possible mechanism to explain the absolute configuration obtained in the process (Scheme 9.16) [55].

In this mechanism, the imine **I** is firstly formed between benzaldehyde and the urea **A**. The imine **I** is further activated through hydrogen bonds with the protonated quinuclidine backbone of catalyst **45**. This activation helps to approach the imine **I** closer to the reaction center of the enamine **II** and drives its possible orientations. On the other hand, the primary amine group on the side chain of the catalyst activates the acetoacetate through the formation of the mentioned enamine **II**. Further attack of the enamine **II** onto the *si* face of the imine **I** gives intermediate **III** (**B**), which, after hydrolysis (**C**), intramolecular cyclization (**D**), and dehydration reaction (**E**), affords products **46** with *R* configuration.

More recently, Moorthy and Saha have reported a highly enantioselective Biginelli reaction catalyzed with organocatalyst **47** containing a chiral secondary amine scaffold and a strong hydrogen-bonding moiety (Scheme 9.17) [56].

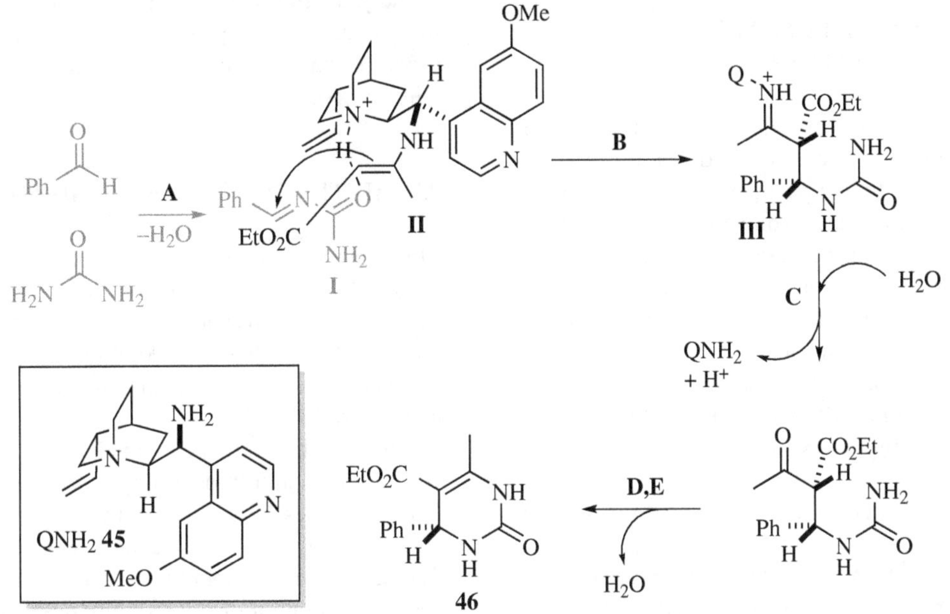

SCHEME 9.16 Mechanism to explain the formation of the *R* absolute configuration observed using catalyst **45**.

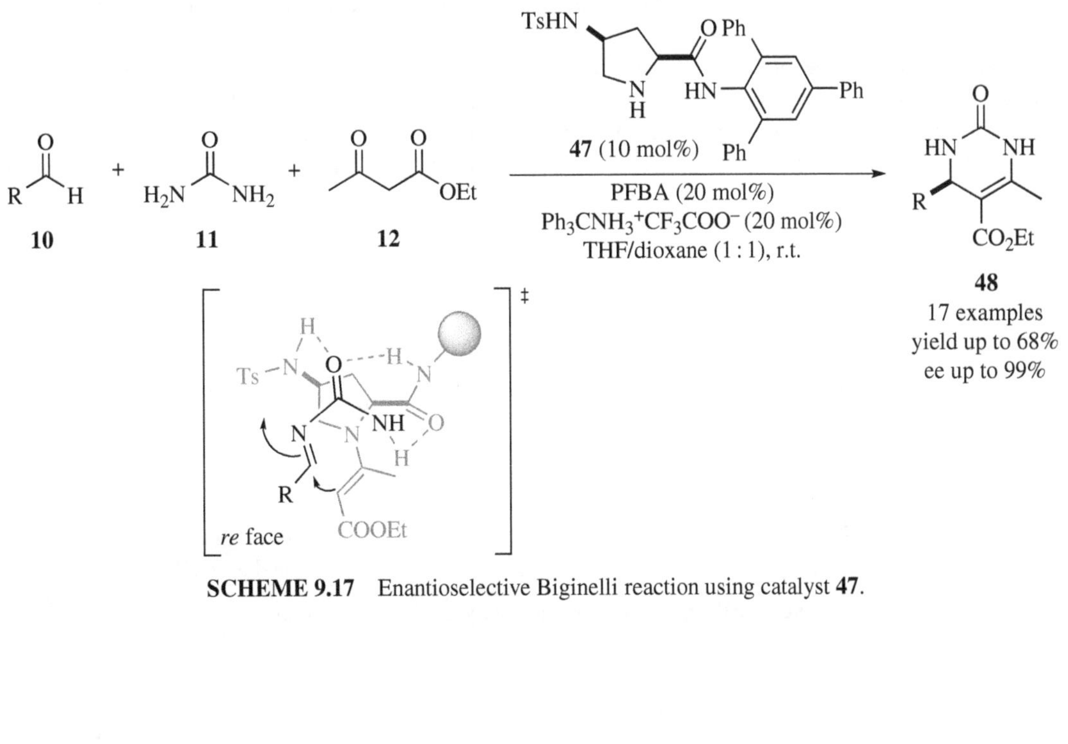

SCHEME 9.17 Enantioselective Biginelli reaction using catalyst **47**.

The protocol was efficient for aromatic and aliphatic aldehydes **10**, rendering final products **48** with high enantioselectivities. The catalyst **47** is a modification of the catalyst **38**, previously reported by Feng and coworkers [51], and the authors suggest similar TS as that shown in Figure 9.3 to support the absolute configuration obtained in the final products (Scheme 9.17). As previously invoked, the acid additive could play a dual role, on one hand catalyzing the imine formation between aldehyde **10** and urea **11** and on the other hand promoting the enamine formation between catalyst and ethyl acetoacetate **12**.

9.6 VARIATIONS OF THE TRADITIONAL BIGINELLI CONDENSATION

In general, aldehydes are more volatile, toxic, or unstable than their corresponding alcohols. And if these are used without previous distillation/purification, the presence of other byproducts could affect not only the real concentration of the active aldehyde in the reaction, but the impurities could also interfere in the course of the reactions reacting with other species present in the process. For this reason, the use of alcohols, with their subsequent oxidation, instead of the corresponding aldehydes is an interesting improvement in this and in other MCRs, and it would widen significantly the versatility and scope of the aldehyde-based MCRs [57].

After the first reported example starting from alcohols [58], Yadav and coworkers developed a new one-pot oxidation methodology of aromatic alcohols **49** using a Brønsted acidic ionic liquid [Hmim]HSO_4 catalyst with $NaNO_3$ followed by their condensation with urea and dicarbonyl compounds in the same vessel at 80°C, giving access to 3,4-dihydropyrimidin-2(1*H*)-ones **51** with moderate to high yield (55–97%) (Scheme 9.18) [59].

The authors explored the scope of this reaction for a variety of aromatic alcohols **49** and also demonstrated the compatibility of this method with a variety of functional groups in the alcohols such as ethers, nitro, hydroxyl, and halides under the best reaction conditions. A plausible mechanism for the present one-pot oxidative cyclocondensation of aromatic alcohols **49**, urea **11**, and 1,3-dicarbonyl compounds **50** was suggested by the authors, and it is disclosed in Scheme 9.19.

Interestingly, the authors envisioned a dual role for [Hmim]HSO_4 catalyst. On one hand, it could act as an acid catalyst as depicted in the Scheme 9.19 and, on the other hand, as a solvent for both the oxidation of alcohols and the subsequent condensation with urea **11** and 1,3-dicarbonyl compounds **50**.

Other interesting variations based on environmentally friendly alternatives, such as the use of microwaves [60], ionic liquids [61], aqueous media [62], or solvent-free [63] processes have also been reported to carry out the strategies with success.

9.7 HETEROCYCLES BEYOND THE DHPMs

This reaction also allows the synthesis of other interesting scaffolds with a wide range of biological activities, among them are coumarins and thiocoumarins [64] and their derivatives that display a remarkable array of biochemical and pharmacological actions and are abundant in numerous natural products.

The Singh group developed a facile route to afford 5-methylmercaptothiocarbonyl-4-aryl-3,4-dihydropyrimidin-2(1*H*)-ones **53** and substituted 2*H*-chromene-2-thiones **55** depending on the starting aldehyde (Scheme 9.20). The explored methodology involves a $SnCl_2$-catalyzed cyclocondensation of *β*-oxodithioesters **52** through a three-component Biginelli reaction and under solvent-free conditions (Scheme 9.21) [65]. It should be mentioned that compounds **55** are not reached through a multicomponent procedure.

On the basis of all the experimental results, an acceptable mechanism for the synthesis of 5-methylmercaptothiocarbonyl-4-aryl-3,4-dihydropyrimidin-2(1*H*)-ones **53** and substituted 2*H*-chromen-2-thiones **55** is disclosed in

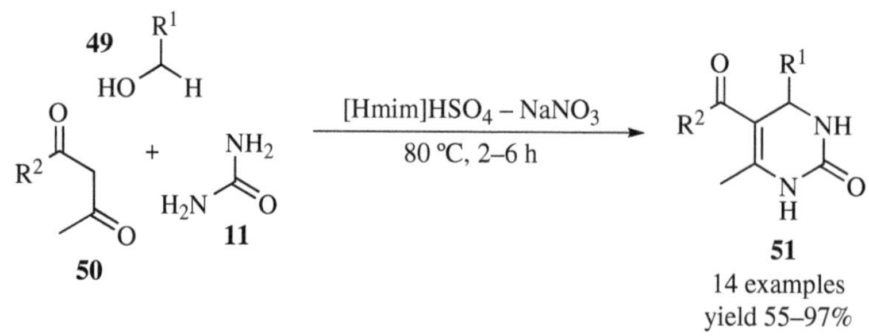

SCHEME 9.18 Ionic liquid-catalyzed Biginelli reaction.

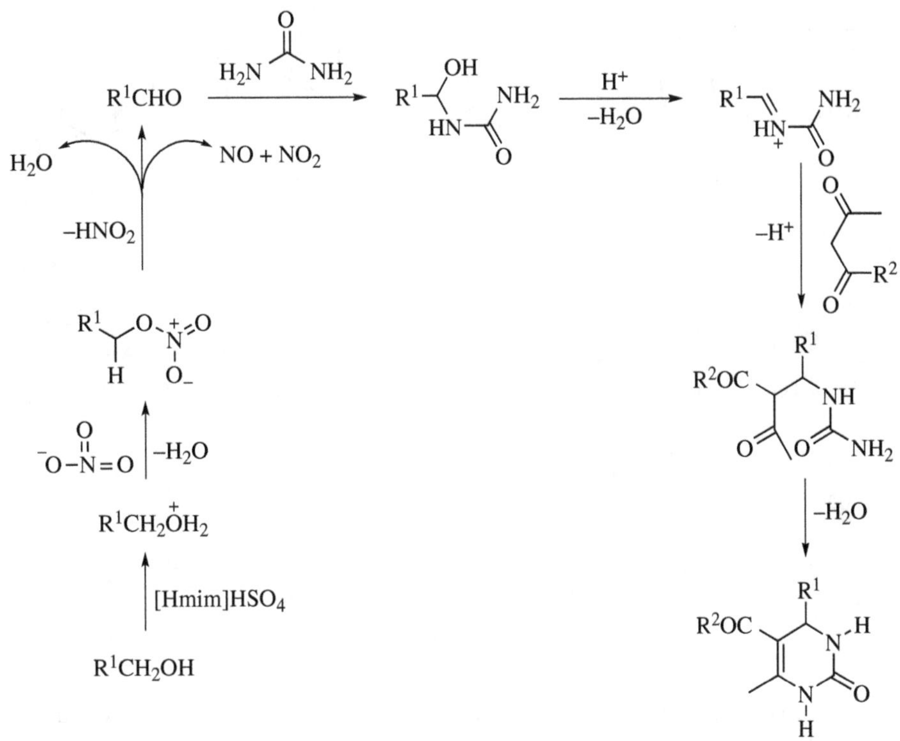

SCHEME 9.19 Invoked mechanism for one-pot oxidative multicomponent Biginelli reaction.

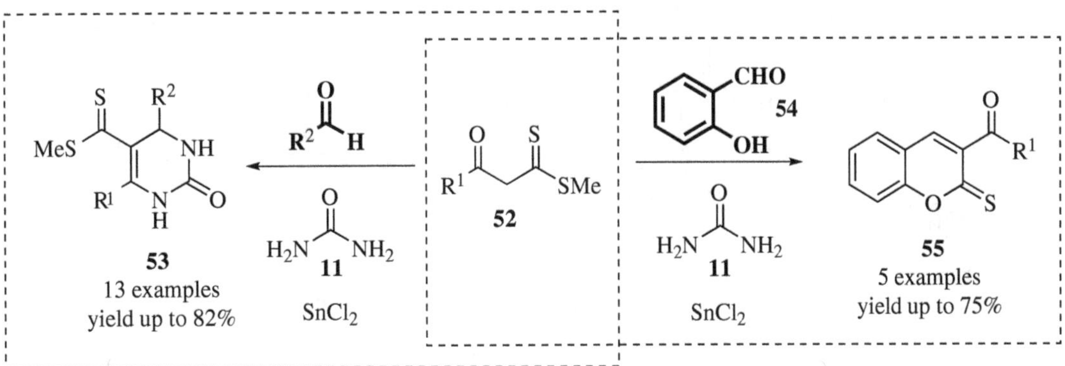

SCHEME 9.20 $SnCl_2$-catalyzed Biginelli reaction using β-oxodithioesters as dicarbonyl compound.

Scheme 9.21. To explain the formation of dihydropyrimidines, the authors propose the general accepted Biginelli mechanism (route A). For the synthesis of coumarins (route B), the overall transformation starts from the condensation between compound **52** and salicylaldehyde **54**, giving intermediate **I**. Further generation of enolate **II**, intramolecular aldol condensation, and dehydration give easy access to coumarins **55**. In an extension of this chemistry, the same group reported a convergent route to allow three interesting bioactive heterocyclic frameworks starting from the same β-oxodithioester **52** under similar reaction conditions (Scheme 9.22) [66]. Interestingly, the reactions were catalyzed using recyclable SiO_2–H_2SO_4.

The compatibility of this methodology with a variety of different substituted reactants and the stability shown by them under these acidic conditions, together with the recyclability of the catalyst without losing catalytic activity, make this strategy highly useful.

9.8 IMPORTANT TARGETS

Many efforts have been invested in the discovery of new active dihydropyrimidones (DHPMs), which have been found to possess a wide range of biological activity (Fig. 9.5) [6]. Although racemic DHPMs are easy accessible

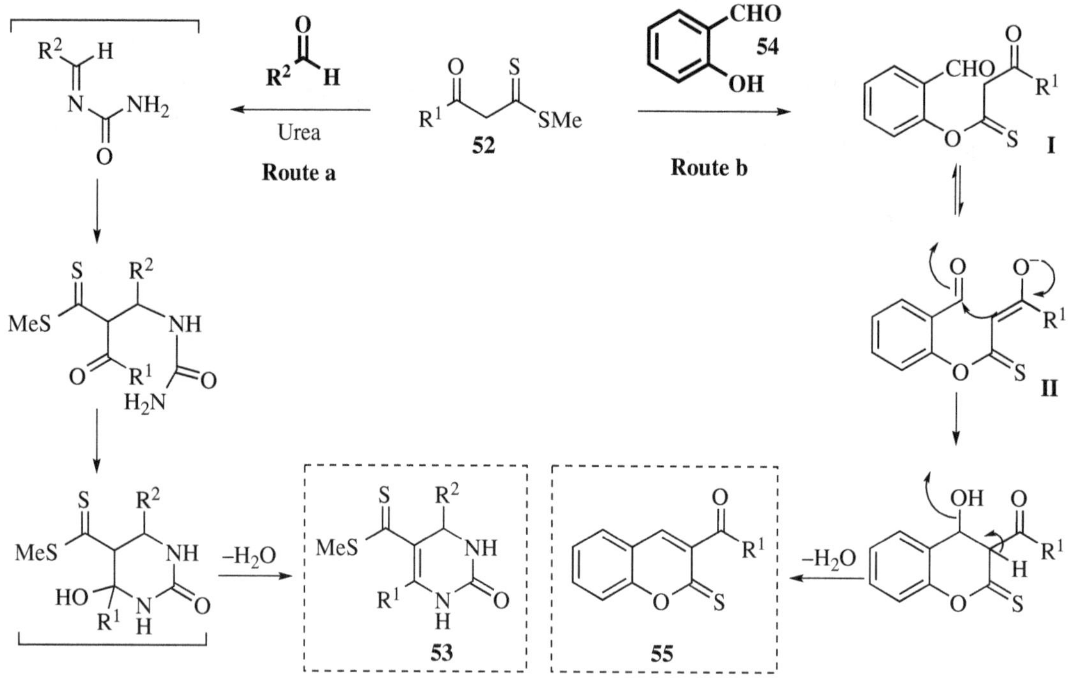

SCHEME 9.21 Mechanistic hypothesis for the synthesis of 5-methylmercaptothiocarbonyl-4-aryl-3,4-dihydropyrimidin-2(1*H*)-ones **53** and substituted 2*H*-chromen-2-thiones **55**.

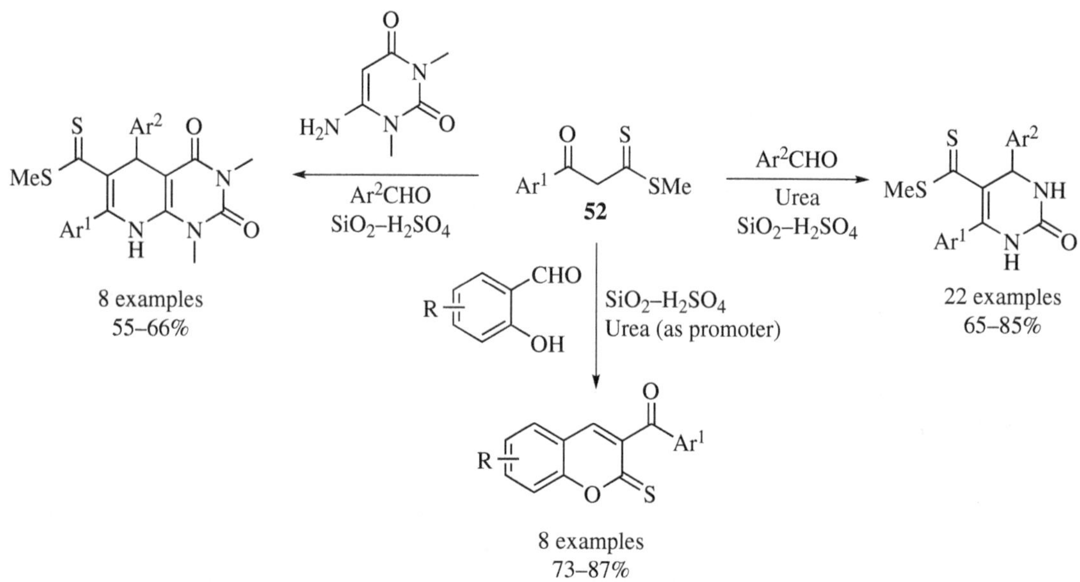

SCHEME 9.22 Synthesis of important heterocycles compounds.

via Biginelli reaction, surprisingly, a scarce number of asymmetric methods have been developed until now for the construction of these chiral heterocycles in high enantiomeric purity (see Sections 9.3 and 9.5). In this part, we want to disclose some selected examples of the preparation of DHPM structures, in which their biological activities have been evaluated [71].

Schaus and Goss reported an enantioselective synthesis of SNAP-7941 **62**, a potent melanin-concentrating hormone receptor antagonist, by using a chiral phosphoric acid-catalyzed Biginelli reaction (Scheme 9.23) [72].

In this process, the authors disclose two organocatalytic routes focus on the preparation of the dihydropyrimidone core **61** of the final product **62** as the key enantioselective

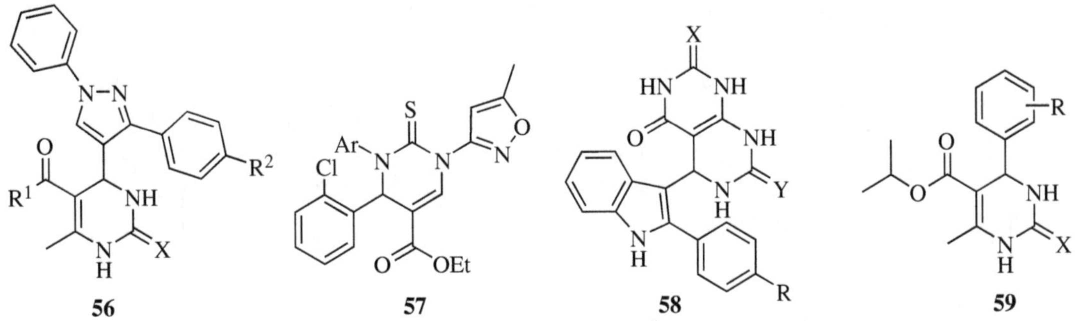

FIGURE 9.5 Selection of active DHPMs: **56** [67], **57** [68], **58** [69], and **59** [70].

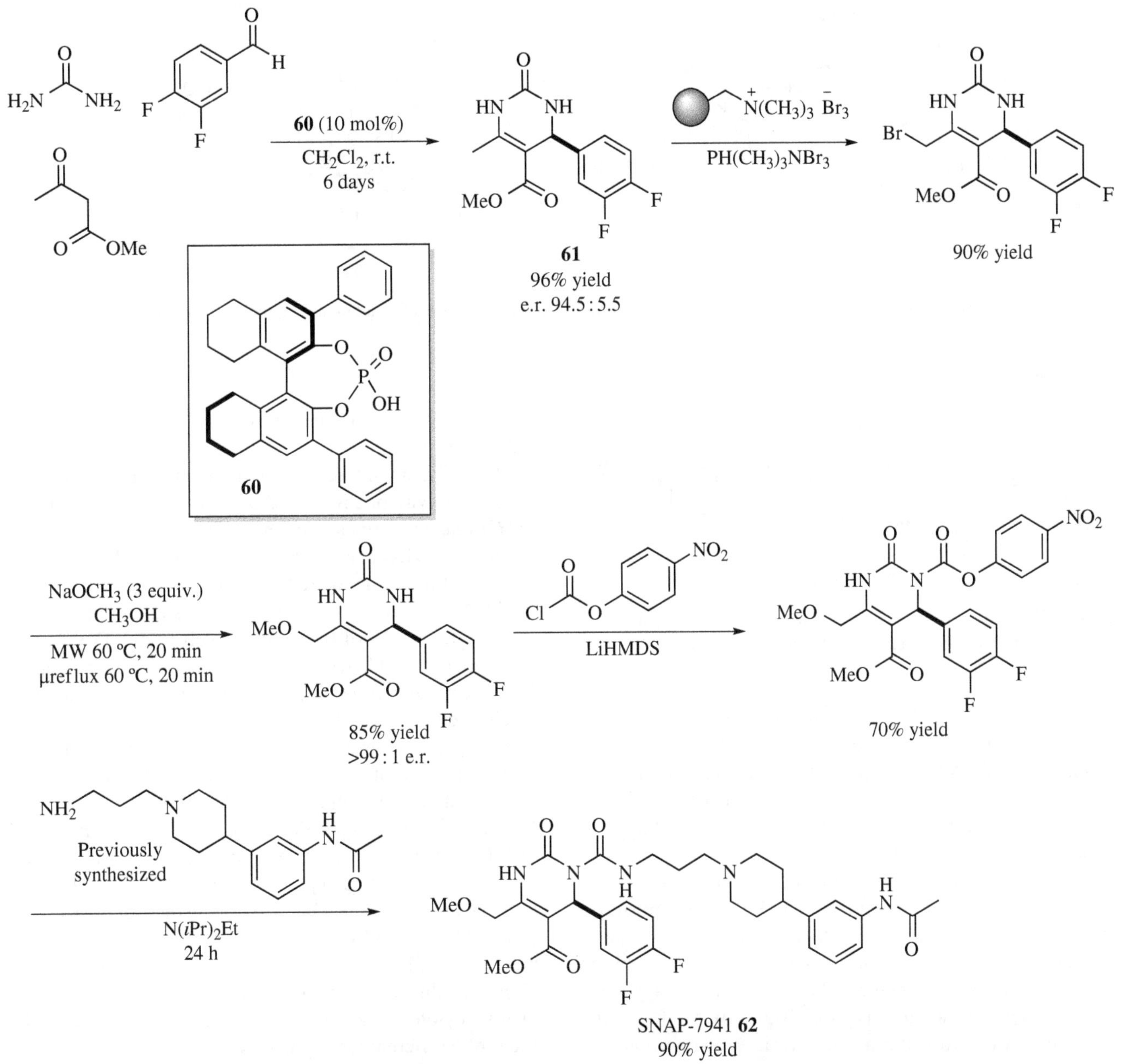

SCHEME 9.23 Enantioselective synthesis of the SNAP-7941 **62** through a key Biginelli reaction.

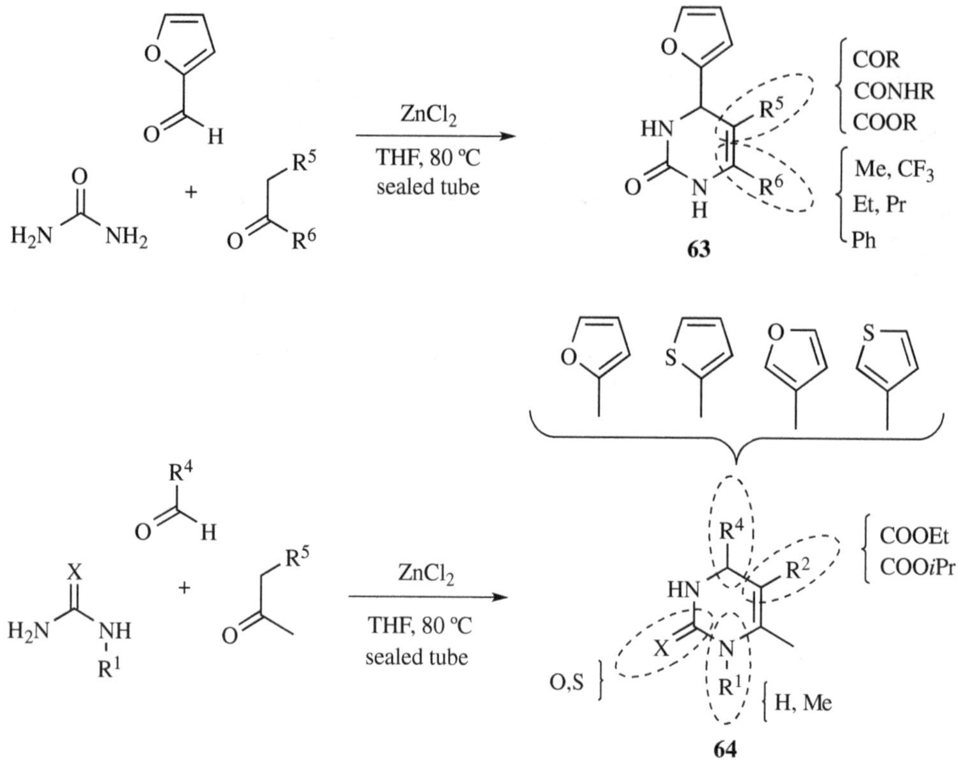

SCHEME 9.24 Collection of 3,4-dihydropyrimidin-2(1*H*)-one structures **63** and **64** as A_{2B} receptor antagonists.

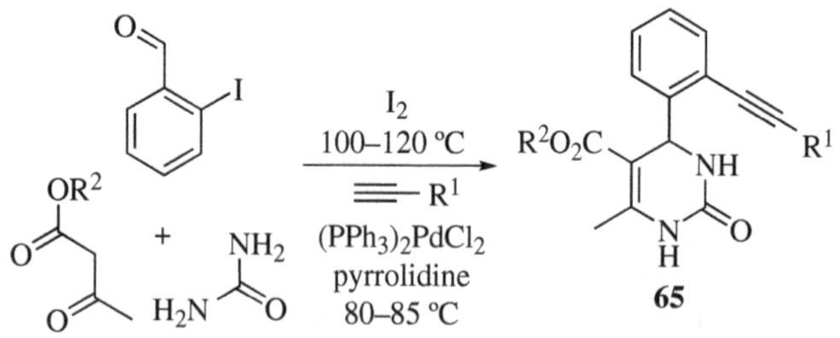

SCHEME 9.25 Synthesis of *o*-alkynylphenyl-substituted dihydropyrimidin-2(1*H*)-one derivatives **65** *via* Biginelli MCR.

step. In Scheme 9.23, the route following a chiral phosphoric acid-catalyzed Biginelli reaction, affording the right configuration of final SNAP-7941 **62** with good results, is shown. This is also one of the scarce examples for the enantioselective synthesis of a target compound using a Biginelli protocol as the key reaction step.

Sotelo and coworkers explored the activity of an extensive collection of optimized 3,4-dihydropyrimidin-2(1*H*)-one structures as a novel family of (nonxanthine) A_{2B} receptor antagonists with an unusually high selectivity profile (Scheme 9.24) [73]. The library was synthesized following a typical Biginelli MCR using ZnCl catalyst, affording a great variety of compounds **63** and **64** and combining simplicity with excellent hA_{2B} AdoR affinity.

Recently, an interesting *in vitro* evaluation study of novel *o*-alkynylphenyl-substituted 3,4-dihydropyrimidin-2(1*H*)-one derivatives as potential inhibitors of chorismate mutase (CM) has been reported [74]. Four MCRs involving sequential iodine-mediated solvent-free Biginelli reaction [75] followed by a copper-free Sonogashira coupling [76] in a single pot afford final alkynyl-substituted DHPMs **65** with good results, which were further screened against *Mycobacterium tuberculosis* H37Rv CM (Scheme 9.25 and Table 9.1).

Some of these structures were found to exhibit good inhibitory activities of CM when tested at 30 μM, and one compound showed dose-dependent inhibition of CM with IC_{50} value of 14.76 ± 0.54 μM. Thus, these alkynyl-substituted DHPM structures could be appealing antitubercular agents.

Another interesting application is the synthesis of a model tricyclic core of the cylindrospermopsin alkaloids **66–68** (Fig. 9.6) reported by Murphy's group following a Biginelli reaction as a key step of the process (Scheme 9.26) [77]. These alkaloids are marine natural products that exhibit diverse cytotoxic activities [78], and they have been the focus of a different total synthesis.

TABLE 9.1 Inhibition of chorismate mutase (CM)

Entry	R¹	R²	Products **65**	Yield (%)	% Inhibition @ 30 μM
1	*n*-Hexyl	Et	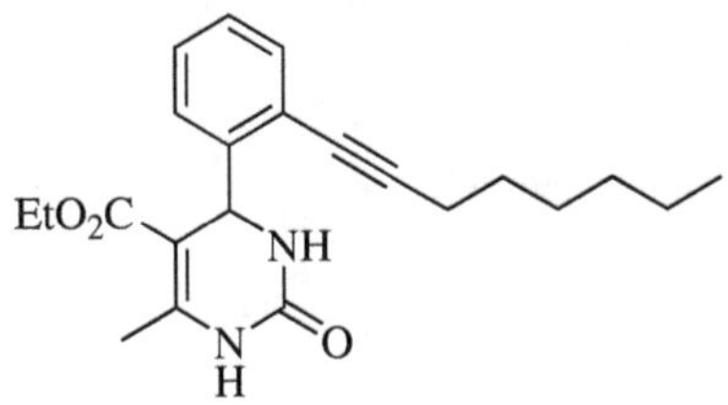	85	31
2	*n*-Pentyl	Et	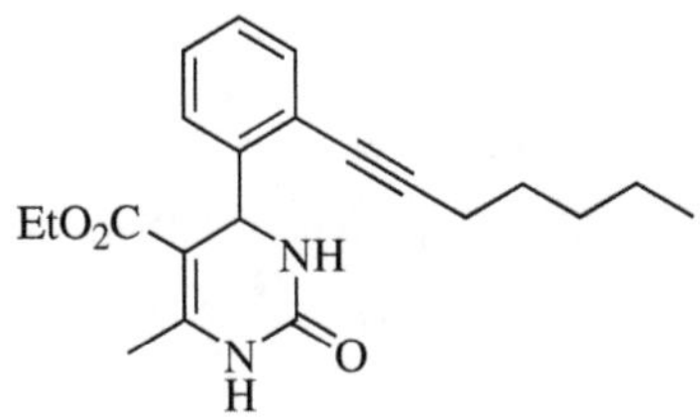	87	29
3	$–CH_2OH$	Et	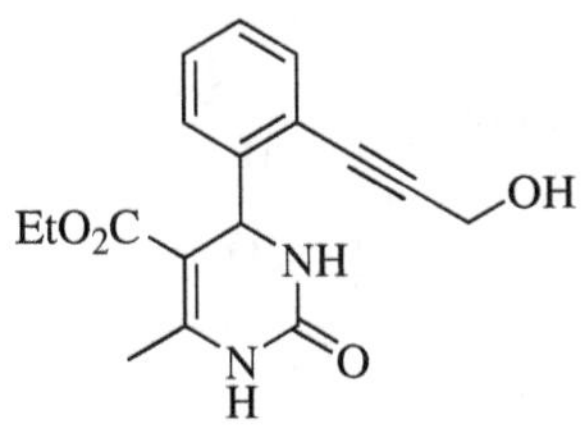	84	60
4	$–CH(OH)CH_3$	Et	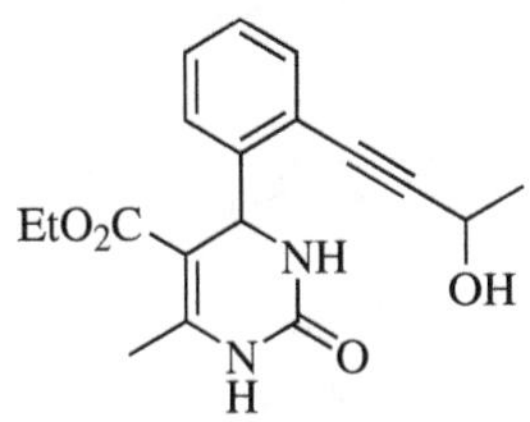	80	41
5	$–(CH_2)_2OH$	Et	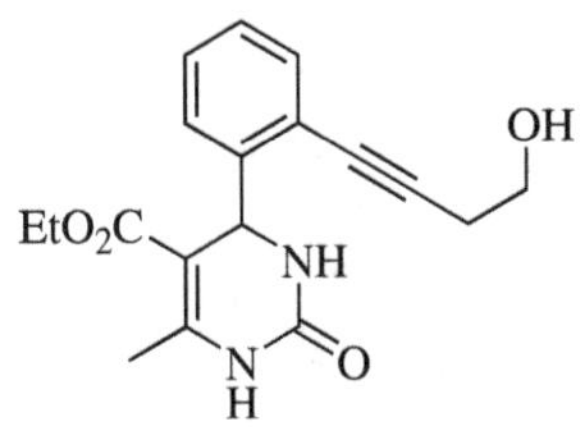	83	63
6	$–(CH_2)_3OH$	Et	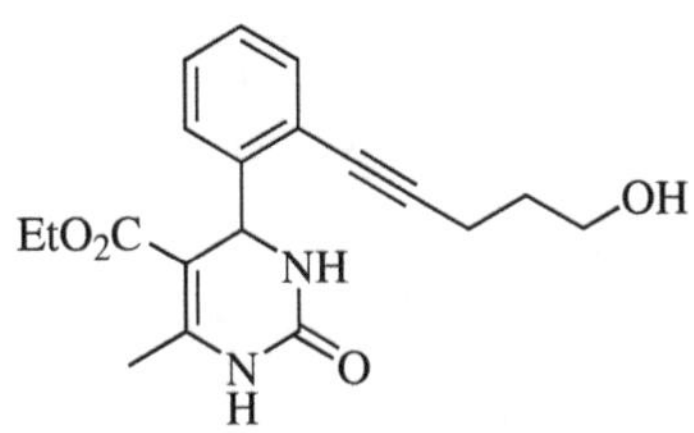	85	57
7	1-Cyclohexenyl	Et	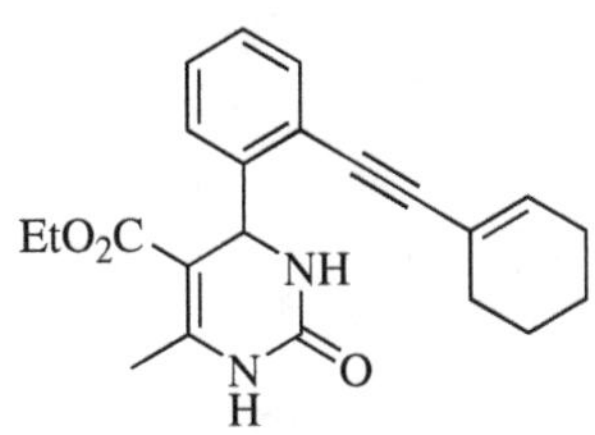	90	27
8	1-Hydroxycyclohexanyl	Et	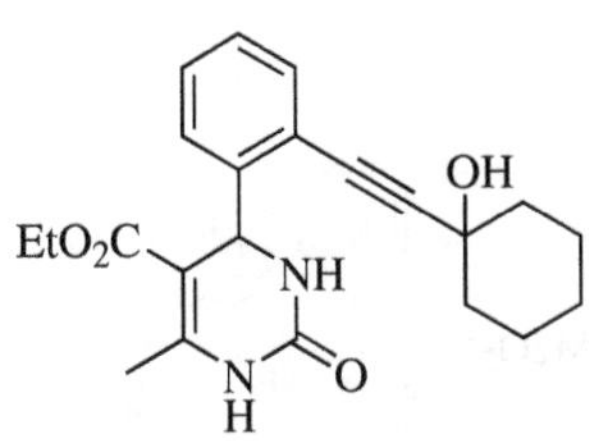	86	20

TABLE 9.1 (*Continued*)

Entry	R^1	R^2	Products **65**	Yield (%)	% Inhibition @ 30 μM
9	Phenyl	Et	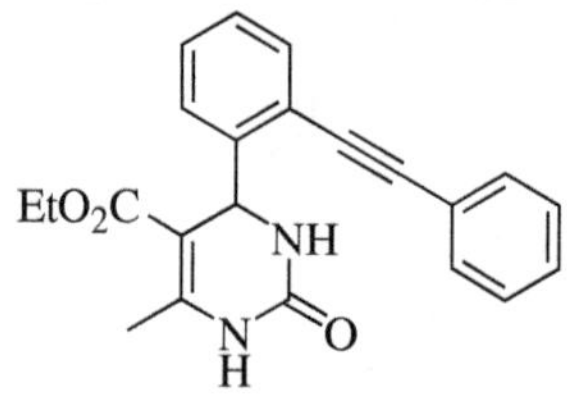	81	17
10	4-(*n*-Pentyl)phenyl	Et	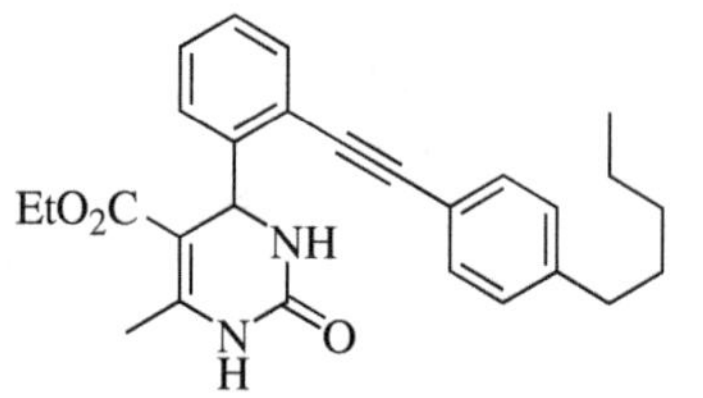	85	19
11	*n*-Pentyl	*t*Bu	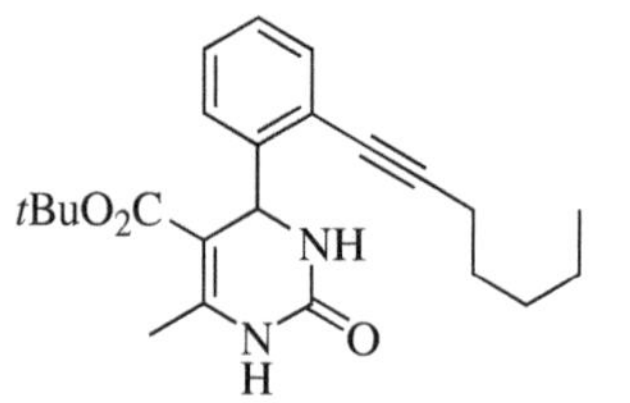	87	30
12	1-Cyclohexenyl	*t*Bu	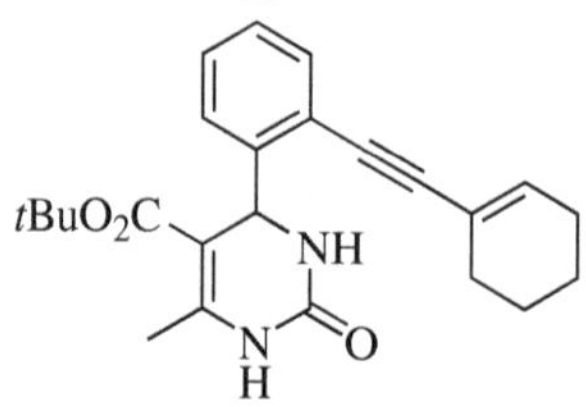	82	22
13	1-Hydroxycyclohexanyl	*t*Bu	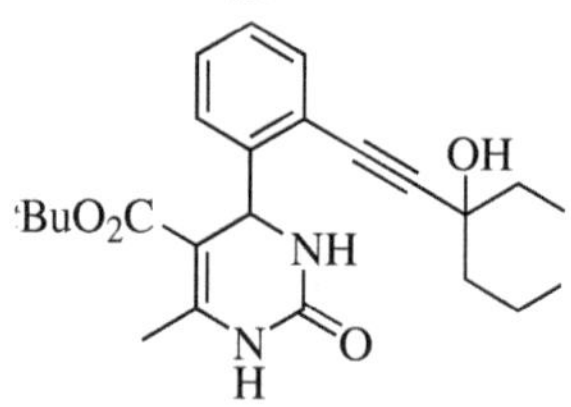	84	18
14	1-Hydroxycyclohexanyl	Me	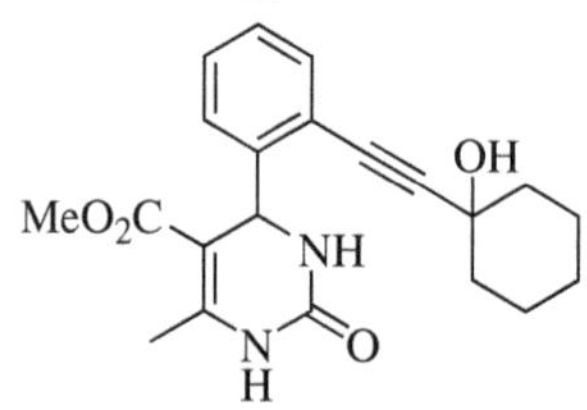	82	25
15	1-Cyclohexenyl	Me	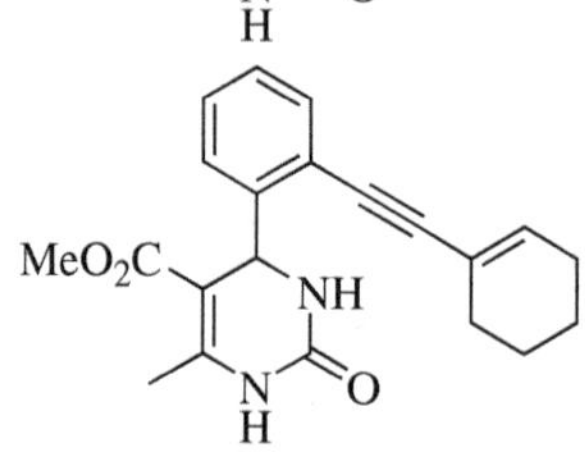	81	21
16	*n*-Hexyl	Me	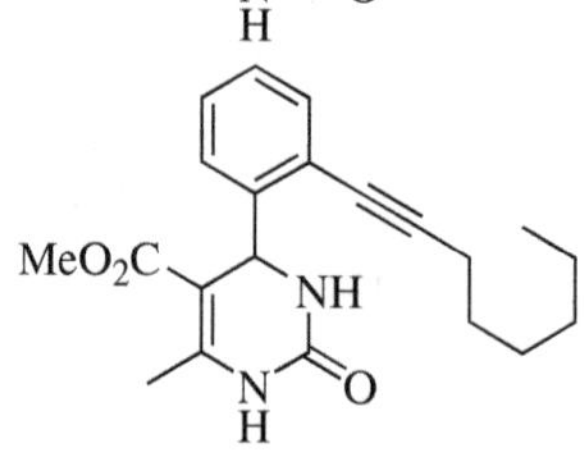	84	28
17	4-Methyl phenyl	Me	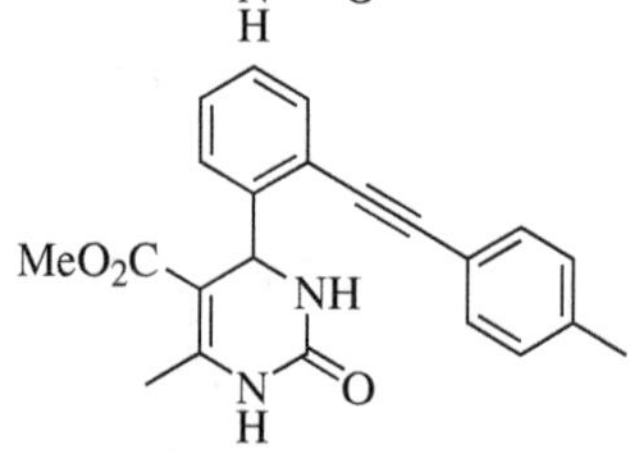	83	16

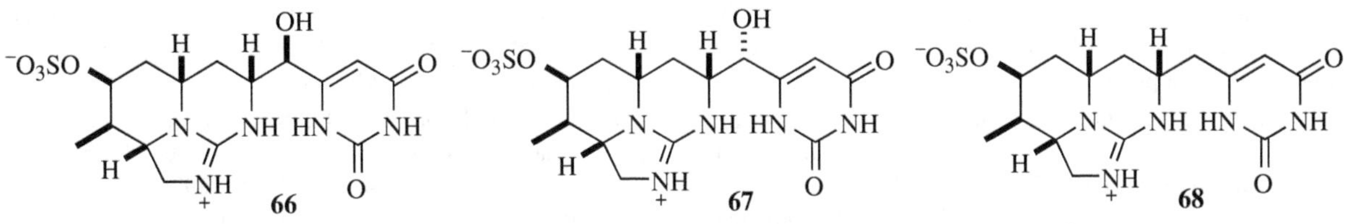

FIGURE 9.6 Cylindrospermopsin alkaloids **66–68**.

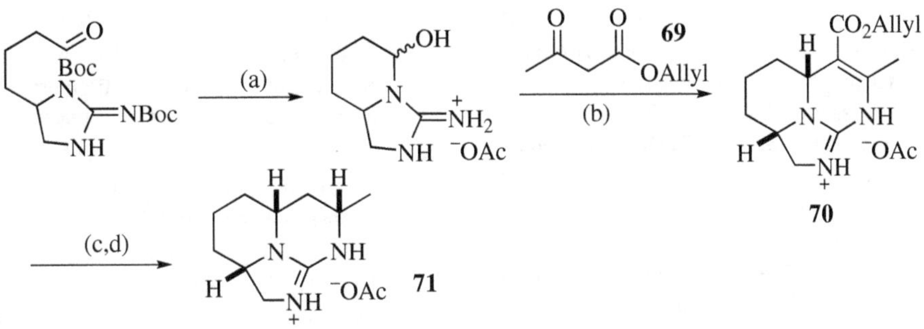

SCHEME 9.26 Reagents and conditions: (a) AcOH, 24 h. (b) Morpholine acetate, **69**, Na_2SO_4, CF_3CH_2OH, 70°C, 12 d, 43%. (c) $Pd(PPh_3)_4$, pyrrolidine, THF/MeOH, 1 h 30 min. (d) $NaBH_3CN$, AcOH/MeOH, 16 h, 57% (two steps from **70**).

9.9 CONCLUSION

In spite of all this incredible number of contributions made by several scientific groups, the search for better promoters still continues being desirable for the development of this reaction. Although success has been achieved in these works, the design and synthesis of new catalysts remain an active and interesting challenge mostly due to the ambiguity of the exact mechanism of the Biginelli reaction and the mode of chiral induction with the aim of achieving optically pure scaffolds. Improvements in more sustainable processes are also required and appealing. In the next years, pivotal outcomes will be achieved with the aim of contributing to the development of this field.

REFERENCES

1 (a) P. Biginelli, *Ber. Dtsch. Chem. Ges.* 1891, 24, 1317–1319. Ueber Aldehyduramide des Acetessigäthers. (b) P. Biginelli, *Ber. Dtsch. Chem. Ges.* 1891, 24, 2962–2967. Ueber Aldehyduramide des Acetessigäthers II. (c) G. C. Tron, A. Minassi, G. Appendino, *Eur. J. Org. Chem.* 2011, 5541–5550. Pietro Biginelli: the man behind the reaction.

2 (a) C. O. Kappe in *Multicomponent reactions* (J. Zhu, H. Bienaymé, Eds.). Wiley-VCH Verlag GmbH, Weinheim, Germany, 2005, pp. 95–120. The Biginelli reaction. (b) A. Dondoni, A. Massi, *Acc. Chem. Res.* 2006, 39, 451–464. Design and synthesis of new classes of heterocyclic *C*-glycoconjugates and carbon-linked sugar and heterocyclic amino acids by asymmetric multicomponent reactions (AMCRs). (c) C. Simon, T. Constantieux, J. Rodriguez, *Eur. J. Org. Chem.* 2004, 4957–4980. Utilisation of 1,3-dicarbonyl derivatives in multicomponent reactions.

3 Suresh, J. S. Sandhu, *Arkivoc* 2012, (i), 66–133. Past, present and future of the Biginelli reaction: a critical perspective.

4 J.-P. Wan, Y. Liu, *Synthesis* 2010, 3943–3953. Synthesis of dihydropyrimidinones and thiones by multicomponent reactions: strategies beyond the classical Biginelli reaction.

5 (a) C. O. Kappe, *Tetrahedron* 1993, 49, 6937–6963. 100 Years of the Biginelli dihydropyrimidine synthesis. (b) C. O. Kappe, *Acc. Chem. Res.* 2000, 33, 879–888. Recent advances in the Biginelli dihydropyrimidine synthesis. New tricks from an old dog. (c) M. Matache, C. Dobrota, N. D. Bogdan, D. P. Funeriu, *Curr. Org. Synth.* 2011, 8, 356–373. Recent developments in the reactivity of the Biginelli compounds. (d) S. S. Panda, P. Khanna, L. Khanna, *Curr. Org. Chem.* 2012, 16, 507–520. Biginelli reaction: a green perspective.

6 (a) C. O. Kappe, *Eur. J. Med. Chem.* 2000, 35, 1043–1052. Biologically active dihydropyrimidones of the Biginelli-type a literature survey. (b) K. Singh, D. Arora, K. Singh, S. Singh, *Mini Rev. Med. Chem.* 2009, 1, 95–106. Genesis of dihydropyrimidinone calcium channel blockers: recent progress in structure-activity relationships and other effects.

7 A. D. Patil, N. V. Kumar, W. C. Kokke, M. F. Bean, A. J. Freyer, C. de Brosse, S. Mai, A. Truneh, D. J. Faulkner, B. Carte, A. L. Breen, R. P. Hertzberg, R. K. Johnson, J. W. Wesley, B. C. M. Potts, *J. Org. Chem.* 1995, 60, 1182–1188. Novel alkaloids from the sponge *Batzella* sp.: inhibitors of HIV gpl2O-human CD4 binding.

8 G. C. Rovnyak, K. S. Atwal, A. Hedberg, S. D. Kimball, S. Moreland, J. Z. Gougoutas, B. C. O'Reilly, J. Schwartz, M. F. Malley, *J. Med. Chem.* 1992, 35, 3254–3263. Dihydropyrimidine calcium channel blockers. 4. Basic 3-substituted-4-aryl-1,4-dihydropyrimidine-5-carboxylic acid esters. Potent antihypertensive agents.

9 E. W. Hurst, *Ann. N. Y. Acad. Sci.* 1962, 98, 275–286. Experimental chemotherapy of infection with agents of the psittacosis-lymphogranuloma-trachoma group.

10 B. Borowsky, M. M. Durkin, K. Ogozalek, M. R. Marzabadi, J. DeLeon, R. Heurich, H. Lichtblau, Z. Shaposhnik, I. Daniewska, T. P. Blackburn, T. A. Branchek, C. Gerald, P. J. Vaysse, C. Forray, *Nat. Med.* 2002, 8, 825–830. Antidepressant, anxiolytic and anorectic effects of a melanin-concentrating hormone-1 receptor antagonist.

11 J. C. Barrow, P. G. Nantermet, H. G. Selnick, K. L. Glass, K. E. Rittle, K. F. Gilbert, T. G. Steele, C. F. Homnick, R. M. Freidinger, R. W. Ransom, P. Kling, D. Reiss, T. P. Broten, T. W. Schorn, R. S. L. Chang, S. S. O'Malley, T. V. Olah, J. D. Ellis, A. Barrish, K. Kassahun, P. Leppert, D. Nagarathnam, C. Forray, *J. Med. Chem.* 2000, 43, 2703–2718. *In vitro* and *in vivo* evaluation of dihydropyrimidinone C-5 amides as potent and selective α_{1A} receptor antagonists for the treatment of benign prostatic hyperplasia.

12 K. Deres, C. H. Schröder, A. Paessens, S. Goldmann, H. J. Hacker, O. Weber, T. Krämer, U. Niewoehner, U. Pleiss, J. Stoltefuss, E. Graef, D. Koletzki, R. N. A. Masantschek, A. Reimann, R. Jaeger, R. Groβ, B. Beckermann, K.-H. Schlemmer, D. Haebich, H. Rübsamen-Waigmann, *Science* 2003, 299, 893–896. Inhibition of hepatitis B virus replication by drug-induced depletion of nucleocapsids.

13 T. U. Mayer, T. M. Kapoor, S. J. Haggarty, R. W. King, S. L. Schreiber, T. J. Mitchison, *Science* 1999, 286, 971–974. Small molecule inhibitor of mitotic spindle bipolarity identified in a phenotype-based screen.

14 I. Garcia-Saez, S. DeBonis, R. Lopez, F. Trucco, B. Rousseau, P. Thuéry, F. Kozielski, *J. Biol. Chem.* 2007, 282, 9740–9747. Structure of human Eg5 in complex with a new monastrol-based inhibitor bound in the *R* configuration.

15 K. S. Atwal, G. C. Rovnyak, S. D. Kimball, D. M. Floyd, S. Moreland, B. N. Swanson, J. Z. Gougoutas, J. Schwartz, K. M. Smillie, M. F. Malley, *J. Med. Chem.* 1990, 33, 2629–2635. Dihydropyrimidine calcium channel blockers. 2. 3-Substituted-4-aryl-1,4-dihydro-6-methyl-5-pyrimidinecarboxylic acid esters as potent mimics of dihydropyridines.

16 K. S. Atwal, B. N. Swanson, S. E. Unger, D. M. Floyd, S. Moreland, A. Hedberg, B. C. O'Reilly, *J. Med. Chem.* 1991, 34, 806–811. Dihydropyrimidine calcium channel blockers. 3. 3-Carbamoyl-4-aryl-1,2,3,4-tetrahydro-6-methyl-5-pyrimidinecarboxylic acid esters as orally effective antihypertensive agents.

17 D. Russowsky, R. F. S. Canto, S. A. A. Sanches, M. G. M. D'Oca, Â. de Fátima, R. A. Pilli, L. K. Kohn, M. A. Antônio, J. E. de Carvalho, *Bioorg. Chem.* 2006, 34, 173–182. Synthesis and differential antiproliferative activity of Biginelli compounds against cancer cell lines: monastrol, oxo-monastrol and oxygenated analogues.

18 P. Biginelli, *Gazz. Chim. Ital.* 1893, 23, 360–416. Synthesis of 3,4-dihydropyrimidin-2(1*H*)-ones.

19 K. Folkers, T. B. Johnson, *J. Am. Chem. Soc.* 1933, 55, 3784–3791. Researches on pyrimidines. CXXXVI. The mechanism of formation of tetrahydropyrimidines by the Biginelli reaction.

20 C. O. Kappe, *J. Org. Chem.* 1997, 62, 7201–7204. Reexamination of the mechanism of the Biginelli dihydropyrimidine synthesis. Support for an *N*-acyliminium ion intermediate.

21 O. M. A. Rodrigo, T. De Souza, E. T. da Penha, H. M. S. Milagre, S. J. Garden, P. M. Esteves, M. N. Eberlin, O. A. C. Antunes, *Chem.–Eur. J.* 2009, 15, 9799–9804. The three-component Biginelli reaction: a combined experimental and theoretical mechanistic investigation.

22 F. Sweet, J. D. Fissekis, *J. Am. Chem. Soc.* 1973, 95, 8741–8749. On the synthesis of 3,4-dihydro-2(1*H*)-pyrimidones and the mechanism of the Biginelli reaction.

23 M.-I. Lannou, F. Hélion, J.-L. Namy, *Synlett* 2008, 105–107. Applications of lanthanide trichloride hydrates, prepared from mischmetall, in the Biginelli reaction.

24 C.-J. Liu, J.-D. Wang, *Molecules* 2009, 14, 763–770. Copper(II) sulfamate: an efficient catalyst for the one-pot synthesis of 3,4-dihydropyrimidine-2(1*H*)-ones and thiones.

25 A. Kamal, T. Krishnaji, M. A. Azhar, *Catal. Commun.* 2007, 8, 1929–1933. Copper(II) tetrafluoroborate as a mild and efficient catalyst for the one-pot synthesis of 3,4-dihydropyrimidin-2(1*H*)-ones under solvent-free conditions.

26 B. Piqani, W. Zhang, *Beilstein J. Org. Chem.* 2011, 7, 1294–1298. Synthesis of diverse dihydropyrimidine-related scaffolds by fluorous benzaldehyde-based Biginelli reaction and post-condensation modifications.

27 N. Ahmed, J. E. van Lier, *Tetrahedron Lett.* 2007, 48, 5407–5409. $TaBr_5$-catalyzed Biginelli reaction: one-pot synthesis of 3,4-dihydropyrimidin-2-(1*H*)-ones/thiones under solvent-free conditions.

28 S. K. De, R. A. Gibbs, *Synthesis* 2005, 1748–1750. Ruthenium(III) chloride-catalyzed one-pot synthesis of 3,4-dihydropyrimidin-2-(1*H*)-ones under solvent-free conditions.

29 N. S. Nandurkar, M. J. Bhanushali, M. D. Bhor, B. M. Bhanage, *J. Mol. Catal A: Chem.* 2007, 271, 14–17. $Y(NO_3)_3{\cdot}6H_2O$: a novel and reusable catalyst for one pot synthesis of 3,4-dihydropyrimidin-2(1*H*)-ones under solvent-free conditions.

30 A. Kumar, R. A. Maurya, *J. Mol. Catal A: Chem.* 2007, 272, 53–56. Synthesis of 3,4-dihydropyrimidin-2(1*H*)-ones using Ziegler–Natta catalyst system under solvent free conditions.

31 P. Salehi, M. Dabiri, M. Koohshari, S. K. Movahed, M. Bararjanian, *Mol. Divers.* 2011, 15, 833–837. One-pot synthesis of 1,2,3-triazole linked dihydropyrimidinones *via* Huisgen 1,3-dipolar/Biginelli cycloaddition.

32 (a) S. El Hankari, B. Motos-Pérez, P. Hesemann, A. Bouhaouss, J. J. E. Moreau, *Chem. Commun.* 2011, 47, 6704–6706. Periodic mesoporous organosilica from zwitterionic precursors. (b) O. Rosati, M. Curini, F. Montanari, M. Nocchetti, S. Genovese, *Catal. Lett.* 2011, 141, 850–853. α-Zirconium sulfophenylphosphonate as a catalyst for the synthesis of 3,4-dihydropyrimidin-2(1*H*)-one derivatives under solvent free conditions. (c) P. Li, S. Regati, R. J. Butcher, H. D. Arman, Z. Chen, S. Xiang, B. Chen, C.-G. Zhao, *Tetrahedron Lett.* 2011, 52, 6220–6222. Hydrogen-bonding 2D metal–organic solids as highly robust and efficient heterogeneous green catalysts for Biginelli reaction.

33 (a) R. Srivastava, *Catal. Lett.* 2010, 139, 17–25. Assessment of the catalytic activities of novel Brönsted acidic ionic liquid catalysts. (b) N. Sharma, U. K. Sharma, R. Kumar, R Richa, A. K. Sinha, *RSC Adv.* 2012, 2, 10648–10651. Green and recyclable glycine nitrate ($GlyNO_3$) ionic liquid triggered multicomponent Biginelli reaction for the efficient synthesis of dihydropyrimidinones.

34 (a) B. K. Banik, A. T. Reddy, A. Dattab, C. Mukhopadhyay, *Tetrahedron Lett.* 2007, 48, 7392–7394. Microwave-induced bismuth nitrate-catalyzed synthesis of dihydropyrimidones *via* Biginelli condensation under solventless conditions. (b) K. S. Niralwad, B. B. Shingate, M. S. Shingare, *Tetrahedron Lett.* 2010, 51, 3616–3618. Microwave-assisted one-pot synthesis of octahydroquinazolinone derivatives using ammonium metavanadate under solvent-free condition. (c) K. K. Pasunooti, H. Chai, C. N. Jensen, B. K. Gorityala, S. Wang, X.-W. Liu, *Tetrahedron Lett.* 2011, 52, 80–84. A microwave-assisted, copper-catalyzed three-component synthesis of dihydropyrimidinones under mild conditions.

35 C.-J. Liu, J.-D. Wang, *Molecules* 2010, 15, 2087–2095. Ultrasound-assisted synthesis of novel 4-(2-phenyl-1,2,3-triazol-4-yl)-3,4-dihydropyrimidin-2(1*H*)-(thio)ones catalyzed by $Sm(ClO_4)_3$.

36 For selected examples, see: (a) S. L. Jain, S. Singhal, B. Sain, *Green Chem.* 2007, 9, 740–741. PEG-assisted solvent and catalyst free synthesis of 3,4-dihydropyrimidinones under mild reaction conditions. (b) Z.-J. Quan, Y.-X. Da, Z. Zhang, X.-C. Wang, *Catal. Commun.* 2009, 10, 1146–1148. PS–PEG–SO_3H as an efficient catalyst for 3,4-dihydropyrimidones *via* Biginelli reaction. (c) H. Murata, H. Ishitani, M. Iwamoto, *Org. Biomol. Chem.* 2010, 8, 1202–1211. Synthesis of Biginelli dihydropyrimidinone derivatives with various substituents on aluminium-planted mesoporous silica catalyst. (d) S. D. Salim, K. G. Akamanchi, *Catal. Commun.* 2011, 12, 1153–1156. Sulfated tungstate: an alternative, eco-friendly catalyst for Biginelli reaction. (e) I. Couto, I. Tellitu, E. Domínguez, *Arkivoc* 2011, (ii), 115–126. Searching for a direct preparation of dihydropyrimidine-5-carboxamides under Biginelli reaction conditions.

37 Y. Huang, F. Yang, C. Zhu, *J. Am. Chem. Soc.* 2005, 127, 16386–16387. Highly enantioselective Biginelli reaction using a new chiral ytterbium catalyst: asymmetric synthesis of dihydropyrimidines.

38 For the pioneering enantioselective work, see: (a) O. Muñoz-Muñiz, E. Juaristi, *Arkivoc* 2003, (xi), 16–26. An enantioselective approach to the Biginelli dihydropyrimidinone condensation reaction using $CeCl_3$ and $InCl_3$ in the presence of chiral ligands. *Corrigendum*: (b) R. González-Olvera, R. Melgar-Fernández, E. Juaristi, *Arkivoc* 2003, (xi), 227.

39 (a) I. Suzuki, Y. Iwata, K. Takeda, *Tetrahedron Lett.* 2008, 49, 3238–3241. Biginelli reactions catalyzed by hydrazine type organocatalyst. (b) A. Debache, M. Amimour, A. Belfaitah, S. Rhouati, B. Carboni, *Tetrahedron Lett.* 2008, 49, 6119–6121. A one-pot Biginelli synthesis of 3,4-dihydropyrimidin-2-(1*H*)-ones/thiones catalyzed by triphenylphosphine as Lewis base. (c) Z.-L. Shen, X.-P. Xu, S.-J. Ji, *J. Org. Chem.* 2010, 75, 1162–1167. Brønsted base-catalyzed one-pot three-component Biginelli-type reaction: an efficient synthesis of 4,5,6-triaryl-3,4-dihydropyrimidin-2(1*H*)-one and mechanistic study. (d) F. Tamaddon, Z. Razmi, A. A. Jafari, *Tetrahedron Lett.* 2010, 51, 1187–1189. Synthesis of 3,4-dihydropyrimidin-2(1*H*)-ones and 1,4-dihydropyridines using ammonium carbonate in water.

40 M. K. Raj, H. S. P. Rao, S. G. Manjunatha, R. Sridharan, S. Nambiar, J. Keshwan, J. Rappai, S. Bhagat, B. S. Shwetha, D. Hegde, U. Santhosh, *Tetrahedron Lett.* 2011, 52, 3605–3609. A mechanistic investigation of Biginelli reaction under base catalysis.

41 For a recent pivotal review, see: M. M. Heravi, S. Asadi, B. M. Lashkariani, *Mol. Divers* 2013, 17, 389–407. Recent progress in asymmetric Biginelli reaction.

42 For a pivotal review, see: (a) L.-Z. Gong, X.-H. Chen, X.-Y. Xu, *Chem.–Eur. J.* 2007, 13, 8920–8926. Asymmetric organocatalytic Biginelli reactions: a new approach to quickly access optically active 3,4-dihydropyrimidin-2-(1*H*)-ones. For more recent examples, see: (b) F. Xu, D. Huang, X. Lin, Y. Wang, *Org. Biomol. Chem.* 2012, 10, 4467–4470. Highly enantioselective Biginelli reaction catalyzed by SPINOL-phosphoric acids. (c) D. An, Y.-S. Fan, Y. Gao, Z.-Q. Zhu, L.-Y. Zheng, S.-Q. Zhang, *Eur. J. Org. Chem.* 2014, 301–306. Highly enantioselective Biginelli reaction catalyzed by double axially chiral bisphosphorylimides.

43 For few non-chiral organocatalytic examples, see: (a) B. Ahmed, R. A. Khan, H Habibullah, M. Keshari, *Tetrahedron Lett.* 2009, 50, 2889–2892. An improved synthesis of Biginelli-type compounds *via* phase-transfer catalysis. (b) S. Verma, S. L. Jain, B. Sain, *Tetrahedron Lett.* 2010, 51, 6897–6900. PEG-embedded thiourea dioxide (PEG.TUD) as a novel organocatalyst for the highly efficient synthesis of 3,4-dihydropyrimidinones. (c) Y.-F. Cai, H.-M. Yang, L. Li, K.-Z. Jiang, G.-Q. Lai, J.-X. Jiang, L.-W. Xu, *Eur. J. Org. Chem.* 2010, 4986–4990. Cooperative and enantioselective $NbCl_5$/primary amine catalyzed Biginelli reaction. (d) M. Kargar, R. Hekmatshoar, A. Mostashari, Z. Hashemi, *Catal. Commun.* 2011, 15, 123–126. Efficient and green synthesis of 3,4-dihydropyrimidin-2(1*H*)-ones/thiones using imidazol-1-yl-acetic acid as a novel, reusable and water-soluble organocatalyst. (e) D. L. da Silva, S. A. Fernandes, A. A. Sabino, Â. de Fátima, *Tetrahedron Lett.* 2011, 52, 6328–6330. *p*-Sulfonic acid calixarenes as efficient and reusable organocatalysts for the synthesis of 3,4-dihydropyrimidin-2(1*H*)-ones/-thiones. (f) D.-Z. Xu, H. Li, Y. Wang, *Tetrahedron* 2012, 68, 7867–7872. Highly enantioselective Biginelli reaction catalyzed by a simple chiral primary amine catalyst: asymmetric synthesis of dihydropyrimidines.

44 X.-H. Chen, X.-Y. Xu, H. Liu, L.-F. Cun, L.-Z. Gong, *J. Am. Chem. Soc.* 2006, 128, 14802–14803. Highly enantioselective organocatalytic Biginelli reaction.

45 For selected pioneering examples, see: (a) D. Uraguchi, K. Sorimachi, M. Terada, *J. Am. Chem. Soc.* 2004, 126, 11804–11805. Organocatalytic asymmetric aza-Friedel-Crafts alkylation of furan. (b) T. Akiyama, H. Morita, J. Itoh, K. Fuchibe, *Org. Lett.* 2005, 7, 2583–2585. Chiral Brønsted acid catalyzed enantioselective hydrophosphonylation of imines: asymmetric synthesis of α-amino phosphonates.

46 (a) N. Li, X.-H. Chen, J. Song, S.-W. Luo, W. Fan, L.-Z. Gong, *J. Am. Chem. Soc.* 2009, 131, 15301–15310. Highly enantioselective organocatalytic Biginelli and Biginelli-like condensations:

reversal of the stereochemistry by tuning the 3,3′-disubstituents of phosphoric acids. *Corrigenda:* (b) *J. Am. Chem. Soc.* 2010, 132, 10953.

47 For previous racemic examples, see: (a) M. M. Abelman, S. C. Smith, D. R. James, *Tetrahedron Lett.* 2003, 44, 4559–4562. Cyclic ketones and substituted α-keto acids as alternative substrates for novel Biginelli-like scaffold syntheses. (b) Z.-T. Wang, L.-W. Xu, C.-G. Xia, H.-Q. Wang, *Tetrahedron Lett.* 2004, 45, 7951–7953. Novel Biginelli-like three-component cyclocondensation reaction: efficient synthesis of 5-unsubstituted 3,4-dihydropyrimidin-2(1*H*)-ones. (c) G. Sabitha, K. B. Reddy, R. Srinivas, J. Yadav, *Helv. Chim. Acta* 2005, 88, 2996–2999. Iodotrimethylsilane-accelerated one-pot synthesis of 5-unsubstituted 3,4-dihydropyrimidin-2(1*H*)-ones: a novel procedure for the Biginelli-like cyclocondensation reaction at room temperature. (d) Y.-L. Zhu, S.-L. Huang, Y.-J. Pan, *Eur. J. Org. Chem.* 2005, 2354–2367. Highly chemoselective multicomponent Biginelli-type condensations of cycloalkanones, urea or thiourea and aldehydes.

48 (a) J. S. Yadav, S. P. Kumar, G. Kondaji, R. S. Rao, K. Nagaiah, *Chem. Lett.* 2004, 33, 1168–1169. A novel L-proline catalyzed Biginelli reaction: one-pot synthesis of 3,4-dihydropyrimidin-2(1*H*)-ones under solvent-free conditions. (b) J. Mabry, B. Ganem, *Tetrahedron Lett.* 2006, 47, 55–56. Studies on the Biginelli reaction: a mild and selective route to 3,4-dihydropyrimidin-2(1*H*)-ones *via* enamine intermediates.

49 R. González-Olvera, P. Demare, I. Regla, E. Juaristi, *Arkivoc* 2008, (vi), 61–72. Application of (1*S*,4*S*)-2,5-diazabicyclo[2.2.1]heptane derivatives in asymmetric organocatalysis: the Biginelli reaction.

50 (a) B. List, *Synlett* 2001, 1675–1686. Asymmetric aminocatalysis. (b) B. List, *Chem. Commun.* 2006, 819–824. The ying and yang of asymmetric aminocatalysis.

51 J. Xin, L. Chang, Z. Hou, D. Shang, X. Liu, X. Feng, *Chem.–Eur. J.* 2008, 14, 3177–3181. An enantioselective Biginelli reaction catalyzed by a simple chiral secondary amine and achiral Brønsted acid by a dual-activation route.

52 J.-H. Sohn, H.-M. Choi, S. Lee, S. Joung, H.-Y. Lee, *Eur. J. Org. Chem.* 2009, 3858–3862. Probing the mode of asymmetric induction of Biginelli reaction using proline ester salts.

53 Y.-Y. Wu, Z. Chai, X.-Y. Liu, G. Zhao, S.-W. Wang, *Eur. J. Org. Chem.* 2009, 904–911. Synthesis of substituted 5-(pyrrolidin-2-yl)tetrazoles and their application in the asymmetric Biginelli reaction.

54 (a) Y. Wang, H. Yang, J. Yu, Z. Miao, R. Chen, *Adv. Synth. Catal.* 2009, 351, 3057–3062. Highly enantioselective Biginelli reaction promoted by chiral bifunctional primary amine-thiourea catalysts: asymmetric synthesis of dihydropyrimidines. (b) Y. Wang, J. Yu, Z. Miao, R. Chen, *Org. Biomol. Chem.* 2011, 9, 3050–3054. Bifunctional primary amine-thiourea–TfOH (BPAT·TfOH) as a chiral phase-transfer catalyst: the asymmetric synthesis of dihydropyrimidines.

55 D. Ding, C.-G. Zhao, *Eur. J. Org. Chem.* 2010, 3802–3805. Primary amine catalyzed Biginelli reaction for the enantioselective synthesis of 3,4-dihydropyrimidin-2(1*H*)-ones.

56 S. Saha, J. N. Moorthy, *J. Org. Chem.* 2011, 76, 396–402. Enantioselective organocatalytic Biginelli reaction: dependence of the catalyst on sterics, hydrogen bonding, and reinforced chirality.

57 For other interesting variations of the Biginelli process, see also: (a) N. Y. Gorobets, Y. V. Sedash, K. S. Ostras, O. V. Zaremba, S. V. Shishkina, V. N. Baumer, O. V. Shishkin, S. M. Kovalenko, S. M. Desenko, E. V. Van der Eycken, *Tetrahedron Lett.* 2010, 51, 2095–2098. Unexpected alternative direction of a Biginelli-like multicomponent reaction with 3-amino-1,2,4-triazole as the urea component. (b) S. V. Ryabukhin, A. S. Plaskon, S. S. Bondarenko, E. N. Ostapchuk, O. O. Grygorenko, O. V. Shishkin, A. A. Tolmachev, *Tetrahedron Lett.* 2010, 51, 4229–4232. Acyl pyruvates as synthons in the Biginelli reaction. (c) I. Essid, S. Touil, *Arkivoc* 2013, (iv) 98–106. β-Ketophosphonates as substrates in the Biginelli multicomponent reaction: an efficient and straightforward synthesis of phosphorylated dihydropyrimidinones.

58 A. R. Khosropour, M. M. Khodaei, M. Beygzadeh, M. Jokar, *Heterocycles* 2005, 65, 767–773. A one-pot synthesis of 3,4-dihydropyrimidin-2-(1*H*)-ones from primary alcohols promoted by $Bi(NO_3)_3{\cdot}5H_2O$ in two different media: organic solvent and ionic liquid.

59 Garima, V. P. Srivastava, L. D. S. Yadav, *Tetrahedron Lett.* 2010, 51, 6436–6438. Biginelli reaction starting directly from alcohols.

60 For selected examples, see: (a) M. Matloobi, C. O. Kappe, *Comb. Chem.* 2007, 9, 275–284. Microwave-assisted solution- and solid-phase synthesis of 2-amino-4-arylpyrimidine derivatives. (b) L. Pisani, H. Prokopcová, J. M. Kremsner, C. O. Kappe, *J. Comb. Chem.* 2007, 9, 415–421. 5-Aroyl-3,4-dihydropyrimidin-2-one library generation *via* automated sequential and parallel microwave-assisted synthesis techniques. (c) L. D. S. Yadav, A. Rai, V. K. Rai, C. Awasthi, *Synlett* 2007, 1905–1908. Biorenewable resources in the Biginelli reaction: cerium(III)-catalyzed synthesis of novel iminosugar-annulated perhydropyrimidines.

61 For selected examples, see: (a) J. C. Legeay, J. J. V. Eynde, L. Toupet, J. P. Bazureau, *Arkivoc* 2007, (iii), 13–28. A three-component condensation protocol based on ionic liquid phase bound acetoacetate for the synthesis of Biginelli 3,4-dihydropyrimidine-2(1*H*)-ones. (b) J. C. Legeay, J. J. V. Eyndeb, J. P. Bazureau, *Tetrahedron Lett.* 2007, 48, 1063–1068. A new approach to N-3 functionalized 3,4-dihydropyrimidine-2(1*H*)-ones with 1,2,4-oxadiazole group as amide isostere *via* ionic liquid-phase technology.

62 For selected examples, see: (a) V. Polshettiwar, R. S. Varma, *Tetrahedron Lett.* 2007, 48, 7343–7346. Biginelli reaction in aqueous medium: a greener and sustainable approach to substituted 3,4-dihydropyrimidin-2(1*H*)-ones. (b) S. Takale, S. Parab, K. Phatangare, R. Pisala, A. Chaskar, *Catal. Sci. Technol.* 2011, 1, 1128–1132. IBX in aqueous medium: a green protocol for the Biginelli reaction.

63 For selected examples, see: (a) H. Adibi, H. A. Samimi, M. Beygzadeh, *Catal. Commun.* 2007, 8, 2119–2124. Iron(III) trifluoroacetate and trifluoromethanesulfonate: recyclable Lewis acid catalysts for one-pot synthesis of 3,4-dihydropy-

rimidinones or their sulfur analogues and 1,4-dihydropyridines *via* solvent-free Biginelli and Hantzsch condensation protocols. (b) M. G. Kulkarni, S. W. Chavhan, M. P. Shinde, D. D. Gaikwad, A. S. Borhade, A. P. Dhondge, Y. B. Shaikh, V. B. Ningdale, M. P. Desai, D. R. Birhade, *Beilstein J. Org. Chem.* 2009, 5(4). Zeolite catalyzed solvent-free one-pot synthesis of dihydropyrimidin-2(1*H*)-ones—a practical synthesis of monastrol. (c) D. Prajapati, D. Bhuyan, M. Gohain, W. Hu, *Mol. Divers.* 2011, 15, 257–261. Green chemistry approaches to the regioselective synthesis of spiroheterobicyclic rings using iodine as a new and efficient catalyst under solvent-free conditions.

64 (a) K. C. Fylaktakidou, D. J. Hadjipavlou-Litina, K. E. Litinas, D. N. Nicolaides, *Curr. Pharm. Des.* 2004, 10, 3813–3826. Natural and synthetic coumarin derivatives with anti-inflammatory/antioxidant activities. (b) S. B. Bodendiek, C. Mahieux, W. Hansel, H. Wulff, *Eur. J. Med. Chem.* 2009, 44, 1838–1852. 4-Phenoxybutoxy-substituted heterocycles—a structure—activity relationship study of blockers of the lymphocyte potassium channel Kv1.3. (c) S. Stanchev, V. Hadjimitova, T. Traykov, T. Boyanov, I. Manolova, *Eur. J. Med. Chem.* 2009, 44, 3077–3082. Investigation of the antioxidant properties of some new 4-hydroxycoumarin derivatives. (d) T. Symeonidis, K. C. Fylaktakidou, D. J. Hadjipavlou-Litina, K. E. Litinas, *Eur. J. Med. Chem.* 2009, 44, 5012–5017. Synthesis and anti-inflammatory evaluation of novel angularly or linearly fused coumarins. (e) C. Xiao, Z.-G. Song, Z.-Q. Liu, *Eur. J. Med. Chem.* 2010, 45, 2559–2566. Synthesis of methyl-substituted xanthotoxol to clarify prooxidant effect of methyl on radical-induced oxidation of DNA. (f) P. O. Patil, S. B. Bari, S. D. Firke, P. K. Deshmukh, S. T. Donda, D. A. Patil, *Bioorg. Med. Chem.* 2013, 21, 2434–2450. A comprehensive review on synthesis and designing aspects of coumarin derivatives as monoamine oxidase inhibitors for depression and Alzheimer's disease.

65 O. M. Singh, N. S. Devi, *J. Org. Chem.* 2009, 74, 3141–3144. Application of β-oxodithioesters in domino and multicomponent reactions: facile route to dihydropyrimidines and coumarins.

66 G. C. Nandi, S. Samai, M. S. Singh, *J. Org. Chem.* 2010, 75, 7785–7795. Biginelli and Hantzsch-type reactions leading to highly functionalized dihydropyrimidinone, thiocoumarin, and pyridopyrimidinone frameworks *via* ring annulation with β-oxodithioesters.

67 A. R. Trivedi, V. R. Bhuva, B. H. Dholariya, D. K. Dodiya, V. B. Kataria, V. H. Shah, *Bioorg. Med. Chem. Lett.* 2010, 20, 6100–6102. Novel dihydropyrimidines as a potential new class of antitubercular agents.

68 E. Rajanarendar, M. N. Reddy, K. R. Murthy, K. G. Reddy, S. Raju, M. Srinivas, B. Praveen, M. S. Rao, *Bioorg. Med. Chem. Lett.* 2010, 20, 6052–6055. Synthesis, antimicrobial, and mosquito larvicidal activity of 1-aryl-4-methyl-3,6-bis-(5-methylisoxazol-3-yl)-2-thioxo-2,3,6,10*b*-tetrahydro-1*H*-pyrimido[5,4-*c*]quinolin-5-ones.

69 R. Gupta, A. Jain, R. Joshi, M. Jain, *Bull. Korean Chem. Soc.* 2011, 32, 899–904. Eco-friendly solventless synthesis of 5-indolylpyrimido[4,5-*d*]pyrimidinones and their antimicrobial activity.

70 S. Chitra, D. Devanathan, K. Pandiarajan, *Eur. J. Med. Chem.* 2010, 45, 367–371. Synthesis and *in vitro* microbiological evaluation of novel 4-aryl-5-isopropoxycarbonyl-6-methyl-3, 4-dihydropyrimidinones.

71 For some recent examples exploring the biological activities, see: (a) N. K. Ladani, M. P. Patel, R. G. Patel, *Arkivoc* 2009, (vii), 292–302. An efficient three component one-pot synthesis of some new octahydroquinazolinone derivatives and investigation of their antimicrobial activities. (b) K. Singh, K. Singh, B. Wan, S. Franzblau, K. Chibale, J. Balzarini, *Eur. J. Med. Chem.* 2011, 46, 2290–2294. Facile transformation of Biginelli pyrimidin-2(1*H*)-ones to pyrimidines. *In vitro* evaluation as inhibitors of mycobacterium tuberculosis and modulators of cytostatic activity. (c) A. Kamal, M. S. Malik, S. Bajee, S. Azeeza, S. Faazil, S. Ramakrishna, V. G. M. Naidu, M. V. P. S. Vishnuwardhan, *Eur. J. Med. Chem.* 2011, 46, 3274–3281. Synthesis and biological evaluation of conformationally flexible as well as restricted dimers of monastrol and related dihydropyrimidones. (d) T. N. Akhaja, J. P. Raval, *Eur. J. Med. Chem.* 2011, 46, 5573–5579. 1,3-Dihydro-2*H*-indol-2-ones derivatives: design, synthesis, *in vitro* antibacterial, antifungal and antitubercular study. (e) H. J. M. Gijsen, D. Berthelot, M. A. J. De Cleyn, I. Geuens, B. Brône, M. Mercken, *Bioorg. Med. Chem. Lett.* 2012, 22, 797–800. Tricyclic 3,4-dihydropyrimidine-2-thione derivatives as potent TRPA1 antagonists. (f) D. L. da Silva, F. S. Reis, D. R. Muniz, A. L. T. G. Ruiz, J. E. de Carvalho, A. A. Sabino, L. V. Modolo, Â. de Fátima, *Bioorg. Med. Chem. Lett.* 2012, 20, 2645–2650. Free radical scavenging and antiproliferative properties of Biginelli adducts. (g) R. K. Yadlapalli, O. P. Chourasia, K. Vemuri, M. Sritharan, R. S. Perali, *Bioorg. Med. Chem. Lett.* 2012, 22, 2708–2711. Synthesis and *in vitro* anticancer and antitubercular activity of diarylpyrazole ligated dihydropyrimidines possessing lipophilic carbamoyl group. (h) A. B. Siddiqui, A. R. Trivedi, V. B. Kataria, V. H. Shah, *Bioorg. Med. Chem. Lett.* 2014, 24, 1493–1495. 4,5-Dihydro-1*H*-pyrazolo[3,4-*d*]pyrimidine containing phenothiazines as antitubercular agents.

72 J. M. Goss, S. E. Schaus, *J. Org. Chem.* 2008, 73, 7651–7656. Enantioselective synthesis of SNAP-7941: chiral dihydropyrimidone inhibitor of MCH1-R.

73 A. Crespo, A. El Maatougui, P. Biagini, J. Azuaje, A. Coelho, J. Brea, M. I. Loza, M. I. Cadavid, X. García-Mera, H. Gutiérrez-de-Terán, E. Sotelo, *ACS Med. Chem. Lett.* 2013, 4, 1031–1036. Discovery of 3,4-dihydropyrimidin-2(1*H*)-ones as a novel class of potent and selective A_{2B} adenosine receptor antagonists.

74 V. M. Rao, P. M. Kumar, D. Rambabu, R. Kapavarapu, S. S. Rani, P. Misra, M. Pal, *Bioorg. Chem.* 2013, 51, 48–53. Novel alkynyl substituted 3,4-dihydropyrimidin-2(1*H*)-one derivatives as potential inhibitors of chorismate mutase.

75 R. S. Bhosale, S. V. Bhosale, S. V. Bhosale, T. Wang, P. K. Zubaidha, *Tetrahedron Lett.* 2004, 45, 9111–9113. An efficient, high yield protocol for the one-pot synthesis of dihydropyrimidin-2(1*H*)-ones catalyzed by iodine.

76 P. M. Kumar, K. S. Kumar, S. R. Poreddy, P. K. Mohakhud, K. Mukkanti, M. Pal, *Tetrahedron Lett.* 2011, 52, 1187–1191.

Biginelli reaction beyond three-component limit: synthesis of functionalized pyrimidinones *via* a one-pot Biginelli-Pd mediated C–C coupling strategy.

77 D. M. Evans, P. J. Murphy, *Chem. Commun.* 2011, 47, 3225–3226. A biomimetic approach to the cylindrospermopsin alkaloids.

78 For reviews concerning the toxicological activity see: (a) D. J. Griffiths, M. L. Saker, *Environ. Toxicol.* 2003, 18, 78–93. The Palm island mystery disease 20 years on: a review of research on the cyanotoxin cylindrospermopsin. (b) S. Kinnear, *Mar. Drugs* 2010, 8, 542–564. Cylindrospermopsin: a decade of progress on bioaccumulation research.

10

BUCHERER–BERGS AND STRECKER MULTICOMPONENT REACTIONS

EUGENIA MARQUÉS-LÓPEZ AND RAQUEL P. HERRERA

Departamento de Química Orgánica, Instituto de Síntesis Química y Catálisis Homogénea (ISQCH), CSIC-Universidad de Zaragoza, Zaragoza, Spain

10.1 BUCHERER–BERGS REACTION

10.1.1 Introduction

The Strecker [1] and Bucherer–Bergs [2] reactions are the most important strategies and powerful tools for the synthesis of α-amino acids from a carbonyl compound and cyanide. The classical first approach uses ammonium hydroxide and potassium cyanide to obtain the corresponding aminonitrile compound **1**, whereas the second one is a four-component condensation between an aldehyde or ketone derivative reacting with potassium cyanide and ammonium carbonate as source of ammonia and carbon dioxide affording hydantoin compounds **2** [3]. In both cases, final hydrolysis would deliver the corresponding α,α-disubstituted α-amino acid derivatives **3** (Scheme 10.1) [4].

Following the Bucherer–Bergs procedure, cyclic α-amino acids can be successfully obtained starting from cyclic ketones [5]. Although the Bucherer–Bergs method provides an easy access to many polycyclic α-amino acids from the corresponding keto derivatives, the classic Bucherer–Bergs reactions require harsh conditions for the hydrolysis of hydantoin intermediates **2**, long reaction times, and elevated temperatures, limiting the scope of this methodology. However, ultrasound irradiation has been proved to improve the methodology compared with conventional heating conditions and proving higher yield with shorter reaction time being tolerant to different functional groups such as ether, ester, and acetals [6]. It is worth noting that Bucherer–Bergs approach has been mainly used for ketones, while Strecker reaction has been more extensively explored using aldehydes.

10.1.2 Comparative Stereochemical Course

Although both methodologies afford final α-amino acid **3** after hydrolysis, the relative stereochemistry of the products in cyclic compounds has been found different. In fact, in 1961, Munday found predominantly one of the possible stereoisomeric hydantoins in the Bucherer–Bergs reaction of 4-*tert*-butylcyclohexanone [7]. Later, Edward and Jitrangsri [8] proved that the high steric hindrance was a determinant factor in the stereochemical course of the Bucherer–Bergs reaction affording final α-amino acid **3** in which the NH and the carboxylic acid groups are preferentially in a *cis*-disposition. However, in the Strecker process, this hindrance is not a main factor in the key step of the mechanistic route, and the attack of the CN would furnish the thermodynamically most stable product, which has a *trans*-disposition between the resulting NH and the carboxylic acid groups (Scheme 10.2).

This mechanistic fact has been successfully employed and demonstrated by some authors in the synthesis of different compounds with *cis*- or *trans*-configuration, as depicted in Scheme 10.2 [9].

10.1.3 Synthesis of Five-Membered Heterocycles

Although the Bucherer–Bergs reaction is mainly employed for the synthesis of α-amino acids, the synthesis of interesting hydantoins can be also addressed (Fig. 10.1) [17]. Curiously, hydantoins were initially observed as undesired by-products in the synthesis of peptides [18]. However, nowadays, this motif displays many important biological and pharmacological properties and is commonly used in drug discovery research lines [4a,b, 19]. These features have

Multicomponent Reactions: Concepts and Applications for Design and Synthesis, First Edition. Edited by Raquel P. Herrera and Eugenia Marqués-López.

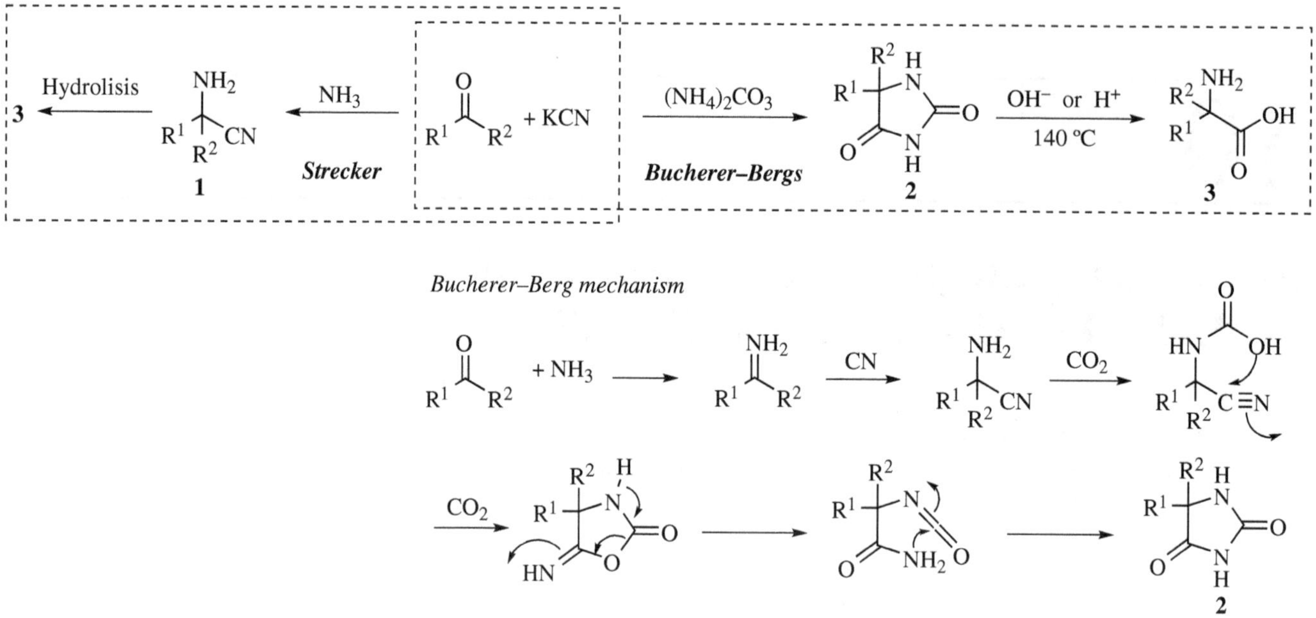

SCHEME 10.1 General Bucherer–Bergs reaction and mechanism.

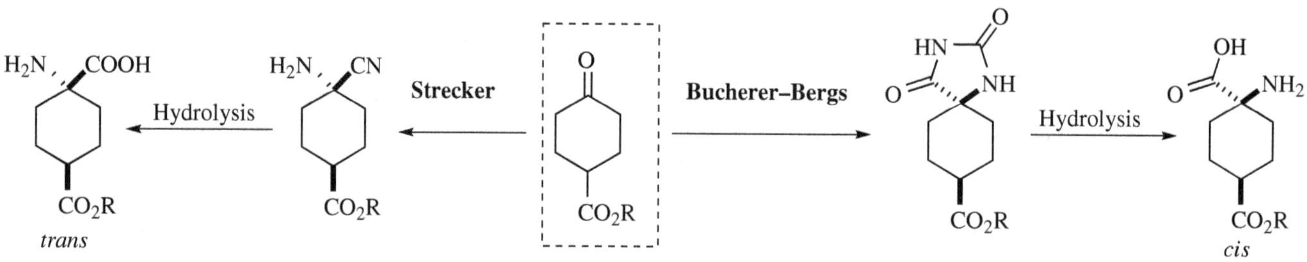

SCHEME 10.2 Relative stereochemistry of final amino acids through a Strecker and Bucherer–Bergs synthesis.

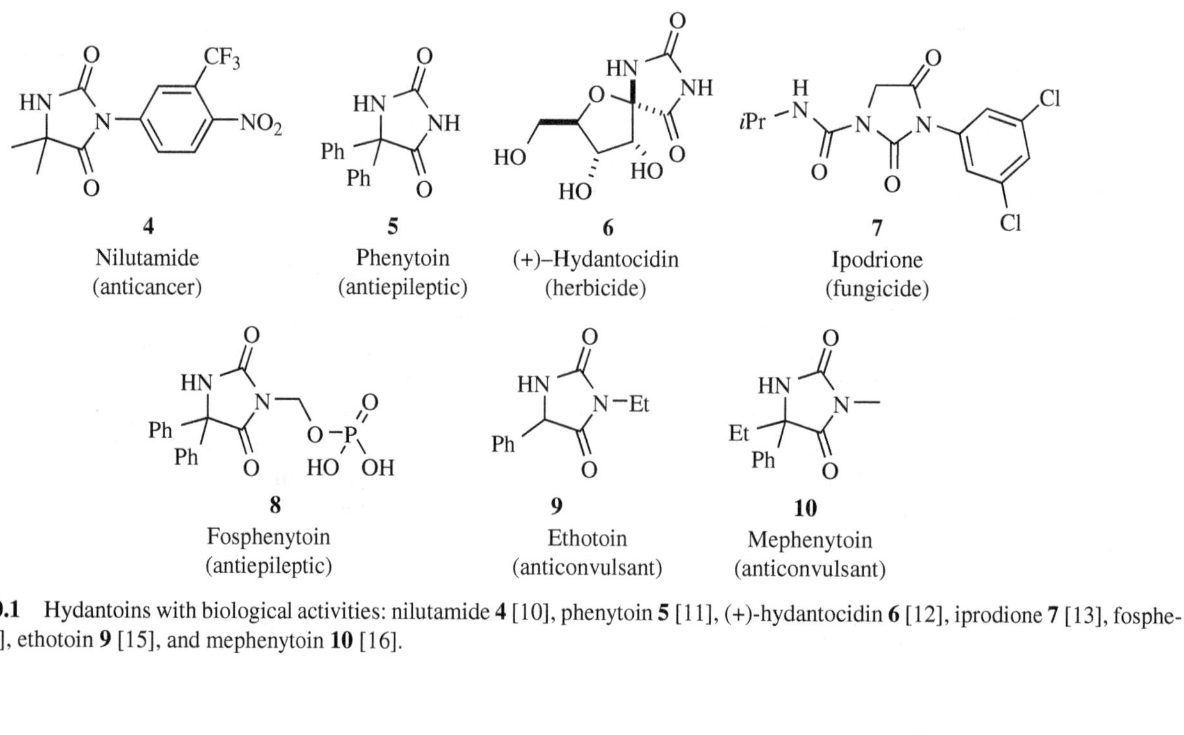

FIGURE 10.1 Hydantoins with biological activities: nilutamide **4** [10], phenytoin **5** [11], (+)-hydantocidin **6** [12], iprodione **7** [13], fosphenytoin **8** [14], ethotoin **9** [15], and mephenytoin **10** [16].

resulted an attractive area of research for the development of new methodologies, providing easy access to new drug candidates [20].

In the last decade, interesting hydantoin scaffolds have been prepared following a Bucherer–Bergs approach (Fig. 10.2). In this part of the chapter, three remarkable examples will be illustrated.

Hameed and coworkers have recently synthesized a variety of spirobicycloimidazolidine-2,4-diones **17** following a Bucherer–Bergs approach as the key reaction step (Scheme 10.3) [26]. The synthesized products **17** were tested for *in vivo* hypoglycemic activity on male albino rats and inhibition of ALR1 and ALR2 extracted from the kidney and eye of the calf, respectively, showing interesting activities in many cases. Some of these molecules could serve as the drug to treat diabetes and diabetic complications in a single medication.

Interestingly, Thomas and coworkers have recently developed a series of 1,3,4,4a,10,10a-hexahydropyrano[4,3-*b*] chromene analogues using for their synthesis a Bucherer–Bergs strategy, which allows the obtainment of the hydantoin cycle intermediate **18** necessary for the preparation of the final products **19** (Scheme 10.4) [27]. Compounds **19** were analyzed as BACE1 inhibitors for Alzheimer's disease with the aim of obtaining efficient inhibitors able to cross the

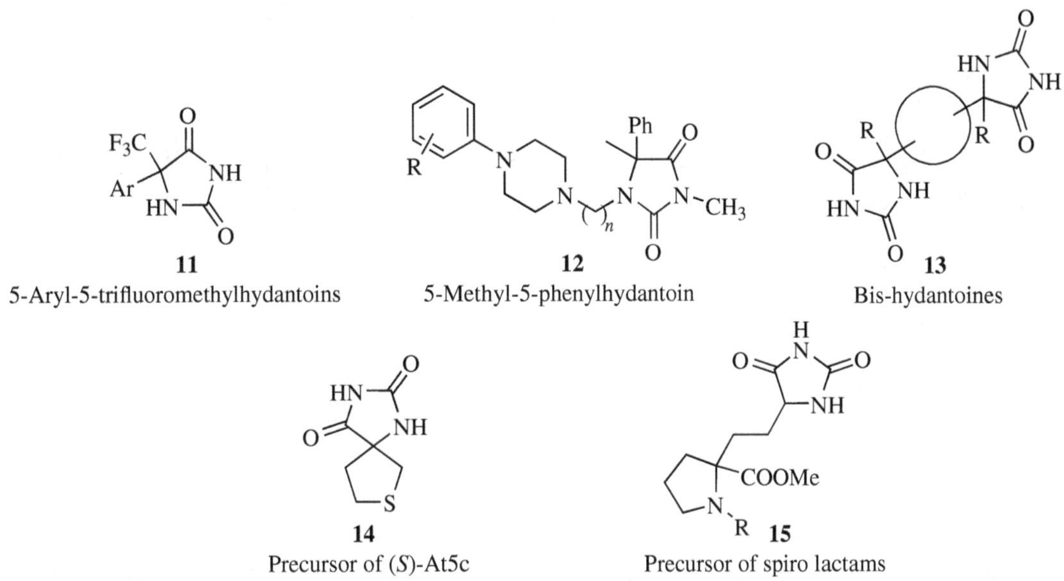

FIGURE 10.2 Synthesis of interesting hydantoins through a Bucherer–Bergs reaction: **11** [21], **12** [22], **13** [23], **14** [24], and **15** [25].

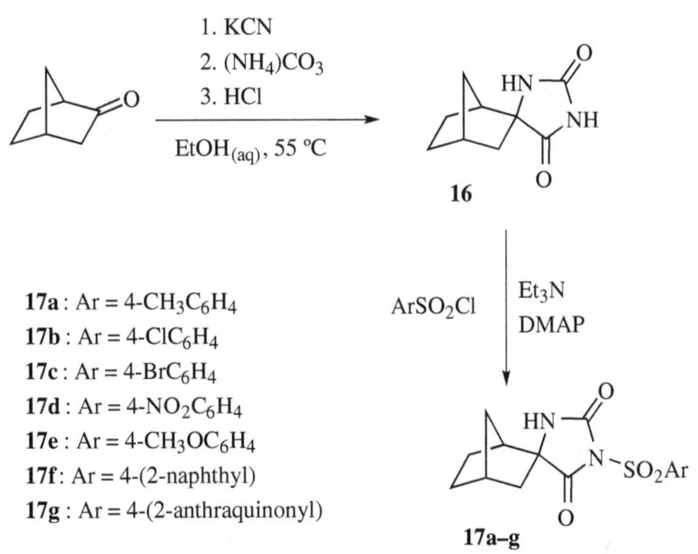

17a–g
51–71% yield

SCHEME 10.3 Synthesis of active spirobicycloimidazolidine-2,4-diones **17**.

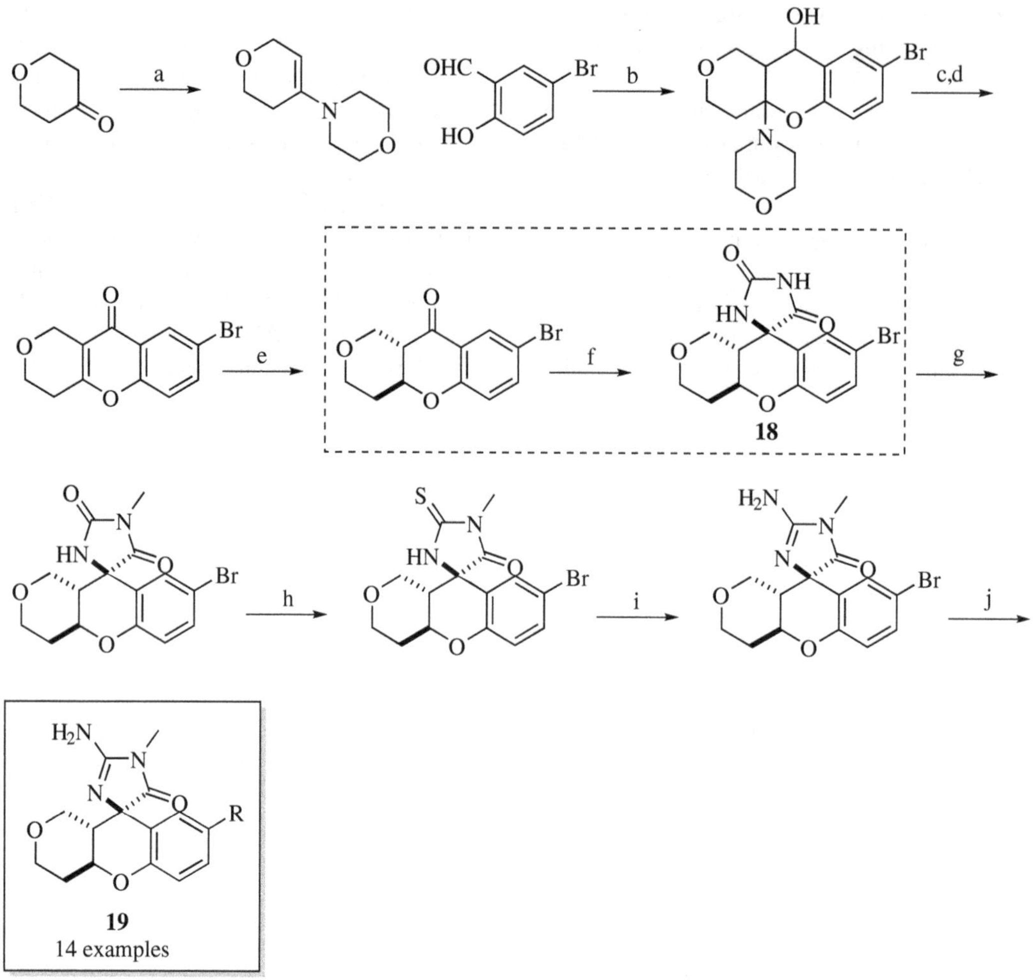

SCHEME 10.4 Synthesis of spirocyclic acyl guanidine derivatives **19**: (a) morpholine, toluene, reflux, Dean–Stark trap. (b) Toluene, r.t. (c) Oxalyl chloride, DMSO, Et_3N, CH_2Cl_2, −78 °C. (d) HOAc, r.t. (e) L-Selectride, THF, −78 °C. (f) KCN, $(NH_4)_2CO_3$, $NaHSO_3$, EtOH, 130 °C. (g) MeI, K_2CO_3, DMF, r.t. (h) Lawesson's reagent, toluene, 90 °C. (i) *t*BuOOH, $NH_4OH_{(aq.)}$, MeOH, 50 °C. (j) $RB(OH)_2$, $Pd[P(Ph)_3]_4$, $Na_2CO_{3(aq.)}$, dioxane, 90 °C.

blood–brain barrier, an interesting and important challenge in this field of research.

Bucherer–Bergs adduct **18** was the major diastereomer of the process, which was enriched to greater than 80% diastereomeric purity by trituration with hot *i*PrOH. The authors also prepared additional analogue compounds using the same Bucherer–Bergs approach, and interesting activity values were found for BACE1 and CatD enzyme assays, a cell assay for the production of Aβ1–40, and for efflux ratio and permeability in an MDR1-transfected LLC-PK1 cell line.

Daïch and coworkers developed an efficient methodology for the synthesis of 4,5-fused imidazolidin-2-ones **25** and **26** from bicyclic and tricyclic ketones **20** and **21** in a four-step sequence (Scheme 10.5) [28]. For such a purpose, the authors carried out the synthesis of a spirohydantoin Bucherer–Bergs intermediate, a successive mono- and dialkylation of the nitrogen atom of the hydantoin ring affording **22** and **23**, a next regioselective reduction of one carbonyl function, and finally a cationic cyclization associated with ring expansion. This methodology gave easy access to polyheterocyclic systems **25** and **26** containing an imidazolidin-2-one nucleus in good to high yields (67–99%) and was compatible with a large-scale production (up to 3 g of product).

Interestingly, novel heterocycles **25** and **26** could have pharmaceutically interest. Additionally, products **26** bearing a secondary amide function could be subsequently transformed, giving access to more interesting scaffolds.

10.1.4 Metal-Catalyzed Synthesis of Hydantoin Derivatives

Beller's group described a pioneering palladium-catalyzed one-pot synthesis of 5-substituted hydantoins **27**, 3,5-substituted hydantoins **28**, and 1,3,5-substituted hydantoins **30** based on the carbonylation of aldehydes in the presence of urea derivatives as an interesting alternative to Bucherer–Bergs reactions (Scheme 10.6) [29].

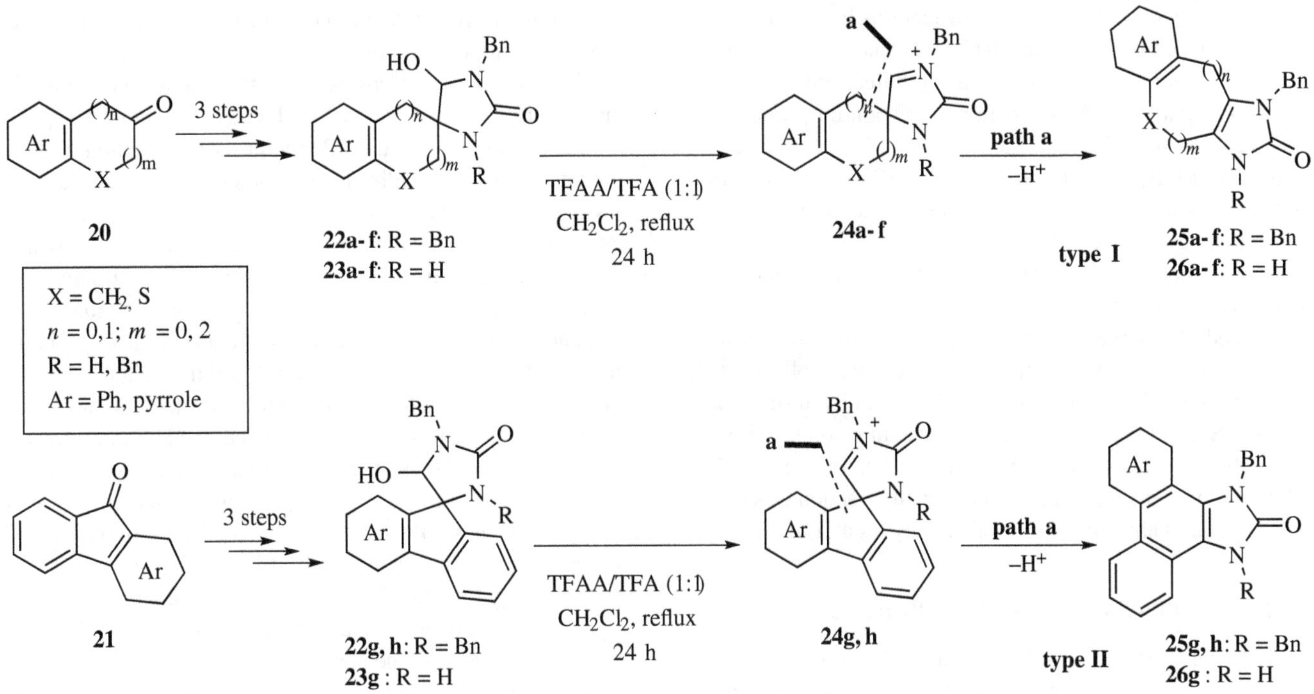

SCHEME 10.5 Synthesis of polyheterocyclic systems **25** and **26**.

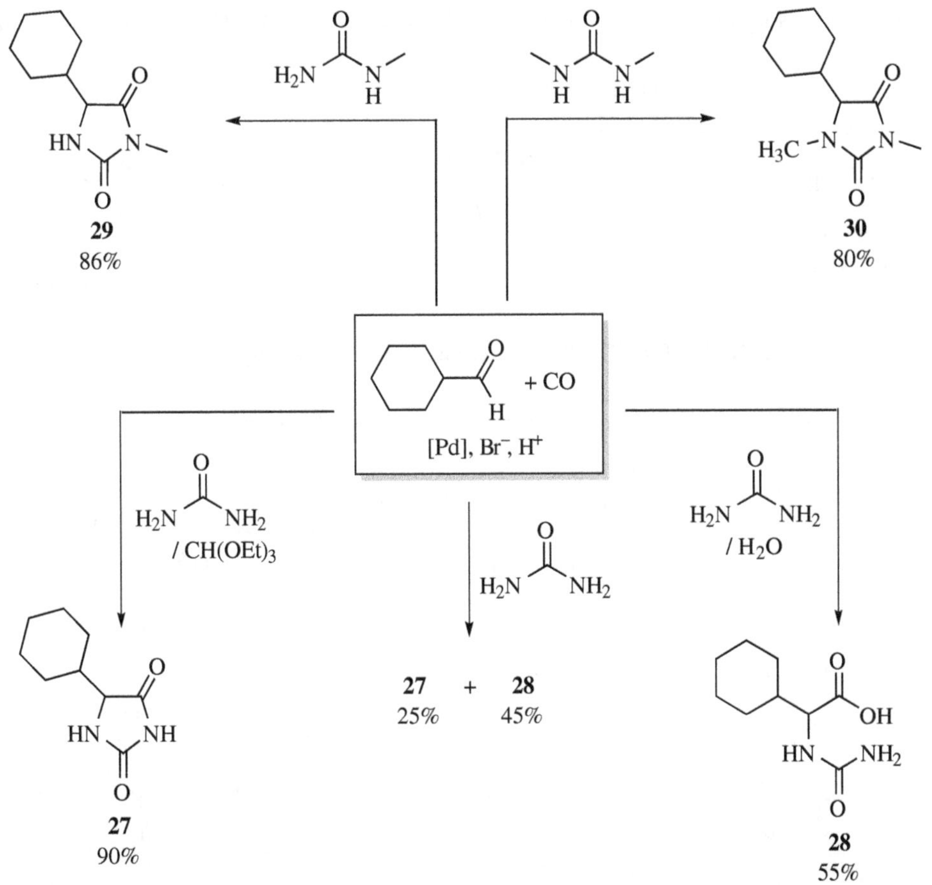

SCHEME 10.6 Pd-catalyzed one-pot synthesis of hydantoins.

More recently, based on a Bucherer–Bergs protocol, Conway and coworkers have developed a one-pot gallium(III) triflate-catalyzed synthesis of 5-substituted and 5,5-disubstituted hydantoins **31** from the corresponding aldehydes and ketones in CH_2Cl_2 (Scheme 10.7) [30]. This one-pot synthesis of hydantoins avoids the difficulty associated with the isolation and purification of the α-aminonitrile intermediate in these processes.

The methodology presents some advantages over the existing classical Bucherer–Bergs reaction. Firstly, the authors developed the strategy using organic solvents, rather than aqueous or ethanol, which makes the procedure applicable to a broader range of substrates favoring the solubility of such compounds. Secondly, the reaction is carried out at lower temperature, which helps to avoid complications related to the volatility of the ammonium carbonate. Moreover, the reaction was found to be operationally simple for both aldehydes and ketones.

10.1.5 Modified Bucherer–Bergs Reaction

Despite the wide scope and potential of the multicomponent (MC) Bucherer–Bergs reaction, this procedure exhibits a limitation since only changes in the structure of the corresponding ketones can be addressed, leading to variations in the skeleton of the final hydantoin [31]. However, with the aim in mind of increasing the utility and scope of this chemistry in the development of new drug, the Shipman group envisioned the possibility of the *in situ* generation of the desired ketones affording a greater structural diversity in the final hydantoin adducts **2** (Schemes 10.8 and 10.9) [32].

The key of this modified Bucherer–Bergs reaction was the use of a nitrile **32** with an organometallic reagent such as RMgX or RLi. The choice of this strategy was based on the following aspects: (i) there is a large number of nitriles, which are commercially available or readily accessible through other methods; (ii) it is also known that common organometallic reagents including RMgX and RLi are able to add to alkyl-, aryl-, and heteroaryl-substituted nitriles in high yields; and (iii) subsequent protonation of the intermediate metalated imine **33** directly leads to the NH imine, the key intermediate in the Bucherer–Bergs process [2]. In this context, the authors successfully developed a simple and practical one-pot strategy for the synthesis of 5,5′-disubstituted hydantoins **2** as represented in Scheme 10.9. It is worth noting that this methodology tolerates larger variation in the structure of the organometallic reagent and the nitrile, showing a broader scope compared with the classical version.

The same group demonstrated the applicability of this strategy in the efficient seven-step synthesis of an amide-functionalized monothiohydantoin **34**, a novel inhibitor of fatty acid amide hydrolase (FAAH) ($IC_{50} = 23.4 \pm 1.1\ \mu M$) (Scheme 10.10) [33], which is a membrane-bound enzyme,

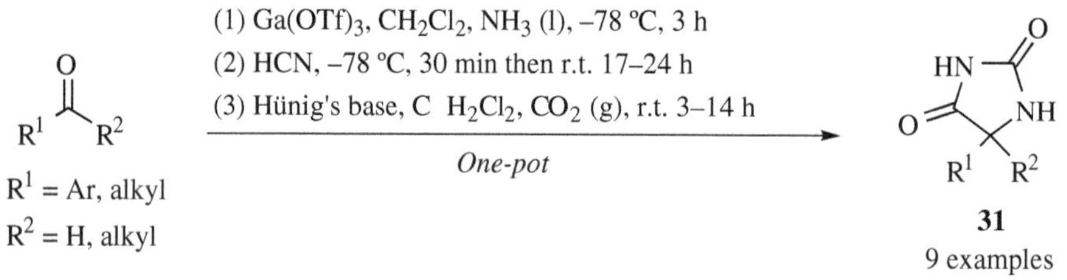

SCHEME 10.7 Ga(III)-catalyzed Bucherer–Bergs one-pot reaction.

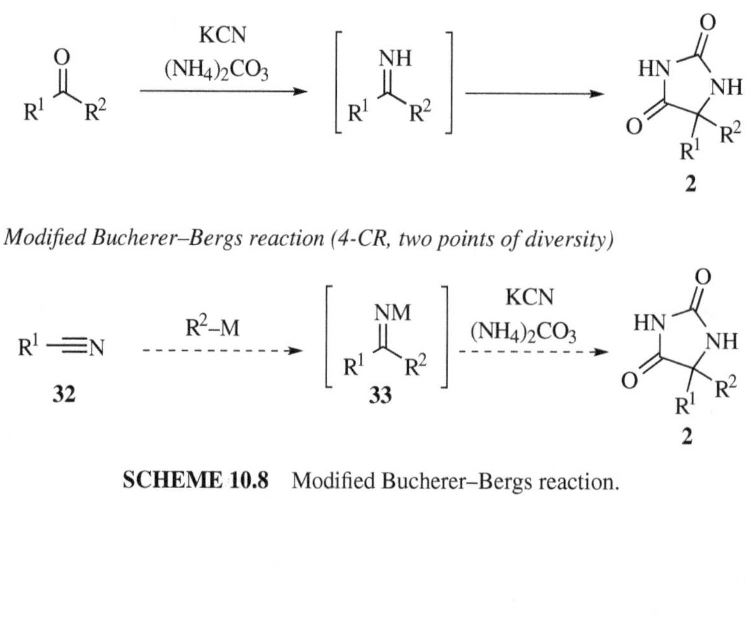

SCHEME 10.8 Modified Bucherer–Bergs reaction.

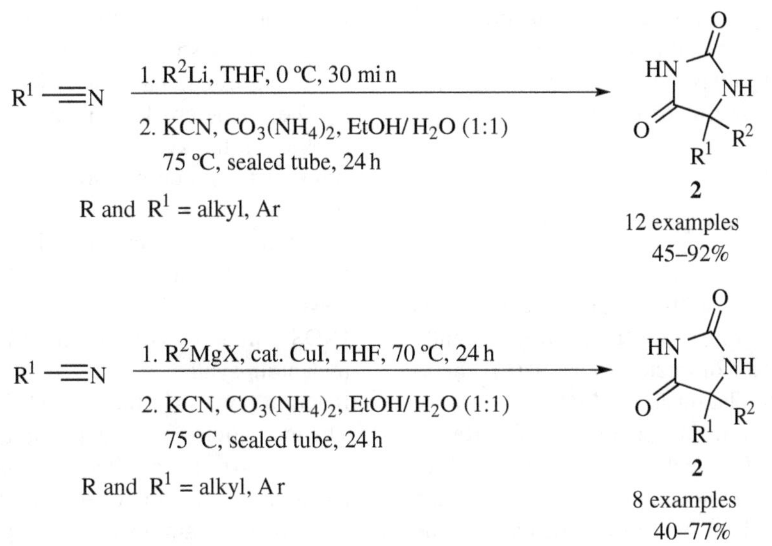

SCHEME 10.9 Synthesis of hydantoins **2** using organolithium compound and Grignard reagents.

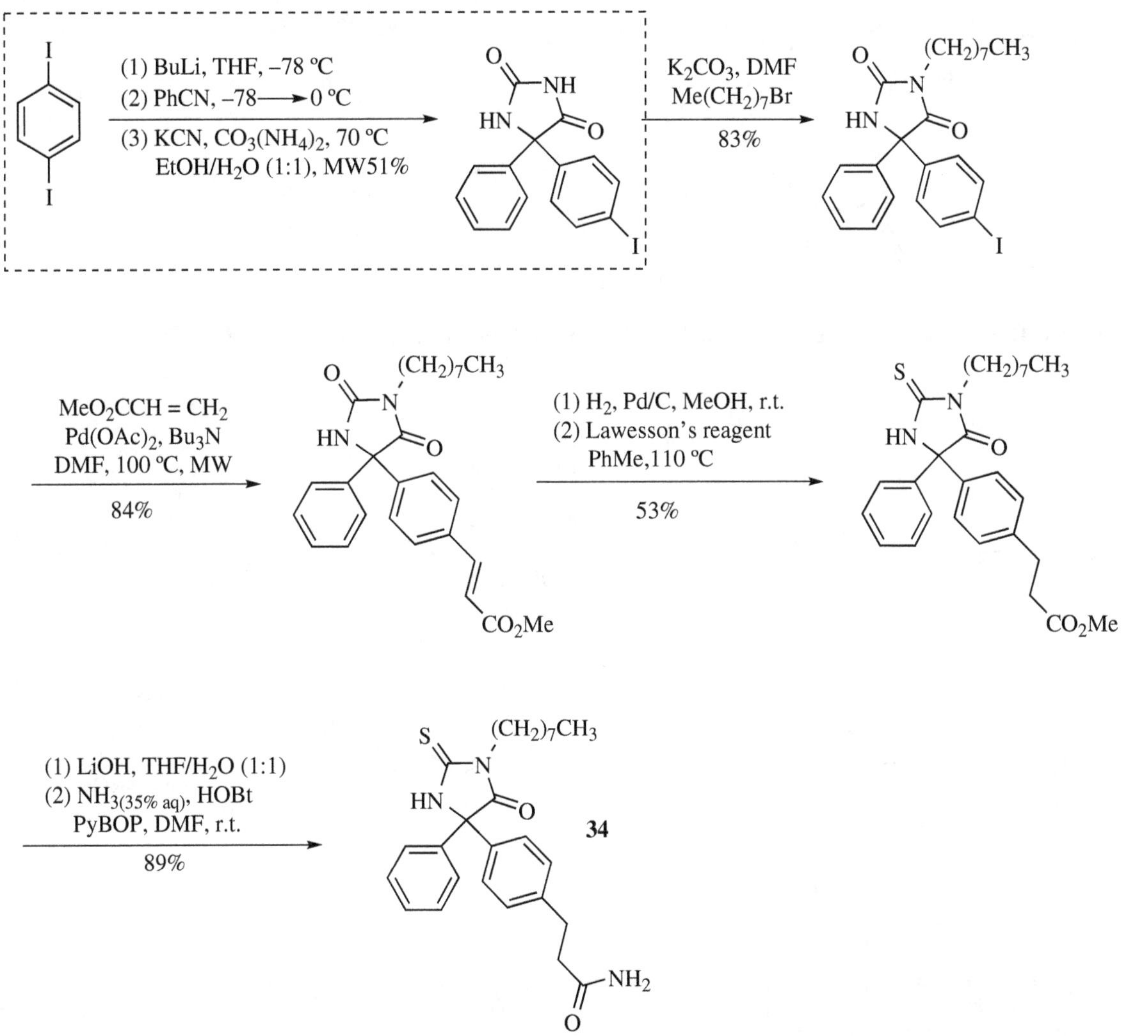

SCHEME 10.10 Synthesis of amide-functionalized monothiohydantoin **34**.

responsible for the hydrolysis of bioactive lipids, including endogenous ligands of the cannabinoid receptors such as anandamide [34]. The synthesis was achieved in 17% overall yield, and it is a useful example of microwave-assisted 4CR.

10.1.6 Synthesis of α-Amino Acids *via* Hydantoin Intermediate

The design and synthesis of a variety of isomeric compounds that possess functional groups on a conformationally restricted or rigid scaffold are of interest in medicinal chemistry for structure–activity relationship (SAR) studies, since the rigidity can be involved in the origin of different biological properties [35]. This feature can be employed for the search of active sites of biological targets, or to explore most efficient ligands for them, since the efficiency of the interaction with a receptor or enzyme could depend on the rigidity of the structure being higher than that for the analogous conformationally more flexible [35].

The Bucherer–Bergs methodology has been successfully applied on the synthesis of diverse amino acid derivatives, some of them conformationally constrained as represented in Figure 10.3.

An interesting and useful example of synthesis of α-amino acids has been recently reported by Wuts and coworkers, describing a scalable process to prepare the INOS inhibitor PHA-399733 **42** using the Bucherer–Bergs hydantoin synthesis as key step to introduce the amino acid function present in the final skeleton (Scheme 10.11) [41]. The methodology was applicable to the isolation of 39.9 kg (75% yield) of the racemic hydantoin **41**, and the desired enantiomer was then opened in basic conditions giving the corresponding amino acid moiety.

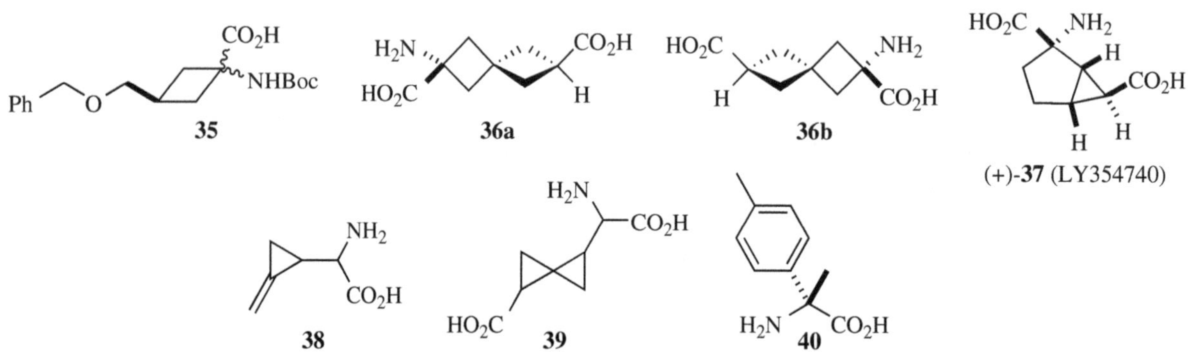

FIGURE 10.3 Representative α-amino acid derivatives: **35** [36], **36a** and **36b** [37], **37** [38], **38** and **39** [39], and **40** [40].

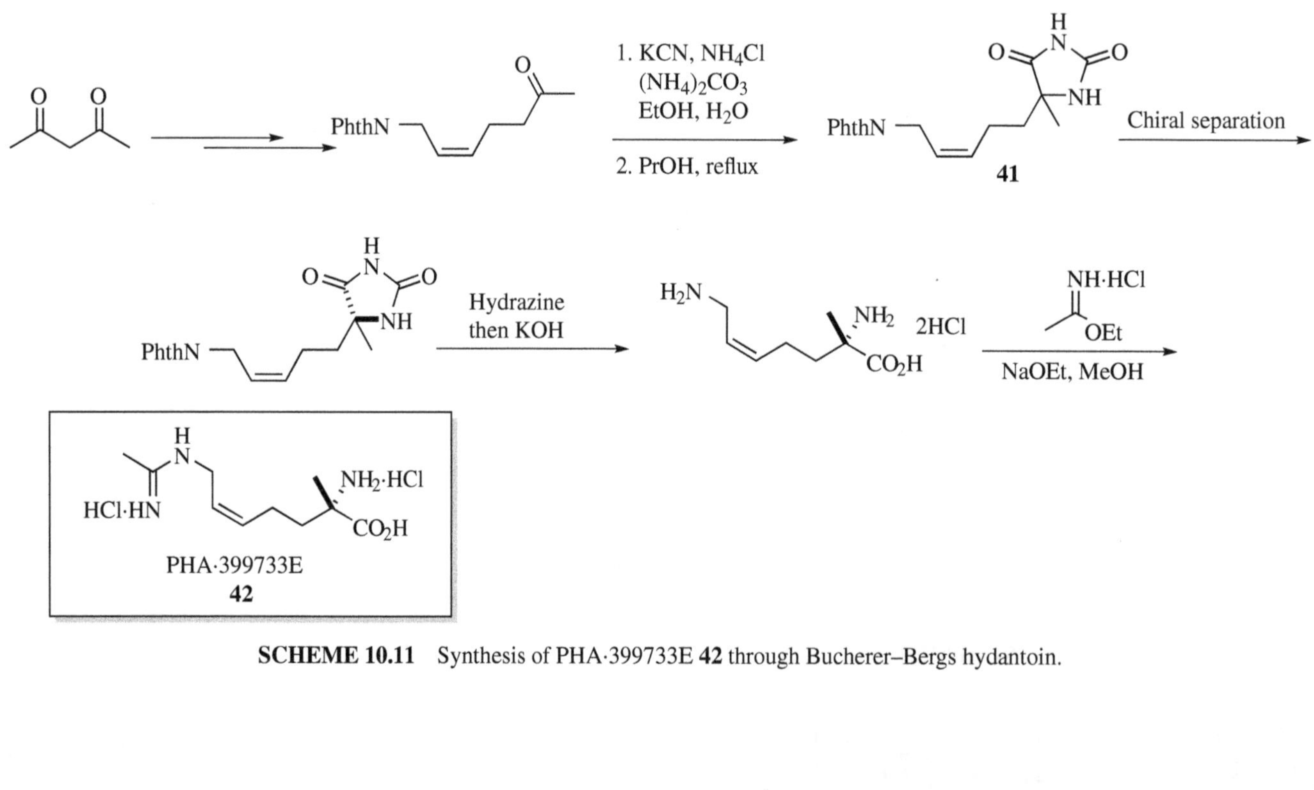

SCHEME 10.11 Synthesis of PHA-399733E **42** through Bucherer–Bergs hydantoin.

The Bucherer–Bergs reaction was also employed to install the masked amino acid functionality present in the diester intermediate **45**, key for the synthesis of methionine amide **46** (LY2140023) (Scheme 10.12) [42].

The authors developed a scalable route for the synthesis of the intermediate core **45** in 12 steps and 0.4% yield and successfully implemented at a pilot plant scale. In this multistep synthesis, racemic hydantoin **43** was obtained after crystallization in 4:1 mixture of diastereomers. Separation of enantiomers by diastereomeric salt formation with (*R*)-2-phenylglycerol and further crystallization followed by salt break gave single isomer (–)-**44**, with the targeted L-glutamic acid configuration.

10.1.7 Synthesis of Diaminodicarboxylic Acids

The chemistry of diaminodicarboxylic acids has gained great interest since these are important biological agents (Fig. 10.4). These compounds have inspired diverse syntheses of a number of structurally related unnatural diaminodicarboxylic acids intended to enhance the bioactivity and/or control the peptide secondary structure [45].

More recently, other authors have also applied the Bucherer–Bergs methodology for the synthesis of conformationally constrained diaminodicarboxylic acid derivatives (Fig. 10.5).

In fact, Iverson and coworkers used the diaminodicarboxylic acid **52** for the synthesis of two novel, rigid, *cis*-**55** and

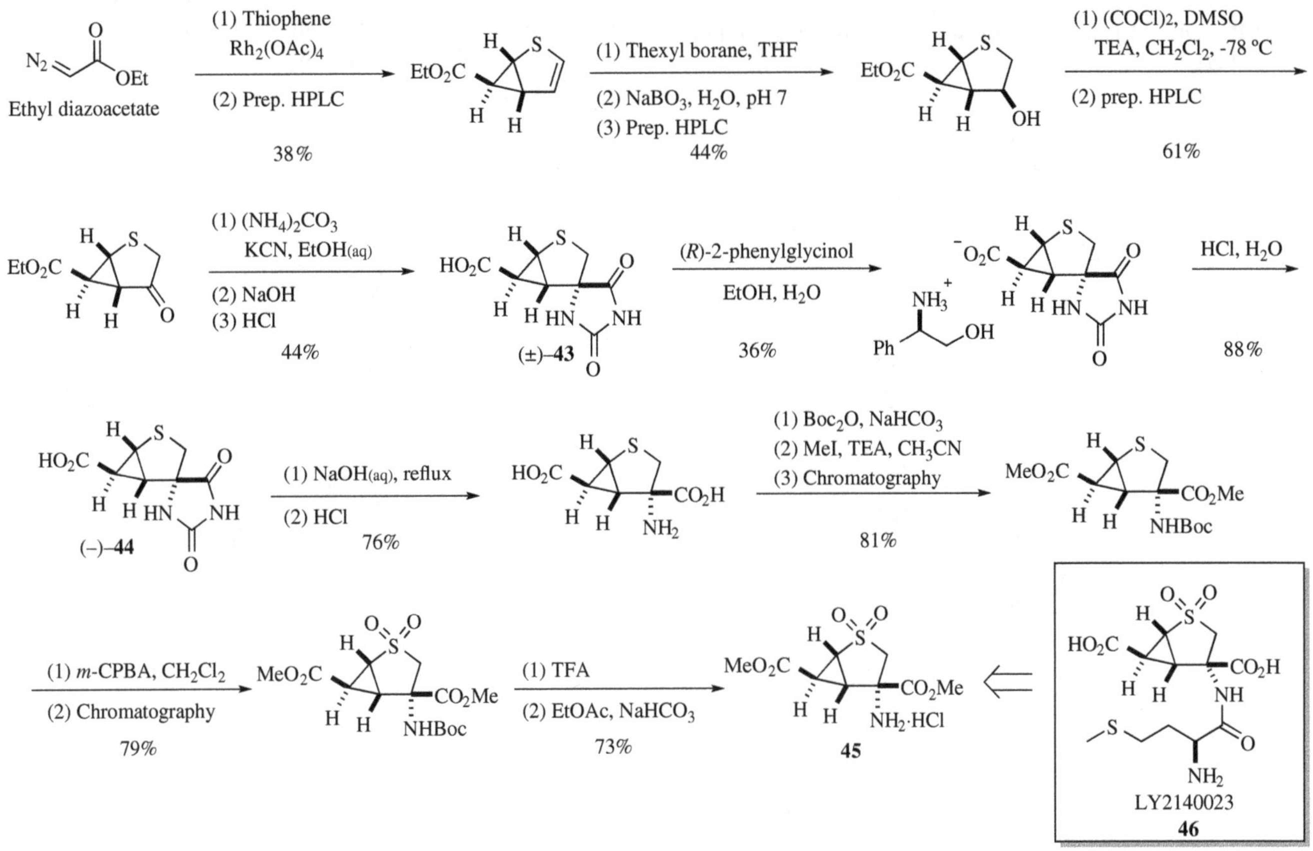

SCHEME 10.12 First-generation synthesis of LY2140023 **46** through Bucherer–Bergs hydantoin.

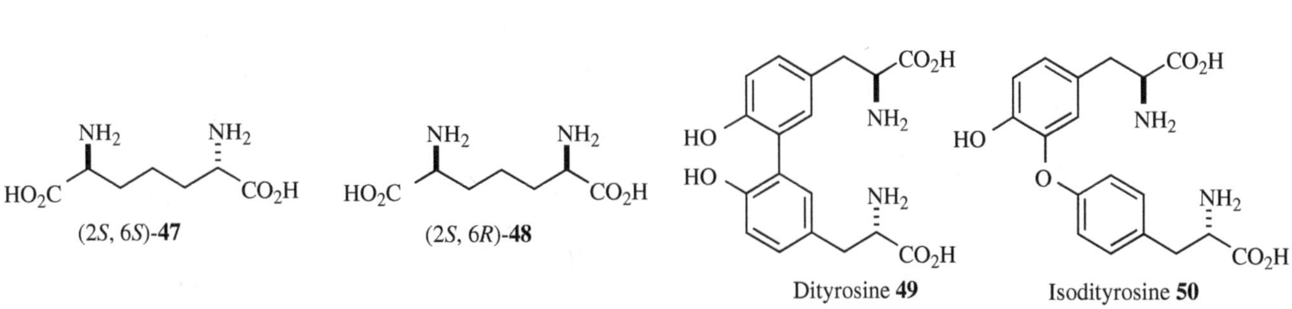

FIGURE 10.4 Diaminodicarboxylic acids biologically active: **47** [43], **48** [43], **49** [44], and **50** [44].

trans-**56** oriented spirocyclic linkers as potential groove binding elements (Scheme 10.13). Dimers **53** and **54** were further explored in preliminary DNA binding studies related with threading bis-intercalation.

In this strategy, the Bucherer–Bergs reaction was crucial for the synthesis of two bis-spirohydantoins with *trans*-geometry represented in Scheme 10.14 in excellent yields. The resulting diamino dimethyl carboxylate **57** is following converted in the desired *trans*-linker **56**.

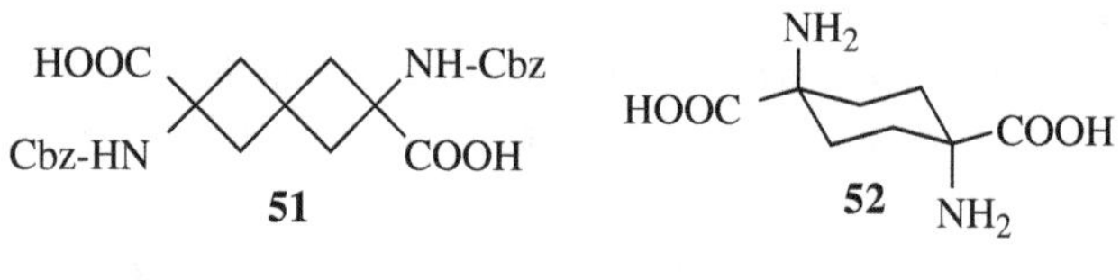

FIGURE 10.5 Diaminodicarboxylic acid derivatives: **51** [46] and **52** [47].

For the preparation of the *cis*-linker **55**, a slight variation in the synthesis was required since the *cis*-oriented bis-hydantoins cannot be obtained *via* Bucherer–Bergs reaction for this structure.

10.2 MC STRECKER REACTION

10.2.1 Introduction

The Strecker reaction [48], discovered in 1850 [1], is the most practical and efficient tool for the preparation of α-amino acids [49] *via* the formation of α-aminonitriles (Scheme 10.15) [50]. This three-component coupling between an amine or its equivalent, a carbonyl compound (generally an aldehyde), and hydrogen cyanide or its alkaline metal cyanides can be also made enantioselectively through the use of both metal-based catalysts and chiral organocatalysts.

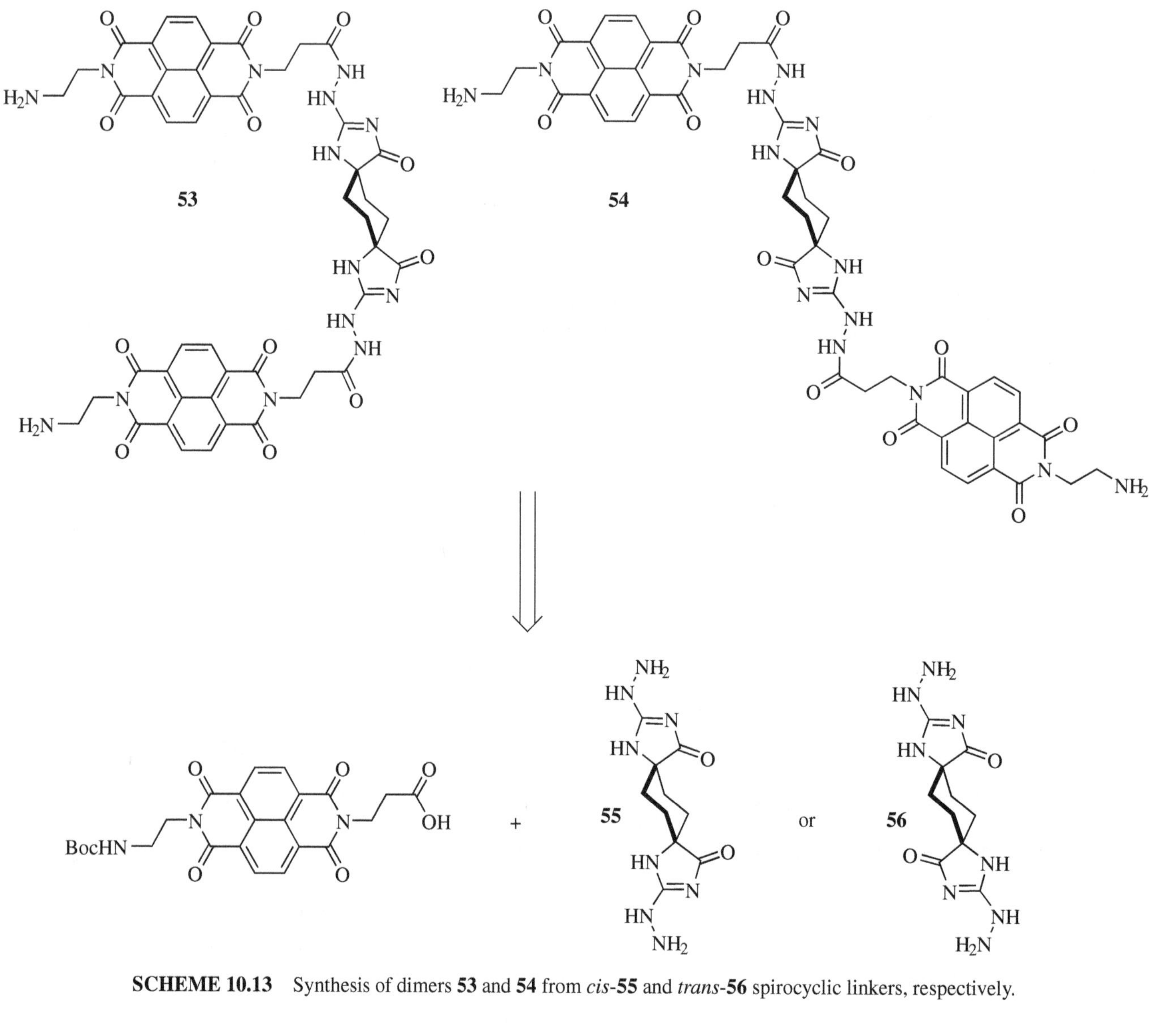

SCHEME 10.13 Synthesis of dimers **53** and **54** from *cis*-**55** and *trans*-**56** spirocyclic linkers, respectively.

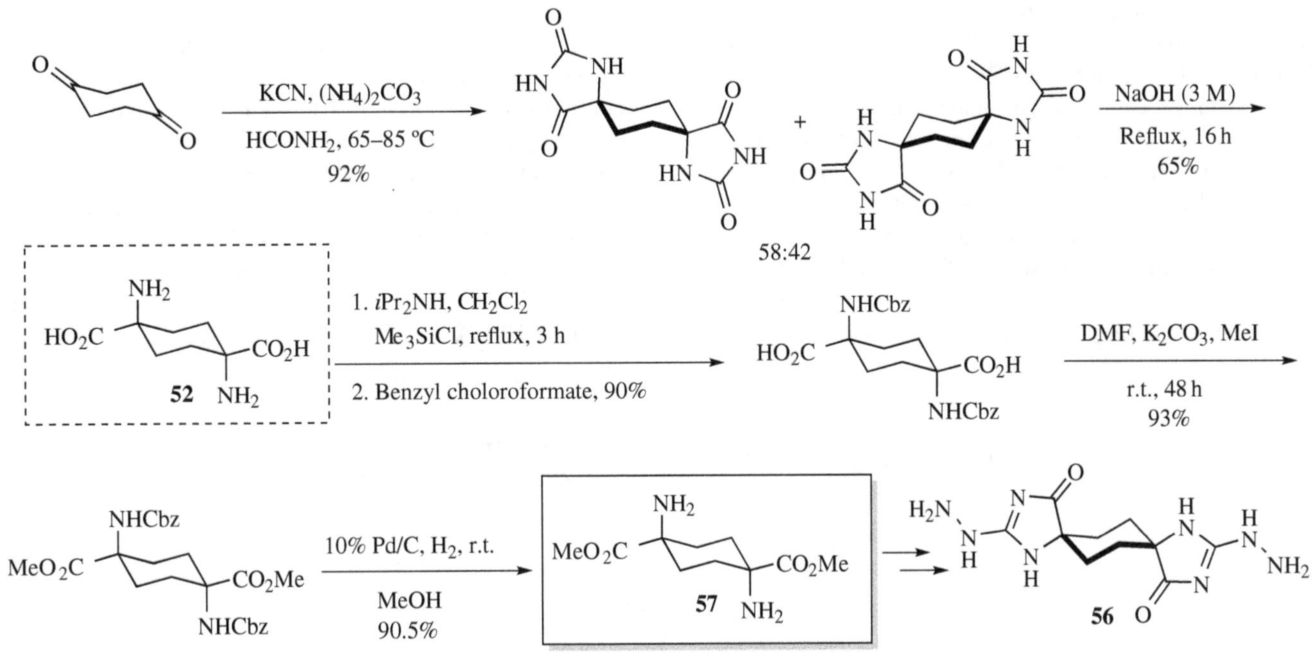

SCHEME 10.14 Synthesis of intermediates **52** and **57**.

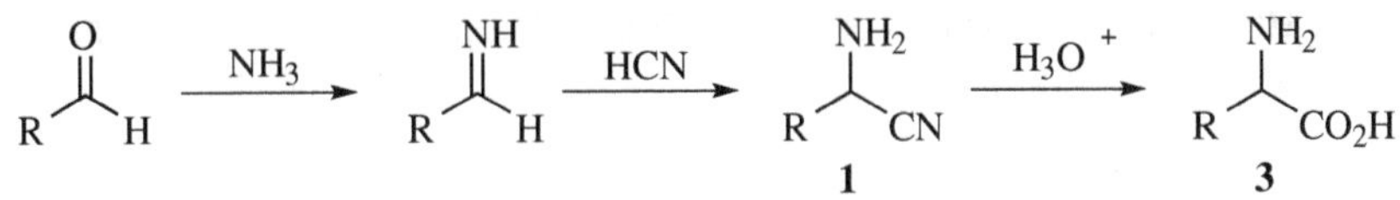

SCHEME 10.15 Strecker synthesis of α-amino acid *via* α-aminonitriles.

Due to the importance of obtaining α-aminonitriles [50], which can be subsequently converted in α-amino acids, many efforts have been engaged to the development of catalytic asymmetric approaches [51] and for applying this methodology in the synthesis of many natural products [52] with biological applications. Although metal catalysis has been the preferred tool of choice for the hydrocyanation of imines, in the last years, asymmetric organocatalysis has gained more interest as an attractive and alternative strategy [48c, 53].

Amino acids are the basic building blocks of proteins in all living organisms, constituting one of the five most important families of natural products [54]. These compounds have registered a growing demand over the years since they are considered key scaffolds in many scientific areas. Currently, they are mainly requested as optically active compounds, which are used for research as chiral catalysts and as chiral pool in the ligand design and in total synthesis, and numerous are the applications of these substrates. For this reason, the development of new and enantioselective synthesis is a continuous challenge. The importance of this process is also reflected in the work reported by Miller supporting the probable key role of the Strecker reaction in the origin of life under primitive Earth conditions for what Miller used a mimic prebiotic synthesis of amino acids from a mixture of CH_4, NH_3, H_2, and H_2O [55].

Although some interesting reviews have appeared in the last years covering this appealing area of research in the field of organocatalysis [48c, 53] and metal catalysis [48d, 51], we will try to remark in this section the most important examples since 2005.

10.2.2 MC Strecker Reaction Using Aldehyde

Although the Strecker reaction has some drawbacks such as the use of highly toxic cyanide derivatives, the use of other milder cyanide sources has been introduced in order to avoid partially this inconvenience. Moreover, instead of the initial aqueous media, this has been changed for other common organic solvents (toluene, methylene chloride, acetonitrile, etc.), improving the solubility of the organic reagents as well as the reaction conditions. In some cases, the methodologies require tedious work-up generating a large amount of waste, or some of the Lewis acids employed are toxic as well as the products resulting from their hydrolysis. Therefore, there is a continuous interest to explore and develop milder, safer, and more efficient protocols for this reaction.

A new MCR was reported using an ammonium salt as new Lewis base catalyst **58** as effective catalysts in the

Strecker reaction using trimethylsilyl cyanide (TMSCN) as cyanide source under mild conditions, affording good yield for the final α-aminonitriles **59**. The small catalytic charge used in this process is also notable (Scheme 10.16) [56].

Based on a previously developed work by Yus and coworkers [57], Zhu's group described a novel three-component synthesis of substituted α-iminonitriles **60** using 2-iodoxybenzoic acid (IBX) as oxidant and a catalytic amount of tetrabutylammonium bromide (TBAB) and affording final products in good to excellent yields under mild conditions (Scheme 10.17) [58].

Electron-rich and electron-poor aromatic aldehydes and α,β-unsaturated aldehydes were all compatible with the oxidative conditions used in this approach. Additionally, *Z* configuration was assigned to all final α-iminonitriles **60** according to NMR studies, except for R^1 = Ph. Interestingly, the authors demonstrated the synthetic applicability of this developed methodology in a two-step construction of appealing indolizidine skeletons **61** (Scheme 10.18).

A plausible mechanism is proposed to explain the role of both TBAB and IBX based on previous precedents (Scheme 10.19).

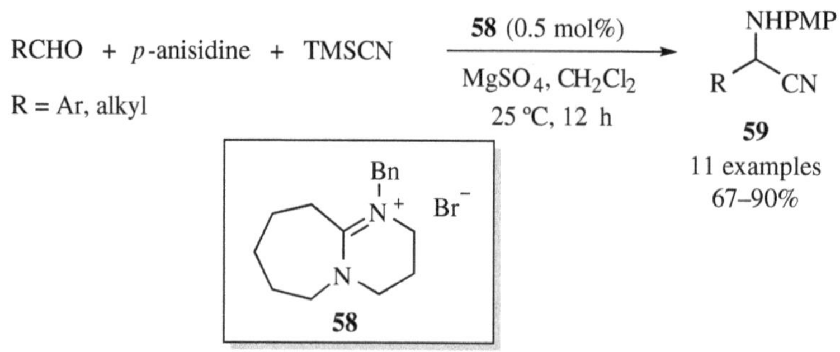

SCHEME 10.16 Synthesis of α-aminonitriles **59** using Lewis base catalyst **58**.

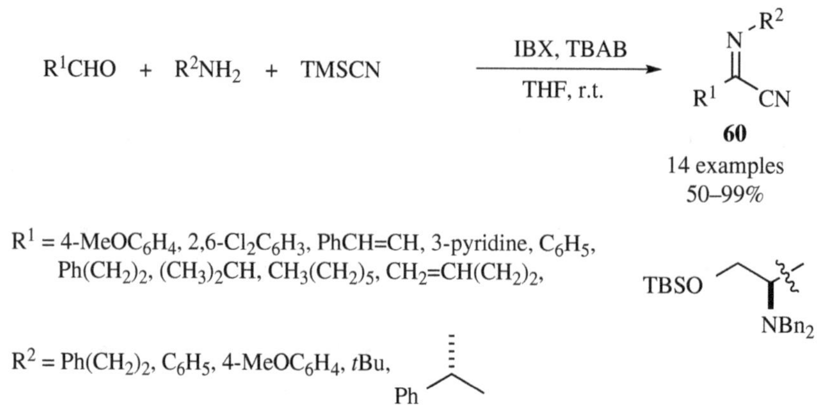

SCHEME 10.17 Multicomponent Strecker synthesis for the obtainment of α-iminonitriles **60**.

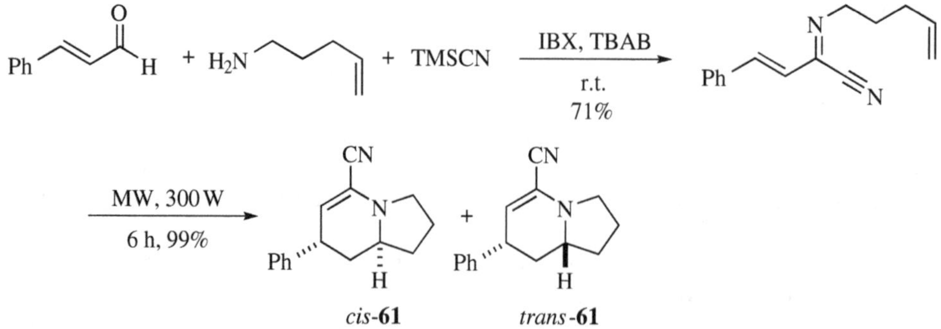

SCHEME 10.18 Synthetic application of α-iminonitriles for the preparation of indolizidines **61**.

The IBX would be activated by the ammonium salt promoting the formation of a pentacoordinated species **62**. Further, the resulting aminonitrile would attack to this intermediate leading to complex **63**. Finally, the elimination of a molecule of water in the *syn*-conformation *via* transition state **64b**, which is less sterically congested than **64a**, would give rise to the formation of final Z-iminonitriles and IBA as by-product (Scheme 10.19). However, the authors did not discard an alternative mechanism demonstrated by Nicolaou and coworkers [59], which would operate through a single-electron transfer (SET) providing intermediate **65** followed by a dehydration step affording the same final adducts.

The same group has also applied a similar methodology to produce carboxamides **66** through an oxidative Strecker reaction of aldehydes, amines, and TMSCN by simply modifying the work-up procedure (Scheme 10.20) [60].

More interesting was the tandem oxidative amidation of alcohols for which the authors used benzyl alcohols bearing electron-withdrawing and electron-donating groups as well as aliphatic ones as precursor of aldehydes. All these reagents tolerated well the conditions employed and led in a one-pot procedure the corresponding carboxamides **67** in very good yields (74–84%) (Scheme 10.21).

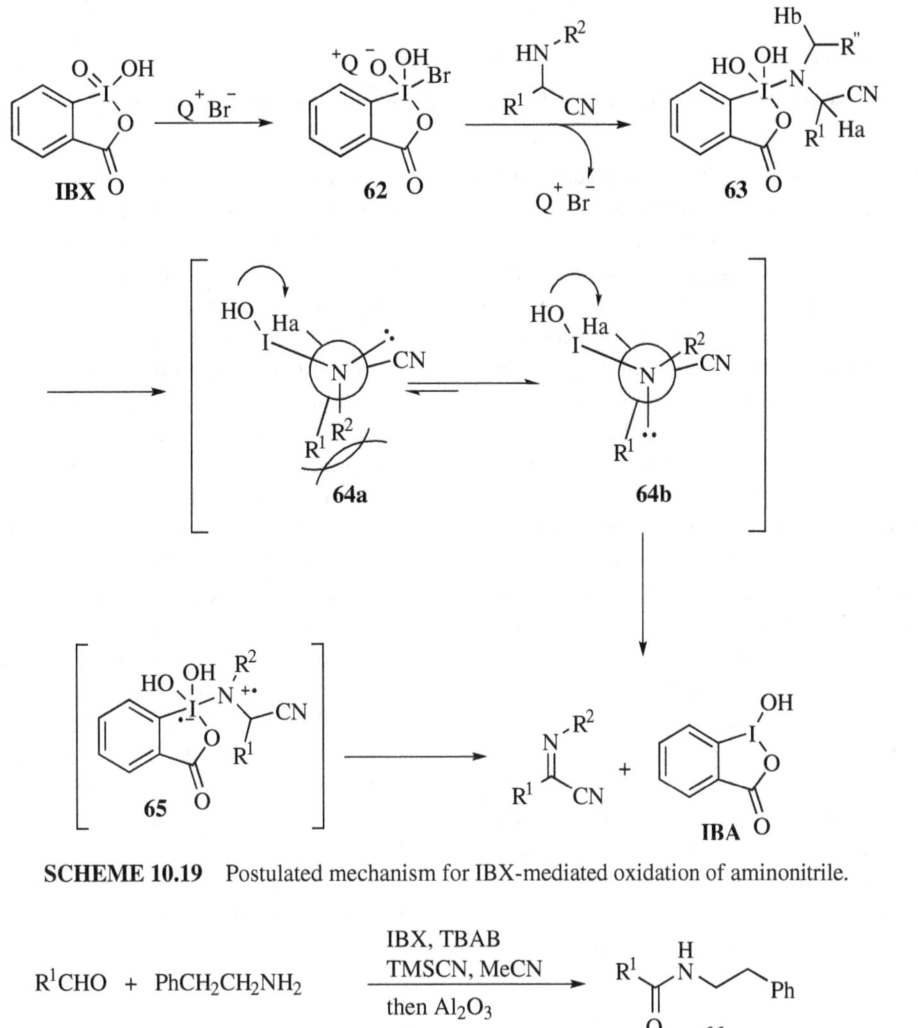

SCHEME 10.19 Postulated mechanism for IBX-mediated oxidation of aminonitrile.

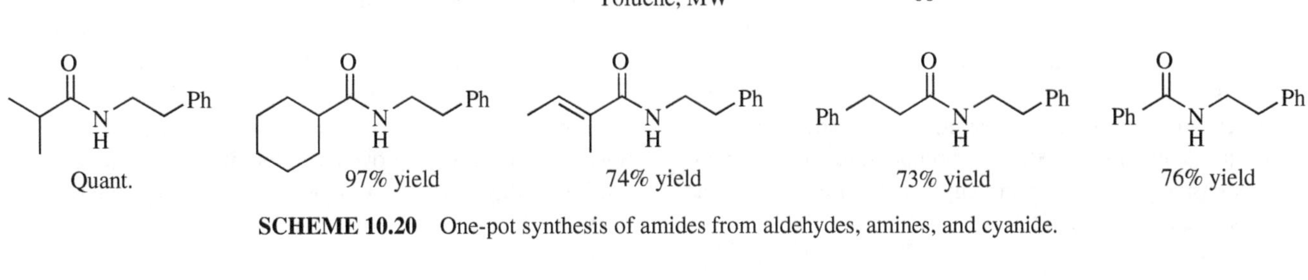

SCHEME 10.20 One-pot synthesis of amides from aldehydes, amines, and cyanide.

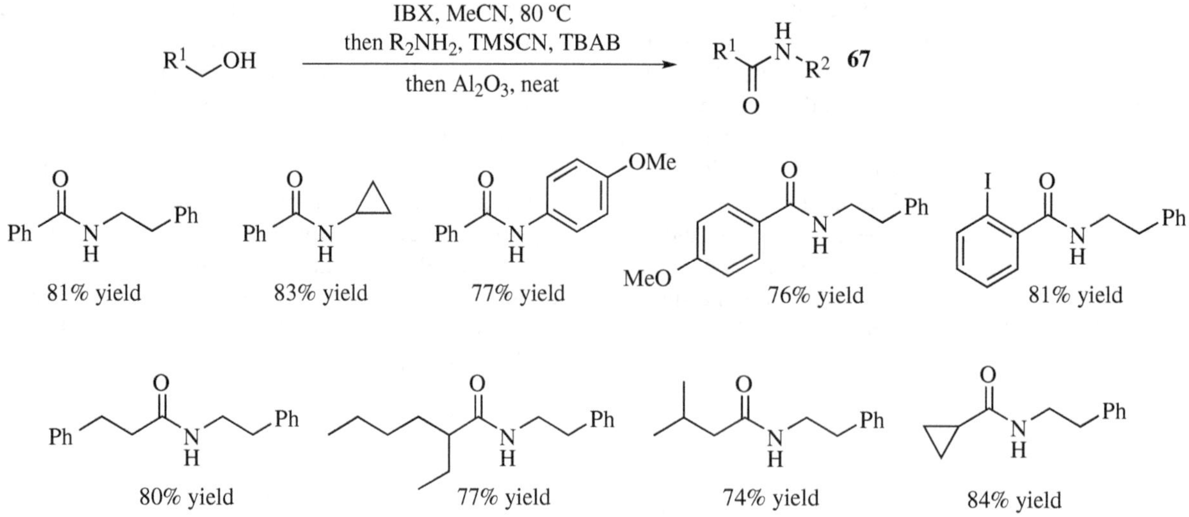

SCHEME 10.21 One-pot synthesis of amides **67** from alcohols, amines, and cyanide.

In all cases, the final addition of Al_2O_3 was crucial for the obtainment of the corresponding carboxamides after formation of the corresponding α-iminonitrile intermediates.

More recently, Liao and coworkers developed an interesting tandem organocatalytic Strecker–allylic alkylation (SAA) reaction for the construction of highly functionalized α-aminonitriles **68** bearing an *N*-substituted quaternary carbon center. The scope and generality of this novel tandem SAA reaction are represented in Scheme 10.22 [61].

These α-aminonitriles were converted into α-methylene-γ-butyrolactams **69** separately through cyclization of adducts **68** in the presence of a catalytic amount of DBU. However, a more complex application of this strategy was illustrated through a Strecker–allylic alkylation–cyclization (SAAC) reaction in a one-pot procedure to afford α-methylene-γ-butyrolactam **69a** as depicted in Scheme 10.23.

The authors envisioned the generation first of the α-aminonitrile **1** and the following α-deprotonation of **1** and *C*-selective allylic alkylation, which would give highly functionalized α-aminonitriles **I**. These compounds could be converted into α-methylene-γ-butyrolactams **69** separately or in a one-pot fashion. In this manner, a great pool of valuable densely functionalized α-aminonitriles **1** and α-methylene-γ-butyrolactams **69** containing quaternary carbon centers could be easily accessed (Scheme 10.24).

10.2.3 Strecker-Type Reaction Using Ketones

Although the Strecker reaction has been extensively studied with aldehydes using various kinds of catalysts and types of catalysis, the MC version using ketones has been less explored, maybe due to the less reactivity shown by these substrates, and only in 2007 it was reported the first efficient and extensively developed strategy using gallium triflate as catalyst (Scheme 10.25) [62].

Various fluorinated and nonfluorinated ketones were successfully employed in this MC Strecker reaction, and interestingly, the proper choice of the reaction conditions (catalyst and solvent) was crucial for the achievement of the process with very good results. Later, this research group developed a similar procedure using TMSOTf [63] or Nafion[R]-H [64] as efficient catalysts of this reaction.

Matsumoto, Kotsuki, Jenner, and coworkers also reported the use of a wide scope of ketones in an MC Strecker process using high pressure [65], and some scarce examples have been also disclosed in other works mainly dedicated to aldehydes [66].

A clean and environmentally friendly protocol was reported by Khan and coworkers using $Fe(Cp)_2PF_6$ (5 mol%) as efficient catalyst of this reaction and performing the process in very short reaction times (20 min) with excellent yields under solvent-free conditions (Scheme 10.26) [67].

A plausible mechanism was envisioned by the authors to explain the role of the catalyst in this process (Scheme 10.27).

The first step would be the *in situ* formation of the imine **71** between the ketone and the amine. Then, the catalyst would polarize the imine favoring the attack of the CN over the imine carbon and producing the TMS derivative **72**, which would be later hydrolyzed with water to give the observed α-aminonitrile **70**.

10.2.4 Catalyst-Free Strecker Reactions in Water

The original protocol of a one-pot Strecker reaction was performed in water as reaction solvent, but in subsequent procedures, water has been replaced by organic solvents, such as toluene, dichloromethane, and acetonitrile, or nonconventional solvent, such as ionic liquid [68], especially with TMSCN as cyanide source, to improve the solubility of organic reagents. Although there are also examples performed in aqueous media [69], including water-containing DMF [70], polyethylene

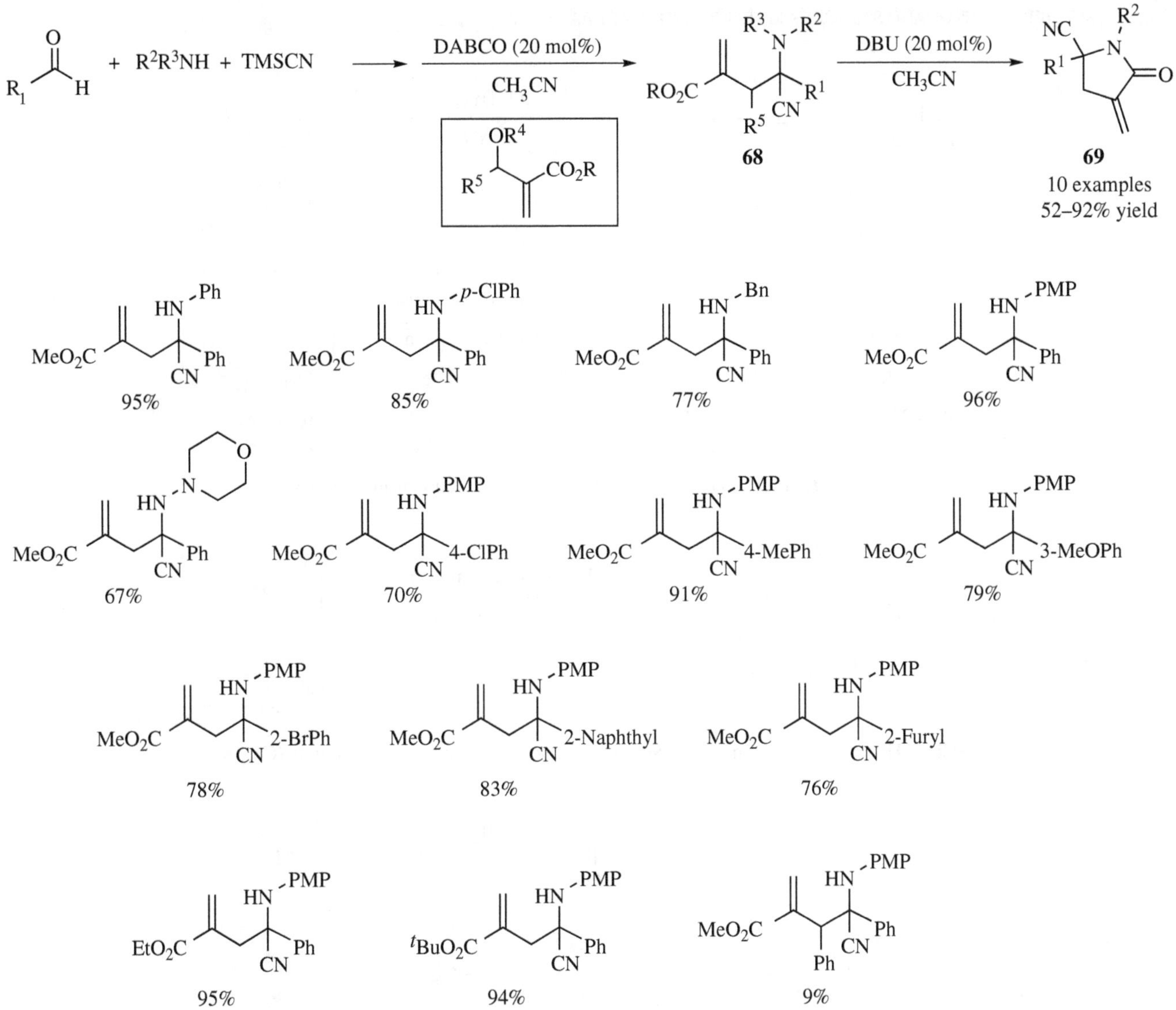

SCHEME 10.22 Synthesis of functionalized **α**-aminonitriles **68** *via* multicomponent tandem SAA reaction.

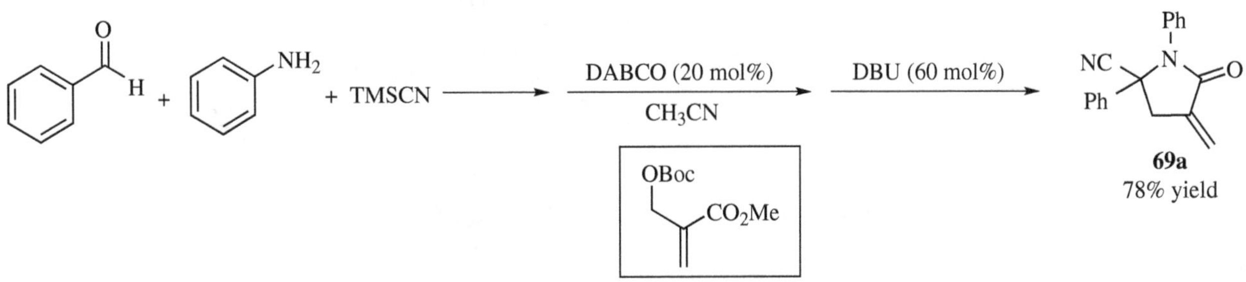

SCHEME 10.23 Synthesis of **α**-methylene-**γ**-butyrolactam **69a** *via* SAAC strategy.

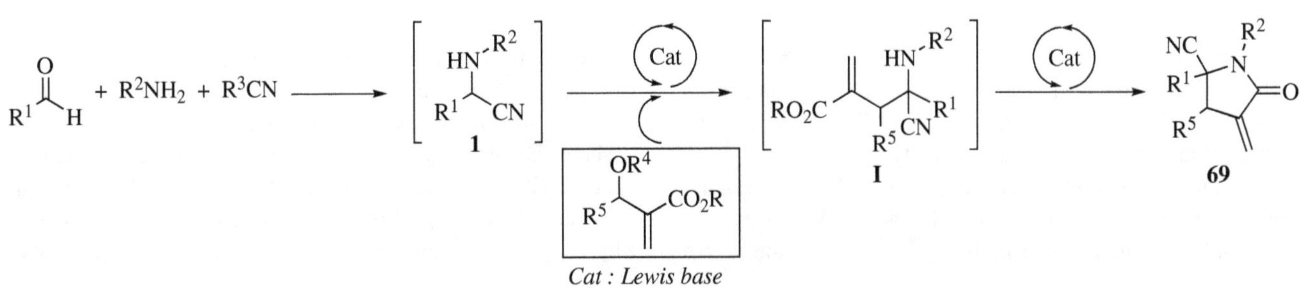

SCHEME 10.24 Mechanism proposal.

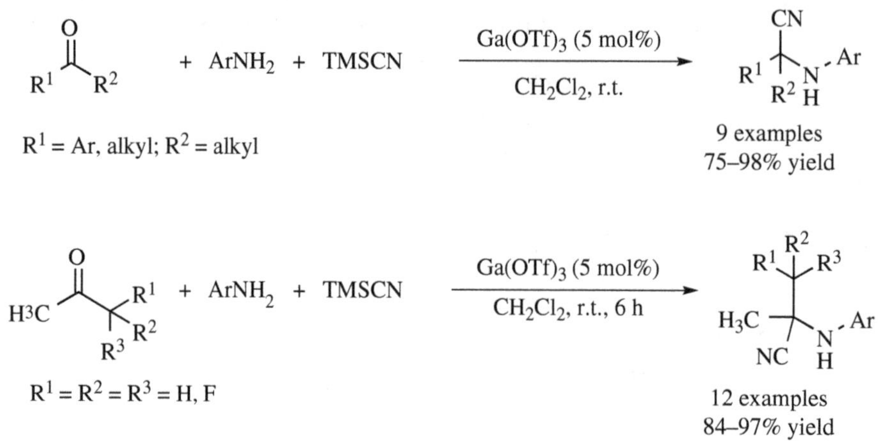

SCHEME 10.25 Ga(OTf)$_3$-catalyzed Strecker reaction using ketones.

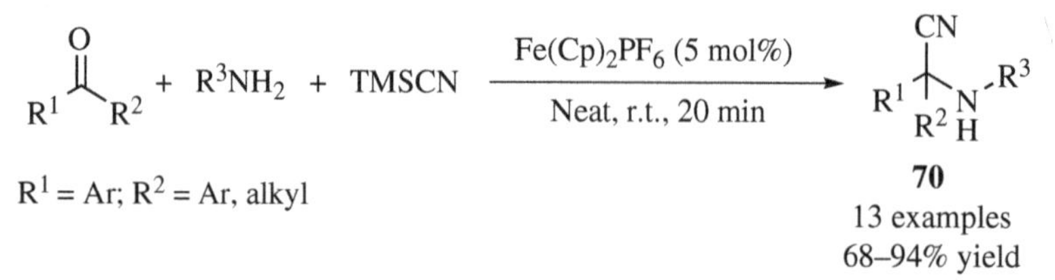

SCHEME 10.26 Multicomponent Strecker synthesis of ketones catalyzed by Fe(Cp)$_2$PF$_6$.

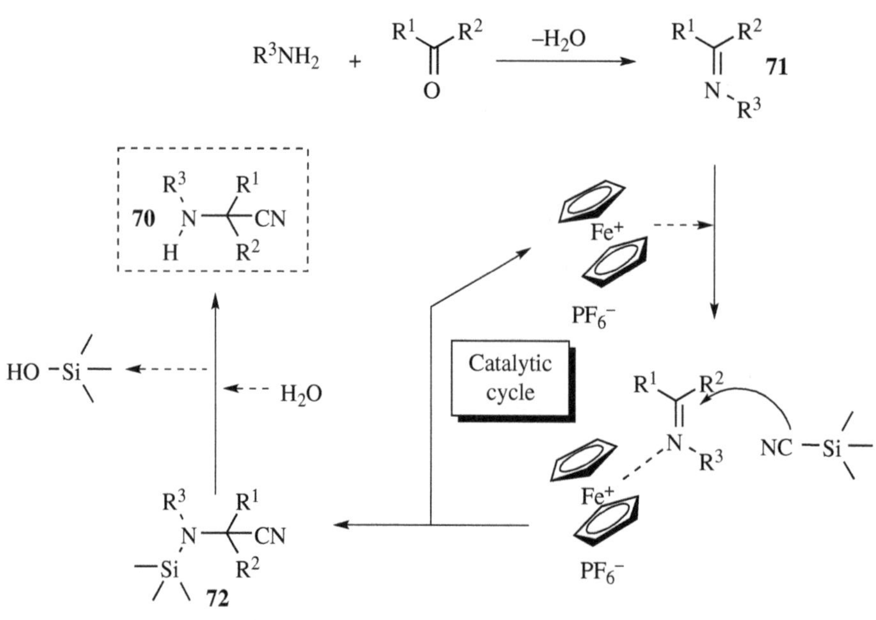

SCHEME 10.27 Plausible Strecker one-pot mechanism.

glycol/water (2:1) [71], β-cyclodextrin in water/methanol (9:1) [72], Sc(OTf)$_3$-catalyzed one-pot reaction in pure water using toxic tributyltin cyanide [73], or K_2PdCl_4-catalyzed MC Strecker reaction [74], this field is clearly less explored.

Fernández and Lassaletta group reported an interesting one-pot Strecker-type procedure between aldehydes and *N,N*-dialkylhydrazine with *in situ* generation of the active species HCN in pure water and in the absence of catalyst (Scheme 10.28) [75].

A plausible mechanism was envisaged by the authors, where the first step would be the hydrolysis of TMSCN to afford HCN that would be activated by the NR_2 lone pair (Scheme 10.29). Additional experiments were also carried out by using a 1:1 mixture of AcOH/KCN for the *in situ* generation

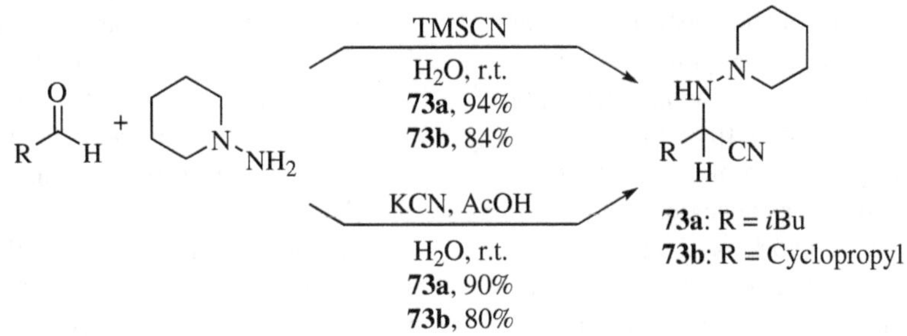

SCHEME 10.28 Strecker-type reaction in pure water.

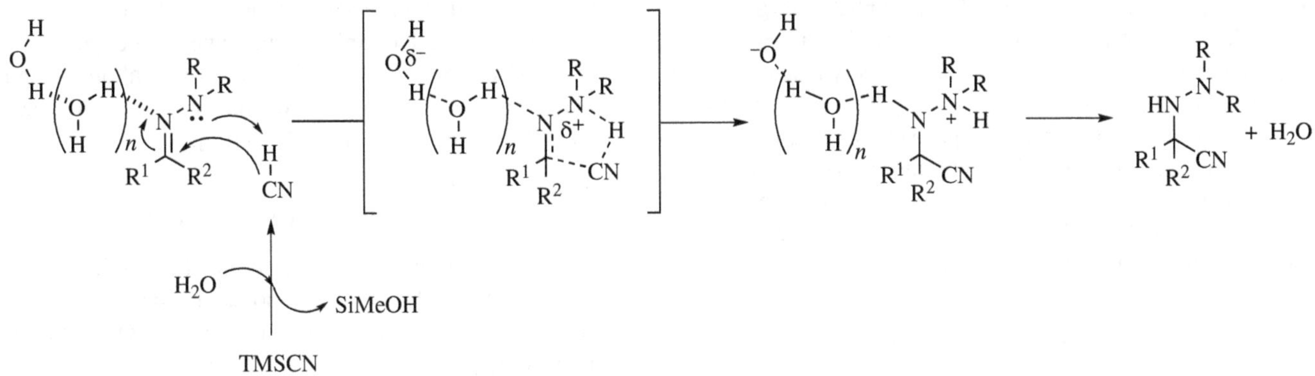

SCHEME 10.29 Plausible reaction mechanism.

of HCN in the reaction media. The reaction also proceeds to give the same adducts in similar yields and with the same reaction rates, while no reaction took place when KCN alone was added to the preformed hydrazone in water. This fact strongly suggests that a previous hydrolysis of TMSCN must occur being the HCN the active species supporting this mechanism.

Since the toxicity and volatility of HCN are high, these facts limit its extensive and practical application in organic synthesis. In this respect, a number of cyanating agents have been developed to avoid the use of toxic HCN, such as TMSCN [76], $(EtO)_2P(O)CN$ [77], Et_2AlCN [78], Bu_3SnCN [79], MeCOCN [80], $K_4[Fe\text{-}(CN)_6]$ [81], and acetone cyanohydrin [82]. Although TMSCN has been the most widely used in the Strecker reaction, this often requires a Brønsted or Lewis acids or bases as catalysts [83].

Although the new developed protocols significantly improved the performance of the reaction in terms of yields and reaction time, sometimes, the new strategies require tedious work-up, leading in some cases to the generation of a large amount of waste. Therefore, there is still field for the discovery of milder and more attractive approaches for this reaction.

Recently, atom-economical cyanation process under catalyst-free conditions has been developed using acetone cyanohydrin [84] in water, as an example of mild and alternative cyanide source (Scheme 10.30) [85].

This method is a simple and practical method for the synthesis of racemic α-aminonitriles through a one-pot, three-component Strecker reaction under mild reaction conditions. Reactions proceeded very efficiently at room

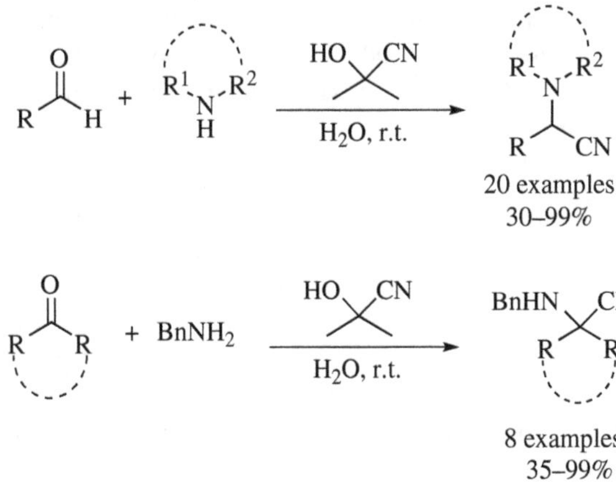

SCHEME 10.30 Strecker reaction using acetone cyanohydrin as cyanide source.

temperature with high selectivity, and in some cases, the expected α-aminonitrile was purified by direct separation from water [86]. This is an example of clean and environmentally friendly processes for the synthesis of α-aminonitriles of synthetic importance, and it represents an interesting precedent. Recently, the same authors have reported the diastereoselective approach of this process [87].

10.2.5 Catalyst-Free Strecker Reactions under Solvent-Free Conditions

The absence of solvent in organic synthesis and the employment of a small excess of reagent make procedures simpler, save energy, and prevent solvent waste, hazards, and toxicity. In this context, few solvent-free syntheses of α-aminonitriles have been reported. However, the pioneering one cannot be considered as MC process [88], and in the second example, the presence of magnesium bromide–diethyl ether complex as a Lewis acid catalyst was necessary [89, 90].

In this field, Najera and coworkers developed more recently a three-component Strecker reaction with TMSCN in the absence of solvent and catalyst, which could potentially be applied at an industrial level (Scheme 10.31) [91].

As expected, aromatic and aliphatic aldehydes reacted faster than ketones, against aromatic and aliphatic amines, and the authors only observed reactivity with cyclic ketones under these conditions. This environmentally friendly methodology proceeds at room temperature and in very short reaction time, without solvent, with almost one equivalent of reagents, in the absence of catalyst, and avoiding the final work-up since the reaction product did not require any further purification [92]. It was considered that the active catalyst of the reaction was the amine employed in this three-component reaction.

10.2.6 Metal-Catalyzed Strecker-Type Reaction

Metal catalysis has been successfully applied in one-pot Strecker reaction using hafnium [93], iron [67], and zirconium catalysts [94]. However, most of the methods involve the use of strong acidic conditions, expensive reagents, long reaction times, harsh conditions, fast hydrolysis, and tedious work-up leading to the generation of a large amount of waste. On the other hand, the use of aldehydes in MC Strecker reactions has been more extensively explored than ketones, which still remains as a more difficult task.

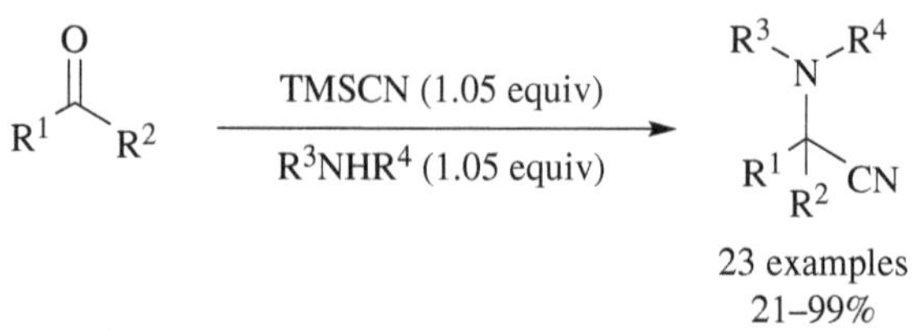

SCHEME 10.31 Strecker reaction under solvent-free conditions.

As mentioned in Section 10.2.3, the application of $Ga(OTf)_3$ as an effective and reusable catalyst for the Strecker reaction has been demonstrated [62]. This is a simple and alternative route to the stepwise methodologies for the Strecker reaction of ketones. Moreover, the use of fluorinated ketones is a significant aspect of this procedure due to the increase interest in organofluorated compounds.

More recently, Jung and coworkers have reported a simple and efficient palladium-catalyzed three-component reaction for the synthesis of α-aminonitriles using aldehydes and ketones. Also interesting was the use of both aliphatic and aromatic amines in this protocol, under mild conditions, reaching high yields (Scheme 10.32) [95].

10.2.7 Organocatalytic MC Strecker Reaction

The progresses reached in the organocatalytic field have been recently illustrated [48c, 53]. Although interesting racemic strategies using nonchiral organocatalysts have been employed such as $(CH_3)_3SiOSO_2CF_3$, NH_2SO_3H, guanidine hydrochloride, *o*-benzenedisulfonimide, thiourea, 2,4,6-trichloro-1,3,5-triazine, and ammonium salt [56, 96], significant asymmetric catalytic results have been achieved and will be described here.

The first organocatalytic asymmetric three-component Strecker reaction was reported by List and Pan using Jacobsen's thiourea catalyst **75**. Diverse α-aminonitriles **76** and **77** were formed in excellent yields and enantioselectivities from aliphatic and aromatic aldehydes, different amines, and acyl cyanides by using 5 mol% of catalyst **75** (Scheme 10.33) [97].

Contemporaneously, an alternative organocatalytic asymmetric three-component Strecker reaction was described using a *C*2-symmetric chiral bisformamide **78** as efficient catalyst, furnishing α-aminonitriles **79** in excellent yields and good enantioselectivities (Scheme 10.34) [98]. However, some final adducts were obtained optically pure (>99% ee)

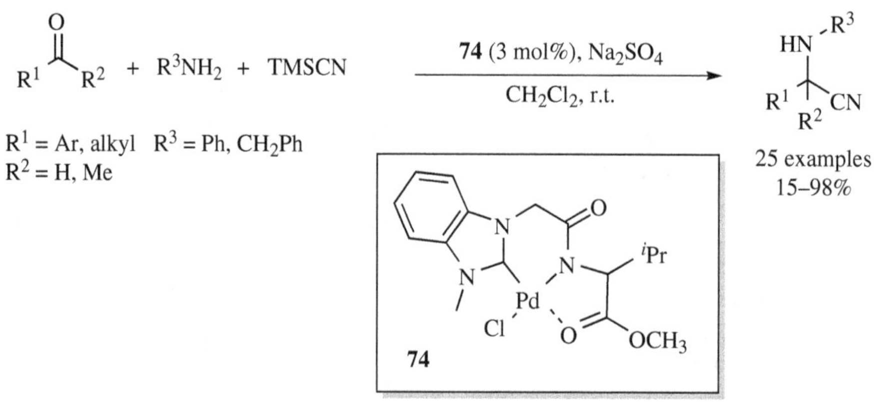

SCHEME 10.32 Palladium-catalyzed one-pot Strecker reaction.

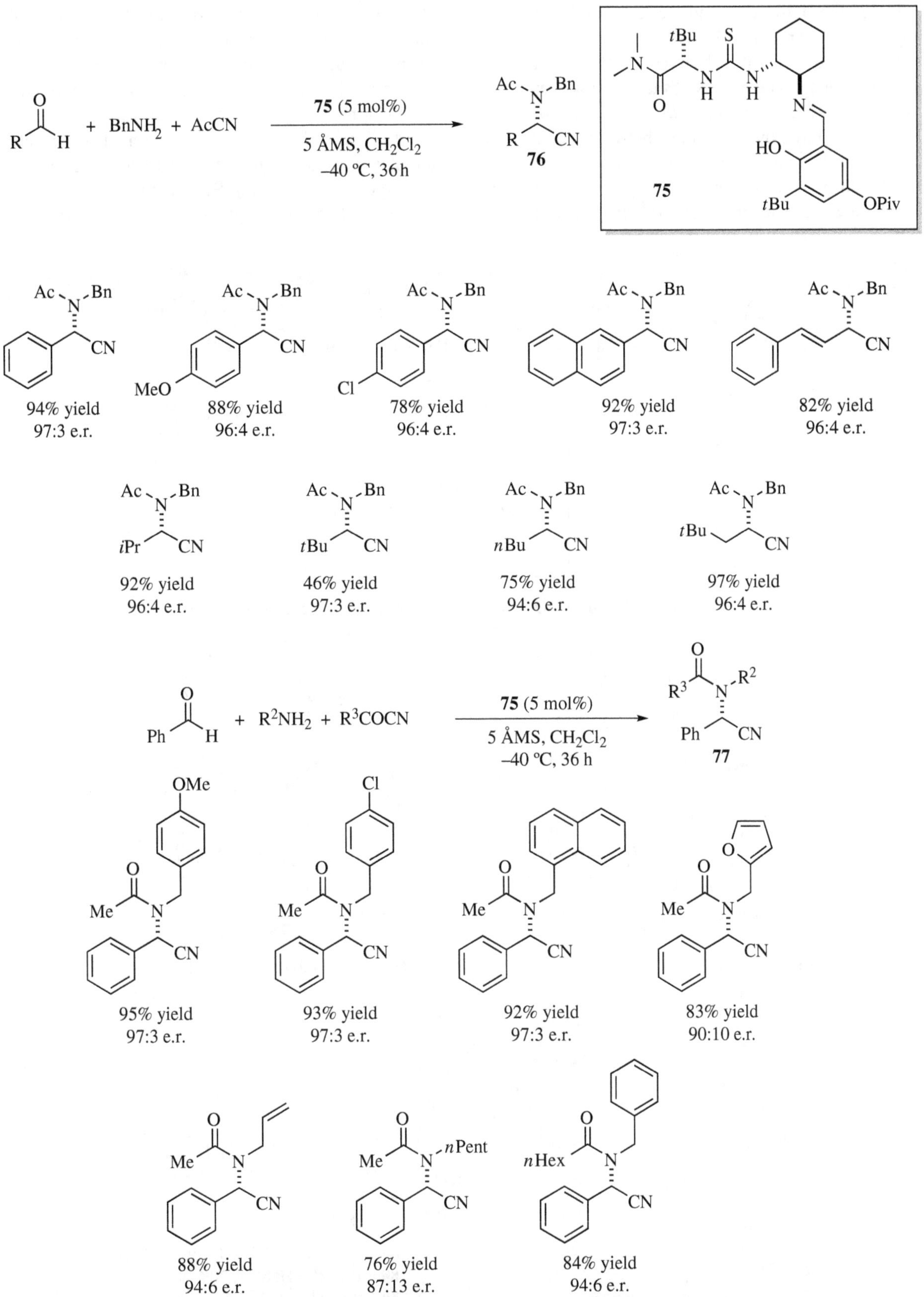

SCHEME 10.33 Thiourea-organocatalyzed asymmetric Strecker 3CR.

after a single recrystallization, and the scope of the reaction is mainly limited to aromatic aldehydes.

The authors explained the observed absolute configuration of final products on the basis of the plausible transition states **TS-1** and **TS-2** (Fig. 10.6). *In situ* generated imine would be activated by a hydrogen bond between the nitrogen of the imine coordinated to the hydrogen of amide. Moreover, TMSCN would be activated by two oxygen atoms coordinated to silicon as depicted in Figure 10.6. In these transition states, the *si* face attack would be much more accessible than the *re* face, in which would there be repulsion between the phenyl group of the imine and one pyrrole unit. The role played by this bifunctional catalyst **78** is supported by DFT calculations at the B3LYP/6-31G(d) level.

More recently, Ma and coworkers have developed an alternative route for the one-pot three-component Strecker reaction of ketones using Brønsted acids as organocatalysts (Scheme 10.35). Although the methodology was extended to

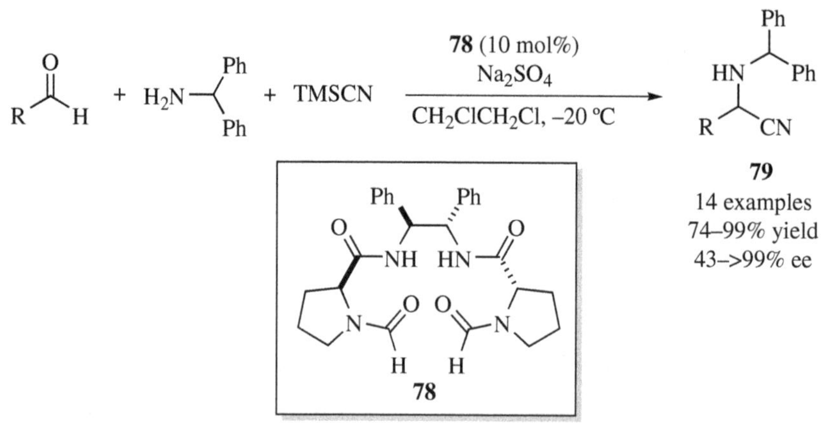

SCHEME 10.34 Organocatalytic asymmetric Strecker 3CR.

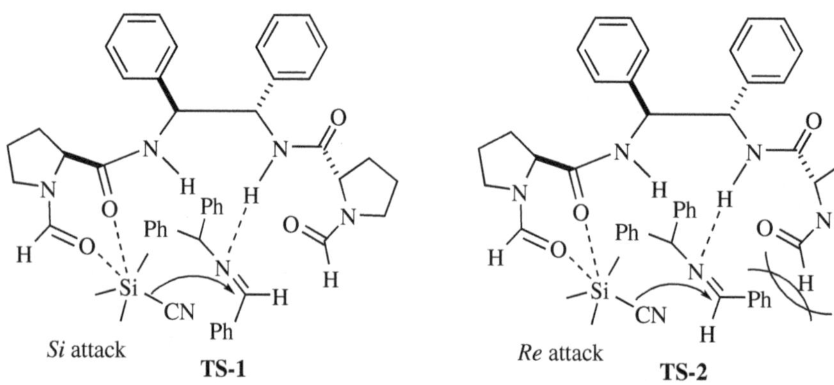

FIGURE 10.6 Proposed transition states for the hydrocyanation catalyzed by bisformamide **78**.

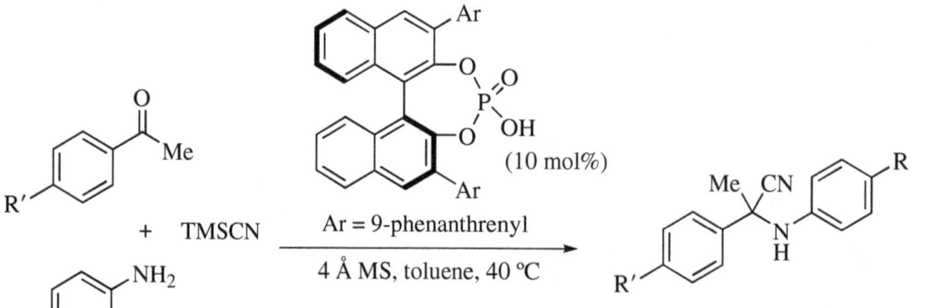

80a: R = 4-MeO, R′ = H, 73%, 40% ee
80b: R = R′ = H, 81%, 20% ee
80c: R = 4-MeO, R′ = 4-F, 78%, 20% ee

SCHEME 10.35 Chiral phosphoric acid-catalyzed three-component Strecker reaction of ketones.

a number of α-aminonitriles **80** with good to excellent yields for the racemic version, the authors have also shown the preliminary results of the catalytic enantioselective version although with poor ee% (up to 40% ee) [99].

10.2.8 Efficient Heterogeneous Catalysis for the Synthesis of α-Aminonitriles

Interestingly, the MC approach for this reaction has been also developed using heterogeneous catalysts such as $K_5CoW_{12}O_{40}·3H_2O$ [100], supported heteropoly acids [101], montmorillonite [102], sulfuric acid-modified PEG-6000 [103], or organometallic hollow spheres bearing bis(*N*-heterocyclic carbene)-palladium species [104], among others.

Prakash, Olah, and coworkers developed an efficient strategy using Nafion^R-H (perfluoroalkanesulfonic acid polymer) and Nafion^R SAC-13 (10–20% Nafion^R-H polymer on amorphous silica, porous nanocomposite) as heterogeneous catalysts for one-pot Strecker reaction using aldehydes, ketones, and fluorinated ketones, affording final α-aminonitriles in high yields [64].

Recently, Heydari and coworkers have employed dehydroascorbic acid capped magnetite (DHAA-Fe_3O_4) catalyst for the one-pot synthetic approach of synthesis of α-aminonitriles. Final adducts are obtained with high yields employing different aldehydes or ketones and amines. A plausible mechanism is proposed to explain this one-pot strategy (Scheme 10.36) [105]

According to this mechanism, DHAA-Fe_3O_4 nanoparticles would be involved in the *in situ* formation of the imine intermediate and in the activation of the imine for the subsequent attack of the CN, as shown in Scheme 10.36.

10.2.9 Synthetic Utility

The α-aminonitrile intermediates resulting from these MC Strecker strategies have been found useful for the synthesis of more complex and interesting structures [106]. This is the case of Fitch and coworkers, who developed an efficient two-pot synthesis for the asymmetric preparation of benzothiadiazine-substituted tetramic acids **83** (Scheme 10.37) [107].

The MC Strecker reaction is the key step in this synthesis for the preparation of intermediate aminonitrile **81**; further acylation of that compound with benzothiadiazine acid **82** under the standard conditions followed by *in situ* addition of triethylamine provided the desired aminotetramic acid **83**.

More recently, Zhu and coworkers have employed the α-aminonitrile intermediate resulting of this process to generate in two steps polysubstituted 2-amino-5-cyanopyrroles **85** through a [4+1] cycloaddition of α,β-unsaturated imidoyl cyanide (2-cyano-1-azadienes) with isocyanides, in combination with a previous *in situ* oxidation of the Strecker adduct (Scheme 10.38) [108, 109].

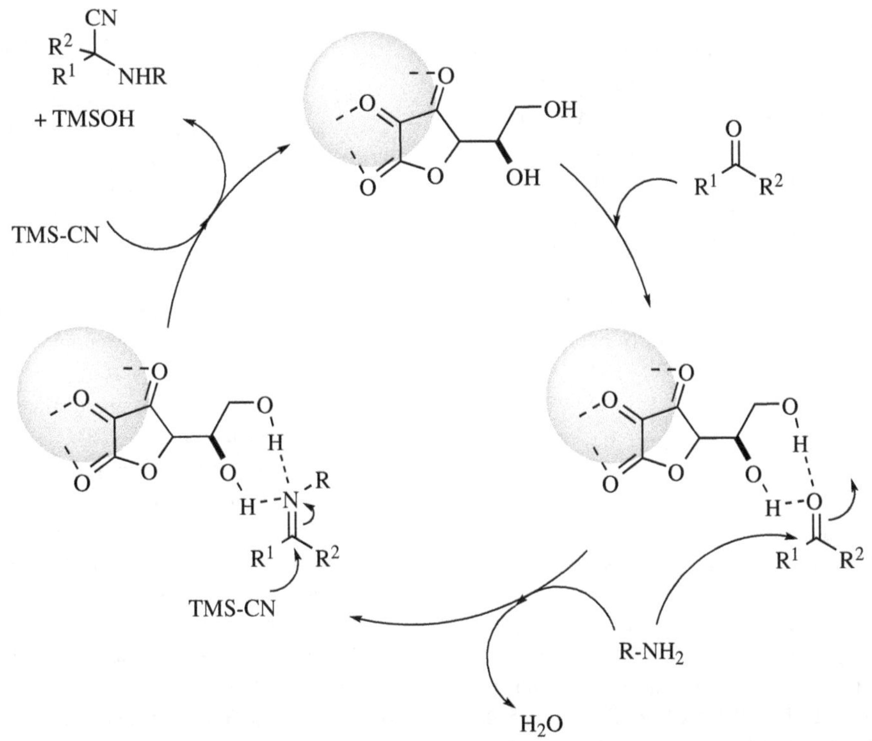

SCHEME 10.36 Plausible catalytic mechanism.

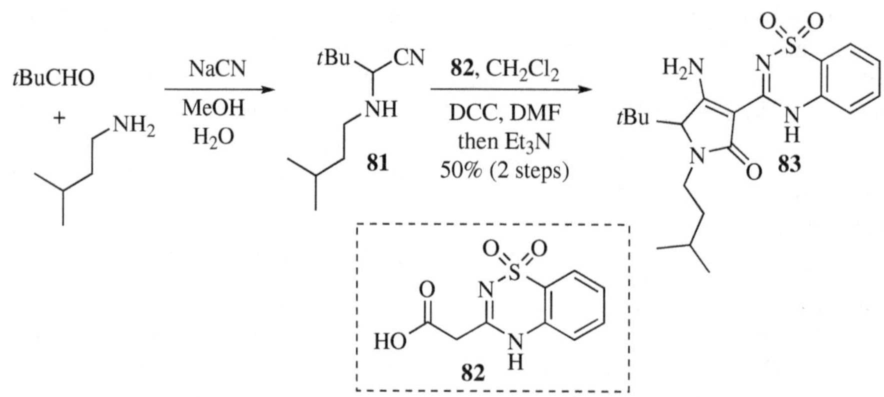

SCHEME 10.37 Synthesis of **83** through α-aminonitrile intermediate **81**.

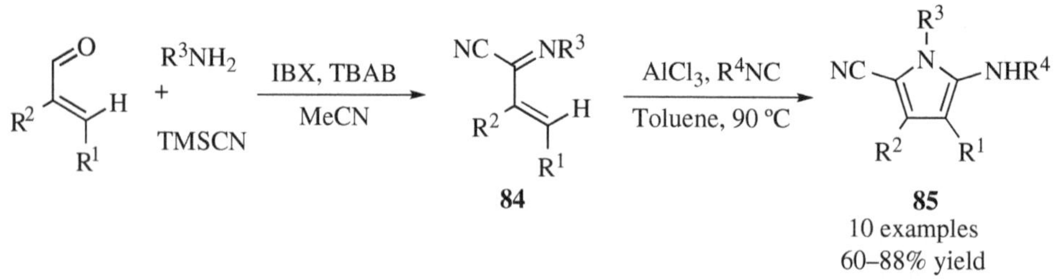

SCHEME 10.38 Sequential MCR/[4+1] cycloaddition strategy for the preparation of **85**.

10.3 CONCLUSIONS

This chapter is an overview of the great work reported during the last 10 years in these two complementing fields, which are two crucial methods for the synthesis of α-aminonitriles. As it has been shown in these pages, the resulting α-aminonitrile adducts could give rise to the synthesis of interesting molecules, such as the corresponding α-amino acids. Although the developed methodologies are nowadays numerous, as in previous chapters, we expect improvements in the next future regarding overall chiral strategies, since the synthesis of enantiomerically enriched α-amino acids is an active aim in organic synthesis, because the biological properties of the molecules can be given by the chirality of the structure. More efficient environmentally benign procedures are also desired.

REFERENCES

1 A. Strecker, *Justus Liebigs Ann. Chem.* 1850, 75, 27–45. Ueber die künstliche Bildung der Milchsäure und einen neuen, dem Glycocoll homologen Körper.

2 (a) H. Bergs, *Ger. Pat.* 566, 094, 1929. (b) H. T. Bucherer, W. Steiner, *J. Prakt. Chem.* 1934, 140, 291–316. Syntheses of hydantoins. I. Reactions of α-hydroxy and α-amino nitriles. (c) H. T. Bucherer, V. A. Lieb, *J. Prakt. Chem.* 1934, 141, 5–43. Syntheses of hydantoins. II. Formation of substituted hydantoins from aldehydes and ketones.

3 A. Rousset, M. Lasperas, J. Taillades, A. Commeyras, *Tetrahedron* 1980, 36, 2649–2661. Systemes de strecker et apparentes-XI: formation et stabilité de l'α-carboxyaminonitrile. Intermédiaire essentiel dans la synthèse des hydantoïnes selon Bucherer-Bergs.

4 (a) E. Ware, *Chem. Rev.* 1950, 46, 403–470. The chemistry of the hydantoins. (b) M. Meusel, M. Guetschow, *Org. Prep. Proced. Int.* 2004, 36, 391–443. Recent developments in hydantoin chemistry. A review. (c) V. A. Soloshonok, A. E. Sorochinsky, *Synthesis* 2010, 2319–2344. Practical methods for the synthesis of symmetrically α,α-disubstituted α-amino acids. (d) S. Kotha, S. Halder, *Synlett* 2010, 337–354. Ethyl isocyanoacetate as a useful glycine equivalent.

5 C. Cativiela, M. D. Diaz-de-Villegas, *Tetrahedron: Asymmetry* 2000, 11, 645–732. Stereoselective synthesis of quaternary α-amino acids. Part 2: cyclic compounds.

6 J.-T. Li, S.-X. Wang, G.-F. Chen, T.-S. Li, *Curr. Org. Synth.* 2005, 2, 415–436. Some applications of ultrasound irradiation in organic synthesis.

7 L. Munday, *J. Chem. Soc.* 1961, 4372–4379. Amino-acids of the cyclohexane series. Part I.

8 J. T. Edward, C. Jitrangsri, *Can. J. Chem.* 1975, 53, 3339–3350. Stereochemistry of the Bucherer-Bergs and Strecker reactions of 4-*tert*-butylcyclohexanone.

9 D. Casabona, C. Cativiela, *Tetrahedron* 2006, 62, 10000–10004. Efficient synthesis of a new pipecolic acid analogue with a bicyclic structure.

10 A. U. Decensi, F. Boccardo, D. Guarneri, N. Positano, M. C. Paoletti, M. Costantini, G. Martorana, L. Giuliani, *J. Urol.* 1991, 146, 377–381. Monotherapy with nilutamide, a pure

nonsteroidal antiandrogen, in untreated patients with metastatic carcinoma of the prostate. The Italian prostatic cancer project.

11 P. Bac, P. Maurois, C. Dupont, N. Pages, J. P. Stables, P. Gressens, P. Evrard, J. Vamecq, *J. Neurosci.* 1998, 18, 4363–4373. Magnesium deficiency-dependent audiogenic seizures (MDDASs) in adult mice: a nutritional model for discriminatory screening of anticonvulsant drugs and original assessment of neuroprotection properties.

12 M. Nakajima, K. Itoi, Y. Takamatsu, T. Kinoshita, T. Okazaki, K. Kawakubo, M. Shindo, T. Honma, M. Tohjigamori, T. Haneishi, *J. Antibiot.* 1991, 44, 293–300. Hydantocidin: a new compound with herbicidal activity from *Streptomyces hygroscopicus*.

13 M. Sauli (Rhône–Poulenc S. A.), Ger. Offen 2, 149, 923, 1972.

14 M. S. Luer, *Neurol. Res.* 1998, 20, 178–182. Fosphenytoin.

15 A. P. Kent, R. A. Webster, *Br. J. Pharmacol.* 1983, 78, 553–559. The evaluation of different types of anticonvulsant drug activity against leptazol induced epilogenic activity in anaesthetized rat.

16 A. Kiipfer, R. Preisig, *Eur. J. Clin. Pharmacol.* 1984, 26, 753–759. Pharmacogenetics of mephenytoin: a new drug hydroxylation polymorphism in man.

17 For recent examples of active hydantoins, see: (a) J. E. Sheppeck, II, J. L. Gilmore, A. Yang, X.-T. Chen, C.-B. Xue, J. Roderick, R.-Q. Liu, M. B. Covington, C. P. Decicco, J. J.-W. Duan, *Bioorg. Med. Chem. Lett.* 2007, 17, 1413–1417. Discovery of novel hydantoins as selective non-hydroxamate inhibitors of tumor necrosis factor-α converting enzyme (TACE). (b) I. Peretto, R. Forlani, C. Fossati, G. A. M. Giardina, A. Giardini, M. Guala, E. La Porta, P. Petrillo, S. Radaelli, L. Radice, L. F. Raveglia, E. Santoro, R. Scudellaro, F. Scarpitta, C. Bigogno, P. Misiano, G. M. Dondio, A. Rizzi, E. Armani, G. Amari, M. Civelli, G. Villetti, R. Patacchini, M. Bergamaschi, M. Delcanale, C. Salcedo, A. G. Fernández, B. P. Imbimbo, *J. Med. Chem.* 2007, 50, 1571–1583. Discovery of diaryl imidazolidin-2-one derivatives, a novel class of muscarinic M3 selective antagonists (Part 1).

18 (a) I. Schön, T. Szirtes, T. Üerhardt, *J. Chem. Soc., Chem. Comm.* 1982, 639–640. Formation of by-products during sodium-liquids ammonia reduction in peptide chemistry. (b) W. Voelter, A. Altenburg, *Liebigs Ann. Chem.* 1983, 1641–1655. Nebenreaktion bei der Darstellung von Methionin-Enkephalin.

19 C. A. López, G. G. Trigo, *Adv. Heterocycl. Chem.* 1985, 38, 177–228. The chemistry of hydantoins.

20 For some recent selected examples, see: (a) G. G. Muccioli, D. Martin, G. K. E. Scriba, W. Poppitz, J. H. Poupaert, J. Wouters, D. M. Lambert, *J. Med. Chem.* 2005, 48, 2509–2517. Substituted 5,5′-diphenyl-2-thioxoimidazolidin-4-one as CB1 cannabinoid receptor ligands: synthesis and pharmacological evaluation. (b) X. Zhang, G. F. Allan, T. Sbriscia, O. Linton, S. G. Lundeen, Z. Sui, *Bioorg. Med. Chem. Lett.* 2006, 16, 5763–5766. Synthesis and SAR of novel hydantoin derivatives as selective androgen receptor modulators. (c) M. Waser, E. D. Moher, S. S. K. Borders, M. M. Hansen, D. W. Hoard, M. E. Laurila, M. E. LeTourneau, R. D. Miller, M. L. Phillips, K. A. Sullivan, J. A. Ward, C. Xie, C. A. Bye, T. Leitner, B. Herzog-Krimbacher, M. Kordian, M. Müllner, *Org. Process Res. Dev.* 2011, 15, 1266–1274. Process development for a key synthetic intermediate of LY2140023, a clinical candidate for the treatment of schizophrenia.

21 T. Martin, C. Massif, N. Wermester, J. Linol, S. Tisse, P. Cardinael, G. Coquerel, J.-P. Bouillon, *Tetrahedron: Asymmetry* 2011, 22, 12–21. Efficient preparation of enantiomerically pure α-aryl-α-trifluoromethylglycines *via* auto seeded programmed polythermic preferential crystallization of 5-aryl-5-trifluoromethylhydantoins.

22 J. Handzlik, A. J. Bojarski, G. Satała, M. Kubacka, B. Sadek, A. Ashoor, A. Siwek, M. Więcek, K. Kucwaj, B. Filipek, K. Kieć-Kononowicz, *Eur. J. Med. Chem.* 2014, 78, 324–339. SAR-studies on the importance of aromatic ring topologies in search for selective 5-HT7 receptor ligands among phenylpiperazine hydantoin derivatives.

23 Z. Khodaee, A. Yahyazadeh, N. O. Mahmoodi, *J. Heterocycl. Chem.* 2013, 50, 288–292. One-pot synthesis and characterization of some new types of 5,5′-disubstituted bis(imidazolidine-2,4-diones).

24 M. Oba, A. Shimabukuro, M. Ono, M. Doi, M. Tanaka, *Tetrahedron: Asymmetry* 2013, 24, 464–467. Synthesis of both enantiomers of cyclic methionine analogue: (*R*)- and (*S*)-3-aminotetrahydrothiophene-3-carboxylic acids.

25 W. A. Loughlin, S. S. Schweiker, I. D. Jenkins, L. C. Henderson, *Tetrahedron* 2013, 69, 1576–1582. Synthesis and evaluation of C8-substituted 4.5-spiro lactams as glycogen phosphorylase a inhibitors.

26 Z. Iqbal, S. Ali, J. Iqbal, Q. Abbas, I. Z. Qureshi, S. Hameed, *Bioorg. Med. Chem. Lett.* 2013, 23, 488–491. Dual action spirobicycloimidazolidine-2,4-diones: antidiabetic agents and inhibitors of aldose reductase-an enzyme involved in diabetic complications.

27 A. A. Thomas, K. W. Hunt, M. Volgraf, R. J. Watts, X. Liu, G. Vigers, D. Smith, D. Sammond, T. P. Tang, S. P. Rhodes, A. T. Metcalf, K. D. Brown, J. N. Otten, M. Burkard, A. A. Cox, M. K. G. Do, D. Dutcher, S. Rana, R. K. DeLisle, K. Regal, A. D. Wright, R. Groneberg, K. Scearce-Levie, M. Siu, H. E. Purkey, J. P. Lyssikatos, I. W. Gunawardana, *J. Med. Chem.* 2014, 57, 878–902. Discovery of 7-tetrahydropyran-2-yl chromans: *β*-site amyloid precursor protein cleaving enzyme 1 (BACE1) inhibitors that reduce amyloid *β*-protein (A*β*) in the central nervous system.

28 A. Pesquet, A. Daïch, L. Van Hijfte, *J. Org. Chem.* 2006, 71, 5303–5311. General and versatile entry to 4,5-fused polycyclic imidazolones systems. Use of the tandem transposition/π-cyclization of *N*-acyliminium species.

29 M. Beller, M. Eckert, W. A. Moradi, H. Neumann, *Angew. Chem. Int. Ed.* 1999, 38, 1454–1457. Palladium-catalyzed synthesis of substituted hydantoins—a new carbonylation reaction for the synthesis of amino acid derivatives.

30 R. G. Murray, D. M. Whitehead, F. Le Strat, S. J. Conway, *Org. Biomol. Chem.* 2008, 6, 988–991. Facile one-pot synthesis of 5-substituted hydantoins.

31 For recent applications, see: (a) J. Mičová, B. Steiner, M. Koóš, V. Langer, D. Gyepesová, *Synlett* 2002, 1715–1717. Synthesis of 4-carbamoyl-2-oxazolidinones C-4-linked with a saccharide

moiety *via* Bucherer-Bergs reaction of hexofuranos-5-uloses. (b) J. Mičová, B. Steiner, M. Koóš, V. Langer, D. Gyepesová, *Carbohydr. Res.* 2003, 338, 1917–1924. Characterisation and X-ray crystallography of products from the Bucherer-Bergs reaction of methyl 2,3-*O*-isopropylidene-α-D-*lyxo*-pentodialdo-1,4-furanoside. (c) U. D. Wermuth, I. D. Jenkins, R. C. Bott, K. A. Byriel, G. Smith, *Aust. J. Chem.* 2004, 57, 461–465. Some stereochemical aspects of the Strecker synthesis and the Bucherer-Bergs reaction. (d) C. Montagne, J. J. Shiers, M. Shipman, *Tetrahedron Lett.* 2006, 47, 9207–9209. Rapid generation of molecular complexity using "sequenced" multicomponent reactions: one-pot synthesis of 5,5′-disubstituted hydantoins from methyleneaziridines. (e) R. G. Murray, D. M. Whitehead, F. Le Strat, S. J. Conway, *Org. Biomol. Chem.* 2008, 6, 988–991. Facile one-pot synthesis of 5-substituted hydantoins.

32 C. Montagne, M. Shipman, *Synlett* 2006, 2203–2206. Modified Bucherer-Bergs reaction for the one-pot synthesis of 5,5′-disubstituted hydantoins from nitriles and organometallic reagents.

33 E. Gallienne, G. G. Muccioli, D. M. Lambert, M. Shipman, *Tetrahedron Lett.* 2008, 49, 6495–6497. Microwave-assisted four-component reaction for the synthesis of a monothiohydantoin inhibitor of a fatty acid amide hydrolase.

34 (a) G. Labar, C. Michaux, *Chem. Biodiversity* 2007, 4, 1882–1902. Fatty acid amide hydrolase: from characterization to therapeutics. (b) G. G. Muccioli, N. Fazio, G. K. E. Scriba, W. Poppitz, F. Cannata, J. H. Poupaert, J. Wouters, D. M. Lambert, *J. Med. Chem.* 2006, 49, 417–425. Substituted 2-thioxoimidazolidin-4-ones and imidazolidine-2,4-diones as fatty acid amide hydrolase inhibitors templates.

35 G. L. Patrick, *An introduction to medicinal chemistry*. 4th ed. Oxford University Press, New York, 2009.

36 L. Martarello, J. McConathy, V. M. Camp, E. J. Malveaux, N. E. Simpson, C. P. Simpson, J. J. Olson, G. D. Bowers, M. M. Goodman, *J. Med. Chem.* 2002, 45, 2250–2259. Synthesis of *syn*- and *anti*-1-amino-3-[^{18}F]fluoromethyl-cyclobutane-1-carboxylic acid (FMACBC), potential PET ligands for tumor detection.

37 D. S. Radchenko, O. O. Grygorenko, I. V. Komarov, *Tetrahedron: Asymmetry* 2008, 19, 2924–2930. Synthesis of conformationally restricted glutamic acid analogs based on the spiro[3.3]heptane scaffold.

38 J. Krysiak, W. H. Midura, W. Wieczorek, L. Sierón, M. Mikołajczyk, *Tetrahedron: Asymmetry* 2010, 21, 1486–1493. Constrained cycloalkyl analogues of glutamic acid: stereocontrolled synthesis of (+)-2-aminobicyclo[3.1.0]hexane-2,6-dicarboxylic acid (LY354740) and its 6-phosphonic acid analogue.

39 E. B. Averina, N. V. Yashin, Y. K. Grishin, T. S. Kuznetsova, N. S. Zefirov, *Synthesis* 2006, 880–884. Synthesis of (±)-(2-methylenecyclopropyl)glycine and (±)-4-[(amino(carboxymethyl)] spiro[2.2]pentane-1-carboxylic acid.

40 L. Ferron, F. Guillen, S. Coste, G. Coquerel, P. Cardinaël, J. Schwartz, J.-M. Paris, J.-C. Plaquevent, *Tetrahedron* 2011, 67, 6036–6044. Design and scalable synthesis of new chiral selectors. Part 1: synthesis and characterization of a new constrained cyclopeptide from unnatural bulky amino acids.

41 P. G. M. Wuts, S. W. Ashford, B. Conway, J. L. Havens, B. Taylor, B. Hritzko, Y. Xiang, P. S. Zakarias, *Org. Process Res. Dev.* 2009, 13, 331–335. A scalable synthesis of the INOS inhibitor PHA-399733.

42 M. Waser, E. D. Moher, S. S. K. Borders, M. M. Hansen, D. W. Hoard, M. E. Laurila, M. E. LeTourneau, R. D. Miller, M. L. Phillips, K. A. Sullivan, J. A. Ward, C. Xie, C. A. Bye, T. Leitner, B. Herzog-Krimbacher, M. Kordian, M. Müllner, *Org. Process Res. Dev.* 2011, 15, 1266–1274. Process development for a key synthetic intermediate of LY2140023, a clinical candidate for the treatment of schizophrenia.

43 R. J. Cox, *Nat. Prod. Rep.* 1996, 13, 29–43. The DAP pathway to lysine as a target for antimicrobial agents.

44 (a) S. O. Andersen, *Biochim. Biophys. Acta* 1964, 93, 213–215. The cross-link in resilin identified as dityrosine and trityrosine. (b) S. C. Fry, *Biochem. J.* 1982, 204, 449–455. Isodityrosine, a new cross-linking acid from plant cell-wall glycoprotein.

45 (a) R. M. Williams, G. J. Fegley, R. Gallegos, F. Schaefer, D. L. Pruess, *Tetrahedron* 1996, 52, 1149–1164. Asymmetric syntheses of (2*S*,3*S*,6*S*)-, (2*S*,3*S*,6*R*)-, and (2*R*,3*R*,6*S*)-2,3-methano-2,6-diaminopimelic acids. Studies directed to the design of novel substrate-based inhibitors of L,L-diaminopimelate epimerase. (b) R. J. Cox, A. Sutherland, J. C. Vederas, *Bioorg. Med. Chem.* 2000, 8, 843–871. Bacterial diaminopimelate metabolism as a target for antibiotic design. (c) B. Lygo, J. Crosby, J. A. Peterson, *Tetrahedron* 2001, 57, 6447–6453. An enantioselective approach to bis-α-amino acids.

46 R. A. Weatherhead, M. D. Carducci, E. A. Mash, *J. Org. Chem.* 2009, 74, 8773–8778. Synthesis of conformationally constrained diaminodicarboxylic acid derivatives.

47 Y. Chu, V. Lynch, B. L. Iverson, *Tetrahedron* 2006, 62, 5536–5548. Synthesis and DNA binding studies of bis-intercalators with a novel spiro-cyclic linker.

48 (a) H. Gröger, *Chem. Rev.* 2003, 103, 2795–2827. Catalytic enantioselective Strecker reactions and analogous syntheses. (b) J. Gawronski, N. Wascinska, J. Gajewy, *Chem. Rev.* 2008, 108, 5227–5242. Recent progress in Lewis base activation and control of stereoselectivity in the additions of trimethylsilyl nucleophiles. (c) P. Merino, E. Marqués-López, T. Tejero, R. P. Herrera, *Tetrahedron* 2009, 65, 1219–1234. Organocatalyzed Strecker reactions. (d) J. Wang, X. Liu, X. Feng, *Chem. Rev.* 2011, 111, 6947–6983. Asymmetric Strecker reactions.

49 (a) J.-A. Ma, *Angew. Chem. Int. Ed.* 2003, 42, 4290–4299. Recent developments in the catalytic asymmetric synthesis of α- and β-amino acids. (b) C. Najera, J. M. Sansano, *Chem. Rev.* 2007, 107, 4584–4671. Catalytic asymmetric synthesis of α-amino acids. (c) A. Perdih, M. S. Dolenc, *Curr. Org. Chem.* 2007, 11, 801–832. Recent advances in the synthesis of unnatural α-amino acids.

50 (a) D. Enders, J. P. Shilvock, *Chem. Soc. Rev.* 2000, 29, 359–373. Some recent applications of α-amino nitrile chemistry. (b) M. North in *Science of synthesis* (S.-L. Murahashi, Ed.). Thieme, Stuttgart, Germany, 2004, Vol. 19, pp. 285–310.

51 (a) L. Yet, *Angew. Chem. Int. Ed.* 2001, 40, 875–877. Recent developments in catalytic asymmetric Strecker-type reactions. (b) C. Spino, *Angew. Chem. Int. Ed.* 2004, 43, 1764–1766.

Recent developments in the catalytic asymmetric cyanation of ketimines. (c) T. Vilaivan, W. Bhanthumnauin, Y. Sritana-Anant, *Curr. Org. Chem.* 2005, 9, 1315–1392. Recent advances in catalytic asymmetric addition to imines and related C=N systems. (d) S. J. Connon, *Angew. Chem. Int. Ed.* 2008, 47, 1176–1178. The catalytic asymmetric Strecker reaction: ketimines continue to join the fold.

52 For recent examples, see: (a) J. F. Sardina, H. Rapoport, *Chem. Rev.* 1996, 96, 1825–1872. Enantiospecific synthesis of heterocycles from α-amino acids. (b) F. P. J. T. Rutjes, L. B. Wolf, H. E. Schoemaker, *J. Chem. Soc., Perkin Trans.* 1 2000, 4197–4212. Applications of aliphatic unsaturated non-proteinogenic α-H-α-amino acids. (c) J. Kaiser, S. S. Kinderman, B. C. J. van Esseveldt, F. L. van Delft, H. E. Schoemaker, R. H. Blaauw, F. P. J. T. Rutjes, *Org. Biomol. Chem.* 2005, 3, 3435–3467. Synthetic applications of aliphatic unsaturated α-H-α-amino acids. (d) U. Kazmaier, *Angew. Chem. Int. Ed.* 2005, 44, 2186–2188. Amino acids-valuable organocatalysts in carbohydrate synthesis. (e) E. R. Jarvo, S. J. Miller, *Tetrahedron* 2002, 58, 2481–2495. Amino acids and peptides as asymmetric organocatalysts.

53 J. Martens, *ChemCatChem* 2010, 2, 379–381. Enantioselective organocatalytic Strecker reactions in the synthesis of α-amino acids.

54 S. V. Bhat, B. A. Nagasampagi, M. Sivakumar (Eds.), *Chemistry of natural products*. Springer, Berlin, 2005.

55 (a) S. L. Miller, *Science* 1953, 117, 528–529. A production of amino acids under possible primitive earth conditions. (b) J. L. Bada, A. Lazcano, *Science* 2003, 300, 745–746. Prebiotic soup-revisiting the Miller experiment. (c) J. L. Bada, *Earth Planet. Sci. Lett.* 2004, 226, 1–15. How life began on Earth: a status report.

56 I. V. P. Raj, G. Suryavanshi, A. Sudalai, *Tetrahedron Lett.* 2007, 48, 7211–7214. Organocatalytic activation of TMSCN by basic ammonium salts for efficient cyanation of aldehydes and imines.

57 R. Martinez, D. J. Ramon, M. Yus, *Tetrahedron Lett.* 2005, 46, 8471–8474. Catalyst-free multicomponent Strecker reaction in acetonitrile.

58 P. Fontaine, A. Chiaroni, G. Masson, J. Zhu, *Org. Lett.* 2008, 10, 1509–1512. One-pot three-component synthesis of α-iminonitriles by IBX/TBAB-mediated oxidative Strecker reaction.

59 (a) K. C. Nicolaou, Y.-L. Zhong, P. S. Baran, *Angew. Chem. Int. Ed.* 2000, 39, 625–628. New synthetic technology for the rapid construction of novel heterocycles-part 2. The reaction of IBX with anilides and related compounds. (b) K. C. Nicolaou, P. S. Baran, R. Kranich, Y.-L. Zhong, K. Sugita, N. Zou, *Angew. Chem. Int. Ed.* 2001, 40, 202–206. Mechanistic studies of periodinane-mediated reactions of anilides and related systems.

60 J.-B. Gualtierotti, X. Schumacher, P. Fontaine, G. Masson, Q. Wang, J. Zhu, *Chem.–Eur. J.* 2012, 18, 14812–14819. Amidation of aldehydes and alcohols through α-iminonitriles and a sequential oxidative three-component Strecker reaction/thio-Michael addition/alumina-promoted hydrolysis process to access β-mercaptoamides from aldehydes, amines, and thiols.

61 F. Pan, J.-M. Chen, Z. Zhuang, Y.-Z. Fang, S. X.-A. Zhang, W.-W. Liao, *Org. Biomol. Chem.* 2012, 10, 2214–2217. Construction of highly functional α-amino nitriles *via* a novel multicomponent tandem organocatalytic reaction: a facile access to α-methylene γ-lactams.

62 G. K. S. Prakash, T. Mathew, C. Panja, S. Alconcel, H. Vaghoo, C. Do, G. A. Olah, *Proc. Natl. Acad. Sci. U. S. A.* 2007, 104, 3703–3706. Gallium (III) triflate catalyzed efficient Strecker reaction of ketones and their fluorinated analogs.

63 G. K. S. Prakash, C. Panja, C. Do, T. Mathew, G. A. Olah, *Synlett* 2007, 2395–2399. Trimethylsilyl trifluoromethanesulfonate as a metal-free, homogeneous and strong Lewis acid catalyst for efficient one-pot synthesis of α-aminonitriles and their fluorinated analogues.

64 G. K. S. Prakash, T. E. Thomas, I. Bychinskaya, A. G. Prakash, C. Panja, H. Vaghoo, G. A. Olah, *Green Chem.* 2008, 10, 1105–1110. Efficient green synthesis of α-aminonitriles, precursors of α-amino acids.

65 K. Matsumoto, J. C. Kim, H. Iida, H. Hamana, K. Kumamoto, H. Kotsuki, G. Jenner, *Helv. Chim. Acta* 2005, 88, 1734–1753. Multicomponent Strecker reaction under high pressure.

66 For selected recent examples, see: (a) M. Suginome, A. Yamamoto, Y. Ito, *Chem. Commun.* 2002, 1392–1393. Bis(dialkylamino)cyanoboranes: highly efficient reagents for the Strecker-type aminative cyanation of aldehydes and ketones. (b) J. McConathi, L. Martarello, E. J. Malveaux, V. M. Camp, N. E. Simpson, C. P. Simpson, G. D. Bowers, J. J. Olson, M. M. Goodman, *J. Med. Chem.* 2002, 45, 2240–2249. Radiolabeled amino acids for tumor imaging with PET: radiosynthesis and biological evaluation of 2-amino-3-[^{18}F]fluoro-2-methylpropanoic acid and 3-[^{18}F]fluoro-2-methyl-2-(methylamino)propanoic acid. (c) G. Jenner, R. B. Salen, J. C. Kim, K. Matsumoto, *Tetrahedron Lett.* 2003, 44, 447–449. Effect of pressure on the Strecker synthesis of hindered α-aminonitriles from ketones and aromatic amines. (d) K. Surendra, N. S. Krishnaveni, A. Mahesh, K. R. Rao, *J. Org. Chem.* 2006, 71, 2532–2534. Supramolecular catalysis of Strecker reaction in water under neutral conditions in the presence of *β*-cyclodextrin.

67 N. H. Khan, S. Agrawal, R. I. Kureshy, S. H. R. Abdi, S. Singh, E. Suresh, R. V. Jasra, *Tetrahedron Lett.* 2008, 49, 640–644. $Fe(Cp)_2PF_6$ catalyzed efficient Strecker reactions of ketones and aldehydes under solvent-free conditions.

68 J. S. Yadav, B. V. S. Reddy, B. Eshwaraiah, M. Srinivas, P. Vishnumurthy, *New J. Chem.* 2003, 27, 462–465. Three-component coupling reactions in ionic liquids: a facile synthesis of α-aminonitriles.

69 For a critical essay about "in water" or "in the presence of water," see: D. G. Blackmond, A. Armstrong, V. Coombe, A. Well, *Angew. Chem. Int. Ed.* 2007, 46, 3798–3800. Water in organocatalytic processes: debunking the myths.

70 E. Takahashi, H. Fujisawa, T. Yanai, T. Mukaiyama, *Chem. Lett.* 2005, 34, 318–319. Lewis base-catalyzed Strecker-type reaction between trimethylsilyl cyanide and *N*-tosylimines in water-containing DMF.

71 M. A. Kumar, M. F. S. Babu, K. Srinivasulu, Y. B. Kiran, C. S. Reddy, *J. Mol. Catal. A: Chem.* 2007, 265, 268–271.

Polyethylene glycol in water: a simple and environment friendly media for Strecker reaction.

72 K. Surendra, N. S. Krishnaveni, A. Mahesh, K. R. Rao, *J. Org. Chem.* 2006, 71, 2532–2534. Supramolecular catalysis of Strecker reaction in water under neutral conditions in the presence of β-cyclodextrin.

73 S. Kobayashi, T. Busujima, S. Nagayama, *Chem. Commun.* 1998, 981–982. Scandium triflate-catalyzed Strecker-type reactions of aldehydes, amines and tributyltin cyanide in both organic and aqueous solutions. Achievement of complete recovery of the tin compounds toward environmentally-friendly chemical processes.

74 B. Karmakar, J. Banerji, *Tetrahedron Lett.* 2010, 51, 2748–2750. K_2PdCl_4 catalyzed efficient multicomponent synthesis of α-aminonitriles in aqueous media.

75 E. Marqués-López, R. P. Herrera, R. Fernández, J. M. Lassaletta, *Eur. J. Org. Chem.* 2008, 3457–3460. Uncatalyzed Strecker-type reaction of *N,N*-dialkylhydrazones in pure water.

76 K. Shen, X. Liu, Y. Cai, L. Lin, X. Feng, *Chem.–Eur. J.* 2009, 15, 6008–6014. Facile and efficient enantioselective Strecker reaction of ketimines by chiral sodium phosphate.

77 S. Harusawa, Y. Hamada, T. Shioiri, *Tetrahedron Lett.* 1979, 20, 4663–4666. Diethyl phosphorocyanidated (DEPC). A novel reagent for the classical Strecker's α-amino nitrile synthesis.

78 S. Nakamura, N. Sato, M. Sugimoto, T. Toru, *Tetrahedron: Asymmetry* 2004, 15, 1513–1516. A new approach to enantioselective cyanation of imines with Et_2AlCN.

79 Z. Xie, G. Li, G. Zhao, J. Wang, *Synthesis* 2009, 2035–2039. Strecker-type reaction catalyzed carboxylic acids in aqueous media.

80 F. Cruz-Acosta, A. Santos-Expósito, P. de Armas, F. García-Tellado, *Chem. Commun.* 2009, 6839–6841. Lewis base-catalyzed three-component Strecker reaction on water. An efficient manifold for the direct α-cyanoamination of ketones and aldehydes.

81 Z. Li, Y. Ma, J. Xu, J. Shi, H. Cai, *Tetrahedron Lett.* 2010, 51, 3922–3926. One-pot three-component synthesis of α-aminonitriles using potassium hexacyanoferrate(II) as an eco-friendly cyanide source.

82 S. Sipos, I. Jablonkai, *Tetrahedron Lett.* 2009, 50, 1844–1846. One-pot synthesis of α-aminonitriles from alkyl and aryl cyanides: a Strecker reaction *via* aldimine alanes.

83 K. Iwanami, H. Seo, J.-C. Choi, T. Sakakura, H. Yasuda, *Tetrahedron* 2010, 66, 1898–1901. Al-MCM-41 catalyzed three-component Strecker-type synthesis of α-aminonitriles.

84 For previous use of acetone cyanohydrin in a one-pot Strecker reaction, see: A. S. Paraskar, A. Sudalai, *Tetrahedron Lett.* 2006, 47, 5759–5762. $Cu(OTf)_2$ or Et_3N-catalyzed three-component condensation of aldehydes, amines and cyanides: a high yielding synthesis of α-aminonitriles.

85 P. Galletti, M. Pori, D. Giacomini, *Eur. J. Org. Chem.* 2011, 3896–3903. Catalyst-free Strecker reaction in water: a simple and efficient protocol using acetone cyanohydrin as cyanide source.

86 For other recent work performed in water, see: S. Ramesh, K. Sivakumar, C. Panja, P. N. Arunachalam, A. Lalitha, *Synth. Commun.* 2010, 40, 3544–3551. Water-mediated Strecker reaction. An efficient and environmentally friendly approach for the synthesis of α-aminonitriles *via* a three-component condensation.

87 M. Pori, P. Galletti, R. Soldati, D. Giacomini, *Eur. J. Org. Chem.* 2013, 1683–1695. Asymmetric Strecker reaction with chiral amines: a catalyst-free protocol using acetone cyanohydrin in water.

88 K. Mai, G. Patil, *Synth. Commun.* 1985, 15, 157–163. A fast *N*-substituted α-aminonitrile synthesis.

89 M. M. Mojtahedi, M. Saed Abaee, H. Abbasi, *Can. J. Chem.* 2006, 84, 429–432. One-pot, solvent-free synthesis of α-aminonitriles under catalysis by magnesium bromide ethyl etherate.

90 For an additional example under solvent-free conditions, see: D. Chaturvedi, A. K. Chaturvedi, N. Mishra, V. Mishra, *Tetrahedron Lett.* 2012, 53, 5398–5401. A novel approach for the synthesis of α-aminonitriles using Mitsunobu's reagent under solvent-free conditions.

91 A. Baeza, C. Najera, J. M. Sansano, *Synthesis* 2007, 1230–1234. Solvent-free synthesis of racemic α-aminonitriles.

92 For other example, see: J. Wang, Y. Masui, M. Onaka, *Eur. J. Org. Chem.* 2010, 1763–1771. Synthesis of α-amino nitriles from carbonyl compounds, amines, and trimethylsilyl cyanide: comparison between catalyst-free conditions and the presence of tin ion-exchanged montmorillonite.

93 X.-L. Zhang, Q.-P. Wu, Q.-S. Zhang, *J. Chem. Res.* 2013, 37, 690–693. Efficient three-component Strecker reaction of acetals and aromatic amines catalyzed by hafnium tetrachloride at room temperature.

94 (a) H. Ishitani, S. Komiyama, Y. Hasegawa, S. Kobayashi, *J. Am. Chem. Soc.* 2000, 122, 762–766. Catalytic asymmetric Strecker synthesis. Preparation of enantiomerically pure α-amino acid derivatives from aldimines and tributyltin cyanide or achiral aldehydes, amines, and hydrogen cyanide using a chiral zirconium catalyst. (b) G. Brahmachari, B. Banerjee, *Asian J. Org. Chem.* 2012, 1, 251–258. Comparison between catalyst-free and $ZrOCl_2{\cdot}8H_2O$-catalyzed Strecker reactions for the rapid and solvent-free one-pot synthesis of racemic α-aminonitrile derivatives.

95 J. Jarusiewicz, Y. Choe, K. S. Yoo, C. P. Park, K. W. Jung, *J. Org. Chem.* 2009, 74, 2873–2876. Efficient three-component Strecker reaction of aldehydes/ketones *via* NHC-amidate palladium(II) complex catalysis.

96 For selected non chiral examples, see: (a) Z. Li, Y. Sun, X. Ren, P. Wei, Y. Shi, P. Ouyang, *Synlett* 2007, 803–805. Sulfamic acid catalyzed direct condensation of aldehydes, amines, and TMSCN to α-aminonitriles at ambient temperature. (b) B. Das, R. A. Kumar, P. Thirupathi, *Helv. Chim. Acta* 2007, 90, 1206–1210. One-Pot three-component synthesis of α-amino nitriles catalyzed by 2,4,6-trichloro-1,3,5-triazine (cyanuric acid). (c) S. C. Pan, B. List, *Synlett* 2007, 318–320. Catalytic one-pot, three-component acyl-Strecker reaction. (d) A. Heydari, A. Arefi, S. Khaksar, R. K. Shiroodi, *J. Mol. Catal. A: Chem.* 2007, 271, 142–144. Guanidine hydrochloride: an active and

simple catalyst for Strecker type reaction. (e) M. Barbero, S. Cadamuro, S. Dughera, G. Ghigo, *Org. Biomol. Chem.* 2012, 10, 4058–4068. *o*-Benzenedisulfonimide and its chiral derivative as Brønsted acids catalysts for one-pot three-component Strecker reaction. Synthetic and mechanistic aspects. (f) J.-M. Chen, Y.-Z. Fang, Z.-L. Wei, W.-W. Liao, *Synthesis* 2012, 44, 1849–1853. Multicomponent tandem phosphine-catalyzed umpolung reaction: facile access to highly functionalized α-aminonitriles.

97 (a) S. C. Pan, B. List, *Org. Lett.* 2007, 9, 1149–1151. Catalytic asymmetric three-component acyl-Strecker reaction. (b) S. C. Pan, B. List, *Chem. Asian J.* 2008, 3, 430–437. The catalytic acylcyanation of imines.

98 (a) Y. Wen, Y. Xiong, L. Chang, J. Huang, X. Liu, X. Feng, *J. Org. Chem.* 2007, 72, 7715–7719. Chiral bisformamides as effective organocatalysts for the asymmetric one-pot, three-component Strecker reaction. (b) Y. Wen, B. Gao, Y. Fu, S. Dong, X. Liu, X. Feng, *Chem.–Eur. J.* 2008, 14, 6789–6795. Asymmetric three-component Strecker reactions catalyzed by *trans*-4-hydroxy-L-proline-derived *N,N′*-dioxides.

99 G.-W. Zhang, D.-H. Zheng, J. Nie, T. Wang, J.-A. Ma, *Org. Biomol. Chem.* 2010, 8, 1399–1405. Brønsted acid-catalyzed efficient Strecker reaction of ketones, amines and trimethylsilyl cyanide.

100 E. Rafiee, A. Azad, M. Joshaghania, *Lett. Org. Chem.* 2007, 4, 60–63. $K_5CoW_{12}O_{40}·3H_2O$: highly efficient heterogeneous catalyst for the synthesis of α-aminonitriles.

101 E. Rafiee, S. Rashidzadeh, A. Azada, *J. Mol. Catal. A: Chem.* 2007, 261, 49–52. Silica-supported heteropoly acids: highly efficient catalysts for synthesis of α-aminonitriles, using trimethylsilyl cyanide or potassium cyanide.

102 J. S. Yadav, B. V. S. Reddy, B. Eeshwaraian, M. Srinivas, *Tetrahedron* 2004, 60, 1767–1771. Montmorillonite KSF clay catalyzed one-pot synthesis of α-aminonitriles.

103 M. Shekouhy, *Catal. Sci. Technol.* 2012, 2, 1010–1020. Sulfuric acid-modified PEG-6000 (PEG-OSO_3H): an efficient Brønsted acid-surfactant combined catalyst for the one-pot three component synthesis of α-aminonitriles in water.

104 J. Choi, Y. Y. Yang, H. J. Kim, S. U. Son, *Angew. Chem. Int. Ed.* 2010, 49, 7718–7722. Organometallic hollow spheres bearing bis(N-heterocyclic carbene)-palladium species: catalytic application in three-component Strecker reactions.

105 D. Saberi, S. Cheraghi, S. Mahdudi, J. Akbari, A. Heydari, *Tetrahedron Lett.* 2013, 54, 6403–6406. Dehydroascorbic acid (DHAA) capped magnetite nanoparticles as an efficient magnetic organocatalyst for the one-pot synthesis of α-aminonitriles and α-aminophosphonates.

106 For some selected examples, see: (a) Y. Pérez-Fuertes, J. E. Taylor, D. A. Tickell, M. F. Mahon, S. D. Bull, T. D. James, *J. Org. Chem.* 2011, 76, 6038–6047. Asymmetric Strecker synthesis of α-arylglycines. (b) J. R. Donald, R. R. Wood, S. F. Martin, *ACS Comb. Sci.* 2012, 14, 135–143. Application of a sequential multicomponent assembly process/Huisgen cycloaddition strategy to the preparation of libraries of 1,2,3-triazole-fused 1,4-benzodiazepines. (c) L. Y. V. Mendez, V. V. Kouznetsov, *Curr. Org. Chem.* 2013, 10, 969–973. First girgensohnine analogs prepared through $InCl_3$-catalyzed Strecker reaction and their bioprospection.

107 D. M. Fitch, K. A. Evans, D. Chai, K. J. Duffy, *Org. Lett.* 2005, 7, 5521–5524. A highly efficient, asymmetric synthesis of benzothiadiazine-substituted tetramic acids: potent inhibitors of hepatitis C virus RNA-dependent RNA polymerase.

108 P. Fontaine, G. Masson, J. Zhu, *Org. Lett.* 2009, 11, 1555–1558. Synthesis of pyrroles by consecutive multicomponent reaction/[4+1] cycloaddition of α-iminonitriles with isocyanides.

109 For a recent related work, see: J.-B. Gualtierotti, X. Schumacher, Q. Wang, J. Zhu, *Synthesis* 2013, 45, 1380–1386. Synthesis of iminonitriles by oxone/TBAB-mediated one-pot oxidative three-component Strecker reaction.

11

UNUSUAL APPROACH FOR MULTICOMPONENT REACTIONS

EUGENIA MARQUÉS-LÓPEZ[1], RAQUEL P. HERRERA[1] AND RAMON RIOS[2]

[1] *Departamento de Química Orgánica, Instituto de Síntesis Química y Catálisis Homogénea (ISQCH), CSIC-Universidad de Zaragoza, Zaragoza, Spain*

[2] *Department of Organic Chemistry, University of Southampton, Highfield Campus, Southampton, UK*

The most famous multicomponent reactions (MCRs) have been disclosed extensively in the previous chapters. In this chapter, some remarkable examples of reactions performed under unusual reaction conditions or by using uncommon catalysts will be discussed [1]. The growing interest for simple and green synthetic procedures has promoted the use of unconventional solvents and catalysts; MCRs using them have become a new research area in green organic syntheses [2]. Among these alternative systems, we will describe some remarkable examples for each, including zeolites, microwave (MW), aqueous media, and ionic liquids.

11.1 ZEOLITE-CATALYZED MCRs

One of the less common catalysts used for promoting MCRs are the zeolites. In the recent years, zeolites have gained considerable importance, owing to not only the possibility of conducting environmentally benign syntheses but also good yields, achieved using these heterogeneous catalytic systems [3]. Zeolites have received increasing attention because of their tunable acidity and diverse structures and pore dimensions, thus exhibiting exceptional properties in terms of both activity and selectivity. Other attractive properties of zeolites include their low cost and excellent thermal stability, which make them economically and environmentally feasible, thus providing green alternatives to homogeneous catalysis. Further, zeolites are widely used to synthesize specialty and fine chemicals and also to develop clean technologies, because the applications of zeolites help overcome the drawbacks of the conventional technologies, mainly environmental pollution and corrosion of equipment.

The interesting structure of zeolites allows the incorporation and immobilization of suitable cations or chiral catalysts on the surface of zeolitic pores. These processes not only enhance the available surface area but also provide highly structured and confined cavities that can increase stereoselectivity. Cation-exchanged zeolites, such as KY and CsY, have received much attention as solid base catalysts owing to the following advantages: (i) easy separation from the reaction mixture, (ii) reusability, (iii) easy modification of their surface and pore size, and (iv) use of nonpolluting natural minerals [4].

Although the first examples on the use of synthetic zeolites as the catalysts show the syntheses of many compounds with interesting structures, the application of synthetic zeolites in MCRs is still an open challenge (Fig. 11.1).

11.1.1 Heterogeneous Hybrid Catalyst

A new concept of immobilization affording novel green catalytic systems is to covalently anchor organocatalysts on solid carriers; this approach can afford efficient and recyclable catalytic systems. An interesting example of this type of structure was recently reported by Arya and coworkers, where they explored this approach in a Mannich MCR by immobilizing proline, an organocatalyst, on pure silica mesoporous materials, thus generating diverse easily recyclable catalysts (Scheme 11.1) [12]. The proline acts as both the catalyst and reaction media, and the solid matrix acts as the nanoscale reactor, thus allowing the reactants and products

Multicomponent Reactions: Concepts and Applications for Design and Synthesis, First Edition. Edited by Raquel P. Herrera and Eugenia Marqués-López.

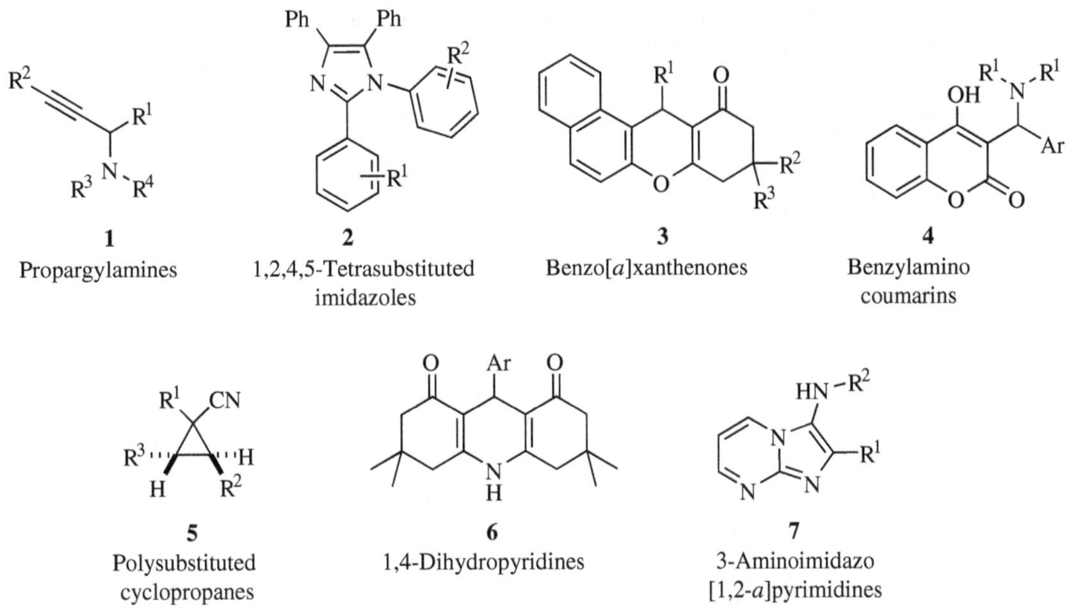

FIGURE 11.1 Interesting syntheses of functionalized structures: **1** [5], **2** [6], **3** [7], **4** [8], **5** [9], **6** [10], and **7** [11].

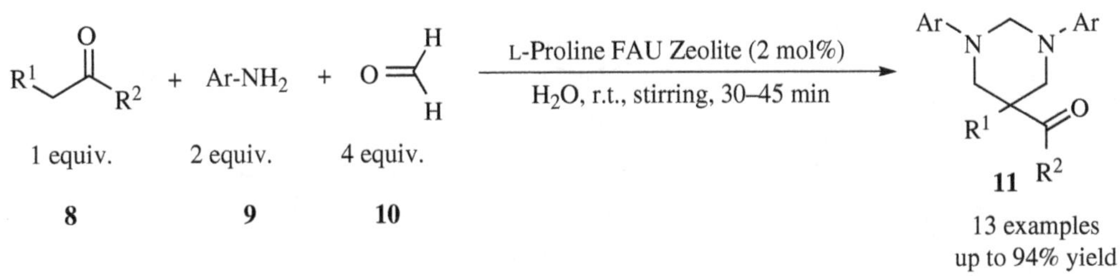

SCHEME 11.1 Mannich multicomponent reaction for the synthesis of spirocyclic compounds.

to be transported in or out. However, some prerequisites are needed for a suitable catalytic system: on one hand, the pore size of the zeolite must be large enough to securely hold the organocatalyst in place, while the channel size of the microporous materials should be small enough to prevent the organocatalyst from leaching. On the other hand, the channel size must also be large enough for the easy transportation of reactants and products. These required properties of this new concept are different from those of the previously reported supported/immobilized organocatalysts [13].

This novel approach was applied to the synthesis of a series of novel spiro[indoline-3,5′-pyrimidin]-2-one and spiro[indene-2,5′-pyrimidin]-1(3*H*)-one derivatives **11** in an aqueous medium (Table 11.1). In this developed procedure, six molecules of reactants are involved, and six new covalent bonds are generated, making the reaction highly atom economic and affording final products in good to excellent yields (68–94%) and in short reaction times (20–45 min). The recovery and reusability of the catalyst were also investigated in one model reaction, and the results showed no loss of catalytic activity even after five cycles.

A plausible mechanism for the reaction is shown in Scheme 11.2. The carbonyl group undergoes two α-aminomethylation reactions on the same α-carbon of the ketone **12**, which is catalyzed by L-proline through enamine catalysis. The cyclocondensation of the resulting substituted amines **19** with formaldehyde affords the desired spiro[indoline-3,5′-pyrimidin]-2-one and spiro[indene-2,5′-pyrimidin]-1(3*H*)-one (Scheme 11.2).

11.2 MW-ASSISTED THREE-COMPONENT REACTIONS

MW-assisted organic transformations have gained considerable popularity, particularly in the past decade, for its ability to accelerate diverse synthetic reactions, thus becoming a powerful tool in synthetic chemistry [14]. MW heating is advantageous over conventional heating because it can decrease the reaction time [15], increase the yield of the final products, provide better selectivity, and simplify the work-up of the reaction, in most cases without the use of catalysts,

TABLE 11.1 Synthesis of spiroheterocycles *via* the one-pot Mannich reaction of ketone, amine, and formaldehyde

Entry	Ketone **8**	Amine **9**	Time (min)	Product **11**
1	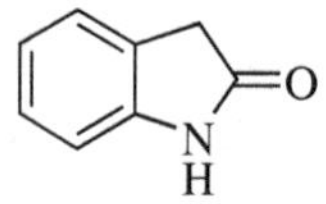	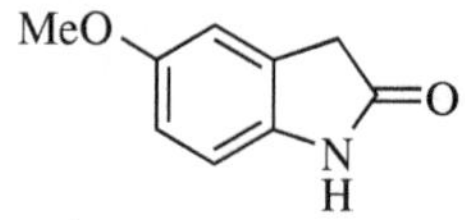	30	94
2	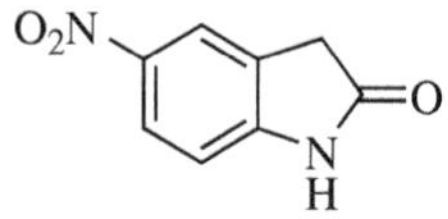	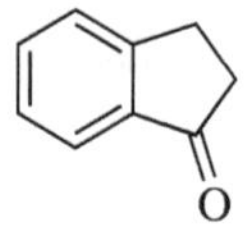	20	90
3	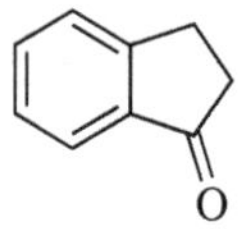	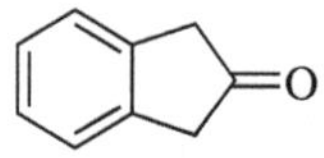	35	88
4	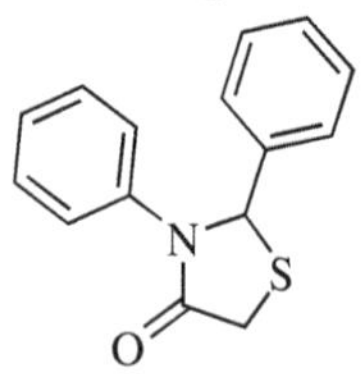	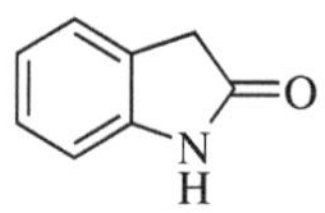	30	87
5	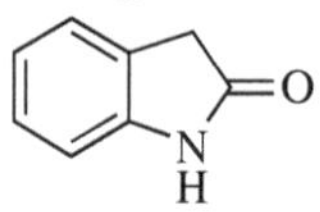	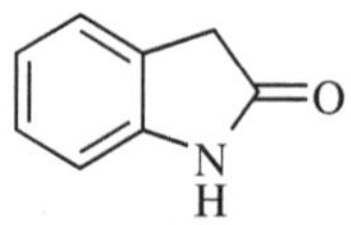	40	90
6	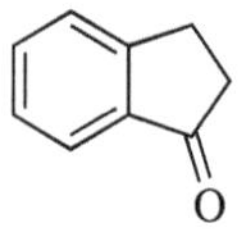	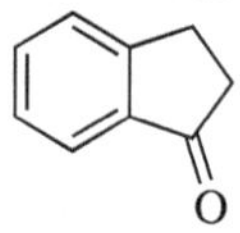	35	93
7	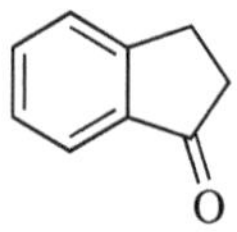	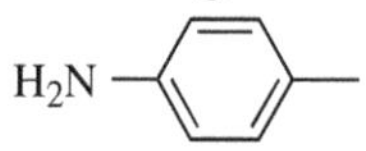	45	90
8	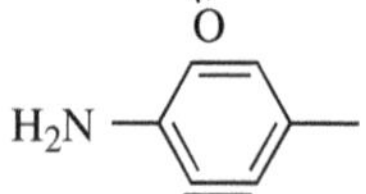	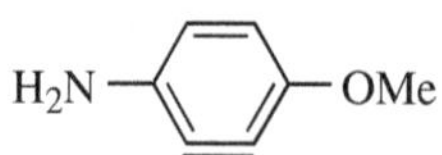	45	65
9	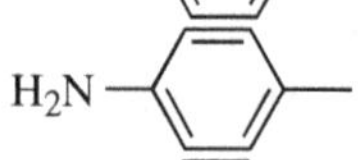	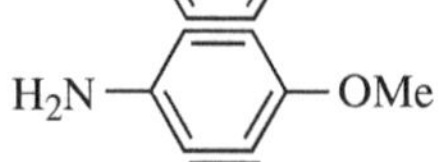	35	80
10	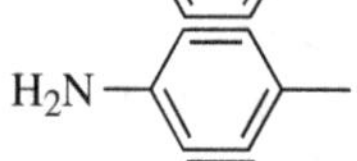	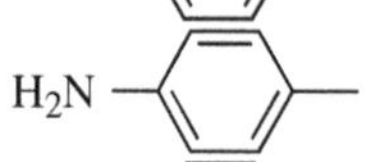	30	88
11	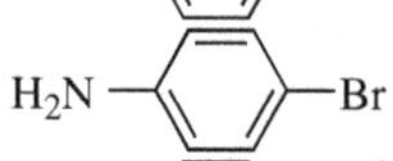	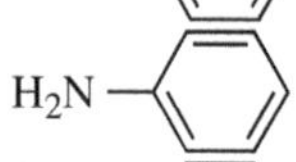	40	78
12	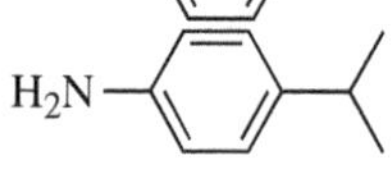	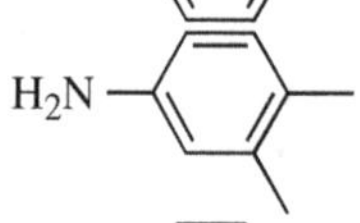	35	68
13	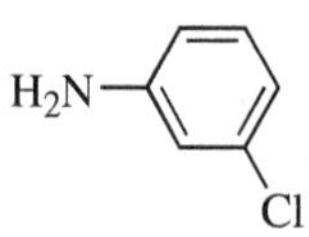	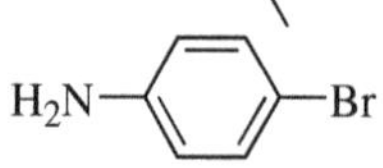	40	70

thus constituting an eco-friendly protocol in the frame of *green chemistry*. Only in the last decade, MW has received special attention in MCRs and currently occupies a central place in research.

MW has been used in the multicomponent syntheses of privileged structures by MCRs (Fig. 11.2). Pyridine nucleus **20** [16] forms the core of diverse natural and synthetic bioactive compounds [17]. *N*-substituted-2-quinolone **21** [18] represents the structural motif of many biologically active compounds, such as protein kinase inhibitors, immunomodulators, antiulcer agents, hypoglycemics, farnesyltransferase inhibitors, and antiviral agents [19]. Pyrido[2,3-*d*]pyrimidine

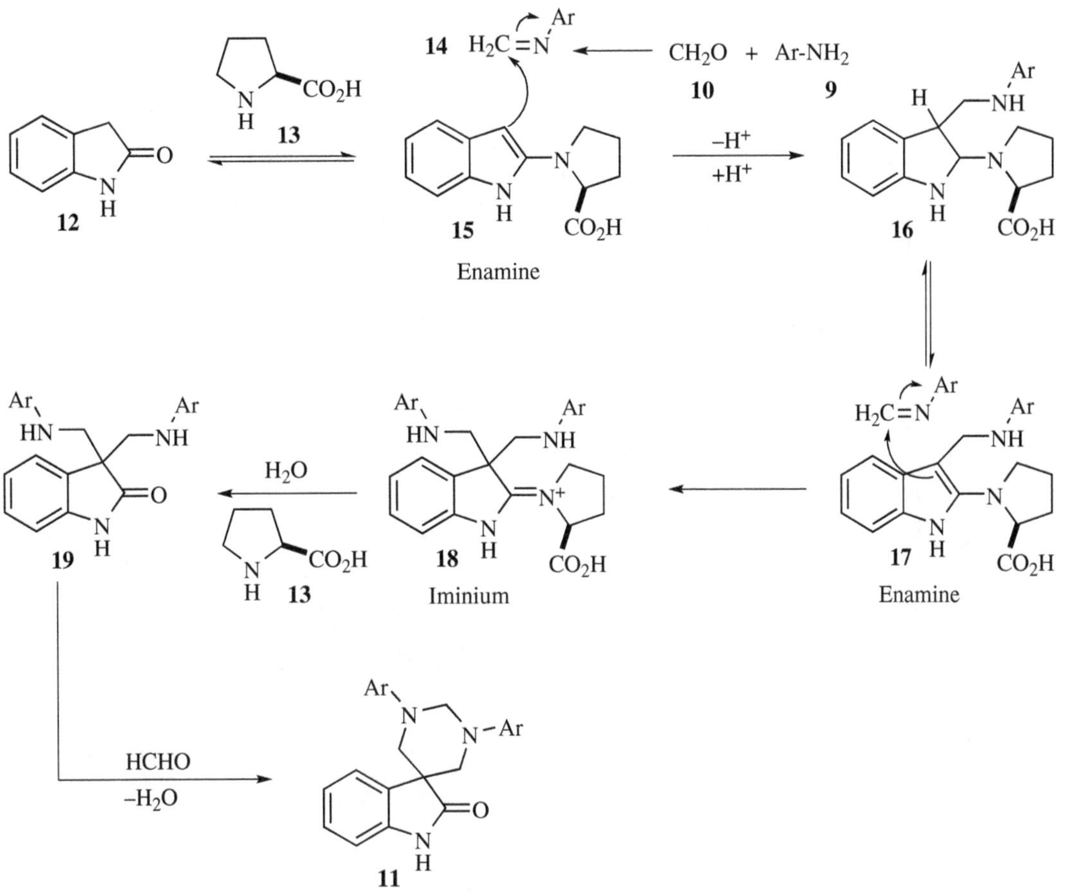

SCHEME 11.2 Mannich-type mechanism through enamine activation.

core **22** [20] has been identified previously as a privileged structure for the inhibition of ATP-dependent kinases, and good potency against Cdks has been reported for representative examples [21]. Thiazolo[3,2-*a*]pyridines **23** [22] contain two fused heterocyclic motifs in one molecule and exhibit many important bioactivities such as inhibiting β-amyloid production, potent CDK2–cyclin A inhibitor, α-glucosidase inhibitor, potential uterus stimulant, and antibacterial and antifungal activities. Quinoline moiety **24** is present in numerous naturally occurring and biologically active compounds [23] and exhibits interesting photochemical properties [24, 25]. Substituted chromenes (benzopyrans) **25** are important compounds because of their applications in pigments, cosmetics, and agrochemicals; moreover, they are an integral part of many natural products [26] and therapeutic agents such as antibacterial [27], antiviral [28], mutagenic [29], antitumor [30], and CNS-active agents [31, 32]. Quinazolines **26** [33] are important scaffolds for drug discovery in pharmaceutical industry [34]. Thus, a facile and environmentally benign approach for the syntheses of these highly functionalized heterocycles has been developed by one-pot multicomponent condensation reactions. Most of the cases involved solvent-free conditions, and MW played an important role in developing these procedures.

A large number of other attractive examples with molecular diversity under the influence of MWs have also been reported [35].

11.2.1 Synthesis of Natural Products

MW-assisted MCR has been used for the syntheses of interesting natural products. Liu and coworkers developed a MCR strategy for the total syntheses of natural alkaloids, glyantrypine (**28a**), fumiquinazoline F (**28b**), and fiscalin B (**28d**) (Scheme 11.3) [36]. These heterocycles contain a pyrazino[2,1-*b*]quinazoline-3,6-dione **27** core [37], which is also present in other natural products and exhibits interesting biological properties [38] (Fig. 11.3).

This method is based on a previously developed method by the same research group [39]: in this case, the method was utilized to synthesize natural products. Notably, the crucial hurdles in these syntheses were the necessity of only a single reagent $P(OPh)_3$ **38**, the identification of an optimal protecting group for the α-amino acid (Boc), and the optimization of thermal conditions for the ring closure to afford the target molecules in maximum yields and without side reactions. The same research group developed a very similar

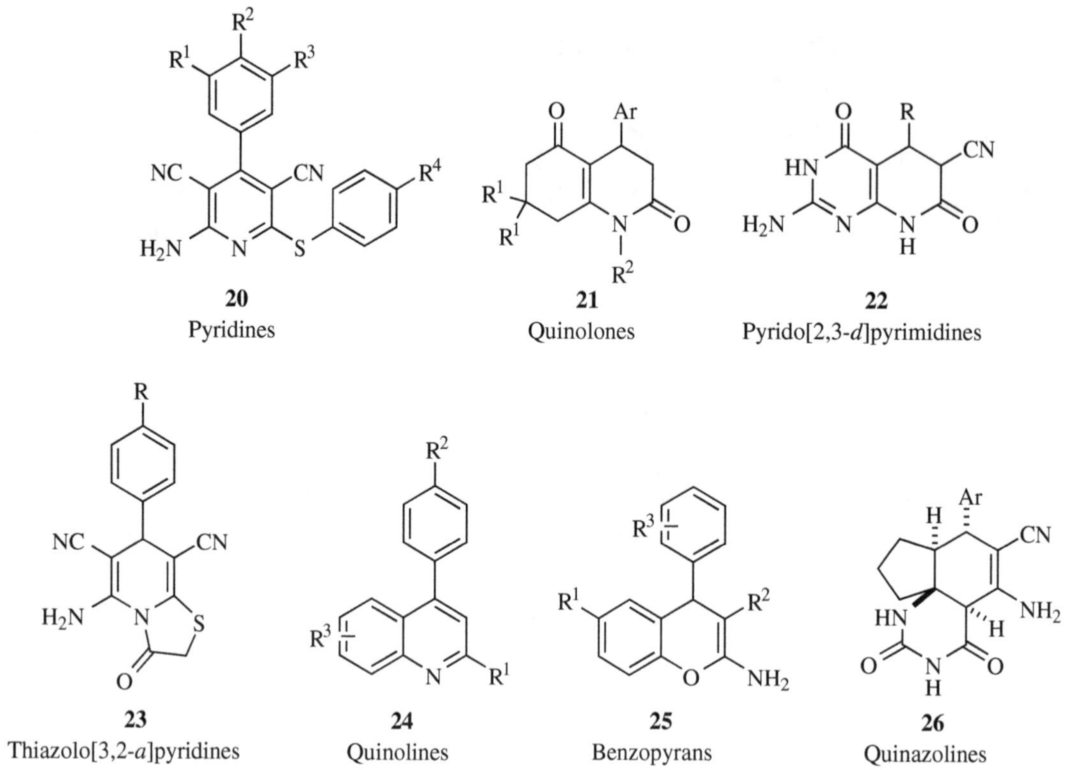

FIGURE 11.2 Representative heterocycles synthesized by MW-assisted MCRs.

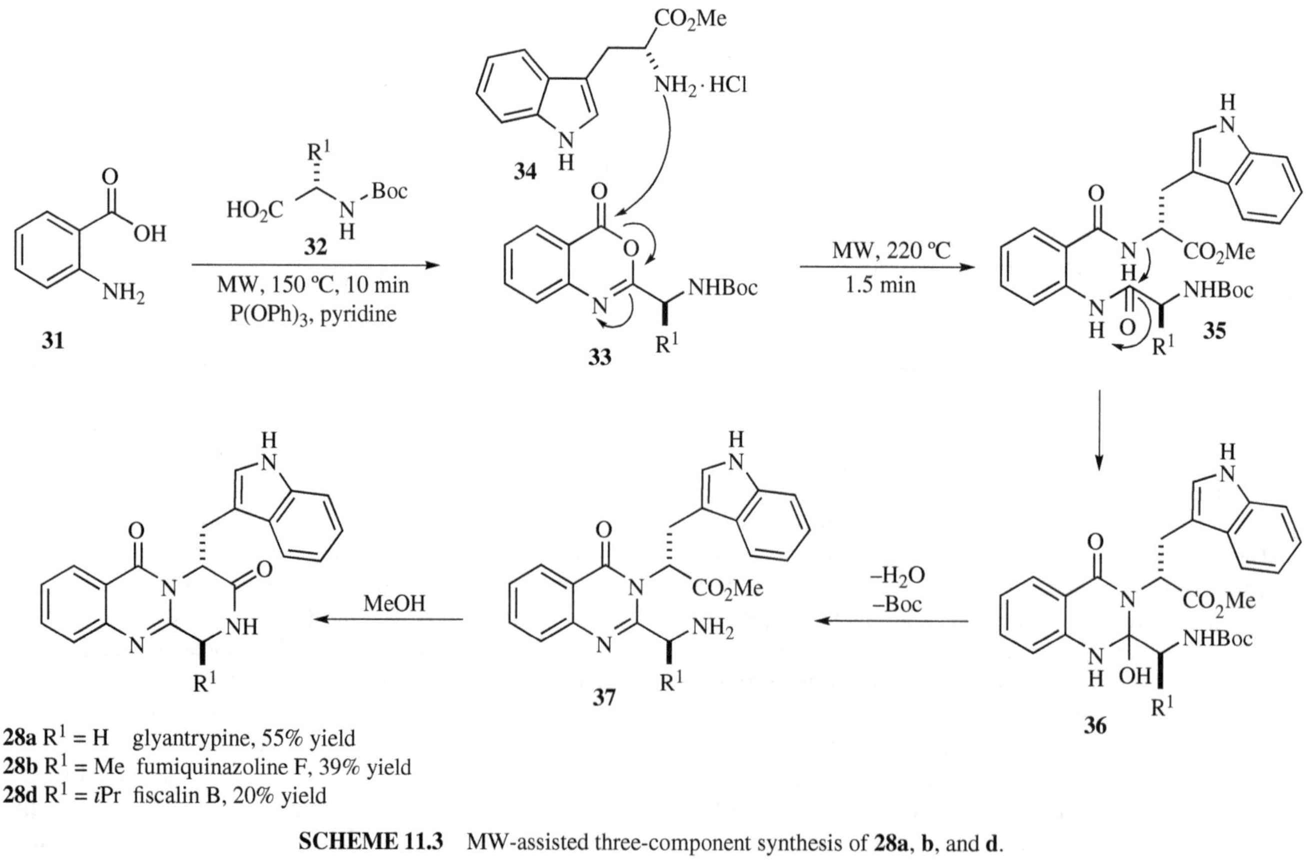

28a R^1 = H glyantrypine, 55% yield
28b R^1 = Me fumiquinazoline F, 39% yield
28d R^1 = *i*Pr fiscalin B, 20% yield

SCHEME 11.3 MW-assisted three-component synthesis of **28a**, **b**, and **d**.

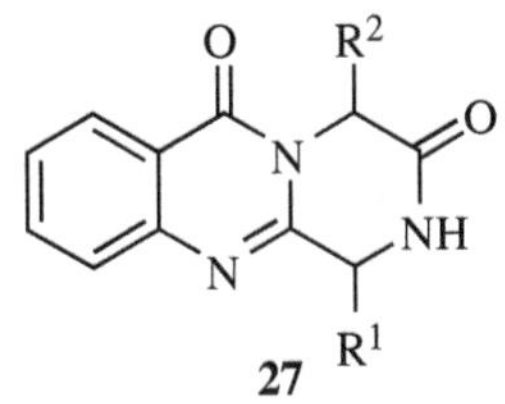

27

Pyrazino[2,1-*b*]quinazoline-3,6-dione core

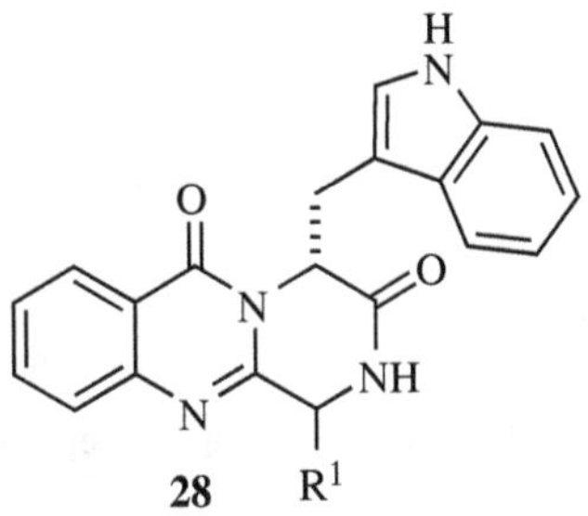

28

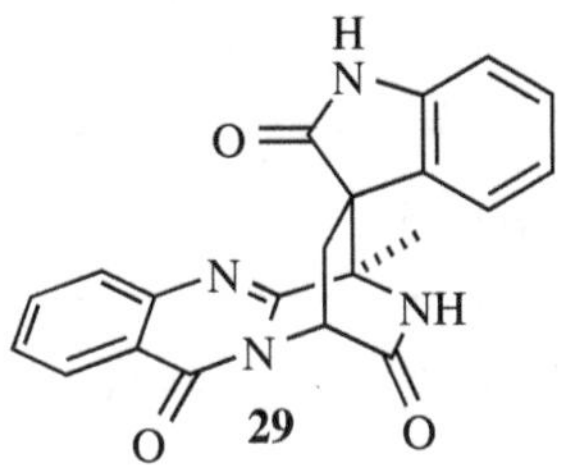

29

Alantrypinone

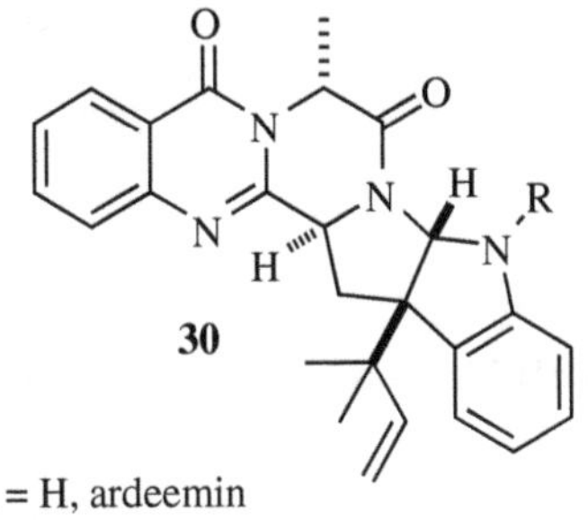

30

R = H, ardeemin
R = Ac, *N*-acetylardeemin

28a R^1 = H glyantrypine
28b R^1 = Me (1*S*)-fumiquinazoline F
28c R^1 = Me (1*R*)-fumiquinazoline G
28d R^1 = *i*Pr (1*S*)-fiscalin B

FIGURE 11.3 Biological active compounds containing the pyrazino[2,1-*b*]quinazoline-3,6-dione core **27**.

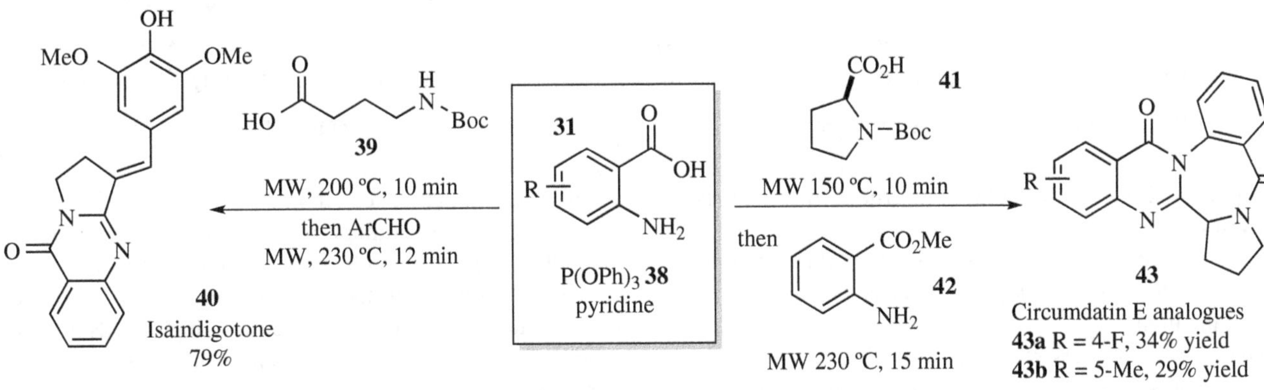

SCHEME 11.4 MW-assisted three-component synthesis of **40** and **43**.

approach for the synthesis of the substituted analogues of circumdatin E **43** [40] and isaindigotone **40** [41] with biological activities (Scheme 11.4).

Eycken and coworkers developed an unprecedented MW-assisted MCR strategy *via* an intramolecular Ugi four-component reaction (4CR) for the synthesis of dibenzo[*c*,*e*] azepinone scaffold **46** (Scheme 11.5) [42]. The core of these products resembles the framework of γ-secretase inhibitor LY411575 **47**, reported by Eli Lilly [43]. The final products were also evaluated for their bioactivity; some of them showed antiproliferative activity in the lower micromolar range against tumor cell lines. The MW irradiation was crucial for the success of the reactions affording the products in excellent yields and with high diastereoselectivity.

11.3 IONIC LIQUID-PROMOTED MCRs

Ionic liquids have attracted much attention in the last decade [44] because they represent an efficient and practical alternative to conventional organic solvents for many important organic transformations. Ionic liquids are advantageous as environmentally benign solvents; they are nonvolatile, nonflammable, nonexplosive, and easily recycled and have great potential to enhance reaction rates [45]. A large number of reactions have been successfully accomplished using ionic liquids as the solvents or catalysts, and many of them utilize the synthetic potential of MCR strategies [46]. In the last decade, a large number of compounds with interesting structures have been synthesized, such as those shown in

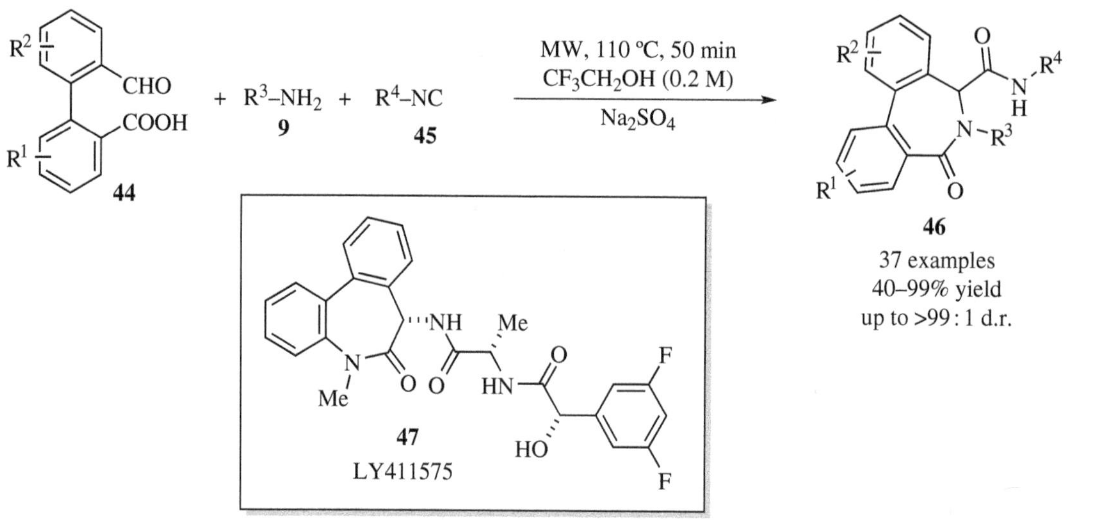

SCHEME 11.5 Microwave-assisted diastereoselective multicomponent reaction to access dibenzo[*c*,*e*]azepinones **46**.

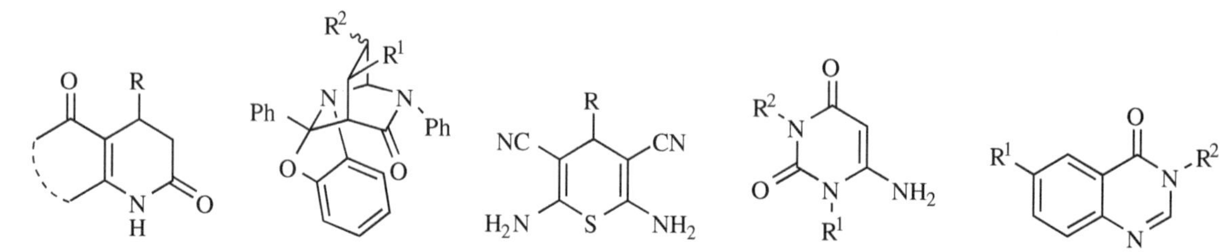

48	**49**	**50**	**51**	**52**
1,4-Dihydropyridines	DABCO core	Thiopyrans	6-Aminouracils	4(3*H*)-Quinazolinones

FIGURE 11.4 Ionic liquid-assisted synthesis of interesting molecules: **48** [47], **49** [48], **50** [49], **51** [50], and **52** [51].

Figure 11.4. Ionic liquids played a crucial role in these syntheses [52].

In this section, we will discuss some important examples where ionic liquids promote the MCRs as the catalysts and solvent medium.

Wang and coworkers reported an interesting strategy for the efficient syntheses of highly substituted cyclohexa-1, 3-diene (Scheme 11.6), polyhydroindene (Scheme 11.7), polyhydronaphthalene (Scheme 11.7), isochromene (Scheme 11.8), isothiochromene (Scheme 11.8), and isoquinoline derivatives (Scheme 11.9) [53]. The syntheses were achieved using MCR procedures in ionic liquid media, with the following advantages: mild reaction conditions, high yields, one-pot procedures, operational simplicity, and environmentally benign conditions.

Although, at this point, the mechanism of this reaction is not clear, the authors proposed a reasonable mechanism as shown in Scheme 11.6. The formation of cyclohexa-1, 3-dienes was explained as follows: first, the Knoevenagel condensation of aryl aldehyde **53** with malononitrile **54** gives malononitrile **58**. Next, the Michael addition reaction of **55** with malononitrile **58** affords intermediate **59**, followed by the intramolecular cyclization to render intermediate **60**. Finally, the isomerization of intermediate **60** affords final products **61**. The authors obtained intermediate **60** by crystallization to support their proposed mechanism.

The reactions led to the corresponding highly substituted products under mild reaction conditions, high yields, one-pot procedures, operational simplicity, and environmentally benign conditions.

11.4 MCRs UNDER SOLVENT-FREE CONDITIONS

The development of new MCRs under solvent-free conditions has emerged as an attractive alternative to conventional syntheses in the recent years. The main advantages of solvent-free processes are cost savings, decreased energy consumption, and reduced reaction times. The E factor is a measure of the "greenness" of a chemical process and is defined as the ratio of the weight of waste to the weight of product. One of the best ways to reduce the E factor of the reactions and their effect on the environment is to carry out the reactions under solvent-free conditions because the best solvent is no solvent.

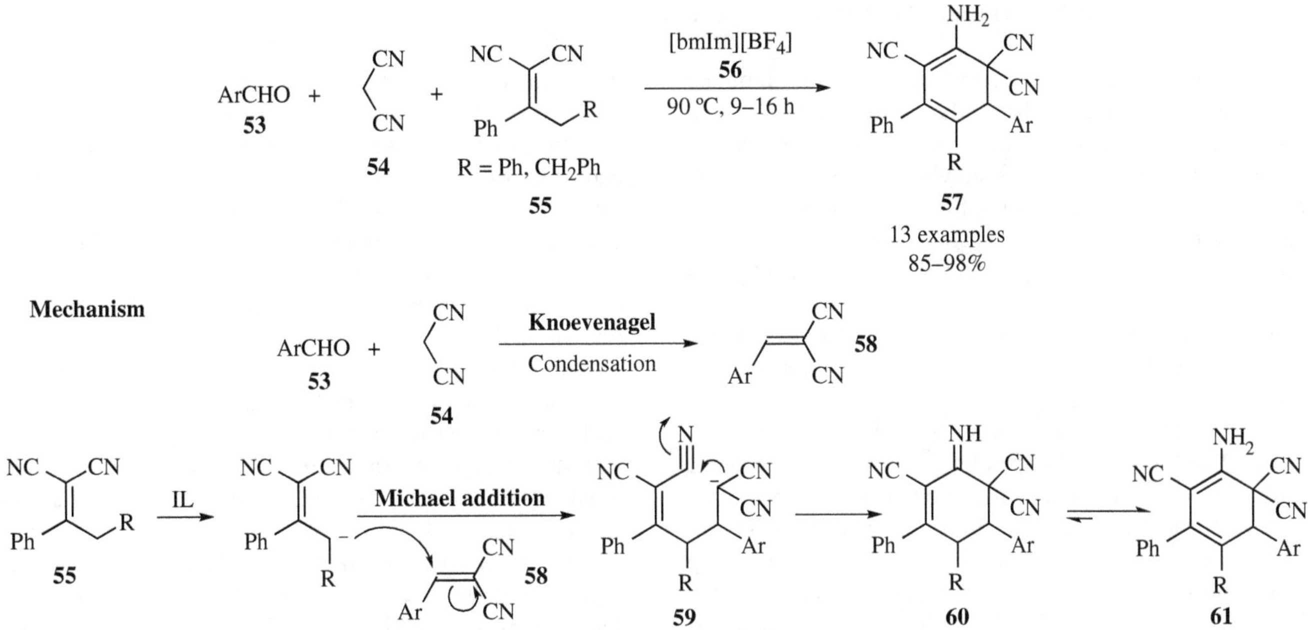

SCHEME 11.6 Synthesis of substituted cyclohexa-1,3-diene.

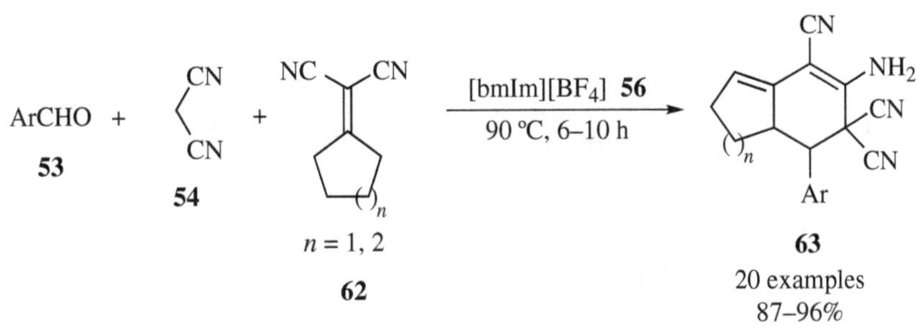

SCHEME 11.7 Synthesis of polyhydroindenes and polyhydronaphthalenes **63**.

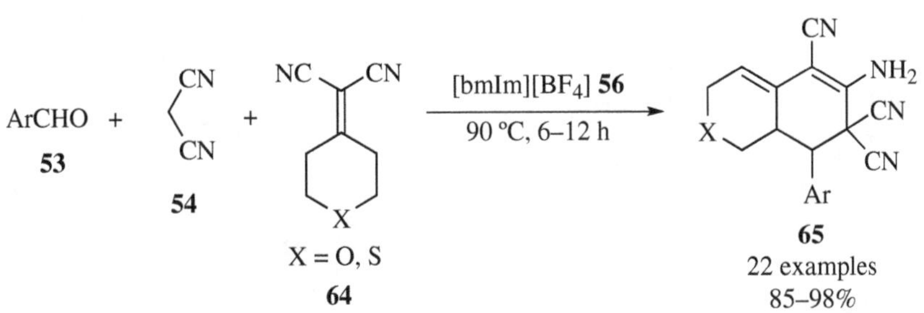

SCHEME 11.8 Synthesis of isochromene and isothiochromene derivatives **65**.

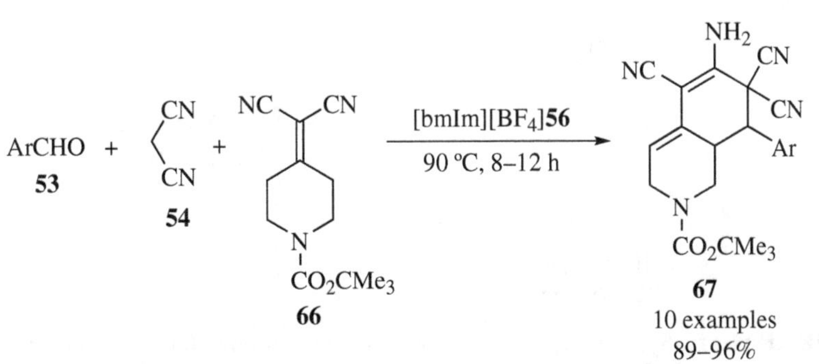

SCHEME 11.9 Synthesis of isoquinoline derivatives **67**.

Based on this concept, Murthy and coworkers reported an intriguing reaction for the synthesis of amidoalkyl naphthols under solvent-free conditions [54]. As shown in Scheme 11.10, β-naphthol **69**, alkyl/aryl aldehydes **53**, and amides/thioureas **68** were stirred at room temperature and catalyzed by silica sulfuric acid (SSA). The three-component reaction afforded amidoalkyl naphthols **70** in excellent yields (79–92%).

One of the limitations of this reaction is that the reaction of 1-naphthol led to a mixture of *ortho-* and *para*-addition products in 1:1.3 ratio.

Later, Adib and coworkers reported the synthesis of 3-amino-2-arylimidazo[1,2-*a*]pyridines **79** by a solvent-free three-component reaction of 2-aminopyridines **71**, benzaldehydes **53**, and imidazoline-2,4,5-triones **72** (Scheme 11.11). The MCR required high temperatures (200°C). The first step of the mechanism of the reaction involves the condensation between benzaldehyde **53** and 2-aminopyridine **71** and between benzaldehyde **53** and imidazoline-2,4,5-trione **72**. Next, the imidazolium ion **74** is attacked by the pyridine *N*-atom of the aldimine **73** to generate adduct **75**, followed by the intramolecular nucleophilic addition of the alkoxide to the adjacent carbonyl group in **75** to afford an epoxide intermediate **76** that undergoes ring opening to give ylide **77**. The intramolecular cyclization of **77** and loss of CO_2 and isocyanic acid (HNCO) afford imidazo[1,2-*a*]pyridine **79** (Scheme 11.11) [55].

In 2010, Pitchumani and Kanagaraj reported the solvent-free multicomponent synthesis of pyranopyrazoles. The reaction of hydrazine, ethyl acetoacetate, malononitrile, and aromatic aldehydes catalyzed by per-6-amino-β-cyclodextrin at room temperature afforded the corresponding pyranopyrazoles in quantitative yields [56].

In the same year, Wang reported the solvent-free uncatalyzed Petasis reaction of salicylaldehydes, boronic acids, and secondary amines [57]. The reaction required high temperatures (80°C) and afforded the corresponding amine derivatives in good to excellent yields (45–96% yield). In 2011, Khosropour and coworkers synthesized pyrazoles and bispyrazoles under solvent-free conditions catalyzed by Zn triflate [58]. The reaction of aldehydes, aryl hydrazones, and ethyl acetate at 120°C

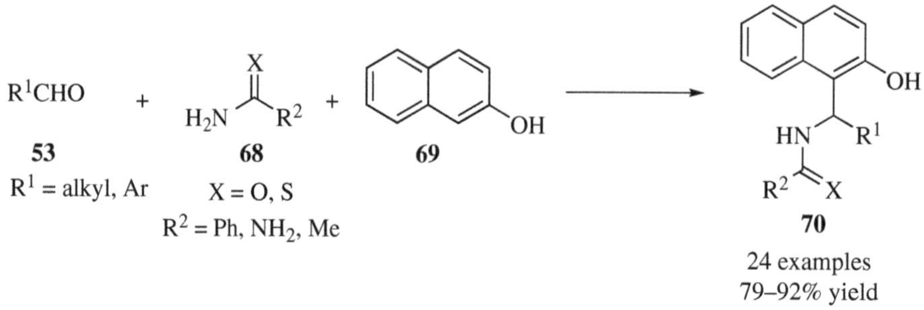

SCHEME 11.10 Synthesis of amidoalkyl naphthols **70**.

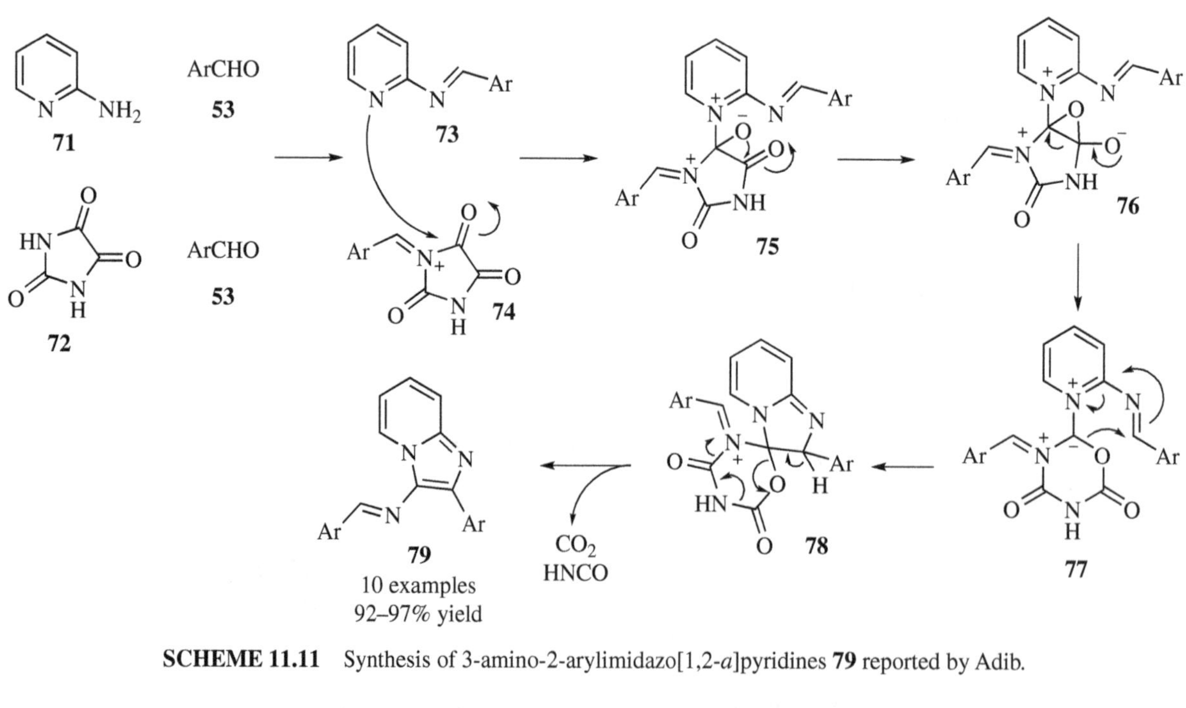

SCHEME 11.11 Synthesis of 3-amino-2-arylimidazo[1,2-*a*]pyridines **79** reported by Adib.

afforded the corresponding pyrazole derivatives in good to excellent yields (83–93%, 12 examples).

Romanelli and coworkers reported the solvent-free multicomponent synthesis of pyridines using the Wells–Dawson (WD) heteropoly acid as the catalyst [59]. The reaction of 3-formylchromones **80** with acetoacetic esters **81** and ammonium acetate gave the corresponding pyridines **83** in good yields (64–99%). As shown in Scheme 11.12, Hantzsch esters **82** were obtained in variable yields (0–36%) as the by-products.

Singh and coworkers reported a method for the synthesis of 2*H*-indazolo[2,1-*b*]phthalazine-1,6,11-triones and 1*H*-pyrazolo[1,2-*b*]phthalazine-5,10-diones under solvent-free and sonochemical conditions [60]. The three-component reaction of phthalhydrazide, 1,3-diketones, and aldehydes catalyzed by (*S*)-camphorsulfonic acid (20 mol%) under ultrasound irradiation at room temperature afforded the corresponding products **86** in good to excellent yields (50–94%) (Scheme 11.13).

Later, a similar reaction was reported by the same research group based on the $Sc(OTf)_3$-catalyzed MCR of hydrazines **87**, aldehydes **53**, and ketoesters **88** under the activation of MWs (Scheme 11.14) [61]. The reactions were very fast (3–6 min) and provided the functionalized pyrazoles **89** in good to excellent yields (74–96%).

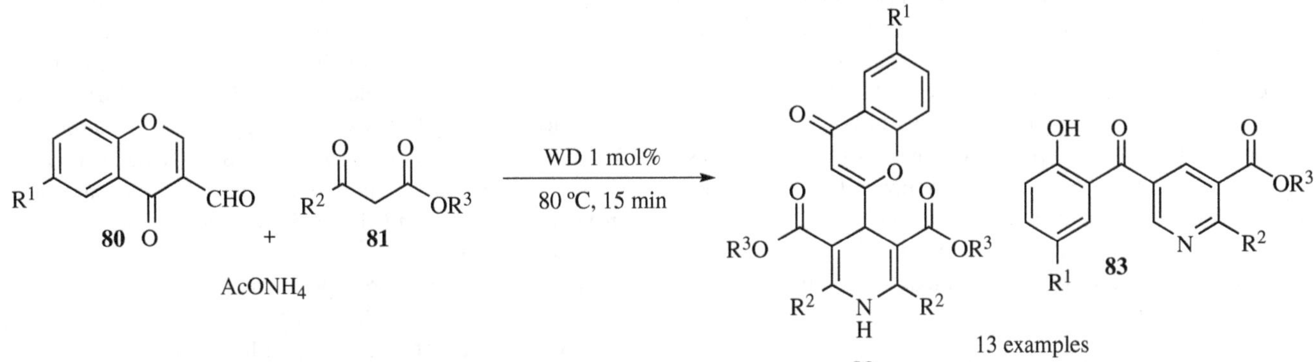

13 examples
83: 64–99% yield
82: 36–0% yield

SCHEME 11.12 Solvent-free multicomponent synthesis of pyridines **83** reported by Romanelli.

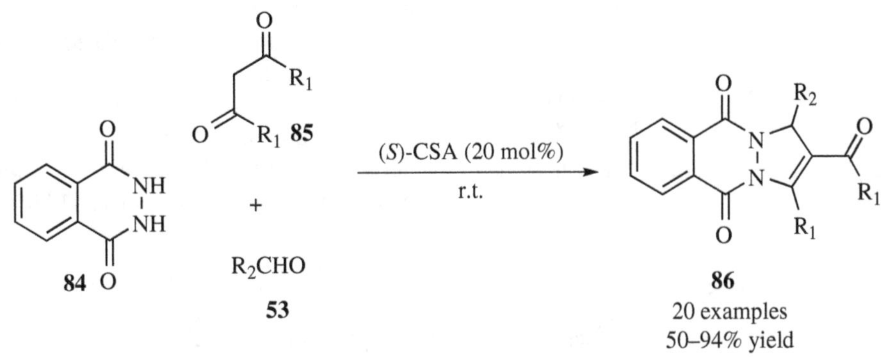

20 examples
50–94% yield

SCHEME 11.13 Synthesis of 2*H*-indazolo[2,1-*b*]phthalazine-1,6,11-triones **86**.

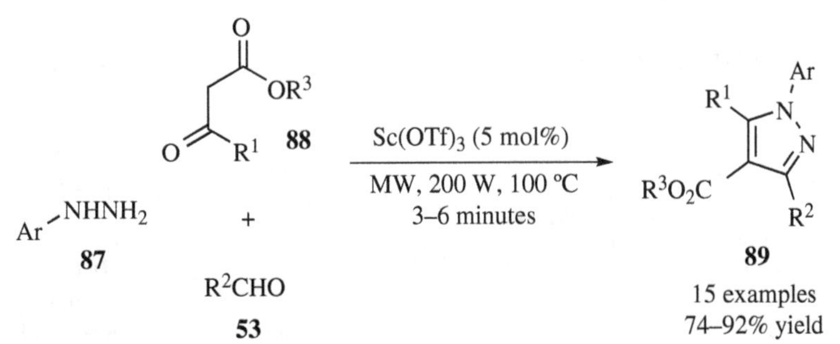

15 examples
74–92% yield

SCHEME 11.14 Synthesis of pyrazoles **89**.

A three-component solvent-free Mannich reaction was developed by Nayak in 2011 [62]. The reaction was catalyzed by camphorsulfonic acid, giving the Mannich adducts in good yields. The authors applied this method to synthesize aminochromanols. A similar reaction was developed by Larsen based on cyclohexanone [63]. As shown in Scheme 11.15, the reaction of cyclohexanone **90** with primary/secondary amines **92** afforded the corresponding imines that reacted with an alkyne, *n*-hexylacetylene **91**, catalyzed by $CuCl_2$. The reactions were carried out under solvent-free conditions at 110°C to afford the corresponding amine derivatives **93** in good to excellent yields (61–98%).

Almost at the same time, Nguyen and coworkers reported the multicomponent oxidative coupling of amines **94** with thioamides **96** catalyzed by elemental sulfur [64]. The reactions were carried out under solvent-free and harsh conditions (130°C) to afford the final thioamides **97** in good to excellent yields (62–95%). The proposed mechanism of this reaction is shown in Scheme 11.16. First, the oxidation of benzylamine **94** with sulfur affords the corresponding imine **95**. Next, the oxidation of imine **95** with sulfur affords the corresponding thioamide **96** that undergoes *trans*thioamidation to afford the final products **97**.

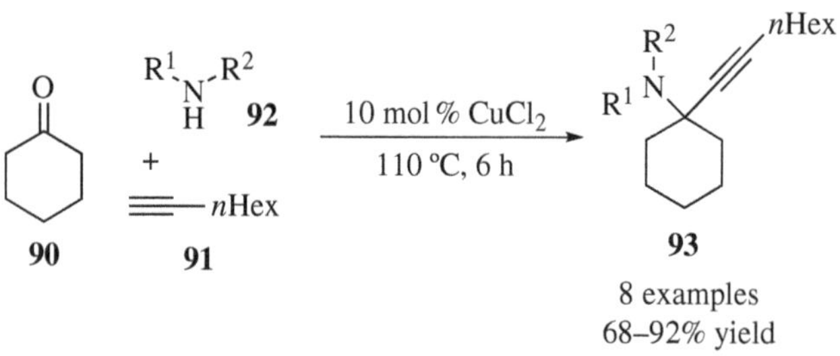

SCHEME 11.15 Three-component solvent-free Mannich reaction developed by Nayak.

Ponnuswamy and coworkers reported an interesting approach for the multicomponent synthesis of spiro compounds **101** under solvent-free conditions based on the Staudinger/aza-Wittig coupling/cyclization [65]. The reaction of azides **98**, triphenylphosphine, ketones **99**, and 2-sulfanylacetic acid **100** under solvent-free conditions and MW irradiation afforded the corresponding spirothiazolidin-4-ones **101** in excellent yields (92–96%) (Scheme 11.17).

Solvent-free grinding methods have been used for the synthesis of pyrano[2,3-*a*]carbazoles [66]. As shown in Scheme 11.18, the reaction starts with the condensation of aldehyde **53** with malononitrile **54** by a Knoevenagel condensation using $NaHCO_3$ as the base. Next, the Michael addition of 1-hydroxy carbazole **102** to the resulting benzylidenemalononitrile **58** followed by the intramolecular nucleophilic addition of the hydroxyl of the carbazole to the cyano group affords intermediate **104** that after the air oxidation and hydrolysis afford pyrano[2,3-*a*]carbazoles **106** in moderate to good yields (55–88%) (Scheme 11.19).

The authors also reported the use of Meldrum's acid instead of malononitrile and obtained similar results (Scheme 11.19).

Karami and coworkers reported multicomponent, solvent-free reactions for the synthesis of benzofurans (Scheme 11.20) [67]. The reaction of arylglyoxals **108** with benzamide derivatives **109** afforded the corresponding iminium intermediate that after the nucleophilic attack by phenol derivatives **110** and intramolecular cyclization gave the corresponding benzofurans **111** in good yields (75–85%). The reaction was efficiently catalyzed by tungstate sulfuric acid (TSA) that after the reaction could be isolated and reused without loss of its catalytic activity.

A similar reaction was reported by Salama and coworkers. The reaction of aldehydes **53**, amides **109**, and phenol derivatives **110** under solvent-free conditions afforded the corresponding amide derivatives **112** in very good to excellent

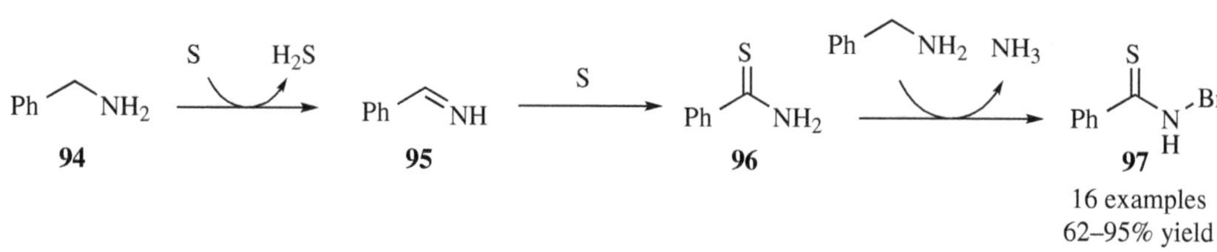

SCHEME 11.16 Multicomponent oxidative coupling reported by Nguyen.

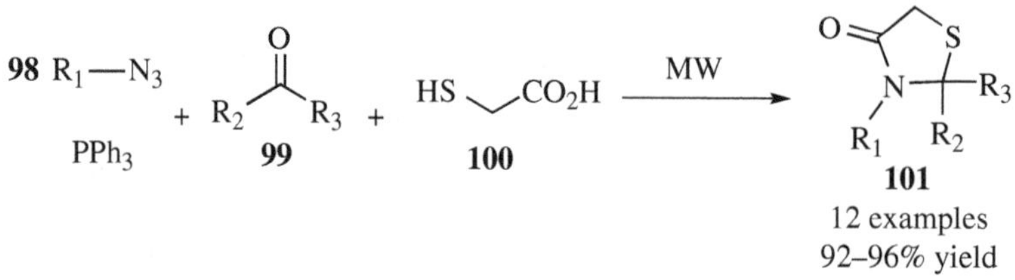

SCHEME 11.17 Multicomponent synthesis reported by Ponnuswamy.

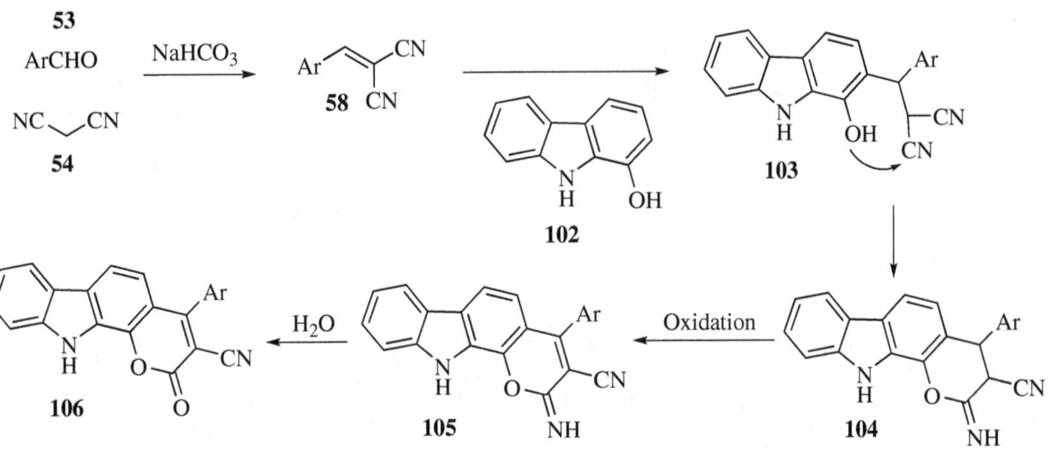

SCHEME 11.18 Proposed mechanism for the synthesis of pyrano[2,3-*a*]carbazoles **106**.

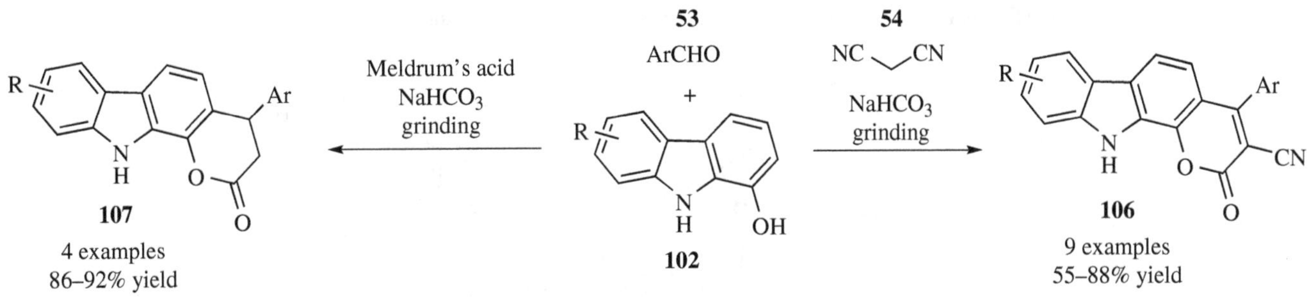

SCHEME 11.19 Synthesis of pyrano[2,3-*a*]carbazoles **107** and **108**.

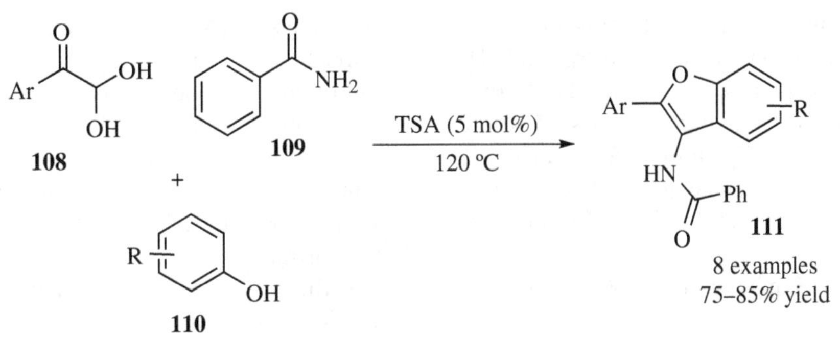

SCHEME 11.20 Synthesis of benzofurans **111**.

yields (85–96%) (Scheme 11.21). The reactions were catalyzed by ethylammonium nitrate (EAN), a reusable ionic liquid that facilitated the reactions at room temperature. Other ionic liquid catalysts required longer reaction times and higher temperatures [68].

Recently, Li and coworkers developed the multicomponent solvent-free domino heteroannulation of heterocyclic ketene aminals for the synthesis of benzo[γ]imidazo[1,2-*a*]quinolinediones [69]. This scaffold is found in diverse natural products and pharmaceuticals with important biological activities ranging from antimicrobial to cytotoxicity.

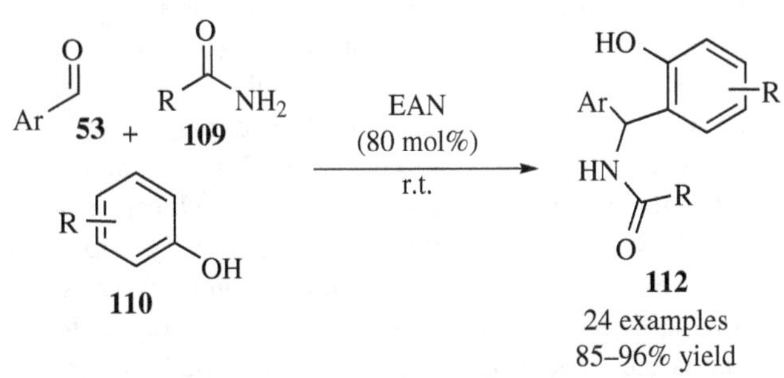

SCHEME 11.21 Synthesis of amides **112**.

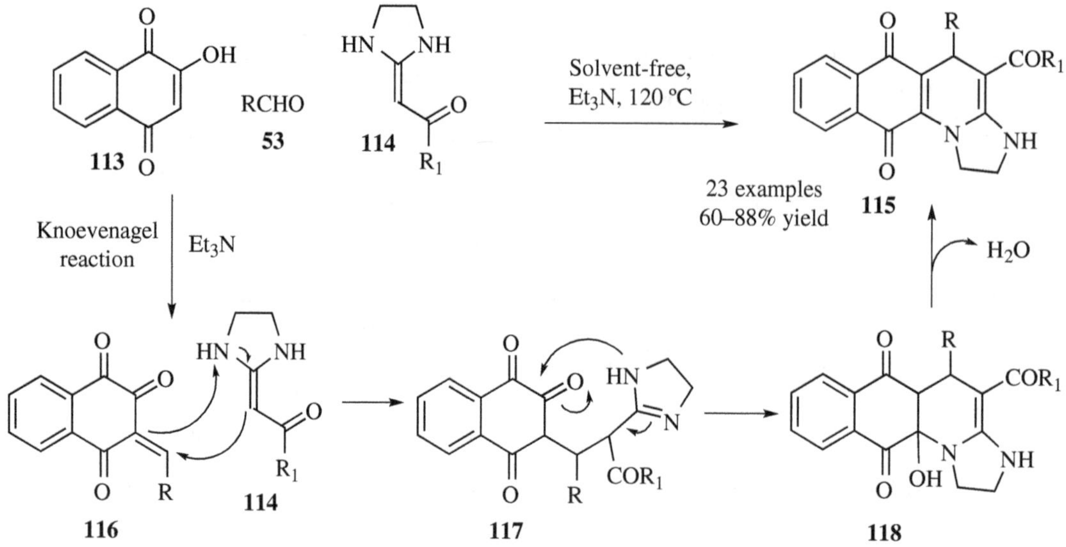

SCHEME 11.22 Synthesis of benzo[γ]imidazo[1,2-*a*]quinolinediones reported by Li.

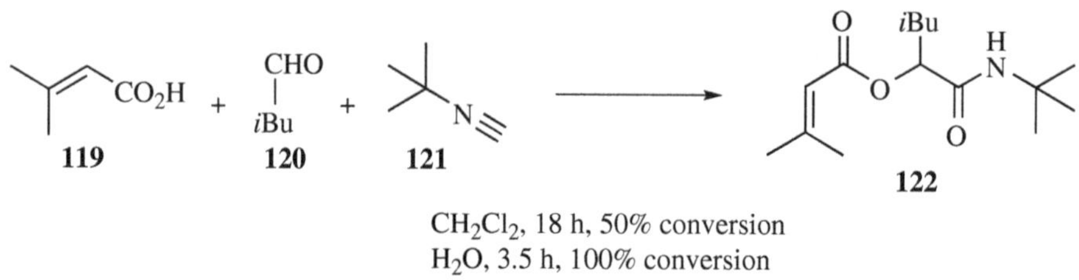

SCHEME 11.23 Enhancement of Passerini reaction in water.

This method affords an easy access to this privileged scaffold by a [3+2+1] annulation catalyzed by Et_3N. The mechanism of the reaction is shown in Scheme 11.22. First, the Knoevenagel reaction of aromatic aldehydes **53** with 2-hydroxy-1,4-naphthoquinone **113** affords **116** that undergoes the aza–ene reaction with the ketene aminal **114** rendering **117**. Next, the intramolecular hemiaminal **118** formation and dehydration afford the corresponding benzo[γ]imidazo[1,2-*a*]quinolinediones **115** in good yields (60–88%).

11.5 MCRs IN AQUEOUS MEDIA

Although chemical transformations in aqueous solutions are known, in general, the poor solubility of organic compounds in water avoids the use of aqueous solvents in chemical reactions. Some early examples of reactions in aqueous medium can be traced back to the Wohler's urea synthesis in 1828 [70] and the Baeyer–Drewsen indigo synthesis in 1882 [71]. MCR protocols, as already stated, are very important for developing new pharmaceuticals and libraries of medicinal scaffolds. However, few examples for MCRs in water as the solvent have been reported in the literature. In this section, we will discuss some selected MCRs "in water" including some MCRs "on water." Sharpless and coworkers defined "on water reactions" as "when insoluble reactants are stirred as aqueous emulsions/suspensions without adding any organic co-solvents" [72].

Classical Passerini and Ugi reactions are the most common MCRs on water. One early example of these reactions was reported by Pirrung and Sarma that clearly demonstrated a significant enhancement in the reaction rate and efficiency of the reaction on water compared to the same reaction in an organic solvent (Scheme 11.23) [73].

An interesting example of the use of MCRs on water is the synthesis of seleno cysteines by the Ugi reaction developed by Wessjohann and coworkers [74]. The reactions of seleno-derived aldehydes **123** with amines **9**, carboxylic acids **125**, and isocyanides **124** afforded the corresponding seleno amino acids **126** in variable yields (24–87%) (Scheme 11.24).

The synthesis of thiophenes by a MCR on water was recently studied by Moghaddam and coworkers [75]. In this study, 3-morpholino-3-thiooxopropanenitrile (**127**), cyclohexyl isocyanide (**128**), and α-haloketones **129** in water were heated at 60°C to afford the corresponding highly functionalized thiophenes **130** in good yields (59–74%) (Scheme 11.25).

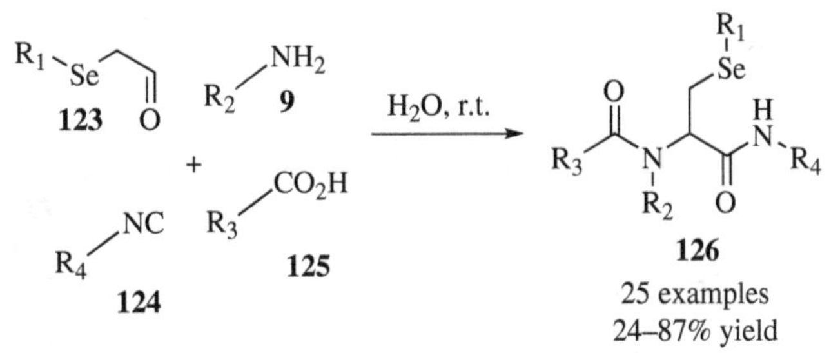

SCHEME 11.24 Synthesis of seleno cysteines **126** reported by Wessjohann.

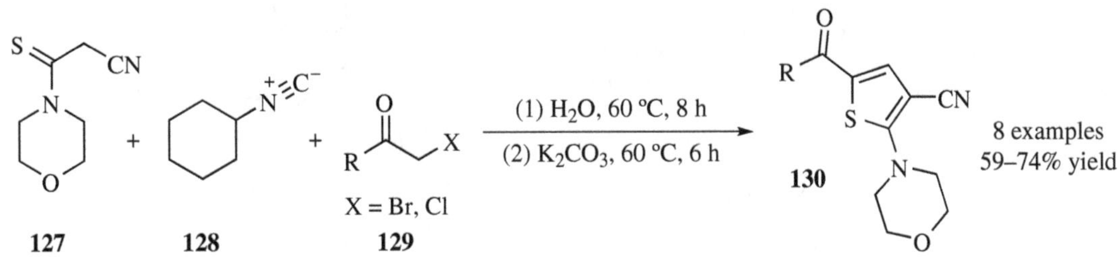

SCHEME 11.25 Synthesis of highly functionalized thiophenes **130**.

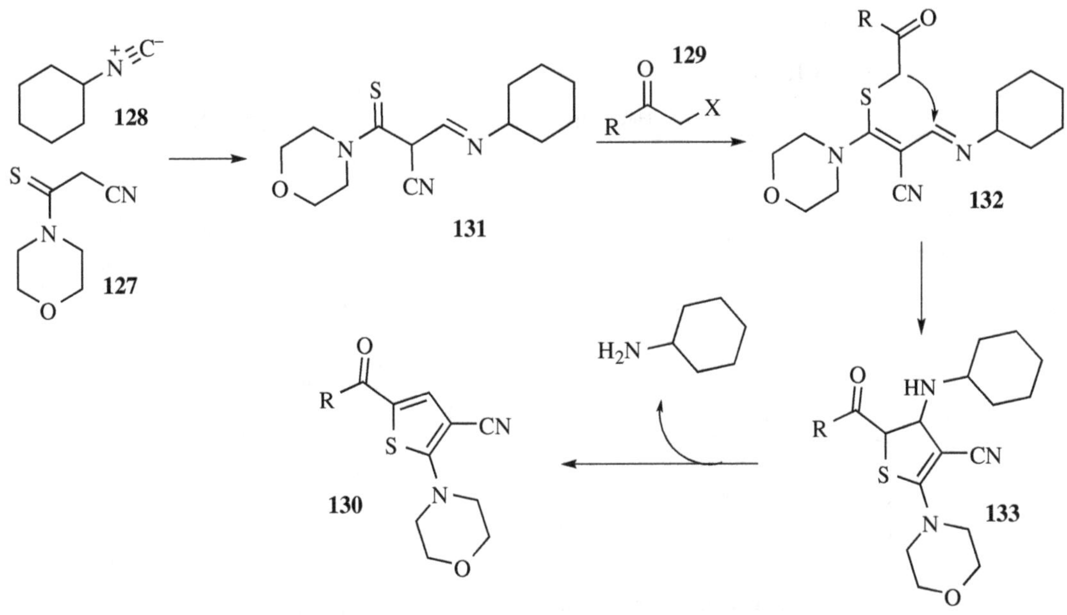

SCHEME 11.26 Proposed mechanism.

The reaction starts with the addition of 3-morpholino-3-thiooxopropanenitrile **127** to cyclohexyl isocyanide **128**. Next, the nucleophilic addition of sulfur in compound **131** to α-haloketones **129**, followed by the base-catalyzed cyclization in compound **132** and elimination of cyclohexylamine, gives the corresponding thiophenes **130** (Scheme 11.26).

A novel MCR on water for the synthesis of spirooxindoles was reported by Dandia in 2012 [76]. The reaction of easily available isatins **134**, malononitrile **54**/α-cyanoacetic ester **137**, and 5-amino-3-methylpyrazole **135** catalyzed by NaCl in water afforded the corresponding spirooxindoles **136** and **138**, respectively, in very good to excellent yields. When α-cyanoacetic ester **137** was used, the reaction afforded spirooxindole **138** as a mixture of diastereomers (3:1 ratio); the relative configuration was ascertained by the X-ray diffraction analysis of a single crystal of the major diastereomer (Scheme 11.27).

Vasuki and Rajarathinam reported a related reaction for the synthesis of 2-azapyrrolidine analogues [77]. The reaction of hydantoins **139**, aldehydes **53**, and

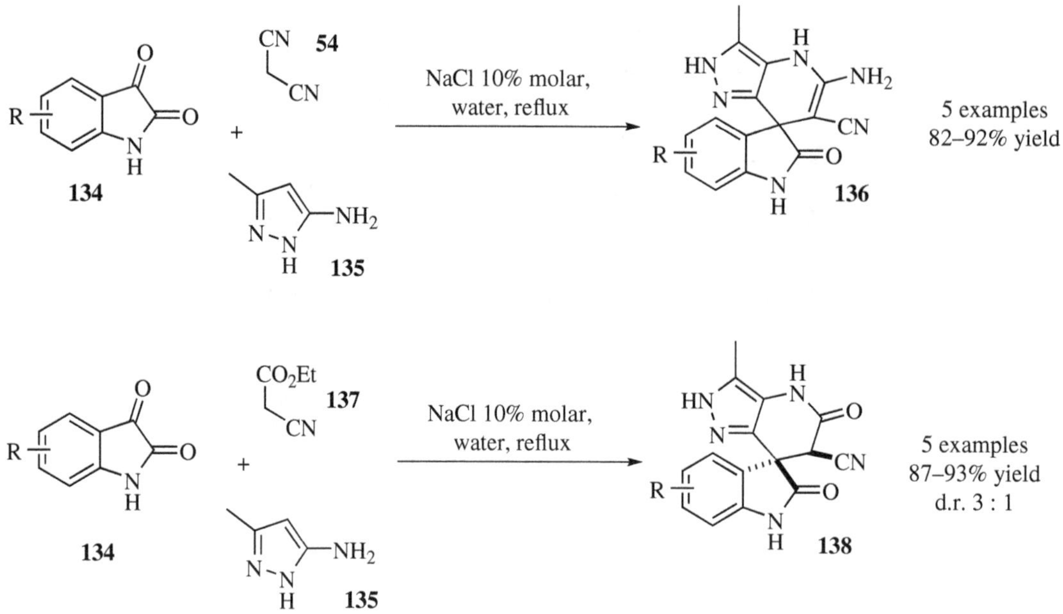

SCHEME 11.27 Synthesis of spirooxindoles reported by Dandia.

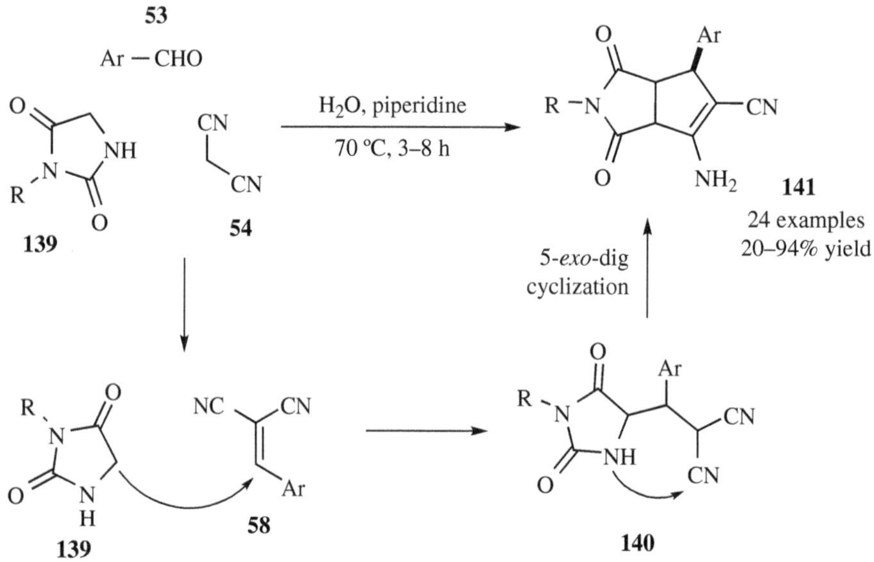

SCHEME 11.28 Synthesis of 2-azapyrrolidines **141**.

malononitrile **54** in water catalyzed by piperidine afforded the corresponding 2-azapyrrolidines **141** in variable yields and as a single diastereomer. The mechanism of the reaction is shown in Scheme 11.28. First, the Knoevenagel condensations of aldehydes **53** with malononitrile **54** afford the corresponding Knoevenagel adducts **58** that undergo the Michael nucleophilic additions to hydantoin **139**. Finally, the intramolecular 5-*exo*-dig cyclizations of the resulting adducts **140** give the corresponding 2-azapyrrolidines **141** in moderate to good yields (20–94%) and as single diastereomers (the relative configurations were ascertained by the X-ray diffraction analyses of single crystal).

Rostami-Charati and Hossaini reported the multicomponent synthesis of phosphonates on water by the reaction of activated acetylenes **143**, 4-hydroxycoumarin/1-(6-hydroxy-2-isopropenyl-1-benzofuran-yl)-1-ethanone (**142**), and trialkylphosphites **144** [78]. The reaction mixtures were heated at 70°C to render the corresponding phosphate derivatives **145** in good yields (Scheme 11.29).

In the same year, Myrboh and coworkers reported the synthesis of β-functionalized 5-methyl-1*H*-pyrazol-3-ol derivatives *via* an

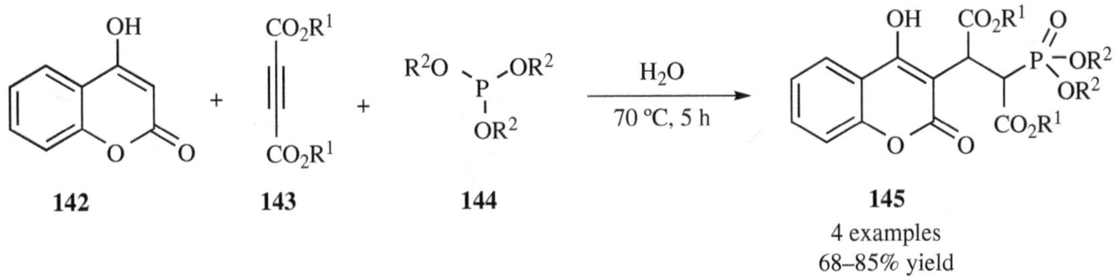

SCHEME 11.29 Multicomponent synthesis of phosphonates on water.

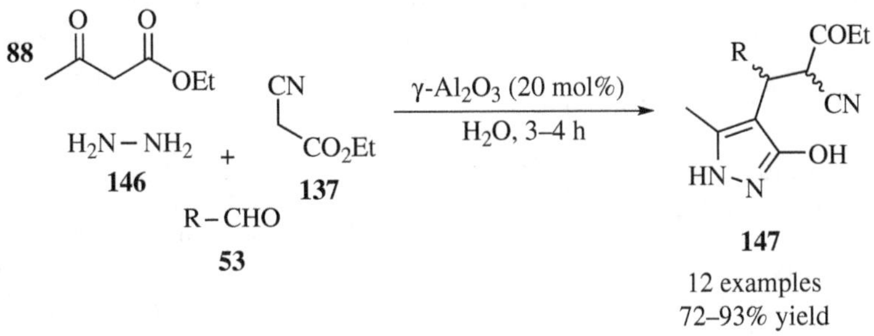

SCHEME 11.30 Synthesis of **β**-functionalized 5-methyl-1*H*-pyrazol-3-ol derivatives **147**.

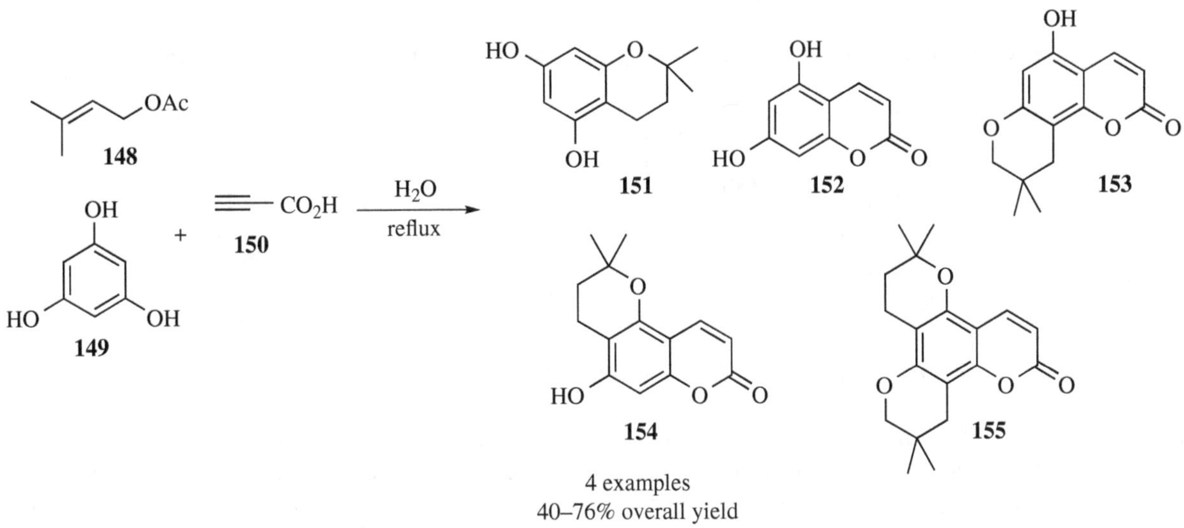

SCHEME 11.31 One-pot synthesis of 5-methoxyseselin and alloxanthoxyletin skeletons.

MCR [79]. The MCR of aldehydes **53**, ethyl cyanoacetate **137**, and *in situ* generated 3-methyl-1*H*-pyrazol-5(4*H*)-one in aqueous media catalyzed by γ-alumina afforded the corresponding pyrazol-3-ol derivatives **147** as 1:1 diastereomeric mixtures in good to excellent yields (72–93%) (Scheme 11.30).

Recently, Qu and coworkers reported the one-pot synthesis of 5-methoxyseselin and alloxanthoxyletin skeletons by MCRs on water [80]. The reaction of isoprenyl acetate **148**, propiolic acid **150**, and phloroglucinol **149** did not require any catalyst and provided a mixture of 5-methoxyseselin, alloxanthoxyletin scaffolds, and other products depending of the reagent's molar ratios (Scheme 11.31).

11.6 HIGH-PRESSURE PROMOTED MCRs

One of the areas that have attracted more attention lately is the development of new reactions using high pressure. However, despite the efforts of chemical community to develop new methods, their use in MCRs is rare. Almost all the reactions developed focused on the bimolecular reactions such as Diels–Alder and cycloadditions [81].

In this section, we will discuss the most important methodologies developed in MCRs under high-pressure conditions in the recent years.

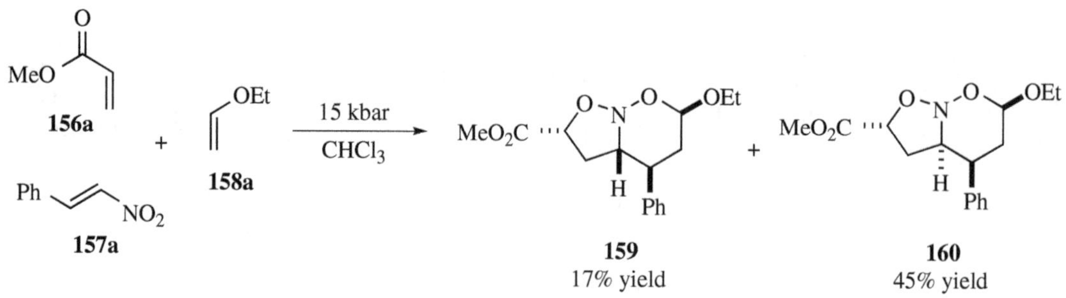

SCHEME 11.32 [4+2]/[3+2] sequential cycloaddition.

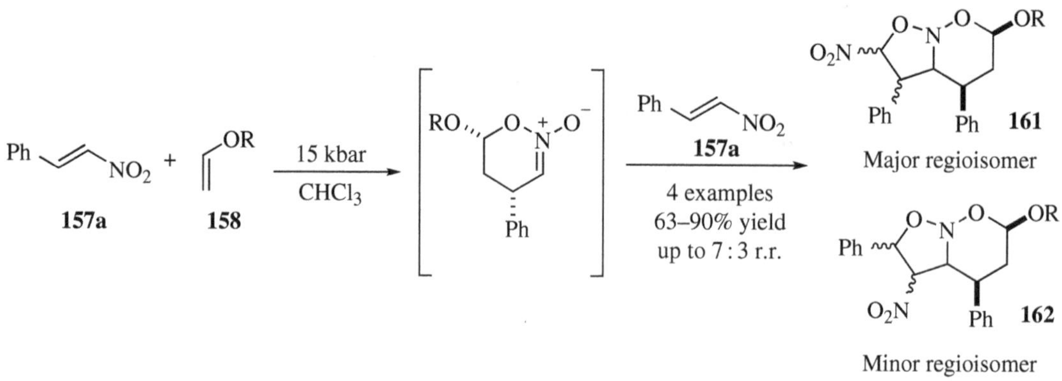

SCHEME 11.33 [4+2]/[3+2] sequential cycloaddition using 2 molecules of nitrostyrene **157a**.

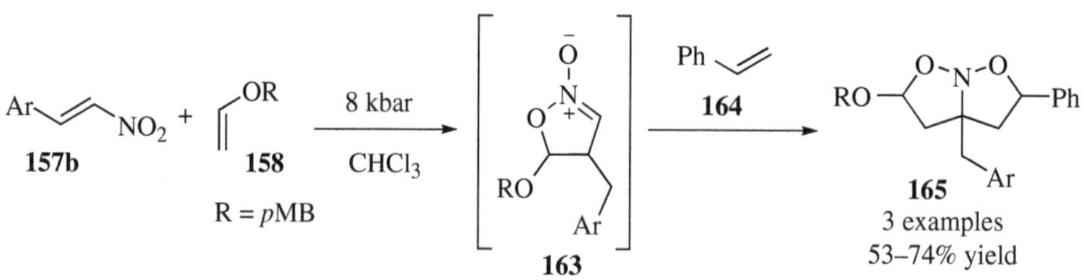

SCHEME 11.34 [4+2]/[3+2] sequential cycloaddition using simple styrenes.

One of the most studied reactions under high pressure is [4+2]/[3+2] sequential cycloaddition of nitroalkenes to electron-rich alkenes such as vinyl ethers. It is well known that cycloaddition reactions are accelerated by high pressure. Scheeren and coworkers used high-pressure conditions to develop a [4+2]/[3+2] sequential cycloaddition affording nitrosoacetals **159** and **160** [82]. The reaction of nitrostyrenes **157a**, acrylates **156a**, and ethyl vinyl ether **158a** at 15 kbar pressure afforded the corresponding cyclic products (**159,160**) in variable yields as a mixture of diastereomers (Scheme 11.32).

Few years later, the same research group reported a high-pressure [4+2]/[3+2] sequential cycloaddition using two molecules of nitrostyrene **157a** instead of acrylates **156** [83]. The reaction was carried out at 15 kbar pressure, affording the corresponding cycloadducts **161** and **162** in good yields as a mixture of regioisomers. Next, the major regioisomer was easily transformed to the corresponding β-lactam by a base-catalyzed rearrangement (Scheme 11.33).

Later, they expanded the substrate scope of the reaction using different electron-deficient alkenes such as maleimides [84]. Moreover, the reaction of 3-[(*E*)-2-nitroeth-1-enyl]pyridine **157b** with alkyl vinyl ethers **158** at a low pressure (8 kbar) afforded five-membered nitronates **163** that reacted with an activated alkene **164** (styrene, acrylates, or maleimides) to afford the corresponding five/five-membered bicyclic nitrosoacetals **165** in moderate yields (53–74%) as shown in Scheme 11.34.

Chataigner and Piettre reported a different approach for the multicomponent domino [4+2]/[3+2] cycloadditions under high pressures, where 3-nitroindole **166** or 3-nitropyrrole derivatives were used as the electron-deficient heterodienes [85]. As shown in Scheme 11.35, the first [4+2] inverse-demand cycloaddition of electron-deficient heterodienes **166** to vinyl ether **158** was completely *endo*-selective; however, the subsequent [3+2] cycloaddition between the Diels–Alder adduct and acrylate **156** showed a complete facial selectivity,

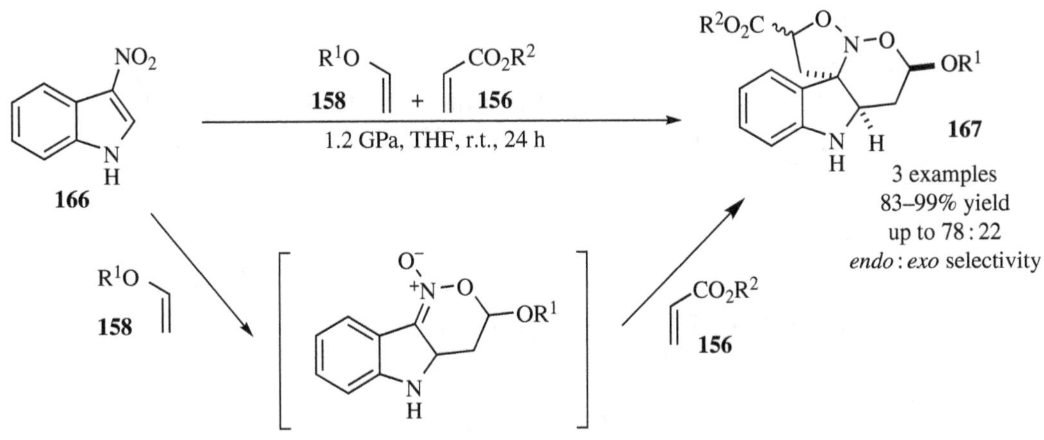

SCHEME 11.35 [4+2]/[3+2] sequential cycloaddition using 3-nitroindole.

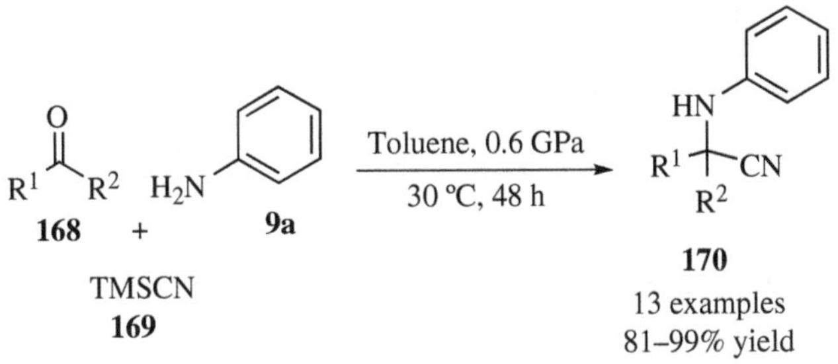

SCHEME 11.36 Strecker reaction.

but slight or no selective *endo*/*exo*-orientation of the acrylate. The reaction was carried out at 1.2 GPa pressure in tetrahydrofuran (THF) at room temperature, affording the final nitrosoacetal **167** in good yields and moderate diastereoselectivity. The Raney nickel-catalyzed hydrogenation of the final compounds **167** under 15 bar pressure for 48 h afforded pyrrolizidines in good to excellent yields.

The Strecker reaction has been widely studied for the synthesis of α-amino acids. Matsumoto and coworkers reported the multicomponent uncatalyzed Strecker reaction under high pressure in 2005 for the synthesis of quaternary α-amino acids **170** [86]. The reaction of ketones **168**, aniline **9a**, and TMSCN (**169**) at 0.6 GPa pressure and 30°C afforded the corresponding α-aminonitriles **170** in very good to excellent yields (81–99%) (Scheme 11.36).

11.7 THREE-COMPONENT REACTIONS USING SUPPORTED REAGENTS

Supported reagents have several advantages in organic chemistry such as the production of resin-bound products, increased stability, and easy purification. Despite these advantages, very few supported reagents have been used in MCRs, in front of the most common approach of using supported catalysts. In this chapter, we will discuss some of the most attractive examples for the use of supported reagents in MCRs.

As stated in Section 11.6, Scheeren and Kuster developed a three-component high-pressure [4+2]/[3+2] sequential reaction. Years later, they developed a solid-support version of the same reaction [87]. They decided to use resin-bound acrylates **171** using Wang's resin. The reaction afforded the corresponding cycloadducts **173** as a mixture of diastereomers in moderate yields (33–52%) (Scheme 11.37).

Later, the same research group reported a similar [4+2]/[3+2] sequential cycloaddition of resin-bound nitroalkenes, ethyl vinyl ether, and styrene under high-pressure conditions [88]. The cycloadducts were obtained in moderate yields (29–51%) as a mixture of diastereomers.

Kennedy and coworkers reported the use of resin-bound isonitriles in the Ugi MCR to afford valuable 2,5-diketopiperazines and 1,4-benzodiazepine-2-5-diones [89]. They developed a resin-bound carbonate convertible isonitrile based on a hydroxymethyl polystyrene resin. As shown in Scheme 11.38, the Ugi reaction with these resin-bound convertible isonitriles afforded, after several derivatizations, the desired 2,5-diketopiperazines in good yields **178**.

Vaultier and coworkers reported Grieco's MCR in water using ammonium chloride as the water-solubilizing moiety [90]. The one-pot acid-catalyzed Grieco's MCR of anilines **9** and aldehydes **53** with electron-rich alkenes afforded the corresponding tetrahydroquinolines **181** and **183**, respectively, in good yields. They supported the reagents (aldehydes or anilines) using ammonium chloride salts **179** and **182** as the soluble support (Scheme 11.39), thus allowing easier purifications after the reaction. The final pure products **181** and **183** were obtained after filtration and recrystallization.

Keshavarz and Albadi used an azide-supported reagent in a macroporous polymer for developing click chemistry [91]. The reaction involves the nucleophilic addition of the supported azide to α-bromo carbonyl compounds (ketones, amides, or esters) followed by the 1,3-dipolar cycloaddition of the preformed azide to an alkyne catalyzed by a polymer-supported

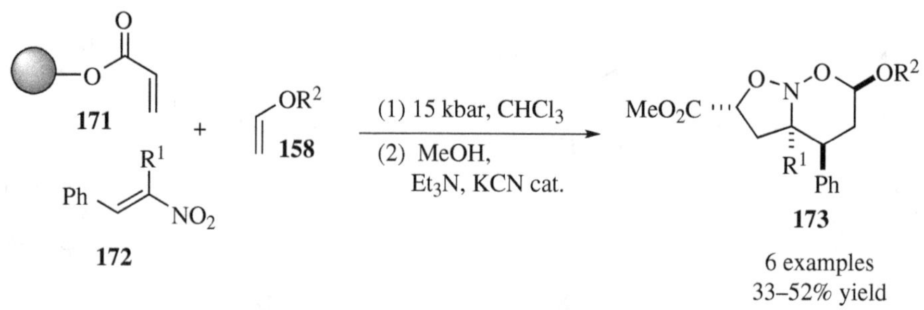

SCHEME 11.37 Supported [4 + 2]/[3 + 2] sequential cycloaddition.

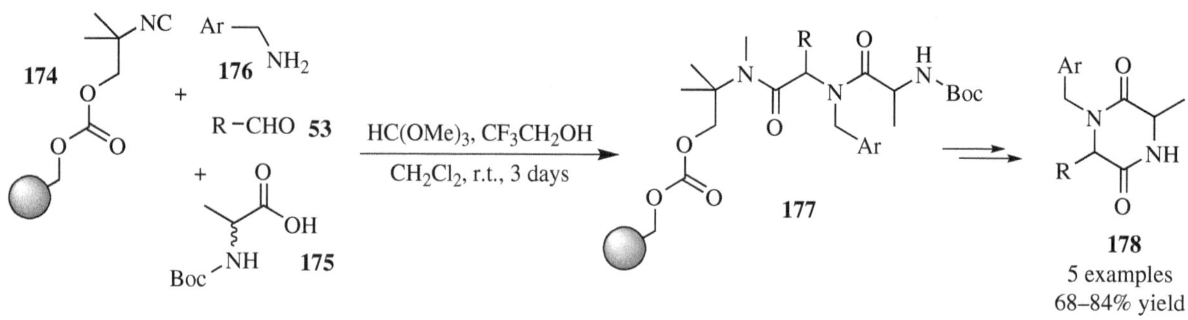

SCHEME 11.38 Supported Ugi reaction.

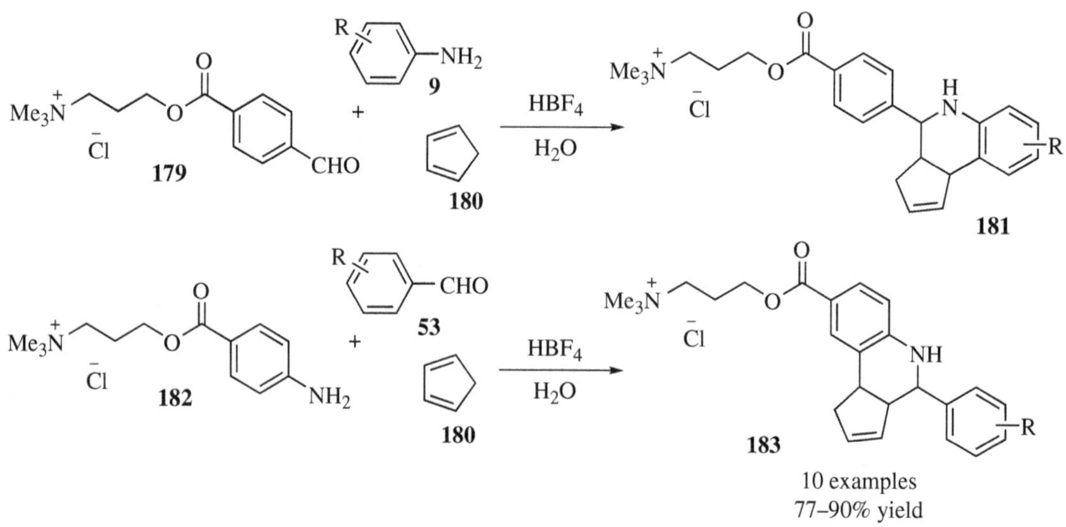

SCHEME 11.39 Supported Grieco's reaction.

Cu(I) complex. This process afforded the desired 1,4-disubstituted 1*H*-1,2,3-triazoles in good yields and with complete regioselectivity (78–92% yields, 16 examples).

11.8 CONCLUSION

MCRs have emerged as one of the most important tools in organic chemistry. The possibility to use several reagents in a one-pot procedure allows the formation of multiple bonds in high yields and often with high regio- and stereoselectivity. In this chapter, we discussed some of the most uncommon approaches to MCRs, for example, using zeolites as the catalysts, supported reagents, MW-assisted reactions, and high-pressure reactions. All these approaches enrich the chemistry knowledge and allow the syntheses of interesting and highly complex scaffolds with some additional benefits using the usual MCRs, such as easy purification, higher yields, and waste-free reactions.

However, despite the large number of studies conducted, yet many opportunities remain to be explored in this area. For example, very few examples on the development of asymmetric methods have been reported; a broad application of high-pressure reactions in difficult cycloaddition reactions is yet to be achieved. Therefore, we expect that a plethora of new methods that complement and improve the existing methods would be developed in the future.

REFERENCES

1 Y. Gu, *Green. Chem.* 2012, 14, 2091–2128. Multicomponent reactions in unconventional solvents: state of the art.

2 (a) P. Tundo, P. Anastas, D. S. Black, J. Breen, T. Collins, S. Memoli, J. Miyamoto, M. Polyakoff, W. Tumas, *Pure Appl. Chem.* 2000, 72, 1207–1228. Synthetic pathways and processes in green chemistry. Introductory overview. (b) K. Alfonsi, J. Colberg, P. J. Dunn, T. Fevig, S. Jennings, T. A. Johnson, H. P. Kleine, C. Knight, M. A. Nagy, D. A. Perry, M. Stefaniak, *Green Chem.* 2008, 10, 31–36. Green chemistry tools to influence a medicinal chemistry and research chemistry based organization.

3 (a) A. Corma, H. Garcia, *Catal. Today* 1997, 38, 257–308. Organic reactions catalyzed over solid acids. (b) R. A. Sheldon, R. S. Downing, *Appl. Catal. A: Gen.* 1999, 189, 163–183. Heterogeneous catalytic transformations for environmentally friendly production.

4 (a) D. Tichit, M. J. M. Ortiz, D. Francova, C. Gerardin, B. Coq, R. Durand, F. Prinetto, G. Ghiotti, *Appl. Catal. A: Gen.* 2007, 318, 170–177. Design of nanostructured multifunctional Pd-based catalysts from layered double hydroxides precursors. (b) P. McMorn, G. J. Hutchings, *Chem. Soc. Rev.* 2004, 33, 108–122. Heterogeneous enantioselective catalysts: strategies for the immobilization of homogeneous catalysts. (c) G. J. Hutchings, *Ann. Rev. Mat. Res.* 2005, 35, 143–166. Heterogeneous asymmetric catalysts: strategies for achieving high enantioselection. (d) X. Zhang, E. S. M. Lai, R. Martin-Aranda, K. L. Yeung, *Appl. Catal. A: Gen.* 2004, 261, 109–118. An investigation of Knoevenagel condensation reaction in microreactors using a new zeolite catalyst. (e) A. Corma, R. M. Martin-Aranda, *Appl. Catal. A: Gen.* 1993, 105, 271–279. Application of solid base catalysts in the preparation of prepolymers by condensation of ketones and malononitrile.

5 (a) R. Maggi, A. Bello, C. Oro, G. Sartori, L. Soldi, *Tetrahedron* 2008, 64, 1435–1439. AgY zeolite as catalyst for the effective three-component synthesis of propargylamines. (b) M. K. Patil, M. Keller, B. M. Reddy, P. Pale, J. Sommer, *Eur. J. Org. Chem.* 2008, 4440–4445. Copper zeolites as green catalysts for multicomponent reactions of aldehydes, terminal alkynes and amines: an efficient and green synthesis of propargylamines.

6 (a) K. Sivakumar, A. Kathirvel, A. Lalitha, *Tetrahedron* 2010, 51, 3018–3021. Simple and efficient method for the synthesis of highly substituted imidazoles using zeolite-supported reagents. (b) A. Teimouria, A. N. Chermahini, *J. Mol. Catal. A: Chem.* 2011, 346, 39–45. An efficient and one-pot synthesis of 2,4,5-trisubstituted and 1,2,4,5-tetrasubstituted imidazoles catalyzed *via* solid acid nano-catalyst.

7 V. Rama, K. Kanagaraj, K. Pitchumani, *Tetrahedron Lett.* 2012, 53, 1018–1024. A multicomponent, solvent-free, one-pot synthesis of benzoxanthenones catalyzed by HY zeolite: their antimicrobial and cell imaging studies.

8 P. P. Ghosh, A. R. Das, *Tetrahedron Lett.* 2012, 53, 3140–3143. Nano crystalline ZnO: a competent and reusable catalyst for one pot synthesis of novel benzylamino coumarin derivatives in aqueous media.

9 V. Rama, K. Kanagaraj, T. Subramanian, P. Suresh, K. Pitchumani, *Catal. Commun.* 2012, 26, 39–43. Pyridinium ylide-assisted KY zeolite catalyzed tandem synthesis of polysubstituted cyclopropanes.

10 M. Nikpassand, M. Mamaghani, K. Tabatabaeian, *Molecules* 2009, 14, 1468–1474. An efficient one-pot three-component synthesis of fused 1,4-dihydropyridines using HY-zeolite.

11 M. J. Thompson, J. M. Hurst, B. Chen, *Synlett* 2008, 3183–3187. Regioselective, solvent-free synthesis of 3-aminoimidazo[1,2-*a*]pyrimidines under microwave irradiation promoted by zeolite HY.

12 K. Arya, U. C. Rajesh, D. S. Rawat, *Green Chem.* 2012, 14, 3344–3351. Proline confined FAU zeolite: heterogeneous hybrid catalyst for the synthesis of spiroheterocycles *via* a Mannich type reaction.

13 (a) F. Calderón, R. Fernández, F. Sánchez, A. Fernández-Mayoralas, *Adv. Synth. Catal.* 2005, 347, 1395–1403. Asymmetric aldol reaction using immobilized proline on mesoporous support. (b) E. G. Doyaguez, F. Calderón, F. Sánchez, A. Fernández-Mayoralas, *J. Org. Chem.* 2007, 72, 9353–9356. Asymmetric aldol reaction catalyzed by a heterogenized proline on a mesoporous support. The role of the nature of solvents. (c) L. Zhong, J. Xiao, C. J. Li, *J. Catal.* 2006, 243, 442–445. An unexpected inversion of enantioselectivity in direct asymmetric aldol reactions on a unique L-proline/γ-Al_2O_3 catalyst. (d) D. Dhar, I. Beadham, S. Chandrasekaran, *Proc. Indian Acad. Sci., Chem. Sci.* 2003, 115, 365–372. Proline and benzylpenicillin derivatives grafted into mesoporous MCM-41: novel organic-inorganic hybrid catalysts for direct aldol reaction.

14 For pivotal reviews, see: (a) P. Lidström, J. Tierney, B. Wathey, J. Westman, *Tetrahedron* 2001, 57, 9225–9283. Microwave assisted organic synthesis—a review. (b) C. O. Kappe, *Angew. Chem. Int. Ed.* 2004, 43, 6250–6284. Controlled microwave heating in modern organic synthesis. (c) B. L. Hayes, *Aldrichim. Acta* 2004, 37, 66–76. Recent advances in microwave assisted synthesis. (d) C. O. Kappe, A. Stadler, *Microwaves in organic and medicinal chemistry*. Wiley-VCH Verlag GmbH, Weinheim, Germany, 2005, pp. 29–55. (e) A. de la Hoz, A. Loupy (Eds.), *Microwaves in organic synthesis*. 2nd ed. Wiley-VCH Verlag GmbH, Weinheim, Germany, 2006. (f) D. Dallinger, C. O. Kappe, *Chem. Rev.* 2007, 107, 2563–2591. Microwave-assisted synthesis in water as solvent. (g) Y. Coquerel, J. Rodriguez, *Eur. J. Org. Chem.* 2008, 1125–1132. Microwave-assisted olefin metathesis. (h) P. Appukkuttan, E. Van der Eycken, *Eur. J. Org. Chem.* 2008, 1133–1155. Recent developments in microwave-assisted, transition-metal-catalysed C–C and C–N bond-forming reactions. (i) S. Caddick, R. Fitzmaurice, *Tetrahedron* 2009, 65, 3325–3355. Microwave enhanced synthesis. (j) H. M. Hügel, *Molecules* 2009, 14, 4936–4972. Microwave multicomponent synthesis.

15 R. Gedye, F. Smith, K. Westawaym, A. Humera, L. Baldisera, L. Laberge, J. Rousell, *Tetrahedron Lett.* 1986, 27, 279–282. The use of microwave ovens for rapid organic synthesis.

16 K. N. Singh, S. K. Singh, *Arkivoc* 2009, (xiii), 153–160. Microwave-assisted, one-pot multicomponent synthesis of highly substituted pyridines using KF/alumina.

17 (a) D. L. Boger, S. Nakahara, *J. Org. Chem.* 1991, 56, 880–884. Diels-Alder reactions of N-sulfonyl-1-aza-1,3-butadiene: development of a synthetic approach to the streptonigrone C ring. (b) D. L. Boger, A. M. Kasper, *J. Am. Chem. Soc.* 1989, 111, 1517–1519. A general solution to implementing the 4.pi. participation of 1-aza-1,3-butadienes in Diels-Alder reactions: inverse electron demand Diels-Alder reactions of α,β-unsaturated N-benzenesulfonyl imines. (c) T. R. K. Reddy, R. Mutter, W. Heal, K. Guo, V. J. Gillet, S. Pratt, B. Chen, *J. Med. Chem.* 2006, 49, 607–615. Library design, synthesis, and screening: pyridine dicarbonitrile as potential prion disease therapeutics. (d) T. Y. Zhang, J. R. Stout, J. G. Keay, E. F. V. Scriven, J. E. Toomey, G. L. Goe, *Tetrahedron* 1995, 51, 13177–13184. Regioselective synthesis of 2-chloro-3-pyridinecarboxylates.

18 S. Tu, X. Zhu, J. Zhang, J. Xu, Y. Zhang, Q. Wang, R. Jia, B. Jiang, J. Zhang, C. Yao, *Bioorg. Med. Chem. Lett.* 2006, 16, 2925–2928. New potential biologically active compounds: design and an efficient synthesis of N-substituted 4-aryl-4,6,7,8-tetrahydroquinoline-2,5(1*H*,3*H*)-diones under microwave irradiation.

19 (a) J. B. Doherty, J. E. Stelmach, M.-H. Chen, L. Liu, J. A. Hunt, R. D. Ruzek, J. L. Goulet, D. D. Wisnoski, S. R. Natarajan, K. M. Rupprecht, J. Bao, S. Miao, X. Hong, WO 0258695, 2002. Preparation of 1,5-diaryl-7-heterocyclyl(alkyl)-2-quinolinones as p38 protein kinase inhibitors. (b) E. Dahlen, M. Andersson, K. Dawe, A. C. Tellander, C. Brunmark, A. Björk, G. Hedlung, *Autoimmunity* 2000, 32, 199–211. Inhibition of autoimmune disease by the immunomodulator linomide correlates with the ability to activate macrophages. (c) Y. Sada, S. Adegawa, K. Mogi, H. Honda, H. Eto, S. Morimoto, J. Okawa, N. Umehara, S. Sato, JP 97-234547, 1999. Preparation and anti-ulcer activity of substituted quinolone derivatives.; (d) N. Shibutani, K. Hashimoto, Y. Inoue, K. Sato, S. Miki, JP 97-183870, 1999. Dihydroquinoline derivatives as hypoglycemics.; (e) D. W. End, M. G. Venet, P. R. Angibaud, G. C. Sanz, WO 96-EP4661, 1997. Farnesyl transferase inhibiting 2-quinolone derivatives. (f) A. Afonso, J. Weinstein, M. J. Gentles, WO 91-US6251, 1995. Preparation of alkyl and acyl substituted quinolines as antiviral agents.

20 S. Tu, J. Zhang, X. Zhu, J. Xu, Y. Zhang, Q. Wang, R. Jia, B. Jiang, J. Zhang, *Bioorg. Med. Chem. Lett.* 2006, 16, 3578–3581. New potential inhibitors of cyclin-dependent kinase 4: design and synthesis of pyrido[2,3-*d*]pyrimidine derivatives under microwave irradiation.

21 P. L. Toogood, *Med. Res. Rev.* 2001, 21, 487–498. Cyclin-dependent kinase inhibitors for treating cancer.

22 F. Shi, C. Li, M. Xia, K. Miao, Y. Zhao, S. Tu, W. Zheng, G. Zhang, N. Maa, *Bioorg. Med. Chem. Lett.* 2009, 19, 5565–5568. Green chemoselective synthesis of thiazolo[3,2-*a*]pyridine derivatives and evaluation of their antioxidant and cytotoxic activities.

23 (a) V. R. Solomon, W. Haq, K. Srivastava, S. K. Puri, S. B. Katti, *J. Med. Chem.* 2007, 50, 394–398. Synthesis and antimalarial activity of side chain modified 4-aminoquinoline derivatives. (b) P. M. S. Chauhan, S. K. Srivastava, *Curr. Med. Chem.* 2001, 8, 1535–1542. Present trends and future strategy in chemotherapy of malaria. (c) J. P. Michael, *Nat. Prod. Rep.* 2004, 21, 650–668. Quinoline, quinazoline and acridone alkaloids. (d) J. P. Michael, *Nat. Prod. Rep.* 2007, 24, 223–246. Quinoline, quinazoline and acridone alkaloids.

24 J. I. Kim, I.-S. Shin, H. Kim, J.-K. Lee, *J. Am. Chem. Soc.* 2005, 127, 1614–1615. Efficient electrogenerated chemiluminescence from cyclometalated iridium(III) complexes.

25 A. Kulkarni, B. Török, *Green Chem.* 2010, 12, 875–878. Microwave-assisted multicomponent domino cyclization–aromatization: an efficient approach for the synthesis of substituted quinolones.

26 (a) E. A. Hafez, M. H. Elnagdi, A. A. Elagamey, F. A. M. El-Taweel, *Heterocycles* 1987, 26, 903–907. Nitriles in heterocyclic synthesis: novel synthesis of benzo[*c*]coumarin and of benzo[c]pyrano[3,2-*c*]quinoline derivatives. (b) M. A. Sofan, F. M. El-Taweel, A. G. A. Elagamey, M. H. Elnagdi, *Liebigs Ann. Chem.* 1989, 935–936. Studies on cinnamonitriles: the reaction of cinnamonitriles with cyclopentanone. (c) F. M. Abdel-Galil, B. Y. Riad, S. M. Sherif, M. H. Elnagdi, *Chem. Lett.* 1982, 11, 1123–1126. Activated nitriles in heterocyclic synthesis: a novel synthesis of 4-azoloyl-2-aminoquinolines. (d) R. S. Varma, R. Dahiya, *J. Org. Chem.* 1998, 63, 8038–8041. An expeditious and solvent-free synthesis of 2-amino-substituted isoflav-3-enes using microwave irradiation.

27 M. Kidwai, S. Saxena, M. K. R. Khan, S. S. Thukral, *Bioorg. Med. Chem. Lett.* 2005, 15, 4295–4298. Aqua mediated synthesis of substituted 2-amino-4*H*-chromenes and *in vitro* study as antibacterial agents.

28 P.W. Smith, S. L. Sollis, P. D. Howes, P. C. Cherry, I. D. Starkey, K. N. Cobley, H. Weston, J. Scicinski, A. Merritt, A. Whittington, P. Wyatt, N. Taylor, D. Green, R. Bethell, S. Madar, R. J. Fenton, P. J. Morley, T. Pateman, A. Beresford, *J. Med. Chem.* 1998, 41, 787–797. Dihydropyrancarboxamides related to zanamivir: a new series of inhibitors of influenza virus sialidases. 1. Discovery, synthesis, biological activity, and structure-activity relationships of 4-guanidino- and 4-amino-4*H*-pyran-6-carboxamides.

29 K. Hiramoto, A. Nasuhara, K. Michikoshi, T. Kato, K. Kikugawa, *Mutat. Res. Fundam. Mol. Mech. Mutagen.* 1997, 395, 47–56. DNA strand-breaking activity and mutagenicity of 2,3-dihydro-3,5-dihydroxy-6-methyl-4*H*-pyran-4-one (DDMP), a Maillard reaction product of glucose and glycine.

30S. J. Mohr, M. A. Chirigos, F. S. Fahrman, J. W. Pryor, *Cancer Res.* 1975, 35, 3750–3754. Pyran copolymer as an effective adjuvant to chemotherapy against a murine leukemia and solid tumor.

31 F. Eiden, F. Denk, *Arch. Pharm.* 1991, 324, 353–354. Pyran derivatives. Part 133. Synthesis and CNS-activity of pyran derivatives: 6,8-dioxabicyclo[3.2.1]octanes.

32 A. C. Shekhar, A. R. Kumar, G. Sathaiah, K. Raju, P. S. Rao, M. Sridhar, B. Narsaiah, P. V. S. S. Srinivas, B. Sridhar, *Helv. Chim. Act.* 2012, 95, 502–508. An efficient one-pot synthesis of substituted 1*H*-naphtho[2,1-*b*]pyrans and 4*H*-1-benzopyrans (chromenes) under solvent-free microwave-irradiation conditions.

33 B. Jiang, S.-J. Tu, P. Kaur, W. Wever, G. Li, *J. Am. Chem. Soc.* 2009, 131, 11660–11661. Four-component domino reaction leading to multifunctionalized quinazolines.

34 (a) S. J. Tu, B. Jiang, R. H. Jia, J. Y. Zhang, Y. Zhang, C. S. Yao, F. Shi, *Org. Biomol. Chem.* 2006, 4, 3664–3668. An efficient one-pot, three-component synthesis of indeno[1,2-*b*] quinoline-9,11(6*H*,10*H*)-dione, acridine-1,8(2*H*,5*H*)-dione and quinoline-3-carbonitrile derivatives from enaminones. (b) S. J. Tu, B. Jiang, Y. Zhang, R. H. Jia, J. Y. Zhang, C. S. Yao, F. Shi, *Org. Biomol. Chem.* 2007, 5, 355–359. An efficient and chemoselective synthesis of N-substituted 2-aminopyridines *via* a microwave-assisted multicomponent reaction. (c) S. J. Tu, C. M. Li, G. G. Li, L. J. Cao, Q. Q. Shao, D. X. Zhou, B. Jiang, J. F. Zhou, M. Xia, *J. Comb. Chem.* 2007, 9, 1144–1148. Microwave-assisted combinatorial synthesis of polysubstituent imidazo[1,2-*a*]quinoline, pyrimido[1,2-*a*] quinoline and quinolino[1,2-*a*]quinazoline derivatives.

35 See, for instances: (a) S. Santra, P. R. Andreana, *Org. Lett.* 2007, 9, 5035–5038. A one-pot, microwave-influenced synthesis of diverse small molecules by multicomponent reaction cascades. (b) M. Presset, Y. Coquerel, J. Rodriguez, *Org. Lett.* 2009, 11, 5706–5709. Microwave-assisted domino and multicomponent reactions with cyclic acylketenes: expeditious syntheses of oxazinones and oxazindiones. (c) W.-J. Hao, B. Jiang, S.-J. Tu, X.-D. Cao, S.-S. Wu, S. Yan, X.-H. Zhang, Z.-G. Han, F. Shi, *Org. Biomol. Chem.* 2009, 7, 1410–1414. A new mild base-catalyzed Mannich reaction of heteroarylamines in water: highly efficient stereoselective synthesis of β-aminoketones under microwave heating. (d) P. Nun, J. Martinez, F. Lamaty, *Synthesis* 2010, 2063–2068. Microwave-assisted neat procedure for the Petasis reaction.

36 J.-F. Liu, P. Ye, B. Zhang, G. Bi, K. Sargent, L. Yu, D. Yohannes, C. M. Baldino, *J. Org. Chem.* 2005, 70, 6339–6345. Three-component one-pot total synthesis of glyantrypine, fumiquinazoline F, and fiscalin B promoted by microwave irradiation.

37 J. P. Michael, *Nat. Prod. Rep.* 2004, 21, 650–668. Quinoline, quinazoline and acridone alkaloids.

38 (a) J. Penn, P. G. Mantle, J. N. Bilton, R. N. Sheppard, *J. Chem. Soc., Perkin Trans.* 1 1992, 1495–1496. Glyantrypine, a novel anthranilic acid-containing metabolite of *Aspergillus clavatus*. (b) A. Numata, C. Takahashi, T. Matsushita, T. Miyamoto, K. Kawai, Y. Usami, E. Matsumura, M. Inoue, H. Ohishi, T. Shingu, *Tetrahedron Lett.* 1992, 33, 1621–1624. Fumiquinazolines, novel metabolites of a fungus isolated from a saltfish. (c) C. Takahashi, T. Matsushita, M. Doi, K. Minoura, T. Shingu, Y. Kumeda, A. Numata, *J. Chem. Soc., Perkin Trans.* 1 1995, 2345–2353. Fumiquinazolines A-G, novel metabolites of a fungus separated from a Pseudolabrus marine fish. (d) S.-M. Wong, L. L. Musza, G. C. Kydd, R. Kullnig, A. M. Gillum, R. Cooper, *J. Antibiot.* 1993, 46, 545–553. Fiscalins: new substance P inhibitors produced by the fungus Neosartorya fischeri. Taxonomy, fermentation, structures, and biological properties. (e) H. Fujimoto, E. Negishi, K. Yamaguchi, N. Nishi, M. Yamazaki, *Chem. Pharm. Bull.* 1996, 44, 1843–1848. Isolation of new tremorgenic metabolites from an ascomycete, corynascus setosus. Some examples include:(f) J. P. Karwowski, M. Jackson, R. R. Rasmussen, P. E. Humphrey, J. B. Poddig, W. L. Kohl, M. H. Scherr, S. Kadam, J. B. McAlpine, *J. Antibiot.* 1993, 46, 374–379. 5-N-acetylardeemin, a novel heterocyclic compound which reverses multiple drug resistance in tumor cells. I. Taxonomy and fermentation of the producing organism and biological activity. (g) J. E. Hochlowski, M. M. Mullally, S. G. Spanton, D. N. Whittern, P. Hill, J. B. McAlpine, *J. Antibiot.* 1993, 46, 380–386. 5-N-acetylardeemin, a novel heterocyclic compound which reverses multiple drug resistance in tumor cells. II. Isolation and elucidation of the structure of 5-N-acetylardeemin and two congeners.

39 J.-F. Liu, J. Lee, A. M. Dalton, G. Bi, L. Yu, C. M. Baldino, E. McElory, M. Brown, *Tetrahedron Lett.* 2005, 46, 1241–1244. Microwave-assisted one-pot synthesis of 2,3-disubstituted 3*H*-quinazolin-4-ones.

40 J.-F. Liu, M. Kaselj, Y. Isome, J. Chapnick, B. Zhang, G. Bi, D. Yohannes, L. Yu, C. M. Baldino, *J. Org. Chem.* 2005, 70, 10488–10493. Microwave-assisted concise total syntheses of quinazolinobenzodiazepine alkaloids.

41 J.-F. Liu, P. Ye, K. Sprague, K. Sargent, D. Yohannes, C. M. Baldino, C. J. Wilson, S.-C. Ng, *Org. Lett.* 2005, 7, 3363–3366. Novel one-pot total synthesis of deoxyvasicinone, mackinazolinone, isaindigotone, and their derivatives promoted by microwave irradiation.

42 V. P. Mehta, S. G. Modha, E. Ruijter, K. Van Hecke, L. Van Meervelt, C. Pannecouque, J. Balzarini, R. V. A. Orru, E. Van der Eycken, *J. Org. Chem.* 2011, 76, 2828–2839. A microwave-assisted diastereoselective multicomponent reaction to access dibenzo[*c,e*]azepinones: synthesis and biological evaluation.

43 J. E. Audia, P. A. Hyslop, J. S. Nissen, R. C. Thompson, J. S. Tung, L. I. Tanner, Patent WO 00/19210, 2000. Biological reagents and methods for determining the mechanism in the generation of β-amyloid peptide.

44 (a) J. Dupont, R. F. de Souza, P. A. Z. Suarez, *Chem. Rev.* 2002, 102, 3667–3691. Ionic liquid (molten salt) phase organometallic catalysis. (b) P. Wasserscheid, W. Keim, *Angew. Chem. Int. Ed. Engl.* 2000, 39, 3772–3789. Ionic liquids—new "solutions" for transition metal catalysis. (c) T. L. Greaves, C. J. Drummond, *Chem. Rev.* 2008, 108, 206–237. Protic ionic liquids: properties and applications. (d) T. Welton, *Chem. Rev.* 1999, 99, 2071–2084. Room-temperature ionic liquids. Solvents for synthesis and catalysis. (e) V. I. Pârvulescu, C. Hardacre, *Chem. Rev.* 2007, 107, 2615–2665. Catalysis in ionic liquids. (f) P. Wasserscheid, T. Welton (Eds.), *Ionic liquids in synthesis*. Wiley-VCH Verlag GmbH, Weinheim, Germany, 2008.

45 (a) T. Welton, *Chem. Rev.* 1999, 99, 2071–2083. Room-temperature ionic liquids. Solvents for synthesis and catalysis. (b) J. P. Hallett, T. Welton, *Chem. Rev.* 2011, 111, 3508–3576. Room-temperature ionic liquids: solvents for synthesis and catalysis. 2. (c) J. S. Wilkes, *Green Chem.* 2002, 4, 73–80. A short history of ionic liquids—from molten salt to neoteric solvents.

46 For a pivotal review, see: N. Isambert, M. M. Sanchez-Duque, J.-C. Plaquevent, Y. Génisson, J. Rodriguez, T. Constantieux, *Chem. Soc. Rev.* 2011, 40, 1347–1357. Multicomponent reactions and ionic liquids: a perfect synergy for eco-compatible heterocyclic synthesis.

47 X. Y. Zhang, Y. Z. Li, X. S. Fan, G. R. Qu, X. Y. Hu, J. J. Wang, *Chin. Chem. Lett.* 2006, 17, 150–152. Multicomponent reaction

in ionic liquid: a novel and green synthesis of 1,4-dihydropyridine derivatives.

48 Z. El Asri, Y. Génisson, F. Guillen, O. Baslé, N. Isambert, M. M. Sanchez-Duque, S. Ladeira, J. Rodriguez, T. Constantieux, J.-C. Plaquevent, *Green Chem.* 2011, 13, 2549–2552. Multicomponent reactions in ionic liquids: convenient and ecocompatible access to the 2,6-DABCO core.

49 X. Zhang, X. Li, X. Fan, X. Wang, D. Li, G. Qu, J. Wang, *Mol. Divers.* 2009, 13, 57–61. Ionic liquid promoted preparation of 4*H*-thiopyran and pyrimidine nucleoside-thiopyran hybrids through one-pot multi-component reaction of thioamide.

50 S. S. Chavan, M. S. Degani, *Green Chem.* 2012, 14, 296–299. Ionic liquid mediated one-pot synthesis of 6-aminouracils.

51 A. Dongamantia, R. Gadiparthia, R. Redamalaa, J. Anireddyb, N. Burrib, J. Vantikommua, *J. Chem. Pharm. Res.* 2012, 4, 3991–4000. Synthesis, anti-bacterial, anti-asthmatic and anti-diabetic activities of novel 3-substituted quinazolin-4-ones using 1-butyl-3-methyl-imidazoliumtetrafluoro borate [bmim+] [BF_4−] as a green, efficient and reusable catalyst under solvent free conditions.

52 For some examples, see: (a) A. Shaabani, A. Rahmati, *Catal. Lett.* 2005, 100, 177–179. Ionic liquid promoted efficient synthesis of 3,4-dihydropyrimidin-2-(1*H*)-ones. (b) S. A. Siddiqui, T. M. Potewar, R. J. Lahoti, K. V. Srinivasan, *Synthesis* 2006, 2849–2854. Ionic liquid promoted facile one-pot synthesis of 1-pyridylimidazo[1,5-*a*]pyridines from dipyridylketone and aryl aldehydes. (c) X.-S. Wang, M.-M. Zhang, H. Jiang, D.-Q. Shi, S.-J. Tu, X.-Y. Wei, Z.-M. Zong, *Synthesis* 2006, 4187–4199. An improved and benign synthesis of 9,10-diarylacridine-1,8-dione and indenoquinoline derivatives from 3-anilino-5,5-dimethylcyclohex-2-enones, benzaldehydes, and 1,3-dicarbonyl compounds in an ionic liquid medium.

53 X.-S. Wang, J.-R. Wu, J. Zhou, S.-J. Tu, *J. Comb. Chem.* 2009, 11, 1011–1022. Green method for the synthesis of highly substituted cyclohexa-1,3-diene, polyhydroindene, polyhydronaphthalene, isochromene, isothiochromene, and isoquinoline derivatives in ionic liquids.

54 G. Srihari, M. Nagaraju, M. M. Murthy, *Helv. Chim. Acta* 2007, 90, 1497–1504. Solvent-free one-pot synthesis of amidoalkyl naphthols catalyzed by silica sulfuric acid.

55 M. Adib, E. Sheibani, L.-G. Zhu, P. Mirzaei, *Tetrahedron Lett.* 2008, 49, 5108–5110. An efficient synthesis of 3-amino-2-arylimidazo[1,2-*a*]pyridines.

56 K. Kanagaraj, K. Pitchumani, *Tetrahedron Lett.* 2010, 51, 3312–3316. Solvent-free multicomponent synthesis of pyranopyrazoles; per-6-amino-β-cyclodextrin as a remarkable catalyst and host.

57 Y. Liu, L. Wang, Y. Sui, J. Yu, *Chin. J. Chem.* 2010, 28, 2039–2044. Solvent-free synthesis of alkylaminophenols *via* Petasis boronic Mannich reaction in one pot without catalysts.

58 S. Safaei, I. Mohammadpoor-Baltork, A. R. Khosropour, M. Moghadam, S. Tangestaninejad, V. Mirkhani, *Synlett* 2011, 2214–2222. Regioselective multicomponent synthesis of fully substituted pyrazoles and bispyrazoles catalyzed by zinc triflate.

59 L. M. Sanchez, A. G. Sathicq, J. L. Jios, G. T. Baronetti, H. J. Thomas, G. P. Romanelli, *Tetrahedron Lett.* 2011, 52, 4412–4416. Solvent-free synthesis of functionalized pyridine derivatives using Wells-Dawson heteropolyacid as catalyst.

60 G. Shukla, R. K. Verma, G. K. Verma, M. S. Singh, *Tetrahedron Lett.* 2011, 52, 7195–7198. Solvent-free sonochemical one-pot three-component synthesis of 2*H*-indazolo[2,1-*b*]phthalazine-1,6,11-triones and 1*H*-pyrazolo[1,2-*b*]phthalazine-5,10-diones.

61 K. Kumari, D. S. Raghuvanshi, V. Jouikov, K. N. Singh, *Tetrahedron Lett.* 2012, 53, 1130–1133. $Sc(OTf)_3$-catalyzed, solvent-free domino synthesis of functionalized pyrazoles under controlled microwave irradiation.

62 K. Kundu, S. K. Nayak, *RSC Adv.* 2012, 2, 480–486. (±)-Camphor-10-sulfonic acid catalyzed direct one-pot three-component Mannich type reaction of alkyl (hetero)aryl ketones under solvent-free conditions: application to the synthesis of aminochromans.

63 C. J. Pierce, C. H. Larsen, *Green Chem.* 2012, 14, 2672–2676. Copper(II) catalysis provides cyclohexanone-derived propargylamines free of solvent or excess starting materials: sole by-product is water.

64 T. B. Nguyen, L. Ermolenko, A. Al-Mourabit, *Org. Lett.* 2012, 14, 4274–4277. Efficient and selective multicomponent oxidative coupling of two different aliphatic primary amines into thioamides by elemental sulfur.

65 A. Ponnuswamy, P. Shanmugavelan, S. Nagarajan, M. Sathishkumar, *Helv. Chim. Acta* 2012, 95, 922–928. The first one-pot, solvent-free, microwave-accelerated, three-component synthesis of spirothiazolidin-4-ones *via* Staudinger/aza-Wittig coupling/cyclization.

66 E. Yamuna, K. Prabakaran, M. Zeller, P. K. J. Rajendra, *Synth. Commun.* 2012, 42, 1330–1340. Convergent synthetic route to pyrano[2,3-*a*]carbazole by multicomponent reaction.

67 B. Karami, S. Khodabakhshi, F. Hashemi, *Tetrahedron Lett.* 2013, 54, 3583–3585. Synthesis of a novel class of benzofurans *via* a three-component, regiospecific intramolecular heterocylization reaction.

68 S. A. R. Mulla, T. A. Salama, M. Y. Pathan, S. M. Inamdar, S. S. Chavan, *Tetrahedron Lett.* 2013, 54, 672–675. Solvent-free, highly efficient one-pot multicomponent synthesis of 1-amido- and 1-carbamato-alkyl naphthols/phenols catalyzed by ethylammonium nitrate as reusable ionic liquid under neat reaction condition at ambient temperature.

69 L.-R. Wen, Q.-C. Sun, H.-L. Zhang, M. Li, *Org. Biomol. Chem.* 2013, 11, 781–786. A new rapid multicomponent domino heteroannulation of heterocyclic ketene aminals: solvent-free regioselective synthesis of functionalized benzo[*g*]imidazo[1,2-*a*]quinolinediones.

70 F. Wohler, *Ann. Phys.* 1828, 2, 253–256. Ueber künstliche Bildung des Harnstoffs.

71 A. Baeyer, V. Drewsen, *Ber. Dtsch. Chem. Ges.* 1882, 15, 2856–2864. Darstellung von Indigblau aus Orthonitrobenzaldehyd.

72 S. Narayan, J. Muldoon, M. G. Finn, V. V. Fokin, H. C. Kolb, K. B. Sharpless, *Angew. Chem. Int. Ed.* 2005, 44, 3275–3279. "On water": unique reactivity of organic compounds in aqueous suspension.

73 M. C. Pirrung, K. D. Sarma, *Tetrahedron* 2005, 61, 11456–11472. Aqueous medium effects on multi-component reactions.

74 M. Abbas, J. Bethke, L. A. Wessjohann, *Chem. Commun.* 2006, 541–543. One pot synthesis of selenocysteine containing peptoid libraries by Ugi multicomponent reactions in water.

75 F. M. Moghaddam, G. R. Bardajee, M. Dolabi, *J. Sulfur Chem.* 2010, 31, 387–393. An efficient one-pot synthesis of tri-substituted thiophenes *via* a multicomponent reaction in water.

76 A. Dandia, A. K. Laxkar, R. Singh, *Tetrahedron Lett.* 2012, 53, 3012–3017. New multicomponent domino reaction on water: highly diastereoselective synthesis of spiro[indoline-3,4′-pyrazolo[3,4-*b*]pyridines] catalyzed by NaCl.

77 B. Rajarathinam, G. Vasuki, *Org. Lett.* 2012, 14, 5204–5206. Diastereoselective multicomponent reaction in water: synthesis of 2-azapyrrolizidine alkaloid analogues.

78 F. Rostami-Charati, Z. Hossaini, *Synlett* 2012, 23, 2397–2399. Facile synthesis of phosphonates *via* catalyst-free multicomponent reactions in water.

79 R. M. Rumum, H. Mecadon, A. T. Khan, B. Myrboh, *Tetrahedron Lett.* 2012, 53, 5261–5264. Synthesis of important β-functionalized 5-methyl-1*H*-pyrazol-3-ol derivatives in the presence of γ-alumina catalyst in aqueous medium.

80 J.-L. Cao, S.-L. Shen, P. Yang, J. Qu, *Org. Lett.* 2013, 15, 3856–3859. A catalyst-free one-pot construction of skeletons of 5-methoxyseselin and alloxanthoxyletin in water.

81 For an excellent review about Diels-Alder reactions see: K. C. Nicolau, S. A. Snyder, T. Montagnon, G. Vassilikogiannakis, *Angew. Chem. Int. Ed.* 2002, 41, 1668–1698. The Diels-Alder reaction in total synthesis.

82 R. M. Uittenbogaard, J.-P. G. Seerden, H. W. Scheeren, *Tetrahedron* 1997, 53, 11929–11936. High-pressure promoted stereoselective tandem [4+2]/[3+2] cycloadditions of nitro alkenes and enol ethers.

83 (a) G. J. Kuster, F. Kalmoua, H. W. Scheeren, R. de Gelder, *Chem. Commun.* 1999, 855–856. A simple entry towards novel bi- and tricyclic *N*-oxy-β-lactams by high pressure promoted tandem [4+2]/[3+2] cycloadditions of enol ethers and β-nitrostyrene. (b) L. W. A. van Berkom, G. J. T. Kuster, R. de Gelder, H. W. Scheeren, *Eur. J. Org. Chem.* 2004, 4397–4404. Synthesis and rearrangement of *N*-organyloxy β-lactams derived from a (4+2)/(3+2) sequential cycloaddition reaction involving enol ethers and nitro alkenes.

84 G. J. T. Kuster, R. H. J. Steeghs, H. W. Scheeren, *Eur. J. Org. Chem.* 2001, 553–560. Novel five/five- and six/five-membered bicyclic nitroso acetals from high-pressure-promoted cyclisation reactions of *p*-methoxybenzyl vinyl ether, 1-nitro-2-heteroaryl ethenes, and mono- and di-substituted olefins.

85 (a) I. Chataigner, S. R. Piettre, *Org. Lett.* 2007, 9, 4159–4162. Multicomponent domino [4+2]/[3+2] cycloadditions of nitroheteroaromatics: an efficient synthesis of fused nitrogenated polycycles. (b) H. Gerard, I. Chataigner, *J. Org. Chem.* 2013, 78, 9233–9242. Selectivities of multicomponent [4+2]/[3+2] cycloadditions of 3-nitroindole with substituted alkenes: a DFT analysis.

86 K. Matsumoto, J. C. Kim, H. Iida, H. Hamana, K. Kumamoto, H. Kotsuki, G. Jenner, *Helv. Chim. Acta* 2005, 88, 1734–1753. Multicomponent Strecker reaction under high pressure.

87 G. J. Kuster, H. W. Scheeren, *Tetrahedron Lett.* 1998, 39, 3613–3616. High pressure promoted tandem [4+2]/[3+2] cycloadditions on a solid phase.

88 G. J. Kuster, H. W. Scheeren, *Tetrahedron Lett.* 2000, 41, 515–519. The preparation of resin-bound nitro alkenes and some applications in high-pressure promoted cycloadditions.

89 A. L. Kennedy, A. M. Fryer, J. A. Josey, *Org. Lett.* 2002, 4, 1167–1170. A new resin-bound universal isonitrile for the Ugi 4CC reaction: preparation and applications to the synthesis of 2,5-diketopiperazines and 1,4-benzodiazepine-2,5-diones.

90 A. Ouach, S. Gmouh, M. Pucheault, M. Vaultier, *Tetrahedron* 2008, 64, 1962–1970. Onium salt supported organic synthesis in water: application to Grieco's multicomponent reaction.

91 J. Albadi, M. Keshavarz, *Synth. Commun.* 2013, 43, 2019–2030. Polymer-supported azide and copper(I): green reusable reagent and catalyst for click cyclization.

12

ESSENTIAL MULTICOMPONENT REACTIONS I

EUGENIA MARQUÉS-LÓPEZ AND RAQUEL P. HERRERA

Departamento de Química Orgánica, Instituto de Síntesis Química y Catálisis Homogénea (ISQCH), CSIC-Universidad de Zaragoza, Zaragoza, Spain

12.1 RADZISZEWSKI REACTIONS (IMIDAZOLE SYNTHESIS)

12.1.1 Introduction

In general, Radziszewski reaction is the synthesis of an imidazole derivative **4** by the condensation of an α-dicarbonyl compound **1** (e.g., glyoxal, pyruvaldehyde, porphyrin-2,3-diones, and benzil), an aldehyde **2**, and two equivalents of dry ammonia in alcohol. Interestingly, the replacement of one equivalent of ammonia with a primary amine **3** results in the formation of 1-substituted imidazoles **4** (Scheme 12.1).

The so-called Radziszewski reaction (occasionally referred as Debus–Radziszewski reaction) [1] is a multicomponent reaction (MCR) published first by Debus in 1858 [2]. In this early work, the condensation of glyoxal with ammonia to give both imidazole and 1*H*,1′*H*-(2,2′)biimidazolyle is described. More than two decades later, Radziszewski repeated this reaction [3], and according to Hofmann, the method carries Radziszewski's name because he extended it to the preparation of simpler imidazoles and demonstrated its general applicability [4]. However, Japp worked on the imidazole-forming reaction before Radziszewski's first article [5], and Plater requested recognition for him for his crucial contributions to the general synthesis of imidazole derivatives [6].

A plausible mechanistic hypothesis for this reaction is that the two molecules of ammonia (or primary amine) react with the α-dicarbonyl compound **1** to form an α-diimine **5**, which condenses with the aldehyde **2** giving an intermediate (**6**) that cyclizes to give imidazole derivative **4**, as a main product, but it can render also oxazole **7** as a by-product (Scheme 12.2) [1].

An example of application of this reaction employing NH_4OAc as a source of ammonia is depicted in Scheme 12.3. It is a key step in the preparation of a pyrimidine–imidazole-based library, where their biological activities were under study [7–9]. Thus, using glyoxal dimethyl acetal **11**, ammonium acetate, and the required asymmetrical 1,2-diketone **10**, prepared from 4-fluorobenzoate **9** and 2-mercapto-4-methylpyrimidine **8**, the key intermediate **12** was obtained in good yield (86%).

12.1.2 Modified Radziszewski Reactions: Efficient Tool for the Synthesis of Substituted Imidazoles

The imidazole moiety is very important in heterocyclic chemistry [10]. It is present in numerous natural products and also in many synthetic compounds. There are imidazole-based drugs such as ketoconazole **13** (antifungal properties) and losartan **14** (a drug against hypertension) (Fig. 12.1) [11]. Some of the different biological activities that this structural motif shows are, for instance, antiallergic activity [12] and anti-inflammatory activity, inhibiting cytokine release or inhibiting the p38 MAP kinase [13]. Some imidazole derivatives are pesticides [14], sodium-channel modulators [15], or antibiotics [16]. In addition, in the last years, these molecules are used as green solvents (ionic liquids (IL)) [17], as N-heterocyclic carbenes [18], and as chemiluminescent [19] and chromotropic materials [20].

Among other reported methods, Radziszewski reaction has demonstrated to be an efficient tool for the synthesis of substituted imidazoles. Regioselectivity in this kind of reaction is a very difficult task; however, playing with the

Multicomponent Reactions: Concepts and Applications for Design and Synthesis, First Edition. Edited by Raquel P. Herrera and Eugenia Marqués-López.

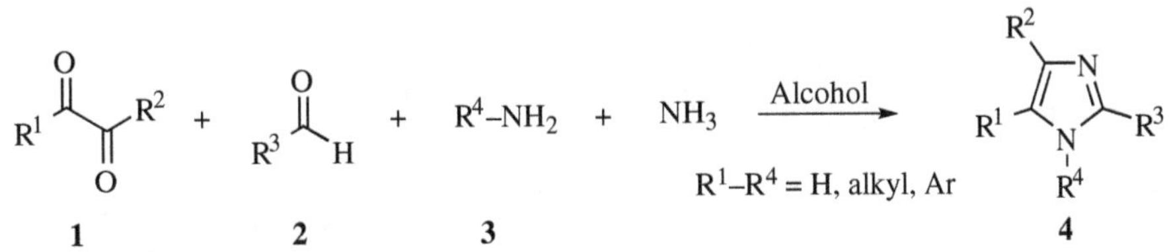

SCHEME 12.1 General Radziszewski reaction.

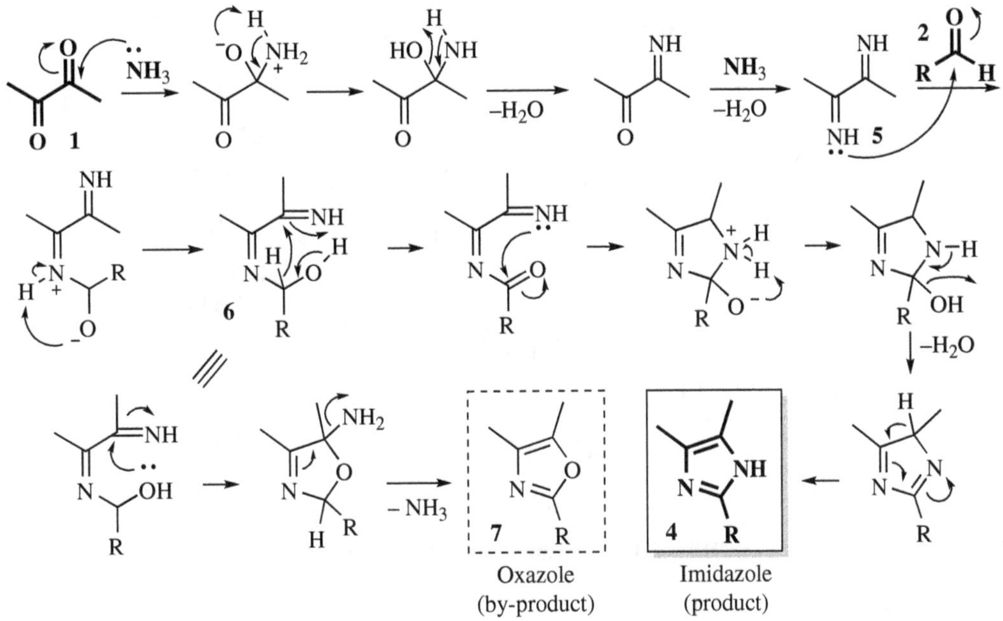

SCHEME 12.2 General mechanistic hypothesis for Radziszewski reaction.

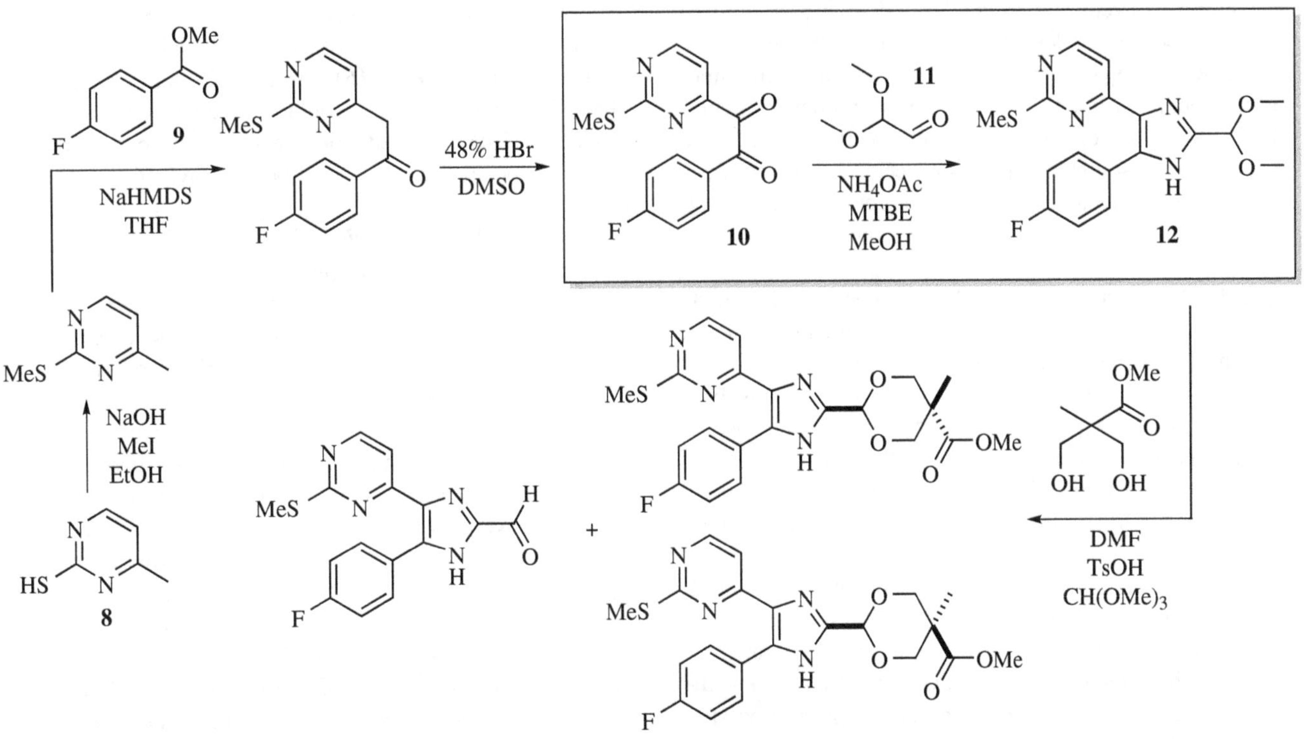

SCHEME 12.3 An example of application of Radziszewski reaction in the preparation of a pyrimidine–imidazole-based library.

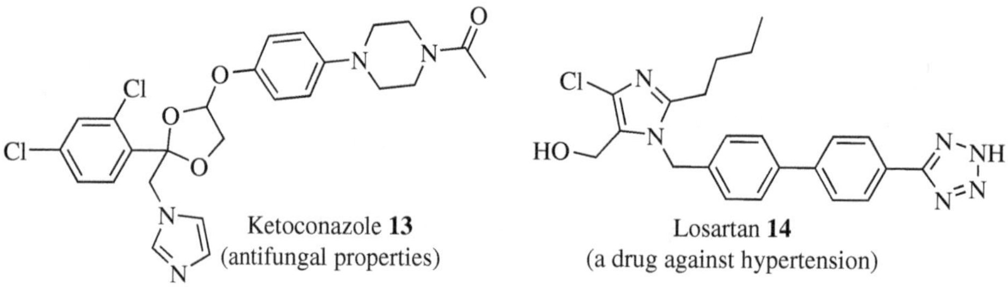

FIGURE 12.1 Imidazole-based drugs.

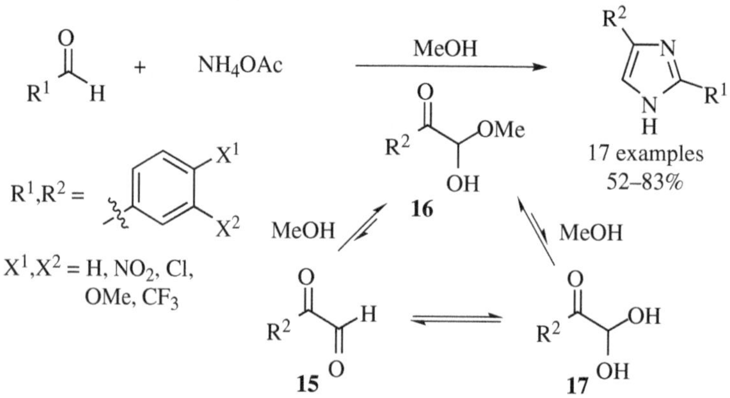

SCHEME 12.4 Synthesis of 2,4-diarylimidazoles.

employed substrates, as well as with the reaction conditions, the substitution of the core can be efficiently controlled in a large number of reported protocols. In this chapter, we will describe some examples of the modified versions of the original Radziszewski reaction that consist in a multicomponent cyclocondensation of 1,2-dicarbonylcompound (generally 1,2-diketone) (or α-hydroxy ketone) with aldehydes and an ammonium source (ammonium acetate) (and amines), involving different catalytic systems and/or modern technologies such as microwaves (MW) or microreactors.

12.1.2.1 2,4-Diarylimidazoles In the original Radziszewski reaction, a 1,2-diketone is used; thus, substituents in C4 and C5 in the imidazole core are given by substituents in that substrate. However, if the goal is the synthesis of C5-unsubstituted imidazoles, a possible method would be the use of monoketone derivative of glyoxal, with the extra difficulty of suppressing competition between aldehyde and glyoxal for the C2 position of the heterocycle. Cocconcelli took advantage of this possibility and reported in 2007 an elegant procedure for the synthesis of C5-unsubstituted 2,4-diaryl-1*H*-imidazoles [21]. They stated that the success of the method strongly depends on the reaction conditions, and they found that the solvent used is especially important. Thus, the use of methanol in the cyclocondensation of phenylglyoxals, benzaldehydes, and ammonium acetate appears to be the solution to the regioselectivity problem. The authors carried out NMR experiments to support the hypothesis that in methanol the equilibrium between the hydrated (**17**), hemiacetal (**16**), and aldehyde (**15**) forms of phenylglyoxal is shifted toward the hemiacetal (**16**) form, and in this case, a formal benzaldehyde excess would effectively be present (Scheme 12.4). In this situation, incorporation of the aldehyde at the C2 position of the core should be favored.

12.1.2.2 2,4,5-Triarylimidazoles This pattern of substitution is the most deeply investigated and the most successfully developed. There are numerous examples in the literature where the target products are achieved efficiently by means of suitable catalyst [22] or MW irradiation [23]. Herein, only some representative examples will be mentioned.

A considerable number of methods base on the Radziszewski reaction employing heterogeneous catalytic systems have come out, allowing the synthesis of 2,4,5-trisubstituted imidazoles in a very efficient way, in terms of yields and time reactions. The authors proclaimed simplicity of the operations and environmental friendliness as additional advantages of the methods. In many cases, the reactions are carried out under solvent-free conditions, and in some cases, it is possible to perform them at room temperature. Some selected examples are those using as catalysts silica sulfuric acid (using water as solvent [24] or without solvent [25]) [$NiCl_2$·$6H_2O$], supported on acidic alumina [26], and uranyl nitrate hexahydrate [$UO_2(NO_3)_2$·$6H_2O$], also supported on acidic alumina [27].

Other catalysts employed in the Radziszewski reaction for the synthesis of 2,4,5-triarylimidazoles are $Yb(OTf)_3$ [28], $ZrCl_4$ [29], I_2 [30], $Eu(OTf)_3$ [31], oxalic acid [32], $NaHSO_3$ [33], phosphomolybdic acid (PMA) [34], $Zr(acac)_4$ [35], L-proline [36], ceric(IV) ammonium nitrate (CAN) and ultrasonication [37], indium trifluoride (InF_3) [38], $Cu(TFA)_2$ [39], $(NH_4)_6Mo_7O_{24}$·$4H_2O$ [40], nanocrystalline MgO [41], anhydrous $FePO_4$ [42], potassium dihydrogenphosphate (KH_2PO_4) [43], and $SnCl_2$·$2H_2O$ and $CeCl_3$·$7H_2O$ [44].

On the other hand, several protocols have been developed using IL as recyclable medium as well as promoters. Thus, in 2005, Srinivasan published a pioneer work using 1-butyl imidazolium tetrafluoroborate ([Hbim]BF_4), without any additional catalyst, in the reaction of 1,2-diketones (or α-hydroxy ketone), aromatic aldehydes, and ammonium. He obtained by this method a variety of 2,4,5-triarylimidazoles in very good yields (85–98%) in short reaction times (25–120 min) [45]. The authors gave mechanistic hypothesis for these reactions, involving 1,2-diketones or α-hydroxy ketone, as starting material, trying to explain the role of the IL (Scheme 12.5). They justified the promotion of the heterocyclization by the inherent Brønsted acidity of the IL, conferred by the most acidic –N–H hydrogen. This makes the IL capable of increasing the reactivities of the carbonyl compounds. In both cases, the IL promotes the generation of ammonia from ammonia acetate. In the case of the 1,2-diketones, the IL may help in the formation of the intermediate **18**, which, under Brønsted acid catalysis of the IL, condenses with the dicarbonyl compound to give intermediate **19**, which then evolves to the final product by rearrangement. On the other hand, in the case of the α-hydroxy ketones, the IL may promote the formation of intermediates **20** and **21**, which then condense and cyclize to give intermediate **22** that provides the final product after dehydrogenation.

An example of the use of a neutral IL (tetrabutylammonium bromide: TBAB) was published in 2008 by Kantevari. TBAB, in refluxing isopropanol, catalyzed efficiently the one-pot three-component synthesis of trisubstituted imidazoles [46]. The same year, Khosropour published his results using 10 mol% of 1-methylimidazolium hydrogen sulfate ([Hmim]HSO_4) as Brønsted acidic IL [47]. A more recent example of the use of IL in this reaction has been reported by Lu [48]. They combined nonchloroaluminate acidic IL ([BMIm][H_2PO_4]) with MW irradiation, developing a three-component condensation with several advantages like high yields, short reaction times, low pollution, and simple experimental procedure.

MW In general, harsh conditions are required in these reactions to accomplish high yields. In recent years, several groups have seen the opportunity of minimizing them by using MW irradiation, drastically shortening the reaction times [49, 50]. In this sense, Wolkenberg reported in 2004 a very effective method for the preparation of imidazoles in yields ranging from 80 to 99% after 5 min of reaction at 180 °C using acetic acid [51]. The method seems to be general for the synthesis of alkyl-, aryl-, and heteroaryl-substituted imidazoles; however, only an example of asymmetric 1,2-diketone is shown. Interestingly, the authors illustrated the utility of the developed methodology by the synthesis of two biologically active compounds: lepidiline B **25** [52], which exhibits micromolar cytotoxicity against several human cancer cell lines, and trifenagrel **28** [53], a drug that exhibits a potent 2,4,5-triarylimidazole arachidonate cyclooxygenase inhibitor that reduces platelet aggregation in several animal species and humans. Molecule **25** was prepared in two steps starting from 2,3-butanedione **23** and acetaldehyde **24** in 43% overall yield (Scheme 12.6a), while compound **28** was obtained in 99% yield after only 5 min by condensation of aldehyde **27** and 1,2-diketone **26** (Scheme 12.6b).

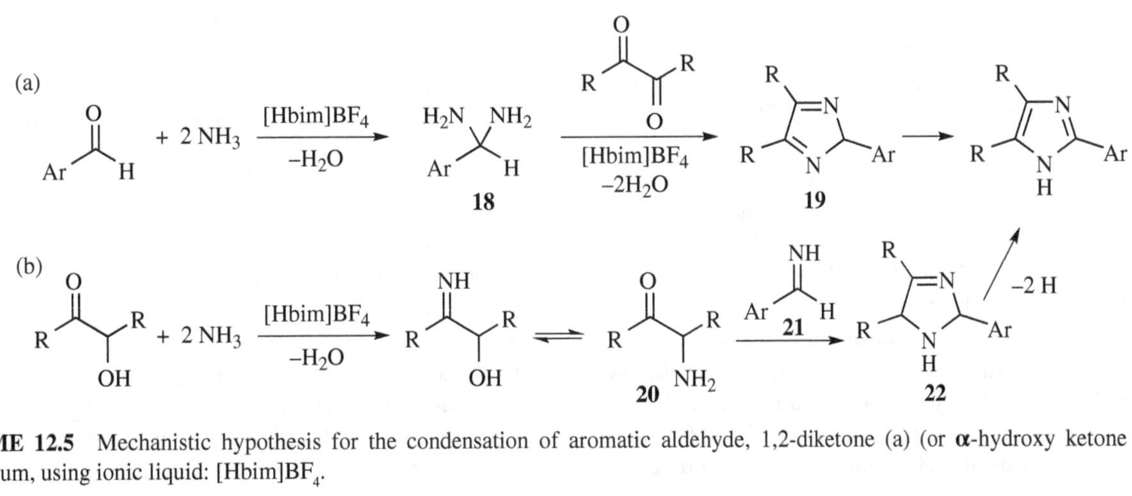

SCHEME 12.5 Mechanistic hypothesis for the condensation of aromatic aldehyde, 1,2-diketone (a) (or α-hydroxy ketone (b)), and ammonium, using ionic liquid: [Hbim]BF_4.

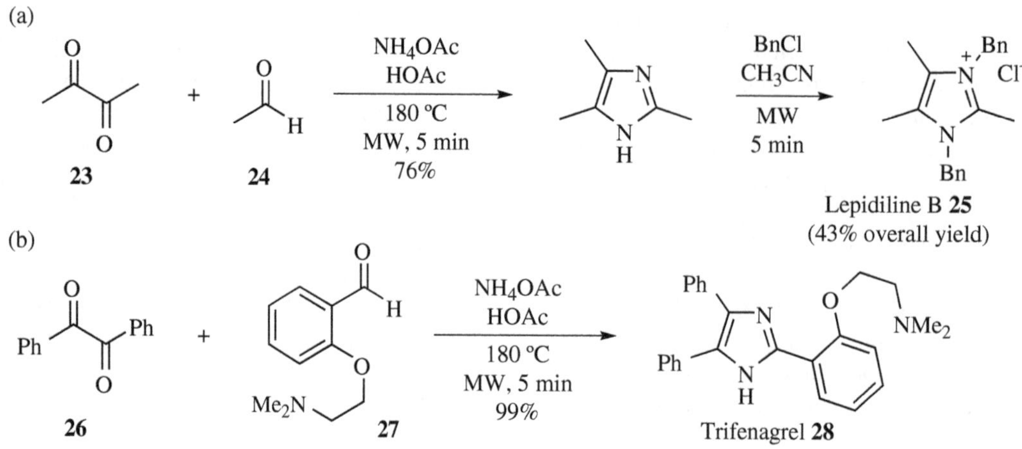

SCHEME 12.6 Synthesis of biologically active compounds, lepidiline B **25** and trifenagrel **28**, *via* Wolkenberg method.

12.1.2.3 1,2,4,5-Tetrasubstituted Imidazoles As well as for trisubstituted imidazoles, the synthesis of 1,2,4,5-tetrasubstituted imidazoles has aroused great interest in organic chemistry. In this case, reactions involve a primary amine in addition to the standard reagents. Thus, reported protocols are based on a four-component condensation of a 1,2-diketone (or α-hydroxyketone or α-ketomonoxime) with an aldehyde, primary amine, and ammonium acetate. In particular, researchers have paid attention to heterogeneous catalysis that provides Brønsted surface sites able to activate the substrates. Here, few efficient and green examples are mentioned, like zeolite HY or silica gel (under MW heating) [54], silica gel-supported sodium bisulfate [$NaHSO_4{\cdot}SiO_2$] (under MW and conventional heating) [55], perchloric acid adsorbed on silica gel [$HClO_4{\cdot}SiO_2$] [56], indium trichloride ($InCl_3{\cdot}3H_2O$) (polar solvent is required (methanol) at room temperature) [57], silica-supported boron trifluoride ($BF_3{\cdot}SiO_2$) (thermal and solvent-free conditions) [58], and Keggin-type heteropoly acid (HPA) $H_4[PMo_{11}VO_{40}]$ (in refluxing ethanol) [59]. In all cases, the cleanness of reactions is highlighted, with almost no side product formation. In general, these methodologies are tolerable to both aromatic and aliphatic primary amines, while the kind of substituent on diones and aldehydes is restricted to aromatic ones (for some examples of aliphatic aldehydes (lower yield than aromatic aldehydes), see Ref. [57]).

Microreactor Technology In the last years, some examples of Radziszewski reaction have been reported using microreactor technology, which allows the development of more sustainable methods thanks to the miniaturization and increased safety and efficiency [60, 61]. In particular, the Stevens and the Orru groups developed in cooperation a procedure for the continuous production of highly substituted imidazoles [62, 63]. They were motivated by the possibility of making the method previously reported by Orru [64] more interesting for industrial applications. The latest procedure was based on MW heating, decreasing substantially the *conventional* reaction time, but it is known that MW-assisted methods are not easy to scale up. Thus, after an exhaustive search for the optimal reaction conditions, the author concluded that the use of *N*-methylpyrrolidinone(NMP)/*n*-pentanol (v/v 1:1) as a solvent at 120 °C and an α-diketone–aldehyde–ammonium acetate–primary amine–acetic acid ratio of 1:1:1:1.2:10, with a final concentration of 0.25 M of the α-diketone in the microreactor, led to the best results, allowing to obtain a variety of tetra- and trisubstituted imidazoles in moderate to good yields (Scheme 12.7).

In general, when these methods (microreaction and MW reaction) are compared, similar regioselectivities are achieved in both cases, while the obtained conversions differ without a clear trend (being higher only in some cases using a microreactor). However, the relatively new microreactor technology has the main advantage of allowing the multigram-scale production of final imidazoles in a continuous way. This possibility is attractive from the standpoint of industrial applicability.

The authors' proposal for the mechanism of this reaction consists in two possible pathways (Scheme 12.8) [62, 64]. In *path I*, the condensation reaction between aldehyde and amine takes place, forming an imine **29**, which then reacts with ammonia to give a diamine **30**. This bulky intermediate adds the primary amine to a carbonyl group of the diketone (the less hindered one in the case of asymmetrical substrate), resulting in intermediate **31**, common to both paths. After that, cyclization occurs by means of an intramolecular attack of the other nucleophilic nitrogen atom to the remaining carbonyl group. Final aromatization of the resulting cycle leads to the final imidazole. In *path II*, ammonia reacts with diketone, giving a geminal aminoalcohol **32**, which reacts by the nitrogen atom with the *in situ* formed imine **29**, generating the aforementioned common intermediate **31**. In the case of an asymmetrical diketone, ammonia (less hindered than diamine intermediate **30**) can add to both carbonyl groups,

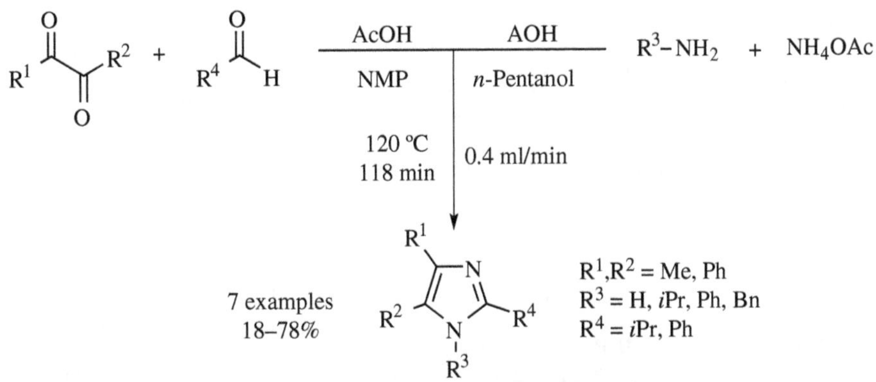

SCHEME 12.7 Modified Radziszewski reaction performed under microreactor conditions.

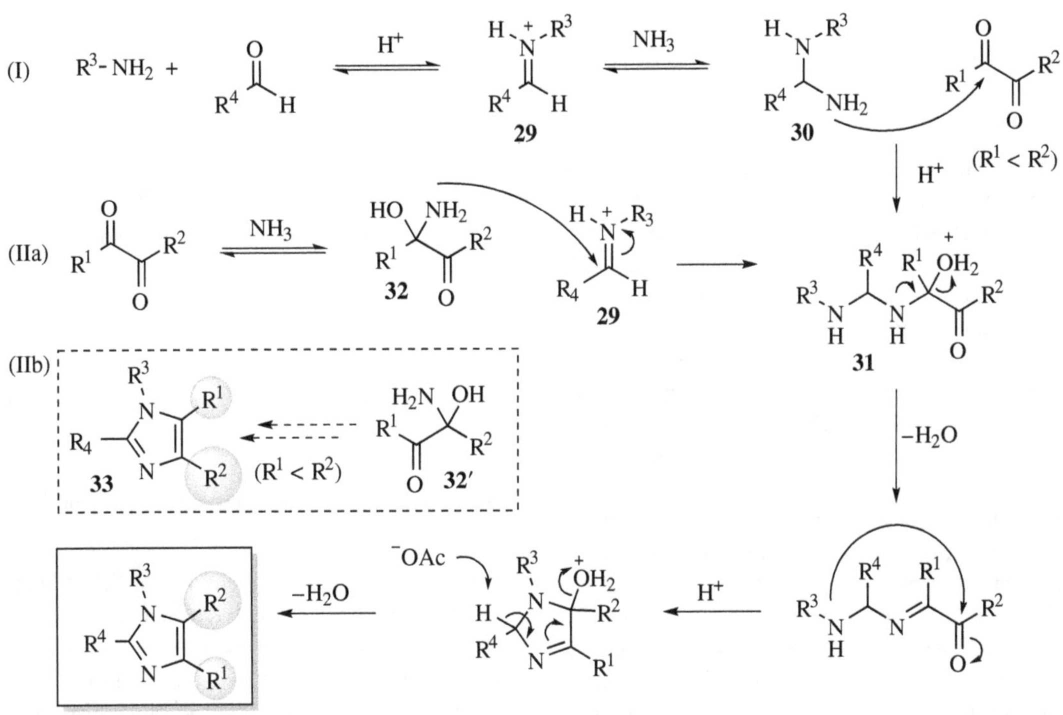

SCHEME 12.8 Mechanistic proposal for the Radziszewski synthesis of imidazoles given by Orru and Stevens.

and if it adds to the more hindered one, the final regioisomer **33** would be obtained, through the formation of geminal aminoalcohol **32** (*path II.b*). Reasonably, the preferred path would depend on the reactivity of the initially formed imine **29** in comparison with the reactivity of the diketone. Thus, depending on the type of amine and aldehyde involved (as well as the diketone), the regioselectivity varies from modest to good. Better results were achieved by using an aromatic aldehyde and an aliphatic amine, with a preference for incorporation of the phenyl group (of the diketone) next to the substituted nitrogen.

12.1.2.4 Weidenhagen Synthesis

The Weidenhagen synthesis of imidazoles is a modification of the Radziszewski reaction [65] and was first published in 1935 [66]. In this method, a mixture of α-hydroxy ketones, aldehydes, and ammonia is treated with cupric acetate followed by hydrogen sulfide or sodium sulfide (Scheme 12.9). It is known that the copper(II) acetate acts as oxidant, converting the hydroxy ketone to the transitory dicarbonyl compound, which couples then with the aldehyde and the ammonia.

The recent applications of Weidenhagen reaction conditions have allowed the synthesis of 4(5)-(2′-hetaryl)imidazoles **38** that can be transformed into potential luminophores [67]. The polycyclic compounds **38** were obtained from the corresponding 2-furomethyl and 2-thenoylmethyl acetates (**34** and **35**, respectively) by condensation with formaldehyde **36** [67a] and benzaldehyde **37** [67b] in the presence of ammonia and copper acetate (Scheme 12.10).

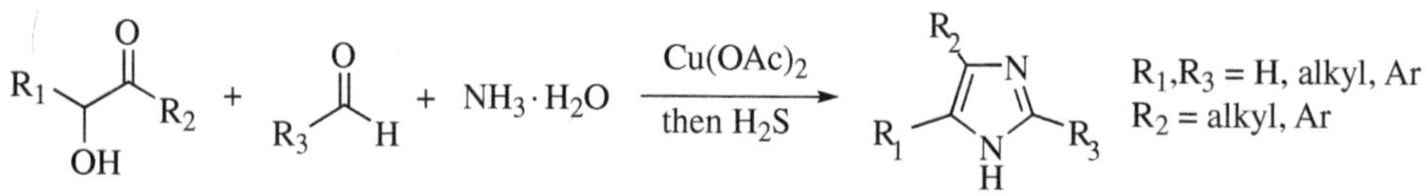

SCHEME 12.9 General Weidenhagen reaction.

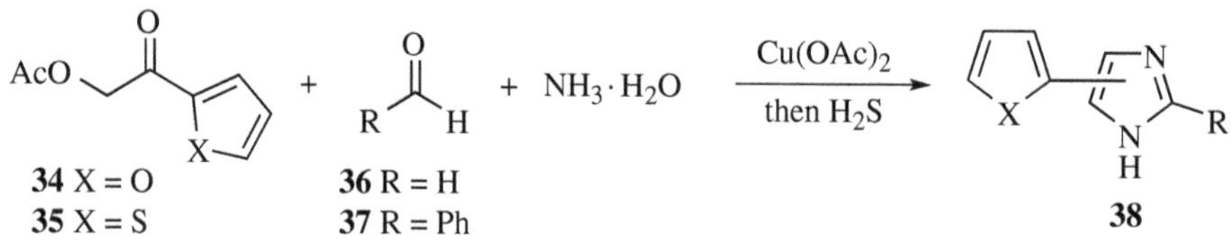

SCHEME 12.10 Synthesis of 2-phenyl-4(5)-(2′-hetaryl)imidazoles **38** by Weidenhagen reaction.

12.2 SAKURAI MCRs

12.2.1 Introduction

The allylation of a carbonyl compound, or equivalent thereof (aldehydes, ketones, acetals, ketals, enones, acid chlorides, epoxides, etc.), performed with an allylsilane **39** promoted by a Lewis acid ($TiCl_4$, $SnCl_4$, $BF_3{\cdot}OEt_2$, $AlCl_3$, Et_2AlCl, etc.) is known as Hosomi–Sakurai reaction (Scheme 12.11, Eq. 1) [68, 69]. This reaction was first described in 1976, allowing the synthesis of homoallylic alcohols **43** from aldehydes and ketones **40** [70]. A year later, the authors extended the method to ketals **41**, which provide homoallylic ethers **44** (Scheme 12.11, Eq. 2) [71], and to α,β-unsaturated ketones **42** (Scheme 12.11, Eq. 3) [72].

A general mechanism of this reaction involves globally a regiospecific transposition of the allylic moiety (Scheme 12.12). The reaction is initiated by the nucleophilic addition of an allylsilane **39** over an electrophilic carbon, which is normally activated by a strong Lewis acid. Thus, a secondary carbocation intermediate (β-silyl cation, **46**) stabilized by hyperconjugation is formed. Then it evolves into a compound with a new double bond by elimination of the silyl group.

In this chapter, only the most recent multicomponent versions of the Hosomi–Sakurai reaction will be covered [68b], in particular those that allowed the synthesis of homoallylic ethers through the combination of an aldehyde (or ketone) with a silyl ether or the corresponding alcohol and an allylsilane (Scheme 12.13, Eq. 1). And also, the synthesis of homoallylic amines where the allylsilane reacts with an imine *in situ* formed from the carbonyl compound and an amine is known as aza-Sakurai reaction (Scheme 12.13, Eq. 2).

12.2.2 Synthesis of Homoallylic Ethers

12.2.2.1 Silyl-Modified Sakurai Reaction Secondary and tertiary homoallylic ethers are important building blocks in the synthesis of biologically active compounds. In 1991, Markó et al. developed an efficient method for the synthesis of those

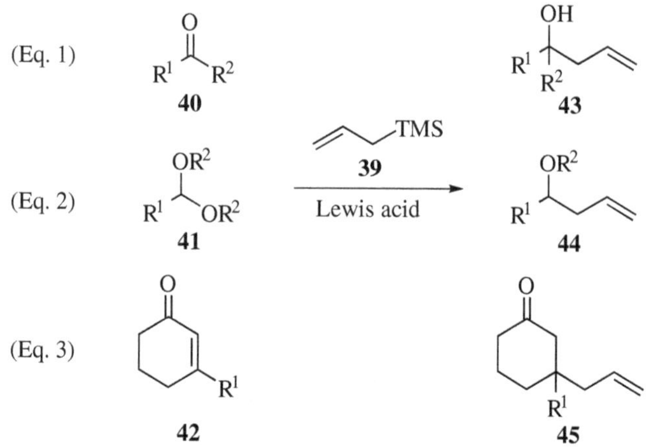

SCHEME 12.11 The first described Hosomi–Sakurai reactions.

compounds based on a variation of the Sakurai reaction, consisting in a three-component reaction that involves a carbonyl compound (aldehydes or ketones), a silyl ether, and an allylsilane catalyzed by trimethylsilyl triflate [73]. This was named by the authors as "silyl-modified Sakurai" (SMS) reaction and has demonstrated to be useful for the preparation of complex natural products [74].

More recently, they developed the first chiral aldehyde-based Sakurai MCR, allowing a rapid, versatile, and highly stereoselective access to homoallylic ethers **49** in good to excellent yields (67–87%) with complete diastereocontrol in the case of secondary and tertiary silyl ethers **48a,e–h** (Scheme 12.14) [75].

The proposed mechanism starts with the activation of the aldehyde (*S*)-**47** by the catalyst TMSOTf, followed by the formation of ketal intermediate **50** after reaction with the silylated alcohol **48**, which is in equilibrium with the oxocarbenium ion **51** (Scheme 12.15). Then the allyl silane **39** adds *via* an S_N1-type mechanism, according to Felkin–Anh transition state **52**, providing the major 2,3-*syn* products **49**.

The synthetic utility of this method was exemplified by the efficient and stereocontrolled synthesis of the pyran **53**, starting by the product **49a**, which results from the use of silyl

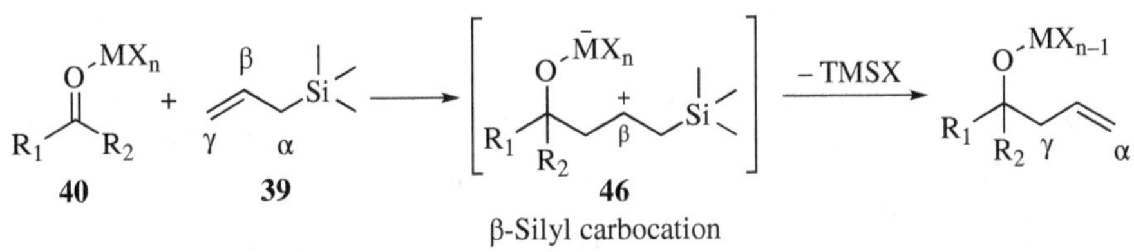

SCHEME 12.12 General mechanism for Hosomi–Sakurai reaction.

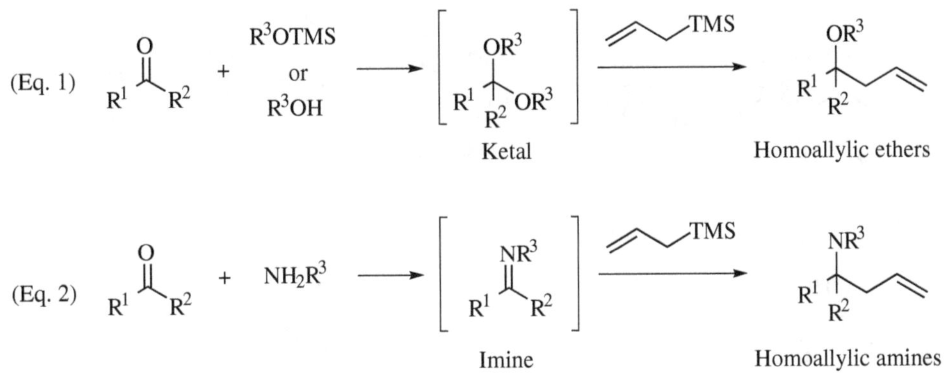

SCHEME 12.13 Multicomponent (aza-)Hosomi–Sakurai reactions: synthesis of homoallylic ethers (Eq. 1) and synthesis of homoallylic amines (Eq. 2).

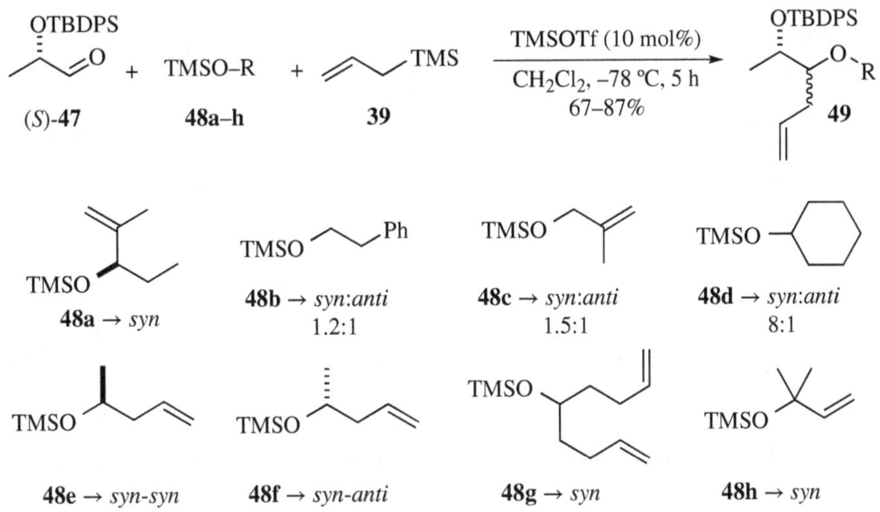

SCHEME 12.14 The first chiral aldehyde-based Sakurai MCR reported by Markó et al.

ether **48a** in the previously explained chiral aldehyde-based Sakurai MCR (6 steps, 59% overall yield), the right-hand subunit of (+)-ambruticin **54**. Later, the authors applied the same methodology for the synthesis of a similar antifungal, jerangolid D **55** (Fig. 12.2) [76].

A year later, after Markó's first article about SMS reaction, Tietze et al. published a stereoselective version of the same reaction using trimethylsilyl ether of (1*R*,2*R*)-*N*-trifluoro-acetylnorpseudoephedrin, which acts as chiral auxiliary [77]. Since then, Tietze's group has been working extensibly in the domino multicomponent allylation reaction (MCAR) for the stereoselective synthesis of homoallylic ethers (Scheme 12.16) [78–80]. The corresponding alcohols can be obtained from the resulting ethers by reductive methods.

The mechanism of this reaction proceeds in a similar way that it was proposed before, *via* a mixed acetal species **50** that losses trimethylsilyloxide ($TMSO^-$) anion and gives an oxocarbenium ion **51**. This intermediate is then attacked by the trimethylallyl silane providing a secondary carbocation **56** that losses TMS group by the attack of the previously formed $TMSO^-$, generating the homoallylic compound and TMS_2O as by-product. The proposed mechanism is supported by the results of experimental and theoretical mechanistic studies [81, 82].

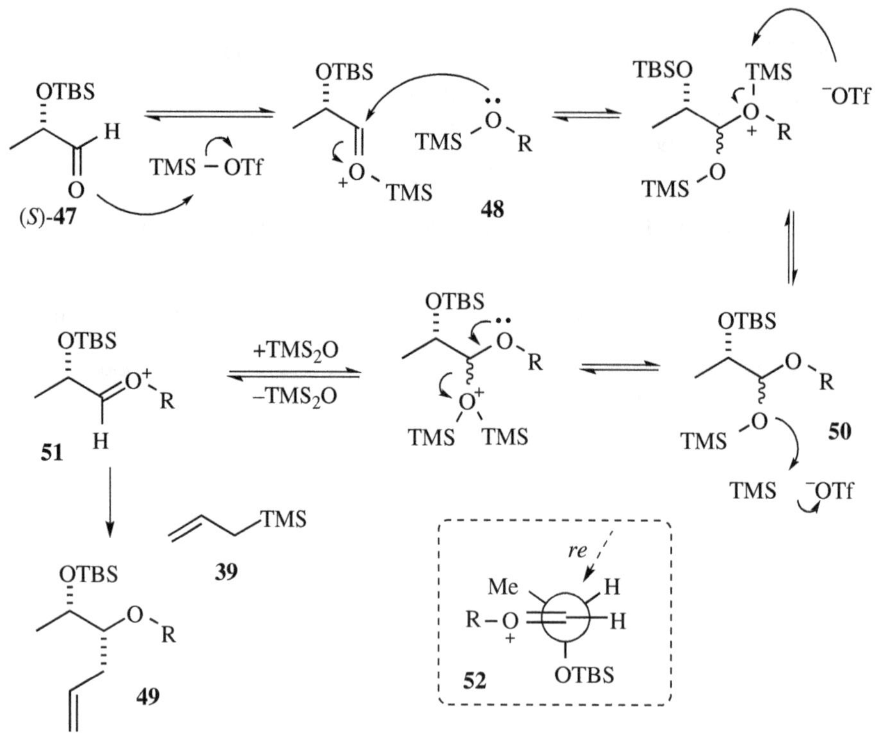

SCHEME 12.15 Proposed mechanism.

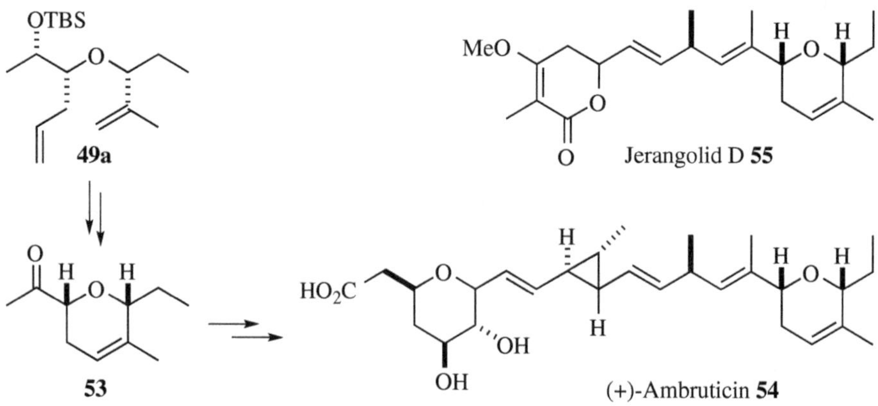

FIGURE 12.2 Antifungal agents: (+)-ambruticin (**54**) [75] and jerangolid D (**55**) [76].

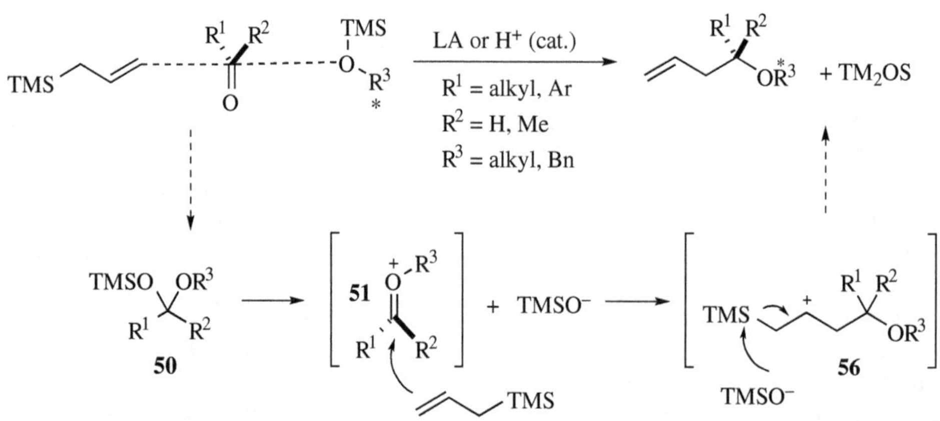

SCHEME 12.16 Domino multicomponent allylation reaction (MCAR) for the stereoselective synthesis of homoallylic ethers and alcohols, a proposed mechanism.

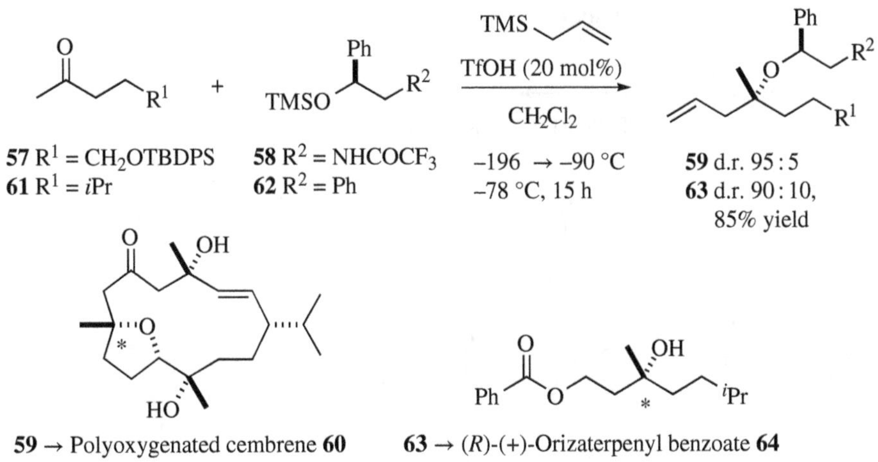

SCHEME 12.17 MCAR for the stereoselective synthesis of natural products.

Tietze has also applied his method for the synthesis of natural products [83]. Recently, he has prepared precursors for the synthesis of polyoxygenated cembrene **60** [84] and (*R*)-(+)-orizaterpenyl benzoate **64** [85] by using the auxiliaries **58** and **62**, respectively, under trifluoromethane-sulfonic acid catalysis (Scheme 12.17).

Mohan has also worked in the multicomponent synthesis of nonchiral homoallyl ethers starting from aldehydes, alkoxytrimethylsilanes, and trimethylallyl silanes using bismuth triflate ($Bi(OTf)_3 \cdot xH_2O$) [86] and iron(III) tosylate ($Fe(OTs)_3 \cdot 6H_2O$) [87] as inexpensive and relatively noncorrosive catalysts.

Other eco-friendly catalysts that has been employed recently in this three-component Hosomi–Sakurai reaction are iodine [88], a cesium salt nanocrystal of tungstosilicic acid ($Cs_{1.8}H_{2.2}SiW_{12}O_{40}$) [89], and HPA/Al-KIT-6 (HPA = $H_3PW_{12}O_{40}$ and $H_3PMo_{12}O_{40}$) [90].

Sakurai reactions are generally mediated by Lewis acid; however, Brønsted acids can be also used as catalysts. In 2008, List and coworkers reported a three-component synthesis of a broad scope of homoallylic ethers in moderate to high yields, starting from aldehydes, silyl ethers, and allylsilanes catalyzed by 2,4-dinitrobenzenesulfonic acid (DNBA, **66**) (Scheme 12.18, *path I*) [91]. Less common is the use of alcohols **65**, instead of alkoxysilane, in these MCRs. Those systems are generally less reactive, which can be due to the lower acetalization of aldehydes with the alcohol, than with the more reactive silyl ether. In addition, the amount of allylsilane can be lower because of its possible reaction with the alcohol (ROH) to form *in situ* ROTMS. However, in this article, the authors developed the first three-component variants in which aldehydes react with allylsilanes and either silyl ethers or their corresponding alcohols, obtaining always poor results in the last case (Scheme 12.18, *path II*). In the proposed mechanism, both paths drive to oxocarbenium ion intermediate **51**, which is attacked by allylsilane **39**, furnishing the desired homoallylic ether in good yield in most cases and TMSX (which could initiate further catalytic cycles).

Finally, Chen's work on the direct allylation of carbonyl compounds using benzyl alcohol in the multicomponent Sakurai reaction catalyzed by selective and green solid acids, such as silicomolybdic acid (SMA–SiO_2) [92] or perchloric acid ($HClO_4$–SiO_2) [93], both supported on silica gel, should be mentioned. In some cases, the use of preformed acetals as substrates provided better results. These methods allowed the synthesis of a broad number of homoallylic ethers in moderate to high yields in a short reaction time. Significantly, catalyst loading of $HClO_4$–SiO_2 is only 2 mol%.

12.2.3 Synthesis of Homoallylic Amines: Aza-Sakurai

Homoallylic amines are useful intermediates in the synthesis of biologically active compounds such as statin, conine, lobeline, sedamine, SKF-100330-A, epothilone, β-lactams, β-amino acids, and many natural products. The synthesis of those interesting precursors is also possible starting from an imine instead of a carbonyl compound, through this kind of reactivity, known in this case as aza-Sakurai reaction [94, 95]. Its multicomponent version involves the *in situ* formation of that imine, and few recent examples are shown here.

In 2007, Tian and coworkers developed an efficient four-component synthesis of Cbz-protected homoallylic amines catalyzed by an environmentally friendly catalyst, iron(II) sulfate heptahydrate ($FeSO_4 \cdot 7H_2O$) (Scheme 12.19) [96]. The authors extended this method to a variety of substrates, obtaining the corresponding products in good yields (43–87%) using only 5 mol% of catalyst, at room temperature.

They envisioned this synthesis as starting from alkyl chloroformate (CbzCl) and 1,1,1,3,3,3-hexamethyldisilazane (HMDS) that provided easily *N*-silylcarbamate

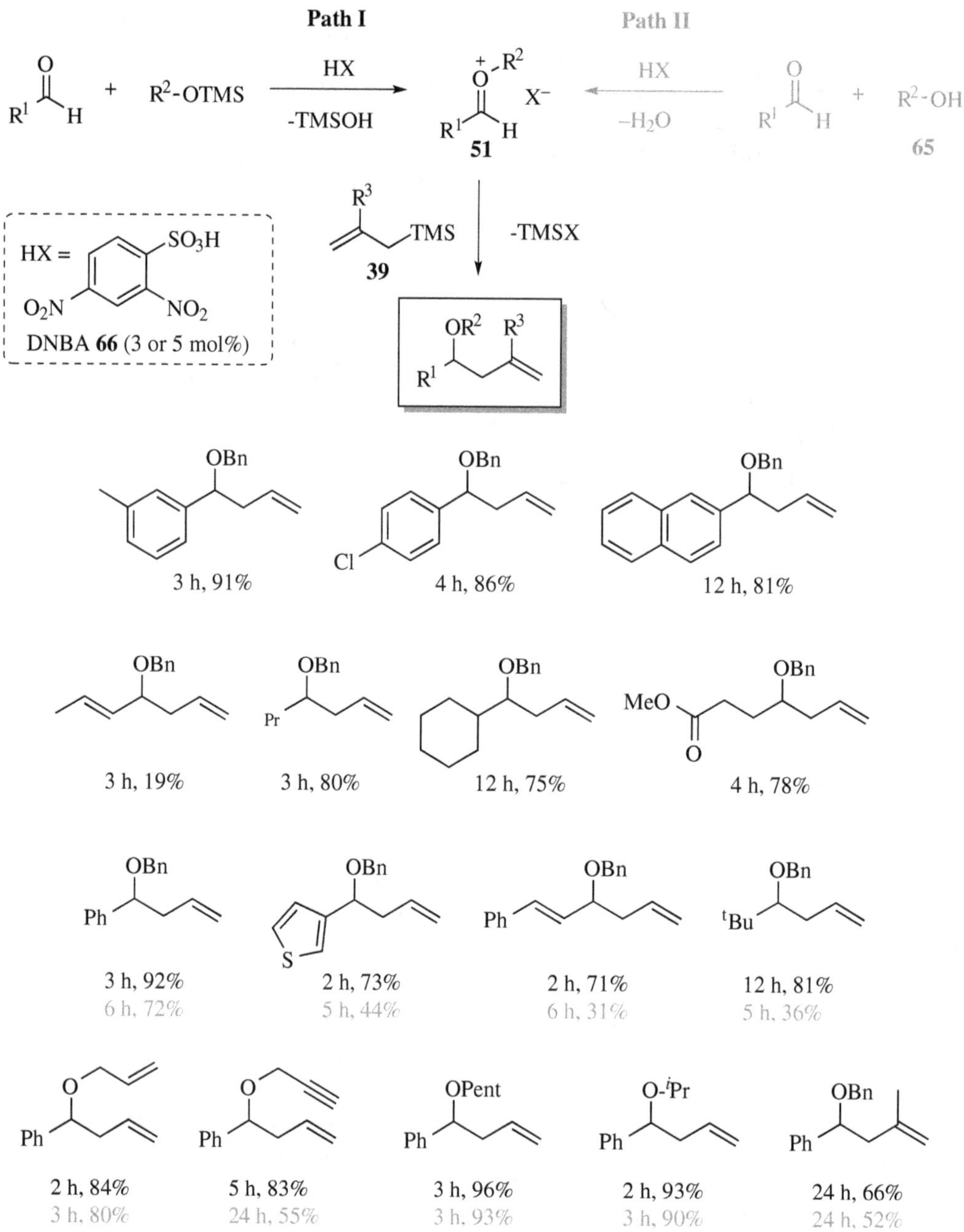

SCHEME 12.18 Brønsted acid-catalyzed three-component Sakurai reaction using either silyl ether (*path I*) or the corresponding alcohol (*path II*).

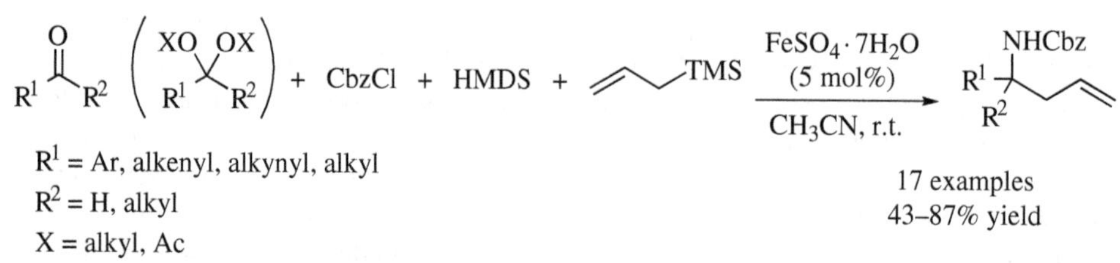

SCHEME 12.19 Four-component aza-Sakurai reaction catalyzed by $FeSO_4{\cdot}7H_2O$.

CbzNH(TMS) that reacts with the carbonyl compound (or acetals/ketals), generating an imine **67**, that finally reacts with allyltrimethylsilane **39**, leading to the final Cbz-protected homoallylic amine after hydrolysis (Scheme 12.20).

Since from the beginning two electrophiles (carbonyl compound and alkyl chloroformate) and two nucleophiles (HMDS and allyltrimethylsilane) are present in the media, the challenge of this reaction is to drive it for the previously described pathway. Side reactions such as allylation of the carbonyl compound and reaction between electrophilic alkyl chloroformate and nucleophilic allyltrimethylsilane were efficiently avoided using the proper Lewis acid as catalyst, being in this case the inexpensive $FeSO_4{\cdot}7H_2O$.

A year later, Yus and coworkers employed another iron-based Lewis acid as catalyst in a similar four-component

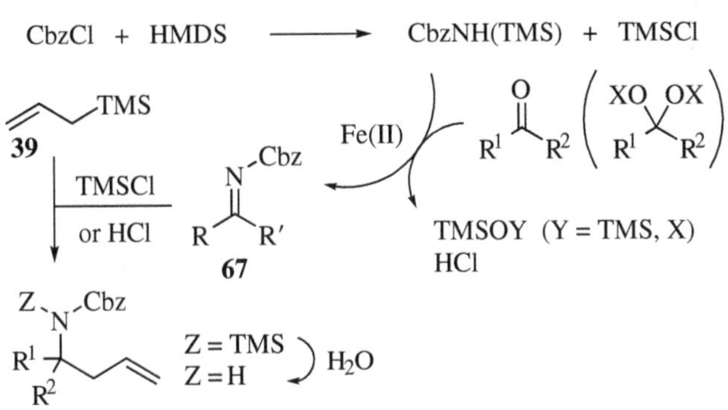

SCHEME 12.20 Mechanistic proposal for the four-component aza-Sakurai reaction.

aza-Sakurai reaction [97, 98]. They showed for the first time the use of commercial nanopowder magnetite (Fe_3O_4) as a heterogeneous Lewis acid catalyst in an organic reaction with a very broad scope, including aromatic and aliphatic acyl chlorides, aldehydes, and ketones, as well as allyl silane derivatives (Scheme 12.21). The catalyst acts selectively, activating first the higher electrophilic aldehydes versus ketone functionalities. They also studied the easy (magnet) recyclability of this material, which allows the possibility to repeat the reaction efficiently up to 15 times, thus making it suitable for continuous industrial processes.

Another elegant example of a multicomponent aza-Sakurai reaction was reported by Jarvo in 2009 [99]. In this case, $Pd(OAc)_2$ was used as catalyst both for the generation *in situ* of the allylsilane from allyl trifluoroacetate and HMDS and also for the allylation of the *in situ* formed imine from the corresponding aldehydes and anisidine (H_2NPMP) (Scheme 12.22). This was concluded after a study base on competition experiments. This procedure allows the synthesis of a variety of homoallylic amines in good to excellent yields (67–96%), starting from different substrates including electron-rich aromatic and aliphatic aldehydes.

Liu and coworkers reported in the same year a three-component allylation of *in situ* generated imines [100, 101]. In this case, 5 mol% of copper(II) triflate ($Cu(OTf)_2$) was used as a catalyst furnishing the corresponding Cbz-protected homoallylic amines in good to excellent yields (75–95%) from an aldehyde, a carbamate, and an allyltrimethyl silane.

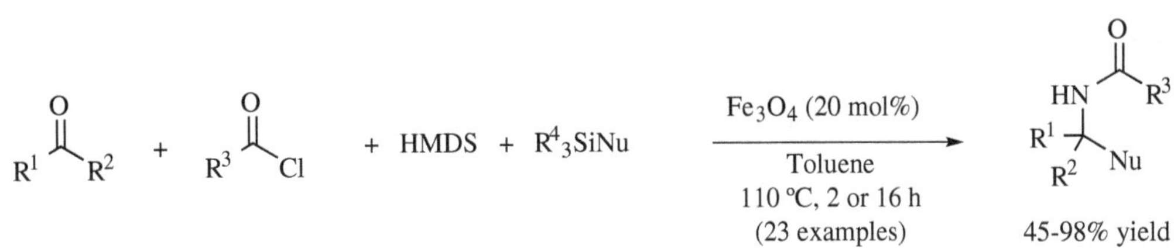

R^1 = Ph, 4-ClC_6H_4, 4-$MeOC_6H_4$, PhCH=CH,*i*Pr, *c*-Hex, *n*Bu
R^2 = H (2 h of reaction), Me (16 h of reaction)
R^3 = OCH_2Ph, $OCH_2CH{=}CH_2$, OCH_2CCH, OPh, OEt, Ph, *t*Bu, (–)-menthyloxy
R^4 = Me, Et
Nu = H, $CH_2CH{=}CH_2$

SCHEME 12.21 Four-component aza-Sakurai reaction catalyzed by Fe_3O_4.

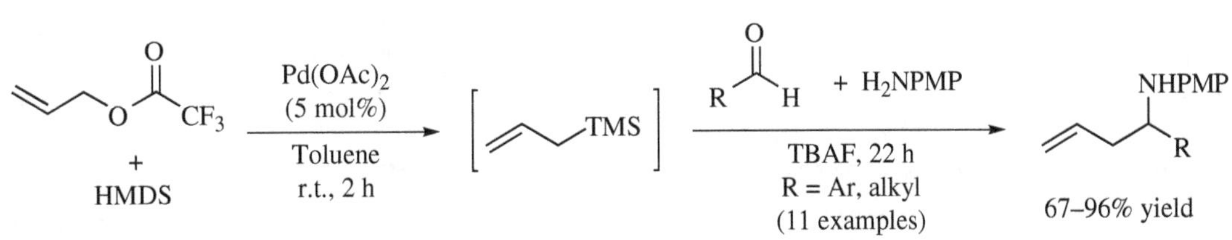

SCHEME 12.22 Three-component one-pot, two-step transformation aza-Sakurai reaction catalyzed by $Pd(OAc)_2$.

12.3 GEWALD MCRs

12.3.1 Introduction

Gewald reaction was first reported in 1961 [102], allowing the synthesis of 2-aminothiophenes **70** from an α-mercaptoaldehydes or ketone **68** (prepared by the reaction of sulfur and carbonyl compound) and an activated nitrile **69** in the presence of base (triethylamine or pyridine) (Scheme 12.23, Eq. 1). After this *first version* [102, 103], three improved versions were published (*second* [104], *third* [105], and *fourth* [106] *versions*) (Eqs. 2, 3 and 4, respectively), trying to solve the problem of the instability of the used carbonylic substrates.

Thus, the *second version* of Gewald reaction is a multicomponent one-pot procedure that involves an α-methylene carbonyl compound **71**, an α-activated acetonitrile **69** (malonodinitrile, cyanoacetic ester, or primary cyanoacetamide), and sulfur (Scheme 12.23, Eq. 2). It is believed that the last one reacts with the α,β-unsaturated nitrile *in situ* generated from the other two reagents. In the case of the *third version*, the aforementioned intermediate **72** is previously synthesized and isolated and later mixed with sulfur and base. This two-pot method is efficient for alkyl aryl ketones and some cycloalkyl ketones, substrates that are not active enough for the *first and second versions* (Scheme 12.23, Eq. 3). In the last basic version of Gewald's protocols, substituted 1,4-dithiane-2,5-diols **73** is employed instead of the α-sulfanylcarbonyl compound **68**. The use of this dimer allows the synthesis of mono- or disubstituted 2-aminothiophenes **70** with R^2=H in acceptable yields (Scheme 12.23, Eq. 4).

In general, the initial formation of the α,β-unsaturated nitrile (**72**) *via* Knoevenagel condensation of an activated nitrile (**69**) with a carbonyl compound (**71**) mediated by a base (:B) is accepted as a proposed mechanism for the three-component Gewald reaction (G-3CR) (Scheme 12.24). Then, after deprotonation of that intermediate (**72**), it reacts with elemental sulfur (S_8) following an S_NX mechanism, generating sulfated compound (**74**), which cyclizes to **75** *via* an intramolecular nucleophilic attack of the sulfur anion to

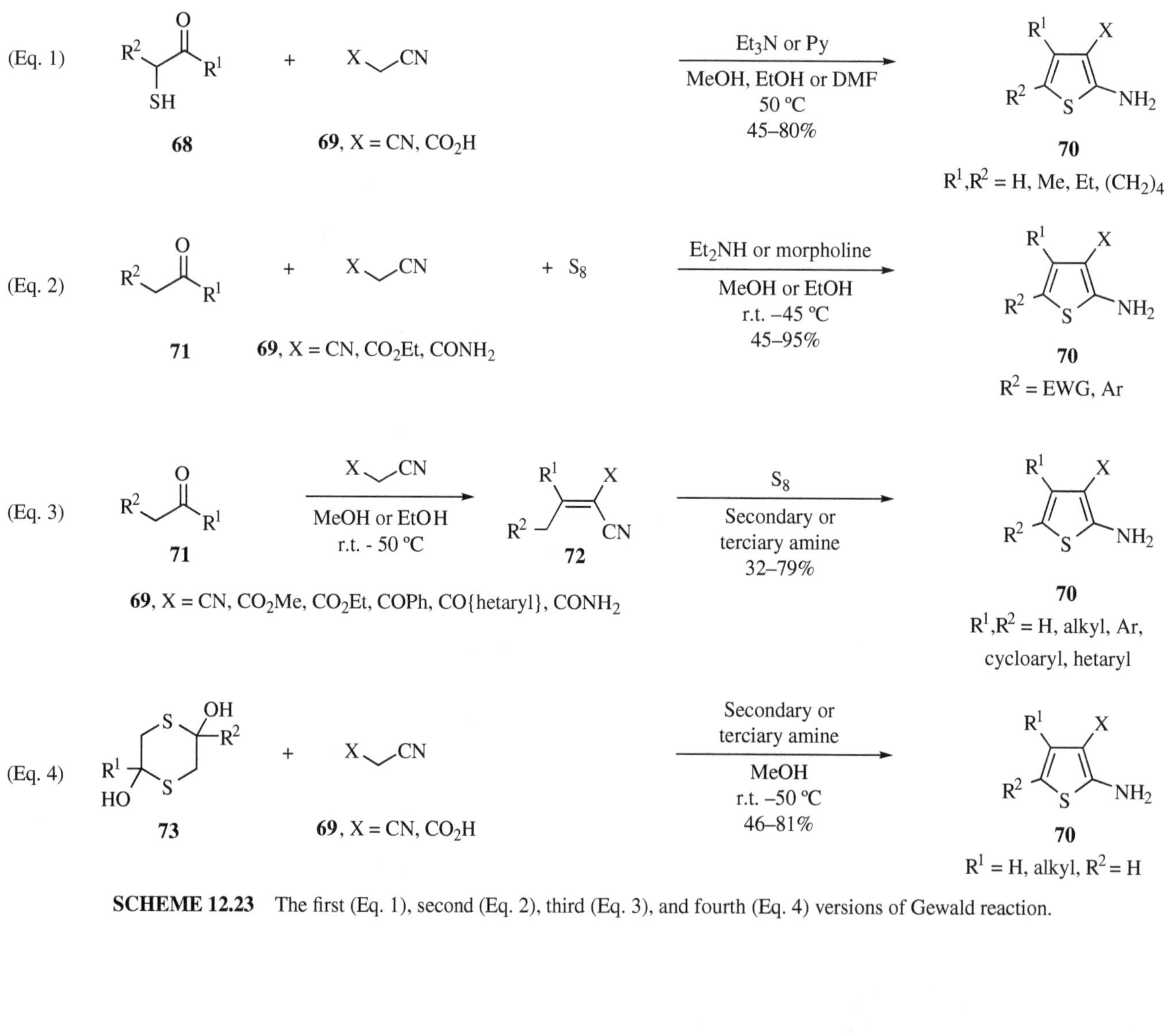

SCHEME 12.23 The first (Eq. 1), second (Eq. 2), third (Eq. 3), and fourth (Eq. 4) versions of Gewald reaction.

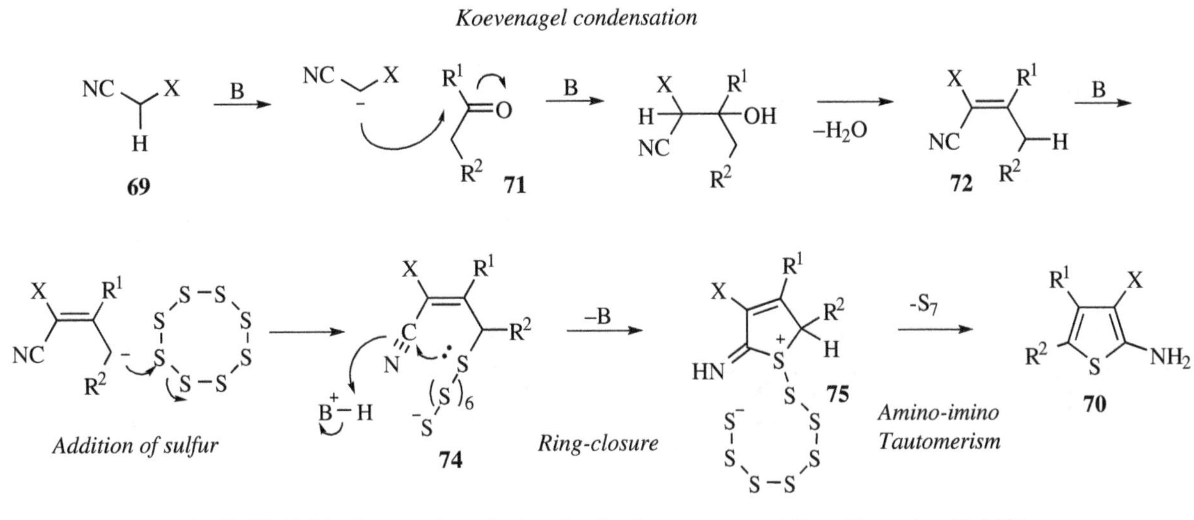

SCHEME 12.24 Proposed mechanism for the three-component Gewald reaction (G-3CR).

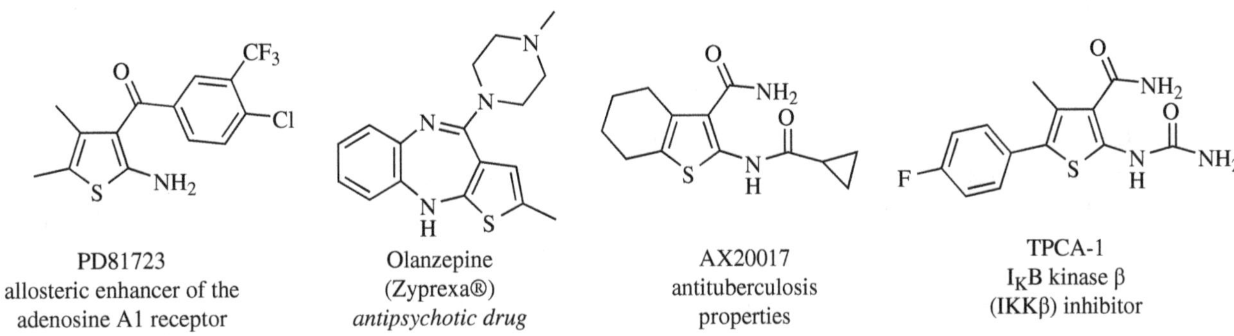

FIGURE 12.3 Biologically active compounds containing 2-aminothiophenes. PD81723 [110], olanzapine [111], AX20017 [112], and TPCA-1 [113].

the cyano group. Elemental sulfur has to be activated to react with compound **72**, either by a base [107] or through an electrochemical activation [108]. The final rearrangement leads to the desired product (**70**) [109].

12.3.2 Easy Protocol for Synthesizing 2-Aminothiophene Derivatives

Substituted thiophene derivatives are important heterocycles present in numerous biologically active compounds (Fig. 12.3) [114]. Aminothiophenes have demonstrated to be of great importance in combinatorial and medicinal chemistry, finding applications, among others [115], in drug discovery [116], agriculture [117], and dyes [118]. Its chemistry has been broadly summarized [119]; in particular, the synthesis of those compounds *via* Gewald protocols has been reviewed [120] first in 1999 by Sabnis [121], later by Végh [122], and very recently by Dömling [123]. Therefore, herein, only few examples of multicomponent Gewald reaction reported in the last years are shown.

In general, G-3CR is considered a MCR with low dimensionality, that is, there are not many possible variations of the different components, with carbonyl compound being the one with more alternatives [124]. Some authors have worked on extending this reaction by varying significantly also the nitrile compound. In this sense, Dömling employed a large number of cyanoacetamides as starting materials (Fig. 12.4), developing an efficient method of synthesizing more than 40 highly substituted 3-amido-2-aminothiophenes, obtained in most cases in good yields with high purity after simple precipitation in water and filtration [125, 126].

In addition to varying the substrates, other authors have worked on the conditions in order to extend the scope of the reaction. In particular, MW [127] and ultrasound [128] irradiations have been applied successfully, decreasing considerably the reactions times. On the other hand, different catalysts have been used substituting the classical secondary or tertiary amines [129], for example, L-proline [130], imidazole [131], cesium carbonate (Cs_2CO_3) [132], amine-functional polysiloxanes (AFPs) [133], Mg/La mixed oxide as a heterogeneous solid base catalyst [134], ZnO (under solvent-free conditions) [135], bovine serum albumin (BSA) [136], and guanidine-based IL [137].

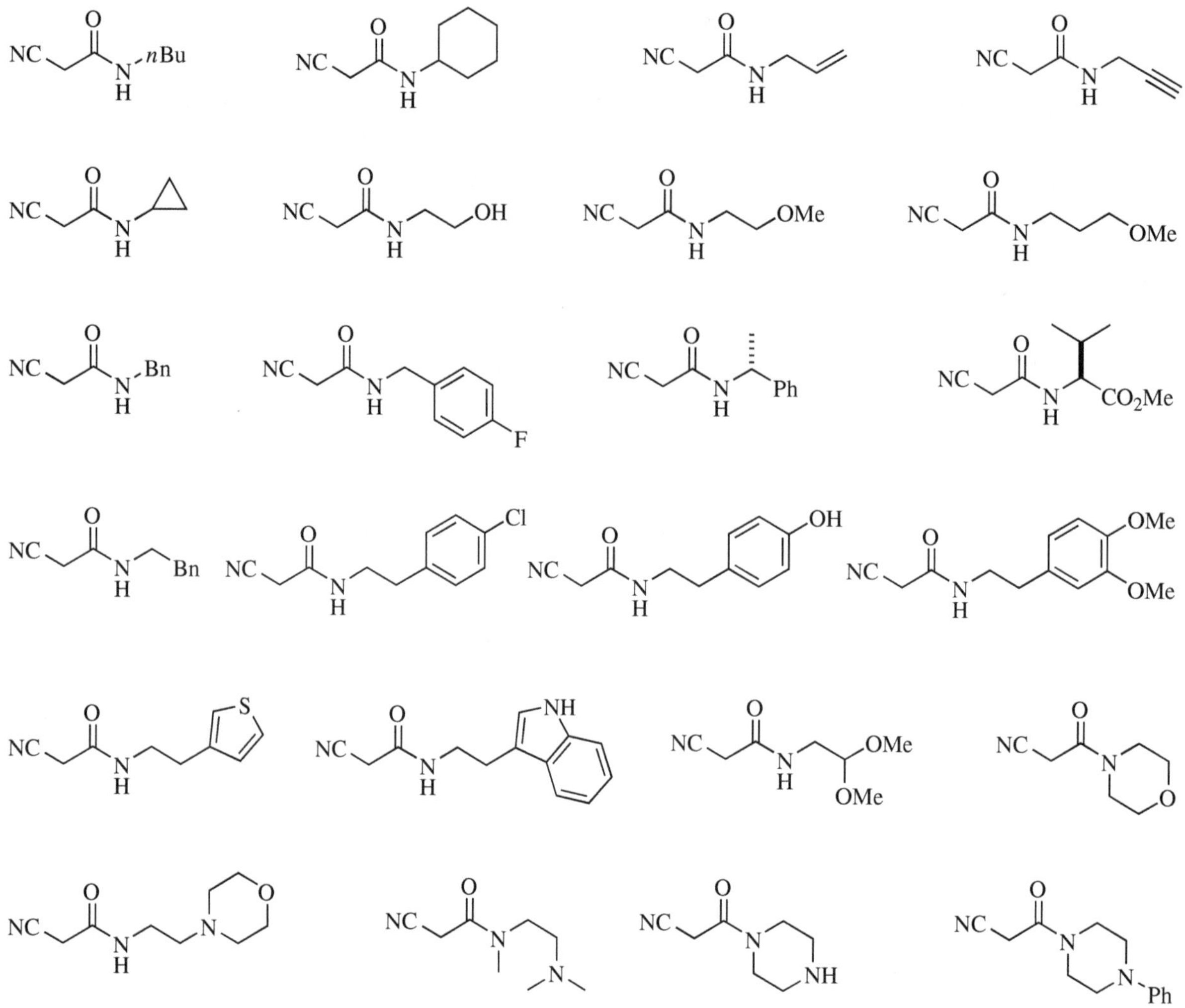

FIGURE 12.4 Cyanoacetamides used as a nitrile component in the G-3CR.

Thus, Xu and Zeng developed in 2010 a method employing 0.1 mol% of L-proline as a catalyst, allowing G-3CR of cyclic ketones in DMF at 60 °C to give the corresponding products in good yields (64–98%) after moderately short times (10–36 h) [130]. The same group published a year later similar reaction for aldehydes and ketones, employing this time imidazole as a catalyst (10 mol%) [131]. The products were achieved in moderate to good yields in shorter reaction times (10–18 h). More recently, another Chinese group reported the use of a polymeric organocatalyst in this reaction [133]. They showed after some optimization work that 5 mol% of the amine-functional polysiloxane (AFP) **76** was active enough in the Gewald reaction of different ketones, using also TsOH (20 mol%) (Scheme 12.25).

The MCR covered herein has found broad applicability in the synthesis of heterocycles containing Gewald moiety (2-amino-thiophene) (Fig. 12.5), with interesting properties [138] and overall antimicrobial activities, such as phthalazinediones [139], thienopyridines [140], thienothiadiazines [141], triazolothienopyrimidines [142], thienopyrimidines [140–142], and thiazolidinones [143].

12.4 KABACHNIK–FIELDS REACTIONS

12.4.1 Introduction

Kabachnik–Fields reaction is a three-component reaction between a carbonyl compound (ketone and aldehydes), an amine, and a dialkyl phosphite [144] to form α-aminoalkylphosphonates **79** (Scheme 12.26). It was first described in 1952 independently by Kabachnik [145] and Fields [146]. This reaction constitutes a general and effective method for the synthesis of α-amino phosphonates, which, in the last five decades, have demonstrated to be a kind of compounds of relevance.

The general mechanism of the reaction is believed to go by the *in situ* formation of an imine intermediate **77**,

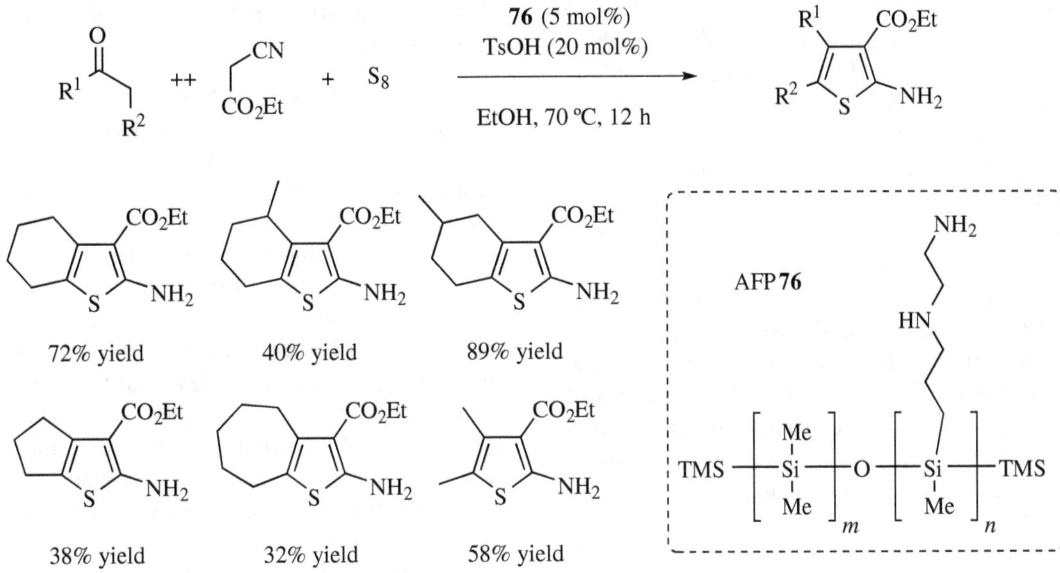

SCHEME 12.25 G-3CR catalyzed by amine-functional polysiloxane **76**.

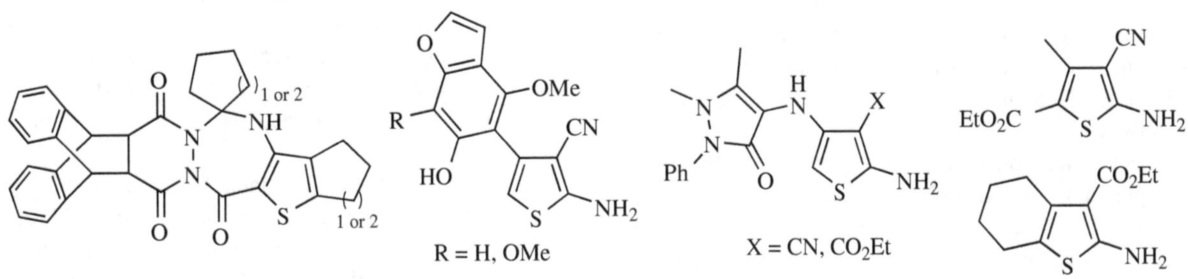

FIGURE 12.5 Thiophenes synthesized by G-3CR as a precursor of biologically active compounds.

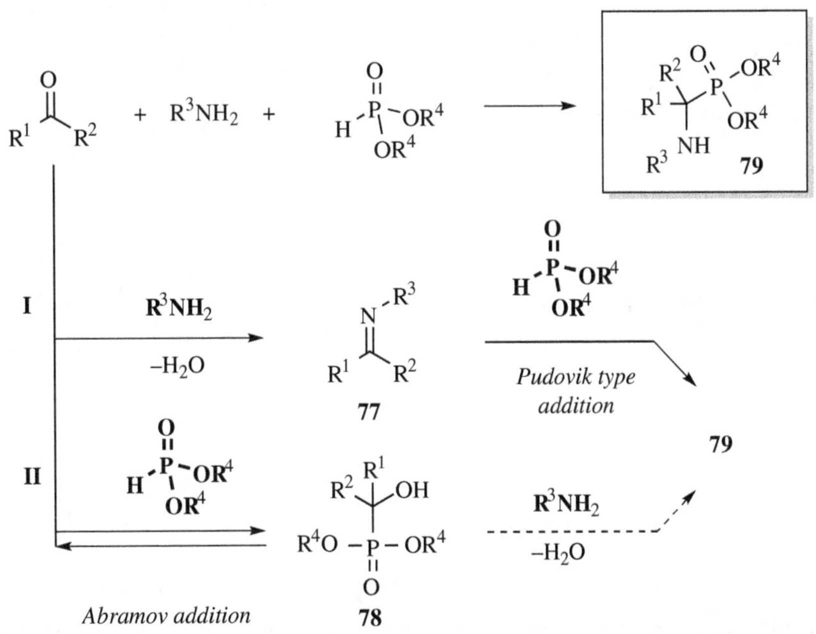

SCHEME 12.26 General scheme and mechanistic pathways for Kabachnik–Fields reaction.

which then reacts with the dialkyl phosphite following a Pudovik-type addition to give the final α-aminoalkylphosphonate **79** (Scheme 12.26, *path I*). However, some authors propose the initial reaction between the carbonyl compound and the dialkyl phosphite (Abramov addition), forming an α-hydroxyphosphonate **78** (Scheme 12.26, *path II*), which then undergoes direct nucleophilic substitution of the hydroxyl group by amine, leading to the corresponding α-amino phosphonates **79** [147].

There is competition between both nucleophiles, dialkyl phosphite (soft) and amine (hard), for the electrophilic carbonyl component; in this sense, the reaction with the P-component will be favored as softer than the C-component [148]. Thus, in general, the mechanism of the reaction depends on starting materials, as well as reaction conditions, but it seems that *path I* (imine route) is more general than *path II* [149]. For instance, a recent mechanistic study of this reaction based on its monitoring by IR spectroscopy and carried out by Keglevich et al. showed imine pathway as the most probable of both of them, in the case of the reaction between benzaldehyde, cyclohexylamine, and dialkyl phosphites [150]. The authors observed, in addition, the formation in low proportion (13%) of alternative α-hydroxyphosphonate, but this route was considered as a "dead-end" one. Theoretical calculations supported all these experimental observations.

This reaction has found numerous synthetic applications. An example of its use in polymer chemistry is shown in Scheme 12.27 [151]. The Kabachnik–Fields reaction is involved, together with a reversible addition fragmentation chain transfer (RAFT) polymerization in a one-pot system that allows the efficient preparation of poly(aminophosphonate)s under controlled radical polymerization (CRP) [152].

Since Kabachnik–Fields reaction and its synthetic applications have been intensively reviewed [147, 149, 153], herein, only some important examples developed in the last years will be shown.

12.4.2 Straightforward Synthesis of α-Amino Phosphonates

α-Aminophosphonic acids may be considered as phosphorus analogues of α-amino acids, as well as their carboxylic counterparts, and exhibit diverse biological activities [154]. Among other organophosphorus compounds, α-amino phosphonates are of special relevance, not only because they are synthetic analogous of α-aminophosphonic acids but also because of their biological interest. For example, they are potent inhibitors of dialkylglycine decarboxylase [155], leucine aminopeptidase [156], serine proteases [157], and tyrosine phosphatases [158]. Certain derivatives are interesting from a pharmacological point of view, thanks to their antimicrobial activities [159] and membrane transport properties (for the design of continuous drug release devices) [160]. Furthermore, they also may find applications in medicinal chemistry [161].

Therefore, α-aminophosphonic acids and their derivatives represent a synthetic goal in modern chemistry [162]. Among other approaches, Kabachnik–Fields reaction represents a straightforward method for the synthesis of α-amino phosphonates. There are a great number of reported racemic versions of Kabachnik–Fields reaction, and some examples will be only mentioned in the following, while scarce enantioselective cases will be treated in detail.

This one-pot three-component reaction can be promoted by acid or base catalysts, external magnetic field [163], MW [164], or ultrasonic [165] irradiation or by heating [166]. Numerous acid catalysts have been applied to this reaction, in particular Lewis acids (e.g., $InCl_3$ [167], $BiCl_3$ [168], $FeCl_3$ [169], $YbCl_3$ [170], $SiO_2/ZnCl_2$ [171], ($CeCl_3{\cdot}7H_2O$) [172], $In(OTf)_3$ [173], $Bi(OTf)_3$ [174], $Al(OTf)_3$ [175], $Sn(OTf)_2$ [176], $Mg(ClO_4)_2$ [177], $Yb(PFO)_3$ [178], $ZrOCl_2{\cdot}5H_2O$ [179], SmI_2 [180], CdI_2 [181], $TaCl_5$–SiO_2 [182], and $NbCl_5$ [183]) and solid acids (e.g., montmorillonite KSF [184], sulfamic acid [185], amberlite IR 120

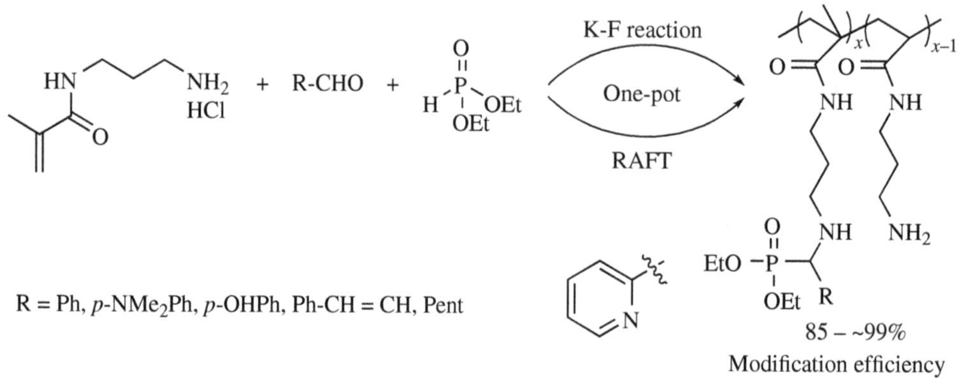

SCHEME 12.27 One-pot synthesis of poly(aminophosphonate)s through K-F–RAFT system.

[186], amberlite IRC-748 [187], silica sulfuric acid [188], and xanthan sulfuric acid [189]). Base catalysts have been also used in this reaction, such as $CaCl_2$ [190], PPh_3 [191], tetramethylguanidine (TMG) [192], and 1,4-dimethylpiperazine [193]. Additionally, other different catalysts have been employed to promote this reaction, for example, ZnO nanoparticles [194], Cu nanoparticles [195], CeO_2 nanoparticles [196], nano-$BF_3{\cdot}SiO_2$ [197], dehydroascorbic acid capped magnetite nanoparticles (DHAA-Fe_3O_4) [198], TiO_2 [199], tetramethyltetra-3,4-pyridinoporphyrazinato copper(II) methyl sulfate [Cu(3,4-tmtppa)$(MeSO_4)_4$] [200], cupric acetate monohydrate ($Cu(OAc)_2$ H_2O) [201], metal tetra-*tert*-butylphthalocyanine complexes (^{t}PcM) [202], gold–bipy complex [$AubpyCl_2$]Cl [203], $Cd(ClO_4)_2{\cdot}xH_2O$ [204], β-cyclodextrin (β-CD) [205], NBS [206], mesoporous aluminosilicate nanocage [207], H-beta zeolite catalyst [208], bifunctional acid–base heterogeneous nanocatalyst IRMOF-3 ($Zn_4O(H_2N\text{-}TA)_3$) [209], choline chloride·$2ZnCl_2$ IL [210], and the HPA PMA ($H_3PMo_{12}O_{40}$) [211].

It worth to mention that despite the importance of the Kabachnik–Fields reaction, stereoselective versions for the synthesis of enantioenriched α-aminophosphonates are scarce [212, 213], and only few enantioselective examples have been published to date (for reviews on enantioselective catalytic direct hydrophosphonylations of imines, see Refs. [162a–c]). Organocatalytic examples use well-known chiral binol-derived phosphoric acid organocatalysts (Fig. 12.6, **80** and **81**) [214], and regarding metal catalysis, chiral scandium(III)-*N,N'*-dioxide and zinc(II)-bis(imidazoline) complexes have been employed (Fig. 12.6, **82** and **83**) [215].

Thus, in 2008, List and coworkers published the first enantioselective version of Kabachnik–Fields reaction [214a]. They found that chiral binol-derived phosphoric acids were efficient in inducing chirality under dynamic kinetic resolution conditions when bulky α-branched aldehydes (isopropyl, cyclopentyl and cyclohexyl) were used as substrate in their reaction with *p*-anisidine and di(3-pentyl) phosphite (Scheme 12.28). In particular, 3,3'-bis(4-anthracenyl-2,6-diisopropylphenyl)-1,1'-8-binaphtyl-2, 2'-diyl hydrogenphosphate **80** allowed the synthesis of the corresponding β,β-disubstituted-α-aminophosphonates **84** in high yields with very good stereocontrol (dr up to 28:1 and er up to 97:3).

In 2010, another example of the use of phosphoric acid catalyst was reported [214]. In this case, 10 mol% of 3,3'-bis(3,5-bis(trifluoromethyl)phenyl)-1,1'-8-binaphtyl-2,2'-diyl hydrogenphosphate **81**, in toluene at 40 °C, allowed the synthesis of the corresponding α-amino phosphates with modest to good enantioselectivities (31–87% ee) and good yields (61–91%).

Moreover, Feng and coworkers reported in 2009 an enantioselective variant of this three-component reaction using metal catalysis [215]. After screening several ligands and lanthanides (Ln), *N,N'*-dioxide-Sc(III) complex **82** was found to be the most efficient in the reaction of aromatic aldehydes, 2-aminophenol, and diphenyl phosphite (Scheme 12.29). The corresponding α-amino

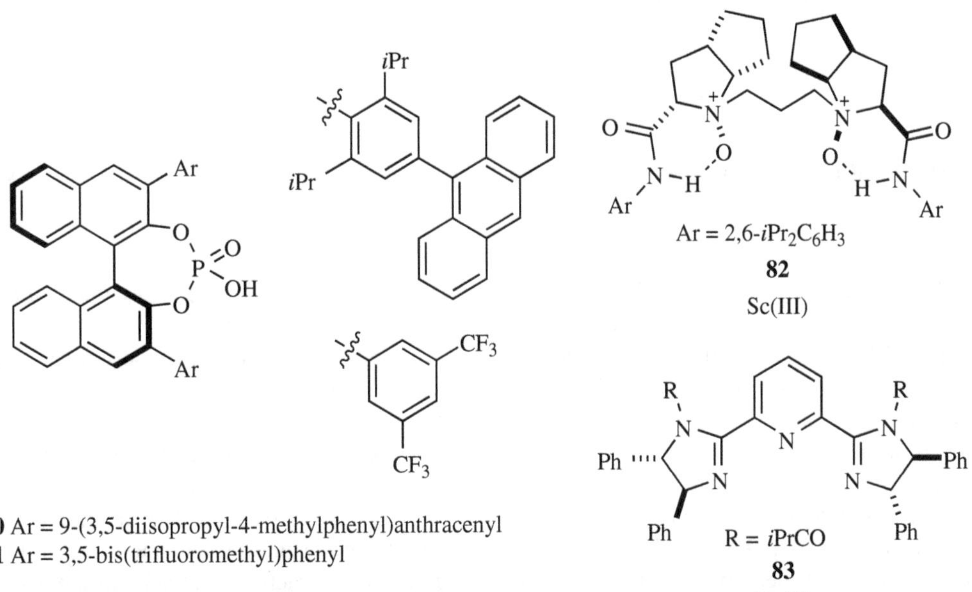

80 Ar = 9-(3,5-diisopropyl-4-methylphenyl)anthracenyl
81 Ar = 3,5-bis(trifluoromethyl)phenyl

82 Sc(III)

83 Zn(II)

FIGURE 12.6 Catalysts **80–83** used in enantioselective versions of Kabachnik–Fields reaction.

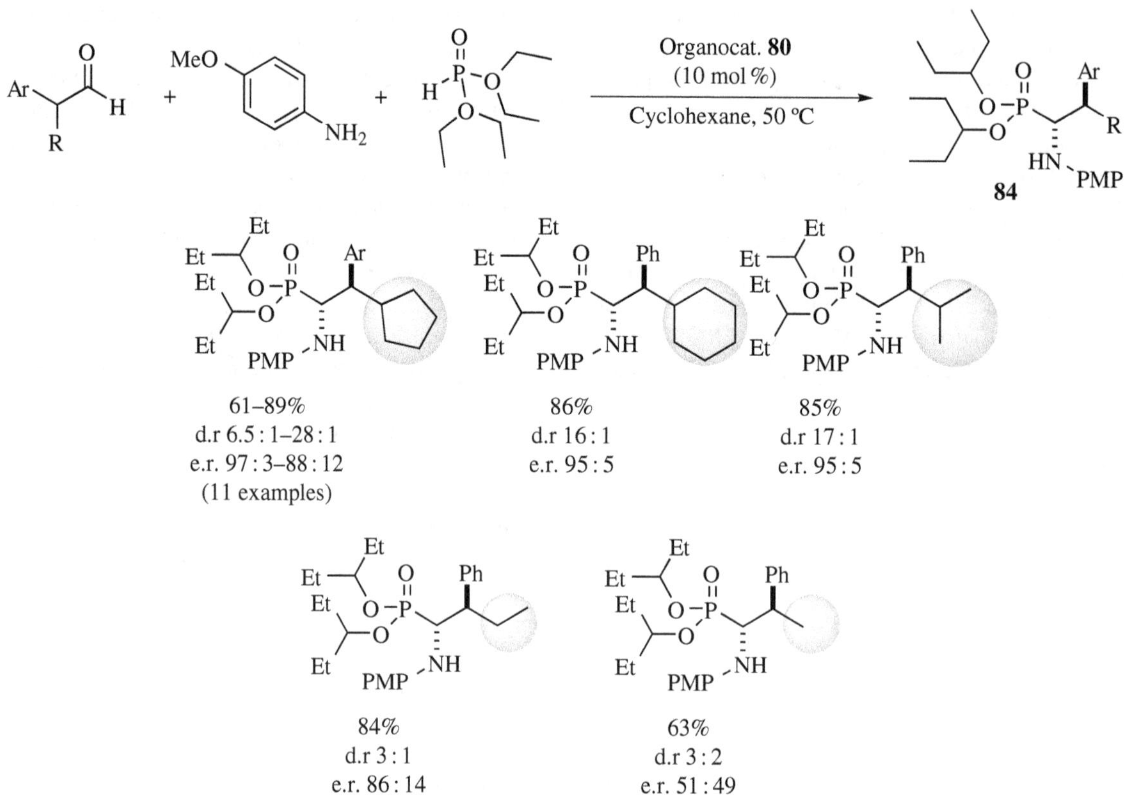

SCHEME 12.28 Enantioselective Kabachnik–Fields reaction catalyzed by chiral binol-derived phosphoric acid **80**.

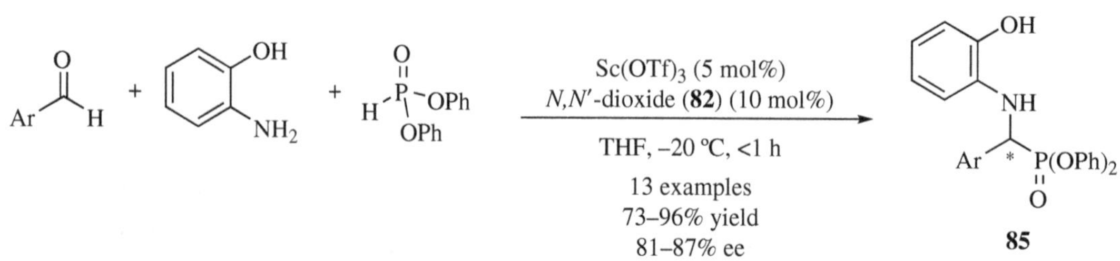

SCHEME 12.29 Enantioselective Kabachnik–Fields reaction catalyzed by chiral Sc(III) complex **82**.

phosphonates **85** were obtained in high yield (73–96%) and with good enantioselectivity (81–87%) after only 1 h of reaction. It should be mentioned that in this one-pot method, aldehyde and amine are allowed to stir for 1 h prior to the addition of the phosphite in order to generate *in situ* the required imine [216].

It was not until 2011 when an enantioselective metal-catalyzed version of a real direct three-component one-pot Kabachnik–Fields reaction was published. Nakamura and Shibata developed an efficient method that allows the synthesis of chiral α-amino phosphonates **86** in good yield (up to 99%) with good enantioselectivity (up to 93% ee) using a zinc(II) complex of 1,3-bis(imidazolin-2-ly)pyridine **83** (Scheme 12.30) [215].

They carried out some experiments in order to clarify the mechanism of the reaction that allowed them to propose the transition state depicted in the Figure 12.7, where the *in situ* generated imine coordinates to the catalyst in the equatorial position, avoiding the steric repulsion of the ligand. In addition, after coordination of the oxygen atom of the P-component to the metallic center, the active phosphite is formed, which then attacks to the azomethine carbon of the imine by its *si* face, thus providing the experimentally observed (*S*)-enantiomer.

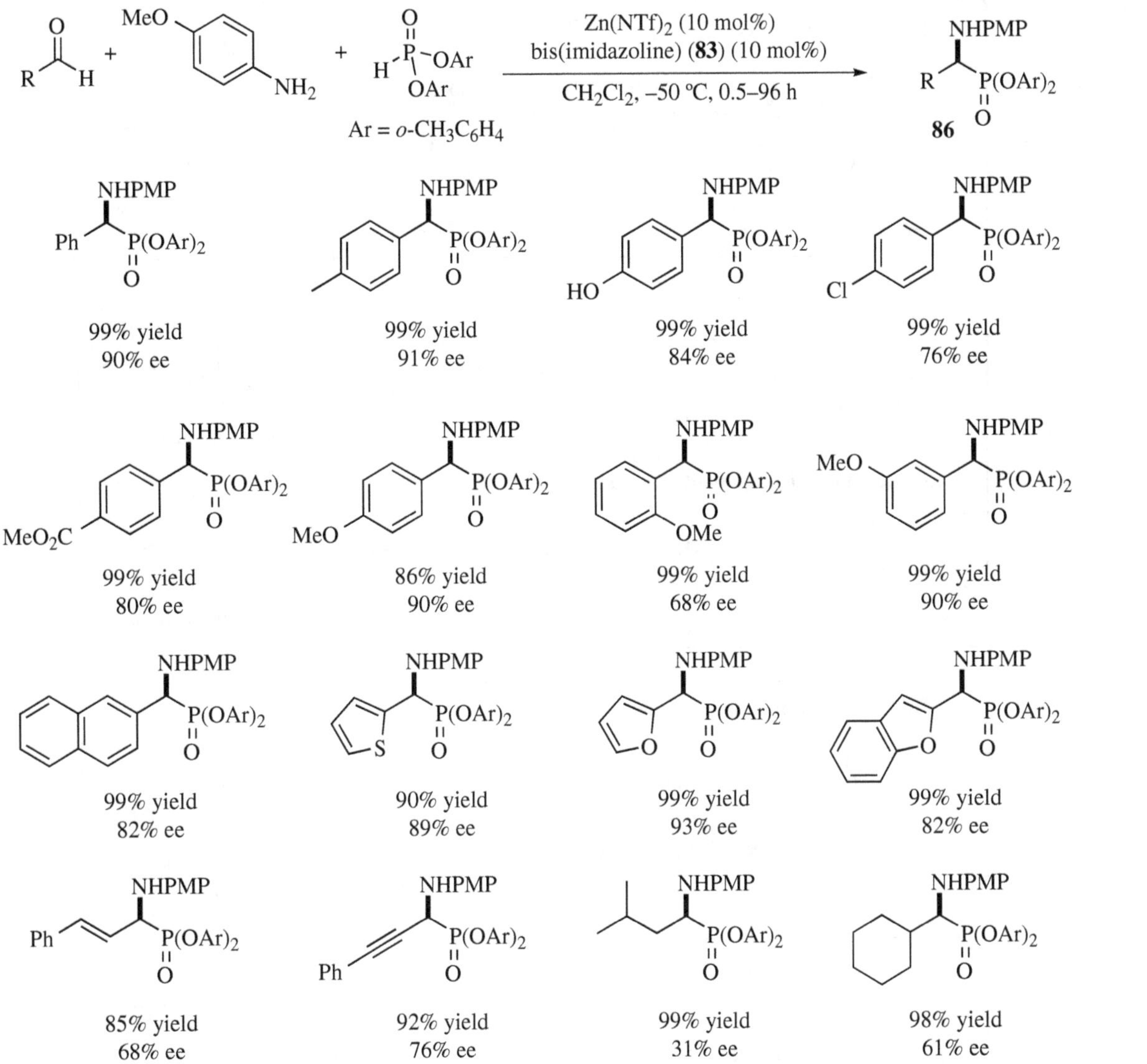

SCHEME 12.30 Enantioselective Kabachnik–Fields reaction catalyzed by chiral bis(imidazoline)-zinc(II) complex **83**.

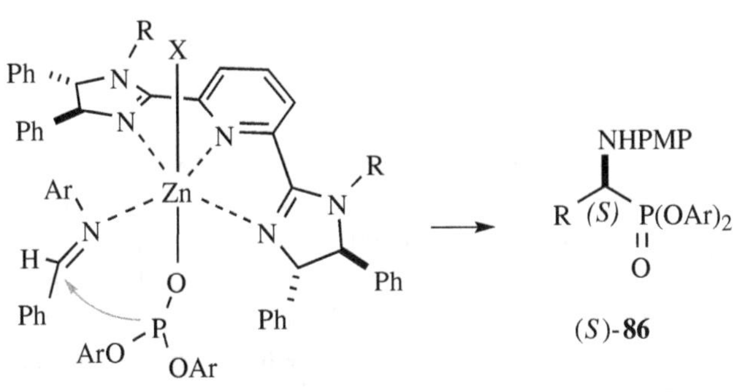

FIGURE 12.7 Proposed transition state for Zn(II)-bis(imidazoline) complex **83** catalyzed Kabachnik–Fields reaction.

12.5 CONCLUSION

In this chapter, four different essential MCRs have been compiled, in particular Radziszewski, (aza-)Sakurai, Gewald, and Kabachnik–Fields reactions (Scheme 12.31). These reactions give access to five types of products: substituted imidazoles, homoallylic ethers(amines), 2-aminothiophene derivatives, and α-aminophosphonates. The interest in these moieties is broadly recognized in the scientific community, since, among other properties, they are found in the structure of numerous biologically active natural or synthesized molecules.

Summarizing in a very general way, all these synthetic methods involve at least a carbonyl compound, which

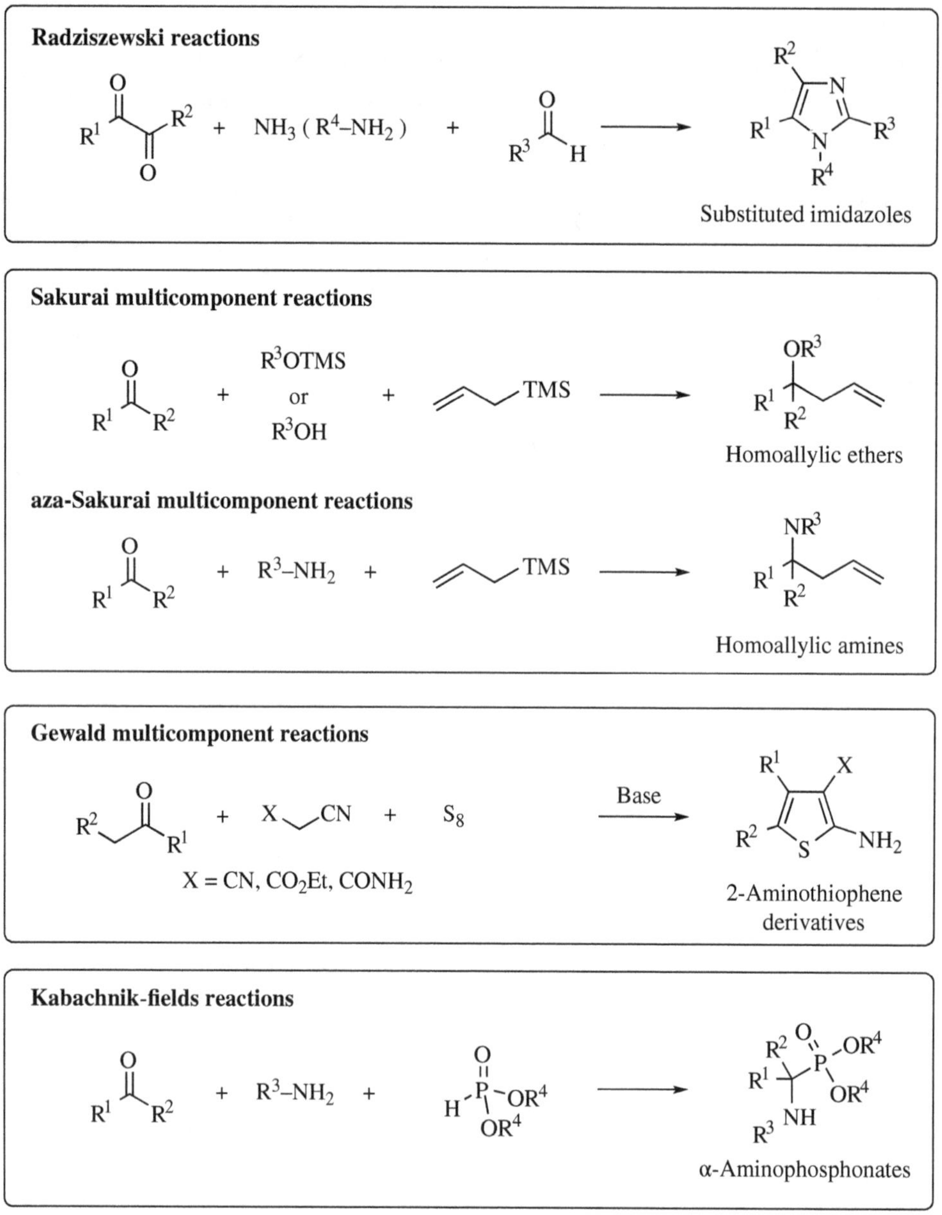

SCHEME 12.31 General scheme for the reactions compiled in this chapter: Radziszewski, Sakurai, Gewald, and Kabachnik–Fields reactions.

reacts normally with an amine to give an intermediate, which interacts with a third component: an aldehyde in the Radziszewski reaction, an allyltrimethylsilane in the aza-Sakurai reaction, and a dialkyl phosphite in the Kabachnik–Fields reaction. However, in the Sakurai and Gewald reactions, the mentioned carbonyl compound reacts with a silyl ether (or an alcohol) and an α-activated acetonitrile, respectively, forming usually an intermediate that interacts then with allyltrimethylsilane, in the case of Sakurai reaction, and with sulfur, in the case of the Gewald reaction.

There are numerous examples in the literature where tri- and tetrasubstituted imidazoles are achieved efficiently by means of suitable catalyst for the Radziszewski reaction. In order to increase the efficiency and greenness, different methods have been applied, such as MW irradiation, sonication, microreactor technology, etc. Some of these methods, as well as a great number of different catalysts, have been used in the other protocol which leads to a heterocycle (2-aminothiophenes): the Gewald reaction. With respect to the other methodologies covered in this chapter, many works have come out since they were reported for the first time. In general, Lewis acids have been broadly used to catalyze them, but in addition, other methods of activation are being more and more employed in the last years. In particular, Kabachnik–Fields reaction can be promoted by acid or base catalysts, external magnetic field, MW, or ultrasonic irradiation and by heating. However, the scarce examples of their stereoselective versions, being those described here more in detail, are remarkable.

As a conclusion, much research has been done in these particular three (four)-multicomponent methods, which allows the efficient synthesis of very interesting products. But there is still room to improve them, especially in the case of the asymmetric versions of the two of them.

REFERENCES

1 Z. Wang, *518. Radziszewski*, in *Comprehensive organic name reactions and reagents*. John Wiley & Sons, Inc., Hoboken, NJ, 2009, Vol. 3, pp. 2293–2297. Radziszewski reaction.

2 H. Debus, *Ann. Chem. Pharm.* 1858, 107, 199–208. Über die Einwirkung des Ammoniaks auf Glyoxal.

3 (a) B. Radziszewski, *Ber. Dtsch. Chem. Ges.* 1882, 15, 1493–1496. Über die Constitution des Lophins und verwandter Verbindungen. (b) B. Radziszewski, *Ber. Dtsch. Chem. Ges.* 1882, 15, 2706–2708. Über Glyoxalin und seine Homologue. (c) B. Radziszewski, *Ber. Dtsch. Chem. Ges.* 1883, 16, 487–494. Über die Synthese der Oxalinbasen. (d)B. Radziszewski, *Ber. Dtsch. Chem. Ges.* 1883, 16, 747–749. Über einige neue Glyoxaline.

4 K. Hofmann, *Chemistry of heterocyclic compounds: imidazole and its derivatives* (A. Weissberger, Ed.). Interscience, New York, 1953, Part I, Vol. 6.

5 (a) F. R. Japp, *J. Chem. Soc., Trans.* 1881, 39, 225–228. On the action of aldehydes on phenanthraquinone in presence of ammonia. (Second Notice). (b) F. R. Japp, F. W. Streatfeild, *J. Chem. Soc., Trans.* 1882, 41, 146–156. On the action of aldehydes on phenanthraquinone in presence of ammonia. (Third Notice). (c) F. R. Japp, H. H. Robinson, *Ber. Dtsch. Chem. Ges.* 1882, 15, 1268–1270. Constitution des Lophins und des Amarins. (d) F. R. Japp, *J. Chem. Soc., Trans.* 1882, 41, 323–329. L.—on the constitution of amarine and lophine.

6 M. J. Plater, *Bull. Hist. Chem.* 2008, 33, 76–81. The crucial early contributions of F. R. Japp to a general synthesis of imidazole derivatives.

7 J. M. McKenna, F. Halley, J. E. Souness, I. M. McLay, S. D. Pickett, A. J. Collis, K. Page, I. Ahmed, *J. Med. Chem.* 2002, 45, 2173–2184. An algorithm-directed two-component library synthesized *via* solid-phase methodology yielding potent and orally bioavailable p38 MAP kinase inhibitors.

8 For an example of the application of Radziszewski reaction in the synthesis of Ru(II) complexes, see: B. Pedras, R. M. F. Batista, L. Tormo, S. P. G. Costa, M. M. M. Raposo, G. Orellana, J. L. Capelo, C. Lodeiro, *Inorg. Chim. Acta* 2012, 381, 95–103. Synthesis, characterization, photophysical studies and interaction with DNA of a new family of Ru(II) furyl- and thienyl-imidazo-phenanthroline polypyridyl complexes.

9 For an example of the application of Radziszewski reaction for intra- and intermolecular polymer coupling, see: K.-S. Krannig, D. Esposito, M. Antonietti, *Macromolecules* 2014, 47, 2350–2353. Highly efficient transfer of amino groups to imidazolium entities for polymer coupling and cross-linking.

10 (a) L. Yet in *Progress in heterocyclic chemistry* (G. W. Gribble, J. A. Joule, Eds.). Elsevier, Oxford, UK, 2013, Vol. 25, pp. 217–256. Five-membered ring systems: with more than one N atom. (b) A. B. Dounay, T. T. Curran, *Imidazoles and Benzimidazoles*, in *Heterocyclic chemistry in drug discovery* (J. J. Li, Ed.). John Wiley & Sons, Inc., Hoboken, NJ, 2013, pp. 333–372.

11 L. Zhang, X.-M. Peng, G. L. V. Damu, R.-X. Geng, C.-H. Zhou, *Med. Res. Rev.* 2014, 34, 340–437. Comprehensive review in current developments of imidazole-based medicinal chemistry.

12 J. W. Black, G. J. Durant, J. C. Emmett, C. R. Ganellin, *Nature* 1974, 248, 65–67. Sulphur-methylene isosterism in the development of metiamide, a new histamine H2-receptor antagonist.

13 (a) P. P. Graczyk, A. Khan, G. S. Bhatia, V. Palmer, D. Medland, H. Numata, H. Oinuma, J. Catchick, A. Dunne, M. Ellis, C. Smales, J. Whitfield, S. J. Neame, B. Shah, D. Wilton, L. Morgan, T. Patel, R. Chung, D. Desmond, J. M. Staddon, N. Sato, A. Inoue, *Bioorg. Med. Chem. Lett.* 2005, 15, 4666–4670. The neuroprotective action of JNK3 inhibitors based on the 6,7-dihydro-5*H*-pyrrolo[1,2-*a*]imidazole scaffold. (b) S. A. Laufer, W. Zimmermann, K. J. Ruff, *J. Med. Chem.* 2004, 47, 6311–6325. Tetrasubstituted imidazole inhibitors of cytokine release: probing substituents in the N-1 position. (c) G. M. Sperandio da Silva, C. M. R. Sant'Anna, E. J. Barreiro, *Bioorg. Med. Chem.* 2004, 12, 3159–3166. A novel 3D-QSAR comparative molecular field analysis (CoMFA) model of imidazole and quinazolinone functionalized p38 MAP kinase inhibitors. (d) J. A. Murry, *Curr. Opin. Drug Discov. Dev.* 2003, 6, 945–965. Synthetic methodology utilized to prepare substituted imidazole p38 MAP kinase inhibitors. (e) J. C. Lee, J. T. Laydon, P. C. McDonnell, T. F. Gallagher, S. Kumar, D. Green, D. McNulty, M. J. Blumenthal, J. R. Keys, S. W. L. Vatter, J. E. Strickler, M. M. McLaughlin, I. R. Siemens, S. M. Fisher, G. P. Livi, J. R. White, J. L. Adams, P. R. Young, *Nature* 2002, 372, 739–742. A protein kinase involved in the regulation of inflammatory cytokine biosynthesis. (f) S. Laufer, G. Wagner, D. Kotschenreuther, *Angew. Chem. Int. Ed.* 2002, 41, 2290–2293. Ones, thiones, and N-oxides: an exercise in imidazole chemistry.

14 Y. Niwano, H. Koga, H. Kodama, K. Kanai, T. Miyazaki, H. Yamaguchi, *Med. Mycol.* 1999, 37, 351–355. Inhibition of sterol 14 α-demethylation of *Candida albicans* with NND-502, a novel optically active imidazole antimycotic agent.

15 (a) A. Liberatore, J. Schulz, J. Pommier, M. Barthelemy, M. Huchet, P. Chabrier, D. Bigg, *Bioorg. Med. Chem. Lett.* 2004, 14, 3521–3523. 2-Alkyl-4-arylimidazoles: structurally novel sodium channel modulators. (b) D. W. Cheung, E. E. Daniel, *Nature* 1980, 283, 485–486. Imidazole inhibits a temperature-dependent component of mammalian skeletal muscle action potential.

16 M. Antolini, A. Bozzoli, C. Ghiron, G. Kennedy, T. Rossi, A. Ursini, *Bioorg. Med. Chem. Lett.* 1999, 9, 1023–1028. Analogs of 4,5-bis(3,5-dichlorophenyl)-2-trifluoromethyl-1*H*-imidazole as potential antibacterial agents.

17 (a) P. Wasserscheid, T. Welton (Eds.), *Ionic liquids in synthesis*. 2nd ed. Wiley-VCH Verlag GmbH, Weinheim, Germany, 2008. (b) T. Welton, *Chem. Rev.* 1999, 99, 2071–2084. Room-temperature ionic liquids. Solvents for synthesis and catalysis.

18 (a) H. L. Lee, M. Bang, C. S. Pak, *Tetrahedron Lett.* 2005, 46, 7139–7142. Efficient synthesis of arylsulfamides by reaction of amines with (arylsulfamoyl)imidazolium triflate. (b) J. M. D. Storey, C. Williamson, *Tetrahedron Lett.* 2005, 46, 7337–7339. Imidazole-based solid-supported catalysts for the benzoin condensation. For reviews, see: (c) W. A. Herrmann, *Angew. Chem. Int. Ed.* 2002, 41, 1290–1309. *N*-heterocyclic carbenes: a new concept in organometallic catalysis. (d) D. Bourissou, O. Guerret, F. P. Gabba, W. G. Bertrand, *Chem. Rev.* 2000, 100, 39–92. Stable carbenes. (e) W. A. Herrmann, C. Kccher, *Angew. Chem. Int. Ed. Eng.* 1997, 36, 2162–2187. Essays on organometallic chemistry. 9. N-Heterocyclic carbenes.

19 K. Yagi, C. F. Soong, M. Irie, *J. Org. Chem.* 2001, 66, 5419–5423. Synthesis of fluorescent diarylethenes having a 2,4,5-triphenylimidazole chromophore.

20 N. Fridman, S. Speiser, M. Kaftory, *Cryst. Growth Des.* 2006, 6, 2281–2288. Chromotropic behavior of lophine nitro-derivatives.

21 V. Zuliani, G. Cocconcelli, M. Fantini, C. Ghiron, M. Rivara, *J. Org. Chem.* 2007, 72, 4551–4553. A practical synthesis of 2,4(5)-diarylimidazoles from simple building blocks.

22 S. Kamijo, Y. Yamamoto, *Chem. Asian J.* 2007, 2, 568–578. Recent progress in the catalytic synthesis of imidazoles.

23 G. K. Gupta, N. Rani, V. Kumar, *Mini Rev. Org. Chem.* 2012, 9, 270–284. Microwave-assisted synthesis of imidazoles.

24 A. Shaabani, A. Rahmati, *J. Mol. Catal. A: Chem.* 2006, 249, 246–248. Silica sulfuric acid as an efficient and recoverable catalyst for the synthesis of trisubstituted imidazoles.

25 B. Maleki, H. K. Shirvan, F. Taimazi, E. Akbarzadeh, *Int. J. Org. Chem.* 2012, 2, 93–99. Sulfuric acid immobilized on silica gel as highly efficient and heterogeneous catalyst for the one-pot synthesis of 2,4,5-triaryl-1*H*-imidazoles.

26 M. M. Heravi, K. Bakhtiari, H. A. Oskooie, S. Taheri, *J. Mol. Catal. A: Chem.* 2007, 263, 279–281. Synthesis of 2,4,5-triaryl-imidazoles catalyzed by $NiCl_2·6H_2O$ under heterogeneous system.

27 V. S. V. Satyanarayana, A. Sivakumar, *Chem. Pap.* 2011, 65, 519–526. An efficient and novel one-pot synthesis of 2,4,5-triaryl-1*H*-imidazoles catalyzed by $UO_2(NO_3)_2·6H_2O$ under heterogeneous conditions.

28 L. M. Wang, Y. H. Wang, H. Tian, Y. Yao, J. Shao, B. Liu, *J. Fluorine Chem.* 2006, 127, 1570–1573. Ytterbium triflate as an efficient catalyst for one-pot synthesis of substituted imidazoles through three-component condensation of benzil, aldehydes and ammonium acetate.

29 G. V. M. Sharma, Y. Jyothi, P. S. Lakshmi, *Synth. Commun.* 2006, 36, 2991–2996. Efficient room-temperature synthesis of tri- and tetrasubstituted imidazoles catalyzed by $ZrCl_4$.

30 (a) A. Parveen, M. D. Ahmed, S. K. Rafi, K. A. Shaikh, S. P. Deshmukh, R. P. Pawar, *Arkivoc* 2007, (xvi), 12–18. Efficient synthesis of 2,4,5-triaryl substituted imidazoles under solvent free conditions at room temperature. (b) M. Kidwai, P. Mothsra, V. Bansal, R. K. Somvanshi, A. S. Ethayathulla, S. Dey, T. P. Singh, *J. Mol. Catal. A: Chem.* 2007, 265, 177–182. One-pot synthesis of highly substituted imidazoles using molecular iodine: a versatile catalyst.

31 C. Yu, M. Lei, W. Su, Y. Xie, *Synth. Commun.* 2007, 37, 3301–3308. Europium triflate-catalyzed one-pot synthesis of 2,4,5-trisubstituted-1*H*-imidazoles *via* a three-component condensation.

32 J. Sangshetti, N. Kokare, S. Kotharkara, D. J. Shinde, *Synthesis* 2007, 2829–2834. One-pot efficient synthesis of 2-aryl-1-arylmethyl-1*H*-benzimidazoles and 2,4,5-triaryl-1*H*-imidazoles using oxalic acid catalyst.

33 J. N. Sangshetti, D. B. Shinde, N. D. Kokare, S. A. Kotharkar, *Monatsh. Chem.* 2008, 139, 125–127. Sodium bisulfite as an efficient and inexpensive catalyst for the one-pot synthesis of 2,4,5-tri-aryl-1*H*-imidazoles from benzil or benzoin and aromatic aldehydes.

34 S. D. Yadhav, N. D. Kokare, S. D. Jadhav, *J. Heterocycl. Chem.* 2008, 45, 1461–1464. Phosphomolybdic acid catalyzed facile one-pot synthesis of 2,4,5-triaryl-1*H*-imidazoles from benzil and aromatic aldehydes.

35 A. R. Khosropour, *Ultrason. Sonochem.* 2008, 15, 659–664. Ultrasound-promoted greener synthesis of 2,4,5-trisubstituted imidazoles catalyzed by $Zr(acac)_4$ under ambient conditions.

36 (a) S. Samai, G. C. Nandi, P. Singh, M. S. Singh, *Tetrahedron* 2009, 65, 10155–10161. L-Proline: an efficient catalyst for the one-pot synthesis of 2,4,5-trisubstituted and 1,2,4,5-tetrasubstituted imidazoles. (b) N. V. Shitole, K. F. Shelke, S. S. Sonar, S. A. Sadaphal, B. B. Shingate, M. S. Shingare, *Bull. Korean Chem. Soc.* 2009, 30, 1963–1966. L-Proline as an efficient catalyst for the synthesis of 2,4,5-triaryl-1*H*-imidazoles.

37 K. F. Shelke, S. B. Sapkal, M. S. Shingare, *Chin. Chem. Lett.* 2009, 20, 283–286. Ultrasound-assisted one-pot synthesis of 2,4,5-triarylimidazole derivatives catalyzed by ceric (IV) ammonium nitrate in aqueous media.

38 M. V. Reddy, Y. T. Jeong, *J. Fluorine Chem.* 2010, 142, 45–51. Indium trifluoride: a highly efficient catalyst for the synthesis of fluorine-containing 2,4,5-trisubstituted imidazoles under solvent-free conditions.

39 D. Song, C. Liu, S. Zhang, D. Luo, *Synth. React. Inorg. Met.-Org. Chem.* 2010, 40, 145–147. One-pot synthesis of 2,4,5-triarylimidazoles catalyzed by copper (II) trifluoroacetate under solvent-free conditions.

40 J. Safari, S. D. Khalili, S. H. Banitaba, *J. Chem. Sci.* 2010, 122, 437–441. A novel and an efficient catalyst for one-pot synthesis of 2,4,5-trisubstituted imidazoles by using microwave irradiation under solvent-free conditions.

41 J. Safari, S. D. Khalili, M. Rezaei, S. H. Banitaba, F. Meshkani, *Monatsh. Chem.* 2010, 141, 1339–1345. Nanocrystalline magnesium oxide: a novel and efficient catalyst for facile synthesis of 2,4,5-trisubstitutedimidazole derivatives.

42 F. K. Behbahani, T. Yektanezhad, A. R. Khorrami, *Heterocycles* 2010, 81, 2313–2321. Anhydrous $FePO_4$: a green and cost-effective catalyst for the one-pot three component synthesis of 2,4,5-triarylated.

43 R. S. Joshi, P. G. Mandhane, M. U. Shaikh, R. S. Kale, C. H. Gill, *Chin. Chem. Lett.* 2010, 21, 429–432. Potassium dihydrogen phosphate catalyzed one-pot synthesis of 2,4,5-triaryl-1*H*-imidazoles.

44 M. V. Marques, M. M. Ruthner, L. A. M. Fontoura, D. Russowsky, *J. Braz. Chem. Soc.* 2012, 23, 171–179. Metal chloride hydrates as Lewis acid catalysts in multicomponent synthesis of 2,4,5-triarylimidazoles or 2,4,5-triaryloxazoles.

45 S. A. Siddiqui, U. C. Narkhede, S. S. Palimkar, T. Daniel, R. J. Lahoti, K. V. Srinivasan, *Tetrahedron* 2005, 61, 3539–3546. Room temperature ionic liquid promoted improved and rapid synthesis of 2,4,5-triaryl imidazoles from aryl aldehydes and 1,2-diketones or α-hydroxyketone.

46 M. V. Chary, N. C. Keerthysri, S. V. N. Vupallapati, N. Lingaiah, S. Kantevari. *Catal. Commun.* 2008, 9, 2013–2017. Tetrabutylammonium bromide (TBAB) in isopropanol: an efficient, novel, neutral and recyclable catalytic system for the synthesis of 2,4,5-trisubstituted imidazoles.

47 A. R. Khosropour, *Can. J. Chem.* 2008, 86, 264–269. Synthesis of 2,4,5-trisubstituted imidazoles catalyzed by $[Hmim]HSO_4$ as a powerful Brønsted acidic ionic liquid.

48 X. Liu, M. Lu, Y. Wang, *Chiang Mai J. Sci.* 2011, 38, 601–607. Non-chloroaluminate ionic liquids supported the synthesis of substituted imidazoles using microwave irradiation.

49 For some examples of the reaction under catalyst and solvent-free, microwave irradiation conditions, see: (a) J. F. Zhou, X. G. Gong, H. Q. Zhu, F. X. Zhu, *Chin. Chem. Lett.* 2009, 20, 1198–1200. Solvent-free and catalyst-free method for the synthesis of 2,4,5-triarylimidazoles under microwave irradiation. (b) J. F. Zhou, G. X. Gong, X. J. Sun, Y. L. Zhu, *Synth. Commun.* 2010, 40, 1134–1141. Facile method for one-step synthesis of 2,4,5-triarylimidazoles under catalyst-free, solvent-free, and microwave-irradiation conditions.

50 For a non-catalytic method of microwaved-assisted protocol using polyethylene glycol (PEG) as solvent, see: (a) S. V. Nalage, M. B. Kalyankar, V. S. Patil, S. V. Bhosale, S. U. Deshmukh, R. P. Pawar, *TOCATJ* 2010, 3, 58–61. An efficient noncatalytic protocol for the synthesis of trisubstituted imidazole in polyethylene glycol using microwaves. For other example using PEG-400 as reaction medium, under classical thermal conditions, see: (b) X. C. Wang, H. P. Gong, Z. J. Quan, L. Li, H. L. Ye, *Chin. Chem. Lett.* 2009, 20, 44–47. PEG-400 as an efficient reaction medium for the synthesis of 2,4,5-triaryl-1*H*-imidazoles and 1,2,4,5-tetraaryl-1*H*-Imidazoles.

51 S. E. Wolkenberg, D. D. Wisnoski, W. H. Leister, Y. Wang, Z. Zhao, C. W. Lindsley, *Org. Lett.* 2004, 6, 1453–1456. Efficient synthesis of imidazoles from aldehydes and 1,2-diketones using microwave irradiation.

52 B. Cui, B. L. Zheng, K. He, Q. Y. Zheng, *J. Nat. Prod.* 2003, 66, 1101–1103. Imidazole alkaloids from *Lepidium meyenii*.

53 (a) A. P. Phillips, H. L. White, S. Rosen, Eur. Patent 58890(A1), 1982. (b) S. L. Abrahams, R. J. Hazen, A. G. Batson, A. P. Phillips, *J. Pharmacol. Exp. Ther.* 1989, 249, 359–365. Trifenagrel: a chemically novel platelet aggregation inhibitor.

54 (a) S. Balalaie, A. Arabanian, *Green Chem.* 2000, 2, 274–276. One-pot synthesis of tetrasubstituted imidazoles catalyzed by zeolite HY and silica gel under microwave irradiation. (b) K. Sivakumar, A. Kathirvel, A. Lalitha, *Tetrahedron Lett.* 2010, 51, 3018–3021. Simple and efficient method for the synthesis of highly substituted imidazoles using zeolite-supported reagents.

55 A. R. Karimi, Z. Alimohammadi, J. Azizian, A. A. Mohammadi, M. R. Mohammadizadeh, *Catal. Commun.* 2006, 7, 728–732. Solvent-free synthesis of tetrasubstituted imidazoles on silica gel/$NaHSO_4$ support.

56 S. Kantevari, S. V. N. Vuppalapati, D. O. Biradar, L. Nagarapu, *J. Mol. Catal. A: Chem.* 2007, 266, 109–113. Highly efficient, one-pot, solvent-free synthesis of tetrasubstituted imidazoles using $HClO_4$–SiO_2 as novel heterogeneous catalyst.

57 S. D. Sharma, P. Hazarika, D. Konwar, *Tetrahedron Lett.* 2008, 49, 2216–2220. An efficient and one-pot synthesis of 2,4,5-trisubstituted and 1,2,4,5-tetrasubstituted imidazoles catalyzed by $InCl_3{\cdot}3H_2O$.

58 B. Sadeghi, B. B. F. Mirjalili, M. M. Hashemi, *Tetrahedron Lett.* 2008, 49, 2575–2577. $BF_3{\cdot}SiO_2$: an efficient reagent system for the one-pot synthesis of 1,2,4,5-tetrasubstituted imidazoles.

59 M. M. Heravi, F. Derikvand, F. F. Bamoharram, *J. Mol. Catal. A: Chem.* 2007, 263, 112–114. Highly efficient, four-component one-pot synthesis of tetrasubstituted imidazoles using Keggin-type heteropolyacids as green and reusable catalysts.

60 A. Cukalovic, J.-C. M. R. Monbaliu, C. V. Stevens, *Top. Heterocycl. Chem.* 2010, 23, 161–198. Microreactor technology as an efficient tool for multicomponent reactions.

61 For an example of continuous Radziszewski reaction in a microreactor setup, to prepare 1,3-dialkylimidazolium-based ionic liquids, see: J. Zimmermann, B. Ondruschka, A. Stark, *Org. Process Res. Dev.* 2010, 14, 1102–1109. Efficient synthesis of 1,3-dialkylimidazolium-based ionic liquids: the modified continuous Radziszewski reaction in a microreactor setup.

62 D. R. J. Acke, R. V. A. Orru, C. V. Stevens. *QSAR Comb. Sci.* 2006, 25, 474–483. Continuous synthesis of tri- and tetrasubstituted imidazoles *via* a multicomponent reaction under microreactor conditions.

63 For the synthesis of tetrasubstituted imidazoles *via* a one-pot modified Radziszewski reaction carried out by means of a continuous flow process, through *a priori* imine formation, see: C. Maton, N. De Vos, B. I. Roman, E. Vanecht, N. R. Brooks, K. Binnemans, S. Schaltin, J. Fransaer, C. V. Stevens, *ChemPhysChem* 2012, 13, 3146–3157. Continuous synthesis of peralkylated imidazoles and their transformation into ionic liquids with improved (electro)chemical stabilities.

64 E. Gelens, F. J. J. De Kanter, R. F. Schmitz, L. A. J. M. Sliedregt, B. J. Van Steen, C. G. Kruse, R. Leurs, M. B. Groen, R. V. A. Orru, *Mol. Divers.* 2006, 10, 17–22. Efficient library synthesis of imidazoles using a multicomponent reaction and microwave irradiation.

65 Z. Wang, *659, Weidenhagen*, in *Comprehensive organic name reactions and reagents*. John Wiley & Sons, Inc., Hoboken, NJ, 2009, Vol. 3, pp. 2951–2955.

66 (a) R. Weidenhagen, R. Herrmann, *Ber. Dtsch. Chem. Ges.* 1935, 68, 1953–1961. Eine neue Synthese von Imidazol-Derivaten. (b) R. Weidenhagen, R. Herrmann, *Ber. Dtsch. Chem. Ges.* 1935, 68, 2205–2209. Über das 4(5)-Amino-5(4)-methyl-imidazol. (c) R. Weidenhagen, R. Herrmann, *Angew. Chem.* 1935, 48, 596–597. Eine neue Synthese von Imidazolderivaten. (d) R. Weidenhagen, *Ber. Dtsch. Chem. Ges.* 1936, 69, 2263–2272. Eine neue Synthese von Benzimidazol-Derivaten. (e) R. Weidenhagen, R. Herrmann, H. Wegner, *Ber. Dtsch. Chem. Ges.* 1937, 70, 570–583. Über neue Abkömmlinge des Imidazols (IV. Mitteil. über Imidazole). (f) R. Weidenhagen, H. Wegner, *Ber. Dtsch. Chem. Ges.* 1937, 70, 2309–2318. Über Alkylamide der Imidazol-carbonsäure-(4 bzw. 5) (V. Mitteil.) über Imidazole. (g) R. Weidenhagen, H. Wegner, *Ber. Dtsch. Chem. Ges.* 1938, 71, 2124–2134. Über die Synthese des Tetrahydro-benzimidazols (4.5-Cyclotetramethylen-imidazols) und seiner Derivate (VI. Mitteil. über Imidazole). (h) R. Weidenhagen, U. Weeden, *Ber. Dtsch. Chem. Ges.* 1938, 71, 2347–2360. Über neue Imidazoloverbindungen der heterocyclischen Reihe (VII. Mitteil. über Imidazole). (i) R. Weidenhagen, H. Rienäcker, *Ber. Dtsch. Chem. Ges.* 1939, 72, 57–67. Die Synthese von Arsonophenyl-imidazolen. (VIII. Mitteil. über Imidazole).

67 (a) E. V. Vlasova, V. M. Stoyanov, M. M. El'chaninov, B. S. Lukyanov, *Chem. Heterocycl. Compd.* 2010, 46, 681–683. Synthesis and methylation products of 4(5)-(2-furyl) imidazole. (b) E. V. Vlasova, A. A. Aleksandrov, M. M. El'chaninov, *Russ. J. App. Chem.* 2010, 83, 1027–1031. Synthesis and properties of 1-methyl-2-phenyl-5-(2-furyl)- and 1-methyl-2-phenyl-5-(2-thienyl)imidazoles.

68 (a) M. A. Biamonte, *Sakurai allylation reaction*, in *Name reactions for homologations*. John Wiley & Sons, Inc., Hoboken, NJ, 2009, Part 1, pp. 539–575. (b) T. Jacques, I. E. Markó, J. Pospíšil, *The modified Sakurai and related reactions*, in *Multicomponent reactions* (J. Zhu, H. Bienaymé, Eds.). Wiley-VCH Verlag GmbH, Weinheim, Germany, 2005, pp. 398–452. (c) A. Hosomi, K. Miura, *Bull. Chem. Soc. Jpn.* 2004, 77, 835–851. Development of new reagents containing silicon and related metals and application to practical organic syntheses. (d) A. Hosomi, *Acc. Chem. Res.* 1988, 21, 200–206. Characteristics in the reactions of allylsilanes and their applications to versatile synthetic equivalents. For Lewis base-mediated allylations, see: (e) S. Rendler, M. Östreich, *Synthesis* 2005, 1727–1747. Hypervalent silicon as a reactive site in selective bond-forming processes. (f) P. Kočovský, A. I. Malkov, *Chiral Lewis bases as catalysts*, in *Enantioselective catalysis* (P. Dalko, Ed.). Wiley-VCH Verlag GmbH, Weinheim, Germany, 2007, pp. 255–286.

69 For some reviews about Sakurai reaction, see: (a) Y. Yamamoto, N. Sasaki in *Chemical bonds—better ways to make them and break them* (I. Bernal, Ed.). Elsevier, Amsterdam, The Netherlands, 1989, pp. 363–441. The stereochemistry of the Sakurai reaction. (b) D. Schinzer, *Synthesis* 1988, 263–273. Intramolecular addition reactions of allylic and propargylic silanes. (c) Y. Yamamoto, N. Asao, *Chem Rev.* 1993, 93, 2207–2293. Selective reactions using allylic metals. (d) S. E. Denmark, J. Fu, *Chem. Rev.* 2003, 103, 2763–2794. Catalytic enantioselective addition of allylic organometallic reagents to aldehydes and ketones. (e) J. Otera (Ed.) *Modern carbonyl chemistry*. Wiley-VCH Verlag GmbH, Weinheim, Germany, 2000.

70 A. Hosomi, H. Sakurai, *Tetrahedron Lett.* 1976, 17, 1295–1298. Syntheses of γ,δ-unsaturated alcohols from allylsilanes and carbonyl compounds in the presence of titanium tetrachloride.

71 A. Hosomi, M. Endo, H. Sakurai, *Chem. Lett.* 1976, 5, 941–942. Allylsilanes as synthetic intermediates. II. Syntheses of homoallyl ethers from allylsilanes and acetals promoted by titanium tetrachloride.

72 A. Hosomi, H. Sakurai, *J. Am. Chem. Soc.* 1977, 99, 1673–1675. Chemistry of organosilicon compounds. 99. Conjugate addition of allylsilanes to α,β-enones. A new method of stereoselective introduction of the angular allyl group in fused cyclic α,β-enones.

73 A. Mekhalfia, I. E. Markó, *Tetrahedron Lett.* 1991, 32, 4779–4782. The silyl modified Sakurai (SMS) reaction. An efficient and versatile one-pot synthesis of homoallylic ethers.

74 I. E. Markó, A. Mekhalfia, F. Murphy, D. J. Bayston, M. Bailey, Z. Janousek, S. Dolan, *Pure Appl. Chem.* 1997, 69, 565–570. Multicomponent cyclisations. Efficient methodologies for the preparation of complex natural products.

75 J. Pospíšil, T. Kumamoto, I. E. Markó, *Angew. Chem. Int. Ed.* 2006, 45, 3357–3360. Highly diastereoselective silyl-modified Sakurai. Multicomponent reaction.

76 J. Pospíšil, I. E. Markó, *J. Am. Chem. Soc.* 2007, 129, 3516–3517. Total synthesis of Jerangolid D.

77 L. F. Tietze, A. Dölle, K. Schiemann, *Angew. Chem. Int. Ed. Engl.* 1992, 31, 1372–1373. Diastereoselective addition of allylsilanes to aldehydes: synthesis of enantiomerically pure homoallylic alcohols.

78 L. F. Tietze, T. Kinzel, C. C. Brazel, *Acc. Chem. Res.* 2009, 42, 367–378. The domino multicomponent allylation reaction for the stereoselective synthesis of homoallylic alcohols.

79 For other approach, see: T. Mukaiyama, M. Ohshima, N. Miyoshi, *Chem. Lett.* 1987, 1121–1124. An efficient and convenient method for the asymmetric allylation of aldehydes with allyltrimethylsilane catalyzed by diphenylboryl triflate.

80 For general reviews on allylations, see: (a) S. E. Denmark, J. Fu, *Chem. Rev.* 2003, 103, 2763–2793. Catalytic enantioselective addition of allylic organometallic reagents to aldehydes and ketones. (b) C. E. Masse, J. S. Panek, *Chem. Rev.* 1995, 95, 1293–1316. Diastereoselective reactions of chiral allyl- and allenylsilanes with activated C=X π bonds. (c) Y. Yamamoto, N. Asao, *Chem. Rev.* 1993, 93, 2207–2293. Selective reactions using allylic metals.

81 (a) L. F. Tietze, T. Kinzel, S. Schmatz, *J. Am. Chem. Soc.* 2006, 128, 11483–11495. Origin of *syn/anti* diastereoselectivity in aldehyde and ketone crotylation reactions: a combined theoretical and experimental study. (b) L. F. Tietze, T. Kinzel, S. Schmatz, *Chem.–Eur. J.* 2009, 15, 1706–1712. Stereoselective allylation of ketones: explanation for the unusual inversion of the induced stereochemistry in the auxiliary-mediated crotylation and pentenylation of butanone by DFT calculations.

82 For several experimental and computational investigations on similar reactions, see: (a) M. N. Paddon-Row, N. G. Rondan, K. N. Houk, *J. Am. Chem. Soc.* 1982, 104, 7162–7166. Staggered models for asymmetric induction: attack trajectories and conformations of allylic bonds from ab initio transition structures of addition reactions. (b) S. E. Denmark, E. J. Weber, *Helv. Chim. Acta* 1983, 159, 1655–1660. On the stereochemistry of allylmetal-aldehyde condensations. (c) S. E. Denmark, E. J. Weber, *J. Am. Chem. Soc.* 1984, 106, 7970–7971. Stereochemistry of allylmetal-aldehyde condensations. 2. Allylstannanes. (d) S. E. Denmark, T. M. Willson, *J. Am. Chem. Soc.* 1989, 111, 3475–3476. Mechanistic and stereochemical divergence in the allylsilane-acetal addition reaction. (e) J. L. Broeker, R. W. Hoffmann, K. N. Houk, *J. Am. Chem. Soc.* 1991, 113, 5006–5017. Conformational analysis of chiral alkenes and oxonium ions: ab initio molecular orbital calculations and an improved MM2 force field. (f) S. E. Denmark, H. Hosoi, *J. Org. Chem.* 1994, 59, 5133–5135. Stereochemical studies on the addition of allylstannanes to aldehydes. The SE' component. (g) S. E. Denmark, N. G. Almstead, Stereochemical studies on the addition of allylsilanes to aldehydes. The SE' component. *J. Org. Chem.* 1994, 59, 5130–5132. (h) A. Bottoni, A. L. Costa, D. Di Tommaso, I. Rossi, E. Tagliavini, *J. Am. Chem. Soc.* 1997, 119, 12131–12135. New computational and experimental evidence for the mechanism of the Sakurai reaction. (i) K. Manju, S. Trehan, *Chem. Commun.* 1999, 1929–1930. Origin of diastereoselectivity in the addition of allylsilane to chiral acyclic mixed acetals. (j) P. S. Mayer, T. H. Morton, *J. Am. Chem. Soc.* 2002, 124, 12928–12929. Gas-phase electrophilic attack of a double bond exhibits stereoselectivity.

83 Vitamin E: (a) L. F. Tietze, J. Görlitzer, A. Schuffenhauer, M. Hübner, *Eur. J. Org. Chem.* 1999, 1075–1084. Enantioselective synthesis of the chromane moiety of vitamin E. (b) L. F. Tietze, J. Görlitzer, *Synlett* 1997, 1049–1050. Preparation of enantiopure precursors for the vitamin E synthesis. A comparison of the asymmetric allylation of ketones and the sharpless bishydroxylation. (+)-Hydroxymyoporone: (c) L. F. Tietze, C. Wegner, C. Wulff, *Chem.–Eur. J.* 1999, 5, 2885–2889. First total synthesis and determination of the absolute configuration of the stress factor (+)-hydroxymyoporone. 5,6-Dihydrocineromycin B: (d) L. F. Tietze, L. Völkel, *Angew. Chem. Int. Ed.* 2001, 40, 901–902. Total synthesis of the macrolide antibiotic 5,6-dihydrocineromycin B.

84 L. F. Tietze, C. C. Brazel, S. Hölsken, J. Magull, A. Ringe, *Angew. Chem. Int. Ed.* 2008, 47, 5246–5249. Total synthesis of polyoxygenated cembrenes.

85 L. F. Tietze, S. Biller, T. Wolfram, *Synlett* 2010, 2130–2132. First enantioselective total synthesis of the reported structure of (*R*)-(+)-orizaterpenyl benzoate using an asymmetric allylation of a prochiral ketone in a domino process.

86 P. W. Anzalone, A. R. Baru, E. M. Danielson, P. D. Hayes, M. P. Nguyen, A. F. Panico, R. C. Smith, R. S. Mohan, *J. Org. Chem.* 2005, 70, 2091–2096. Bismuth compounds in organic synthesis. A one-pot synthesis of homoallyl ethers and homoallyl acetates from aldehydes catalyzed by bismuth triflate.

87 M. J. Spafford, E. D. Anderson, J. R. Lacey, A. C. Palma, R. S. Mohan, *Tetrahedron Lett.* 2007, 48, 8665–8667. Iron(III) *p*-toluenesulfonate catalyzed synthesis of homoallyl ethers from acetals and aldehydes.

88 D. Kataki, P. Phukan, *Tetrahedron Lett.* 2009, 50, 1958–1960. Iodine-catalyzed one-pot three-component synthesis of homoallyl benzyl ethers from aldehydes.

89 R. Fazaeli, H. Aliyan, M. J. Bagi, *Phosphorus, Sulfur Silicon Relat. Elem.* 2012, 187, 677–688. Cesium salt of tungstosilicic acid fabricated by nano-casting strategy: an efficient catalyst for the three-component Hosomi-Sakurai reaction.

90 R. Fazaeli, H. Aliyan, E. Naderi, *Phosphorus, Sulfur Silicon Relat. Elem.* 2013, 188, 745–754. Host (nanocavity of aluminium-incorporated mesoporous kit-6)–guest (12-tungesto(molybdato) phosphoric acid) nanocomposite material: efficient and reusable catalysts for the Hosomi-Sakurai three-component coupling reaction.

91 D. Kampen, A. Ladépêche, G. Clasen, B. List, *Adv. Synth. Catal.* 2008, 350, 962–966. Brønsted acid-catalyzed three-component Hosomi-Sakurai reactions.

92 K. Murugan, S. Srimurugan, C. Chen, *Tetrahedron* 2011, 67, 5621–5629. Silicomolybdic acid supported on silica gel: an efficient catalyst for Hosomi-HOSOMI-SAKURAI Sakurai reactions.

93 K. Murugan, C. Chen, *Tetrahedron Lett.* 2011, 52, 5827–5830. $HClO_4$-supported on silica gel: a mild and efficient catalyst for Hosomi-Sakurai reaction.

94 For some examples of the direct three-component preparation of homoallylic amines, see: (a) G. Smitha, B. Miriyala, J. S. Williamson, *Synlett* 2005, 839–841. Phosphomolybdic acid-catalyzed efficient three-component reaction: a facile synthesis of protected homoallylic amines. (b) P. Phukan, *J. Org. Chem.* 2004, 69, 4005–4006. Iodine as a very powerful catalyst for three-component synthesis of protected homoallylic amines. (c) J.-R. Ella-Menye, W. Dobbs, M. Billet, P. Klotz, A. Mann, *Tetrahedron Lett.* 2005, 46, 1897–1900. Unexpected 1,2 *syn* diastereoselectivity in the three-component "aza Sakurai–Hosomi" reaction. (d) M. Sugiura, K. Hirano, S. Kobayashi, *J. Am. Chem. Soc.* 2004, 126, 7182–7183. α-Aminoallylation of aldehydes with ammonia: stereoselective synthesis of homoallylic primary amines. (e) M. Billet, P. Klotz, A. Mann, *Tetrahedron Lett.* 2001, 42, 631–634. *Syn* diastereoselectivity in the synthesis of homoallylamine using crotylsilane in the three-component reaction. (f) L. Niimi, K.-I. Serita, S. Hiraoka, T. Yokozawa, *Tetrahedron Lett.* 2000, 41, 7075–7078. Lewis acid-catalyzed three-component condensation reactions of aldehydes, *N*-silylcarbamates, and allylsilane: synthesis of *N*-homoallylcarbamates.

95 For multicomponent synthesis of homoallylic amines using allyltributylstannane instead of allyltrimethylsilane and catalyzed by different species, see: Polymer-bound-Yb salts (polyfluoroalkyl sulfonic acids): (a) Y. Yin, G. Zhao, G.-L. Li, *Tetrahedron* 2005, 61, 12042–12052. Synthesis of polystyrene-bound perfluoroalkyl sulfonic acids and the application of their ytterbium salts in multicomponent reactions (MCRs). (Bromodimethyl) sulfonium bromide ($Me_2S^+BrBr^-$): (b) B. Das, B. Ravikanth, P. Thirupathi, B. Vittal Rao, *Tetrahedron Lett.* 2006, 47, 5041–5044. (Bromodimethyl)sulfonium bromide catalyzed efficient multicomponent one-pot synthesis of homoallylic amines.

Sulphamic acid (NH_2SO_3H): (c) J. S. Yadav, H. Ather, P. P. Rao, R. S. Rao, K. Nagaiah, A. R. Prasad, *Cat. Commun.* 2006, 7, 797–801. Sulphamic acid: an efficient, cost-effective and recyclable solid acid catalyst for three-component synthesis of homoallylic amines. Zirconyl chloride ($ZrOCl_2$): (d) B. Das, B. Ravikanth, K. R. Reddy, B. V. Rao, *Helv. Chim. Acta* 2007, 90, 105–109. Zirconyl chloride (=dichlorooxozirconium) as an efficient catalyst for one-pot multicomponent synthesis of homoallylic amines. Solid-supported perchloric acid ($HClO_4$–SiO_2): (e) L. Nagarapu, V. Paparaju, G. Pathuri, S. Kantevari, R. R. Pakkiru, R. Kamalla, *J. Mol. Cat. A: Chem.* 2007, 267, 53–56. $HClO_4$–SiO_2 as a novel and recyclable catalyst for the one-pot synthesis of homoallylic amines. Cuprous iodide (CuI): (f) P. K. Kalita, P. Phukan, *Tetrahedron Lett.* 2008, 49, 5495–5497. A three-component synthesis of homoallylic amines catalyzed by CuI. Indium (starting directly from nitro compounds in water): (g) B. Das, G. Satyalakshmi, K. Suneel, B. Shashikanth, *Tetrahedron Lett.* 2008, 49, 7209–7212. A distinct approach for the rapid synthesis of homoallylic amines starting directly from nitro compounds in water. Phosphomolybdic acid (PMA) in water: (h) Z. Xie, G. Li, G. Zhao, J. Wang, *Chin. J. Chem.* 2009, 27, 925–929. Three-component synthesis of homoallylic amines catalyzed by phosphomolybdic acid in water. Indium triflate ($In(OTf)_3$): (i) A. V. Narsaiah, J. K. Kumar, P. Narsimha, *Synthesis* 2010, 1609–1612. A simple and efficient method for the three-component synthesis of homoallylic amines catalyzed by indium triflate. MgI_2 etherate: (j) X. Zhang, S. Ye, S. Hu, *Lett. Org. Chem.* 2011, 8, 427–430. MgI_2 etherate-catalyzed three-component allylation: a facile and efficient synthesis of homoallylic amines. (k)Y. Wang, Y. Liu, S. Hu, X. Zhang, *J. Chem. Res.* 2012, 36, 21–24. An efficient three-component synthesis of homoallylic amines catalysed by MgI_2 etherate.

96 Q.-Y. Song, B.-L. Yang, S.-K. Tian. *J. Org. Chem.* 2007, 72, 5407–5410. $FeSO_4{\cdot}7H_2O$-catalyzed four-component synthesis of protected homoallylic amines.

97 R. Martínez, D. J. Ramón, M. Yus, *Adv. Synth. Catal.* 2008, 350,1235–1240. Unmodified nano-powder magnetite catalyzes a four-component aza-Sakurai reaction.

98 F. Alonso, F. Foubelo, J. C. González-Gómez, R. Martínez, D. J. Ramón, P. Riente, M. Yus, *Mol. Divers.* 2010, 14, 411–424. Efficiency in chemistry: from hydrogen autotransfer to multicomponent catalysis.

99 R. E. Grote, E. R. Jarvo, *Org. Lett.* 2009, 11, 485–488. Palladium-catalyzed, one-pot, three-component synthesis of homoallylic amines from aldehydes, anisidine, and allyl trifluoroacetate.

100 K. K. Pasunooti, M. L. Leow, S. Vedachalam, B. K. Gorityala, X.-W. Liu, *Tetrahedron Lett.* 2009, 50, 2979–2981. A general and mild copper-catalyzed three-component synthesis of protected homoallyl amines.

101 For a similar example, using 1 mol% of bismuth triflate ($Bi(OTf)_3$) as catalyst, see: T. Ollevier, T. Ba, *Tetrahedron Lett.* 2003, 44, 9003–9005. Highly efficient three-component synthesis of protected homoallylic amines by bismuth triflate-catalyzed allylation of aldimines.

102 K. Gewald, *Angew. Chem.* 1961, 73, 114–116. Reaction of (±)-oxomercaptans with nitriles.

103 (a) K. Gewald, *Chem. Ber.* 1965, 98, 3571–3577. Heterocycles from CH-acidic nitriles. VII. 2-Aminothiophene from a-oxo mercaptans and methylene-active nitriles. (b) R. Gompper, E. Kutter, *Angew. Chem.* 1962, 74, 251–252. New thiophenes and thieno[2,3-*b*]thiophenes. (c) K. Gewald, E. Schinke, *Chem. Ber.* 1966, 99, 2712–2715. Heterocyclen aus CH-aciden Nitrilen, X. Notiz zur Reaktion von Aceton mit Cyanessigester und Schwefel. (d) K. Gewald, G. Neumann, *Chem. Ber.* 1968, 101, 1933–1939. Heterocyclen aus CH-aciden Nitrilen, XIV. 2-Amino-thionaphthene.

104 K. Gewald, E. Schinke, H. Böttcher, *Chem. Ber.* 1966, 99, 94–100. 2-Amino-thiophene aus methylenaktiven nitrilen carbonylverbindungen und schwefel. (Heterocycles from C–H acidic nitriles VIII. 2-Aminothiophenes from methylene active nitriles, carbonyl compounds, and sulfur).

105 (a) R. W. Sabnis, *Sulfur Rep.* 1994, 16, 1–17. The Gewald synthesis. (b) I. Adachi, T. Yamamori, Y. Hiramatsu, K. Sakai, S.-I. Mihara, M. Kawakami, M. Masui, O. Uno, M. Ueda, *Chem. Pharm. Bull.* 1988, 36, 4389–4402. Studies on dihydropyridines. III. Synthesis of 4,7-dihydrothieno [2,3-*b*]-pyridines with vasodilator and antihypertensive activities. (c) B. Naumann, R. Böhm, F. Fülöp, G. Bernáth, *Pharmazie* 1996, 51, 4–6. On thieno compounds. Part 13. Preparation of new trifunctional thiophene derivatives.

106 (a) S. Gronowitz, J. Fortea-Laguna, S. Ross, B. Sjöberg, N. E. Stjernström, *Acta Pharm. Suec.* 1968, 5, 563–578. Thiophene analogs of methaqualone-like compounds. (b) G. Hallas, J.-H. Choi, *Dyes Pigm.* 1999, 42, 249–265. Synthesis and spectral properties of azo dyes derived from 2-aminothiophenes and 2-aminothiazoles.

107 (a) V. Zavarzin, V. N. Yarovenko, A. V. Shirokov, N. G. Smirnova, A. A. Es'kov, M. M. Krayushkin, *Arkivoc* 2003, (xiii), 205–223. Synthesis and reactivity of monothiooxamides. (b) D. M. Barnes, A. R. Haight, T. Hameury, M. A. McLaughlin, J. Mei, J. S. Tedrow, J. D. Riva Toma, *Tetrahedron* 2006, 62, 11311–11319. New conditions for the synthesis of thiophenes *via* the Knoevenagel/Gewald reaction sequence. Application to the synthesis of a multitargeted kinase inhibitor. (c) Z. Puterová, A. Andicsová, D. Végh, *Tetrahedron* 2008, 64, 11262–11269. Synthesis of π-conjugated thiophenes starting from substituted 3-oxopropanenitriles *via* Gewald reaction.

108 (a) G. Le Guillanton, Q. T. Do, J. Simonet, *Bull. Soc. Chim. Fr.* 1989, 126, 433–440. Use of carbon-sulfur cathodes in organic electrochemistry. I. Application to the reduction of vinylic compounds with leaving groups of Z-3-chloro (or 3-tosyloxy or 3-benzoyloxy)-2-phenylpropeninitrile. (b) G. Le Guillanton, Q. T. Do, J. Simonet, *Bull. Soc. Chim. Fr.* 1990, 127, 427–439. Use of carbon-sulfur cathodes in electroorganic chemistry. Part 2. Reactions with activated alkenes. Evidence for a vicarious substitution specific of this type of electrode. (c) M. Feroci, I. Chiarotto, L. Rossi, A. Inesi, *Adv. Synth. Catal.* 2008, 350, 2740–2746. Activation of elemental sulfur by electrogenerated cyanomethyl anion: synthesis of substituted 2-aminothiophenes by the Gewald reaction.

109 N. P. Peet, S. Sunder, R. J. Barbuch, A. P. Vinogradoff, *J. Heterocycl. Chem.* 1986, 23, 129–134. Mechanistic observations in the Gewald syntheses of 2-aminothiophenes.

110 R. V. Mudumbi, S. C. Montamat, R. F. Bruns, R. E. Vestal, *Am. J. Physiol.* 1993, 264, H1017–H1022. Cardiac functional-responses to adenosine by PD-81723, an allosteric enhancer of the adenosine-A1-receptor.

111 H. Y. Meltzer, H. C. Fibiger, *Neuropsychopharmacology* 1996, 14, 83–85. Olanzapine: a new atypical antipsychotic drug.

112 A. Walburger, A. Koul, G. Ferrari, L. Nguyen, C. Prescianotto-Baschong, K. Huygen, B. Klebl, C. Thompson, G. Bacher, J. Pieters, *Science* 2004, 304, 1800–1804. Protein kinase G from pathogenic mycobacteria promotes survival within macrophages.

113 P. L. Podolin, J. F. Callahan, B. J. Bolognese, Y. H. Li, K. Carlson, T. G. Davis, G. W. Mellor, C. Evans, A. K. Roshak, *J. Pharmacol. Exp. Ther.* 2005, 312, 373–381. Attenuation of murine collagen-induced arthritis by a novel, potent, selective small molecule inhibitor of I kappa B kinase 2, TPCA-1 (2-[(aminocarbonyl)amino]-5-(4-fluorophenyl)-3-thiophenecarboxamide), occurs *via* reduction of proinflammatory cytokines and antigen-induced T cell proliferation.

114 (a) C. Heyde, I. Zug, H. Hartmann, *Eur. J. Org. Chem.* 2000, 3273–3278. A simple route to *N,N*-dialkyl derivatives of 2-amino-5-thiophenecarboxylates. (b) A. Noack, H. Hartmann, *Tetrahedron* 2002, 58, 2137–2146. Synthesis and characterisation of *N,N*-disubstituted 2-amino-5-acylthiophenes and 2-amino-5-acylthiazoles. (c) T. S. Jagodzinski, *Chem. Rev.* 2003, 103, 197–227. Thioamides as useful synthons in the synthesis of heterocycles. (d) O. Zeika, H. Hartmann, *Synthesis* 2004, 377–380. Preparation and characterization of 3,4-diaryl-substituted 2-diarylaminothiophenes. (e) O. Zeika, H. Hartmann, *Tetrahedron* 2004, 60, 8213–8219. On the oxidative coupling of *N,N*-disubstituted 2-aminothiophenes—synthesis of *N,N′*-persubstituted 5,5′-diamino-2,2′-bithiophenes.

115 Biodiagnostics: (a) K. Doré, S. Dubus, H. A. Ho, I. Lévesque, M. Brunette, G. Corbeil, M. Boissinot, G. Boivin, M. G. Bergeron, D. Boudreau, M. Leclerc, *J. Am. Chem. Soc.* 2004, 126, 4240–4244. Fluorescent polymeric transducer for the rapid, simple, and specific detection of nucleic acids at the zeptomole level. Electronic and optoelectronic devices: (b) C. Rost, S. Karg, W. Riess, M. A. Loi, M. Murgia, M. Muccini, *Appl. Phys. Lett.* 2004, 85, 1613–1615. Ambipolar light-emitting organic field-effect transistor. Conductivity-based sensors: (c) D. M. Vriezema, J. Hoogboom, K. Velonia, K. Takazawa, P. C. M. Christianen, J. C. Maan, A. E. Rowan, R. J. M. Nolte, *Angew. Chem. Int. Ed.* 2003, 42, 772–776. Vesicles and polymerized vesicles from thiophene-containing rod-coil block copolymers. Self-assembled superstructures: (d) H. Yu, A. E. Pullen, M. G. Büschel, T. M. Swager, *Angew. Chem. Int. Ed.* 2004, 43, 3700–3703. Charge-specific interactions in segmented conducting polymers: an approach to selective ionoresistive responses.

116 C. Wu, E. R. Decker, N. Blok, H. Bui, T. J. You, J. Wang, A. R. Bourgoyne, V. Knowles, K. L. Berens, G. W. Holland, T. A. Brock, R. A. F. Dixon, *J. Med. Chem.* 2004, 47, 1969–1986. Discovery, modeling, and human pharmacokinetics of N-(2-acetyl-4,6-dimethylphenyl)-3-(3,4-dimethylisoxazol-5-ylsulfamoyl)thiophene-2-carboxamide (TBC3711), a second generation, ETA selective, and orally bioavailable endothelin antagonist.

117 J. Swanston, Thiophene in *Ullmann's encyclopedia of industrial chemistry*. Wiley-VCH Verlag GmbH, Weinheim, Germany, 2006.

118 M. S. Yen, I. J. Wang, *Dyes Pigm.* 2004, 61, 243–250. Synthesis and absorption spectra of hetarylazo dyes derived from coupler 4-aryl-3-cyano-2-aminothiophenes.

119 (a) R. K. Norris, *Aminothiophenes and their derivatives*, in *The chemistry of heterocyclic compounds* (S. Gronowitz, Ed.). Interscience, 1986, Part 2, Vol. 44, pp. 631–799. (b) J. Schatz in *Science of synthesis, Original edition* (G. Maas, Ed.). Thieme Chemistry, Stuttgart, Germany, 2000, Vol. 9, pp. 287–422. Reactions involving a carbonyl compound as electrophilic component. Using a nitrile and sulfur as nucleophiles (Gewald reaction). (c) J. Schatz, M. Selßer, *Thiophenes, thiophene 1,1-dioxides, and thiophene 1-oxides,* in *Science of synthesis, knowledge updates 2011/2* (A. Fuerstner, Ed.). Thieme Chemistry, Stuttgart, Germany, 2011, Vol. 9, pp. 371–444. (d) E. R. Biehl, *Top. Heterocycl. Chem.* 2012, 29, 347–380. Recent advances in the synthesis of thiophenes and benzothiophenes. (e) A. El-Mekabaty, *Synth. Commun.* 2014, 44, 1–31. Chemistry of 2-amino-3-carbethoxythiophene and related compounds.

120 For Gewald reaction in general multicomponent revision, see: (a) V. Zavarzin, V. N. Yarovenko, A. V. Shirokov, N. G. Smirnova, A. A. Es'kov, M. M. Krayushkin, *Arkivoc* 2003, (xiii), 205–223. Synthesis and reactivity of monothio-oxamides. (b) J. M. Tinsley, *Gewald aminothiophene synthesis,* in *Name reactions in heterocyclic chemistry* (J. J. Li, E. J. Corey, Eds.). John Wiley & Sons, Inc., Hoboken, NJ, 2005, pp. 193–198. (c) C. Hulme, *Applications of multicomponent reactions in drug discovery – Lead generation to process development,* in *Multicomponent reactions* (J. Zhu, H. Bienaymé, Eds.). Wiley-VCH Verlag GmbH, Weinheim, Germany, 2005, pp. 311–341. (d) A. Dömling, *Chem. Rev.* 2006, 106, 17–89. Recent developments in isocyanide based multicomponent reactions in applied chemistry. (e) Z. Wang, *267, Gewald reaction,* in *Comprehensive organic name reactions and reagents*. John Wiley & Sons, Inc., Hoboken, NJ, 2009, Vol. 1, pp. 1213–1216.

121 R. W. Sabnis, D. W. Rangnekar, N. D. Sonawane, *J. Heterocycl. Chem.* 1999, 36, 333–345. 2-Aminothiophenes from the Gewald reaction.

122 Z. Puterová, A. Krutošíková, D. Végh, *Arkivoc* 2010, (i), 209–246. Gewald reaction: synthesis, properties and applications of substituted 2-aminothiophenes.

123 (a) Y. Huang, A. Dömling, *Mol. Divers.* 2011, 15, 3–33. The Gewald multicomponent reaction. (b) Y. Huang, A. Dömling, *Reactions involving a carbonyl compound as electrophilic component. Using a nitrile and sulfur as nucleophiles (Gewald reaction),* in *Science of synthesis, multicomponent reactions* (T. J. J. Mueller, Ed.). Thieme Chemistry, Stuttgart, Germany, 2014, Vol. 1, pp. 531–552.

124 For an example of the use of 1,5-diketone under Gewald conditions to give enamines and thioamides, see: N. Paul, R. Sathishkumar, C. Anuba, S. Muthusubramanian, *RSC Adv.* 2013, 3, 7445–7451. Reactions of diphenacylaniline and diphenacyl sulfide under Gewald conditions: generation of enamines and thioamides.

125 K. Wang, D. Kim, A. Dömling, *J. Comb. Chem.* 2010, 12, 111–118. Cyanoacetamide MCR (III): three-component Gewald reactions revisited.

126 For examples of variation of the nitrile compound (cyanoacetamides) in the preparation of the starting thiophenes used in the synthesis of heterocycles with potential biological

activities, see: (a) J. Saravanan, S. Mohan, J. J. Roy, *Eur. J. Med. Chem.* 2010, 45, 4365–4369. Synthesis of some 3-substituted amino-4,5-tetramethylene thieno[2,3-*d*][1,2,3]-triazin-4(3*H*)-ones as potential antimicrobial agents. (b) R. Srinath, D. V. Pinkal, J. Saravanan, S. J. Pravin, R. Prashant, S. Aravind, *Int. J. Res. Pharm. Sci.* 2010, 1, 143–150. Synthesis and evaluation of anti-depressant like activity of some novel thieno 1,2,3-triazine-4-ones.

127 G. Revelant, S. Dunand, S. Hesse, G. Kirsch, *Synthesis* 2011, 2935–2940. Microwave-assisted synthesis of 5-substituted 2-aminothiophenes starting from arylacetaldehydes.

128 (a) M. M. Mojtahedi, M. S. Abaee, P. Mahmoodi, M. Adib, *Synth. Commun.* 2010, 40, 2067–2074. Convenient synthesis of 2-aminothiophene derivatives by acceleration of Gewald reaction under ultrasonic aqueous conditions. (b) B. D. d. C. F. d. Santos, J. S. Bello Forero, E. M. de Carvalho, J. Jones Junior, F. M. da Silva, *Heterocycl. Lett.* 2012, 2, 31–36. A solventless synthesis of 2-aminothiophenes *via* the Gewald reaction under ultrasonic conditions.

129 For some recent examples of use of classical amines, see: Et_3N in aqueous conditions, 4-CR: (a) M. S. Abaee, S. Cheraghi, *Arkivoc* 2014, (iv), 1–10. A facile four-component Gewald reaction under organocatalyzed aqueous conditions. Et_3N in aqueous conditions, 3-CR: (b) M. S. Abaee, S. Cheraghi, *J. Sulfur Chem.* 2014, 35, 261–269. Efficient three-component Gewald reactions under Et_3N/H_2O conditions. Et_2NH: (c) R. Mishra, I. Tomer, S. Kumar, *Pharm. Sin.* 2012, 3, 332–336. Synthesis and antimicrobial evaluation of novel thiophene derivatives. Piperinidina: (d) L.-H. Chen, Y.-S. Chuang, B. D. Narhe, C.-M. Sun, *RSC Adv.* 2013, 3, 13934–13943. A concise synthesis of 2-(2-aminothiophene)-benzimidazoles by one-pot multicomponent reaction.

130 T. Wang, X.-G. Huang, J. Liu, B. Li, J.-J. Wu, K.-X. Chen, W.-L. Zhu, X.-Y. Xu, B.-B. Zeng, *Synlett* 2010, 1351–1354. An efficient one-pot synthesis of substituted 2-aminothiophenes *via* three-component Gewald reaction catalyzed by L-proline.

131 X.-G. Huang, J. Liu, J. Ren, T. Wang, W. Chen, B.-B. Zeng, *Tetrahedron* 2011, 67, 6202–6205. A facile and practical one-pot synthesis of multisubstituted 2-aminothiophenes *via* imidazole-catalyzed Gewald reaction.

132 F. Moeinpour, R. Omidinia, N. Dorostkar-Ahmadi, B. Khoshdeli, *Bull. Korean Chem. Soc.* 2011, 32, 2091–2092. Cesium carbonate as a heterogeneous base catalyst for synthesis of 2-aminothiophenes *via* Gewald reaction.

133 Z.-J. Zheng, L.-X. Liu, G. Gao, H. Dong, J.-X. Jiang, G.-Q. Lai, L.-W. Xu, *RSC Adv.* 2012, 2, 2895–2901. Amine-functional polysiloxanes (AFPs) as efficient polymeric organocatalyst for amino catalysis: efficient multicomponent Gewald reaction, α-allylic alkylation of aldehydes, and Knoevenagel condensation.

134 F. Moeinpour, N. Dorostkar, M. Vafaei, *Synth. Commun.* 2012, 42, 2367–2374. Mg/La mixed oxide as an efficient heterogeneous basic catalyst for synthesis of 2-aminothiophenes under microwave irradiation.

135 (a) R. Tayebee, S. J. Ahmadi, E. R. Seresht, F. Javadi, M. A. Yasemi, M. Hosseinpour, B. Maleki, *Ind. Eng. Chem. Res.* 2012, 51, 14577–14582. Commercial zinc oxide: a facile, efficient, and eco-friendly catalyst for the one-pot three-component synthesis of multisubstituted 2-aminothiophenes *via* the gewald reaction. For an example of use of nano ZnO under solvent-free conditions, see: (b) R. Tayebee, F. Javadi, G. Argi, *J. Mol. Catal. A: Chem.* 2013, 368–369, 16–23. Easy single-step preparation of ZnO nano-particles by sedimentation method and studying their catalytic performance in the synthesis of 2-aminothiophenes *via* Gewald reaction.

136 D.-D. Zhao, L. Li, F. Xu, Q. Wu, X.-F. Lin, *J. Mol. Catal. B: Enzym.* 2013, 95, 29–35. Bovine serum albumin-catalyzed one-pot synthesis of 2-aminothiophenes *via* Gewald reaction.

137 P. K. Kunda, J. V. Rao, K. Mukkanti, M. Induri, G. D. Reddy, *Trop. J. Pharm. Res.* 2013, 12, 566–576. Synthesis, anticonvulsant activity and *in silico* studies of Schiff bases of 2-aminothiophenes *via* guanidine-catalyzed Gewald reaction.

138 For the synthesis of aminobithiophenes as fluorescent stable fluorophores, see: (a) Y. Dong, A. Bolduc, N. McGregor, W. G. Skene, *Org. Lett.* 2011, 13, 1844–1847. Push-pull aminobithiophenes; highly fluorescent stable fluorophores. For the synthesis of macrocycles containing a 2-aminothiophene unit with potent antituberculosis activity, see: (b) D. Kim, Y. Huang, K. Wang, A. Dömling, *Chem. Heterocycl. Compd.* 2013, 49, 849–859. New macrocycles with potent antituberculosis activity accessed by one-pot multicomponent reactions. (c) C. Valant, L. Aurelio, S. M. Devine, T. D. Ashton, J. M. White, P. M. Sexton, A. Christopoulos, P. J. Scammells, *J. Med. Chem.* 2012, 55, 2367–2375. Synthesis and characterization of novel 2-amino-3-benzoylthiophene derivatives as biased allosteric agonists and modulators of the adenosine A1 receptor. For the synthesis of thieno[3,4-*c*]pyrrole-4,6-dione (TPD)-based polymers, see: (d) P. Berrouard, S. Dufresne, A. Pron, J. Veilleux, M. Leclerc, *J. Org. Chem.* 2012, 77, 8167–8173. Low-cost synthesis and physical characterization of thieno[3,4-*c*]pyrrole-4,6-dione-based polymers. For the synthesis of thieno[2,3-*d*]pyrimidines as potential inhibitors of PDE4, see: (e) R. Adepu, D. Rambabu, B. Prasad, C. Lakshmi T. Meda, A. Kandale, G. R. Krishna, C. M. Reddy, L. N. Chennuru, K. V. L. Parsa, M. Pal, *Org. Biomol. Chem.* 2012, 10, 5554–5569. Novel thieno[2,3-*d*]pyrimidines: their design, synthesis, crystal structure analysis and pharmacological evaluation. For the synthesis of thieno[3,2-*c*]pyran-4-one as potential anticancer agents, see: (f) A. Nakhi, R. Adepu, D. Rambabu, R. Kishore, G. R. Vanaja, A. M. Kalle, M. Pal, *Bioorg. Med. Chem. Lett.* 2012, 22, 4418–4427. Thieno[3,2-*c*]pyran-4-one based novel small molecules: their synthesis, crystal structure analysis and in vitro evaluation as potential anticancer agents. For the synthesis of cycloalkylthiophene–imine derivatives with antiviral properties, see: (g) S. Ke, Y. Wei, Z. Yang, K. Wang, Y. Liang, L. Shi, *Bioorg. Med. Chem. Lett.* 2013, 23, 5131–5134. Novel cycloalkylthiophene–imine derivatives bearing benzothiazole scaffold: synthesis, characterization and antiviral activity evaluation.

139 A. E.-G. M. Khalil, M. A. Berghot, M. A. Gouda, *J. Serb. Chem. Soc.* 2011, 76, 329–339. Design, synthesis and antibacterial activity of new phthalazinedione derivatives.

140 E. M. Ahmed, N. A. Marzouk, S. A. Hessien, A. M. Ali, *World J. Chem.* 2011, 6, 25–31. Synthesis, reactions and antimicrobial activity of some new thienopyridine and thienopyrimidine derivatives.

141 H. M. Aly, N. M. Saleh, H. A. Elhady, *Eur. J. Med. Chem.* 2011, 46, 4566–4572. Design and synthesis of some new thiophene, thienopyrimidine and thienothiadiazine derivatives of antipyrine as potential antimicrobial agents.

142 S. I. Panchamukhi, A. K. Mohammed Iqbal, A. Y. Khan, M. B. Kalashetti, I. M. Khazi, *Pharm. Chem. J.* 2011, 44, 694–696. Synthesis, characterization, antibacterial and antifungal activity of thienopyrimidines and triazolothienopyrimidines.

143 R. Mishra, I. Tomer, S. Kumar, *Pharm. Sin.*, 2012, 3, 332–336. Synthesis and antimicrobial evaluation of novel thiophene derivatives.

144 For some examples of the use of other phosphorus components, see: (a) N. Azizi, M. R. Saidi, *Eur. J. Org. Chem.* 2003, 4630–4633. Lithium perchlorate-catalyzed three-component coupling: a facile and general method for the synthesis of α-aminophosphonates under solvent-free conditions. (b) K. C. K. Swamy, S. Kumaraswamy, K. S. Kumar, C. Muthiah, *Tetrahedron Lett.* 2005, 46 3347–3351. Cyclic chlorophosphites as scaffolds for the one-pot synthesis of α-aminophosphonates under solvent-free conditions. (c) S. Kudrimoti, V. R. Bommena, *Tetrahedron Lett.* 2005, 46, 1209–1210. (Bromodimethyl)sulfonium bromide: an inexpensive reagent for the solvent-free, one-pot synthesis of α-aminophosphonates. (d) H. Liu, S. Cai, J. Xu, *J. Peptide Sci.* 2006, 12, 337–340. Asymmetric synthesis of *N*-protected chiral 1-aminoalkylphosphonic acids and synthesis of side chain-functionalized depsiphosphonopeptides. (e) A. Heydari, A. Arefi, *Cat. Commun.* 2007, 8, 1023–1026. One-pot three-component synthesis of α-amino phosphonate derivatives. (f) M. Kasthuraiah, K. A. Kumar, C. S. Reddy, C. D. Reddy, *Heteroat. Chem.* 2007, 18, 2–8. Syntheses, spectral property, and antimicrobial activities of 6-α-amino dibenzo [*d,f*][1,3,2] dioxaphosphepin 6-oxides. (g) A. Heydari, H. Hamadi, M. Pourayoubi, *Catal. Commun.* 2007, 8, 1224–1226. A new one-pot synthesis of α-amino phosphonates catalyzed by $H_3PW_{12}O_{40}$. (h) S. Ramalingam, P. Kumar, *Catal. Lett.* 2008, 125, 315–319. Synthesis of α-amino phosphonates by three component condensation of carbonyl compound, amine, and dialkyl phosphite using yttria-zirconia based Lewis acid catalyst. (i) S. M. Vahdat, R. Baharfar, M. Tajbakhsh, A. Heydari, S. M. Baghbanian, S. Haksar, *Tetrahedron Lett.* 2008, 49, 6501–6504. Organocatalytic synthesis of α-hydroxy and α-aminophosphonates. (j) M. Tajbakhsh, A. Heydari, H. Alinezhad, M. Ghanei, S. Khaksar, *Synthesis* 2008, 352–354. Coupling of aldehydes, amines, and trimethyl phosphite promoted by amberlyst-15: highly efficient synthesis of α-aminophosphonates. (k) S. S. Sonar, K. F. Shelke, G. K. Kakade, B. B. Shingate, M. S. Shingare, *Chin. Chem. Lett.* 2009, 20, 1042–1046. Alum: an efficient catalyst for one-pot synthesis of α-aminophosphonates. (l) M. T. Maghsoodlou, S. M. Habibi-Khorassani, R. Heydari, N. Hazeri, S. S. Sajadikhah, M. Rostamizadeh, *Chin. J. Chem.* 2010, 28, 285–288. $Al(H_2PO_4)_3$ as an efficient and reusable catalyst for one-pot three-component synthesis of α-amino phosphonates under solvent-free conditions. (m) K. S. Niralwad, B. B. Shingate, M. S. Shingare, *Ultrason. Sonochem.* 2010, 17, 760–763. Solvent-free sonochemical preparation of α-aminophosphonates catalyzed by 1-hexanesulphonic acid sodium salt. (n) N. Gangwar, V. K. Kasana, *Synth. Commun.* 2011, 41, 2800–2804. Tartaric acid-catalyzed synthesis of α-aminophosphonates under solvent-free conditions. (o) P. G. Mandhane, R. P. Joshi, D. R. Nagargoje, G. H. Gill, *Chin. Chem. Lett.* 2011, 22, 563–566. Thiamine hydrochloride (VB1): an efficient catalyst for one-pot synthesis of α-aminophosphonates under ultrasonic irradiation. (p) M. Wu, R. Liu, D. Wan, *Heteroat. Chem.* 2013, 24, 110–115. Convenient one-pot synthesis of α-amino phosphonates in water using *p*-toluenesulfonic acid as catalyst for the Kabachnik-Fields reaction. (q) G. Keglevich, E. Bálint, R. Kangyal, M. Bálint, M. Milen, *Heteroat. Chem.* 2014, 25, 282–289. A critical overview of the Kabachnik-Fields reactions utilizing trialkyl phosphites in water as the reaction medium: a study of the benzaldehyde-benzylamine triethyl phosphite/diethyl phosphite models. (r) M. Milen, P. Ábrányi-Balogh, A. Dancsó, D. Frigyes, L. Pongó, G. Keglevich, *Tetrahedron Lett.* 2013, 54, 5430–5433. T3P®-promoted Kabachnik-Fields reaction: an efficient synthesis of α-aminophosphonates. (s) M. Milen, P. Ábrányi-Balogh, R. Kangyal, A. Dancsó, D. Frigyes, G. Keglevich, *Heteroat. Chem.* 2014, 25, 245–255. T3P®-mediated one-pot synthesis of bis(α-aminophosphonates). (t) M. Lashkari, N. Hazeri, M. T. Maghsoodlou, S. M. H. Khorassani, N. A. Torbati, A. Hosseinian, S. García-Granda, L. Torre-Fernández, *Heteroat. Chem.* 2013, 24, 58–65. Synthesis and crystal structure study of diethyl aryl(benzo[*d*]thiazol-2-ylamino)methyl phosphonates. (u) C. B. Reddy, K. S. Kumar, M. A. Kumar, M. V. N. Reddy, B. S. Krishna, M. Naveen, M. K. Arunasree, C. S. Reddy, C. N. Raju, C. D. Reddy, *Eur. J. Med. Chem.* 2012, 47, 553–559. PEG-SO_3H catalyzed synthesis and cytotoxicity of α-aminophosphonates.

145 M. I. Kabachnik, T. Y. Medved, *Dokl. Akad. Nauk SSSR* 1952, 83, 689–692. [Dokl. Chem. Technol., Dokl. Chem. (Engl. Transl.)]. New method for preparation of α-aminophosphonic acids.

146 E. K. Fields, *J. Am. Chem. Soc.* 1952, 74, 1528–1531. The synthesis of esters of substituted amino phosphonic acids.

147 R. A. Cherkasov, V. I. Galkin, *Russ. Chem. Rev.* 1998, 67, 857–882. The Kabachnik-Fields reaction: synthetic potential and the problem of the mechanism.

148 R. Gancarz, *Tetrahedron* 1995, 51, 10627–10632. Nucleophilic addition to carbonyl compounds. Competition between hard (amine) and soft (phosphite) nucleophile.

149 N. S. Zefirov, E. D. Matveeva, *Arkivoc* 2008, (i), 1–17. Catalytic Kabachnik-Fields reaction: new horizons for old reaction.

150 (a) G. Keglevich, N. Z. Kiss, D. K. Menyhárd, A. Fehérvári, I. Csontos, *Heteroat. Chem* 2012, 23, 171–178. A study on the Kabachnik-Fields reaction of benzaldehyde, cyclohexylamine, and dialkyl phosphites. (b) G. Keglevich, E. Bálint, *Molecules* 2012, 17, 12821–12835. The Kabachnik-Fields reaction: mechanism and synthetic use.

151 Y. Zhang, Y. Zhao, B. Yang, C. Zhu, Y. Wei, L. Tao, *Polym. Chem.* 2014, 5, 1857–1862. "One pot" synthesis of well-defined poly(aminophosphonate)s: time for the Kabachnik-Fields reaction on the stage of polymer chemistry.

152 For an example of the use of Kabachnik-Fields reaction in post-polymerization modification, see: R. Kakuchi, P. Theato, *ACS Macro Lett.* 2014, 3, 329–332. Efficient multicomponent postpolymerization modification based on Kabachnik-Fields reaction.

153 (a) Z. Wang, *353, Kabachnik–Fields reaction,* in *Comprehensive organic name reactions and reagents*. John Wiley & Sons, Inc., Hoboken, NJ, 2009, Vol. 2, pp. 1588–1592. (b) N. Zefirov, E. D. Matveeva, M. V. Shuvalov in *Science of synthesis, multicomponent reactions* (T. J. J. Mueller, Ed.). Thieme Chemistry, Stuttgart, Germany, 2014, Vol. 1, pp. 273–295. The chapter title is: reactions involving a carbonyl compound as electrophilic component. Third component phosphonate (Kabachnik-Fields reaction).

154 (a) V. P. Kukhar, H. R. Hudson (Eds.), *Aminophosphonic and aminophosphinic acids: chemistry and biological activity*. John Wiley & Sons, Ltd, Chichester, UK, 2000. (b) B. Lejczak, P. Kafarski, *Top. Heterocycl. Chem.* 2009, 20, 31–63. Biological activity of aminophosphonic acids and their short peptides. (c) F. Orsini, G. Sello, M. Sisti, *Curr. Med. Chem.* 2010, 17, 264–289. Aminophosphonic acids and derivatives. Synthesis and biological applications.

155 W. Liu, C. J. Rogers, A. J. Fisher, M. D. Toney, *Biochemistry* 2002, 41, 12320–12328. Aminophosphonate inhibitors of dialkylglycine decarboxylase: structural basis for slow binding inhibition.

156 J. Grembecka, A. Mucha, T. Cierpicki, P. Kafarski, *J. Med. Chem.* 2003, 46, 2641. The most potent organophosphorus inhibitors of leucine aminopeptidase. Structure-based design, chemistry, and activity.

157 M. Sienczyk, J. Oleksyszyn, *Curr. Med. Chem.* 2009, 16, 1673–1687. Irreversible inhibition of serine proteases—design and in vivo activity of diaryl α-aminophosphonate derivatives.

158 Q. Wang, M. Zhu, R. Zhu, L. Lu, C. Yuan, S. Xing, X. Fu, Y. Mei, Q. Hang, *Eur. J. Med. Chem.* 2012, 49, 354–364. Exploration of α-aminophosphonate *N*-derivatives as novel, potent and selective inhibitors of protein tyrosine phosphatases.

159 (a) C. H. Mohan, B. Hari Babu, C. Naga Raju, R. Usha Nagalakshmi, *E-J. Chem.* 2008, 5, 679–687. A convenient synthesis and antibacterial activity of novel α-aminophosphonic acid esters from amino acids/esters (Kabachnik-Fields reaction). (b) N. Long, X. J. Cai, B. A. Song, S. Yang, Z. Chen, P. S. Bhadury, D. Y. Hu, L. H. Jin, W. Xue, *J. Agric. Food Chem.* 2008, 56, 5242–5246. Synthesis and antiviral activities of cyanoacrylate derivatives containing an α-aminophosphonate moiety. (c) D. Y. Hu, Q. Q. Wan, S. Yang, B. A. Song, P. S. Bhadury, L. H. Jin, K. Yan, F. Liu, Z. Chen, W. Xue, *J. Agric. Food Chem.* 2008, 56, 998–1001. Synthesis and antiviral activities of amide derivatives containing the α-aminophosphonate moiety. (d) B. Song, L. Jin, S. Yang, P. S. Bhadury, *Environment-friendly antiviral agents for plants*. Springer, Berlin, Germany, 2010, pp. 7–93.

160 (a) R. A. Cherkasov, A. S. Talan, A. V. Tarasov, A. P. Garifzyanov, *Russ. J. Gen. Chem.* 2008, 78, 1330–1333. Synthesis of novel mono- and bisaminophosphoryl compounds and their membrane transport properties for acidic substrates. (b) R. A. Cherkasov, A. R. Garifzyanov, A. S. Talan, R. R. Davletshin, N. V. Kurnosova, *Russ. J. Gen. Chem.* 2009, 79, 1835–1849. Synthesis of new liophilic functionalized aminomethylphosphine oxides and their acid-base and membrane-transport properties toward acidic substrates. (c) R. A. Cherkasov, A. R. Garifzyanov, S. A. Koshkin, *Russ. J. Gen. Chem.* 2011, 81, 773–774. Synthesis of optically active α-aminophosphine oxides and enantioselective membrane transport of acids with their participation. (d) D. C. Danila, X. Y. Wang, H. Hubble, I. S. Antipin, E. Pinkhassik, *Bioorg. Med. Chem. Lett.* 2008, 18, 2320–2323. Increasing permeability of phospholipid bilayer membranes to alanine with synthetic α-aminophosphonate carriers.

161 C. A. Boswell, C. A. S. Regino, K. E. Baidoo, K. J. Wong, D. E. Milenic, J. A. Kelley, C. C. Lai, M. W. Brechbiel, *Bioorg. Med. Chem.* 2009, 17, 548–552. A novel side-bridged hybrid phosphonate/acetate pendant cyclam: synthesis, characterization, and ^{64}Cu small animal PET imaging.

162 (a) M. Ordóñez, H. Rojas-Cabrera, C. Cativiela, *Tetrahedron* 2009, 65, 17–49. An overview of stereoselective synthesis of α-aminophosphonic acids and derivatives. (b) P. Merino, E. Marqués-López, R. P. Herrera, *Adv. Synth. Catal.* 2008, 350, 1195–1208. Catalytic enantioselective hydrophosphonylation of aldehydes and imines. (c) Ł. Albrecht, A. Albrecht, H. Krawczyk, K. A. Jørgensen, *Chem.–Eur. J.* 2010, 16, 28–48. Organocatalytic asymmetric synthesis of organophosphorus compounds. (d)I. L. Odinets, E. V. Matveeva, *Russ. Chem. Rev.* 2012, 81, 221–238. The application of green chemistry methods in organophosphorus synthesis. (e) M. Ordóñez, F. J. Sayago, C. Cativiela, *Tetrahedron* 2012, 68, 6369–6412. Synthesis of quaternary α-aminophosphonic acids.

163 R. Karpowicz, J. Lewkowski, E. Miękoś, M. Zieliński, *Heteroat. Chem* 2014, 25, 163–170. The Kabachnik-Fields reaction accelerated in external magnetic field.

164 (a) Z.-P. Zhan, R.-F. Yang, J.-P. Li, *Chem. Lett.* 2005, 34, 1042–1043. Microwave-assisted one-pot synthesis of α-amino phosphonates *via* three component coupling on a silica gel support. (b) M. M. Kabachnik, E. V. Zobnina, V. Y. Pavlov, I. O. Konstantinov, I. P. Beletskaya, *Russ. Chem. Bull.* 2005, 54, 262–265. First synthesis of α-aminophosphonates from natural porphyrin derivatives by the Kabachnik-Fields reaction. (c) A. K. Bhattacharya, T. Kaur, *Synlett* 2007, 745–748. An efficient one-pot synthesis of α-amino phosphonates catalyzed by bismuth nitrate pentahydrate. (d) S.-G. Lee, J. K. Lee, C. E. Song, D.-C. Kim, *Bull. Korean Chem. Soc.* 2002, 23, 667–668. Microwave-assisted Kabachnik-Fields reaction in ionic liquid. (e) B. Kaboudin, N. Rahman, *Tetrahedron Lett.* 2001, 42, 8211–8213. Microwave-assisted synthesis of 1-aminoalkyl phosphonates under solvent-free conditions. (f) K. V. Ramana, S. Rasheed, K. C. Sekhar, S. Adam, C. N. Raju, *Pharm. Lett.* 2012, 4, 456–463. One-pot and catalyst-free synthesis of novel α-aminophosphonates under microwave irradiation and their biological activity.

165 For synthesis of α-aminophosphonates under catalyst and ultrasonic irradiation, see: M. Xia, Y.-D. Lu, *Ultrason. Sonochem.* 2007, 14, 235–240. Ultrasound-assisted one-pot approach to α-amino phosphonates under solvent-free and catalyst-free conditions.

166 For solvent and catalyst-free conditions, see: (a) B. C. Ranu, A. Hajra, *Green Chem.* 2002, 4, 551–554. A simple and green procedure for the synthesis of α-aminophosphonate by a one-pot three-component condensation of carbonyl compound, amine and diethyl phosphite without solvent and catalyst. (b) S. Chandrasekhar, C. Narshmulu, S. S. Sultana, B. Saritha, S. J. Prakash, *Synlett* 2003, 505–506. Solvent and catalyst free three-component coupling of carbonyl compounds, amines and triethylphosphite; a new synthesis of α-aminophosphonates. (c) M. M. Kabachnik, E. V. Zobnina, I. P. Beletskaya, *Synlett* 2005, 1393–1396. Catalyst-free microwave-assisted synthesis of α-aminophosphonates in a three-component system: $R_1C(O)R_2$–$(EtO)_2P(O)H$–RNH_2. (d) X.-J. Mu, M.-Y. Lei, J.-P. Zou, W. Zhang, *Tetrahedron Lett.* 2006, 47, 1125–1127. Microwave-assisted solvent-free and catalyst-free Kabachnik-Fields reactions for α-amino phosphonates. (e) G. Keglevich, A. Szekrenyi, *Lett. Org. Chem.* 2008, 5, 616–622. Eco-friendly accomplishment of the extended Kabachnik-Fields reaction; a solvent- and catalyst-free microwave-assisted synthesis of α-aminophosphonates and α-aminophosphine oxides. (f) K. U. Maheswara Rao, G. R. Devi, N. J. Reddy, P. Santhipriya, C. S. Reddy, *Phar. Chem.* 2010, 2, 51–57. Uncatalyzed three component synthesis and anti-microbial activity of α-amino phosphonates.

167 B. C. Ranu, A. Hajra, U. Jana, *Org. Lett.* 1999, 1, 1141–1143. General procedure for the synthesis of α-amino phosphonates from aldehydes and ketones using indium(III) chloride as a catalyst.

168 Z. P. Zhan, J. P. Li, *Synth. Commun.* 2005, 35, 2501–2508. Bismuth(III) chloride–catalyzed three- component coupling: synthesis of α-amino phosphonates.

169 (a) Z. Rezaei, H. Firouzabadi, N. Iranpoor, A. Ghaderi, M. R. Jafari, A. A. Jafari, H. R. Zare, *Eur. J. Med. Chem.* 2009, 44, 4266–4275. Design and one-pot synthesis of α-aminophosphonates and bis(α-aminophosphonates) by iron(III) chloride and cytotoxic activity. (b) W. M. Abdou, R. F. Barghash, M. S. Bekheit, *Monatsh Chem.* 2011, 142, 649–656. Multicomponent reactions in a one-pot synthesis of α-aminophosphonates and α-aminophosphonic diamides with anti-inflammatory properties.

170 F. Xu, Y. Q. Luo, J. T. Wu,Q. Shen, H. Chen, *Heteroat. Chem* 2006, 17, 389–392. Facile one-pot synthesis of α-amino phosphonates using lanthanide chloride as catalyst.

171 D. Subba Rao, S. Rasheed, S. K. Thaslim Basha, C. Naga Raju, K. Naresh, *Pharma Chem.* 2013, 5, 61–74. $SiO_2/ZnCl_2$ catalyzed α-aminophosphonates and phosphonated *N*-(substituted phenyl) sulfonamides of 2-aminothiophene: synthesis and biological evaluation.

172 (a) A. Ali Jafari, M. Nazarpour, M. Abdollahi-Alibeik, *Heteroat. Chem.* 2010, 21, 397–403. $CeCl_3{\cdot}7H_2O$-catalyzed one-pot Kabachnik-Fields reaction: a green protocol for three-component synthesis of α-aminophosphonates. (b) M. Varalakshmi, D. Srinivasulu, D. Rajasekhar, C. Naga Raju, S. Sreevani, *Phosphorus Sulfur Silicon Relat. Elem.* 2014, 189, 106–112. $CeCl_3{\cdot}7H_2O$ catalyzed, microwave-assisted high-yield synthesis of α-aminophosphonates and their biological studies.

173 R. Ghosh, S. Maiti, A. Chakraborty, D. K. Maiti, *J. Mol. Catal. A: Chem.* 2004, 210, 53–57. $In(OTf)_3$ catalysed simple one-pot synthesis of α-amino phosphonates.

174 A. Banik, S. Batta, D. Bandyopadhyay, B. K. Banik, *Molecules* 2010, 15, 8205–8213. A highly efficient bismuth salts-catalyzed route for the synthesis of α-aminophosphonates.

175 (a) S. Sobhani, Z. Tashrifi, *Synth. Commun.* 2009, 39, 120–131. $Al(OTf)_3$ as an efficient catalyst for one-pot synthesis of primary diethyl 1-aminophosphonates under solvent-free conditions. (b) S. Sobhani, Z. Tashrifi, *Heteroat. Chem.* 2009, 20, 109–115. One-pot synthesis of primary 1-aminophosphonates: coupling reaction of carbonyl compounds, hexamethyldisilazane, and diethyl phosphite catalyzed by $Al(OTf)_3$. For a cross-linked polystyrene supported aluminium triflate (Ps-$Al(OTf)_3$) under solvent-free conditions, see: (c) K. P. Boroujeni, *Synth. React. Inorg., Metal-Org., Nano-Metal Chem.* 2011, 41, 173–176. Synthesis of α-aminophosphonates using polystyrene supported $Al(OTf)_3$ as a heterogeneous catalyst.

176 R. Gallardo-Macias, K. Nakayama, *Synthesis* 2010, 57–62. Tin(II) compounds as catalysts for the Kabachnik-Fields reaction under solvent-free conditions: facile synthesis of α-aminophosphonates.

177 S. Bhagat, A. K. Chakraborti, *J. Org. Chem.* 2007, 72, 1263–1270. An extremely efficient three-component reaction of aldehydes/ketones, amines, and phosphites (Kabachnik-Fields reaction) for the synthesis of α-aminophosphonates catalyzed by magnesium perchlorate.

178 J. Tang, L. Wang, W. Wang, L. Zhang, S. Wu, D. Mao, *J. Fluor. Chem.* 2011, 132, 102–106. A facile synthesis of α-aminophosphonates catalyzed by ytterbium perfluorooctanoate under solvent-free conditions.

179 S. Bhagat, A. K. Chakraborti, *J. Org. Chem.* 2008, 73, 6029–6032. Zirconium(IV) compounds as efficient catalysts for synthesis of α-aminophosphonates.

180 F. Xu, Y. Q. Luo, M. Y. Deng, Q. Shen, *Eur. J. Org. Chem.* 2003, 4728–4730. One-pot synthesis of α-amino phosphonates using samarium diiodide as a catalyst precursor.

181 (a) N. V. Yashin, E. V. Villemson, A. V. Chemagin, E. B. Averina, M. M. Kabachnik, T. S. Kuznetsova, *Synthesis* 2008, 464–468. Synthesis of novel α-aminophosphonates containing small rings. (b) M. M. Kabachnik, L. I. Minaeva, I. P. Beletskaya, *Synthesis* 2009, 2357–2360. Synthesis of novel α-aminophosphonates containing adamantyl fragment.

182 S. Chandrasekhar, S. J. Prakash, V. Jagadeshwar, C. Narsihmulu, *Tetrahedron Lett.* 2001, 42, 5561–5563. Three component coupling catalyzed by $TaCl_5$–SiO_2: synthesis of α-amino phosphonates.

183 J.-T. Hou, J.-W. Gao, Z.-H. Zhang, *Appl. Organomet. Chem.* 2011, 25, 47–53. $NbCl_5$: an efficient catalyst for one-pot synthesis of α-aminophosphonates under solvent-free conditions.

184 J. S. Yadav, B. V. S. Reddy, C. Madan, *Synlett* 2001, 1131–1133. Montmorillonite clay-catalyzed one-pot synthesis of α-amino phosphonates.

185 S. D. Mitragotri, D. M. Pore, U. V. Desai, P. P. Wadgaonkar, *Catal. Commun.* 2008, 9, 1822–1826. Sulfamic acid: an efficient and cost-effective solid acid catalyst for the synthesis of α-aminophosphonates at ambient temperature.

186 A. K. Bhattacharya, K. C. Rana, *Tetrahedron Lett.* 2008, 49, 2598–2601. Amberlite-IR 120 catalyzed three-component synthesis of α-amino phosphonates in one-pot.

187 N. D. Shashikumar, *J. Chem.* 2013, 8. Preparation of new α-aminophosphonate derivatives by Kabachnik-Fields reaction using a recyclable catalyst.

188 J. J. Yang, N. Dang, Y. W. Chang, *Lett. Org. Chem.* 2009, 6, 470–473. Silica sulfuric acid as a recyclable catalyst for a one-pot synthesis of α-aminophosphonates in solvent-free conditions.

189 G.-Y. Sun, J.-T. Hou, J.-J. Dou, J. Lu, Y.-J. Hou, T. Xue, Z.-H. Zhang, *J. Chin. Chem. Soc.* 2010, 57, 1315–1320. Xanthan sulfuric acid as an efficient biodegradable and recyclable catalyst for the one-pot synthesis of α-amino phosphonates.

190 B. Kaboudin, H. Zahedi, *Chem. Lett.* 2008, 37, 540–541. Calcium chloride as an efficient Lewis base catalyst for the one-pot synthesis of α-aminophosphonic esters.

191 Y. P. Tian, F. Xu, Y. Wang, J. J. Tang, H. L. Li, *J. Chem. Res.* 2009, 78–80. PPh_3-catalysed one-pot three-component syntheses of α-aminophosphonates under solvent-free conditions.

192 B. S. Kumar, *Arch. Appl. Sci. Res.* 2011, 3, 37–43. TMG catalysed one pot-synthesis of α-aminophosphonates.

193 C. Sampath, K. Veena Vani, Y. Kotaiah, N. Hari Krishna, C. Naga Raju, C. Venkata Rao, *J. Chem. Pharm. Res.* 2012, 4, 1375–1382. A facile and efficient one-pot three component reaction (Kabachnik-Fields reaction) for the synthesis of novel α-aminophosphonates by 1,4-dimethylpiperazine as a new catalyst.

194 M. Z. Kassaee, F. Movahedi, H. Masrouri, *Synlett* 2009, 1326–1330. ZnO nanoparticles as an efficient catalyst for the one-pot synthesis of α-amino phosphonates.

195 M. Kidwai, S. Bhardwaj, N. K. Mishra, A. Jain, A. Kumar, S. Mozzumdar, *Catal. Sci. Technol.* 2011, 1, 426–430. Application of mobilized Cu-nanoparticles as heterogeneous catalyst for the synthesis of α-amino phosphonates *via* A2-P coupling.

196 S. M. Agawane, J. M. Nagarkar, *Tetrahedron Lett.* 2011, 52, 3499–3504. Nano ceria catalyzed synthesis of α-aminophosphonates under ultrasonication.

197 D. Rajasekhar, D. S. Rao, D. Srinivasulu, C. N. Raju, M. Balaji, *Phosphorus, Sulfur Silicon Relat. Elem.* 2013, 188, 1017–1025. Microwave assisted synthesis of biologically active α-aminophosphonates catalyzed by nano-$BF_3{\cdot}SiO_2$ under solvent-free conditions.

198 D. Saberi, S. Cheraghi, S. Mahdudi, J. Akbari, A. Heydari, *Tetrahedron Lett.* 2013, 54, 6403–6406. Dehydroascorbic acid (DHAA) capped magnetite nanoparticles as an efficient magnetic organocatalyst for the one-pot synthesis of α-aminonitriles and α-aminophosphonates.

199 M. Hosseini-Sarvari, *Tetrahedron* 2008, 64, 5459–5466. TiO_2 as a new and reusable catalyst for one-pot three-component syntheses of α-aminophosphonates in solvent-free conditions.

200 S. Sobhani, E. Safaei, M. Asadi, F. Jalili, *J. Organomet. Chem.* 2008, 693, 3313–3317. An eco-friendly procedure for the efficient synthesis of dialkyl α-aminophosphonates in aqueous media.

201 S. S. Prasad, K. S. Kumar, S. H. Jayaprakash, B. S. Krishna, C. S. Sundar, P. V. Rao, T. M. Babu, W. Rajendra, C. S. Reddy, *Arch. Pharm. Chem. Life Sci.* 2013, 346, 380–391. Design, synthesis, antioxidant, and anti-breast cancer activities of novel diethyl(alkyl/aryl/heteroarylamino)(4-(pyridin-2-yl) phenyl)methylphosphonates.

202 (a) E. D. Matveeva, T. A. Podrugina, E. V. Tishkovskaya, N. S. Zefirov, *Synlett* 2003, 2321–2324. A novel catalytic three-component synthesis (Kabachnick–Fields reaction) of α-aminophosphonates from ketones. (b) E. D. Matveeva, T. A. Podrugina, M. V. Prisyazhnoi, N. S. Zefirov, *Russ. Chem. Bull. Int. Ed.* 2006, 55, 1209–1214. Ketones in the catalytic three-component "one-pot" Kabachnik-Fields synthesis of α-amino phosphonates. (c) E. D. Matveeva, N. S. Zefirov, *Russ. J. Org. Chem.* 2006, 42, 1237–1238. Amino acids in a three-component synthesis of α-aminophosphonates derivatives. (d) E. D. Matveeva, T. A. Podrugina, M. V. Prisyazhnoi, I. N. Rusetskaya, N. S. Zefirov, *Russ. Chem. Bull. Int. Ed.* 2007, 56, 798–805. Three-component catalytic method for synthesis of α-amino phosphonates with the use of α-amino acids as amine component. (e) E. D. Matveeva, T. A. Podrugina, I. N. Kolesnikova, A. A. Borisenko, N. S. Zefirov, *Russ. Chem. Bull. Int. Ed.* 2009, 58, 119–125. Aminopyridines as amino components in the catalytic synthesis of α-aminophosphonates. For an example of silica gel-immobilized aluminum phthalocyanine complex, see: (f) E. D. Matveeva, M. V. Shuvalov, N. S. Zefirov, *Russ. Chem. Bull. Int. Ed.* 2011, 60, 242–247. Silica gel-immobilized aluminum phthalocyanine complex as a heterogeneous catalyst for the synthesis of α-amino phosphonates.

203 Y. Zhang, C. Zhu, *Catal. Commun.* 2012, 134–137. Gold complex-catalyzed C–P bond formation by Kabachnik-Fields reactions.

204 S. D. Dindulkar, M. V. Reddy, Y. T. Jeong, *Catal. Commun.* 2012, 114–117. $Cd(ClO_4)_2{\cdot}xH_2O$ as a novel catalyst for the synthesis of α-aminophosphonates under solvent-free conditions.

205 B. Kaboudin, M. Sorbiun, *Tetrahedron Lett.* 2007, 48, 9015–9017. β-Cyclodextrin as an efficient catalyst for the one-pot synthesis of 1-aminophosphonic esters in water.

206 J. Wu, W. Sun, X. Y. Sun, H. G. Xia, *Green Chem.* 2006, 8, 365–367. Expeditious approach to α-amino phosphonates *via* three-component solvent-free reactions catalyzed by NBS or CBr_4.

207 A. Vinu, P. Kalita, V. V. Balasubramanian, H. Oveisi, T. Selvan, A. Mano, M. A. Chari, B. V. S. Reddy, *Tetrahedron Lett.* 2009, 50, 7132–7136. Mesoporous aluminosilicate

nanocage-catalyzed three-component coupling reaction: an expedient synthesis of α-aminophosphonates.

208 V. H. Tillu, D. K. Dumbre, R. D. Wakharkar, V. R. Choudhary, *Tetrahedron Lett.* 2011, 52, 863–866. One-pot three-component Kabachnik-Fields synthesis of α-aminophosphonates using H-beta zeolite catalyst.

209 S. Rostamnia, H. Xin, N. Nouruzi, *Microporous Mesoporous Mater.* 2013, 179, 99–103. Metal–organic frameworks as a very suitable reaction inductor for selective solvent-free multicomponent reaction: IRMOF-3 as a heterogeneous nanocatalyst for Kabachnik-Fields three-component reaction.

210 S. T. Disale, S. R. Kale, S. S. Kahandal, T. G. Srinivasan, R. V. Jayaram, *Tetrahedron Lett.* 2012, 53, 2277–2279. Choline chloride·2$ZnCl_2$ ionic liquid: an efficient and reusable catalyst for the solvent free Kabachnik-Fields reaction.

211 P. S. Reddy, P. V. G. Reddy, S. M. Reddy, *Tetrahedron Lett.* 2014, 55, 3336–3339. Phosphomolybdic acid promoted Kabachnik-Fields reaction: an efficient one-pot synthesis of α-aminophosphonates from 2-cyclopropylpyrimidine-4-carbaldehyde.

212 For catalyzed diastereoselective synthesis of α-aminophosphonates, see: (a) Y. Song, G. Xing-Wen, D. Chun-Ling, S. Bao-An, J. Lin-Hong, X. Guang-Fang, Z. Guo-Ping, W. Wei, H. De-Yu, X. Wei, Z. Xia, L. Ping, *Chin. J. Chem.* 2006, 24, 1581–1588. Synthesis and antifungal activity of novel chiral α-aminophosphonates containing fluorine moiety. (b) Ambica, S. Kumar, S. C. Taneja, M. S. Hundal, K. K. Kapoor, *Tetrahedron Lett.* 2008, 49, 2208–2212. One-pot synthesis of α-aminophosphonates catalyzed by antimony trichloride adsorbed on alumina.

213 For the diastereoselective synthesis of α-aminophosphonates under solvent and catalyst free-conditions, see: (a) G. D. Tibhe, S. Lagunas-Rivera, E. Vargas-Díaz, O. García-Barradas, M. Ordoñez, *Eur. J. Org. Chem.* 2010, 6573–6581. Uncatalyzed one-pot diastereoselective synthesis of α-amino phosphonates under solvent-free conditions. (b) G. D. Tibhe, V. Labastida-Galván, M. Ordóñez, *Rapid Commun. Mass Spectrom.* 2011, 25, 951–959. Study of the fragmentation pathway of α-aminophosphonates by chemical ionization and fast atom bombardment mass spectrometry. (c) J. L. Viveros-Ceballos, C. Cativiela, M. Ordóñez, *Tetrahedron: Asymmetry* 2011, 22, 1479–1484. One-pot three-component highly diastereoselective synthesis of isoindolin-1-one-3-phosphonates under solvent and catalyst free-conditions. (d) G. D. Tibhe, M. A. Reyes-González, C. Cativiela, M. Ordóñez, *J. Mex. Chem. Soc.* 2012, 56, 183–187. Microwave-assisted high diastereoselective synthesis of α-aminophosphonates under solvent and catalyst free-conditions.

214 (a) X. Cheng, R. Goddard, G. Buth, B. List, *Angew. Chem. Int. Ed.* 2008, 47, 5079–5081. Direct catalytic asymmetric three-component Kabachnik-Fields reaction. (b) L. Wang, S. Cui, W. Meng, G. Zhang, J. Nie, J. Ma, *Chin. Sci. Bull.* 2010, 55, 1729–1731. Asymmetric synthesis of α-aminophosphonates by means of direct organocatalytic three-component hydrophosphonylation.

215 (a) X. Zhou, D. Shang, Q. Zhang, L. Lin, X. Liu, X. Feng, *Org. Lett.* 2009, 11, 1401–1404. Enantioselective three-component Kabachnik-Fields reaction catalyzed by chiral scandium(III)-*N,N′*-dioxide complexes. (b) M. Ohara, S. Nakamura, N. Shibata, *Adv. Synth. Catal.* 2011, 353, 3285–3289. Direct enantioselective three-component Kabachnik-Fields reaction catalyzed by chiral bis(imidazoline)-zinc(II) catalysts.

216 For pioneering work on the highly enantioselective one-pot reaction of *in situ* preformed imines using a chiral salalen-aluminum catalyst, see: B. Saito, H. Egami, T. Katsuki, *J. Am. Chem. Soc.* 2007, 129, 1978–1986. Synthesis of an optically active Al(salalen) complex and its application to catalytic hydrophosphonylation of aldehydes and aldimines.

13

ESSENTIAL MULTICOMPONENT REACTIONS II

DANIEL KÖNNING[1] AND JÖRG SWATSCHEK[2]

[1] Department of Medicinal Chemistry, Helmholtz Centre for Infection Research, Braunschweig, Germany

[2] TU Dortmund, Fakultät für Chemie und Chemische Biologie, Lehrbereich Anorganische Chemie, Dortmund, Germany

13.1 KNOEVENAGEL REACTIONS IN MULTICOMPONENT SYNTHESES

13.1.1 Introduction

At the end of the nineteenth century, Emil Knoevenagel reported the condensation of aldehydes with malonic acid derivatives under the catalysis of several amines to yield 2-alkylidenemalonic acid derivatives [1]. Today, this and closely related reactions are commonly referred to as the Knoevenagel condensation [2, 3]. In multicomponent syntheses, this essential reaction often is the initial step followed by a wide variety of reactions in a one-pot fashion to give rise to complex molecular structures starting from rather simple molecules. The α,β-unsaturated carbonyl compound generated is capable to undergo Michael additions, hetero-Diels–Alder reactions [4], 1,3-dipolar cycloadditions, and many more, providing access to a large range of substance classes. The wide application of the Knoevenagel condensation makes the categorization not easy. So this subchapter will confine itself to the Knoevenagel condensation-induced multicomponent syntheses of variably substituted 4*H*-pyrans, 3,4-dihydro-2*H*-pyrans, dihydropyridines, diazines, and various N-, O-, S-, and P-heterocyclic compounds published since 2005. At the end of this subchapter, other Knoevenagel combinations that did not fit into these categories will be discussed. The topic of the Knoevenagel-initiated multicomponent reactions (MCRs), by far, cannot be exhausted by this subchapter, but it gives a deeper insight into this matter.

The reaction mechanism of the initial Knoevenagel condensation strongly depends on the conditions applied [3]. The base-mediated reaction is shown in Scheme 13.1. Bases such as acetates or mixed metal oxides (MMO) are believed to facilitate the Knoevenagel condensation following this so-called Hann–Lapworth mechanism [5]. Initially, the base forms an enolate of the most CH-acidic component **1**. This undergoes a 1,2-addition to the aldehyde with subsequent protonation of the generated alkoxide. A base-mediated E1cb elimination pathway then generates the desired condensation product **2**.

While tertiary amines act as bases as well, primary and secondary amines undergo a different mechanism, as outlined in Scheme 13.2 with piperidine **3** as a representative example [3]. Here, the aldehyde condenses with the secondary amine to provide an iminium intermediate. This is attacked by the preformed enolate of the most CH-acidic component. Thereafter, a similar base-mediated E1cb-elimination takes place with release of piperidine, which leads to the formation of the desired condensation product **2**.

Lewis- and Brønsted acid-mediated Knoevenagel condensations are thought to follow a diverse type of mechanism (Scheme 13.3). The acid-mediated enol formation with subsequent 1,2-addition to the activated aldehyde component provides a β-hydroxy carbonyl **4**. Dehydration of this compound leads to the Knoevenagel condensation product **2**.

With the Knoevenagel condensation product in hand, the second step to complete the MCR is the Michael addition of the other CH-acidic component with subsequent cyclization of the generated intermediate. Differentially substituted CH acids provide diversely substituted 4*H*-pyrans, so the mechanistic pathways will be discussed for the three mainly applied types of CH acids (nitriles **5**, carbonyls **6**, and enamines **7**) separately (Schemes 13.4 and 13.5).

Multicomponent Reactions: Concepts and Applications for Design and Synthesis, First Edition. Edited by Raquel P. Herrera and Eugenia Marqués-López.

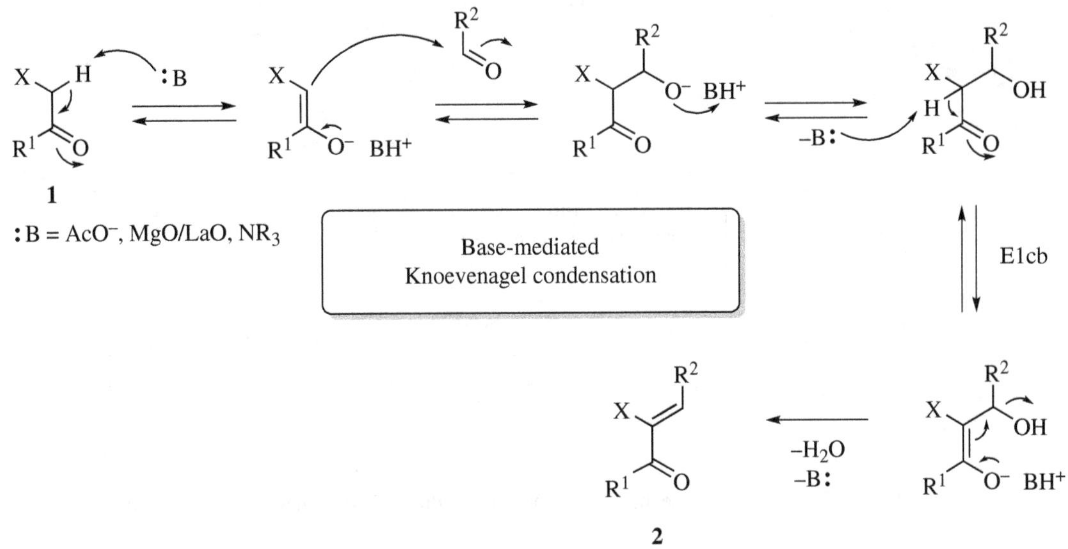

SCHEME 13.1 Proposed mechanism of the base-mediated Knoevenagel condensation.

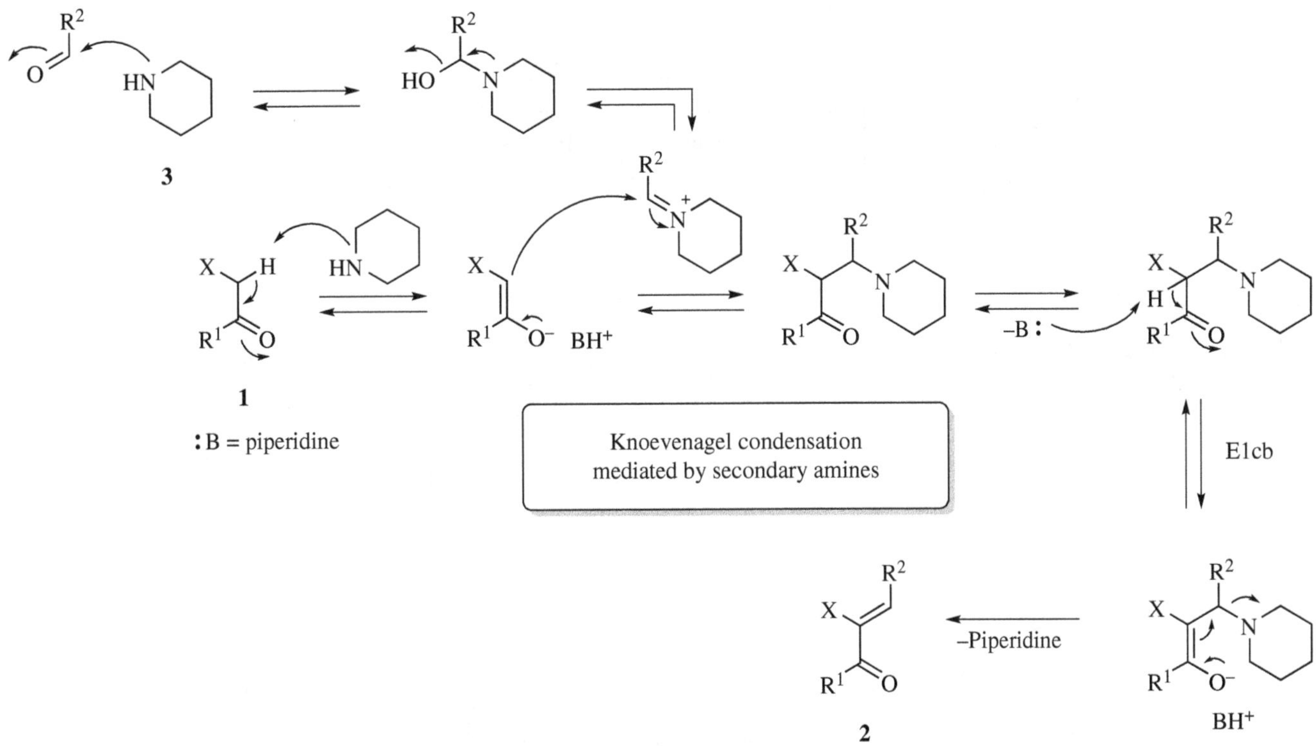

SCHEME 13.2 Proposed mechanism of the Knoevenagel condensation mediated by secondary amines.

Under basic conditions, both carbonyl- and nitrile-bearing CH acids will undergo deprotonation first to give the corresponding enolate and enolate analogue, respectively (Scheme 13.4). Those add to the Knoevenagel condensation product in a Michael addition fashion, creating an enolate that undergoes direct cyclization with the carbonyl or nitrile. In the case of a carbonyl group, the generated hemiacetal undergoes a base-mediated dehydration to provide the 4*H*-pyran **9**. In the case of a nitrile group, an imine/enamine tautomerism provides the 2-amino-4*H*-pyran **8**. For a mechanistic proposal under Lewis acid catalysis, see Perumal et al. [6].

Enamines **7** directly add to the Michael acceptor without the need of deprotonation (Scheme 13.5). Cyclization is taking place by substitution of a good leaving group present

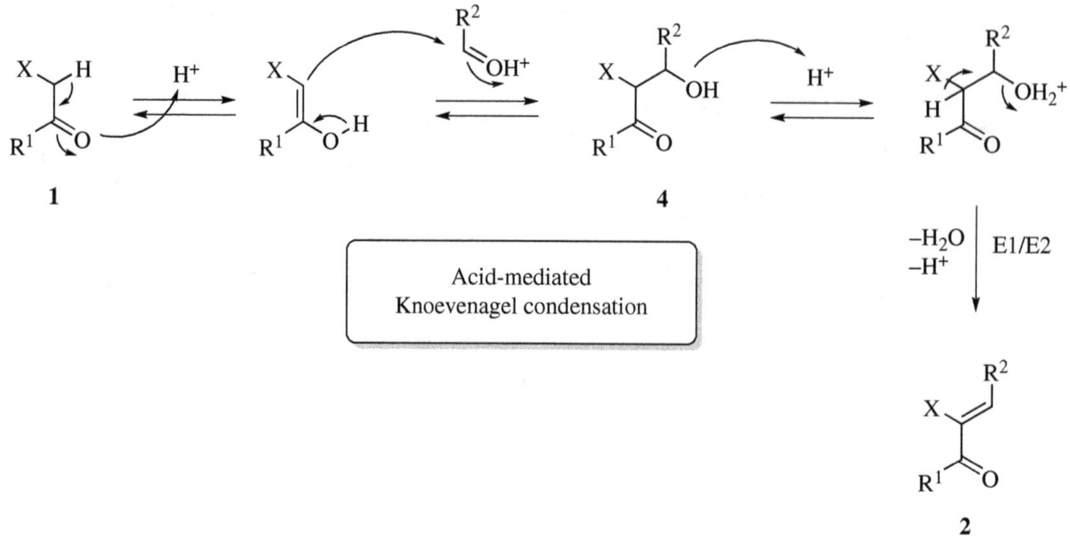

SCHEME 13.3 Proposed mechanism of the acid-mediated Knoevenagel condensation.

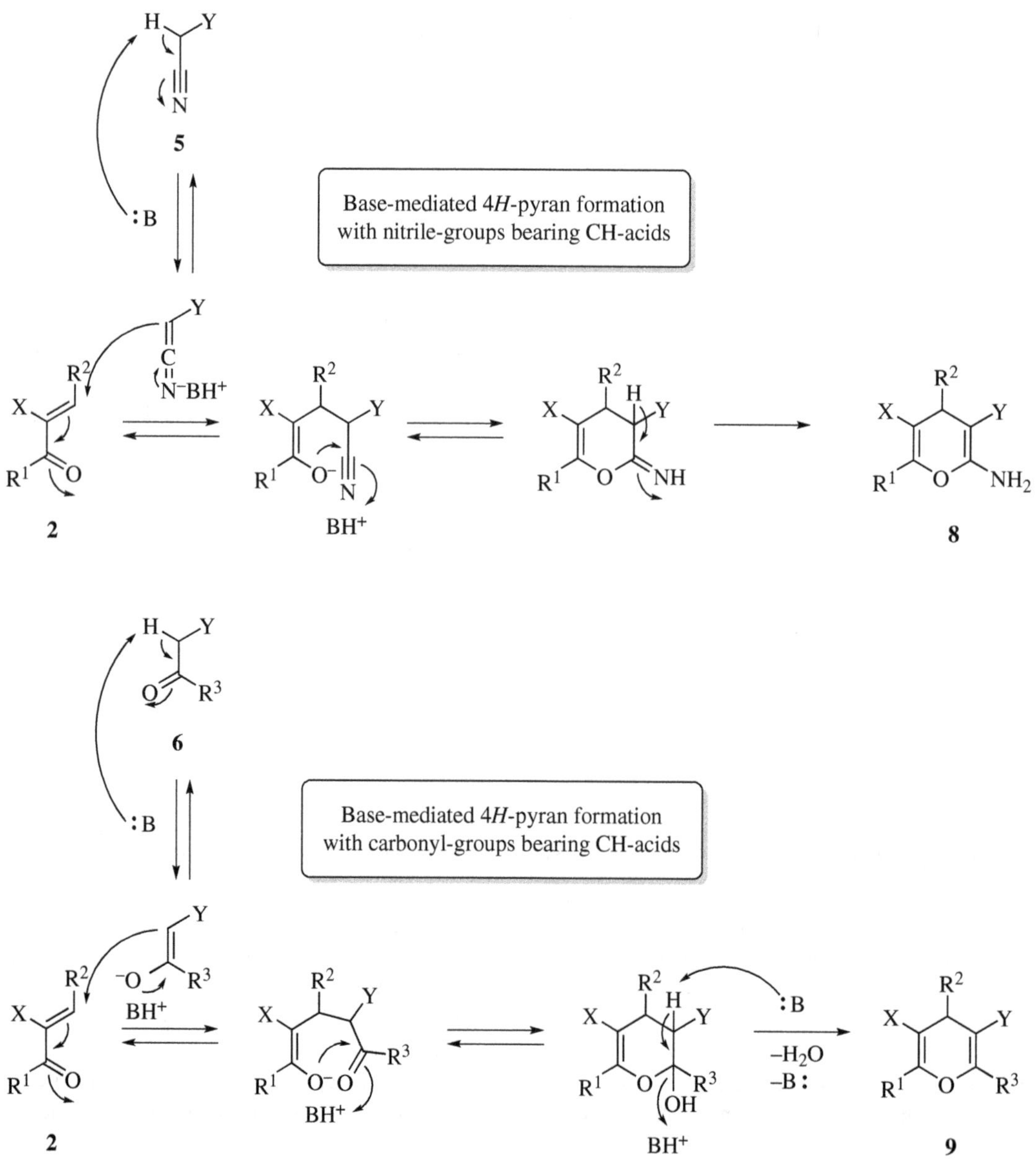

SCHEME 13.4 Proposed mechanism of the base-mediated MCR completion for nitrile-**5** and carbonyl-**6** bearing CH-acids.

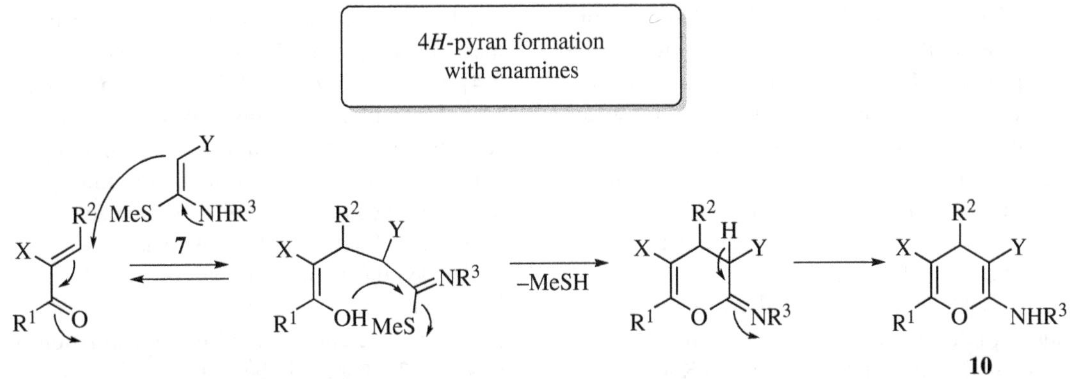

SCHEME 13.5 Proposed mechanism of the enamine MCR.

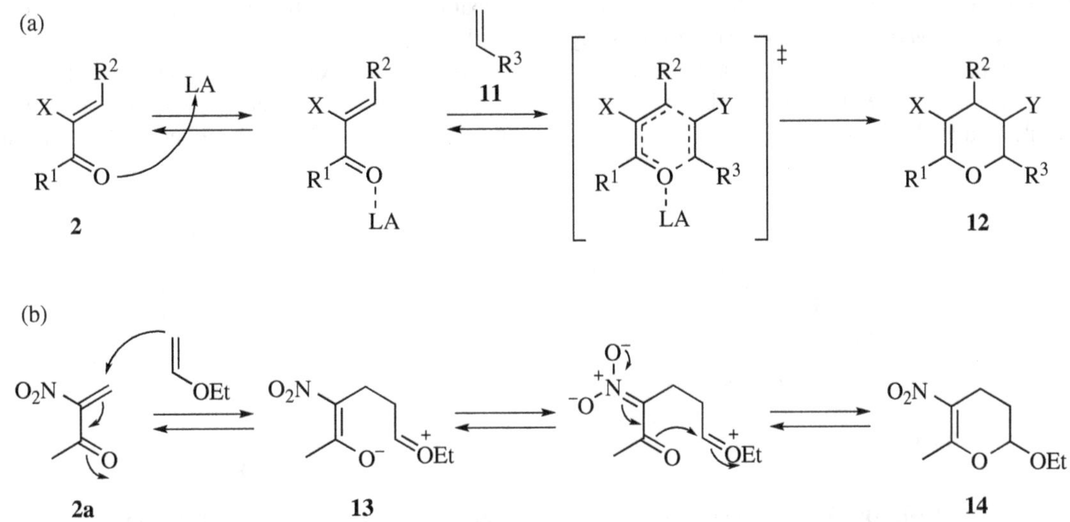

SCHEME 13.6 (a) General concerted mechanism for the IEDHDA reaction to generate dihydropyran derivatives **12**; (b) stepwise alternative for certain substrates such as **2a** [3, 8, 9].

in the enamine moiety. Imine/enamine tautomerism then leads to the corresponding N-substituted 2-amino-4*H*-pyran **10** [7].

3,4-Dihydro-2*H*-pyrans **12** are generally created by an inverse electron demand hetero-Diels–Alder reaction (IEDHDA). This requires an electron-deficient heteroatom-bearing diene **2** (Knoevenagel condensation product) and an electron-rich dienophile **11**, often represented by alkyl vinyl ethers. A Lewis-acidic catalyst coordinated to the hetero-diene lowers its electron density, making the reaction even more likely to occur (Scheme 13.6a). Commonly, a concerted mechanism with an aromatic transition state is assumed for such Diels–Alder reactions, but in certain cases, the mechanism can deviate from this assumption [3]. Tietze et al. have shown the high electrophilicity of 3-nitro-3-buten-2-one **2a** being responsible for a switch in the reaction mechanism. The concerted reaction pathway is shifted to a stepwise process where the vinyl ether first adds to the Michael acceptor. The oxocarbenium intermediate **13** then undergoes cyclization to the desired 3,4-dihydro-2*H*-pyran **14** (Scheme 13.6b) [8]. The formation of this relatively long-living oxocarbenium species explains the observed generation of diastereomeric mixtures in cases where 2-substituted vinyl ethers are used [9].

The MCR mechanism to provide dihydropyridine derivatives starting from enamines is similar to the one discussed for the synthesis of 4*H*-pyrans **10** (Scheme 13.5) [10]. In contrast to the discussed mechanism, the cyclization in the syntheses of dihydropyridine derivatives takes place by addition/dehydration instead of nucleophilic displacement. As an amidine acts like an enamine as well, the reaction pathway for diazine syntheses can be formulated in analogy to the one discussed for the enamine MCR to create N-substituted 2-amino-4*H*-pyrans **10** (Scheme 13.5) [11].

13.1.2 Domino Knoevenagel/Hetero-Diels–Alder Reaction and Pyran Syntheses

13.1.2.1 4H-Pyrans In 2007, Balalaie et al. reported the synthesis of several 4*H*-tetrahydro benzo[*b*]pyrans **18** starting from a number of aryl aldehydes **16**, CH acids **17**, and dimedone **15** (Scheme 13.7) [12]. The diammonium hydrogenphosphate-catalyzed reactions provided the corresponding products **18** in water at ambient temperature in good to excellent yields. Due to the reduced water solubility of alkyl cyanoacetates, water/EtOH mixtures (1:1) were applied in those cases. Lower reactivity of the alkyl cyanoacetates in comparison with malonodinitrile made the use of elevated reaction temperatures and elongated reaction times unavoidable.

The products could be easily isolated by filtration, and recrystallization from ethanol led to the pure desired compounds. The authors suggested an iminium ion-catalyzed Knoevenagel condensation mechanism, followed by enamine–Michael addition and intramolecular cyclization. One year later, the same group reported the diammonium hydrogenphosphate-catalyzed synthesis of various pyrano-pyrimidinones by exchanging dimedone by barbituric and thiobarbituric acid in aqueous ethanol at ambient temperature [13]. The desired compounds could be afforded in good yields.

Zheng et al. applied an ionic liquid catalyst to the synthesis of 4*H*-pyrans **18** of the type shown in Scheme 13.7 [14]. By using 5 mol% of the basic ionic liquid *N*-benzyl-2-(dimethylamino)-*N*,*N*-dimethylethanaminium chloride under solvent-free conditions at elevated temperatures, the desired compounds were isolated in good to high yields after short reaction times (up to 80 min). The reuse of the catalyst was possible without loss of catalytic activity.

The synthesis of a similar compound class was reported by Lingaiah et al. in 2008 [15]. The application of MMO as strongly basic catalysts led to the formation of the desired products **22** that could be isolated in moderate to high yields after refluxing in methanol for a short reaction time (Scheme 13.8). Special features of this reaction are the easy separation of the catalyst by centrifugation and its recycling without loss of catalytic activity.

Magnetic Fe_3O_4 nanoparticles have been reported by Kidwai et al. to efficiently catalyze the formation of 4*H*-pyran derivatives **25** in ethanol at 40 °C (Scheme 13.9) [16].

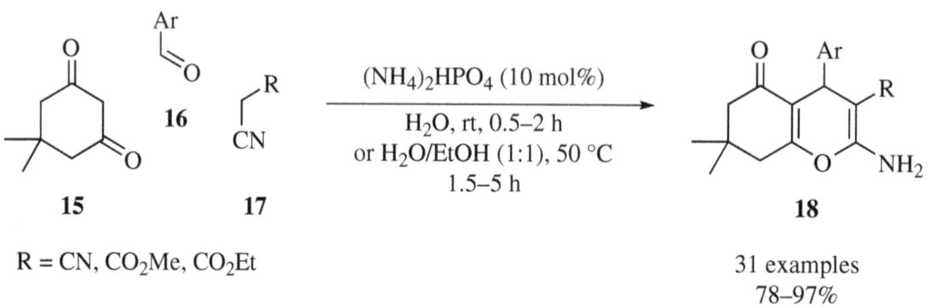

SCHEME 13.7 Diammonium hydrogenphosphate-catalyzed synthesis of 4*H*-pyrans **18**.

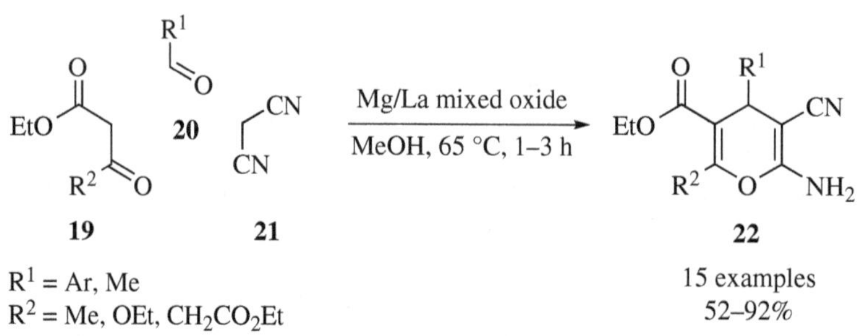

SCHEME 13.8 Mixed metal oxides (MMO) catalyze the formation of 4*H*-pyran derivatives **22**.

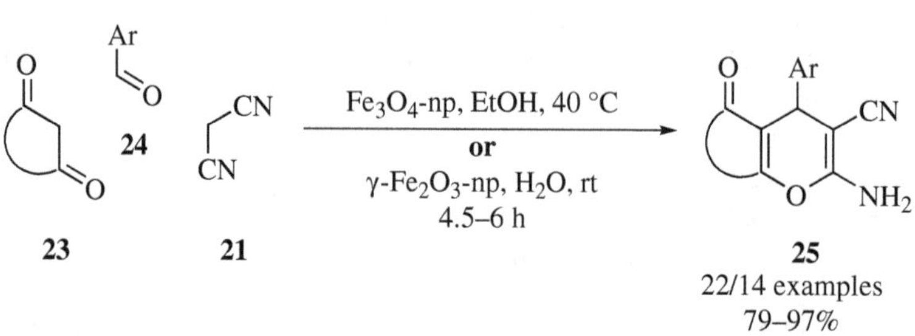

SCHEME 13.9 Iron nanoparticles in the synthesis of 2-amino-4*H*-pyrans **25**.

The removal of the catalyst was possible by the application of an external magnetic field, and the catalyst was proven to be reusable up to four times with only slight loss of catalytic activity. A similar procedure by Rostamnia et al. uses γ-Fe_2O_3 nanoparticles in water at rt with excellent yields of the desired products (Scheme 13.9) [17]. The catalyst was isolated in a similar fashion by the application of an external magnetic field and reused up to five times.

4*H*-Pyran derivatives with extended cyclic systems have been reported. Perumal et al. used 20 mol% of $InCl_3$ to catalyze the pyrano-ring formation between 3-cyanoacetylindole **26**, malonodinitrile **21**, and various aromatic aldehydes **27** (Scheme 13.10) [6]. This method led to the formation of the desired products **28** in short reaction times in refluxing ethanol, which could be isolated in good to high yields by a filtration/recrystallization procedure.

Li et al. applied the ionic liquid 1-butyl-3-methyl-1*H*-imidazol-3-ium bromide ([bmim]Br, **32**) for the reaction of 2-hydroxy-1,4-naphthoquinone **29** with dimedone derivatives **31** and several aromatic aldehydes **30** to afford a tetracyclic 4*H*-pyran **33** (Scheme 13.11) [18]. The ionic liquid was proven to be reusable four times, and the product **33** could be isolated in good yield (80–89%) by simple filtration/recrystallization.

The synthesis of a different type of tetracyclic 4*H*-pyran compounds **36** was reported by Shestopalov et al. (Scheme 13.12) [19]. The application of a catalytic amount of *N*-methylmorpholine (NMM) in a mixture of ethanol and DMF at elevated temperatures provided the desired products in short reaction times and good yields (75–88%).

Elinson et al. used a catalytic amount of sodium acetate for the synthesis of bicyclic 4*H*-pyran derivatives of

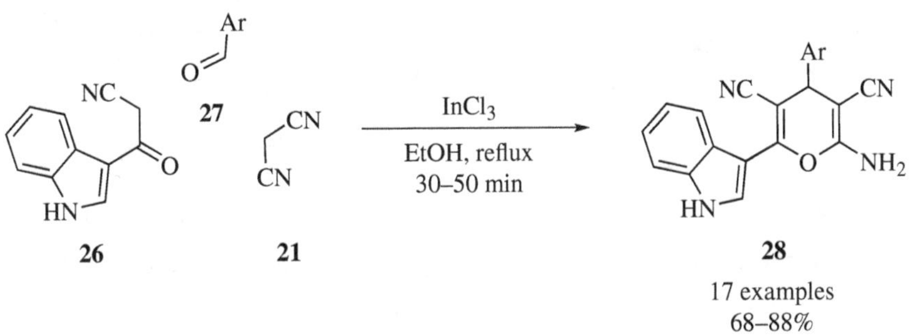

SCHEME 13.10 Indium(III) chloride-catalyzed 2-amino-4*H*-pyran **28** formation [6].

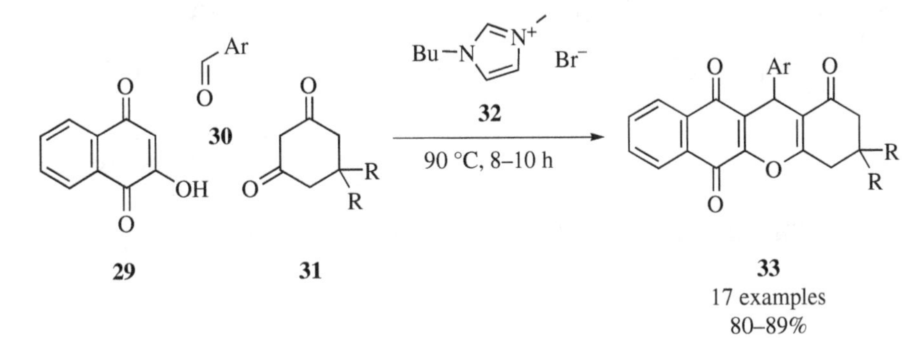

SCHEME 13.11 Synthesis of a tetracyclic 4*H*-pyran derivative **33** using [bmim]Br ionic liquid **32**.

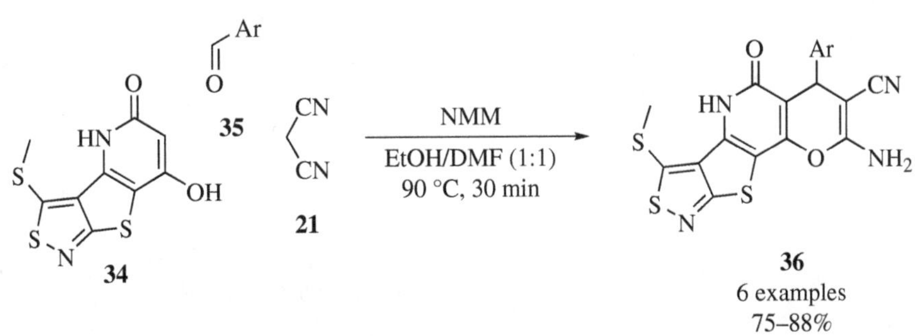

SCHEME 13.12 *N*-Methylmorpholine (NMM)-catalyzed 4*H*-pyran **36** formation.

2-pyrazolin-5-ones [20]. Short reaction times in alcoholic solvents at rt provided good to excellent yields of the desired products.

Two reports have been published by Naimi-Jamal et al. and Singh et al., making use of 2-naphthol for the construction of variably substituted tricyclic 4*H*-pyrans **39** (Schemes 13.13 and 13.14, respectively) [21, 22]. Naimi-Jamal et al. used 10 mol% of sodium carbonate under solvent-free conditions at high temperatures to afford the desired products **39** in excellent yields (88%-quant.) (Scheme 13.13).

The procedure by Singh et al. actually is a four-component reaction (4CR) catalyzed by 10 mol% $InCl_3$ in alcoholic solvent at elevated temperature. By using alcohols as the solvent, the product of the three-component reaction undergoes subsequent transesterification to provide the shown tricyclic products **42** in good yields (58–83%) (Scheme 13.14).

The reaction between isatin derivatives **44**, malonodinitrile **21**, and CH acids **43** led to the formation of spiropyran compounds **46** by the application of a variety of catalysts. Yuan et al. reported the use of 10 mol% cupreine **45** as an organocatalyst that led to spiro compounds with enhanced enantiomeric excess (Scheme 13.15) [23]. The reactions are conducted in dichloromethane or 1,2-dichloroethane as solvents at 0 °C, while the addition of 4 Å molecular sieves provides shorter reaction times and higher ee's.

A very similar, though not enantioselective, reaction was reported by Mahdavinia et al. in 2013 [24]. DABCO is used as the organocatalyst in refluxing ethanol for 10–15 min, and

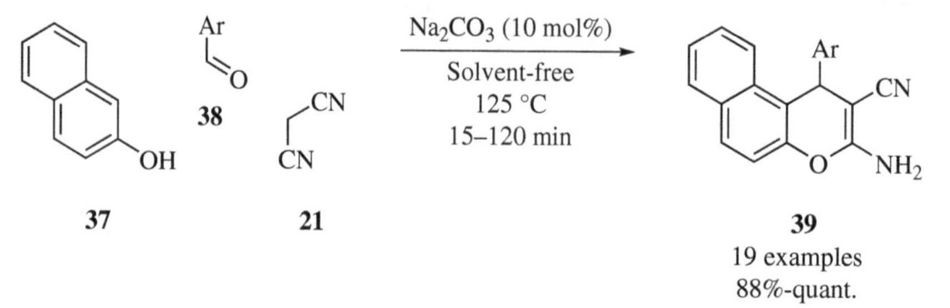

SCHEME 13.13 Sodium carbonate-catalyzed formation of a tricyclic pyran **39**.

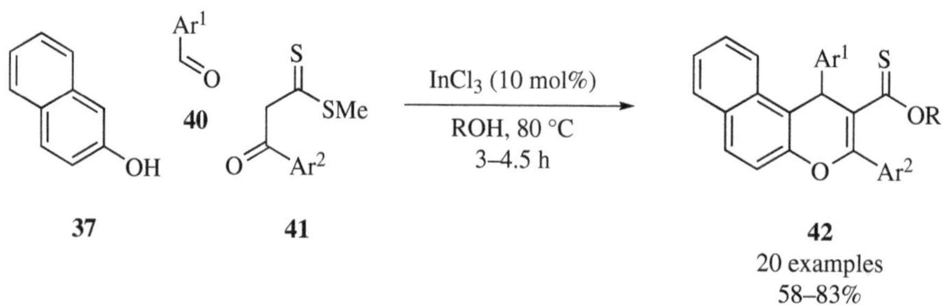

SCHEME 13.14 Indium(III) chloride-catalyzed four-component reaction.

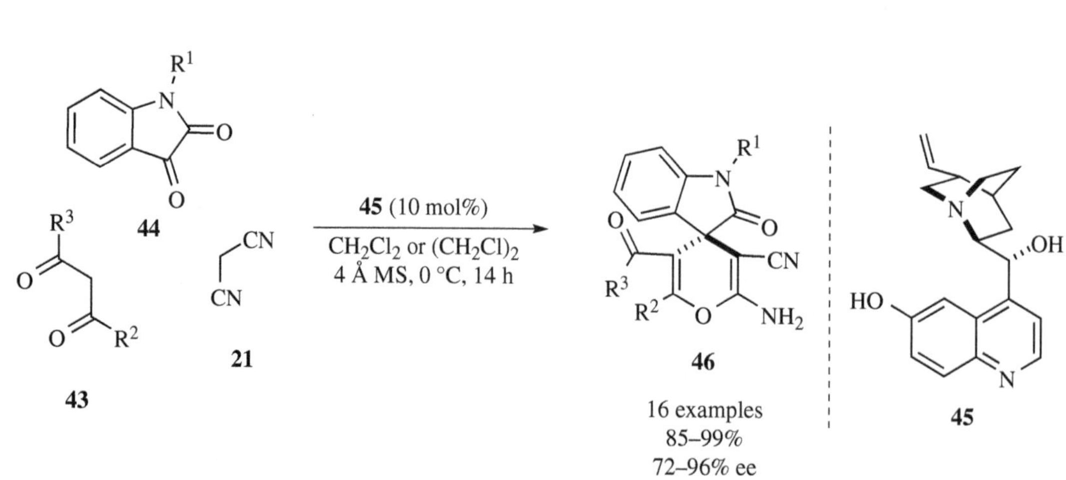

SCHEME 13.15 Cupreine **45**-catalyzed enantioselective synthesis of spiropyrans **46**.

the desired products were isolated in excellent yields. Sachdeva et al. applied NiO_{100} nanoparticles/piperidine to catalyze the formation of the desired spiro compounds [25]. In ethanol under microwave irradiation, the desired products were formed in short reaction times of 8–10 min and could be isolated in good yields of up to 90%. The catalyst could be recycled three times without loss of catalytic activity.

The use of special enamines in the three-component Knoevenagel–Michael addition/cyclization sequence directly led to *N*-alkyl- or *N*-aryl-substituted 2-amino-4*H*-pyrans. Perumal et al. therefore applied catalytic amounts of piperidine in ethanol at ambient temperature to add (*E*)-*N*-methyl-1-(methylthio)-2-nitroethenamine **49** to the previously formed Michael acceptor in a one-pot fashion (Scheme 13.16) [7]. This protocol provides the desired *N*-methyl-2-amino-4*H*-pyrans **50/51** in good yields. The methylthio substituent serves as a good leaving group facilitating the cyclization subsequent to the Michael addition reaction by nucleophilic substitution.

Analogous compounds could be obtained by Kumar et al. by applying the same enamine to different Michael acceptors derived from cyanoacetylindoles or cyanoacetylbenzene and aryl aldehydes [26]. A solvent-free procedure was published by Singh et al. making use of α-oxoketene-*N*,*S*-arylaminoacetals as the enamine component [27]. A stoichiometric amount of DABCO at elevated temperatures led to a rapid conversion of the three components to give the desired *N*-aryl-2-amino-4*H*-pyrans in good yields.

In 2006, Fadda et al. reported the piperidine-catalyzed reaction of 1-chloro-3,4-dihydronaphthalene-2-carboxaldehyde **53** with pyrazole **52** and malonodinitrile **21** to provide an access to a chromeno[2,3-*c*]pyrazole **55** at high temperatures (Scheme 13.17) [28]. Exchange of the piperidine catalyst by an excess of ammonium acetate led to the formation of a pyrazolo[3,4-*b*]quinoline derivative **54** instead.

In 2012, Bodwell and Nandaluru applied a pyrrolidine-mediated six-step domino reaction process to the synthesis of 6*H*-dibenzo[*b*,*d*]pyran-6-ones **59** (Scheme 13.18) [29].

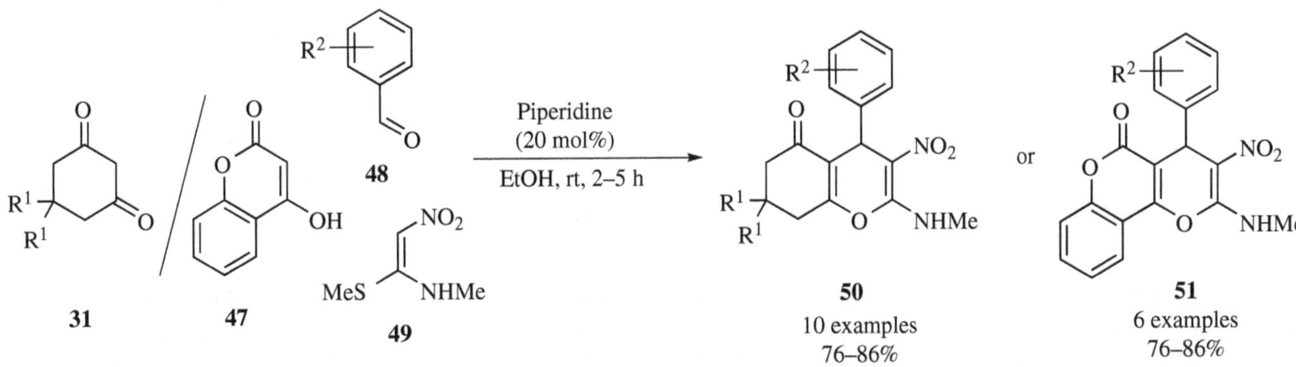

SCHEME 13.16 Piperidine-catalyzed Knoevenagel/enamine–Michael addition/cyclization sequence.

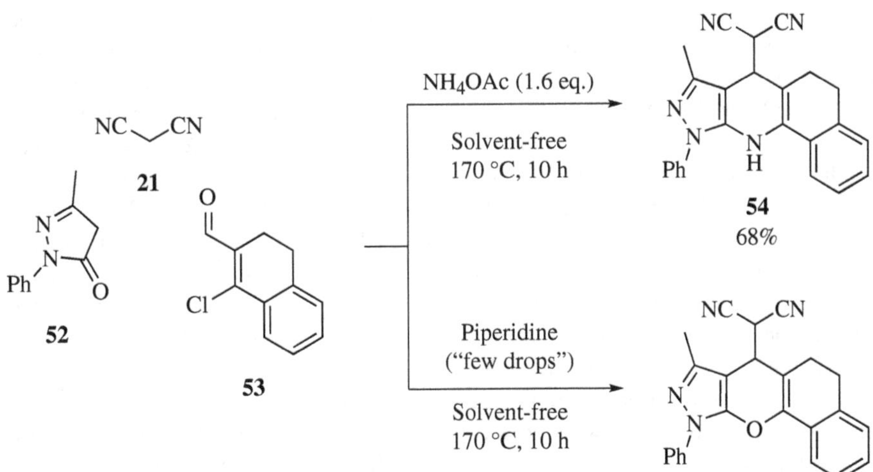

SCHEME 13.17 Pyrazolo[3,4-*b*]quinoline **54** and chromeno[2,3-*c*]pyrazole **55** syntheses.

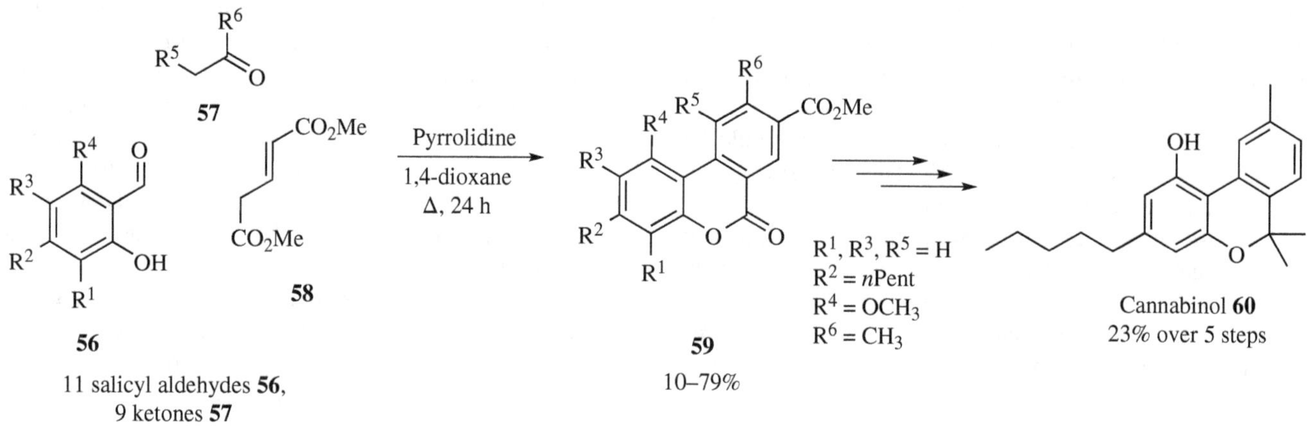

SCHEME 13.18 Application of a three-component Knoevenagel-induced six-step domino process in the synthesis of cannabinol **60**.

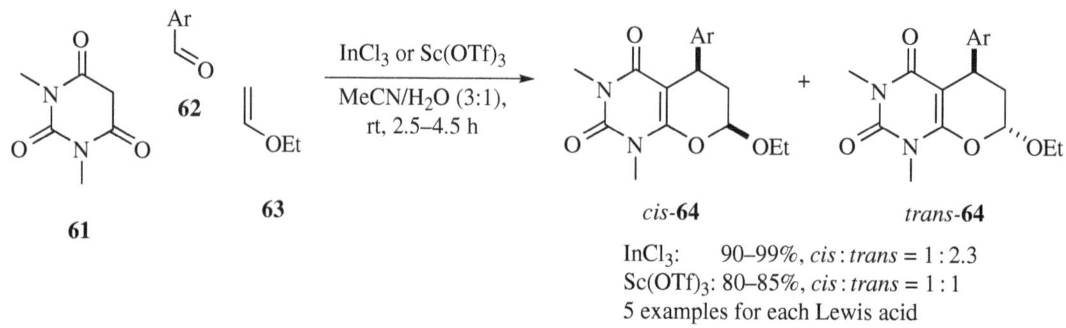

SCHEME 13.19 Indium(III) chloride (or scandium(III) triflate)-catalyzed Knoevenagel/IEDHDA reaction.

The tricyclic pyranones are provided in moderate to good yields with an excess of pyrrolidine in 1,4-dioxane at elevated temperatures for 24 h. One of the desired products is shown to be further transformed into cannabinol **60**.

13.1.2.2 ***3,4-Dihydro-2H-pyrans and the Knoevenagel/ IEDHDA Reaction*** In 2006, Prajapati and Gohain reported the reaction of 1,3-dimethylbarbituric acid **61** with various aromatic aldehydes **62** and two types of vinyl ethers (ethyl vinyl ether **63** and 2,3-dihydrofuran) to give a racemic *cis-/trans-*mixture of the corresponding Knoevenagel/ hetero-Diels–Alder products **64** (Scheme 13.19) [30]. In the case of 2,3-dihydrofuran as the dienophile, the product showed to be exclusively *cis*.

The authors tested $InCl_3$ and $Sc(OTf)_3$ as Lewis acids to catalyze the domino process. Scandium(III) triflate mostly gave a 1:1 mixture of *cis*- and *trans*-**64**, while indium(III) chloride was able to provide mostly a 1:2.3 ratio. The yields showed to be higher with $InCl_3$. The reaction could also be conducted without any catalyst, but the yields were considerably lower (65–70%, *cis*/*trans* = 1:1).

Four years later, Pałasz showed the catalyst-free reaction in aqueous suspension without acetonitrile to be mostly *cis*-selective at ambient temperature [31]. The reactions of 1,3-dimethyl barbituric acid **61** or thiobarbituric acid **65** with various aromatic aldehydes **66** and ethyl vinyl ether **63** in water at ambient temperature led to the desired products **68** in excellent yields and good *cis*-selectivity. Using styrenes and *N*-vinyl-1,3-oxazolidin-2-one **67** as the dienophile component, the reactions resulted in exclusive *cis*-selectivity (Scheme 13.20). The shown by-product **69** was formed in the cases of ethyl vinyl ether **63** as the dienophile due to acidic hydrolysis of the enol and reaction with the resulting acetaldehyde. In 2012, the work was published with a broader screening of reaction conditions and the examination of terephthaldehyde [32].

Formaldehyde **72** is, among aryl aldehydes, the most widely used aldehyde component in the MC synthesis of 3,4-dihydro-2*H*-pyrans **73** (Scheme 13.21). It serves as a methylene-dication equivalent for the creation of 4-unsubstituted pyrans. Gu et al. examined the MCR of dimedone **15**, formaldehyde **72**, and styrenes in glycerol at higher temperatures to yield 2-aryl-3,4-dihydro-2*H*-pyrans **73** [33]. Hilt et al. reported the reaction of cyclic 1,3-diketones **70**, formaldehyde **72**, and several alkenes **71** utilizing the catalyst system hydroquinone (5 mol%)/KOAc (10 mol%) in acetic acid at elevated temperatures [34]. Jiménez-Alonso et al. used 2,5-dihydroxybenzoquinone **74** and formaldehyde **72** to react them with several dienophiles at high temperature in 1,4-dioxane as well as microwave irradiation conditions in 1,2-dichloroethane. In comparison to the high temperature

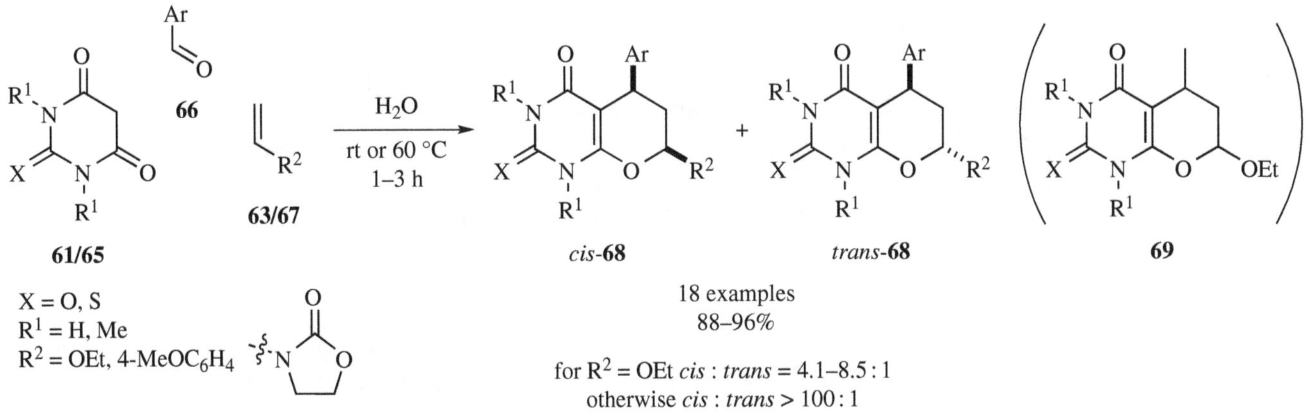

SCHEME 13.20 Knoevenagel/IEDHDA reaction in aqueous suspension.

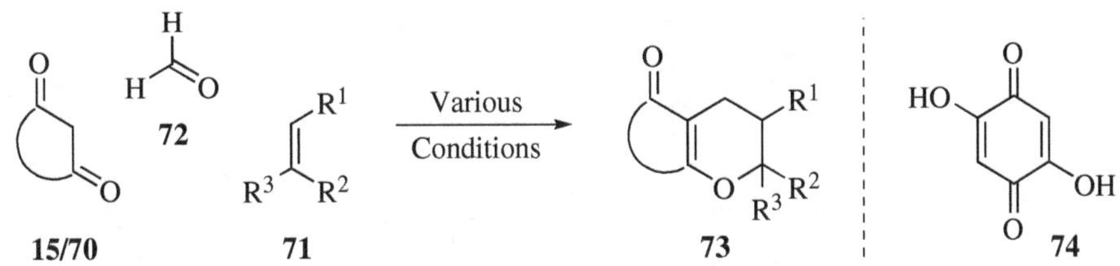

SCHEME 13.21 Formaldehyde **72** in the synthesis of 4-unsubstituted 3,4-dihydro-2*H*-pyrans **73** [33–36].

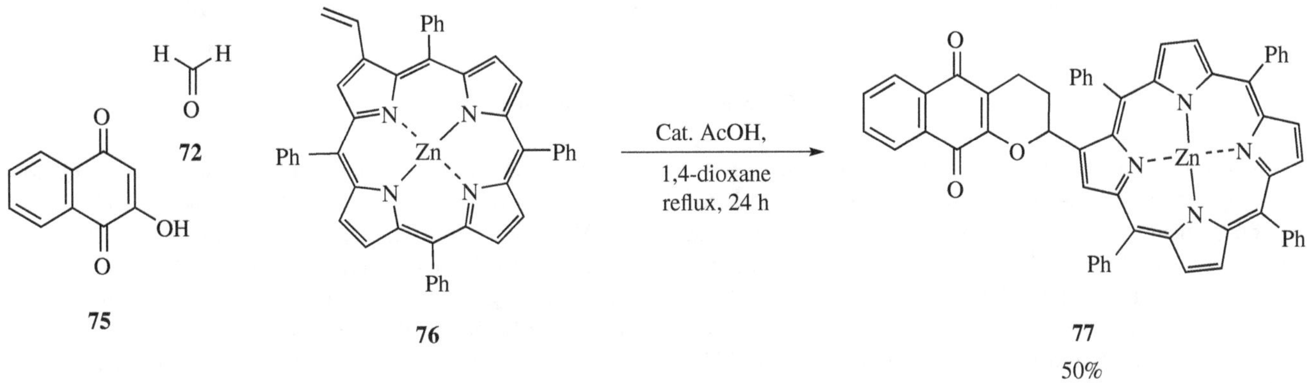

SCHEME 13.22 Formaldehyde **72** in the multicomponent synthesis of substituted porphyrins **77**.

conditions, the MWI reaction led to very high yields of the desired *bis*-Knoevenagel/IEDHDA product in short reaction times of only 15–20 min [35]. Substitution at the 6-position of the benzoquinone exclusively provided the monoreaction product [36].

Cavaleiro et al. reported the reaction of vinyl-substituted zinc porphyrins **76** with Knoevenagel products of formaldehyde **72** with several quinones **75** in refluxing 1,4-dioxane and *ortho*-dichlorobenzene (Scheme 13.22) [37]. The desired products **77** could be obtained in moderate to high yields where the exclusive formation of the *para*-benzoquinone is noteworthy (see also [36]). A catalytic amount of acetic acid reduces the reaction time and increases the product yield.

Xu et al. examined the reaction of 2-hydroxynaphthalene-1,4-dione with various aldehydes, including formaldehyde **72**, to give the corresponding Knoevenagel products that underwent smooth one-pot reaction with several dienophiles providing the pyrano products in moderate yields [38]. Ferreira examined the same reaction with aryl aldehydes and styrenes in a refluxing mixture of EtOH and water (1:1) [39]. The desired products could be obtained in moderate to good yields. Here, the *ortho*-benzoquinones were found in up to 33% yield.

Soman and Patel reported the reaction of an extended hydroxychromenone **78** with aryl aldehydes **79** and 2-methoxypropene **80** in the presence of a catalytic amount of ethylenediammonium diacetate (EDDA) to provide an intermediate **81** for the synthesis of new furowarfarins (Scheme 13.23) [40]. The reaction is conducted in dry 1,4-dioxane at 85–90 °C in a screw-cap pressure tube.

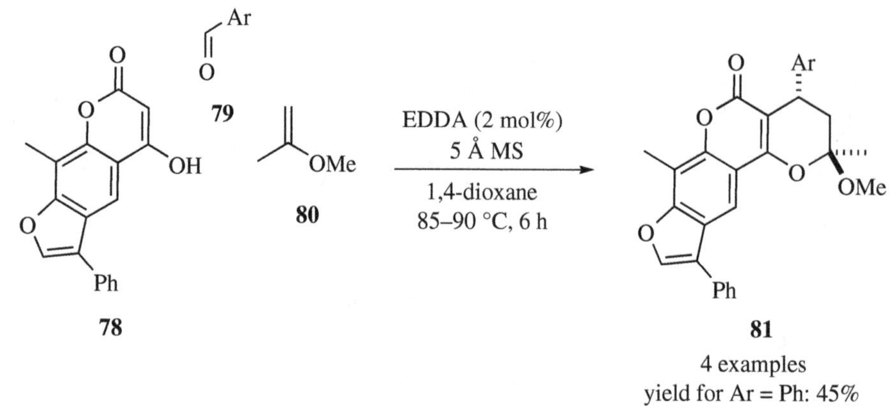

SCHEME 13.23 Application of a Knoevenagel/hetero-Diels–Alder MCR in the synthesis of new furowarfarins.

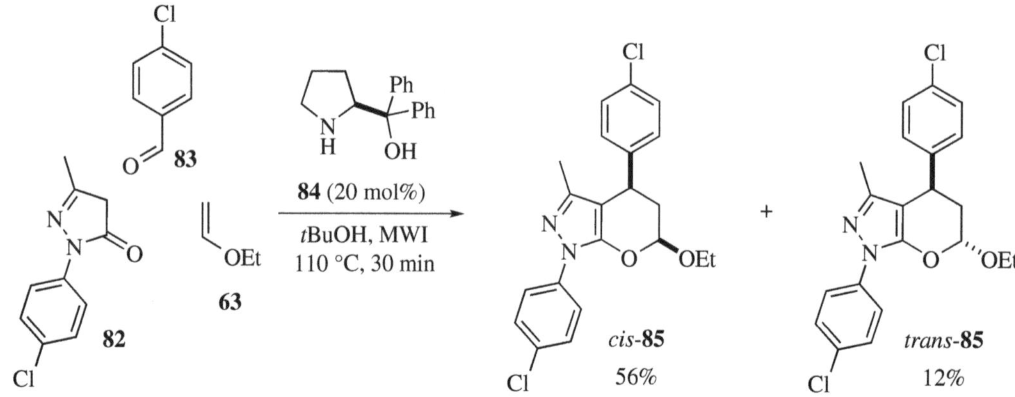

SCHEME 13.24 MCR synthesis of new potential antitubercular agents.

In the research on new potential antitubercular agents, Botta et al. applied a Knoevenagel–HDA MCR, which was catalyzed by α,α-diphenylprolinol **84** under microwave irradiation at 110 °C (Scheme 13.24) [41]. The organocatalyst activates the carbonyl group in a double hydrogen bond donation motif.

Very similar products could be obtained by Parmar et al. utilizing the ionic liquid tetraethylammonium acetate [42]. This reaction was completed under solvent-free reaction conditions within 5 h at ambient temperature or within 8–12 min under microwave irradiation. The yields were comparably high in both cases (86–92%).

Jérôme et al. reported the catalyst-free reaction of mostly acyclic 1,3-dicarbonyl compounds with formaldehyde **72** and styrene derivatives in water at elevated temperatures [43]. As the authors complemented one year later, the reaction does not take place at all under anhydrous conditions [44]. A DFT calculation study revealed that the water plays an essential role in the dehydration step of the initial Knoevenagel addition product (Scheme 13.25).

In the water-unassisted reaction, a four-membered cyclic transition state was assumed for the dehydration. The water-assisted pathway passes a six-membered transition state instead. This is more likely as can be deduced from the difference in the calculated Gibbs free energy for each transition state. A pseudo-five-component reaction could be demonstrated to give a tricyclic product **88** (Scheme 13.26) [43].

Mostly acyclic 1,3-dicarbonyl compounds were also reacted with aldehydes and alkyl vinyl ethers under solvent-free conditions at 80 °C by Lin et al. [45]. Several catalysts have been screened, though EDDA gave the best results.

In 2009, Tietze et al. reported the synthesis of D-(+)-forosamine sugars *via* a racemic Knoevenagel–HDA reaction (Scheme 13.27a) [8]. The resolution of the enantiomers was carried out by chiral HPLC at the end of the synthesis. The authors suggest a stepwise zwitterionic reaction mechanism for this reaction. DFT calculations supported this mechanism two years later, where an improved procedure for the synthesis of sugars of the forosamine and ossamine series was published (Scheme 13.27b) [9]. The zwitterionic intermediates explain the *cis-/trans*-selectivity, as mentioned before.

In 2012, Rueping et al. reported the proline-mediated reaction of 1,3-diketones with aldehydes to provide 2-hydroxy-3,4-dihydro-2*H*-pyran derivatives in good to excellent yields [46]. The reaction mechanism involves a Knoevenagel–Michael addition sequence with subsequent hemiacetalization. The hemiacetal was oxidized with TPAP/NMO or PCC to give the corresponding lactones **96**. An enantioselective variant utilizing stoichiometric amounts of an

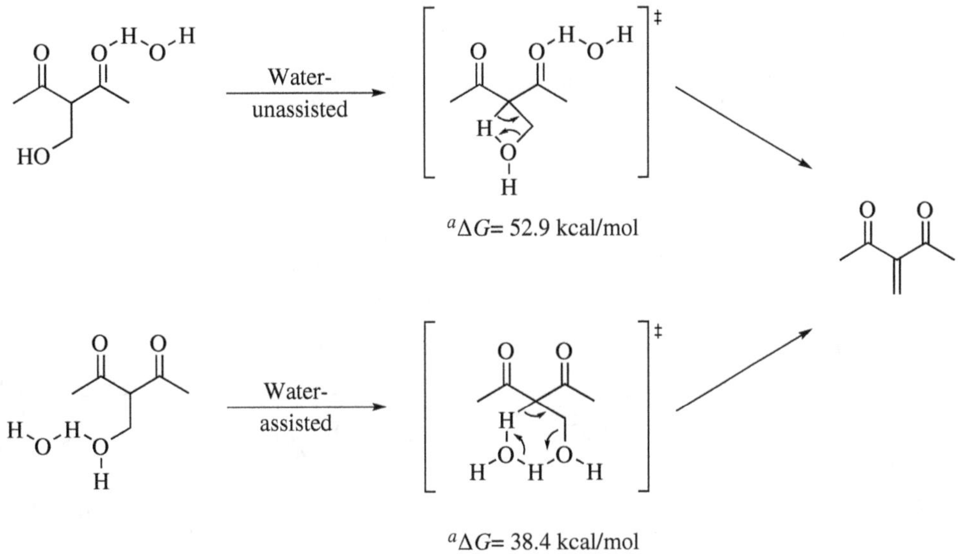

SCHEME 13.25 Assumed transition states for the dehydration step of the initial Knoevenagel addition product and the corresponding Gibbs free energies based on DFT calculations of monohydrated compounds [44].

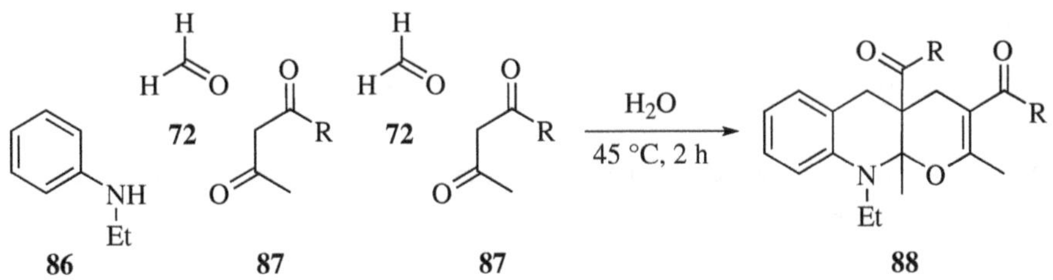

SCHEME 13.26 Pseudo-5CR provides a tricyclic product **88**.

α,α-diarylprolinol-based organocatalyst **95** was developed aswell, which was coupled with the following oxidation step. This afforded the desired lactones **96** in moderate yields (49–52%), with good enantioselectivity (83–89% ee) (Scheme 13.28).

A similar reaction was published by Song et al. in 2013 (Scheme 13.29) [47]. 2-Hydroxynaphthoquinone **75** was reacted with aromatic aldehydes **97** and ethyl 4,4,4-trifluoro-3-oxobutanoate **98** catalyzed by a mixture of ammonium acetate and acetic acid (25 mol% each). A Knoevenagel–Michael addition sequence was followed by hemiketal formation to give the desired product **99** in moderate to good yields. Dehydration of the product yielded the 4*H*-pyran derivatives.

13.1.3 Useful Syntheses of Heterocycles: 1,4-Dihydropyridine and Diazine Syntheses

13.1.3.1 1,4-Dihydropyridines and 3,4-Dihydropyridones

In 2006, Wang et al. reacted the dimedone-derived enamine **101** with various benzaldehydes **100** and cyclic 1,3-dicarbonyl compounds **31** to afford the tricyclic 1,4-dihydropyridine products **102** in excellent yields (Scheme 13.30) [10]. The reactions were carried out in the ionic liquid 1-butyl-3-methyl-imidazolium bromide ([bmim]Br) at elevated temperatures and were completed within 3–5 h. The ionic liquid has been shown to be reusable up to four times. The scope of this reaction was further extended one year later to include several active methylene compounds (such as malonodinitrile and 1*H*-indene-1,3(2*H*)-dione), utilizing the ionic liquid [bmim]BF_4 [48].

Similar enamines **104** were used by Ji et al. in 2009 to afford spirodihydropyridines **105** from isatin derivatives **103** and malonodinitrile **21** (Scheme 13.31) [49]. The reaction smoothly led to the desired products under microwave irradiation at elevated temperatures within 7–10 min, which could be isolated in moderate to high yields (66–85%).

An acid-catalyzed version of this reaction was published by Wang et al. in 2013 [50]. A substoichiometric amount of acetic acid in refluxing ethanol mediates the Knoevenagel/enamine–Michael addition/cyclization sequence to provide the spirodihydropyridines in high yields.

The $InCl_3$-catalyzed reaction of barbituric and thiobarbituric acid-derived enamines **107** with aryl aldehydes **106** and 3-(2-cyanoacetyl)indole **26** was described by Bhuyan and

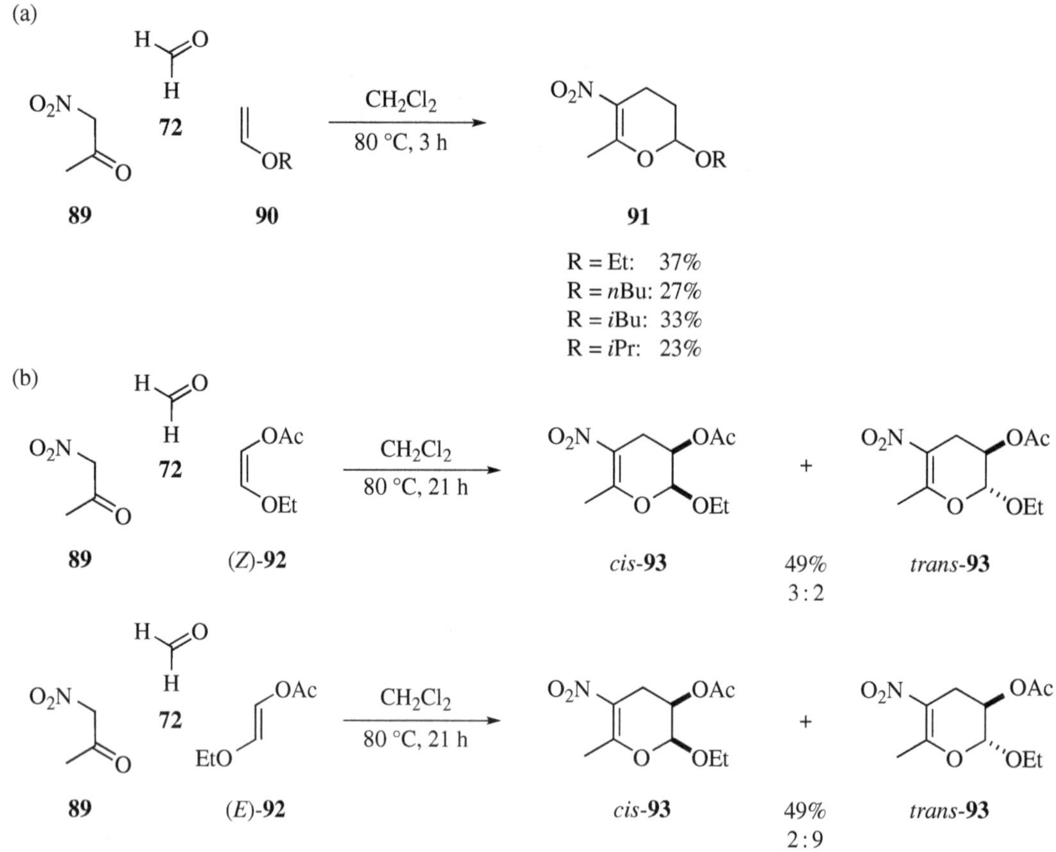

SCHEME 13.27 (a) Multicomponent Knoevenagel/IEDHDA reaction in Tietze's forosamine synthesis (2009) [8]; (b) key step in the improved synthesis of the forosamine- and ossamine-type sugars (2011) [9].

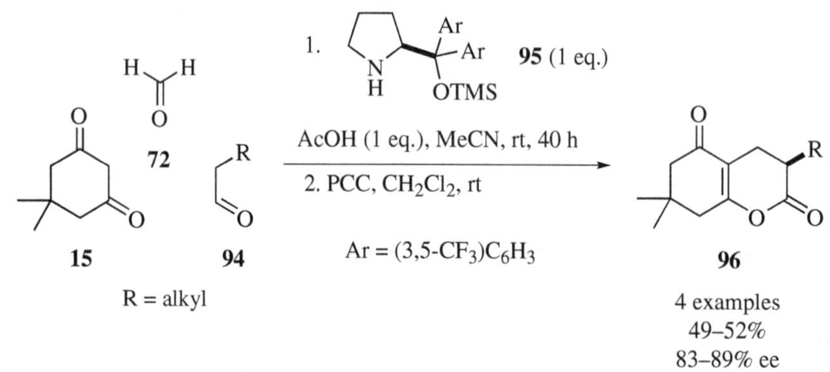

SCHEME 13.28 Organocatalyst-mediated construction of 2-hydroxy-3,4-dihydro-2*H*-pyrans **96** followed by lactol oxidation.

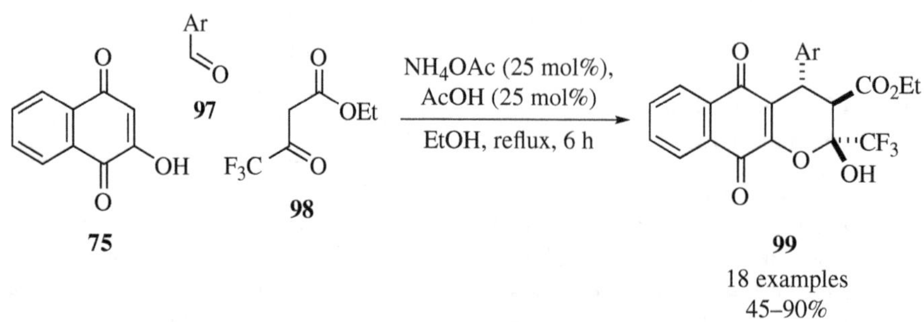

SCHEME 13.29 Ammonium acetate-/acetic acid-catalyzed synthesis of 2-hydroxy-2,3-dihydro-4*H*-pyran **99**.

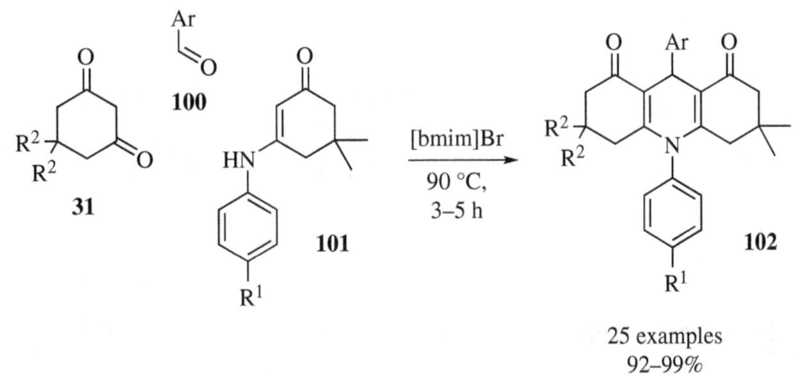

SCHEME 13.30 Ionic liquid synthesis of tricyclic dihydropyridines **102** by Wang et al.

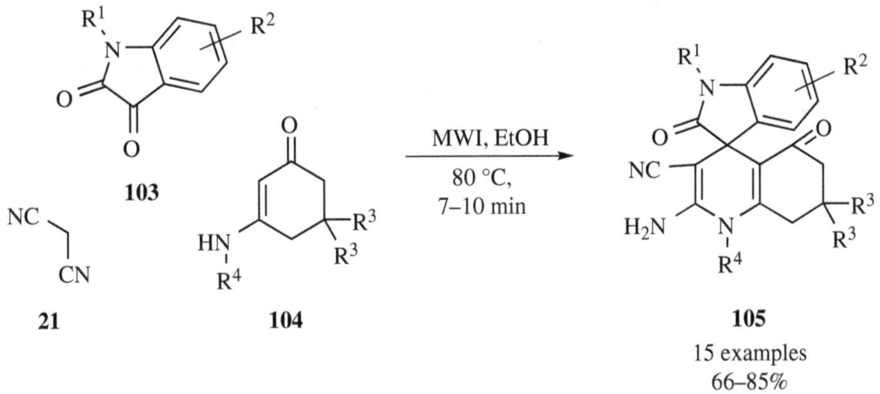

SCHEME 13.31 Synthesis of spirodihydropyridine derivatives **105** by Ji et al.

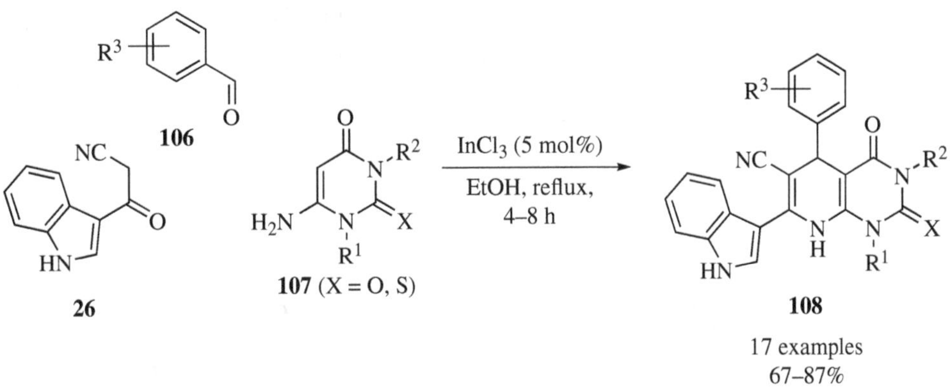

SCHEME 13.32 Indium(III) chloride-catalyzed synthesis of dihydropyrido[2,3-*d*]pyrimidines **108**.

coworkers in 2012 (Scheme 13.32) [51]. This reaction led to several 1,4-dihydropyrido[2,3-*d*]pyrimidine derivatives **108** in moderate to high yields (67–87%) within four to eight hours in refluxing ethanol.

The reaction of dimedone-derived enamines **110**, methyl (2-cyano)acetate **109**, and formaldehyde **72** mediated by lithium perchlorate led to the formation of a Knoevenagel/enamine–Michael addition product (Scheme 13.33) [52]. This could be cyclized in a one-pot procedure with substoichiometric amounts of triphenyl phosphine to provide dihydropyridone derivatives **111** in moderate to high yields (67–85%).

A direct cyclization to the corresponding pyridones is possible by the use of Meldrum's acid **112** as the CH-acidic component (Scheme 13.34) [53–55]. This serves as a synthetic "CH_2-C(O)" -equivalent, as discussed in Section 13.3 (*vide infra*). Various reaction conditions have been applied for those reactions, including neutral and acidic ionic liquids as well as piperidine as a secondary amine catalyst in refluxing ethanol.

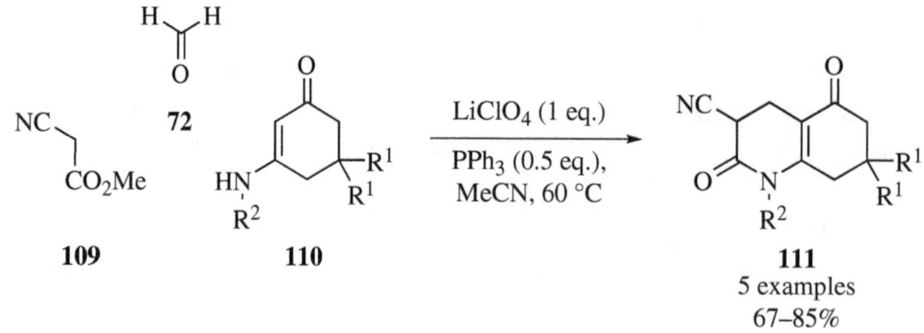

SCHEME 13.33 Dihydropyridone **111** synthesis mediated by lithium perchlorate and triphenyl phosphine by Georg and Gu.

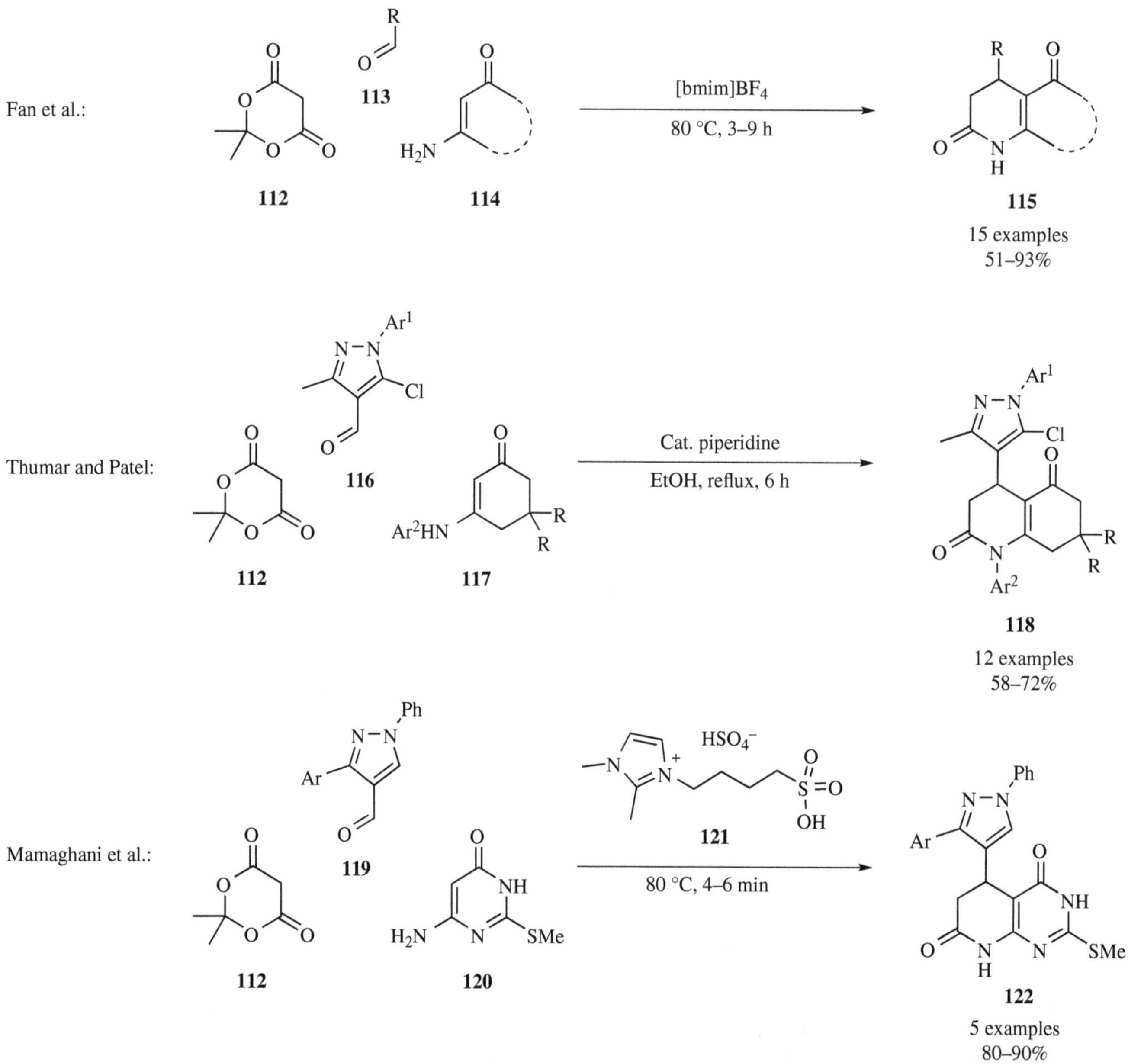

SCHEME 13.34 Direct cyclization of Knoevenagel/enamine–Michael addition products to the corresponding pyridone derivatives **115**, **118**, and **122** [53–55].

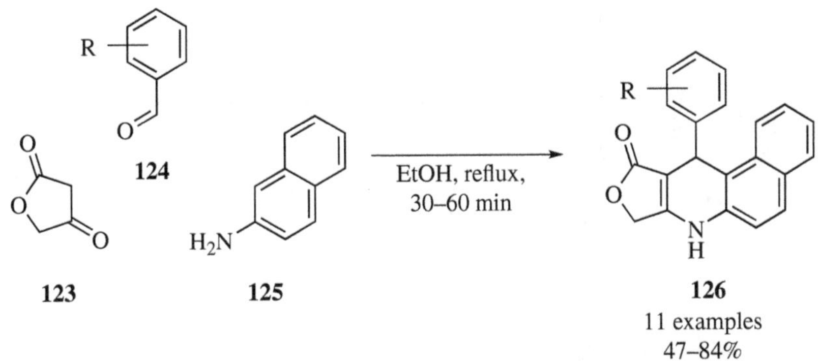

SCHEME 13.35 Synthesis of a tetracyclic 1,4-dihydropyridine derivative **126** by Pashkovskii et al.

As 2-aminonaphthalene **125** can provide an enamine-like reactivity, Pashkovskii et al. used this compound for the synthesis of tetracyclic 1,4-dihydropyridine derivatives **126** starting from tetronic acid **123** and several aryl aldehydes **124** (Scheme 13.35) [56]. The reactions were conducted in refluxing ethanol within 30–60 min and the desired products **126** could be obtained in moderate to good yields (47–84%).

The enamine can be formed *in situ* in a 4CR as well (Scheme 13.36). A primary amine **128** [57] or, in most cases, ammonium acetate [58–61] is reacted with a 1,3-dicarbonyl compound to yield the corresponding enamine. This undergoes nucleophilic addition to the formed Michael acceptor (by reaction of Meldrum's acid **112** with the corresponding aldehydes) followed by cyclization with the Meldrum's acid moiety present in the Michael acceptor molecule. Thus, pyridones are accessible in a 4CR starting from simple, readily available substances. Mostly, acidic conditions are applied for their syntheses, but microwave irradiation under neutral reaction conditions works fine as well.

Jiang et al. described the proline-catalyzed reaction of several amines **143** with alkynes **141**, various aldehydes **142**, and 1,3-dicarbonyl compounds **144** to afford 1,4-dihydropyridines **145** in moderate to good yields (65–85%) (Scheme 13.37) [62]. Mainly three reactions are involved in the production of those products: the first one is a proline-catalyzed Knoevenagel reaction between the aldehydes **142** and the 1,3-dicarbonyl compounds **144** to give Michael acceptors. The second one is a hydroamination reaction of the alkyne **141** to yield enamines, which in the third reaction undergo an enamine–Michael addition/cyclization sequence to provide the desired products.

Bicyclic 1,4-dihydropyridines were successfully synthesized by Li et al. by the reaction of imidazolidine-based enamines **148** with CH acids **146/112** and aldehydes **147/151** in the presence of a weak base (Scheme 13.38) [63]. Cyclization of those products under basic conditions afforded the tetracyclic 1,4-dihydropyridine derivatives **150/153**. One year later, this reactions scope was extended by the group utilizing malonodinitrile as the CH-acidic component to yield 2-amino-3-cyano-1,4-dihydropyridines [64].

Thiazolidine-derived enamines **155** led to thio-substituted bicyclic 1,4-dihydropyridine derivatives **156** in good to excellent yields [65]. A 4CR using thiazolidinone-derived enamines **158** gave the arylidene-substituted products **159/160** (Scheme 13.39) [66].

Monocyclic thio-substituted 1,4-dihydropyridines could be obtained by the reaction of α-CH-acidic thioamides with aldehydes and CH acids (Scheme 13.40) [67]. Here, first, the pyridothiones are afforded under microwave irradiation that can be converted to the desired products via S-alkylation. Also tricyclic compounds **163/166** of this type can be accessed by the application of thioamides **161/164** that carry certain halogenoaryl substituents [68]. The alkylation step is displaced by an intramolecular S-arylation. Potassium fluoride on neutral alumina catalyzed this reaction in PEG6000 under microwave conditions to provide the desired products **163** in high yields (73–88%). Triethylamine instead has been shown to be a good catalyst for this reaction if Meldrum's acid is used as the CH-acidic component [69].

Imidazole efficiently catalyzed the 1,4-dihydropyridine formation between 2-cyanoacetamide, aromatic aldehydes, and 1,3-dicarbonyl compounds [70].

The preceding reactions represent 1,4-dihydropyridine formations utilizing two dissimilar CH acids that readily contain a nitrogen atom. Other methods have been described where the CH acids do not bear any nitrogen. Instead, this is provided by nitrogen sources like ammonium acetate [71].

An efficient approach to pyrido[2,3-*d*]pyrimidines **171** was reported by Singh et al. in 2011 [72]. The Knoevenagel condensation is catalyzed by L-proline **170** in refluxing ethanol to give an *ortho*-iminemethide intermediate, which forms the desired products by subsequent HDA reaction in moderate to good yields (46–88%) (Scheme 13.41).

Rostamnia et al. successfully applied a sulfonic acid-modified mesoporous silica nanoreactor with ultrasonic irradiation in ethanol at ambient temperature for the synthesis of 1,4-dihydropyridine derivatives [73]. The desired products were isolated in excellent yields after short reaction times,

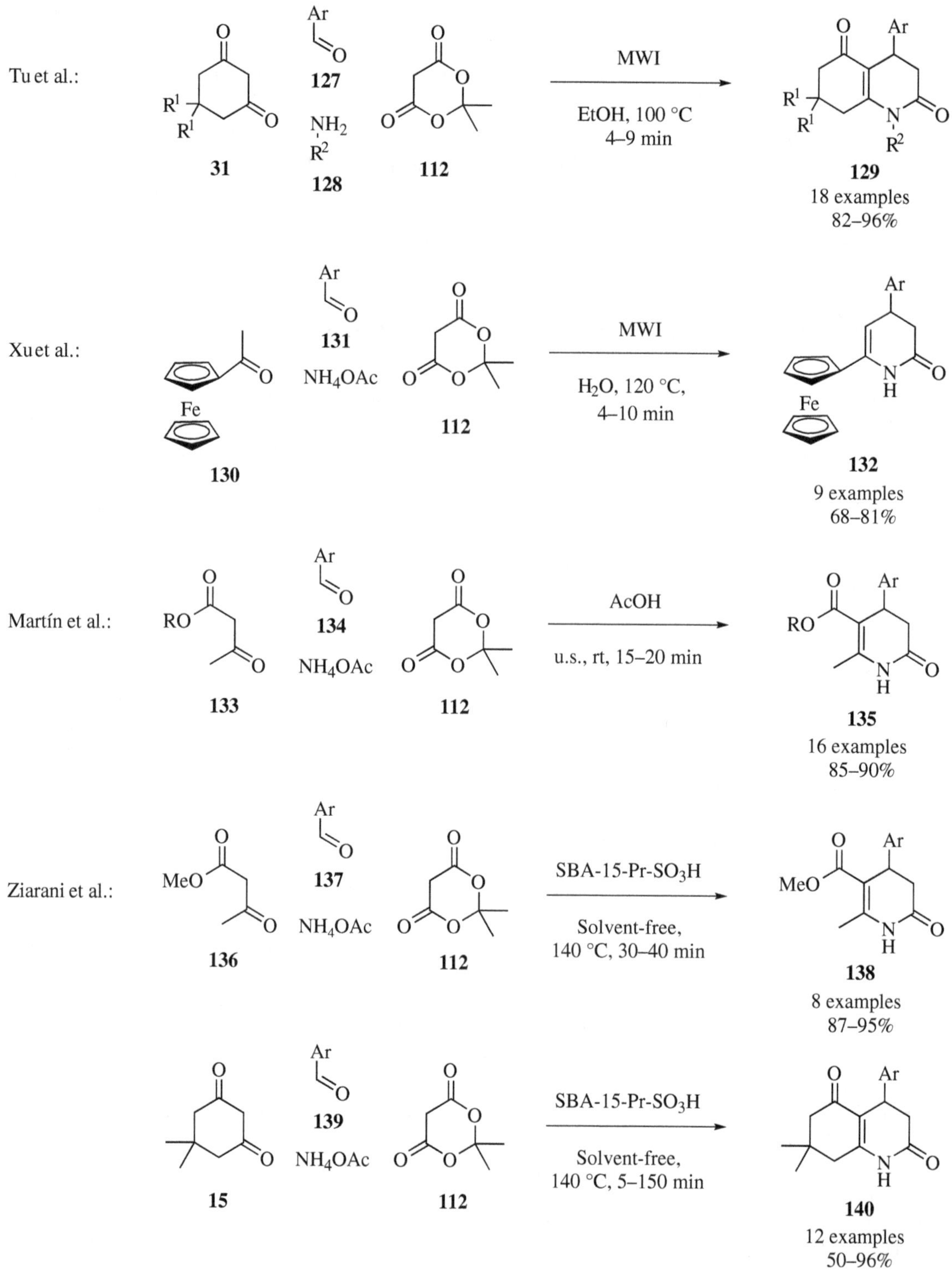

SCHEME 13.36 Four-component syntheses of pyridone derivatives **129**, **132**, **135**, **138**, and **140** [57–61].

while the catalyst was shown to be recyclable up to five times with only little loss of catalytic activity.

*13.1.3.2 **Diazines*** Variably substituted 4-amino-5-pyrimidinecarbonitriles **174** were accessed by the base-mediated Biginelli-type reaction of phenylamidine or guanidine **173** with several aromatic aldehydes **172** and malonodinitrile **21** (Scheme 13.42) [11]. The best results could be obtained with substoichiometric amounts of triethylamine in toluene under microwave irradiation for up to 2 min. These results were

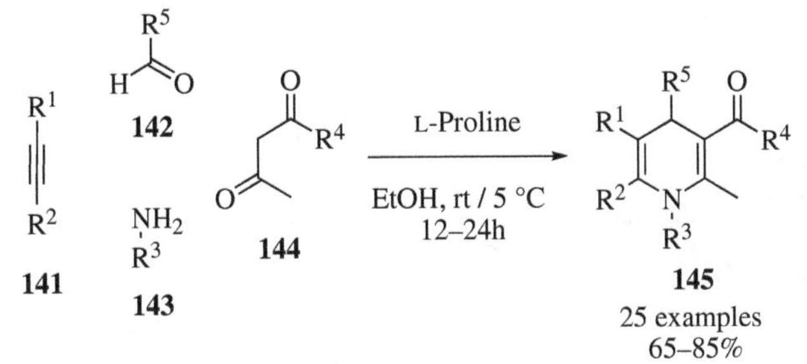

SCHEME 13.37 Four-component 1,4-dihydropyridine **145** synthesis by Jiang et al.

compared to thermal reaction conditions in a water/ethanol (10:1) mixture with stoichiometric sodium acetate. In this case, the reaction medium had to be heated for up to 12 h to complete the reactions, probably not least because of the higher dilution with respect to the toluene/MWI conditions.

Later, the authors extended the scope of this reaction by utilizing MgO as an effective mediator in refluxing acetonitrile [74]. The reaction times could be reduced to 45–120 min, while the yield was improved. The application of ethyl 2-cyanoacetate **175** instead of malonodinitrile **21** led to pyrimidinone derivatives **177**. Using phenylhydrazonopropane-2-one **179**,

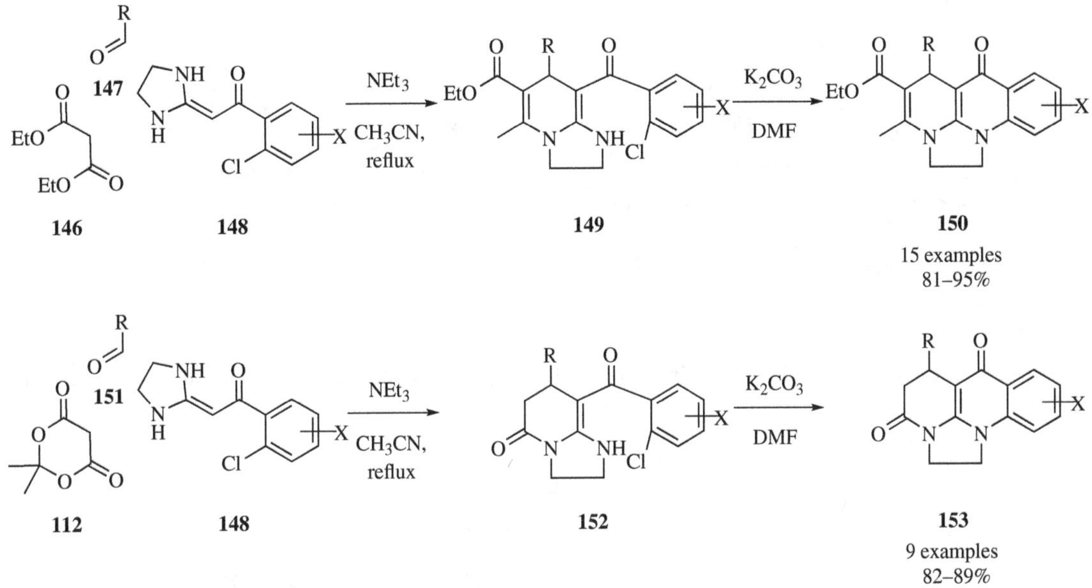

SCHEME 13.38 Synthesis of tetracyclic 1,4-dihydropyridine derivatives **150** and **153** by Li et al.

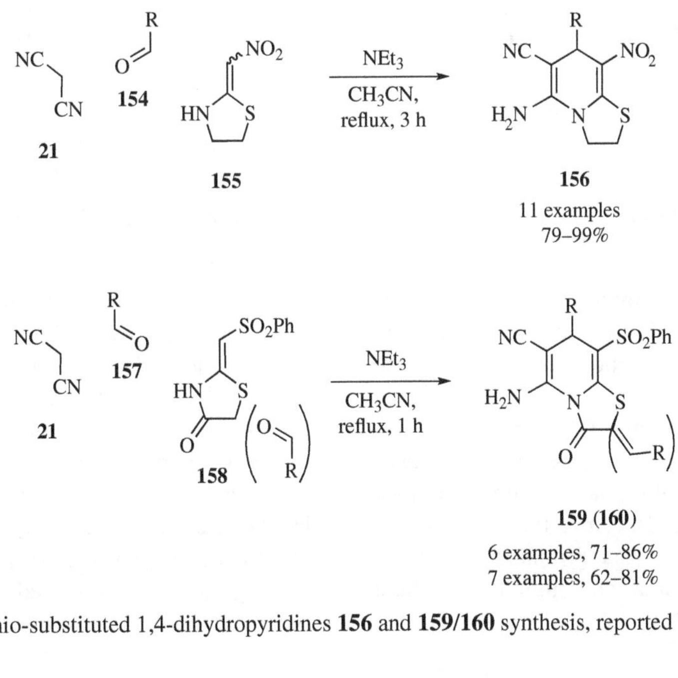

SCHEME 13.39 Thio-substituted 1,4-dihydropyridines **156** and **159/160** synthesis, reported by Altuğ et al. [65, 66].

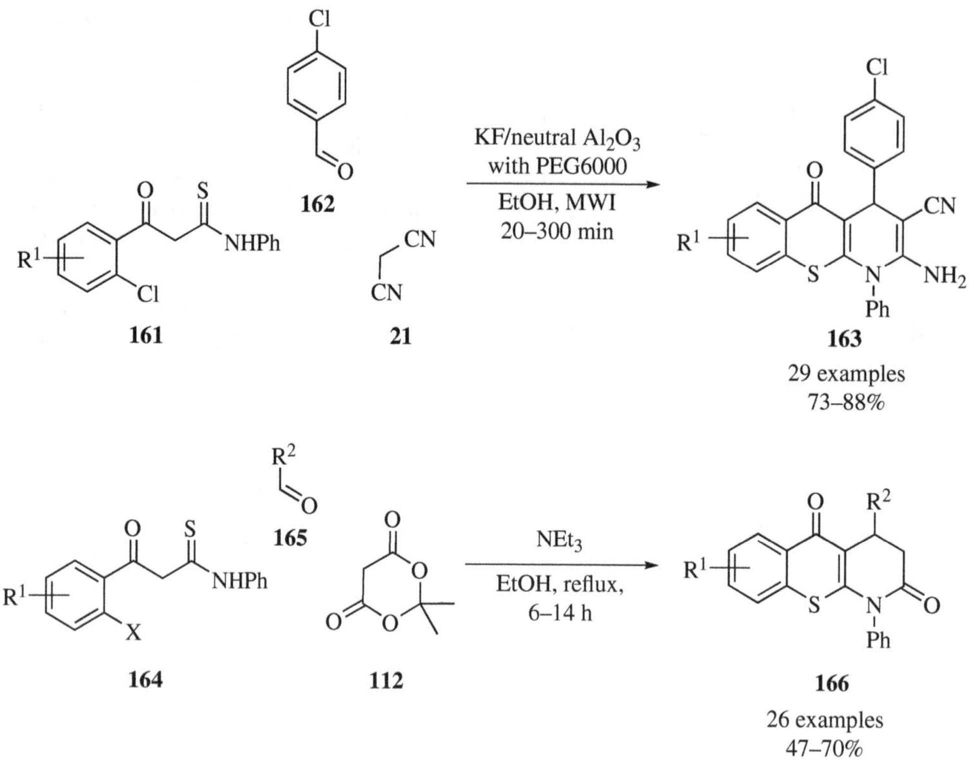

SCHEME 13.40 Tricyclic 2-amino-1,4-dihydropyridines **163** and 1,4-dihydropyridones **166** cyclized by nucleophilic aromatic substitution [68, 69].

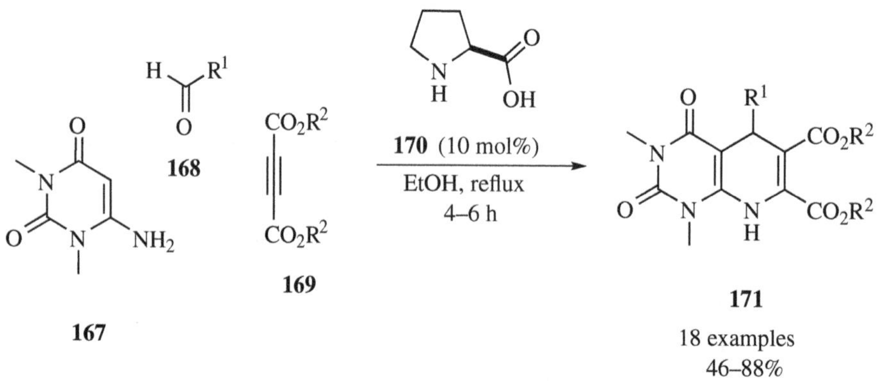

SCHEME 13.41 Knoevenagel/hetero-Diels–Alder approach to pyrido[2,3-*d*]pyrimidines **171**.

the 2,5-dihydropyridazines **180** could be obtained in high yields (50–78%) [75]. The use of high-surface-area magnesium oxide significantly reduced the reaction time and increased the yields.

An access to spiro-2-aminopyrimidinones **184** was reported by Balalaie et al. in 2008 by reacting N-substituted 4-piperidinones **182** with guanidinium carbonate **183** and alkyl 2-cyanoacetates **181** in refluxing methanol (Scheme 13.43) [76]. The desired products **184** were obtained in short reaction times and high to excellent yields (70–96%).

As mentioned before for the synthesis of pyridones, 5-unsubstituted pyrimidinones **186** can be accessed by utilizing Meldrum's acid **112** as the CH-acidic component (Scheme 13.44) [77]. In refluxing ethanol, the desired product could be obtained under catalyst-free conditions in moderate to good yields (51–83%). In DMF at high temperatures with the use of an amino-functionalized highly porous MCM-41 catalyst, the yields could be increased, as reported by Mirza-Aghayan et al. [78].

Singh and Devi reported the tin(II)-mediated Biginelli-type reaction of 3-substituted methyl β-ketodithioates with

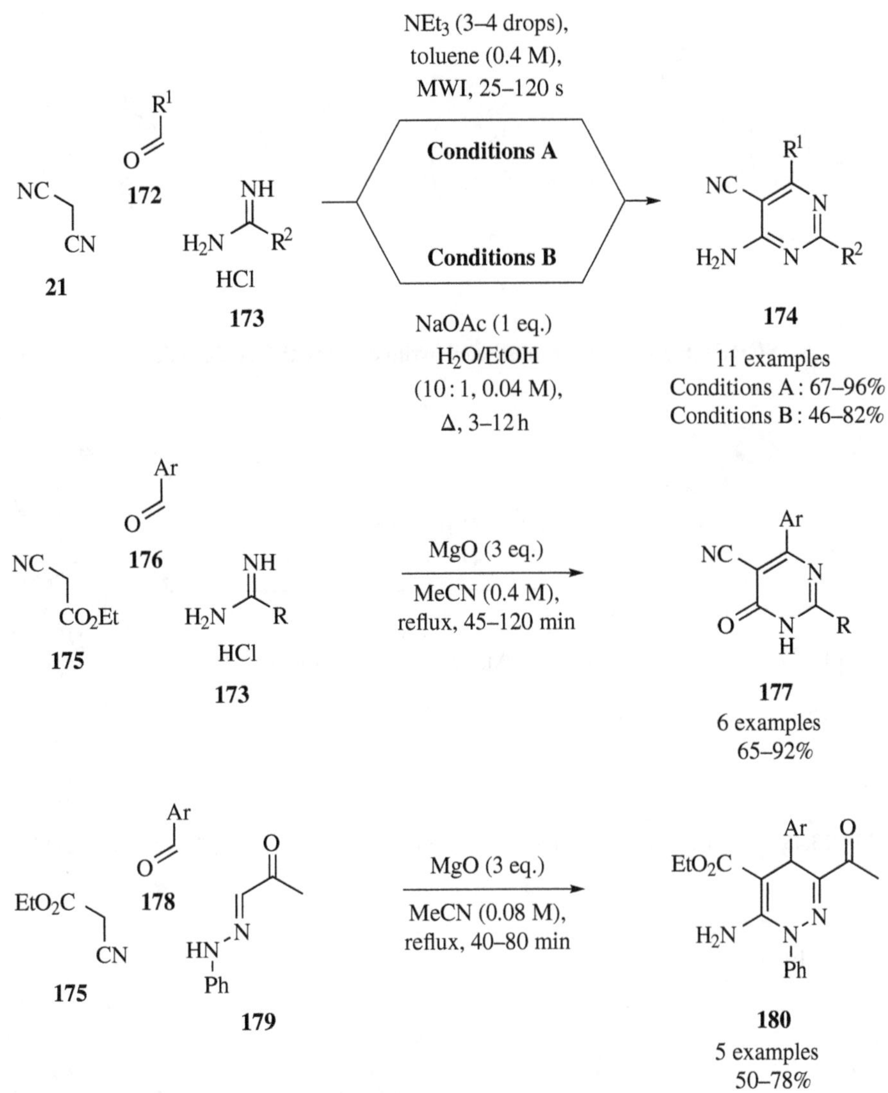

SCHEME 13.42 Base-mediated pyrimidine and pyrimidinone formation under thermal and MWI conditions [11, 74, 75].

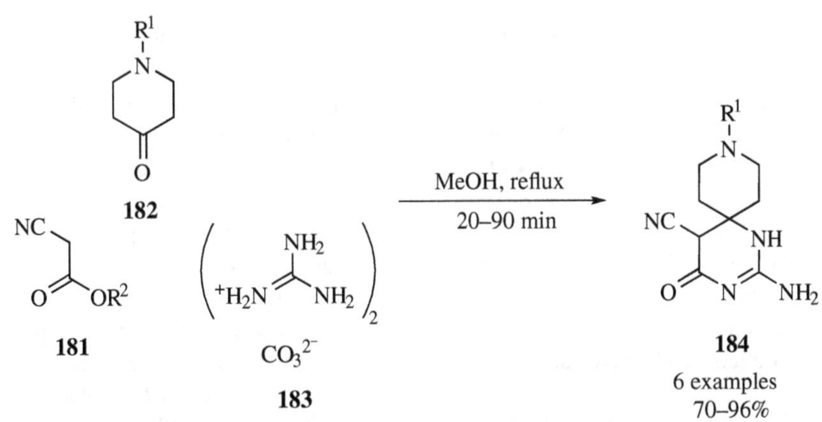

SCHEME 13.43 Spiro-2-aminopyrimidinone **184** synthesis.

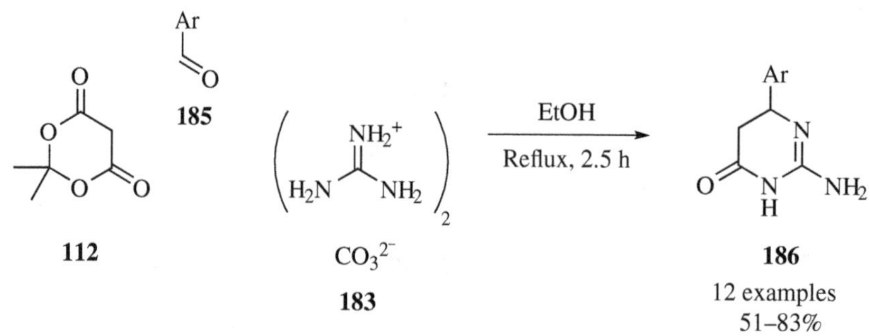

SCHEME 13.44 5-Unsubstituted pyrimidinones **186** by Balalaie et al.

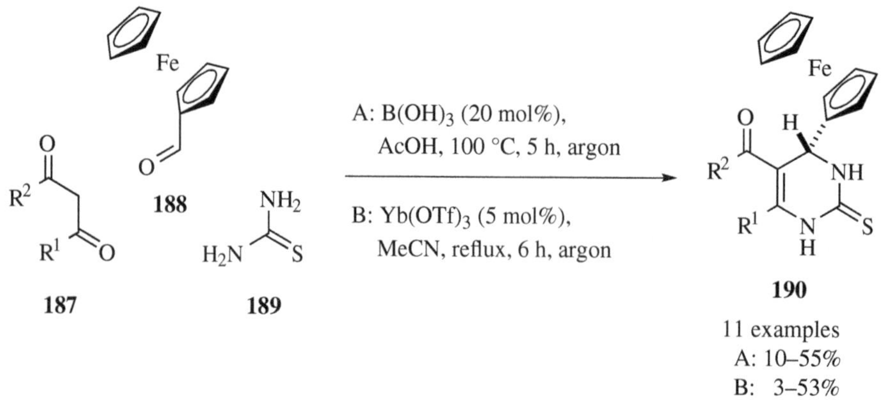

SCHEME 13.45 Biginelli reaction for the synthesis of ferrocenyl-dihydro-1,3-pyrimidine-2-thiones **190**.

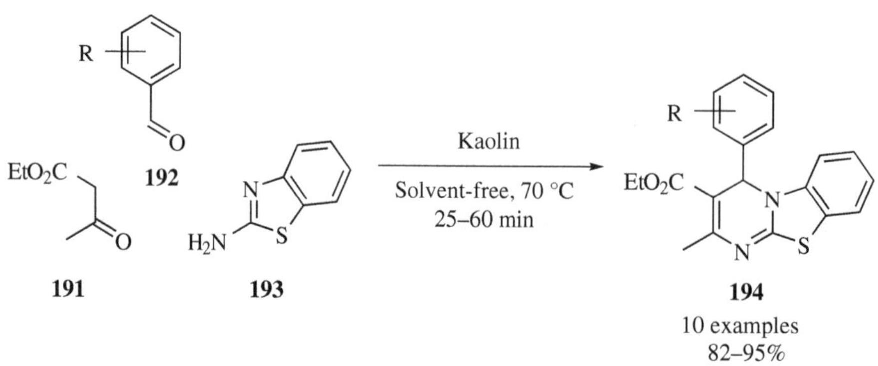

SCHEME 13.46 Kaolin-catalyzed Biginelli-type reaction.

urea and a set of aldehydes to obtain dihydro-1,3-pyrimidine-2-ones at high temperatures in good yields [79]. A Biginelli-type reaction of formylferrocene **188** with thiourea **189** and 1,3-dicarbonyl compounds **187** led to the formation of ferrocenyl-dihydro-1,3-pyrimidine-2-thiones **190** (Scheme 13.45) [80]. This reaction was catalyzed by boronic acid in acetic acid at high temperatures under inert atmosphere or by ytterbium(III) triflate in refluxing acetonitrile, also under inert atmosphere. The products **190** were obtained in moderate yields.

Analogous compounds have been prepared by the kaolin-catalyzed reaction of ethyl acetoacetate **191** with various aldehydes **192** and urea or thiourea [81]. This reaction is carried out without any solvent at elevated temperature. The kaolin catalyst has been recycled up to four times with only slight loss of activity. By means of changing the thiourea component to 2-aminobenzothiazole **193**, tricyclic pyrimidine derivatives **194** could be achieved in good yields (82–95%) (Scheme 13.46).

Structurally related spiro compounds have been described by Arya and Kumar in 2011 [82].

5-Aminotetrazole **197** contains a structurally implemented guanidine moiety that has been applied by Cai et al. in the multicomponent synthesis of dihydrotetrazolopyrimidines **198** [83].

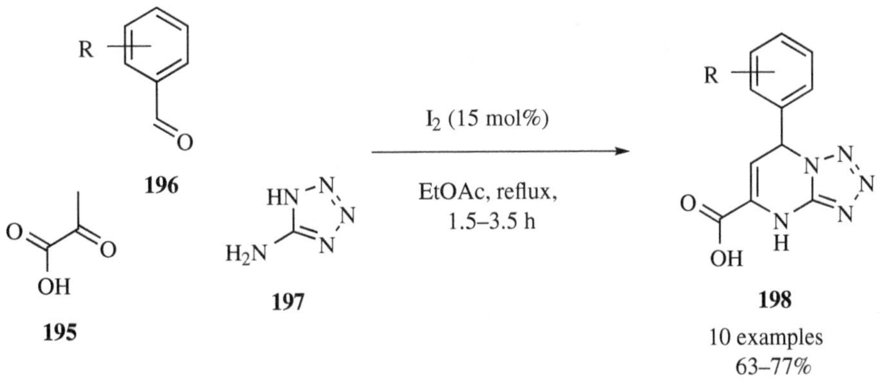

SCHEME 13.47 Iodine-catalyzed Biginelli-type reaction of an implemented guanidine moiety.

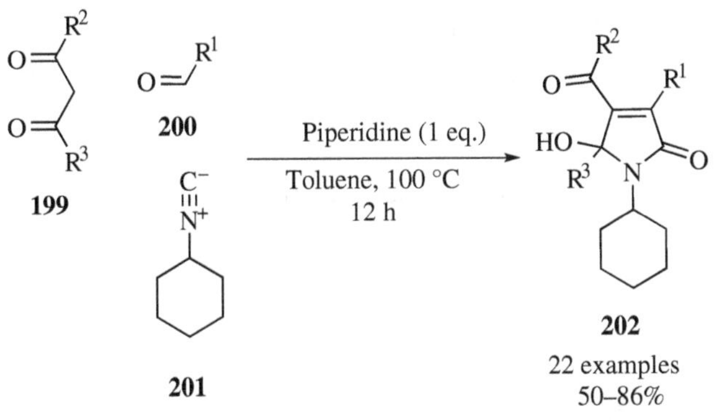

SCHEME 13.48 5-Hydroxy-2*H*-pyrrol-2-one **202** synthesis by Liang et al. [85].

Various aryl aldehydes **196** have been cyclized with 5-aminotetrazole **197** and pyruvic acid **195** catalyzed by iodine in refluxing ethyl acetate (Scheme 13.47).

An enzyme-catalyzed Biginelli reaction has been reported by Zhang et al. [84]. Porcine pancreatic trypsin showed to be the most effective enzyme among 13 enzymes tested. The reactions were conducted in anhydrous ethanol at 37 °C for 72 h and the desired products could be isolated in good to excellent yields.

13.1.4 Useful Syntheses of Heterocycles: Various Heterocyclic Scaffolds

13.1.4.1 N-Heterocycles Liang et al. reported the synthesis of several substituted 5-hydroxy-2*H*-pyrrol-2-one derivatives **202** by piperidine-mediated reaction of aromatic and aliphatic aldehydes **200** with 1,3-dicarbonyl compounds **199** followed by cyclization with cyclohexyl isocyanide **201** (Scheme 13.48) [85]. The reaction is believed to follow a sequence consisting of Knoevenagel condensation, [4+1] cycloaddition, and oxidative rearrangement as suggested by Quai et al. (Scheme 13.49) [86]. The desired products could be obtained in moderate to good yields after 12 h at high temperature. Several aliphatic aldehydes could be converted to the desired products in a one-pot fashion.

In 2007, Renuga et al. investigated the solvent-free reaction of bis(aroylmethyl) sulfides **208** with aromatic aldehydes **207** and ammonium acetate for the synthesis of various thiazole derivatives **209** under microwave irradiation (Scheme 13.50a) [87]. These reactions led to the formation of the desired products in moderate yields (62–73%), in short reaction times (10 min). The suggested mechanism is shown in Scheme 13.50b.

A set of rather complex dispiropyrrolidines **218** has been synthesized by Li et al. (Scheme 13.51a) [88]. Their 4CR strategy involves a Knoevenagel condensation/1,3-dipolar cycloaddition sequence to access the desired products in high yields (63–92%) and short reaction times (1–2.5 h). Sarcosine **215** serves in this case as a Knoevenagel condensation mediator as well as a reactant for the following 1,3-dipolar cycloaddition (Scheme 13.51b).

In 2013, a related reaction was published by Dandia et al. in a three-component fashion with a preformed Knoevenagel condensation product **223** (Scheme 13.52) [89]. This reaction follows a similar mechanism as discussed in Scheme 13.51 and provides the desired products **226** in high yields up to 92% in refluxing aqueous methanol.

Several 5-substituted tetrazoles have been synthesized by Bazgir et al. in 2012 starting from aryl aldehydes **227** (or isatin derivatives **229**) and malonodinitrile **21** (Scheme 13.53) [90]. The Knoevenagel condensation was followed by a 1,3-dipolar cycloaddition of sodium azide with one of the cyano groups to afford the tetrazoles **228** and **230** in moderate to good yields. The reactions were conducted in aqueous solution at elevated temperatures and were mostly finished within one day of reaction time.

Spiro-isoxazolines **234** have been prepared by a Knoevenagel condensation followed by 1,3-dipolar cycloaddition by Perumal and Uma Maheswari (Scheme 13.54) [91]. The desired products were obtained at rt after a few hours in high yields (72–89%). The basic conditions generate 1,3-dipoles from several aryl hydroximinoyl chlorides **232**, which smoothly undergo 1,3-dipolar cycloaddition with the previously formed Knoevenagel condensation product.

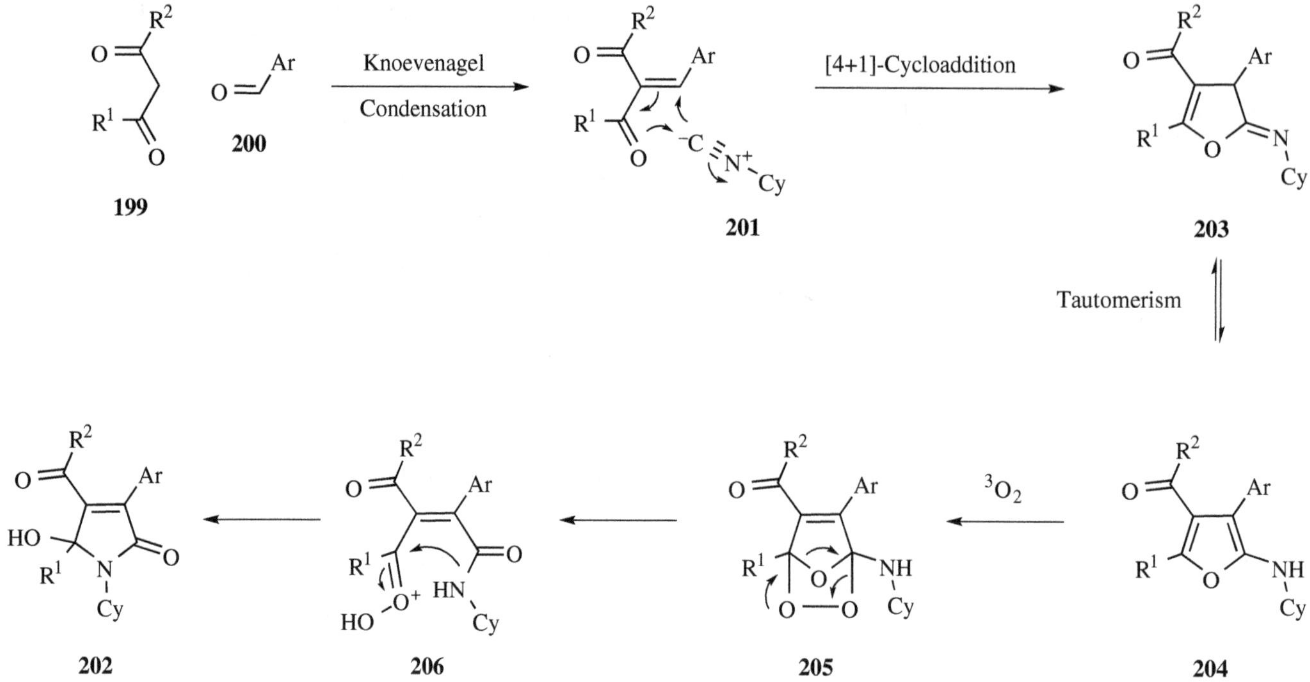

SCHEME 13.49 Proposed mechanism for the formation of 5-hydroxy-2*H*-pyrrol-2-ones **202** by Quai et al. [86].

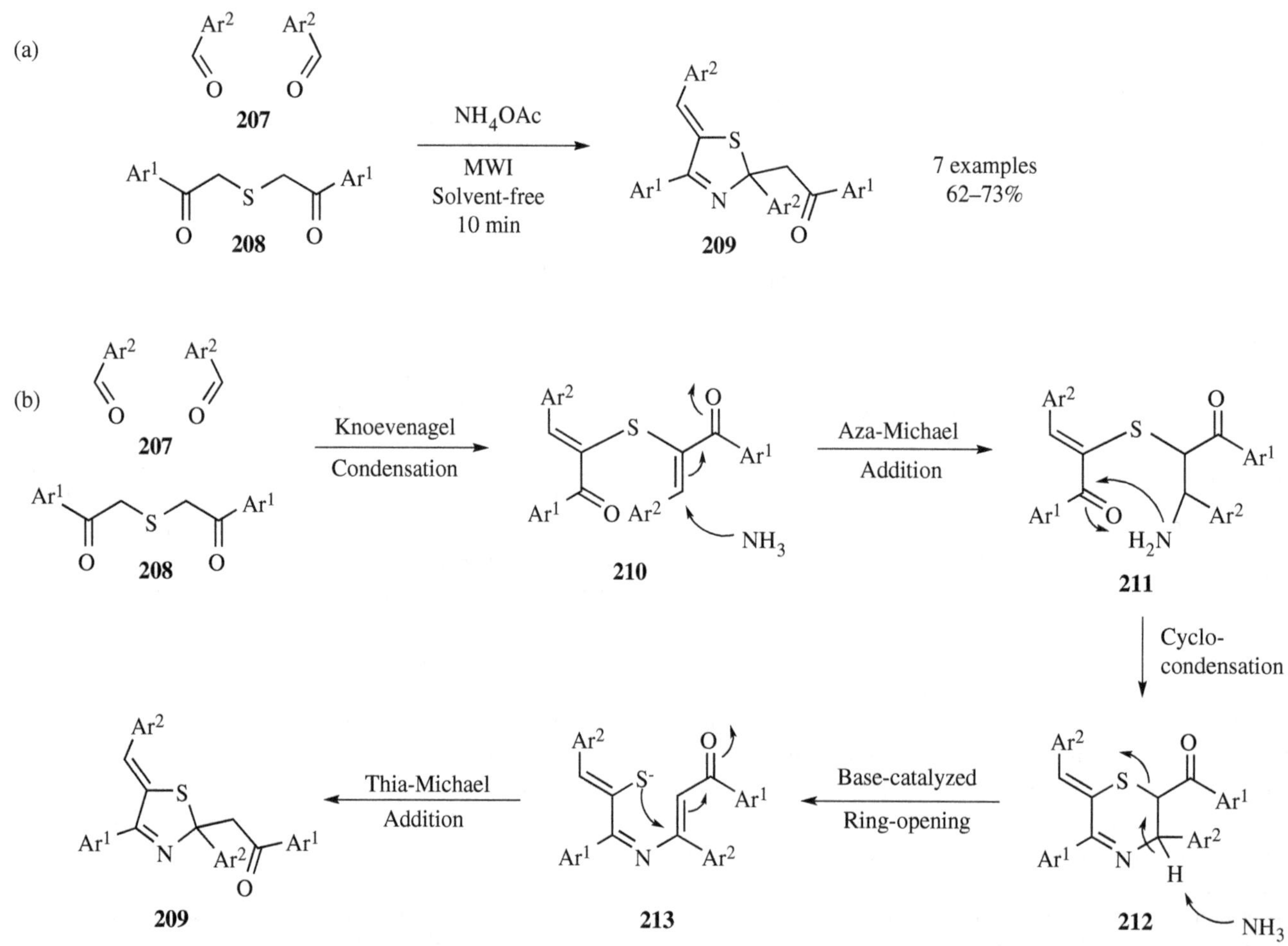

SCHEME 13.50 (a) Three-component thiazole **209** synthesis by Renuga et al. and (b) the proposed mechanism for its formation [87].

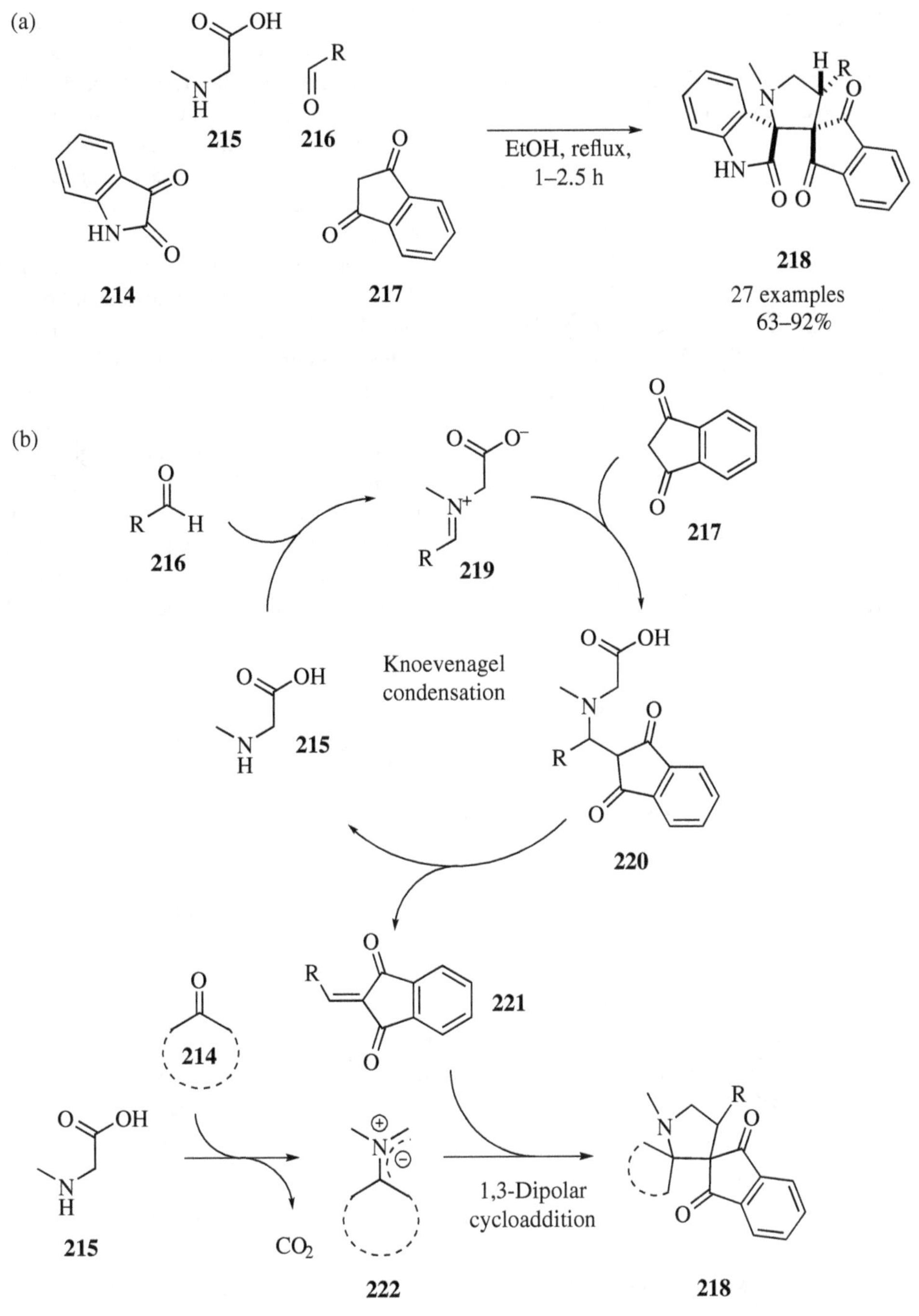

SCHEME 13.51 (a) Four-component dispiropyrrolidine **218** synthesis by Li et al. and (b) the proposed mechanism for its formation [88].

1*H*-Pyrazolo[1,2-*b*]phthalazine-5,10-dione derivatives **238** have been prepared by Bazgir et al. *via* an acid-catalyzed Knoevenagel condensation/aza-Michael addition/cyclization sequence in the ionic liquid [bmim]Br [92]. The reactions were carried out at high temperature within 2–5 h using 30 mol% *p*-toluenesulfonic acid (PTSA) as the catalyst. They provide the desired products **238** in good to excellent yields (73–97%) (Scheme 13.55).

An analogous reaction has been examined by Rostamnia et al. [73]. The authors applied a sulfonic acid-modified mesoporous silica nanoreactor with ultrasonic irradiation for the three-component reaction of aryl aldehydes **239** with dimedone **15** and phthalhydrazide **235** to give analogous products **240** to the ones discussed in Scheme 13.55 in excellent yields (92–96%), after short reaction times of 15–20 min (Scheme 13.56). The catalyst has been reused up to seven times without significant loss of activity.

Heydari and Tahamipour reported a Knoevenagel condensation/cyclization sequence of certain 1,10-phenanthrolinium N-ylides (prepared from their corresponding pyridinium salts **241**) with aromatic aldehydes **242** and malonodinitrile **21** to give tetracyclic 1,10-phenanthroline derivatives **243** (Scheme 13.57) [93]. The reaction is complete within 15 min and provides the desired products in high yields over 80%. The 4CR of phenanthridine **245**, aromatic aldehydes **244**, malonodinitrile **21**, and several isonitriles **246** gives tetracyclic phenanthridine derivatives **247** in excellent yields over 90% and reaction times between 16 and 48 h at rt, as published by Marandi et al. [94]. Both reactions follow a Knoevenagel condensation/Michael addition/cyclization mechanism.

Ionic liquid-bound pyrrolo-[1,2-*a*]-benzimidazoles have been synthesized by Sun et al. by reacting IL-supported 2-(cyanomethyl)benzimidazole with aldehydes and isonitriles under microwave irradiation at high temperature [95]. After precipitation and cleavage from the IL support, the desired products could be obtained in good yields and moderate to high purity without chromatographic purification.

A silica tungstic acid (STA)-catalyzed Knoevenagel condensation/Michael addition/double Mannich reaction

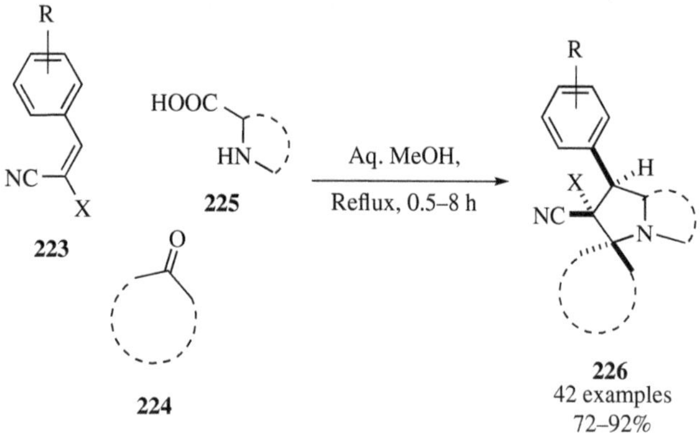

SCHEME 13.52 Three-component spiropyrrolidine **226** synthesis by Dandia et al. [89].

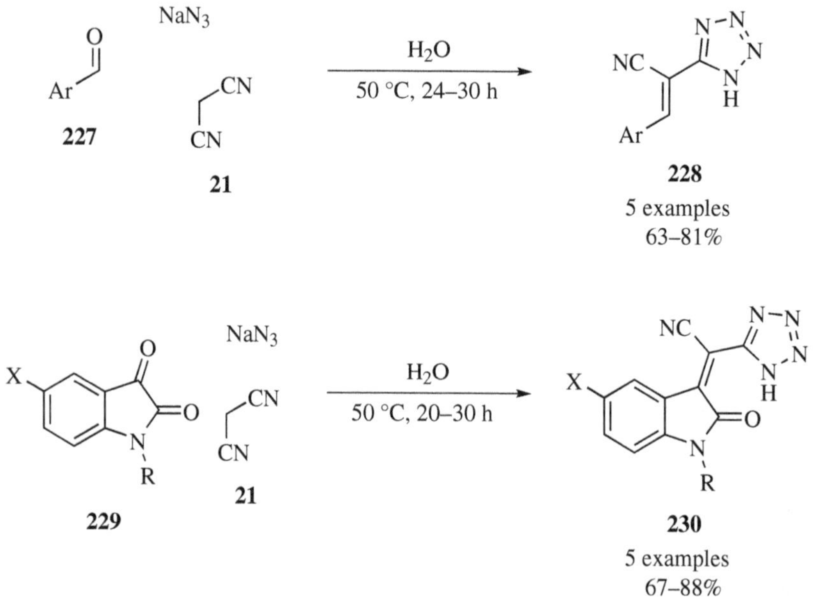

SCHEME 13.53 MCR to afford 5-substituted thiazoles **228** and **230** as reported by Bazgir et al. [90].

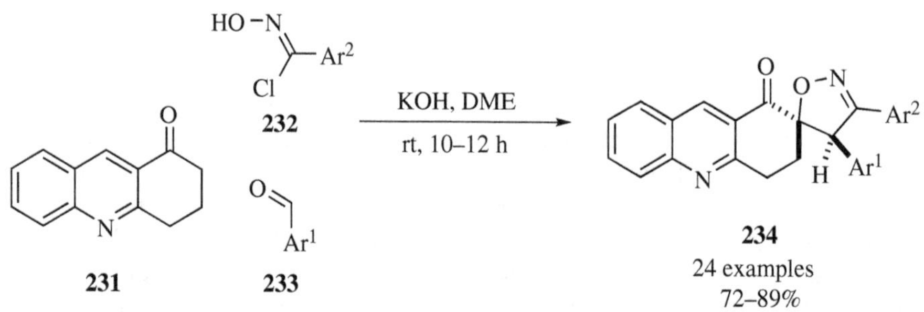

SCHEME 13.54 Knoevenagel condensation/1,3-dipolar cycloaddition to afford spiro-isoxazolines **234** [91].

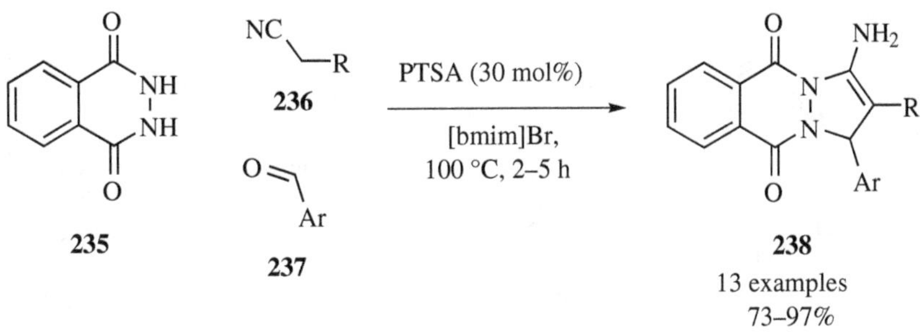

SCHEME 13.55 Synthesis of 1*H*-pyrazolo[1,2-*b*]phthalazine-5,10-dione derivatives **238** by Bazgir et al. [92].

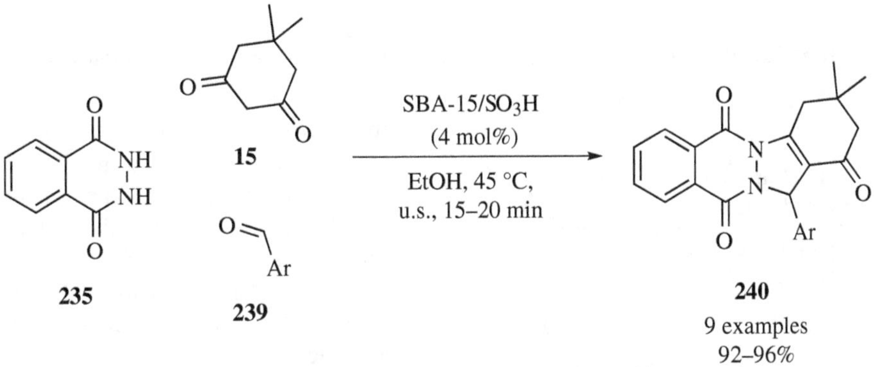

SCHEME 13.56 Synthesis of 1*H*-pyrazolo[1,2-*b*]phthalazine-5,10-dione derivatives **240** by Rostamnia et al. [73].

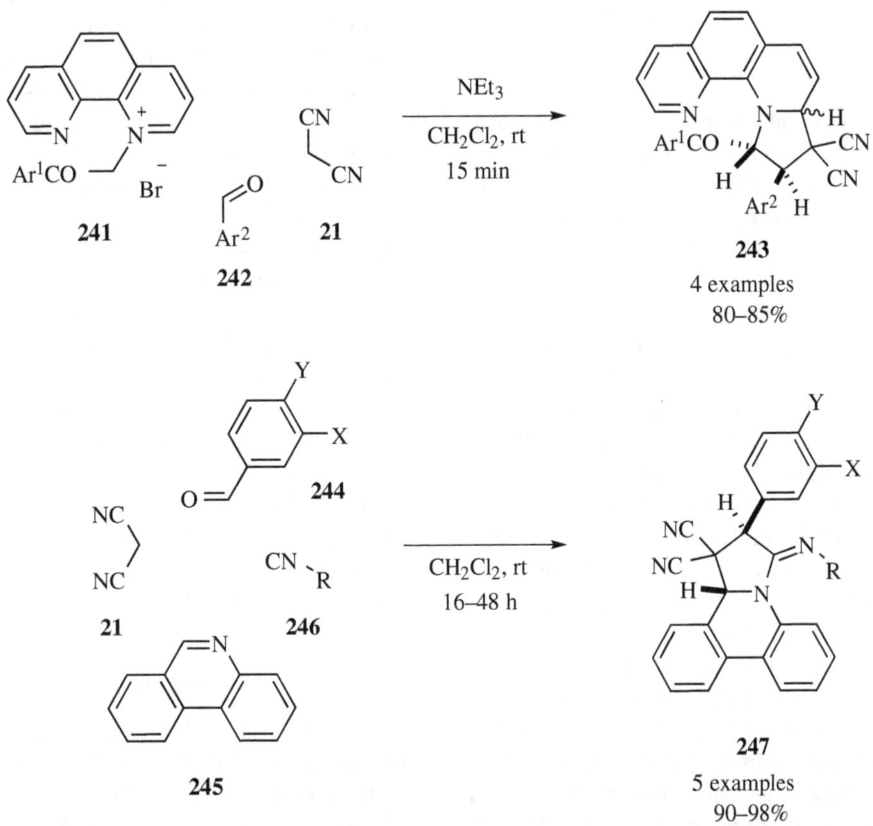

SCHEME 13.57 Syntheses of tetracyclic derivatives of 1,10-phenanthroline **243** and phenanthridine **247** [93, 94].

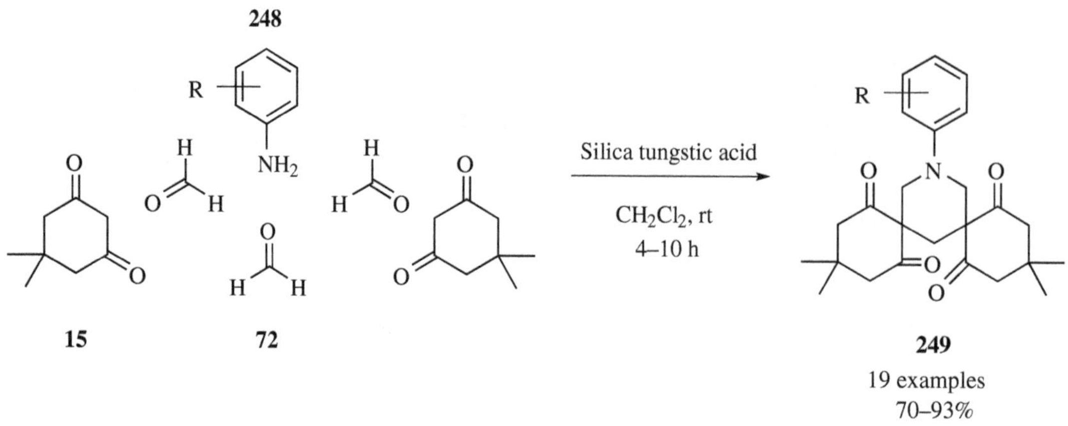

SCHEME 13.58 Silica tungstic acid (STA)-catalyzed dispiro-compound **249** formation [96].

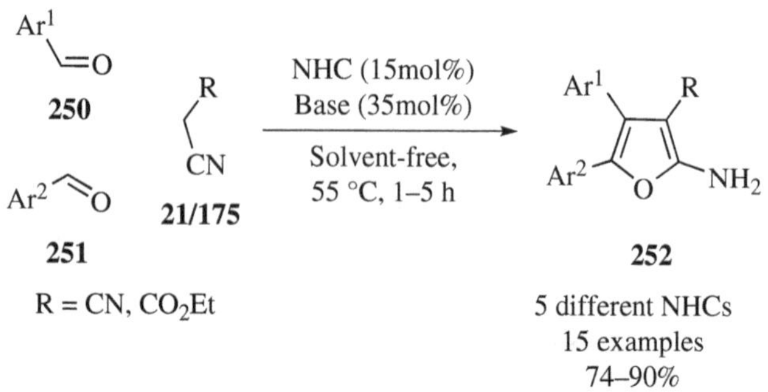

SCHEME 13.59 NHC umpolung to deliver fully substituted furan derivatives **252** [97].

has been reported by Atar and Jeong (Scheme 13.58) [96]. The reaction provides the desired dispiro compounds **249** in good to high yields (70–93%), after a few hours at ambient temperature. The STA catalyst has been shown to be reusable up to six times with only slight loss of catalytic activity.

13.1.4.2 O-, P-, and S-Heterocycles The reaction of two similar or dissimilar aryl aldehydes **250/251** with malononitrile **21** or ethyl 2-cyanoacetate **175** catalyzed by *N*-heterocyclic carbenes (NHCs) has been demonstrated to provide fully substituted furans **252** in good to high yields (74–90%), within short reaction times up to 5 h under solvent-free conditions (Scheme 13.59) [97]. This transformation is based on the umpolung of one of the aldehydes by the NHC, while the other one undergoes a Knoevenagel condensation with the CH-acidic reaction partner. The Breslow intermediate then attacks the condensation product in fashion of a Michael addition. After elimination of the NHC, base-catalyzed cyclization provides the desired products. Five different NHCs have been tested catalyzing this reaction.

Tetracyclic 2,3,4,11*b*-tetrahydro-1*H*,6*H*-6λ^5-[1,2]benzoxaphospholo[2,3-*b*][1,2]benzoxaphosphol-1-ones **254** could be accessed by the reaction of several salicylic aldehydes **253**, cyclic 1,3-dicarbonyl compounds **31** and triethyl phosphite under solvent-free conditions at high temperatures [98]. The desired compounds **254** were achieved in high yields and short reaction times. The Knoevenagel condensation product **255**, which is formed in the first reaction step, undergoes a subsequent phospha-Michael addition/double cyclization sequence. The mechanism is shown in Scheme 13.60 by means of salicylic aldehyde **253a**, dimedone **15**, and trimethyl phosphite.

Variably substituted 2-aminothiophenes **260**, **262**, and **264** have been synthesized by several groups in the last years by the application of modified Gewald reactions (Scheme 13.61). Those reactions were catalyzed by secondary amines like piperidine [99] and proline [100] and by polysiloxane-bound primary amines [101]. In the first case, an equimolar amount of piperidine has been used; in the last two cases, only catalytic amounts of the amines were necessary. All reactions were conducted in polar solvents at elevated temperatures and the desired products could be isolated in moderate to high yields.

A series of 2-amino-4*H*-thiopyrans **268** has been synthesized by one-pot Knoevenagel condensation/Michael addition/cyclization of several β-oxodithioesters **265**, active methylene compounds **267**, and various aldehydes **266** (Scheme 13.62) [102]. The products could be isolated in good to excellent yields (70–93%), after stirring for a few hours in refluxing dichloromethane. Analogous 4-spirothiopyrans **272** have been prepared by Majumdar et al. [103]. These reactions were conducted without any catalyst in refluxing ethanol, and the products could also be isolated in excellent yields (89–99%).

13.1.5 Other Knoevenagel Combinations

Ramachary et al. reported an organocatalyzed two-carbon homologation that directly led to saturated Knoevenagel condensation products **277** by utilizing Hantzsch esters **275** as *in situ* reducing agents (Scheme 13.63) [104–109]. The Knoevenagel reaction is catalyzed by proline in

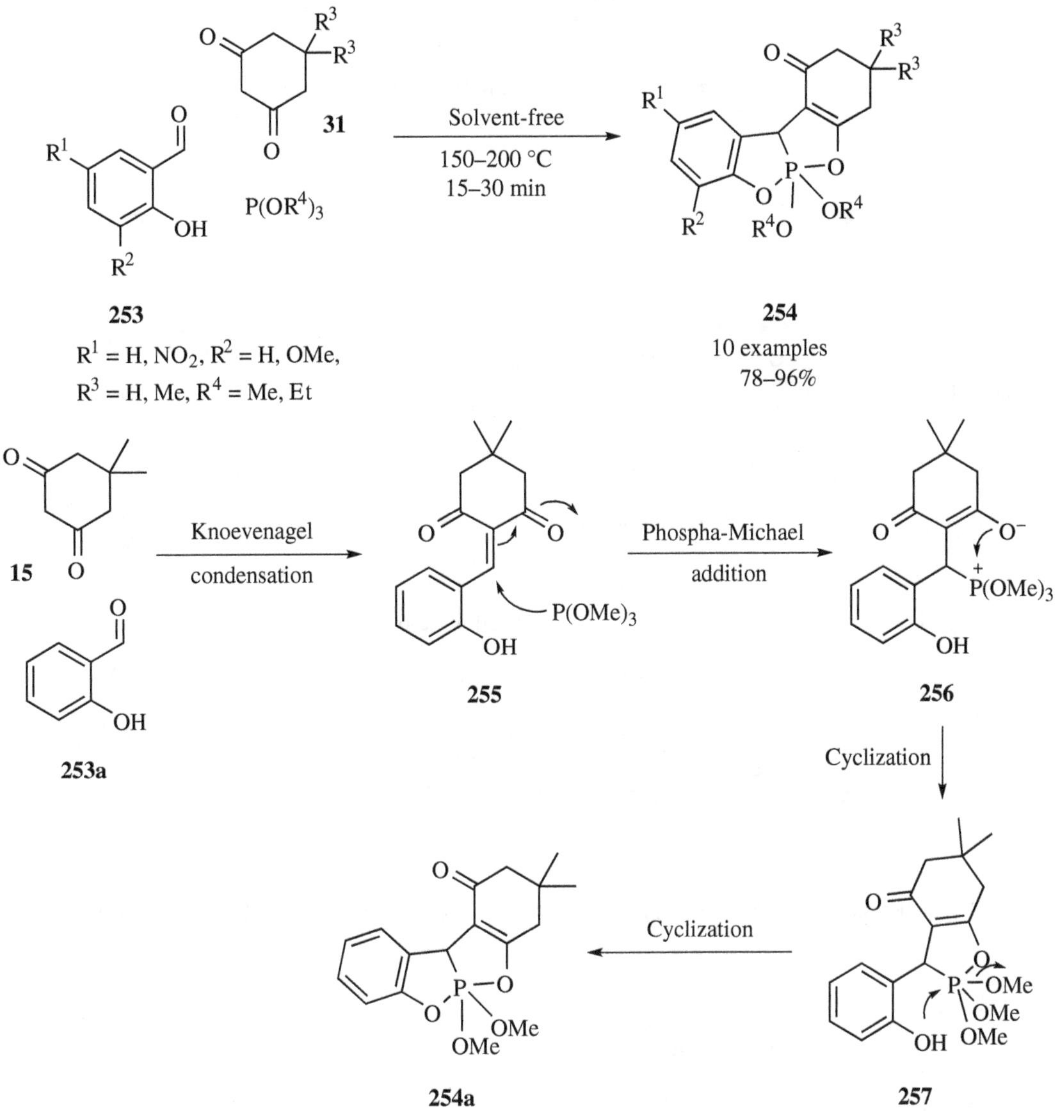

SCHEME 13.60 Knoevenagel condensation/phospha-Michael addition/double cyclization sequence for the synthesis of tetracyclic 2,3,4,11*b*-tetrahydro-1*H*,6*H*-6λ⁵-[1, 2]benzoxaphospholo[2,3-*b*][1,2]benzoxa-phosphol-1-ones **254** [98].

ethanol, acetonitrile, DMSO, dichloromethane, or water at ambient temperature. The presence of Hantzsch ester derivatives directly led to conjugate reduction of the newly formed double bond to provide the desired products in good to excellent yields. As the carbonyl component for this reaction, aliphatic and aromatic aldehydes, as well as ketones, have been shown to give good results. CH acids like 1,3-cyclohexanedione, dimedone, Meldrum's acid, barbituric acid, malonodinitrile, and alkyl 2-cyanoacetates have been used, while the reducing agents either could be Hantzsch ester derivatives or a combination of aldehydes with 1,2-diaminobenzene **276**. In this case, the generated 2,3-dihydro-1*H*-benzo[*d*]imidazole derivative fills the role as the reducing agent.

This homologation has been extended to a five-component reaction sequence consisting of a Knoevenagel condensation/Diels–Alder reaction/epimerization/Knoevenagel condensation/hydrogenation (Scheme 13.64) [110]. The desired products **283** are generated in a highly diastereospecific fashion and could be isolated in good yields (60–75%).

A three-component reaction for the synthesis of β-acetamido ketones **287** starting from aldehydes **286**, enolizable ketones **284**, and nitrile compounds **285**, in the presence of acetyl chloride in the corresponding nitrile as a reaction medium, is catalyzed by iodine at ambient temperature (Scheme 13.65) [111]. The desired products **287** are produced in good yields (75–89%), in about 4–6 h. This type of reaction has also been shown to give good

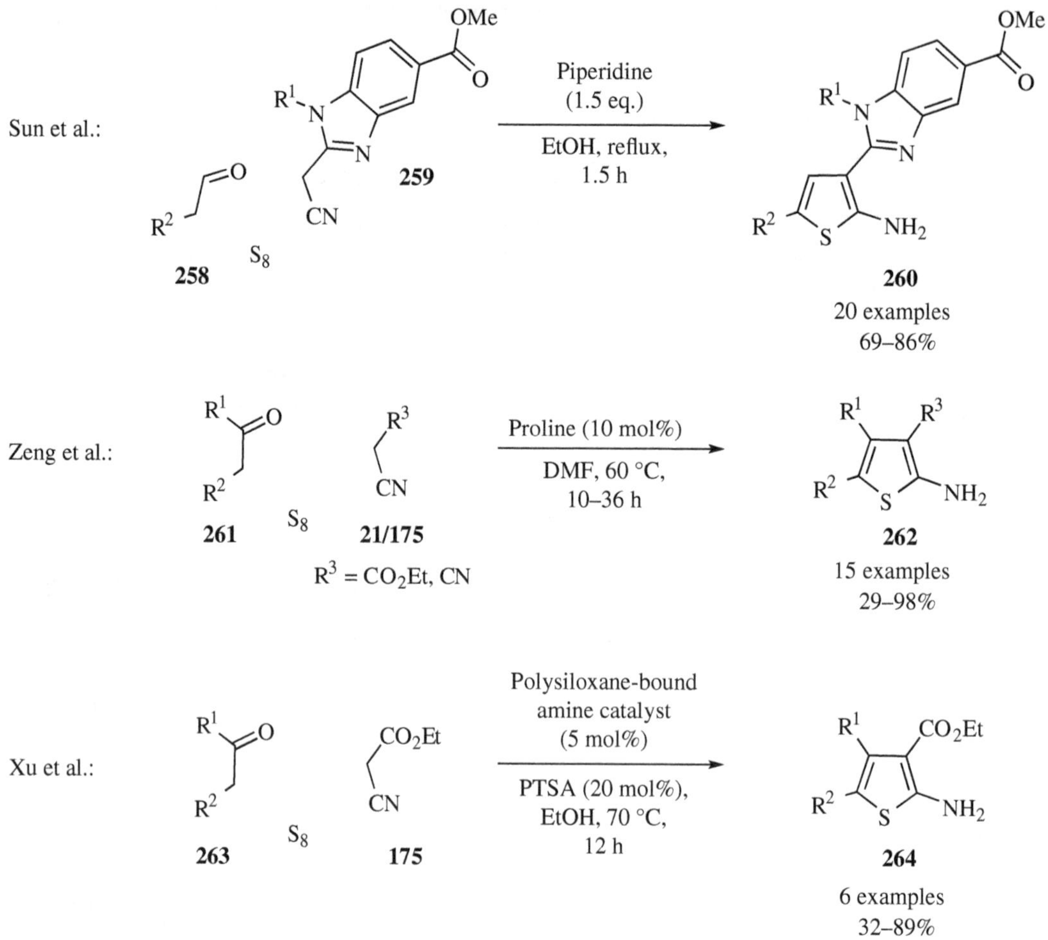

SCHEME 13.61 Variably substituted 2-aminothiophenes by modified Gewald reactions [99–101].

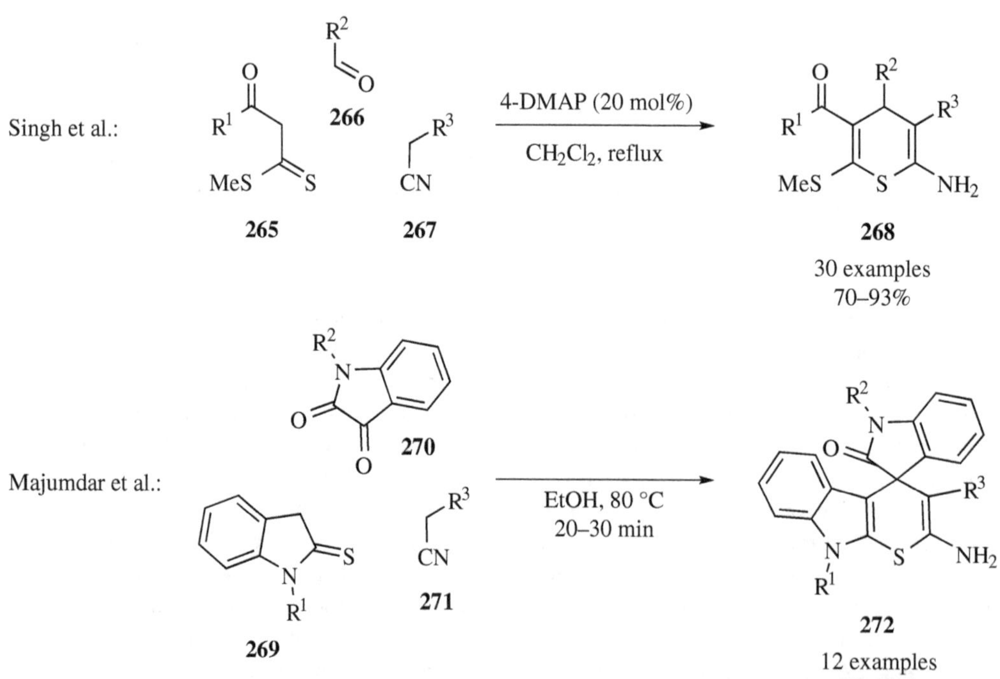

SCHEME 13.62 2-Aminothiopyran **268** and **272** syntheses [102, 103].

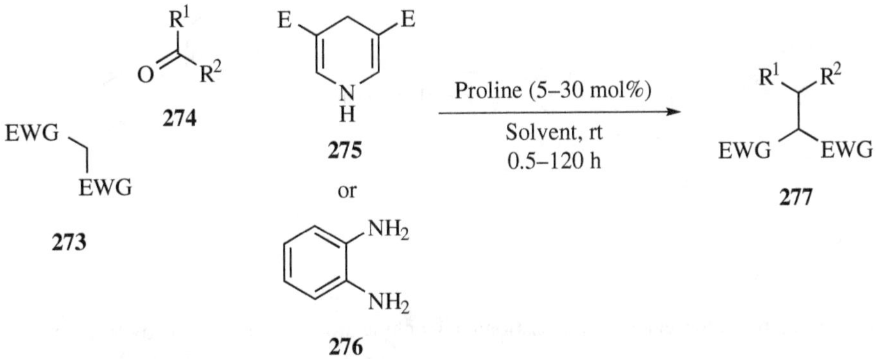

SCHEME 13.63 Two-carbon homologation by Ramachary et al. [104–109].

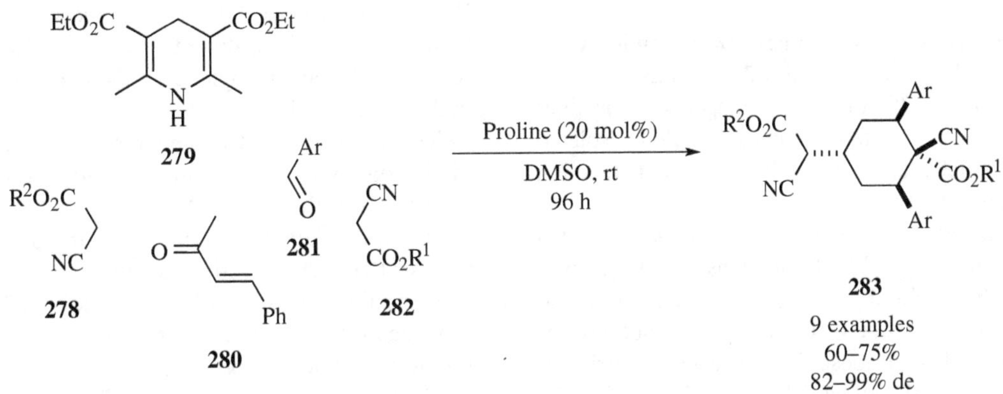

SCHEME 13.64 Five-component Knoevenagel condensation/Diels–Alder reaction/epimerization/Knoevenagel condensation/hydrogenation sequence [110].

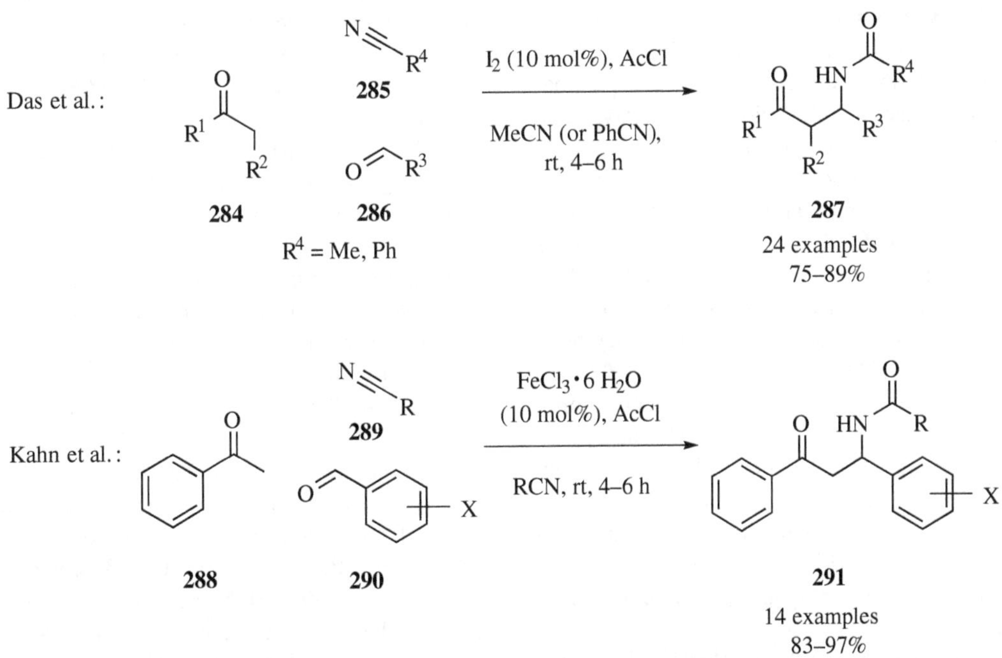

SCHEME 13.65 Syntheses of β-acetamido ketones **287** and **291** [111, 112].

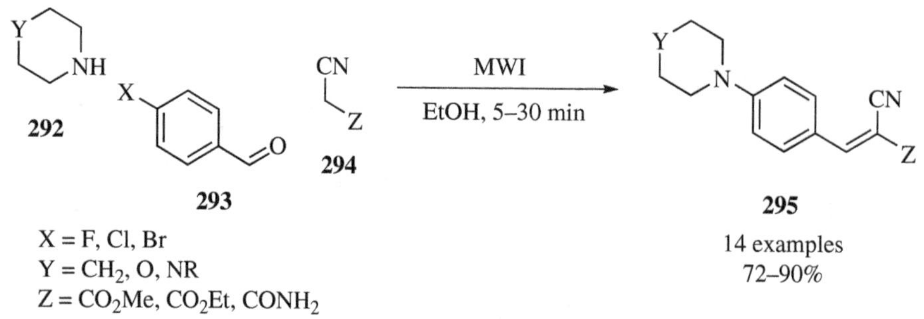

SCHEME 13.66 Knoevenagel condensation/nucleophilic aromatic substitution by Wang et al. [113].

results under iron catalysis [112]. The exchange of the ketone by β-ketoesters as enolizable carbonyl component was possible.

By reacting cyclic secondary amines **292** under microwave irradiation with 4-halobenzaldehydes **293** and 2-cyanoacetates **294**, the products **295** of a Knoevenagel condensation/nucleophilic aromatic substitution sequence could be isolated in good to high yields (72–90%), after very short reaction times [113] (Scheme 13.66).

2-Thioxothiazolidin-4-one derivatives **298/302** have been functionalized under various MCR conditions (Scheme 13.67). Yadav et al. applied a chiral ionic liquid for the enantio- and diastereoselective generation of 6-mercaptopyranothiazoles **300** [114]. Anderluh et al. describe the reaction of rhodanine **302** with several aldehydes **301** and amines **303** under microwave irradiation conditions to generate 2-amino-5-alkylidene-thiazol-4-ones **304** [115]. The reactions were performed at high temperatures using catalytic amounts of acetic acid and give the desired products mostly in high yields (11–92%), within short reaction times (20 min). An analogous functionalization of rhodanine has been performed by Mukhopadhyay and Ray [116]. Their methodology involves a surface-modified silica gel catalyst that makes it possible to conduct the reaction at ambient temperature by cooperative acid–base catalysis. The conversions are complete at ambient temperature within 2–2.5 h in a 1:1-mixture of water and EtOH, and the desired products **308** have been isolated in moderate to high yields. With ketones instead of aldehydes, heating the reactions to reflux is necessary, which provides the products in moderate to high yields within 1–1.5 h [117].

The synthesis of highly functionalized cyclohexanones has been published by Liu et al. in 2009 [118]. The utilization of α-alkenoyl ketene *S,S*-acetals in this three-component reaction with aldehydes and CH-acidic compounds provides the cyclohexanone derivatives in moderate to good yields under acidic conditions in refluxing THF. With cyclic CH acids such as dimedone, the corresponding spiro compounds have been prepared.

Wang et al. reported the synthesis of several tetrahydroisoquinolines **311** by the reaction of N-protected 4-(dicyanomethylene)piperidines **309** with aromatic aldehydes **310** and malonodinitrile **21** in the ionic liquid [bmim]BF_4 (Scheme 13.68) [119]. The products could be isolated in high yields (83–95%), after several hours at 50 °C. A screening of various dicyanomethylene compounds has been published in the same year [120].

The morpholine-promoted reaction of *N*-alkyl piperidine-4-one with malonodinitrile and β-nitrostyrene derivatives in refluxing EtOH directly led to substituted tetrahydroisoquinoline derivatives in moderate to good yields [121]. The reaction is initiated by a Knoevenagel condensation, in which a Michael addition/Thorpe–Ziegler sequence follows. Oxidation by air is the last step to occur providing the desired products.

Various stilbene derivatives **315** have been synthesized via a tandem double Knoevenagel condensation/decarboxylation/Heck reaction by Sinha et al. (Scheme 13.69) [122]. This rather complex reaction sequence is promoted by a catalyst system consisting of tetrakis(triphenylphosphine)palladium(0), piperidine, potassium carbonate, and lithium chloride. It is conducted under microwave irradiation at 150 °C and provides the desired products **315** in moderate to good yields.

The asymmetric Michael addition of acetone **317** to *in situ* generated isatylidine malononitriles is efficiently catalyzed by a cooperative acid–base organocatalyst consisting of a quinidine-derived primary amine **318** and phosphoric acid BINOL ester **319** [123]. This Knoevenagel condensation/asymmetric Michael addition sequence works well in THF at ambient temperature giving the desired products **320** in good to excellent yields (85–97%) and high enantiomeric excesses (86–99% ee) (Scheme 13.70).

A Knoevenagel condensation/Diels–Alder cycloaddition approach has been developed by Delgado et al. [124]. By infrared light irradiation at 50 °C and solvent-free conditions, the desired products **324/325** were isolated in moderate yields (40–65%), with also moderate to good *endo*-selectivity (Scheme 13.71). Frontier molecular orbital calculations rationalized the stereoselectivity of this process. In the cases of diethyl malonate as the

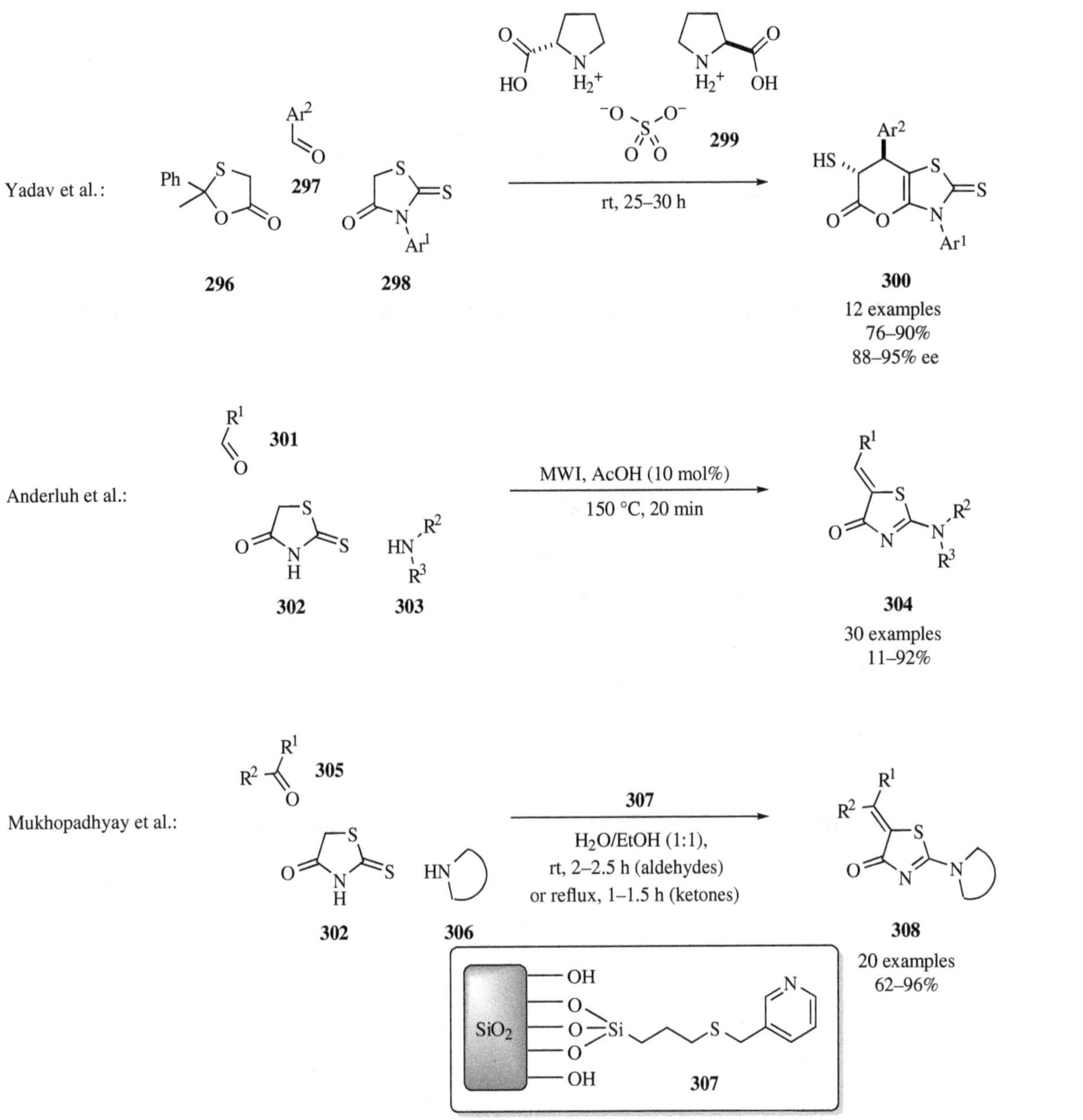

SCHEME 13.67 Functionalization of 2-thioxothiazolidin-2-one derivatives [114–117].

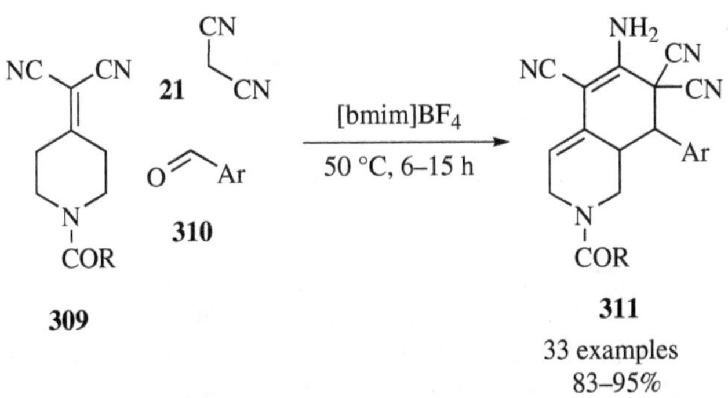

SCHEME 13.68 Tetrahydroisoquinoline **311** synthesis by Wang et al. [119].

CH-acidic component ($R^1=R^2=CO_2Et$), none of the desired products could be isolated, even within extended reaction times of up to 240 min; instead, only the Knoevenagel condensation product with the aldehyde and the shown by-product **326** were formed. The latter one arises from a dimerization of compound **323** in a homo-Diels–Alder reaction.

Another Knoevenagel condensation/hetero-Diels–Alder cycloaddition strategy led to *ortho*-amido alkylation of phenols in good yields [125]. These reactions are promoted by silica-immobilized $HClO_4$ in refluxing acetonitrile with phenols, formaldehyde **72**, and various lactams as the reacting components.

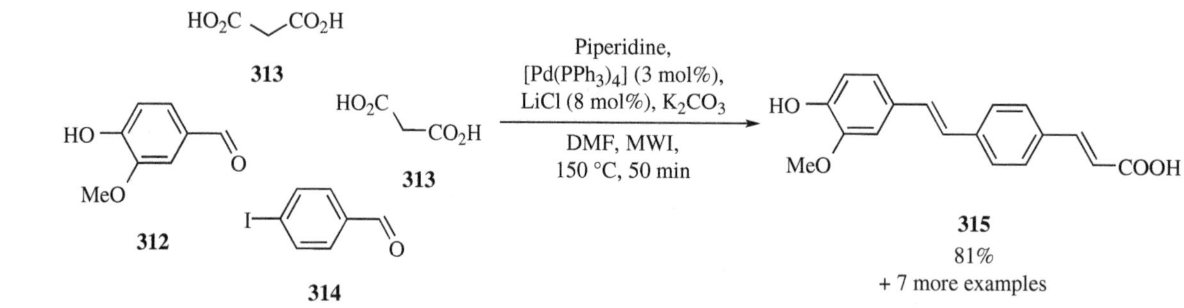

SCHEME 13.69 Model reaction for the multicomponent synthesis of stilbene derivatives **315** [122].

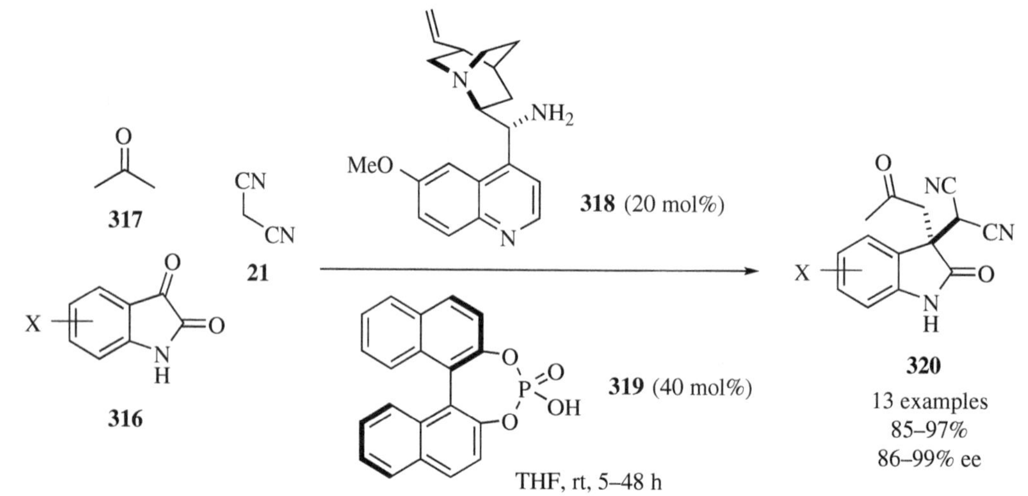

SCHEME 13.70 Knoevenagel condensation/asymmetric Michael addition by Wang et al. [123].

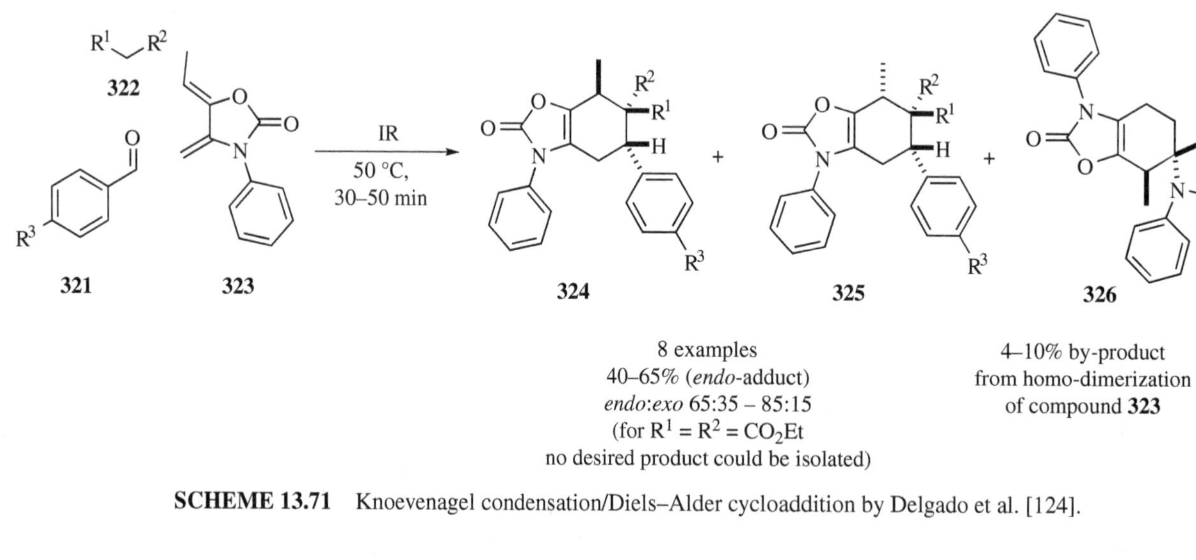

SCHEME 13.71 Knoevenagel condensation/Diels–Alder cycloaddition by Delgado et al. [124].

13.2 YONEMITSU-TYPE TRIMOLECULAR CONDENSATIONS

13.2.1 Introduction and Mechanistic Aspects

In 1978, Yonemitsu et al. published the three-component reaction of Meldrum's acid **112** with various aldehydes **328** and indole **327** in acetonitrile at slightly elevated temperatures providing their condensation products **329** in high yields (80–98%) (Scheme 13.72) [126]. This reaction has been shown to be accelerated by the addition of 5 mol% proline. Subsequent decarboxylative ethanolysis gave variably substituted ethyl indolepropionates **330** in good yields (62–87%).

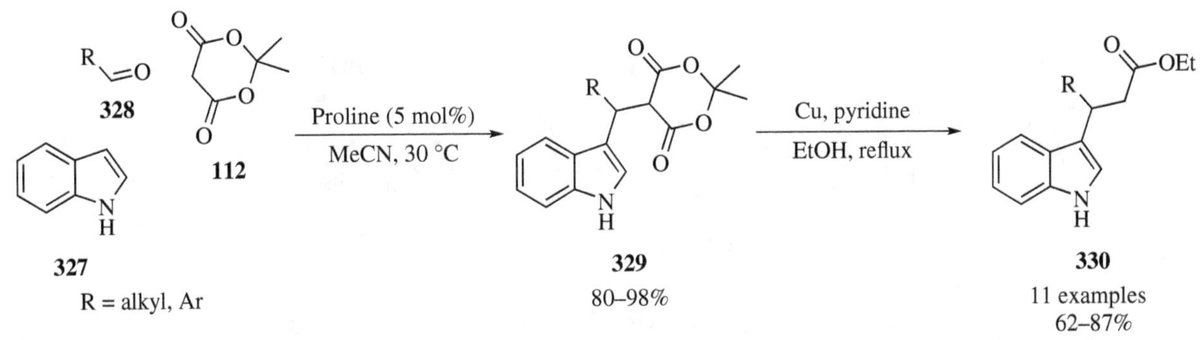

SCHEME 13.72 Multicomponent reaction of Meldrum's acid **112** and indole **327** with several aldehydes **328** by Yonemitsu et al.

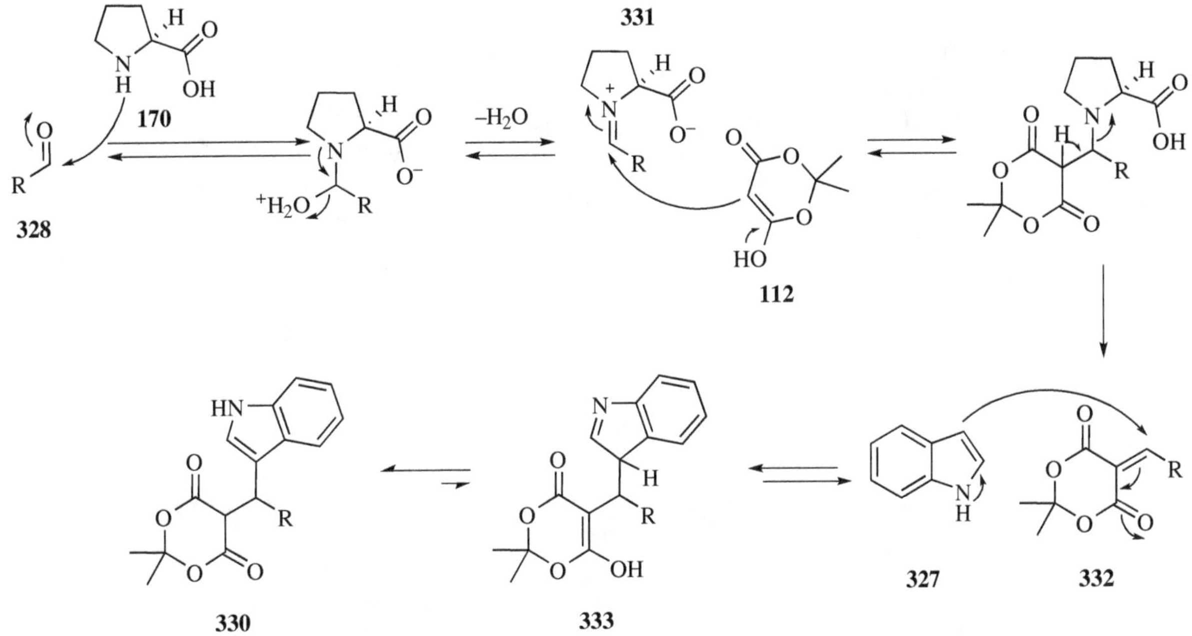

SCHEME 13.73 Proposed mechanism of the Yonemitsu reaction.

The reaction combines a proline-catalyzed Knoevenagel condensation with the Michael addition of the electron-rich heterocycle (Scheme 13.73). Proline **170** initially condenses with the aldehyde **328** to form an iminium intermediate **331**. This is attacked by the CH-acidic Meldrum's acid **112** with subsequent elimination of proline to provide the Knoevenagel condensation product **332**. Nucleophilic 1,4-addition of the indole **327** followed by tautomerizations of the rather unstable intermediate **333** liberates the desired products **330**.

The original Yonemitsu reaction is found relatively rare in contemporary literature, but many modifications by exchange of the CH-acidic or the heterocyclic component have been published. These modifications make the Yonemitsu reaction a broadly applied MCR in organic synthesis, though their classification being a modification of the Yonemitsu reaction is not always mentioned in the literature.

The recent literature since 2005 dealing with the original Yonemitsu reaction will separately be discussed from the modified Yonemitsu reactions in this chapter. An interesting variation of the mechanism under alkaline conditions will be presented, and at the end, the tetramolecular version of the Yonemitsu reaction will be discussed.

13.2.2 Applications of the Original Yonemitsu Trimolecular Condensation

In 2005, Sapi et al. reported the Yonemitsu reaction as the key step in their diastereoselective synthesis of 3,4-heterocycle-annulated tetrahydro-β-carbolines **336** (Scheme 13.74) [127]. The configuration of D-glyceraldehyde **334** or Garner's aldehyde (not shown) dictated the stereochemistry of the newly generated stereocenters. D,L-Proline efficiently catalyzed the trimolecular condensation step and provided the

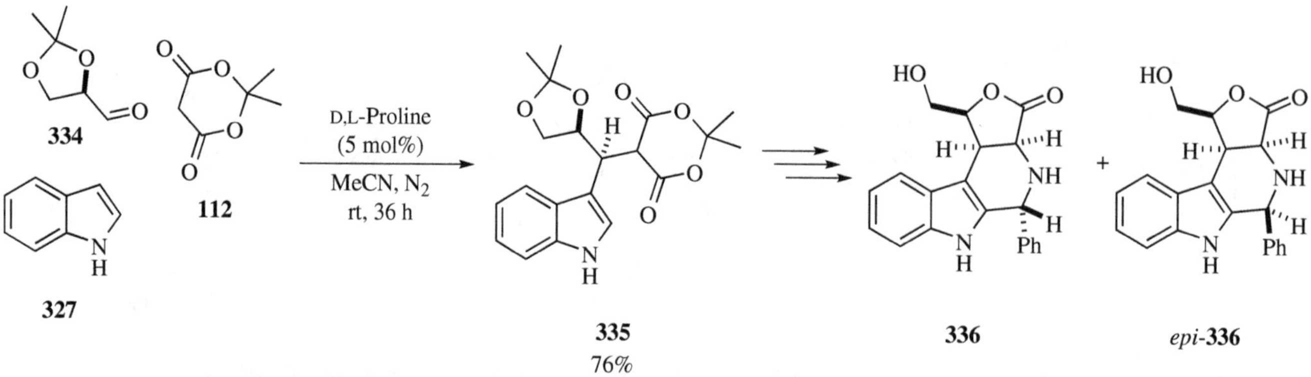

SCHEME 13.74 Yonemitsu reaction for the synthesis of 3,4-furanone annulated tetrahydro-β-carbolines **336**.

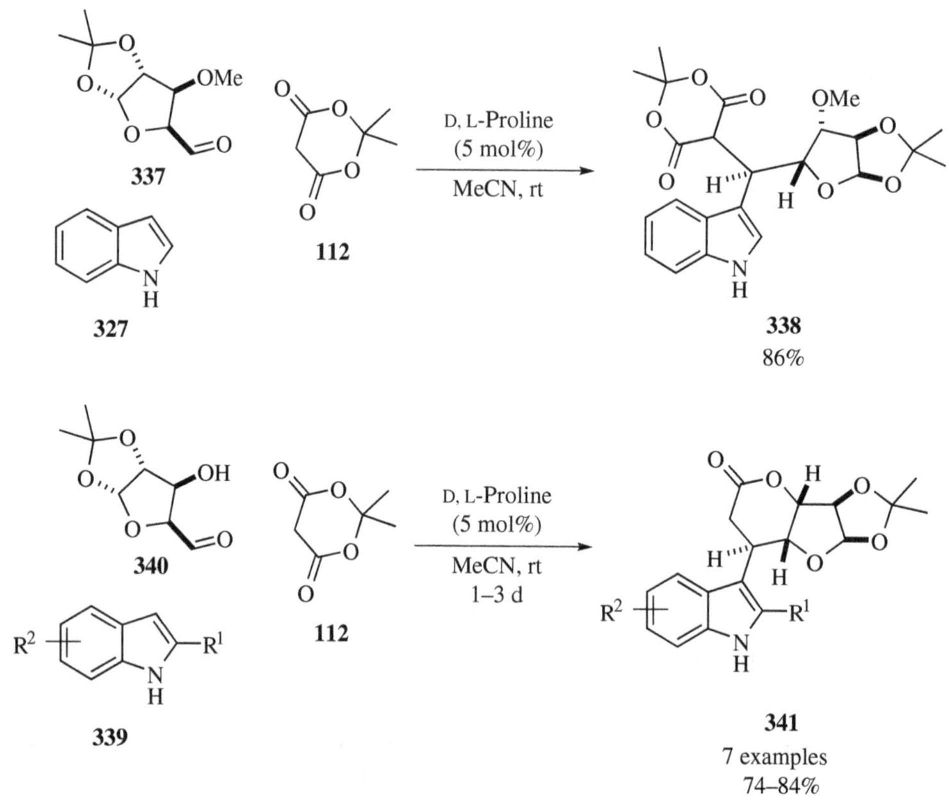

SCHEME 13.75 Proline-catalyzed construction of compounds **338** and **341**.

desired synthesis intermediates **335** in good yields. Subsequent transformations led to the desired products **336** in only a few more steps.

Sabitha et al. published an analogous reaction utilizing a D-glucose-derived aldehyde **337** for the construction of the Yonemitsu reaction product **338** (Scheme 13.75) [128]. The aldehyde **340** containing a free hydroxyl group directly led to a decarboxylative cyclization step with the Meldrum's acid moiety, subsequent to the formation of the Yonemitsu product providing 7-(1*H*-3-indolyl)-2,3-dimethoxy-perhydrofuro[3,2-*b*]pyran-5-ones **341** in good yields (74–84%).

A similar reaction procedure with various sugar-derived aldehydes was published by Sapi et al. in 2010 [129]. Deprotection of the oxygen-substituents present in the aldehyde moiety led to the previously discussed decarboxylative cyclization in one additional step providing the desired products in good yields over two steps with high diastereoselectivity.

In 2012, Kumar et al. reported the Yonemitsu condensation/decarboxylative cyclization sequence in a saccharine-derived ionic liquid at rt, which provides the products in high yields around 90% [130]. The ionic liquid was shown to be recyclable up to four times with only small decrease in the products yield. One year later, a different group applied an ionic liquid derived from a defined composition of choline chloride and urea for the Yonemitsu reaction (Scheme 13.76) [131]. The IL could be used in substoichiometric amounts of 20 mol% facilitating the reaction efficiently. Subsequent aminolysis of the Yonemitsu product **344** in a one-pot fashion led to 3-substituted 3-indolepropionic amides **345** in high yields (74–92%).

The Yonemitsu reaction-derived products **348** have been shown to undergo a Lewis acid-catalyzed nucleophilic substitution by Kerr et al. (Scheme 13.77) [132]. This method uses the nitrogen-supported easy elimination of the Meldrum's acids anion under Lewis-acidic conditions to generate a conjugated iminium ion **350** that is trapped by the nucleophile in a conjugate addition. As nucleophiles mostly aromatic compounds have been tested.

13.2.3 Yonemitsu-Type Reactions and Tetramolecular Condensations

13.2.3.1 Variation of the CH-Acidic Component In 2005, Zeng et al. revealed the hydroxide-promoted Yonemitsu-type reaction of indoles **351** with deoxybenzoin **353** under ultrasonic irradiation conditions (Scheme 13.78) [133]. With this approach, a novel access to β-indolyl phenyl ketones **356** has been developed, which provides the desired products in moderate to good yields (51–82%), in short reaction times (1–3 h). It is noteworthy that the authors suggested a reaction mechanism that is fundamentally different from the mechanism of the proline-catalyzed Yonemitsu reaction. In this mechanistic suggestion, the indole first undergoes a base-assisted condensation with the aldehyde **352** to furnish a conjugate imine **355**. This is attacked by the anion of the CH-acidic component to generate the desired product **356**.

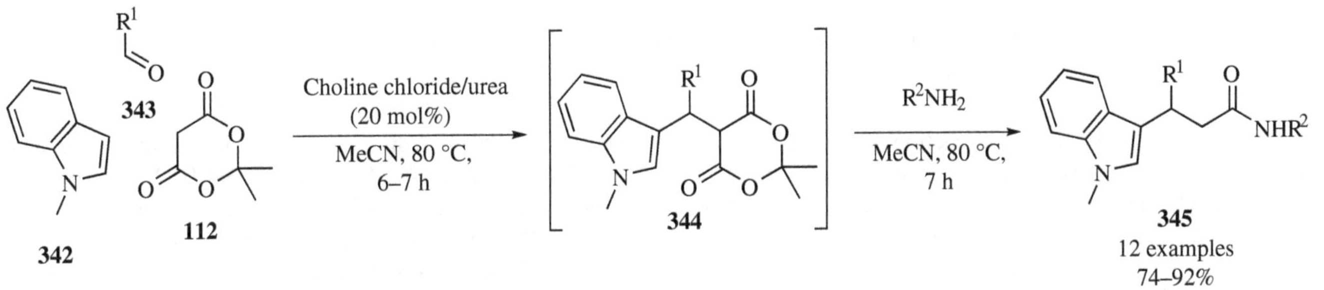

SCHEME 13.76 One-pot Yonemitsu reaction/aminolysis to 3-substituted 3-indolepropionic amides **345**.

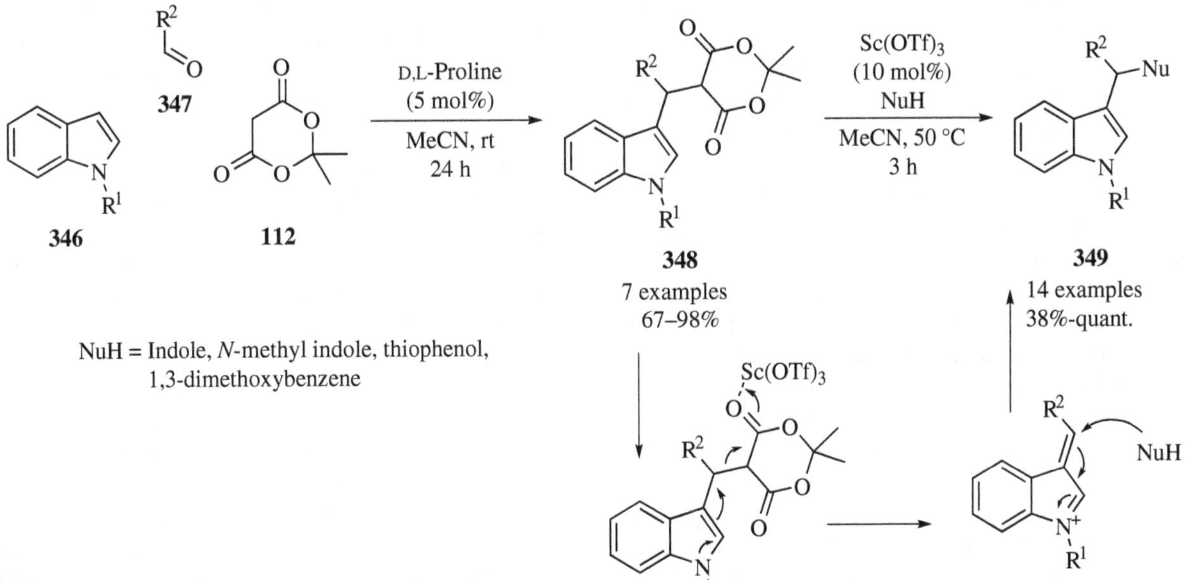

SCHEME 13.77 Proline-catalyzed Yonemitsu reaction with subsequent nucleophilic substitution of the Meldrum's acid moiety and its proposed mechanism.

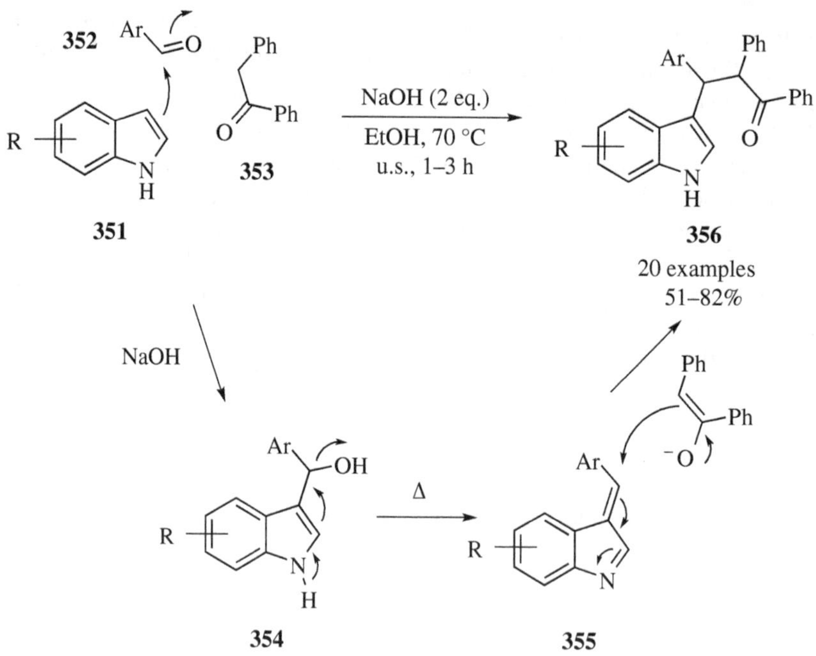

SCHEME 13.78 Base-mediated Yonemitsu-type reaction and its proposed mechanism.

The thermal Yonemitsu-type reaction of ethyl 2-nitroacetate with variably substituted indoles and formaldehyde in toluene led to the formation of various tryptophan precursors in moderate to high yields [134]. Molecular sieves supported the Knoevenagel condensation step. A closely related catalyst-free version was published by Jérôme et al. in 2009 [43]. β-Ketoesters represent the CH-acidic component that undergoes a Yonemitsu-type reaction at slightly elevated temperatures to provide the desired products in moderate yields within 6–20h.

A few Lewis acids have been shown to catalyze the Yonemitsu-type reactions of indoles, aldehydes, and several CH acids (Scheme 13.79). Dimethylmalonate **358** was successfully reacted with indole **327** and several aldehydes **357** under solvent-free ultrasound irradiation conditions to provide the desired products **359** in moderate yields [135]. Ytterbium triflate was used as the Lewis-acidic catalyst in this case. A copper–salen complex was utilized in water at elevated temperature to facilitate the Yonemitsu-type reaction of indoles **360**, aldehydes **361**, and malonodinitrile **21** [136]. In close similarity copper(II) acetate was used in polyethylene glycol at elevated temperatures to provide Yonemitsu-type products **366** in moderate to high yields (48–98%) [137].

13.2.3.2 Variation of the Heterocyclic Component Exchanging the indole as the heterocyclic component in the Yonemitsu reaction by imidazo[1,2-*a*]pyridines **367** led to a multicomponent synthesis pathway for the preparation of a library of imidazo[1,2-*a*]pyridin-3-yl carboxylic acids and amides (Scheme 13.80) [138]. Heterocycles bearing a primary amine group tend to decarboxylative cyclization subsequent to the Yonemitsu reaction providing dihydropyridinone derivatives **373/376/378/380/382**. In particular, 2,6-diaminopyrimidin-4-one [144], 5-aminopyrazoles **370/371/374** [139–141, 145], their corresponding isoxazoles [139], 1,2-diamino-4-phenylimidazole [146], 2-amino-4-iminothiazolidines [147], its selenium analogues [148], 4-aminoisothiazoles [149], 2-amino-4-cyanopyrrols [150], and variably substituted anilines [142, 151] have been shown to undergo this reaction sequence. With aromatic dialdehydes **379**, linked dihydropyridinones **380** have been prepared [143]. In comparison to the amine analogues, hydroxyl-substituted heterocycles and phenol derivatives give 4*H*-pyran-2-ones [152].

An elegant method for the utility of unstable 3-unsubstituted 2-aminothiophene derivatives **385** in the Yonemitsu-type reaction has been elaborated by Krayushkin et al. (Scheme 13.81) [153]. Starting from stable 4,5-disubstituted ethyl 2-aminothiophene-3-carboxylates **383**, the ester functionality is hydrolyzed to yield the corresponding alkali metal carboxylates **384**. These are directly subjected to the three-component Yonemitsu-type reaction in acetic acid as the reaction medium. Protonation of the carboxylate followed by decarboxylation generates the 2-aminothiophene derivatives **385** that directly undergo Yonemitsu reaction without the need of isolation of those intermediates. This method has been applied to methyl 2,4-diaminothiazole-5-carboxylates **388** [154], ethyl 5-aminopyrazole-4-carboxylates **389** [155], dimethyl 3-aminopyrrol-2,4-dicarboxylates **390** [156], 4,5-disubstituted methyl 3-aminothiophene-2-carboxylates **391** [157], dimethyl 3-amino-4-phenylthiophene-2,5-dicarboxylate **392** [156], and ethyl 5-aminoimidazol-4-carboxylates **393** [158].

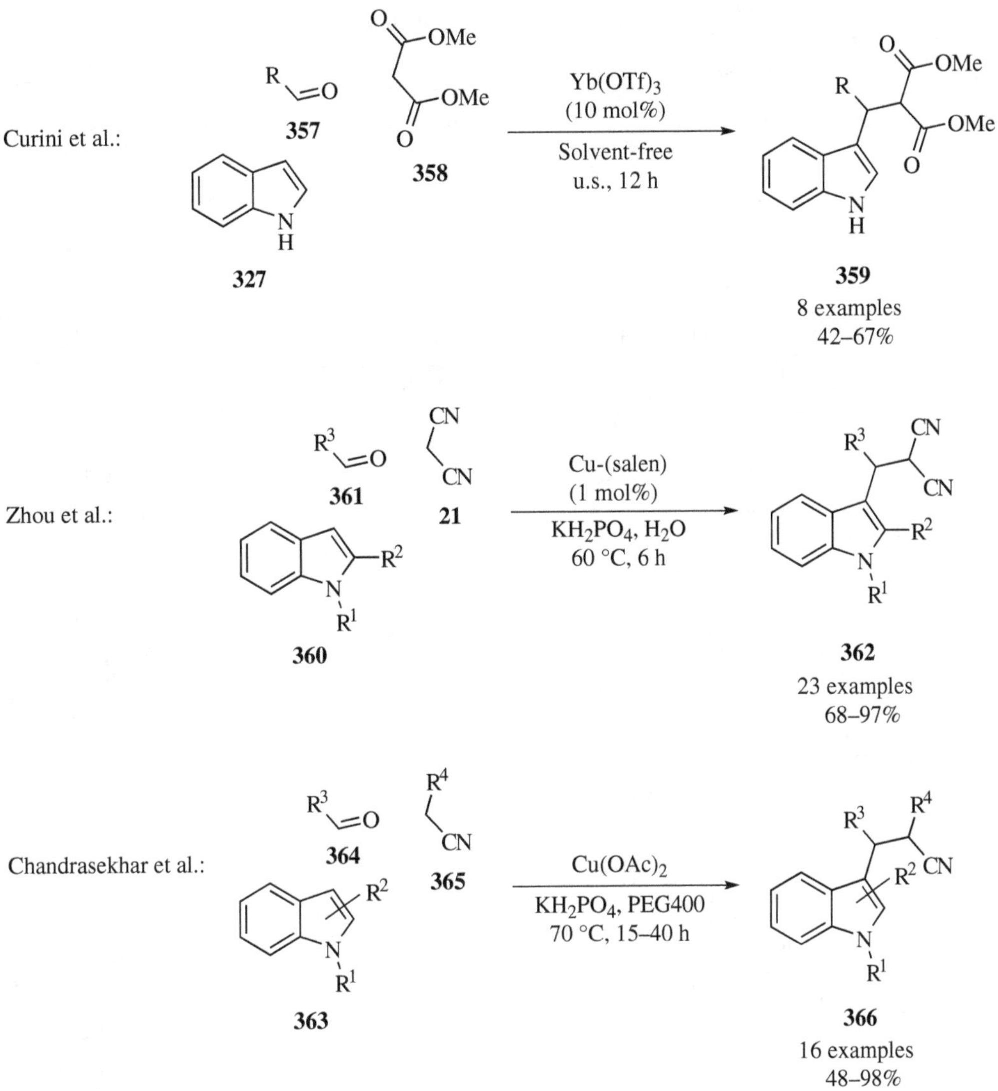

SCHEME 13.79 Lewis acid-catalyzed Yonemitsu-type reactions [135–137].

The decarboxylative cyclization could be prevented by using two equivalents of the corresponding aldehyde component **395**, which led to spirodihydropyridines **396** as shown by Li et al. (Scheme 13.82) [139], or by exchanging the amino group by a hydroxyl substituent and adding ammonium acetate as the amine source [159]. In this 4CR, the Meldrum's acid moiety is hydrolyzed to give the carboxylic acid **399** that is stabilized by an intramolecular hydrogen bond.

13.2.3.3 Variation of Both the Heterocyclic and CH-Acidic Component and Tetramolecular Yonemitsu-Type Reactions Heterocyclic components bearing an exocyclic amino group have also been shown to undergo reactions with various CH-acidic components different from Meldrum's acid (Scheme 13.83). In the cases of cyclic CH acids like dimedone **15**, tricyclic dihydropyridines **401** were obtained [140, 160, 161], and with aromatic dialdehydes linked tricyclic dihydropyridines have been prepared as well [143].

Under electrolytic reaction conditions, pyrazol-5(4*H*)-one derivatives undergo a Yonemitsu-type reaction with sodium bromide as electrolyte in ethanol at ambient temperature (Scheme 13.84) [162]. Under those reductive conditions, the base is generated *in situ* from the alcoholic solvent. After deprotonation of the CH-acidic compound **413**, the regular Yonemitsu mechanism is passed through, giving the desired products **414** in good to excellent yields (75–97%).

Recently, a four-component Yonemitsu-type reaction has been reported that generates both the Michael acceptor and the heterocyclic nucleophile *in situ* from readily available starting materials (Scheme 13.85) [163]. A substoichiometric amount of alumina acts as the mediator for this reaction and provides the desired products **417** in high yields in water at slightly elevated temperature. The alumina has been shown to be recyclable four times with only slight loss in activity.

Another four-component Yonemitsu-type reaction was published in 2013 (Scheme 13.85) [164]. Copper iodide

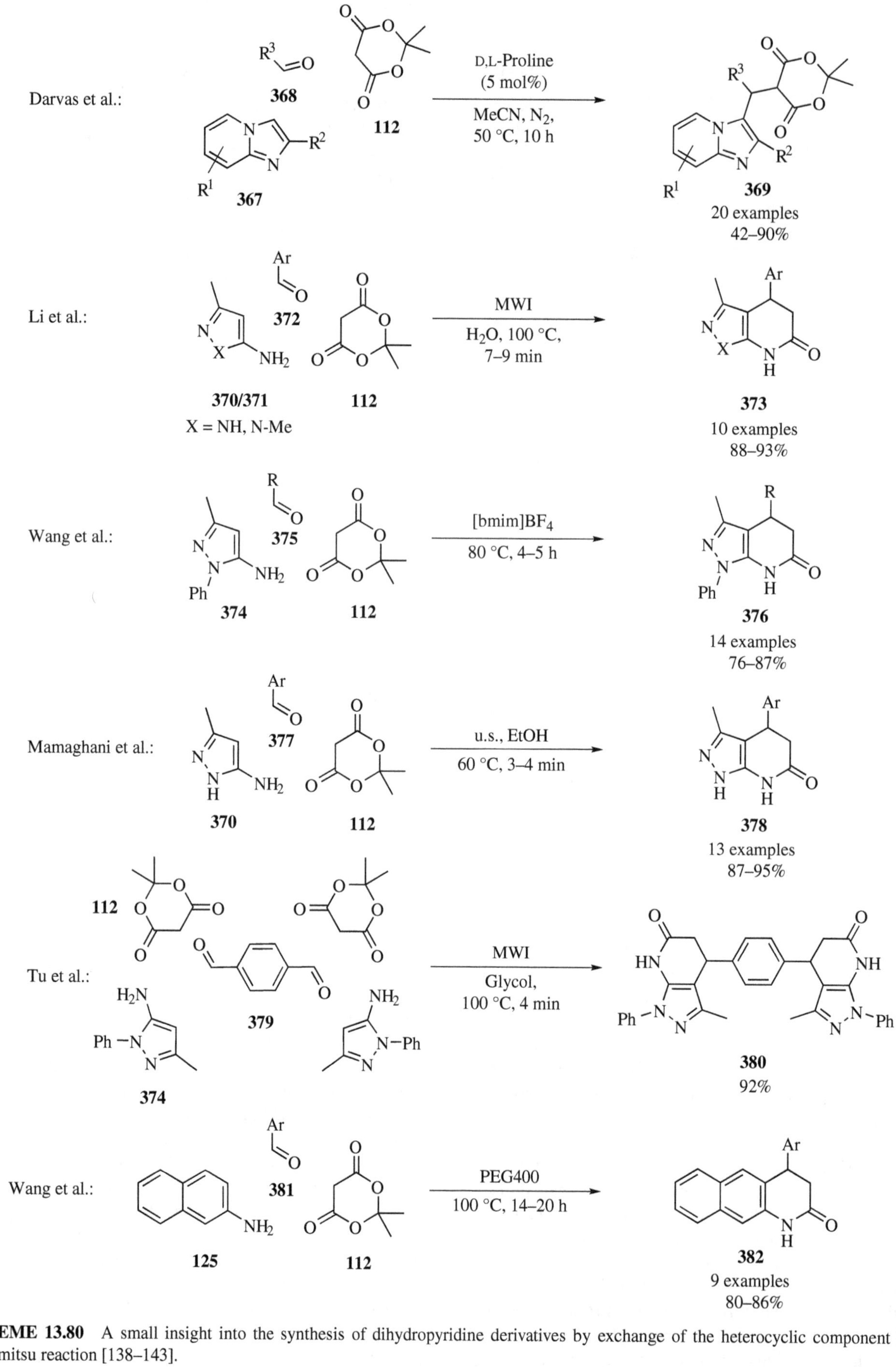

SCHEME 13.80 A small insight into the synthesis of dihydropyridine derivatives by exchange of the heterocyclic component in the Yonemitsu reaction [138–143].

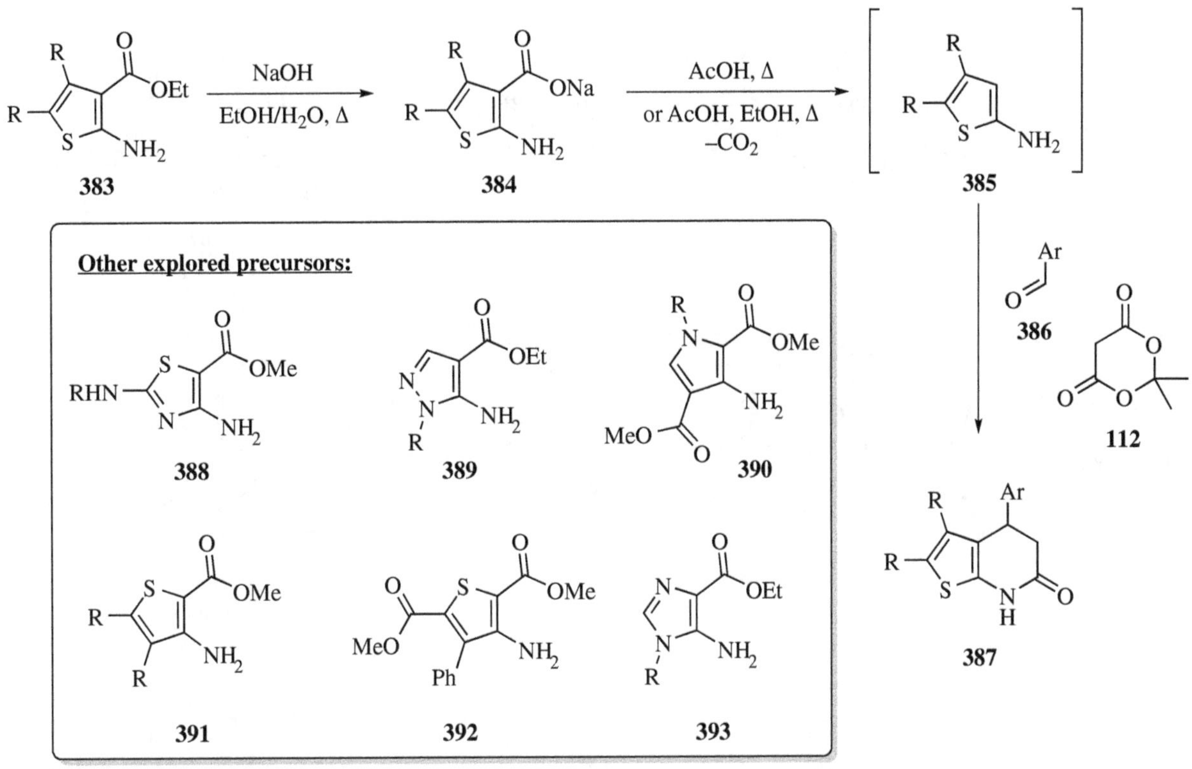

SCHEME 13.81 Yonemitsu-type trimolecular condensation with *in situ* generation of unstable heterocyclic compounds **385** from stable precursors by Krayushkin et al. [153–158].

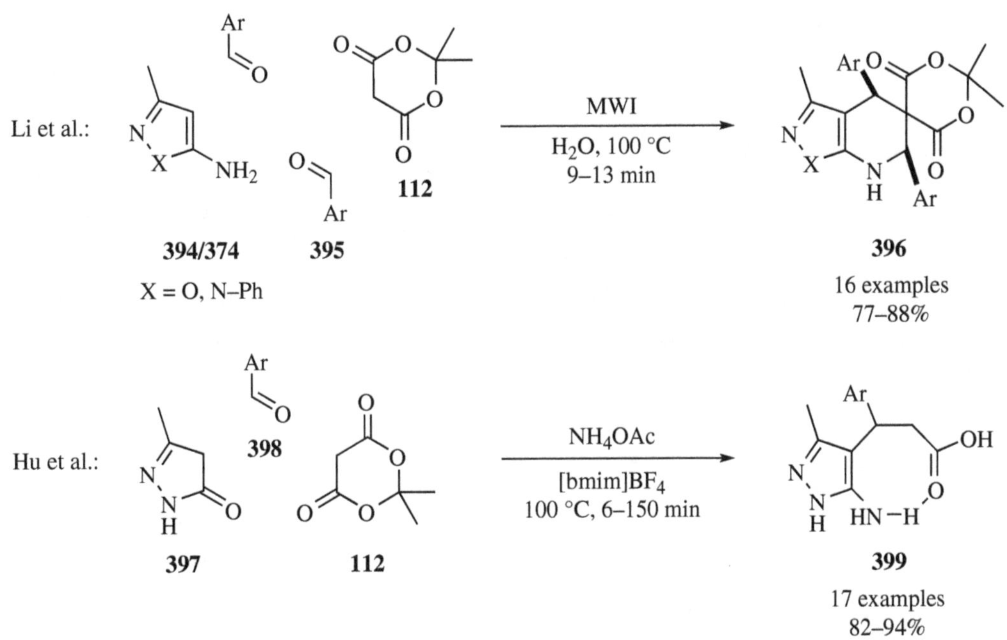

SCHEME 13.82 Prevention of the decarboxylative cyclization [139, 159].

nanoparticles mediated the formation of the actually reacting CH-acidic component from acyl chlorides **418** and Meldrum's acid **112** before the other reaction components are added. In short reaction times and refluxing water, the desired products **420** were isolated in good to excellent yields. The copper nanoparticles have been shown to be recyclable up to four times with only little loss of activity.

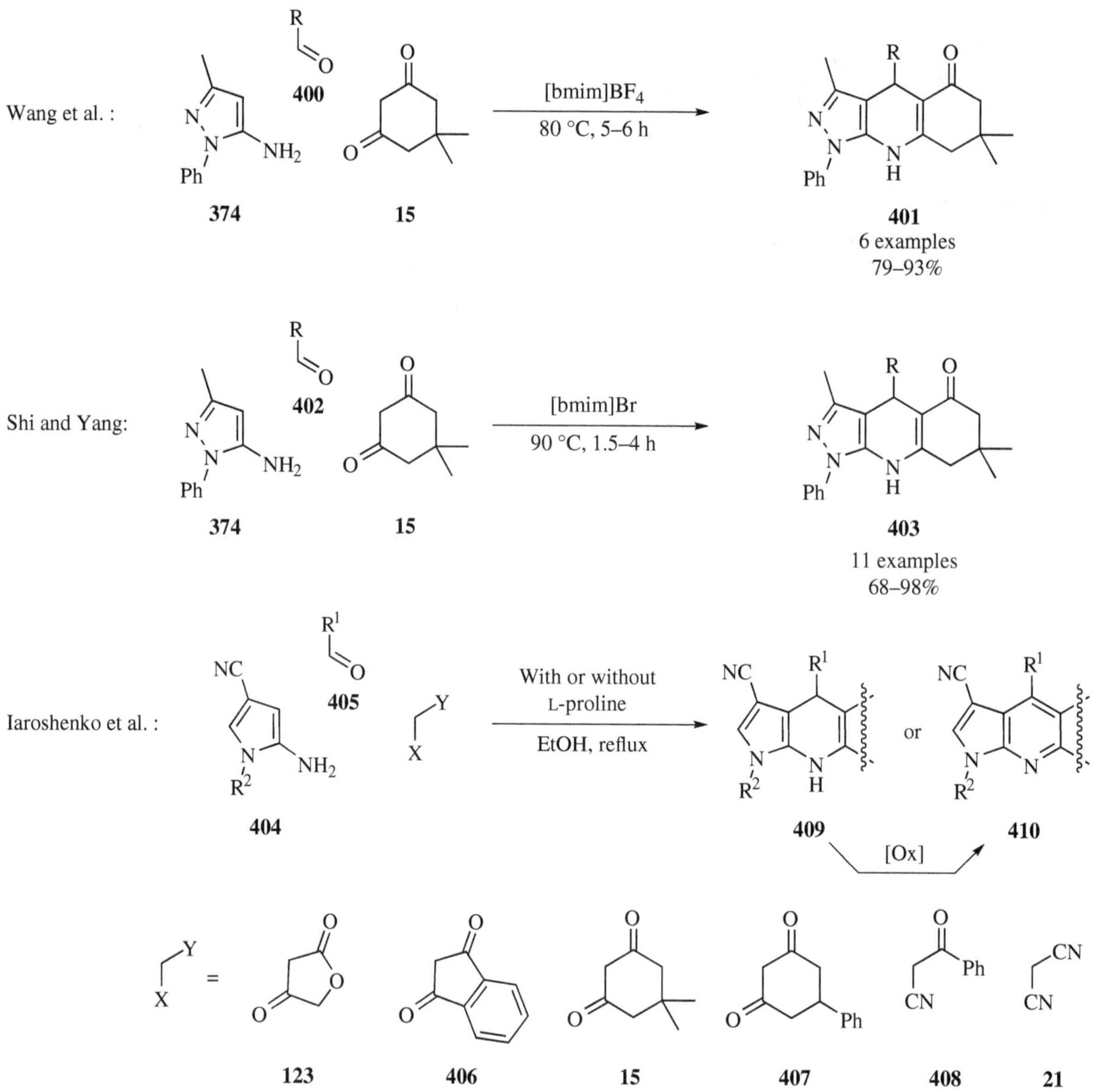

SCHEME 13.83 Preparation of tricyclic dihydropyridine derivatives by amino-heterocycles and various cyclic CH acids different from Meldrum's acid [140, 143, 160, 161].

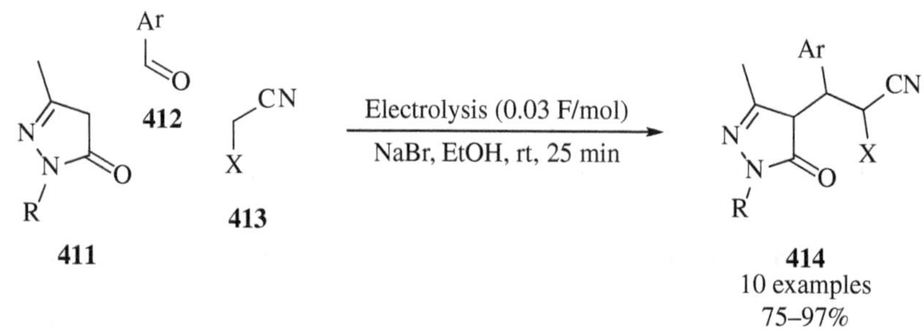

SCHEME 13.84 Yonemitsu-type reaction under electrolytic reaction conditions.

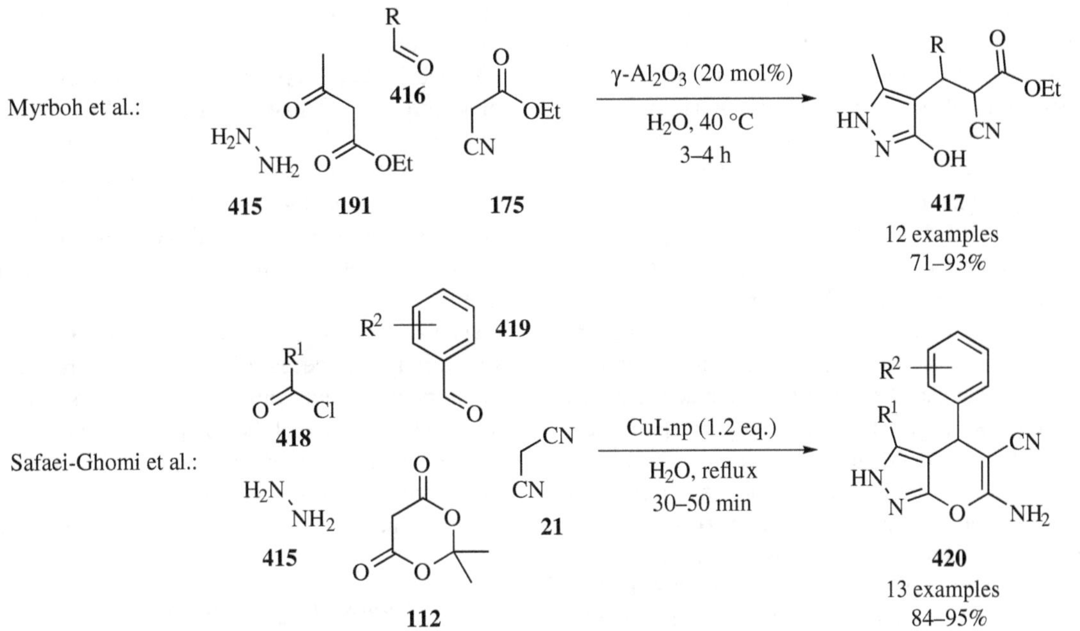

SCHEME 13.85 Four-component Yonemitsu-type reactions [163, 164].

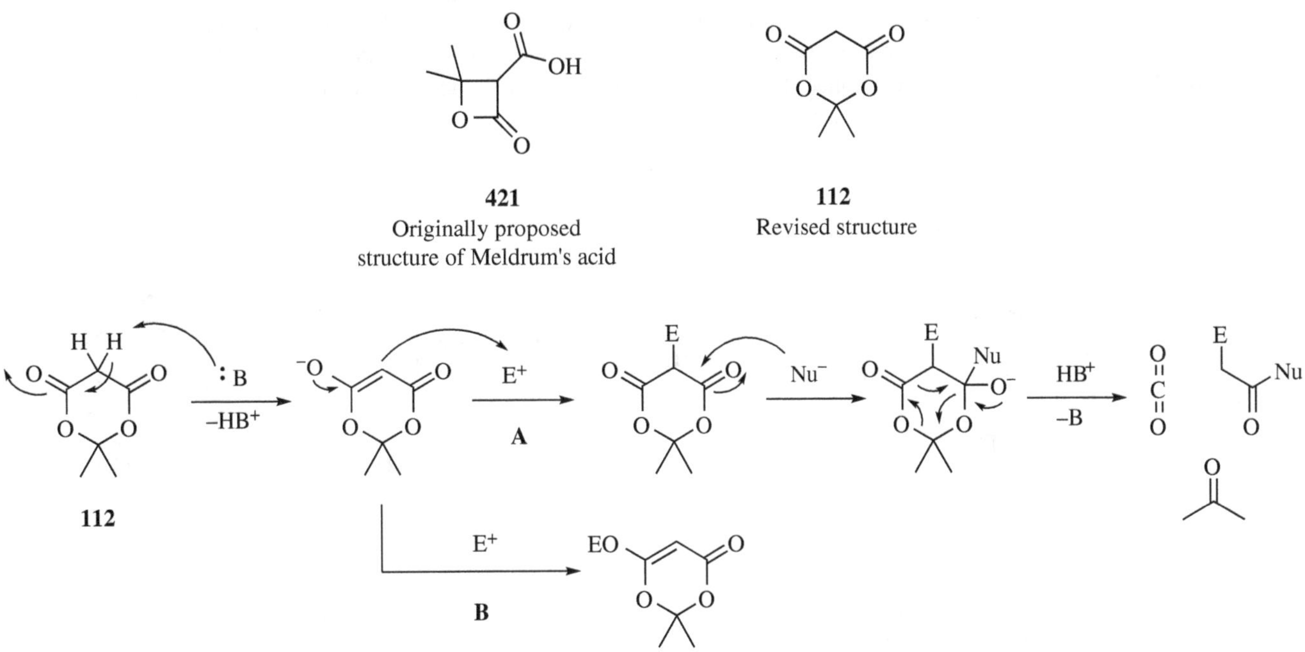

SCHEME 13.86 Proposed structures **421** and **112** of Meldrum's acid and its acidic, nucleophilic, and electrophilic properties.

13.3 MCRs INVOLVING MELDRUM'S ACID

13.3.1 Introduction

In 1908, Andrew N. Meldrum prepared a white crystalline substance from malonic acid, acetone, acetic anhydride, and a catalytic amount of conc. sulfuric acid that he assigned as the β-lactonic acid **421** (Scheme 13.86) [165]. Nowadays, the substance is widely known as Meldrum's acid. Caused by the fact that its first structural assignment as β-lactonic acid **421** was incorrect, its structure was revised about four decades later as 2,2-dimethyl-1,3-dioxane-4,6-dione **112** [166]. Due to its combination of interesting chemical properties, Meldrum's acid became a widely applied compound in organic syntheses [167, 168]. The comparably high CH acidity (pK_a in DMSO=7.325) [168] helps to generate a nucleophilic species even in the presence of weak bases. This ability and the two electrophilic ester functionalities give rise to a broad spectrum of possible transformations that makes Meldrum's acid a versatile tool in synthetic applications.

C-2 modification of the 1,3-dicarbonyl compound is possible by deprotonation in the 2-position followed by electrophilic capture (path **A**). Depending on the type of electrophile used, oxygen modification is also possible (path **B**). Nucleophilic attack on one of the Meldrum's acids carbonyl groups leads to a tetrahedral intermediate, which decomposes easily to generate carbon dioxide, acetone, and an enolate product, which is protonated either by acidic reaction media or in a separate work-up step. This nucleophilic displacement reaction equals the use of Meldrum's acid as a $-CH_2C(O)$ equivalent. Indeed, this is a very common application of Meldrum's acid in multicomponent syntheses (*vide infra*).

The following chapter will give an overview on the literature dealing with the application of Meldrum's acid in multicomponent syntheses since 2005. The literature will be summarized in three main categories depending on the products formed. Reactions in which the Meldrum's acid moiety is intact in the product will be followed by reactions in which **112** serves as a $-CH_2C(O)$ equivalent. At the end, the literature that does not fit into the first two categories will be discussed. As Meldrum's acid is an extensively applied CH-acidic component in MCRs and as it is one of the three components in the Yonemitsu reaction, the reader is referred to Sections 13.1 and 13.2 at this point, in which some examples of the use of Meldrum's acid have already been discussed.

Demand for huge libraries of structurally diverse compounds in short times, especially in the screening process for new drug candidates, has led to the development of strategies in combinatorial chemistry commonly referred to as diversity-oriented synthesis (DOS) [169]. Starting from a mutual synthetic intermediate, the application of a variety of reagents and/or reaction conditions leads to a broad set of structurally (not necessarily) related compounds. Especially in multicomponent syntheses, this can easily be achieved by the variation of one of the components (e.g., an aldehyde, an amine, an isonitrile, or a CH acid). A lot of the examples in this chapter will use the techniques of DOS to create a large library of new compounds and potential drug candidates.

13.3.2 Applications and DOS

Wang et al. found in 2005 that the reaction of certain imines **422** with Meldrum's acid **112** led to significant amounts of a spirocyclic by-product **424**, beneath the desired benzo[*f*]-quinolin-3-ones **423** (Scheme 13.87) [170]. Those findings encouraged Tu et al. to develop a three-component reaction selectively generating dispiro[4.2.5.2]pentadecanes **427**

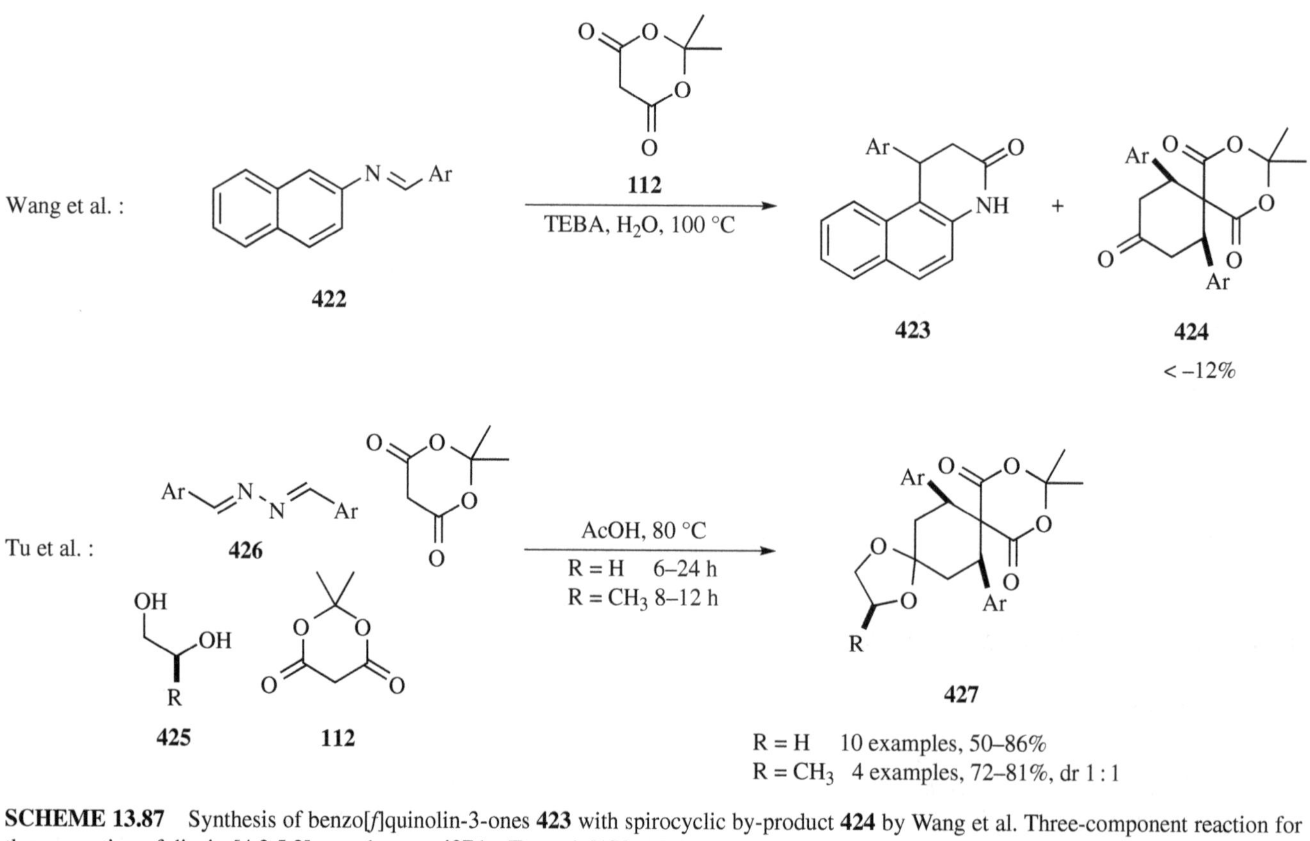

SCHEME 13.87 Synthesis of benzo[*f*]quinolin-3-ones **423** with spirocyclic by-product **424** by Wang et al. Three-component reaction for the generation of dispiro[4.2.5.2]pentadecanes **427** by Tu et al. [170, 171].

from Meldrum's acid **112**, 1,2-diols **425**, and symmetric double imines **426** (Scheme 13.87) [171]. This reaction is conducted in acetic acid at elevated temperatures providing the desired products in moderate to good yields. Interestingly, the first reaction step, which is a Knoevenagel-type condensation, generates acetohydrazide *in situ* from the double imine and Meldrum's acid. This autocatalyst mediates the following Diels–Alder reaction forming the first spirocyclic substructure. Ketone acetalization with the 1,2-diol then creates the second spirocyclic substructure.

The synthesis of the spiro[5.5]undecane substructure was possible by the reaction of Meldrum's acid with aromatic aldehydes in acetone, as described by An et al. in 2010 [172]. This reaction is mediated by sodium acetate, which is added in excess to the reaction. Within 48 h, the desired products could be isolated in 43–85% yield. Analogous products **430** containing a urea structural motif have been prepared by Jetti et al. by the reaction of urea **428** with several aromatic aldehydes **429** and Meldrum's acid **112** at 80 °C under solvent-free conditions (Scheme 13.88) [173]. This reaction is mediated by a combination of stoichiometric *N*-bromosuccinimide (NBS) and azobisisobutyronitrile (AIBN). The NBS could be recycled up to three times with only slight loss of activity.

The preparation of ring-contracted variants of such spiro compounds was possible by reacting Meldrum's acid **112** with aromatic aldehydes **432** and α-thiocyanato ketones **431** under microwave irradiation in ethanol at higher temperatures (Scheme 13.89) [174]. Substoichiometric sodium ethanolate catalyzed this reaction efficiently to give the desired *trans*-cyclopropanes **433** in good yields (74–87%) and short reaction times (13–17 min).

Zwitterionic salts with Meldrum's acid as the anionic structural element are easily accessible by reacting a CH-acidic *N*-alkyl pyridinium salt **434** with 2-alkylidene Meldrum's acid derivatives (Scheme 13.90) [175]. The 2-alkylidene Meldrum's acids are generated *in situ* from Meldrum's acid **112** and an aldehyde **435** under basic conditions. Other heterocycles have been reported to undergo the same reaction [176]. In a 4CR, the pyridinium salts can be formed *in situ* [177]. The generation of zwitterionic salts from Meldrum's acid, several heterocycles and dimethyl acetylenedicarboxylate has been reported [178].

The exchange of the heterocyclic component by a phosphine led to phosphonium based zwitterionic salts [179]. Catalytic pyrrolidine and benzoic acid act as mediators for the Knoevenagel reaction generating the 2-alkylidene Meldrum's acids. The formed zwitterionic salts can easily be transformed into highly substituted furan derivatives utilizing carboxylic acid chlorides.

An intermolecular salt formation has been reported by the reaction of DBU with Meldrum's acid (and other CH-acidic compounds) and pyrimidine-tetraone in refluxing chloroform [180]. As cyanide ions readily add to 2-alkylidene Meldrum's acid derivatives, an intermolecular salt is formed between sodium cyanide and *in situ* generated 2-alkylidene Meldrum's acids in a three-component reaction [181].

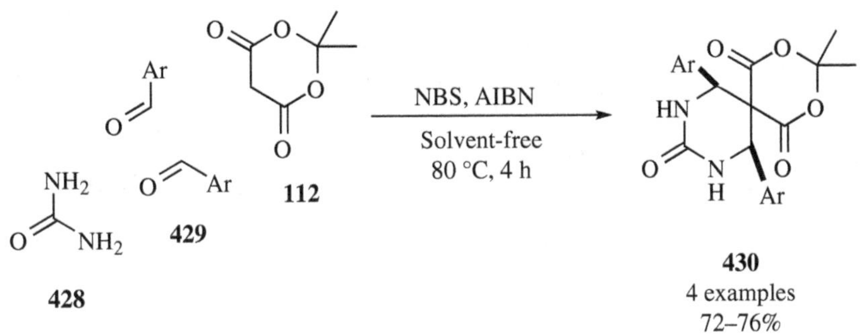

SCHEME 13.88 Synthesis of diazaspiro[5.5]undecanes **430** by Jetti et al.

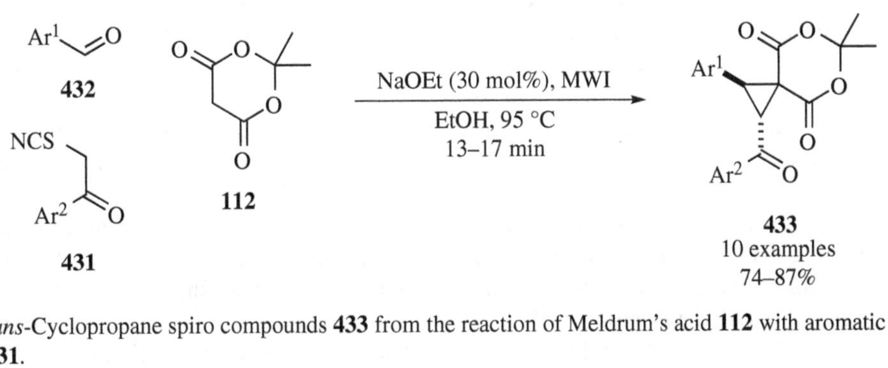

SCHEME 13.89 *trans*-Cyclopropane spiro compounds **433** from the reaction of Meldrum's acid **112** with aromatic aldehydes **432** and α-thiocyanato ketones **431**.

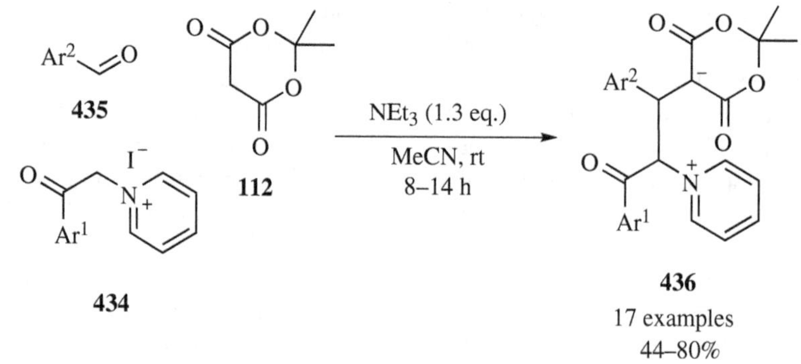

SCHEME 13.90 An example of zwitterionic salts **436** with Meldrum's acid as the anionic structural element.

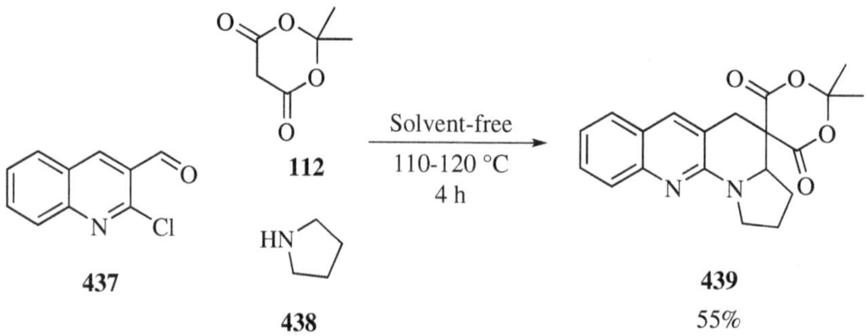

SCHEME 13.91 Synthesis of novel pentacyclic quinoline derivatives **439** under solvent-free conditions.

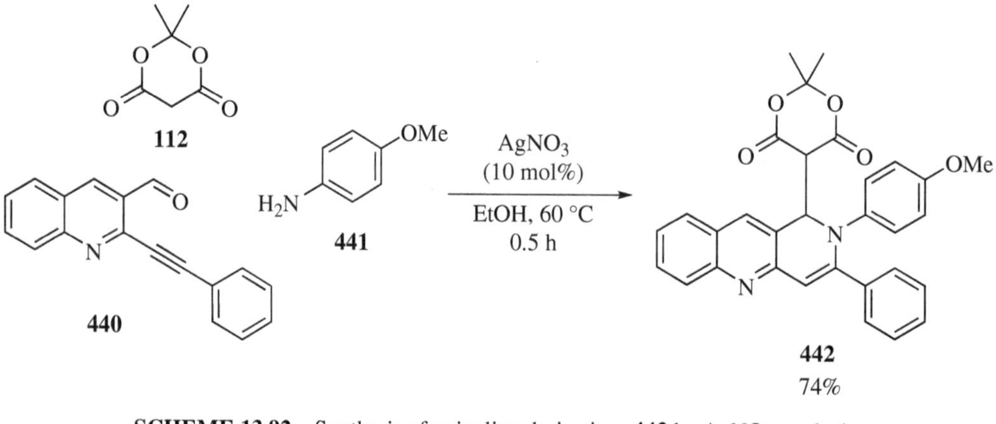

SCHEME 13.92 Synthesis of quinoline derivatives **442** by $AgNO_3$ catalysis.

In 2006, an MCR pathway to pentacyclic quinoline derivatives **439** was established by Bhuyan et al. (Scheme 13.91) [182]. Novel annelated quinolines have been prepared in a three-component reaction under solvent-free conditions at high temperatures. In addition to Meldrum's acid **112** and pyrrolidine **438**, a variety of other CH acids and amines have successfully been applied in this reaction.

A different pathway to substituted quinoline derivatives **442** involving Meldrum's acid **112** was reported by Verma et al. [183]. This report utilizes a huge library of amines and CH acids that undergo reaction with the *ortho*-alkynyl aldehyde **440** by cooperative catalysis of $AgNO_3$ and proline or by the catalysis of $AgNO_3$ alone as in the case of the example shown in Scheme 13.92.

A catalytic amount of tributylphosphine in toluene at 80 °C catalyzes the formation of highly substituted tetrahydrofuran derivatives **445**, as reported by Pedduri and Williamson in 2008 (Scheme 13.93) [184].

The syntheses of variably substituted 4(1*H*)-quinolones **449**, 5*H*-thiazolo[3,2-*a*]pyrimidin-5-one (not shown), and 4*H*-pyrimido[2,1-*b*]benzothiazol-4-ones **452** could be achieved by reacting substituted anilines **446** and the corresponding 2-aminothiazoles **450**, respectively, with trimethylorthoformate **447** and Meldrum's acid **112** in the

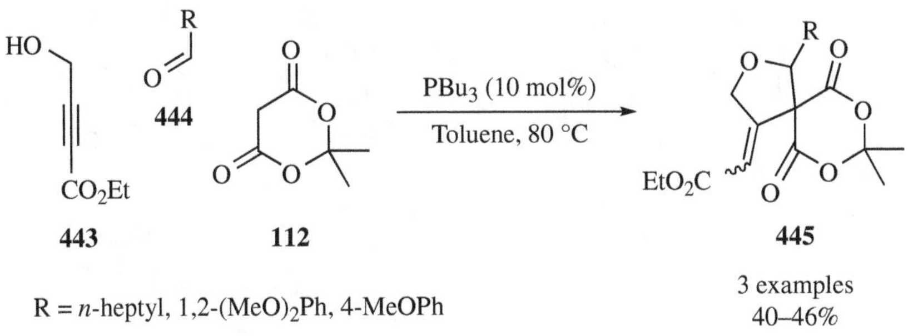

SCHEME 13.93 Tributylphosphine-catalyzed synthesis of highly substituted tetrahydrofuran derivatives **445**.

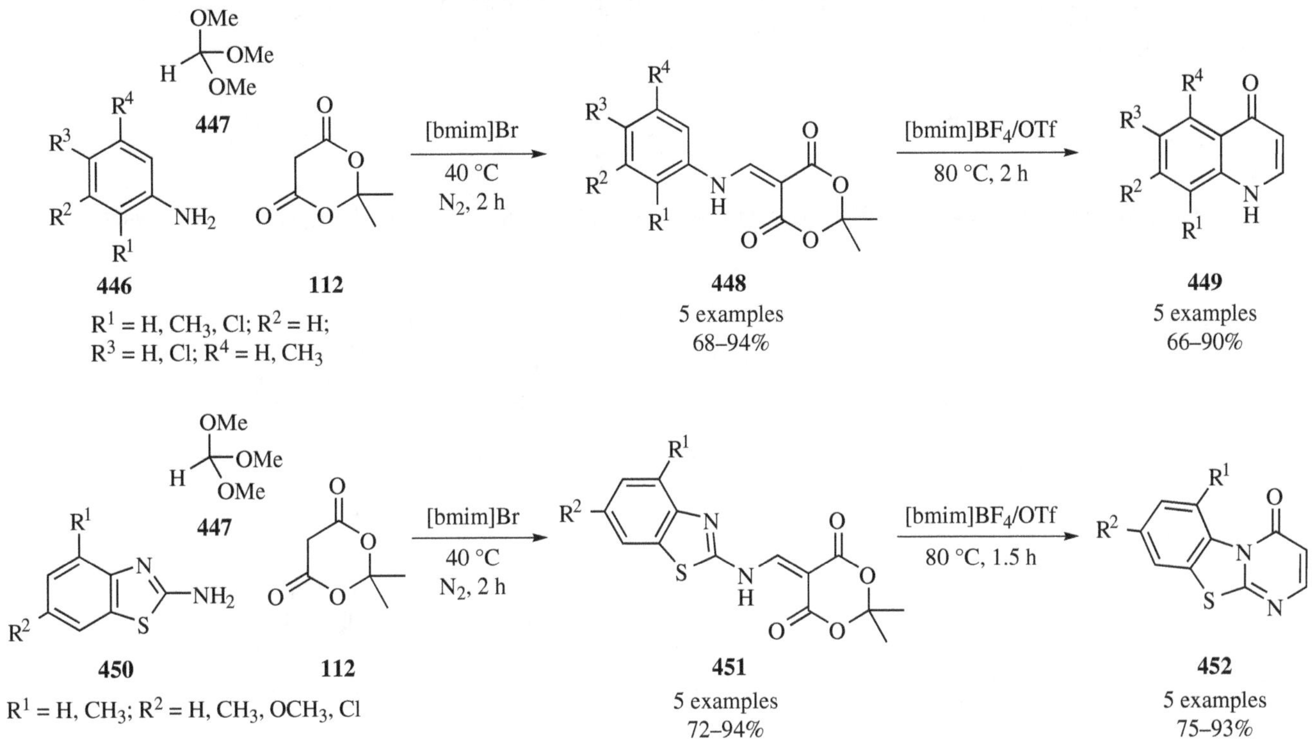

SCHEME 13.94 Synthesis of 4(1*H*)-quinolones **449** and 4*H*-pyrimido[2,1-*b*]benzothiazol-4-ones **452** in a two-step fashion introduced by a three-component reaction step.

ionic liquid [bmim]Br at slightly elevated temperature (Scheme 13.94) [185]. This three-component reaction led to an intermediate (**448**/**451**) with intact Meldrum's acid structural element that can be cyclized to give the desired compounds mentioned earlier in only one additional step.

13.3.3 Meldrum's Acid as Synthetic Equivalent

Within a sequence of Knoevenagel condensation, Michael addition and condensation, Meldrum's acid can be used as a "CH_2-C(O)" equivalent, through the elimination of acetone and subsequent decarboxylation (Scheme 13.86). In this sequence, other 1,3-dicarbonyl compounds within the reaction mixture serve as nucleophile in the Michael addition. By addition of a nitrogen source to this reaction mixture, for example, ammonium acetate or primary amines, the synthesis of 3,4-dihydropyridinones is possible. Furthermore, the use of amines and alcohols instead of 1,3-dicarbonyl compounds can cause solvolysis of the Meldrum's acid fragment, inhibiting the decarboxylation and thus, allowing the use of Meldrum's acid as a malonic acid equivalent.

13.3.3.1 3,4-Dihydropyranones The use of Meldrum's acid as a " CH_2-C(O)" equivalent was successfully applied in the synthesis of 3,4-dihydropyranones. The reaction of Meldrum's acid **112** with several aldehydes **453** and dimedone **15** under grinding conditions to form 3,4-dihydropyranones **454** was reported by Rong in 2007 (Scheme 13.95) [186].

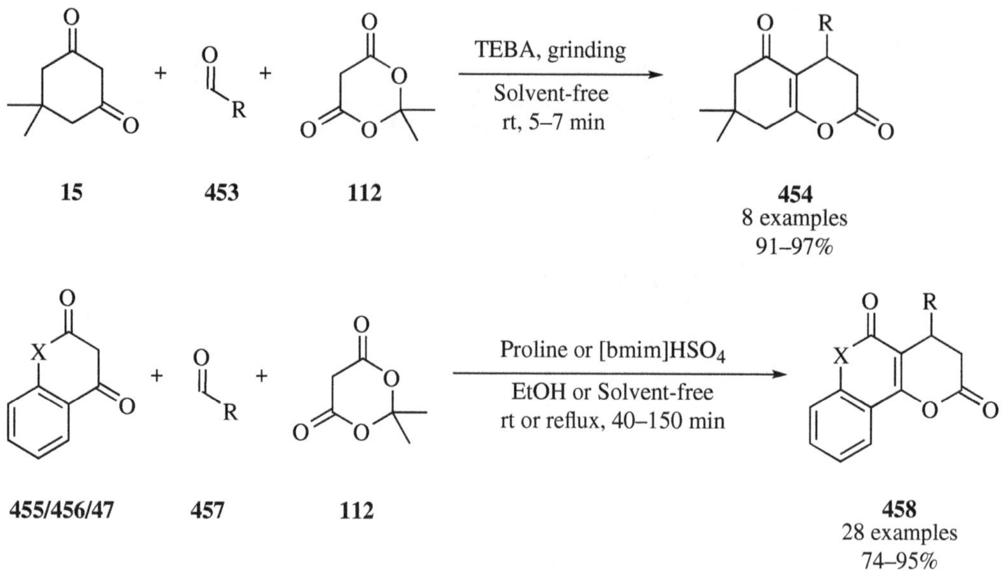

SCHEME 13.95 Synthesis of 3,4-dihydropyranones **454/458** starting from 1,3-dicarbonyl compounds **15/47/455/456**.

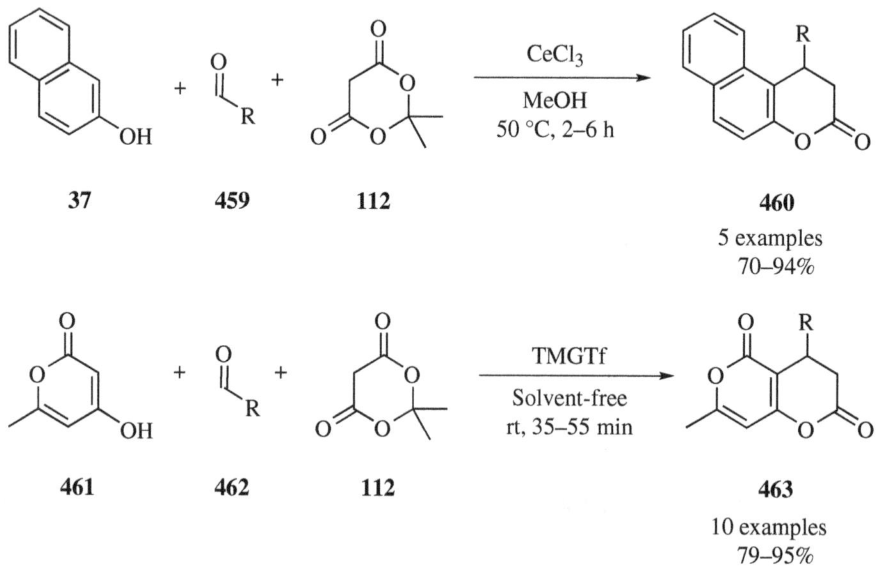

SCHEME 13.96 Synthesis of 3,4-dihydropyranones **460/463** starting from stable enols **37/461**.

Analogous to the enolizable 1,3-dicarbonyls, stable enols like 2-naphthol **37** [187a] or **461** [187b], can also be used in the same manner (Scheme 13.96).

The barbituric acids **464/61** represent a special class of substrates, which enable highly functionalized polyheterocyclic products, such as 3,4-dihydropyranones **466** (Scheme 13.97) [188].

The introduction of an amide at C4 is possible through the addition of an isonitrile **468/471** to the reaction. Depending on the solvent or further addition of an alcohol, Meldrum's acid can be solvolized by an alcohol, inhibiting the decarboxylation (Scheme 13.98) [189]. Thereby obtained products in general show good *cis*-selectivity according to Shaabani and Rezayan.

13.3.3.2 3,4-Dihydropyridinones Similar to the Hantzsch 1,4-dihydropyridine synthesis (see Section 13.5.3), the addition of ammonium acetate [190] or primary amines [57] led to the incorporation of a nitrogen atom and therefore to a broad range of 3,4-dihydropyridinones **474** and **477** (Scheme 13.99). Both cyclic and acyclic 1,3-dicarbonyl compounds were found to give the desired products in good yields by electromagnetic irradiation.

In the case of a highly electron-withdrawing group like trifluoromethyl in the β-ketoester, the final dehydration is circumvented and tetrahydropyridinones **480** are obtained (Scheme 13.100) [190a].

Comparable to the synthesis of 3,4-dihydropyranones, preformed enamines **104/485** [191] and arylamines **125** [192]

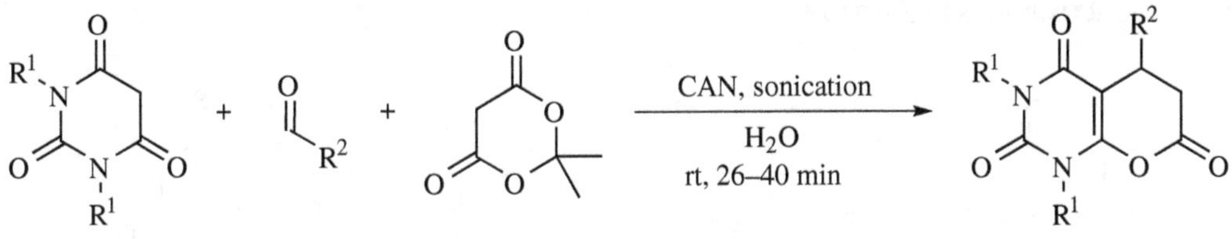

61/464 **465** **112** **466**

8 examples
83–96%

SCHEME 13.97 Synthesis of 3,4-dihydropyranones **466** starting from barbituric acids **61/464**.

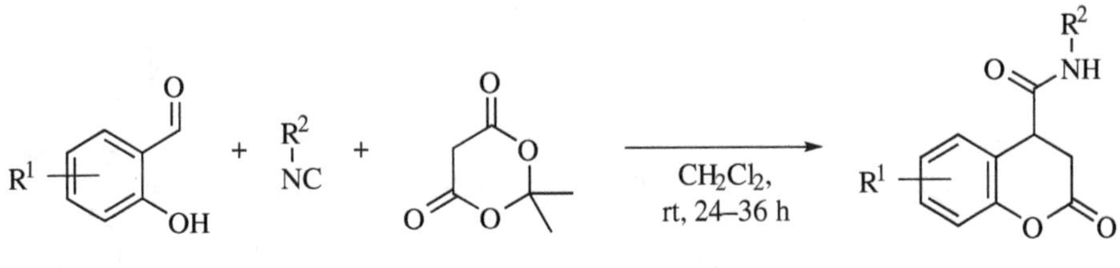

253/467 **468** **112** **469**

19 examples
65–81%

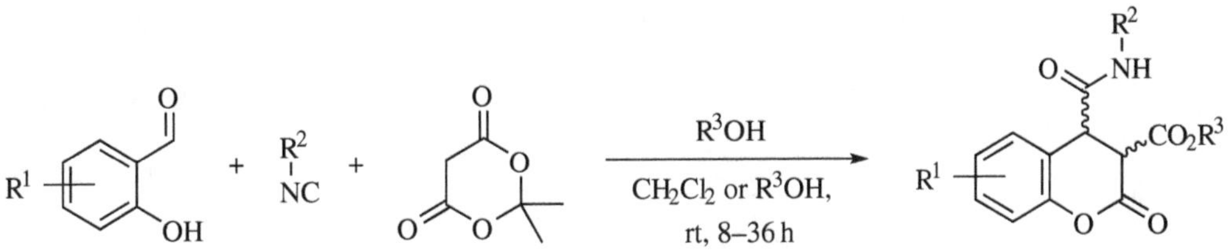

253/470 **471** **112** **472**

42 examples
63–96%
cis:trans 1:1–1:0

SCHEME 13.98 Synthesis of C4-amide-substituted 3,4-dihydropyranones **469/472**.

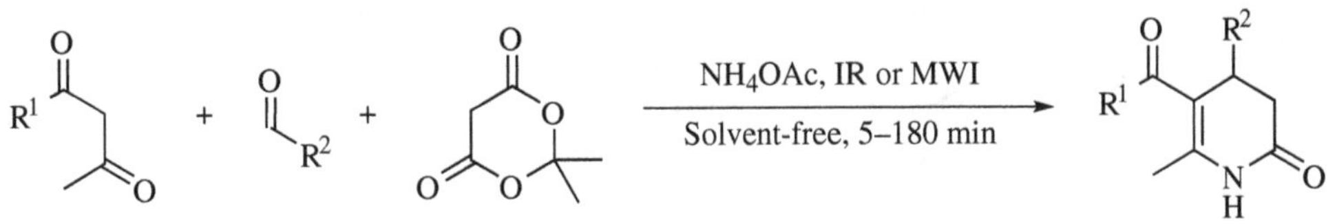

136/191 **473** **112** **474**

25 examples
50–75%

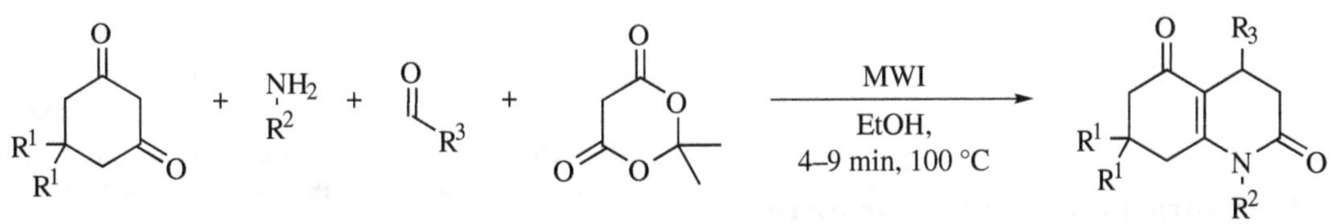

15/71 **475** **476** **112** **477**

18 examples
82–96%

SCHEME 13.99 Synthesis of 3,4-dihydropyridinones **474/477** using ammonium acetate or primary amines **475** [57, 190].

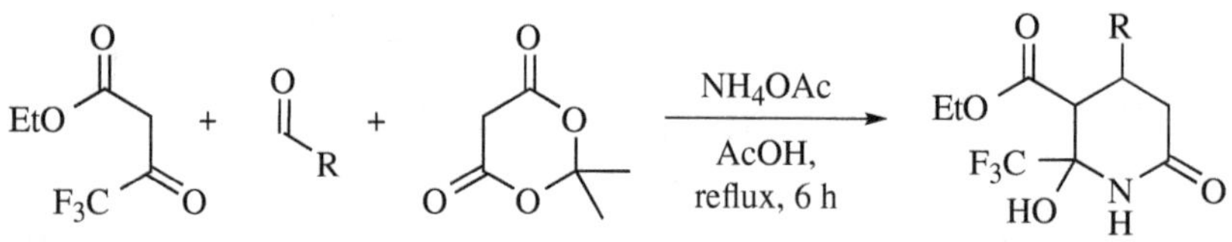

478 479 112 480

13 examples
61–74%

SCHEME 13.100 Synthesis of tetrahydropyridinones **480** [190a].

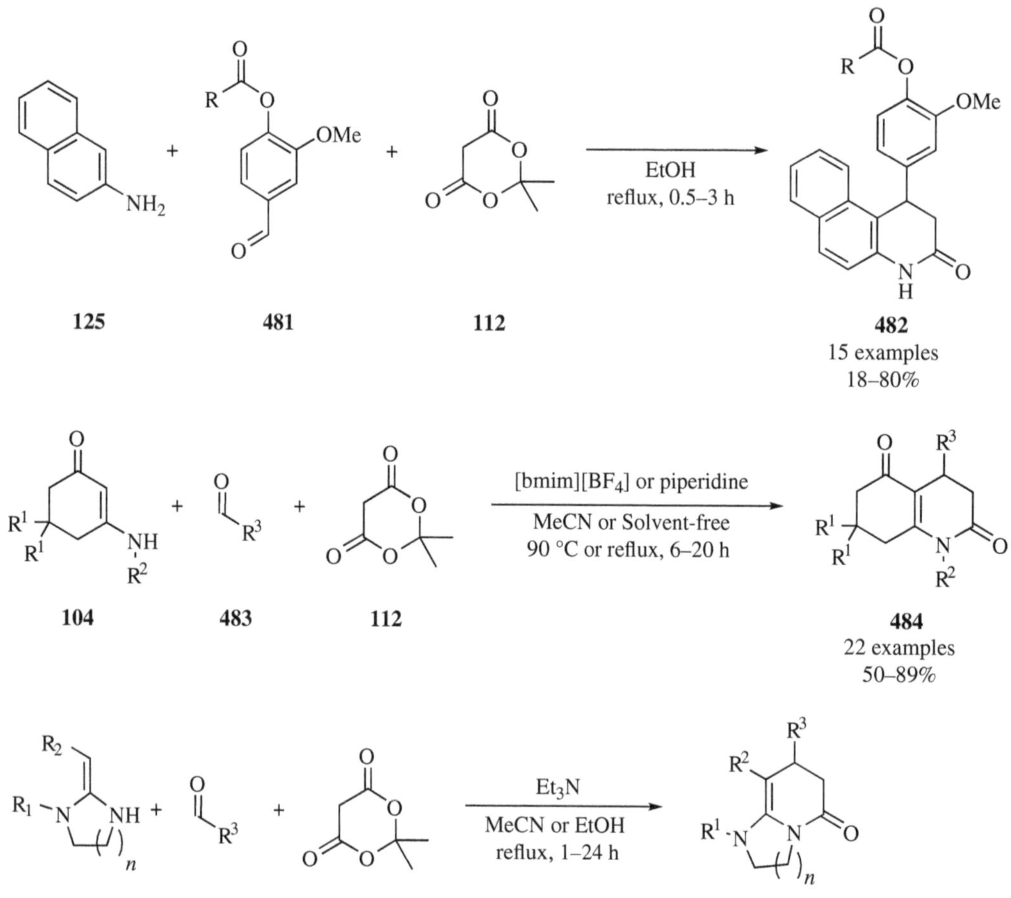

125 481 112 482
15 examples
18–80%

104 483 112 484
22 examples
50–89%

485 486 112 487
39 examples
15–96%

SCHEME 13.101 Synthesis of 3,4-dihydropyridinones **482/484/487** starting from **125/104/485**, respectively [191, 192].

are suitable for the preparation of 3,4-dihydropyridinones and therefore enable complex substitution patterns like tetrahydroimidazopyridinones or tetrahydropyridopyrimidinones (Scheme 13.101).

13.3.4 Meldrum's Acid as Malonic Acid Equivalent

Another application of Meldrum's acid is its use as a malonic acid equivalent. Through substitution of the 1,3-dicarbonyl compound by amines or alcohols, solvolysis of the Meldrum's acid fragment occurs and inhibits the decarboxylation. For example, after Knoevenagel condensation of Meldrum's acid **112** with the ferrocenylcarbaldehyde **188**, an isonitrile **488** can attack the condensation product, leading to the corresponding amide **489** (Scheme 13.102). On account of the solvolysis and the absence of an appropriate intramolecular nucleophile, the isolation of acyclic products is possible [193].

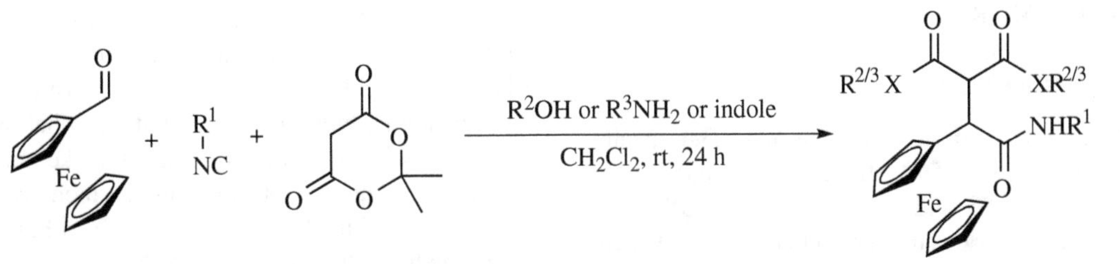

188 **488** **112** **489** 18 examples 55–90% (X = O or NH)

SCHEME 13.102 Meldrum's acid as malonic acid equivalent—I.

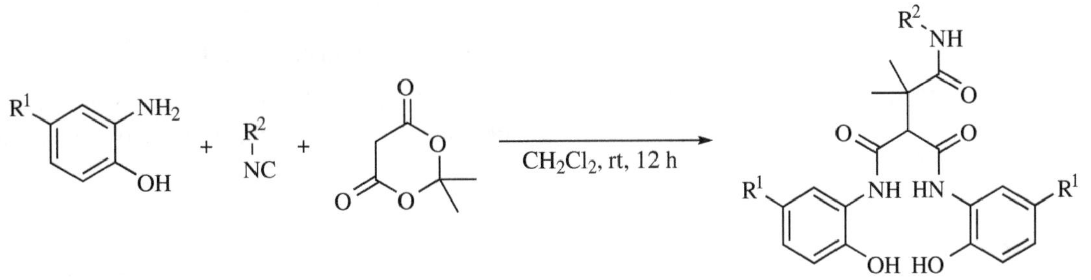

490 **491** **112** **492** 6 examples 75–87%

SCHEME 13.103 Meldrum's acid as malonic acid equivalent—II.

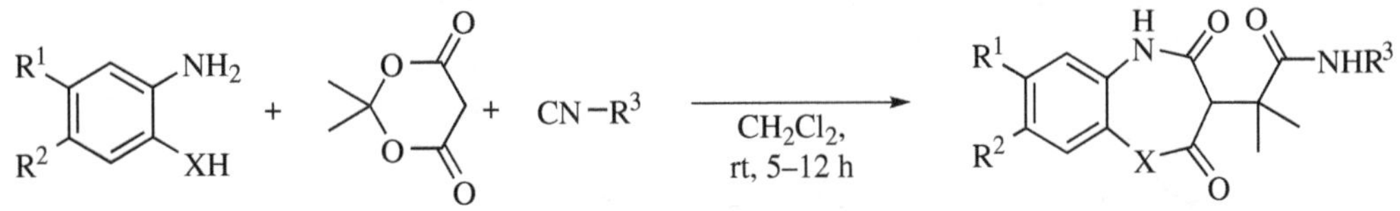

276/493 X = NH, O **112** **494** **495** 22 examples 70–90%

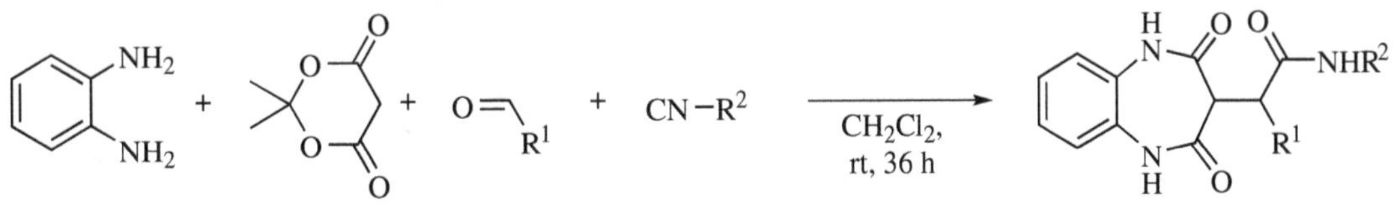

276 **112** **496** **497** **498** 13 examples 55–88%

SCHEME 13.104 Synthesis of benzodiazepinedions and benzooxazepinedions (**495**, **498**) [194, 195].

Without any aldehyde or ketone being present, the acetone from Meldrum's acid cleavage reacts instead, leading to α-dimethyl substituted amides **492** (Scheme 13.103) [194].

According to this reactivity, 1,2-diaminobenzenes **276** or 2-aminophenols **493** led to substituted benzodiazepinedions or benzooxazepinedions **495** and **498** (Scheme 13.104) [194, 195].

13.4 POVAROV MCRs

13.4.1 Introduction

In 1963, Povarov observed the formation of 1,2,3,4-tetrahydroquinoline derivatives **500** through the reaction of *N*-benzylideneaniline **499** and vinyl alkyl ethers **90** in the presence of boron trifluoride etherate (Scheme 13.105) [196].

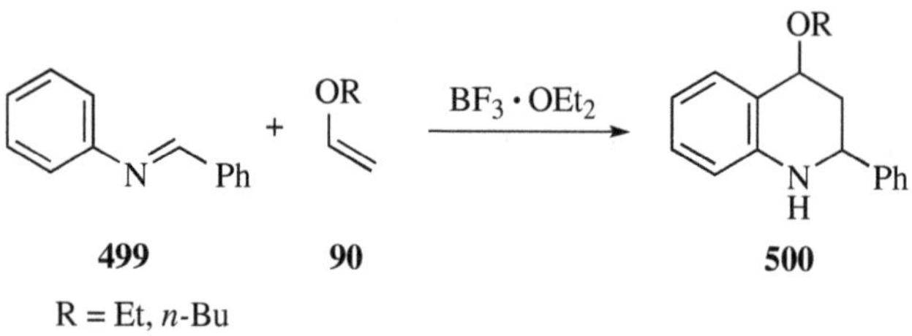

SCHEME 13.105 1,2,3,4-Tetrahydroquinoline synthesis reported by Povarov in 1963.

Substituted 1,2,3,4-tetrahydroquinolines are present in several natural products, and the synthesis of some of those have been achieved starting with a Povarov reaction [197]. For example, luotonin A **502**, with a rather simple structure but promising cytotoxicity against leukemia cells, was the target of a four-step total synthesis, which involves a Povarov reaction combined with an oxidation to build the quinoline intermediate **501** (Scheme 13.106) [197c].

A slightly more complex application of a Povarov reaction in total synthesis is the preparation of martinellic acid **504**, reported by Batey in 2002 (Scheme 13.107) [197b]. Using the Povarov reaction, it was possible to build the highly substituted 1,2,3,4-tetrahydroquinoline core **503** in only two steps with good diastereoselectivity.

13.4.2 Mechanistic Aspects

Although the mechanism of the Povarov reaction is still debated, a stepwise mechanism via ionic intermediates is the

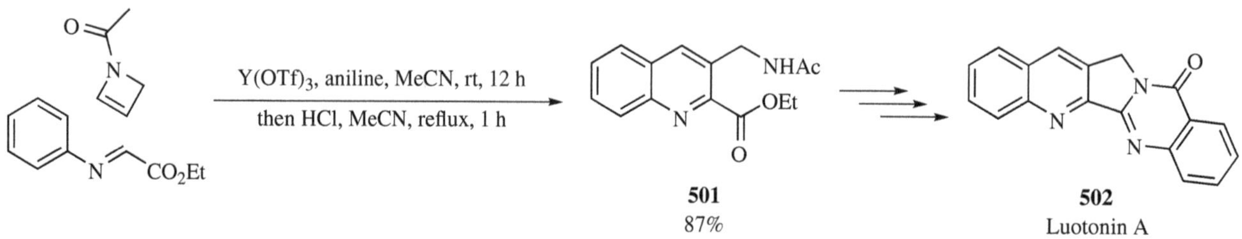

SCHEME 13.106 Use of Povarov reaction in total synthesis of luotonin A **502**.

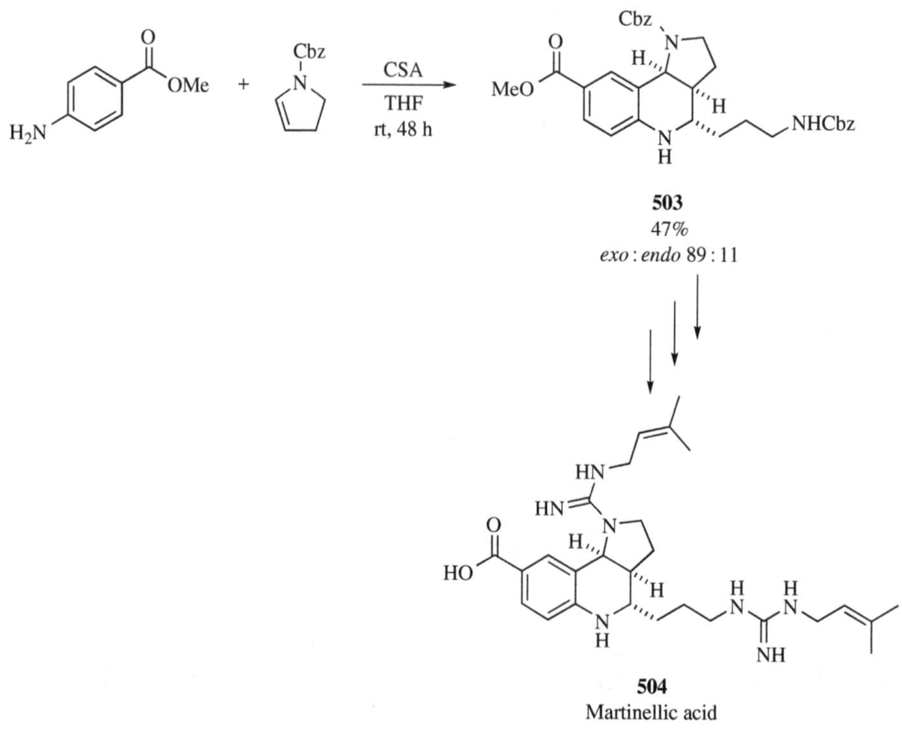

SCHEME 13.107 Use of the Povarov reaction in total synthesis of martinellic acid **504**.

most probable one [198]. Therefore, the condensation of a primary aromatic amine **I** with an aldehyde **II** to give the corresponding Schiff base **III** is the initial step (Scheme 13.108).

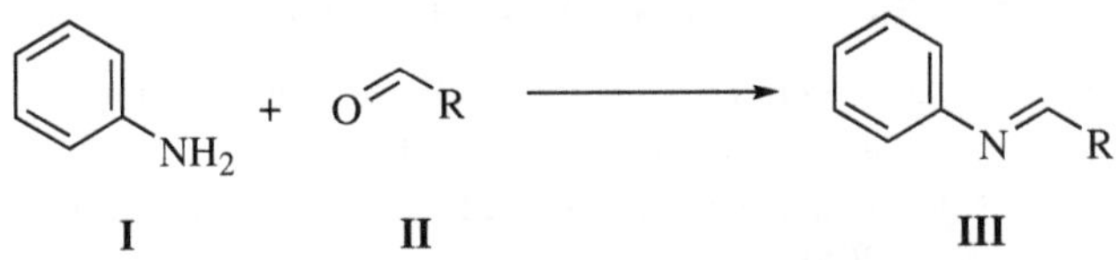

SCHEME 13.108 Mechanism of the Povarov reaction—part I.

The present Brønsted or Lewis acid (A) activates the Schiff base **III** for an electrophilic addition of an electron-rich alkene, generating a carbenium ion **IV**, which is stabilized by an electron-donating group. A final intramolecular electrophilic aromatic substitution gives the 1,2,3,4-tetrahydroquinolines **V** (Scheme 13.109).

Depending on the stability of the electron-donating group, elimination can take place to give 1,2-dihydroquinoline **VI**, which can also directly be obtained by the reaction of alkynes instead of alkenes (Scheme 13.110).

Various systems were examined and proven as applicable for the activation of the Schiff base **III**. Among them are

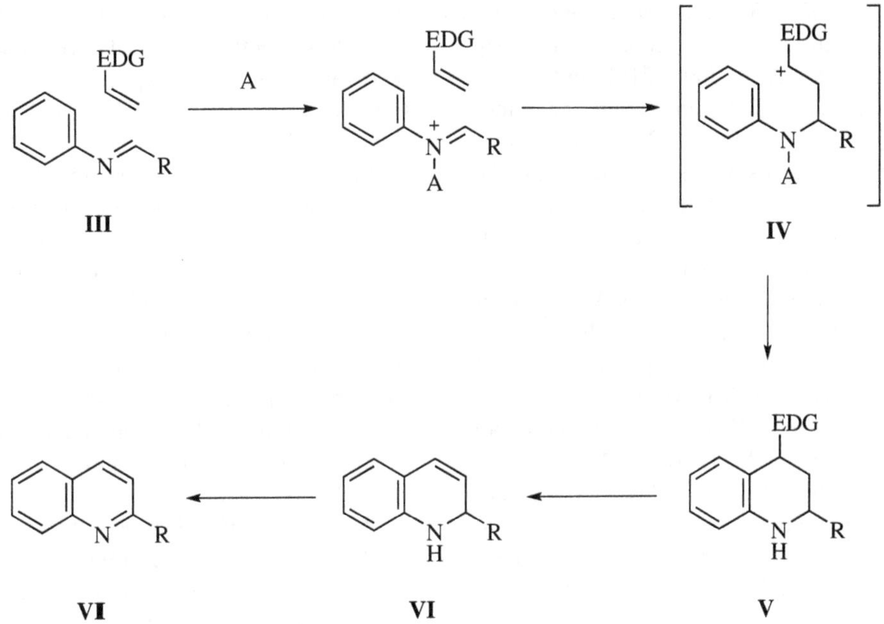

SCHEME 13.109 Mechanism of the Povarov reaction—part II.

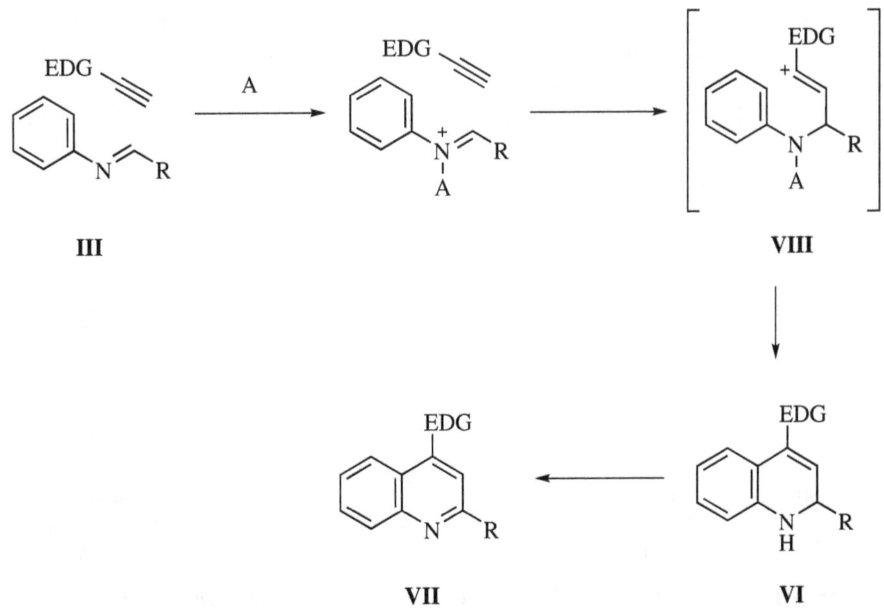

SCHEME 13.110 Mechanism of the Povarov reaction—part III.

simple proton sources like trifluoroacetic acid (TFA) [199], trifluoromethanesulfonic acid (TfOH) [200], trifluoromethanesulfonimid (Tf_2NH) [201], or trifluoroethanol (TFE) [202], Lewis acids like ceric ammonium nitrate (CAN) [203], bismuth chloride [204], boron trifluoride etherate [205], niobium chloride [206], ytterbium [207], or scandium [208] trifluoromethanesulfonate. Also, molecular iodine [209] and materials with a complex structure like a nanoporous aluminosilicate (AlKIT-5) [210] and perchloric acid on montmorillonite-K10 [211] are able to activate the Schiff base **III**.

The range of possible residues R includes nearly each conceivable alkyl chain, branched ones, aromatic and heteroaromatic systems. The most commonly used alkenes are vinyl ethers [202, 203, 203d, 209c], vinyl amides [204, 212], 2,3-dihydrofuran [203c, 206, 207d, 210], 3,4-dihydro-2*H*-pyran [206, 207d, 209b, 210], and acrylic acid derivatives [212], but also, *in situ* formed enamines [213], norbornene [205], *N*-protected 2,3-dihydropyrroles [208], and 1,2-disubstituted alkenes [199, 207c] can be applied in the Povarov reaction. In the case of alkynes, terminal ones [200, 207b, 209a, 211] and esters of acetylenedicarboxylic acid [207a] are favored.

The obtained 1,2,3,4-tetrahydroquinolines **V** are rather stable, but the 1,2-dihydroquinolines **VI** are highly reactive and usually oxidized to the corresponding quinolines **VII** (Scheme 13.110) on exposure to air [200, 202, 211, 212] or by a hydrogen transfer reaction [207b] in a domino one-pot process with the elimination [202, 212] or the Povarov reaction itself [202, 207a, b, 209a, 211]. Also direct oxidation of the 1,2,3,4-tetrahydroquinolines can be accomplished stoichiometrically with 2,3-dichloro-5,6-dicyano-1,4-benzoquinone (DDQ) [201] or catalytically with palladium combined with air [212].

The stereochemical result of the Povarov reaction is closely linked to the applied alkene species. In the case of an acyclic alkene, the 2,4-*cis* product is highly favored because of a bis-equatorial arrangement of both residues in the transition state. Cyclic alkenes generate predominantly the *exo* product. However, 2,3-dihydrofuranes and dihydropyrroles give poor diastereomeric ratios (*endo*/*exo* 1:1–1:2), while 3,4-dihydro-2*H*-pyranes give excellent ratios under the same reaction conditions [206, 207d]. The competition between a possible coordination and steric repulsion in the transition state is proposed as a reason for this observation [198a].

13.4.3 Efficient Synthesis of 1,2,3,4-Tetrahydroquinolines

The simplest access to 2,4-*cis* 1,2,3,4-tetrahydroquinolines **506** via Povarov reaction is the use of vinyl ethers or vinyl amides **67** under Lewis acid catalysis (Scheme 13.111) [203b, 204, 205].

In contrast to the usual substitution pattern, the 2,4-*trans* product **509** can be obtained by *in situ* substitution prior to the intramolecular electrophilic aromatic substitution. This product is selectively formed during the reaction and cannot be obtained by simple exposure of the 4-ethoxy compound (not shown in Scheme 13.112) to CAN and an arylamine (Scheme 13.112) [203a].

After condensation of the substrates, the imine can tautomerize to the enamine and thereby act as the electrophile, as well as the alkene, substituting the widely used vinyl ethers, in this case even without a catalyst (Scheme 13.113) [213].

Cyclic vinyl ethers like 3,4-dihydro-2*H*-pyranes and 2,3-dihydrofurans **512** are also suitable for the Povarov reaction, providing the corresponding tricyclic compounds **514** (Scheme 13.114). As mentioned earlier, the diastereoselectivity for the products obtained from 3,4-dihydro-2*H*-pyrans is much better than for those from 2,3-dihydrofuranes [206, 209b].

Besides simple arylamines, even more complex aromatic amines like substituted 3-aminocoumarins **515** can be subjected to the cyclization conditions (Schemes 13.115) [207d].

The nanoporous alumosilicate (AlKIT-5) with its complex structure facilitates not only the opening of 2,3-dihydrofurans and 3,4-dihydro-2*H*-pyrans to give the corresponding aldehydes but also opens access to the *endo* products **518** with good selectivity (Scheme 13.116) [210].

Bicyclic alkenes, like norbornene **519**, can be also used in the Povarov reaction and give the *exo*–*exo* product **521** with good yields and excellent selectivity, in general (Scheme 13.117) [205].

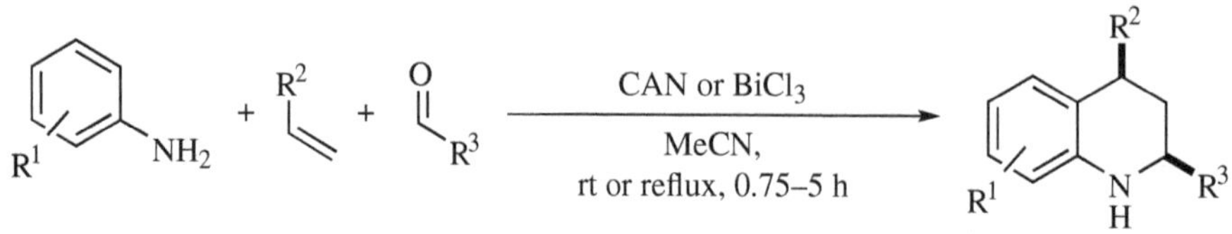

248 **67** **505** **506**
23 examples
47–98%

SCHEME 13.111 Synthesis of 2,4-*cis* 1,2,3,4-tetrahydroquinolines **506** [203b, 204, 205].

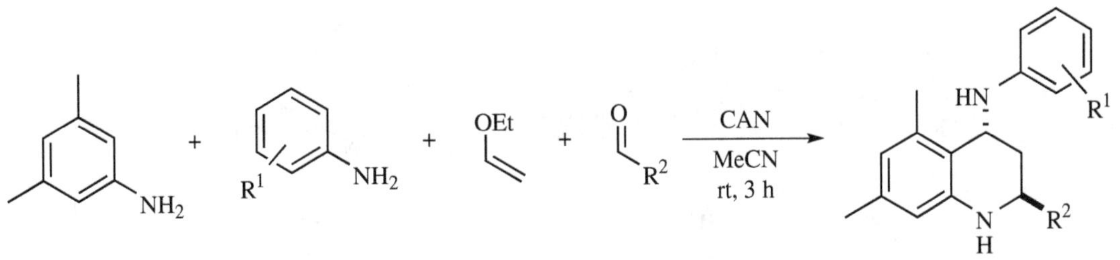

507 248 63 508 509
3 examples
72–87%

SCHEME 13.112 Synthesis of 2,4-*trans* 1,2,3,4-tetrahydroquinolines **509**.

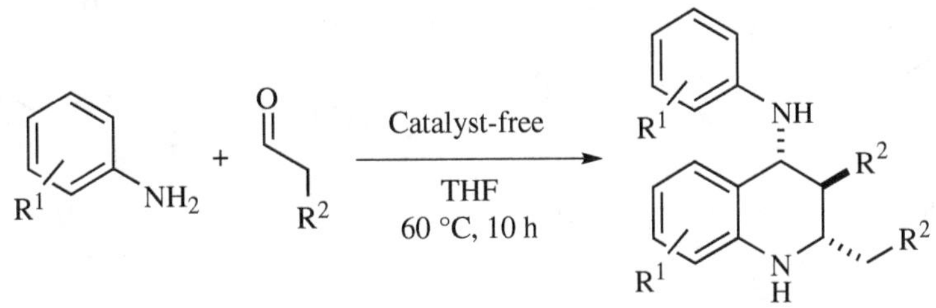

248 510 511
12 examples
24–98%

SCHEME 13.113 Synthesis of 2,3,4-substituted 1,2,3,4-tetrahydroquinolines **511**.

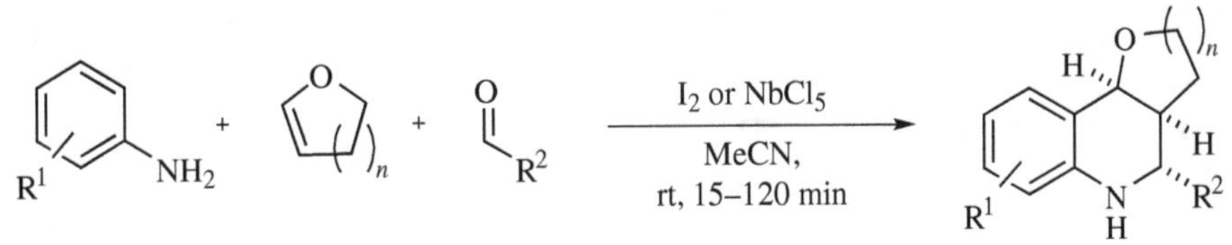

248 512 $n = 1, 2$ 513 514
22 examples
57–90%
endo : *exo* 1 : 2–1 : 49

SCHEME 13.114 Synthesis of 1,2,3,4-tetrahydroquinolines **514** starting from 2,3-dihydrofurans and 3,4-dihydro-2*H*-pyrans **512** [206, 209b].

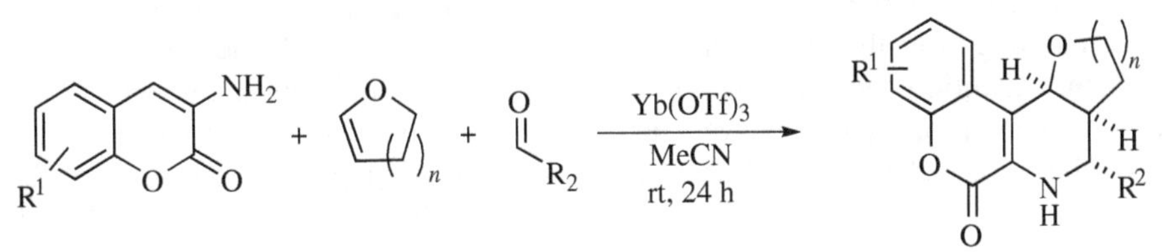

515 512 $n = 1, 2$ 516 517
13 examples
26–71%
endo : *exo* 1 : 1–1 : 19

SCHEME 13.115 Synthesis of 1,2,3,4-tetrahydroquinolines **517** starting from 3-aminocoumarins **515** [207d].

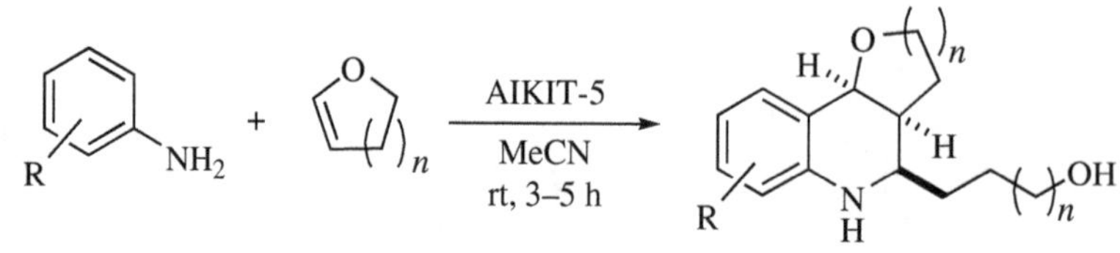

248 512 $n = 1, 2$ 518 10 examples 89–92% *endo* : *exo* 8 : 1–19 : 1

SCHEME 13.116 *Endo*-selective synthesis of 1,2,3,4-tetrahydroquinolines **518**.

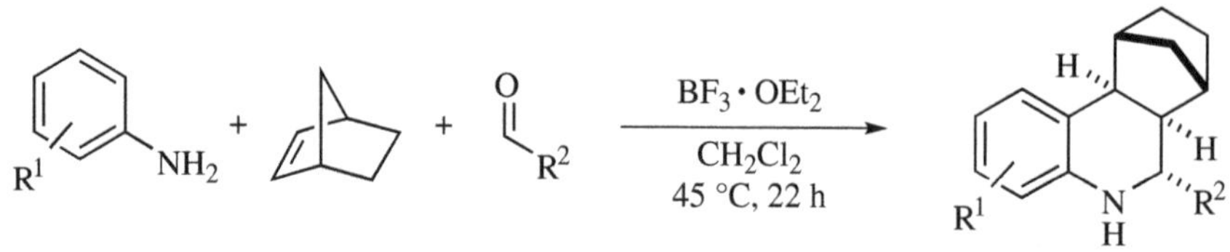

248 519 520 521 17 examples 55–91% *exo–endo* : *exo–exo* 1 : 1–1 : 49

SCHEME 13.117 Synthesis of 1,2,3,4-tetrahydroquinolines **521** starting from norbornene **519**.

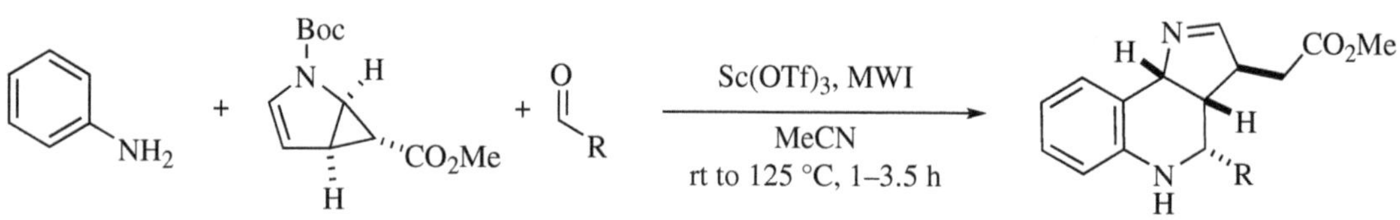

522 523 524 525 7 examples 53–70% *endo* : *exo* 1.5 : 1–2.6 : 1

SCHEME 13.118 Synthesis of 1,2,3,4-tetrahydroquinolines **525** starting from dihydropyrrole **523**.

The cyclopropane substructure in the *N*-protected dihydropyrrole derivative **523** collapses after the cyclization and delivers a highly functionalized product **525** (Scheme 13.118) [208].

By introducing chiral information into the substrate, a diastereo- and enantioselective preparation of 1,2,3,4-tetrahydroquinolines is possible. Jørgensen therefore combined an organocatalytic Michael addition (not shown) with a Povarov reaction in a one-pot process to achieve good selectivities in the synthesis of octahydroacridines **527** (Scheme 13.119) [199].

13.4.4 Efficient Synthesis of Quinolines

There are different approaches to synthesize quinolines utilizing the Povarov reaction. The first one is the most simple, a one-pot process of the Povarov reaction followed by final oxidation of the obtained 1,2,3,4-tetrahydroquinoline (Scheme 13.120) [201].

In the case of a suitable leaving group, a domino one-pot process can be applied, consisting of Povarov reaction, subsequent elimination/decarboxylation (including the loss of R^2), and the final oxidation of the obtained 1,2-dihydroquinolines (Scheme 13.121) [202a, 212]. On account of a slightly different mechanism and no leaving group being present, the unusual 2,3-substitution pattern and a direct oxidation of 1,2,3,4-tetrahydroquinolines to quinolines by palladium catalysis can be achieved (Schemes 13.121) [212].

An appropriate strategy, containing a semiprotected diamine **534** as substrate, also allows the synthesis of asymmetric 1,10-phenantrolines **536** by repeating a sequence of Povarov reaction and subsequent oxidation (Scheme 13.122) [202b].

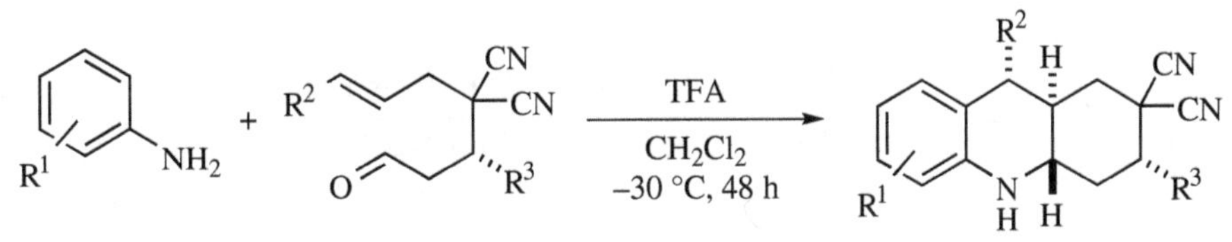

248 526 527
18 examples
59–93%
89–92% ee
dr 12 : 1–20 : 1

SCHEME 13.119 Enantioselective synthesis of octahydroacridines **527**.

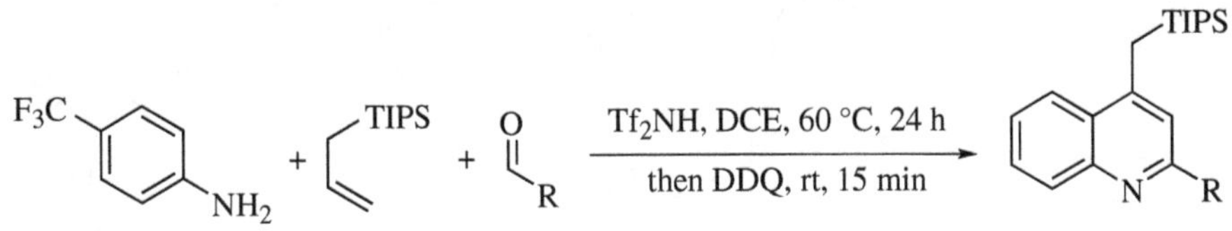

528 529 530 531
2 examples
52–86%

SCHEME 13.120 One-pot synthesis of quinolines **531** by a Povarov reaction and subsequent oxidation with DDQ.

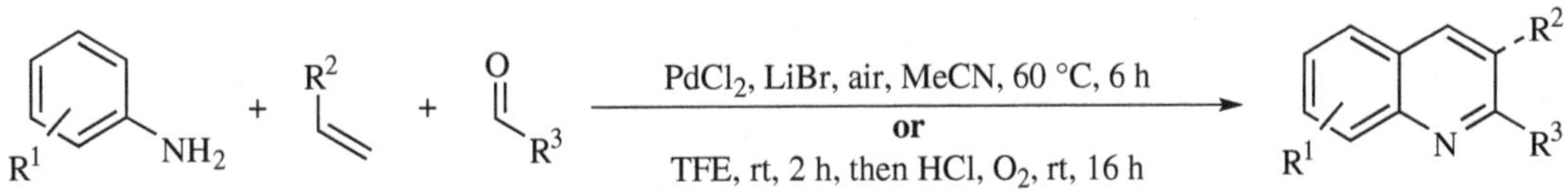

248 67 532 533
54 examples
35–93%

SCHEME 13.121 Domino one-pot synthesis of quinolines **533** by Povarov reaction and subsequent oxidation by O_2 or air [202a, 212].

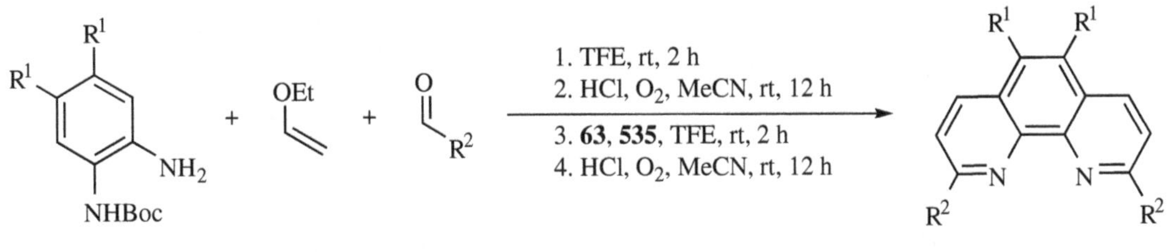

534 63 535 536
8 examples
15–30%

SCHEME 13.122 Synthesis of 1,10-phenantrolines **536** by a Povarov reaction.

The other possible synthesis of quinolines by Povarov reaction requires an alkyne **537** instead of an alkene and produces directly 1,2-dihydroquinolines, which easily undergo oxidation (Scheme 13.123) [200, 207b].

By using methyl or ethyl acetylenedicarboxylates, a highly substituted pyridine core can be obtained, and additionally, in combination with 2-aminonaphthalene **125**, there is an access to benzo[*f*]quinolines **542** (Scheme 13.124) [207a].

As mentioned before, not only simple arylamines are suitable for the Povarov reaction but also more complex aromatic amines like substituted 3-aminocoumarins **515** along

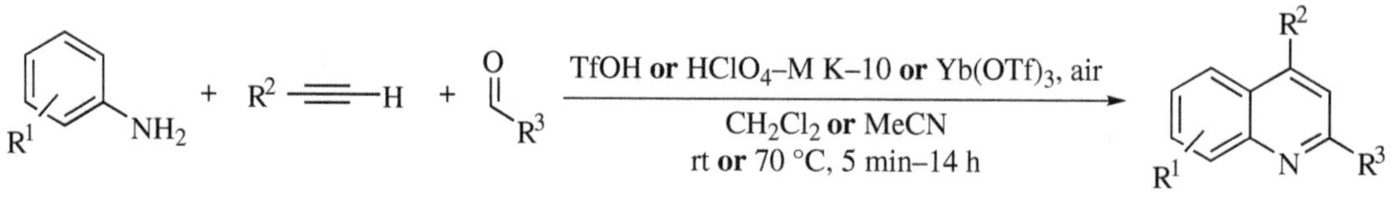

248 537 538 539 57 examples 26–81%

SCHEME 13.123 Synthesis of quinolines **539** using terminal alkynes **537** [200, 207b].

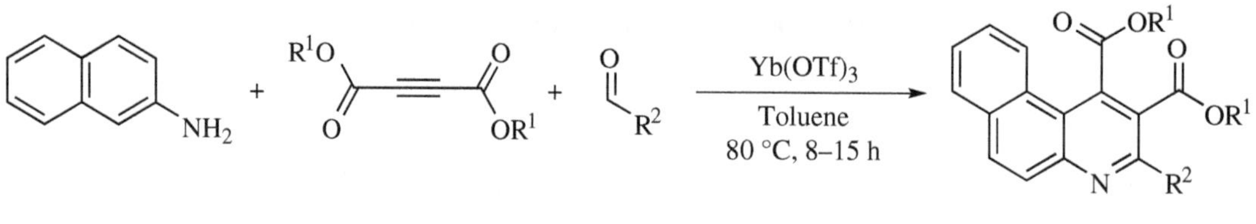

125 540 541 542 16 examples 76–92%

SCHEME 13.124 Synthesis of benzoquinolines **542** using acetylenedicarboxylates **540** and 2-aminonaphthalene **125**.

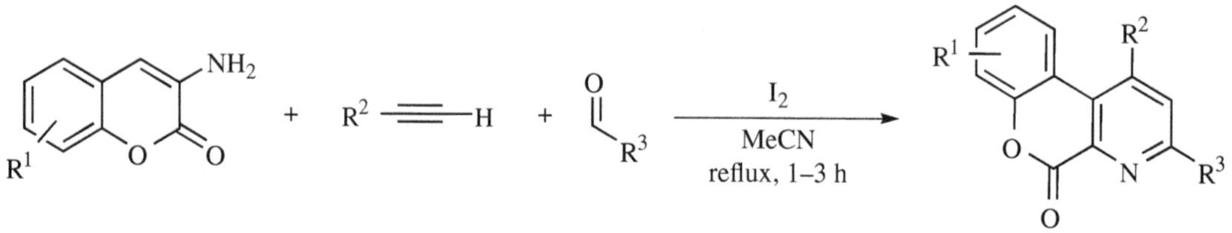

515 537 543 544 22 examples 72–94%

SCHEME 13.125 Synthesis of chromenopyridinones **544** using 3-aminocoumarins **515**.

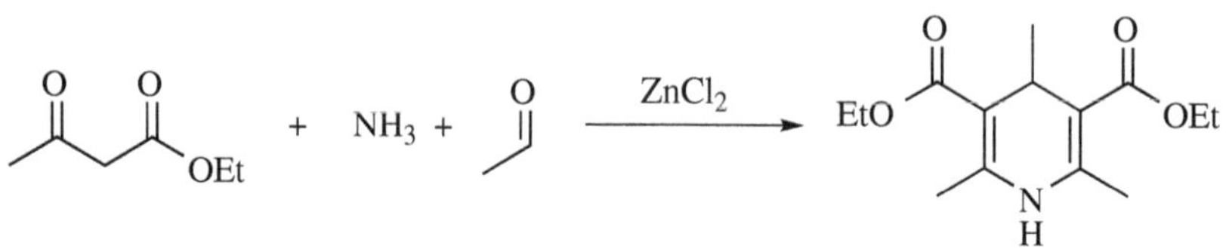

191 545 546 547

SCHEME 13.126 Hantzsch 1,4-dihydropyridine synthesis from 1881.

with terminal alkynes **537** leading to 5*H*-chromeno[3,4-*b*]pyridin-5-ones **544** (Scheme 13.125) [209a].

13.5 HANTZSCH MULTICOMPONENT SYNTHESIS OF HETEROCYCLES

13.5.1 Introduction

In 1881, Hantzsch first noted the formation of 1,4-dihydropyridines **547** through the reaction of ethyl acetoacetate **191**, ammonia **545**, and acetaldehyde **546** in the presence of zinc chloride as the Lewis acid [214]. After oxidation and decarboxylation, he was able to isolate 2,4,6-collidine (Scheme 13.126).

A few years later, in 1890, he observed the reaction between chloroacetone **548**, ethyl acetoacetate **191**, and ammonia **545**, forming pyrrole **549** (Scheme 13.127) [215]. The properties and reactivity of the obtained product were the same as for a compound obtained by Knorr in the same year.

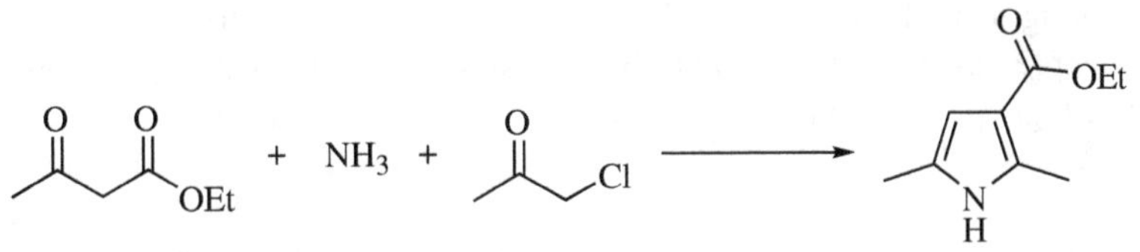

191 **545** **548** **549**

SCHEME 13.127 Hantzsch pyrrole synthesis from 1890.

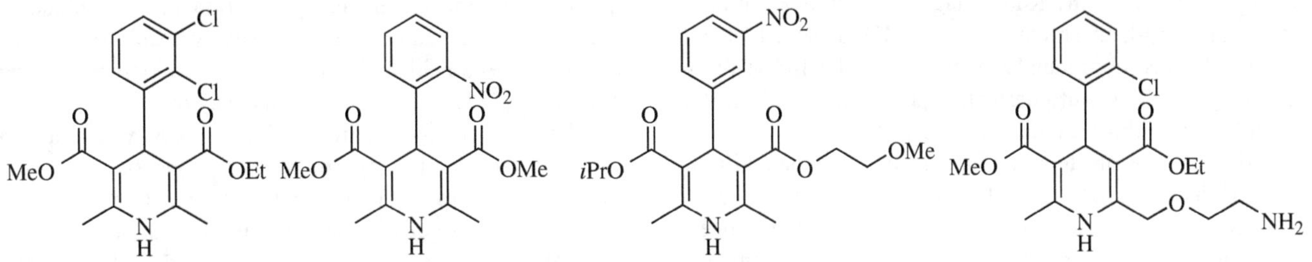

550 Felodipine | **551** Nifedipine | **552** Nimodipine | **553** Amlodipine

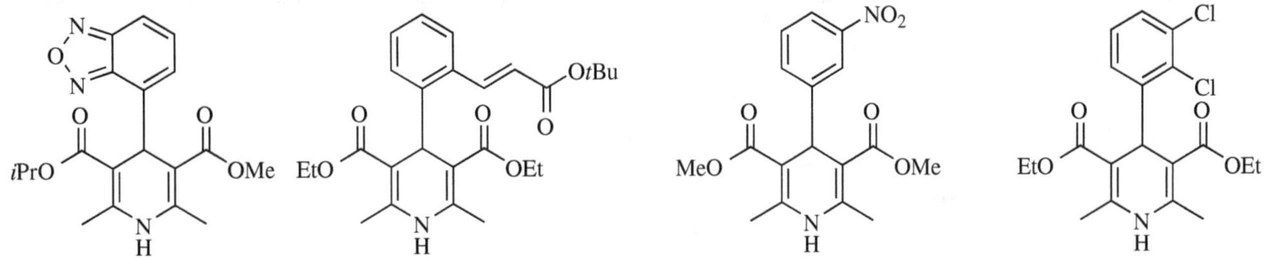

554 Isradipine | **555** Lacidipine | **556** Nitrendipine | **557** Nemadipine B

FIGURE 13.1 Pharmaceutically active 1,4-dihydropyridines.

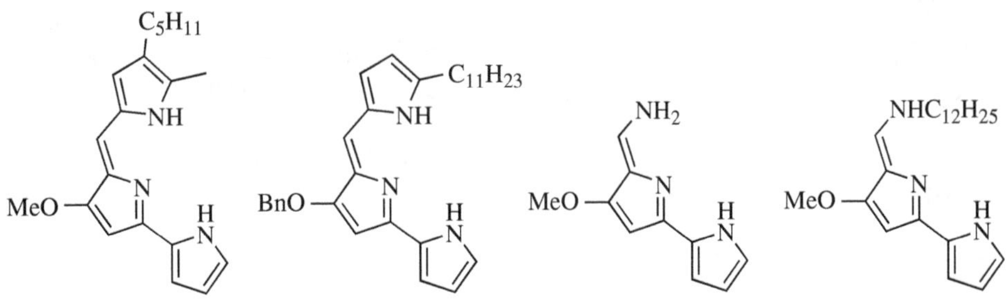

558 Prodigiosin | **559** PNU-156804 | **560** Tambjamine A | **561** BE-18591

FIGURE 13.2 Pharmaceutically active pyrroles.

Both 1,4-dihydropyridines and pyrroles are widely spread motives in natural products and drugs (Figs. 13.1 and 13.2). Several calcium channel blockers **550–557** with their additional use in the treatment of, for example, hypertension, angina pectoris, vasospasm, and coronary artery disease are all examples of 1,4-dihydropyridine drugs obtainable by Hantzsch or Hantzsch-type reactions (Fig. 13.1).

Among the pyrrole-containing natural products and therapeutics, PNU-156804 **559** expresses promising activity in the suppression of heart allograft rejection [216], and BE18591 **561** do so against tumors [217] (Fig. 13.2).

13.5.2 Catalysis and Mechanism

In the Hantzsch 1,4-dihydropyridine synthesis, the β-ketocarbonyl compound **II** condenses on the one hand with a nitrogen source forming an enamine **I** and, on the other hand, with the appropriate aldehyde in a Knoevenagel condensation, forming an unsaturated β-ketocarbonyl compound **III** (Scheme 13.128).

Both formed compounds react in a Michael-type reaction to give, after tautomerization and condensation, the 1,4-dihydropyridines **IV** (Scheme 13.129).

All condensation steps can be facilitated by the addition of a Brønsted or Lewis acid, which increases the reaction rate. Thus, various catalyst systems have been examined and proven as applicable. Among them are simple proton sources like montmorillonite K-10 (MK-10) [218], cellulose [219], silica [220], sulfuric acid, or several Lewis acids [221]. Also, *in situ* generated HCl, from 2,4,6-trichloro[1,3,5]triazine (TCT) [222], and materials with a complex structure like tungsten phosphoric acid on silica [223], hybrid tin phosphonate nanoparticles (HSnP-2) [224], a molybdenum coordination polymer [225], nickel nanoparticles [226], or magnesium oxide nanotubes [227], or hydrotalcites ($MgAl_2$-HT) [228] were examined for their catalytic activity. Even molecular iodine [229], amino acids [230], urotropin [231], and thiamine [232] are able to catalyze the 1,4-dihydropyridine formation.

The range of possible residues R^2 includes nearly each conceivable alkyl chain, branched ones, aromatic and heteroaromatic systems. Even highly functionalized or hindered aldehydes like carbohydrate derivatives [233], linear-extended biscarbaldehydes [234], or even polymer-bound aldehydes [235] are suitable for this reaction.

The obtained 1,4-dihydropyridines can easily be oxidized to the corresponding pyridines by various oxidants (see Section 13.5.3) or in a hydrogen transfer reaction [236].

At first, analogous to the 1,4-dihydropyridine synthesis, in the Hantzsch pyrrole synthesis, a β-ketocarbonyl compound reacts with ammonia or a secondary amine, forming an enamine **II**. This enamine **II** reacts with the α-haloketone **I** in a nucleophilic substitution and gives, after tautomerization and condensation, the pyrroles **III** (Scheme 13.130).

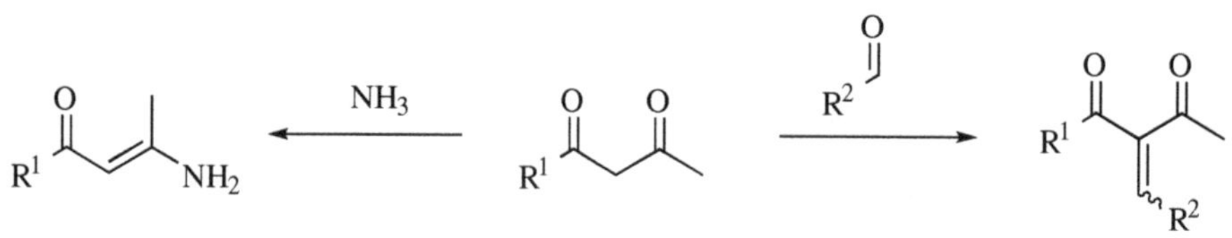

I II III

SCHEME 13.128 Mechanism of the Hantzsch 1,4-dihydropyridine synthesis—part I.

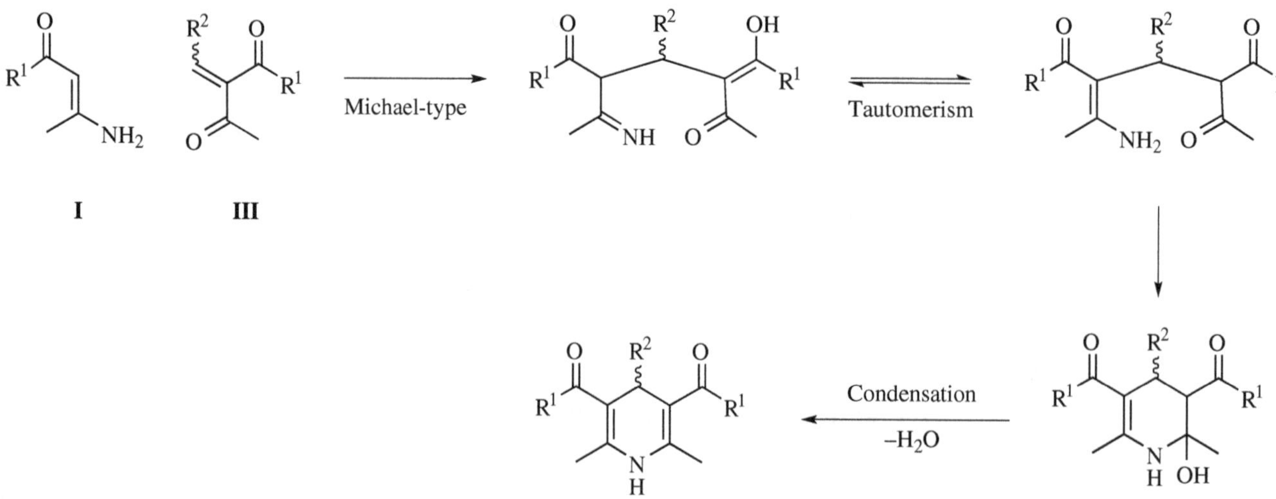

SCHEME 13.129 Mechanism of the Hantzsch 1,4-dihydropyridine synthesis—part II.

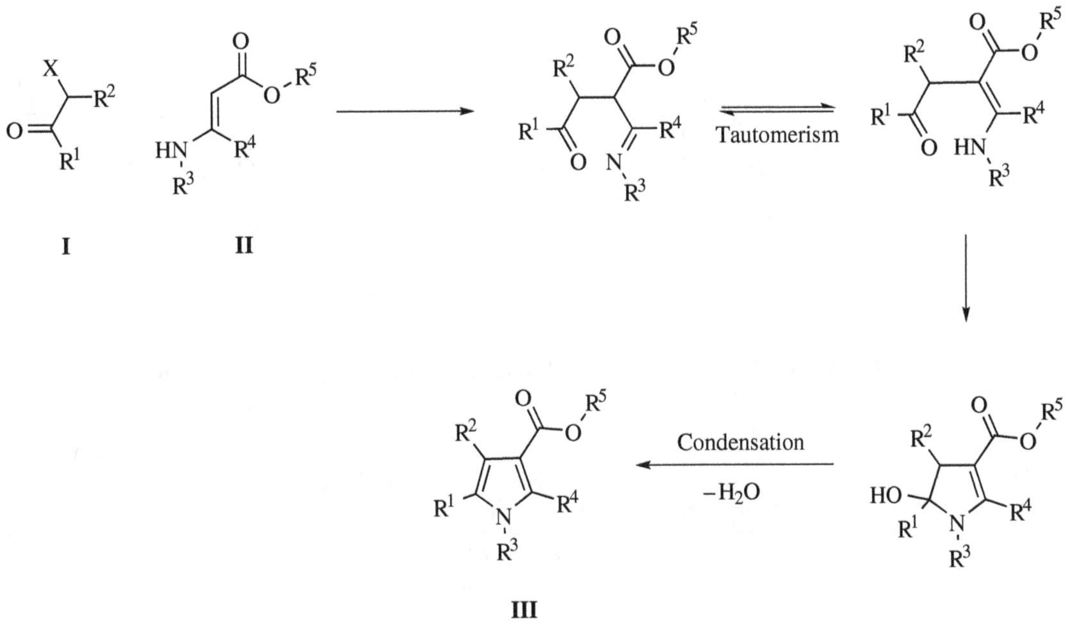

SCHEME 13.130 Mechanism of the Hantzsch pyrrole synthesis.

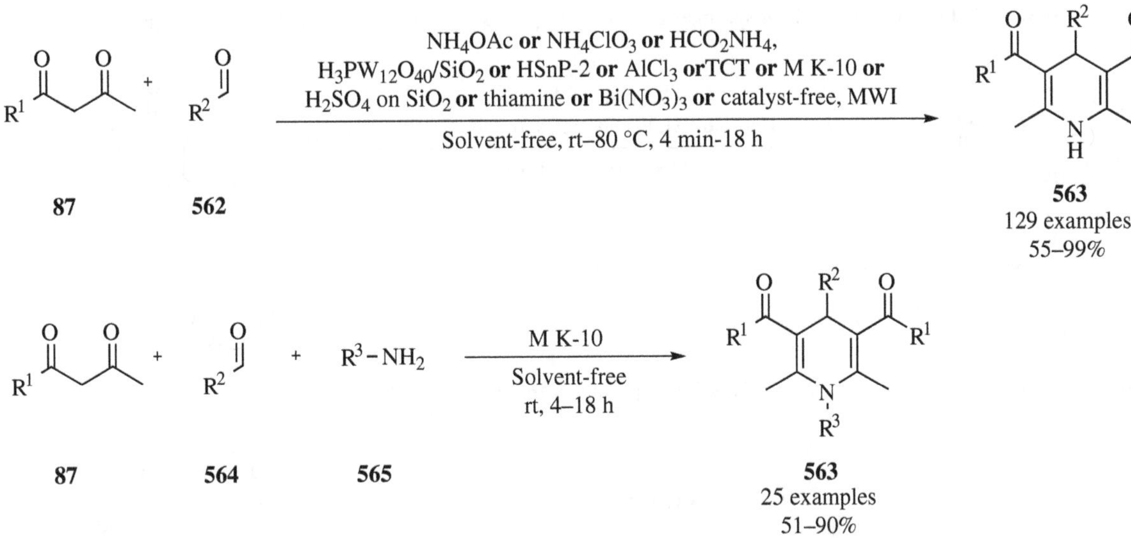

SCHEME 13.131 Synthesis of symmetric 1,4-dihydropyridines **563** under solvent-free conditions [220–224, 232, 235, 237, 238].

13.5.3 Syntheses of 1,4-Dihydropyridines and Their Oxidation to Pyridines

13.5.3.1 Symmetric 1,4-Dihydropyridines The most common Hantzsch synthesis produces symmetrical 1,4-dihydropyridines by the use of two equivalents of the β-ketocarbonyl compound. For the solvent-free performance, the nitrogen sources are limited to solid or high boiling ones, like ammonium acetate [220–224, 232, 235], ammonium chlorate [237], ammonium formate [238], or primary aryl amines (Scheme 13.131) [218b].

A possibility to reduce the reaction time to only minutes is the use of microwave irradiation [221c, 238]. In some cases, not even a catalyst is necessary to achieve good results (Scheme 13.131) [238]. Under microwave conditions, a synthesis of solid-supported 1,4-dihydropyridines **567** with resin-bound aldehydes **566** is also possible (Scheme 13.132) [235].

In the case of a reaction in solution, all nitrogen sources mentioned earlier are suitable too [218c, 221b, d, e, 227, 228, 239], but furthermore, ammonium carbonate [240] and ammonium hydroxide [241] are used. Besides specially developed bifunctional (Lewis acid and nitrogen source) reagents, like aluminum and magnesium nitride [234b, 242], the modest nitrogen source ammonia is also

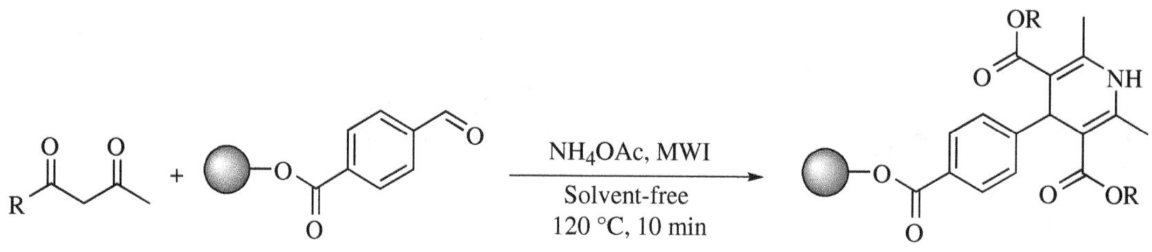

87 566 567
2 examples
96–97%

SCHEME 13.132 Synthesis of resin-bound symmetric 1,4-dihydropyridines **567**.

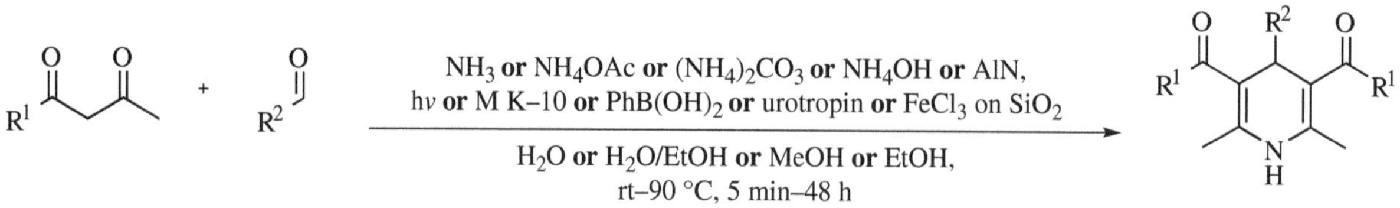

87 568 563
119 examples
29–99%

SCHEME 13.133 Synthesis of symmetric 1,4-dihydropyridines **563** in water, methanol, and ethanol.

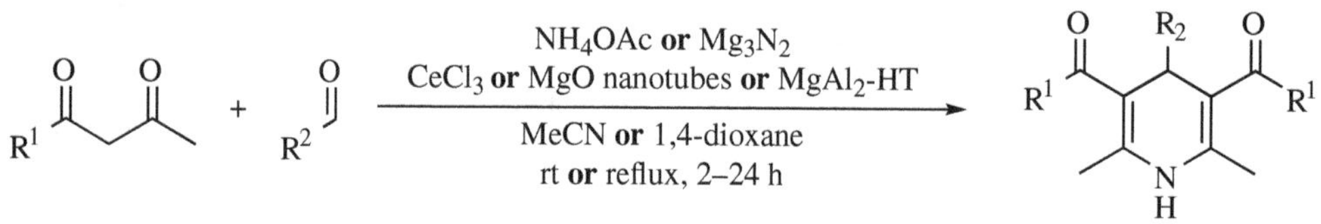

87 569 563
56 examples
22–97%

SCHEME 13.134 Synthesis of symmetric 1,4-dihydropyridines **563** in acetonitrile and 1,4-dioxane [221e, 227, 228, 242].

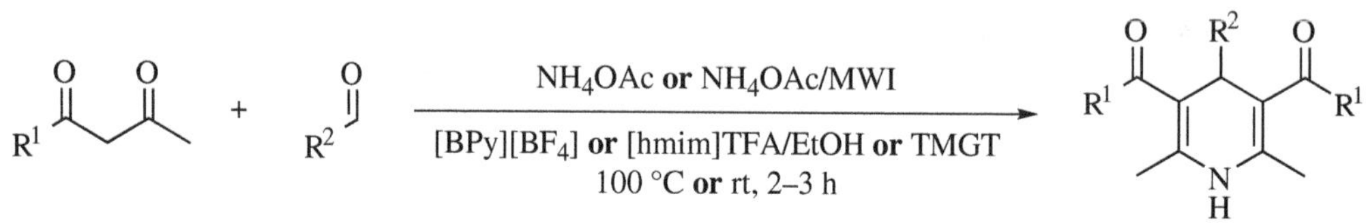

87 570 563
30 examples
87–97%

SCHEME 13.135 Synthesis of symmetric 1,4-dihydropyridines **563** in ionic liquids [239b–d].

suitable [218a, 243]. The most common solvents for the Hantzsch 1,4-dihydropyridine synthesis are water [218a, 234b, 240], methanol [221d, 239a], ethanol [218c, 221b], or a mixture of water and ethanol [231, 241, 243] (Scheme 13.133).

In addition to the common polar solvents water and alcohols, acetonitrile [221e, 227, 228] and dioxane [242] have also proven as utilizable solvents in the synthesis of 1,4-dihydropyridines (Scheme 13.134).

Ionic liquids with their unique properties and recyclability were developed into an environmentally friendly solvent during the last decades. Therefore, it is not surprising that they were studied in this context as well (Scheme 13.135) [239b–d].

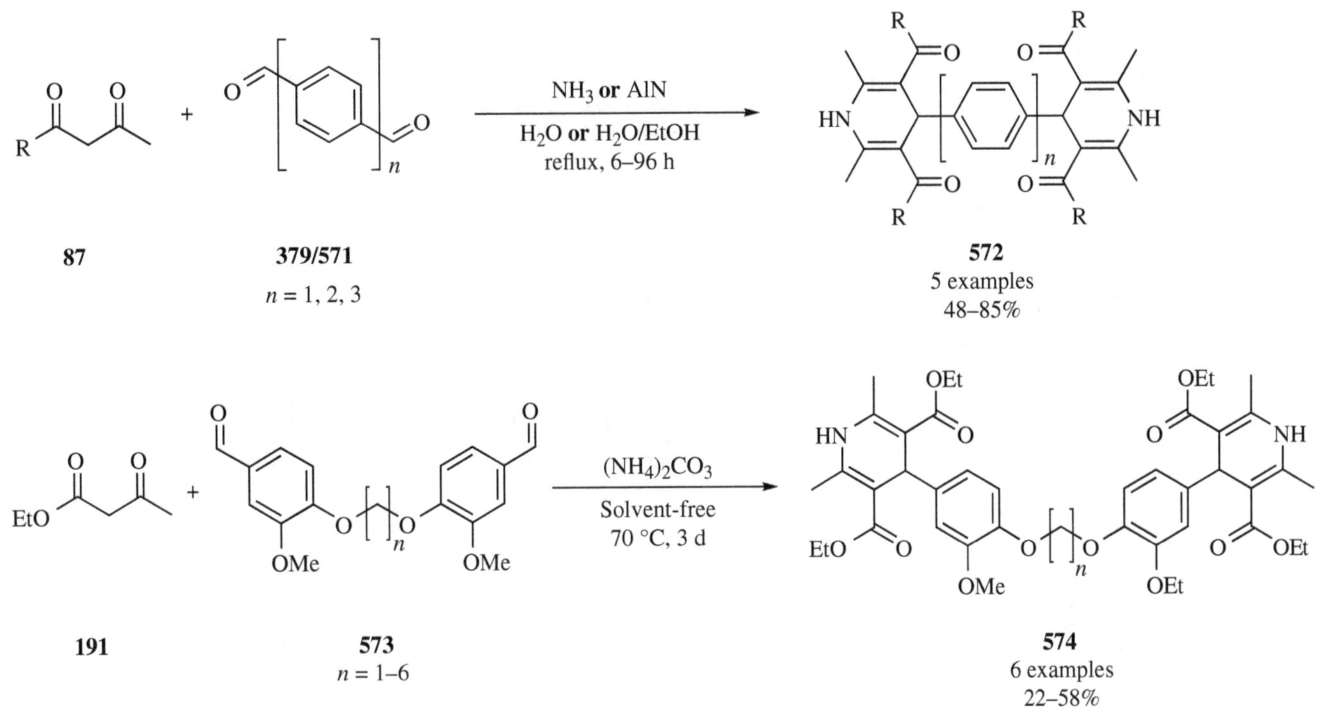

SCHEME 13.136 Synthesis of bridged bis-1,4-dihydropyridines **572**/**574**.

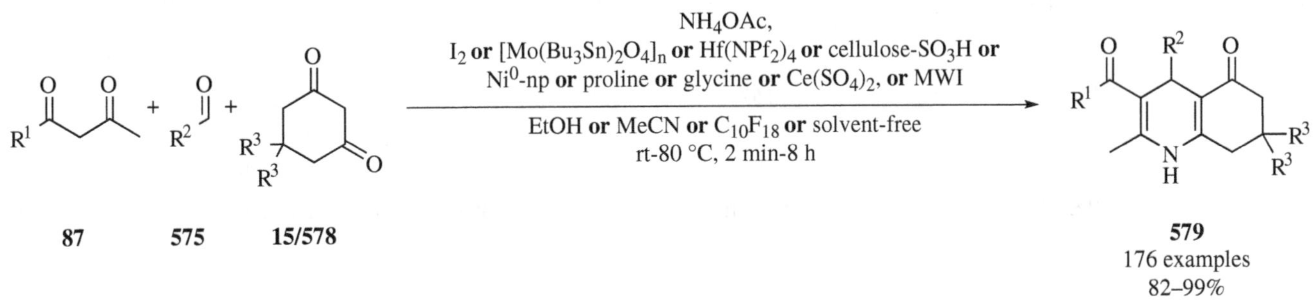

SCHEME 13.137 Synthesis of asymmetric 1,4-dihydropyridines **579** using dimedone **15** or cyclohexane-1,3-dione **578**.

Through the highly selective reaction pathway, even the synthesis of bridged bis-1,4-dihydropyridines **572** and **574** is possible by the use of linear-extended biscarbaldehydes (Scheme 13.136) [234].

13.5.3.2 Asymmetric 1,4-Dihydropyridines Instead of two equivalents of one β-ketocarbonyl compound, two different equimolar β-ketocarbonyl compounds with different reactivities can be subjected to the Hantzsch 1,4-dihydropyridine synthesis. The products of those reactions are asymmetric 1,4-dihydropyridines **579**.

In most examples, the second β-ketocarbonyl is a 1,3-dioxocycloalkane, like cyclohexane-1,3-dione **578** [219a, 221g, 225, 229], dimedone **15** (Scheme 13.137) [219a, 221f, g, 225, 226, 229, 230, 230e, 244], or 1*H*-indene-1,3(2*H*)-dione **217** (Scheme 13.138) [219a, 230d, 245].

It is possible to use preformed α,β-unsaturated carbonyls **582** [219b] or enamines **167** [246] in the Hantzsch 1,4-dihydropyridine synthesis to obtain asymmetric 1,4-dihydropyridines **583** and **586** too (Schemes 13.139 and 13.140, respectively). Especially in cases of C5-unsubstituted dihydropyridines, it is necessary to perform the Knoevenagel condensation in an upstream reaction (Scheme 13.139).

In contrast, preformed enamines are sometimes the better choice to introduce a special substitution pattern at C5 and C6, for example, a pyrimidinedione **167** (Scheme 13.140) [246].

13.5.3.3 Catalyzed Oxidations of 1,4-Dihydropyridines to Pyridines As mentioned earlier, the obtained 1,4-dihydropyridines can easily be oxidized to the corresponding pyridines by various oxidants. The most economic one certainly is molecular oxygen (Scheme 13.141)

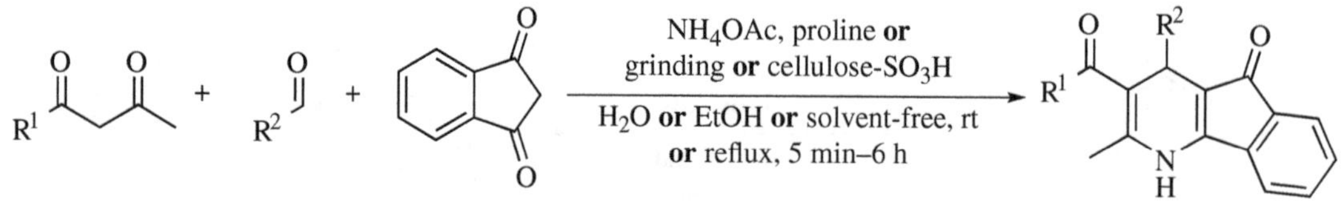

87 **580** **217** **581** 30 examples 67–95%

SCHEME 13.138 Synthesis of asymmetric 1,4-dihydropyridines **581** starting from 1*H*-indene-1,3(2*H*)-dione **217** [219a, 230d, 245].

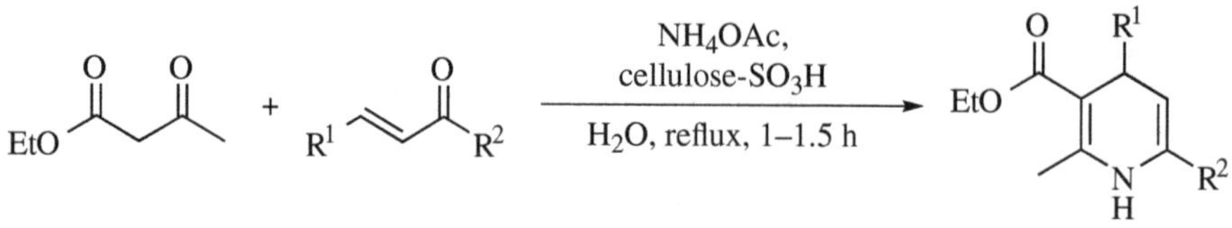

191 **582** **583** 12 examples 87–98%

SCHEME 13.139 Synthesis of asymmetric 1,4-dihydropyridines **583** with preformed α,β-unsaturated carbonyls **582** [219b].

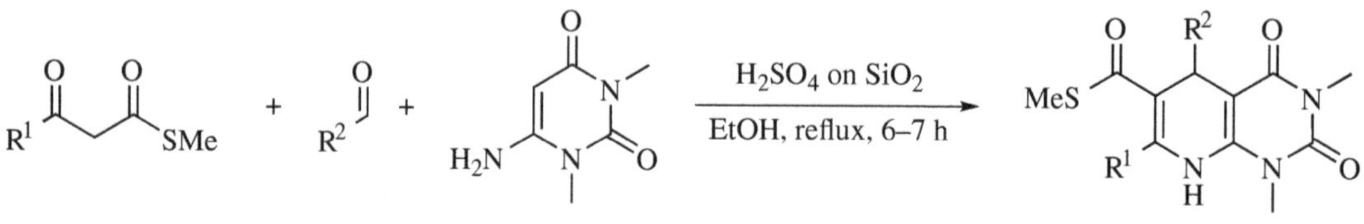

584 **585** **167** **586** 8 examples 55–66%

SCHEME 13.140 Synthesis of asymmetric 1,4-dihydropyridines **586** with preformed enamines **167**.

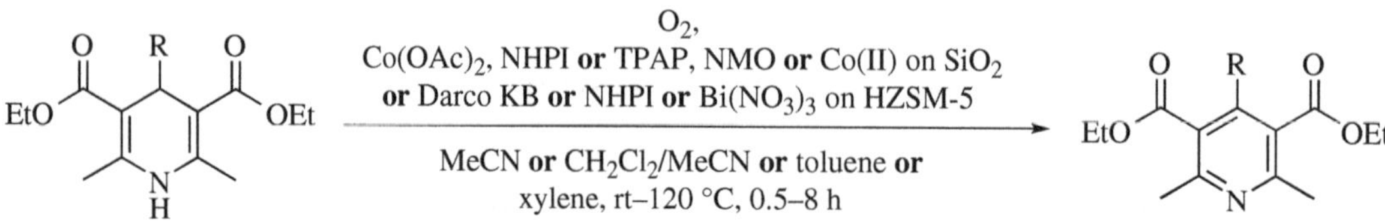

587 **588** 69 examples 65–99%

SCHEME 13.141 Catalytic oxidation of 1,4-dihydropyridines **587** by molecular oxygen.

[247], but for this kind of oxidation, usually, a catalyst or catalyst system is necessary. The same is true for hydrogen peroxide (Scheme 13.142) [221a, 248] and sodium periodate (Scheme 13.143) [249].

The catalysts for the oxidation with molecular oxygen and hydrogen peroxide vary from cobalt complexes [247a, c, 248a], aluminum chloride [221a], *N*-hydroxyphthalimide (NHPI) [247e], and vanadium oxide [248b] to tetra-*N*-propylammonium perruthenate (TPAP) [247b], activated carbon (Darco KB) [247d], and bismuth nitrate on zeolite (HZSM-5) [247g] (Schemes 13.141 and 13.142). A catalyst-free version of the oxidation is also known and proceeds smoothly in dimethylsulfoxide at elevated temperature and exposure to air [250].

In the case of sodium periodate as oxidant, manganese(III) complexes with a bis(salicylaldehyde)-4-methyl-1,2-phenylenediimine (BSMP) [249b] or tetrapyridylporphyrin supported on chloromethylated polystyrene (TPyP)-CMP [249a] have proven as the catalysts of choice (Scheme 13.143).

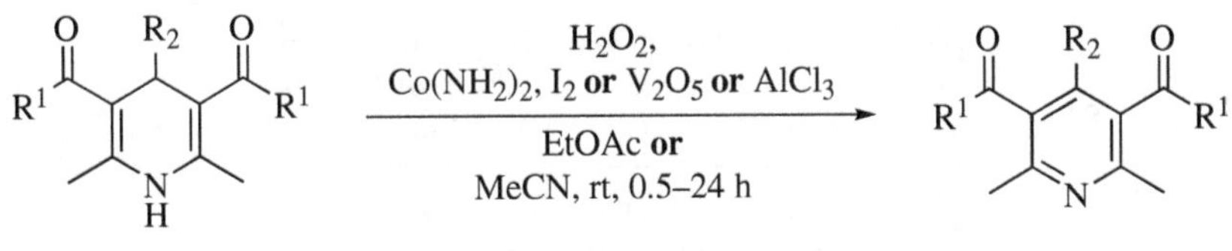

563

589
45 examples
73–99%

SCHEME 13.142 Catalytic oxidation of 1,4-dihydropyridines **563** by hydrogen peroxide [221a, 248].

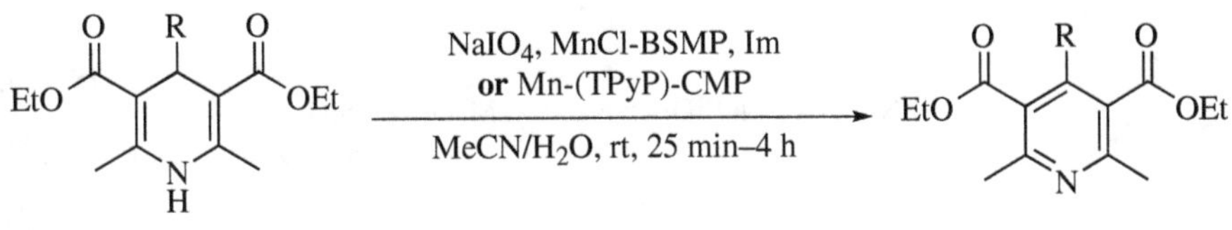

587

588
23 examples
70–95%

SCHEME 13.143 Catalytic oxidation of 1,4-dihydropyridines **587** by sodium periodate [249].

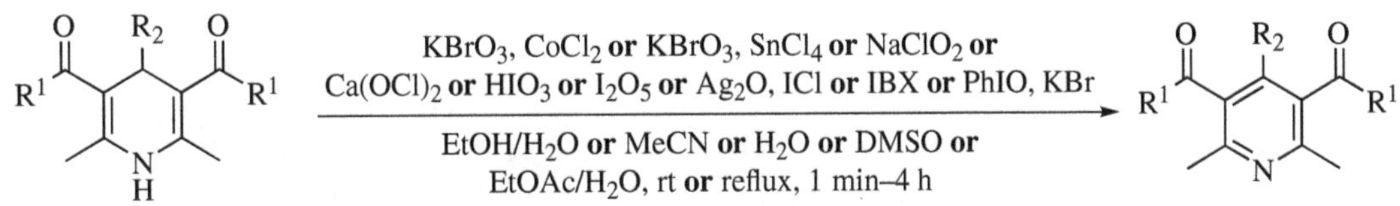

563

589
127 examples
80–99%

SCHEME 13.144 Stoichiometric oxidation of 1,4-dihydropyridines **563** by halogen-based oxidants [251–257].

It has to be noted that some catalytic oxidations also cleave the carbon–carbon bond between C4 and the corresponding residue. This cleavage takes place if the residue can form a stabilized leaving group, like *iso*-propyl [247e, 248a, 249b] or benzyl [248a, 249b].

13.5.3.4 Oxidation of 1,4-Dihydropyridines to Pyridines with Stoichiometric Oxidants Many of the known stoichiometric oxidants were examined in the oxidation of 1,4-dihydropyridines to the corresponding pyridines. Thus, halogen-based oxidants, such as potassium bromate [251], sodium chlorite [252], calcium hypochlorite [253], iodic acid [254], its anhydride [255], iodine chloride [256], or hypervalent iodine species [257], give good results (Scheme 13.144).

Metal [258], metalloid [259], and nonmetal [260] -based oxidants are very efficient but rather toxic agents; thus, vanadium oxytrichloride [258f], 2,6-dicarboxypyridinum chlorochromate (DCPCC) [258a], bis(trimethylsilyl) chromate (BTSC) [258e], chromic acid [261], and selenium dioxide [260] have been used in the oxidations of 1,4-dihydropyridines **563** (Scheme 13.145).

The salts of nitrous [262] and nitric [263] acid with their well-known oxidative capacity are also able to convert 1,4-dihydropyridines into the corresponding pyridines (Scheme 13.146).

Even peroxo species like cetyltrimethylammonium (CTA) peroxodisulfate [264] or the hydrogen peroxide–urea complex (UHP) [265] were found to be suitable oxidants. The CTA counterion acts as a phase-transfer catalyst and is necessary to solubilize the peroxodisulfate ion in organic solvents, while the hydrogen peroxide–urea complex represents a concentrated and crystalline version of hydrogen peroxide.

Although all oxidants provide good results in the oxidation of 1,4-dihydropyridines to pyridines, many of them also cleave the carbon–carbon bond between C4 and the corresponding residue. The cleavage occurs in cases of residues that can form a stabilized leaving group, like *iso*-propyl [251a, 254, 256, 257b, 258b, f, 259, 264–266], benzyl [258f, 259, 264, 266], 1-phenylethyl [258b, 264, 265], or

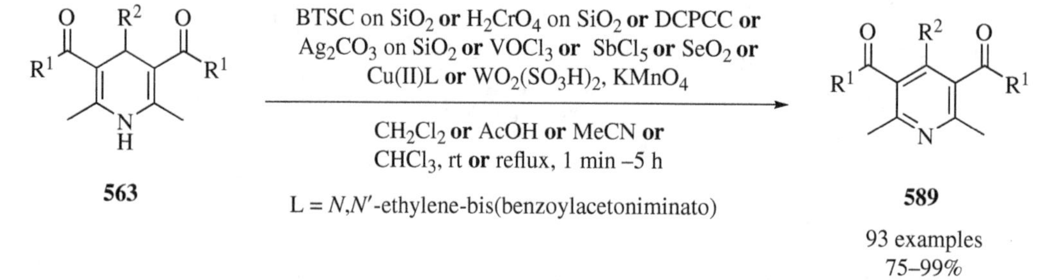

SCHEME 13.145 Stoichiometric oxidation of 1,4-dihydropyridines **563** by metal, metalloid, and nonmetal-based oxidants [258–260].

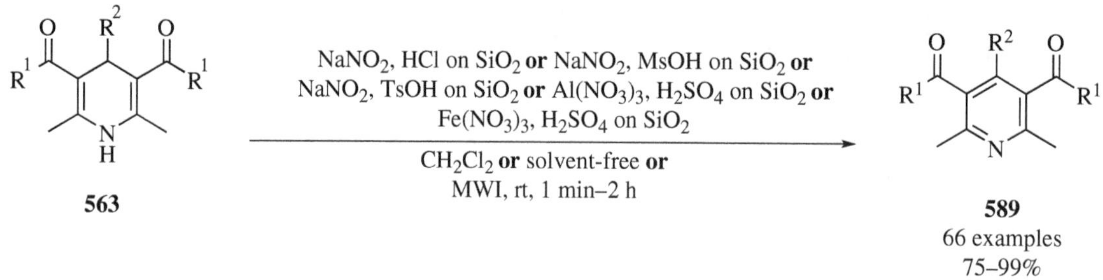

SCHEME 13.146 Stoichiometric oxidation of 1,4-dihydropyridines **563** by nitrates and nitrites [262, 263].

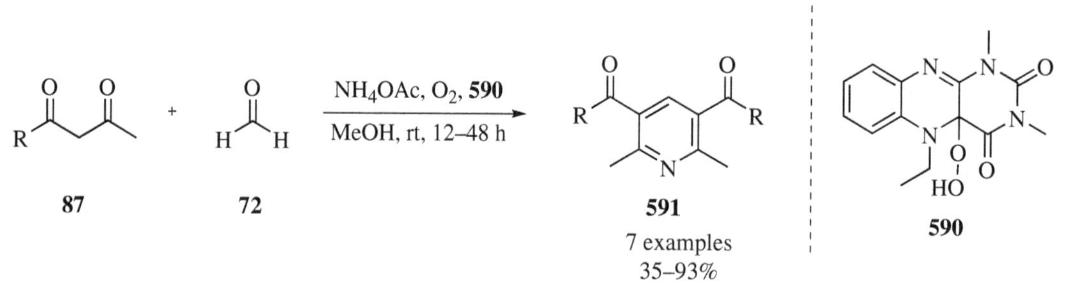

SCHEME 13.147 Domino Hantzsch synthesis and subsequent oxidation—I [267].

cyclohexyl [264] but in some examples also with an ethyl [259] or *n*-propyl [251a, 254, 258b, f] residue.

13.5.3.5 Syntheses of Pyridines via Domino/One-Pot Hantzsch Synthesis–Oxidation Combination of the Hantzsch 1,4-dihydropyridine synthesis with subsequent oxidation in a domino process enables the direct access to substituted pyridines.

For example, a bioinspired oxidation with hydroperoxyflavins **590** combined with a catalyst-free 1,4-dihydropyridine synthesis effectively produces C4-unsubstituted pyridines **591** (Scheme 13.147) [267], whereas a palladium on carbon oxidation combined with an MK-10 catalyzed 1,4-dihydropyridine formation under microwave assistance reduces the reaction time (Scheme 13.148) [268]. A one-pot performance of the pyridine synthesis is also possible in water with stoichiometric soluble oxidants like iron chloride or potassium permanganate (Scheme 13.148) [266]. Nevertheless, both examples (Scheme 13.148) cleave off the C4 substituent in the case of an *iso*-propyl, benzyl, and 1-phenylethyl residue [266, 268].

13.5.4 Multicomponent Pyrrole Syntheses

In the last years, only some examples of multicomponent pyrrole syntheses have been published, but those are highly interesting in terms of reaction conditions and substitution patterns of the obtained pyrroles. One example is the conversion of α-iodoketones **594** in a Hantzsch pyrrole synthesis under high-speed vibration milling conditions (HSVM) and CAN catalysis, with silver nitrate as scavenger for the produced hydroiodic acid (Scheme 13.149) [269].

Another example deals with the direct synthesis of pyrrol-2-carbaldehydes **598** and **601** to avoid installation of these moieties into preformed pyrroles. Therefore, it is necessary to change from a normal β-ketocarbonyl compound to a stabilized ketonitrile **596** to ensure only one possible reaction pathway toward the enamine (Scheme 13.150) [270].

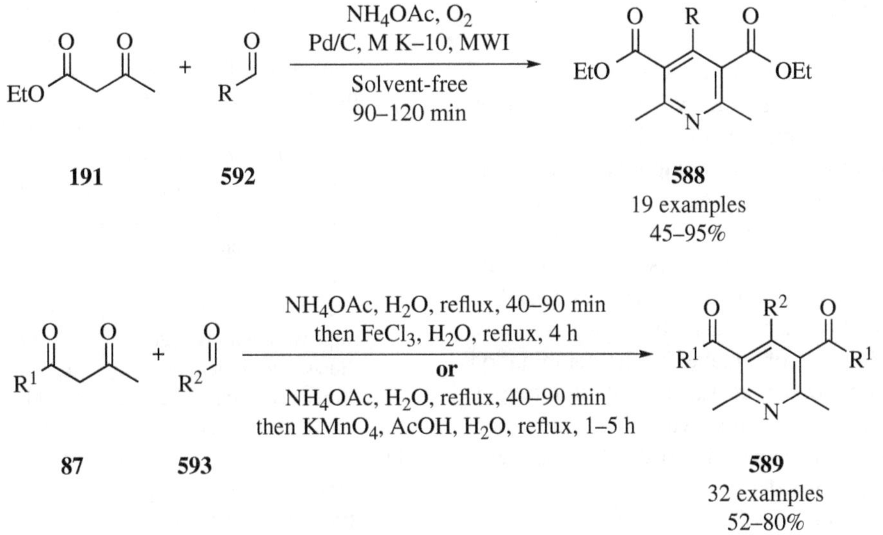

SCHEME 13.148 Domino Hantzsch synthesis and subsequent oxidation—II [266, 268].

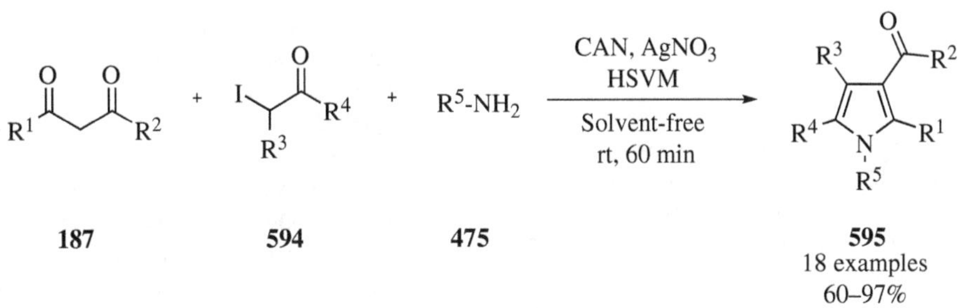

SCHEME 13.149 Hantzsch pyrrole synthesis under HSVM conditions.

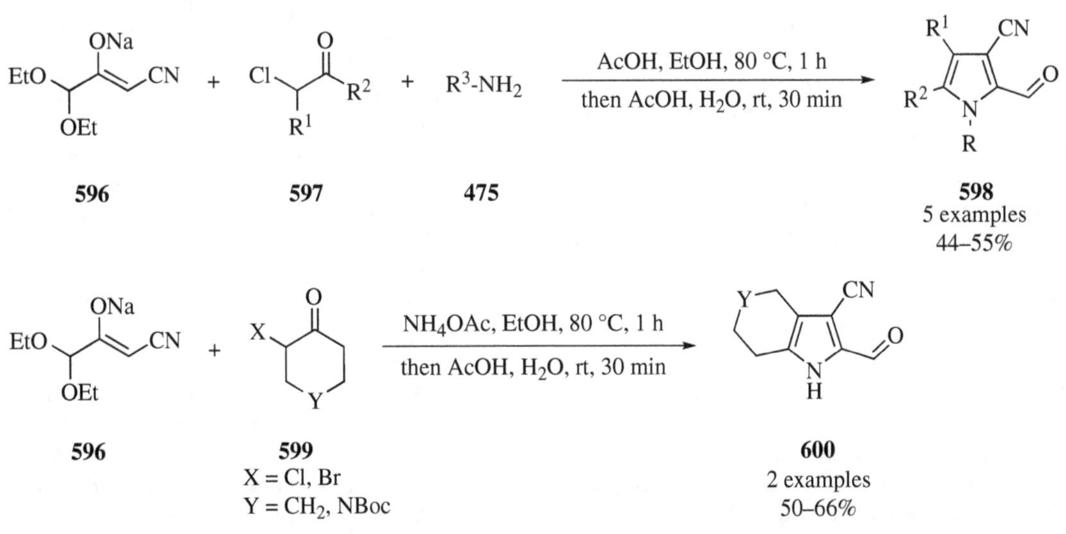

SCHEME 13.150 Hantzsch synthesis of pyrrol-2-carbaldehydes **598/600**.

13.6 CONCLUSIONS

The preceding chapter illustrated versatile applications of the discussed MCRs as tools in the syntheses of complex molecular scaffolds, containing various N-, O-, S-, and P-heterocyclic or heteropolycyclic compounds. The most common examples of the Knoevenagel, Yonemitsu-type, Povarov, and Hantzsch reactions and reactions containing Meldrum's acid as the key substrate have been covered along with motivating examples of their use in the total synthesis of natural products or pharmaceuticals. The shown examples highlighted the good results obtainable by the combination of simple catalysts (Brønsted acid, Lewis acid, amino acid, etc.) or catalytic systems combined with simple commercially available molecules in a one-pot process. Nevertheless, there are only a few examples demonstrating high diastereo-/enantioselectivity. Especially, with regard to the use of those reactions in total synthesis, this is an aspect that requires further investigation, as well as several reaction mechanisms. As a result, this will cause a development culminating in more complex products, new scaffolds for total synthesis, and augmented industrial applications.

REFERENCES

1 (a) E. Knoevenagel, *Ber. Dtsch. Chem. Ges.* 1894, 27, 2345–2346. Ueber eine Darstellungsweise der Glutarsäure. (b) E. Knoevenagel, *Ber. Dtsch. Chem. Ges.* 1898, 31, 2596–2619. Condensation von Malonsäure mit aromatischen Aldehyden durch Ammoniak und Amine.

2 (a) L. F. Tietze, U. Beifuss in *Comprehensive organic synthesis. Selectivity, strategy, and efficiency in modern organic chemistry.* (B. M. Trost, I. Fleming, Eds.) Pergamon Press, Oxford, England, 1991. (b) G. Jones in *Organic reactions.* John Wiley & Sons, Inc., Hoboken, NJ, 2011, Vol. 15, pp. 204–599.

3 L. Kürti, B. Czakó, *Strategic applications of named reactions in organic synthesis. Background and detailed mechanisms.* Elsevier Academic Press, Amsterdam, The Netherlands, 2005.

4 L. F. Tietze, N. Rackelmann in *Multicomponent reactions.* (J. Zhu, H. Bienaymé, Eds.) Wiley-VCH Verlag GmbH, Weinheim, Germany, 2005, pp. 121–168.

5 A. C. O. Hann, A. Lapworth, *J. Chem. Soc., Trans.* 1904, 85, 46–56. Optically active esters of β-ketonic and β-aldehydic acids. Part IV. Condensation of aldehydes with menthyl acetoacetate.

6 N. V. Lakshmi, P. Thirumurugan, K. Noorulla, P. Perumal, *Bioorg. Med. Chem. Lett.* 2010, 20, 5054–5061. $InCl_3$ mediated one-pot multicomponent synthesis, anti-microbial, antioxidant and anticancer evaluation of 3-pyranyl indole derivatives.

7 J. Kamalraja, D. Muralidharan, P. Perumal, *Synlett* 2012, 23, 2894–2898. An efficient, one-pot regioselective synthesis of highly functionalized chromen-5-ones and pyrano[3,2-*c*]chromen-5-ones *via* a tandem Knoevenagel-Michael-cyclization sequence.

8 L. F. Tietze, N. Böhnke, S. Dietz, *Org. Lett.* 2009, 11, 2948–2950. Synthesis of the deoxyaminosugar (+)-D-forosamine *via* a novel domino-Knoevenagel-hetero-Diels–Alder reaction.

9 L. F. Tietze, S. Dietz, N. Böhnke, M. A. Düfert, I. Objartel, D. Stalke, *Eur. J. Org. Chem.* 2011, 6574–6580. Three-component domino Knoevenagel/hetero-Diels–Alder reaction for the synthesis of the amino sugars 2-acetoxyforosamine and 2-acetoxyossamine—experimental and theoretical results.

10 X.-S. Wang, M.-M. Zhang, H. Jiang, D.-Q. Shi, S.-J. Tu, X.-Y. Wei, Z.-M. Zong, *Synthesis* 2006, 4187–4199. An improved and benign synthesis of 9,10-diarylacridine-1,8-dione and indenoquinoline derivatives from 3-anilino-5,5-dimethylcyclohex-2-enones, benzaldehydes, and 1,3-dicarbonyl compounds in an ionic liquid medium.

11 H. Sheibani, A. S. Saljoogi, A. Bazgir, *Arkivoc* 2008, (ii), 115–123. Three-component process for the synthesis of 4-amino-5-pyrimidinecarbonitriles under thermal aqueous conditions or microwave irradiation.

12 S. Balalaie, M. Bararjanian, M. Sheikh-Ahmadi, S. Hekmat, P. Salehi, *Synth. Commun.* 2007, 37, 1097–1108. Diammonium hydrogen phosphate: an efficient and versatile catalyst for the one-pot synthesis of tetrahydrobenzo[*b*]pyran derivatives in aqueous media.

13 S. Balalaie, S. Abdolmohammadi, H. R. Bijanzadeh, A. M. Amani, *Mol. Divers.* 2008, 12, 85–91. Diammonium hydrogen phosphate as a versatile and efficient catalyst for the one-pot synthesis of pyrano[2,3-*d*]pyrimidinone derivatives in aqueous media.

14 L. Chen, Y.-Q. Li, X.-J. Huang, W.-J. Zheng, *Heteroatom Chem.* 2009, 20, 91–94. *N,N*-Dimethylamino-functionalized basic ionic liquid catalyzed one-pot multicomponent reaction for the synthesis of 4*H*-benzo[*b*]pyran derivatives under solvent-free condition.

15 N. Seshu Babu, N. Pasha, K. Venkateswara Rao, P. Sai Prasad, N. Lingaiah, *Tetrahedron Lett.* 2008, 49, 2730–2733. A heterogeneous strong basic Mg/La mixed oxide catalyst for efficient synthesis of polyfunctionalized pyrans.

16 M. Kidwai, A. Jain, S. Bhardwaj, *Mol. Divers.* 2012, 16, 121–128. Magnetic nanoparticles catalyzed synthesis of diverse *N*-heterocycles.

17 S. Rostamnia, A. Nuri, H. Xin, A. Pourjavadi, S. H. Hosseini, *Tetrahedron Lett.* 2013, 54, 3344–3347. Water dispersed magnetic nanoparticles (H_2O-DMNPs) of γ-Fe_2O_3 for multicomponent coupling reactions: a green, single-pot technique for the synthesis of tetrahydro-4*H*-chromenes and hexahydroquinoline carboxylates.

18 B. Du, G. Cai, B. Zhao, X. Meng, X. Wang, Y. Li, *Res. Chem. Intermed.* 2013, 39, 1323–1333. Efficient one-pot three-component synthesis of 3,4-dihydro-12-phenyl-2*H*-benzo[*b*]xanthene-1,6,11(12*H*)-trione derivatives in ionic liquid.

19 A. M. Shestopalov, N. A. Larionova, A. E. Fedorov, L. A. Rodinovskaya, V. Y. Mortikov, A. A. Zubarev, I. S. Bushmarinov, *ACS Comb. Sci.* 2013, 15, 541–545. Synthesis of isomeric isothiazolo[4′,3′:4,5]- and isothiazolo[4′,5′:4,5]thieno[3,2-*b*]pyrano[2,3-*d*]pyridines by combination of domino reactions.

20 M. N. Elinson, R. F. Nasybullin, G. I. Nikishin, *C. R. Chim.* 2013, 16, 789–794. Sodium acetate catalyzed tandem Knoevenagel-Michael multicomponent reaction of aldehydes, 2-pyrazolin-5-ones, and cyano-functionalized C-H acids: facile and efficient way to 3-(5-hydroxypyrazol-4-yl)-3-aryl-propionitriles.

21 M. R. Naimi-Jamal, S. Mashkouri, A. Sharifi, *Mol. Divers.* 2010, 14, 473–477. An efficient, multicomponent approach for solvent-free synthesis of 2-amino-4*H*-chromene scaffold.

22 S. Samai, G. C. Nandi, M. S. Singh, *Tetrahedron* 2012, 68, 1247–1252. Highly convergent one-pot four-component regioselective synthesis of 4*H*-benzo[*f*]chromenes *via* annulation of β-oxodithioesters.

23 W.-B. Chen, Z.-J. Wu, Q.-L. Pei, L.-F. Cun, X.-M. Zhang, W.-C. Yuan, *Org. Lett.* 2010, 12, 3132–3135. Highly enantioselective construction of spiro[4*H*-pyran-3,3′-oxindoles] through a domino Knoevenagel/Michael/cyclization sequence catalyzed by cupreine.

24 G. H. Mahdavinia, M. Mirzazadeh, B. Notash, *Tetrahedron Lett.* 2013, 54, 3487–3492. A rapid and simple diversity-oriented synthesis of novel 3-amino-2′-oxospiro[benzo[*c*]pyrano[3,2-*a*]phenazine-1,3′-indoline]-2-carbonitrile/carboxylate derivatives *via* a one-pot, four-component domino reaction.

25 H. Sachdeva, D. Dwivedi, R. R. Bhattacharjee, S. Khaturia, R. Saroj, *J. Chem.* 2013, 1–10. NiO nanoparticles: an efficient catalyst for the multicomponent one-pot synthesis of novel spiro and condensed indole derivatives.

26 S. Sivakumar, S. Kanchithalaivan, R. R. Kumar, *RSC Adv.* 2013, 3, 13357–13364. A one-pot three-component domino protocol for the synthesis of penta-substituted 4*H*-pyrans.

27 M. S. Singh, G. C. Nandi, S. Samai, *Green Chem.* 2012, 14, 447–455. DABCO-promoted three-component regioselective synthesis of functionalized chromen-5-ones and pyrano[3,2-*c*]chromen-5-ones *via* direct annulation of α-oxoketene-*N*,*S*-arylaminoacetals under solvent-free conditions.

28 S. Bondock, W. Khalifa, A. A. Fadda, *Synth. Commun.* 2006, 36, 1601–1612. Utility of 1-chloro-3,4-dihydronaphthalene-2-carboxaldehyde in the synthesis of novel heterocycles with pharmaceutical interest.

29 P. R. Nandaluru, G. J. Bodwell, *Org. Lett.* 2012, 14, 310–313. Multicomponent synthesis of 6*H*-dibenzo[*b*,*d*]pyran-6-ones and a total synthesis of cannabinol.

30 D. Prajapati, M. Gohain, *Beilstein J. Org. Chem.* 2006, 2, (No. 11). An efficient synthesis of novel pyrano[2,3-*d*]- and furopyrano[2,3-*d*]pyrimidines *via* indium-catalyzed multicomponent domino reaction.

31 A. Pałasz, *Synthesis* 2010, 4021–4032. A green approach to the synthesis of fused uracils: pyrano[2,3-*d*]pyrimidines. "On-water" one-pot synthesis by domino Knoevenagel/Diels–Alder reactions.

32 A. Pałasz, *Monatsh. Chem.* 2012, 143, 1175–1185. Synthesis of fused uracils: pyrano[2,3-*d*]pyrimidines and 1,4-bis(pyrano[2,3-*d*]pyrimidinyl)benzenes by domino Knoevenagel/Diels–Alder reactions.

33 M. Li, C. Chen, F. He, Y. Gu, *Adv. Synth. Catal.* 2010, 352, 519–530. Multicomponent reactions of 1,3-cyclohexanediones and formaldehyde in glycerol: stabilization of paraformaldehyde in glycerol resulted from using dimedone as substrate.

34 F. Erver, G. Hilt, K. Harms, *Synthesis* 2011, 972–978. Cobalt(I) catalysis in the diastereoselective two-step synthesis of tricyclic systems.

35 S. Jiménez-Alonso, A. Estévez-Braun, Á. G. Ravelo, R. Zárate, M. López, *Tetrahedron* 2007, 63, 3066–3074. Double domino Knoevenagel hetero Diels–Alder strategy towards bis-pyrano-1,4-benzoquinones.

36 S. Jiménez-Alonso, H. Chávez, A. Estévez-Braun, Á. G. Ravelo, G. Feresin, A. Tapia, *Tetrahedron* 2008, 64, 8938–8942. An efficient synthesis of embelin derivatives through domino Knoevenagel hetero Diels–Alder reactions under microwave irradiation.

37 J. Menezes, A. Gomes, A. Silva, M. Faustino, M. Neves, A. Tomé, F. d. C. d. Silva, V. Ferreira, J. Cavaleiro, *Synlett* 2011, 1841–1844. Reaction of β-vinyl-*meso*-tetraphenylporphyrin with *o*-quinone methides.

38 D.-Q. Peng, Y. Liu, Z.-F. Lu, Y.-M. Shen, J.-H. Xu, *Synthesis* 2008, 1182–1192. Synthesis of siloxy-α-lapachone derivatives by chemo- and regioselective Diels–Alder reactions of 3-methylene-1,2,4-naphthotriones with silyl enol ethers.

39 S. B. Ferreira, F. de Carvalho da Silva, F. A. F. M. Bezerra, M. C. S. Lourenço, C. R. Kaiser, A. C. Pinto, V. F. Ferreira, *Arch. Pharm. Chem. Life Sci.* 2010, 343, 81–90. Synthesis of α- and β-pyran naphthoquinones as a new class of antitubercular agents.

40 S. S. Soman, J. M. Patel, *Synth. Commun.* 2013, 43, 775–783. Facile synthesis of furowarfarins *via* hetero-Diels–Alder cycloaddition reaction.

41 M. Radi, V. Bernardo, B. Bechi, D. Castagnolo, M. Pagano, M. Botta, *Tetrahedron Lett.* 2009, 50, 6572–6575. Microwave-assisted organocatalytic multicomponent Knoevenagel/hetero Diels–Alder reaction for the synthesis of 2,3-dihydropyran[2,3-*c*]pyrazoles.

42 N. J. Parmar, H. A. Barad, B. R. Pansuriya, N. P. Talpada, *RSC Adv.* 2013, 3, 8064–8070. A highly efficient, rapid one-pot synthesis of some new heteroaryl pyrano[2,3-*c*]pyrazoles in ionic liquid under microwave-irradiation.

43 Y. Gu, R. de Sousa, G. Frapper, C. Bachmann, J. Barrault, F. Jérôme, *Green Chem.* 2009, 11, 1968–1972. Catalyst-free aqueous multicomponent domino reactions from formaldehyde and 1,3-dicarbonyl derivatives.

44 G. Frapper, C. Bachmann, Y. Gu, R. Coval De Sousa, F. Jérôme, *Phys. Chem. Chem. Phys.* 2011, 13, 628–636. Mechanisms of the Knoevenagel hetero Diels–Alder sequence in multicomponent reactions to dihydropyrans: experimental and theoretical investigations into the role of water.

45 L. Li, Q. Wu, B.-K. Liu, X.-F. Lin, *Synthesis* 2011, 563–570. Facile multicomponent synthesis of novel 2,3-dihydropyran derivatives under solvent-free conditions.

46 M. Rueping, E. Merino, M. Bolte, *Org. Biomol. Chem.* 2012, 10, 6201–6210. Efficient proline and prolinol ether mediated 3-component synthesis of 3- and 3,4-substituted chromenone derivatives.

47 Y. Duan, X. Wang, X. Xu, Z. Kang, M. Zhang, L. Song, H. Deng, *Synthesis* 2013, 45, 2193–2200. Convenient one-pot

three-component synthesis of trifluoromethylated tetrahydrobenzo[*g*]-chromene derivatives.

48 X.-S. Wang, M.-M. Zhang, H. Jiang, C.-S. Yao, S.-J. Tu, *Tetrahedron* 2007, 63, 4439–4449. Three-component green synthesis of *N*-arylquinoline derivatives in ionic liquid [bmim$^+$] [BF_4^-]: reactions of arylaldehyde, 3-arylamino-5,5-dimethylcyclohex-2-enone, and active methylene compounds.

49 S.-L. Zhu, K. Zhao, X.-M. Su, S.-J. Ji, *Synth. Commun.* 2009, 39, 1355–1366. Microwave-assisted synthesis of new spiro[indoline-3,4′-quinoline] derivatives *via* a one-pot multicomponent reaction.

50 P. H. Yang, C. T. Qu, W. Z. Wang, *Res. Chem. Intermed.* 2013, 39, 463–468. AcOH catalyzed three-component synthesis of spirooxindole derivatives.

51 P. Seetham Naidu, P. Borah, P. J. Bhuyan, *Tetrahedron Lett.* 2012, 53, 4015–4017. Synthesis of some novel functionalized dihydropyrido[2,3-*d*]pyrimidines *via* an one-pot three-component reaction catalysed by $InCl_3$.

52 X. Gu, G. I. Georg, *Tetrahedron* 2013, 69, 9406–9416. Lithium perchlorate-, acetic anhydride-, and triphenylphosphine-assisted multicomponent syntheses of 4-unsubstituted 2,5-dioxo-octahydroquinoline-3-carboxylates and 3-carbonitriles.

53 X. S. Fan, Y. Z. Li, X. Y. Zhang, G. R. Qu, J. J. Wang, X. Y. Hu, *Heteroatom Chem.* 2006, 17, 382–388. An efficient and green synthesis of 1,4-dihydropyridine derivatives through multicomponent reaction in ionic liquid.

54 N. J. Thumar, M. P. Patel, *Saudi Pharm. J.* 2011, 19, 75–83. Synthesis, characterization, and antimicrobial evaluation of carbostyril derivatives of 1*H*-pyrazole.

55 R. H. Nia, M. Mamaghani, K. Tabatabaeian, F. Shirini, M. Rassa, *Acta Chim. Slov.* 2013, 60, 889–895. A rapid one-pot synthesis of pyrido[2,3-*d*]pyrimidine derivatives using Brønsted-acidic ionic liquid as catalyst.

56 N. G. Kozlov, S. L. Bondarev, A. P. Kadutskii, L. I. Basalaeva, F. S. Pashkovskii, *Russ. J. Org. Chem.* 2008, 44, 1031–1037. Tetronic acid in reaction with aromatic aldehydes and 2-naphthylamine. Investigation of fluorescent and nonlinear-optical characteristics of compounds obtained.

57 S. Tu, X. Zhu, J. Zhang, J. Xu, Y. Zhang, Q. Wang, R. Jia, B. Jiang, J. Zhang, C. Yao, *Bioorg. Med. Chem. Lett.* 2006, 16, 2925–2928. New potential biologically active compounds: design and an efficient synthesis of *N*-substituted 4-aryl-4,6,7,8-tetrahydroquinoline-2,5(1*H*,3*H*)-diones under microwave irradiation.

58 E. Ruiz, H. Rodríguez, J. Coro, E. Salfrán, M. Suárez, R. Martínez-Alvarez, N. Martín, *Ultrason. Sonochem.* 2011, 18, 32–36. Ultrasound-assisted one-pot, four component synthesis of 4-aryl 3,4-dihydropyridone derivatives.

59 J.-P. Zhang, J. Ding, N. Ma, B. Jiang, L.-C. Xu, S.-J. Tu, *J. Hetercycl. Chem.* 2013, 50, 66–70. Microwave-assisted aqueous synthesis of 6-ferrocenyl pyridin-2(1*H*)-one derivative.

60 G. M. Ziarani, S. Asadi, A. Badiei, S. Mousavi, P. Gholamzadeh, *Res. Chem. Intermed.* 2015, 41, 637–645. One-pot four-component synthesis of 2,5-dioxo-1,2,3,4,5,6,7,8-octahydroquinolines catalyzed by silica-based sulfonic acid.

61 G. M. Ziarani, S. Mousavi, N. Lashgari, A. Badiei, *J. Chem. Sci.* 2013, 125, 1359–1364. Mesostructured SBA-15-Pr-SO_3H: an efficient solid acid catalyst for one-pot and solvent-free synthesis of 3,4-dihydro-2-pyridone derivatives.

62 H. Jiang, R. Mai, H. Cao, Q. Zhu, X. Liu, *Org. Biomol. Chem.* 2009, 7, 4943–4953. L-Proline-catalyzed synthesis of highly functionalized multisubstituted 1,4-dihydropyridines.

63 L.-R. Wen, C. Liu, M. Li, L.-J. Wang, *J. Org. Chem.* 2010, 75, 7605–7614. Modulating the reactivity of heterocyclic ketene aminals in MCR: selective construction of tetrahydrobenzo[*b*]-imidazo[3,2,1-*ij*][1,8]naphthyridines.

64 L.-R. Wen, C.-Y. Jiang, M. Li, L.-J. Wang, *Tetrahedron* 2011, 67, 293–302. Application of 2-(2-chloroaroyl)methyleneimidazolidines in domino and multicomponent reaction: new entries to imidazo[1,2-*a*]pyridines and benzo[*b*]imidazo[1,2,3-*ij*][1,8]naphthyridines.

65 C. Altuğ, A. K. Burnett, E. Caner, Y. Dürüst, M. C. Elliott, R. P. Glanville, C. Guy, A. D. Westwell, *Tetrahedron* 2011, 67, 9522–9528. An efficient one-pot multicomponent approach to 5-amino-7-aryl-8-nitrothiazolo[3,2-*a*]pyridines.

66 C. Altuğ, E. Caner, *C. R. Chim.* 2013, 16, 217–221. A facile strategy for the annulation of 2-phenylsulfonyl methylene thiazolidin-4-one *via* multicomponent reactions.

67 M. Li, Z. Zuo, L. Wen, S. Wang, *J. Comb. Chem.* 2008, 10, 436–441. Microwave-assisted combinatorial synthesis of hexa-substituted 1,4-dihydropyridines scaffolds using one-pot two-step multicomponent reaction followed by a *S*-alkylation.

68 L. Wen, C. Ji, Y. Li, M. Li, *J. Comb. Chem.* 2009, 11, 799–805. Application of β-(2-chloroaroyl)thioacetanilide in synthesis(III): an efficient three-component synthesis of thiochromeno[2,3-*b*] pyridines catalyzed by KF/neutral Al_2O_3 co-operated with PEG 6000 under microwave irradiation.

69 L.-R. Wen, C. Ji, M. Li, H.-Y. Xie, *Tetrahedron* 2009, 65, 1287–1293. Application of *ortho*-chloro-β-aroylthioamides in synthesis(II): an efficient one-pot, three-component synthesis of tricyclic thiochromeno[2,3-*b*]pyridine derivatives.

70 Z. Liu, L. Tan, Q. Wu, X. Lin, *Chin. J. Chem.* 2012, 30, 2343–2348. Imidazole-catalyzed three-component cascade reaction for the facile synthesis of highly substituted 3,4-dihydropyridin-2-one derivatives.

71 (a) S. Gupta, P. Gupta, A. Sachar, R. L. Sharma, *J. Heterocycl. Chem.* 2010, 47, 334–349. Facile and one pot synthetic routes for various novel, differently fused and promising heteropolycycles. (b) M. Saha, A. K. Pal, *Tetrahedron Lett.* 2011, 52, 4872–4877. Palladium(0) nanoparticles: an efficient catalyst for the one-pot synthesis of polyhydroquinolines.

72 S. Samai, G. Chandra Nandi, S. Chowdhury, M. S. Singh, *Tetrahedron* 2011, 67, 5935–5941. L-Proline catalyzed synthesis of densely functionalized pyrido[2,3-*d*]pyrimidines *via* three-component one-pot domino Knoevenagel aza-Diels–Alder reaction.

73 S. Rostamnia, H. Xin, X. Liu, K. Lamei, *J. Mol. Catal. A: Chem.* 2013, 374–375, 85–93. Simultaneously application of SBA-15 sulfonic acid nanoreactor and ultrasonic irradiation as a very useful novel combined catalytic system: an ultra-fast, selective, reusable and waste-free green approach.

74 H. Sheibani, M. Seifi, A. Bazgir, *Synth. Commun.* 2009, 39, 1055–1064. Three-component synthesis of pyrimidine and pyrimidinone derivatives in the presence of high-surface-area MgO, a highly effective heterogeneous base catalyst.

75 H. Sheibani, M. A. Amrollahi, Z. Esfandiarpoor, *Mol. Divers.* 2010, 14, 277–283. A convenient one-pot synthesis of 3-amino-2,5-dihydropyridazine and pyrimidine derivatives in the presence of high surface area MgO as a highly effective heterogeneous base catalyst.

76 S. Ramezanpour, M. S. Hashtroudi, H. R. Bijanzadeh, S. Balalaie, *Tetrahedron Lett.* 2008, 49, 3980–3982. A novel and efficient domino reaction for the one-pot synthesis of spiro-2-aminopyrimidinones.

77 M. Mohammadnejad, M. S. Hashtroudi, S. Balalaie, *Heterocycl. Commun.* 2009, 15, 459–465. Efficient synthesis of 2-amino-6-aryl-5,6-dihydro-3*H*-pyrimidin-4-one building blocks *via* domino reaction.

78 M. Mirza-Aghayan, T. Baie Lashaki, M. Rahimifard, R. Boukherroub, A. A. Tarlani, *J. Iran. Chem. Soc.* 2011, 8, 280–286. Amino-functionalized MCM-41 base-catalyzed one-pot synthesis of 2-amino-5,6-dihydropyrimidin-4(3*H*)-ones.

79 O. M. Singh, N. S. Devi, *J. Org. Chem.* 2009, 74, 3141–3144. Application of β-oxodithioesters in domino and multicomponent reactions: facile route to dihydropyrimidines and coumarins.

80 K. Kiss, A. Csámpai, P. Sohár, *J. Organomet. Chem.* 2010, 695, 1852–1857. New ferrocenyl-substituted heterocycles. Formation under Biginelli conditions, DFT modelling, and structure determination.

81 P. K. Sahu, P. K. Sahu, D. D. Agarwal, *RSC Adv.* 2013, 3, 9854–9864. Efficient and facile synthesis of heterocycles and their mechanistic consideration using kaolin.

82 A. K. Arya, M. Kumar, *Mol. Divers.* 2011, 15, 781–789. Base catalyzed multicomponent synthesis of spiroheterocycles with fused heterosystems.

83 L.-Y. Zeng, Y.-M. Ren, C. Cai, *Synth. Commun.* 2011, 41, 3635–3643. Iodine-catalyzed, multicomponent, one-pot synthesis of 5-aryl-5,8-dihydrotetrazolo[1,5-*a*]pyrimidine-7-carboxylic acids.

84 P. Zhang, W. Li, G. Zhou, Y. Lai, S. Xu, *Heterocycles* 2011, 83, 2067–2077. One-pot synthesis of dihydropyrimidiones *via* environmentally friendly enzyme-catalyzed Biginelli reaction.

85 M.-J. Fan, B. Qian, L.-B. Zhao, Y.-M. Liang, *Tetrahedron* 2007, 63, 8987–8992. A facile reaction involving zwitterionic intermediates for the synthesis of 5-hydroxy-2*H*-pyrrol-2-one derivatives.

86 M. Quai, S. Frattini, U. Vendrame, M. Mondoni, S. Dossena, E. Cereda, *Tetrahedron Lett.* 2004, 45, 1413–1416. 5-Hydroxy-2*H*-pyrrol-2-ones and not 2-aminofurans are the cycloaddition products between alkyl isocyanides and benzyliden-1,3-diketones.

87 S. Renuga, M. Gnanadeebam, B. M. Vinosha, S. Perumal, *Tetrahedron* 2007, 63, 10054–10058. A novel four-component tandem protocol for the stereoselective synthesis of highly functionalised thiazoles.

88 M. Li, W.-L. Yang, L.-R. Wen, F.-Q. Li, *Eur. J. Org. Chem.* 2008, 2008, 2751–2758. A first resource-efficient and highly flexible procedure for a four-component synthesis of dispiro-pyrrolidines.

89 A. Dandia, A. K. Jain, A. K. Laxkar, D. S. Bhati, *Tetrahedron Lett.* 2013, 54, 3180–3184. A highly efficient protocol for the regio- and stereo-selective synthesis of spiro pyrrolidine and pyrrolizidine derivatives by multicomponent reaction.

90 Z. N. Tisseh, M. Dabiri, M. Nobahar, H. R. Khavasi, A. Bazgir, *Tetrahedron* 2012, 68, 1769–1773. Catalyst-free, aqueous and highly diastereoselective synthesis of new 5-substituted 1*H*-tetrazoles *via* a multi-component domino Knoevenagel condensation/1,3-dipolar cycloaddition reaction.

91 S. Uma Maheswari, S. Perumal, *Tetrahedron Lett.* 2012, 53, 6885–6888. A facile sequential three-component regio- and stereoselective synthesis of novel spiro-isoxazoline/acridinone hybrids.

92 R. Ghahremanzadeh, G. Shakibaei, A. Bazgir, *Synlett* 2008, 1129–1132. An efficient one-pot synthesis of 1*H*-pyrazolo[1,2-*b*]phthalazine-5,10-dione derivatives.

93 R. Heydari, B. Tahamipour, *Chin. Chem. Lett.* 2011, 22, 1281–1284. Highly regioselective synthesis of dicyano-8*a*,10,11-trihydropyrrolo[1,2-*a*][1,10]phenanthrolines *via* a domino-Knoevenagel-cyclization.

94 G. Marandi, M. T. Maghsoodlou, N. Hazeri, S. M. Habibi-Khorassani, N. A. Torbati, F. Rostami-Charati, B. W. Skelton, M. Makha, *Mol. Divers.* 2011, 15, 197–201. Synthesis of cyano-2,3-dihydropyrrolo[1,2-*f*]phenanthridine derivatives *via* a domino-Knoevenagel-cyclization.

95 W.-S. Hsu, V. Paike, C.-M. Sun, *Mol. Divers.* 2013, 17, 285–294. One-pot three component reaction for the rapid synthesis of pyrrolo[1,2-*a*]benzimidazoles.

96 A. B. Atar, Y. T. Jeong, *Tetrahedron Lett.* 2013, 54, 1302–1306. Silica supported tungstic acid (STA): an efficient catalyst for the synthesis of bis-spiro piperidine derivatives under milder condition.

97 C. Yu, J. Lu, T. Li, D. Wang, B. Qin, H. Zhang, C. Yao, *Synlett* 2011, 2420–2424. A NHC-involved, cascade, metal-free, and three-component synthesis of 2,3-diarylated fully substituted furans under solvent-free conditions.

98 M. Adib, E. Sheikhi, M. Bagheri, H. R. Bijanzadeh, M. Amanlou, *Tetrahedron* 2012, 68, 3237–3242. Solvent-free and three-component synthesis of 1*H*,6*H*-6λ^5-[1,2]benzoxaphospholo[2,3-*b*][1,2]benzoxaphosphol-1-ones.

99 L.-H. Chen, Y.-S. Chuang, B. D. Narhe, C.-M. Sun, *RSC Adv.* 2013, 3, 13934–13943. A concise synthesis of 2-(2-aminothiophene)-benzimidazoles by one-pot multicomponent reaction.

100 T. Wang, X.-G. Huang, J. Liu, B. Li, J.-J. Wu, K.-X. Chen, W.-L. Zhu, X.-Y. Xu, B.-B. Zeng, *Synlett* 2010, 1351–1354. An efficient one-pot synthesis of substituted 2-aminothiophenes *via* three-component Gewald reaction catalyzed by L-proline.

101 Z.-J. Zheng, L.-X. Liu, G. Gao, H. Dong, J.-X. Jiang, G.-Q. Lai, L.-W. Xu, *RSC Adv.* 2012, 2, 2895–2901. Amine-functional polysiloxanes (AFPs) as efficient polymeric

organocatalyst for amino catalysis: efficient multicomponent Gewald reaction, α-allylic alkylation of aldehydes, and Knoevenagel condensation.

102 R. K. Verma, G. K. Verma, G. Shukla, A. Nagaraju, M. S. Singh, *ACS Comb. Sci.* 2012, 14, 224–230. 4-Dimethylamino pyridine-promoted one-pot three-component regioselective synthesis of highly functionalized 4*H*-thiopyrans *via* hetero-annulation of β-oxodithioesters.

103 K. Majumdar, S. Ponra, R. K. Nandi, *Tetrahedron Lett.* 2012, 53, 1732–1737. One-pot efficient green synthesis of spirooxindole-annulated thiopyran derivatives *via* Knoevenagel condensation followed by Michael addition.

104 D. B. Ramachary, G. B. Reddy, *Org. Biomol. Chem.* 2006, 4, 4463–4468. Towards organo-click reactions: development of pharmaceutical ingredients by using direct organocatalytic bio-mimetic reductions.

105 D. B. Ramachary, M. Kishor, K. Ramakumar, *Tetrahedron Lett.* 2006, 47, 651–656. A novel and green protocol for two-carbon homologation: a direct amino acid/K_2CO_3-catalyzed four-component reaction of aldehydes, active methylenes, Hantzsch esters and alkyl halides.

106 D. B. Ramachary, M. Kishor, G. B. Reddy, *Org. Biomol. Chem.* 2006, 4, 1641–1646. Development of drug intermediates by using direct organocatalytic multi-component reactions.

107 D. B. Ramachary, M. Kishor, *J. Org. Chem.* 2007, 72, 5056–5068. Organocatalytic sequential one-pot double cascade asymmetric synthesis of Wieland–Miescher ketone analogues from a Knoevenagel/hydrogenation/Robinson annulation sequence: scope and applications of organo-catalytic biomimetic reductions.

108 D. B. Ramachary, M. Kishor, *Org. Biomol. Chem.* 2008, 6, 4176–4187. Direct amino acid-catalyzed cascade biomimetic reductive alkylations: application to the asymmetric synthesis of Hajos–Parrish ketone analogues.

109 D. B. Ramachary, C. Venkaiah, Y. V. Reddy, M. Kishor, *Org. Biomol. Chem.* 2009, 7, 2053–2062. Multi-catalysis cascade reactions based on the methoxycarbonylketene platform: diversity-oriented synthesis of functionalized non-symmetrical malonates for agrochemicals and pharmaceuticals.

110 D. B. Ramachary, Y. V. Reddy, B. V. Prakash, *Org. Biomol. Chem.* 2008, 6, 719–726. Double cascade reactions based on the Barbas dienamine platform: highly stereoselective synthesis of functionalized cyclohexanes for cardiovascular agents.

111 B. Das, K. Reddy, R. Ramu, P. Thirupathi, B. Ravikanth, *Synlett* 2006, 2006, 1756–1758. Iodine as an efficient catalyst for one-pot multicomponent synthesis of β-acetamido ketones.

112 A. T. Khan, T. Parvin, L. H. Choudhury, *Tetrahedron* 2007, 63, 5593–5601. Iron(III) chloride-catalyzed convenient one-pot synthesis of β-acetamido carbonyl compounds.

113 H. Xu, X. Yu, L. Sun, J. Liu, W. Fan, Y. Shen, W. Wang, *Tetrahedron Lett.* 2008, 49, 4687–4689. Microwave-assisted three-component Knoevenagel-nucleophilic aromatic substitution reactions.

114 L. D. S. Yadav, B. S. Yadav, V. K. Rai, *J. Heterocycl. Chem.* 2008, 45, 1315–1319. Multicomponent reactions in chiral ionic liquids: a stereocontrolled route to mercapto-pyranothiazoles.

115 M. Anderluh, M. Jukič, R. Petrič, *Tetrahedron* 2009, 65, 344–350. Three-component one-pot synthetic route to 2-amino-5-alkylidene-thiazol-4-ones.

116 C. Mukhopadhyay, S. Ray, *Tetrahedron Lett.* 2011, 52, 6431–6438. Rapid and straightforward one-pot expeditious synthesis of 2-amino-5-alkylidene-thiazol-4-ones at room temperature.

117 C. Mukhopadhyay, S. Ray, *Tetrahedron* 2011, 67, 7936–7945. Synthesis of 2-amino-5-alkylidenethiazol-4-ones from ketones, rhodanine, and amines with the aid of re-usable heterogeneous silica-pyridine based catalyst.

118 Y. Ma, M. Wang, D. Li, B. Bekturhun, J. Liu, Q. Liu, *J. Org. Chem.* 2009, 74, 3116–3121. α-Alkenoyl ketene *S,S*-acetal-based multicomponent reaction: an efficient approach for the selective construction of polyfunctionalized cyclohexanones.

119 X.-S. Wang, J.-R. Wu, Q. Li, M.-M. Zhang, *J. Heterocycl. Chem.* 2009, 46, 1355–1363. A novel and green method for the synthesis of highly substituted isoquinoline derivatives in ionic liquid.

120 X.-S. Wang, J.-R. Wu, J. Zhou, S.-J. Tu, *J. Comb. Chem.* 2009, 11, 1011–1022. Green method for the synthesis of highly substituted cyclohexa-1,3-diene, polyhydroindene, polyhydro-naphthalene, isochromene, isothiochromene, and isoquinoline derivatives in ionic liquids.

121 K. Balamurugan, V. Jeyachandran, S. Perumal, J. C. Menéndez, *Tetrahedron* 2011, 67, 1432–1437. A three-component domino protocol for the facile synthesis of highly functionalized tetrahydroisoquinolines by creation of their benzene ring.

122 N. Sharma, A. Sharma, A. Shard, R. Kumar, Saima, A. K. Sinha, *Chem.–Eur. J.* 2011, 17, 10350–10356. Pd-catalyzed orthogonal Knoevenagel/Perkin condensation-decarboxylation-Heck/Suzuki sequences: tandem transformations of benzaldehydes into hydroxy-functionalized antidiabetic stilbene-cinnamoyl hybrids and asymmetric distyrylbenzenes.

123 L. Liu, D. Wu, X. Li, S. Wang, H. Li, J. Li, W. Wang, *Chem. Commun.* 2012, 48, 1692–1694. Organocatalytic enantioselective conjugate addition of ketones to isatylidine malononitriles.

124 M. I. Flores-Conde, L. Reyes, R. Herrera, H. Rios, M. A. Vazquez, R. Miranda, J. Tamariz, F. Delgado, *Int. J. Mol. Sci.* 2012, 13, 2590–2617. Highly regio- and stereoselective Diels–Alder cycloadditions *via* two-step and multicomponent reactions promoted by infrared irradiation under solvent-free conditions.

125 R. Mudududdla, S. K. Jain, J. B. Bharate, A. P. Gupta, B. Singh, R. A. Vishwakarma, S. B. Bharate, *J. Org. Chem.* 2012, 77, 8821–8827. *ortho*-Amidoalkylation of phenols *via* tandem one-pot approach involving oxazine intermediate.

126 Y. Oikawa, H. Hirasawa, O. Yonemitsu, *Tetrahedron Lett.* 1978, 19, 1759–1762. Meldrum's acid in organic synthesis. 1. A convenient one-pot synthesis of ethyl indolepropionates.

127 E. Dardennes, Á. Kovács-Kulyassa, M. Boisbrun, C. Petermann, J.-Y. Laronze, J. Sapi, *Tetrahedron: Asymmetry* 2005, 16, 1329–1339. Diastereocontrolled multicomponent pathway to 3,4-heterocycle-annulated tetrahydro-β-carbolines.

128 G. Sabitha, M. Raj Kumar, M. Shashi Kumar Reddy, J. Yadav, K. Rama Krishna, A. Kunwar, *Tetrahedron Lett.* 2005, 46, 1659–1661. A D,L-proline catalyzed diastereoselective trimolecular condensation: an approach to the one-pot synthesis of perhydrofuro[3,2-*b*]pyran-5-ones.

129 E. Dardennes, S. Gérard, C. Petermann, J. Sapi, *Tetrahedron: Asymmetry* 2010, 21, 208–215. Diastereoselective trimolecular condensation between indole, Meldrum's acid and chiral sugar-derived aldehydes.

130 A. Kumar, P. Kumar, V. D. Tripathi, S. Srivastava, *RSC Adv.* 2012, 2, 11641–11644. A novel access to indole-3-substituted dihydrocoumarins in artificial sweetener saccharin based functional ionic liquids.

131 E. Siddalingamurthy, K. M. Mahadevan, T. O. S. Kumar, *Synth. Commun.* 2013, 43, 3153–3162. Choline chloride/urea ionic liquid catalyzed a convenient one-pot synthesis of indole-3-propanamide derivatives.

132 E. L. Armstrong, H. K. Grover, M. A. Kerr, *J. Org. Chem.* 2013, 78, 10534–10540. Scandium triflate-catalyzed nucleophilic additions to indolylmethyl Meldrum's acid derivatives *via* a gramine-type fragmentation: synthesis of substituted indolemethanes.

133 Z.-L. Shen, S.-J. Ji, S.-Y. Wang, X.-F. Zeng, *Tetrahedron* 2005, 61, 10552–10558. A novel base-promoted synthesis of β-indolylketones *via* a three-component condensation under ultrasonic irradiation.

134 Y. Sui, L. Liu, J.-L. Zhao, D. Wang, Y.-J. Chen, *Tetrahedron Lett.* 2007, 48, 3779–3782. An efficient one-pot reaction of indoles, nitroacetate, and paraformaldehyde for the synthesis of tryptophan derivatives.

135 F. Epifano, S. Genovese, O. Rosati, S. Tagliapietra, C. Pelucchini, M. Curini, *Tetrahedron Lett.* 2011, 52, 568–571. Ytterbium triflate catalyzed synthesis of β-functionalized indole derivatives.

136 Y. Qu, F. Ke, L. Zhou, Z. Li, H. Xiang, D. Wu, X. Zhou, *Chem. Commun.* 2011, 47, 3912–3914. Synthesis of 3-indole derivatives by copper sulfonato Salen catalyzed three-component reactions in water.

137 S. Chandrasekhar, V. Patro, G. P. K. Reddy, R. Grée, *Tetrahedron Lett.* 2012, 53, 6223–6225. A ligand-free copper(II)-catalyzed three-component reaction in poly(ethylene glycol) medium: a versatile protocol for the preparation of selected 3-indole derivatives.

138J. Gerencsér, G. Panka, T. Nagy, O. Egyed, G. Dormán, L. Ürge, F. Darvas, *J. Comb. Chem.* 2005, 7, 530–538. Procedure for the parallel preparation of 3-imidazo[1,2-*a*]pyridin-3-yl-propionic acid derivatives involving Meldrum's acid.

139 N. Ma, B. Jiang, G. Zhang, S.-J. Tu, W. Wever, G. Li, *Green Chem.* 2010, 12, 1357–1361. New multicomponent domino reactions (MDRs) in water: highly chemo-, regio- and stereoselective synthesis of spiro{[1,3]dioxanopyridine}-4,6-diones and pyrazolo[3,4-*b*]pyridines.

140 X. Zhang, D. Li, X. Fan, X. Wang, X. Li, G. Qu, J. Wang, *Mol. Divers.* 2010, 14, 159–167. Ionic liquid-promoted multi-component reaction: novel and efficient preparation of pyrazolo[3,4-*b*] pyridinone, pyrazolo[3,4-*b*]-quinolinone and their hybrids with pyrimidine nucleoside.

141 A. A. Roshan, M. Mamaghani, N. O. Mahmoodi, F. Shirini, *Chin. Chem. Lett.* 2012, 23, 399–402. An efficient regioselective sonochemical synthesis of novel 4-aryl-3-methyl-4,5-dihydro-1*H*-pyrazolo[3,4-*b*]pyridin-6(7*H*)-ones.

142 X.-M. Wang, H.-L. Ye, Z.-J. Qaun, X.-C. Wang, *Res. Chem. Intermed.* 2013, 39, 2357–2367. One-pot synthesis of benzoquinoline and coumarin derivatives using Meldrum's acid in three-component reactions.

143 S. Tu, Q. Wang, Y. Zhang, J. Xu, J. Zhang, X. Zhu, F. Shi, *J. Heterocycl. Chem.* 2007, 44, 811–814. Design and synthesis of new and significative bifunctional compounds containing two pyrazolo[3,4-*b*]pyridine nucleus through multicomponent reaction under microwave irradiation.

144 (a) S. Tu, Q. Wang, J. Xu, X. Zhu, J. Zhang, B. Jiang, R. Jia, Y. Zhang, J. Zhang, *J. Heterocycl. Chem.* 2006, 43, 855–858. An efficient one-pot synthesis of 5-aryl substituted 2-amino-5,8-dihydropyrido[2,3-*d*]pyrimidin-4,7-diones under microwave irradiation without catalyst. (b) D.-Q. Shi, J.-W. Shi, S.-F. Rong, *J. Heterocycl. Chem.* 2009, 46, 1331–1334. An efficient and clean synthesis of pyrido[2,3-*d*]pyrimidine-4,7-dione derivatives in aqueous media.

145 (a) V. V. Lipson, V. V. Borodina, M. G. Shirobokova, S. M. Desenko, O. V. Shishkin, R. I. Zubatyuk, *Chem. Heterocycl. Compd.* 2007, 43, 490–495. Interaction of aminoazoles with Meldrum's acid and dialkyl ketones or cyclohexanone. (b) C.-L. Shi, H. Chen, D.-Q. Shi, *J. Heterocycl. Chem.* 2011, 48, 351–354. An efficient one-pot synthesis of pyrazolo[3,4-*b*] pyridinone derivatives catalyzed by L-proline. (c) L.-M. Yang, P.-L. Sun, L.-Q. Wu, *J. Chin. Chem. Soc.* 2012, 59, 1500–1503. One-pot three-component synthesis of spiro[pyrazolo[3,4-*b*] pyridine-4,3′-indoline] derivatives catalyzed by melamine trisulfonic acid. (d) M. Mamaghani, F. Shirini, N. O. Mahmoodi, A. Azimi-Roshan, H. Hashemlou, *J. Mol. Struct.* 2013, 1051, 169–176. A green, efficient and recyclable Fe^{+3}@K10 catalyst for the synthesis of bioactive pyrazolo[3,4-*b*] pyridin-6(7*H*)-ones under "on water" conditions. (e) X. Zhong, G. Dou, D. Wang, *Molecules* 2013, 18, 13139–13147. Polyethylene glycol (PEG-400): an efficient and recyclable reaction medium for the synthesis of pyrazolo[3,4-*b*]pyridin-6(7*H*)-one derivatives.

146 V. V. Lipson, N. V. Svetlichnaya, S. V. Shishkina, O. V. Shishkin, *Mendeleev Commun.* 2008, 18, 141–143. Cascade cyclization of 1,2-diamino-4-phenylimidazole with aromatic aldehydes and Meldrum's acid.

147 A. A. Dudinov, B. V. Lichitsky, I. A. Antonov, A. N. Komogortsev, P. A. Belyakov, M. M. Krayushkin, *Russ. Chem. Bull.* 2008, 57, 1740–1743. Three-component condensation of 2,4-diaminothiazoles with aldehydes and Meldrum's acid. The synthesis of 7-aryl- and 7-alkyl-6,7-dihydro-4*H*-thiazolo[4,5-*b*]pyridin-5-ones.

148 A. A. Dudinov, A. N. Komogortsev, B. V. Lichitsky, M. M. Krayushkin, *Phosphorus, Sulfur Silicon Relat. Elem.* 2010,

185, 1230–1234. Three-component condensation of 2,4-diaminoselenazole with aldehydes and Meldrum's acid: synthesis of 7-aryl(alkyl)-substituted 2-amino-6,7-dihydro-4*H*-selenazolo[4,5-*b*]pyridin-5-ones.

149 B. V. Lichitsky, A. N. Komogortsev, R. M. Belyi, A. A. Dudinov, M. M. Krayushkin, *Russ. Chem. Bull.* 2009, 58, 1538–1541. Three-component condensation of 4-amino-isothiazole derivatives with aldehydes and Meldrum's acid. Synthesis of 6,7-dihydro-4*H*-isothiazolo[4,5-*b*] pyridin-5-ones.

150 M. Vilches-Herrera, A. Spannenberg, P. Langer, V. O. Iaroshenko, *Tetrahedron* 2013, 69, 5955–5967. Novel and efficient synthesis of 4,7-dihydro-1*H*-pyrrolo[2,3-*b*]pyridine derivatives *via* one-pot, three-component approach from *N*-substituted 5-amino-3-cyanopyrroles, various carbonyl and active methylene compounds.

151 (a) J. Peng, R. Jia, Y. Xu, S. Tu, N. Ma, G. Zhang, *Chin. J. Chem.* 2009, 27, 2392–2396. A simple one-pot synthesis of indeno[2,1-*f*]quinolin-3-one derivatives under microwave irradiation. (b) D. Azarifar, D. Sheikh, *Synth. Commun.* 2013, 43, 2517–2526. $ZrOCl_2{\cdot}8H_2O$: an efficient, eco-friendly, and recyclable catalyst for ultrasound-accelerated, one-pot, solvent-free synthesis of 8-aryl-7,8-dihydro-[1,3] dioxolo[4,5-*g*]quinolin-6-(5*H*)-one and 4-aryl-3,4-dihydro-quinolin-2(1*H*)-one derivatives.

152 (a) S. Gao, C. Tsai, C.-F. Yao, *Synlett* 2009, 949–954. A simple and green approach for the synthesis of tetrahydrobenzo[*a*]-xanthen-11-one derivatives using tetra-butyl ammonium fluoride in water. (b) L.-Q. Wu, W.-L. Li, F.-L. Yan, *Collect. Czech. Chem. Commun.* 2011, 76, 235–241. $CeCl_3{\cdot}7H_2O$ as mild and efficient catalyst for the one-pot multicomponent synthesis of 8-aryl-7,8-dihydro[1,3] dioxolo[4,5-*g*]chromen-6-ones. (c) E. Yamuna, K. Prabakaran, M. Zeller, K. J. Rajendra Prasad, *Synth. Commun.* 2012, 42, 1330–1340. Convergent synthetic route to pyrano[2,3-*a*]carbazole by multicomponent reaction. (d) B. Karami, K. Eskandari, S. Khodabakhshi, *Arkivoc* 2012, (ix), 76–84. One-pot three-component approach to synthesis of fused heterocyclic compounds: synthesis of fused pyran-2-ones.

153 B. V. Lichitsky, A. N. Komogortsev, A. A. Dudinov, M. M. Krayushkin, *Russ. Chem. Bull.* 2008, 57, 2175–2179. Three-component condensation of 2-aminothiophene-3-carboxylic acid derivatives with aldehydes and Meldrum's acid.

154 A. A. Dudinov, B. V. Lichitsky, A. N. Komogortsev, M. M. Krayushkin, *Mendeleev Commun.* 2009, 19, 87–88. Three-component condensation of 2,4-diaminothiazoles with aldehydes and Meldrum's acid: synthesis of 7-aryl(alkyl)-substituted 6,7-dihydro-4*H*-thiazolo[4,5-*b*]pyridin-5-ones.

155 B. V. Lichitsky, A. N. Komogortsev, A. A. Dudinov, M. M. Krayushkin, *Russ. Chem. Bull.* 2009, 58, 1504–1508. Three-component condensation of 5-aminopyrazole derivatives with isatins and Meldrum's acid. Synthesis of 1,7-dihydrospiro[pyrazolo[3,4-*b*]-pyridine-4,3′-indole]-2′,6(1′*H*,5*H*)-diones.

156 B. V. Lichitsky, A. A. Dudinov, A. N. Komogortsev, M. M. Krayushkin, *Mendeleev Commun.* 2010, 20, 255–256. Synthesis of 5-oxo-4,5,6,7-tetrahydro-1*H*-pyrrolo-[3,2-*b*] pyridine-3-carboxylic acids by three-component condensation of 3-aminopyrrole derivatives.

157 (a) B. V. Lichitsky, R. M. Belyi, A. N. Komogortsev, A. A. Dudinov, M. M. Krayushkin, *Russ. Chem. Bull.* 2009, 58, 387–391. Three-component condensation of 3-aminothio-phene-2-carboxylic acid derivatives with aldehydes and Meldrum's acid. (b) B. V. Lichitsky, R. M. Belyi, A. N. Komogortsev, A. A. Dudinov, M. M. Krayushkin, *Russ. Chem. Bull.* 2013, 62, 1026–1031. Three-component condensation of 3-aminothiophene derivatives with isatines and Meldrum's acid. Synthesis of 2,5′-dioxo-3′-phenyl-5′,6′-dihydro-4′*H*-spiro[indoline-3,7′-thieno[3,2-*b*]pyridine]-2′-carboxylic acids.

158 B. V. Lichitsky, A. N. Komogortsev, A. A. Dudinov, M. M. Krayushkin, *Russ. Chem. Bull.* 2012, 61, 1591–1595. Three-component condensation of 5-aminoimidazole derivatives with aldehydes and Meldrum's acid. Synthesis of 3,4,6,7-tetrahydroimidazo[4,5-*b*]pyridin-5-ones.

159 Z. Xiao, M. Lei, L. Hu, *Tetrahedron Lett.* 2011, 52, 7099–7102. An unexpected multi-component reaction to synthesis of 3-(5-amino-3-methyl-1*H*-pyrazol-4-yl)-3-arylpropanoic acids in ionic liquid.

160 D.-Q. Shi, F. Yang, *J. Heterocycl. Chem.* 2011, 48, 308–311. An efficient synthesis of pyrazolo[3,4-*b*]quinolin-5(6*H*)-one derivatives in ionic liquid.

161 M. Vilches-Herrera, I. Knepper, N. de Souza, A. Villinger, V. Y. Sosnovskikh, V. O. Iaroshenko, *ACS Comb. Sci.* 2012, 14, 434–441. One-pot, three-component synthesis of 7-azaindole derivatives from *N*-substituted 2-amino-4-cyanopyrroles, various aldehydes, and active methylene compounds.

162 M. N. Elinson, A. S. Dorofeev, R. F. Nasybullin, S. K. Feducovich, G. I. Nikishin, *Electrochim. Acta* 2008, 53, 5033–5038. Electrocatalytic tandem Knoevenagel–Michael reaction of 3-methyl-2-pyrazolin-5-ones, aryl aldehydes and cyano-functionalized C-H acids: facile and convenient multicomponent way to substituted 3-(5-hydroxy-3-methylpyrazol-4-yl)-3-arylpropionitriles.

163 M. Rumum Rohman, H. Mecadon, A. T. Khan, B. Myrboh, *Tetrahedron Lett.* 2012, 53, 5261–5264. Synthesis of important β-functionalized 5-methyl-1*H*-pyrazol-3-ol derivatives in the presence of γ-alumina catalyst in aqueous medium.

164 J. Safaei-Ghomi, A. Ziarati, M. Tamimi, *Acta Chim. Slov.* 2013, 60, 403–410. A novel method for the one-pot five-component synthesis of highly functionalized pyranopyrazoles catalyzed by CuI nanoparticles.

165 A. N. Meldrum, *J. Chem. Soc., Trans.* 1908, 93, 598–601. A β-lactonic acid from acetone and malonic acid.

166 D. Davidson, S. A. Bernhard, *J. Am. Chem. Soc.* 1948, 70, 3426–3428. The structure of Meldrum's supposed β-lactonic acid.

167 (a) B.-C. Chen, *Heterocycles* 1991, 32, 529–597. Meldrum's acid in organic synthesis. (b) A. S. Ivanov, *Chem. Soc. Rev.* 2008, 37, 789–811. Meldrum's acid and related compounds in the synthesis of natural products and analogs. (c) V. V. Lipson, N. Y. Gorobets, *Mol. Divers.* 2009, 13, 399–419. One hundred years of Meldrum's acid: advances in the synthesis

of pyridine and pyrimidine derivatives. (d) W.-T. Gao, M.-R. Zheng, W.-D. Hou, *Chin. J. Org. Chem.* 2010, 30, 958–967. Progress in application of Meldrum's acid to synthesis of heterocyclic compounds.

168 J. Gerencsér, G. Dormán, F. Darvas, *QSAR Comb. Sci.* 2006, 25, 439–448. Meldrum's acid in multicomponent reactions: applications to combinatorial and diversity-oriented synthesis.

169 (a) D. R. Spring, *Org. Biomol. Chem.* 2003, 1, 3867–3870. Diversity-oriented synthesis; a challenge for synthetic chemists. (b) M. D. Burke, S. L. Schreiber, *Angew. Chem. Int. Ed. Engl.* 2004, 43, 46–58. A planning strategy for diversity-oriented synthesis. (c) R. J. Spandl, A. Bender, D. R. Spring, *Org. Biomol. Chem.* 2008, 6, 1149–1158. Diversity-oriented synthesis; a spectrum of approaches and results. (d) R. J. Spandl, M. Díaz-Gavilán, K. M. G. O'Connell, G. L. Thomas, D. R. Spring, *Chem. Rec.* 2008, 8, 129–142. Diversity-oriented synthesis.

170 X.-S. Wang, M.-M. Zhang, Z.-S. Zeng, D.-Q. Shi, S.-J. Tu, X.-Y. Wei, Z.-M. Zong, *Tetrahedron Lett.* 2005, 46, 7169–7173. A simple and clean procedure for the synthesis of polyhydroacridine and quinoline derivatives: reaction of Schiff base with 1,3-dicarbonyl compounds in aqueous medium.

171 B. Jiang, W.-J. Hao, J.-P. Zhang, S.-J. Tu, F. Shi, *Org. Biomol. Chem.* 2009, 7, 2195–2201. A new domino autocatalytic reaction leading to polyfunctionalized spiro[5.5]undecanes and dispiro[4.2.5.2]pentadecanes.

172 L. An, F. Yang, R. Yao, C. Yan, *Chin. J. Chem.* 2010, 28, 2451–2454. Sodium acetate catalyzed multicomponent cyclization of aromatic aldehydes, acetone and Meldrum acid.

173 S. R. Jetti, D. Verma, S. Jain, *J. Chem. Pharm. Res.* 2012, 4, 2373–2379. NBS/AIBN promoted one-pot multi component regioselective synthesis of spiro heterobicyclic rings *via* Biginelli-like condensation reaction.

174 F.-Y. Wu, Y. Li, H. Feng, Q. Wu, B. Jiang, F. Shi, S.-J. Tu, *Synthesis* 2011, 2459–2465. Stereoselective synthesis of functionalized cyclopropane derivatives *via* α-thiocyanate ketone-based three-component reaction.

175 Y. Han, J. Chen, L. Hui, C.-G. Yan, *Tetrahedron* 2010, 66, 7743–7748. Synthesis of zwitterionic salts *via* three-component reactions of pyridacylpyridinum iodide, aromatic aldehydes, and Meldrum acid or *N,N*-dimethylbarbituric acid.

176 (a) L. Hui, Y.-X. Jin, J.-X. Wang, C.-G. Yan, *Eur. J. Org. Chem.* 2011, 7194–7198. Molecular diversity of the three-component reactions of *N*-benzylbenzimidazolium salts, aromatic aldehydes, and active methylene compounds. (b) A. Kumar, G. Gupta, S. Srivastava, *Org. Lett.* 2011, 13, 6366–6369. Synthesis of new class of alkyl azarene pyridinium zwitterions *via* iodine mediated sp^3 C-H bond activation. (c) H. Hou, Y. Zhang, C.-G. Yan, *Chem. Commun.* 2012, 48, 4492–4494. One-pot synthesis of 4-substituted isoquinolinium zwitterionic salts by metal-free C-H bond activation.

177 Q.-F. Wang, L. Hui, H. Hou, C.-G. Yan, *J. Comb. Chem.* 2010, 12, 260–265. Synthesis of zwitterionic salts of pyridinium-Meldrum acid and barbiturate through unique four-component reactions.

178 E.-Y. Xia, J. Sun, R. Yao, C.-G. Yan, *Tetrahedron* 2010, 66, 3569–3574. Synthesis of zwitterionic salts *via* three component reactions of nitrogen-containing heterocycles, acetylenedicarboxylate and cyclic 1,3-dicarbonyl compounds.

179 C.-J. Lee, Y.-J. Jang, Z.-Z. Wu, W. Lin, *Org. Lett.* 2012, 14, 1906–1909. Preparation of functional phosphorus zwitterions from activated alkanes, aldehydes, and tributylphosphine: synthesis of polysubstituted furo[3,2-*c*]coumarins.

180 S. Ahadi, M. Abaszadeh, H. R. Khavasi, A. Bazgir, *J. Iran. Chem. Soc.* 2013, 10, 1145–1149. An efficient three-component synthesis of new amidinium salts.

181 E. Soleimani, M. M. Khodaei, A. Taheri Kal Koshvandi, *C. R. Chim.* 2012, 15, 273–277. One-pot three-component reaction: synthesis of substituted β-cyanocarbonyls in aqueous media.

182 I. Devi, B. Baruah, P. Bhuyan, *Synlett* 2006, 2593–2596. α-Cyclisation of tertiary amines: synthesis of some novel annelated quinolines *via* a three-component reaction under solvent-free conditions.

183 A. K. Verma, S. K. Reddy Kotla, D. Choudhary, M. Patel, R. K. Tiwari, *J. Org. Chem.* 2013, 78, 4386–4401. Silver-catalyzed tandem synthesis of naphthyridines and thienopyridines *via* three-component reaction.

184 Y. Pedduri, J. S. Williamson, *Tetrahedron Lett.* 2008, 49, 6009–6012. One-pot synthesis of highly substituted tetrahydrofurans from activated propargyl alcohols using Bu_3P.

185 A. K. Yadav, G. R. Sharma, P. Dhakad, T. Yadav, *Tetrahedron Lett.* 2012, 53, 859–862. A novel ionic liquid mediated synthesis of 4(1*H*)-quinolones, 5*H*-thiazolo[3,2-*a*]pyrimidin-5-one and 4*H*-pyrimido[2,1-*b*]benzothiazol-4-ones.

186 (a) L. Rong, X. Li, H. Wang, D. Shi, S. Tu, Q. Zhuang, *Synth. Commun.* 2007, 37, 183–189. Efficient green procedure for the synthesis of coumarin derivatives by a one-pot, three-component reaction under solvent-free conditions. (b) I. Yavari, M. Sabbaghan, Z. Hossaini, *Synlett* 2008, 1153–1154. Proline-promoted efficient synthesis of 4-aryl-3,4-dihydro-2*H*,5*H*-pyrano[3,2-*c*] chromene-2,5-diones in aqueous media. (c) K. Rad-Moghadam, S. C. Azimi, E. Abbaspour-Gilandeh, *Tetrahedron Lett.* 2013, 54, 4633–4636. Synthesis of novel pyrano[3,2-*c*]quinoline-2,5-diones using an acidic ionic liquid catalyst. (d) S. L. Zhu, J. Wang, Z. Xu, J. Li, *Molecules* 2012, 17, 13856–13863. An efficient one-pot synthesis of pyrano[3,2-*c*]quinolin-2, 5-dione derivatives catalyzed by L-proline.

187 (a) M. Kidwai, A. Jahan, N. K. Mishra, *C. R. Chim.* 2012, 15, 324–330. A novel method for the synthesis of tetrahydrobenzo[*a*]-xanthen-11-one derivatives using cerium(III) chloride as a highly efficient catalyst. (b) K. Rad-Moghadam, M. Sharifi-Kiasaraie, S. C. Azimi, *Tetrahedron* 2012, 68, 6472–6476. Synthesis of 4-substituted pyrano[4,3-*b*]pyran-2,5-diones in an ionic liquid.

188 A. Dandia, S. L. Gupta, S. Bhaskaran, *Eur. Chem. Bull.* 2013, 2, 836–841. Cerric ammonium nitrate (CAN) catalysed synthesis of pyrano[2,3-*d*]pyrimidine-2,4,7-triones in aqueous medium under sonication *via* one-pot three-component reaction.

189 (a) A. H. Rezayan, *C. R. Chim.* 2012, 15, 499–503. Stereoselective synthesis of 3,4-dihydro-7-nitrocoumarins *via* isocyanide-based multicomponent reaction. (b) A. H.

Rezayan, P. Azerang, S. Sardari, A. Sarvary, *Chem. Biol. Drug Des.* 2012, 80, 929–936. Synthesis and biological evaluation of coumarin derivatives as inhibitors of mycobacterium bovis (BCG). (c) A. Shaabani, A. Sarvary, E. Soleimani, A. H. Rezayan, M. Heidary, *Mol. Divers.* 2008, 12, 197–202. A novel method for the synthesis of substituted 3,4-dihydrocoumarin derivatives *via* isocyanide-based three-component reaction. (d) A. Shaabani, E. Soleimani, A. H. Rezayan, A. Sarvary, H. R. Khavasi, *Org. Lett.* 2008, 10, 2581–2584. Novel isocyanide-based four-component reaction: a facile synthesis of fully substituted 3,4-dihydrocoumarin derivatives.

190 (a) P. Wang, L. Song, H. Yi, M. Zhang, S. Zhu, H. Deng, M. Shao, *Tetrahedron Lett.* 2010, 51, 3975–3977. Convenient one-pot synthesis of fluorinated DHPs derivatives and their further transformations. (b) M. O. Noguez, V. Marcelino, H. Rodríguez, O. Martín, J. O. Martínez, G. A. Arroyo, F. J. Pérez, M. Suárez, R. Miranda, *Int. J. Mol. Sci.* 2011, 12, 2641–2649. Infrared assisted production of 3,4-dihydro-2(1*H*)-pyridones in solvent-free conditions. (c) G. Y. Fu, X. L. Zhang, S. R. Sheng, M. H. Wei, X. L. Liu, *Synth. Commun.* 2008, 38, 1249–1258. Rapid microwave-assisted liquid-phase synthesis of 4-substituted-5-methoxycarbonyl-6-methyl-3,4-dihydropyridones on poly(ethylene glycol) support.

191 (a) C.-Y. Yu, P.-H. Yang, M.-X. Zhao, Z.-T. Huang, *Synlett* 2006, 1835–1840. A novel one-pot reaction of heterocyclic ketene aminals: synthesis of a small library of tetrahydropyridinone-fused 1,3-diazaheterocycles. (b) B. Hu, Y.-Y. Zhou, X. Xu, Z. Li, W.-P. Deng, W. Du, *Synth. Commun.* 2011, 41, 1112–1118. Facile synthesis of tetrahydroimidazolpyridinones *via* an MCR involving 6-Cl-PMNI, aldehydes, and Meldrum's acid. (c) S. Kazemi Movahed, M. Dabiri, A. Bazgir, *Helv. Chim. Acta* 2013, 96, 525–532. An efficient one-pot four-component synthesis of functionalized imidazo[1,2-*a*]pyridines. (d) X.-S. Wang, M.-M. Zhang, H. Jiang, C.-S. Yao, S.-J. Tu, *Tetrahedron* 2007, 63, 4439–4449. Three-component green synthesis of N-arylquinoline derivatives in ionic liquid [bmim$^+$][BF_4^-]: reactions of arylaldehyde, 3-arylamino-5,5-dimethylcyclohex-2-enone, and active methylene compounds. (e) N. J. Thumar, M. P. Patel, *Med. Chem. Res.* 2012, 21, 1751–1761. Synthesis, characterization and biological activity of some new carbostyril bearing 1*H*-pyrazole moiety. (f) H. Sheibani, K. Saidi, M. Lakaei, *J. Heterocycl. Chem.* 2012, 49, 1386–1390. Three-component one-pot synthesis of 4-aryl-2, 3-dihydropyrimido[1,2-*a*]benzimidazol-2-ones catalyzed by L-proline.

192 N. G. Kozlov, L. I. Basalaeva, *Russ. J. Org. Chem.* 2006, 42, 509–514. Three-component condensation of Meldrum's acid with 2-naphthylamine and esters derived from vanillin.

193 R. Akbarzadeh, T. Amanpour, P. Mirzaei, A. Bazgir, *J. Organomet. Chem.* 2011, 696, 3421–3424. A simple four-component synthesis of ferrocenyl amidodiesters and ferrocenyl triamides.

194 A. Shaabani, H. Mofakham, A. Maleki, F. Hajishaabanha, *J. Comb. Chem.* 2010, 12, 630–632. Novel isocyanide-based one-pot multicomponent syntheses of tetrahydrobenzo[*b*][1,4]oxazepine and malonamide derivatives.

195 (a) A. Shaabani, A. H. Rezayan, S. Keshipour, A. Sarvary, S. W. Ng, *Org. Lett.* 2009, 11, 3342–3345. A novel one-pot three-(*in situ* five-)component condensation reaction: an unexpected approach for the synthesis of tetrahydro-2,4-dioxo-1*H*-benzo[*b*][1,5]diazepine-3-yl-2-methylpropanamide derivatives. (b) R. Akbarzadeh, T. Amanpour, A. A. Soorki, A. Bazgir, *Helv. Chim. Acta* 2012, 95, 483–490. Isocyanide-based five-component synthesis of 2-aryl-2-(2,3,4,5-tetrahydro-2,4-dioxo-1*H*-1,5-benzodiazepin-3-yl)acetamides (=α-aryl-2,3,4,5-tetrahydro-2,4-dioxo-1*H*-1,5-benzodiazepine-3-acetamides).

196 (a) L. S. Povarov, B. M. Mikhailov, *Russ. Chem. Bull.* 1963, 12, 871–871. A new type of diene condensation reaction. (b) L. S. Povarov, B. M. Mikhailov, *Izv. Akad. Nauk SSSR, Ser. Khim.* 1963, 5, 955–956. A new type of diene condensation reaction. (c) L. S. Povarov, V. I. Grigos, B. M. Mikhailov, *Russ. Chem. Bull.* 1963, 12, 1878–1880. Reaction of benzylideneaniline with some unsaturated compounds. (d) L. S. Povarov, V. I. Grigos, B. M. Mikhailov, *Izv. Akad. Nauk SSSR, Ser. Khim.* 1963, 11, 2039–2041. Reaction of benzylideneaniline with some unsaturated compounds.

197 (a) R. N. Asolkar, D. Schroder, R. Heckmann, S. Lang, I. Wagner-Dobler, H. Laatsch, *J. Antibiot.* 2004, 57, 17–23. Helquinoline, a new tetrahydroquinoline antibiotic from *Janibacter limosus* Hel 1. (b) D. A. Powell, R. A. Batey, *Org. Lett.* 2002, 4, 2913–2916. Total synthesis of the alkaloids martinelline and martinellic acid *via* a hetero Diels–Alder multicomponent coupling reaction. (c) D. Osborne, P. J. Stevenson, *Tetrahedron Lett.* 2002, 43, 5469–5470. A concise formal synthesis of luotonin A. (d) S. Omura, A. Nakagawa, H. Hashimoto, R. Oiwa, Y. Iwai, A. Hirano, N. Shibukawa, Y. Kojima, *J. Antibiot.* 1980, 33, 1395–1396. Virantmycin, a potent anti-viral antibiotic produced by a strain of streptomyces. (e) I. Jacquemond-Collet, F. Benoit-Vical, A. Valentin, E. Stanislas, M. Mallie, I. Fouraste, *Planta Med.* 2002, 68, 68–69. Antiplasmodial and cytotoxic activity of galipinine and other tetrahydroquinolines from *Galipea officinalis*.

198 (a) M. J. Alves, N. G. Azoia, A. G. Fortes, *Tetrahedron* 2007, 63, 727–734. Regio- and stereo-selective aza-Diels–Alder reaction of ethyl glyoxylate 4-methoxyphenylimine with 1,3-dienes in the presence of $BF_3{\cdot}Et_2O$. Evidence for a non-concerted mechanism. (b) J. M. Mellor, G. D. Merriman, *Tetrahedron* 1995, 51, 6115–6132. Reaction of electron-rich alkenes with anilines and formaldehyde-syntheses of tetrahydroquinolines. (c) S. Hermitage, J. A. K. Howard, D. Jay, R. G. Pritchard, M. R. Probert, A. Whiting, *Org. Biomol. Chem.* 2004, 2, 2451–2460. Mechanistic studies on the formal aza-Diels–Alder reactions of *N*-aryl imines: evidence for the non-concertedness under Lewis-acid catalysed conditions. (d) O. Jiménez, G. de la Rosa, R. Lavilla, *Angew. Chem. Int. Ed.* 2005, 44, 6521–6525. Straightforward access to a structurally diverse set of oxacyclic scaffolds through a four-component reaction. (e) S. Hermitage, D. A. Jay, A. Whiting, *Tetrahedron Lett.* 2002, 43, 9633–9636.

Evidence for the non-concerted [4+2]-cycloaddition of *N*-aryl imines when acting as both dienophiles and dienes under Lewis acid-catalysed conditions.

199 G. Dickmeiss, K. L. Jensen, D. Worgull, P. T. Franke, K. A. Jørgensen, *Angew. Chem. Int. Ed.* 2011, 50, 1580–1583. An asymmetric organocatalytic one-pot strategy to octahydroacridines.

200 Y. L. Zhao, W. Zhang, S. Wang, Q. Liu, *J. Org. Chem.* 2007, 72, 4985–4988. Ethynyl ketene-*S,S*-acetals: the highly reactive electron-rich dienophiles and applications in the synthesis of 4-functionalized quinolines *via* a one-pot three-component reaction.

201 N. Shindoh, H. Tokuyama, Y. Takemoto, K. Takasu, *J. Org. Chem.* 2008, 73, 7451–7456. Auto-tandem catalysis in the synthesis of substituted quinolines from aldimines and electron-rich olefins: cascade Povarov-hydrogen-transfer reaction.

202 (a) C. Venkateswarlu, P. V. Balaji, K. De, B. Crousse, B. Figadere, J. Legros, *J. Fluor. Chem.* 2013, 152, 94–98. Straightforward synthesis of 2-propylquinolines under multicomponent conditions in fluorinated alcohols. (b) K. De, J. Legros, B. Crousse, S. Chandrasekaran, D. Bonnet-Delpona, *Org. Biomol. Chem.* 2011, 9, 347–350. Synthesis of substituted 8-aminoquinolines and phenanthrolines through a Povarov approach.

203 (a) P. Ribelles, V. Sridharan, M. Villacampa, M. T. Ramos, J. C. Menendez, *Org. Biomol. Chem.* 2013, 11, 569–579. Diastereoselective, multicomponent access to *trans*-2-aryl-4-arylamino-1,2,3,4-tetrahydroquinolines *via* an AA'BC sequential four-component reaction and their application to 2-arylquinoline synthesis. (b) V. Sridharan, C. Avendano, J. C. Menendez, *Synthesis* 2008, 1039–1044. New findings on the cerium(IV) ammonium nitrate catalyzed Povarov reaction: stereoselective synthesis of 4-alkoxy-2-aryl-1,2,3,4-tetrahydroquinoline derivatives. (c) V. Sridharan, C. Avendano, J. C. Menendez, *Synlett* 2007, 1079–1082. CAN-catalyzed vinylogous Povarov reactions: the first three-component synthesis of 2-functionalized tetrahydroquinolines from anilines, cinnamaldehyde and vinyl ethers. (d) M. Viji, R. Nagarajan, *Synthesis* 2012, 44, 253–258. CAN-catalyzed regioselective synthesis of pyrido[2,3-c]carbazoles by the Povarov reaction.

204 (a) V. V. Kouznetsov, C. M. M. Gomez, J. H. B. Jaimes, *J. Heterocycl. Chem.* 2010, 47, 1148–1152. Transformations of 2-aryl-4-(2-oxopyrrolidinyl-1)-1,2,3,4-tetrahydroquinolines, cycloadducts of the $BiCl_3$-catalyzed three-component Povarov reaction: oxidation and reduction processes towards new potentially bioactive 2-arylquinoline derivatives. (b) V. V. Kouznetsov, C. M. M. Gomez, F. A. R. Ruiz, E. del Olmo, *Tetrahedron Lett.* 2012, 53, 3115–3118. Simple entry to new 2-alkyl-1,2,3,4-tetrahydroquinoline and 2,3-dialkylquinoline derivatives using $BiCl_3$-catalyzed three component reactions of anilines and aliphatic aldehydes in the presence (or lack) of *N*-vinyl amides.

205 C. D. Smith, J. I. Gavrilyuk, A. J. Lough, R. A. Batey, *J. Org. Chem.* 2010, 75, 702–715. Lewis acid catalyzed three-component hetero-Diels–Alder (Povarov) reaction of *N*-arylimines with strained norbornene-derived dienophiles.

206 B. da Silva, L. M. Martins, L. C. da Silva, *Synlett* 2012, 1973–1977. Niobium pentachloride catalyzed multicomponent Povarov reaction.

207 (a) X. S. Wang, J. Zhou, K. Yang, C. S. Yao, *Tetrahedron Lett.* 2010, 51, 5721–5723. $Yb(OTf)_3$: an efficient catalyst for the synthesis of 3-arylbenzo[*f*]quinoline-1,2-dicarboxylate derivatives *via* imino-Diels–Alder reaction. (b) K. Pericherla, A. Kumar, A. Jha, *Org. Lett.* 2013, 15, 4078–4081. Povarov-reductive amination cascade to access 6-aminoquinolines and anthrazolines. (c) A. A. Kudale, D. O. Miller, L. N. Dawe, G. J. Bodwell, *Org. Biomol. Chem.* 2011, 9, 7196–7206. Intramolecular Povarov reactions involving 3-amino-coumarins. (d) A. A. Kudale, J. Kendall, D. O. Miller, J. L. Collins, G. J. Bodwell, *J. Org. Chem.* 2008, 73, 8437–8447. Povarov reactions involving 3-aminocoumarins: synthesis of 1,2,3,4-tetrahydropyrido[2,3-*c*]coumarins and pyrido[2,3-*c*] coumarins.

208 S. Roy, O. Reiser, *Angew. Chem. Int. Ed.* 2012, 51, 4722–4725. A catalytic multicomponent approach for the stereoselective synthesis of *cis*-4,5-disubstituted pyrrolidinones and tetrahydro-3*H*-pyrrolo[3,2-*c*]quinolines.

209 (a) A. T. Khan, D. K. Das, K. Islam, P. Das, *Tetrahedron Lett.* 2012, 53, 6418–6422. A simple and expedient synthesis of functionalized pyrido[2,3-*c*]coumarin derivatives using molecular iodine catalyzed three-component reaction. (b) N. P. Rai, S. Shashikanth, P. N. Arunachalam, *Synth. Commun.* 2009, 39, 2125–2136. Iodine-catalyzed aza-Diels–Alder reactions of aliphatic *N*-arylaldimines. (c) M. Y. Yin, M. M. Zhang, W. Wang, Y. L. Li, X. S. Wang, *J. Heterocycl. Chem.* 2012, 49, 1239–1242. An efficient method for the synthesis of 3-aryl-4,7-phenanthroline derivatives catalyzed by iodine.

210 S. Chauhan, G. P. Mane, C. Anand, D. S. Dhawale, B. V. S. Reddy, S. M. J. Zaidi, S. S. Al-Deyab, V. V. Balasubramanian, T. Mori, A. Vinu, *Synlett* 2012, 2237–2240. Low-temperature synthesis of pyrano- and furo[3,2-*c*]quinolines *via* Povarov reaction using a highly ordered 3D nanoporous catalyst with a high acidity.

211 S. K. Guchhait, K. Jadeja, C. Madaan, *Tetrahedron Lett.* 2009, 50, 6861–6865. A new process of multicomponent Povarov reaction-aerobic dehydrogenation: synthesis of polysubstituted quinolines.

212 X. C. Ji, H. W. Huang, Y. B. Li, H. J. Chen, H. F. Jiang, *Angew. Chem. Int. Ed.* 2012, 51, 7292–7296. Palladium-catalyzed sequential formation of C-C bonds: efficient assembly of 2-substituted and 2,3-disubstituted quinolines.

213 Z. T. Rong, Q. J. Li, W. H. Lin, Y. X. Jia, *Tetrahedron Lett.* 2013, 54, 4432–4434. Reagent-free synthesis of 2,3,4-polysubstituted tetrahydroquinolines: application to the formal synthesis of (+/–)-martinellic acid and martinelline.

214 A. Hantzsch, *Ber. Dtsch. Chem. Ges.* 1881, 14, 1637–1638. Condensationsprodukte aus Aldehydammoniak und ketonartigen Verbindungen.

215 A. Hantzsch, *Ber. Dtsch. Chem. Ges.* 1890, 23, 1474–1476. Neue Bildungsweise von Pyrrolderivaten.

216(a) S. M. Stepkowski, Z. S. Nagy, M. E. Wang, F. Behbod, R. Erwin-Cohen, B. D. Kahan, R. A. Kirken, *Transplant. Proc.* 2001, 33, 3272–3273. PNU156804 inhibits Jak3 tyrosine kinase and rat heart allograft rejection. (b) S. M. Stepkowski, R. A. Erwin-Cohen, F. Behbod, M.-E. Wang, X. Qu, N. Tejpal, Z. S. Nagy, B. D. Kahan, R. A. Kirken, *Blood* 2002, 99, 680–689. Selective inhibitor of Janus tyrosine kinase 3, PNU156804, prolongs allograft survival and acts synergistically with cyclosporine but additively with rapamycin.

217 K. Kojiri, S. Nakajima, H. Suzuki, A. Okura, H. Suda, *J. Antibiot.* 1993, 46, 1799–1803. A new antitumor substance, BE-18591, produced by a streptomycete. I. Fermentation, isolation, physico-chemical and biological properties.

218 (a) A. M. Zonouz, N. Sahranavard, *E-J. Chem.* 2010, 7, S372–S376. Synthesis of 1,4-dihydropyridine derivatives under aqueous media. (b) Y. P. Liu, J. M. Liu, X. Wang, T. M. Cheng, R. T. Li, *Tetrahedron* 2013, 69, 5242–5247. Multicomponent reactions leading to symmetric and asymmetric multi-substituted 1,4-dihydropyridines on montmorillonite. (c) A. M. Zonouz, S. B. Hosseini, *Synth. Commun.* 2008, 38, 290–296. Montmorillonite K10 clay: an efficient catalyst for Hantzsch synthesis of 1,4-dihydropyridine derivatives.

219 (a) M. Mamaghani, K. Tabatabaeian, M. Mohammadi, A. Khorshidi, *J. Chem.* 2013. Cellulose-sulfuric acid as an efficient biosupported catalyst in one-pot synthesis of novel heteroaryl substituted 1,4-dihydropyridines. (b) J. Safari, S. H. Banitaba, S. D. Khalili, *J. Mol. Catal. A: Chem.* 2011, 335, 46–50. Cellulose sulfuric acid catalyzed multicomponent reaction for efficient synthesis of 1,4-dihydropyridines *via* unsymmetrical Hantzsch reaction in aqueous media.

220 E. Kolvari, M. A. Zolfigol, N. Koukabi, B. Shirmardi-Shaghasemi, *Chem. Pap.* 2011, 65, 898–902. A simple and efficient one-pot synthesis of Hantzsch 1,4-dihydropyridines using silica sulphuric acid as a heterogeneous and reusable catalyst under solvent-free conditions.

221 (a) S. Das Sharma, P. Hazarika, D. Konwar, *Catal. Commun.* 2008, 9, 709–714. A simple, green and one-pot four-component synthesis of 1,4-dihydropyridines and their aromatization. (b) J. Safaei-Ghomi, A. Ziarati, S. Zahedi, *J. Chem. Sci.* 2012, 124, 933–939. Silica (NPs) supported Fe (III) as a reusable heterogeneous catalyst for the one-pot synthesis of 1,4-dihydropyridines under mild conditions. (c) D. Bandyopadhyay, S. Maldonado, B. K. Banik, *Molecules* 2012, 17, 2643–2662. A microwave-assisted bismuth nitrate-catalyzed unique route toward 1,4-dihydropyridines. (d) A. Debache, R. Boulcina, A. Belfaitah, S. Rhouati, B. Carboni, *Synlett* 2008, 509–512. One-pot synthesis of 1,4-dihydropyridines *via* a phenylboronic acid catalyzed Hantzsch three-component reaction. (e) G. Sabitha, K. Arundhathi, K. Sudhakar, B. S. Sastry, J. S. Yadav, *Synth. Commun.* 2009, 39, 2843–2851. $CeCl_3{\cdot}7H_2O$-catalyzed one-pot synthesis of Hantzsch 1,4-dihydropyridines at room temperature. (f) E. Mosaddegh, A. Hassankhani, *Bull. Chem. Soc. Ethiop.* 2012, 26, 461–465. One-pot synthesis of polyhydropyridine derivatives *via* Hantzsch four component condensation in water medium: use of a recyclable Lewis acid $Ce(SO_4)_2$ catalyst. (g) M. Hong, C. Cai, W. B. Yi, *J. Fluor. Chem.* 2010, 131, 111–114. Hafnium (IV) bis(perfluorooctanesulfonyl)imide complex catalyzed synthesis of polyhydro-quinoline derivatives *via* unsymmetrical Hantzsch reaction in fluorous medium.

222 G. V. M. Sharma, K. L. Reddy, P. S. Lakshmi, P. R. Krishna, *Synthesis* 2006, 55–58. "*In situ*" generated "HCl"–an efficient catalyst for solvent-free Hantzsch reaction at room temperature: synthesis of new dihydropyridine glycoconjugates.

223 E. Rafiee, S. Eavani, S. Rashidzadeh, M. Joshaghani, *Inorg. Chim. Acta* 2009, 362, 3555–3562. Silica supported 12-tungstophosphoric acid catalysts for synthesis of 1,4-dihydropyridines under solvent-free conditions.

224 M. Pramanik, A. Bhaumik, *J. Mater. Chem. A* 2013, 1, 11210–11220. Self-assembled hybrid tinphosphonate nanoparticles with bimodal porosity: an insight towards the efficient and selective catalytic process for the synthesis of bioactive 1,4-dihydropyridines under solvent-free conditions.

225 B. Das, M. Srilatha, B. Veeranjaneyulu, B. S. Kanth, *Helv. Chim. Acta* 2011, 94, 885–891. Molybdenum- and tungsten-based coordination polymers as catalysts for an efficient and rapid synthesis of hexahydro-5-oxoquinoline-3-carboxylates and 1,4-dihydropyridine-3,5-dicarboxylates.

226 L. Saikia, D. Dutta, D. K. Dutta, *Catal. Commun.* 2012, 19, 1–4. Efficient clay supported Ni-0 nanoparticles as heterogeneous catalyst for solvent-free synthesis of Hantzsch polyhydroquinoline.

227 R. Murugan, K. Ramamoorthy, S. Sundarrajan, S. Ramakrishna, *Tetrahedron* 2012, 68, 7196–7201. Magnesium oxide nanotubes: synthesis, characterization and application as efficient recyclable catalyst for pyrazolyl 1,4-dihydropyridine derivatives.

228 C. A. Antonyraj, S. Kannan, *Appl. Catal., A* 2008, 338, 121–129. Hantzsch pyridine synthesis using hydrotalcites or hydrotalcite-like materials as solid base catalysts.

229 S. Ko, M. N. V. Sastry, C. C. Lin, C. F. Yao, *Tetrahedron Lett.* 2005, 46, 5771–5774. Molecular iodine-catalyzed one-pot synthesis of 4-substituted-1,4-dihydropyridine derivatives *via* Hantzsch reaction.

230 (a) A. Kumar, R. A. Maurya, *Tetrahedron* 2007, 63, 1946–1952. Synthesis of polyhydroquinoline derivatives through unsymmetric Hantzsch reaction using organocatalysts. (b) N. N. Karade, V. H. Budhewar, S. V. Shinde, W. N. Jadhav, *Lett. Org. Chem.* 2007, 4, 16–19. L-Proline as an efficient organocatalyst for the synthesis of polyhydroquinoline *via* multicomponent Hantzsch reaction. (c) M. Z. Kassaee, H. Masrouri, F. Movahedi, *Monatsh. Chem.* 2010, 141, 317–322. ZnO-nanoparticle-promoted synthesis of polyhydroquinoline derivatives *via* multicomponent Hantzsch reaction. (d) F. K. Behbahani, H. S. Alaei, *J. Chem. Sci.* 2013, 125, 623–626. L-Proline-catalysed synthesis of functionalized unsymmetrical dihydro-1*H*-indeno[1,2-*b*]pyridines. (e) S. K. Singh, K. N. Singh, *J. Heterocycl. Chem.* 2010, 47, 194–198. Glycine-catalyzed easy and efficient one-pot synthesis of polyhydroquinolines through hantzsch multicomponent condensation under controlled microwave.

231 K. Garkushenko, M. A. Makarova, O. P. Sorokina, N. V. Poendaev, M. A. Vorontsova, G. P. Sagitullina, *Chem. Heterocycl. Compd.* 2011, 47, 482–486. Synthesis of 3,5-diarylpyridines.

232 M. Lei, L. Ma, L. H. Hu, *Synth. Commun.* 2011, 41, 1969–1976. Thiamine hydrochloride-catalyzed one-pot synthesis of 1,4-dihydropyridine derivatives under solvent-free conditions.

233 (a) A. Dondoni, A. Massi, *Acc. Chem. Res.* 2006, 39, 451–463. Design and synthesis of new classes of heterocyclic *C*-glycoconjugates and carbon-linked sugar and heterocyclic amino acids by asymmetric multicomponent reactions (AMCRs). (b) D. R. B. Ducatti, A. Massi, M. D. Noseda, M. E. R. Duarte, A. Dondoni, *Org. Biomol. Chem.* 2009, 7, 1980–1986. Dihydropyridine *C*-glycoconjugates by organocatalytic Hantzsch cyclocondensation. Stereoselective synthesis of α-threofuranose *C*-nucleoside enantiomers. (c) A. Dondoni, A. Massi, M. Aldhoun, *J. Org. Chem.* 2007, 72, 7677–7687. Hantzsch-type three-component approach to a new family of carbon-linked glycosyl amino acids. Synthesis of C-glycosylmethyl pyridylalanines.

234 (a) Z. B. Gan, A. Okui, Y. Kawashita, M. Hayashi, *Chem. Lett.* 2008, 37, 1302–1303. Convenient synthesis of linear-extended bipyridines involving a central phenyl linking group. (b) A. Ghorbani-Choghamarani, M. A. Zolfigol, M. Hajjami, H. Goudarziafshar, M. Nikoorazm, S. Yousefi, B. Tahmasbi, *J. Braz. Chem. Soc.* 2011, 22, 525–531. Nano aluminium nitride as a solid source of ammonia for the preparation of Hantzsch 1,4-dihydropyridines and *bis*-(1,4-dihydropyridines) in water *via* one pot multicomponent reaction. (c) D. M. Liu, L. T. Du, J. Sun, C. G. Yan, *Synth. Commun.* 2010, 40, 1333–1338. Efficient synthesis of alkylene bridging bis-dihydropyridines.

235 J. C. Legeay, J. J. Vanden Eynde, J. P. Bazureau, *Tetrahedron* 2005, 61, 12386–12397. Ionic liquid phase technology supported the three component synthesis of Hantzsch 1,4-dihydropyridines and Biginelli 3,4-dihydropyrimidin-2(1*H*)-ones under microwave dielectric heating.

236 M. A. Zolfigol, A. G. Choghamarani, M. Shahamirian, M. Safaiee, I. Mohammadpoor-Baltork, S. Mallakpour, M. Abdollahi-Alibeik, *Tetrahedron Lett.* 2005, 46, 5581–5584. 4-Phenyl-1,2,4-triazole-3,5-dione as a novel and reusable reagent for the aromatization of 1,4-dihydropyridines under mild conditions.

237 M. Nasr-Esfahani, B. Karami, M. Behzadi, *J. Heterocycl. Chem.* 2009, 46, 931–935. A simple, efficient, one-pot three-component domino synthesis of Hantzsch pyridines under solvent-free condition.

238 M. Saha, S. Roy, S. K. Chaudhuri, S. Bhar, *Green Chem. Lett. Rev.* 2008, 1, 99–102. Microwave-assisted ammonium formate-mediated synthesis of Hantzsch dihydropyridines under solvent-free conditions—a green protocol.

239 (a) A. Samzadeh-Kermani, H. Shafaroodi, R. Miri, H. Mirkhani, M. Vosooghi, A. Shafiee, *Med. Chem. Res.* 2009, 18, 112–126. Lipophilic 2-(4-chlorophenyl)-4-thiazolyl-1,4-dihydropyridines: synthesis, calcium channel antagonist activity, and protection against pentylenetetrazole-induced seizure. (b) X. Y. Wu, *Synth. Commun.* 2012, 42, 454–459. Facile and green synthesis of 1,4-dihydropyridine derivatices in *n*-butyl pyridinium tetrafluoroborate. (c) J. R. Avalani, D. S. Patel, D. K. Raval, *J. Chem. Sci.* 2012, 124, 1091–1096. 1-Methylimidazolium trifluoroacetate [Hmim][TFA]: mild and efficient Brønsted acidic ionic liquid for Hantzsch reaction under microwave irradiation. (d) A. Shaabani, A. H. Rezayan, A. Rahmati, M. Sharifi, *Monatsh. Chem.* 2006, 137, 77–81. Ultrasound-accelerated synthesis of 1,4-dihydropyridines in an ionic liquid.

240 F. Tamaddon, Z. Rami, A. A. Jafari, *Tetrahedron Lett.* 2010, 51, 1187–1189. Synthesis of 3,4-dihydropyrimidin-2(1*H*)-ones and 1,4-dihydropyridines using ammonium carbonate in water.

241 S. Ghosh, F. Saikh, J. Das, A. K. Pramanik, *Tetrahedron Lett.* 2013, 54, 58–62. Hantzsch 1,4-dihydropyridine synthesis in aqueous ethanol by visible light.

242 S. Long, M. Panunzio, A. Petroli, W. L. Qin, Z. N. Xia, *Synthesis* 2011, 1071–1078. The use of magnesium nitride for the synthesis of enantiomerically pure 1,4-dihydropyridines *via* the Hantzsch reaction.

243 B. R. P. Kumar, P. Masih, E. Karthikeyan, A. Bansal, S., P. Vijayan, *Med. Chem. Res.* 2010, 19, 344–363. Synthesis of novel Hantzsch dihydropyridines and Biginelli dihydropyrimidines of biological interest: a 3D-QSAR study on their cytotoxicity.

244 S. B. Sapkal, K. F. Shelke, B. B. Shingate, M. S. Shingare, *Tetrahedron Lett.* 2009, 50, 1754–1756. Nickel nanoparticle-catalyzed facile and efficient one-pot synthesis of polyhydroquinoline derivatives *via* Hantzsch condensation under solvent-free conditions.

245 S. Samai, G. C. Nandi, R. Kumar, M. S. Singh, *Tetrahedron Lett.* 2009, 50, 7096–7098. Multicomponent one-pot solvent-free synthesis of functionalized unsymmetrical dihydro-1*H*-indeno[1,2-*b*]pyridines.

246 G. C. Nandi, S. Samai, M. S. Singh, *J. Org. Chem.* 2010, 75, 7785–7795. Biginelli and Hantzsch-type reactions leading to highly functionalized dihydropyrimidinone, thiocoumarin, and pyridopyrimidinone frameworks *via* ring annulation with β-oxodithioesters.

247 (a) B. Han, Z. G. Liu, Q. Liu, L. Yang, Z. L. Liu, W. Yu, *Tetrahedron* 2006, 62, 2492–2496. An efficient aerobic oxidative aromatization of Hantzsch 1,4-dihydropyridines and 1,3,5-trisubstituted pyrazolines. (b) R. Murugan, K. Ramamoorthy, S. Sundarrajan, S. Ramakrishna, *Tetrahedron* 2011, 67, 2998–3002. Simple and efficient synthesis of 2,6-dialkyl-3,5-dialkoxycarbonyl-4-(3-aryl-1-phenyl-pyrazol-4-yl)pyridines using TPAP/NMO as a catalyst under mild conditions. (c) T. Shamim, M. Gupta, S. Paul, *J. Mol. Catal. A: Chem.* 2009, 302, 15–19. The oxidative aromatization of Hantzsch 1,4-dihydropyridines by molecular oxygen using surface functionalized silica supported cobalt catalysts. (d) Y. Kawashita, M. Hayashi, *Molecules* 2009, 14, 3073–3093. Synthesis of heteroaromatic compounds by oxidative aromatization using an activated carbon/molecular oxygen system. (e) B. Han, Q. Liu, Z. G. Liu, R. Z. Mu, W. Zhang, Z. L. Liu, W. Yu, *Synlett* 2005, 2333–2334. Metal-free catalytic aerobic aromatization of Hantzsch 1,4-dihydropyridines by *N*-hydroxyphthalimide. (f) M. Hayashi, *Chem. Rec.* 2008, 8, 252–267. Oxidation using activated carbon and molecular oxygen system. (g) M. M. Heravi, K. Bakhtiari, H. A. Oskooie, R. Hekmatshoa, *Heterocycl. Commun.* 2006, 12, 209–212. Zeobis, a versatile reagent for the fast aromatization of Hantzsch 1,4-dihydropyridines.

248 (a) M. Filipan-Litvic, M. Litvic, V. Vinkovic, *Tetrahedron* 2008, 64, 5649–5656. An efficient, metal-free, room temperature aromatization of Hantzsch-1,4-dihydropyridines with urea-hydrogen peroxide adduct, catalyzed by molecular iodine. (b) J. L. Su, C. Zhang, D. Lin, Y. Q. Duan, X. K. Fu, R. Z. Mu, *Synth. Commun.* 2010, 40, 595–600. Oxidative aromatization of Hantzsch 1,4-dihydro-pyridines by H_2O_2/V_2O_5 at room temperature.

249 (a) M. Moghadam, M. Nasr-Esfahani, S. Tangestaninejad, V. Mirkhani, M. A. Zolfigol, *Can. J. Chem.* 2006, 84, 1–4. Biomimetic aromatization of Hantzsch 1,4-dihydropyridines with sodium periodate catalyzed by a new polystyrene-bound manganese porphyrin. (b) M. Montazerozohori, M. Nasr-Esfahani, S. Joohari, N. Haghighat, *Asian J. Chem.* 2010, 22, 4249–4254. Biomimetic catalytic oxidation of Hantsch 1,4-dihydropyridines using *bis*-(salicylaldehyde)-4-methyl-1,2-phenylenediimine -Mn(III) chloride/sodium periodate under mild conditions.

250 A. Saini, S. Kumar, J. S. Sandhu, *Synth. Commun.* 2007, 37, 2317–2324. New strategy for the oxidation of Hantzsch 1,4-dihydropyridines and dihydropyrido[2,3-*d*]pyrimidines catalyzed by DMSO under aerobic conditions.

251 (a) K. A. Dilmaghani, B. Zeynizadeh, M. Mirzaei, *Heterocycl. Commun.* 2007, 13, 139–142. Oxidative-aromatization of Hantzsch ester 1,4-dihydropyridines by $KBrO_3$/$CoCl_2·6H_2O$ under mild condition. (b) B. Zeynizadeh, K. A. Dilmaghani, A. Roozijoy, *Synth. Commun.* 2005, 35, 557–562. Oxidative-aromatization of Hantzsch ester 1,4-dihydropyridines by $KBrO_3$/$SnCl_4·5H_2O$ under mild condition.

252 X. L. Liao, W. B. Lin, J. Lu, C. Wang, *Tetrahedron Lett.* 2010, 51, 3859–3861. Oxidative aromatization of Hantzsch 1,4-dihydropyridines by sodium chlorite.

253 F. Tamaddon, Z. Razmi, *Synth. Commun.* 2011, 41, 485–492. Oxidation of 1,4-dihydropyridines and 3,4-dihydropyrimidin-2(1*H*)-ones to substituted pyridines and pyrimidinones using $Ca(OCl)_2$ in aqueous media.

254 B. Zeynizadeh, K. A. Dilmaghani, M. Mirzaei, *Asian J. Chem.* 2009, 21, 2969–2972. Mild and efficient method for aromatization of Hantzsch esters of 1,4-dihydropyridines with HIO_3.

255 L. Z. Chai, Y. K. Zhao, Q. J. Sheng, Z. Q. Liu, *Tetrahedron Lett.* 2006, 47, 9283–9285. Aromatization of Hantzsch 1,4-dihydropyridines and 1,3,5-trisubstituted pyrazolines with HIO_3 and I_2O_5 in water.

256 M. Montazerozohori, B. Karami, M. Nasr-Esfabani, S. A. Musavi, *Heterocycl. Commun.* 2007, 13, 289–294. Silver salts/iodine monochloride as a new oxidation system for the oxidative aromatization of 1,4-dihydropyridines.

257 (a) J. S. Yadav, B. V. S. Reddy, A. K. Basak, G. Baishya, A. V. Narsaiah, *Synthesis* 2006, 451–454. Iodoxybenzoic acid (IBX): an efficient and novel oxidizing agent for the aromatization of 1,4-dihydropyridines. (b) P. Kumar, *J. Heterocycl. Chem.* 2010, 47, 1429–1433. A novel, facile, simple and convenient oxidative aromatization of Hantzsch 1,4-dihydropyridines to pyridines using polymeric iodosobenzene with KBr. (c) J. M. Chen, X. M. Zeng, *Synth. Commun.* 2009, 39, 3521–3526. β-Cyclodextrin-catalyzed mild aromatization of Hantzsch 1,4-dihydropyridines with o-iodoxy-benzoic acid in water/acetone.

258 (a) M. M. Heravi, K. Bakhtiari, H. A. Oskooie, *Arkivoc* 2007, (xiii), 190–194. 2,6-Dicarboxypyridinium chlorochromate: an efficient oxidizing agent for the very fast oxidation of Hantzsch 1,4-dihydro-pyridines. (b) A. R. Momeni, T. Sameh, H. Golmohammadi, H. J. Naghash, H. Aliyan, A. R. Massah, S. Solati, *Bull. Korean Chem. Soc.* 2006, 27, 355–356. An efficient oxidation of 1,4-dihydropyridines to pyridines using silver carbonate on silica gel and celite. (c) S. Dehghanpour, M. M. Heravi, F. Derikvand, *Molecules* 2007, 12, 433–438. *N,N'*-ethylene-bis(benzoylacetoniminato) copper(II), $Cu(C_{22}H_{22}N_2O_2)$, a new reagent for aromatization of Hantzsch 1,4-dihydropyridines. (d) B. Karami, M. Montazerozohori, M. H. Habibi, M. A. Zolfigol, *Heterocycl. Commun.* 2005, 11, 513–516. Tungstate sulfuric acid/$KMnO_4$ as a novel heterogeneous system for the rapid aromatization of Hantzsch 1,4-dihydropyridines under mild conditions. (e) M. M. Heravi, F. Derikvand, H. A. Oskooie, R. H. Shoar, *Synth. Commun.* 2006, 36, 77–82. Silica gel-supported *bis*(trimethylsilyl) chromate: oxidation of 1,4-dihydropyridines to pyridines. (f) M. Filipan-Litvic, M. Litvic, V. Vinkovic, *Tetrahedron* 2008, 64, 10912–10918. Rapid, efficient, room temperature aromatization of Hantzsch-1,4-dihydropyridines with vanadium(V) salts: superiority of classical technique *versus* microwave promoted reaction.

259 M. Filipan-Litvic, M. Litvic, I. Cepanec, V. Vinkovic, *Arkivoc* 2008, (xi), 96–103. Antimony(V) chloride-promoted room temperature aromatization of 1,4-dihydropyridines in non-acidic solvent.

260 X. H. Cai, H. J. Yang, G. L. Zhang, *Can. J. Chem.* 2005, 83, 273–275. Aromatization of 1,4-dihydropyridines with selenium dioxide.

261 M. A. Zolfigol, P. Salehi, A. Ghorbani-Choghamarani, M. Safaiee, M. Shahamirian, *Synth. Commun.* 2007, 37, 1817–1823. Silica chromate as a novel oxidizing agent for the oxidation of 1,4-dihydropyridines.

262 (a) K. Niknam, M. A. Zolfigol, *Heterocycl. Commun.* 2006, 12, 183–186. Aromatization of 1,4-dihydropyridines in the presence of toluenesulfonyl chloride/$NaNO_2$/wet-SiO_2 under microwave irradiation. (b) K. Niknam, M. A. Zolfigol, S. M. Razavian, I. Mohammahpoor-Baltork, *J. Heterocycl. Chem.* 2006, 43, 199–202. Aromatization of 1,4-dihydropyridines in the presence of methanesulfonic acid/$NaNO_2$/wet SiO_2 under both heterogeneous and solvent free conditions. (c) M. M. Hashemi, H. Ghafuri, Z. Karimi-Jaberi, *Monatsh. Chem.* 2006, 137, 197–200. Room temperature catalytic aromatization of Hantzsch 1,4-dihydropyridines by sodium nitrite in the presence of acidic silica gel.

263 A. Ghorbani-Choghamarani, J. Zeinivand, *Synth. Commun.* 2010, 40, 2457–2463. Aromatization of Hantzsch 1,4-dihydropyridines with $Al(NO_3)_3·9H_2O$ and/or $Fe(NO_3)_3·9H_2O$ in the presence of silica sulfuric acid under mild and heterogeneous conditions.

264 P. Kumar, A. Kumar, *Bull. Korean Chem. Soc.* 2010, 31, 2299–2303. An expeditious oxidative aromatization of

Hantzsch 1,4-dihydropyridines to pyridines using cetyltrimethylammonium peroxodisulfate: a phase transferring oxidant.

265 A. R. Momeni, H. Aliyan, H. Mombeini, A. R. Massah, H. J. Naghash, *Z. Naturforsch.* 2006, 61b, 331–333. Aromatization of Hantzsch 1,4-dihydropyridines with urea-hydrogen peroxide/maleic anhydride.

266 J. J. Xia, G. W. Wang, *Synthesis* 2005, 2379–2383. One-pot synthesis and aromatization of 1,4-dihydropyridines in refluxing water.

267 S. Chen, M. S. Hossain, F. W. Foss, *ACS Sustainable Chem. Eng.* 2013, 1, 1045–1051. Bioinspired oxidative aromatizations: one-pot syntheses of 2-substituted benzothiazoles and pyridines by aerobic organocatalysis.

268 O. de Paolis, J. Baffoe, S. M. Landge, B. Torok, *Synthesis* 2008, 3423–3428. Multicomponent domino cyclization-oxidative aromatization on a bifunctional Pd/C/K-10 catalyst: an environmentally benign approach toward the synthesis of pyridines.

269 V. Estevez, M. Villacampa, J. C. Menendez, *Chem. Commun.* 2013, 49, 591–593. Three-component access to pyrroles promoted by the CAN-silver nitrate system under high-speed vibration milling conditions: a generalization of the Hantzsch pyrrole synthesis.

270 T. A. Moss, T. Nowak, *Tetrahedron Lett.* 2012, 53, 3056–3060. Synthesis of 2,3-dicarbonylated pyrroles and furans *via* the three-component Hantzsch reaction.

INDEX

Multicomponent Reactions: Concepts and Applications for Design and Synthesis, First Edition. Edited by Raquel P. Herrera and Eugenia Marqués-López.

www.ingramcontent.com/pod-product-compliance
Lightning Source LLC
LaVergne TN
LVHW062136200926
848484LV00007B/1820
9781118016008